Forms Involving $\sqrt{u^2 - a^2}$

39 $\displaystyle\int \sqrt{u^2 - a^2}\,du = \frac{u}{2}\sqrt{u^2 - a^2} - \frac{a^2}{2}\ln|u + \sqrt{u^2 - a^2}| + C$

40 $\displaystyle\int u^2\sqrt{u^2 - a^2}\,du = \frac{u}{8}(2u^2 - a^2)\sqrt{u^2 - a^2} - \frac{a^4}{8}\ln|u + \sqrt{u^2 - a^2}| + C$

41 $\displaystyle\int \frac{\sqrt{u^2 - a^2}}{u}\,du = \sqrt{u^2 - a^2} - a\cos^{-1}\frac{a}{u} + C$

42 $\displaystyle\int \frac{\sqrt{u^2 - a^2}}{u^2}\,du = -\frac{\sqrt{u^2 - a^2}}{u} + \ln|u + \sqrt{u^2 - a^2}| + C$

43 $\displaystyle\int \frac{du}{\sqrt{u^2 - a^2}} = \ln|u + \sqrt{u^2 - a^2}| + C$

44 $\displaystyle\int \frac{u^2\,du}{\sqrt{u^2 - a^2}} = \frac{u}{2}\sqrt{u^2 - a^2} + \frac{a^2}{2}\ln|u + \sqrt{u^2 - a^2}| + C$

45 $\displaystyle\int \frac{du}{u^2\sqrt{u^2 - a^2}} = \frac{\sqrt{u^2 - a^2}}{u^2 u} + C$

46 $\displaystyle\int \frac{du}{(u^2 - a^2)^{3/2}} = -\frac{u}{a^2\sqrt{u^2 - a^2}} + C$

Forms Involving $a + bu$

47 $\displaystyle\int \frac{u\,du}{a + bu} = \frac{1}{b^2}(a + bu - a\ln|a + bu|) + C$

48 $\displaystyle\int \frac{u^2\,du}{a + bu} = \frac{1}{2b^3}[(a + bu)^2 - 4a(a + bu) + 2a^2\ln|a + bu|] + C$

49 $\displaystyle\int \frac{du}{u(a + bu)} = \frac{1}{a}\ln\left|\frac{u}{a + bu}\right| + C$

50 $\displaystyle\int \frac{du}{u^2(a + bu)} = -\frac{1}{au} + \frac{b}{a^2}\ln\left|\frac{a + bu}{u}\right| + C$

51 $\displaystyle\int \frac{u\,du}{(a + bu)^2} = \frac{a}{b^2(a + bu)} + \frac{1}{b^2}\ln|a + bu| + C$

52 $\displaystyle\int \frac{du}{u(a + bu)^2} = \frac{1}{a(a + bu)} - \frac{1}{a^2}\ln\left|\frac{a + bu}{u}\right| + C$

53 $\displaystyle\int \frac{u^2\,du}{(a + bu)^2} = \frac{1}{b^3}\left(a + bu - \frac{a^2}{a + bu} - 2a\ln|a + bu|\right) + C$

54 $\displaystyle\int u\sqrt{a + bu}\,du = \frac{2}{15b^2}(3bu - 2a)(a + bu)^{3/2} + C$

55 $\displaystyle\int \frac{u\,du}{\sqrt{a + bu}} = \frac{2}{3b^2}(bu - 2a)\sqrt{a + bu}$

56 $\displaystyle\int \frac{u^2\,du}{\sqrt{a + bu}} = \frac{2}{15b^3}(8a^2 + 3b^2u^2 - 4abu)\sqrt{a + bu}$

57 $\displaystyle\int \frac{du}{u\sqrt{a + bu}} = \frac{1}{\sqrt{a}}\ln\left|\frac{\sqrt{a + bu} - \sqrt{a}}{\sqrt{a + bu} + \sqrt{a}}\right| + C, \quad \text{if } a > 0$

$\displaystyle\qquad = \frac{2}{\sqrt{-a}}\tan^{-1}\sqrt{\frac{a + bu}{-a}} + C, \quad \text{if } a < 0$

58 $\displaystyle\int \frac{\sqrt{a + bu}}{u}\,du = 2\sqrt{a + bu} + a\int \frac{du}{u\sqrt{a + bu}}$

59 $\displaystyle\int \frac{\sqrt{a + bu}}{u^2}\,du = -\frac{\sqrt{a + bu}}{u} + \frac{b}{2}\int \frac{du}{u\sqrt{a + bu}}$

60 $\displaystyle\int u^n\sqrt{a + bu}\,du = \frac{2u^n(a + bu)^{3/2}}{b(2n + 3)} - \frac{2na}{b(2n + 3)}\int \frac{u^{n-1}}{\sqrt{a + bu}}\,du$

61 $\displaystyle\int \frac{u^n\,du}{\sqrt{a + bu}} = \frac{2u^n\sqrt{a + bu}}{b(2n + 1)} - \frac{2na}{b(2n + 1)}\int \frac{u^{n-1}\,du}{\sqrt{a + bu}}$

62 $\displaystyle\int \frac{du}{u^n\sqrt{a + bu}} = -\frac{\sqrt{a + bu}}{a(n - 1)u^{n-1}} - \frac{b(2n - 3)}{2a(n - 1)}\int \frac{du}{u^{n-1}\sqrt{a + bu}}$

Trigonometric Forms

63 $\displaystyle\int \sin^2 u\,du = \tfrac{1}{2}u - \tfrac{1}{4}\sin 2u + C$

64 $\displaystyle\int \cos^2 u\,du = \tfrac{1}{2}u + \tfrac{1}{4}\sin 2u + C$

65 $\displaystyle\int \tan^2 u\,du = \tan u - u + C$

66 $\displaystyle\int \cot^2 u\,du = -\cot u - u + C$

67 $\displaystyle\int \sin^3 u\,du = -\tfrac{1}{3}(2 + \sin^2 u)\cos u + C$

68 $\displaystyle\int \cos^3 u\,du = \tfrac{1}{3}(2 + \cos^2 u)\sin u + C$

69 $\displaystyle\int \tan^3 u\,du = \tfrac{1}{2}\tan^2 u + \ln|\cos u| + C$

70 $\displaystyle\int \cot^3 u\,du = -\tfrac{1}{2}\cot^2 u - \ln|\sin u| + C$

71 $\displaystyle\int \sec^3 u\,du = \tfrac{1}{2}\sec u\tan u + \tfrac{1}{2}\ln|\sec u + \tan u| + C$

72 $\displaystyle\int \csc^3 u\,du = -\tfrac{1}{2}\csc u\cot u + \tfrac{1}{2}\ln|\csc u - \cot u| + C$

73 $\displaystyle\int \sin^n u\,du = -\frac{1}{n}\sin^{n-1} u\cos u + \frac{n - 1}{n}\int \sin^{n-2} u\,du$

74 $\displaystyle\int \cos^n u\,du = \frac{1}{n}\cos^{n-1} u\sin u + \frac{n - 1}{n}\int \cos^{n-2} u\,du$

(Continued inside back cover)

Prindle, Weber & Schmidt is a division of Wadsworth, Inc.

Library of Congress Cataloging in Publication Data

Swokowski, Earl William
Calculus with analytic geometry.

Includes index.
1. Calculus. 2. Geometry, Analytic. I. Title.
QA303.S94 1979 515'.15 78-31267
ISBN 0-87150-268-2

Printed in the United States of America.
Fifth printing: August, 1981

Production by Nancy Blodget and Elizabeth Thomson, under the direction of David Chelton and Michael Michaud. Text design by Elizabeth Thomson. Text composition in Monophoto Times Roman and Plantin by Technical Filmsetters Europe Limited. Technical art by Phil Carver & Friends, Inc. Cover printed by Lehigh Press. Text printed and bound by Rand McNally & Co.

Preface to Second Edition

This edition has benefited greatly from comments of users of the first edition, and from the constructive criticism of those who reviewed the manuscript. Following their suggestions, many exercise sets were expanded by adding drill-type problems, together with some challenging ones for highly motivated students. There are now approximately 5500 exercises, almost 1500 more than appeared previously. The number of solved examples has also increased significantly. For more specific information regarding the differences between this edition and the previous edition the following will be useful.

Chapter 2 was partially rewritten, keeping in mind the goals of presenting limits and continuity of functions in a mathematically sound manner, while providing students with a strong intuitive feeling for these important concepts. A noteworthy change is that the sections on limits at infinity and functions which become infinite have been moved to Chapter 4, where they are used in conjunction with graphs of rational functions. This move makes it possible to introduce the derivative earlier than before, and thus maintain student interest by using the limit notion in an important way very early in the course.

The proof of the Chain Rule in Chapter 3 has been simplified by avoiding the case where it is necessary to introduce an auxiliary function of some type. However, those who are interested in a complete proof will find one in Appendix II.

The discussion of extrema in Chapter 4 has been improved through the addition of examples and figures. Additional emphasis has been placed on end-point extrema. The section on applications to economics was rewritten.

In Chapter 5 the proof of the Fundamental Theorem of Calculus has been amplified, and error estimates for numerical integration are stressed more than in the earlier edition. The definite integral as a limit of a sum is strongly emphasized in Chapter 6; however, solutions to examples are constructed so that after this important idea is thoroughly understood, the student may bypass the subscript part of the procedure and merely "set up" the required integrals.

Chapter 8 on exponential and logarithmic functions was completely rewritten. The present version should make the development of these important functions much easier to follow than before.

The principal changes in Chapters 9 and 10 consist of the introduction of many new examples and exercises, a better discussion of hyperbolic functions, and a new section on the use of tables of integrals.

In Chapter 11, two physical examples are introduced to help motivate improper integrals. The section on Taylor's Formula has been clarified by graphically illustrating what happens if the number of terms of the approximating polynomial is increased.

Some of the material on infinite series in Chapter 12 has been rewritten and rearranged. Several new theorems, including the Root Test, were added, together with many new exercises. The discussion of power series representations of functions has also been enlarged.

The concept of the *direction* of a two-dimensional vector has been replaced by the simpler (and more easily generalizable) notions of the *same* or *opposite* direction of vectors. This, together with a reorganization of topics, leads to a smoother development in Chapter 14. A stronger emphasis has been placed on geometric problems. This is especially true with applications of the vector product.

The discussion of derivatives and integrals of vector-valued functions in Chapter 15 has been modified to help unify this material. A new section on Kepler's Laws was added to illustrate the power of vector techniques.

The changes in Chapters 16 and 17 consist primarily of additional examples, figures, exercises, and minor rewriting. In Chapter 18, two new sections have been added on transformations of coordinates, Jacobians, and change of variables in multiple integrals.

The order of topics is flexible. For instance, some users of the first edition introduced trigonometric functions very early in the course. A natural place to do so is after the discussion of the Chain Rule in Chapter 3. As a matter of fact, a remark at the end of Section 3.5 refers to the derivative of the sine function. This could be extended by stating formulas for derivatives of the remaining trigonometric functions, and then selecting appropriate exercises from Chapter 9 as students progress through subsequent sections. In like manner, integrals of the trigonometric functions may be introduced in Chapter 5.

Chapter 7, on analytic geometry, can be covered immediately after Chapter 1. Of course, in this event, exercises involving derivatives and integrals cannot be assigned. Chapter 19, on differential equations, may be discussed upon the completion of techniques of integration in Chapter 10, provided that the sections on exact equations and series solutions are omitted. It is also possible to discuss the material on vectors, in Chapter 14, prior to Chapter 13. If desired, Chapter 12, on infinite series, may be postponed until later in the text.

There is a review section at the end of each chapter consisting of a list of important topics together with pertinent exercises. The review exercises are similar in scope to those which appear throughout the text and may be used by students to prepare for examinations. Answers to odd-numbered exercises are given at the end of the text. Instructors may obtain an answer booklet for the even-numbered exercises from the publisher.

I wish to thank the following individuals, who reviewed all, or parts of, the manuscript for the second edition, and offered many helpful suggestions: Phillip W. Bean, Mercer University; Daniel D. Benice, Montgomery College; Delmar L. Boyer, University of Texas–El Paso; Ronald E. Bruck, University of Southern California; David C. Buchthal, The University of Akron; John E. Derwent, University of Notre Dame; William R. Fuller, Purdue University; Gary Haggard, University of Maine–Orono; Douglas Hall, Michigan State University; George Johnson, University of South Carolina; Andy Karantinos, University of South Dakota; G. Otis Kenny, Boise State University; Eleanor Killam, University of Massachusetts–Amherst;

Prem K. Kulshrestha, University of New Orleans; Margaret Lial, American River College; Phil Locke, University of Maine–Orono; Burnett Meyer, University of Colorado; Joseph Miles, University of Illinois–Urbana; Charles D. Miller, American River College; John Nohel, University of Wisconsin–Madison; Neal C. Raber, The University of Akron; John T. Scheick, The Ohio State University; Richard D. Semmler, Northern Virginia Community College–Annandale; Ray C. Shifflet, California State University–Fullerton; David Shochat, Santa Monica College; Carol M. Smith, Birmingham–Southern College; William M. Snyder, University of Maine–Orono; Eugene Speer, Rutgers University; John Tung, Miami University; Dale E. Walston, University of Texas–Austin; Frederick R. Ward, Boise State University.

In addition many instructors were kind enough to respond to a survey conducted by my publisher. Heron S. Collins, Louisiana State University-Baton Rouge; Karl Peterson, University of North Carolina; M. Evans Munroe, University of New Hampshire; John Berglund, Virginia Commonwealth University; George Johnson, University of South Carolina; Frank Quinn, University of Virginia; Lawrence Runyan, Shoreline College; Robert W. Owens, Lewis and Clark College; Karl Gentry, University of North Carolina–Greensboro; George Haborak, College of Charleston; James M. Sobota, University of Wisconsin–La Crosse; Gene A. de Both, St. Norbert College; David Greenstein, Northeastern Illinois University; Jerry Wagenblast, Valparaiso University; Gene Vanden Boss, Adrian College; Duane E. Deal, Ball State University; Carol Smith, Birmingham Southern University; Gary Eichelsdorfer, University of Scranton; Robert E. Spencer, Virginia Polytechnic Institute and State University; Albert L. Rabenstein, Washington and Jefferson College; Charles A. Grobe, Jr., Bowdoin College; Teisuke Ito, Northern Michigan University; Stanley R. Samsky, University of Delaware, were among those who sent in valuable information and ideas.

I also wish to acknowledge the advice of my colleagues at Marquette University, who offered numerous suggestions for improvements. Additional recognition is due Drs. Thomas Bronikowski and Michael Ziegler, for their careful work with solutions to the exercises.

I am grateful for the valuable assistance of the staff of Prindle, Weber & Schmidt. In particular, Elizabeth Thomson, Nancy Blodget, and Mary Le Quesne have been very helpful in the production of the text. As usual, Executive Editor John Martindale was a constant source of advice and encouragement.

Special thanks are due to my wife Shirley and the members of our family: Mary, Mark, John, Steven, Paul, Thomas, Robert, Nancy, and Judy. All have had an influence on the book—either directly, by working exercises, proofreading, or typing—or indirectly, through continued interest and moral support.

In addition to all of the persons named here, I express my sincere appreciation to the many unnamed students and teachers who have helped shape my views on how calculus should be presented in the classroom.

Earl W. Swokowski

is placed near the end of the book, where portions may be omitted without interrupting the continuity of the text. The same is true for Chapter 19 on differential equations.

I wish to thank the following individuals, who reviewed the manuscript and offered many helpful suggestions: James Cornette, Iowa State University; August Garver, University of Missouri–Rolla; Douglas Hall, Michigan State University; Alan Heckenbach, Iowa State University; Simon Hellerstein, University of Wisconsin; David Mader, Ohio State University; William Meyers, California State University, San Jose; David Minda, University of Cincinnati; Chester Miracle, University of Minnesota; Ada Peluso, Hunter College of the City University of New York; Leonard Shapiro, University of Minnesota; Donald Sherbert, University of Illinois; Charles Van Gorden, Millersville State College; Dale Walston, University of Texas.

Special thanks are due to Dr. Thomas Bronikowski of Marquette University, who carefully read the entire manuscript, worked every exercise, and was responsible for many improvements in the text. In addition, he has written a student supplement which contains detailed solutions for approximately one-third of the exercises.

I am grateful to Carolyn Meitler for an excellent job of typing the manuscript, and to the staff of Prindle, Weber & Schmidt, for their painstaking work in the production of this book. In particular, John Martindale, a fine editor and friend, has been a constant source of encouragement during my association with the company. Above all, I owe a debt of thanks to my family, for their patience and understanding over long periods of writing.

Earl W. Swokowski

Table of Contents

What is Calculus?

Calculus was invented in the seventeenth century to provide a tool for solving problems involving motion. The subject matter of geometry, algebra, and trigonometry is applicable to objects which move at constant speeds; however, methods introduced in calculus are required to study the orbits of planets, to calculate the flight of a rocket, to predict the path of a charged particle through an electromagnetic field and, for that matter, to deal with all aspects of motion.

In order to discuss objects in motion it is essential first to define what is meant by *velocity* and *acceleration.* Roughly speaking, the velocity of an object is a measure of the rate at which the distance traveled changes with respect to time. Acceleration is a measure of the rate at which velocity changes. Velocity may vary considerably, as is evident from the motion of a drag-strip racer or the descent of a space capsule as it reenters the Earth's atmosphere. In order to give precise meanings to the notions of velocity and acceleration it is necessary to use one of the fundamental concepts of calculus, the *derivative.*

Although calculus was introduced to help solve problems in physics, it has been applied to many different fields. One of the reasons for its versatility is the fact that the derivative is useful in the study of rates of change of many entities other than objects in motion. For example, a chemist may use derivatives to forecast the outcome of various chemical reactions. A biologist may employ it in the investigation of the rate of growth of bacteria in a culture. An electrical engineer uses the derivative to describe the change in current in an electric circuit. Economists have applied it to problems involving corporate profits and losses.

The derivative is also used to find tangent lines to curves. Although this has some independent geometric interest, the significance of tangent lines is of major importance in physical problems. For example, if a particle moves along a curve, then the tangent line indicates the direction of motion. If we restrict our attention to a sufficiently small portion of the curve, then in a certain sense the tangent line may be used to approximate the position of the particle.

Many problems involving maximum and minimum values may be attacked with the aid of the derivative. Some typical questions that can be answered are: At what angle of elevation should a projectile be fired in order to achieve its maximum range? If a tin can is to hold one gallon of a liquid, what dimensions require the least amount of tin? At what point between two light sources will the illumination be greatest? How can certain corporations maximize their revenue? How can a manufacturer minimize the cost of producing a given article?

Another fundamental concept of calculus is known as the *definite integral.* It, too, has many applications in the sciences. A physicist uses it to find the work required to stretch or compress a spring. An engineer may use it to find the center of mass or moment of inertia of a solid. The definite integral can be used by a biologist to calculate the flow of blood through an arteriole. An economist may employ it to estimate depreciation of equipment in a manufacturing plant. Mathematicians use definite integrals to investigate such concepts as areas of surfaces, volumes of geometric solids, and lengths of curves.

All the examples we have listed, and many more, will be discussed in detail as we progress through this book. There is literally no end to the applications of calculus. Indeed, in the future perhaps *you,* the reader, will discover new uses for this important branch of mathematics.

The derivative and the definite integral are defined in terms of certain limiting processes. The notion of limit is the initial idea which separates calculus from the more elementary branches of mathematics. Sir Isaac Newton (1642–1727) and Gottfried Wilhelm Leibniz (1646–1716) discovered the connection between derivatives and integrals. Because of this, and their other contributions to the subject, they are credited with the invention of calculus. Many other mathematicians have added a great deal to its development.

The preceding discussion has not answered the question "What is calculus?" Actually, there is no simple answer. Calculus could be called the study of limits, derivatives, and integrals; however, this statement is meaningless if definitions of the terms are unknown. Although we have given a few examples to illustrate what can be accomplished with derivatives and integrals, neither of these concepts has been given any meaning. Defining them will be one of the principal objectives of our early work in this text.

Prerequisites for Calculus

This chapter contains topics necessary for the study of calculus. After a brief review of real numbers, coordinate systems, and graphs in two dimensions, we turn our attention to one of the most important concepts in mathematics—the notion of function.

1.1 REAL NUMBERS

Real numbers are used considerably in precalculus mathematics and it will be assumed that the reader is familiar with the fundamental properties of addition, subtraction, multiplication, division, exponents, radicals, and so on. Throughout this chapter, unless otherwise specified, lower-case letters $a, b, c, \ldots$ will denote real numbers.

The **positive integers** $1, 2, 3, 4, \ldots$ may be obtained by adding the real number 1 successively to itself. The **integers** consist of all positive and negative integers together with the real number 0. A **rational number** is a real number that can be expressed as a quotient a/b, where a and b are integers and $b \neq 0$. Real numbers that are not rational are called **irrational**. The ratio of the circumference of a circle to its diameter is irrational. This real number is denoted by π and the notation $\pi \approx 3.1416$ is used to indicate that π is *approximately equal* to 3.1416. Another example of an irrational number is $\sqrt{2}$.

Real numbers may be represented by nonterminating decimals. For example, the decimal representation for the rational number 7434/2310 is found by long division to be 3.2181818..., where the digits 1 and 8 repeat indefinitely. Rational numbers may always be represented by repeating decimals. Decimal representations for irrational numbers may also be obtained; however, they are nonterminating and nonrepeating.

It is possible to associate real numbers with points on a line l in such a way that to each real number a there corresponds one and only one point, and conversely, to each point P there corresponds precisely one real number. Such an association between two sets is referred to as a **one-to-one correspondence**. We first choose an arbitrary point O, called the **origin**, and associate with it the real number 0. Points associated with the integers are then determined by laying off successive line segments of equal length on either side of O as illustrated in Figure 1.1. The points corresponding to rational numbers such as 23/5 and $-1/2$ are obtained by subdividing the equal line segments. Points associated with certain irrational numbers, such as $\sqrt{2}$, can be found by geometric construction. For other irrational numbers such as π, no construction is possible. However, the point corresponding to π can be approximated to any degree of accuracy by locating successively the points corresponding to 3, 3.1, 3.14, 3.141, 3.1415, 3.14159, and so on. It can be shown that to every irrational number there corresponds a unique point on l and, conversely, every point that is not associated with a rational number corresponds to an irrational number.

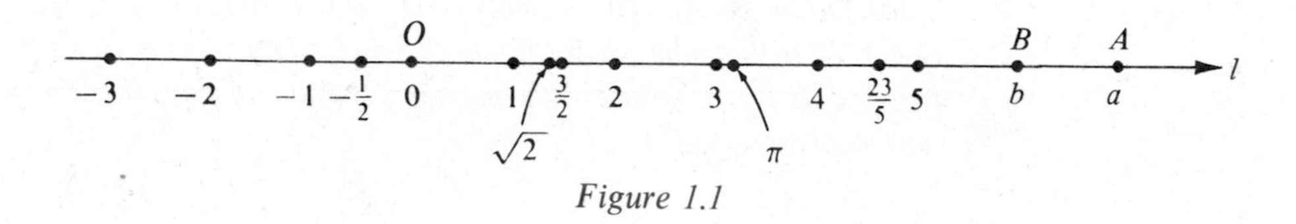

Figure 1.1

The number a that is associated with a point A on l is called the **coordinate** of A. An assignment of coordinates to points on l is called a **coordinate system** for l, and l is called a **coordinate line**, or a **real line**. A direction can be assigned to l by taking the **positive direction** to the right and the **negative direction** to the left. The positive direction is noted by placing an arrowhead on l as shown in Figure 1.1.

The real numbers which correspond to points to the right of O in Figure 1.1 are called **positive real numbers**, whereas those which correspond to points to the left of O are **negative real numbers**. The real number 0 is neither positive nor negative. The collection of positive real numbers is **closed** relative to addition and multiplication; that is, if a and b are positive, then so is the sum $a + b$ and the product ab.

If a and b are real numbers, and $a - b$ is positive, we say that ***a* is greater than *b*** and write $a > b$. An equivalent statement is ***b* is less than *a***, written $b < a$. The symbols $>$ or $<$ are called **inequality signs** and expressions such as $a > b$ or $b < a$ are called **inequalities**. From the manner in which we constructed the coordinate line l in Figure 1.1, we see that if A and B are points with coordinates a and b, respectively, then $a > b$ (or $b < a$) *if and only if A lies to the right of B*. Since $a - 0 = a$, it follows that $a > 0$ if and only if a is positive. Similarly, $a < 0$ means that a is negative. The following rules can be proved.

(1.1)

If $a > b$ and $b > c$, then $a > c$.

If $a > b$, then $a + c > b + c$.

If $a > b$ and $c > 0$, then $ac > bc$.

If $a > b$ and $c < 0$, then $ac < bc$.

Analogous rules for "less than" can also be established.

The symbol $a \geq b$, which is read ***a* is greater than or equal to *b***, means that either $a > b$ or $a = b$. The symbol $a < b < c$ means that $a < b$ and $b < c$, in which case we say that ***b* is *between* *a* and *c***. The notations $a \leq b$, $a < b \leq c$, $a \leq b < c$, $a \leq b \leq c$, and so on, have similar meanings.

Another property, called **completeness**, is needed to characterize the real numbers. This property will be discussed in Chapter 12.

If a is a real number, then it is the coordinate of some point A on a coordinate line l, and the symbol $|a|$ is used to denote the number of units (or distance) between A and the origin, without regard to direction. Referring to Figure 1.2 we see that for the point with coordinate -4 we have $|-4| = 4$. Similarly, $|4| = 4$. In general, if a is negative we change its sign to find $|a|$, whereas if a is nonnegative, then $|a| = a$. Thus we have the following definition.

(1.2) **Definition**

$$|a| = \begin{cases} a & \text{if } a \geq 0 \\ -a & \text{if } a < 0 \end{cases}$$

The nonnegative number $|a|$ is called the **absolute value** of a.

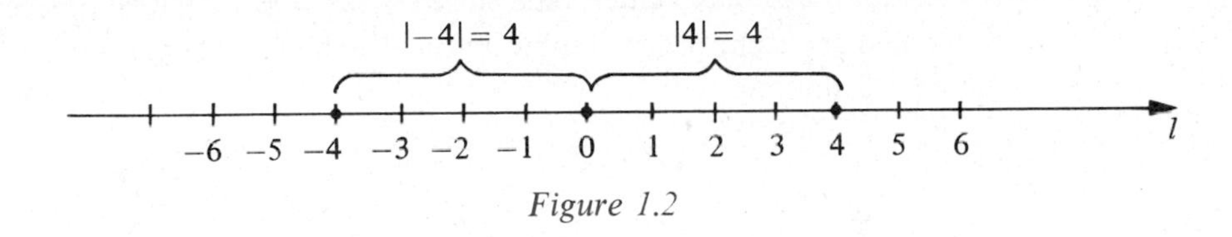

Figure 1.2

Example 1 Find $|3|, |-3|, |0|, |\sqrt{2} - 2|$, and $|2 - \sqrt{2}|$.

Solution Since $3, 2 - \sqrt{2}$, and 0 are nonnegative, we have by (1.2),

$$|3| = 3, \quad |2 - \sqrt{2}| = 2 - \sqrt{2}, \quad \text{and} \quad |0| = 0.$$

Since -3 and $\sqrt{2} - 2$ are negative, we use the formula $|a| = -a$ of (1.2) to obtain

$$|-3| = -(-3) = 3 \quad \text{and} \quad |\sqrt{2} - 2| = -(\sqrt{2} - 2) = 2 - \sqrt{2}.$$

It can be proved that for all real numbers a and b,

(1.3)
$$|a| = |-a|$$
$$|ab| = |a|\,|b|$$
$$-|a| \leq a \leq |a|.$$

It can also be shown that if b is any positive real number, then

(1.4)
$$|a| < b \quad \text{if and only if } -b < a < b$$
$$|a| > b \quad \text{if and only if } a > b \text{ or } a < -b$$
$$|a| = b \quad \text{if and only if } a = b \text{ or } a = -b.$$

It follows from (1.4) that if b is positive, then $|a| \leq b$ means $-b \leq a \leq b$. This fact is used in the proof of the next statement.

(1.5) Triangle Inequality

> If a and b are real numbers, then
>
> $$|a + b| \leq |a| + |b|.$$

Proof. From (1.3), $-|a| \leq a \leq |a|$ and $-|b| \leq b \leq |b|$. Adding corresponding sides we obtain

$$-(|a| + |b|) \leq a + b \leq |a| + |b|$$

and hence, from the remark following (1.4), $|a + b| \leq |a| + |b|$.

The concept of absolute value may be used to define the distance between any two points on a coordinate line. Let us begin by noting that the points with coordinates 2 and 7 shown in Figure 1.3 are 5 units apart on l, and that 5 is the difference, $7 - 2$, obtained by subtracting the smaller coordinate from the larger. If we employ absolute values, then since $|7 - 2| = |2 - 7|$, it is unnecessary to be concerned about the order of subtraction. We shall use this as our motivation for the next definition.

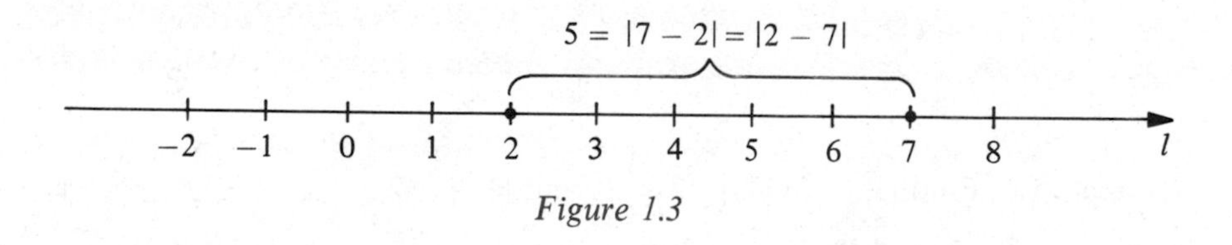

Figure 1.3

(1.6) Definition

> Let a and b be the coordinates of two points A and B, respectively, on a coordinate line l. The **distance between A and B**, denoted by $d(A, B)$, is given by
>
> $$d(A, B) = |b - a|.$$

The nonnegative number $d(A, B)$ in Definition (1.6) is also called the **length of the line segment AB**. Observe that since $d(B, A) = |a - b|$ and $|b - a| = |a - b|$, we have $d(A, B) = d(B, A)$. Also note that the distance between the origin O and the point A is

$$d(O, A) = |a - 0| = |a|$$

which agrees with the geometric interpretation of absolute value illustrated in Figure 1.2.

Example 2 If A, B, C, and D have coordinates $-5, -3, 1$, and 6 respectively, find $d(A, B)$, $d(C, B)$, $d(O, A)$, and $d(C, D)$.

Solution The points are indicated in Figure 1.4. By Definition (1.6),

$$d(A, B) = |-3 - (-5)| = |-3 + 5| = |2| = 2.$$
$$d(C, B) = |-3 - 1| = |-4| = 4.$$
$$d(O, A) = |-5 - 0| = |-5| = 5.$$
$$d(C, D) = |6 - 1| = |5| = 5.$$

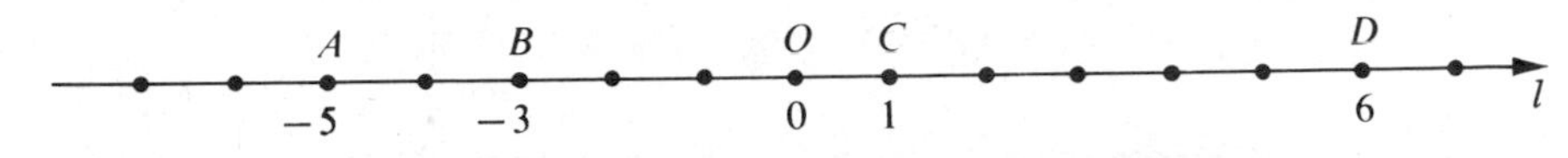

Figure 1.4

The concept of absolute value has uses other than that of finding distances between points. Generally, it is employed whenever one is interested in the magnitude or numerical value of a real number without regard to its sign.

In order to shorten explanations it is sometimes convenient to use the notation and terminology of sets. A **set** may be thought of as a collection of objects of some type. The objects are called **elements** of the set. Throughout our work $\mathbb{R}$ will denote the set of real numbers, $\mathbb{N}$ the set of positive integers, and $\mathbb{Z}$ the integers. If S is a set, then $a \in S$ means that a is an element of S, whereas $a \notin S$ signifies that a is not an element of S. If every element of a set S is also an element of a set T, then S is called a **subset** of T. Two sets S and T are said to be **equal**, written $S = T$, if S and T contain precisely the same elements. The notation $S \neq T$ means that S and T are not equal. If S and T are sets, their **union** $S \cup T$ consists of the elements which are either in S, in T, or in *both* S and T. The **intersection** $S \cap T$ consists of the elements which the sets have in common.

If the elements of a set S have a certain property, then we write $S = \{x: \ldots\}$ where the property describing the arbitrary element x is stated in the space after the colon. For example, $\{x: x > 3\}$ may be used to represent the set of all real numbers greater than 3.

Of major importance in calculus are certain subsets of $\mathbb{R}$ called **intervals**. If $a < b$, the symbol (a, b) will sometimes be used for the set of all real numbers between a and b; that is,

$$(a, b) = \{x: a < x < b\}.$$

The set (a, b) is called an **open interval**. The **graph** of a set S of real numbers is defined as the totality of points on a coordinate line which correspond to the numbers in S. In particular, the graph of the open interval (a, b) consists of all points between the points corresponding to a and b. In Figure 1.5 we have sketched the graphs of a general open interval (a, b) and the special open intervals $(-1, 3)$ and $(2, 4)$. The parentheses in the figure are used to indicate that the **end-points** are not to be included.

Figure 1.5. Graphs of the open intervals (a, b), $(-1, 3)$, and $(2, 4)$

Closed intervals denoted by $[a, b]$, and **half-open intervals** denoted by $[a, b)$ or $(a, b]$, are defined as follows.

$$[a, b] = \{x : a \leq x \leq b\}$$
$$[a, b) = \{x : a \leq x < b\}$$
$$(a, b] = \{x : a < x \leq b\}$$

Typical graphs are sketched in **Figure 1.6**. A bracket in the figure indicates that the corresponding end-point is included in the graph.

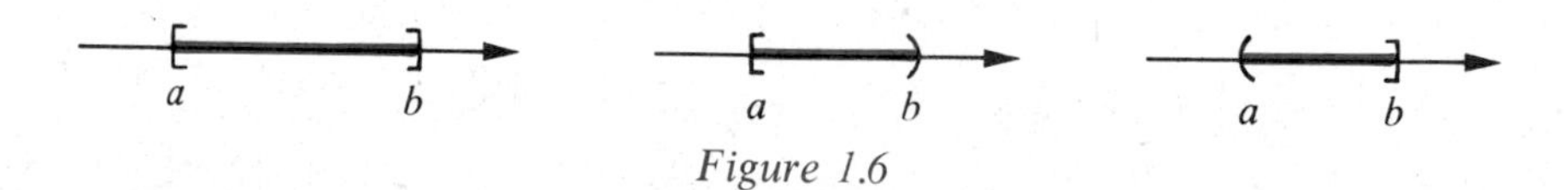

Figure 1.6

For convenience we sometimes use the terms *interval* and *graph of an interval* interchangeably. In future discussions of intervals, whenever the magnitudes of a and b are not stated explicitly it will always be assumed that $a < b$. If an interval is a subset of another interval I it is called a **subinterval** of I. For example, the closed interval $[2, 3]$ is a subinterval of $[0, 5]$.

Infinite intervals are defined as follows.

$$(a, \infty) = \{x : x > a\}$$
$$[a, \infty) = \{x : x \geq a\}$$
$$(-\infty, a) = \{x : x < a\}$$
$$(-\infty, a] = \{x : x \leq a\}$$
$$(-\infty, \infty) = \mathbb{R}$$

For example, $(1, \infty)$ represents all real numbers greater than 1. The symbol ∞ denotes "infinity" and is merely a notational device. It is not to be interpreted as representing a real number.

As indicated in this section, we frequently make use of symbols to denote arbitrary elements of a set. For example, we may use x to denote a real number, although no *particular* real number is specified. A letter which is used to represent any element of a given set is sometimes called a **variable**. Throughout this text, unless otherwise specified, variables will represent real numbers. The **domain of a variable** is the set of real numbers represented by the variable. To illustrate, given the expression $\sqrt{x}$, we note that in order to obtain a real number we must have $x \geq 0$, and hence in this case the domain of x is assumed to be the set of nonnegative real numbers. Similarly, when working with the expression $1/(x - 2)$ we must exclude $x = 2$ (Why?) and consequently we take the domain of x as the set of all real numbers different from 2.

Inequalities which involve variables occur frequently in the study of calculus. One illustration is

$$x^2 - 3 < 2x + 4.$$

If certain numbers such as 4 or 5 are substituted for x, we obtain the false statements $13 < 12$ or $22 < 14$, respectively. Other numbers such as 1 or 2 produce the true statements $-2 < 6$ or $1 < 8$, respectively. In general, if we are given an inequality in x and if a true statement is obtained when x is replaced by a real number a, then a is called a **solution** of the inequality. Thus 1 and 2 are solutions of the inequality $x^2 - 3 < 2x + 4$, whereas 4 and 5 are not solutions. To **solve** an inequality means to find all solutions. We say that two inequalities are **equivalent** if they have exactly the same solutions.

A standard method for solving an inequality is to replace it with a chain of equivalent inequalities, terminating in one for which the solutions are obvious. The main tools used in applying this method are properties such as those listed in (1.1), (1.3), and (1.4). For example, if x represents a real number, then adding the same expression in x to both sides leads to an equivalent inequality. We may multiply both sides of an inequality by an expression containing x if we are certain that the expression is positive for all values of x under consideration. If we multiply both sides of an inequality by an expression that is always negative, such as $-7 - x^2$, then the inequality sign is reversed.

The reader should supply reasons for the solutions of the following inequalities.

Example 3 Solve the inequality $4x + 3 > 2x - 5$.

Solution The following inequalities are equivalent.

$$4x + 3 > 2x - 5$$

$$4x > 2x - 8$$

$$2x > -8$$

$$x > -4$$

Hence the solutions consist of all real numbers greater than -4, that is, the numbers in the infinite interval $(-4, \infty)$.

Example 4 Solve the inequality $-5 < \dfrac{4 - 3x}{2} < 1$.

Solution We may proceed as follows:

$$-5 < \frac{4 - 3x}{2} < 1$$

$$-10 < 4 - 3x < 2$$

$$-14 < -3x < -2$$

$$\frac{14}{3} > x > \frac{2}{3}$$

$$\frac{2}{3} < x < \frac{14}{3}$$

Hence the solutions are the numbers in the open interval $(2/3, 14/3)$.

Example 5 Solve $x^2 - 7x + 10 > 0$.

Solution Since the inequality may be written

$$(x - 5)(x - 2) > 0,$$

it follows that x is a solution if and only if both factors $x - 5$ and $x - 2$ are positive, or both are negative. The diagram in Figure 1.7 indicates the signs of these factors for various real numbers. Evidently, both factors are positive if x is in the interval $(5, \infty)$ and both are negative if x is in $(-\infty, 2)$. Hence the solutions consist of all real numbers in the union $(-\infty, 2) \cup (5, \infty)$.

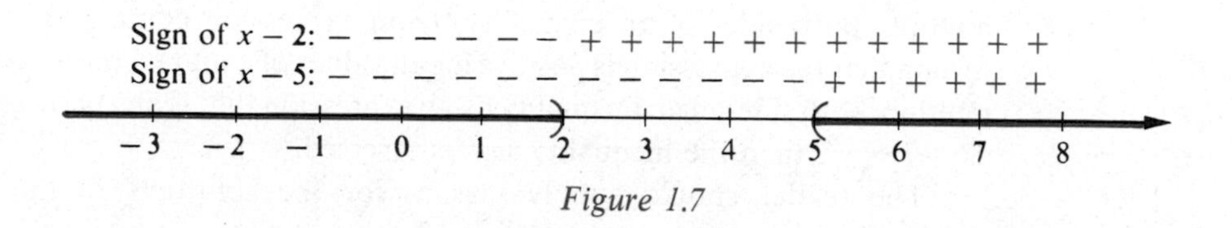

Figure 1.7

Among the most important inequalities occurring in calculus are those containing absolute values of the type illustrated in the next example.

Example 6 Solve the inequality $|x - 3| < 0.1$.

Solution Using (1.4), the given inequality is equivalent to

$$-0.1 < x - 3 < 0.1$$

and hence to

$$-0.1 + 3 < (x - 3) + 3 < 0.1 + 3$$

or

$$2.9 < x < 3.1.$$

Thus the solutions are the real numbers in the open interval $(2.9, 3.1)$.

Example 7 Solve $|2x - 7| > 3$.

Solution By (1.4), x is a solution of $|2x - 7| > 3$ if and only if either

$$2x - 7 > 3 \quad \text{or} \quad 2x - 7 < -3.$$

The first of these two inequalities is equivalent to $2x > 10$, or $x > 5$. The second is equivalent to $2x < 4$, or $x < 2$. Hence the solutions of $|2x - 7| > 3$ are the numbers in the union $(-\infty, 2) \cup (5, \infty)$.

EXERCISES 1.1

In Exercises 1 and 2 replace the comma between each pair of real numbers with the appropriate symbol $<$, $>$, or $=$.

1 (a) $-2, -5$ (b) $-2, 5$ (c) $6-1, 2+3$
(d) $2/3, 0.66$ (e) $2, \sqrt{4}$ (f) $\pi, 22/7$

2 (a) $-3, 0$ (b) $-8, -3$ (c) $8, -3$
(d) $\frac{3}{4} - \frac{2}{3}, \frac{1}{15}$ (e) $\sqrt{2}, 1.4$ (f) $\frac{4053}{1110}, 3.6513$

Rewrite the expressions in Exercises 3 and 4 without using symbols for absolute values.

3 (a) $|2-5|$ (b) $|-5|+|-2|$
(c) $|5|+|-2|$ (d) $|-5|-|-2|$
(e) $|\pi - 22/7|$ (f) $(-2)/|-2|$
(g) $|1/2 - 0.5|$ (h) $|(-3)^2|$

4 (a) $|4-8|$ (b) $|3-\pi|$
(c) $|-4|-|-8|$ (d) $|-4+8|$
(e) $|-3|^2$ (f) $|2-\sqrt{4}|$
(g) $|-0.67|$ (h) $-|-3|$

5 If A, B, and C are points on a coordinate line with coordinates -5, -1, and 7, respectively, find the following distances.

(a) $d(A, B)$ (b) $d(B, C)$
(c) $d(C, B)$ (d) $d(A, C)$

6 Rework Exercise 5 if A, B, and C have coordinates 2, -8, and -3, respectively.

Solve the inequalities in Exercises 7–34 and express the solutions in terms of intervals.

7 $5x - 6 > 11$

8 $3x - 5 < 10$

9 $2 - 7x \leq 16$

10 $7 - 2x \geq -3$

11 $|2x + 1| > 5$

12 $|x + 2| < 1$

13 $3x + 2 < 5x - 8$

14 $2 + 7x < 3x - 10$

15 $12 \geq 5x - 3 > -7$

16 $5 > 2 - 9x > -4$

17 $-1 < \frac{3 - 7x}{4} \leq 6$

18 $0 \leq 4x - 1 \leq 2$

19 $\frac{5}{7 - 2x} > 0$

20 $\frac{4}{x^2 + 9} > 0$

21 $|x - 10| < 0.3$

22 $\left|\frac{2x + 3}{5}\right| < 2$

23 $\left|\frac{7 - 3x}{2}\right| \leq 1$

24 $|3 - 11x| \geq 41$

25 $|25x - 8| > 7$

26 $|2x + 1| < 0$

27 $3x^2 + 5x - 2 < 0$

28 $2x^2 - 9x + 7 < 0$

29 $2x^2 + 9x + 4 \geq 0$

30 $x^2 - 10x \leq 200$

31 $1/x^2 < 100$

32 $5 + \sqrt{x} < 1$

33 $\dfrac{3x+2}{2x-7} \leq 0$

34 $\dfrac{3}{x-9} > \dfrac{2}{x+2}$

35 Prove that $|a - b| \geq |a| - |b|$. (*Hint*: Write $|a| = |(a - b) + b|$ and apply (1.5).)

36 If n is any positive integer and $a_1, a_2, \ldots, a_n$ are real numbers, prove that

$$|a_1 + a_2 + \cdots + a_n| \leq |a_1| + |a_2| + \cdots + |a_n|.$$

(*Hint*: By (1.5), $|a_1 + a_2 + \cdots + a_n| \leq |a_1| + |a_2 + \cdots + a_n|$.)

37 If $0 < a < b$, prove that $(1/a) > (1/b)$. Why is the restriction $0 < a$ necessary?

38 If $0 < a < b$, prove that $a^2 < b^2$. Why is the restriction $0 < a$ necessary?

39 If $a < b$ and $c < d$, prove that $a + c < b + d$.

40 If $a < b$ and $c < d$ is it always true that $ac < bd$? Explain.

1.2 COORDINATE SYSTEMS IN TWO DIMENSIONS

In Section 1 we discussed how coordinates may be assigned to points on a straight line. Coordinate systems can also be introduced in planes by means of *ordered pairs*. The term **ordered pair** refers to two real numbers, where one is designated as the "first" number and the other as the "second." The symbol (a, b) is used to denote the ordered pair consisting of the real numbers a and b where a is first and b is second. There are many uses for ordered pairs. They were used in Section 1 to denote open intervals. In this section they will represent points in a plane. Although ordered pairs are employed in different situations, there is little chance for confusion, since it should always be clear from the discussion whether the symbol (a, b) represents an interval, a point, or some other mathematical object. We consider two ordered pairs (a, b) and (c, d) **equal**, and write

$$(a, b) = (c, d) \quad \text{if and only if} \quad a = c \text{ and } b = d.$$

This implies, in particular, that $(a, b) \neq (b, a)$ if $a \neq b$. The set of all ordered pairs will be denoted by $\mathbb{R} \times \mathbb{R}$.

A **rectangular**, or **Cartesian**,* **coordinate system** may be introduced in a plane by considering two perpendicular coordinate lines in the plane which intersect in the origin O on each line. Unless specified otherwise, the same unit of length is chosen on each line. Usually one of the lines is horizontal with positive direction to the right, and the other line is vertical with positive direction upward, as indicated by the arrowheads in Figure 1.8. The two lines are called **coordinate axes** and the point O is called the **origin**. The horizontal line is often referred to as the ***x*-axis** and the vertical line as the ***y*-axis**, and they are labeled x and y, respectively. The plane is

* The term "Cartesian" is used in honor of the French mathematician and philosopher René Descartes (1596–1650), who was one of the first to employ such coordinate systems.

then called a **coordinate plane** or, with the preceding notation for coordinate axes, an ***xy*-plane**. In certain applications different labels such as *d*, *t*, etc., are used for the coordinate lines. The coordinate axes divide the plane into four parts called the **first, second, third,** and **fourth quadrants** and labeled I, II, III, and IV, respectively, as shown in (i) of Figure 1.8.

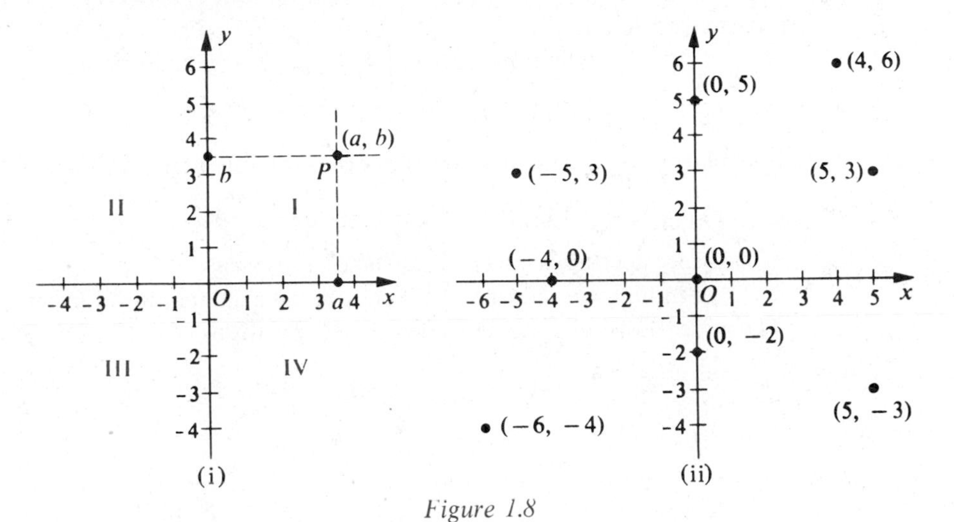

Figure 1.8

Each point P in an xy-plane may be assigned a unique ordered pair. If vertical and horizontal lines through P intersect the x- and y-axes at points with coordinates a and b, respectively (see (i) of Figure 1.8), then P is assigned the ordered pair (a, b). The number a is called the **x-coordinate** (or **abscissa**) of P, and b is called the **y-coordinate** (or **ordinate**) of P. We sometimes say that *P has coordinates* (a, b). Conversely, every ordered pair (a, b) determines a point P in the xy-plane with coordinates a and b. Specifically, P is the point of intersection of lines perpendicular to the x-axis and y-axis at the points having coordinates a and b, respectively. This establishes a one-to-one correspondence between the set of all points in the xy-plane and the set of all ordered pairs. It is sometimes convenient to refer to the *point* (a, b) meaning the point with abscissa a and ordinate b. The symbol $P(a, b)$ will denote the point P with coordinates (a, b). To **plot a point** $P(a, b)$ means to locate, in a coordinate plane, the point P with coordinates (a, b). This point is represented by a dot in the appropriate position, as illustrated in (ii) of Figure 1.8.

The next statement provides a formula for finding the distance $d(P, Q)$ between two points P and Q in a coordinate plane.

(1.7) Distance Formula

> The distance between any two points $P_1(x_1, y_1)$ and $P_2(x_2, y_2)$ in a coordinate plane is given by
>
> $$d(P_1, P_2) = \sqrt{(x_2 - x_1)^2 + (y_2 - y_1)^2}.$$

Proof. If $x_1 \neq x_2$ and $y_1 \neq y_2$, then as illustrated in Figure 1.9, the points P_1, P_2, and $P_3(x_2, y_1)$ are vertices of a right triangle. By the Pythagorean Theorem,

$$[d(P_1, P_2)]^2 = [d(P_1, P_3)]^2 + [d(P_3, P_2)]^2.$$

Using the fact that $d(P_1, P_3) = |x_2 - x_1|$ and $d(P_3, P_2) = |y_2 - y_1|$ gives us the desired formula.

If $y_1 = y_2$, the points P_1 and P_2 lie on the same horizontal line and

$$d(P_1, P_2) = |x_2 - x_1| = \sqrt{(x_2 - x_1)^2}.$$

Similarly, if $x_1 = x_2$, the points are on the same vertical line and

$$d(P_1, P_2) = |y_2 - y_1| = \sqrt{(y_2 - y_1)^2}.$$

These are special cases of the Distance Formula.

Although we referred to Figure 1.9, the argument used in this proof of the Distance Formula is independent of the positions of the points P_1 and P_2.

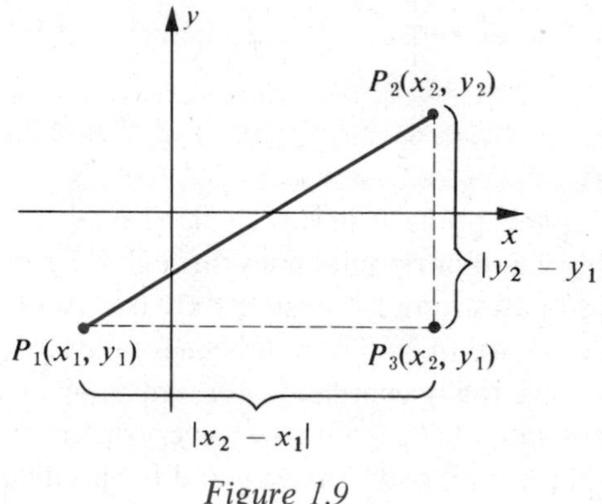

Figure 1.9

Example 1 Prove that the triangle with vertices $A(-1, -3)$, $B(6, 1)$, and $C(2, -5)$ is a right triangle and find its area.

Solution By the Distance Formula,

$$d(A, B) = \sqrt{(-1-6)^2 + (-3-1)^2} = \sqrt{49 + 16} = \sqrt{65}$$
$$d(B, C) = \sqrt{(6-2)^2 + (1+5)^2} = \sqrt{16 + 36} = \sqrt{52}$$
$$d(A, C) = \sqrt{(-1-2)^2 + (-3+5)^2} = \sqrt{9 + 4} = \sqrt{13}.$$

Hence $[d(A, B)]^2 = [d(B, C)]^2 + [d(A, C)]^2$; that is, the triangle is a right triangle with hypotenuse AB. The area is $(1/2)\sqrt{52}\sqrt{13} = 13$ square units.

It is easy to obtain a formula for the midpoint of a line segment. Let $P_1(x_1, y_1)$ and $P_2(x_2, y_2)$ be two points in a coordinate plane and let M be the midpoint of the segment P_1P_2. The lines through P_1 and P_2 parallel to the y-axis intersect the x-axis at $A_1(x_1, 0)$ and $A_2(x_2, 0)$ and, from plane geometry, the line through M

parallel to the y-axis bisects the segment A_1A_2 (see Figure 1.10). If $x_1 < x_2$, then $x_2 - x_1 > 0$, and hence $d(A_1, A_2) = x_2 - x_1$. Since M_1 is halfway from A_1 to A_2, the abscissa of M_1 is

$$\begin{aligned} x_1 + \tfrac{1}{2}(x_2 - x_1) &= x_1 + \tfrac{1}{2}x_2 - \tfrac{1}{2}x_1 \\ &= \tfrac{1}{2}x_1 + \tfrac{1}{2}x_2 \\ &= \frac{x_1 + x_2}{2}. \end{aligned}$$

It follows that the abscissa of M is also $(x_1 + x_2)/2$. It can be shown in similar fashion that the ordinate of M is $(y_1 + y_2)/2$. Moreover, these formulas hold for all positions of P_1 and P_2. This gives us the following result.

(1.8) Midpoint Formula

The midpoint of the line segment from $P_1(x_1, y_1)$ to $P_2(x_2, y_2)$ is

$$\left(\frac{x_1 + x_2}{2}, \frac{y_1 + y_2}{2}\right).$$

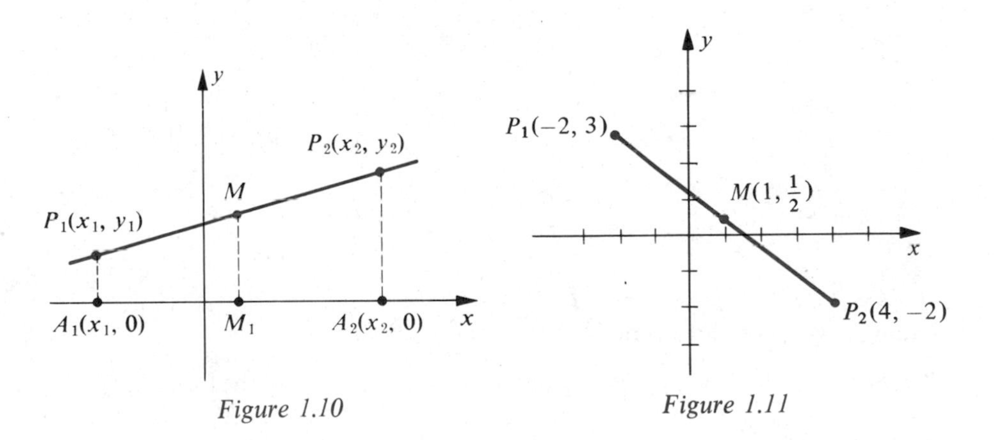

Figure 1.10

Figure 1.11

Example 2 Find the midpoint M of the line segment from $P_1(-2, 3)$ to $P_2(4, -2)$. Plot the points P_1, P_2, M and verify that $d(P_1, M) = d(P_2, M)$.

Solution Applying (1.8), the coordinates of M are

$$\left(\frac{-2+4}{2}, \frac{3+(-2)}{2}\right) \quad \text{or} \quad \left(1, \frac{1}{2}\right).$$

The three points P_1, P_2, and M are plotted in Figure 1.11. Using the Distance Formula we obtain

$$d(P_1, M) = \sqrt{(-2-1)^2 + (3-\tfrac{1}{2})^2} = \sqrt{9 + (\tfrac{25}{4})} = \sqrt{61}/2$$
$$d(P_2, M) = \sqrt{(4-1)^2 + (-2-\tfrac{1}{2})^2} = \sqrt{9 + (\tfrac{25}{4})} = \sqrt{61}/2$$

Hence $d(P_1, M) = d(P_2, M)$.

If W is a set of ordered pairs, then we may speak of the point $P(x, y)$ in a coordinate plane which corresponds to the ordered pair (x, y) in W. The **graph** of W is the set of all points which correspond to the ordered pairs in W. The phrase "sketch the graph of W" means to illustrate the significant features of the graph geometrically on a coordinate plane.

Example 3 Describe the graph of $W = \{(x, y): |x| \leq 2, |y| \leq 1\}$.

Solution From (1.4), the indicated inequalities are equivalent to $-2 \leq x \leq 2$ and $-1 \leq y \leq 1$. Hence the graph of W consists of all points within and on the boundary of the rectangular region shown in Figure 1.12.

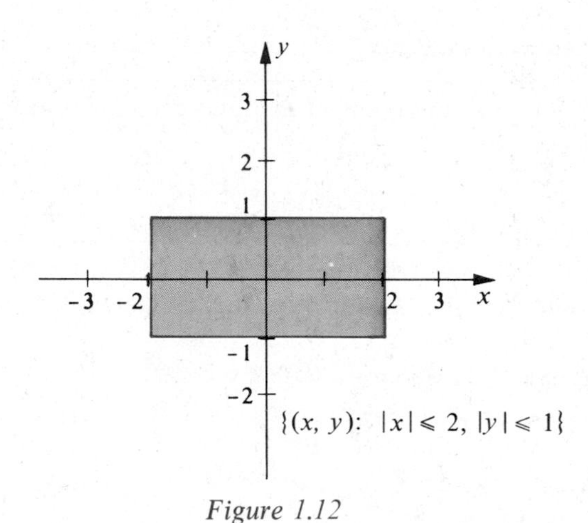

Figure 1.12

Example 4 Sketch the graph of $W = \{(x, y): y = 2x - 1\}$.

Solution We begin by finding points with coordinates of the form (x, y) where the ordered pair (x, y) is in W. It is convenient to list these coordinates in tabular form as shown below, where for each real number x the corresponding value for y is $2x - 1$.

x	-2	-1	0	1	2	3
y	-5	-3	-1	1	3	5

After plotting, it appears that the points with these coordinates all lie on a line and we sketch the graph accordingly (see Figure 1.13). Ordinarily the few points we have plotted would not be enough to illustrate the graph; however, in this elementary case we can be reasonably sure that the graph is a line. In the next section we will prove that our conjecture is correct.

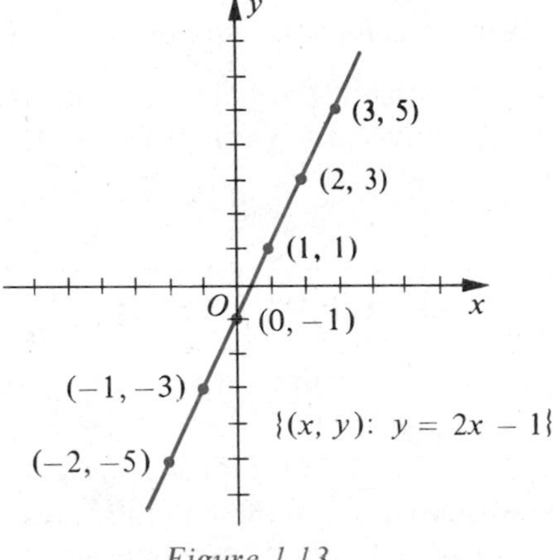

Figure 1.13

The abscissas of points at which a graph intersects the x-axis are called the **x-intercepts** of the graph. Similarly, the ordinates of points at which a graph intersects the y-axis are called the **y-intercepts**. In Example 4, there is one x-intercept $1/2$ and one y-intercept -1.

It is impossible to sketch the entire graph in Example 4 since x may be assigned values which are numerically as large as desired. Nevertheless, we often call a drawing of the type given in Figure 1.13 *the graph of W* or *a sketch of the graph*, where it is understood that the drawing is only a device for visualizing the actual graph and the line does not terminate as shown in the figure. In general, the sketch of a graph should illustrate enough of the graph so that the remaining parts are evident.

The graph in Example 4 is determined by the equation $y = 2x - 1$ in the sense that for every real number x, the equation can be used to find a number y such that (x, y) is in W. Given an equation in x and y, we say that an ordered pair (a, b) is a **solution** of the equation if equality is obtained when a is substituted for x and b for y. For example, $(2, 3)$ is a solution of $y = 2x - 1$ since substitution of 2 for x and 3 for y leads to $3 = 4 - 1$, or $3 = 3$. Two equations in x and y are said to be **equivalent** if they have exactly the same solutions. The solutions of an equation in x and y determine a set S of ordered pairs, and we define the **graph of the equation** as the graph of S. Notice that the solutions of the equation $y = 2x - 1$ are the pairs (a, b) such that $b = 2a - 1$, and hence the solutions are identical with the set W given in Example 4. Consequently the graph of the equation $y = 2x - 1$ is the same as the graph of W (see Figure 1.13).

For some of the equations we shall encounter in this chapter the technique used for sketching the graph will consist of plotting a sufficient number of points until some pattern emerges, and then sketching the graph accordingly. This is obviously a crude (and often inaccurate) way to arrive at the graph; however, it is a method often employed in elementary courses. As we progress through this text, techniques will be introduced which will enable us to sketch accurate graphs without plotting many points.

Example 5 Sketch the graph of the equation $y = x^2$.

Solution In order to obtain the graph, it is necessary to plot more points than in the previous example. Increasing successive abscissas by $\frac{1}{2}$, we obtain the following table.

x	-3	$-\frac{5}{2}$	-2	$-\frac{3}{2}$	-1	$-\frac{1}{2}$	0	$\frac{1}{2}$	1	$\frac{3}{2}$	2	$\frac{5}{2}$	3
y	9	$\frac{25}{4}$	4	$\frac{9}{4}$	1	$\frac{1}{4}$	0	$\frac{1}{4}$	1	$\frac{9}{4}$	4	$\frac{25}{4}$	9

Larger numerical values of x produce even larger values of y. For example, the points $(3, 9)$, $(4, 16)$, $(5, 25)$, and $(6, 36)$ are on the graph, as are $(-3, 9)$, $(-4, 16)$, $(-5, 25)$, and $(-6, 36)$. Plotting the points given by the table and drawing a smooth curve through these points gives us the sketch in Figure 1.14, where we have labeled only points with integer coordinates.

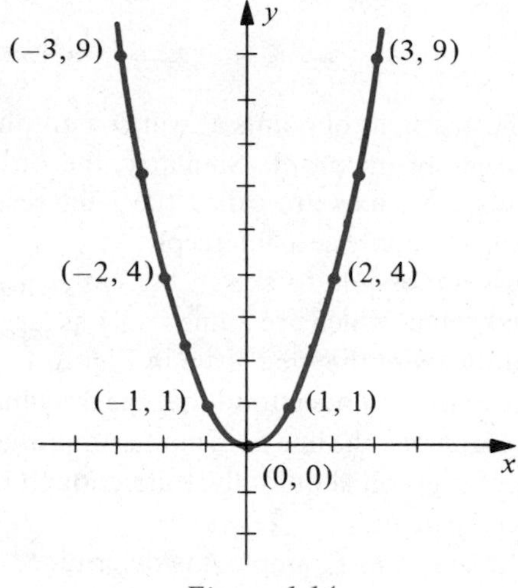

Figure 1.14

The graph in Example 5 is called a **parabola**. The lowest point $(0, 0)$ is called the **vertex** of the parabola and we say that the parabola **opens upward**. If the graph were inverted, as would be the case for $y = -x^2$, then the parabola **opens downward**. The y-axis is called the **axis of the parabola**. Parabolas and their properties will be discussed in detail in Chapter 7, where it is shown that the graph of every equation of the form $y = ax^2 + bx + c$, with $a \neq 0$, is a parabola. The next example illustrates the special case of this equation where $a = -1$, $b = 2$, and $c = 0$.

Example 6 Sketch the graph of the equation $y = 2x - x^2$.

Solution Tabulating coordinates of several points (x, y) on the graph, we obtain the table shown below.

x	-2	-1	0	1	2	3	4
y	-8	-3	0	1	0	-3	-8

The graph is sketched in Figure 1.15.

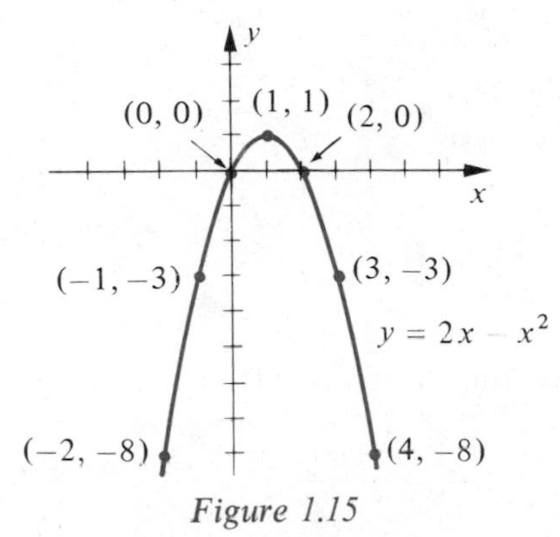

Figure 1.15

If $C(h, k)$ is a point in a coordinate plane, then a circle with center C and radius r may be defined as the collection of all points in the plane that are r units from C. As shown in (i) of Figure 1.16, a point $P(x, y)$ is on the circle if and only if $d(C, P) = r$ or, by the Distance Formula, if and only if

$$\sqrt{(x-h)^2 + (y-k)^2} = r.$$

The equivalent equation

(1.9) $$(x-h)^2 + (y-k)^2 = r^2, \quad r > 0,$$

is called the **standard equation of a circle of radius r and center $C(h, k)$**. If $h = 0$ and $k = 0$, this equation reduces to $x^2 + y^2 = r^2$, which is an equation of a circle of radius r with center at the origin (see (ii) of Figure 1.16). If $r = 1$, the graph of (1.9) is called a **unit circle**.

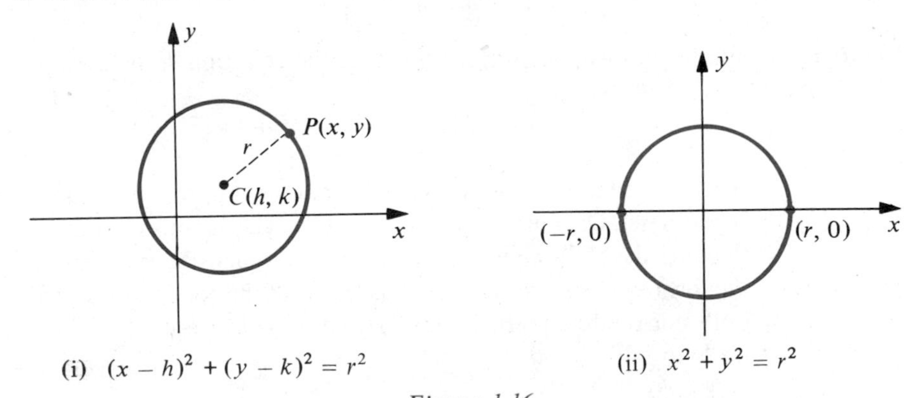

Figure 1.16

Example 7 Find an equation of the circle with center $C(-2, 3)$ which passes through the point $D(4, 5)$.

Solution Since D is on the circle, the radius r is $d(C, D)$. By the Distance Formula,

$$r = \sqrt{(-2-4)^2 + (3-5)^2} = \sqrt{36 + 4} = \sqrt{40}.$$

Applying (1.9) with $h = -2$ and $k = 3$ gives us

$$(x + 2)^2 + (y - 3)^2 = 40$$

or equivalently

$$x^2 + y^2 + 4x - 6y - 27 = 0.$$

Squaring the terms in (1.9) and simplifying leads to an equation of the form

(1.10)
$$x^2 + y^2 + ax + by + c = 0.$$

Conversely, given (1.10) we may complete the squares in x and y, to obtain an equation of the form

$$(x - h)^2 + (y - k)^2 = l$$

where h, k, l are real numbers. The method is illustrated in Example 8. If $l > 0$, then the graph is a circle with center (h, k) and radius $r = \sqrt{l}$. If $l = 0$, the only solution of the equation is (h, k) and hence the graph consists of only one point. Finally, if $l < 0$ there are no solutions and consequently there is no graph. This proves that *the graph of* (1.10), *if it exists, is either a circle or a point.*

Example 8 Find the center and radius of the circle with equation

$$x^2 + y^2 - 4x + 6y - 3 = 0.$$

Solution We begin by arranging the terms of the equation as follows:

$$(x^2 - 4x \qquad) + (y^2 + 6y \qquad) = 3.$$

Next we complete the squares by adding appropriate numbers within the parentheses. Of course, to obtain equivalent equations we must add the numbers to *both* sides of the equation. In order to complete the square for an expression of the form $x^2 + ax$, we add the square of half the coefficient of x, that is, $(a/2)^2$, to both sides of the equation. Similarly, for $y^2 + by$ we add $(b/2)^2$ to both sides. This leads to

$$(x^2 - 4x + 4) + (y^2 + 6y + 9) = 3 + 4 + 9$$

or

$$(x - 2)^2 + (y + 3)^2 = 16.$$

By (1.9) the center is $(2, -3)$ and the radius is 4.

EXERCISES 1.2

In Exercises 1–6, find (a) the distance $d(A, B)$ between the points A and B, and (b) the midpoint of the segment AB.

1 $A(6, -2)$, $B(2, 1)$

2 $A(-4, -1)$, $B(2, 3)$

3 $A(0, -7)$, $B(-1, -2)$

4 $A(4, 5)$, $B(4, -4)$

5 $A(-3, -2)$, $B(-8, -2)$

6 $A(11, -7)$, $B(-9, 0)$

In Exercises 7 and 8 prove that the triangle with vertices A, B, and C is a right triangle and find its area.

7 $A(-3, 4)$, $B(2, -1)$, $C(9, 6)$

8 $A(7, 2)$, $B(-4, 0)$, $C(4, 6)$

In each of Exercises 9–14 sketch the graph of the given set W of ordered pairs.

9 $W = \{(x, y): x = 4\}$

10 $W = \{(x, y): y = -3\}$

11 $W = \{(x, y): xy < 0\}$

12 $W = \{(x, y): xy = 0\}$

13 $W = \{(x, y): |x| < 2, |y| > 1\}$

14 $W = \{(x, y): |x| > 1, |y| \leq 2\}$

In each of Exercises 15–32 sketch the graph of the equation.

15 $y = 3x + 1$

16 $y = 4x - 3$

17 $y = -2x + 3$

18 $y = 2 - 3x$

19 $y = 2x^2 - 1$

20 $y = -x^2 + 2$

21 $4y = x^2$

22 $3y + x^2 = 0$

23 $y = -\frac{1}{2}x^3$

24 $y = \frac{1}{2}x^3$

25 $y = x^3 - 2$

26 $y = 2 - x^3$

27 $y = \sqrt{x}$

28 $y = \sqrt{x} - 1$

29 $y = \sqrt{-x}$

30 $y = \sqrt{x - 1}$

31 $x^2 + y^2 = 16$

32 $4x^2 + 4y^2 = 25$

In each of Exercises 33–40 find an equation of a circle satisfying the stated conditions.

33 Center $C(3, -2)$, radius 4

34 Center $C(-5, 2)$, radius 5

35 Center at the origin, passing through $P(-3, 5)$

36 Center $C(-4, 6)$, passing through $P(1, 2)$

37 Center $C(-4, 2)$, tangent to the x-axis

38 Center $C(3, -5)$, tangent to the y-axis

39 Endpoints of a diameter $A(4, -3)$ and $B(-2, 7)$

40 Tangent to both axes, center in the first quadrant, radius 2

In each of Exercises 41–46 find the center and radius of the circle with the given equation.

41 $x^2 + y^2 + 4x - 6y + 4 = 0$

42 $x^2 + y^2 - 10x + 2y + 22 = 0$

43 $x^2 + y^2 + 6x = 0$

44 $x^2 + y^2 + x + y - 1 = 0$

45 $2x^2 + 2y^2 - x + y - 3 = 0$

46 $9x^2 + 9y^2 - 6x + 12y - 31 = 0$

1.3 LINES

The following concept is fundamental for the study of lines. All lines referred to are considered to be in some fixed coordinate plane.

(1.11) Definition

If l is a line which is not parallel to the y-axis and if $P_1(x_1, y_1)$ and $P_2(x_2, y_2)$ are distinct points on l, then the **slope** m of l is given by

$$m = \frac{y_2 - y_1}{x_2 - x_1}.$$

If l is parallel to the y-axis, then the slope is undefined.

The numerator $y_2 - y_1$ in the formula for m measures the vertical change in direction when proceeding from P_1 to P_2 and the denominator $x_2 - x_1$ measures the amount of horizontal change when going from P_1 to P_2. In finding the slope of a line, it is immaterial which point is labeled P_1 and which is labeled P_2, since

$$\frac{y_2 - y_1}{x_2 - x_1} = \frac{y_1 - y_2}{x_1 - x_2}.$$

Consequently, we may as well assume that the points are labeled such that $x_1 < x_2$, as in Figure 1.17. In this event, $x_2 - x_1 > 0$ and hence the slope is positive, negative, or zero according as $y_2 > y_1$, $y_2 < y_1$, or $y_2 = y_1$. The slope of

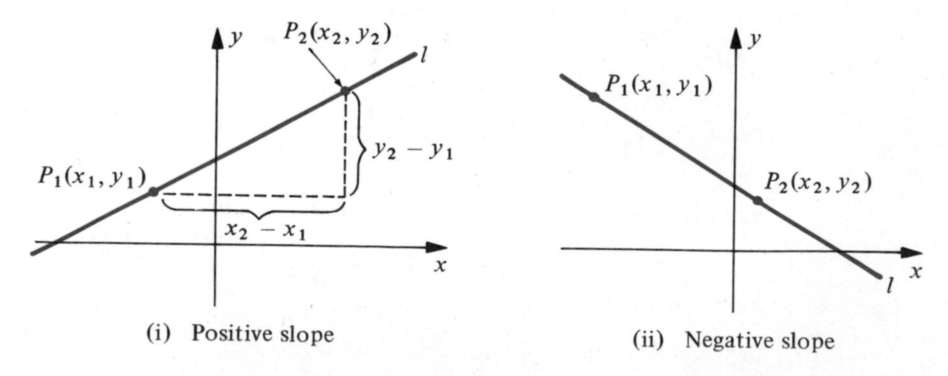

Figure 1.17

the line shown in (i) of Figure 1.17 is positive, whereas the slope of the line shown in (ii) of the figure is negative. The slope is zero if and only if the line is horizontal. If the slope is positive, then as the abscissas of points on the line increase so do the ordinates, and we say that the line *rises*. If the slope is negative, then as the abscissas increase the corresponding ordinates decrease, and we say that the line *falls*.

It is important to note that the definition of slope is independent of the two points that are chosen on l, for if other points $P_1'(x_1', y_1')$ and $P_2'(x_2', y_2')$ are used, then as in Figure 1.18, the triangle with vertices P_1', P_2', and $P_3'(x_2', y_1')$ is similar to the triangle with vertices P_1, P_2, and $P_3(x_2, y_1)$. Since the ratios of corresponding sides are equal it follows that

$$\frac{y_2 - y_1}{x_2 - x_1} = \frac{y_2' - y_1'}{x_2' - x_1'}.$$

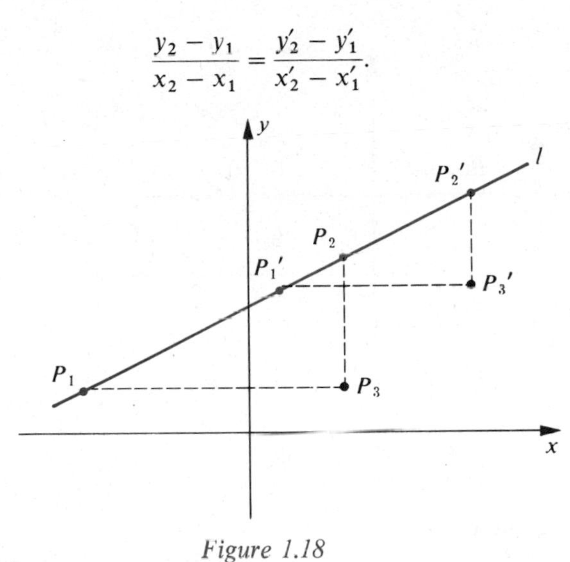

Figure 1.18

Example 1 Sketch the lines through the following pairs of points and find their slopes.

(a) $A(-1, 4)$ and $B(3, 2)$
(b) $A(2, 5)$ and $B(-2, -1)$
(c) $A(4, 3)$ and $B(-2, 3)$
(d) $A(4, -1)$ and $B(4, 4)$.

Solution The lines are sketched in Figure 1.19 on the next page. Using Definition (1.11) gives us the slopes for parts (a)–(c).

(a) $$m = \frac{2 - 4}{3 - (-1)} = \frac{-2}{4} = -\frac{1}{2}$$

(b) $$m = \frac{5 - (-1)}{2 - (-2)} = \frac{6}{4} = \frac{3}{2}$$

(c) $$m = \frac{3 - 3}{-2 - 4} = \frac{0}{-6} = 0$$

(d) The slope is undefined since the line is vertical. This is also seen by noting that if (1.11) is used, then the denominator $x_2 - x_1$ is zero.

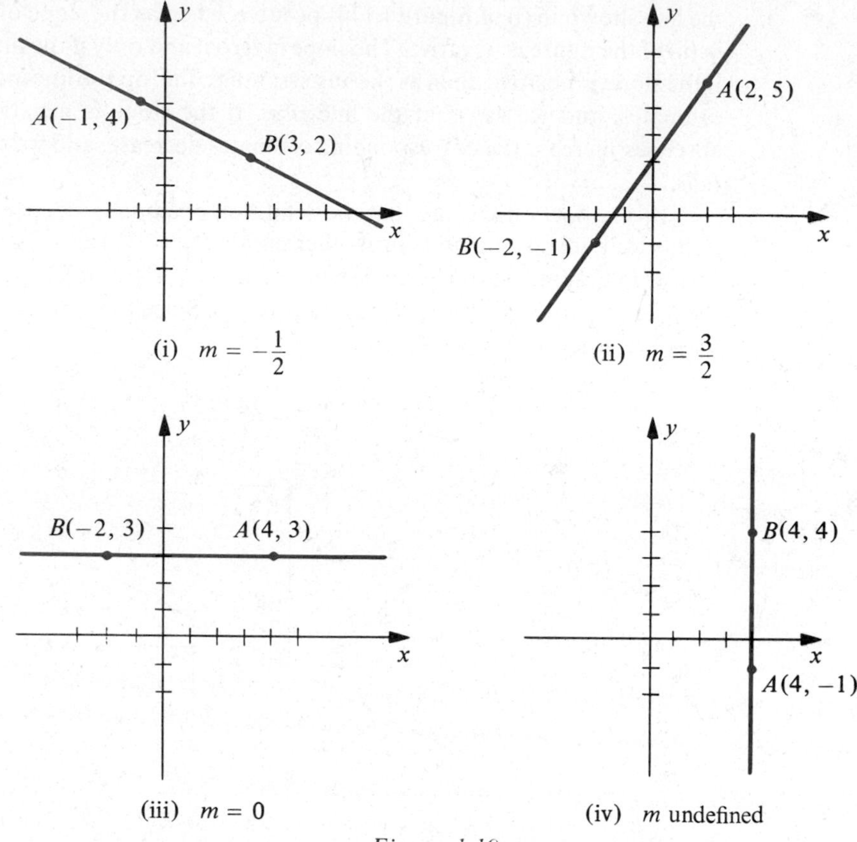

Figure 1.19

(1.12) Definition

> If l is a line which is not parallel to the x-axis and P is the point of intersection of l and the x-axis, then the **inclination** of l is the smallest angle α through which the x-axis must be rotated in a counterclockwise direction about P in order to coincide with l. If l is parallel to the x-axis, then $\alpha = 0°$.

If l is not horizontal, then $0° < \alpha < 180°$. The line shown in (i) of Figure 1.20 illustrates the case $0° < \alpha < 90°$ and that in (ii) illustrates $90° < \alpha < 180°$. Two lines are parallel if and only if they have the same inclination.

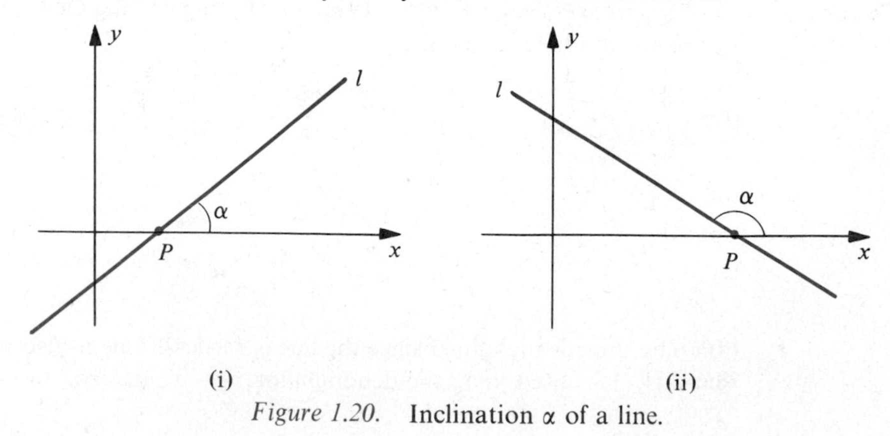

Figure 1.20. Inclination α of a line.

The next theorem exhibits the connection between the slope and inclination of a line. In the proof we shall use the definition of *tangent of an angle*. A review of this and other trigonometric concepts will be found in Appendix III.

(1.13) Theorem

> If a line has slope m and inclination α, then $m = \tan\alpha$.

Proof. If the line is horizontal, then $m = 0$, $\alpha = 0°$, and since $0 = \tan 0°$, the theorem is true.

A typical line l whose inclination α is acute is sketched in Figure 1.21. As illustrated in the figure, let l' be a line through the origin which is parallel to l, and choose any point (x, y) on l' with $y > 0$. If (x_1, y_1) and (x_2, y_2) are distinct points on l with $y_2 > y_1$, then using similar triangles and the definition of tangent of an angle, we obtain

$$m = \frac{y_2 - y_1}{x_2 - x_1} = \frac{y}{x} = \tan\alpha.$$

A similar argument can be given if α is not acute.

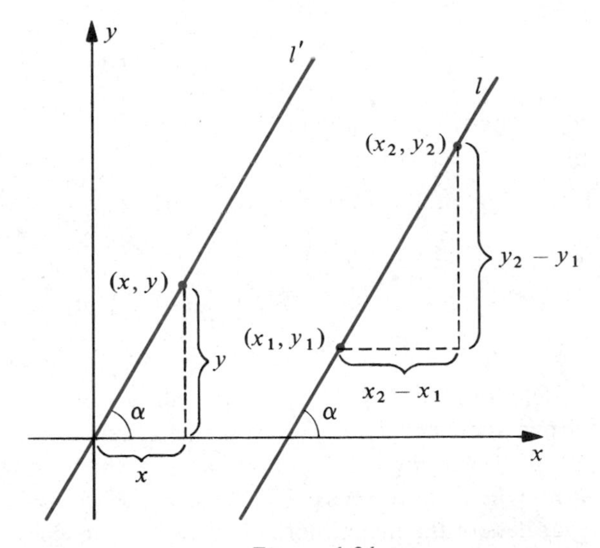

Figure 1.21

(1.14) Corollary

> Two lines with slopes m_1 and m_2 are parallel if and only if $m_1 = m_2$.

Proof. If the lines have inclinations α_1 and α_2 then they are parallel if and only if $\alpha_1 = \alpha_2$ or, equivalently, $\tan\alpha_1 = \tan\alpha_2$. The corollary now follows from Theorem (1.13).

(1.15) Theorem

Two lines with slopes m_1 and m_2 are perpendicular if and only if $m_1 m_2 = -1$.

Proof. If α_1 and α_2 denote the inclinations of the lines, then by Theorem (1.13),

$$m_1 = \tan \alpha_1 \quad \text{and} \quad m_2 = \tan \alpha_2.$$

We may assume, without loss of generality, that $\alpha_2 > \alpha_1$ as illustrated in Figure 1.22. If θ is the angle indicated in the figure, then

$$\alpha_2 = \alpha_1 + \theta, \quad \text{or} \quad \theta = \alpha_2 - \alpha_1.$$

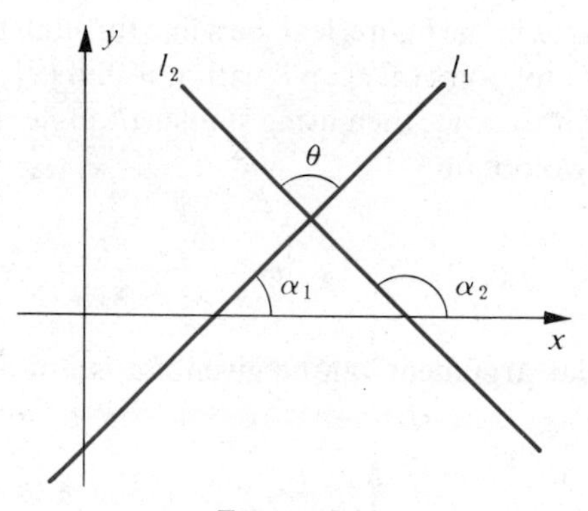

Figure 1.22

Employing a trigonometric identity,

$$\tan \theta = \tan(\alpha_2 - \alpha_1) = \frac{\tan \alpha_2 - \tan \alpha_1}{1 + \tan \alpha_1 \tan \alpha_2}$$

or

$$\tan \theta = \frac{m_2 - m_1}{1 + m_1 m_2}.$$

By definition, the lines are perpendicular if and only if $\theta = 90°$, or equivalently, $\tan \theta$ is undefined. However, by the preceding formula, $\tan \theta$ is undefined if and only if $1 + m_1 m_2 = 0$, or $m_1 m_2 = -1$, which is what we wished to prove. Although we have referred to Figure 1.22, a similar argument may be used regardless of the magnitudes of α_1 and α_2, or where the lines intersect.

Example 2 Prove that the triangle with vertices $A(-1, -3)$, $B(6, 1)$, and $C(2, -5)$ is a right triangle.

Solution If m_1 is the slope of the line through B and C, and m_2 the slope of the line through A and C, then by Definition (1.11)

$$m_1 = \frac{1 - (-5)}{6 - 2} = \frac{3}{2}; \quad m_2 = \frac{-3 - (-5)}{-1 - 2} = -\frac{2}{3}.$$

Since $m_1 m_2 = -1$, the angle at C is a right angle.

The equation $y = b$, where b is a real number, may be considered the same as the equation $0 \cdot x + y = b$ in two variables x and y. The solutions consist of all ordered pairs of the form (x, b), where x has *any* value and b is fixed. It follows that the graph of $y = b$ is a straight line parallel to the x-axis with y-intercept b. Similarly, the graph of the equation $x = a$ is a line parallel to the y-axis with x-intercept a. The graphs are sketched in Figure 1.23.

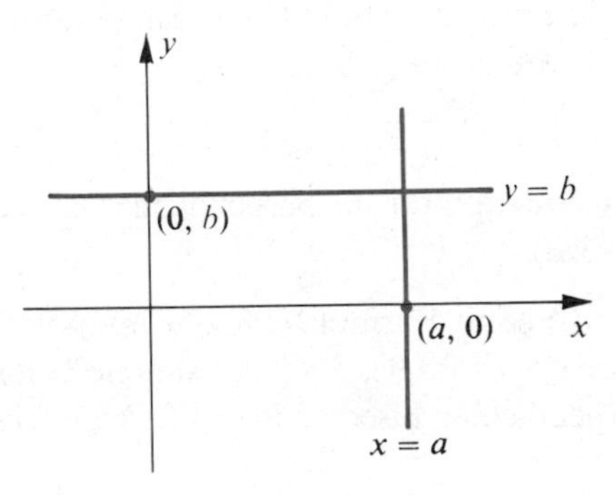

Figure 1.23

Let us next find an equation of a line l through a point $P_1(x_1, y_1)$ with slope m (only one such line exists). If $P(x, y)$ is any point with $x \neq x_1$, then P is on l if and only if the slope of the line through P_1 and P is m; that is,

$$\frac{y - y_1}{x - x_1} = m.$$

This equation may be written in the form

$$y - y_1 = m(x - x_1).$$

Note that (x_1, y_1) is also a solution of the latter equation and hence the points on l are precisely the points which correspond to the solutions. This equation for l is referred to as the **point-slope form**. Our discussion may be summarized as follows:

(1.16) Point-Slope Form for the Equation of a Line

> An equation for the line through the point $P(x_1, y_1)$ with slope m is
>
> $$y - y_1 = m(x - x_1).$$

Example 3 Find an equation of the line through the points $A(1, 7)$ and $B(-3, 2)$.

Solution By Definition (1.11) the slope m of the line is

$$m = \frac{7 - 2}{1 - (-3)} = \frac{5}{4}.$$

Using the coordinates of A in the Point-Slope Form (1.16) gives us

$$y - 7 = \frac{5}{4}(x - 1)$$

which is equivalent to

$$4y - 28 = 5x - 5 \quad \text{or} \quad 5x - 4y + 23 = 0.$$

The same equation would have been obtained if the coordinates of point B had been substituted in (1.16).

Example 4 Find an equation for the perpendicular bisector of the line segment from $A(1, 7)$ to $B(-3, 2)$.

Solution By the Midpoint Formula (1.8), the midpoint M of the segment AB is $(-1, 9/2)$. Since the slope of AB is $5/4$ (see Example 3), it follows from (1.15) that the slope of the perpendicular bisector is $-4/5$. Applying the Point-Slope Form,

$$y - \frac{9}{2} = \frac{-4}{5}(x + 1).$$

Multiplying both sides by 10 and simplifying leads to $8x + 10y - 37 = 0$.

The equation in (1.16) may be rewritten as $y = mx - mx_1 + y_1$, which is of the form

$$y = mx + b$$

where $b = -mx_1 + y_1$. The real number b is the y-intercept of the graph, as may be seen by setting $x = 0$. Since the equation $y = mx + b$ displays the slope m and y-intercept b of l, it is called the **slope-intercept form** for the equation of a line. Conversely, given an equation of that form we may change it to

$$y - b = m(x - 0).$$

Comparing with (1.16) we see that the graph is a line with slope m and passing through the point $(0, b)$. This gives us the next result.

(1.17) Slope-Intercept Form for the Equation of a Line

The graph of the equation $y = mx + b$ is a line having slope m and y-intercept b.

An equation of the form $ax + by + c = 0$, where a and b are not both zero, is called a **linear equation**. It follows from the discussion in this section that every line is the graph of a linear equation and, conversely, the graph of a linear equation is a line. For simplicity we shall often use the terminology *the line*

$ax + by + c = 0$ instead of the more accurate phrase *the line with equation* $ax + by + c = 0$.

Example 5 Sketch the graph of the equation $2x - 5y = 8$.

Solution Since the graph is a straight line, it is sufficient to find two points on the graph. Substituting $y = 0$ in the given equation we obtain the x-intercept 4, whereas substituting $x = 0$ gives us the y-intercept $-8/5$. This leads to the graph sketched in Figure 1.24.

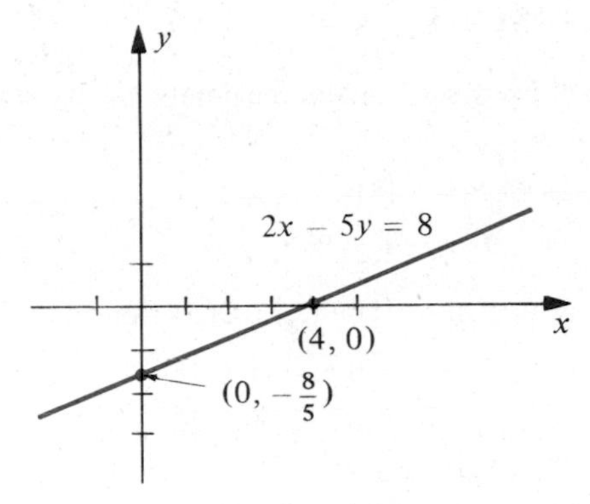

Figure 1.24

Another method of solution is to express the given equation in Slope-Intercept Form. To do this we begin by isolating the term involving y on one side of the equals sign, obtaining

$$5y = 2x - 8.$$

Next, dividing both sides by 5 gives us

$$y = \frac{2}{5}x + \left(\frac{-8}{5}\right).$$

By comparison with (1.17) we obtain the slope $m = 2/5$ and the y-intercept $b = -8/5$. We may then sketch a line through the point $(0, -8/5)$ with slope $2/5$.

Example 6 Find an equation of a line through the point $(5, -7)$ which is parallel to the line $6x + 3y - 4 = 0$.

Solution Let us express the given equation in Slope-Intercept Form. We begin by writing

$$3y = -6x + 4$$

and then divide both sides by 3, obtaining

$$y = -2x + \frac{4}{3}.$$

The last equation is in Slope-Intercept Form (1.17) with $m = -2$, and hence the slope is -2. From Corollary (1.14) the required line also has slope -2. Applying the Point-Slope Form (1.16) gives us

$$y + 7 = -2(x - 5)$$

or equivalently,

$$y + 7 = -2x + 10 \quad \text{or} \quad 2x + y - 3 = 0.$$

EXERCISES 1.3

In each of Exercises 1–4 plot the points A and B and find the slope of the line through A and B.

1 $A(-4, 6), B(-1, 18)$

2 $A(6, -2), B(-3, 5)$

3 $A(-1, -3), B(-1, 2)$

4 $A(-3, 4), B(2, 4)$

5 Show that $A(-3, 1)$, $B(5, 3)$, $C(3, 0)$, and $D(-5, -2)$ are vertices of a parallelogram.

6 Show that $A(2, 3)$, $B(5, -1)$, $C(0, -6)$, and $D(-6, 2)$ are vertices of a trapezoid.

7 Prove that the following points are vertices of a rectangle: $A(6, 15)$, $B(11, 12)$, $C(-1, -8)$, $D(-6, -5)$.

8 Prove that the points $A(1, 4)$, $B(6, -4)$, and $C(-15, -6)$ are vertices of a right triangle.

9 If three consecutive vertices of a parallelogram are $A(-1, -3)$, $B(4, 2)$, and $C(-7, 5)$, find the fourth vertex.

10 Let $A(x_1, y_1)$, $B(x_2, y_2)$, $C(x_3, y_3)$, and $D(x_4, y_4)$ denote the vertices of an arbitrary quadrilateral. Prove that the line segments joining midpoints of adjacent sides form a parallelogram.

In each of Exercises 11–24 find an equation for the line satisfying the given conditions.

11 Through $A(2, -6)$, slope $1/2$

12 Slope -3, y-intercept 5

13 Through $A(-5, -7)$, $B(3, -4)$

14 x-intercept -4, y-intercept 8

15 Through $A(8, -2)$, y-intercept -3

16 Slope 6, x-intercept -2

17 Through $A(10, -6)$, parallel to (a) the y-axis; (b) the x-axis.

18 Through $A(-5, 1)$, perpendicular to (a) the y-axis; (b) the x-axis.

19 Through $A(7, -3)$, perpendicular to the line with equation $2x - 5y = 8$.

20 Through $(-3/4, -1/2)$, parallel to the line with equation $x + 3y = 1$.

21 Given $A(3, -1)$ and $B(-2, 6)$, find an equation for the perpendicular bisector of the line segment AB.

22 Find an equation for the line which bisects the second and fourth quadrants.

23 Find equations for the altitudes of the triangle with vertices $A(-3, 2)$, $B(5, 4)$, $C(3, -8)$, and find the point at which they intersect.

24 Find equations for the medians of the triangle in Exercise 23, and find their point of intersection.

In each of Exercises 25–34 use the Slope-Intercept Form (1.17) to find the slope and y-intercept of the line with the given equation and sketch the graph of each line.

25 $3x - 4y + 8 = 0$

26 $2y - 5x = 1$

27 $x + 2y = 0$

28 $8x = 1 - 4y$

29 $y = 4$

30 $x + 2 = (1/2)y$

31 $5x + 4y = 20$

32 $y = 0$

33 $x = 3y + 7$

34 $x - y = 0$

35 Find a real number k such that the point $P(-1, 2)$ is on the line $kx + 2y - 7 = 0$.

36 Find a real number k such that the line $5x + ky - 3 = 0$ has y-intercept -5.

37 If a line l has nonzero x- and y-intercepts a and b, respectively, prove that an equation for l is

$$\frac{x}{a} + \frac{y}{b} = 1.$$

(This is called the **intercept form** for the equation of a line.) Express the equation $4x - 2y = 6$ in intercept form.

38 Prove that an equation of the line through $P_1(x_1, y_1)$ and $P_2(x_2, y_2)$ is

$$(y - y_1)(x_2 - x_1) = (y_2 - y_1)(x - x_1).$$

(This is called the **two-point form** for the equation of a line.) Use the two-point form to find an equation of the line through $A(7, -1)$ and $B(4, 6)$.

39 Find all values of r such that the slope of the line through the points $(r, 4)$ and $(1, 3 - 2r)$ is less than 5.

40 Find all values of t such that the slope of the line through $(t, 3t + 1)$ and $(1 - 2t, t)$ is greater than 4.

1.4 FUNCTIONS

One of the most useful concepts in mathematics is that of *function.* A function may be thought of as a rule, or correspondence, that associates with each element of a set X *one and only one element* of a set Y. As an illustration, let X denote the set of books in a library and Y the set of integers. If with each book we associate the number of pages in the book, a function from X to Y is obtained. Note that there may be elements of Y which are not associated with elements of X. For example, the negative integers are in this category, since a book cannot have a negative number of pages. As another illustration, let X and Y both denote the set $\mathbb{R}$ of real numbers. With each real number x let us associate its square, x^2. Thus with 3 we associate 9, with $-5/4$ we associate $25/16$, with $\sqrt{2}$ the number 2, and so on. This determines a function from $\mathbb{R}$ to $\mathbb{R}$. The next definition summarizes the preceding remarks and introduces some new terminology.

(1.18) Definition

> A **function** f from a set X to a set Y is a correspondence that associates with each element x of X a unique element y of Y. The element y is called the **image** of x under f and is denoted by $f(x)$. The set X is called the **domain** of the function. The **range** of the function consists of all images of elements of X.

The symbol $f(x)$ used for the element associated with x is read "f of x." Sometimes $f(x)$ is called the **value** of f at x.

Functions may be represented pictorially by diagrams of the type shown in Figure 1.25. The curved arrows indicate that the elements $f(x), f(w), f(z)$, and $f(a)$ of Y are associated with the elements x, w, z, and a, respectively, of X. We might imagine a whole family of arrows of this type, where each arrow connects an element of X to some specific element of Y. Although the sets X and Y have been pictured as having no elements in common, this is not required by Definition (1.18). As a matter of fact, we often take $X = Y$. It is important to note that with each x in X there is associated precisely one image $f(x)$; however, different elements such as w and z in Figure 1.25 may have the same image in Y.

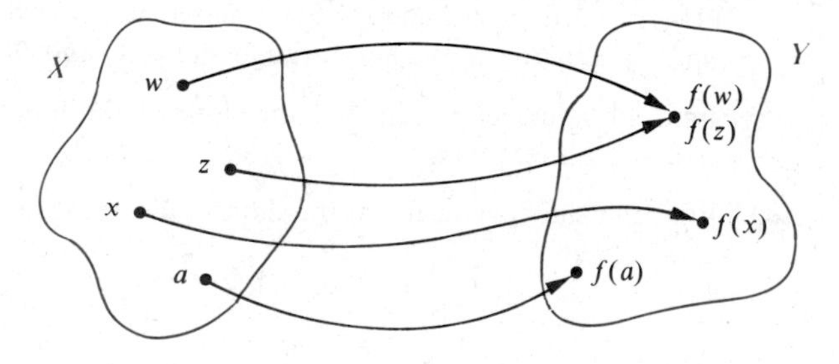

Figure 1.25

If the sets X and Y of Definition (1.18) are intervals, or other sets of real numbers, then instead of using points within regions to represent elements, as in Figure 1.25, we may use two coordinate lines l and l'. This technique is illustrated in Figure 1.26, where two images for a function f are represented graphically.

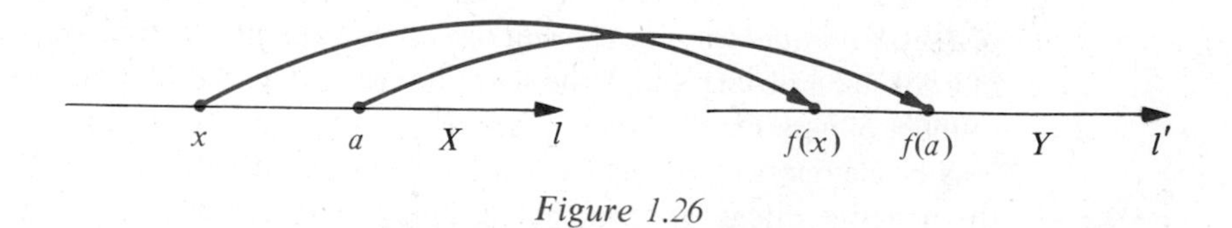

Figure 1.26

Beginning students are sometimes confused by the symbols f and $f(x)$ in Definition (1.18). Remember that f is used to represent the function. It is neither in X nor in Y. However, $f(x)$ is an element of Y, namely the element which f assigns to x.

Two functions f and g from X to Y are said to be **equal**, and we write $f = g$, provided $f(x) = g(x)$ for every x in X. For example, if $g(x) = (1/2)(2x^2 - 6) + 3$ and $f(x) = x^2$ for all x in $\mathbb{R}$, then $g = f$.

Example 1 Let f be the function with domain $\mathbb{R}$ such that $f(x) = x^2$ for every x in $\mathbb{R}$. Find $f(-6), f(\sqrt{3})$, and $f(a + b)$, where a, b are real numbers. What is the range of f?

Solution Values of f (or images under f) may be found by substituting for x in the equation $f(x) = x^2$. Thus

$$f(-6) = (-6)^2 = 36, \quad f(\sqrt{3}) = (\sqrt{3})^2 = 3$$

and

$$f(a + b) = (a + b)^2 = a^2 + 2ab + b^2.$$

If T denotes the range of f, then by Definition (1.18) T consists of all numbers of the form $f(a)$ where a is in $\mathbb{R}$. Hence T is the set of all squares a^2, where a is a real number. Since the square of any real number is nonnegative, T is contained in the set of all nonnegative real numbers. Moreover, every nonnegative real number c is an image under f, since $f(\sqrt{c}) = (\sqrt{c})^2 = c$. Hence the range of f is the set of all nonnegative real numbers.

In order to describe a function f it is necessary to specify the image $f(x)$ of *each* element x of the domain. A common method for doing this is to use an equation, as in Example 1. In this event, the symbol used for the variable is immaterial. Thus, expressions such as $f(x) = x^2, f(s) = s^2, f(t) = t^2$, and so on, all define the same function f. This is true because if a is any number in the domain of f, then the image a^2 is obtained no matter which expression is employed.

Occasionally one of the notations

$$X \xrightarrow{f} Y, \quad f: X \to Y \quad \text{or} \quad f: x \to f(x)$$

is used to signify that f is a function from X to Y. It is not unusual in this event to say ***f* maps *X* into *Y*** or ***f* maps *x* into *f*(*x*)**. If f is the function in Example 1, then f maps x into x^2 and we may write $f: x \to x^2$.

Many formulas which occur in mathematics and the sciences determine functions. As an illustration, the formula $A = \pi r^2$ for the area A of a circle of radius r associates with each positive real number r a unique value of A and hence determines a function f where $f(r) = \pi r^2$. The letter r, which represents an arbitrary number from the domain of f, is often called an **independent variable**. The letter A, which represents a number from the range of f, is called a **dependent variable**, since its value depends on the number assigned to r. When two variables r and A are related in this manner, it is customary to use the phrase *A is a function of r*. To cite another example, if an automobile travels at a uniform rate of 50 miles per hour, then the distance d (miles) traveled in time t (hours) is given by $d = 50t$ and we say that d is a function of t.

We have seen that different elements in the domain of a function may have the same image. If images are always different, then the function is called *one-to-one*.

(1.19) Definition

> A function f from X to Y is a **one-to-one function** if, whenever $a \neq b$ in X, then $f(a) \neq f(b)$ in Y.

If f is one-to-one, then each $f(x)$ in the range is the image of *precisely one* x in X. The function illustrated in Figure 1.25 is not one-to-one since two different elements w and z of X have the same image in Y. If the range of f is Y and f is one-to-one, then sets X and Y are said to be in **one-to-one correspondence**. In this case each element of Y is the image of precisely one element of X. The association between real numbers and points on a coordinate line is an example of a one-to-one correspondence.

Example 2 (a) If $f(x) = 3x + 2$, where x is real, prove that f is one-to-one.

(b) If $g(x) = x^2 + 5$, where x is real, prove that g is not one-to-one.

Solution (a) If $a \neq b$, then $3a \neq 3b$ and hence $3a + 2 \neq 3b + 2$, or $f(a) \neq f(b)$. Hence f is one-to-one by Definition (1.19).

(b) The function g is not one-to-one since different numbers in the domain may have the same image. For example, although $-1 \neq 1$, both $g(-1)$ and $g(1)$ are equal to 6.

If f is a function from X to X and if $f(x) = x$ for every x, that is, every element x maps into itself, then f is called the **identity function** on X. A function f is a **constant function** if there is some (fixed) element c such that $f(x) = c$ for every x in the domain. If a constant function is represented by a diagram of the type shown in Figure 1.25, then every arrow from X terminates at the same point in Y.

In the remainder of our work, unless specified otherwise, the phrase *f is a function* will mean that the domain and range are sets of real numbers. If a function is defined by means of some expression as in Examples 1 or 2, and the domain X is not stated explicitly, then X is considered to be the totality of real numbers for which the given expression is meaningful. To illustrate, if $f(x) = \sqrt{x}/(x - 1)$, then the domain is assumed to be the set of nonnegative real numbers different from 1. (Why?) If x is the domain we sometimes say that **f is defined at x**, or that **$f(x)$ exists**. If a set S is contained in the domain we often say that **f is defined on S**. The terminology **f is undefined at x** means that x is not in the domain of f.

The concept of ordered pair can be used to obtain an alternate approach to functions. We first observe that a function f from X to Y determines the following set W of ordered pairs:

$$W = \{(x, f(x)) : x \text{ is in } X\}.$$

Thus W is the totality of ordered pairs for which the first number is in X and the second number is the image of the first. In Example 1, where $f(x) = x^2$, W consists of all pairs of the form (x, x^2) where x is any real number. It is important to note that for each x there is exactly one ordered pair (x, y) in W having x in the first position.

Conversely, if we begin with a set W of ordered pairs such that each x in X appears exactly once in the first position of an ordered pair, and numbers from Y

appear in the second position, then W determines a function from X to Y. Specifically, for any x in X there is a unique pair (x, y) in W, and by letting y correspond to x, we obtain a function from X to Y.

It follows from the preceding discussion that the statement given below could also be used as a definition of function. We prefer, however, to think of it as an alternate approach to this concept.

(1.20) Alternate Definition of a Function

> A function with domain X is a set W of ordered pairs such that for each x in X, there is exactly one ordered pair (x, y) in W having x in the first position.

The **graph of a function** f is defined as the set of all points $(x, f(x))$ in a coordinate plane, where x is in the domain of f. Graphs are very useful for describing the behavior of $f(x)$ as x varies. The graph of f can also be described as the set of all points $P(x, y)$ such that $y = f(x)$. Thus the graph of f is the same as the graph of the equation $y = f(x)$, and if $P(x, y)$ is on the graph of f, then the ordinate y is the functional value $f(x)$. It is important to note that since there is a unique $f(x)$ for each x in the domain, there is only *one* point on the graph with abscissa x.

Example 3 Sketch the graph of f if (a) $f(x) = 2x - 1$; (b) $f(x) = x^2$; (c) $f(x) = 2x - x^2$.

Solution The graph in (a) consists of all points $(x, 2x - 1)$ and hence is identical with that considered in Example 4 of Section 2 (see Figure 1.13). Note that the graph is the same as the graph of the equation $y = 2x - 1$. Similarly, the graphs in (b) and (c) are the same as the graphs of the equations $y = x^2$ and $y = 2x - x^2$ sketched in Figures 1.14 and 1.15, respectively.

Example 4 Sketch the graph of f if $f(x) = x^3$.

Solution Coordinates $(x, f(x))$ of some points on the graph are listed in the following table.

x	-2	-1	$-\frac{1}{2}$	0	$\frac{1}{2}$	1	2
$f(x)$	-8	-1	$-\frac{1}{8}$	0	$\frac{1}{8}$	1	8

Plotting points, we find that the graph has the shape illustrated in Figure 1.27.

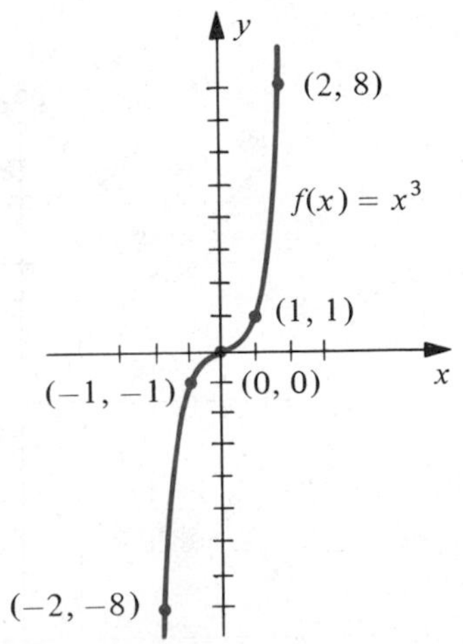

Figure 1.27

Example 5 Sketch the graph of f if $f(x) = \sqrt{x - 1}$.

Solution The domain of f does not include values of x such that $x - 1 < 0$, since $f(x)$ is not a real number in this case. Consequently there are no points (x, y) with $x < 1$ on the graph of f. The following table lists some points $(x, f(x))$ on the graph.

x	1	2	3	4	5	6
$f(x)$	0	1	$\sqrt{2}$	$\sqrt{3}$	2	$\sqrt{5}$

Plotting points leads to the sketch shown in Figure 1.28.

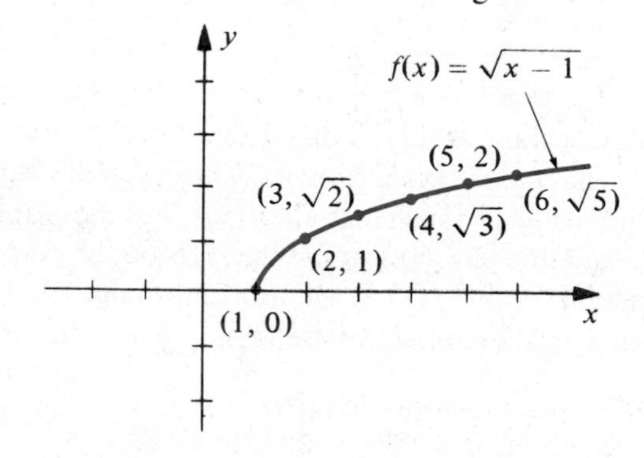

Figure 1.28

The x-intercepts of the graph of a function f are the solutions of the equation $f(x) = 0$. These numbers are called the **zeros** of the function. In the preceding example, the number 1 is a zero of f since $f(1) = 0$. In Example 4, $f(0) = 0$ and hence 0 is a zero of f.

Example 6 Sketch the graph of f if $f(x) = |x|$.

Solution If $x \geq 0$, then $f(x) = x$ and hence the part of the graph to the right of the y-axis is identical to the graph of $y = x$, which is a line through the origin with slope 1. If $x < 0$, then by (1.2), $f(x) = |x| = -x$, and hence the part of the graph to the left of the y-axis is the same as the graph of $y = -x$. The graph is sketched in Figure 1.29.

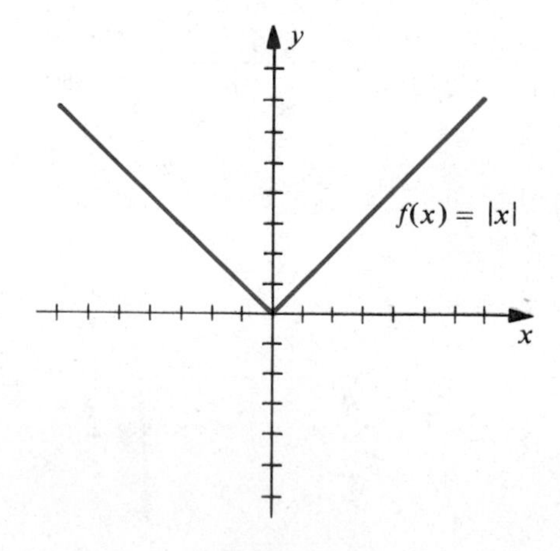

Figure 1.29

Example 7 Sketch the graph of f if $f(x) = 1/x$.

Solution The domain of f is the set of all nonzero real numbers. If x is positive, so is $f(x)$, and hence no part of the graph lies in quadrant IV. Quadrant II is also excluded, since if $x < 0$ then $f(x) < 0$. If x is close to zero, the ordinate $1/x$ is very large numerically. As x increases through positive values, $1/x$ decreases and is close to zero when x is large. Similarly, if we let x take on numerically large negative values, the ordinate $1/x$ is close to zero. Using these remarks and plotting several points gives us the sketch in Figure 1.30.

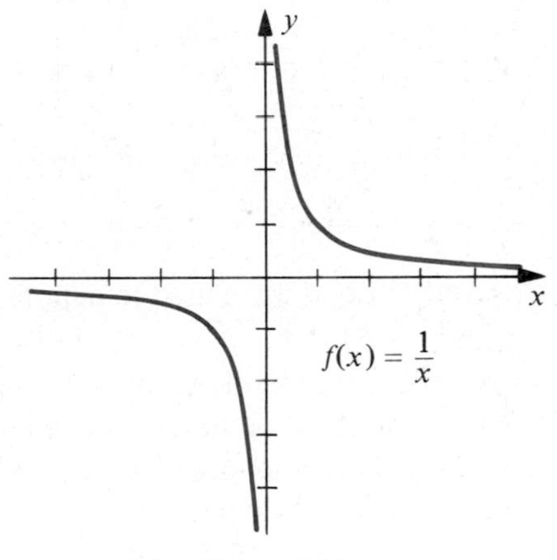

Figure 1.30

Example 8 Describe the graph of a constant function.

Solution If $f(x) = c$, where c is a real number, then the graph of f is the same as the graph of the equation $y = c$ and hence is a horizontal line with y-intercept c.

Example 9 If x is any real number, then there exist consecutive integers n and $n + 1$ such that $n \leq x < n + 1$. Let f be the function from $\mathbb{R}$ to $\mathbb{R}$ defined as follows: If $n \leq x < n + 1$, then $f(x) = n$. Sketch the graph of f.

Solution Abscissas and ordinates of points on the graph may be listed as follows:

Values of x	$f(x)$
...	...
$-2 \leq x < -1$	-2
$-1 \leq x < 0$	-1
$0 \leq x < 1$	0
$1 \leq x < 2$	1
$2 \leq x < 3$	2
...	...

Since f behaves in the same manner as a constant function when x is between integral values, the corresponding part of the graph is a segment of a horizontal line. Part of the graph is sketched in Figure 1.31 on the next page.

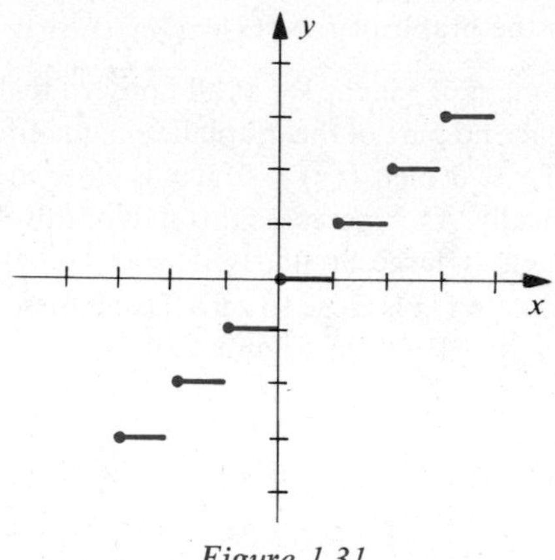

Figure 1.31

The symbol $[x]$ is often used to denote the largest integer z such that $z \leq x$. For example $[1.6] = 1$, $[\sqrt{5}] = 2$, $[\pi] = 3$, and $[-3.5] = -4$. Using this notation, the function f of Example 9 may be defined by $f(x) = [x]$. It is customary to refer to f as the **greatest integer function**.

Sometimes functions are described in terms of several expressions, as illustrated in the next example.

Example 10 Sketch the graph of the function f which is defined as follows:

$$f(x) = \begin{cases} 2x + 3 & \text{if } x < 0 \\ x^2 & \text{if } 0 \leq x < 2 \\ 1 & \text{if } x \geq 2. \end{cases}$$

Solution If $x < 0$, then $f(x) = 2x + 3$. This means that when x is negative, the expression $2x + 3$ should be used to find functional values. Consequently, if $x < 0$, then the graph of f coincides with the line $y = 2x + 3$ and we sketch that portion of the graph to the left of the y-axis as indicated in Figure 1.32.

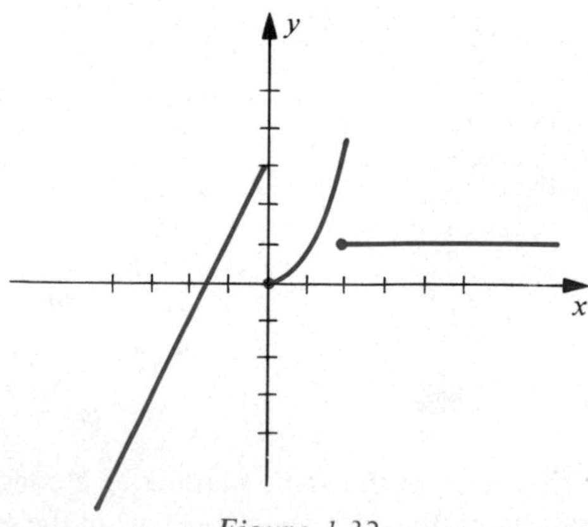

Figure 1.32

If $0 \leq x < 2$, we use x^2 to find functional values of f, and therefore this part of the graph of f coincides with the graph of the equation $y = x^2$ (see Figure 1.14). We then sketch the part of the graph of f between $x = 0$ and $x = 2$ as indicated in Figure 1.32.

Finally, if $x \geq 2$, the graph of f coincides with the constant function having values equal to 1. That part of the graph is the horizontal half-line illustrated in Figure 1.32.

EXERCISES 1.4

1 If $f(x) = x^3 + 4x - 3$, find $f(1)$, $f(-1)$, $f(0)$, and $f(\sqrt{2})$.

2 If $f(x) = \sqrt{x - 1} + 2x$, find $f(1)$, $f(3)$, $f(5)$, and $f(10)$.

In Exercises 3 and 4 find each of the following, where a, b, and h are real numbers:

(a) $f(a)$ (b) $f(-a)$

(c) $-f(a)$ (d) $f(a + h)$

(e) $f(a) + f(h)$ (f) $\dfrac{f(a + h) - f(a)}{h}$ provided $h \neq 0$

3 $f(x) = 3x^2 - x + 2$ **4** $f(x) = 1/(x^2 + 1)$

In Exercises 5 and 6 find each of the following:

(a) $g(1/a)$ (b) $1/g(a)$ (c) $g(a^2)$

(d) $[g(a)]^2$ (e) $g(\sqrt{a})$ (f) $\sqrt{g(a)}$

5 $g(x) = 1/(x^2 + 4)$ **6** $g(x) = 1/x$

In each of Exercises 7–12 find the largest subset of $\mathbb{R}$ that can serve as the domain of the function f.

7 $f(x) = \sqrt{3x - 5}$ **8** $f(x) = \sqrt{7 - 2x}$

9 $f(x) = \sqrt{4 - x^2}$ **10** $f(x) = \sqrt{x^2 - 9}$

11 $f(x) = \dfrac{x + 1}{x^3 - 9x}$ **12** $f(x) = \dfrac{4x + 7}{6x^2 + 13x - 5}$

In each of Exercises 13–18 find the number that maps into 4. If $a > 0$, what number maps into a? Find the range of f.

13 $f(x) = 7x - 5$ **14** $f(x) = 3x$

15 $f(x) = \sqrt{x - 3}$ **16** $f(x) = 1/x$

17 $f(x) = x^3$ **18** $f(x) = \sqrt[3]{x - 4}$

In each of Exercises 19–26 determine if the function f is one-to-one.

19 $f(x) = 2x + 9$ **20** $f(x) = 1/(7x + 9)$

21 $f(x) = 5 - 3x^2$ **22** $f(x) = 2x^2 - x - 3$

23 $f(x) = \sqrt{x}$ **24** $f(x) = x^3$

25 $f(x) = |x|$ **26** $f(x) = 4$

A function f with domain X is termed (i) **even** if $f(-a) = f(a)$ for every a in X, or (ii) **odd** if $f(-a) = -f(a)$ for every a in X. In each of Exercises 27–36 determine whether f is even, odd, or neither even nor odd.

27 $f(x) = 3x^3 - 4x$

28 $f(x) = 7x^4 - x^2 + 7$

29 $f(x) = 9 - 5x^2$

30 $f(x) = 2x^5 - 4x^3$

31 $f(x) = 2$

32 $f(x) = 2x^3 + x^2$

33 $f(x) = 2x^2 - 3x + 4$

34 $f(x) = \sqrt{x^2 + 1}$

35 $f(x) = \sqrt[3]{x^3 - 4}$

36 $f(x) = |x| + 5$

In Exercises 37–60 sketch the graph and determine the domain and range of f.

37 $f(x) = -4x + 3$

38 $f(x) = 4x - 3$

39 $f(x) = -3$

40 $f(x) = 3$

41 $f(x) = 4 - x^2$

42 $f(x) = -(4 + x^2)$

43 $f(x) = \sqrt{4 - x^2}$

44 $f(x) = \sqrt{x^2 - 4}$

45 $f(x) = 1/(x - 4)$

46 $f(x) = 1/(4 - x)$

47 $f(x) = 1/(x - 4)^2$

48 $f(x) = -1/(x - 4)^2$

49 $f(x) = |x - 4|$

50 $f(x) = |x| - 4$

51 $f(x) = x/|x|$

52 $f(x) = x + |x|$

53 $f(x) = \sqrt{4 - x}$

54 $f(x) = 2 - \sqrt{x}$

55 $f(x) = \begin{cases} -1 & \text{if } x < 0 \\ 1 & \text{if } x \geq 0 \end{cases}$

56 $f(x) = \begin{cases} 1 & \text{if } x \text{ is an integer} \\ 0 & \text{if } x \text{ is not an integer} \end{cases}$

57 $f(x) = \begin{cases} -5 & \text{if } x < -5 \\ x & \text{if } -5 \leq x \leq 5 \\ 5 & \text{if } x > 5 \end{cases}$

58 $f(x) = \begin{cases} -x & \text{if } x < 0 \\ 2 & \text{if } 0 \leq x < 1 \\ x^2 & \text{if } x \geq 1 \end{cases}$

59 $f(x) = \begin{cases} x^2 & \text{if } x \leq -1 \\ x^3 & \text{if } |x| < 1 \\ 2x & \text{if } x \geq 1 \end{cases}$

60 $f(x) = \begin{cases} x & \text{if } x \leq 1 \\ -x^2 & \text{if } 1 < x < 2 \\ x & \text{if } x \geq 2 \end{cases}$

61 Explain why the graph of the equation $x^2 + y^2 = 1$ is not the graph of a function.

62 (a) Define a function f whose graph is the upper half of a circle with center at the origin and radius 1.

(b) Define a function f whose graph is the lower half of a circle with center at the origin and radius 1.

1.5 COMBINATIONS OF FUNCTIONS

In calculus and its applications it is common to encounter functions that are defined in terms of sums, differences, products, and quotients of various expressions. For example, if $h(x) = x^2 + \sqrt{5x + 1}$, then we may regard $h(x)$ as a sum of values of the simpler functions f and g defined by $f(x) = x^2$ and $g(x) = \sqrt{5x + 1}$. It is natural to refer to the function h as the *sum* of f and g. More generally, if f and g are any functions and *D is the intersection of their domains*, then the **sum** of f and g is the function s defined by

$$s(x) = f(x) + g(x)$$

where x is in D.

It is convenient to denote s by the symbol $f + g$. Since f and g are functions, not numbers, the $+$ used between f and g is not to be considered as addition of real numbers. It is used to indicate that the image of x under $f + g$ is $f(x) + g(x)$, that is,

$$(f + g)(x) = f(x) + g(x).$$

Similarly, the **difference** $f - g$ and the **product** fg of f and g are defined by

$$(f - g)(x) = f(x) - g(x) \quad \text{and} \quad (fg)(x) = f(x)g(x)$$

where x is in D. Finally, the **quotient** f/g of f by g is given by

$$\left(\frac{f}{g}\right)(x) = \frac{f(x)}{g(x)}$$

where x is in D and $g(x) \neq 0$.

Example 1 If $f(x) = \sqrt{4 - x^2}$ and $g(x) = 3x + 1$, find the sum, difference, and product of f and g, and the quotient of f by g.

Solution The domain of f is the closed interval $[-2, 2]$ and the domain of g is $\mathbb{R}$. Consequently, the intersection of their domains is $[-2, 2]$ and the required functions are given by

$$\begin{aligned}
(f + g)(x) &= \sqrt{4 - x^2} + (3x + 1), && -2 \le x \le 2 \\
(f - g)(x) &= \sqrt{4 - x^2} - (3x + 1), && -2 \le x \le 2 \\
(fg)(x) &= \sqrt{4 - x^2}(3x + 1), && -2 \le x \le 2 \\
(f/g)(x) &= \sqrt{4 - x^2}/(3x + 1), && -2 \le x \le 2, x \neq -1/3.
\end{aligned}$$

If g is a constant function such that $g(x) = c$ for every x, and if f is any function, then cf will denote the product of g and f; that is, $(cf)(x) = cf(x)$ for all x in the domain of f. To illustrate, if f is the function of Example 1, then we have $(cf)(x) = c\sqrt{4 - x^2}$, $-2 \le x \le 2$.

A function f is called a **polynomial function** if

$$f(x) = a_n x^n + a_{n-1} x^{n-1} + \cdots + a_1 x + a_0 \tag{1.21}$$

for every x, where the coefficients $a_0, a_1, \ldots, a_n$ are real numbers and the exponents are nonnegative integers. The expression on the right in (1.21) is called a **polynomial in x** (with real coefficients) and each $a_k x^k$ is called a **term** of the polynomial. We often use the phrase *the polynomial* $f(x)$ when referring to expressions of this type. If $a_n \neq 0$, then a_n is called the **leading coefficient** of $f(x)$ and we say that f (or $f(x)$) has **degree n**.

If a polynomial function f has degree 0, then $f(x) = c$, where $c \neq 0$, and hence f is a constant function. If a coefficient a_i is zero we often abbreviate (1.21) by deleting the term $a_i x^i$. If *all* the coefficients of a polynomial are zero it is called the **zero polynomial** and is denoted by 0. It is customary not to assign a degree to the zero polynomial.

If $f(x)$ is a polynomial of degree 1, then $f(x) = ax + b$, where $a \neq 0$. From Section 3, the graph of f is a straight line and, accordingly, f is called a **linear function**.

Any polynomial $f(x)$ of degree 2 may be written

$$f(x) = ax^2 + bx + c,$$

where $a \neq 0$. In this case f is called a **quadratic function**. The graph of f or, equivalently, of the equation $y = ax^2 + bx + c$, is a *parabola*.

A **rational function** is a quotient of two polynomial functions. Thus q is rational if, for every x in its domain,

$$q(x) = \frac{f(x)}{h(x)}$$

where $f(x)$ and $h(x)$ are polynomials. The domain of a polynomial function is $\mathbb{R}$, whereas the domain of a rational function consists of all real numbers except the zeros of the polynomial in the denominator.

A function f is called **algebraic** if it can be expressed in terms of sums, differences, products, quotients, or roots of polynomial functions. For example, if

$$f(x) = 5x^4 - 2\sqrt[3]{x} + \frac{x(x^2 + 5)}{\sqrt{x^3 + \sqrt{x}}}$$

then f is an algebraic function. Functions which are not algebraic are termed **transcendental**. The trigonometric, exponential, and logarithmic functions considered later in this book are examples of transcendental functions.

We shall conclude this section by describing an important method of using two functions f and g to obtain a third function. Suppose X, Y, and Z are sets of real numbers and let f be a function from X to Y and g a function from Y to Z. In terms of the arrow notation we have

$$X \xrightarrow{f} Y \xrightarrow{g} Z$$

that is, f maps X into Y and g maps Y into Z. A function from X to Z may be defined in a natural way. For every x in X, the number $f(x)$ is in Y. Since the domain of g is Y, we may then find the image of $f(x)$ under g. Of course, this element of Z is written as $g(f(x))$. By associating $g(f(x))$ with x, we obtain a function from X to Z called the **composite function** of g by f. This is illustrated geometrically in Figure 1.33 where we have represented the domain X of f on a coordinate line l, the domain Y of g (and the range of f) on a coordinate line l', and the range of g on a coordinate line l''. The dashes indicate the correspondence defined from X to Z. We sometimes use an operation symbol $\circ$ and denote the latter function $g \circ f$. The next definition summarizes this discussion.

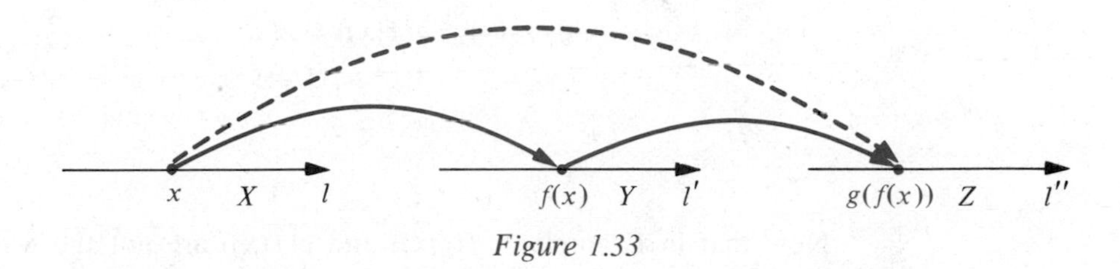

Figure 1.33

(1.22) Definition

> If f is a function from X to Y and g is a function from Y to Z, then the **composite function** $g \circ f$ is the function from X to Z defined by
>
> $$(g \circ f)(x) = g(f(x)),$$
>
> for every x in X.

Actually, it is not essential that the domain of g be all of Y but merely that it *contain* the range of f. In certain cases we may wish to restrict x to some subset of X so that $f(x)$ is in the domain of g. This is illustrated in the next example.

Example 2 If $f(x) = x - 2$ and $g(x) = 5x + \sqrt{x}$, find $(g \circ f)(x)$.

Solution Using the definitions of $g \circ f, f$, and g,

$$\begin{aligned}(g \circ f)(x) &= g(f(x)) = g(x - 2)\\ &= 5(x - 2) + \sqrt{x - 2}\\ &= 5x - 10 + \sqrt{x - 2}.\end{aligned}$$

The domain X of f is the set of all real numbers; however, the last equality implies that $(g \circ f)(x)$ is a real number only if $x \geq 2$. Thus, when working with the composite function $g \circ f$ it is necessary to restrict x to the interval $[2, \infty)$.

If $X = Y = Z$ in Definition (1.22), then it is possible to find $f(g(x))$. In this case we first obtain the image of x under g and then apply f to $g(x)$. This gives us a composite function from Z to X denoted by $f \circ g$. Thus, by definition,

$$(f \circ g)(x) = f(g(x)),$$

for all x in Z.

Example 3 If $f(x) = x^2 - 1$ and $g(x) = 3x + 5$, find $(f \circ g)(x)$ and $(g \circ f)(x)$.

Solution By the definitions of $f \circ g$, g, and f,

$$\begin{aligned}(f \circ g)(x) &= f(g(x)) = f(3x + 5)\\ &= (3x + 5)^2 - 1\\ &= 9x^2 + 30x + 24.\end{aligned}$$

Similarly,

$$\begin{aligned}(g \circ f)(x) &= g(f(x)) = g(x^2 - 1)\\ &= 3(x^2 - 1) + 5\\ &= 3x^2 + 2.\end{aligned}$$

Note that in Example 3, $f(g(x))$ and $g(f(x))$ are not the same; that is, $f \circ g \neq g \circ f$.

EXERCISES 1.5

In each of Exercises 1–6, find the sum, difference, and product of f and g, and the quotient of f by g.

1 $f(x) = 3x^2, g(x) = 1/(2x - 3)$

2 $f(x) = \sqrt{x + 3}, g(x) = \sqrt{x + 3}$

3 $f(x) = x + 1/x, g(x) = x - 1/x$

4 $f(x) = x^3 + 3x, g(x) = 3x^2 + 1$

5 $f(x) = 2x^3 - x + 5, g(x) = x^2 + x + 2$

6 $f(x) = 7x^4 + x^2 - 1, g(x) = 7x^4 - x^3 + 4x$

In Exercises 7–20 find $(f \circ g)(x)$ and $(g \circ f)(x)$.

7 $f(x) = 2x^2 + 5, g(x) = 4 - 7x$

8 $f(x) = 1/(3x + 1), g(x) = 2/x^2$

9 $f(x) = x^3, g(x) = x + 1$

10 $f(x) = \sqrt{x^2 + 4}, g(x) = 7x^2 + 1$

11 $f(x) = 3x^2 + 2, g(x) = 1/(3x^2 + 2)$

12 $f(x) = 7, g(x) = 4$

13 $f(x) = \sqrt{2x + 1}, g(x) = x^2 + 3$

14 $f(x) = 6x - 12, g(x) = \frac{1}{6}x + 2$

15 $f(x) = |x|, g(x) = -5$

16 $f(x) = \sqrt[3]{x^2 + 1}, g(x) = x^3 + 1$

17 $f(x) = x^2, g(x) = 1/x^2$

18 $f(x) = \dfrac{1}{x + 1}, g(x) = x + 1$

19 $f(x) = 2x - 3, g(x) = \dfrac{x + 3}{2}$

20 $f(x) = x^3 - 1, g(x) = \sqrt[3]{x + 1}$

21 If $f(x)$ and $g(x)$ are polynomials of degree 5 in x, does it follow that $f(x) + g(x)$ has degree 5? Explain.

22 Prove that the degree of the product of two nonzero polynomials equals the sum of the degrees of the polynomials.

23 Using the terminology of Exercise 27 in Section 4, prove that (a) the product of two odd functions is even; (b) the product of two even functions is even; and (c) the product of an even function and an odd function is odd.

24 Which parts of Exercise 23 are true if the word "product" is replaced by "sum"?

25 Prove that every function with domain $\mathbb{R}$ can be written as the sum of an even function and an odd function.

26 Show that there exist an infinite number of rational functions f and g such that $f + g = fg$.

1.6 INVERSE FUNCTIONS

In Example 3 of the previous section we considered two functions f and g such that $f(g(x))$ and $g(f(x))$ are not always the same, that is, $f \circ g \neq g \circ f$. In certain cases it may happen that equality *does* occur. Of major importance is the case in which $f(g(x))$ and $g(f(x))$ are not only identical, but both are equal to x. Needless to say, f and g must be very special functions in order for this to happen. In the following discussion we indicate the manner in which they will be restricted.

Suppose f is a *one-to-one function* with domain X and range Y (see Definition (1.19)). This implies that each element y of Y is the image of precisely one element x of X. Another way of phrasing this is to say that *each element y of Y can be written in one and only one way in the form $f(x)$*, where x is in X. We may then define a function g from Y to X by demanding that

$$g(y) = g(f(x)) = x \quad \text{for every } x \text{ in } X.$$

This amounts to *reversing* the correspondence given by f. If f is represented geometrically by drawing arrows as in (i) of Figure 1.34, then g can be represented by simply *reversing* these arrows as illustrated in (ii) of the figure. It follows that g is a one-to-one function with domain Y and range X. (See Exercise 16.)

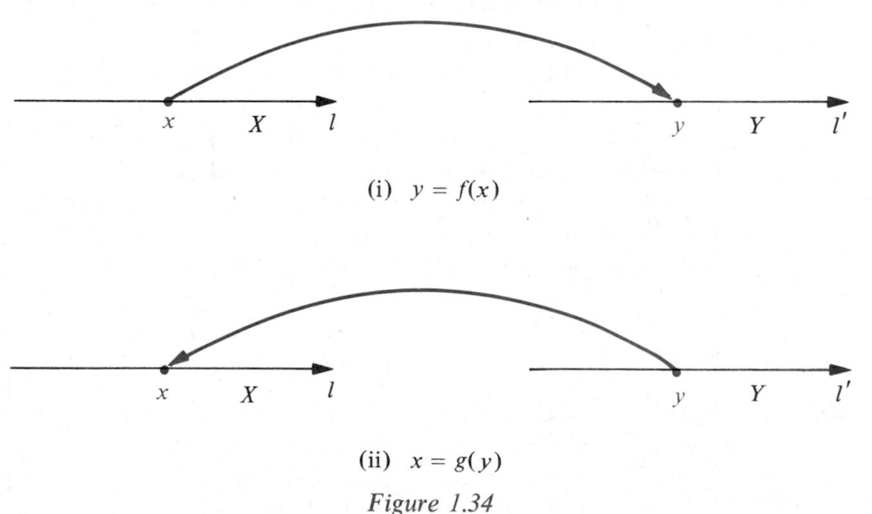

Figure 1.34

As illustrated in Figure 1.34, if $f(x) = y$, then $x = g(y)$. This means that

$$f(g(y)) = y \quad \text{for every } y \text{ in } Y.$$

Since the notation used for the variable is immaterial, we may write

$$f(g(x)) = x \quad \text{for every } x \text{ in } Y.$$

The functions f and g are called *inverse functions* of one another, according to the following definition.

(1.23) Definition

If f is a one-to-one function with domain X and range Y, then a function g with domain Y and range X is called the **inverse function of f** if

$$f(g(x)) = x \quad \text{for every } x \text{ in } Y \text{ and}$$
$$g(f(x)) = x \quad \text{for every } x \text{ in } X.$$

There can be only one inverse function of f (see Exercise 18). Moreover, if g is the inverse function of f, then by Definition (1.23), $(g \circ f)(x) = x$ for every x in X, that is, $g \circ f$ is the identity function on X. Similarly, since $(f \circ g)(x) = x$, for all x in Y, $f \circ g$ is the identity function on Y. For this reason the symbol f^{-1} is often used to denote the inverse function of f. Employing this notation,

$$f^{-1}(f(x)) = x \quad \text{for every } x \text{ in } X$$
$$f(f^{-1}(x)) = x \quad \text{for every } x \text{ in } Y. \qquad (1.24)$$

The symbol -1 in (1.24) should not be mistaken for an exponent; that is, $f^{-1}(x)$ does not mean $1/f(x)$. The reciprocal $1/f(x)$ may be denoted by $[f(x)]^{-1}$.

Inverse functions are very important in the study of trigonometry. In Chapter 8 we shall discuss two other important classes of inverse functions. It is important to remember that in order to define the inverse of a function f *it is absolutely essential that f be one-to-one.*

An algebraic method can sometimes be used to find the inverse of a one-to-one function f with domain X and range Y. Given any x in X, its image y in Y may be found by means of the equation $y = f(x)$. In order to determine the inverse function f^{-1} we wish to *reverse* this procedure, in the sense that given y, the element x may be found. Since x and y are related by means of $y = f(x)$, it follows that if the latter equation can be solved for x in terms of y, we may arrive at the inverse function f^{-1}. This technique is illustrated in the following examples.

Example 1 If $f(x) = 3x - 5$ for every real number x, find the inverse function of f.

Solution It is not difficult to show that f is a one-to-one function with domain and range $\mathbb{R}$, and hence the inverse function g exists. If we let

$$y = 3x - 5$$

and then solve for x in terms of y, we get

$$x = \frac{y+5}{3}.$$

The last equation enables us to find x when given y. Letting

$$g(y) = \frac{y+5}{3}$$

gives us a function g from Y to X that reverses the correspondence determined by f. Since the symbol used for the independent variable is immaterial, we may replace y by x in the expression for g, obtaining

$$g(x) = \frac{x+5}{3}.$$

To verify that g is actually the inverse function of f, we must verify that the two conditions stated in Definition (1.23) are fulfilled. Thus

$$f(g(x)) = f\left(\frac{x+5}{3}\right) = 3\left(\frac{x+5}{3}\right) - 5 = x.$$

Similarly,

$$g(f(x)) = g(3x-5) = \frac{(3x-5)+5}{3} = x.$$

This proves that g is the inverse function of f. Using the notation of (1.24),

$$f^{-1}(x) = \frac{x+5}{3}.$$

Example 2 Find the inverse function of f if the domain X is the interval $[0, \infty)$ and $f(x) = x^2 - 3$ for all x in X.

Solution The domain has been restricted so that f is one-to-one. The range of f is the interval $[-3, \infty)$. As in Example 1 we begin by considering the equation

$$y = x^2 - 3.$$

Solving for x gives us

$$x = \pm\sqrt{y+3}.$$

Since x is nonnegative we reject $x = -\sqrt{y+3}$ and, as in the preceding example, we let

$$g(y) = \sqrt{y+3}$$

or equivalently,

$$g(x) = \sqrt{x+3}.$$

We now check the two conditions in Definition (1.23), obtaining

$$f(g(x)) = f(\sqrt{x+3}) = (\sqrt{x+3})^2 - 3 = (x+3) - 3 = x$$

and

$$g(f(x)) = g(x^2 - 3) = \sqrt{(x^2 - 3) + 3} = x.$$

This proves that

$$f^{-1}(x) = \sqrt{x+3}, \quad \text{where } x \geq -3.$$

There is an interesting relationship between the graphs of a function f and its inverse function f^{-1}. We first note that f maps a into b if and only if f^{-1} maps b into a; that is, $b = f(a)$ means the same thing as $a = f^{-1}(b)$. These equations imply that the point (a, b) is on the graph of f if and only if the point (b, a) is on the graph of f^{-1}. As an illustration, in Example 2 we found that the functions f and f^{-1} given by

$$f(x) = x^2 - 3 \quad \text{and} \quad f^{-1}(x) = \sqrt{x+3}$$

are inverse functions of one another, provided x is suitably restricted. Some points on the graph of f are $(0, -3)$, $(1, -2)$, $(2, 1)$, and $(3, 6)$. Corresponding points on the graph of f^{-1} are $(-3, 0)$, $(-2, 1)$, $(1, 2)$, and $(6, 3)$. The graphs of f and f^{-1} are sketched on the same coordinate axes in Figure 1.35. If the page is folded along the line l which bisects quadrants I and III (as indicated by the dashes in the figure), then the graphs of f and f^{-1} coincide. The two graphs are said to be *reflections* of one another through the line with equation $y = x$. This is typical of the graphs of all functions f which have inverse functions f^{-1}.

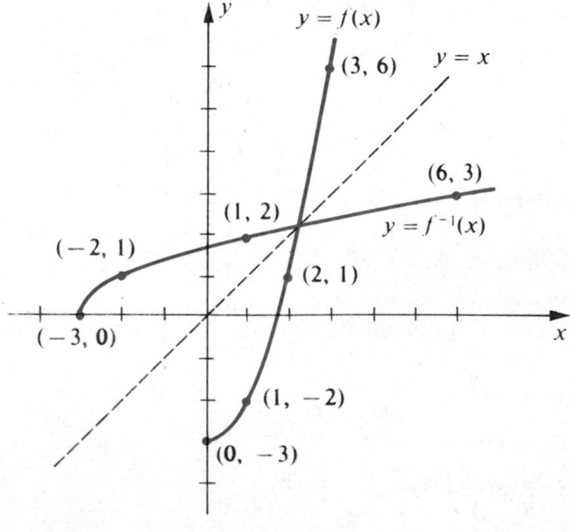

Figure 1.35

EXERCISES 1.6

In Exercises 1–4 prove that f and g are inverse functions of one another.

1 $f(x) = 9x + 2, g(x) = (1/9)x - 2/9$

2 $f(x) = x^3 + 1, g(x) = \sqrt[3]{x - 1}$

3 $f(x) = \sqrt{2x + 1}, x \geq -1/2;\ g(x) = (1/2)x^2 - 1/2, x \geq 0$

4 $f(x) = 1/(x - 1), x > 1;\ g(x) = (1 + x)/x, x > 0$

In Exercises 5–14 find the inverse function of f.

5 $f(x) = 8 + 11x$

6 $f(x) = 1/(8 + 11x), x > -8/11$

7 $f(x) = 6 - x^2, 0 \leq x \leq \sqrt{6}$

8 $f(x) = 2x^3 - 5$

9 $f(x) = \sqrt{7x - 2}, x \geq 2/7$

10 $f(x) = \sqrt{1 - 4x^2}, 0 \leq x \leq 1/2$

11 $f(x) = 7 - 3x^3$

12 $f(x) = x$

13 $f(x) = (x^3 + 8)^5$

14 $f(x) = x^{1/3} + 2$

15 Sketch the graphs of f and g given in Exercise 1 on the same coordinate plane. Do the same for the functions defined in Exercise 3.

16 If f is a one-to-one function with domain X and range Y, prove that f^{-1} is a one-to-one function with domain Y and range X.

17 Prove that the linear function f defined by $f(x) = ax + b$ has an inverse function if $a \neq 0$. Does a constant function have an inverse? Does the identity function have an inverse?

18 Prove that a function f can have at most one inverse function.

19 Prove that not every polynomial function has an inverse function.

20 If f has an inverse function f^{-1}, prove that f^{-1} has an inverse function and that $(f^{-1})^{-1} = f$.

1.7 REVIEW

Concepts

Define or discuss each of the following.

1 Rational and irrational numbers

2 A real number a is greater than a real number b

3 Coordinate line

4 Intervals (open, closed, half-open, infinite)

5 Absolute value of a real number

6 Ordered pair

7 Rectangular coordinate system in a plane

8 The abscissa and ordinate of a point

9 The Distance Formula

10 The Midpoint Formula

11 The graph of an equation in x and y

12 The standard equation of a circle

13 The slope of a line

14 The inclination of a line

15 The Point-Slope Form

16 The Slope-Intercept Form

17 Function

18 The domain and range of a function

19 One-to-one function

20 Constant function

21 The graph of a function

22 The sum, difference, product, and quotient of two functions

23 Polynomial function

24 Rational function

25 The composite function of two functions

26 The inverse of a function

Exercises

Solve the inequalities in Exercises 1–8 and express the solutions ir [illegible] terval

1 $4 - 3x > 7 + 2x$

2 $\frac{7}{2} > \frac{1-4x}{5} > \frac{3}{2}$

3 $|2x - 7| \leq 0.01$

4 $|6x - 7| > 1$

5 $2x^2 < 5x - 3$

6 $\frac{2x^2 - 3x - 20}{x + 3}$ [illegible]

7 $\frac{1}{3x - 1} < \frac{2}{x + 5}$

8 $x^2 + 4 \geq 4x$

9 Given the points $A(2, 1)$, $B(-1, 4)$, and $C(-2, -3)$,

(a) prove that A, B, and C are vertices of a right triangle and find its area.
(b) find the coordinates of the midpoint of AB.
(c) find the slope of the line through B and C.

Sketch the graphs of the equations in Exercises 10–13.

10 $3x - 5y = 10$

11 $x^2 + y = 4$

12 $x = y^3$

13 $|x + y| = 1$

In Exercises 14–17 sketch the graph of the set W.

14 $W = \{(x, y): x > 0\}$

15 $W = \{(x, y): y > x\}$

16 $W = \{(x, y): x^2 + y^2 < 1\}$

17 $W = \{(x, y): |x - 4| < 1, |y + 3| < 2\}$

In each of Exercises 18–20 find an equation of the circle satisfying the given conditions.

18 Center $C(4, -7)$ and passing through the origin

19 Center $C(-4, -3)$ and tangent to the line with equation $x = 5$

20 Passing through the points $A(-2, 3)$, $B(4, 3)$, and $C(-2, -1)$

21 Find the center and radius of the circle which has equation

$$x^2 + y^2 - 10x + 14y - 7 = 0.$$

Given the points $A(-4, 2)$, $B(3, 6)$, and $C(2, -5)$, solve the problems stated in Exercises 22–26.

22 Find an equation for the line through B which is parallel to the line through A and C.

23 Find an equation for the line through B which is perpendicular to the line through A and C.

24 Find an equation for the line through C and the midpoint of the line segment AB.

25 Find an equation for the line through A which is parallel to the y-axis.

26 Find an equation for the line through C which is perpendicular to the line with equation $3x - 10y + 7 = 0$.

In Exercises 27–30 find the largest subset of $\mathbb{R}$ that can serve as the domain of f.

27 $f(x) = \dfrac{2x - 3}{x^2 - x}$

28 $f(x) = \dfrac{x}{\sqrt{16 - x^2}}$

29 $f(x) = \dfrac{1}{\sqrt{x - 5}\sqrt{7 - x}}$

30 $f(x) = \dfrac{1}{\sqrt{x(x - 2)}}$

31 If $f(x) = 1/\sqrt{x + 1}$ find each of the following.

(a) $f(1)$ (b) $f(3)$ (c) $f(0)$ (d) $f(\sqrt{2} - 1)$
(e) $f(-x)$ (f) $-f(x)$ (g) $f(x^2)$ (h) $(f(x))^2$

In each of exercises 32–35, sketch the graph of f.

32 $f(x) = 1 - 4x^2$

33 $f(x) = 100$

34 $f(x) = -1/(x + 1)$

35 $f(x) = |x + 5|$

In Exercises 36–38 find $(f + g)(x)$, $(f - g)(x)$, $(fg)(x)$, $(f/g)(x)$, $(f \circ g)(x)$, and $(g \circ f)(x)$.

36 $f(x) = x^2 + 3x + 1$, $g(x) = 2x - 1$

37 $f(x) = x^2 + 4$, $g(x) = \sqrt{2x + 5}$

38 $f(x) = 5x + 2$, $g(x) = 1/x^2$

In Exercises 39 and 40, prove that f is one-to-one and find the inverse function of f.

39 $f(x) = 5 - 7x$

40 $f(x) = 4x^2 + 3, x \geq 0$

Limits and Continuity of Functions

The concept of limit of a function *is one of the fundamental ideas that distinguishes calculus from areas of mathematics such as algebra or geometry. The student should be warned that the notion of limit is not easily mastered. Indeed, it is usually necessary for the beginner to study the definition many times, looking at it from various points of view, before the meaning becomes clear. In spite of the complexity of the definition, it is easy to develop an intuitive feeling for limits. With this in mind, the discussion in the first section is not rigorous. The mathematically precise description of limit of a function will be presented in Section 2. The remainder of the chapter contains important theorems and further concepts pertaining to limits.*

2.1 INTRODUCTION

In calculus and its applications we are often interested in the values $f(x)$ of a function f when x is *very close* to a number a, *but not necessarily equal to* a. As a matter of fact, in many instances the number a is not in the domain of f; that is, $f(a)$ is undefined. Roughly speaking, we ask the following question: As x gets closer and closer to a (but $x \neq a$), does $f(x)$ get closer and closer to some number L? If the answer is *yes*, we say that *the limit of* $f(x)$, *as* x *approaches* a, *equals* L, and we write

$$\lim_{x \to a} f(x) = L. \tag{2.1}$$

As an illustration of (2.1), suppose a physicist wishes to obtain a certain physical measurement when the air pressure is zero. Since it is impossible to achieve a perfect vacuum in a laboratory, a natural way to attack the problem is to obtain measurements at smaller and smaller pressures. If, as the pressure approaches zero, the corresponding measurements approach some number L, then it may be assumed that the measurement in a vacuum is also L. Moreover, if it is found that at a pressure of x lbs/in.^{2} the measurement is given by $f(x)$, where f is a function, then this experimental result may be expressed as

$$\lim_{x \to 0} f(x) = L.$$

Note that in this experiment the pressure x *never equals* 0; however, modern vacuum equipment can achieve pressures extremely close to 0.

A chemist uses a similar limiting process to measure the *equivalent conductibility* of a sodium chloride solution in an electric cell. By definition, the *conductivity* k of the solution is given by $k = b/R$, where R is the cell resistance and b is a constant which depends on the cell. The conductivity is a measure of the ability of a solution to conduct electric current. If the resistance R is large, then the conductivity k is small; if R is small, then k is large. If the solution has a concentration of C moles per liter, then the *equivalent conductibility* E_C is defined by

$$E_C = \frac{k}{C} = \frac{b}{RC}.$$

In order to learn about the motion of ions in the solution, it is necessary to study the values of E_C in extremely dilute solutions. This leads the physical chemist to investigate the *limiting equivalent conductibility* E of the solution, as given by

$$E = \lim_{C \to 0} E_C.$$

This number is also called the *equivalent conductibility at infinite dilution.* Again note that we cannot let $C = 0$, since then no sodium chloride would be present and E_C would not exist.

Let us next consider a mathematical illustration of (2.1). In plane geometry, the tangent line l at a point P on a circle may be defined as the line which has only the point P in common with the circle, as illustrated in (i) of Figure 2.1. This definition cannot be extended to arbitrary graphs, since a tangent line may intersect a graph several times, as shown in (ii) of Figure 2.1.

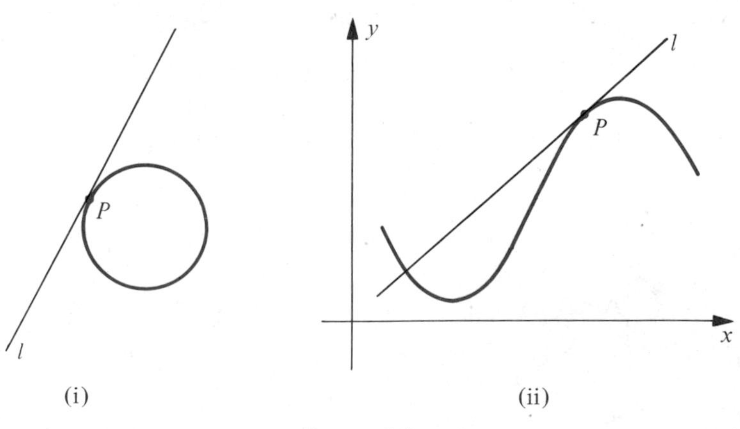

Figure 2.1

In order to define the tangent line l at a point P on the graph of an equation it is sufficient to state the slope m of l, since this completely determines the line. To arrive at m we begin by choosing any other point Q on the graph and considering the line through P and Q, as in (i) of Figure 2.2. A line which cuts through a graph in this way is called a **secant line** for the graph. Next, we study the variation of the secant line as Q gets closer and closer to P, as illustrated by the dashes in (ii) of Figure 2.2. It appears that if Q is close to P, then the slope m_{PQ} of the secant line should be close to the slope of l. For this reason, if the slope m_{PQ} has a limiting value as Q approaches P, we define this value as the slope m of the tangent line l. If a is the abscissa of P and x is the abscissa of Q (see (i) of Figure 2.2), then for many graphs the phrase "Q approaches P" may be replaced by "x approaches a" and we have

(2.2)
$$m = \lim_{x \to a} m_{PQ}.$$

Once again it is important to observe that $x \neq a$ throughout this limiting process. Indeed, if we let $x = a$, then $P = Q$ and m_{PQ} does not exist!

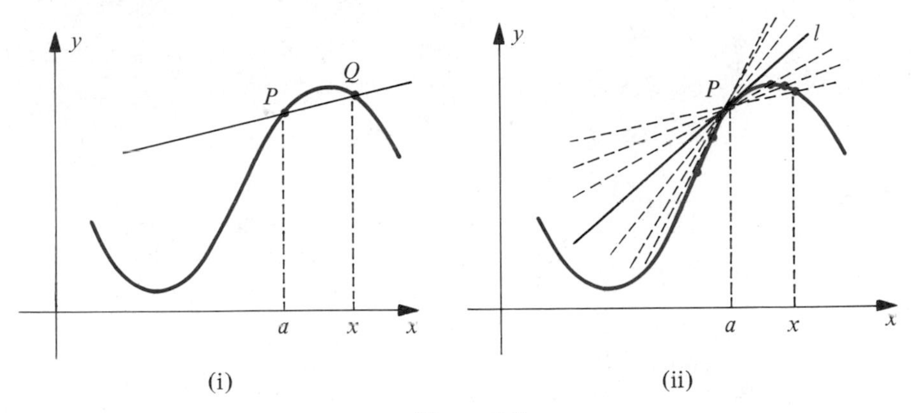

Figure 2.2

Example 1 If a is any real number, use (2.2) to find the slope of the tangent line to the graph of $y = x^2$ at the point $P(a, a^2)$. Find an equation of the tangent line to the graph at the point $(3/2, 9/4)$.

Solution The graph of $y = x^2$ and typical points $P(a, a^2)$ and $Q(x, x^2)$ are illustrated in Figure 2.3. By the Slope Formula (1.11),

$$m_{PQ} = \frac{x^2 - a^2}{x - a} = \frac{(x + a)(x - a)}{x - a} = x + a.$$

Applying (2.2), the slope m of the tangent line at P is

$$m = \lim_{x \to a} m_{PQ} = \lim_{x \to a} (x + a).$$

As x gets closer and closer to a, the expression $x + a$ gets close to $a + a$, or $2a$. Consequently $m = 2a$.

Since the slope of the tangent line at the point $(3/2, 9/4)$ is the special case in which $a = 3/2$, we have $m = 2(3/2) = 3$. Using the point-slope form (1.16), an equation of the tangent line is

$$y - \tfrac{9}{4} = 3(x - \tfrac{3}{2}).$$

The reader may verify that this equation simplifies to

$$12x - 4y - 9 = 0.$$

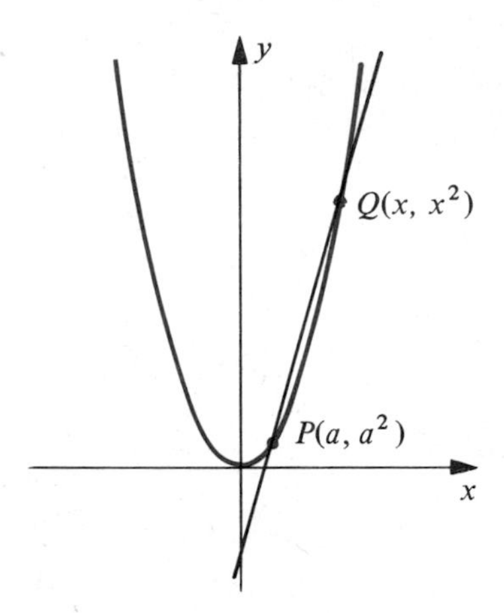

Figure 2.3

The preceding discussion and example lack precision because of the haziness of the phrases "very close" and "closer and closer." This will be remedied later, when a formal definition of limit is stated. In the remainder of this section we shall continue in an intuitive manner. As a simple illustration of the limit formula (2.1) suppose $f(x) = \frac{1}{2}(3x - 1)$ and consider $a = 4$. Although 4 is in the domain of the function f, we are primarily interested in the behavior of $f(x)$ when x is *close* to 4, but *not necessarily equal* to 4. The following values give some indication of this behavior.

$$f(3.9) = 5.35 \qquad f(4.1) = 5.65$$
$$f(3.99) = 5.485 \qquad f(4.01) = 5.515$$
$$f(3.999) = 5.4985 \qquad f(4.001) = 5.5015$$
$$f(3.9999) = 5.49985 \qquad f(4.0001) = 5.50015$$
$$f(3.99999) = 5.499985 \qquad f(4.00001) = 5.500015$$

It appears that the closer x is to 4, the closer $f(x)$ is to 5.5. This can also be verified by observing that if x is close to 4, then $3x - 1$ is close to 11, and hence $\frac{1}{2}(3x - 1)$ is close to 5.5. Consequently we write

$$\lim_{x \to 4} \tfrac{1}{2}(3x - 1) = 5.5.$$

In this illustration the number 4 could actually have been substituted for x, thereby obtaining 5.5. The next two examples show that it is not always possible to find the number L in (2.1) by merely substituting a for x.

Example 2 If $f(x) = \dfrac{x - 9}{\sqrt{x} - 3}$, find $\lim_{x \to 9} f(x)$.

Solution The number 9 is not in the domain of f since the denominator $\sqrt{x} - 3$ is zero for this value of x. However, writing

$$f(x) = \frac{(\sqrt{x} - 3)(\sqrt{x} + 3)}{(\sqrt{x} - 3)}$$

it is evident that for all nonnegative values of x, *except* $x = 9$, $f(x)$ is given by $\sqrt{x} + 3$. Thus, the closer x is to 9 (but $x \neq 9$), the closer $f(x)$ is to $\sqrt{9} + 3$, or 6. Using the notation of (2.1),

$$\lim_{x \to 9} \frac{x - 9}{\sqrt{x} - 3} = 6.$$

Example 3 If $f(x) = \dfrac{2x^2 - 5x + 2}{5x^2 - 7x - 6}$, find $\lim_{x \to 2} f(x)$.

Solution Note that 2 is not in the domain of f since 0/0 is obtained when 2 is substituted for x. Factoring the numerator and denominator gives us

$$f(x) = \frac{(x - 2)(2x - 1)}{(x - 2)(5x + 3)}$$

and, if $x \neq 2$, the values of $f(x)$ are the same as those of $(2x - 1)/(5x + 3)$. It follows that if x is close to 2 (but $x \neq 2$), then $f(x)$ is close to $(4 - 1)/(10 + 3)$ or 3/13. Thus it appears that

$$\lim_{x \to 2} \frac{2x^2 - 5x + 2}{5x^2 - 7x - 6} = \lim_{x \to 2} \frac{2x - 1}{5x + 3} = \frac{3}{13}.$$

The preceding examples demonstrate that algebraic manipulations can sometimes be used to simplify the task of finding limits. In other cases a considerable amount of ingenuity is necessary in order to determine whether or not a limit exists. This will be especially true when limits of trigonometric, exponential, and logarithmic functions are discussed. For example, it will be shown later that

$$\lim_{x \to 0} \frac{\sin x}{x} = 1.$$

This important formula cannot be obtained algebraically.

The function f defined by $f(x) = 1/x$ provides an illustration in which no limit exists as x approaches 0. If x is assigned values closer and closer to 0 (but $x \neq 0$), $f(x)$ increases without bound numerically, as illustrated in Figure 1.30. We shall have more to say about this function in Example 2 of the next section.

EXERCISES 2.1

In Exercises 1–16 proceed in an intuitive manner and find the limits, if they exist.

1 $\lim_{x \to 2} \dfrac{x^2 - 4}{x - 2}$

2 $\lim_{x \to 3} \dfrac{2x^3 - 6x^2 + x - 3}{x - 3}$

3 $\lim_{x \to 1} \dfrac{x^2 - x}{2x^2 + 5x - 7}$

4 $\lim_{r \to -3} \dfrac{r^2 + 2r - 3}{r^2 + 7r + 12}$

5 $\lim_{x \to 5} \dfrac{3x^2 - 13x - 10}{2x^2 - 7x - 15}$

6 $\lim_{x \to 25} \dfrac{\sqrt{x} - 5}{x - 25}$

7 $\lim_{k \to 4} \dfrac{k^2 - 16}{\sqrt{k} - 2}$

8 $\lim_{h \to 0} \dfrac{(x + h)^3 - x^3}{h}$

9 $\lim_{h \to 0} \dfrac{(x + h)^2 - x^2}{h}$

10 $\lim_{h \to 2} \dfrac{h^3 - 8}{h^2 - 4}$

11 $\lim_{h \to -2} \dfrac{h^3 + 8}{h + 2}$

12 $\lim_{z \to 10} \dfrac{1}{z - 10}$

13 $\lim_{x \to -3/2} \dfrac{2x + 3}{4x^2 + 12x + 9}$

14 $\lim_{s \to -1} \dfrac{1}{s^2 + 2s + 1}$

15 $\lim_{x \to 0} \dfrac{1}{x^2}$

16 $\lim_{t \to 1} \dfrac{(1/t) - 1}{t - 1}$

In each of Exercises 17–20 find (a) the slope of the tangent line to the graph of f at the point $P(a, f(a))$; (b) the equation of the tangent line at the point $P(2, f(2))$.

17 $f(x) = 5x^2 - 4x$

18 $f(x) = 3 - 2x^2$

19 $f(x) = x^3$

20 $f(x) = x^4$

In Exercises 21–24, use (2.2) to find the slope of the tangent line at the point with abscissa a on the graph of the given equation. Also find the equation of the tangent line at the indicated point P. Sketch the graph and the tangent line at P.

21 $y = 3x + 2,\ P(1, 5)$

22 $y = \sqrt{x},\ P(4, 2)$

23 $y = 1/x,\ P(2, \frac{1}{2})$

24 $y = x^{-2},\ P(2, 1/4)$

25 Give a geometric argument to show that the graph of $y = |x|$ has no tangent line at the point $(0, 0)$.

26 Give a geometric argument to show that the graph of the greatest integer function (see Example 9 in Section 1.4) has no tangent line at the point $P(1, 1)$.

27 Refer to Exercise 19. Show that the tangent line to the graph of $y = x^3$ at the point $P(0, 0)$ crosses the curve at that point.

28 If $f(x) = ax + b$, prove that the tangent line to the graph of f at any point coincides with the graph of f.

29 Refer to Example 1. Sketch the graph of $y = x^2$ together with tangent lines at the points having abscissas $-3, -2, -1, 0, 1, 2$, and 3. At what point on the graph is the slope of the tangent line equal to 6?

30 Sketch a graph which has three horizontal tangent lines and one vertical tangent line.

2.2 DEFINITION OF LIMIT

Let us return to the illustration $\lim_{x\to 4}\frac{1}{2}(3x-1) = 5.5$ discussed in Section 1 and consider in more detail the variation of $f(x) = \frac{1}{2}(3x-1)$ when x is close to 4. Using the functional values on page 54 we arrive at the following statements.

If	$3.9 < x < 4.1$	then	$5.35 < f(x) < 5.65$.
If	$3.99 < x < 4.01$	then	$5.485 < f(x) < 5.515$.
If	$3.999 < x < 4.001$	then	$5.4985 < f(x) < 5.5015$.
If	$3.9999 < x < 4.0001$	then	$5.49985 < f(x) < 5.50015$.
If	$3.99999 < x < 4.00001$	then	$5.499985 < f(x) < 5.500015$.

Each of these statements has the following form, where the Greek letters ε (epsilon) and δ (delta) are used to denote small positive real numbers:

(2.3) $$\text{If}\quad 4-\delta < x < 4+\delta, \quad \text{then}\quad 5.5-\varepsilon < f(x) < 5.5+\varepsilon.$$

For example, the first statement follows from (2.3) by letting $\delta = 0.1$ and $\varepsilon = 0.15$; the second is the case $\delta = 0.01$ and $\varepsilon = 0.015$; for the third let $\delta = 0.001$ and $\varepsilon = 0.0015$; and so on.

We may rewrite (2.3) in terms of intervals as follows:

(2.4) If x is in the open interval $(4-\delta, 4+\delta)$, then $f(x)$ is in the open interval $(5.5-\varepsilon, 5.5+\varepsilon)$.

A geometric interpretation is given in Figure 2.4, where the curved arrow indicates the correspondence between x and $f(x)$.

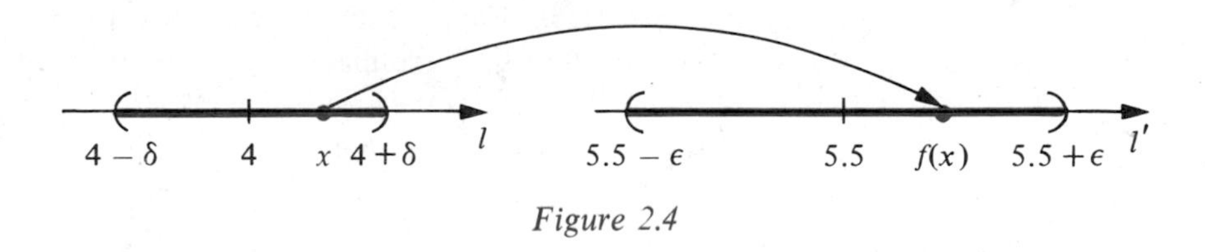

Figure 2.4

Evidently (2.3) is equivalent to:

$$\text{If}\quad -\delta < x-4 < \delta, \quad \text{then}\quad -\varepsilon < f(x)-5.5 < \varepsilon.$$

Employing absolute values, this may be written:

$$\text{If}\quad |x-4| < \delta, \quad \text{then}\quad |f(x)-5.5| < \varepsilon.$$

If we do not wish to consider what happens at $x = 4$, then it is necessary to add the condition $0 < |x-4|$. This gives us:

(2.5) $$\text{If}\quad 0 < |x-4| < \delta, \quad \text{then}\quad |f(x)-5.5| < \varepsilon.$$

A statement of this type will appear in the definition of limit; however, it is necessary to change our point of view to some extent. To arrive at (2.5), we first considered the domain of f and assigned values to x which were close to 4. We then noted the closeness of $f(x)$ to 5.5. In the definition of limit we shall reverse this process by first considering an open interval $(5.5 - \varepsilon, 5.5 + \varepsilon)$ and then, second, determining whether there is an open interval $(4 - \delta, 4 + \delta)$ in the domain of f such that (2.5) is true.

Henceforth, the fact that ε and δ are positive real numbers will not always be mentioned explicitly. The next definition is patterned after the previous remarks.

(2.6) Definition

Let f be a function that is defined on an open interval containing a, except possibly at a itself, and let L be a real number. The statement

$$\lim_{x \to a} f(x) = L$$

means that for every $\varepsilon > 0$ there exists a $\delta > 0$ such that

$$\text{if} \quad 0 < |x - a| < \delta, \quad \text{then} \quad |f(x) - L| < \varepsilon.$$

If $\lim_{x \to a} f(x) = L$, we say that **the limit of $f(x)$, as x approaches a, is L**. Since ε can be arbitrarily small, the last two inequalities in this definition are sometimes phrased *$f(x)$ can be made arbitrarily close to L by choosing x sufficiently close to a.* The last part of (2.6) may be stated in terms of open intervals as follows:

(2.7) If x is in the open interval $(a - \delta, a + \delta)$, and $x \neq a$, then $f(x)$ is in the open interval $(L - \varepsilon, L + \varepsilon)$.

In order to get a better understanding of the relationship between the positive numbers ε and δ in (2.6) and (2.7), let us consider a geometric interpretation similar to that in Figure 1.26, where the domain of f is represented by certain points on a coordinate line l, and the range by other points on a coordinate line l'. The limit process may be outlined as follows.

To prove that $\lim_{x \to a} f(x) = L$:

Step 1. For any $\varepsilon > 0$ consider the open interval $(L - \varepsilon, L + \varepsilon)$ (see Figure 2.5).

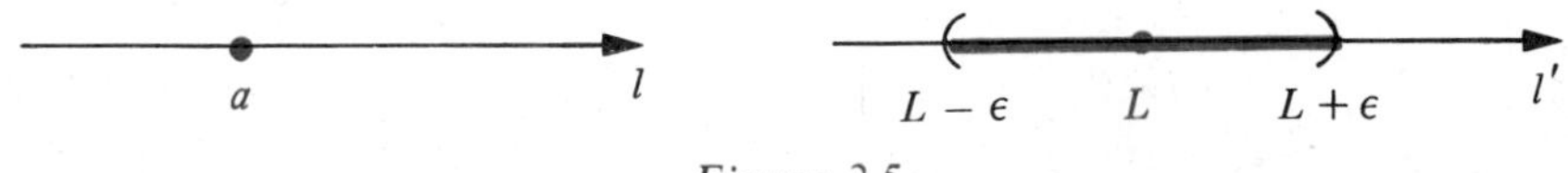

Figure 2.5

Step 2. Show that there exists an open interval $(a - \delta, a + \delta)$ in the domain of f such that (2.7) is true (see Figure 2.6).

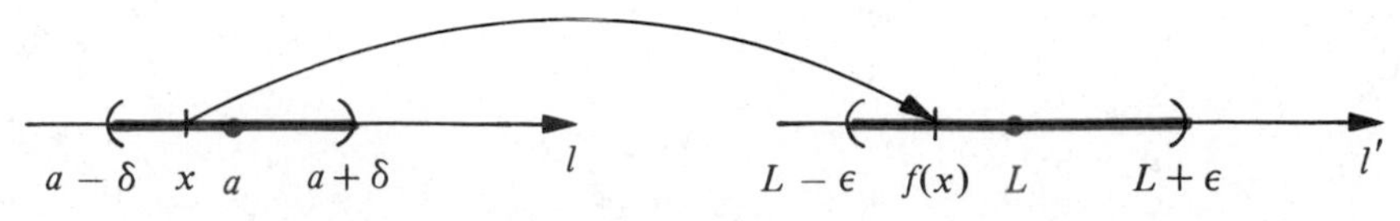

Figure 2.6

It is extremely important to remember that *first* we consider the arbitrary interval $(L - \varepsilon, L + \varepsilon)$ and then, *second*, we show that an interval $(a - \delta, a + \delta)$ of the required type exists in the domain of f. One scheme for remembering the proper sequence of events is to think of the function f as a cannon which shoots a cannonball from the point on l with coordinate x to the point on l' with coordinate $f(x)$, as illustrated by the curved arrow in Figure 2.6. Step 1 may then be regarded as setting up a target of radius ε with bull's eye at L. To apply Step 2 we must find an open interval containing a in which to place the cannon such that the cannonball hits the target. Incidentally, there is no guarantee that it will hit the bull's eye; however, if $\lim_{x \to a} f(x) = L$ we can make the cannonball land as close as we please to the bull's eye.

It should be clear that the number δ in the limit definition is not unique, for if a specific δ can be found, then any *smaller* positive number δ' will also satisfy the requirements.

Since a function may be described geometrically by means of a graph on a rectangular coordinate system, it is of interest to interpret Definition (2.6) graphically. Figure 2.7 illustrates the graph of a function f where, for any x in the domain of $f, f(x)$ is the ordinate of the point on the graph with abscissa x. Given any $\varepsilon > 0$, we consider the open interval $(L - \varepsilon, L + \varepsilon)$ on the y-axis, and the horizontal lines $y = L \pm \varepsilon$ shown in the figure. If there exists an open interval $(a - \delta, a + \delta)$ such that for all x in $(a - \delta, a + \delta)$, with the possible exception of $x = a$, the point $P(x, f(x))$ lies between the horizontal lines, that is, within the shaded rectangle shown in Figure 2.7, then

$$L - \varepsilon < f(x) < L + \varepsilon$$

and hence $\lim_{x \to a} f(x) = L$.

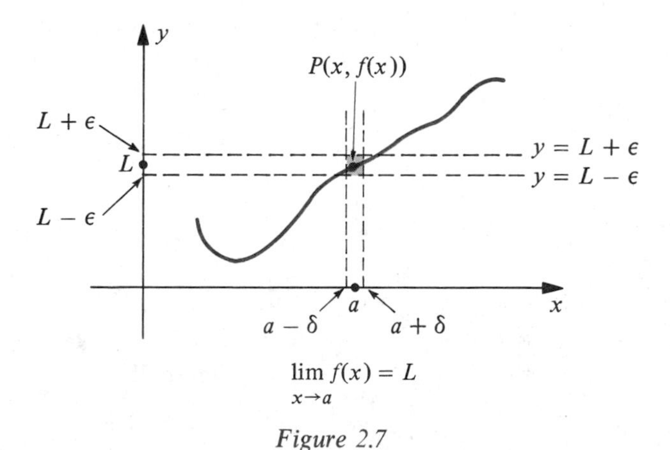

Figure 2.7

To shorten explanations, whenever the notation $\lim_{x \to a} f(x) = L$ is used we shall often assume that all the conditions given in Definition (2.6) are satisfied. Thus, it will not always be pointed out that f is defined on an open interval containing a, and so on. Moreover, we shall not always specify L but merely write "$\lim_{x \to a} f(x)$ exists," or "$f(x)$ has a limit as x approaches a." The phrase "find

$\lim_{x\to a} f(x)$" means "find a number L such that $\lim_{x\to a} f(x) = L$." If no such L exists we write "$\lim_{x\to a} f(x)$ does not exist." It can be proved that if $f(x)$ has a limit as x approaches a, then that limit is unique (see Appendix II).

In the following example we return to the function considered at the beginning of this section and *prove* that the limit exists.

Example 1 Use Definition (2.6) to prove that

$$\lim_{x\to 4} \frac{1}{2}(3x - 1) = \frac{11}{2}.$$

Solution Let $f(x) = \frac{1}{2}(3x - 1)$, $a = 4$, and $L = 11/2$. According to Definition (2.6) we must show that for every $\varepsilon > 0$, there exists a $\delta > 0$ such that

$$\text{if} \quad 0 < |x - 4| < \delta, \quad \text{then} \quad \left|\frac{1}{2}(3x - 1) - \frac{11}{2}\right| < \varepsilon.$$

A clue to the choice of δ can be found by examining the last inequality involving ε. The following is a list of equivalent inequalities.

$$\begin{aligned} |\tfrac{1}{2}(3x - 1) - \tfrac{11}{2}| &< \varepsilon \\ \tfrac{1}{2}|(3x - 1) - 11| &< \varepsilon \\ |3x - 1 - 11| &< 2\varepsilon \\ |3x - 12| &< 2\varepsilon \\ 3|x - 4| &< 2\varepsilon \\ |x - 4| &< \tfrac{2}{3}\varepsilon \end{aligned}$$

The final inequality gives us the needed clue. If we let $\delta = \frac{2}{3}\varepsilon$, then if $0 < |x - 4| < \delta$, the last inequality in the list is true and consequently so is the first. Hence by Definition (2.6), $\lim_{x\to 4} \frac{1}{2}(3x - 1) = 11/2$.

It was relatively easy to use the definition of limit in the previous example because the function f was linear. Limits of more complicated functions may also be verified by direct applications of the definition; however, the task of showing that for every $\varepsilon > 0$ there exists a suitable $\delta > 0$ often requires a great deal of ingenuity. In Section 3 we shall introduce theorems which can be used to find many limits without resorting to a search for the general number δ which appears in Definition (2.6).

The next two examples indicate how the geometric interpretation illustrated in Figure 2.7 may be used to show that certain limits do not exist.

Example 2 Show that $\lim_{x\to 0} \left(\frac{1}{x}\right)$ does not exist.

Solution Let us proceed in an indirect manner. Thus, suppose it *were* true that

$$\lim_{x\to 0} \frac{1}{x} = L$$

for some number L. Let us consider any pair of horizontal lines $y = L \pm \varepsilon$ as illustrated in Figure 2.8. Since we are assuming that the limit exists, it should be possible to find an open interval $(0 - \delta, 0 + \delta)$ or equivalently, $(-\delta, \delta)$, containing 0, such that whenever $-\delta < x < \delta$ and $x \neq 0$, the point $(x, 1/x)$ lies between the horizontal lines. However, since $1/x$ can be made as large as desired by choosing x close to 0, not every point $(x, 1/x)$ with nonzero abscissa in $(-\delta, \delta)$ has this property. Consequently our supposition is false; that is, the limit does not exist.

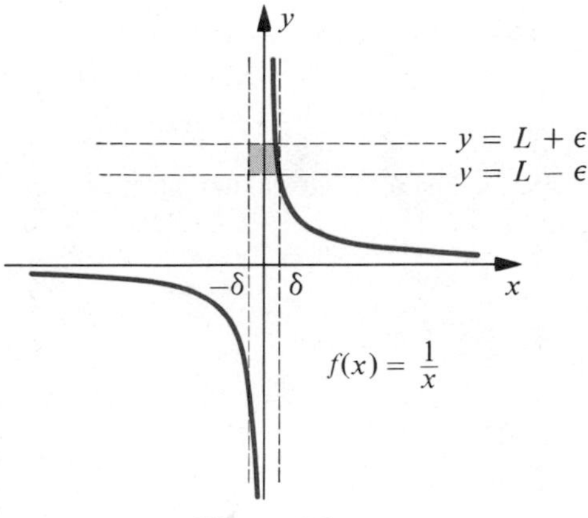

Figure 2.8

Example 3 If $f(x) = |x|/x$, show that $\lim_{x \to 0} f(x)$ does not exist.

Solution If $x > 0$, then $|x|/x = x/x = 1$ and hence, to the right of the y-axis, the graph of f coincides with the line $y = 1$. If $x < 0$, then $|x|/x = -x/x = -1$, which means that to the left of the y-axis the graph of f coincides with the line $y = -1$. If it were true that $\lim_{x \to 0} f(x) = L$ for some L, then the preceding remarks imply that $-1 \leq L \leq 1$. As shown in Figure 2.9, if we consider any pair of horizontal lines $y = L \pm \varepsilon$, where $0 < \varepsilon < 1$, then there exist points on the graph which are not between these lines for some nonzero x in *every* interval $(-\delta, \delta)$ containing 0. It follows that the limit does not exist.

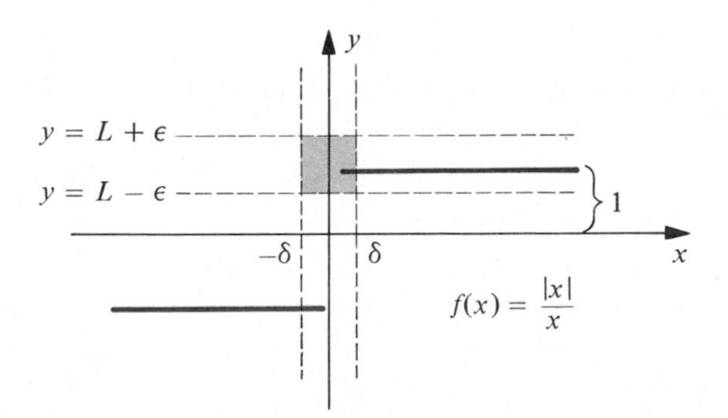

Figure 2.9

EXERCISES 2.2

Establish the limits in Exercises 1–10 by means of Definition (2.6).

1 $\lim_{x \to 2} (5x - 3) = 7$

2 $\lim_{x \to -3} (2x + 1) = -5$

3 $\lim_{x \to -6} (10 - 9x) = 64$

4 $\lim_{x \to 4} (8x - 15) = 17$

5 $\lim_{x \to 5} \left(\frac{x}{4} + 2\right) = \frac{13}{4}$

6 $\lim_{x \to 6} \left(9 - \frac{x}{6}\right) = 8$

7 $\lim_{x \to 3} 5 = 5$

8 $\lim_{x \to 5} 3 = 3$

9 $\lim_{x \to \pi} x = \pi$

10 $\lim_{x \to a} 2 = 2$

Use the method illustrated in Examples 2 and 3 to show that the limits in Exercises 11–14 do not exist.

11 $\lim_{x \to 3} \frac{|x - 3|}{x - 3}$

12 $\lim_{x \to -2} \frac{x + 2}{|x + 2|}$

13 $\lim_{x \to -5} \frac{1}{x + 5}$

14 $\lim_{x \to 1} \frac{1}{(x - 1)^2}$

15 Give an example of a function f which is defined at a and $\lim_{x \to a} f(x)$ exists, but $\lim_{x \to a} f(x) \neq f(a)$.

16 If f is the greatest integer function and a is any integer, show that $\lim_{x \to a} f(x)$ does not exist.

17 Let f be defined by the following conditions: $f(x) = 0$ if x is rational and $f(x) = 1$ if x is irrational. Prove that for every real number a, $\lim_{x \to a} f(x)$ does not exist.

18 Why is it impossible to investigate $\lim_{x \to 0} \sqrt{x}$ by means of Definition (2.6)?

19 In Section 3 it will be shown that $\lim_{x \to a} x^2 = a^2$. Illustrate this fact graphically as in Figure 2.7, by showing that for every pair of horizontal lines $y = a^2 \pm \varepsilon$, there exists an open interval $(a - \delta, a + \delta)$ such that if x is in $(a - \delta, a + \delta)$, then the point (x, x^2) lies between the horizontal lines.

20 Demonstrate graphically that $\lim_{x \to a} x^3 = a^3$. (See Exercise 19.)

2.3 THEOREMS ON LIMITS

It would be an excruciating task to solve each problem on limits by means of Definition (2.6). The purpose of this section is to introduce theorems which may be used to simplify the process. In order to prove the theorems it is necessary to employ (2.6); however, once they are established it will be possible to determine many limits without referring to an ε or a δ. Several theorems are proved in this section; the remaining proofs will be found in Appendix II.

The simplest limit to consider involves the constant function defined by $f(x) = c$, where c is a real number. If f is represented geometrically by means of coordinate lines l and l', then *every* arrow from l terminates at the same point on l', namely, the point with coordinate c, as indicated in Figure 2.10.

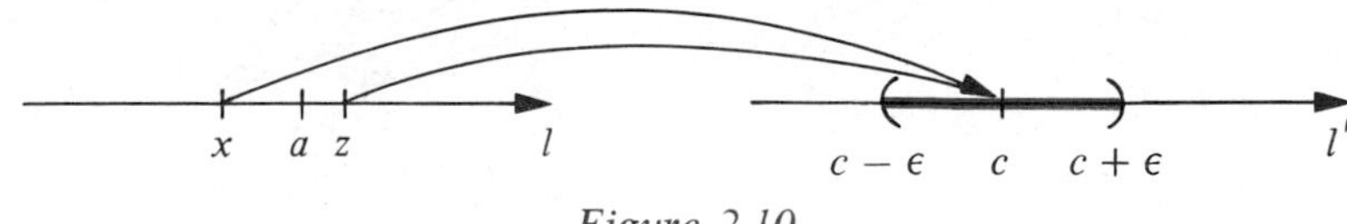

Figure 2.10

It is easy to prove that for every real number a, $\lim_{x\to a} f(x) = c$. Thus, if $\varepsilon > 0$, consider the open interval $(c - \varepsilon, c + \varepsilon)$ on l' as illustrated in the figure. Since $f(x) = c$ is always in this interval, *any* number δ will satisfy the conditions of Definition (2.6); that is, for *every* $\delta > 0$,

$$\text{if} \quad 0 < |x - a| < \delta, \quad \text{then} \quad |f(x) - c| < \varepsilon.$$

We have proved the following theorem.

(2.8) Theorem

If a and c are any real numbers, then $\lim_{x\to a} c = c$.

Sometimes Theorem (2.8) is phrased "the limit of a constant is the constant." To illustrate,

$$\lim_{x\to 3} 8 = 8, \quad \lim_{x\to 8} 3 = 3, \quad \lim_{x\to \pi} \sqrt{2} = \sqrt{2}, \quad \lim_{x\to a} 0 = 0$$

and so on.

The next result tells us that in order to find the limit of a linear function as x approaches a, we merely substitute a for x.

(2.9) Theorem

If a, b, and m are real numbers, then

$$\lim_{x\to a} (mx + b) = ma + b.$$

Proof. If $m = 0$, then $mx + b = b$ and the statement of the theorem reduces to $\lim_{x\to a} b = b$, which was proved in Theorem (2.8).

Next suppose $m \neq 0$. If we let $f(x) = mx + b$ and $L = ma + b$, then according to Definition (2.6) we must show that for every $\varepsilon > 0$ there exists a $\delta > 0$ such that

$$\text{if} \quad 0 < |x - a| < \delta, \quad \text{then} \quad |(mx + b) - (ma + b)| < \varepsilon.$$

As in the solution of Example 1 in the previous section, a clue to the choice of δ can be found by examining the inequality involving ε. All inequalities in the following list are equivalent.

$$|(mx + b) - (ma + b)| < \varepsilon$$
$$|mx - ma| < \varepsilon$$
$$|m|\,|x - a| < \varepsilon$$
$$|x - a| < \frac{\varepsilon}{|m|}.$$

The last inequality suggests that we choose $\delta = \varepsilon/|m|$. Thus, given any $\varepsilon > 0$,

$$\text{if} \quad 0 < |x - a| < \delta, \quad \text{where} \quad \delta = \varepsilon/|m|$$

then the last inequality in the above list is true, and hence, so is the first inequality, which is what we wished to prove.

As special cases of Theorem (2.9), we have

$$\lim_{x \to a} x = a$$
$$\lim_{x \to 4} (3x - 5) = 3 \cdot 4 - 5 = 7$$
$$\lim_{x \to \sqrt{2}} (13x + \sqrt{2}) = 13\sqrt{2} + \sqrt{2} = 14\sqrt{2}.$$

The next theorem states that if a function f has a positive limit as x approaches a, then $f(x)$ is positive throughout some open interval containing a, with the possible exception of a.

(2.10) Theorem

> If $\lim_{x \to a} f(x) = L$ and $L > 0$, then there exists an open interval $(a - \delta, a + \delta)$ such that $f(x) > 0$ for all x in $(a - \delta, a + \delta)$, except possibly $x = a$.

Proof. If $\varepsilon = L/2$, then the interval $(L - \varepsilon, L + \varepsilon)$ contains only positive numbers. By (2.7) there exists a $\delta > 0$ such that whenever x is in the open interval $(a - \delta, a + \delta)$ and $x \neq a$, then $f(x)$ is in $(L - \varepsilon, L + \varepsilon)$, and hence $f(x) > 0$.

In like manner, it can be shown that if f has a *negative* limit as x approaches a, then there is an open interval I containing a such that $f(x) < 0$ for all x in I, with the possible exception of $x = a$.

Many functions may be expressed as sums, differences, products, and quotients of other functions. In particular, suppose a function s is a sum of two functions f and g, so that $s(x) = f(x) + g(x)$ for every x in the domain of s. If $f(x)$ and $g(x)$ have limits L and M, respectively, as x approaches a, it is natural to conclude that $s(x)$ has the limit $L + M$ as x approaches a. The fact that this and analogous statements are true for products and quotients is a consequence of the next theorem. The proofs are rather technical and have been placed in Appendix II.

(2.11) Theorem

If $\lim_{x \to a} f(x) = L$ and $\lim_{x \to a} g(x) = M$, then

(i) $\lim_{x \to a} [f(x) + g(x)] = L + M.$

(ii) $\lim_{x \to a} f(x) \cdot g(x) = L \cdot M.$

(iii) $\lim_{x \to a} \dfrac{f(x)}{g(x)} = \dfrac{L}{M}$, provided $M \neq 0$.

The preceding theorem is often written as follows:

(2.12)

(i) $\lim_{x \to a} [f(x) + g(x)] = \lim_{x \to a} f(x) + \lim_{x \to a} g(x)$

(ii) $\lim_{x \to a} f(x) \cdot g(x) = \lim_{x \to a} f(x) \cdot \lim_{x \to a} g(x)$

(iii) $\lim_{x \to a} \dfrac{f(x)}{g(x)} = \dfrac{\lim_{x \to a} f(x)}{\lim_{x \to a} g(x)}$

where it is assumed that the indicated limits exist and $\lim_{x \to a} g(x) \neq 0$ in (iii). The results in (i) and (ii) may be extended to sums and products of more than two functions. In words, (2.12) may be stated as follows:

(i) *The limit of a sum is the sum of the limits.*
(ii) *The limit of a product is the product of the limits.*
(iii) *The limit of a quotient is the quotient of the limits.*

Example 1 Find $\lim_{x \to 2} \dfrac{3x + 4}{5x + 7}$.

Solution The numerator and denominator of the indicated quotient define linear functions whose limits exist as in Theorem (2.9). Applying (iii) of (2.12) together with (2.9) gives us

$$\lim_{x \to 2} \frac{3x + 4}{5x + 7} = \frac{\lim_{x \to 2} (3x + 4)}{\lim_{x \to 2} (5x + 7)} = \frac{3(2) + 4}{5(2) + 7} = \frac{10}{17}.$$

Notice the simple manner in which the limit in Example 1 was found. It would be a lengthy task to verify the limit by means of Definition (2.6).

Example 2 Prove that for every real number a,

$$\lim_{x\to a} x^2 = a^2.$$

Solution Using (ii) of (2.12) with $f(x) = g(x) = x$, together with Theorem (2.9), we obtain

$$\lim_{x\to a} x^2 = \lim_{x\to a} (x \cdot x) = (\lim_{x\to a} x) \cdot (\lim_{x\to a} x) = a \cdot a = a^2.$$

The techniques used in Example 2 can be extended to the expression x^n, where n is any positive integer. We merely write x^n as a product $x \cdot x \cdot \cdots \cdot x$ of n factors and then take the limit of each factor. This gives us

(2.13) $$\lim_{x\to a} x^n = a^n$$

for every positive integer n and every real number a. Similarly, if $\lim_{x\to a} f(x)$ exists we can use (ii) of (2.11) to prove that

(2.14) $$\lim_{x\to a} [f(x)]^n = [\lim_{x\to a} f(x)]^n$$

for every positive integer n.

Example 3 Find $\lim_{x\to 2} (3x + 4)^5$.

Solution Applying (2.14) and Theorem (2.9),

$$\begin{aligned}\lim_{x\to 2} (3x + 4)^5 &= [\lim_{x\to 2} (3x + 4)]^5 \\ &= [3(2) + 4]^5 \\ &= 10^5 = 100{,}000.\end{aligned}$$

If we are given an expression of the form $cf(x)$, where c is a real number and the function f has a limit as x approaches a, we may use (ii) of (2.12) to write

$$\lim_{x\to a} [c \cdot f(x)] = [\lim_{x\to a} c] \cdot [\lim_{x\to a} f(x)].$$

Applying Theorem (2.8) we see that

(2.15) $$\lim_{x\to a} [cf(x)] = c[\lim_{x\to a} f(x)]$$

for every real number c.

Since $f(x) - g(x) = f(x) + (-1)g(x)$ we may use (i) of (2.12) and then (2.15) with $c = -1$, to obtain

(2.16) $$\lim_{x\to a} [f(x) - g(x)] = \lim_{x\to a} f(x) - \lim_{x\to a} g(x)$$

provided the limits exist.

An important special case of (2.15) is

(2.17) $$\lim_{x\to a} cx^n = c \lim_{x\to a} x^n = ca^n.$$

Example 4 Find $\lim_{x \to -2} (3x^4 + 7x)$.

Solution First employing (i) of (2.12) and then (2.17) gives us

$$\begin{aligned} \lim_{x \to -2} (3x^4 + 7x) &= \lim_{x \to -2} 3x^4 + \lim_{x \to -2} 7x \\ &= 3(-2)^4 + 7(-2) \\ &= 48 - 14 = 34. \end{aligned}$$

Note that (2.17) states that the indicated limit may be found by substituting a for x. We now prove that the same is true for limits involving polynomial and rational functions.

(2.18) Theorem

> If f is a polynomial function, then
>
> $$\lim_{x \to a} f(x) = f(a)$$
>
> for every real number a.

Proof. We may write $f(x)$ in the form

$$f(x) = b_n x^n + b_{n-1} x^{n-1} + \cdots + b_0$$

where the b_i are real numbers. Employing (2.12) and (2.17),

$$\begin{aligned} \lim_{x \to a} f(x) &= \lim_{x \to a} (b_n x^n) + \lim_{x \to a} (b_{n-1} x^{n-1}) + \cdots + \lim_{x \to a} b_0 \\ &= b_n a^n + b_{n-1} a^{n-1} + \cdots + b_0 \\ &= f(a). \end{aligned}$$

(2.19) Corollary

> If q is a rational function and a is in the domain of q, then
>
> $$\lim_{x \to a} q(x) = q(a).$$

Proof. We may write $q(x) = f(x)/h(x)$ where f and h are polynomial functions. If a is in the domain of q, then $h(a) \neq 0$. Using (2.12) and (2.18),

$$\lim_{x \to a} q(x) = \frac{\lim_{x \to a} f(x)}{\lim_{x \to a} h(x)} = \frac{f(a)}{h(a)} = q(a).$$

Example 5 Find $\lim_{x \to 3} \dfrac{5x^2 - 2x + 1}{6x - 7}$.

Solution Applying (2.19),

$$\lim_{x \to 3} \frac{5x^2 - 2x + 1}{6x - 7} = \frac{5(3)^2 - 2(3) + 1}{6(3) - 7}$$
$$= \frac{45 - 6 + 1}{18 - 7} = \frac{40}{11}.$$

The next result shows that for positive integral roots of x, we may again determine a limit by a simple substitution. The proof makes use of Definition (2.6) and will be found in Appendix II.

(2.20) Theorem

> If $a > 0$ and n is a positive integer, or if $a \leq 0$ and n is an odd positive integer, then
>
> $$\lim_{x \to a} \sqrt[n]{x} = \sqrt[n]{a}.$$

If m and n are positive integers and $a > 0$, then using (2.14) and Theorem (2.20),

$$\lim_{x \to a} (\sqrt[n]{x})^m = (\lim_{x \to a} \sqrt[n]{x})^m = (\sqrt[n]{a})^m.$$

In terms of fractional exponents this may be expressed as

(2.21) $$\lim_{x \to a} x^{m/n} = a^{m/n}$$

where $a > 0$ and m and n are positive integers. Extensions to negative exponents may also be made.

Example 6 Find $\displaystyle\lim_{x \to 8} \frac{x^{2/3} + 3\sqrt{x}}{4 - (16/x)}$.

Solution The reader should supply reasons for each of the following steps.

$$\lim_{x \to 8} \frac{x^{2/3} + 3\sqrt{x}}{4 - (16/x)} = \frac{\lim_{x \to 8} (x^{2/3} + 3\sqrt{x})}{\lim_{x \to 8} (4 - (16/x))}$$
$$= \frac{\lim_{x \to 8} x^{2/3} + \lim_{x \to 8} 3\sqrt{x}}{\lim_{x \to 8} 4 - \lim_{x \to 8} (16/x)}$$
$$= \frac{8^{2/3} + 3\sqrt{8}}{4 - (16/8)}$$
$$= \frac{4 + 6\sqrt{2}}{4 - 2} = 2 + 3\sqrt{2}$$

(2.22) Theorem

If a function f has a limit as x approaches a, then

$$\lim_{x \to a} \sqrt[n]{f(x)} = \sqrt[n]{\lim_{x \to a} f(x)}$$

provided either n is an odd positive integer or n is an even positive integer and $\lim_{x \to a} f(x) > 0$.

The preceding theorem will be proved in Section 5. In the meantime we shall use it whenever applicable to gain experience in finding limits which involve roots of algebraic expressions.

Example 7 Find $\lim_{x \to 5} \sqrt[3]{3x^2 - 4x + 9}$.

Solution Using Theorems (2.22) and (2.18),

$$\lim_{x \to 5} \sqrt[3]{3x^2 - 4x + 9} = \sqrt[3]{\lim_{x \to 5} (3x^2 - 4x + 9)}$$
$$= \sqrt[3]{75 - 20 + 9} = \sqrt[3]{64} = 4.$$

The beginning student should not be misled by the preceding examples. It is not always possible to find limits merely by substitution. As illustrated in Section 2, it may be necessary to perform some type of algebraic manipulation before a limit can be obtained. Sometimes other devices must be employed. The next theorem concerns three functions f, h, and g, where $h(x)$ is always "sandwiched" between $f(x)$ and $g(x)$. If f and g have a common limit L as x approaches a, then as stated below, h must have the same limit. A proof will be found in Appendix II.

(2.23) The Sandwich Theorem

If $f(x) \leq h(x) \leq g(x)$ for all x in an open interval containing a, except possibly at a, and if

$$\lim_{x \to a} f(x) = L = \lim_{x \to a} g(x),$$

then

$$\lim_{x \to a} h(x) = L.$$

Example 8 The sine function has the property that $-1 \leq \sin t \leq 1$ for every real number t. Use this fact and the Sandwich Theorem (2.23) to prove that

$$\lim_{x \to 0} x \sin \frac{1}{x} = 0.$$

Solution The limit cannot be found by substituting 0 for x, or by using an algebraic manipulation. However, since all values of the sine function are between -1 and 1, $|\sin(1/x)| \leq 1$ and, therefore,

$$\left|x \sin\frac{1}{x}\right| = |x| \left|\sin\frac{1}{x}\right| \leq |x|$$

for every $x \neq 0$. Consequently

$$0 \leq \left|x \sin\frac{1}{x}\right| \leq |x|.$$

It is not difficult to show that $\lim_{x\to 0} |x| = 0$. Hence, by the Sandwich Theorem (2.23), with $f(x) = 0$ and $g(x) = |x|$, we see that

$$\lim_{x\to 0} \left|x \sin\frac{1}{x}\right| = 0.$$

It now follows from the definition of limit (2.6) that

$$\lim_{x\to 0} x \sin\frac{1}{x} = 0.$$

EXERCISES 2.3

In Exercises 1–36 find the limits, if they exist.

1 $\lim_{x\to -2} (3x^3 - 2x + 7)$

2 $\lim_{x\to 4} (5x^2 - 9x - 8)$

3 $\lim_{x\to \sqrt{2}} (x^2 + 3)(x - 4)$

4 $\lim_{t\to -3} (3t + 4)(7t - 9)$

5 $\lim_{x\to 4} \sqrt[3]{x^2 - 5x - 4}$

6 $\lim_{x\to -2} \sqrt{x^4 - 4x + 1}$

7 $\lim_{x\to 7} 0$

8 $\lim_{x\to 1/2} \dfrac{4x^2 - 6x + 3}{16x^3 + 8x - 7}$

9 $\lim_{x\to \sqrt{2}} 15$

10 $\lim_{x\to 15} \sqrt{2}$

11 $\lim_{x\to 1/2} \dfrac{2x^2 + 5x - 3}{6x^2 - 7x + 2}$

12 $\lim_{x\to -3} \dfrac{x + 3}{(1/x) + (1/3)}$

13 $\lim_{x\to 2} \dfrac{x - 2}{x^3 - 8}$

14 $\lim_{x\to 2} \dfrac{x^2 - x - 2}{(x - 2)^2}$

15 $\lim_{x\to 16} \dfrac{x - 16}{\sqrt{x} - 4}$

16 $\lim_{x\to -2} \dfrac{x^3 + 8}{x^4 - 16}$

17 $\lim_{s\to 4} \dfrac{6s - 1}{2s - 9}$

18 $\lim_{x\to \pi} (x - 3.1416)$

19 $\lim_{x\to 1} \left(\dfrac{x^2}{x - 1} - \dfrac{1}{x - 1}\right)$

20 $\lim_{x\to 1} \left(\sqrt{x} + \dfrac{1}{\sqrt{x}}\right)^6$

21 $\lim_{x\to 16} \dfrac{2\sqrt{x} + x^{3/2}}{\sqrt[4]{x} + 5}$

22 $\lim_{x\to -8} \dfrac{16x^{2/3}}{4 - x^{4/3}}$

23 $\lim_{x\to 3} \sqrt[3]{\dfrac{2 + 5x - 3x^3}{x^2 - 1}}$

24 $\lim_{x\to \pi} \sqrt[5]{\dfrac{x - \pi}{x + \pi}}$

25 $\lim_{h\to 0} \dfrac{4 - \sqrt{16 + h}}{h}$

26 $\lim_{h\to 0} \left(\dfrac{1}{h}\right)\left(\dfrac{1}{\sqrt{1 + h}} - 1\right)$

27 $\lim_{x\to 1} \dfrac{(x - 1)^5}{x^5 - 1}$

28 $\lim_{x\to 6} (x + 4)^3(x - 6)^2$

29 $\lim_{v\to 3} v^2(3v - 4)(9 - v^3)$

30 $\lim_{k\to 2} \sqrt{3k^2 + 4}\sqrt[3]{3k + 2}$

31 $\lim_{t\to -1} \dfrac{(4t^2 + 5t - 3)^3}{(6t + 5)^4}$

32 $\lim_{t\to 7} \dfrac{\sqrt[5]{3 - 5t}}{(t - 5)^3}$

33 $\lim_{x\to 9} \dfrac{x^2 - 81}{3 - \sqrt{x}}$

34 $\lim_{x\to 8} \dfrac{x - 8}{\sqrt[3]{x} - 2}$

35 $\lim_{h\to 0} \dfrac{(2 + h)^{-2} - 2^{-2}}{h}$

36 $\lim_{h\to 0} \dfrac{(9 + h)^{-1} - 9^{-1}}{h}$

37 If r is any rational number and $a > 0$, prove that $\lim_{x\to a} x^r = a^r$. Under what conditions will this be true if $a < 0$?

38 If $\lim_{x\to a} f(x) = L \neq 0$ and $\lim_{x\to a} g(x) = 0$, prove that $\lim_{x\to a} [f(x)/g(x)]$ does not exist. (*Hint:* Assume there is a number M such that $\lim_{x\to a} [f(x)/g(x)] = M$ and consider

$$\lim_{x\to a} f(x) = \lim_{x\to a} \left[g(x)\cdot\frac{f(x)}{g(x)} \right].$$

39 (a) Illustrate the Sandwich Theorem geometrically by sketching graphs of three arbitrary functions f, h, and g which satisfy the conditions of (2.23).

(b) Use the Sandwich Theorem and the fact that $\lim_{x\to 0} (|x| + 1) = 1$ to prove that $\lim_{x\to 0} (x^2 + 1) = 1$.

40 Use the Sandwich Theorem (2.23) with $f(x) = 0$ and $g(x) = |x|$ to prove that

$$\lim_{x\to 0} \frac{|x|}{\sqrt{x^4 + 4x^2 + 7}} = 0.$$

41 If c is a nonnegative real number and $0 \le f(x) \le c$ for every x, use the Sandwich Theorem (2.23) to prove that

$$\lim_{x\to 0} x^2 f(x) = 0.$$

42 Prove that $\lim_{x\to 0} x^4 \sin(1/\sqrt[3]{x}) = 0$. (*Hint:* See Example 8.)

43 Prove that if a function f has a negative limit as x approaches a, then there is some open interval I containing a such that $f(x)$ is negative for every x in I except possibly $x = a$.

2.4 ONE-SIDED LIMITS

If $f(x) = \sqrt{x-2}$ and $a > 2$, then f is defined throughout an open interval containing a and, by Theorem (2.22),

$$\lim_{x \to a} \sqrt{x-2} = \sqrt{a-2}.$$

The case $a = 2$ is not covered by Definition (2.6) since there is no open interval containing 2 throughout which f is defined (note that $\sqrt{x-2}$ is not real if $x < 2$). A natural way to extend the definition of limit to include this exceptional case is to restrict x to values *greater* than 2. Thus, we replace the condition $2 - \delta < x < 2 + \delta$, which arises from Definition (2.6), by the condition $2 < x < 2 + \delta$. The corresponding limit is called *the limit of $f(x)$ as x approaches 2 from the right*. This is a special case of the next definition.

(2.24) Definition

Let f be a function that is defined on an open interval (a, c), and let L be a real number. The statement

$$\lim_{x \to a^+} f(x) = L$$

means that for every $\varepsilon > 0$, there exists $\delta > 0$, such that

$$\text{if} \quad a < x < a + \delta, \quad \text{then} \quad |f(x) - L| < \varepsilon.$$

If $\lim_{x \to a^+} f(x) = L$, we say that **the limit of $f(x)$, as x approaches a from the right, is L.** We also refer to L as the **right-hand limit** of $f(x)$ as x approaches a. The symbol $x \to a^+$ is used to indicate that values of x are always larger than a. Note that the only difference between Definitions (2.24) and (2.6) is that for *right*-hand limits we restrict x to the *right* half $(a, a + \delta)$ of the interval $(a - \delta, a + \delta)$. Definition (2.24) is illustrated geometrically in (i) of Figure 2.11. Intuitively, we think of $f(x)$ getting close to L as x gets close to a, through values *larger* than a.

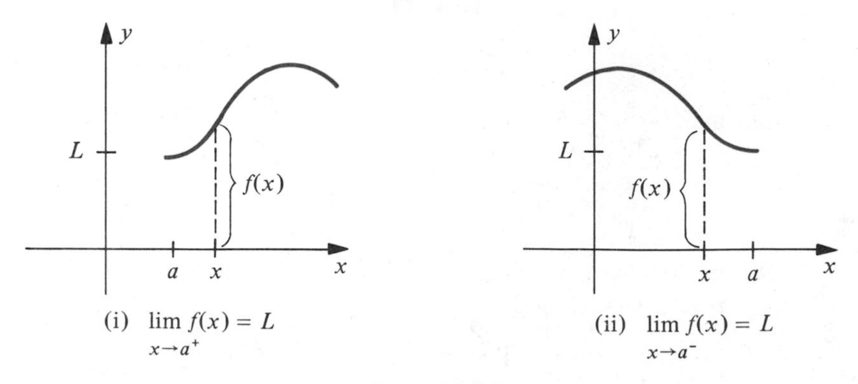

Figure 2.11

The notion of **left-hand limit** is defined in similar fashion. For example, if $f(x) = \sqrt{2 - x}$, then we restrict x to values *less* than 2. The general definition follows.

(2.25) Definition

> Let f be a function that is defined on an open interval (c, a), and let L be a real number. The statement
>
> $$\lim_{x \to a^-} f(x) = L$$
>
> means that for every $\varepsilon > 0$, there exists a $\delta > 0$, such that
>
> $$\text{if} \quad a - \delta < x < a, \quad \text{then} \quad |f(x) - L| < \varepsilon.$$

If $\lim_{x \to a^-} f(x) = L$, we say that **the limit of $f(x)$ as x approaches a from the left, is L**; or that L is the **left-hand limit** of $f(x)$ as x approaches a. The symbol $x \to a^-$ is used to indicate that x is restricted to values *less* than a. A geometric illustration of Definition (2.25) is given in (ii) of Figure 2.11. Note that for the *left*-hand limit, x is in the *left* half $(a - \delta, a)$ of the interval $(a - \delta, a + \delta)$.

Sometimes (2.24) and (2.25) are referred to as **one-sided limits** of $f(x)$ as x approaches a. The relation between one-sided limits and limits is stated in the next theorem. The proof is left as an exercise.

(2.26) Theorem

> If f is defined throughout an open interval containing a, except possibly at a itself, then $\lim_{x \to a} f(x) = L$ if and only if both $\lim_{x \to a^-} f(x) = L$ and $\lim_{x \to a^+} f(x) = L$.

The preceding theorem tells us that the limit of $f(x)$ as x approaches a exists if and only if both the right- and left-hand limits exist and are equal.

Theorems similar to the limit theorems of the previous section can be proved for one-sided limits. For example,

$$\lim_{x \to a^+} [f(x) + g(x)] = \lim_{x \to a^+} f(x) + \lim_{x \to a^+} g(x)$$

and

$$\lim_{x \to a^+} \sqrt[n]{f(x)} = \sqrt[n]{\lim_{x \to a^+} f(x)}$$

with the usual restrictions on the existence of limits and nth roots. Analogous results are true for left-hand limits.

Example 1 Find $\lim_{x \to 2^+} (1 + \sqrt{x - 2})$.

Solution Using (one-sided) limit theorems,

$$\lim_{x\to 2^+} (1 + \sqrt{x-2}) = \lim_{x\to 2^+} 1 + \lim_{x\to 2^+} \sqrt{x-2}$$
$$= 1 + \sqrt{\lim_{x\to 2^+} (x-2)}$$
$$= 1 + 0 = 1.$$

The graph of $f(x) = 1 + \sqrt{x-2}$ is sketched in Figure 2.12. Note that there is no left-hand limit, since $\sqrt{x-2}$ is not a real number if $x < 2$.

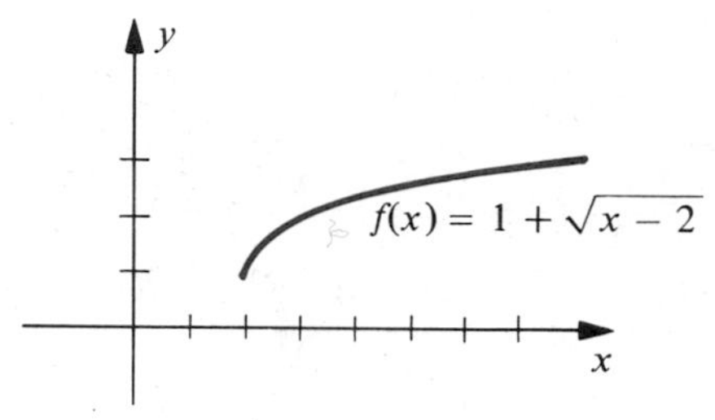

Figure 2.12

Example 2 Suppose $f(x) = |x|/x$ if $x \neq 0$ and $f(0) = 1$. Show that $\lim_{x\to 0^+} f(x) = 1$ and $\lim_{x\to 0^-} f(x) = -1$. What is $\lim_{x\to 0} f(x)$?

Solution If $x > 0$, then $|x| = x$ and $f(x) = x/x = 1$. Consequently,

$$\lim_{x\to 0^+} f(x) = \lim_{x\to 0^+} 1 = 1.$$

If $x < 0$, then $|x| = -x$ and $f(x) = -x/x = -1$. Therefore

$$\lim_{x\to 0^-} f(x) = \lim_{x\to 0^-} (-1) = -1.$$

Since these right- and left-hand limits are unequal, it follows from Theorem (2.26) that $\lim_{x\to 0} f(x)$ does not exist. The graph of f is sketched in Figure 2.13.

Example 3 Suppose $f(x) = x + 2$ if $x \neq 1$ and $f(1) = 0$. Find $\lim_{x\to 1^-} f(x)$, $\lim_{x\to 1^+} f(x)$, and $\lim_{x\to 1} f(x)$.

Solution The graph of f consists of the point $P(1, 0)$ together with all points on the line $y = x + 2$ except the point with coordinates $(1, 3)$ as shown in Figure 2.14. Evidently, $\lim_{x\to 1^+} f(x) = 3 = \lim_{x\to 1^-} f(x)$. Hence by Theorem (2.26), $\lim_{x\to 1} f(x) = 3$. Note that $\lim_{x\to 1} f(x) \neq f(1)$.

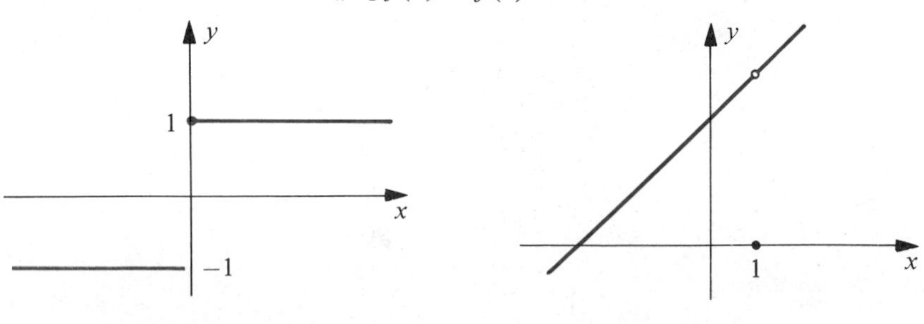

Figure 2.13 *Figure 2.14*

EXERCISES 2.4

In each of Exercises 1–16, find the limit (if it exists).

1 $\lim_{x \to 0^+} (4 + \sqrt{x})$

2 $\lim_{x \to 0^+} (4x^{3/2} - \sqrt{x} + 3)$

3 $\lim_{x \to -6^+} (\sqrt{x+6} + x)$

4 $\lim_{x \to 5/2^-} (\sqrt{5 - 2x} - x^2)$

5 $\lim_{x \to 5^+} (\sqrt{x^2 - 25} + 3)$

6 $\lim_{x \to 3^-} x\sqrt{9 - x^2}$

7 $\lim_{x \to 3^-} \dfrac{|x - 3|}{x - 3}$

8 $\lim_{x \to -10^+} \dfrac{x + 10}{|x + 10|}$

9 $\lim_{x \to 3^+} \dfrac{\sqrt{(x - 3)^2}}{x - 3}$

10 $\lim_{x \to -10^-} \dfrac{x + 10}{\sqrt{(x + 10)^2}}$

11 $\lim_{x \to 5^+} \dfrac{1 + \sqrt{2x - 10}}{x + 3}$

12 $\lim_{x \to 4^+} \dfrac{\sqrt[4]{x^2 - 16}}{x + 4}$

13 $\lim_{x \to -7^+} \dfrac{x + 7}{|x + 7|}$

14 $\lim_{x \to \pi^-} \dfrac{|\pi - x|}{x - \pi}$

15 $\lim_{x \to 0^+} \dfrac{1}{x}$

16 $\lim_{x \to 8^-} \dfrac{1}{x - 8}$

For each f defined in Exercises 17 and 18, find $\lim_{x \to 2^-} f(x)$, $\lim_{x \to 2^+} f(x)$, and sketch the graph of f.

17 $f(x) = \begin{cases} 3x & \text{if } x \le 2 \\ x^2 & \text{if } x > 2 \end{cases}$

18 $f(x) = \begin{cases} x^3 & \text{if } x \le 2 \\ 4 - 2x & \text{if } x > 2 \end{cases}$

In Exercises 19 and 20 find $\lim_{x \to -3^+} f(x)$, $\lim_{x \to -3^-} f(x)$, and $\lim_{x \to -3} f(x)$, if they exist.

19 $f(x) = \begin{cases} 1/(2 - 3x) & \text{if } x < -3 \\ \sqrt[3]{x + 2} & \text{if } x \ge -3 \end{cases}$

20 $f(x) = \begin{cases} 9/x^2 & \text{if } x \le -3 \\ 4 + x & \text{if } x > -3 \end{cases}$

In Exercises 21–23, n denotes an arbitrary integer. For each function f, sketch the graph of f and find $\lim_{x \to n^-} f(x)$ and $\lim_{x \to n^+} f(x)$.

21 $f(x) = (-1)^n \quad \text{if } n \le x < n + 1$

22 $f(x) = \begin{cases} 0 & \text{if } x = n \\ 1 & \text{if } x \ne n \end{cases}$

23 $f(x) = \begin{cases} x & \text{if } x = n \\ 0 & \text{if } x \ne n \end{cases}$

In Exercises 24 and 25, [ ] denotes the greatest integer function and n is an arbitrary integer.

24 Find $\lim_{x \to n^-} [x]$ and $\lim_{x \to n^+} [x]$.

25 If $f(x) = x - [x]$, find $\lim_{x \to n^-} f(x)$ and $\lim_{x \to n^+} f(x)$.

26 If $f(x) = (x^2 - 9)/(x - 3)$ for $x \ne 3$ and $f(3) = 5$, find $\lim_{x \to 3^+} f(x)$, $\lim_{x \to 3^-} f(x)$, and $\lim_{x \to 3} f(x)$.

27 If f is a polynomial function, prove that $\lim_{x\to a^+} f(x) = f(a)$ and $\lim_{x\to a^-} f(x) = f(a)$ for every real number a.

28 Prove Theorem (2.26).

29 Sketch geometric interpretations for right-hand limits which are analogous to those in Figures 2.5 and 2.6. Do the same for left-hand limits.

30 The illustrations of limits from physics and chemistry given in Section 1 are one-sided. Why?

2.5 CONTINUOUS FUNCTIONS

In arriving at the definition of $\lim_{x\to a} f(x)$ we emphasized the restriction $x \neq a$. A number of examples in preceding sections have brought out the fact that $\lim_{x\to a} f(x)$ may exist even though f is undefined at a. Let us now turn our attention to the case in which a is in the domain of f. If f is defined at a and $\lim_{x\to a} f(x)$ exists, then this limit may, or may not, equal $f(a)$. If $\lim_{x\to a} f(x) = f(a)$ then f is said to be *continuous* at a according to the next definition.

(2.27) Definition

A function f is **continuous** at a number a if the following three conditions are satisfied.

(i) f is defined on an open interval containing a.

(ii) $\lim\limits_{x\to a} f(x)$ exists.

(iii) $\lim\limits_{x\to a} f(x) = f(a)$.

If f is not continuous at a, then we say it is **discontinuous** at a, or has a **discontinuity** at a.

Example 1 (a) Prove that a polynomial function is continuous at every real number a.

(b) Prove that a rational function is continuous at every real number in its domain.

Solution (a) A polynomial function f is defined throughout $\mathbb{R}$. Moreover, by Theorem (2.18), $\lim_{x\to a} f(x) = f(a)$ for every real number a. Thus f satisfies conditions (i)–(iii) of Definition (2.27) and hence is continuous at a.

(b) If q is a rational function, then $q = f/h$, where f and h are polynomial functions. Consequently q is defined for all real numbers *except* the zeros of h. It follows that if $h(a) \neq 0$, then q is defined throughout an open interval containing a. Moreover, by (2.19), $\lim_{x\to a} q(x) = q(a)$. Applying Definition (2.27), q is continuous at a.

If f is continuous at a, then by (i) of Definition (2.27) there is a point $(a, f(a))$ on the graph of f. Moreover, since $\lim_{x \to a} f(x) = f(a)$, the closer x is to a, the closer $f(x)$ is to $f(a)$ or, in geometric terms, the closer the point $(x, f(x))$ on the graph of f is to the point $(a, f(a))$. More precisely, in a manner similar to that illustrated in Figure 2.7 (but with the point $(a, f(a))$ included), for every pair of horizontal lines $y = f(a) \pm \varepsilon$, there exist vertical lines $x = a \pm \delta$, such that if $a - \delta < x < a + \delta$, then the point $(x, f(x))$ on the graph of f lies within the shaded rectangular region. For this reason functions that are continuous at every number in a given interval are sometimes thought of as functions whose graphs can be sketched without lifting the pencil from the paper; that is, there are no breaks in the graph. Another way of interpreting a continuous function f is to say that a small change in x produces only a small change in the functional value $f(x)$. These are not accurate descriptions, but rather devices to help develop an intuitive feeling for continuous functions.

Since the notion of continuity involves the fact that $\lim_{x \to a} f(x) = f(a)$, the following result may be obtained by replacing L in Definition (2.6) by $f(a)$.

(2.28) Theorem

If a function f is defined on an open interval containing a, then f is continuous at a if for every $\varepsilon > 0$ there exists a $\delta > 0$ such that

$$|f(x) - f(a)| < \varepsilon \quad \text{whenever} \quad |x - a| < \delta$$

Note that we do not require $0 < |x - a|$ in Theorem (2.28), since f is defined at a. Moreover, if $x = a$, then

$$|f(x) - f(a)| = |f(a) - f(a)| = 0 < \varepsilon.$$

Graphs of several functions which are *not* continuous at a are sketched in Figure 2.15. The function having the graph illustrated in (i) fails to be continuous since f is undefined at a. The functions whose graphs are sketched in (ii) and (iii) are discontinuous at a since $\lim_{x \to a} f(x)$ does not exist. For the function with a graph as in (iv), both $f(a)$ and $\lim_{x \to a} f(x)$ exist but they are unequal and hence f is discontinuous at a by (iii) of Definition (2.27). The last illustration shows the necessity for checking all three conditions of the definition.

As a concrete illustration, consider the function f of Example 3 in the previous section, where $f(x) = x + 2$ if $x \neq 1$ and $f(1) = 0$ (see Figure 2.14). Since

$$\lim_{x \to 1} f(x) = 3 \neq f(1),$$

condition (iii) of Definition (2.27) is not satisfied, and hence f has a discontinuity at $x = 1$.

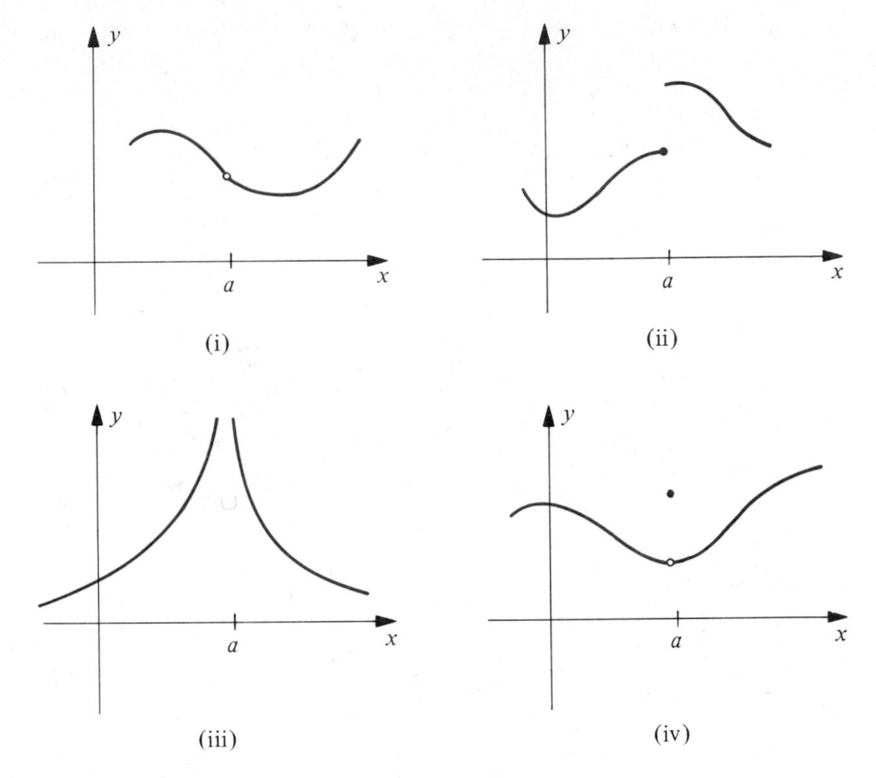

Figure 2.15

If $f(x) = 1/x$, then f has a discontinuity at $x = 0$. In this case *none* of the conditions of Definition (2.27) is satisfied (see Figure 2.8).

The functions whose graphs are sketched in Figure 2.15 appear to be continuous at numbers other than a. Most functions considered in calculus are of this type; that is, they may be discontinuous at certain numbers of their domains and continuous elsewhere.

If a function f is continuous at every number in an open interval (a, b), we say that **f is continuous on the interval (a, b)**. Similarly, a function is said to be continuous on an infinite interval of the form (a, ∞), $(-\infty, b)$, or $(-\infty, \infty)$ if it is continuous at every number in the interval. The next definition covers the case of a closed interval.

(2.29) Definition

If a function f is defined on a closed interval $[a, b]$, then **f is continuous on $[a, b]$** if it is continuous on (a, b) and if, in addition,

$$\lim_{x \to a^+} f(x) = f(a) \quad \text{and} \quad \lim_{x \to b^-} f(x) = f(b).$$

Generally, if any function f has either a right-hand or left-hand limit of the type indicated in Definition (2.29) we say that **f is continuous from the right at a** or that **f is continuous from the left at b**, respectively.

Example 2 If $f(x) = \sqrt{9 - x^2}$, sketch the graph of f and prove that f is continuous on the closed interval $[-3, 3]$.

Solution By (1.9), the graph of $x^2 + y^2 = 9$ or equivalently $y^2 = 9 - x^2$ is a circle with center at the origin and radius 3. It follows that the graph of $y = \sqrt{9 - x^2}$ and, therefore, the graph of f, is the upper half of that circle (see Figure 2.16). If $-3 < c < 3$ then, using Theorem (2.22),

$$\lim_{x \to c} f(x) = \lim_{x \to c} \sqrt{9 - x^2} = \sqrt{9 - c^2} = f(c).$$

Hence, by Definition (2.27), f is continuous at c.

According to Definition (2.29), all that remains is to check the end-points of the interval using one-sided limits. Since

$$\lim_{x \to -3^+} f(x) = \lim_{x \to -3^+} \sqrt{9 - x^2} = \sqrt{9 - 9} = 0 = f(-3)$$

f is continuous from the right at -3. We also have

$$\lim_{x \to 3^-} f(x) = \lim_{x \to 3^-} \sqrt{9 - x^2} = \sqrt{9 - 9} = 0 = f(3)$$

and hence f is continuous from the left at 3. This completes the proof that f is continuous on $[-3, 3]$.

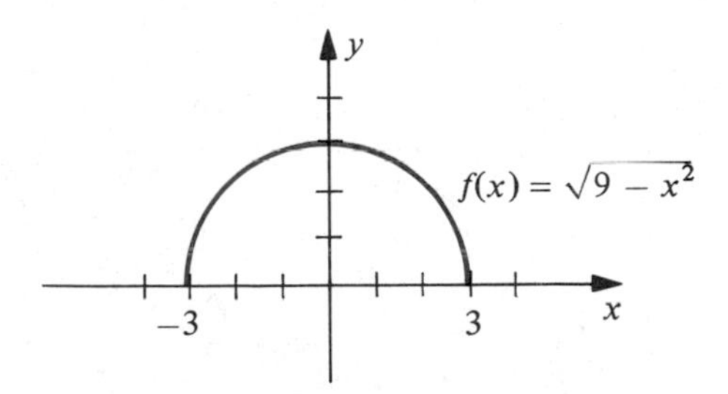

Figure 2.16

It should be evident how continuity on other types of intervals is defined. For example, a function f is said to be continuous on $[a, b)$ or $[a, \infty)$ if it is continuous at every number greater than a in the interval and if, in addition, f is continuous from the right at a. For intervals of the form $(a, b]$ or $(-\infty, b]$ we require continuity at every number less than b in the interval and also continuity from the left at b.

As illustrated in the next example, when asked to discuss the continuity of a function f we shall list the largest intervals on which f is continuous. Of course f will also be continuous on any subinterval of those intervals.

Example 3 Discuss the continuity of f if

$$f(x) = \frac{\sqrt{x^2 - 9}}{x - 4}.$$

Solution The function is undefined if the denominator $x - 4$ is zero (that is, if $x = 4$) or if the radicand $x^2 - 9$ is negative (that is, if $-3 < x < 3$). Any other real number is in one of the intervals $(-\infty, -3]$, $[3, 4)$, or $(4, \infty)$. The proof that f is continuous on each of these intervals is similar to that given in the solution of Example 2. Thus to prove continuity on $[3, 4)$ it is necessary to show that

$$\lim_{x \to c} f(x) = f(c) \quad \text{if } 3 < c < 4$$

and also that

$$\lim_{x \to 3^+} f(x) = f(3).$$

We shall leave the details of the proof for this, and the other intervals, to the reader.

Limit theorems discussed in Section 3 may be used to establish the following important result.

(2.30) Theorem

> If the functions f and g are continuous at a, then so are the sum $f + g$, the difference $f - g$, the product fg, and, if $g(a) \neq 0$, the quotient f/g.

Proof. If f and g are continuous at a, then $\lim_{x \to a} f(x) = f(a)$ and $\lim_{x \to a} g(x) = g(a)$. By the definition of sum $(f + g)(x) = f(x) + g(x)$. Consequently

$$\begin{aligned} \lim_{x \to a} (f + g)(x) &= \lim_{x \to a} [f(x) + g(x)] \\ &= \lim_{x \to a} f(x) + \lim_{x \to a} g(x) \\ &= f(a) + g(a) \\ &= (f + g)(a), \end{aligned}$$

which proves that $f + g$ is continuous at a. The remainder of the theorem is proved in similar fashion.

If f and g are continuous on an interval I it follows that $f + g$, $f - g$, and fg are continuous on I. If, in addition, $g(a) \neq 0$ throughout I, then f/g is continuous on I. These results may be extended to more than two functions; that is, sums, differences, products, or quotients involving any number of continuous functions are continuous (provided zero denominators do not occur).

The next result on limits of composite functions has many applications.

(2.31) Theorem

> If f and g are functions such that $\lim_{x \to a} g(x) = b$ and f is continuous at b, then
>
> $$\lim_{x \to a} f(g(x)) = f(b) = f(\lim_{x \to a} g(x)).$$

Proof. As was pointed out in Chapter 1 (see Figure 1.33), the composite function $f(g(x))$ may be represented geometrically by means of three real lines l, l', and l'' as shown in Figure 2.17, where to each coordinate x on l there corresponds the coordinate $g(x)$ on l' and then, in turn, $f(g(x))$ on l''. We wish to prove that $f(g(x))$ has the limit $f(b)$ as x approaches a. In terms of Definition (2.6) we must show that for every $\varepsilon > 0$ there exists a $\delta > 0$ such that

(2.32) $$\text{if} \quad 0 < |x - a| < \delta, \quad \text{then} \quad |f(g(x)) - f(b)| < \varepsilon.$$

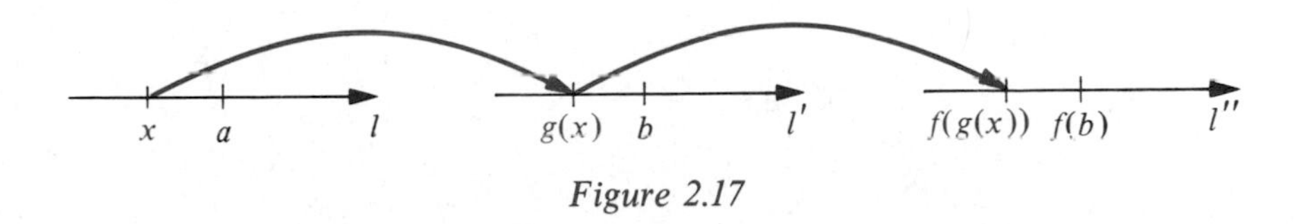

Figure 2.17

Let us begin by considering the interval $(f(b) - \varepsilon, f(b) + \varepsilon)$ on l'' shown in color in Figure 2.18. Since f is continuous at b, $\lim_{z \to b} f(z) = f(b)$ and hence, as illustrated in the figure, there exists a number $\delta_1 > 0$ such that

(2.33) $$\text{if} \quad |z - b| < \delta_1, \quad \text{then} \quad |f(z) - f(b)| < \varepsilon.$$

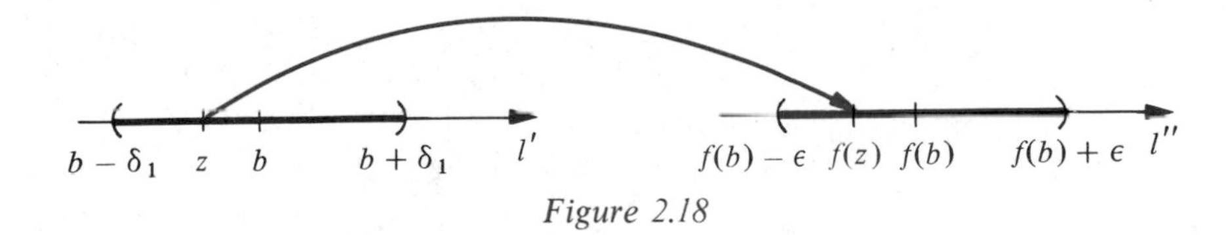

Figure 2.18

In particular, if we let $z = g(x)$ we see that

(2.34) $$\text{if} \quad |g(x) - b| < \delta_1, \quad \text{then} \quad |f(g(x)) - f(b)| < \varepsilon.$$

Next, turning our attention to the interval $(b - \delta_1, b + \delta_1)$ on l' and using the definition of $\lim_{x \to a} g(x) = b$, we obtain the fact illustrated in Figure 2.19, that there exists a $\delta > 0$ such that

(2.35) $$\text{if} \quad 0 < |x - a| < \delta, \quad \text{then} \quad |g(x) - b| < \delta_1.$$

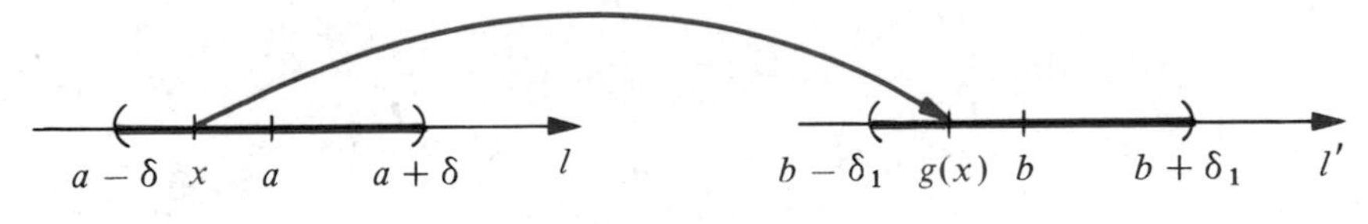

Figure 2.19

Finally, combining (2.35) and (2.34) we see that

$$\text{if} \quad 0 < |x - a| < \delta \quad \text{then} \quad |f(g(x)) - f(b)| < \varepsilon$$

which is the desired conclusion (2.32).

The principal use of Theorem (2.31) is to establish other theorems. To illustrate, if n is a positive integer and $f(x) = \sqrt[n]{x}$, then

$$f(g(x)) = \sqrt[n]{g(x)}$$

and

$$f(\lim_{x\to a} g(x)) = \sqrt[n]{\lim_{x\to a} g(x)}.$$

If we now substitute in the equation

$$\lim_{x\to a} f(g(x)) = f(\lim_{x\to a} g(x))$$

the result stated in Theorem (2.22) is obtained, that is,

$$\lim_{x\to a} \sqrt[n]{g(x)} = \sqrt[n]{\lim_{x\to a} g(x)},$$

where it is assumed that the indicated nth roots exist.

If, in Theorem (2.31), g is continuous at a and f is continuous at $b = g(a)$, then

(2.36) $$\lim_{x\to a} f(g(x)) = f(\lim_{x\to a} g(x)) = f(g(a));$$

that is, the composite function of f by g is continuous at a. This result may be extended to functions which are continuous on intervals. Sometimes this is expressed by the statement "the composite function of a continuous function by a continuous function is continuous."

Example 4 If $f(x) = |x|$, prove that f is continuous at every real number a.

Solution Since $|x| = \sqrt{x^2}$ we have, by (2.22) and (2.13),

$$\lim_{x\to a} f(x) = \lim_{x\to a} |x| = \lim_{x\to a} \sqrt{x^2}$$

$$= \sqrt{\lim_{x\to a} x^2} = \sqrt{a^2} = |a| = f(a).$$

Hence, from Definition (2.27), f is continuous at a.

We shall conclude this section by stating an important property of continuous functions. A proof may be found in more advanced texts on calculus.

(2.37) The Intermediate Value Theorem

> If a function f is continuous on a closed interval $[a, b]$ and if $f(a) \neq f(b)$, then f takes on every value between $f(a)$ and $f(b)$ in the interval $[a, b]$.

Theorem (2.37) states that if w is any number between $f(a)$ and $f(b)$, then there is a number c between a and b such that $f(c) = w$. If the graph of the continuous function f is regarded as extending in an unbroken manner from the point $(a, f(a))$ to the point $(b, f(b))$, as illustrated in Figure 2.20, then for any number w between $f(a)$ and $f(b)$ it appears that a horizontal line with y-intercept w should intersect the graph in at least one point P. The abscissa c of P is a number such that $f(c) = w$.

A corollary of Theorem (2.37) is that if $f(a)$ and $f(b)$ have opposite signs, then there is a number c between a and b such that $f(c) = 0$; that is, f has a zero at c. Geometrically, this implies that if the point $(a, f(a))$ on the graph of a continuous function lies below the x-axis, and the point $(b, f(b))$ lies above the x-axis, or vice versa, then the graph crosses the x-axis at some point $(c, 0)$, where $a < c < b$.

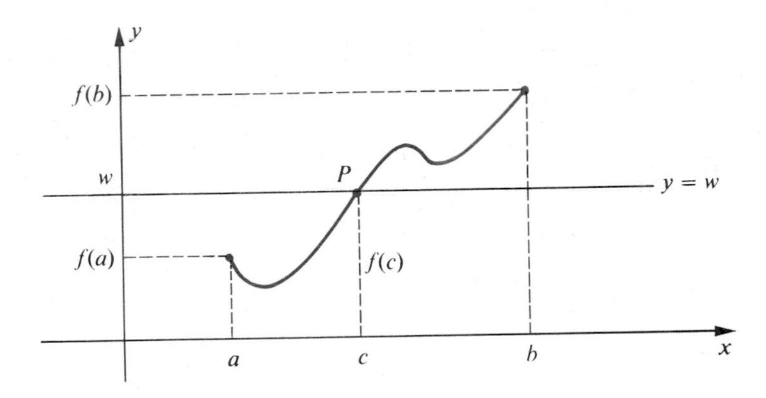

Figure 2.20

Example 5 Verify the Intermediate Value Theorem (2.37) if $f(x) = \sqrt{x + 1}$ and the interval is $[3, 24]$.

Solution The function f is continuous on $[3, 24]$ and $f(3) = 2, f(24) = 5$. If w is any real number between 2 and 5 we must find a number c in the interval $[3, 24]$ such that $f(c) = w$, that is, $\sqrt{c + 1} = w$. Squaring both sides of the last equation and solving for c we obtain $c = w^2 - 1$. This number c is in the interval $[3, 24]$, since if $2 < w < 5$, then

$$4 < w^2 < 25, \quad \text{or} \quad 3 < w^2 - 1 < 24.$$

To check our work we see that

$$f(c) = f(w^2 - 1) = \sqrt{(w^2 - 1) + 1} = w.$$

EXERCISES 2.5

In Exercises 1–6 show that the function f is continuous at the given number a.

1 $f(x) = \sqrt{2x - 5} + 3x, a = 4$

2 $f(x) = 3x^2 + 7 - 1/\sqrt{-x}, a = -2$

3 $f(x) = \dfrac{x}{x^2 - 4}, a = 3$

4 $f(x) = \dfrac{1}{x}, a = 10^{-6}$

5 $f(x) = \sqrt[3]{x^2 + 2}, a = -5$

6 $f(x) = \dfrac{\sqrt[3]{x}}{2x + 1}, a = 8$

In Exercises 7–10, show that f is continuous on the indicated interval.

7 $f(x) = \sqrt{x-4}$; $[4, 8]$

8 $f(x) = \sqrt{16-x}$; $(-\infty, 16]$

9 $f(x) = \dfrac{1}{x^2}$; $(0, \infty)$

10 $f(x) = \dfrac{1}{x-1}$; $(1, 3)$

In each of Exercises 11–22 find all numbers for which the function f is continuous.

11 $f(x) = \dfrac{3x-5}{2x^2-x-3}$

12 $f(x) = \dfrac{x^2-9}{x-3}$

13 $f(x) = \sqrt{2x-3} + x^2$

14 $f(x) = \dfrac{x}{\sqrt[3]{x-4}}$

15 $f(x) = \dfrac{x-1}{\sqrt{x^2-1}}$

16 $f(x) = \dfrac{x}{\sqrt{1-x^2}}$

17 $f(x) = \dfrac{|x+9|}{x+9}$

18 $f(x) = \dfrac{x}{x^2+1}$

19 $f(x) = \dfrac{5}{x^3-x^2}$

20 $f(x) = \dfrac{4x-7}{(x+3)(x^2+2x-8)}$

21 $f(x) = \dfrac{\sqrt{x^2-9}\sqrt{25-x^2}}{x-4}$

22 $f(x) = \dfrac{\sqrt{9-x}}{\sqrt{x-6}}$

23–28 Discuss the discontinuities of the functions defined in Exercises 21–26 of Section 4.

29 Prove that if $f(x) = 1/x$, then f is continuous on every open interval that does not contain the origin. What is true for open intervals containing the origin?

30 If

$$f(x) = \begin{cases} 1 & \text{if } x \neq 3 \\ 0 & \text{if } x = 3 \end{cases}$$

is f continuous at 3? Explain.

31 If

$$f(x) = \begin{cases} \dfrac{|x-3|}{x-3} & \text{if } x \neq 3 \\ 1 & \text{if } x = 3 \end{cases}$$

is f continuous at 3? Explain.

32 Suppose $f(x) = 0$ if x is rational and $f(x) = 1$ if x is irrational. Prove that f is discontinuous at every real number a.

33 Prove that a function f is continuous at a if and only if $\lim_{h \to 0} f(a+h) = f(a)$.

34 If f is continuous on an interval containing c, and if $f(c) > 0$, prove that $f(x)$ is positive throughout an interval containing c. (*Hint:* See Theorem (2.10).)

In each of Exercises 35–38, verify the Intermediate Value Theorem (2.37) for f on the stated interval $[a, b]$ by showing that if w is any number between $f(a)$ and $f(b)$, then there is a number c in $[a, b]$ such that $f(c) = w$.

35 $f(x) = x^3 + 1$; $[-1, 2]$

36 $f(x) = -x^3$; $[0, 2]$

37 $f(x) = x^2 + 4x + 4$; $[0, 1]$

38 $f(x) = x^2 - x$; $[-1, 3]$

39 If $f(x) = x^3 - 5x^2 + 7x - 9$, prove that there is a real number a such that $f(a) = 100$.

40 Prove that the equation

$$x^5 - 3x^4 - 2x^3 - x + 1 = 0$$

has a solution between 0 and 1.

2.6 REVIEW

Concepts

Define or discuss each of the following.

1. The limit of a function as x approaches a
2. The geometric interpretations of $\lim_{x \to a} f(x) = L$
3. Theorems on limits
4. Limits of polynomial and rational functions
5. Right- and left-hand limits of functions
6. Continuous function
7. Discontinuities of a function
8. Continuity on an interval
9. Limit of a composite function
10. The Intermediate Value Theorem

Exercises

In each of Exercises 1–20 find the limit, if it exists.

1 $\displaystyle\lim_{x \to 3} \frac{5x + 11}{\sqrt{x + 1}}$

2 $\displaystyle\lim_{x \to -2} \frac{6 - 7x}{(3 + 2x)^4}$

3 $\displaystyle\lim_{x \to -2} (2x - \sqrt{4x^2 + x})$

4 $\displaystyle\lim_{x \to 4^-} (x - \sqrt{16 - x^2})$

5 $\displaystyle\lim_{x \to 3/2} \frac{2x^2 + x - 6}{4x^2 - 4x - 3}$

6 $\displaystyle\lim_{x \to 2} \frac{3x^2 - x - 10}{x^2 - x - 2}$

7 $\displaystyle\lim_{x \to 2} \frac{x^4 - 16}{x^2 - x - 2}$

8 $\displaystyle\lim_{x \to 3^+} \frac{1}{x - 3}$

9 $\displaystyle\lim_{x \to 0^+} \frac{1}{\sqrt{x}}$

10 $\displaystyle\lim_{x \to 5} \frac{(1/x) - (1/5)}{x - 5}$

11 $\displaystyle\lim_{x \to 1/2} \frac{8x^3 - 1}{2x - 1}$

12 $\displaystyle\lim_{x \to 2} 5$

13 $\lim_{x\to 3^+} \dfrac{3-x}{|3-x|}$

14 $\lim_{x\to 2} \dfrac{\sqrt{x}-\sqrt{2}}{x-2}$

15 $\lim_{h\to 0} \dfrac{(a+h)^4-a^4}{h}$

16 $\lim_{x\to -3} \sqrt[3]{\dfrac{x+3}{x^3+27}}$

17 $\lim_{h\to 0} \dfrac{(2+h)^{-3}-2^{-3}}{h}$

18 $\lim_{x\to 5} (x^2+3)^0$

19 $\lim_{x\to 2^+} \dfrac{|x-2|}{2-x}$

20 $\lim_{x\to 1} \dfrac{x-1}{\sqrt{(x-1)^2}}$

Find the limits in Exercises 21 and 22, where [ ] denotes the greatest integer function.

21 $\lim_{x\to 3^+} ([x]-x^2)$

22 $\lim_{x\to 3^-} ([x]-x^2)$

23 Prove, directly from the definition of limit, that $\lim_{x\to 6}(5x-21)=9$.

24 Suppose $f(x)=1$ if x is rational and $f(x)=-1$ if x is irrational. Prove that $\lim_{x\to a} f(x)$ does not exist for any real number a.

In each of Exercises 25–28, find all numbers for which f is continuous.

25 $f(x)=2x^4-\sqrt[3]{x}+1$

26 $f(x)=\sqrt{(2+x)(3-x)}$

27 $f(x)=\dfrac{\sqrt{9-x^2}}{x^4-16}$

28 $f(x)=\dfrac{\sqrt{x}}{x^2-1}$

In each of Exercises 29–32 find the discontinuities of f.

29 $f(x)=\dfrac{|x^2-16|}{x^2-16}$

30 $f(x)=\dfrac{1}{x^2-16}$

31 $f(x)=\dfrac{x^2-x-2}{x^2-2x}$

32 $f(x)=\dfrac{x+2}{x^3-8}$

33 If $f(x)=1/x^2$, verify the Intermediate Value Theorem (2.37) for f on the interval $[2,3]$.

34 Prove that the equation

$$x^5+7x^2-3x-5=0$$

has a root between -2 and -1.

The Derivative

The derivative of a function *is one of the most powerful tools in mathematics. Indeed, it is indispensable for nonelementary investigations in both the natural and human sciences. We shall begin our work by reformulating the notion of tangent line introduced in the preceding chapter and then discussing the velocity of a moving object. These two concepts serve to motivate the definition of derivative given in Section 2. The remainder of the chapter is concerned primarily with properties of the derivative.*

3.1 INTRODUCTION

Let $P(a, f(a))$ be any point on the graph of a function f. Another point on the graph may be denoted by $Q(a + h, f(a + h))$, where h is the difference between the abscissas of Q and P (see (i) of Figure 3.1). By Definition (1.11), the slope m_{PQ} of the secant line through P and Q is

$$m_{PQ} = \frac{f(a + h) - f(a)}{h}.$$

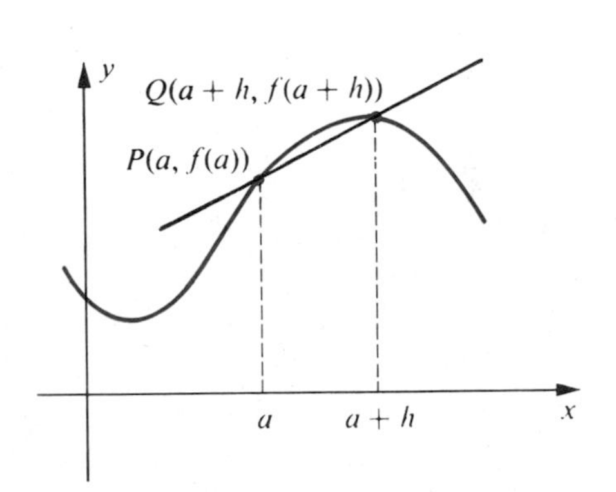

(i) Slope of secant line:

$$m_{PQ} = \frac{f(a + h) - f(a)}{h}$$

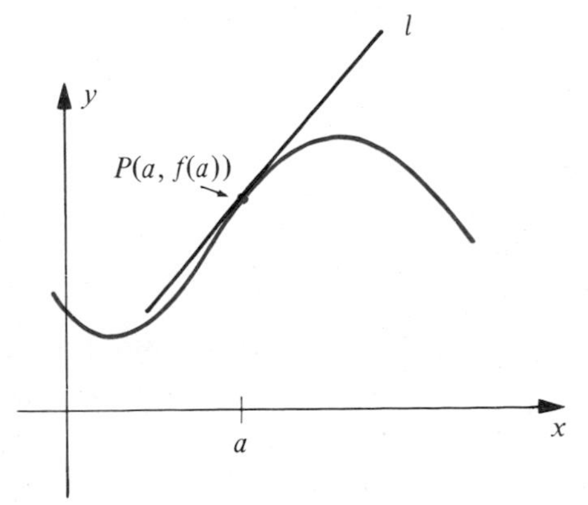

(ii) Slope of tangent line l:

$$m = \lim_{h \to 0} \frac{f(a + h) - f(a)}{h}$$

Figure 3.1

In Chapter 2 the slope m of the tangent line l at P was defined as the limiting value of m_{PQ} as Q approaches P (see (ii) of Figure 3.1). If f is continuous, then we can make Q approach P by letting h approach 0. It is natural, therefore, to define m as follows.

(3.1) Definition

If a function f is defined on an open interval containing a, then the **slope m of the tangent line** to the graph of f at the point $P(a, f(a))$ is given by

$$m = \lim_{h \to 0} \frac{f(a+h) - f(a)}{h}$$

provided the limit exists.

If a tangent line is vertical, its slope is undefined and the limit in (3.1) does not exist. Vertical tangent lines will be studied in the next chapter.

Example 1 If $f(x) = x^2$, find the slope of the tangent line to the graph of f at the point $P(a, a^2)$.

Solution This problem is the same as the one stated for equations in Example 1 of the first section in Chapter 2. Using the quotient in Definition (3.1),

$$\begin{aligned}\frac{f(a+h) - f(a)}{h} &= \frac{(a+h)^2 - a^2}{h}\\ &= \frac{a^2 + 2ah + h^2 - a^2}{h}\\ &= \frac{2ah + h^2}{h}\\ &= 2a + h.\end{aligned}$$

The slope m of the tangent line is, therefore,

$$m = \lim_{h \to 0} (2a + h) = 2a.$$

One of the main reasons for the invention of calculus was the need for a way to study the behavior of objects in motion. Let us consider the problem of arriving at a satisfactory definition for the velocity, or speed, of an object at a given instant. We shall assume, for simplicity, that the object is moving on a straight line. Motion on a straight line is called **rectilinear motion**. It is easy to define the **average velocity** r during an interval of time. We merely use the formula

$$r = \frac{d}{t} \tag{3.2}$$

where t denotes the length of the time interval and d is the distance between the initial position of the object and its position after t units of time.

As an elementary illustration, suppose an automobile leaves city A at 1:00 P.M. and travels along a straight highway, arriving at city B, 150 miles from A, at 4:00 P.M. Employing (3.2) we see that its average velocity r during the indicated time interval is 150/3, or 50 miles per hour. This is the velocity which, if maintained for three hours, would enable the automobile to travel the distance from A to B. The average velocity gives no information whatsoever about the velocity at any instant. For example, at 2:30 P.M. the automobile's speedometer may have registered 40, or 30, or the automobile may not even have been moving. If we wish to determine the rate at which the automobile is traveling at 2:30 P.M., information is needed about its motion or position *near* this time. For example, suppose at 2:30 P.M. the automobile is 80 miles from A and at 2:35 P.M. it is 84 miles from A, as illustrated in Figure 3.2.

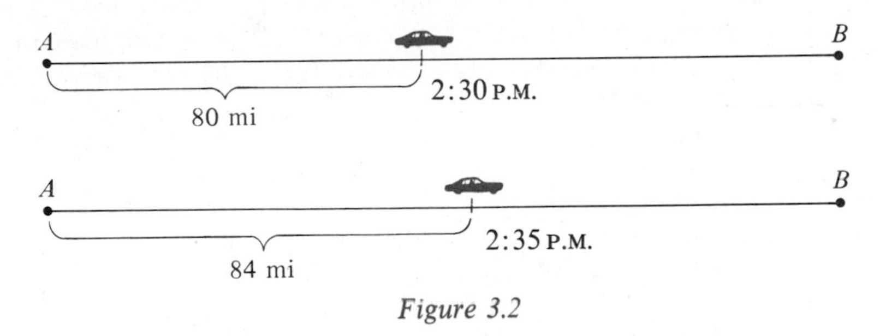

Figure 3.2

For the interval from 2:30 P.M. to 2:35 P.M. the elapsed time t is 5 minutes, or 1/12 hours, and the distance d is 4 miles. Substituting in (3.2), the average velocity r during this time interval is

$$r = \frac{4}{(1/12)} = 48 \text{ miles per hour.}$$

However, this is still not an accurate indication of the velocity at 2:30 P.M. since, for example, the automobile may have been traveling very slowly at 2:30 P.M. and then speeded up considerably so as to arrive at the point 84 miles from A at 2:35 P.M. Evidently, a better approximation of the motion would be obtained by considering the average velocity during a smaller time interval, say from 2:30 P.M. to 2:31 P.M. Indeed, it appears that the best procedure would be to take smaller and smaller time intervals near 2:30 P.M. and study the average velocity in each time interval. This leads us into a limiting process similar to that discussed for tangent lines.

In order to base our discussion on mathematical concepts, let us assume that the position of an object moving rectilinearly may be represented by a point P on a coordinate line l. We shall sometimes refer to the motion of the *point* P on l, or the motion of a *particle* on l whose position is specified by P. We further assume that the position of P is known at every instant in a given interval of time. If $f(t)$ denotes the coordinate of P at time t, then the function f determined in this way is called the **position function** for P. If we keep track of time by means of a clock, then for each t the point P is $f(t)$ units from the origin, as illustrated in Figure 3.3.

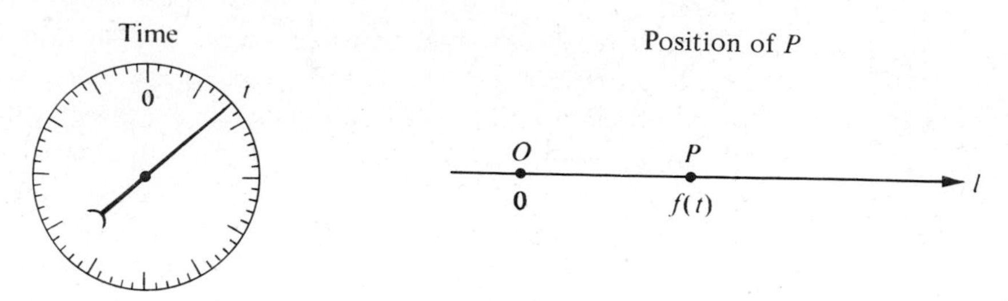

Figure 3.3

In order to define the velocity of P at time a, we begin by investigating the average velocity in a (small) time interval near a. Thus we consider times a and $a + h$ where h may be positive or negative but not zero. The corresponding positions of P on the l-axis are given by $f(a + h)$ and $f(a)$, as illustrated in Figure 3.4, and hence the amount of change in the position of P is $f(a + h) - f(a)$. The latter number may be positive, negative, or zero, depending on whether the position of P at time $a + h$ is to the right, to the left, or the same as, its position at time a. This number is not necessarily the distance traversed by P during the time interval $[a, a + h]$ since, for example, P may have moved beyond the point corresponding to $f(a + h)$ and then returned to that point at time $a + h$.

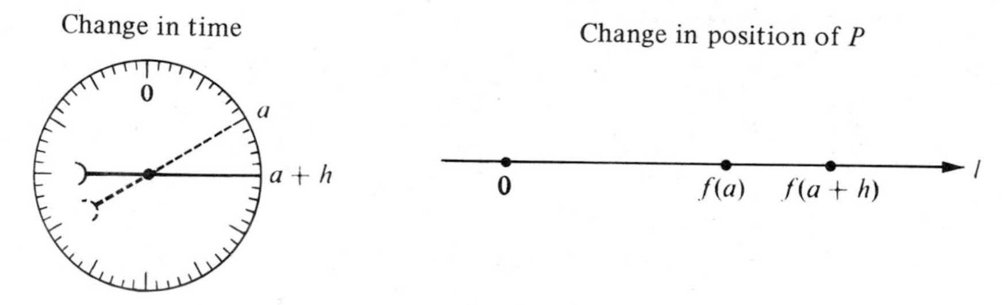

Figure 3.4

By (3.2), the average velocity of P during the time interval $[a, a + h]$ is given by

$$\text{Average Velocity} = \frac{f(a + h) - f(a)}{h}.$$

As in our previous discussion, the smaller h is numerically, the closer this quotient should approximate the velocity of P at time a. Accordingly, we *define* the velocity as the limit, as h approaches zero, of the average velocity, provided the limit exists. This limit is also called the *instantaneous velocity* of P at time a. Summarizing, we have the following definition.

(3.3) **Definition**

If a point P moves on a coordinate line l such that its coordinate at time t is $f(t)$, then the **velocity** $v(a)$ of P at time a is given by

$$v(a) = \lim_{h \to 0} \frac{f(a + h) - f(a)}{h}$$

provided the limit exists.

If $f(t)$ is measured in centimeters and t in seconds, then the unit of velocity is centimeters per second, abbreviated cm/sec. If $f(t)$ is in miles and t in hours, then velocity is in miles per hour (mi/hr). Other units of measurement may, of course, be used.

We shall return to the velocity concept in Chapter 4, where we will see that if the velocity is positive in a given time interval, then the point is moving in the positive direction on l, whereas if the velocity is negative, the point is moving in the negative direction. Although these facts have not been proved we shall use them in the following example.

Example 2 The position of a point P on a coordinate line l is given by $f(t) = t^2 - 6t$, where $f(t)$ is measured in feet and t in seconds. Find the velocity at time a. What is the velocity at $t = 0$? At $t = 4$? Determine time intervals in which (a) P moves in the positive direction on l and (b) P moves in the negative direction. At what time is the velocity 0?

Solution From the formula for $f(t)$ we obtain

$$\begin{aligned}\frac{f(a+h) - f(a)}{h} &= \frac{[(a+h)^2 - 6(a+h)] - [a^2 - 6a]}{h} \\ &= \frac{a^2 + 2ah + h^2 - 6a - 6h - a^2 + 6a}{h} \\ &= \frac{2ah + h^2 - 6h}{h} \\ &= 2a + h - 6.\end{aligned}$$

Consequently, by Definition (3.3), the velocity $v(a)$ at time a is

$$v(a) = \lim_{h \to 0} \frac{f(a+h) - f(a)}{h} = \lim_{h \to 0} (2a + h - 6) = 2a - 6.$$

In particular, the velocity at $t = 0$ is $v(0) = 2(0) - 6 = -6$ ft/sec. At $t = 4$ it is $v(4) = 2(4) - 6 = 2$ ft/sec. According to the remarks preceding this example, P moves to the left when the velocity is negative; that is, when $2a - 6 < 0$, or $a < 3$. The particle moves to the right when $2a - 6 > 0$ or $a > 3$. In terms of intervals, the motion is to the left in the time interval $(-\infty, 3)$ and to the right in $(3, \infty)$. The velocity is zero when $2a - 6 = 0$, that is, when $a = 3$ seconds.

The reader has undoubtedly noted the similarity of the limit in Definition (3.3) to that used in the definition of tangent line in (3.1). Indeed, the two expressions are identical! There are many different mathematical and physical applications which lead to precisely this same limit. Several of these will be discussed in the next chapter.

EXERCISES 3.1

In Exercises 1–4 find the slope of the tangent line at the point $P(a, f(a))$ on the graph of f. Sketch the graph and show the tangent lines at various points.

1 $f(x) = 2 - x^3$

2 $f(x) = 3x - 5$

3 $f(x) = \sqrt{x} + 1$

4 $f(x) = \dfrac{1}{x} - 1$

In each of Exercises 5 and 6, the position of a point P moving on a coordinate line l is given by $f(t)$ where t is measured in seconds and $f(t)$ in centimeters.

(a) Find the average velocity of P in the following time intervals: $[1, 1.2]$; $[1, 1.1]$; $[1, 1.01]$; $[1, 1.001]$.
(b) Find the velocity of P at $t = 1$.
(c) Determine the time intervals in which P moves in the positive direction.
(d) Determine the time intervals in which P moves in the negative direction.

5 $f(t) = 4t^2 + 3t$

6 $f(t) = t^3$

7 If an object is projected directly upward from the ground with an initial velocity of 112 ft/sec, then its distance $f(t)$ above the ground after t seconds is $112t - 16t^2$. What is the velocity of the object at $t = 2$, $t = 3$, and $t = 4$? At what time does the object reach its maximum height? When does the object strike the ground? What is the velocity at the moment of impact?

8 If an object is dropped from a balloon 500 feet above the ground, then its distance $f(t)$ above ground after t seconds is $500 - 16t^2$. Find the velocity at $t = 1$, $t = 2$, and $t = 3$. With what velocity does the object strike the ground?

9 If the position of an object which moves rectilinearly is given by a polynomial function of degree 1, prove that the velocity is constant.

10 If the position function of a rectilinearly moving object is a constant function, prove that the velocity is 0 at all times. Describe the motion of the particle.

3.2 DEFINITION OF DERIVATIVE

In the preceding section the same limiting process occurred in two very different situations. The limit which appears in Definitions (3.1) and (3.3) is one of the fundamental concepts of calculus. In the remainder of this chapter a number of rules pertaining to this concept will be developed. We begin by introducing the terminology and notation associated with this limit.

(3.4) Definition

If a function f is defined on an open interval containing a, then the **derivative of f at a**, written $f'(a)$, is given by

$$f'(a) = \lim_{h \to 0} \frac{f(a+h) - f(a)}{h}$$

provided the limit exists.

The symbol $f'(a)$ is read "f prime of a." The terminology "$f'(a)$ exists" will mean that the limit in Definition (3.4) exists. If $f'(a)$ exists we say that the function f is **differentiable at a** or **f has a derivative at a**.

It is important to observe that if f is differentiable at a, then by Definition (3.1), **$f'(a)$ is the slope of the tangent line to the graph of f at the point $(a, f(a))$.**

We say that **a function f is differentiable on an open interval (a, b)** if it is differentiable at every number c in (a, b). In like manner, we refer to functions which are differentiable on intervals of the form (a, ∞), $(-\infty, a)$, or $(-\infty, \infty)$. For closed intervals we use the following convention, which is analogous to the definition of continuity on a closed interval given in (2.29).

(3.5) Definition

A function **f is differentiable on a closed interval $[a, b]$** if it is differentiable on (a, b) and if the following limits exist:

$$\lim_{h \to 0^+} \frac{f(a+h) - f(a)}{h} \quad \text{and} \quad \lim_{h \to 0^-} \frac{f(b+h) - f(b)}{h}.$$

The one-sided limits specified in Definition (3.5) are sometimes referred to as the **right-hand** and **left-hand derivatives** of f at a and b, respectively.

If f is defined on a closed interval $[a, b]$ and is undefined outside of this interval, then the right-hand and left-hand derivatives allow us to define the slope of the tangent lines at the points $P(a, f(a))$ and $R(b, f(b))$, as illustrated in Figure 3.5. Thus, for the slope of the tangent line at P we take the limiting value of the slope m_{PQ} of the secant line through P and Q as Q approaches P from the right. For the tangent line at R, the point Q approaches R from the left.

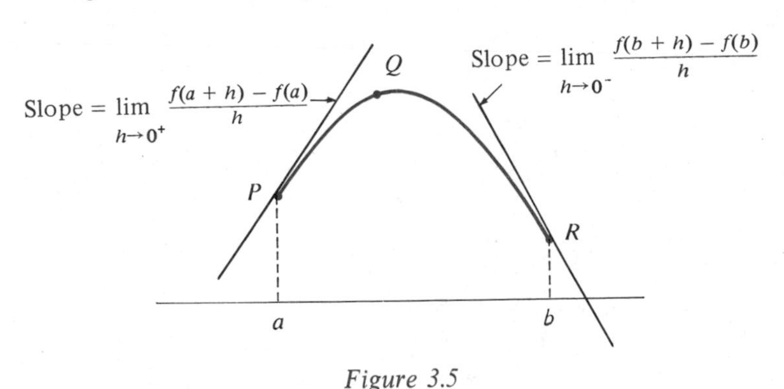

Figure 3.5

Differentiability on an interval of the form $[a, b)$, $[a, \infty)$, $(a, b]$, or $(-\infty, b]$ is defined in the obvious way, using a one-sided limit at an end-point.

It should be clear that if f is defined on an open interval containing a, then *$f'(a)$ exists if and only if both the right-hand and left-hand derivatives exist at a, and* *are equal.* The functions whose graphs are sketched in Figure 3.6 have right-hand and left-hand derivatives at a which give the slopes of the indicated lines l_1 and l_2, respectively. However, since the slopes of l_1 and l_2 are unequal, $f'(a)$ does not exist. Generally, if the graph of f has a corner at $(a, f(a))$, then f is not differentiable at a.

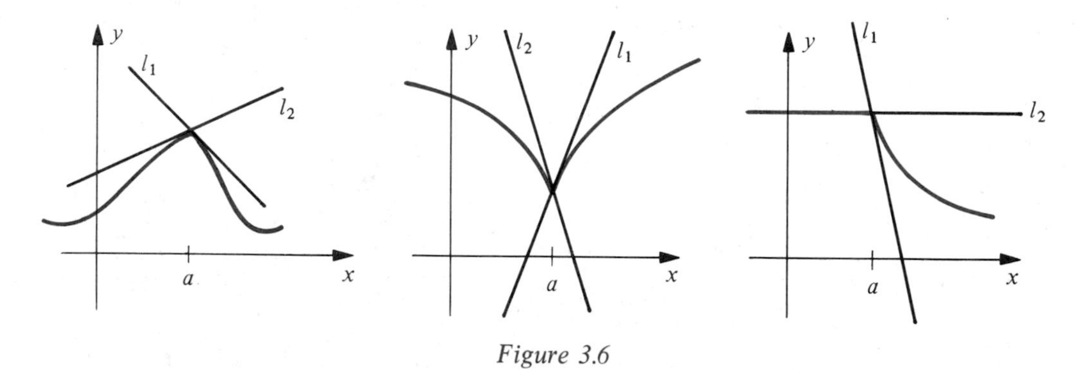

Figure 3.6

Given a function f, let $S = \{a : f'(a) \text{ exists}\}$. If f is defined only on certain intervals, then the terminology "$f'(a)$ exists" may refer to the existence of certain right-hand or left-hand limits of the type specified in Definition (3.5). If, with each x in S, we associate the number $f'(x)$, we obtain a function f' with domain S called the **derivative of f**. The value of f' at x is given by

(3.6)
$$f'(x) = \lim_{h \to 0} \frac{f(x+h) - f(x)}{h}$$

or by an appropriate one-sided limit.

It is important to note that in determining $f'(x)$, the number x is fixed, but arbitrary, and the limit is taken as h approaches zero. The terminology "differentiate $f(x)$" is used interchangeably with "find the derivative of $f(x)$." In statements of definitions and theorems, the expression "f is differentiable" will mean that $f'(x)$ exists for all numbers x in some set of real numbers. The domain of f' will not always be stated explicitly; however, for specific problems, it may often be found by inspection.

Example 1 If $f(x) = 3x^2 - 5x + 4$, find $f'(x)$. What is the domain of f'? Find $f'(2)$, $f'(-\sqrt{2})$, and $f'(a)$.

Solution Employing (3.6),

$$\begin{aligned} f'(x) &= \lim_{h \to 0} \frac{f(x+h) - f(x)}{h} \\ &= \lim_{h \to 0} \frac{[3(x+h)^2 - 5(x+h) + 4] - (3x^2 - 5x + 4)}{h} \\ &= \lim_{h \to 0} \frac{(3x^2 + 6xh + 3h^2 - 5x - 5h + 4) - (3x^2 - 5x + 4)}{h} \\ &= \lim_{h \to 0} \frac{6xh + 3h^2 - 5h}{h} = \lim_{h \to 0} (6x + 3h - 5) \\ &= 6x - 5. \end{aligned}$$

Since $f'(x) = 6x - 5$ for all x, the domain of f' is $\mathbb{R}$. Substituting for x, we have

$$f'(2) = 6(2) - 5 = 7$$
$$f'(-\sqrt{2}) = 6(-\sqrt{2}) - 5 = -(6\sqrt{2} + 5)$$
$$f'(a) = 6a - 5.$$

Example 2 Find $f'(x)$ if $f(x) = \sqrt{x}$. What is the domain of f'?

Solution By (3.6),

$$f'(x) = \lim_{h \to 0} \frac{\sqrt{x+h} - \sqrt{x}}{h}.$$

When working problems we shall always assume that values of variables are chosen so that the given expressions are defined. Consequently, in this example it is assumed that x and h are numbers for which $\sqrt{x+h}$ and $\sqrt{x}$ exist. In particular, $x > 0$. In order to find the limit we begin by multiplying numerator and denominator of the indicated quotient by $\sqrt{x+h} + \sqrt{x}$. Thus

$$\begin{aligned}
f'(x) &= \lim_{h \to 0} \frac{\sqrt{x+h} - \sqrt{x}}{h} \cdot \frac{\sqrt{x+h} + \sqrt{x}}{\sqrt{x+h} + \sqrt{x}} \\
&= \lim_{h \to 0} \frac{(x+h) - x}{h(\sqrt{x+h} + \sqrt{x})} \\
&= \lim_{h \to 0} \frac{1}{\sqrt{x+h} + \sqrt{x}} \\
&= \frac{1}{\sqrt{x} + \sqrt{x}} \\
&= \frac{1}{2\sqrt{x}}
\end{aligned}$$

which is true for every $x > 0$. Let us examine the case $x = 0$ separately. If f' is differentiable at 0, then using (3.6) with $x = 0$,

$$\begin{aligned}
f'(0) &= \lim_{h \to 0^+} \frac{\sqrt{0+h} - \sqrt{0}}{h} \\
&= \lim_{h \to 0^+} \frac{\sqrt{h}}{h} = \lim_{h \to 0^+} \frac{1}{\sqrt{h}}
\end{aligned}$$

where a one-sided limit is necessary since 0 is an end-point of the domain of f. Since the last limit does not exist (see Exercise 38 of Section 2.3), $f'(0)$ does not exist. Hence the domain of f' is the set of positive real numbers.

Example 3 Show that the function f defined by $f(x) = |x|$ is not differentiable at 0.

Solution The graph of f is sketched in Figure 1.29. It is geometrically evident that f has no derivative at 0 since the graph has a corner at the origin. We can prove that $f'(0)$

does not exist by showing that the right-hand and left-hand derivatives of f at 0 are not equal. Using the limits in (3.5) with $a = 0$ and $b = 0$.

$$\lim_{h \to 0^+} \frac{f(0+h) - f(0)}{h} = \lim_{h \to 0^+} \frac{|0+h| - |0|}{h} = \lim_{h \to 0^+} \frac{|h|}{h} = 1$$

$$\lim_{h \to 0^-} \frac{f(0+h) - f(0)}{h} = \lim_{h \to 0^-} \frac{|0+h| - |0|}{h} = \lim_{h \to 0^-} \frac{|h|}{h} = -1.$$

Thus $f'(0)$ does not exist.

It follows from Example 3 that the graph of $y = |x|$ does not have a tangent line at the point $P(0, 0)$.

In (3.4), the derivative was defined as a certain limit. There is another important limit formula for $f'(a)$. To see how it arises geometrically, let us begin by labeling the graph of f as shown in Figure 3.7 (compare with (i) of Figure 3.1).

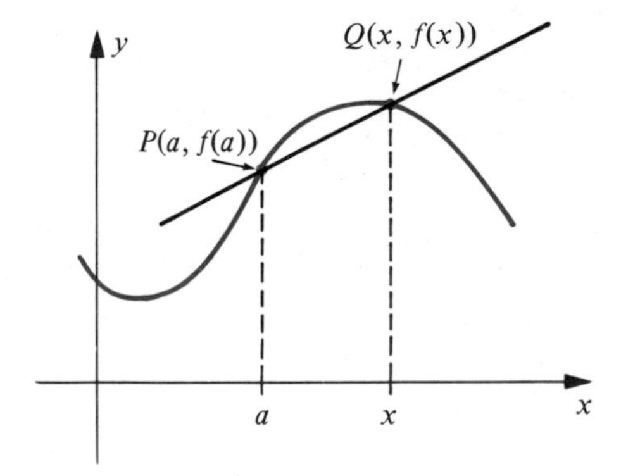

Figure 3.7

By Definition (1.10), the slope m_{PQ} of the secant line through P and Q is given by

$$m_{PQ} = \frac{f(x) - f(a)}{x - a}.$$

If f is continuous at a, we can make Q approach P by letting x approach a. Thus it appears that the slope of the tangent line at P is given by

$$m = \lim_{x \to a} \frac{f(x) - f(a)}{x - a}$$

provided the limit exists. Since $m = f'(a)$ this leads to the alternate formula for the derivative of f stated in the next theorem. A nongeometric proof of the formula will be found in Appendix II.

(3.7) Theorem

If f is defined on an open interval containing a, then

$$f'(a) = \lim_{x \to a} \frac{f(x) - f(a)}{x - a}$$

provided the limit exists.

Example 4 If $f(x) = x^{1/3}$ and $a \neq 0$, calculate $f'(a)$.

Solution By Theorem (3.7),

$$f'(a) = \lim_{x \to a} \frac{x^{1/3} - a^{1/3}}{x - a}$$

provided the limit exists. We now change the form of the indicated quotient as follows:

$$\lim_{x \to a} \frac{x^{1/3} - a^{1/3}}{x - a} = \lim_{x \to a} \frac{x^{1/3} - a^{1/3}}{(x^{1/3})^3 - (a^{1/3})^3}$$

$$= \lim_{x \to a} \frac{x^{1/3} - a^{1/3}}{(x^{1/3} - a^{1/3})(x^{2/3} + x^{1/3}a^{1/3} + a^{2/3})}.$$

Dividing numerator and denominator of the last quotient by $x^{1/3} - a^{1/3}$ and taking thc limit gives us

$$f'(a) = \frac{1}{a^{2/3} + a^{2/3} + a^{2/3}} = \frac{1}{3a^{2/3}}.$$

(3.8) Theorem

If a function f is differentiable at a, then f is continuous at a.

Proof. If x is in the domain of f and $x \neq a$, then $f(x)$ may be written as follows:

$$f(x) = f(a) + \frac{f(x) - f(a)}{x - a}(x - a).$$

Employing limit theorems and Theorem (3.7),

$$\lim_{x \to a} f(x) = \lim_{x \to a} f(a) + \lim_{x \to a} \frac{f(x) - f(a)}{x - a} \cdot \lim_{x \to a} (x - a)$$

$$= f(a) + f'(a) \cdot 0$$

$$= f(a).$$

Thus, by Definition (2.27), f is continuous at a.

The converse of Theorem (3.8) is false; that is, *there exist continuous functions which are not differentiable.* The function in Example 4 is continuous but not differentiable at 0. Similarly, if $f(x) = |x|$, then f is continuous at 0; however, as was proved in Example 3, f is not differentiable at 0.

Various symbols are used for derivatives. Sometimes it is convenient to denote $f'(x)$ by $D_x[f(x)]$. The subscript x on D is employed to designate the independent variable. For example, if the independent variable is t we write $f'(t) = D_t[f(t)]$. The symbols D_x, D_t, and so on, are referred to as **differential operators**. Standing alone, D_x has no practical significance; however, if an expression in x is placed to the right of it, then the derivative is obtained. To illustrate, using Example 1,

$$D_x(3x^2 - 5x + 4) = 6x - 5.$$

We say that D_x *operates* on the expression $3x^2 - 5x + 4$.

If f is defined by an equation $y = f(x)$ we write

$$f'(x) = D_x[f(x)] = D_xy = y'.$$

In this context we sometimes refer to D_xy as **the derivative of y with respect to x**. As indicated, y' may be used as an abbreviation for this derivative. Of course, variables are often denoted by symbols other than x and y. For example, if $s = f(t)$ where f is a function, we may write

$$f'(t) = D_t[f(t)] = D_ts = s'.$$

In Section 4 we shall introduce still another notation for derivatives. All of these representations are used in mathematics and applications, and it is advisable for students to become familiar with the various forms.

EXERCISES 3.2

In Exercises 1–10 use (3.6) to find $f'(x)$.

1 $f(x) = 37$

2 $f(x) = 17 - 6x$

3 $f(x) = 9x - 2$

4 $f(x) = 7x^2 - 5$

5 $f(x) = 2 + 8x - 5x^2$

6 $f(x) = x^3 + x$

7 $f(x) = 1/(x - 2)$

8 $f(x) = (1 + \sqrt{3})^2$

9 $f(x) = \sqrt{3x + 1}$

10 $f(x) = 1/2x$

In Exercises 11–14 find D_xy.

11 $y = 7/\sqrt{x}$

12 $y = (2x + 3)^2$

13 $y = 2x^3 - 4x + 1$

14 $y = x/(3x + 4)$

In Exercises 15–20 find $f'(a)$ by means of Theorem (3.7).

15 $f(x) = x^2$

16 $f(x) = \sqrt{2}x$

17 $f(x) = 6/x^2$

18 $f(x) = 8 - x^3$

19 $f(x) = 1/(x + 5)$

20 $f(x) = \sqrt{x}$

In Exercises 21 and 22 use right-hand and left-hand derivatives to prove that f is not differentiable at $x = 5$.

21 $f(x) = |x - 5|$

22 $f(x) = [x]$ (the greatest integer function)

23 Prove that if n is any integer, then the greatest integer function f is not differentiable at n.

24 Suppose $f(x) = 0$ if x is rational and $f(x) = 1$ if x is irrational. Prove that f is not differentiable at any real number a.

25 If $f(x)$ is a polynomial of degree 1, prove that $f'(x)$ is a polynomial of degree 0. What if $f(x)$ is a polynomial of degree 2 or 3?

26 Prove that for every real number c, $D_x c = 0$.

3.3 RULES FOR FINDING DERIVATIVES

This section contains some general rules which simplify the task of finding derivatives.

(3.9) Theorem

If f is the constant function defined by $f(x) = c$, then $f'(x) = 0$.

Proof. Since every value of f is c we have $f(x) = c$ and $f(x + h) = c$. Hence by (3.6),

$$f'(x) = \lim_{h \to 0} \frac{f(x + h) - f(x)}{h} = \lim_{h \to 0} \frac{c - c}{h} = \lim_{h \to 0} 0 = 0$$

where the last step follows from Theorem (2.8).

In terms of the operator notation introduced in the previous section, the conclusion of the preceding theorem may be written

$$D_x c = 0.$$

This is sometimes referred to by the statement "**the derivative of a constant is zero**."

(3.10) Theorem

If $f(x) = x$, then $f'(x) = 1$.

Proof. Using (3.6) and the definition of f,

$$\begin{aligned} f'(x) &= \lim_{h\to 0} \frac{f(x+h) - f(x)}{h} \\ &= \lim_{h\to 0} \frac{(x+h) - x}{h} \\ &= \lim_{h\to 0} \frac{h}{h} = \lim_{h\to 0} 1 = 1. \end{aligned}$$

Using the operator notation,

$$D_x(x) = 1.$$

In the proof of the next rule we shall make use of the *Binomial Theorem*, which states that the following formula is true for all real numbers a and b and every positive integer n.

(3.11)

$$\begin{aligned} (a+b)^n = a^n + na^{n-1}b + \frac{n(n-1)}{2!}a^{n-2}b^2 \\ + \cdots + \binom{n}{r} a^{n-r}b^r + \cdots + nab^{n-1} + b^n \end{aligned}$$

where the symbol $\binom{n}{r}$ used for the coefficient of the term involving b^r is defined by

$$\binom{n}{r} = \frac{n(n-1)(n-2)\cdots(n-r+1)}{r(r-1)(r-2)\cdots 1} = \frac{n!}{(n-r)!\,r!}.$$

The validity of (3.11) may be established by mathematical induction. The special cases $n = 2$, $n = 3$, and $n = 4$ are

$$\begin{aligned} (a+b)^2 &= a^2 + 2ab + b^2 \\ (a+b)^3 &= a^3 + 3a^2b + 3ab^2 + b^3 \\ (a+b)^4 &= a^4 + 4a^3b + 6a^2b^2 + 4ab^3 + b^4. \end{aligned}$$

(3.12) The Power Rule

If $f(x) = x^n$, where n is a positive integer, then $f'(x) = nx^{n-1}$.

Proof. By (3.6)

$$f'(x) = \lim_{h\to 0} \frac{(x+h)^n - x^n}{h}$$

provided the limit exists. Employing the Binomial Theorem (3.11) with $a = x$ and $b = h$ gives us

$$(x+h)^n = x^n + nx^{n-1}h + \frac{n(n-1)}{2!}x^{n-2}h^2 + \cdots + nxh^{n-1} + h^n.$$

It is important to observe that every term after the first contains h to some positive integral power. If we subtract x^n and divide by h we obtain

$$f'(x) = \lim_{h\to 0}\left[nx^{n-1} + \frac{n(n-2)}{2!}x^{n-2}h + \cdots + nxh^{n-2} + h^{n-1}\right].$$

Since each term within the brackets except the first contains a power of h, we see that

$$f'(x) = nx^{n-1}.$$

If $x \neq 0$, then (3.12) is also true if $n = 0$, for in this case $f(x) = x^0 = 1$ and by Theorem (3.9), $f'(x) = 0 = 0 \cdot x^{0-1}$.

In terms of the operator notation, the Power Rule may be written

(3.13) $$D_x(x^n) = nx^{n-1}.$$

To illustrate,

$$D_x(x^3) = 3x^2 \quad \text{and} \quad D_x(x^8) = 8x^7.$$

Similarly, if $y = x^{100}$, then $D_x y = 100x^{99}$ or $y' = 100x^{99}$.

If symbols other than x are used for the independent variable, then (3.13) may be written

$$D_t(t^n) = nt^{n-1}, \quad D_z(z^n) = nz^{n-1}, \quad D_v(v^n) = nv^{n-1},$$

and so on. In Section 7 we shall prove that the Power Rule is valid for every rational number n.

(3.14) Theorem

> If f is differentiable, then
>
> $$D_x[cf(x)] = cD_x[f(x)].$$

Proof. If we let $g(x) = cf(x)$, then

$$\begin{aligned} D_x[cf(x)] &= D_x[g(x)] \\ &= \lim_{h\to 0}\frac{g(x+h) - g(x)}{h} \\ &= \lim_{h\to 0}\frac{cf(x+h) - cf(x)}{h} \\ &= c\lim_{h\to 0}\frac{f(x+h) - f(x)}{h} \\ &= cf'(x) = cD_x[f(x)]. \end{aligned}$$

The differentiation formula in Theorem (3.14) is stated in terms of an arbitrary value $f(x)$ of f. If we wish to state this rule without referring to functional values we may write

(3.15) $$(cf)' = cf'.$$

For the special case $f(x) = x^n$, (3.14) and (3.13) lead to the next formula, which is true for every real number c and every positive integer n.

(3.16) $$D_x(cx^n) = cnx^{n-1}$$

Example 1 (a) Find $D_x(7x^4)$.
(b) Find $F'(z)$ if $F(z) = -3z^{15}$.

Solution (a) Using (3.16),

$$D_x(7x^4) = (7)4x^3 = 28x^3.$$

(b) Again using (3.16), but with z as the independent variable,

$$F'(z) = (-3)(15)z^{14} = -45z^{14}.$$

The theorems which follow involve several functions. As usual we assume that every number x is chosen from the intersection of the domains of the functions.

(3.17) Theorem If f and g are differentiable, then

$$D_x[f(x) + g(x)] = D_x[f(x)] + D_x[g(x)].$$

Proof. Let $k(x) = f(x) + g(x)$. We wish to show that $k'(x) = f'(x) + g'(x)$. This may be done as follows.

$$\begin{aligned} k'(x) &= \lim_{h\to 0} \frac{k(x+h) - k(x)}{h} \\ &= \lim_{h\to 0} \frac{[f(x+h) + g(x+h)] - [f(x) + g(x)]}{h} \\ &= \lim_{h\to 0} \left[\frac{f(x+h) - f(x)}{h} + \frac{g(x+h) - g(x)}{h}\right] \\ &= \lim_{h\to 0} \frac{f(x+h) - f(x)}{h} + \lim_{h\to 0} \frac{g(x+h) - g(x)}{h} \\ &= f'(x) + g'(x) \end{aligned}$$

To express the rule in Theorem (3.17) without referring to functional values we may write

(3.18) $$(f+g)' = f' + g'.$$

Similar formulas hold for differences. Thus

(3.19) $$D_x[f(x) - g(x)] = D_x[f(x)] - D_x[g(x)], \text{ or } (f-g)' = f' - g'.$$

The preceding results may be extended to sums or differences of any number of functions.

Since a polynomial is a sum of terms of the form cx^n where n is a nonnegative integer, we may use (3.16) and results on sums and differences to obtain the derivative, as illustrated in the next example.

Example 2 Find $f'(x)$ if $f(x) = 2x^4 - 5x^3 + x^2 - 4x + 1$.

Solution

$$\begin{aligned} f'(x) &= D_x(2x^4 - 5x^3 + x^2 - 4x + 1) \\ &= D_x(2x^4) - D_x(5x^3) + D_x(x^2) - D_x(4x) + D_x(1) \\ &= 8x^3 - 15x^2 + 2x - 4. \end{aligned}$$

(3.20) The Product Rule

> If f and g are differentiable, then
>
> $$D_x[f(x)g(x)] = f(x)D_x[g(x)] + g(x)D_x[f(x)].$$

Proof. Let $k(x) = f(x)g(x)$. We wish to show that

$$k'(x) = f(x)g'(x) + g(x)f'(x).$$

If $k'(x)$ exists, then

$$\begin{aligned} k'(x) &= \lim_{h \to 0} \frac{k(x+h) - k(x)}{h} \\ &= \lim_{h \to 0} \frac{f(x+h)g(x+h) - f(x)g(x)}{h}. \end{aligned}$$

In order to change the form of the quotient so that the limit may be evaluated, we subtract and add the expression $f(x+h)g(x)$ in the numerator. Thus

$$k'(x) = \lim_{h \to 0} \frac{f(x+h)g(x+h) - f(x+h)g(x) + f(x+h)g(x) - f(x)g(x)}{h}$$

which may be written

$$\begin{aligned} k'(x) &= \lim_{h \to 0} \left[f(x+h) \cdot \frac{g(x+h) - g(x)}{h} + g(x) \cdot \frac{f(x+h) - f(x)}{h} \right] \\ &= \lim_{h \to 0} f(x+h) \cdot \lim_{h \to 0} \frac{g(x+h) - g(x)}{h} + \lim_{h \to 0} g(x) \cdot \lim_{h \to 0} \frac{f(x+h) - f(x)}{h}. \end{aligned}$$

Since f is differentiable at x, it is continuous and hence $\lim_{h\to 0} f(x+h) = f(x)$. Also, $\lim_{h\to 0} g(x) = g(x)$ since x is fixed in this limiting process. Finally, applying the definition of derivative to $f(x)$ and $g(x)$ we obtain

$$k'(x) = f(x)g'(x) + g(x)f'(x).$$

The Product Rule may be phrased as follows: *The derivative of a product equals the first factor times the derivative of the second factor, plus the second times the derivative of the first.* If we do not wish to refer to functional values we may write

(3.21) $$(fg)' = fg' + gf'.$$

Example 3 Find $f'(x)$ if $f(x) = (x^3+1)(2x^2+8x-5)$.

Solution Using the Product Rule (3.20),

$$\begin{aligned} f'(x) &= (x^3+1)D_x(2x^2+8x-5) + (2x^2+8x-5)D_x(x^3+1) \\ &= (x^3+1)(4x+8) + (2x^2+8x-5)(3x^2) \\ &= 4x^4 + 8x^3 + 4x + 8 + 6x^4 + 24x^3 - 15x^2 \\ &= 10x^4 + 32x^3 - 15x^2 + 4x + 8. \end{aligned}$$

We could also find $f'(x)$ by first multiplying the two factors x^3+1 and $2x^2+8x-5$ and then differentiating the resulting polynomial.

(3.22) The Quotient Rule

> If f and g are differentiable, and $g(x) \neq 0$, then
>
> $$D_x\left[\frac{f(x)}{g(x)}\right] = \frac{g(x)D_x[f(x)] - f(x)D_x[g(x)]}{[g(x)]^2}.$$

Proof. Let $k(x) = f(x)/g(x)$. We wish to show that

$$k'(x) = \frac{g(x)f'(x) - f(x)g'(x)}{[g(x)]^2}.$$

Using the definitions of $k'(x)$ and $k(x)$,

$$\begin{aligned} k'(x) &= \lim_{h\to 0} \frac{k(x+h) - k(x)}{h} \\ &= \lim_{h\to 0} \frac{\dfrac{f(x+h)}{g(x+h)} - \dfrac{f(x)}{g(x)}}{h} \\ &= \lim_{h\to 0} \frac{g(x)f(x+h) - f(x)g(x+h)}{hg(x+h)g(x)}. \end{aligned}$$

Subtracting and adding $g(x)f(x)$ in the numerator of the last quotient

$$k'(x) = \lim_{h \to 0} \frac{g(x)f(x+h) - g(x)f(x) + g(x)f(x) - f(x)g(x+h)}{hg(x+h)g(x)}$$

or equivalently,

$$k'(x) = \lim_{h \to 0} \frac{g(x)\left[\dfrac{f(x+h) - f(x)}{h}\right] - f(x)\left[\dfrac{g(x+h) - g(x)}{h}\right]}{g(x+h)g(x)}.$$

Taking the limit of the numerator and denominator gives us the desired formula.

Another way of stating the Quotient Rule is

(3.23)
$$\left(\frac{f}{g}\right)' = \frac{gf' - fg'}{g^2}$$

which may be stated as follows: *The derivative of a quotient equals the denominator times the derivative of the numerator minus the numerator times the derivative of the denominator, divided by the square of the denominator.*

Example 4 Find y' if $y = \dfrac{3x^2 - x + 2}{4x^2 + 5}$.

Solution By the Quotient Rule (3.22),

$$y' = \frac{(4x^2 + 5)D_x(3x^2 - x + 2) - (3x^2 - x + 2)D_x(4x^2 + 5)}{(4x^2 + 5)^2}$$

$$= \frac{(4x^2 + 5)(6x - 1) - (3x^2 - x + 2)(8x)}{(4x^2 + 5)^2}$$

$$= \frac{(24x^3 - 4x^2 + 30x - 5) - (24x^3 - 8x^2 + 16x)}{(4x^2 + 5)^2}$$

$$= \frac{4x^2 + 14x - 5}{(4x^2 + 5)^2}.$$

It is now a simple matter to extend the Power Rule (3.12) to the case in which the exponent is a negative integer.

(3.24) Corollary

> If n is a positive integer, then
>
> $$D_x(x^{-n}) = -nx^{-n-1}.$$

Proof. Using the definition of x^{-n} together with (3.22) and (3.12),

$$D_x(x^{-n}) = D_x\left(\frac{1}{x^n}\right) = \frac{x^n(0) - 1(nx^{n-1})}{(x^n)^2}$$

$$= \frac{-nx^{n-1}}{x^{2n}} = (-n)x^{(n-1)-2n} = (-n)x^{-n-1}.$$

Thus (3.12) and (3.16) are true for negative as well as positive exponents.

Example 5 Differentiate the following. (a) $g(w) = 1/w^4$. (b) $H(s) = 3/s$.

Solution (a) Writing $g(w) = w^{-4}$ and using the Power Rule (with w as independent variable), $g'(w) = -4w^{-5} = -4/w^5$.

(b) Since $H(s) = 3s^{-1}$, $H'(s) = 3(-1)s^{-2} = -3/s^2$.

EXERCISES 3.3

Differentiate the functions defined in Exercises 1–32.

1 $f(x) = 10x^2 + 9x - 4$

2 $f(x) = 6x^3 - 5x^2 + x + 9$

3 $f(s) = 15 - s + 4s^2 - 5s^4$

4 $f(t) = 12 - 3t^4 + 4t^6$

5 $g(x) = (x^3 - 7)(2x^2 + 3)$

6 $k(x) = (2x^2 - 4x + 1)(6x - 5)$

7 $h(r) = r^2(3r^4 - 7r + 2)$

8 $g(s) = (s^3 - 5s + 9)(2s + 1)$

9 $f(x) = \dfrac{4x - 5}{3x + 2}$

10 $h(x) = \dfrac{8x^2 - 6x + 11}{x - 1}$

11 $h(z) = \dfrac{8 - z + 3z^2}{2 - 9z}$

12 $f(w) = \dfrac{2w}{w^3 - 7}$

13 $f(x) = 3x^3 - 2x^2 + 4x - 7$

14 $g(z) = 5z^4 - 8z^2 + z$

15 $F(t) = t^2 + (1/t^2)$

16 $s(x) = 2x + 1/(2x)$

17 $g(x) = (8x^2 - 5x)(13x^2 + 4)$

18 $H(y) = (y^5 - 2y^3)(7y^2 + y - 8)$

19 $G(v) = \dfrac{v^3 - 1}{v^3 + 1}$

20 $f(t) = \dfrac{8t + 15}{t^2 - 2t + 3}$

21 $f(x) = \dfrac{1}{1 + x + x^2 + x^3}$

22 $p(x) = 1 + \dfrac{1}{x} + \dfrac{1}{x^2} + \dfrac{1}{x^3}$

23 $g(z) = z(2z^3 - 5z - 1)(6z^2 + 7)$

24 $N(v) = 4v(v - 1)(2v - 3)$

25 $K(s) = (3s)^{-4}$

26 $W(s) = (3s)^4$

27 $h(x) = (5x - 4)^2$

28 $g(r) = (5r - 4)^{-2}$

29 $f(t) = \dfrac{(3/5t) - 1}{(2/t^2) + 7}$

30 $S(w) = (2w + 1)^3$

31 $M(x) = (2x^3 - 7x^2 + 4x + 3)/x^2$

32 $f(x) = (3x^2 - 5x + 8)/7$

In each of Exercises 33 and 34 find y' by means of (a) the Quotient Rule (3.22), (b) the Product Rule (3.20), and (c) simplifying algebraically and using the Power Rule (3.12).

33 $y = \dfrac{3x - 1}{x^2}$

34 $y = \dfrac{x^2 + 1}{x^4}$

In each of Exercises 35 and 36 find y' by (a) using the Product Rule (3.20) and (b) first multiplying the two factors.

35 $y = (12x - 17)(5x + 3)$

36 $y = 8x^2(5x - 9)$

37 If f, g, and h are differentiable, use the Product Rule (3.20) to prove that

$$D_x[f(x)g(x)h(x)] = f(x)g(x)h'(x) + f(x)h(x)g'(x) + h(x)g(x)f'(x).$$

As a corollary, let $f = g = h$ to prove that

$$D_x[f(x)]^3 = 3[f(x)]^2 f'(x).$$

38 Extend Exercise 37 to the derivative of a product of four functions and then find a formula for $D_x[f(x)]^4$.

In Exercises 39 and 40, use Exercise 37 to find y'.

39 $y = (8x - 1)(x^2 + 4x + 7)(x^3 - 5)$

40 $y = (3x^4 - 10x^2 + 8)(2x^2 - 10)(6x + 7)$

41 Find an equation of the tangent line to the graph of $y = 5/(1 + x^2)$ at the following points.

(a) $P(0, 5)$ (b) $P(1, 5/2)$ (c) $P(\ \ 2, 1)$

42 Find an equation of the tangent line to the graph of $y = 2x^3 + 4x^2 - 5x - 3$ at the following points.

(a) $P(0, -3)$ (b) $P(-1, 4)$ (c) $P(1, -2)$

43 Find the abscissas of points on the graph of $y = x^3 + 2x^2 - 4x + 5$ at which the tangent line is (a) horizontal; (b) parallel to the line $2y + 8x - 5 = 0$.

44 Find the point P on the graph of $y = x^3$ such that the tangent line at P has x-intercept 4.

In Exercises 45 and 46 the position of a point P moving on a coordinate line l is given by $f(t)$. Determine the time intervals in which P moves in (a) the positive direction; (b) the negative direction. When is the velocity 0?

45 $f(t) = 2t^3 + 9t^2 - 60t + 1$

46 $f(t) = 3t^5 - 5t^3$

If f and g are differentiable functions such that $f(2) = 3$, $f'(2) = -1$, $g(2) = -5$, and $g'(2) = 2$, find the numbers in Exercises 47 and 48.

47 (a) $(f + g)'(2)$ (b) $(f - g)'(2)$ (c) $(4f)'(2)$
(d) $(fg)'(2)$ (e) $(f/g)'(2)$

48 (a) $(g - f)'(2)$ (b) $(g/f)'(2)$
(c) $(4g)'(2)$ (d) $(ff)'(2)$

In Exercises 49 and 50 find the points at which the graphs of f and f' intersect.

49 $f(x) = x^3 - x^2 + x + 1$

50 $f(x) = x^2 + 2x + 1$

51 Find an equation of the tangent line to the graph of $y = 3x^2 + 4x - 6$ which is parallel to the line $5x - 2y - 1 = 0$.

52 Find equations of the tangent lines to the graph of $y = x^3$ which are parallel to the line $16x - 3y + 17 = 0$.

53 Find an equation of the line through the point $P(5, 9)$ which is tangent to the graph of $y = x^2$.

54 Find equations of the lines through $P(3, 1)$ which are tangent to the graph of $xy = 4$.

3.4 INCREMENTS AND DIFFERENTIALS

Let us consider the equation $y = f(x)$ where f is a function. In many applications the independent variable x is subjected to a small change and it is necessary to find the corresponding change in the dependent variable y. If x changes from x_1 to x_2, then the amount of change is often denoted by the symbol Δx (read "delta x"); that is,

$$\Delta x = x_2 - x_1.$$

The number Δx is called an **increment** of x. Note that $x_2 = x_1 + \Delta x$; that is, the new value of x equals the old value plus the change, Δx, in x. Similarly, Δy will denote the change in the dependent variable y which corresponds to the change Δx. Thus

(3.25) $$\Delta y = f(x_2) - f(x_1) = f(x_1 + \Delta x) - f(x_1).$$

The geometric representations for these increments in terms of the graph of f are shown in Figure 3.8.

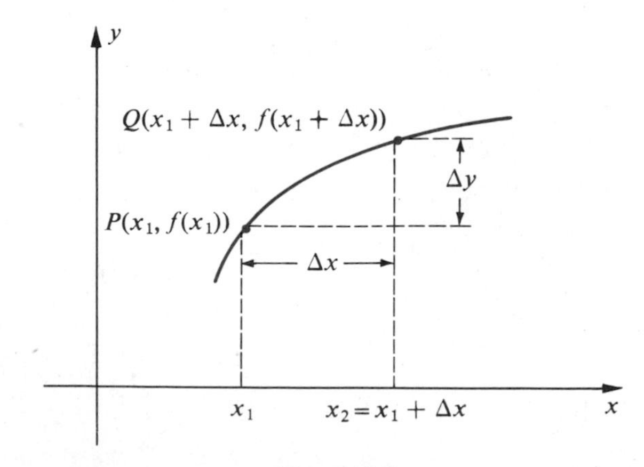

Figure 3.8

Example 1 If $y = 3x^2 - 5$, find the change Δy if the initial value of x is 2 and $\Delta x = 0.1$.

Solution Using (3.25) with $f(x) = 3x^2 - 5$, we obtain

$$\begin{aligned}\Delta y &= f(x_1 + \Delta x) - f(x_1)\\ &= f(2.1) - f(2)\\ &= [3(2.1)^2 - 5] - [3(2)^2 - 5]\\ &= 1.23.\end{aligned}$$

The increment notation may be used in the definition of the derivative of a function. All that is necessary is to substitute Δx for h in (3.6) and regard x as the initial value of the independent variable. Thus

$$f'(x) = \lim_{\Delta x \to 0} \frac{f(x + \Delta x) - f(x)}{\Delta x}.$$

If we use (3.25), with x as initial value, then

(3.26) $$\Delta y = f(x + \Delta x) - f(x).$$

Substitution in the preceding formula for $f'(x)$ gives us

(3.27) $$f'(x) = \lim_{\Delta x \to 0} \frac{\Delta y}{\Delta x}.$$

In words (3.27) may be phrased as follows. "The derivative of f is the limit of the ratio of the increment Δy of the dependent variable to the increment Δx of the independent variable as Δx approaches zero." Notice that if $x = x_1$ in Figure 3.8, then $\Delta y/\Delta x$ is the slope of the secant line through P and Q. It follows that if f is differentiable, then

(3.28) $$\frac{\Delta y}{\Delta x} \approx f'(x) \quad \text{if } \Delta x \approx 0.$$

Geometrically this says that if Δx is close to 0, then the slope $\Delta y/\Delta x$ of the secant line through P and Q is close to the slope $f'(x)$ of the tangent line at P. We may also write

(3.29) $$\Delta y \approx f'(x)\,\Delta x \quad \text{if } \Delta x \approx 0.$$

We give the expression $f'(x)\,\Delta x$ in (3.29) a special name in the next definition.

(3.30) Definition

If $y = f(x)$, where f is differentiable, and if Δx is an increment of x, then
(i) the **differential** $\boldsymbol{dy}$ of the dependent variable y is given by $dy = f'(x)\,\Delta x$
(ii) the **differential** $\boldsymbol{dx}$ of the independent variable x is given by $dx = \Delta x$.

Observe that dy depends, for its value, on *both* x and Δx. From (ii) we also see that as far as the independent variable x is concerned, there is no difference between the increment Δx and the differential dx. Substituting dx for Δx in (i) leads to

(3.31) $$dy = f'(x)\,dx.$$

If both sides of (3.31) are divided by dx (assuming that $dx \neq 0$) we obtain

(3.32) $$\frac{dy}{dx} = f'(x).$$

Thus, the derivative $f'(x)$ may be expressed as a quotient of two differentials. Indeed, this is one of the reasons for defining dx and dy as we do in (3.30).

Example 2 If $y = x^4 - 3x^2 + 5x + 4$, find (a) dy; (b) the value of dy if $x = 2$ and $\Delta x = -0.1$.

Solution (a) Writing $f(x) = x^4 - 3x^2 + 5x + 4$ we have, from (3.31),

$$dy = (4x^3 - 6x + 5)\,dx.$$

(b) By Definition (3.30), $dx = \Delta x$. Thus we let $x = 2$ and $dx = -0.1$ in the formula for dy. This gives us

$$dy = (4 \cdot 2^3 - 6 \cdot 2 + 5)(-0.1) = (25)(-0.1) = -2.5.$$

It is instructive to study the geometric interpretations of dx and dy. Consider the graphs in Figure 3.9, where l is the tangent line at the point P. We have pictured both Δx and Δy as positive; however, they may also take on negative values. If T is the point of intersection of l and the vertical line through Q, and if R is the point with coordinates $(x + \Delta x, f(x))$, then it can be seen from triangle PRT that the slope $f'(x)$ of the tangent line is the ratio $r/\Delta x$, where r denotes the length of the segment RT. Hence, $r/\Delta x = f'(x)$ or $r = f'(x)\Delta x = dy$. A similar result holds if one or both increments are negative. In general, if x is given an increment Δx, then dy is the amount that the *tangent line* rises (or falls) when the independent variable changes from x to $x + \Delta x$. This is in contrast to the amount Δy that the *graph* rises (or falls) between P and Q.

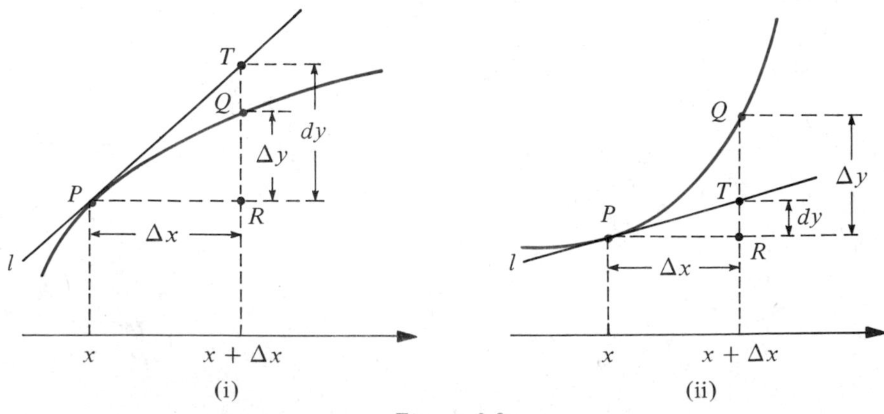

Figure 3.9

It follows from (3.29) and (i) of Definition (3.30) that if Δx is small, then

(3.33) $$\Delta y \approx dy = f'(x)\,dx = (D_x y)\,dx.$$

This is also evident geometrically from Figure 3.9. Consequently, if $y = f(x)$, then dy can be used as an approximation to the exact change Δy of the dependent variable corresponding to a small change Δx in x. This observation is useful in applications where only a rough estimate of the change in y is desired.

Example 3 If $y = 3x^2 - 5$, use dy to approximate Δy if x changes from 2 to 2.1.

Solution If $f(x) - 3x^2 - 5$, then by Example 1 $\Delta y = 1.23$. Using (3.31),

$$dy = f'(x)\,dx = 6x\,dx.$$

In the present example, $x = 2$, $\Delta x = dx = 0.1$, and

$$dy = (6)(2)(0.1) = 1.2.$$

Hence our approximation is correct to the nearest tenth.

The next example illustrates how differentials may be used to find certain approximation formulas.

Example 4 (a) Use differentials to obtain a formula for approximating the volume of a thin cylindrical shell of altitude h, inner radius r, and thickness t.

(b) What error is involved in using the formula?

Solution (a) A typical cylindrical shell is illustrated in Figure 3.10, where the thickness t has been denoted by Δr. We wish to find the volume contained between the "inner" cylinder of radius r and the "outer" cylinder of radius $r + \Delta r$. The volume V of the inner cylinder is given by

$$V = \pi r^2 h.$$

If we increase r by an amount Δr, then the volume of the shell is the change ΔV in V. Using differentials as in (3.33),

$$\Delta V \approx dV = (D_r V)\,\Delta r$$

and hence a formula for approximating the volume of the shell is

$$\Delta V \approx (2\pi r h)\,\Delta r = (2\pi r h)t$$

or, in words,

$$\text{Volume} \approx (\text{Area of inner cylinder wall}) \times (\text{Thickness}).$$

(b) The exact volume of the shell is

$$\Delta V = \pi(r + \Delta r)^2 h - \pi r^2 h$$

which simplifies to

$$\Delta V = (2\pi rh)\,\Delta r + \pi h(\Delta r)^2.$$

$$\Delta V - dV = \pi h(\Delta r)^2.$$

This indicates that the approximation formula is quite accurate if Δr is small in comparison to h.

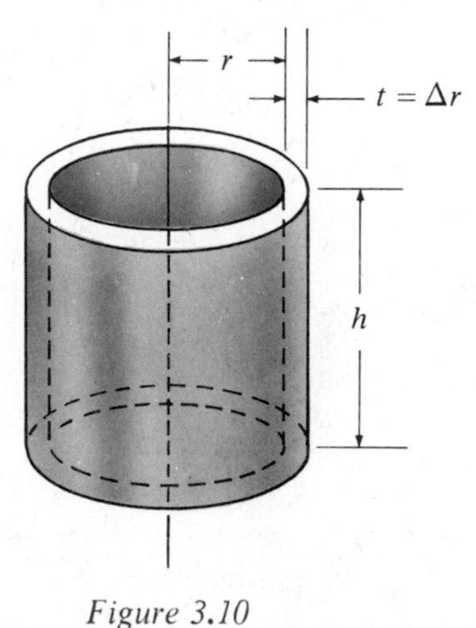

Figure 3.10

Differentials may sometimes be used to estimate errors, as illustrated by the next example.

Example 5 The radius of a spherical balloon is estimated to be 12 inches with a maximum error in measurement of 0.05 inches. Estimate the maximum error in the calculated volume of the sphere.

Solution If x denotes the radius of the balloon and V the volume, then $V = \frac{4}{3}\pi x^3$. If the maximum error in the measured value of x is denoted by dx (or equivalently by Δx), then the exact value of the radius is between $x - dx$ and $x + dx$. If we let ΔV denote the error in V caused by the error dx in x, then we may approximate ΔV by means of the differential

$$dV = (D_x V)\,dx = 4\pi x^2\,dx.$$

Letting $x = 12$ and $dx = \pm 0.05$, we obtain

$$dV = 4\pi(12)^2(\pm 0.05) \approx \pm 90.$$

Thus the maximum error in volume due to the error in measurement of the radius is approximately 90 cubic inches.

If there is an error in the measurement of a quantity, we define

(3.34) $$\textbf{Average error} = \frac{\textbf{error in measurement}}{\textbf{measured value}}.$$

The average error is also called the **relative error**. For example, if the measured length of an object is 25 inches with a possible error of 0.1 inches, then from (3.34) the average error is 0.1/25 or 0.004. The significance of this number is that the error involved is, *on the average*, 0.004 inches per inch. The **percentage error** is defined as the relative error multiplied by 100. In the illustration just given, the percentage error is (0.004)(100) or 0.4%.

In terms of differentials, if w represents a measurement with a possible error of dw, then from our previous remarks, the average error is dw/w. Of course, if dw is an approximation to the error in w, then dw/w is an approximation to the average error. In Example 5, the average error in the radius is dx/x or $0.05/12 = 1/240 \approx 0.0042$ inches per inch. The average error in V is approximately dV/V or 0.012 cubic inches per cubic inch. The percentage errors are approximately 0.4% and 1.2%, respectively.

In (3.32) we arrived at the expression $dy/dx = f'(x)$. The notation dy/dx was employed by Leibniz in his pioneering work on the calculus in the late 1600s and is still widely used in mathematics and the applied sciences. It is not always essential to regard dy/dx as a quotient of two differentials. As a matter of fact dy/dx is often thought of formally as just another symbol for the derivative $f'(x)$. For example,

$$\text{if} \quad y = 7x^3 - 4x^2 + 2, \quad \text{then} \quad \frac{dy}{dx} = 21x^2 - 8x.$$

The symbol d/dx is used in the same manner as the operator D_x. Thus

$$f'(x) = D_x[f(x)] = \frac{d}{dx}[f(x)] = \frac{d[f(x)]}{dx}.$$

In particular, rules for differentiating sums, products, and quotients may be stated as follows, where $u = f(x)$ and $v = g(x)$:

$$\text{(3.35)} \qquad \begin{aligned} \frac{d}{dx}(u + v) &= \frac{du}{dx} + \frac{dv}{dx} \\ \frac{d}{dx}(uv) &= u\frac{dv}{dx} + v\frac{du}{dx} \\ \frac{d}{dx}\left(\frac{u}{v}\right) &= \frac{v\dfrac{du}{dx} - u\dfrac{dv}{dx}}{v^2}. \end{aligned}$$

EXERCISES 3.4

In Exercises 1–4 use (3.25) to find Δy if the initial values of x and Δx are as indicated.

1 $y = 2x^2 - 4x + 5$, $x = 2$, $\Delta x = -0.2$

2 $y = x^3 - 3x^2 + x - 4$, $x = -1$, $\Delta x = 0.1$

3 $y = 1/x^2$, $x = 3$, $\Delta x = 0.3$

4 $y = (x + 2)/(x + 3)$, $x = 0$, $\Delta x = -0.03$

In Exercises 5–12, (a) use (3.26) to find a general formula for Δy; (b) find dy; (c) find $dy - \Delta y$.

5 $y = 3x^2 + 5x - 2$

6 $y = 4 - 7x - 2x^2$

7 $y = 1/x$

8 $y = 7x + 12$

9 $y = 4 - 9x$

10 $y = 8$

11 $y = x^3$

12 $y = x^{-2}$

In Exercises 13 and 14 use differentials to approximate the change in $f(x)$ if x changes from a to b.

13 $f(x) = 4x^5 - 6x^4 + 3x^2 - 5$; $a = 1$, $b = 1.03$

14 $f(x) = -3x^3 + 8x - 7$; $a = 4$, $b = 3.96$

15 If $f(z) = z^3 - 3z^2 + 2z - 7$ and $w = f(z)$, find dw. Use dw to approximate the change in w if z changes from 4 to 3.95.

16 If $F(t) = 1/(2 - t^2)$ and $s = F(t)$, find ds. Use ds to approximate the change in s if t changes from 1 to 1.02.

17 The radius of a circular disc is estimated to be 8 inches with a maximum error in measurement of 0.06 inches. Use differentials to estimate the maximum error in the calculated area of one side of the disc. What is the approximate average error and the approximate percentage error?

18 If the side of a square is estimated as 1 foot with a maximum error in measurement of 1/16 inch, use differentials to estimate the maximum error in the calculated area. What is the approximate average error and the approximate percentage error?

19 Use differentials to approximate the increase in volume of a cube if the length of each edge changes from 10 inches to 10.1 inches. What is the exact change in volume?

20 Use differentials to approximate the increase in surface area of a spherical balloon if the diameter changes from 2 feet to 2.02 feet.

21 One side of a house has the shape of a square surmounted by an equilateral triangle. If the length of the base is measured as 48 feet, with a maximum error in measurement of 1 in., calculate the area of the side and use differentials to estimate the maximum error in the calculation. What is the approximate average error and the approximate percentage error?

22 A silo has the shape of a right circular cylinder surmounted by a hemisphere. The altitude of the cylinder is exactly 50 ft. The circumference of the base is measured as 30 ft, with a maximum error in measurement of 6 in. Calculate the volume of the silo from these measurements and use differentials to estimate the maximum error in the calculation. What is the approximate average error and the approximate percentage error?

23 As sand leaks out of a container, it forms a conical pile whose altitude is always the same as the radius. If, at a certain instant, the radius is 10 cm, use differentials to approximate the change in radius that will increase the volume of the pile by 2 cm^3.

24 Use the technique illustrated in Example 4 of this section to find formulas which can be used to approximate a thin shell-shaped solid whose surface is (a) spherical; (b) cubical.

25 Newton's Law of Gravitation states that the force F of attraction between two particles having masses m_1 and m_2 is given by $F = gm_1m_2/s^2$, where g is a constant and s is the distance between the particles. If $s = 20$ cm, use differentials to approximate the change in s that will increase F by 10%.

26 Boyle's Law states that the pressure p and volume v of a confined gas are related by the formula $pv = c$, where c is a constant, or equivalently, by $p = c/v$, where $v \neq 0$. Show that dp and dv are related by means of the formula $pdv + vdp = 0$.

27 Use differentials to approximate $(0.98)^4$. (*Hint:* Let $y = f(x) = x^4$ and consider $f(x + \Delta x) = f(x) + \Delta y$, where $x = 1$ and $\Delta x = -0.02$.) What is the exact value of $(0.98)^4$?

28 Use differentials to approximate

$$N = (2.01)^4 - 3(2.01)^3 + 4(2.01)^2 - 5.$$

What is the exact value of N?

29 The area A of a square of side s is given by $A = s^2$. If s increases by an amount Δs, give geometric illustrations of dA and $\Delta A - dA$.

30 The volume V of a cube of edge x is given by $V = x^3$. If x increases by an amount Δx, give geometric illustrations of dV and $\Delta V - dV$.

3.5 THE CHAIN RULE

Let us consider two differentiable functions f and g, where

$$y = f(u) \quad \text{and} \quad u = g(x).$$

If $g(x)$ is in the domain of f, then we may write

$$y = f(u) = f(g(x))$$

that is, y is a function of x. Indeed, the function involved is the same as the composite function $f \circ g$ defined in Section 1.5. The next theorem provides a formula which specifies the derivative $D_x y$ of the composite function in terms of the derivatives of f and g. In the statement of the theorem it is assumed that variables are chosen such that the composite function $f \circ g$ is defined, and that if g has a derivative at x, then f has a derivative at $g(x)$.

(3.36) The Chain Rule

If $y = f(u)$, $u = g(x)$, and the derivatives $D_u y$ and $D_x u$ both exist, then the composite function defined by $y = f(g(x))$ has a derivative given by

$$D_x y = (D_u y)(D_x u) = f'(u)g'(x).$$

Partial Proof. Let Δx be an increment such that both x and $x + \Delta x$ are in the domain of the composite function. Since $y = f(g(x))$, the corresponding increment of y is given by

$$\Delta y = f(g(x + \Delta x)) - f(g(x)).$$

If the composite function has a derivative at x, then by definition

$$D_x y = \lim_{\Delta x \to 0} \frac{\Delta y}{\Delta x}.$$

Next consider $u = g(x)$ and let Δu be the increment of u which corresponds to Δx, that is,

$$\Delta u = g(x + \Delta x) - g(x).$$

Since

$$g(x + \Delta x) = g(x) + \Delta u = u + \Delta u$$

we may express the formula for Δy as

$$\Delta y = f(u + \Delta u) - f(u).$$

If $y = f(u)$ is differentiable at u, then by definition

$$D_u y = f'(u) = \lim_{\Delta u \to 0} \frac{\Delta y}{\Delta u}.$$

Similarly, if $u = g(x)$ is differentiable at x, then

$$D_x u = g'(x) = \lim_{\Delta x \to 0} \frac{\Delta u}{\Delta x}.$$

Let us assume that there exists an open interval I containing x such that whenever $x + \Delta x$ is in I and $\Delta x \neq 0$, then $\Delta u \neq 0$. In this case we may write

$$D_x y = \lim_{\Delta x \to 0} \frac{\Delta y}{\Delta x} = \lim_{\Delta x \to 0} \left(\frac{\Delta y}{\Delta u} \frac{\Delta u}{\Delta x} \right) = \left(\lim_{\Delta x \to 0} \frac{\Delta y}{\Delta u} \right) \left(\lim_{\Delta x \to 0} \frac{\Delta u}{\Delta x} \right)$$

provided the limits exist. Since g is differentiable at x, it is continuous at x. Hence if $\Delta x \to 0$, then $g(x + \Delta x)$ approaches $g(x)$ and $\Delta u \to 0$. It follows that the last limit formula displayed above may be written

$$\begin{aligned} D_x y &= \left(\lim_{\Delta u \to 0} \frac{\Delta y}{\Delta u} \right) \left(\lim_{\Delta x \to 0} \frac{\Delta u}{\Delta x} \right) \\ &= (D_u y)(D_x u) = f'(u)g'(x) \end{aligned}$$

which is what we wished to prove.

In most applications of the Chain Rule, $u = g(x)$ has the property assumed in the preceding paragraph. If g does not satisfy this property, then every open interval containing x contains a number $x + \Delta x$, with $\Delta x \neq 0$, such that $\Delta u = 0$. In this case our proof is invalid, since Δu occurs in a denominator. In order to construct a proof which takes functions of this type into account, it is necessary to introduce techniques which are more sophisticated than those we have used. A complete proof of the Chain Rule is given in Appendix II.

An easy way to memorize the Chain Rule is to use the differential notation. If we let $y = f(u)$ and $u = g(x)$, so that $y = f(g(x))$, then (3.36) may be written

(3.37)
$$\frac{dy}{dx} = \frac{dy}{du}\frac{du}{dx}.$$

This formula for dy/dx may be remembered by thinking of canceling du on the right-hand side.

Example 1 If $y = u^3$ and $u = x^2 + 1$, find dy/dx.

Solution The problem could be worked by substituting $x^2 + 1$ for u in the first equation and then calculating dy/dx. Another approach is to use (3.37). Thus

$$\frac{dy}{dx} = \frac{dy}{du}\frac{du}{dx} = (3u^2)(2x).$$

If $x^2 + 1$ is substituted for u we obtain

$$\frac{dy}{dx} = 3(x^2 + 1)^2(2x).$$

One of the main uses for the Chain Rule (3.36) is to establish other differentiation formulas. As a first illustration we shall obtain a formula for the derivative of a power of a function.

(3.38) The Power Rule for Functions

If g is a differentiable function and n is an integer, then

$$D_x[g(x)]^n = n[g(x)]^{n-1}D_x[g(x)].$$

Proof. If we let $y = u^n$ and $u = g(x)$, then $y = [g(x)]^n$ and, by the Chain Rule (3.36),

$$D_xy = (D_uy)(D_xu) = nu^{n-1}D_xu = n[g(x)]^{n-1}D_x[g(x)].$$

This completes the proof.

The Power Rule may be remembered in terms of the formula

(3.39)
$$D_x(u^n) = nu^{n-1}D_xu$$

where $u = g(x)$. Note that if $u = x$, then $D_xu = 1$ and (3.39) reduces to (3.13).

Example 2 Find $f'(x)$ if $f(x) = (x^5 - 4x + 8)^7$.

Solution Using (3.39) with $u = x^5 - 4x + 8$ and $n = 7$,

$$\begin{aligned} f'(x) &= D_x(x^5 - 4x + 8)^7 = 7(x^5 - 4x + 8)^6 D_x(x^5 - 4x + 8) \\ &= 7(x^5 - 4x + 8)^6(5x^4 - 4). \end{aligned}$$

Example 3 Find y' if $y = \dfrac{1}{(4x^2 + 6x - 7)^3}$.

Solution Writing $y = (4x^2 + 6x - 7)^{-3}$ and applying the Power Rule (3.38) with $g(x) = 4x^2 + 6x - 7$ and $n = -3$,

$$\begin{aligned} y' &= -3(4x^2 + 6x - 7)^{-4} D_x(4x^2 + 6x - 7) \\ &= -3(4x^2 + 6x - 7)^{-4}(8x + 6) \\ &= \frac{-6(4x + 3)}{(4x^2 + 6x - 7)^4}. \end{aligned}$$

Example 4 If $F(z) = (2z + 5)^3(3z - 1)^4$, find $F'(z)$.

Solution Using first the Product Rule, second the Power Rule, and then factoring the result, we have

$$\begin{aligned} F'(z) &= (2z + 5)^3 D_z(3z - 1)^4 + (3z - 1)^4 D_z(2z + 5)^3 \\ &= (2z + 5)^3 \cdot 4(3z - 1)^3(3) + (3z - 1)^4 \cdot 3(2z + 5)^2(2) \\ &= 6(2z + 5)^2(3z - 1)^3[2(2z + 5) + (3z - 1)] \\ &= 6(2z + 5)^2(3z - 1)^3(7z + 9). \end{aligned}$$

If g is a function and $y = g(x)$, then (3.39) may be written in any of the forms

(3.40)
$$D_x(y^n) = ny^{n-1}D_x y = ny^{n-1}y' = ny^{n-1}\frac{dy}{dx}.$$

The dependent variable y represents the expression $g(x)$ and consequently when differentiating y^n it is *essential* to multiply ny^{n-1} by the derivative y'. A common error is to write $D_x y^r = ry^{r-1}$. Formula (3.40) will be extremely important in our work with implicit functions in the next section.

The Chain Rule will have far-reaching consequences in our later work with differentiation and integration. For example, when we discuss the trigonometric functions in Chapter 9 it will be shown that for every real number x,

$$D_x \sin x = \cos x.$$

As a corollary, if $y = \sin u$, and $u = g(x)$, then from (3.36),

$$D_x y = (D_u y)(D_x u) = (\cos u)D_x u$$

that is,

$$D_x[\sin g(x)] = [\cos g(x)]g'(x).$$

In like manner, if we have *any* rule for the derivative of a function of x, then we can immediately use the Chain Rule (3.36) to extend it to a function of u, where $u = g(x)$.

EXERCISES 3.5

Differentiate the functions defined in Exercises 1–26.

1 $f(x) = (x^2 - 3x + 8)^3$

2 $f(x) = (4x^3 + 2x^2 - x - 3)^2$

3 $g(x) = (8x - 7)^{-5}$

4 $k(x) = (5x^2 - 2x + 1)^{-3}$

5 $f(x) = \dfrac{x}{(x^2 - 1)^4}$

6 $g(x) = \dfrac{x^4 - 3x^2 + 1}{(2x + 3)^4}$

7 $f(x) = (8x^3 - 2x^2 + x - 7)^5$

8 $g(w) = (w^4 - 8w^2 + 15)^4$

9 $F(v) = (17v - 5)^{1000}$

10 $K(x) = (3x^2 - 5x + 7)^{-1}$

11 $s(t) = (4t^5 - 3t^3 + 2t)^{-2}$

12 $p(s) = 1/(8 - 5s + 7s^2)^{10}$

13 $N(x) = (6x - 7)^3(8x^2 + 9)^2$

14 $f(w) = (2w^2 - 3w + 1)(3w + 2)^4$

15 $g(z) = \left(z^2 - \dfrac{1}{z^2}\right)^6$

16 $S(t) = \left(\dfrac{3t + 4}{6t - 7}\right)^3$

17 $k(u) = \dfrac{(u^2 + 1)^3}{(4u - 5)^5}$

18 $g(x) = (3x - 8)^{-2}(7x^2 + 4)^{-3}$

19 $f(x) = \left(\dfrac{3x^2 - 5}{2x^2 + 7}\right)^2$

20 $M(z) = \dfrac{9z^3 + 2z}{(6z + 1)^3}$

21 $G(s) = (s^{-4} + 3s^{-2} + 2)^{-6}$

22 $F(v) = (v^{-1} - 2v^{-2})^{-3}$

23 $h(x) = [(2x + 1)^{10} + 1]^{10}$

24 $f(t) = \left[\left(1 + \dfrac{1}{t}\right)^{-1} + 1\right]^{-1}$

25 $F(t) = 2t(2t + 1)^2(2t + 3)^3$

26 $N(x) = \dfrac{7x(x^2 + 1)^2}{(3x + 10)^4}$

For each of Exercises 27–30 find (a) an equation of the tangent line to the graph of the given equation at the indicated point, and (b) the points on the graph at which the tangent line is horizontal.

27 $y = (4x^2 - 8x + 3)^4$, $P(2, 81)$

28 $y = \left(x + \dfrac{1}{x}\right)^5$, $P(1, 32)$

29 $y = (2x - 1)^{10}$, $P(1, 1)$

30 $y = (x^2 - 1)^7$, $P(0, -1)$

31 If $y = (x^4 - 3x^2 + 1)^{10}$, find dy and use it to approximate the change in y if x changes from 1 to 1.01.

32 If $w = z^3(z - 1)^5$, find dw and use it to approximate the change in w if z changes from 2 to 1.98.

33 If $w = f(z)$ and $z = g(s)$, express the Chain Rule formula for dw/ds in terms of the differential notation.

34 If $v = F(u)$ and $u = G(t)$, express the Chain Rule formula for dv/dt in terms of the differential notation.

35 Use the result of Exercise 33 to find dw/ds if $w = z^3 - (2/z)$ and $z = (s^2 + 1)^5$.

36 Use the result of Exercise 34 to find dv/dt if $v = (u^4 + 2u^2 + 1)^3$ and $u = 4t^2$.

37 If $f(x) = x^4 - 3x^3 + 3x + 2$ and $g(x) = x^3 - 3x^2 + 2x + 1$, use differentials to approximate the change in $f(g(x))$ if x changes from 1 to 0.99.

38 If $f(x) = x^5 + x^4 + x^3 + x^2 + x + 1$, approximate the change in $f(f(x))$ if x changes from 0 to 0.01.

39 If $k(x) = f(g(x))$ with $f(2) = -4$, $g(2) = 2$, $f'(2) = 3$, and $g'(2) = 5$, find $k'(2)$.

40 If r, s, and t are functions such that $r(x) = s(t(x))$, and if $s(0) = -1$, $t(0) = 0$, $s'(0) = -3$, and $r'(0) = 2$, find $t'(0)$.

41 If $z = k(y)$, $y = f(u)$, and $u = g(x)$, show that under suitable restrictions

$$D_x z = (D_y z)(D_u y)(D_x u).$$

Extend this result to any number of functions.

42 A function f is *even* if $f(-x) = f(x)$ for all x in its domain D, whereas f is *odd* if $f(-x) = -f(x)$ for all x in D. Suppose f is differentiable. Use the Chain Rule to prove that

(a) if f is even, then f' is odd.

(b) if f is odd, then f' is even.

Use polynomial functions to give examples of (a) and (b).

43 The **normal line** at a point $P(x_1, y_1)$ on the graph of a differentiable function f is defined as the line through P which is perpendicular to the tangent line. If $f'(x_1) \neq 0$, prove that an equation of the normal line at $P(x_1, y_1)$ is

$$y - y_1 = -\frac{1}{f'(x_1)}(x - x_1).$$

44 Refer to Exercise 43. What is an equation of the normal line if $f'(x_1) = 0$?

In Exercises 45–48 find an equation of the normal line to the graph of f at the indicated point P. Sketch the graph, showing the normal line at P.

45 $f(x) = x^2 + 1$, $P(1, 2)$

46 $f(x) = 8 - x^3$, $P(1, 7)$

47 $f(x) = 1/(x^2 + 1)$, $P(1, \frac{1}{2})$

48 $f(x) = (x - 1)^4$, $P(2, 1)$

49 Find an equation of the line with slope 4 which is normal to the graph of $y = 1/(8x + 3)^2$.

50 Find the points on the graph of $y = (2x - 1)^3$ at which the normal line is parallel to the line $x + 12y = 36$.

51 In Chapter 8 a function denoted by ln is defined which has the property $D_x \ln x = 1/x$ for all $x > 0$. Use the Chain Rule to find $D_x \ln(x^4 + 2x^2 + 5)$.

52 In Chapter 8 a function f is defined which has the property $D_x[f(x)] = f(x)$ for all x. Find a formula for $D_x[f(g(x))]$, where g is any differentiable function.

3.6 IMPLICIT DIFFERENTIATION

Given an equation of the form

$$y = 2x^2 - 3$$

it is customary to say that y *is a function of* x, since we may write

$$y = f(x), \quad \text{where} \quad f(x) = 2x^2 - 3. \tag{3.41}$$

The equation

$$4x^2 - 2y = 6 \tag{3.42}$$

defines the same function f, since solving for y gives us

$$-2y = -4x^2 + 6, \quad \text{or } y = 2x^2 - 3.$$

In the case of (3.42) we say that y is an **implicit function** of x, or that f is defined *implicitly* by the equation.

Substituting $f(x)$ for y in (3.42) we obtain

$$\begin{aligned} 4x^2 - 2f(x) &= 6 \\ 4x^2 - 2(2x^2 - 3) &= 6 \\ 4x^2 - 4x^2 + 6 &= 6 \end{aligned}$$

which is an identity since it is true for every x in the domain of f. This result is characteristic of every function f defined implicitly by an equation in x and y; that is, f is implicit if and only if substitution of $f(x)$ for y leads to an identity.

Equations in x and y often determine more than one implicit function, as illustrated in the next example.

Example 1 How many different implicit functions are defined by the equation $x^2 + y^2 = 1$?

Solution Solving the equation for y in terms of x we obtain

$$y = \pm\sqrt{1 - x^2}.$$

Two obvious implicit functions are given by

$$f(x) = \sqrt{1 - x^2} \quad \text{and} \quad g(x) = -\sqrt{1 - x^2}.$$

The graphs of f and g are the upper and lower halves, respectively, of the unit circle with center at the origin (see (i) and (ii) of Figure 3.11). To find other implicit functions we may let a be any number between -1 and 1 and then define the function k by

$$k(x) = \begin{cases} \sqrt{1 - x^2} & \text{if} \quad -1 \le x \le a \\ -\sqrt{1 - x^2} & \text{if} \quad a < x \le 1. \end{cases}$$

The graph of k is sketched in (iii) of Figure 3.11. Note that there is a discontinuity at $x = a$. The function k is defined implicitly by $x^2 + y^2 = 1$, since

$$x^2 + [k(x)]^2 = 1$$

for every x in the domain of k. By letting a take on different values, it is possible to obtain as many implicit functions as desired. There are also many other types of functions which are determined implicitly by $x^2 + y^2 = 1$.

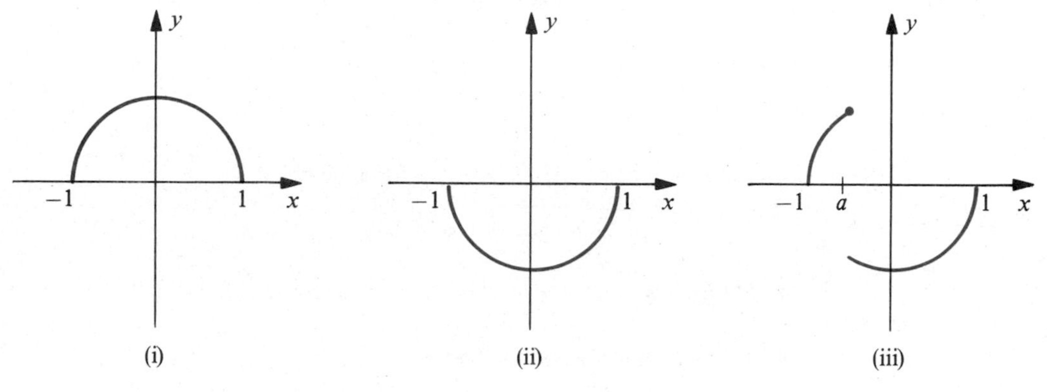

Figure 3.11

It is not obvious that an equation such as

$$y^4 + 3y - 4x^3 = 5x + 1$$

determines an implicit function f in the sense that if $f(x)$ is substituted for y, then

$$[f(x)]^4 + 3[f(x)] - 4x^3 = 5x + 1$$

is an identity for all x in the domain of f. It is possible to state conditions under which an implicit function exists and is differentiable at numbers in its domain; however, the proof requires advanced methods and hence is omitted here. In the examples to follow it will be assumed that a given equation in x and y determines a differentiable function f such that if $f(x)$ is substituted for y, the equation is an identity for all x in the domain of f. The derivative of f may then be found by the method of **implicit differentiation** as illustrated in the following example.

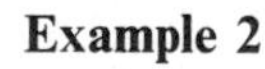

Example 2 Assuming that the equation $y^4 + 3y - 4x^3 = 5x + 1$ determines, implicitly, a differentiable function f such that $y = f(x)$, find its derivative.

Solution We think of y as a symbol which denotes $f(x)$ and regard the equation as an identity for all x in the domain of f. It follows that the derivatives of both sides are equal, that is,

$$D_x(y^4 + 3y - 4x^3) = D_x(5x + 1),$$

or

$$D_x(y^4) + D_x(3y) - D_x(4x^3) = D_x(5x) + D_x(1). \tag{3.43}$$

It is important to remember that in general, $D_x y^4 \neq 4y^3$. Instead, since $y = f(x)$, we use the Power Rule for Functions (3.38), obtaining $D_x(y^4) = 4y^3y'$. In like manner, $D_x(3y) = 3D_x y = 3y'$. Substituting in (3.43),

$$4y^3y' + 3y' - 12x^2 = 5 + 0.$$

We now solve for y'. Thus

$$(4y^3 + 3)y' = 12x^2 + 5$$

and

$$y' = \frac{12x^2 + 5}{4y^3 + 3}$$

provided the denominator is not zero. In terms of f, this becomes

$$f'(x) = \frac{12x^2 + 5}{4(f(x))^3 + 3}.$$

The last two equations bring out a disadvantage in using the method of implicit differentiation, namely that the formula for y' (or $f'(x)$) may contain the expression y (or $f(x)$). However, these formulas can still be extremely useful in analyzing f and its graph.

In the next example, implicit differentiation is used to find the slope of the tangent line at a point $P(a, b)$ on the graph of an equation. In problems of this type we shall assume that the equation defines an implicit function f whose graph coincides with the graph of the equation for all x in some open interval containing a.

Example 3 Find the slope of the tangent line to the graph of $y^4 + 3y - 4x^3 = 5x + 1$ at the point $P(1, -2)$.

Solution By Definition (3.1) the slope m of the tangent line at a point $P(a, b)$ on the graph is the value of the derivative y' when $x = a$ and $y = b$. The given equation is the same as that in Example 3, where it was found that $y' = (12x^2 + 5)/(4y^3 + 3)$. Substituting 1 for x and -2 for y gives us

$$m = \frac{12(1)^2 + 5}{4(-2)^3 + 3} = -\frac{17}{29}.$$

Example 4 Assume that $x^2 + y^2 = 1$ determines a function f such that $y = f(x)$. Find y'.

Solution The given equation was considered in Example 1. Proceeding as in Example 2, we differentiate both sides with respect to x, obtaining

$$D_x(x^2) + D_x(y^2) = D_x(1).$$

Consequently,

$$2x + 2yy' = 0, \quad \text{or} \quad yy' = -x$$

and

$$y' = -\frac{x}{y}, \quad \text{if } y \neq 0.$$

The equation $x^2 + y^2 = 1$ determines many implicit functions (see Example 1). It can be shown that the method of implicit differentiation provides the derivative of *any* differentiable function determined by an equation in two variables. To illustrate, the slope of the tangent line at the point (x, y) on any of the graphs in Figure 3.11 is given by $y' = -x/y$, provided this derivative exists.

Example 5 Find y' if $4xy^3 - x^2y + x^3 - 5x + 6 = 0$.

Solution Differentiating both sides of the equation with respect to x we obtain

$$D_x(4xy^3) - D_x(x^2y) + D_x(x^3) - D_x(5x) + D_x(6) = D_x(0).$$

Since y denotes $f(x)$, where f is a function, the Product Rule must be applied to $D_x(4xy^3)$ and $D_x(x^2y)$. Thus

$$\begin{aligned} D_x(4xy^3) &= 4xD_x(y^3) + y^3D_x(4x) \\ &= 4x(3y^2y') + y^3(4) \\ &= 12xy^2y' + 4y^3 \end{aligned}$$

and

$$\begin{aligned} D_x(x^2y) &= x^2D_xy + yD_x(x^2) \\ &= x^2y' + y(2x). \end{aligned}$$

Substituting these in the first equation of the solution and differentiating the other terms leads to

$$(12xy^2y' + 4y^3) - (x^2y' + 2xy) + 3x^2 - 5 = 0.$$

Collecting the terms containing y' and transposing the remaining terms to the right side of the equation gives us

$$(12xy^2 - x^2)y' = 5 - 3x^2 + 2xy - 4y^3.$$

Consequently,

$$y' = \frac{5 - 3x^2 + 2xy - 4y^3}{12xy^2 - x^2}.$$

EXERCISES 3.6

Find at least one implicit function f determined by each of the equations in Exercises 1–8 and state the domain of f.

1 $3x - 2y + 4 = 2x^2 + 3y - 7x$

2 $x^3 - xy + 4y = 1$

3 $x^2 + y^2 - 16 = 0$

4 $3x^2 - 4y^2 = 12$

5 $x^2 - 2xy + y^2 = x$

6 $3x^2 - 4xy + y^2 = 0$

7 $\sqrt{x} + \sqrt{y} = 1$

8 $|x - y| = 2$

Assuming that each of the equations in Exercises 9–20 determines a function f such that $y = f(x)$, find y'.

9 $8x^2 + y^2 = 10$

10 $4x^3 - 2y^3 = x$

11 $2x^3 + x^2y + y^3 = 1$

12 $5x^2 + 2x^2y + y^2 = 8$

13 $5x^2 - xy - 4y^2 = 0$

14 $x^4 + 4x^2y^2 - 3xy^3 + 2x = 0$

15 $(1/x^2) + (1/y^2) = 1$

16 $x = y + 2y^2 + 3y^3$

17 $x^2y^3 + 4xy + x - 6y = 2$

18 $4 - 7xy = (y^2 + 4)^5$

19 $(y^2 - 9)^4 = (4x^2 + 3x - 1)^2$

20 $(1 + xy)^3 = 2x^2 - 9$

In each of Exercises 21–24 find an equation of the tangent line to the graph of the given equation at the indicated point P.

21 $xy + 16 = 0$, $P(-2, 8)$

22 $y^2 - 4x^2 = 5$, $P(-1, 3)$

23 $2x^3 - x^2y + y^3 - 1 = 0$, $P(2, -3)$

24 $(1/x) + (3/y) = 1$, $P(2, 6)$

25 Show that the equation $x^2 + y^2 + 1 = 0$ does not determine a function f such that $y = f(x)$.

26 Show that the equation $y^2 = x$ determines an infinite number of implicit functions. Sketch the graphs of four such functions.

27 How many implicit functions are determined by the following equations?
(a) $x^4 + y^4 - 1 = 0$ (b) $x^4 + y^4 = 0$

28 Use implicit differentiation to prove that if P is any point on the circle $x^2 + y^2 = a^2$, then the tangent line at P is perpendicular to OP.

29 Suppose that $3x^2 - x^2y^3 + 4y = 12$ determines a differentiable function f such that $y = f(x)$. If $f(2) = 0$, use differentials to approximate the change in $f(x)$ if x changes from 2 to 1.97.

30 Suppose that $x^3 + xy + y^4 = 19$ determines a differentiable function f such that $y = f(x)$. If $P(1, 2)$ is a point on the graph of f, use differentials to approximate the ordinate b of the point $Q(1.10, b)$ on the graph.

3.7 DERIVATIVES OF ALGEBRAIC FUNCTIONS

The Power Rule (3.12) can be extended to the case where the exponent is a rational number. Thus, if

$$y = x^{m/n}$$

where m and n are integers and $n \neq 0$, then

$$y^n = x^m.$$

Assuming that y' exists and no zero denominators occur we have, by implicit differentiation,

$$ny^{n-1}\frac{dy}{dx} = mx^{m-1}.$$

Consequently,

$$\frac{dy}{dx} = \frac{m}{n}x^{m-1}y^{1-n} = \frac{m}{n}x^{m-1}(x^{m/n})^{1-n}.$$

Simplifying the right side by means of laws of exponents we obtain

$$D_x(x^{m/n}) = \frac{m}{n}x^{(m/n)-1} \tag{3.44}$$

which is the Power Rule (3.12) where the exponent is a rational number m/n. Our proof is incomplete in the sense that we assumed the existence of y'. In order to give a complete proof the definition of derivative (3.6) could be used.

Example 1 Find y' if $y = 6\sqrt[3]{x^4} + 4/\sqrt{x}$.

Solution Introducing exponents, $y = 6x^{4/3} + 4x^{-1/2}$. It follows that

$$\begin{aligned} y' &= 6(4/3)x^{(4/3)-1} + 4(-1/2)x^{-(1/2)-1} \\ &= 8x^{1/3} - 2x^{-3/2} \\ &= 8x^{1/3} - 2/x^{3/2} \end{aligned}$$

or, in terms of radicals,

$$y' = 8\sqrt[3]{x} - 2/\sqrt{x^3}$$

for all $x > 0$.

We may employ the proof used to establish (3.38) to show that the Power Rule for Functions is true for every *rational* exponent. The following examples illustrate the extension of (3.39) to the case where n is a rational number.

Example 2 Find $f'(x)$ if $f(x) = \sqrt[3]{5x^2 - x + 4}$.

Solution Writing $f(x) = (5x^2 - x + 4)^{1/3}$ and using (3.39) with $n = 1/3$,

$$\begin{aligned} f'(x) &= \frac{1}{3}(5x^2 - x + 4)^{-2/3} D_x(5x^2 - x + 4) \\ &= \left(\frac{1}{3}\right)\frac{1}{(5x^2 - x + 4)^{2/3}}(10x - 1) \\ &= \frac{10x - 1}{3\sqrt[3]{(5x^2 - x + 4)^2}}. \end{aligned}$$

Example 3 Find dy/dx if $y = (3x + 1)^6\sqrt{2x - 5}$.

Solution Since $y = (3x + 1)^6(2x - 5)^{1/2}$, we have, by the Product and Power Rules,

$$\begin{aligned} \frac{dy}{dx} &= (3x + 1)^6\frac{1}{2}(2x - 5)^{-1/2}(2) + (2x - 5)^{1/2}6(3x + 1)^5(3) \\ &= \frac{(3x + 1)^6}{\sqrt{2x - 5}} + 18(3x + 1)^5\sqrt{2x - 5} \\ &= \frac{(3x + 1)^6 + 18(3x + 1)^5(2x - 5)}{\sqrt{2x - 5}} \\ &= \frac{(3x + 1)^5(39x - 89)}{\sqrt{2x - 5}}. \end{aligned}$$

The next example is of interest since it illustrates the fact that after the Power Rule is applied to $[g(x)]^r$, it may be necessary to apply it again in order to find $g'(x)$.

Example 4 Find $f'(x)$ if $f(x) = (7x + \sqrt{x^2 + 6})^4$.

Solution Applying the Power Rule,

$$\begin{aligned} f'(x) &= 4(7x + \sqrt{x^2 + 6})^3 D_x(7x + \sqrt{x^2 + 6}) \\ &= 4(7x + \sqrt{x^2 + 6})^3 [D_x(7x) + D_x\sqrt{x^2 + 6}]. \end{aligned}$$

Again applying the Power Rule,

$$\begin{aligned} D_x\sqrt{x^2 + 6} &= D_x(x^2 + 6)^{1/2} = \frac{1}{2}(x^2 + 6)^{-1/2} D_x(x^2 + 6) \\ &= \frac{1}{2\sqrt{x^2 + 6}}(2x) = \frac{x}{\sqrt{x^2 + 6}}. \end{aligned}$$

Therefore

$$f'(x) = 4(7x + \sqrt{x^2 + 6})^3 \left[7 + \frac{x}{\sqrt{x^2 + 6}}\right].$$

EXERCISES 3.7

Differentiate the functions defined in Exercises 1–26.

1 $f(x) = \sqrt[3]{x^2} + 4\sqrt{x^3}$

2 $f(x) = 10\sqrt[5]{x^3} + \sqrt[3]{x^5}$

3 $k(r) = \sqrt[3]{8r^3 + 27}$

4 $h(z) = (2z^2 - 9z + 8)^{-2/3}$

5 $F(v) = 5/\sqrt[5]{v^5 - 32}$

6 $k(s) = 1/\sqrt{3s - 4}$

7 $f(x) = \sqrt{2x}$

8 $g(x) = \sqrt[5]{1/x}$

9 $f(z) = 10\sqrt{z^3} + 3/\sqrt[3]{z}$

10 $f(t) = \sqrt[3]{t^2} - (1/\sqrt{t^3})$

11 $g(w) = (w^2 - 4w + 3)/w^{3/2}$

12 $K(x) = 8x^2\sqrt{x} + 3x\sqrt[3]{x}$

13 $M(x) = \sqrt{4x^2 - 7x + 4}$

14 $F(s) = \sqrt[3]{5s - 8}$

15 $f(t) = 4/(9t^2 + 16)^{2/3}$

16 $k(v) = 1/\sqrt{(v^4 + 7v^2)^3}$

17 $H(u) = \sqrt{\dfrac{3u + 8}{2u + 5}}$

18 $G(x) = \left(\dfrac{x}{x^2 + 1}\right)^{5/2}$

19 $k(s) = \sqrt[4]{s^2 + 9}(4s + 5)^4$

20 $g(y) = (15y + 2)(y^2 - 2)^{3/4}$

21 $h(x) = (x^2 + 4)^{5/3}(x^3 + 1)^{3/5}$

22 $f(w) = \sqrt{w^3(9w + 1)^5}$

23 $g(z) = \dfrac{\sqrt[3]{2z + 3}}{\sqrt{3z + 2}}$

24 $H(x) = \dfrac{2x + 3}{\sqrt{4x^2 + 9}}$

25 $f(x) = (7x + \sqrt{x^2 + 3})^6$

26 $p(z) = [1 + (1 + 2z)^{1/2}]^{1/2}$

In each of Exercises 27–30 find an equation of the tangent line to the graph of the equation at the point P.

27 $y = \sqrt{2x^2 + 1}$, $P(-1, \sqrt{3})$

28 $y = (5x - 8)^{1/3}$, $P(7, 3)$

29 $y = 4x^{2/3} + 2x^{-1/3} - 10$, $P(-8, 5)$

30 $y = 4x/\sqrt{x + 1}$, $P(3, 6)$

31 Find the point P on the graph of $y = \sqrt{2x - 4}$ such that the tangent line at P passes through the origin.

32 Find the points on the graph of $y = x^{5/3} + x^{1/3}$ at which the tangent line is perpendicular to the line $2y + x = 7$.

Assuming that each of the equations in Exercises 33–38 determines a function f such that $y = f(x)$, find y'.

33 $\sqrt{x} + \sqrt{y} = 100$

34 $x^{2/3} + y^{2/3} = 4$

35 $6x + \sqrt{xy} - 3y = 4$

36 $xy^2 + \sqrt[3]{xy} + x^4 = 7$

37 $3x^2 + \sqrt[3]{xy} = 2y^2 + 20$

38 $2x - \sqrt{xy} + y^3 = 16$

In Exercises 39 and 40 find dy.

39 $y = \sqrt[3]{6x + 11}$

40 $y = 1/\sqrt{x^2 + 1}$

41 Use differentials to approximate $\sqrt[3]{65}$.

42 Use differentials to approximate $\sqrt{99}$.

43 Use differentials to approximate the change in $f(x) = (4x^2 + 9)^{3/2}$ if x changes from 2 to 1.998.

44 Use differentials to approximate $N = 5(1.01)^{3/5} - 3(1.01)^{1/5} + 7$.

45 The curved surface area S of a right circular cone having altitude h and base radius r is given by $S = \pi r\sqrt{r^2 + h^2}$. For a certain cone, $r = 6$ cm and the altitude is measured as 8 cm, with a maximum error in measurement of 0.1 cm. Calculate S from these measurements and use differentials to estimate the maximum error in the calculation. What is the approximate percentage error?

46 The period T of a simple pendulum of length l may be calculated by means of the formula $T = 2\pi\sqrt{l/g}$, where g is a constant. Use differentials to approximate the change in l that will increase T by 1%.

3.8 HIGHER ORDER DERIVATIVES

The derivative of a function f leads to another function f'. If f' has a derivative, it is denoted by f'' and is called the **second derivative** of f. Thus

$$f''(x) = D_x(f'(x)) = D_x(D_x(f(x))) \tag{3.45}$$

for every x such that the indicated derivatives exist. The expression on the right in (3.45) is abbreviated $D_x^2 f(x)$. In like manner the **third derivative** f''' of f is the derivative of the second derivative. Specifically,

$$f'''(x) = D_x(f''(x)) = D_x(D_x^2 f(x)) = D_x^3 f(x).$$

In general, if n is a positive integer, then $f^{(n)}$ denotes the nth derivative of f and is found by starting with f and differentiating, successively, n times. Using operator notation, $f^{(n)}(x) = D_x^n f(x)$. The integer n is called the **order** of the derivative $f^{(n)}(x)$.

If $y = f(x)$, then the first n derivatives are denoted by

$$D_x y, D_x^2 y, D_x^3 y, \ldots, D_x^n y$$

or

$$y', y'', y''', \ldots, y^{(n)}.$$

If the differential notation is used we write

$$\frac{dy}{dx}, \frac{d^2y}{dx^2}, \frac{d^3y}{dx^3}, \ldots, \frac{d^ny}{dx^n}.$$

In this case, the symbols used for higher derivatives should *not* be interpreted as quotients.

Example 1 If $f(x) = 4x^2 - 5x + 8 - (3/x)$, find the first four derivatives of $f(x)$.

Solution Since $f(x) = 4x^2 - 5x + 8 - 3x^{-1}$,

$$f'(x) = 8x - 5 + 3x^{-2} = 8x - 5 + \frac{3}{x^2}$$

$$f''(x) = 8 - 6x^{-3} = 8 - \frac{6}{x^3}$$

$$f'''(x) = 18x^{-4} = \frac{18}{x^4}$$

$$f^{(4)}(x) = -72x^{-5} = -\frac{72}{x^5}.$$

The next example illustrates a technique for finding higher derivatives of implicit functions.

Example 2 Find y'' if $y^4 + 3y - 4x^3 = 5x + 1$.

Solution The given equation was investigated in Example 3 of Section 6, where we found that

$$y' = (12x^2 + 5)/(4y^3 + 3).$$

Hence,

$$y'' = D_x(y') = D_x\left(\frac{12x^2 + 5}{4y^3 + 3}\right).$$

We now use the quotient rule, differentiating implicitly as follows:

$$\begin{aligned} y'' &= \frac{(4y^3 + 3)D_x(12x^2 + 5) - (12x^2 + 5)D_x(4y^3 + 3)}{(4y^3 + 3)^2} \\ &= \frac{(4y^3 + 3)(24x) - (12x^2 + 5)(12y^2y')}{(4y^3 + 3)^2}. \end{aligned}$$

Substituting for y' yields

$$\begin{aligned} y'' &= \frac{(4y^3 + 3)(24x) - (12x^2 + 5)\cdot 12y^2\left(\dfrac{12x^2 + 5}{4y^3 + 3}\right)}{(4y^3 + 3)^2} \\ &= \frac{(4y^3 + 3)^2(24x) - 12y^2(12x^2 + 5)^2}{(4y^3 + 3)^3}. \end{aligned}$$

There are numerous mathematical and physical applications of higher derivatives. Some of them will be discussed in the next chapter.

EXERCISES 3.8

Find the first and second derivatives of the functions defined in Exercises 1–10.

1 $f(x) = 3x^4 - 4x^2 + x - 2$

2 $g(x) = 3x^8 - 2x^5$

3 $H(s) = \sqrt[3]{s} + \dfrac{2}{s^2}$

4 $F(t) = t^{3/2} - 2t^{1/2} + 4t^{-1/2}$

5 $g(z) = \sqrt{3z + 1}$

6 $k(s) = (s^2 + 4)^{2/3}$

7 $k(r) = (4r + 7)^5$

8 $f(x) = \sqrt[5]{10x + 7}$

9 $f(x) = \sqrt{x^2 + 4}$

10 $h(x) = 1$

In Exercises 11–16 find $D_x^3 y$.

11 $y = 2x^5 + 3x^3 - 4x + 1$

12 $y = \sqrt{2 - 5x}$

13 $y = \dfrac{2x - 3}{3x + 1}$

14 $y = \dfrac{1}{x^2 + 4}$

15 $y = \sqrt[3]{2 - 9x}$

16 $y = (3x + 1)^4$

Assuming that each of the equations in Exercises 17–20 determines a function f, such that $y = f(x)$, find y''.

17 $x^3 - y^3 = 1$

18 $x^2y^3 = 1$

19 $x^2 - 3xy + y^2 = 4$

20 $\sqrt{xy} - y + x = 0$

21 Find all the nonzero derivatives of $f(x) = x^6 - 2x^4 + 3x^3 - x + 2$.

22 Find all the nonzero derivatives of $f(x) = (x^2 - 1)^3$.

23 If $f(x) = 1/x$, find a formula for $f^{(n)}(x)$ where n is any positive integer. What is $f^{(n)}(1)$?

24 If $f(x) = \sqrt{x}$, find a formula for $f^{(n)}(x)$ where n is any positive integer.

25 If $f(x)$ is a polynomial of degree n, show that $f^{(k)}(x) = 0$ if $k > n$.

26 Find a polynomial $f(x)$ of degree 2 such that $f(1) = 5$, $f'(1) = 3$, and $f''(1) = -4$.

27 If $f(x) = 2x^3 + 3x^2 - 12x + 7$, find the value of f'' at each zero of f'.

28 If $f(x) = x^4 - x^3 - 6x^2 + 7x$, find an equation of the tangent line to the graph of f' at the point $P(2, -2)$.

29 If $f(x) = x^4 - 10x^2 + x + 2$, use differentials to approximate the change in $f'(x)$ if x changes from 2 to 2.005.

30 Suppose $u = f(x)$ and $v = g(x)$ where f and g have derivatives of all orders. If $y = uv$, prove that

$$y'' = u''v + 2u'v' + uv''$$
$$y''' = u'''v + 3u''v' + 3u'v'' + uv'''$$
$$y^{(4)} = u^{(4)}v + 4u'''v' + 6u''v'' + 4u'v''' + uv^{(4)},$$

Formulate a conjecture for $y^{(n)}$.

31 If $y = f(g(x))$ and f'' and g'' exist, use the Chain Rule to express $D_x^2 y$ in terms of the first and second derivatives of f and g.

32 Suppose f has a second derivative. Prove that

(a) if f is an even function, then f'' is even.

(b) if f is an odd function, then f'' is odd.

Illustrate these facts by using polynomial functions. (*Hint:* See Exercise 42 of Section 5.)

3.9 REVIEW

Concepts

Define or explain each of the following.

1 The slope of the tangent line to the graph of a function

2 The velocity of a point moving rectilinearly

3 The derivative of a function

4 Differentiable function

5 Increment notation

6 Differentials

7 Right-hand derivative of a function

8 Left-hand derivative of a function

9 Differentiability on an interval

10 The relationship between continuity and differentiability of a function

11 The Product Rule

12 The Quotient Rule

13 The Chain Rule

14 The Power Rule for Functions

15 Implicit differentiation

16 Higher order derivatives of functions

Exercises

In Exercises 1 and 2 find $f'(x)$ directly from the definition of derivative.

1 $f(x) = 4/(3x^2 + 2)$

2 $f(x) = \sqrt{5 - 7x}$

Find the first derivatives in Exercises 3–32.

3 $f(x) = 2x^3 - 7x + 2$

4 $k(x) = 1/(x^4 - x^2 + 1)$

5 $g(t) = \sqrt{6t + 5}$

6 $h(t) = 1/\sqrt{6t + 5}$

7 $F(z) = \sqrt[3]{7z^2 - 4z + 3}$

8 $f(w) = \sqrt[5]{3w^2}$

9 $G(x) = 6/(3x^2 - 1)^4$

10 $H(x) = (3x^2 - 1)^4/6$

11 $F(y) = (y^2 - y^{-2})^{-2}$

12 $h(z) = [(z^2 - 1)^5 - 1]^5$

13 $g(x) = \sqrt[5]{(3x + 2)^4}$

14 $P(x) = (x + x^{-1})^2$

15 $r(s) = \left(\dfrac{8s^2 - 4}{1 - 9s^3}\right)^4$

16 $g(w) = \dfrac{(w - 1)(w - 3)}{(w + 1)(w + 3)}$

17 $F(x) = (x^6 + 1)^5(3x + 2)^3$

18 $k(z) = (z^2 + (z^2 + 9)^{1/2})^{1/2}$

19 $g(y) = (7y - 2)^{-2}(2y + 1)^{2/3}$

20 $p(x) = (2x^4 + 3x^2 - 1)/x^2$

21 $f(x) = (x^2 + 1)(x^2 + 2)(x^2 + 3)$

22 $H(t) = (t^6 - t^{-6})^6$

23 $h(x) = \sqrt{x + \sqrt{x + \sqrt{x}}}$

24 $K(r) = \sqrt{r}\sqrt{r + 1}\sqrt{r + 2}$

25 $f(x) = \sqrt[3]{2x + 3}/\sqrt{3x + 2}$

26 $f(x) = 6x^2 - \dfrac{5}{x} + \dfrac{2}{\sqrt[3]{x^2}}$

27 $g(z) = (9z^{5/3} - 5z^{3/5})^3$

28 $F(t) = (5t^2 - 7)/(t^2 + 2)$

29 $k(s) = (2s^2 - 3s + 1)(9s - 1)^4$

30 $H(x) = (3 + 2x^{-2} + 4x^{-3})^{-5}$

31 $f(w) = \sqrt{\dfrac{2w + 5}{7w - 9}}$

32 $S(t) = \sqrt{t^2 + t + 1}\sqrt[3]{4t - 9}$

Assuming that each of the equations in Exercises 33–36 determines a function f such that $y = f(x)$, find y'

33 $5x^3 - 2x^2y^2 + 4y^3 - 7 = 0$

34 $3x^2 - xy^2 + y^{-1} = 1$

35 $\dfrac{\sqrt{x} + 1}{\sqrt{y} + 1} = y$

36 $y^2 - \sqrt{xy} + 3x = 2$

In each of Exercises 37–39, find an equation of the tangent line to the graph of the given equation at the indicated point P.

37 $y = 2x - (4/\sqrt{x}), P(4, 6)$

38 $y = (x^3 + 2)^5, P(-1, 1)$

39 $x^2y - y^3 = 8, P(-3, 1)$

In Exercises 40–42, find y', y'', and y'''.

40 $y = 5x^3 + 4\sqrt{x}$

41 $y = x^{-2} + x^{-1}$

42 $y = 7$

43 If $x^2 + 4xy - y^2 = 8$, find y'' by implicit differentiation.

44 Given $f(x) = x^3 - x^2 - 5x + 2$.

(a) Find the abscissas of all points on the graph of f at which the tangent line is parallel to the line through $A(-3, 2)$ and $B(1, 14)$.

(b) Find the value of f'' at each zero of f'.

45 If $f(x) = 1/(1 - x)$, find a formula for $f^{(n)}(x)$, where n is any positive integer.

46 If $y = 5x/(x^2 + 1)$, find dy and use it to approximate the change in y if x changes from 2 to 1.98. What is the exact change in y?

47 The side of an equilateral triangle is estimated to be 4 inches with a maximum error of 0.03 inches. Use differentials to estimate the maximum error in the calculated area of the triangle. What is the approximate percentage error?

48 If $s = 3r^2 - 2\sqrt{r + 1}$ and $r = t^3 + t^2 + 1$, use the Chain Rule (3.36) to find the value of ds/dt at $t = 1$.

49 If $f(x) = 2x^3 + x^2 - x + 1$ and $g(x) = x^5 + 4x^3 + 2x$, use differentials to approximate the change in $g(f(x))$ if x changes from -1 to -1.01.

50 Use differentials to approximate $\sqrt[3]{64.2}$.

Applications of the Derivative

In this chapter the derivative is used as a tool to investigate some mathematical and physical problems. Included are methods for determining maximum and minimum values of functions, applications of extrema to applied situations, and the graphical concepts of concavity *and* points of inflection. *We shall also examine how the derivative may be used to find rates at which quantities change, not only with respect to time, but also with respect to other variables. The concept of* antiderivative *is introduced near the end of the chapter and applied to problems on rectilinear motion. The final section contains a discussion of how derivatives may be used in economics.*

4.1 LOCAL EXTREMA OF FUNCTIONS

Suppose that the graph illustrated in Figure 4.1 was made by a recording instrument used to measure the variation of a physical quantity with respect to time. In this case the x-axis represents time and ordinates of points on the graph denote magnitudes of the quantity. For example, y-values might represent measurements such as temperature, pressure, current in an electrical circuit, blood pressure of an individual, the amount of chemical in a solution, the bacteria count in a culture, and so on. From the graph we see that the quantity increased in

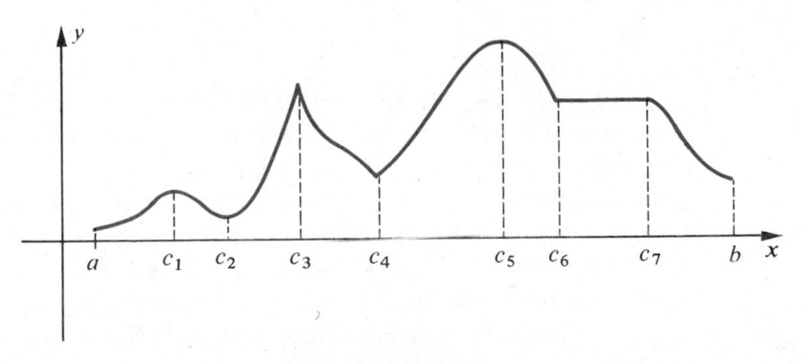

Figure 4.1

the time interval $[a, c_1]$, decreased in $[c_1, c_2]$, increased in $[c_2, c_3]$, and so on. If we restrict our attention to the interval $[c_1, c_4]$, the quantity had its largest (or maximum) value at c_3 and its smallest (or minimum) value at c_2. In other intervals there were different largest or smallest values. For example, over the entire interval $[a, b]$, the maximum value occurred at c_5 and the minimum value at a.

If Figure 4.1 is a sketch of the graph of a function f, it is convenient to use similar terminology to describe the behavior of $f(x)$ as x varies. The mathematical terms which are employed are included in the next two definitions.

(4.1) Definition

If a function f is defined on an interval I, then

(i) f is **increasing** on I if $f(x_1) < f(x_2)$ whenever x_1, x_2 are in I and $x_1 < x_2$.

(ii) f is **decreasing** on I if $f(x_1) > f(x_2)$ whenever x_1, x_2 are in I and $x_1 < x_2$.

(iii) f is **constant** on I if $f(x_1) = f(x_2)$ for every x_1, x_2 in I.

Geometric interpretations of the preceding definition are given in Figure 4.2, where the interval I is not indicated. Of course, if f is constant on I, then the graph is part of a horizontal line.

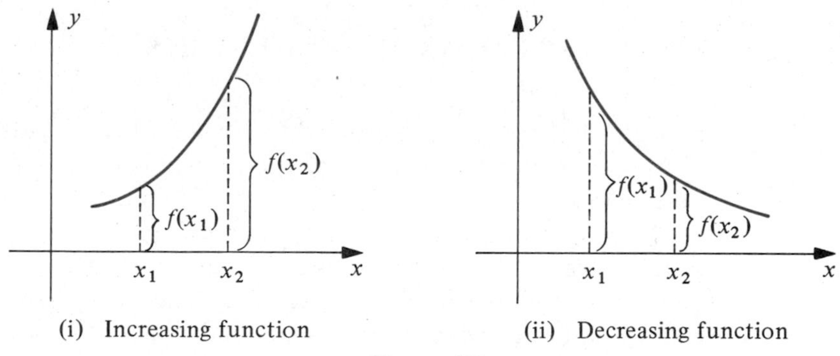

(i) Increasing function (ii) Decreasing function

Figure 4.2

We shall use the phrases "f is increasing" and "$f(x)$ is increasing" interchangeably. This will also be done for the term "decreasing." If a function is increasing, then the graph rises as x increases, as illustrated in (i) of Figure 4.2. If a function is decreasing, then the graph falls as x increases, as in (ii) of the figure. If the sketch in Figure 4.1 represents the graph of a function f, then f is increasing on the intervals $[a, c_1]$, $[c_2, c_3]$, and $[c_4, c_5]$. It is decreasing on $[c_1, c_2]$, $[c_3, c_4]$, $[c_5, c_6]$, and $[c_7, b]$. The function is constant on the interval $[c_6, c_7]$.

The next definition introduces terminology we shall use for largest and smallest values of functions on an interval.

(4.2) Definition

> Suppose a function f is defined on an interval I and c is a number in I. Then
> (i) $f(c)$ is the **maximum value** of f on I if $f(x) \leq f(c)$ for every x in I.
> (ii) $f(c)$ is the **minimum value** of f on I if $f(x) \geq f(c)$ for every x in I.

Maximum and minimum values are illustrated graphically in Figure 4.3. We have pictured I as a closed interval $[a, b]$; however, Definition (4.2) may be applied to any interval. Although the graphs in the figure have horizontal tangent lines at the point $(c, f(c))$, maximum and minimum values can also occur at abscissas of points where graphs have corners or at end-points of domains.

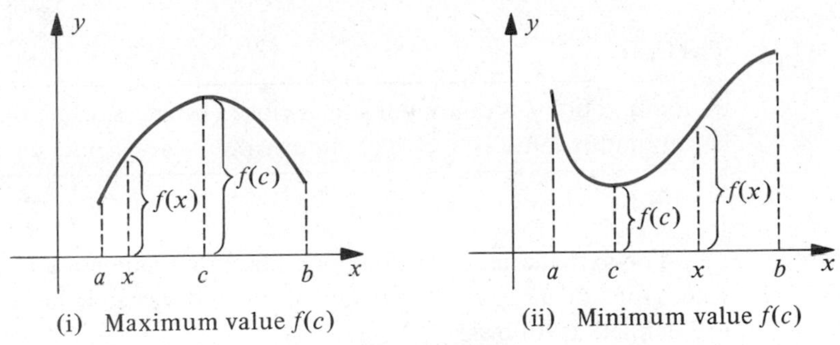

(i) Maximum value $f(c)$ (ii) Minimum value $f(c)$

Figure 4.3

If $f(c)$ is the maximum value of f on I, we say that f *takes on* its maximum value at c. If we restrict our attention to points with abscissas in I, then the point $(c, f(c))$ is a highest point on the graph as illustrated in (i) of Figure 4.3. Similarly, if $f(c)$ is the minimum value of f on I we say that f *takes on* its minimum value at c, and, as in (ii) of the figure, $(c, f(c))$ is a lowest point on the graph. Maximum and minimum values are sometimes called **extreme values** or **extrema** of f. A function can take on a maximum or minimum value more than once. Indeed, if f is a constant function, then $f(c)$ is both a maximum and a minimum value of f for *every* real number c.

Certain functions may have a maximum value but no minimum value on an interval, or vice versa. Other functions may have neither a maximum nor minimum value. To illustrate, suppose $f(x) = 1/x$ if $x \neq 0$ and $f(0) = 0$. The graph of f is sketched in Figure 4.4. Note that f is not continuous at 0. On the

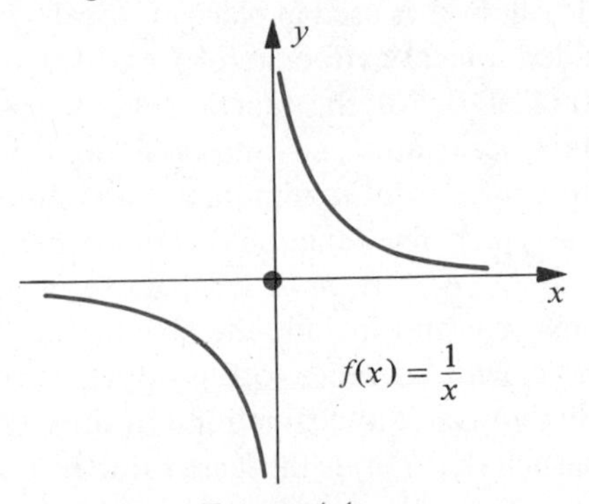

Figure 4.4

closed interval $[-1, 1]$, the function has neither a maximum value nor a minimum value. (Why?) The same is true on the open interval $(0, 1)$, for if $0 < a < 1$, then there always exist numbers x_1 and x_2 in $(0, 1)$ such that $f(x_1) > f(a)$ and $f(x_2) < f(a)$. On the half-open interval $(0, 1]$ there is a minimum value $f(1)$, but no maximum value. On the closed interval $[1, 2]$ f has a maximum value $f(1)$ and a minimum value $f(2)$.

From the preceding illustration it appears that the existence of maximum or minimum values may depend on both the continuity of the function and whether the interval under consideration is open, closed, half-open, and so on. The next theorem provides sufficient conditions under which a function takes on a maximum or minimum value on an interval. The interested reader may consult more advanced texts in calculus for the proof.

(4.3) Theorem

> If a function f is continuous on a closed interval $[a, b]$, then f takes on its minimum value and its maximum value at least once in $[a, b]$.

The extrema are also called the **absolute minimum** and **absolute maximum values** for f on an interval. We shall also be interested in **local extrema** of f which are defined as follows.

(4.4) Definition

> Let c be a number in the domain of a function f.
>
> (i) $f(c)$ is a **local maximum** of f if there exists an open interval (a, b) containing c such that $f(x) \leq f(c)$ for all x in (a, b).
>
> (ii) $f(c)$ is a **local minimum** of f if there exists an open interval (a, b) containing c such that $f(x) \geq f(c)$ for all x in (a, b).

The term "local" is used because we *localize* our attention to a sufficiently small open interval containing c where f takes on a largest (or smallest) value at c. Outside that interval f may take on larger (or smaller) values. Sometimes the word "relative" is used in place of "local." Each local maximum or minimum will be called a **local extremum** of f and the totality of such numbers are the **local extrema** of f. For the function whose graph is sketched in Figure 4.1, local maxima occur at c_1, c_3, and c_5 whereas local minima occur at c_2 and c_4. It is possible for a local minimum such as $f(c_4)$ to be *larger* than a local maximum such as $f(c_1)$. The functional values which correspond to numbers in the open interval (c_6, c_7) are *both* local maxima and local minima. (Why?) The local extrema may not include the absolute minimum or maximum values of f. For example, with reference to Figure 4.1, $f(a)$ is the minimum value of f on $[a, b]$, although it is not a local minimum since there is no *open* interval I contained in $[a, b]$ such that $f(a)$ is the least value of f on I. The number $f(c_5)$ is both a local maximum and the absolute maximum for f on $[a, b]$.

At a point corresponding to a local extremum of the function graphed in Figure 4.1, the tangent line is either horizontal or there is a corner. Abscissas of these points are numbers at which the derivative is zero or does not exist. The next theorem brings out the fact that this is generally true.

(4.5) **Theorem**

If a function f has a local extremum at a number c, then either $f'(c) = 0$ or $f'(c)$ does not exist.

Proof. Suppose f has a local extremum at c. If $f'(c)$ does not exist there is nothing more to prove. If $f'(c)$ exists, then precisely one of the following occurs: $f'(c) > 0$, (ii) $f'(c) < 0$, or (iii) $f'(c) = 0$. We shall arrive at (iii) by proving that neither (i) nor (ii) can occur. Thus, suppose $f'(c) > 0$. Employing Theorem (3.7) for the derivative,

$$\lim_{x \to c} \frac{f(x) - f(c)}{x - c} > 0$$

and hence by Theorem (2.10), there exists an open interval (a, b) containing c such that

$$\frac{f(x) - f(c)}{x - c} > 0$$

for all x in (a, b) different from c. The last inequality implies that if $a < x < b$ and $x \neq c$, then $f(x) - f(c)$ and $x - c$ are either both positive or both negative; that is,

$$\begin{cases} f(x) - f(c) < 0 & \text{whenever } x - c < 0, \text{ and} \\ f(x) - f(c) > 0 & \text{whenever } x - c > 0. \end{cases}$$

Another way of stating these facts is that if x is in (a, b) and $x \neq c$, then

$$\begin{cases} f(x) < f(c) & \text{whenever } x < c, \text{ and} \\ f(x) > f(c) & \text{whenever } x > c. \end{cases}$$

It follows that $f(c)$ is neither a local maximum nor a local minimum for f, contrary to hypothesis. Consequently (i) cannot occur. In like manner, the assumption that $f'(c) < 0$ leads to a contradiction. Hence (iii) must hold and the theorem is proved.

(4.6) **Corollary**

If $f'(c)$ exists and $f'(c) \neq 0$, then $f(c)$ is not a local extremum of the function f.

A result similar to Theorem (4.5) is true for the absolute maximum and minimum values of a function which is continuous on a closed interval $[a, b]$, provided the extrema occur on the *open* interval (a, b). The theorem may be stated as follows.

(4.7) Theorem

> If a function f is continuous on a closed interval $[a, b]$ and has its maximum or minimum value at a number c in the open interval (a, b), then either $f'(c) = 0$ or $f'(c)$ does not exist.

The proof is exactly the same as that of Theorem (4.5) with the word "local" deleted.

It follows from Theorems (4.5) and (4.7) that the numbers at which the derivative is either zero or does not exist play a crucial role in the search for extrema of a function. Because of this, we give these numbers a special name in the next definition.

(4.8) Definition

> A number c in the domain of a function f is a **critical number** of f if either $f'(c) = 0$ or $f'(c)$ does not exist.

Referring to Theorem (4.7) we see that if f is continuous on a closed interval $[a, b]$, then the absolute maximum and minimum values occur either at the end-points a or b of the interval, or at a critical number of f. Thus, to find the *absolute* extrema for such functions we may proceed as follows:

(i) Calculate $f(c)$ for each critical number c.
(ii) Calculate $f(a)$ and $f(b)$.

The absolute maximum and minimum of f on $[a, b]$ will then be the largest and smallest of these functional values. If either $f(a)$ or $f(b)$ is an extremum we call it an **end-point extremum**.

Example 1 If $f(x) = x^3 - 12x$, find the absolute maximum and minimum values of f on the closed interval $[-3, 5]$. Sketch the graph of f.

Solution We begin by finding the critical numbers of f. Differentiating,

$$f'(x) = 3x^2 - 12 = 3(x^2 - 4) = 3(x + 2)(x - 2).$$

Since the derivative exists everywhere, the only critical numbers are those for which the derivative is zero, that is, -2 and 2. Since f is continuous on $[-3, 5]$, it follows from the discussion preceding this example that the absolute maximum and minimum are among the numbers $f(-2), f(2), f(-3)$, and $f(5)$. Calculating these values we obtain

$$\begin{aligned} f(-2) &= (-2)^3 - 12(-2) = -8 + 24 = 16 \\ f(2) &= 2^3 - 12(2) = 8 - 24 = -16 \\ f(-3) &= (-3)^3 - 12(-3) = -27 + 36 = 9 \\ f(5) &= 5^3 - 12(5) = 125 - 60 = 65. \end{aligned}$$

Thus, the minimum value of f on $[-3, 5]$ is $f(2) = -16$ and the maximum value is the end-point extremum $f(5) = 65$.

Using the functional values we have calculated, and plotting several more points, leads to the sketch in Figure 4.5, where for clarity, different scales have been used on the x- and y-axes. It will follow from our work in Section 4 that $f(-2) = 16$ is a local maximum for f, as indicated by the graph.

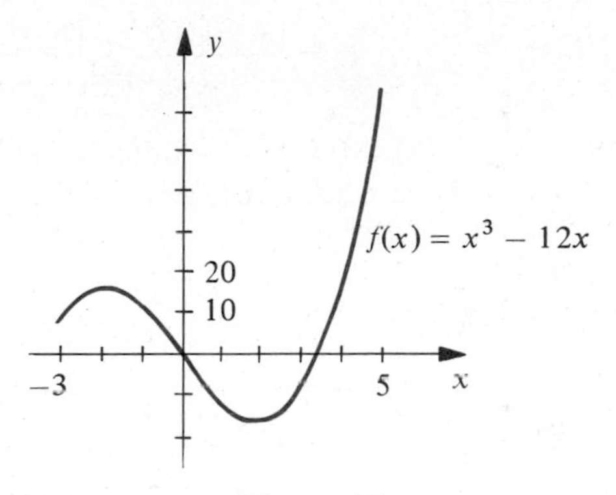

Figure 4.5

We see from Theorem (4.5) that if a function has a *local* extremum, then it *must* occur at a critical number; however, not every critical number leads to a local extremum as illustrated by the following example.

Example 2 If $f(x) = x^3$, prove that f has no local extremum.

Solution The derivative of f is $f'(x) = 3x^2$, which exists for all x and is zero only if $x = 0$. Consequently 0 is the only critical number. However, if $x < 0$, then $f(x)$ is negative whereas if $x > 0$, then $f(x)$ is positive. Hence $f(0)$ is neither a local maximum nor a local minimum. Since a local extremum *must* occur at a critical number (see Theorem (4.5)) it follows that f has no local extrema. The graph of f is sketched in Figure 1.27. Note that the tangent line is horizontal at the point $P(0, 0)$, which has the critical number 0 as abscissa.

In later sections methods are developed for finding local extrema of functions. At that time it may be necessary to obtain critical numbers of functions which are defined by fairly complicated expressions, as illustrated in the next example.

Example 3 Find the critical numbers of f if $f(x) = (x+5)^2\sqrt[3]{x-4}$.

Solution Differentiating $f(x) = (x+5)^2(x-4)^{1/3}$ we obtain

$$f'(x) = (x+5)^2\frac{1}{3}(x-4)^{-2/3} + 2(x+5)(x-4)^{1/3}.$$

As an aid to finding the critical numbers we simplify $f'(x)$ as follows:

$$\begin{aligned} f'(x) &= \frac{(x+5)^2}{3(x-4)^{2/3}} + 2(x+5)(x-4)^{1/3} \\ &= \frac{(x+5)^2 + 6(x+5)(x-4)}{3(x-4)^{2/3}} \\ &= \frac{(x+5)[(x+5)+6(x-4)]}{3(x-4)^{2/3}} \\ &= \frac{(x+5)(7x-19)}{3(x-4)^{2/3}}. \end{aligned}$$

Consequently, $f'(x) = 0$ if $x = -5$ or $x = 19/7$. The derivative $f'(x)$ does not exist at $x = 4$. Thus f has three critical numbers, namely -5, $19/7$, and 4.

EXERCISES 4.1

In each of Exercises 1–4, find the absolute maximum and minimum of f on the indicated closed interval.

1 $f(x) = 5 - 6x^2 - 2x^3$; $[-3, 1]$

2 $f(x) = 3x^2 - 10x + 7$; $[-1, 3]$

3 $f(x) = 1 - x^{2/3}$; $[-1, 8]$

4 $f(x) = x^4 - 5x^2 + 4$; $[0, 2]$

5 (a) If $f(x) = x^{1/3}$, prove that 0 is the only critical number of f, and that $f(0)$ is not a local extremum.

(b) If $f(x) = x^{2/3}$, prove that 0 is the only critical number of f, and that $f(0)$ is a local minimum of f.

6 If $f(x) = |x|$, prove that 0 is the only critical number of f, that $f(0)$ is a local minimum of f, and that the graph of f has no tangent line at the point $(0, 0)$.

In each of Exercises 7 and 8, prove that f has no local extrema. Sketch the graph of f. Prove that f is continuous on the interval $(0, 1)$, but f has neither a maximum nor minimum value on $(0, 1)$. Why doesn't this contradict Theorem (4.3)?

7 $f(x) = x^3 + 1$

8 $f(x) = 1/x^2$

Find the critical numbers of the functions defined in Exercises 9–20.

9 $f(x) = 4x^2 - 3x + 2$

10 $g(x) = 2x + 5$

11 $s(t) = 2t^3 + t^2 - 20t + 4$

12 $K(z) = 4z^3 + 5z^2 - 42z + 7$

13 $F(w) = w^4 - 32w$

14 $k(r) = r^5 - 2r^3 + r - 12$

15 $f(z) = \sqrt{z^2 - 16}$

16 $M(x) = \sqrt[3]{x^2 - x - 2}$

17 $g(t) = t^2\sqrt[3]{2t - 5}$

18 $T(v) = (4v + 1)\sqrt{v^2 - 16}$

19 $G(x) = (2x - 3)/(x^2 - 9)$

20 $f(s) = s^2/(5s + 4)$

21 Prove that a polynomial function of degree 1 has no local or absolute extrema on the interval $(-\infty, \infty)$. What is true on a closed interval $[a, b]$?

22 If f is a constant function and (a, b) is any open interval, prove that $f(c)$ is both a local and an absolute extremum of f for every number c in (a, b).

23 If f is the greatest integer function, prove that every number is a critical number of f.

24 Let f be defined by the following conditions: $f(x) = 0$ if x is rational and $f(x) = 1$ if x is irrational. Prove that every number is a critical number of f.

25 Prove that a quadratic function has exactly one critical number on $(-\infty, \infty)$.

26 Prove that a polynomial function of degree 3 has either two, one, or no critical numbers on $(-\infty, \infty)$. Sketch graphs which illustrate how each of these possibilities can occur.

27 If $f(x) = x^n$, where n is a positive integer, prove that f has either one or no local extrema on $(-\infty, \infty)$ according as n is even or odd, respectively. Sketch typical graphs illustrating each case.

28 Prove that a polynomial function of degree n can have at most $n - 1$ local extrema on $(-\infty, \infty)$.

4.2 ROLLE'S THEOREM AND THE MEAN VALUE THEOREM

Sometimes it is extremely difficult to find the critical numbers of a function. As a matter of fact, there is no guarantee that critical numbers even exist. The next theorem, credited to the French mathematician Michel Rolle (1652–1719), provides sufficient conditions for the existence of a critical number. The theorem is stated for a function f which is continuous on an interval $[a, b]$, differentiable on (a, b), and for which $f(a) = f(b)$. Some typical graphs of functions of this type are sketched in Figure 4.6.

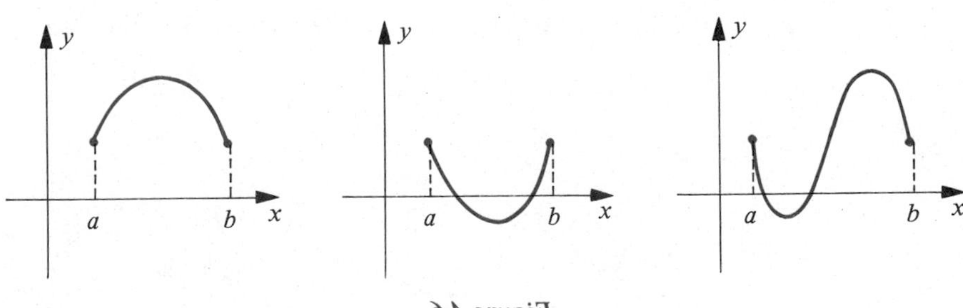

Figure 4.6

Referring to the sketches in Figure 4.6, it seems reasonable to expect that there is at least one number c between a and b such that the tangent line at the point $(c, f(c))$ is horizontal, or equivalently, such that $f'(c) = 0$. This is precisely the conclusion of the following theorem.

(4.9) Rolle's Theorem

> If a function f is continuous on a closed interval $[a, b]$, differentiable on the open interval (a, b), and if $f(a) = f(b)$, then $f'(c) = 0$ for at least one number c in (a, b).

Proof. The function f must fall into at least one of the following three categories.

(i) $f(x) = f(a)$ *for all* x *in* (a, b). In this case f is a constant function and hence $f'(x) = 0$ for all x. Consequently *every* number c in (a, b) is a critical number.

(ii) $f(x) > f(a)$ *for some* x *in* (a, b). In this case the maximum value of f in $[a, b]$ is greater than $f(a)$ or $f(b)$ and, therefore, must occur at some number c in the *open* interval (a, b). Since the derivative exists throughout (a, b), we conclude from Theorem (4.5) that $f'(c) = 0$.

(iii) $f(x) < f(a)$ *for some* x *in* (a, b). In this case the minimum value of f in $[a, b]$ is less than $f(a)$ or $f(b)$ and must occur at some number c in (a, b). As in (ii), $f'(c) = 0$.

(4.10) Corollary

> If a function f is continuous on a closed interval $[a, b]$ and if $f(a) = f(b)$, then f has at least one critical number in the open interval (a, b).

Proof. On the one hand, if f' does not exist at some number c in (a, b), then by Definition (4.8), c is a critical number. On the other hand, if f' exists throughout (a, b), then a critical number exists by Rolle's Theorem.

Before discussing the next theorem, which may be considered a generalization of Rolle's Theorem to the case in which $f(a) \neq f(b)$, let us consider the points $P(a, f(a))$ and $R(b, f(b))$ on the graph of f as illustrated by any of the sketches in Figure 4.7. If f' exists throughout the open interval (a, b), then it appears geometrically evident that there is at least one point $T(c, f(c))$ on the graph at which the tangent line is parallel to the secant line through P and R. This

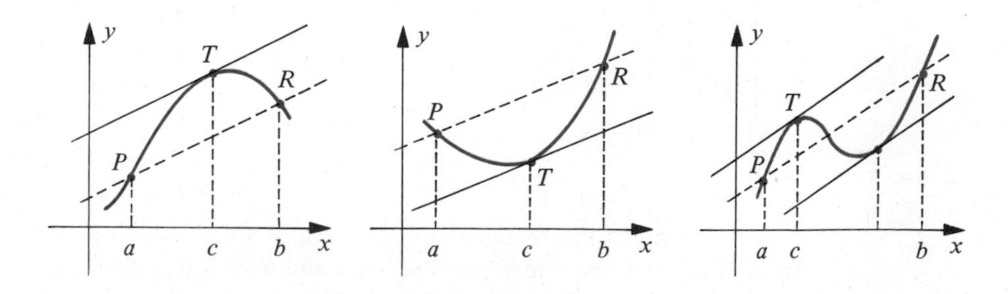

Figure 4.7

fact may be expressed in terms of slopes as follows:

(4.11) $$f'(c) = \frac{f(b) - f(a)}{b - a}$$

where the quotient on the right is obtained by using the formula for the slope of the line through P and R. If we multiply both sides of this equation by $b - a$ we obtain the formula stated in the next theorem.

(4.12) **The Mean Value Theorem**

If a function f is continuous on a closed interval $[a, b]$ and is differentiable on the open interval (a, b), then there exists a number c in (a, b) such that

$$f(b) - f(a) = f'(c)(b - a).$$

Proof. Let us define a function g as follows:

$$g(x) = f(x) - f(a) - \left[\frac{f(b) - f(a)}{b - a}\right](x - a)$$

for all x in $[a, b]$. Although we have seemingly "pulled g out of the air," there is an interesting geometric interpretation for $g(x)$ (see Exercise 22). Since f is continuous on $[a, b]$ and differentiable on (a, b), the same is true for g. Differentiating we obtain

$$g'(x) = f'(x) - \frac{f(b) - f(a)}{b - a}.$$

Moreover, by direct substitution we see that $g(a) = g(b) = 0$, and hence the function g satisfies the hypotheses of Rolle's Theorem. Consequently, there exists a number c in (a, b) such that $g'(c) = 0$, or equivalently,

$$f'(c) - \frac{f(b) - f(a)}{b - a} = 0.$$

The last equation may be written in the form stated in the conclusion of the theorem.

The Mean Value Theorem is also called **The Theorem of the Mean**. It will be employed later to help establish several very important results. The following example provides a numerical illustration of the Mean Value Theorem.

Example Prove that the function f defined by $f(x) = x^3 - 8x - 5$ satisfies the hypotheses of the Mean Value Theorem on the interval $[1, 4]$ and find a number c in the interval $(1, 4)$ which satisfies the conclusion of the theorem.

Solution Since f is a polynomial function it is continuous and differentiable for all real numbers. In particular, it is continuous on $[1, 4]$ and differentiable on the open interval $(1, 4)$. According to the Mean Value Theorem, there exists a number c in $(1, 4)$ such that

$$f(4) - f(1) = f'(c)(4 - 1).$$

Since $f'(x) = 3x^2 - 8$, this is equivalent to

$$27 - (-12) = (3c^2 - 8)(3).$$

We leave it to the reader to show that the last equation implies that $c = \pm\sqrt{7}$. Hence the desired number in the interval $(1, 4)$ is $c = \sqrt{7}$.

EXERCISES 4.2

1 If $f(x) = |x|$, show that $f(-1) = f(1)$ but $f'(c) \neq 0$ for every number c in the open interval $(-1, 1)$. Why doesn't this contradict Rolle's Theorem?

2 If $f(x) = 5 + 3(x - 1)^{2/3}$, show that $f(0) = f(2)$, but $f'(c) \neq 0$ for every number c in the open interval $(0, 2)$. Why doesn't this contradict Rolle's Theorem?

3 If $f(x) = 4/x$, prove that there is no number c such that $f(4) - f(-1) = f'(c)[4 - (-1)]$. Why doesn't this contradict the Mean Value Theorem applied to the interval $[-1, 4]$?

4 If f is the greatest integer function and if a and b are real numbers such that $b - a \geq 1$, prove that there is no number c such that $f(b) - f(a) = f'(c)(b - a)$. Why doesn't this contradict the Mean Value Theorem?

In Exercises 5–8 show that f satisfies the hypotheses of Rolle's Theorem on the indicated interval $[a, b]$ and find all numbers c in (a, b) such that $f'(c) = 0$.

5 $f(x) = 3x^2 - 12x + 11,\ [0, 4]$

6 $f(x) = 5 - 12x - 2x^2,\ [-7, 1]$

7 $f(x) = x^4 + 4x^2 + 1,\ [-3, 3]$

8 $f(x) = x^3 - x,\ [-1, 1]$

In Exercises 9–18 determine whether the function f satisfies the hypotheses of the Mean Value Theorem on the indicated interval $[a, b]$ and if so, find all numbers c in (a, b) such that $f(b) - f(a) = f'(c)(b - a)$.

9 $f(x) = x^3 + 1,\ [-2, 4]$

10 $f(x) = 5x^2 - 3x + 1,\ [1, 3]$

11 $f(x) = x + (4/x),\ [1, 4]$

12 $f(x) = 3x^5 + 5x^3 + 15x,\ [-1, 1]$

13 $f(x) = x^{2/3},\ [-8, 8]$

14 $f(x) = 1/(x - 1)^2,\ [0, 2]$

15 $f(x) = 4 + \sqrt{x - 1},\ [1, 5]$

16 $f(x) = 1 - 3x^{1/3},\ [-8, -1]$

17 $f(x) = x^3 - 2x^2 + x + 3,\ [-1, 1]$

18 $f(x) = |x - 3|,\ [-1, 4]$

19 Prove that if f is a linear function, then f satisfies the hypotheses of the Mean Value Theorem on every closed interval $[a, b]$, and that *every* number c satisfies the conclusion of the theorem.

20 If f is a quadratic function and $[a, b]$ is any closed interval, prove that there is precisely one number c in the interval (a, b) which satisfies the conclusion of the Mean Value Theorem.

21 If f is a polynomial function of degree 3 and $[a, b]$ is any closed interval, prove that there are at most two numbers in (a, b) which satisfy the conclusion of the Mean Value Theorem. Sketch graphs which illustrate the various possibilities. What can be said of a polynomial function of degree 4? Illustrate with sketches. Generalize to polynomial functions of degree n, where n is any positive integer.

22 Prove that if $g(x)$ is the function defined in the proof of the Mean Value Theorem (4.12), then $|g(x)|$ is the distance (measured along a vertical line with x-intercept x) between the graph of f and the line through $P(a, f(a))$ and $R(b, f(b))$.

23 If f is continuous on $[a, b]$ and if $f'(x) = c$ for every x in (a, b), use the Mean Value Theorem to prove that $f(x) = cx + d$ for some real number d.

24 If $f(x)$ is a polynomial of degree 3, use Rolle's Theorem to prove that f has at most three real zeros. Extend this result to polynomials of degree n.

4.3 THE FIRST DERIVATIVE TEST

The following theorem indicates how the derivative may be used to determine intervals on which a function is increasing or decreasing.

(4.13) Theorem

> Let f be a function that is continuous on a closed interval $[a, b]$ and differentiable on the open interval (a, b).
>
> (i) If $f'(x) > 0$ for all x in (a, b), then f is increasing on $[a, b]$.
>
> (ii) If $f'(x) < 0$ for all x in (a, b), then f is decreasing on $[a, b]$.

Proof. (i) Suppose $f'(x) > 0$ for all x in (a, b) and consider any numbers x_1, x_2 in $[a, b]$ such that $x_1 < x_2$. We wish to show that $f(x_1) < f(x_2)$. Applying the Mean Value Theorem (4.12) to the interval $[x_1, x_2]$,

$$f(x_2) - f(x_1) = f'(w)(x_2 - x_1),$$

where w is in the open interval (x_1, x_2). Since $x_2 - x_1 > 0$ and since, by hypothesis, $f'(w) > 0$, the right-hand side of the previous equation is positive; that is, $f(x_2) - f(x_1) > 0$. Hence $f(x_2) > f(x_1)$, which is what we wished to show. The proof of (ii) is similar and is left as an exercise.

Figures 4.8 and 4.9 give geometric illustrations of Theorem (4.13). A typical tangent line l to the graph is shown at a point whose abscissa x is in the interval

(a, b). As illustrated in Figure 4.8, if $f'(x) > 0$ the tangent line rises and, as we have proved, so does the graph of f. If $f'(x) < 0$ both the tangent line and the graph of f fall, as illustrated in Figure 4.9.

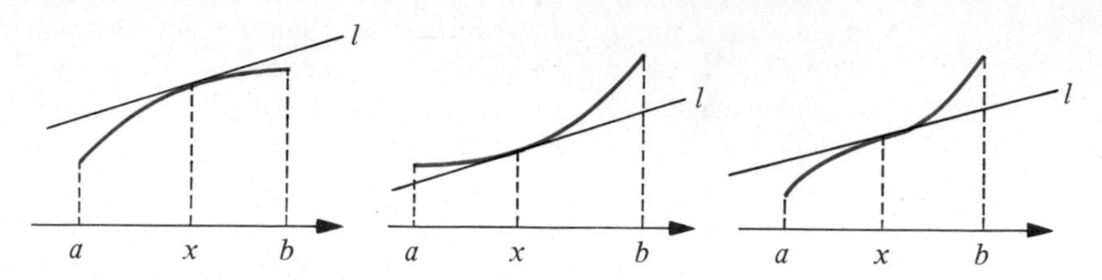

Figure 4.8. $f'(x) > 0$; f increasing on $[a, b]$

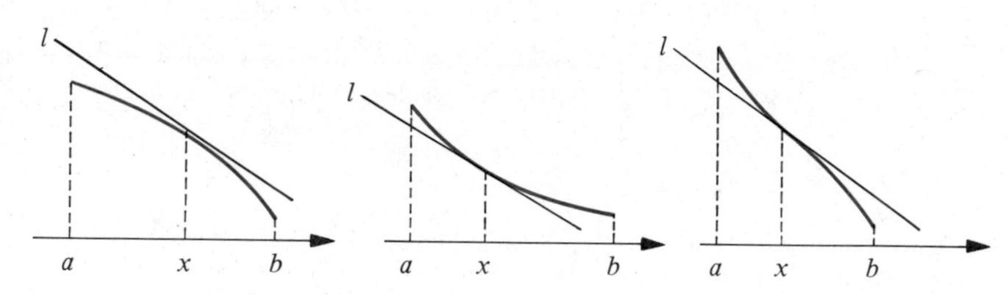

Figure 4.9. $f'(x) < 0$; f decreasing on $[a, b]$

The proof given for Theorem (4.13) may also be used to show that if $f'(x) > 0$ throughout an infinite interval of the form $(-\infty, a)$ or (b, ∞), then f is increasing on $(-\infty, a]$ or $[b, \infty)$, respectively, provided f is continuous on those intervals. An analogous result holds for decreasing functions if $f'(x) < 0$.

Example 1 If $f(x) = x^3 + x^2 - 5x - 5$, find the intervals on which f is increasing and the intervals on which f is decreasing. Sketch the graph of f.

Solution Differentiating, we obtain

$$f'(x) = 3x^2 + 2x - 5 = (3x + 5)(x - 1).$$

By Theorem (4.13) it is sufficient to find the intervals in which $f'(x) > 0$ and those in which $f'(x) < 0$. The factored form of $f'(x)$ and the critical numbers $-5/3$ and 1 suggest that we consider the open intervals $(-\infty, -5/3)$, $(-5/3, 1)$, and $(1, \infty)$. To keep track of signs, it is convenient to arrange our work in tabular form as follows, where the symbol $+$ indicates that the expression under which it appears is positive, whereas $-$ indicates that the expression is negative.

Interval	$(3x + 5)$	$x - 1$	$f'(x)$	f
$(-\infty, -\frac{5}{3})$	$-$	$-$	$+$	increasing on $(-\infty, -\frac{5}{3}]$
$(-\frac{5}{3}, 1)$	$+$	$-$	$-$	decreasing on $[-\frac{5}{3}, 1]$
$(1, \infty)$	$+$	$+$	$+$	increasing on $[1, \infty)$

Note that $f(x)$ may be written as $f(x) = x^2(x + 1) - 5(x + 1) = (x^2 - 5)(x + 1)$ and hence the x-intercepts of the graph are $\sqrt{5}$, $-\sqrt{5}$, and -1. The y-intercept is $f(0) = -5$. The points corresponding to the critical numbers are $(-5/3, 40/27)$ and $(1, -8)$. Plotting these six points and using the information on where f is increasing or decreasing gives us the sketch shown in Figure 4.10.

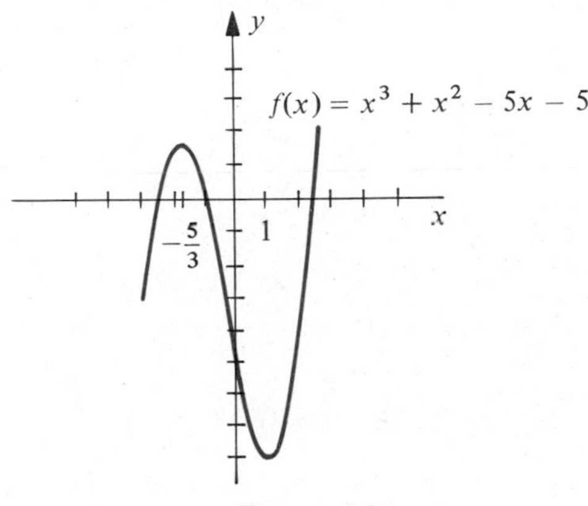

Figure 4.10

As we saw in Section 1, if a function has a local extremum, then it must occur at a critical number; however, not every critical number leads to a local extremum (see Example 2 of Section 1). In order to find the local extrema, we may begin by locating all the critical numbers of the function. Next, each critical number should be tested to determine whether or not a local extremum occurs. There are several methods for conducting this test. The following theorem is based on the sign of the first derivative of f. Roughly speaking, it states that if $f'(x)$ changes sign as x increases through a critical number c, then f has a local maximum or minimum at c. If $f'(x)$ does not change sign, then there is no extremum at c.

(4.14) The First Derivative Test

Suppose c is a critical number of a function f and (a, b) is an open interval containing c. Suppose further that f is continuous on $[a, b]$ and differentiable on (a, b), except possibly at c.

(i) If $f'(x) > 0$ for $a < x < c$ and $f'(x) < 0$ for $c < x < b$, then $f(c)$ is a local maximum of f.

(ii) If $f'(x) < 0$ for $a < x < c$ and $f'(x) > 0$ for $c < x < b$, then $f(c)$ is a local minimum of f.

(iii) If $f'(x) > 0$ or if $f'(x) < 0$ for all x in (a, b) except $x = c$, then $f(c)$ is not a local extremum of f.

Proof. If $f'(x)$ behaves as indicated in (i), then by Theorem (4.13) f is increasing on $[a, c]$ and decreasing on $[c, b]$. It follows that $f(x) < f(c)$ for all x in (a, b) different from c. Hence, by Definition (4.4), $f(c)$ is a local maximum for f. Parts (ii) and (iii) may be proved in like manner.

A device which may be used to remember the first derivative test is to think of graphs of the type sketched in Figure 4.11. In the case of a local maximum, as illustrated in (i) of Figure 4.11, the slope of the tangent line at $P(x, f(x))$ is positive if $x < c$ and negative if $x > c$. In the case of a local minimum, as illustrated in (ii) of Figure 4.11, the opposite situation occurs. Similar drawings can be sketched if the graph has a corner at the point $(c, f(c))$.

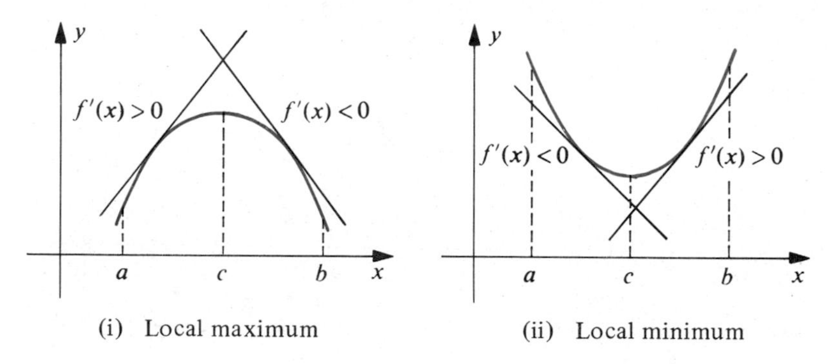

Figure 4.11

Example 2 Find the local extrema of f if $f(x) = x^3 + x^2 - 5x - 5$.

Solution This function is the same as that considered in Example 1. The critical numbers of f are $-5/3$ and 1. We see from the table in Example 1 that the sign of $f'(x)$ changes from positive to negative as x increases through $-5/3$. Hence, by the First Derivative Test, f has a local maximum at $-5/3$. This maximum value is $f(-5/3) = 40/27$. Similarly, a local minimum occurs at 1 since the sign of $f'(x)$ changes from negative to positive as x increases through 1. This minimum value is $f(1) = -8$. The reader should refer to Figure 4.10 for the geometric significance of these local extrema.

Example 3 Find the local maxima and minima of f if $f(x) = x^{1/3}(8 - x)$. Sketch the graph of f.

Solution By the Product Rule

$$f'(x) = x^{1/3}(-1) + (8 - x)\frac{1}{3}x^{-2/3}$$

$$= \frac{-3x + (8 - x)}{3x^{2/3}} = \frac{4(2 - x)}{3x^{2/3}}$$

and hence the critical numbers of f are 0 and 2. The following table indicates the sign of $f'(x)$ in intervals determined by the critical numbers, and where f is increasing or decreasing.

Interval	$2 - x$	$x^{2/3}$	$f'(x)$	f
$(-\infty, 0)$	+	+	+	increasing on $(-\infty, 0]$
$(0, 2)$	+	+	+	increasing on $[0, 2]$
$(2, \infty)$	−	+	−	decreasing on $[2, \infty)$

By the First Derivative Test, f has a local maximum at 2 since the sign of $f'(x)$ changes from + to − as x increases through 2. This local maximum is $f(2) = 2^{1/3}(8 - 2) = 6\sqrt[3]{2} \approx 7.6$. The function does not have an extremum at 0 since the sign of $f'(x)$ does not change as x increases through 0. To sketch the graph we first plot points corresponding to the critical numbers. From the formula for $f(x)$ it is evident that the x-intercepts of the graph are 0 and 8. The graph is sketched in Figure 4.12.

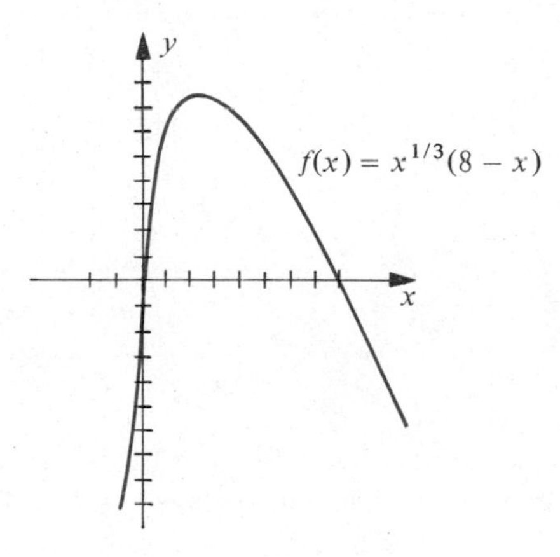

Figure 4.12

Example 4 Find the local extrema of f if $f(x) = x^{2/3}(x^2 - 8)$. Sketch the graph of f.

Solution By the Product Rule

$$f'(x) = x^{2/3}(2x) + (x^2 - 8)\left(\frac{2}{3}x^{-1/3}\right)$$
$$= \frac{6x^2 + 2(x^2 - 8)}{3x^{1/3}} = \frac{8(x^2 - 2)}{3x^{1/3}}.$$

The critical numbers are $-\sqrt{2}, 0$, and $\sqrt{2}$. These suggest an examination of the sign of $f'(x)$ in the intervals $(-\infty, -\sqrt{2})$, $(-\sqrt{2}, 0)$, $(0, \sqrt{2})$, and $(\sqrt{2}, \infty)$. Arranging our work in tabular form we have

Interval	$x^2 - 2$	$x^{1/3}$	$f'(x)$	f
$(-\infty, -\sqrt{2})$	+	−	−	decreasing on $(-\infty, -\sqrt{2}]$
$(-\sqrt{2}, 0)$	−	−	+	increasing on $[-\sqrt{2}, 0]$
$(0, \sqrt{2})$	−	+	−	decreasing on $[0, \sqrt{2}]$
$(\sqrt{2}, \infty)$	+	+	+	increasing on $[\sqrt{2}, \infty)$

By the First Derivative Test, f has local minima at $-\sqrt{2}$ and $\sqrt{2}$ and a local maximum at 0. The corresponding functional values are $f(0) = 0$ and

$$f(\sqrt{2}) = -6\sqrt[3]{2} = f(-\sqrt{2}).$$

Note that the derivative does not exist at 0. The graph is sketched in Figure 4.13.

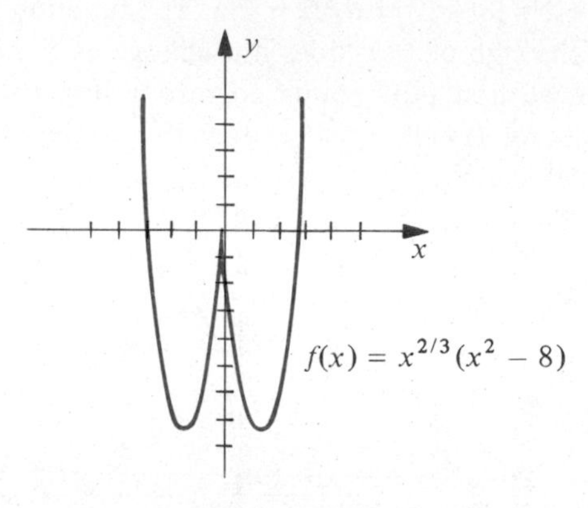

Figure 4.13

It was pointed out in Section 1 that an absolute maximum or minimum of a function may not be included among the local extrema. Recall that if a function f is continuous on a closed interval $[a, b]$ and if it is required to find the absolute maximum and minimum values, all the local extrema should be found and *in addition* the values $f(a)$ and $f(b)$ of f at the end-points a and b of the interval $[a, b]$ should be calculated. The number which is largest among the local extrema and the values $f(a)$ and $f(b)$ is the absolute maximum of f on $[a, b]$, whereas the smallest of these numbers is the absolute minimum of f on $[a, b]$. To illustrate these remarks, let us consider the function discussed in the previous example, but restrict our attention to certain intervals.

Example 5 If $f(x) = x^{2/3}(x^2 - 8)$, find the absolute maximum and minimum values of f in each of the following intervals: (a) $[-1, \frac{1}{2}]$, (b) $[-1, 3]$, (c) $[-3, -2]$.

Solution The graph in Figure 4.13 indicates the local extrema and the intervals in which f is increasing or decreasing. Consequently, all we have to do is find the ordinates of the high and low points corresponding to each of the given intervals. We shall arrange our work in tabular form and leave it to the reader to check that the entries are correct.

Interval	Minimum	Maximum
$[-1, 1/2]$	$f(-1) = -7$	$f(0) = 0$
$[-1, 3]$	$f(\sqrt{2}) = -6\sqrt[3]{2}$	$f(3) = \sqrt[3]{9}$
$[-3, -2]$	$f(-2) = -4\sqrt[3]{4}$	$f(-3) = \sqrt[3]{9}$

Note that on some intervals the maximum or minimum value of f is also a local extremum, whereas on other intervals this is not the case.

EXERCISES 4.3

In Exercises 1–16, find the local extrema of f. Describe the intervals in which f is increasing or decreasing and sketch the graph of f.

1 $f(x) = 5 - 7x - 4x^2$

2 $f(x) = 6x^2 - 9x + 5$

3 $f(x) = 2x^3 + x^2 - 20x + 1$

4 $f(x) = x^3 - x^2 - 40x + 8$

5 $f(x) = x^4 - 8x^2 + 1$

6 $f(x) = x^3 - 3x^2 + 3x + 7$

7 $f(x) = x^{4/3} + 4x^{1/3}$

8 $f(x) = x^{2/3}(8 - x)$

9 $f(x) = x^2\sqrt[3]{x^2 - 4}$

10 $f(x) = x\sqrt{4 - x^2}$

11 $f(x) = x^{2/3}(x - 7)^2 + 2$

12 $f(x) = 4x^3 - 3x^4$

13 $f(x) = x^3 + (3/x)$

14 $f(x) = 8 - \sqrt[3]{x^2 - 2x + 1}$

15 $f(x) = 10x^3(x - 1)^2$

16 $f(x) = (x^2 - 10x)^4$

In Exercises 17–22, find the local extrema of f.

17 $f(x) = \sqrt[3]{x^3 - 9x}$

18 $f(x) = x^2/\sqrt{x + 7}$

19 $f(x) = (x - 2)^3(x + 1)^4$

20 $f(x) = x^2(x - 5)^4$

21 $f(x) = (2x - 5)/(x + 3)$

22 $f(x) = (x^2 + 3)/(x - 1)$

23–26 For the functions defined in Exercises 1–4, find the absolute maximum and minimum valucs on cach of the following intervals:

(a) $[-1, 1]$ (b) $[-4, 2]$ (c) $[0, 5]$

In Exercises 27 and 28 sketch the graph of a differentiable function f which satisfies the given conditions.

27 $f'(-5) = 0$; $f'(0) = 0$; $f'(5) = 0$; $f'(x) > 0$ if $|x| > 5$; $f'(x) < 0$ if $0 < |x| < 5$.

28 $f'(a) = 0$ for $a = 1, 2, 3, 4, 5$, and $f'(x) > 0$ for all other values of x.

In Exercises 29 and 30 find the local extrema of f'. Describe the intervals in which f' is increasing or decreasing. Sketch the graph of f and study the variation of the slope of the tangent line as x increases through the domain of f.

29 $f(x) = x^4 - 6x^2$

30 $f(x) = 4x^3 - 3x^4$

31 If $f(x) = ax^3 + bx^2 + cx + d$, determine values for a, b, c, and d such that f has a local maximum 2 at $x = -1$ and a local minimum -1 at $x = 1$.

32 If $f(x) = ax^4 + bx^3 + cx^2 + dx + e$, determine values of a, b, c, d, and e such that f has a local maximum 2 at $x = 0$, and a local minimum -14 at $x = -2$ and $x = 2$, respectively.

4.4 CONCAVITY AND THE SECOND DERIVATIVE TEST

The concept of *concavity* is useful for describing the graph of a differentiable function f. If $f'(c)$ exists, then f has a tangent line with slope $f'(c)$ at the point $P(c, f(c))$. Figure 4.14 illustrates three possible situations which may occur if $f'(c) > 0$. Similar situations occur if $f'(c) < 0$ or $f'(c) = 0$. Note that in (i) of Figure 4.14 there is an open interval (a, b) containing c such that for all x in (a, b), except $x = c$, the point $Q(x, f(x))$ on the graph of f lies above the corresponding point on the tangent line having abscissa x. In this case we say that on the interval (a, b) the graph of f is *above* the tangent line through P. In (ii) of Figure 4.14 we say that the graph of f is *below* the tangent line through P. In (iii), for every open interval (a, b) containing c the tangent line is neither above nor below the graph but instead *crosses* the graph at that point.

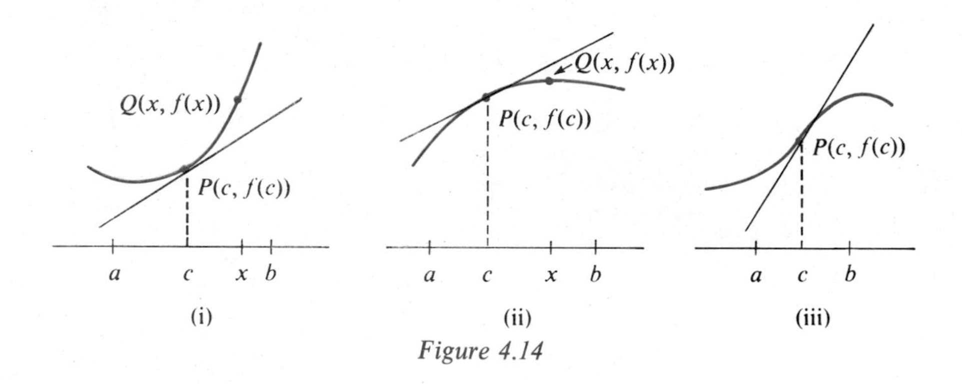

Figure 4.14

The terminology introduced in the next definition is used to describe the graphs illustrated in (i) and (ii) of Figure 4.14.

(4.15) Definition

Let f be a function that is differentiable at a number c.

(i) The graph of f is **concave upward** at the point $P(c, f(c))$ if there exists an open interval (a, b) containing c such that on (a, b) the graph of f is above the tangent line through P.

(ii) The graph of f is **concave downward** at $P(c, f(c))$ if there exists an open interval (a, b) containing c such that on (a, b) the graph of f is below the tangent line through P.

Consider a function f whose graph is concave upward at the point $P(c, f(c))$. Let (a, b) be an open interval containing c such that on (a, b) the graph of f is above the tangent line through P. If f is differentiable on (a, b), then the graph of f has a tangent line at every point $Q(x, f(x))$ with abscissa x in (a, b). If we consider a number of these tangent lines, as indicated in Figure 4.15, it appears that as the abscissa x of the point of tangency increases, then the slope $f'(x)$ of the tangent line increases. Thus in (i) of Figure 4.15, a larger positive slope is obtained as P

moves to the right. In (ii) of the figure, the slope of the tangent line also increases, becoming less negative as P moves to the right.

In like manner, if the graph of f is concave downward at $P(c, f(c))$, then the situations illustrated in Figure 4.16 may occur. Here the slope of the tangent line *decreases* as P moves to the right.

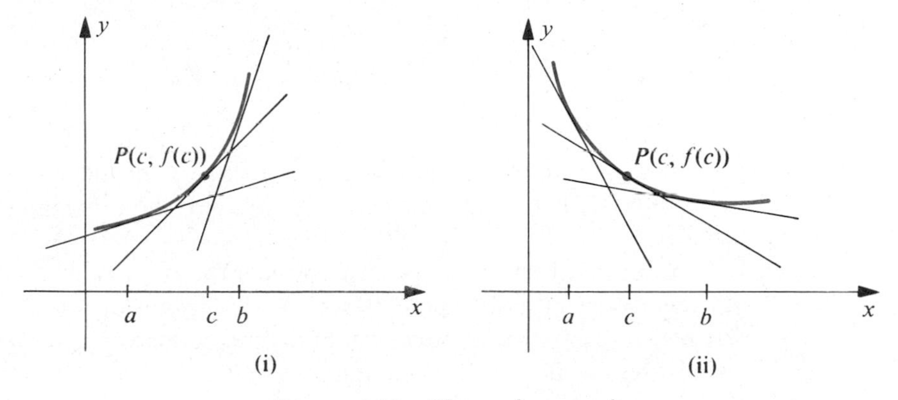

Figure 4.15. Upward concavity

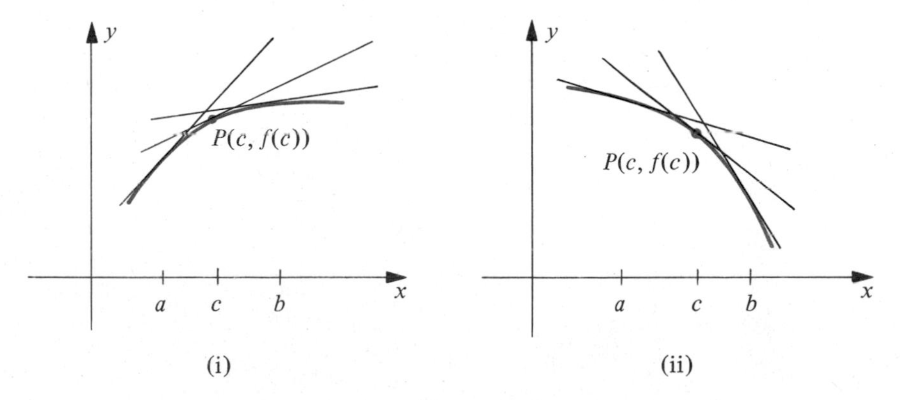

Figure 4.16. Downward concavity

Conversely, we might expect that if the derivative $f'(x)$ increases as x increases through c, then the graph is concave upward at $P(c, f(c))$, whereas if $f'(x)$ decreases, then the graph is concave downward. According to Theorem (4.13), if the derivative of a function is positive on an interval, then the function is increasing. Consequently, if the values of the *second* derivative f'' are positive on an interval, then the *first* derivative f' is increasing. Similarly, if f'' is negative, then f' is decreasing. This suggests that the sign of f'' can be used to test concavity as stated in the next theorem.

(4.16) Test for Concavity

> If a function f is differentiable on an open interval containing c, then at the point $P(c, f(c))$, the graph is
>
> (i) concave upward if $f''(c) > 0$;
>
> (ii) concave downward if $f''(c) < 0$.

Proof. Applying the formula in Theorem (3.7) to the function f',

$$f''(c) = \lim_{x \to c} \frac{f'(x) - f'(c)}{x - c}.$$

If $f''(c) > 0$, then by Theorem (2.10) there is an open interval (a, b) containing c such that

$$\frac{f'(x) - f'(c)}{x - c} > 0$$

for all x in (a, b) different from c. Hence $f'(x) - f'(c)$ and $x - c$ both have the same sign for all x in (a, b) different from c. We next show that this implies upward concavity.

For any x in the interval (a, b), consider the point (x, y) on the tangent line through $P(c, f(c))$ and the point $Q(x, f(x))$ on the graph of f (see Figure 4.17). If we let $g(x) = f(x) - y$, then we can establish upward concavity by showing that $g(x)$ is positive for all x such that $x \neq c$.

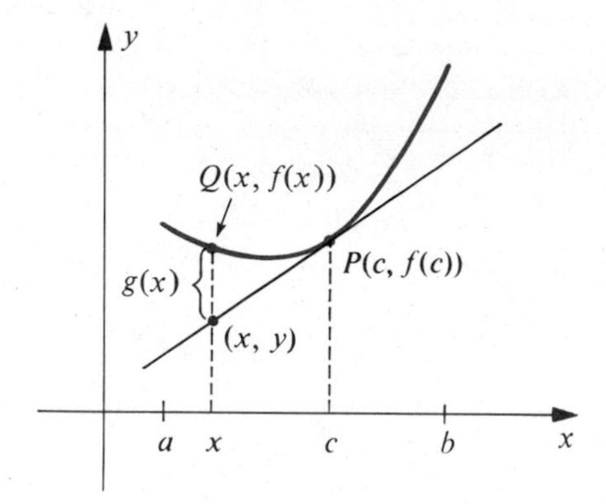

Figure 4.17

Since the equation of the tangent line at P is $y - f(c) = f'(c)(x - c)$, the ordinate y of the point (x, y) on the tangent line is given by $y = f(c) + f'(c)(x - c)$. Consequently,

$$g(x) = f(x) - y = f(x) - f(c) - f'(c)(x - c).$$

Applying the Mean Value Theorem (4.12) to f and the interval $[x, c]$, we see that there exists a number w in the open interval (x, c) such that

$$f(x) - f(c) = f'(w)(x - c).$$

Substituting for $f(x) - f(c)$ in the preceding formula for $g(x)$ leads to

$$g(x) = f'(w)(x - c) - f'(c)(x - c)$$

which can be factored as follows:

$$g(x) = [f'(w) - f'(c)](x - c).$$

Since w is in the interval (a, b) we know, from the first paragraph of the proof, that $f'(w) - f'(c)$ and $w - c$ have the same sign. Moreover, since w is between x and c, $w - c$ and $x - c$ have the same sign. Consequently, $f'(w) - f'(c)$ and $x - c$ have the same sign for all x in (a, b) such that $x \neq c$. It follows from the factored form of $g(x)$ that $g(x)$ is positive if $x \neq c$ which is what we wished to prove. Part (ii) may be established in similar fashion.

There may be points on the graph of a function at which the concavity changes from upward to downward, or from downward to upward. Such points are called *points of inflection* according to the next definition.

(4.17) Definition

A point $P(c, f(c))$ on the graph of a function f is a **point of inflection** if there exists an open interval (a, b) containing c such that one of the following statements holds.

(i) $f''(x) > 0$ if $a < x < c$ and $f''(x) < 0$ if $c < x < b$; or

(ii) $f''(x) < 0$ if $a < x < c$ and $f''(x) > 0$ if $c < x < b$.

The sketch in Figure 4.18 displays typical points of inflection on a graph. A graph is said to be **concave upward** (or **downward**) **on an interval** if it is concave upward (or downward) at every number in the interval. Intervals on which the graph in Figure 4.18 is concave upward or concave downward are abbreviated CU or CD, respectively. Observe that a corner may, or may not, be a point of inflection.

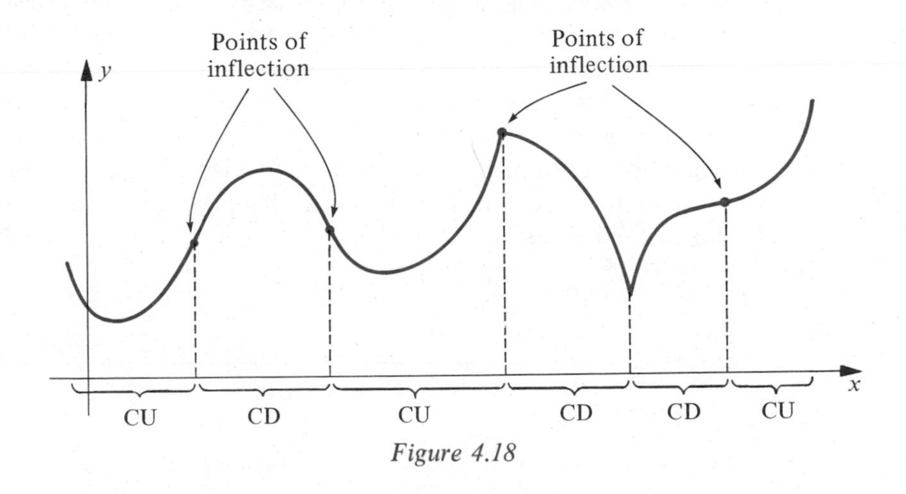

Figure 4.18

Example 1 If $f(x) = x^3 + x^2 - 5x - 5$, determine intervals on which the graph of f is concave upward and intervals on which the graph is concave downward.

Solution The function f was considered in Examples 1 and 2 of the preceding section. Since $f'(x) = 3x^2 + 2x - 5$, we have

$$f''(x) = 6x + 2 = 2(3x + 1).$$

Hence $f''(x) < 0$ if $3x + 1 < 0$, that is, if $x < -1/3$. It follows from the Test for Concavity (4.16) that the graph is concave downward on the infinite interval $(-\infty, -1/3)$. Similarly $f''(x) > 0$ if $x > -1/3$ and, therefore, the graph is concave upward on $(-1/3, \infty)$. By Definition (4.17) the point $P(-1/3, -88/27)$ at which the concavity changes from downward to upward is a point of inflection. The graph of f obtained previously (see Figure 4.10) is sketched again in Figure 4.19 in order to show the point of inflection.

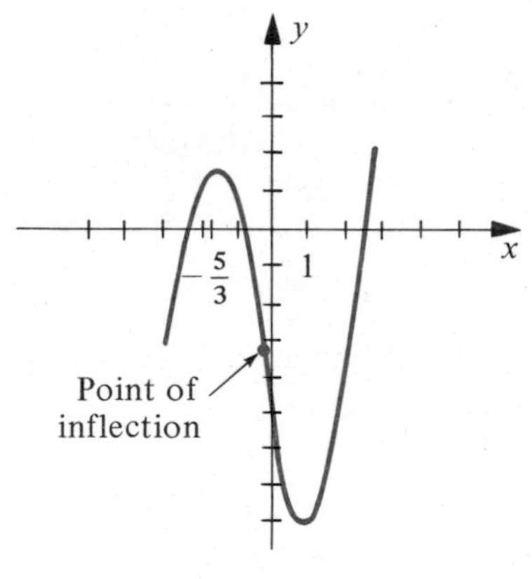

Figure 4.19

If $P(c, f(c))$ is a point of inflection on the graph of f and if f'' is continuous on an open interval containing c, then necessarily $f''(c) = 0$. To prove this, note that if $f''(c) > 0$, then $f''(x) > 0$ throughout an interval (a, b) containing c, contrary to Definition (4.17). We arrive at a similar contradiction if $f''(c) < 0$. Consequently, to locate points of inflection for a function whose second derivative is continuous, we begin by finding all numbers x such that $f''(x) = 0$. Each of these numbers is then tested to determine whether it is an abscissa of a point of inflection. Before giving an example of this technique, we state the following useful test for local maxima and minima.

(4.18) The Second Derivative Test

Suppose a function f is differentiable on an open interval containing c and $f'(c) = 0$.

(i) If $f''(c) < 0$, then f has a local maximum at c.

(ii) If $f''(c) > 0$, then f has a local minimum at c.

Proof. If $f'(c) = 0$, then the tangent line to the graph at $P(c, f(c))$ is horizontal. If, in addition $f''(c) < 0$, then the graph is concave downward at c and hence there is an interval (a, b) containing c such that the graph lies below the tangent line at P. It follows that $f(c)$ is a local maximum for f as illustrated in (i) of Figure 4.20. A similar proof may be given for part (ii), which is illustrated graphically in (ii) of Figure 4.20.

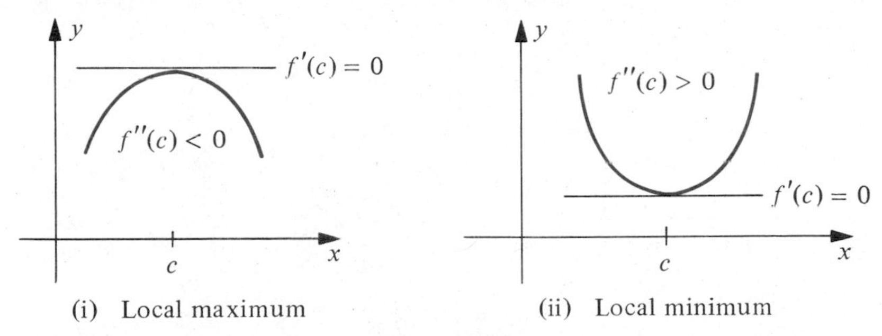

Figure 4.20

If $f''(c) = 0$, the Second Derivative Test is not applicable. In such cases the First Derivative Test should be employed.

Example 2 If $f(x) = 12 + 2x^2 - x^4$, use the Second Derivative Test to find the local maxima and minima of f. Discuss concavity, find the points of inflection, and sketch the graph of f.

Solution We begin by finding the first and second derivatives and factoring them as follows:

$$f'(x) = 4x - 4x^3 = 4x(1 - x^2)$$
$$f''(x) = 4 - 12x^2 = 4(1 - 3x^2).$$

The expression for $f'(x)$ is used to find the critical numbers 0, 1, and -1. The values of f'' at these numbers are

$$f''(0) = 4 > 0, \quad f''(1) = -8 < 0 \quad \text{and} \quad f''(-1) = -8 < 0.$$

Hence, by the Second Derivative Test there is a local minimum at 0 and local maxima at 1 and -1. The corresponding functional values are $f(0) = 12$ and $f(1) = 13 = f(-1)$. To locate the possible points of inflection we solve the equation $f''(x) = 0$, that is, $4(1 - 3x^2) = 0$. Evidently, the solutions are $-\sqrt{3}/3$ and $\sqrt{3}/3$. This suggests an examination of the sign of $f''(x)$ in the intervals $(-\infty, -\sqrt{3}/3)$, $(-\sqrt{3}/3, \sqrt{3}/3)$, and $(\sqrt{3}/3, \infty)$. We arrange our work in tabular form as follows.

Interval	$f''(x)$	Concavity
$(-\infty, -\sqrt{3}/3)$	$-$	downward
$(-\sqrt{3}/3, \sqrt{3}/3)$	$+$	upward
$(\sqrt{3}/3, \infty)$	$-$	downward

Since the sign of $f''(x)$ changes as x increases through $-\sqrt{3}/3$ and $\sqrt{3}/3$, the corresponding points $(\pm\sqrt{3}/3, 113/9)$ on the graph are points of inflection. These are the points at which the sense of concavity changes. Indeed, as shown in the table, the graph is concave upward on the open interval $(-\sqrt{3}/3, \sqrt{3}/3)$ and concave downward outside of $[-\sqrt{3}/3, \sqrt{3}/3]$. The graph is sketched in Figure 4.21.

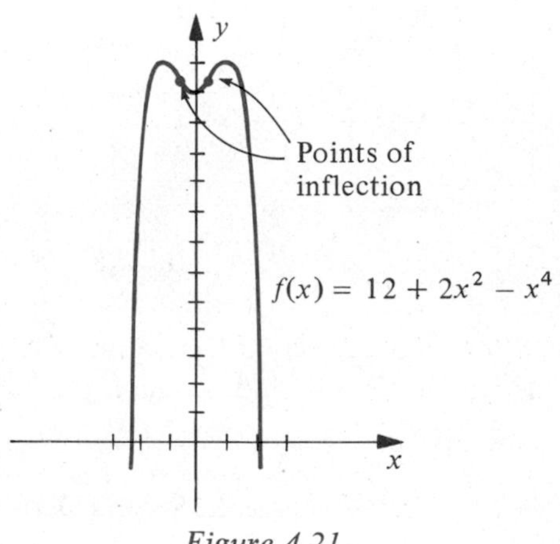

Figure 4.21

Example 3 If $f(x) = x^5 - 5x^3$, use the Second Derivative Test to find the local extrema of f. Discuss concavity, find the points of inflection, and sketch the graph of f.

Solution Following the same pattern used in the solution of Example 2 we have

$$f'(x) = 5x^4 - 15x^2 = 5x^2(x^2 - 3)$$
$$f''(x) = 20x^3 - 30x = 10x(2x^2 - 3).$$

The formula for $f'(x)$ leads to the critical numbers 0, $-\sqrt{3}$, and $\sqrt{3}$. The values of f'' at these numbers are

$$f''(0) = 0$$
$$f''(\sqrt{3}) = 10\sqrt{3}(6 - 3) = 30\sqrt{3} > 0$$
$$f''(-\sqrt{3}) = -30\sqrt{3} < 0.$$

It follows from the Second Derivative Test that f has a local minimum at $\sqrt{3}$ and a local maximum at $-\sqrt{3}$ given by $f(\sqrt{3}) = -6\sqrt{3}$ and $f(-\sqrt{3}) = 6\sqrt{3}$, respectively.

Since $f''(0) = 0$, the Second Derivative Test is not applicable at 0 and hence we turn to the First Derivative Test. If $-\sqrt{3} < x < 0$ then $f'(x) < 0$, and if $0 < x < \sqrt{3}$ then $f'(x) < 0$. Since $f'(x)$ does not change sign, there is neither a maximum nor a minimum at $x = 0$.

To find possible points of inflection we consider the equation $f''(x) = 0$, that is, $10x(2x^2 - 3) = 0$. The solutions of this equation, in order of magnitude, are $-\sqrt{6}/2, 0$, and $\sqrt{6}/2$, and we construct the following table.

Interval	$f''(x)$	Concavity
$(-\infty, -\sqrt{6}/2)$	$-$	downward
$(-\sqrt{6}/2, 0)$	$+$	upward
$(0, \sqrt{6}/2)$	$-$	downward
$(\sqrt{6}/2, \infty)$	$+$	upward

Since the sign of $f''(x)$ changes as x increases through $-\sqrt{6}/2, 0$, and $\sqrt{6}/2$, the points $(0, 0)$, $(-\sqrt{6}/2, 21\sqrt{6}/8)$, and $(\sqrt{6}/2, -21\sqrt{6}/8)$ are points of inflection. The graph is sketched in Figure 4.22 where, for clarity, we have used different scales on the x- and y-axes.

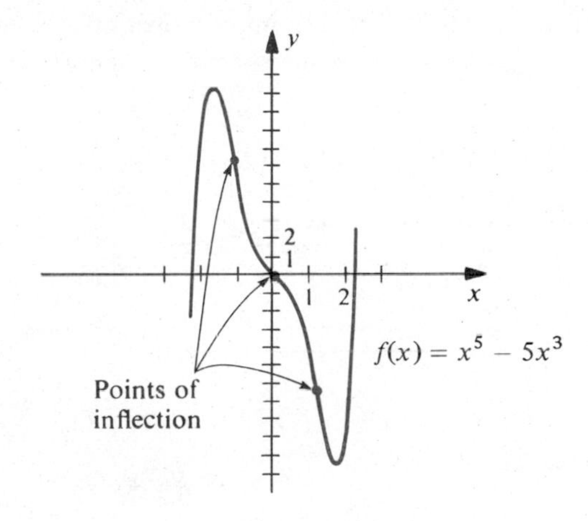

Figure 4.22

In all the preceding examples f'' was continuous. It is also possible for $(c, f(c))$ to be a point of inflection when either $f'(c)$ or $f''(c)$ does not exist, as illustrated in the next example.

Example 4 If $f(x) = 1 - x^{1/3}$, find the local extrema, discuss concavity, find the points of inflection, and sketch the graph of f.

Solution Differentiating we obtain

$$f'(x) = -\frac{1}{3}x^{-2/3} = -\frac{1}{3x^{2/3}}$$

$$f''(x) = \frac{2}{9}x^{-5/3} = \frac{2}{9x^{5/3}}.$$

The first derivative does not exist at $x = 0$, and 0 is the only critical number for f. Since $f''(0)$ is undefined, the Second Derivative Test is not applicable. However, if $x \neq 0$, then $x^{2/3} > 0$ and $f'(x) = -1/3x^{2/3} < 0$, which means that f is decreasing throughout its domain. Consequently $f(0)$ is not a local extremum.

Since f'' is undefined at $x = 0$, the point $(0, 1)$ on the graph of f *may* be a point of inflection. To check we shall apply Definition (4.17) with $c = 0$. If $x < 0$, then $x^{5/3} < 0$. Hence

$$f''(x) = \frac{2}{9x^{5/3}} < 0 \quad \text{if } x < 0$$

which implies that the graph of f is concave downward on the interval $(-\infty, 0)$. If $x > 0$, then $x^{5/3} > 0$. Thus

$$f''(x) = \frac{2}{9x^{5/3}} > 0 \quad \text{if } x > 0$$

which means that the graph is concave upward on the interval $(0, \infty)$. It follows that the point $(0, 1)$ is a point of inflection. Using this information and plotting several points gives us the sketch in Figure 4.23.

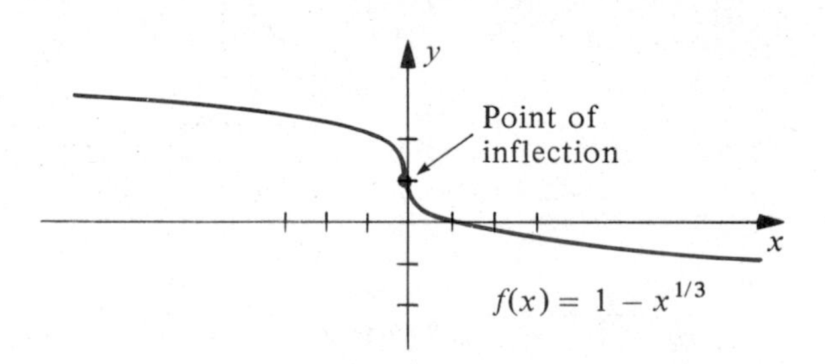

Figure 4.23

EXERCISES 4.4

In Exercises 1–18, use the Second Derivative Test (whenever applicable) to find the local extrema of f. Discuss concavity, find abscissas of points of inflection, and sketch the graph of f.

1 $f(x) = x^3 - 2x^2 + x + 1$

2 $f(x) = x^3 + 10x^2 + 25x - 50$

3 $f(x) = 3x^4 - 4x^3 + 6$

4 $f(x) = 8x^2 - 2x^4$

5 $f(x) = 2x^6 - 6x^4$

6 $f(x) = 3x^5 - 5x^3$

7 $f(x) = (x^2 - 1)^2$

8 $f(x) = x - (16/x)$

9 $f(x) = \sqrt[5]{x} - 1$

10 $f(x) = (x + 4)/\sqrt{x}$

11 $f(x) = x^2 - (27/x^2)$

12 $f(x) = x^{2/3}(1 - x)$

13 $f(x) = x/(x^2 + 1)$

14 $f(x) = x^2/(x^2 + 1)$

15 $f(x) = \sqrt[3]{x^2}(3x + 10)$

16 $f(x) = x^4 - 4x^3 + 10$

17 $f(x) = 8x^{1/3} + x^{4/3}$

18 $f(x) = x\sqrt{4 - x^2}$

In each of Exercises 19–26, sketch the graph of a continuous function f which satisfies all of the stated conditions.

19 $f(0) = 1; f(2) = 3; f'(0) = f'(2) = 0; f'(x) < 0$ if $|x - 1| > 1; f'(x) > 0$ if $|x - 1| < 1$; $f''(x) > 0$ if $x < 1$; $f''(x) < 0$ if $x > 1$.

20 $f(0) = 4; f(2) = 2; f(5) = 6; f'(0) = f'(2) = 0; f'(x) > 0$ if $|x - 1| > 1; f'(x) < 0$ if $|x - 1| < 1; f''(x) < 0$ if $x < 1$ or if $|x - 4| < 1; f''(x) > 0$ if $|x - 2| < 1$ or if $x > 5$.

21 $f(0) = 2; f(2) = f(-2) = 1; f'(0) = 0; f'(x) > 0$ if $x < 0$; $f'(x) < 0$ if $x > 0$; $f''(x) < 0$ if $|x| < 2$; $f''(x) > 0$ if $|x| > 2$.

22 $f(1) = 4; f'(x) > 0$ if $x < 1$; $f'(x) < 0$ if $x > 1$; $f''(x) > 0$ for all $x \neq 1$.

23 $f(-2) = f(6) = -2; f(0) = f(4) = 0; f(2) = f(8) = 3$; f' is undefined at 2 and 6; $f'(0) = 1$; $f'(x) > 0$ throughout $(-\infty, 2)$ and $(6, \infty)$; $f'(x) < 0$ if $|x - 4| < 2$; $f''(x) < 0$ throughout $(-\infty, 0)$, $(4, 6)$, and $(6, \infty)$; $f''(x) > 0$ throughout $(0, 2)$ and $(2, 4)$.

24 $f(0) = 2$; $f(2) = 1$; $f(4) = f(10) = 0$; $f(6) = -4$; $f'(2) = f'(6) = 0$; $f'(x) < 0$ throughout $(-\infty, 4)$, $(4, 6)$, and $(10, \infty)$; $f'(x) > 0$ throughout $(6, 10)$; $f'(4)$ and $f'(10)$ do not exist; $f''(x) > 0$ throughout $(-\infty, 2)$, $(4, 10)$, and $(10, \infty)$; $f''(x) < 0$ throughout $(2, 4)$.

25 If n is an odd integer, then $f(n) = 1$ and $f'(n) = 0$; if n is an even integer, then $f(n) = 0$ and $f'(n)$ does not exist; if n is any integer then

(a) $f'(x) > 0$ whenever $2n < x < 2n + 1$;

(b) $f'(x) < 0$ whenever $2n - 1 < x < 2n$;

(c) $f''(x) < 0$ whenever $2n < x < 2n + 2$.

26 $f(x) = x$ if $x = -1, 2, 4$, or 8; $f'(x) = 0$ if $x = -1, 4, 6$, or 8; $f'(x) < 0$ throughout $(-\infty, -1)$, $(4, 6)$, and $(8, \infty)$; $f'(x) > 0$ throughout $(-1, 4)$ and $(6, 8)$; $f''(x) > 0$ throughout $(-\infty, 0)$, $(2, 3)$, and $(5, 7)$; $f''(x) < 0$ throughout $(0, 2)$, $(3, 5)$, and $(7, \infty)$.

27 Prove that the graph of a quadratic function has no points of inflection. State conditions under which the graph is always (a) concave upward; (b) concave downward. Illustrate with sketches.

28 Prove that the graph of a polynomial function of degree 3 has exactly one point of inflection. Illustrate this fact with sketches.

29 Prove that the graph of a polynomial function of degree $n > 2$ has at most $n - 2$ points of inflection.

30 If $f(x) = x^n$, where $n > 1$, prove that the graph of f has either one or no points of inflection, according to whether n is odd or even. Illustrate with sketches.

4.5 HORIZONTAL AND VERTICAL ASYMPTOTES

Thus far we have used the derivative primarily as a tool for sketching graphs. In particular, we have developed techniques for determining where a graph rises or falls, the high and low points on a graph, concavity, and points of inflection. Before turning to physical applications of derivatives, we shall discuss several other interesting characteristics of some graphs. When applied to the derivative f', our discussion will provide information about vertical tangent lines to the graph of f.

Let us begin by considering the graph of $f(x) = 2 + 1/x$ sketched in Figure 4.24 and concentrate on values of $f(x)$ if x is large and positive. For example, $f(100) = 2.01, f(1000) = 2.001, f(10{,}000) = 2.0001$, and $f(100{,}000) = 2.00001$. It is evident that we can make $f(x)$ as close to 2 as desired or, equivalently, we can make $|f(x) - 2|$ arbitrarily small, by choosing x sufficiently large. This behavior of $f(x)$ is expressed by writing

$$\lim_{x \to \infty} \left(2 + \frac{1}{x}\right) = 2,$$

which may be read "the limit of $2 + 1/x$ as x becomes infinite (or as x increases without bound) is 2."

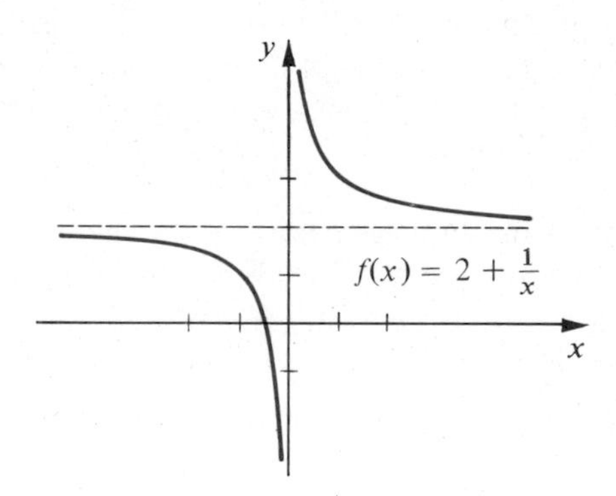

Figure 4.24

The general definition which describes the behavior of $f(x)$ in the preceding illustration may be phrased as follows.

(4.19) Definition

> Let f be defined on an interval (c, ∞). The statement
>
> $$\lim_{x \to \infty} f(x) = L$$
>
> means that for every $\varepsilon > 0$ there corresponds a positive number N such that
>
> $$|f(x) - L| < \varepsilon \quad \text{whenever } x > N.$$

If $\lim_{x \to \infty} f(x) = L$ we say that **the limit of $f(x)$ is L as x becomes infinite** (or **as x increases without bound**). To emphasize the fact that x is large and positive, we often say that the limit of $f(x)$ is L as x becomes *positively* infinite. It is important to remember that the symbol ∞ does not represent a real number. It is used here to denote that x increases without bound. Later, in Definition (4.22), we shall employ it in a different context.

It is of interest to study the graphical significance of the preceding definition. Suppose $\lim_{x \to \infty} f(x) = L$, and consider any horizontal lines $y = L \pm \varepsilon$ (see Figure 4.25). According to Definition (4.19), if x is larger than some number N, then all points $P(x, f(x))$ lie between these horizontal lines. Roughly speaking, the graph of f gets closer and closer to the line $y = L$ as x gets larger and larger. The line $y = L$ is called a **horizontal asymptote** for the graph of f. For example, the line $y = 2$ in Figure 4.24 is a horizontal asymptote for the graph of $f(x) = 2 + 1/x$.

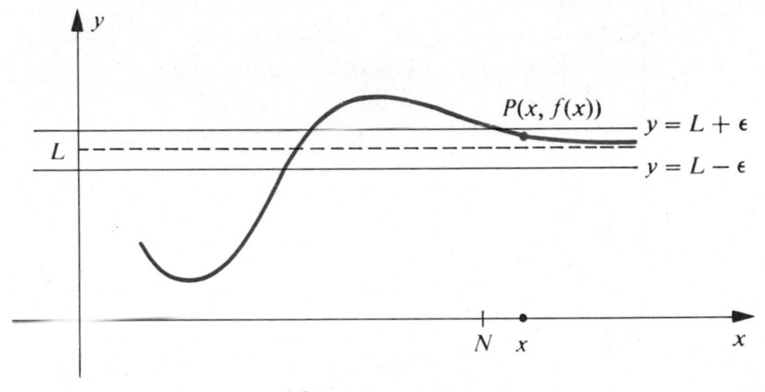

Figure 4.25. $\lim_{x \to \infty} f(x) = L$

In Figure 4.25, the graph of f approaches the asymptote $y = L$ from above, that is, with $f(x) > L$. However, we could also have $f(x) < L$ or other variations.

The situation in which x is numerically large and *negative* is also important. In particular, we write

(4.20)
$$\lim_{x \to -\infty} f(x) = L$$

if $|f(x) - L|$ can be made arbitrarily small by choosing numerically large *negative* values of x. In this case we say that $f(x)$ has the limit L as x becomes *negatively* infinite. If $f(x)$ is defined on $(-\infty, c)$ then the definition of (4.20) may be obtained from (4.19) by taking N negative and replacing $x > N$ by $x < N$. A geometric illustration of (4.20) is given in Figure 4.26. If we consider any lines $y = L \pm \varepsilon$, then all points $P(x, f(x))$ on the graph lie between these lines if x is *less* than some number N. As before, we refer to the line $y = L$ as a horizontal asymptote for the graph of f.

Limit theorems analogous to those in Chapter 2 may be established. In particular, Theorem (2.11) concerning limits of sums, products, and quotients is true for the cases $x \to \infty$ or $x \to -\infty$. Similarly, Theorem (2.22) on the limit of $\sqrt[n]{g(x)}$ holds if $x \to \infty$ or $x \to -\infty$. Finally, it is trivial to show that

$$\lim_{x \to \infty} c = c \quad \text{and} \quad \lim_{x \to -\infty} c = c.$$

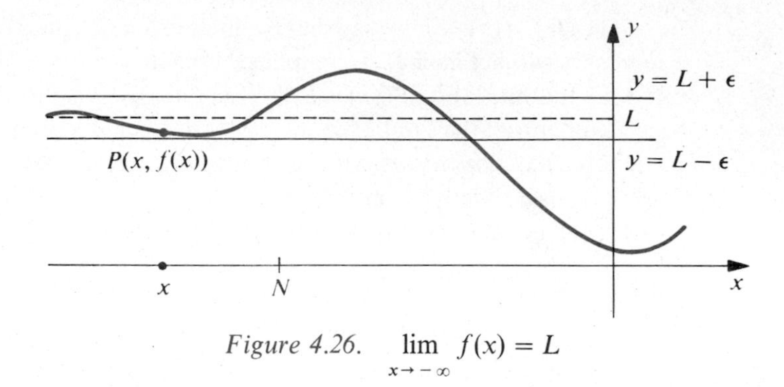

Figure 4.26. $\lim_{x \to -\infty} f(x) = L$

The next theorem provides an important tool for calculating specific limits.

(4.21) Theorem

If k is a positive rational number and c is any nonzero real number, then

$$\lim_{x \to \infty} \frac{c}{x^k} = 0 \quad \text{and} \quad \lim_{x \to -\infty} \frac{c}{x^k} = 0,$$

provided x^k is always defined.

Proof. For any $\varepsilon > 0$, the following four inequalities are equivalent.

$$\left|\frac{c}{x^k} - 0\right| < \varepsilon; \quad \frac{|x|^k}{|c|} > \frac{1}{\varepsilon}; \quad |x|^k > |c|/\varepsilon; \quad x > (|c|/\varepsilon)^{1/k}.$$

If we let $N = (|c|/\varepsilon)^{1/k}$, then whenever $x > N$ the last, and hence the first, inequality is true. Consequently, by Definition (4.19), $\lim_{x \to \infty} c/x^k = 0$. The second part of the theorem may be proved in similar fashion.

If f is a rational function, then limits as $x \to \infty$ or $x \to -\infty$ may be found by first dividing numerator and denominator of $f(x)$ by a suitable power of x and then applying Theorem (4.21), as illustrated in the next two examples.

Example 1 Investigate $\displaystyle\lim_{x \to -\infty} \frac{2x^2 - 5}{3x^2 + x + 2}$.

Solution Since we are interested in numerically large (negative) values of x, we may assume that $x \neq 0$. Dividing numerator and denominator of the given expression by x^2 and using the appropriate limit theorems, we obtain

$$\lim_{x \to -\infty} \frac{2x^2 - 5}{3x^2 + x + 2} = \lim_{x \to -\infty} \frac{2 - (5/x^2)}{3 + (1/x) + (2/x^2)}$$

$$= \frac{\lim_{x \to -\infty} [2 - (5/x^2)]}{\lim_{x \to -\infty} [3 + (1/x) + (2/x^2)]}$$

$$= \frac{\lim_{x \to -\infty} 2 - \lim_{x \to -\infty} (5/x^2)}{\lim_{x \to -\infty} 3 + \lim_{x \to -\infty} (1/x) + \lim_{x \to -\infty} (2/x^2)}$$

$$= \frac{2 - 0}{3 + 0 + 0} = \frac{2}{3}.$$

It follows that the line $y = 2/3$ is a horizontal asymptote for the graph of f.

Example 2 If $f(x) = 4x/(x^2 + 9)$, determine the horizontal asymptotes, find the relative maxima and minima, the points of inflection, discuss concavity, and sketch the graph of f.

Solution We divide numerator and denominator of $f(x)$ by x^2 and use limit theorems as follows:

$$\lim_{x \to \infty} f(x) = \lim_{x \to \infty} \frac{4x}{x^2 + 9}$$

$$= \lim_{x \to \infty} \frac{(4/x)}{1 + (9/x^2)}$$

$$= \frac{\lim_{x \to \infty} (4/x)}{\lim_{x \to \infty} 1 + \lim_{x \to \infty} (9/x^2)}$$

$$= \frac{0}{1 + 0} = \frac{0}{1} = 0.$$

Hence $y = 0$ (the x-axis) is a horizontal asymptote for the graph of f. Similarly, $\lim_{x \to -\infty} f(x) = 0$.

It is left to the reader to show that the first and second derivatives of f are given by

$$f'(x) = \frac{4(9 - x^2)}{(x^2 + 9)^2}; \quad f''(x) = \frac{8x(x^2 - 27)}{(x^2 + 9)^3}.$$

The critical numbers of f are ± 3. (Why?) Using either the First or Second Derivative Test, we can show that $f(-3)$ is a relative minimum and $f(3)$ is a relative maximum. Points of inflection have abscissas 0 and $\pm\sqrt{27} = \pm 3\sqrt{3}$. The reader may verify that the graph is concave downward in the intervals $(-\infty, -3\sqrt{3})$ and $(0, 3\sqrt{3})$ and concave upward in $(-3\sqrt{3}, 0)$ and $(3\sqrt{3}, \infty)$. The graph is sketched in Figure 4.27.

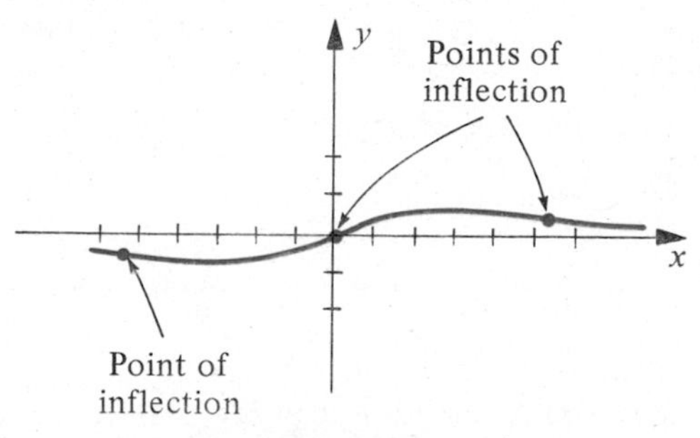

Figure 4.27

Example 3 Investigate $\lim_{x\to\infty} \dfrac{\sqrt{9x^2+2}}{3-4x}$.

Solution Dividing numerator and denominator of the quotient by x and applying limit theorems gives us

$$\begin{aligned}
\lim_{x\to\infty} \frac{\sqrt{9x^2+2}}{3-4x} &= \lim_{x\to\infty} \frac{\sqrt{9+(2/x^2)}}{(3/x)-4} \\
&= \frac{\lim_{x\to\infty} \sqrt{9+(2/x^2)}}{\lim_{x\to\infty} [(3/x)-4]} \\
&= \frac{\sqrt{\lim_{x\to\infty} (9+(2/x^2))}}{\lim_{x\to\infty} (3/x) - \lim_{x\to\infty} 4} \\
&= \frac{\sqrt{\lim_{x\to\infty} 9 + \lim_{x\to\infty} (2/x^2)}}{0-4} \\
&= \frac{\sqrt{9+0}}{-4} = -\frac{3}{4}.
\end{aligned}$$

Let us next consider the function f defined by $f(x) = 1/(x-3)^2$. If x is close to 3 (but $x \neq 3$) the denominator $(x-3)^2$ is close to zero, and hence $f(x)$ is very large. Indeed, $f(x)$ can be made as large as desired by choosing x sufficiently close to 3. This behavior of $f(x)$ is symbolized by writing

$$\lim_{x\to 3} \frac{1}{(x-3)^2} = \infty.$$

(4.22) Definition

Let f be defined on an open interval containing a (except possibly at $x = a$). The statement **$f(x)$ becomes infinite** (or **increases without bound**) as x approaches a, written

$$\lim_{x\to a} f(x) = \infty$$

means that for every positive number N there corresponds a $\delta > 0$ such that

$$f(x) > N \quad \text{whenever } 0 < |x-a| < \delta.$$

Sometimes Definition (4.22) is referred to by stating that $f(x)$ becomes *positively* infinite as x approaches a. A geometric interpretation in terms of the graph of f is shown in Figure 4.28. If we consider any horizontal line $y = N$, then when x is in a suitable interval $(a - \delta, a + \delta)$, but $x \neq a$, the points on the graph of f lie *above* the horizontal line.

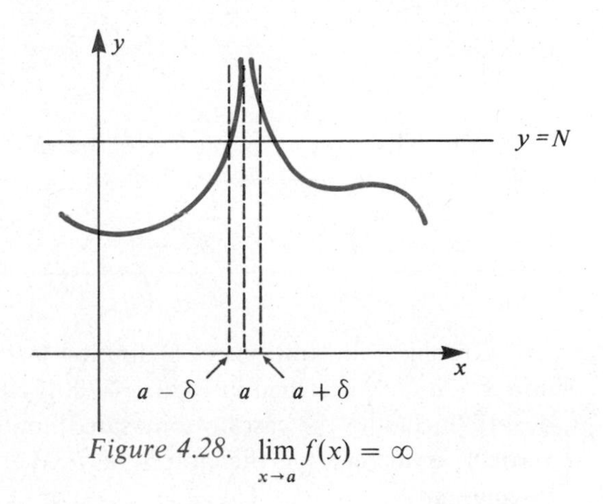

Figure 4.28. $\lim_{x \to a} f(x) = \infty$

As we mentioned earlier, the symbol ∞ is used as a device for describing certain types of functional behavior. It is *not* a real number. In particular, $\lim_{x \to a} f(x) = \infty$ does not mean that the limit *exists* as x approaches a. The limit notation is merely used to denote the behavior of $f(x)$ we have described.

The notion of $f(x)$ increasing without bound as x approaches a from the right or left can also be introduced. This is indicated by writing

(4.23) $$\lim_{x \to a^+} f(x) = \infty \quad \text{or} \quad \lim_{x \to a^-} f(x) = \infty,$$

respectively. Only minor modifications of Definition (4.22) are needed to define these concepts. Thus, for $x \to a^+$, it is assumed that $f(x)$ exists on some open interval (a, c), and the final inequality in Definition (4.22) is changed from $0 < |x - a| < \delta$ to $a < x < a + \delta$; that is, x may only take on values *larger* than a. Similar statements may be made for $x \to a^-$. Graphs illustrating (4.23) are sketched in Figure 4.29.

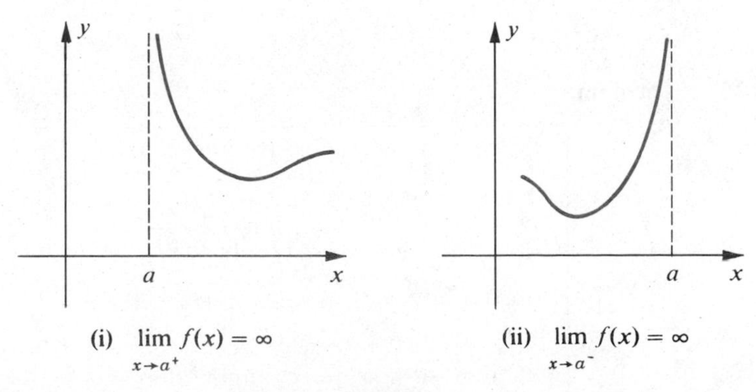

(i) $\lim_{x \to a^+} f(x) = \infty$ (ii) $\lim_{x \to a^-} f(x) = \infty$

Figure 4.29

(4.24) Definition

Let f be defined on an open interval containing a (except possibly at $x = a$). The statement **$f(x)$ becomes negatively infinite** (or **decreases without bound**), written

$$\lim_{x \to a} f(x) = -\infty$$

means that for every negative number N there corresponds a $\delta > 0$ such that

$$f(x) < N \quad \text{whenever } 0 < |x - a| < \delta.$$

Graphical illustrations of Definition (4.24) together with the cases $x \to a^+$ and $x \to a^-$ are sketched in Figure 4.30. If either of Definitions (4.22) or (4.24) occurs (including the cases of one-sided limits), then the line $x = a$ is called a **vertical asymptote** for the graph of f. We also say that f has an **infinite discontinuity** at $x = a$. The line $x = a$ in Figures (4.28)–(4.30) is a vertical asymptote for the given graph.

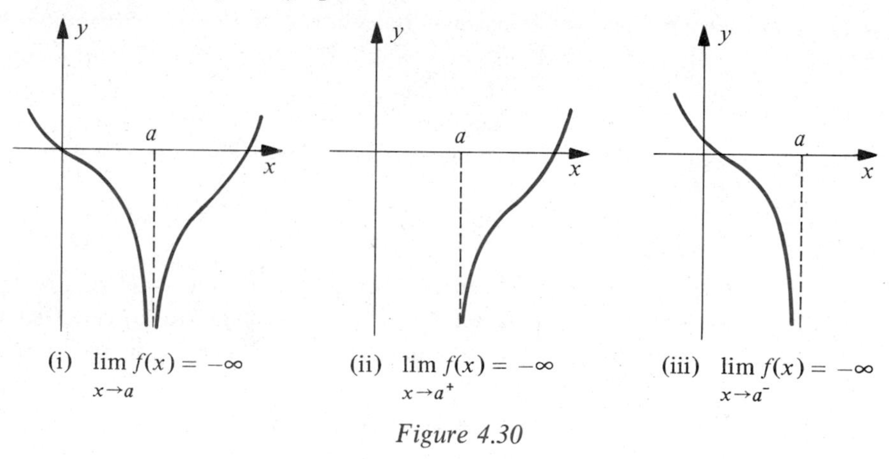

(i) $\lim_{x \to a} f(x) = -\infty$ (ii) $\lim_{x \to a^+} f(x) = -\infty$ (iii) $\lim_{x \to a^-} f(x) = -\infty$

Figure 4.30

The next theorem, stated without proof, is useful in the investigation of certain limits.

(4.25) Theorem

(i) If n is an even positive integer then

$$\lim_{x \to a} \frac{1}{(x - a)^n} = \infty.$$

(ii) If n is an odd positive integer, then

$$\lim_{x \to a^+} \frac{1}{(x - a)^n} = \infty \quad \text{and} \quad \lim_{x \to a^-} \frac{1}{(x - a)^n} = -\infty.$$

It follows that if n is a positive integer and $f(x) = 1/(x - a)^n$, then the graph of f has the line $x = a$ as a vertical asymptote. Moreover, since $\lim_{x\to\infty} 1/(x - a)^n = 0$, the x-axis is a horizontal asymptote.

It is unnecessary to memorize Theorem (4.25), since for any specific problem it is usually possible to determine the answer intuitively. As an illustration, consider $\lim_{x\to -5^-} 1/(x + 5)^3$. If x is close to -5 and $x < -5$, then $x + 5$ is close to 0 and is *negative*. Consequently $(x + 5)^3$ is close to 0 and negative and, therefore, $1/(x + 5)^3$ is a numerically large negative number. This suggests that

$$\lim_{x\to -5^-} \frac{1}{(x + 5)^3} = -\infty.$$

To investigate $x \to -5^+$, we first note that if $x > -5$, then $x + 5 > 0$, and reason that

$$\lim_{x\to -5^+} \frac{1}{(x + 5)^3} = \infty.$$

Example 4 If $f(x) = 1/(x - 2)^3$, discuss $\lim_{x\to 2^+} f(x)$, $\lim_{x\to 2^-} f(x)$, and sketch the graph of f.

Solution If x is very close to 2, but larger than 2, then $x - 2$ is a small positive number and hence $1/(x - 2)^3$ is a large positive number. Thus it is evident that

$$\lim_{x\to 2^+} \frac{1}{(x - 2)^3} = \infty.$$

This fact also follows from (ii) of Theorem (4.25) with $a = 2$ and $n = 3$.

If x is close to 2 but *less* than 2, then $x - 2$ is close to 0 and negative. Hence

$$\lim_{x\to 2^-} \frac{1}{(x - 2)^3} = -\infty.$$

This is also a consequence of (ii) of (4.25). The graph of f is sketched in Figure 4.31.

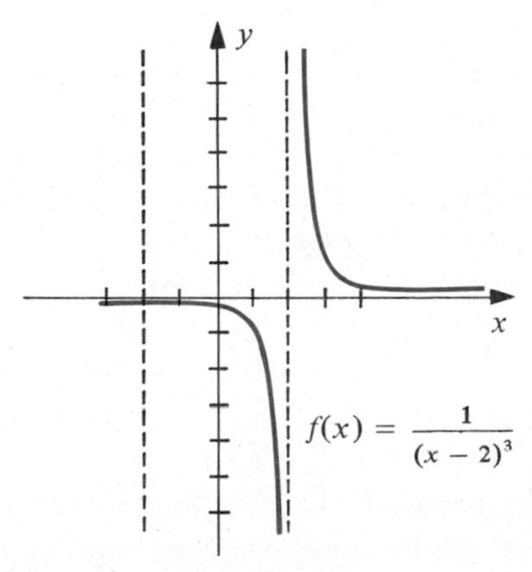

Figure 4.31

The line $x = 2$ is a vertical asymptote. The reader should verify that

$$\lim_{x \to \infty} \frac{1}{(x-2)^3} = 0 = \lim_{x \to -\infty} \frac{1}{(x-2)^3}$$

and hence $y = 0$ (the x-axis) is a horizontal asymptote.

The following are special cases of Theorem (4.25) with $a = 0$:

(i) If n is a positive even integer, then

(4.26) $$\lim_{x \to 0} \frac{1}{x^n} = \infty.$$

(ii) If n is a positive odd integer, then

$$\lim_{x \to 0^+} \frac{1}{x^n} = \infty \quad \text{and} \quad \lim_{x \to 0^-} \frac{1}{x^n} = -\infty.$$

Similar results may be established for positive rational exponents.

Several properties of sums, products, and quotients of functions which become infinite are stated below. Analogous results hold for the cases $x \to a^+$ and $x \to a^-$.

(4.27) Theorem

If $\lim_{x \to a} f(x) = \infty$ and $\lim_{x \to a} g(x) = c$ then:

(i) $\lim_{x \to a} [g(x) + f(x)] = \infty$.

(ii) If $c > 0$, then $\lim_{x \to a} [g(x)f(x)] = \infty$ and $\lim_{x \to a} \frac{f(x)}{g(x)} = \infty$.

(iii) If $c < 0$, then $\lim_{x \to a} [g(x)f(x)] = -\infty$ and $\lim_{x \to a} \frac{f(x)}{g(x)} = -\infty$.

(iv) $\lim_{x \to a} \frac{g(x)}{f(x)} = 0$.

The conclusions of Theorem (4.27) appear intuitively evident. In (i), for example, if $g(x)$ is close to c and if $f(x)$ increases without bound as x approaches a, then it is reasonable to expect that $g(x) + f(x)$ can be made arbitrarily large by choosing x sufficiently close to a. In (iii), if $\lim_{x \to a} g(x) = c < 0$, then $g(x)$ is negative when x is close to a. Consequently, if $f(x)$ is large and positive, then when x is close to a the product $g(x)f(x)$ is numerically large and *negative*. This suggests that $\lim_{x \to a} g(x)f(x) = -\infty$. The reader should supply similar arguments for the remaining parts of Theorem (4.27). We shall not give formal proofs for these results. A similar theorem can be stated for $\lim_{x \to a} f(x) = -\infty$. Theorems involving more than two functions can also be proved.

Graphs of rational functions frequently have vertical and horizontal asymptotes. Illustrations of this are shown in Figures 4.24, 4.27, and 4.31. Another is given in the following example.

Example 5 If $f(x) = 2x^2/(9 - x^2)$, find the vertical and horizontal asymptotes for the graph of f. Sketch the graph of f.

Solution We begin by writing

$$f(x) = \frac{2x^2}{(3 - x)(3 + x)}.$$

The denominator vanishes at $x = 3$ and $x = -3$, and hence the corresponding lines are likely candidates for vertical asymptotes. (Why?) Since

$$\lim_{x \to 3^-} \frac{1}{3 - x} = \infty \quad \text{and} \quad \lim_{x \to 3^-} \frac{2x^2}{3 + x} = 3$$

we see from (ii) of Theorem (4.27) that

$$\lim_{x \to 3^-} f(x) = \lim_{x \to 3^-} \left(\frac{1}{3 - x}\right)\left(\frac{2x^2}{3 + x}\right) = \infty;$$

that is, $f(x)$ becomes positively infinite as x approaches 3 from the left. Moreover, since

$$\lim_{x \to 3^+} \frac{1}{3 - x} = -\infty \quad \text{and} \quad \lim_{x \to 3^-} \frac{2x^2}{3 + x} = 3 > 0$$

it follows that

$$\lim_{x \to 3^+} f(x) = \lim_{x \to 3^+} \left(\frac{1}{3 - x}\right)\left(\frac{2x^2}{3 + x}\right) = -\infty.$$

Thus $f(x)$ becomes negatively infinite as x approaches 3 from the right. This behavior near the vertical asymptote $x = 3$ is illustrated in Figure 4.32.

In like manner, since

$$\lim_{x \to -3^-} \frac{1}{3 + x} = -\infty \quad \text{and} \quad \lim_{x \to -3^-} \frac{2x^2}{3 - x} = 3 > 0$$

we conclude that

$$\lim_{x \to -3^-} f(x) = \lim_{x \to -3^-} \left(\frac{1}{3 + x}\right)\left(\frac{2x^2}{3 - x}\right) = -\infty.$$

However,

$$\lim_{x \to -3^+} \frac{1}{3 + x} = \infty \quad \text{and hence} \quad \lim_{x \to -3^+} f(x) = \infty.$$

This behavior of $f(x)$ near $x = -3$ is illustrated in Figure 4.32.

To find the horizontal asymptotes we consider

$$\lim_{x\to\infty} f(x) = \lim_{x\to\infty} \frac{2x^2}{9 - x^2} = \lim_{x\to\infty} \frac{2}{(9/x^2) - 1} = -2.$$

The same limit is obtained if $x \to -\infty$. Consequently $y = -2$ is a horizontal asymptote. Using the above information and plotting several points gives us the sketch in Figure 4.32.

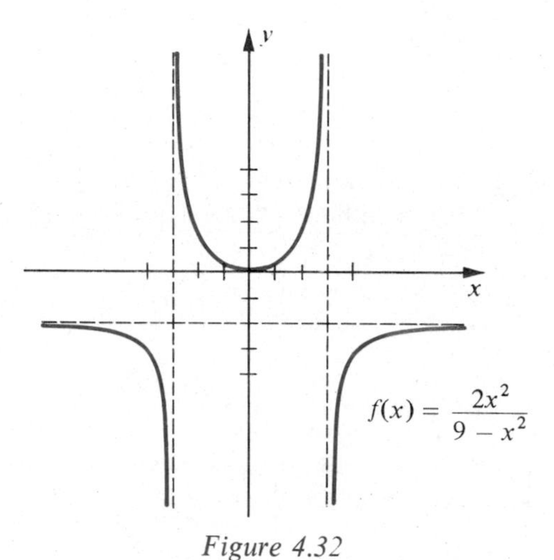

Figure 4.32

If more detailed information about the graph is desired, we could use the first derivative to show that f is decreasing on the intervals $(-\infty, -3)$ and $(-3, 0]$, and increasing on $[0, 3)$ and $(3, \infty)$, with a minimum occurring at $x = 0$ as shown in the figure. The second derivative could be used to investigate concavity.

If we regard a tangent line l as a limiting position of a secant line (see Figure 2.2), then it is possible for l to be vertical. In such cases, as the secant line approaches l its slope increases or decreases without bound. With this in mind we formulate the next definition.

(4.28) Definition

The graph of a function f is said to have a **vertical tangent line** at the point $P(a, f(a))$ if f is continuous at a and

$$\lim_{x\to a} |f'(x)| = \infty.$$

Example 6 If $f(x) = (x - 8)^{1/3} + 1$ and $g(x) = (x - 8)^{2/3} + 1$, show that the graphs of f and g have vertical tangent lines at $P(8, 1)$. Sketch the graphs, showing the tangent lines at P.

Solution Both f and g are continuous at 8. Differentiating we obtain

$$f'(x) = \frac{1}{3}(x-8)^{-2/3} = \frac{1}{3(x-8)^{2/3}}$$

$$g'(x) = \frac{2}{3}(x-8)^{-1/3} = \frac{2}{3(x-8)^{1/3}}.$$

Evidently

$$\lim_{x \to 8} |f'(x)| = \infty = \lim_{x \to 8} |g'(x)|.$$

The graphs are sketched in Figure 4.33. Note that for f the slope of the tangent line becomes positively infinite whether x approaches 8 from the left or from the right. However, for g the slope becomes negatively infinite if $x \to 8^-$ and positively infinite if $x \to 8^+$.

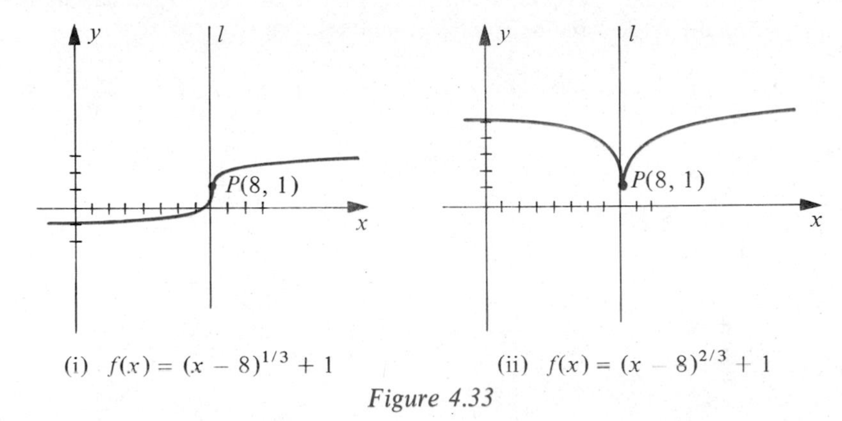

(i) $f(x) = (x-8)^{1/3} + 1$ (ii) $f(x) = (x-8)^{2/3} + 1$

Figure 4.33

EXERCISES 4.5

Find the limits in Exercises 1–12.

1 $\lim_{x \to \infty} \dfrac{5x^2 - 3x + 1}{2x^2 + 4x - 7}$

2 $\lim_{x \to \infty} \dfrac{3x^3 - x + 1}{6x^3 + 2x^2 - 7}$

3 $\lim_{x \to -\infty} \dfrac{4 - 7x}{2 + 3x}$

4 $\lim_{x \to -\infty} \dfrac{(3x+4)(x-1)}{(2x+7)(x+2)}$

5 $\lim_{x \to \infty} \sqrt[3]{\dfrac{8 + x^2}{x(x+1)}}$

6 $\lim_{x \to -\infty} \dfrac{4x - 3}{\sqrt{x^2 + 1}}$

7 $\lim_{x \to \infty} \dfrac{\sqrt{4x+1}}{10 - 3x}$

8 $\lim_{x \to \infty} \dfrac{2x^2 - x + 3}{x^3 + 1}$

9 $\lim_{x \to \infty} \sqrt{x^2 + 5x} - x$

10 $\lim_{x \to \infty} (x - \sqrt{x^2 - 3x})$

11 $\lim_{x \to -\infty} \dfrac{1 + \sqrt[5]{x}}{1 - \sqrt[5]{x}}$

12 $\lim_{x \to \infty} \dfrac{\sqrt[3]{x}}{x^3 + 1}$

13 Determine the horizontal asymptotes and sketch the graph of $y = 4x^2/(1 + x^2)$.

14 Determine the horizontal asymptotes and sketch the graph of $y = 4/(9 + x^2)$.

15 If f and g are polynomial functions of degree n, find $\lim_{x\to\infty} f(x)/g(x)$ and $\lim_{x\to-\infty} f(x)/g(x)$. What is true if the degree of f is less than the degree of g?

16 If f is any polynomial function of degree greater than 0, prove that $\lim_{x\to\infty} 1/f(x) = 0$.

17 Prove, by a direct application of Definition (4.19), that $\lim_{x\to\infty} [2 + (1/x)] = 2$.

18 Define (4.20) in a manner similar to (4.19).

19 If f and g are functions such that $0 < f(x) < g(x)$ for all x, and if $\lim_{x\to\infty} g(x) = 0$, prove that $\lim_{x\to\infty} f(x) = 0$.

20 Prove that if $f(x)$ has a limit as x becomes infinite, then the limit is unique.

Find $\lim_{x\to a^+} f(x)$ and $\lim_{x\to a^-} f(x)$ in Exercises 21–30 and sketch the graph of f. Identify the vertical and horizontal asymptotes.

21 $f(x) = \dfrac{5}{x - 4}, a = 4$

22 $f(x) = \dfrac{5}{4 - x}, a = 4$

23 $f(x) = \dfrac{8}{(2x + 5)^3}, a = -5/2$

24 $f(x) = \dfrac{-4}{7x + 3}, a = -\dfrac{3}{7}$

25 $f(x) = \dfrac{3x}{(x + 8)^2}, a = -8$

26 $f(x) = \dfrac{3x^2}{(2x - 9)^2}, a = \dfrac{9}{2}$

27 $f(x) = \dfrac{2x^2}{x^2 - x - 2}, a = -1, a = 2$

28 $f(x) = \dfrac{4x}{x^2 - 4x + 3}, a = 1, a = 3$

29 $f(x) = \dfrac{1}{x(x - 3)^2}, a = 0, a = 3$

30 $f(x) = \dfrac{x^2}{x + 1}, a = -1$

In each of Exercises 31–34 determine the vertical and horizontal asymptotes and sketch the graph of f.

31 $f(x) = \dfrac{x^2 + 3x + 2}{x^2 + 2x - 3}$

32 $f(x) = \dfrac{x^2 - 5x}{x^2 - 25}$

33 $f(x) = \dfrac{x + 4}{x^2 - 16}$

34 $f(x) = \dfrac{\sqrt[3]{16 - x^2}}{4 - x}$

35 If f and g are polynomial functions and the degree of $f(x)$ is greater than the degree of $g(x)$, what is

$$\lim_{x\to\infty} \frac{f(x)}{g(x)}?$$

36 Prove Theorem (4.25).

In each of Exercises 37–40, find the points where the graph of f has a vertical tangent line.

37 $f(x) = x(x + 2)^{3/5}$

38 $f(x) = \sqrt{x + 2}$

39 $f(x) = \sqrt{16 - 9x^2} + 3$

40 $f(x) = \sqrt[3]{x} - 5$

41 Prove that the graph of a rational function has no vertical tangent lines.

42 Sketch the graph of $y = |x^3 - x|$ and determine where the graph has the following.
(a) A horizontal tangent line.
(b) A vertical tangent line.
(c) No tangent line.

4.6 APPLICATIONS OF EXTREMA

The theory we have developed for finding extrema of functions can be applied to certain practical problems. These problems may either be described orally, or stated in written words, as is the case in textbooks. In order to solve them, it is necessary to convert verbal statements into the language of mathematics by introducing formulas, functions, or equations. Since the types of applications are unlimited, it is difficult to state specific rules for finding solutions. However, it is possible to develop a general strategy for attacking such problems. The following guidelines are often helpful.

Guidelines for Solving Applied Problems Involving Extrema

1. Read the problem carefully several times and think about the given facts, together with the unknown quantities that are to be found.
2. If possible, sketch a picture or diagram and label it appropriately, introducing variables for unknown quantities. Words such as "what," "find," "how much," "how far," or "when" should alert you to the unknown quantities.
3. Make a list of known facts together with any relationships involving the variables. A relationship may often be described by means of an equation of some type.
4. After analyzing the list in Guideline 3, determine which variable is to be maximized or minimized and express this variable as a function of *one* of the other variables.
5. Find the critical numbers of the function obtained in Guideline 4 and test each of them for maxima or minima.
6. Check to see whether extrema occur at the end-points of the domain of the function obtained in Guideline 4.
7. Don't become discouraged if you are unable to solve a given problem. It takes a great deal of effort and practice to become proficient in solving applied problems. Keep trying!

The use of these guidelines is illustrated in the following examples.

Example 1 An open box with a rectangular base is to be constructed from a rectangular piece of cardboard 16 inches wide and 21 inches long by cutting out a square from each corner and then bending up the sides. Find the size of the corner square which will produce a box having the largest possible volume.

We begin by drawing a picture of the cardboard, as shown in (i) of Figure 4.34, where we have introduced a variable x to denote the length of the side of the square to be cut out of each corner of the cardboard. Our goal is to maximize the volume V of the box to be constructed by folding along the dashed lines (see (ii) of Figure 4.34).

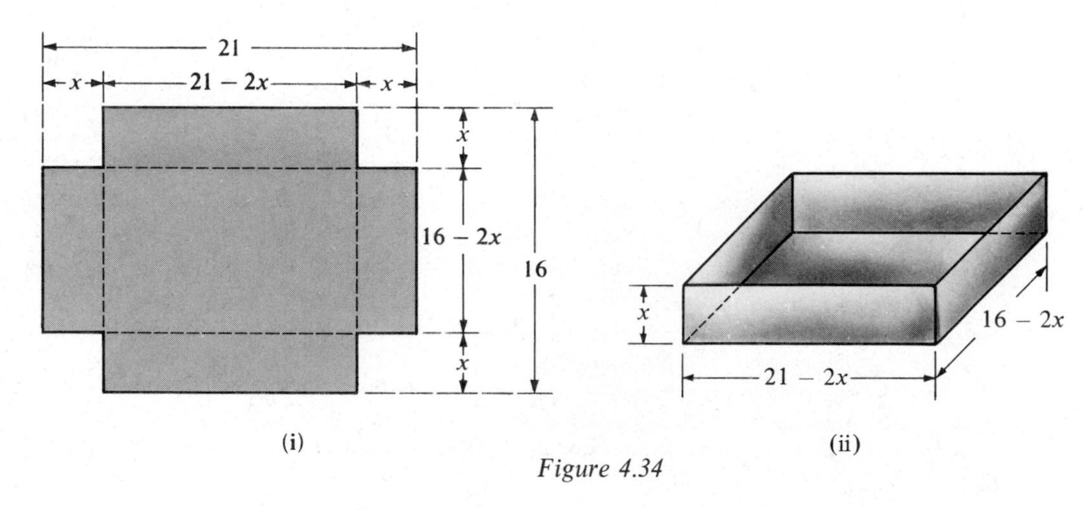

Figure 4.34

The volume V of the box shown in (ii) of the figure is given by

$$V = x(16 - 2x)(21 - 2x) = 2(168x - 37x^2 + 2x^3).$$

This equation expresses V as a function of x and we proceed to find its critical numbers. Differentiating with respect to x we obtain

$$\begin{aligned} D_x V &= 2(168 - 74x + 6x^2) \\ &= 4(3x^2 - 37x + 84) \\ &= 4(3x - 28)(x - 3). \end{aligned}$$

Thus the possible critical numbers are 28/3 and 3. Since 28/3 is outside the domain of x, the only critical number is 3.

The second derivative of V is given by

$$D_x^2 V = 2(-74 + 12x) = 4(6x - 37).$$

Substituting 3 for x,

$$D_x^2 V = 4(18 - 37) = -76 < 0$$

and hence by the Second Derivative Test V has a relative maximum at $x = 3$.

Finally, we check for end-point extrema (see Guideline 6). Since $0 \leq x \leq 8$ (Why?) and since $V = 0$ if either $x = 0$ or $x = 8$, the maximum value of V does not occur at an end-point of the domain of x. Consequently, a three-inch square should be cut from each corner of the cardboard in order to maximize the volume of the resulting box.

Example 2 A circular cylindrical container, open at the top and having a capacity of 24π cubic inches, is to be manufactured. If the cost of the material used for the bottom of the container is three times that used for the curved part and if there is no waste of material, find the dimensions which will minimize the cost.

Solution Begin by drawing a picture as in Figure 4.35, letting r denote the length (in inches) of the radius of the base of the container and h the altitude (in inches). Since the volume is 24π cubic inches, we have

$$\pi r^2 h = 24\pi.$$

This gives us the relationship

$$h = \frac{24}{r^2}$$

between r and h.

Our goal is to minimize the cost C of the material used to manufacture the container.

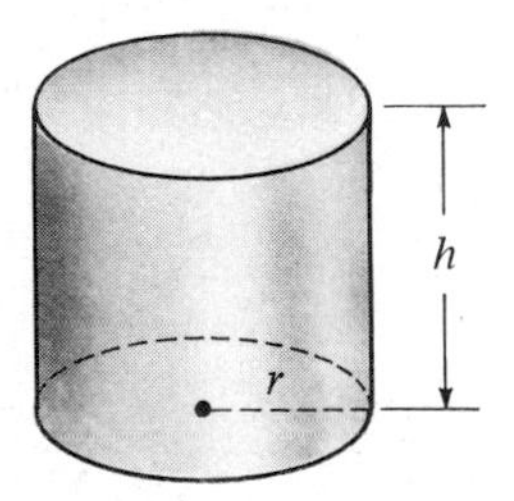

Figure 4.35

If a denotes the cost per square inch of the material to be used for the curved part, then the material to be used for the bottom costs $3a$ per square inch. The cost of the material for the curved part is, therefore, $a(2\pi rh)$ and the cost for the base is $3a(\pi r^2)$. The total cost C for material is given by

$$C = 3a(\pi r^2) + a(2\pi rh) = a\pi(3r^2 + 2rh)$$

or, since $h = 24/r^2$,

$$C = a\pi\left(3r^2 + \frac{48}{r}\right).$$

Since a is fixed, this expresses C as a function of one variable r as recommended in Guideline 4. To find the value of r which will lead to the smallest value for C, we find the critical numbers. Differentiating the last equation with respect to r gives us

$$D_r C = a\pi\left(6r - \frac{48}{r^2}\right) = 6a\pi\left(\frac{r^3 - 8}{r^2}\right).$$

Since $D_r C = 0$ if $r = 2$, we see that 2 is the only critical number. (The number 0 is not a critical number since C is undefined if $r = 0$.) Since $D_r C < 0$ if $r < 2$ and

$D_rC > 0$ if $r > 2$, it follows from the First Derivative Test that C has its minimum value if the radius of the cylinder is 2 inches. The corresponding value for the altitude (obtained from $h = 24/r^2$) is 24/4, or 6 inches.

Since the domain of the variable r is the infinite interval $(0, \infty)$, there can be no end-point extrema.

Example 3 Find the maximum volume of a right circular cylinder that can be inscribed in a cone of altitude 12 cm and base radius 4 cm, if the axes of the cylinder and cone coincide.

Solution The problem is illustrated in Figure 4.36 where (ii) represents a cross-section through the axis of the cone and cylinder. The volume V of the cylinder is given by $V = \pi r^2 h$. In order to express V in terms of one variable (see Guideline 4) we must find a relationship between r and h. Referring to (ii) of Figure 4.36 and using similar triangles we see that

$$\frac{h}{4 - r} = \frac{12}{4} = 3 \quad \text{or} \quad h = 3(4 - r).$$

Consequently,

$$V = \pi r^2 h = \pi r^2 [3(4 - r)] = 3\pi(4r^2 - r^3).$$

If either $r = 0$ or $r = 4$ we see that $V = 0$, and hence the maximum volume is not an end-point extremum. It is sufficient, therefore, to search for relative maximum values. Since

$$D_r V = 3\pi(8r - 3r^2) = 3\pi r(8 - 3r)$$

the critical numbers for V are $r = 0$ and $r = 8/3$. Let us apply the Second Derivative Test for $r = 8/3$. Differentiating $D_r V$ gives us

$$D_r^2 V = 3\pi(8 - 6r).$$

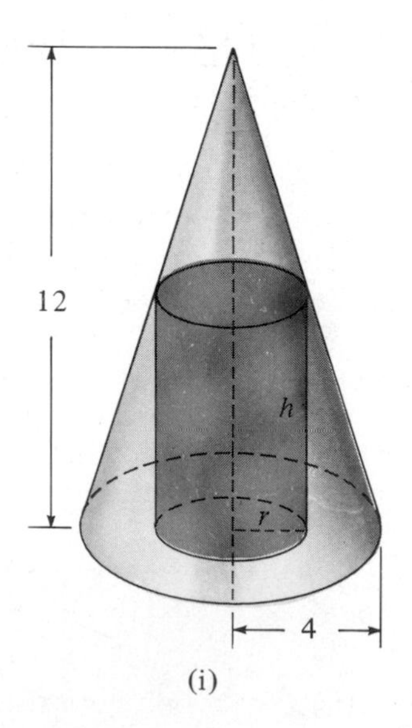

(i)

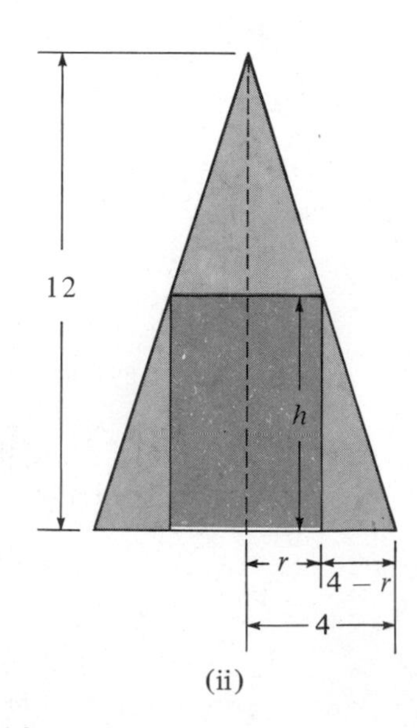

(ii)

Figure 4.36

Substituting 8/3 for r we obtain

$$3\pi[8 - 6(8/3)] = 3\pi(8 - 16) = -24\pi < 0$$

which means that a maximum occurs. The corresponding value for $h = 3(4 - r)$ is

$$h = 3(4 - \tfrac{8}{3}) = 3(\tfrac{4}{3}) = 4.$$

Hence the maximum volume of the inscribed cylinder is

$$V = \pi\left(\frac{8}{3}\right)^2(4) = \pi\left(\frac{64}{9}\right)(4) = \frac{256\pi}{9}\text{ cm}^3.$$

Example 4 A North–South highway intersects an East–West highway at a point P. An automobile crosses P at 10:00 A.M., traveling east at a constant speed of 20 mph. At that same instant another automobile is two miles north of P, traveling south at 50 mph. Find the time at which they are closest to each other and approximate the minimum distance between the automobiles.

Solution For convenience let us represent the highways by means of x- and y-axes, with P at the origin, East toward the right, and North upward. If t denotes the number of hours after 10:00 A.M., then the slower automobile is $20t$ miles east of P, as indicated by point A in Figure 4.37. The faster automobile is at B, $50t$ miles south of its position at 10:00 A.M. The distance d between A and B is given by

$$\begin{aligned} d &= \sqrt{(2 - 50t)^2 + (20t)^2} \\ &= \sqrt{4 - 200t + 2500t^2 + 400t^2} \\ &= \sqrt{4 - 200t + 2900t^2}. \end{aligned}$$

Evidently d has its smallest value when the expression under the radical is minimal. Thus we may simplify our work by letting

$$f(t) = 4 - 200t + 2900t^2$$

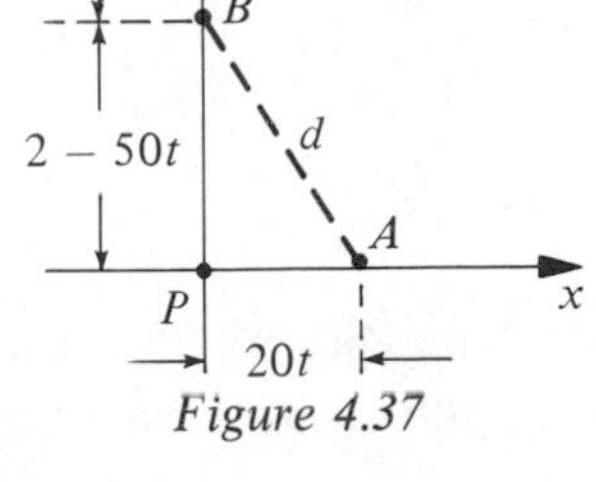

y
2
50t
B
d
2 – 50t
A
P
x
20t

Figure 4.37

and finding the value of t for which f has a minimum. Since

$$f'(t) = -200 + 5800t$$

the only critical number for f is

$$t = \frac{200}{5800} = \frac{1}{29}.$$

Moreover, since $f''(t) = 5800$, the second derivative is always positive and, therefore, f has a relative minimum at $t = 1/29$, and $f(1/29) \approx 0.55$. Since the domain of t is $[0, \infty)$ and since $f(0) = 4$, there is no end-point extremum. Consequently the automobiles will be closest at 1/29 hours (or approximately 2.07 minutes) after 10:00 A.M. The minimal distance is

$$\sqrt{f(1/29)} \approx \sqrt{0.55} \approx 0.74 \text{ miles}.$$

Example 5 A man in a rowboat 2 miles from the nearest point on a straight shoreline wishes to reach a point 6 miles further down the shore. If he can row at a rate of 3 miles per hour and run at a rate of 5 miles per hour, how should he proceed in order to arrive at his destination in the shortest amount of time?

Solution A diagram of the problem is shown in Figure 4.38, where A denotes the position of the boat, B is the nearest point on shore, and C is the man's destination. As shown in the figure, D is the point at which the boat will reach shore and x denotes the distance between B and D. Thus x is restricted to the interval $[0, 6]$.

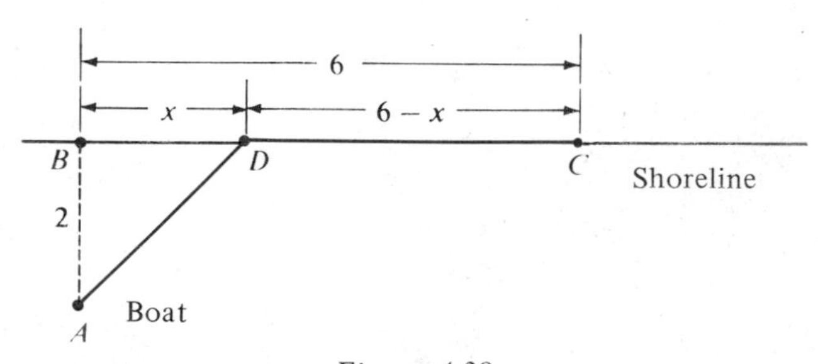

Figure 4.38

By the Pythagorean Theorem, the distance between A and D is $\sqrt{x^2 + 4}$. Using the formula *time* = *distance*/*rate*, the time it takes the man to row from A to D is $\sqrt{x^2 + 4}/3$, whereas the time it takes the man to run from D to C is $(6 - x)/5$. Hence the total time T for the trip is given by

$$T = \frac{\sqrt{x^2 + 4}}{3} + \frac{6 - x}{5}$$

or equivalently,

$$T = \frac{1}{3}(x^2 + 4)^{1/2} + \frac{6}{5} - \frac{1}{5}x.$$

We wish to find the minimum value for T. Note that the case $x = 0$ corresponds to the extreme situation in which the man rows directly to B and then runs the entire distance from B to C. If $x = 6$, then the man rows directly from A to C. These numbers may be considered as end-points of the domain of T. If $x = 0$, then from the above formula for T,

$$T = \frac{\sqrt{4}}{3} + \frac{6}{5} - 0 = \frac{28}{15} \approx 1.87$$

which is approximately 1 hour and 52 minutes. If $x = 6$, then

$$T = \frac{\sqrt{40}}{3} + \frac{6}{5} - \frac{6}{5} = \frac{2\sqrt{10}}{3} \approx 2.11$$

or approximately 2 hours and 7 minutes.

Differentiating the general formula for T we see that

$$\begin{aligned} D_x T &= \frac{1}{3} \cdot \frac{1}{2}(x^2 + 4)^{-1/2}(2x) - \frac{1}{5} \\ &= \frac{x}{3(x^2 + 4)^{1/2}} - \frac{1}{5}. \end{aligned}$$

To find the critical numbers we let $D_x T = 0$. This leads to the following chain of equations:

$$\begin{aligned} 5x &= 3(x^2 + 4)^{1/2} \\ 25x^2 &= 9(x^2 + 4) \\ 16x^2 &= 36 \\ x^2 &= 36/16 \\ x &= 6/4 = 3/2. \end{aligned}$$

Thus 3/2 is the only critical number. We next employ the Second Derivative Test. Applying the quotient rule to the formula for $D_x T$,

$$\begin{aligned} D_x^2 T &= \frac{3(x^2 + 4)^{1/2} - x \cdot 3(1/2)(x^2 + 4)^{-1/2}(2x)}{9(x^2 + 4)} \\ &= \frac{3(x^2 + 4) - 3x^2}{9(x^2 + 4)^{3/2}} \\ &= \frac{4}{3(x^2 + 4)^{3/2}} \end{aligned}$$

which is positive if $x = 3/2$. Consequently T has a local minimum at $x = 3/2$. The time T which corresponds to $x = 3/2$ is

$$T = \frac{1}{3}\left(\frac{9}{4} + 4\right)^{1/2} + \frac{6}{5} - \frac{3}{10} = \frac{26}{15}$$

or equivalently 1 hour and 44 minutes.

We have already examined the values of T at the end-points of the domain, obtaining approximately 1 hour 52 minutes and over 2 hours, respectively. Hence the minimum value of T occurs at $x = 3/2$ and, therefore, the boat should land between B and C, $1\frac{1}{2}$ miles from B. For a similar problem, but one in which the end-points of the domain lead to minimum time, see Exercise 22.

Example 6 A wire 60 inches long is to be cut into two pieces. One of the pieces will be bent into the shape of a circle and the other into the shape of an equilateral triangle. Where should the wire be cut in order that the sum of the areas of the circle and triangle is (a) a maximum; (b) a minimum?

Solution If the wire is cut and x denotes the length of one of the pieces of wire, then the length of the other piece is $60 - x$. Let the piece of length x be bent to form a circle of radius r so that $2\pi r = x$. If the remaining piece is bent into an equilateral triangle of side s, then $3s = 60 - x$ (see Figure 4.39).

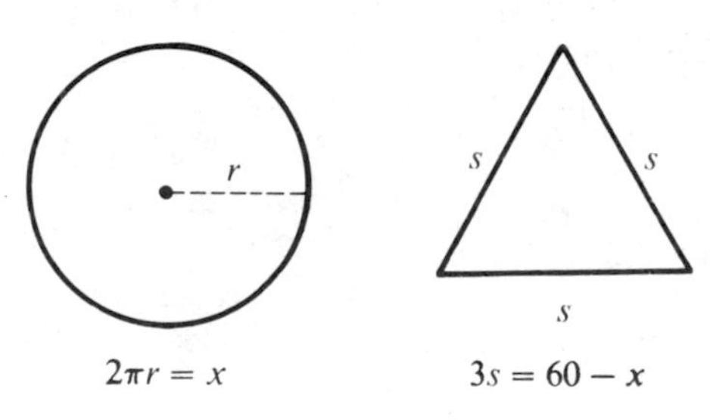

Figure 4.39

The area of the circle in Figure 4.39 is

$$\pi r^2 = \pi\left(\frac{x}{2\pi}\right)^2 = \left(\frac{1}{4\pi}\right)x^2$$

and the area of the triangle is

$$\frac{\sqrt{3}}{4}s^2 = \frac{\sqrt{3}}{4}\left(\frac{60 - x}{3}\right)^2.$$

Consequently, the sum A of the areas can be expressed in terms of one variable x as follows:

$$A = \left(\frac{1}{4\pi}\right)x^2 + \left(\frac{\sqrt{3}}{36}\right)(60 - x)^2.$$

We now find the critical numbers. Differentiating,

$$D_xA = \left(\frac{1}{2\pi}\right)x - \left(\frac{\sqrt{3}}{18}\right)(60 - x)$$

$$= \left(\frac{1}{2\pi} + \frac{\sqrt{3}}{18}\right)x - \frac{10\sqrt{3}}{3}$$

and, therefore, $D_xA = 0$ if and only if

$$x = \frac{10\sqrt{3}/3}{(1/2\pi) + (\sqrt{3}/18)} \approx 22.61.$$

The second derivative

$$D_x^2A = \frac{1}{2\pi} + \frac{\sqrt{3}}{18}$$

is always positive and hence the indicated critical number will yield a minimum value for A. This minimum value is approximated by

$$A \approx \frac{1}{4\pi}(22.61)^2 + \frac{\sqrt{3}}{36}(60 - 22.61)^2 \approx 107.94.$$

Since there are no other critical numbers, the maximum value of A must be an end-point extremum. If $x = 0$, then all the wire is used to form a triangle and

$$A = \frac{\sqrt{3}}{36}(60)^2 \approx 173.21.$$

If $x = 60$, then all the wire is used for the circle, and

$$A = \frac{1}{4\pi}(60)^2 \approx 268.48.$$

Thus the maximum value of A occurs if the wire is not cut and the entire length of wire is bent into the shape of a circle.

EXERCISES 4.6

1 Find two real numbers whose difference is 40 and whose product is a minimum.

2 Find two positive real numbers whose sum is 40 and whose product is a maximum.

3 If a box with a square base and open top is to have a volume of 4 cubic feet, find the dimensions that require the least material (neglect the thickness of the material and waste in construction).

4 Work Exercise 3 if the box has a closed top.

5 A fence 8 feet tall stands on level ground and runs parallel to a tall building. If the fence is 1 foot from the building, find the shortest ladder that will extend from the ground over the fence to the wall of the building.

6 A page of a book is to have an area of 90 square inches, with 1-inch margins at the bottom and sides and a 1/2-inch margin at the top. Find the dimensions of the page which will allow the largest printed area.

7 Find the dimensions of the rectangle of maximum area that can be inscribed in a semicircle of radius a, if two vertices lie on the diameter.

8 Find the dimensions of the rectangle of maximum area that can be inscribed in an equilateral triangle of side a, if two vertices of the rectangle lie on one of the sides of the triangle.

9 Of all possible right circular cones that can be inscribed in a sphere of radius a, find the volume of the one which has maximum volume.

10 Find the dimensions of the right circular cylinder of maximum volume that can be inscribed in a sphere of radius a.

11 A metal cylindrical container with an open top is to hold one cubic foot. If there is no waste in construction, find the dimensions which require the least amount of material.

12 If the circular base of the container in Exercise 11 is cut from a square sheet and the remaining metal is discarded, find the dimensions which require the least amount of material.

13 A long rectangular sheet of metal, 12 inches wide, is to be made into a rain gutter by turning up two sides at right angles to the sheet. How many inches should be turned up to give the gutter its greatest capacity?

14 Work Exercise 13 if the sides of the gutter make an angle of 120° with the base.

15 Prove that the rectangle of largest area having a given perimeter p is a square.

16 A right circular cylinder is generated by rotating a rectangle of perimeter p about one of its sides. What dimensions of the rectangle will generate the cylinder of maximum volume?

17 The strength of a rectangular beam varies jointly as the width and the square of the depth of a cross section. Find the dimensions of the strongest beam that can be cut from a cylindrical log of radius a.

18 A window has the shape of a rectangle surmounted by a semicircle. If the perimeter of the window is 15 feet, find the dimensions which will allow the maximum amount of light to enter.

19 Find the point on the graph of $y = x^2 + 1$ that is closest to the point $(3, 1)$.

20 Find the abscissa of the point on the graph of $y = x^3$ that is closest to the point $(4, 0)$.

21 A manufacturer sells a certain article to dealers at a rate of $20 each if less than 50 are ordered. If 50 or more are ordered (up to 600) the price per article is reduced at a rate of 2 cents times the number ordered. What size order will produce the maximum amount of money for the manufacturer?

22 Refer to Example 5 of this section. If the man is in a motorboat which can travel at an average rate of 15 miles per hour, how should he proceed in order to arrive at his destination in the least time?

23 The illumination from a light source is directly proportional to the strength of the source and inversely proportional to the square of the distance from the source. If two light sources of strengths S_1 and S_2 are d units apart, at what point on the line segment joining the two sources is the illumination minimal?

24 At 1:00 P.M. ship A is 30 miles due south of ship B and is sailing north at a rate of 15 miles per hour. If ship B is sailing west at a rate of 10 miles per hour, at what time will the distance between the ships be minimal?

25 A veterinarian has 100 feet of fencing and wishes to construct six dog kennels by first building a fence around a rectangular region, and then subdividing that region into six smaller rectangles by placing five fences parallel to one of the sides. What dimensions of the region will maximize the total area?

26 A paper cup having the shape of a right circular cone is to be constructed. If the volume desired is 36π in.^{3}, find the dimensions that require the least amount of paper (neglect any waste that may occur in the construction).

27 A steel storage tank for propane gas is to be constructed in the shape of a right circular cylinder with a hemisphere at each end. If the desired capacity is 100 ft^3, what dimensions will require the least amount of steel?

28 A pipeline for transporting oil will connect two points A and B which are 3 miles apart and on opposite banks of a straight river one mile wide. Part of the pipeline will run under water from A to a point C on the opposite bank, and then above ground from C to B. If the cost per mile of running the pipeline under water is four times the cost per mile of running it above ground, find the location of C which will minimize the cost (ignore the slope of the river bed).

29 A wire 36 cm long is to be cut into two pieces. One of the pieces will be bent into the shape of an equilateral triangle and the other into the shape of a rectangle whose length is twice its width. Where should the wire be cut if the combined area of the triangle and rectangle is (a) a minimum? (b) a maximum?

30 An isosceles triangle has base b and equal sides of length a. Find the dimensions of the rectangle of maximum area that can be inscribed in the triangle if one side of the rectangle lies on the base of the triangle.

31 Two vertical poles of lengths 6 ft and 8 ft stand on level ground, with their bases 10 ft apart. Approximate the minimal length of cable that can reach from the top of one pole to some point on the ground between the poles, and then to the top of the other pole.

32 Find the dimensions of the rectangle of maximum area having two vertices on the x-axis and two vertices above the x-axis, on the graph of $y = 4 - x^2$.

33 A window has the shape of a rectangle surmounted by an equilateral triangle. If the perimeter of the window is 12 ft, find the dimensions of the rectangle which will produce the largest area for the window.

34 If a trapezoid has three nonparallel sides of length 8, find the length of the fourth side which will maximize the area.

35 The owner of an apple orchard estimates that if 24 trees are planted per acre, then each mature tree will yield 600 apples per year. For each additional tree planted per acre the number of apples produced by each tree decreases by 12 per year. How many trees should be planted per acre in order to obtain the most apples?

36 A real estate company owns 180 apartments which are fully occupied when the rent is \$300 per month. The company estimates that for each \$10 increase in rent, five apartments will become unoccupied. What rent should be charged in order to obtain the largest gross income?

37 A package can be sent by parcel post only if the sum of its length and girth (the perimeter of the base) is not more than 96 inches. Find the dimensions of the box of maximum volume that can be sent if the base of the box is a square.

38 A North–South highway A and an East–West highway B intersect at a point P. At 10:00 A.M. an automobile crosses P traveling north on highway A at a speed of 50 mph. At that same instant, an airplane flying east at a speed of 200 mph and an altitude of 26,400 ft is directly above the point on highway B which is 100 miles west of P. If the automobile and airplane maintain the same speed and direction, at what time will they be closest to one another?

39 Use the First Derivative Test to prove that the shortest distance from a point (x_1, y_1) to the line $ax + by + c = 0$ is given by

$$d = \frac{|ax_1 + by_1 + c|}{\sqrt{a^2 + b^2}}.$$

40 Prove that the shortest distance from a point (x_1, y_1) to the graph of a differentiable function f is measured along a normal line to the graph, that is, a line perpendicular to the tangent line.

4.7 THE DERIVATIVE AS A RATE OF CHANGE

All quantities encountered in everyday life change with time. This is especially evident in scientific investigations. For example, a chemist may be interested in the rate at which a certain substance dissolves in water. An electrical engineer may wish to know the rate of change of current in part of an electrical circuit. A biologist may be studying the rate at which the bacteria in a culture are increasing or decreasing. Numerous other examples could be cited, including many from fields other than the natural sciences. Let us consider the following general situation, which can be applied to all of the above examples.

Suppose a variable w is a function of time such that at time t, w is given by $w = g(t)$, where g is a differentiable function. The difference between the initial and final values of w in the time interval $[t, t + h]$ is given by $g(t + h) - g(t)$. As in our development of the velocity concept, we formulate the following definition.

(4.29) Definition

The **average rate of change** of $w = g(t)$ in the interval $[t, t + h]$ is

$$\frac{g(t + h) - g(t)}{h}.$$

The **rate of change** of $w = g(t)$ with respect to t is

$$\frac{dw}{dt} = g'(t) = \lim_{h \to 0} \frac{g(t + h) - g(t)}{h}.$$

The units to be used in Definition (4.29) depend on the nature of the quantity represented by w. Sometimes dw/dt is referred to as the **instantaneous rate of change** of w with respect to t.

Example 1 A scientist finds that if a certain substance is heated, the Celsius temperature after t minutes, where $0 \leq t \leq 5$, is given by the formula $g(t) = 30t + 6\sqrt{t} + 8$.

(a) Find the average rate of change of $g(t)$ during the time interval $[4, 4.41]$.

(b) Find the rate of change of $g(t)$ at $t = 4$.

Solution (a) Letting $t = 4$ and $h = 0.41$ in Definition (4.29), the average rate of change of g in $[4, 4.41]$ is

$$\frac{g(4.41) - g(4)}{0.41} = \frac{[30(4.41) + 6\sqrt{4.41} + 8] - [120 + 6\sqrt{4} + 8]}{0.41}$$

$$= \frac{12.9}{0.41} \approx 31.46°\text{C/min}.$$

(b) From Definition (4.29) the rate of change of g at time t is given by $g'(t) = 30 + 3/\sqrt{t}$. In particular,

$$g'(4) = 30 + 3/\sqrt{4} = 31.5\,°\text{C/min}.$$

The velocity of a point P which moves on a coordinate line l was defined in Chapter 3. In future work we shall often use the letter s to denote functions whose values are distances along lines or other graphs. If the coordinate of P at time t is $s(t)$, then by Definition (3.3) the velocity of P at time t is $s'(t)$. Using the terminology introduced at the beginning of this section, the velocity is the rate of change of $s(t)$ with respect to time. We call s the **position function** of P. Denoting the velocity of P at time t by $v(t)$ we have the following alternate definition of velocity.

(4.30) Definition

> If the position of a point P on a coordinate line l at time t is given by $s(t)$, then the **velocity** $v(t)$ of P at time t is $v(t) = s'(t)$.

We shall call v the **velocity function** for P. The function v is the derivative of the position function s; that is,

$$v = D_t s = \frac{ds}{dt}.$$

If t is measured in seconds and $s(t)$ in centimeters, then the unit for $v(t)$ is cm/sec (read "centimeters per second"). Similarly, if t is measured in hours and $s(t)$ in miles, then $v(t)$ is in mi/hr (miles per hour), and so on.

The theory developed in Section 3 may be used to determine the direction of motion of P. If the velocity $v(t)$ is positive in a given time interval I, then $s'(t) > 0$ and hence, by Theorem (4.13), $s(t)$ is increasing; that is, P is moving in the positive direction on l. Similarly, if $v(t)$ is negative, then P is moving in the negative direction. Since $v(t) = s'(t)$, the times at which the velocity is zero are critical numbers for the position function s and hence lead to possible local maxima or minima for s. If a local maximum or minimum for s occurs at t_1, then t_1 is usually a time at which P reverses direction.

The **speed** of P at time t is defined as $|v(t)|$. The speed indicates how fast P is moving without specifying the direction of motion.

The derivative $v'(t)$ is called the **acceleration** of P at time t and is denoted by $a(t)$. Since $v(t) = s'(t)$,

(4.31) $$a(t) = v'(t) = s''(t)$$

The function a is called the **acceleration function** of the point P. As in Definition (4.29), the acceleration $a(t)$ is the rate of change of velocity of P at time t. The units for $a(t)$ are cm/sec^2 (centimeters per second per second), mi/hr^2 (miles per hour per hour), and so on.

Again applying the theory of increasing and decreasing functions, it follows that if $a(t) > 0$, then at time t the velocity is increasing, whereas if $a(t) < 0$, then the velocity is decreasing. Note that if *both* $a(t) < 0$ and $v(t) < 0$, then the velocity is decreasing in the sense of becoming more negative. In this event, even though the velocity is decreasing, the speed $|v(t)|$ is increasing in magnitude. An illustration of this occurs in Example 2 below.

Since the function a is the second derivative of the position function s, it can sometimes be used in conjunction with the Second Derivative Test to find the local extrema of s. For example, if $v(t_1) = 0$ and $a(t_1) < 0$, then $s(t_1)$ is a local maximum for s and hence P changes direction from right to left at time t_1. Similarly, if $v(t_1) = 0$ and $a(t_1) > 0$, then $s(t_1)$ is a local minimum for s and P changes direction from left to right at time t_1.

Example 2 The position function s of a point P on a coordinate line is given by

$$s(t) = t^3 - 12t^2 + 36t - 20,$$

where t is measured in seconds and $s(t)$ in centimeters. Describe the motion of P during the time interval $[-1, 9]$.

Solution Differentiating,

$$v(t) = s'(t) = 3t^2 - 24t + 36 = 3(t - 2)(t - 6)$$

and

$$a(t) = v'(t) = 6t - 24 = 6(t - 4).$$

Consequently the velocity is 0 at $t = 2$ and $t = 6$. The following table describes the direction of motion of P.

Time Interval	$v(t)$	Direction of Motion
$(-1, 2)$	+	right
$(2, 6)$	−	left
$(6, 9)$	+	right

The table shows that the function s has a local maximum at time $t = 2$ and a local minimum at $t = 6$. This can also be verified by noting that $a(2) = -12 < 0$ and $a(6) = 12 > 0$. The next table displays the values of the position, velocity, and acceleration functions at important times, namely the smallest and largest numbers in the time interval $[-1, 9]$ and the times at which the velocity or acceleration is zero.

t	-1	2	4	6	9
$s(t)$	-69	12	-4	-20	61
$v(t)$	63	0	-12	0	63
$a(t)$	-30	-12	0	12	30

It is convenient to represent the motion of P schematically as in Figure 4.40. The curve above the coordinate line is not the path of the point, but is intended to show the manner in which P moves on l. As indicated by the table and Figure 4.40, at $t = -1$ the point is 69 cm to the left of the origin and is moving to the right with a velocity of 63 cm/sec. The negative acceleration -30 cm/sec^2 indicates that the velocity is decreasing at a rate of 30 cm per second each second. The point continues to move to the right and to slow down until it has zero velocity at $t = 2$, 12 cm to the right of the origin. The point P then reverses direction and moves until, at $t = 6$, it is 20 cm to the left of the origin. Then it again reverses direction and moves to the right for the remainder of the time interval, with increasing velocity. The direction of motion is indicated by the arrows on the curve in Figure 4.40.

The time $t = 4$, at which the acceleration is zero, is a critical number for the velocity function. (Why?) The significance of this time is clearer if we consider the speed $|v(t)| = |3(t-2)(t-6)|$. In the time interval $[2, 6]$, the speed increases from 0 at $t = 2$ to a local maximum $|-12| = 12$ at $t = 4$ and then decreases to 0 at $t = 6$.

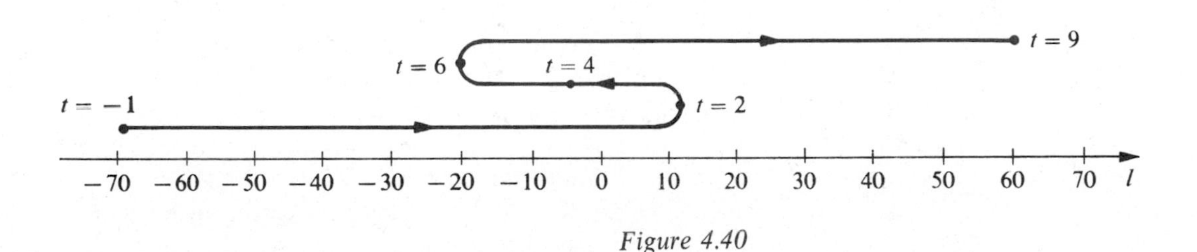

Figure 4.40

Example 3 A projectile is fired straight upward with a velocity of 400 ft/sec. Its distance above the ground t seconds after being fired is given by $s(t) = -16t^2 + 400t$. Find the time and the velocity at which the projectile hits the ground. What is the maximum altitude achieved by the projectile? What is the acceleration at any time t?

Solution The path of the projectile is on a vertical coordinate line with origin at ground level and positive direction upward. The projectile will be on the ground when $-16t^2 + 400t = 0$, that is, when $-16t(t - 25) = 0$. Hence the projectile hits the ground after 25 seconds. The velocity at time t is given by

$$v(t) = s'(t) = -32t + 400.$$

In particular, when $t = 25$,

$$v(25) = -32(25) + 400 = -400 \text{ ft/sec}.$$

The maximum altitude occurs when $s'(t) = 0$, that is, when $-32t + 400 = 0$. Solving for t gives us $t = 400/32 = 25/2$ and hence its maximum altitude is

$$s(25/2) = -16(25/2)^2 + 400(25/2) = 2500 \text{ feet}.$$

Finally, the acceleration at any time t is $a(t) = v'(t) = -32 \text{ ft/sec}^2$.

We may investigate rates of change with respect to variables other than time. Indeed, if $y = f(x)$, where x is *any* variable, then the **average rate of change of y with respect to x in the interval $[x, x + h]$** is defined as the quotient

$$\frac{f(x + h) - f(x)}{h}.$$

The limit of this quotient as h approaches 0, that is, dy/dx, is called the **rate of change of y with respect to x**. Thus, if the variable x changes, then y changes at a rate of dy/dx units per unit change in x. To illustrate, suppose a quantity of gas is enclosed in a spherical balloon. If the gas is heated or cooled and if the pressure remains constant, then the balloon expands or contracts and its volume V is a function of the temperature T. The derivative dV/dT gives us the rate of change of volume with respect to temperature.

Example 4 The current I (in amperes) in a certain electrical circuit is given by $I = 100/R$, where R denotes resistance (in ohms). Find the rate of change of I with respect to R when the resistance is 20 ohms.

Solution Since $dI/dR = -100/R^2$, we see that if $R = 20$ then $dI/dR = -100/400 = -1/4$. Thus, if R is increasing, then when $R = 20$ the current I is decreasing at a rate of 1/4 amperes per ohm.

EXERCISES 4.7

1 As a spherical balloon is being inflated, its radius (in centimeters) after t minutes is given by $r(t) = 3\sqrt[3]{t+8}$, where $0 \le t \le 10$. What is the rate of change with respect to t of each of the following at $t = 8$?

(a) $r(t)$ (b) The volume of the balloon (c) The surface area

2 The volume of a spherical balloon (in ft^3) t hours after 1:00 P.M. is given by $V(t) = \frac{4}{3}\pi(9 - 2t)^3$, where $0 \le t \le 4$. What is the rate of change with respect to t of each of the following at 4:00 P.M.?

(a) $V(t)$ (b) The radius of the balloon (c) The surface area

3 Suppose that t seconds after he starts to run, an individual's pulse rate (in beats/min) is given by $P(t) = 56 + 2t^2 - t$, where $0 \le t \le 7$. Find the rate of change of $P(t)$ with respect to t at:

(a) $t = 2$, (b) $t = 4$, (c) $t = 6$.

4 The temperature T (degrees Celsius) of a solution at time t (minutes) is given by $T(t) = 10 + 4t + 3/(t + 1)$, where $1 \le t \le 10$. Find the rate of change of $T(t)$ with respect to t at:

(a) $t = 2$, (b) $t = 5$, (c) $t = 9$.

5 A stone is dropped into a lake, causing circular waves. If, after t seconds, the radius of one of the waves is $40t$ cm, find the rate of change with respect to t of the area of the circle formed by the wave at:

(a) $t = 1$, (b) $t = 2$, (c) $t = 3$.

6 Boyle's Law for gases states that $pv = c$, where p denotes the pressure, v the volume, and c is a constant. Suppose that at time t (minutes) the pressure is $20 + 2t\ \text{gm/cm}^2$, where $0 \le t \le 10$. If the volume is $60\ \text{cm}^3$ at $t = 0$, find the rate at which the volume is changing with respect to t at $t = 5$.

Position functions of points moving rectilinearly are defined in Exercises 7–16. Find the velocity and acceleration at time t and describe the motion of the point during the indicated time interval. Illustrate the motion by means of a diagram of the type shown in Figure 4.40.

7 $s(t) = 3t^2 - 12t + 1$, $[0, 5]$

8 $s(t) = t^2 + 3t - 6$, $[-2, 2]$

9 $s(t) = t^3 - 9t + 1$, $[-3, 3]$

10 $s(t) = 24 + 6t - t^3$, $[-2, 3]$

11 $s(t) = t + 4/t$, $[1, 4]$

12 $s(t) = 2\sqrt{t} + 1/\sqrt{t}$, $[1, 4]$

13 $s(t) = 2t^4 - 6t^2$, $[-2, 2]$

14 $s(t) = 2t^3 - 6t^5$, $[-1, 1]$

15 $s(t) = t^3 + 1$, $[0, 4]$

16 $s(t) = \sqrt[3]{t}$, $[-8, 0]$

17 An object is thrown directly upward with a velocity of 144 ft/sec. Its height $s(t)$ in feet above the ground after t seconds is given by $s(t) = 144t - 16t^2$. What are the velocity and acceleration after t seconds? After 3 seconds? What is the maximum height? When does the object strike the ground?

18 An object rolls down an inclined plane such that the distance $s(t)$ (in inches) it rolls in t seconds is given by $s(t) = 5t^2 + 2$. What is the velocity after 1 second? 2 seconds? When will the velocity be 28 inches per second?

19 The position function s of a point moving rectilinearly is given by $s(t) = 2t^3 - 15t^2 + 48t - 10$ where t is in seconds and $s(t)$ is in meters. Find the acceleration when the velocity is 12 m/sec. Find the velocity when the acceleration is 10 m/sec^2.

20 Work Exercise 19 if $s(t) = t^5 - (5/3)t^3 - 48t$.

21 The illumination I from a light source is directly proportional to the strength S of the source and inversely proportional to the square of the distance d from the source. If, for a certain light, I is 120 units at a distance of 2 ft, find the rate of change of I with respect to d at a distance of 20 ft.

22 Show that the rate of change of the volume of a sphere with respect to its radius is numerically equal to the surface area of the sphere.

23 Show that the rate of change of the radius of a circle with respect to its circumference is independent of the size of the circle. Illustrate this fact by using great circles on two spheres—one the size of a basketball and the other the size of the Earth.

24 The formula for the adiabatic expansion of air is $pv^{1.4} = c$, where p is the pressure, v the volume, and c is a constant. Find a formula for the rate of change of pressure with respect to volume.

25 The relationship between the temperature F on the Fahrenheit scale and the temperature C on the Celsius scale is given by $C = \frac{5}{9}(F - 32)$. What is the rate of change of F with respect to C?

26 If a sum of money P_0 is invested at an interest rate of $100r$ per cent per year, compounded monthly, then the principal P at the end of one year is $P = P_0(1 + r/12)^{12}$. Find the rate of change of P with respect to r when $P_0 = \$1,000$ and $r = 6\%$.

27 The period T of a simple pendulum, that is, the time required for one complete oscillation, varies directly as the square root of its length l. What can be said about the rate of change of T with respect to l?

28 The electrical resistance R of a copper wire of fixed length is inversely proportional to the square of its diameter d. What is the rate of change of R with respect to d?

29 An open box is to be constructed from a rectangular piece of cardboard 40 cm wide and 60 cm long by cutting out a square of side s cm from each corner and then bending up the cardboard. Express the volume V of the box as a function of s and find the rate of change of V with respect to s.

30 The formula $1/f = (1/p) + (1/q)$ is used in optics, where f is the focal length of a convex lens, and p and q are the distances from the lens to the object and image, respectively. If f is fixed, find a general formula for the rate of change of q with respect to p.

4.8 RELATED RATES

In applications it is not unusual to encounter two variables x and y which are differentiable functions of time t, say $x = f(t)$ and $y = g(t)$. In addition, x and y may be related by means of some equation such as

$$x^2 - y^3 - 2x + 7y^2 - 2 = 0.$$

Differentiating with respect to t and using the Chain Rule (3.36) produces an equation which involves the rates of change dx/dt and dy/dt. As an illustration, the previous equation leads to

$$2x\frac{dx}{dt} - 3y^2\frac{dy}{dt} - 2\frac{dx}{dt} + 14y\frac{dy}{dt} = 0.$$

The derivatives dx/dt and dy/dt in this equation are called **related rates**, since they are related by means of the equation. The last equation can be used to find one of the rates when the other is known. This observation has many practical applications. The examples which follow give several illustrations.

Example 1 A ladder 20 ft long leans against a vertical building. If the bottom of the ladder slides away from the building horizontally at a rate of 2 ft/sec, how fast is the ladder sliding down the building when the top of the ladder is 12 ft above the ground?

Solution We begin by representing a general position of the ladder schematically, as in Figure 4.41, introducing a variable x to denote the distance from the base of the building to the bottom of the ladder, and a variable y to denote the distance from the ground to the top of the ladder.

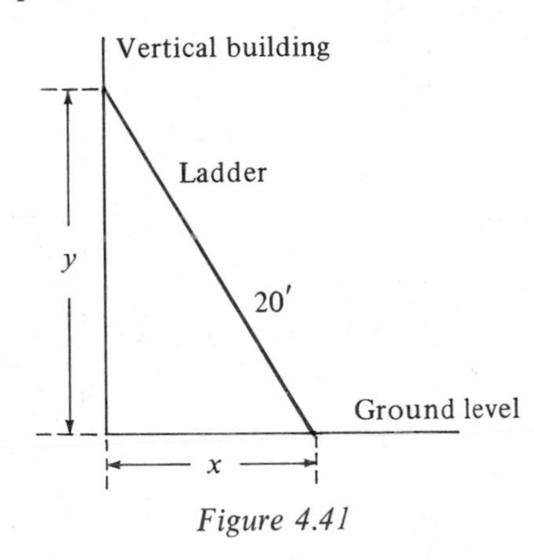

Figure 4.41

Since x is changing at a rate of 2 ft/sec, we may write

$$\frac{dx}{dt} = 2 \text{ ft/sec}.$$

Our objective is to find dy/dt, the rate at which the top of the ladder is sliding down the building, at the instant that $y = 12$ ft.

The relationship between x and y may be determined by applying the Pythagorean Theorem to the triangle in the figure. This gives us

$$x^2 + y^2 = 400.$$

Differentiating with respect to t leads to

$$2x\frac{dx}{dt} + 2y\frac{dy}{dt} = 0$$

from which it follows that

$$\frac{dy}{dt} = -\frac{x}{y}\frac{dx}{dt}.$$

The last equation is a *general* formula which relates the two rates of change under consideration. Let us now consider the special case $y = 12$. The corresponding value of x may be determined from

$$x^2 + 144 = 400, \quad \text{or} \quad x^2 = 400 - 144 = 256.$$

Thus $x = \sqrt{256} = 16$ when $y = 12$. Substituting in the general formula for dy/dt we obtain

$$\frac{dy}{dt} = -\frac{16}{12}(2) = -\frac{8}{3}\text{ ft/sec.}$$

Some of the Guidelines listed in Section 6 for applications of extrema are useful when attempting to solve a related rate problem. In particular, the first three Guidelines—(1) read the problem carefully, (2) sketch and label a suitable diagram, and (3) list known facts and relationships—are strongly recommended. After following these steps, it is essential to formulate a *general* equation which relates the variables involved in the problem, as we did in Example 1. A common error is to introduce specific values for the rates and variable quantities too early in the solution. Always remember to obtain a *general* formula which involves the rates of change at *any* time t. Specific values should not be substituted for variables until the final steps of the solution.

Example 2 A water tank has the shape of an inverted right circular cone of altitude 12 ft and base radius 6 ft. If water is being pumped into the tank at a rate of 10 gal/min, approximate the rate at which the water level is rising when it is 3 ft deep (1 gal $\approx$ 0.1337 ft^3).

Solution We begin by sketching the tank as in Figure 4.42, letting r denote the radius of the surface of the water when the depth is h. Note that r and h are functions of time t. The volume V of water in the tank corresponding to depth h is given by

$$V = \frac{1}{3}\pi r^2 h.$$

Let us express V in terms of one variable. Referring to similar triangles in Figure 4.42 we see that r and h are related by the equations

$$\frac{r}{h} = \frac{6}{12}, \quad \text{or} \quad r = \frac{h}{2}.$$

Consequently, at depth h,

$$V = \frac{1}{3}\pi\left(\frac{h}{2}\right)^2 h = \frac{1}{12}\pi h^3.$$

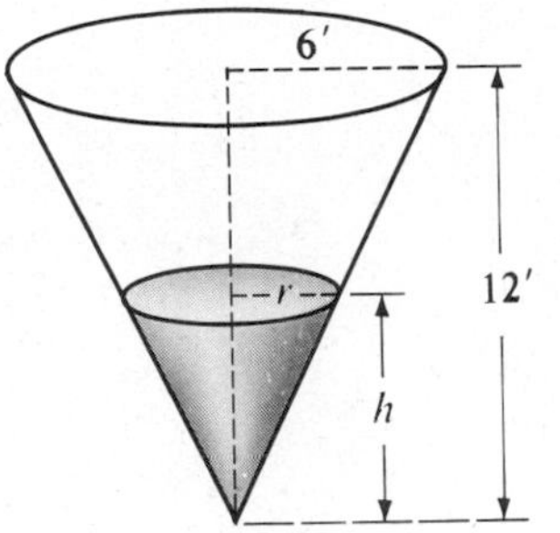

Figure 4.42

Differentiating with respect to t gives us the following general relationship between the rates of change of V and h at any time:

$$\frac{dV}{dt} = \frac{1}{4}\pi h^2 \frac{dh}{dt}.$$

An equivalent formula is

$$\frac{dh}{dt} = \frac{4}{\pi h^2}\frac{dV}{dt}.$$

In particular, if $h = 3$, and $dV/dt = 10$ gal/min ≈ 1.337 ft^3/min, we see that

$$\frac{dh}{dt} \approx \frac{4}{\pi(9)}(1.337) \approx 0.189 \text{ ft/min}.$$

Example 3 At 1:00 P.M., ship A is 25 miles due south of ship B. If ship A is sailing west at a rate of 16 mi/hr and ship B is sailing south at a rate of 20 mi/hr, find the rate at which the distance between the ships is changing at 1:30 P.M.

Solution Let x and y denote the miles covered by ships A and B, respectively, in t hours after 1:00 P.M. We then have the situation sketched in Figure 4.43, where P and Q are their respective positions at 1:00 P.M. If z is the distance between the ships at time t, then by the Pythagorean Theorem,

$$z^2 = x^2 + (25 - y)^2.$$

Differentiation with respect to t gives us

$$2z\frac{dz}{dt} = 2x\frac{dx}{dt} + 2(25 - y)\left(-\frac{dy}{dt}\right)$$

or

$$z\frac{dz}{dt} = x\frac{dx}{dt} + (y - 25)\frac{dy}{dt}.$$

It is given that

$$\frac{dx}{dt} = 16 \text{ mi/hr} \quad \text{and} \quad \frac{dy}{dt} = 20 \text{ mi/hr}.$$

Our objective is to find dz/dt.

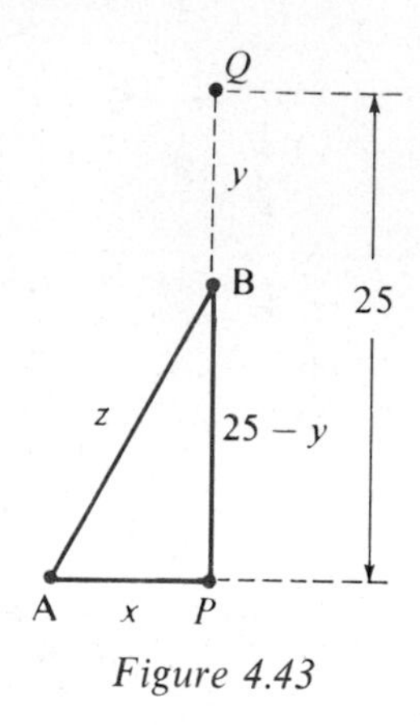

Figure 4.43

At 1:30 P.M. the ships have traveled for half an hour and we have $x = 8$, $y = 10, 25 - y = 15$, and, therefore, $z^2 = 64 + 225 = 289$, or $z = 17$. Substituting in the last displayed equation gives us

$$17\frac{dz}{dt} = 8(16) + (-15)(20)$$

or

$$\frac{dz}{dt} = -\frac{172}{17} \approx -10.12 \text{ mi/hr}.$$

The negative sign indicates that the distance between the ships is decreasing at 1:30 P.M.

Another method of solution is to write $x = 16t$, $y = 20t$, and

$$z = [x^2 + (25 - y)^2]^{1/2} = [256t^2 + (25 - 20t)^2]^{1/2}.$$

The derivative dz/dt may then be found, and substitution of $1/2$ for t produces the desired rate of change.

EXERCISES 4.8

1 A ladder 20 ft long leans against a vertical building. If the bottom of the ladder slides away from the building horizontally at a rate of 3 ft/sec, how fast is the ladder sliding down the building when the top of the ladder is 8 ft from the ground?

2 As a circular metal disc is being heated, its diameter changes at a rate of 0.01 cm/min. When the diameter is 5 cm at what rate is the area of one side changing?

3 Gas is being pumped into a spherical balloon at a rate of 5 ft^3/min. If the pressure is constant, find the rate at which the radius is changing when the diameter is 18 in.

4 A girl starts at a point A and runs east at a rate of 10 ft/sec. One minute later, another girl starts at A and runs north at a rate of 8 ft/sec. At what rate is the distance between them changing 1 minute after the second girl starts?

5 A light is at the top of a 16-ft pole. A boy 5 ft tall walks away from the pole at a rate of 4 ft/sec. At what rate is the tip of his shadow moving when he is 18 ft from the pole? At what rate is the length of his shadow increasing?

6 A man on a dock is pulling in a boat by means of a rope attached to the vertical bow of the boat 1 ft above water level and passing through a simple pulley located on the dock 8 ft above water level. If he pulls in the rope at a rate of 2 ft/sec, how fast is the boat approaching the dock when the bow of the boat is 25 ft from a point on the water directly below the pulley?

7 The top of a silo has the shape of a hemisphere of diameter 20 ft. If it is coated uniformly with a layer of ice, and if the thickness is decreasing at a rate of ¼ in./hr, how fast is the volume of ice changing when the ice is 2 in. thick?

8 As sand leaks out of a hole in a container, it forms a conical pile whose altitude is always the same as its radius. If the height of the pile is increasing at a rate of 6 in./min, find the rate at which the sand is leaking out when the altitude is 10 in.

9 A boy flying a kite pays out string at a rate of 2 ft/sec as the kite moves horizontally at an altitude of 100 ft. Assuming there is no sag in the string, find the rate at which the kite is moving when 125 ft of string have been paid out.

10 A weather balloon is rising vertically at a rate of 2 ft/sec. An observer is situated 100 yds from a point on the ground directly below the balloon. At what rate is the distance between the balloon and the observer changing when the altitude of the balloon is 500 ft?

11 Boyle's Law for gases states that $pv = c$, where p denotes the pressure, v the volume, and c is a constant. At a certain instant the volume is 75 in.^{3}, the pressure is 30 lbs/in.^{2}, and the pressure is decreasing at a rate of 2 lbs/in.^{2} every minute. At what rate is the volume changing at this instant?

12 Suppose a spherical snowball is melting and the radius is decreasing at a constant rate, changing from 12 in. to 8 in. in 45 minutes. How fast was the volume changing when the radius was 10 in.?

13 The ends of a water trough 8 ft long are equilateral triangles whose sides are 2 ft long. If water is being pumped into the trough at a rate of 5 ft^3/min, find the rate at which the water level is rising when the depth is 8 in.

14 Work Exercise 13 if the ends of the trough have the shape of the graph of $y = 2|x|$ between the points $(-1, 2)$ and $(1, 2)$.

15 A point $P(x, y)$ moves on the graph of the equation $y = x^3 + x^2 + 1$, the abscissa changing at a rate of 2 units per second. How fast is the ordinate changing at the point $(1, 3)$?

16 A point $P(x, y)$ moves on the graph of $y^2 = x^2 - 9$ such that $dx/dt = 1/x$. Find dy/dt at the point $(5, 4)$.

17 The area of an equilateral triangle is decreasing at a rate of 4 cm^2/min. Find the rate at which the length of a side is changing when the area of the triangle is 200 cm^2.

18 Gas is escaping from a spherical balloon at a rate of 10 ft^3/hr. At what rate is the radius changing when the volume is 400 ft^3?

19 A stone is dropped into a lake, causing circular waves whose radii increase at a constant rate of 0.5 m/sec. At what rate is the circumference of a wave changing when its radius is 4 m?

20 A softball diamond has the shape of a square with sides 60 ft long. If a player is running from second base to third at a speed of 24 ft/sec, at what rate is her distance from home plate changing when she is 20 ft from third?

21 When two electrical resistances R_1 and R_2 are connected in parallel, the total resistance R is given by $1/R = (1/R_1) + (1/R_2)$. If R_1 and R_2 are increasing at rates of 0.01 ohms/sec and 0.02 ohms/sec, respectively, at what rate is R changing at the instant that $R_1 = 30$ ohms and $R_2 = 90$ ohms?

22 The formula for the adiabatic expansion of air is $pv^{1.4} = c$, where p denotes pressure, v denotes volume, and c is a constant. At a certain instant the pressure is 40 dynes/cm^2 and is increasing at a rate of 3 dynes/cm^2 per second. If, at that same instant, the volume is 60 cm^3, find the rate at which the volume is changing.

23 If a spherical tank of radius a contains water which has a maximum depth h, then the volume V of water in the tank is given by $V = \frac{1}{3}\pi h^2(3a - h)$. Suppose a spherical tank of radius 16 ft is being filled at a rate of 100 gal/min. Approximate the rate at which the water level is rising when $h = 4$ ft (1 gal ≈ 0.1337 ft^3).

24 A spherical tank is coated uniformly with a 2-inch layer of ice. If the volume of ice is melting at a rate which is directly proportional to its surface area, show that the outside diameter is decreasing at a constant rate.

25 From the edge of a cliff which overlooks a lake 200 ft below, a boy drops a stone and then, two seconds later, drops another stone from exactly the same position. Discuss the rate at which the distance between the two stones is changing during the next second. (Assume that the distance an object falls in t seconds is $16t^2$ ft.)

26 A metal rod has the shape of a right circular cylinder. As it is being heated, its length is increasing at a rate of 0.005 cm/min and its diameter is increasing at 0.002 cm/min. At what rate is the volume changing when the rod has length 40 cm and diameter 3 cm?

27 An airplane, flying at a constant speed of 360 mph and climbing at an angle of 45°, passes over a point P on the ground at an altitude of 10,560 ft. Find the rate at which its distance from P is changing one minute later.

28 Refer to Exercise 38 of Section 6. At what rate is the distance between the airplane and the automobile changing at 10:15 A.M.?

29 A paper cup containing water has the shape of a frustum of a right circular cone of altitude 6 in. and lower and upper base radii 1 in. and 2 in., respectively. If water is leaking out of the cup at a rate of 3 in.^{3}/hr, at what rate is the water level decreasing when its depth is 4 in.? (*Note:* The volume V of a frustum of a right circular cone of altitude h and base radii a and b is given by $V = \frac{1}{3}\pi h(a^2 + b^2 + ab)$.)

30 The top part of a swimming pool is a rectangle of length 60 ft and width 30 ft. The depth of the pool varies uniformly from 4 ft to 9 ft through a horizontal distance of 40 ft and then is level for the remaining 20 ft, as illustrated by the cross-sectional view in Figure 4.44. If the pool is being filled with water at a rate of 500 gal/min, approximate the rate at which the water level is rising when the depth at the deep end is 4 ft (1 gal ≈ 0.1337 ft^3).

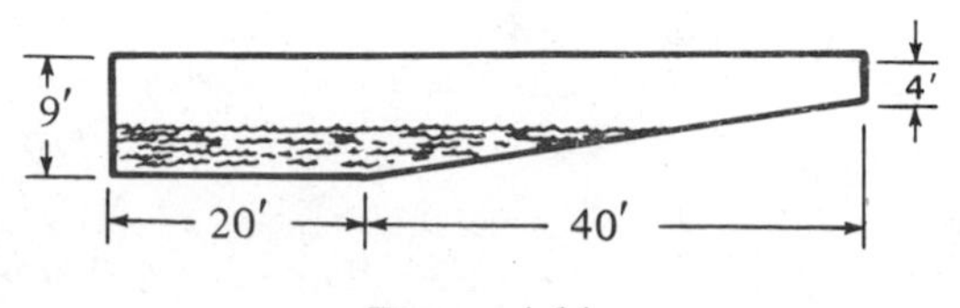

Figure 4.44

4.9 ANTIDERIVATIVES

In Chapter 3 certain problems were stated in the form "*Given a function g, find the derivative g′.*" We shall now consider the converse problem, namely, "*Given the derivative g′, find the function g.*" If g' is denoted by f and g by F, then an equivalent way of stating this converse problem is, "Given a function f, find a function F such that $F' = f$." As a simple illustration, suppose $f(x) = 8x^3$. In this case it is easy to find a function F which has f as its derivative. We know that differentiating a power of x *reduces* the exponent by 1 and therefore to obtain F we must *increase* the given exponent by 1. Thus $F(x) = ax^4$ for some number a. Differentiating, we obtain $F'(x) = 4ax^3$ and, in order for this to equal $f(x)$, a must equal 2. Consequently, the function F defined by $F(x) = 2x^4$ has the desired property. According to the next definition, F is called an *antiderivative* of f.

(4.32) **Definition**

A function F is an **antiderivative** of a function f if $F' = f$.

For convenience, we shall use the phrase "$F(x)$ is an antiderivative of $f(x)$" synonymously with "F is an antiderivative of f." The domain of an antiderivative will not usually be specified. It follows from the Fundamental Theorem of Calculus (5.31) in the next chapter that a suitable domain is any closed interval $[a, b]$ on which f is continuous.

Antiderivatives are never unique. Indeed, since the derivative of a constant is zero, it follows that if F is an antiderivative of f, so is the function G defined by $G(x) = F(x) + C$, for every number C. For example, if $f(x) = 8x^3$, then functions defined by expressions such as $2x^4 + 7$, $2x^4 - \sqrt{3}$, and $2x^4 + (2/5)$ are antiderivatives of f. The next theorem brings out the fact that functions of this type are the only possible antiderivatives of f.

(4.33) **Theorem**

If F_1 and F_2 are differentiable functions such that $F_1'(x) = F_2'(x)$ for all x in a closed interval $[a, b]$, then $F_2(x) = F_1(x) + C$ for some number C and all x in $[a, b]$.

Proof. If we define the function g by

$$g(x) = F_2(x) - F_1(x)$$

it follows that

$$g'(x) = F_2'(x) - F_1'(x) = 0$$

for all x in $[a, b]$. If x is any number such that $a < x \leq b$, then applying the Mean Value Theorem (4.12) to the function g and the closed interval $[a, x]$, there exists a number z in the open interval (a, x) such that

$$g(x) - g(a) = g'(z)(x - a) = 0 \cdot (x - a) = 0.$$

Hence $g(x) = g(a)$ for all x in $[a, b]$. Substitution in the first equation stated in the proof gives us

$$g(a) = F_2(x) - F_1(x).$$

Adding $F_1(x)$ to both sides we obtain the desired conclusion, with $C = g(a)$.

If F_1 and F_2 are antiderivatives of the same function f, then $F_1' = f = F_2'$ and hence, by the theorem just proved, $F_2(x) = F_1(x) + C$ for some C. In other words, if $F(x)$ is an antiderivative of $f(x)$, then every other antiderivative has the form $F(x) + C$ where C is an arbitrary constant (that is, an unspecified real number). We shall refer to $F(x) + C$ as the **most general antiderivative** of $f(x)$.

(4.34) Theorem

> If f is a function such that $f'(x) = 0$ for all x in $[a, b]$, then f is a constant function on $[a, b]$.

Proof. Denote f by F_2 and let the function F_1 be defined by $F_1(x) = 0$ for all x. Since $F_1'(x) = 0$ and $F_2'(x) = f'(x) = 0$, we see that $F_1'(x) = F_2'(x)$ for all x in $[a, b]$. Applying Theorem (4.33), there is a number C such that $F_2(x) = F_1(x) + C$; that is, $f(x) = C$ for all x in $[a, b]$. This completes the proof.

Rules for derivatives may be used to obtain formulas for antiderivatives, as in the proof of the following important result.

(4.35) Power Rule for Antidifferentiation

> Let a be any real number, r any rational number different from -1, and C an arbitrary constant.
>
> $$\text{If} \quad f(x) = ax^r, \quad \text{then} \quad F(x) = \left(\frac{a}{r+1}\right)x^{r+1} + C$$
>
> is the most general antiderivative of $f(x)$.

Proof. It is sufficient to show that $F'(x) = f(x)$. This fact follows readily from the Power Rule for Derivatives (3.39), since

$$F'(x) = \left(\frac{a}{r+1}\right)(r + 1)x^r = ax^r = f(x).$$

Example 1 Find the most general antiderivative of (a) $4x^5$; (b) $7/x^3$; (c) $\sqrt[3]{x^2}$.

Solution (a) Using the Power Rule (4.35) with $a = 4$ and $r = 5$ gives us the antiderivative

$$\tfrac{4}{6}x^6 + C, \quad \text{or} \quad \tfrac{2}{3}x^6 + C.$$

(b) Writing $7/x^3$ as $7x^{-3}$ and using Rule (4.35) with $a = 7$ and $r = -3$ leads to

$$\frac{7}{-2}x^{-2} + C, \quad \text{or} \quad -\frac{7}{2x^2} + C.$$

(c) Since $\sqrt[3]{x^2} = x^{2/3}$ we may apply the Power Rule (4.35), with $a = 1$ and $r = 2/3$, obtaining

$$\frac{1}{(5/3)}x^{5/3} + C, \quad \text{or} \quad \frac{3}{5}x^{5/3} + C.$$

In order to avoid algebraic errors it is a good idea to check the preceding solutions by differentiating the antiderivatives. In each case the given expression should be obtained.

Recall from Chapter 3 that the derivative of a sum of functions is the sum of the derivatives. A similar situation exists for antiderivatives. Thus if F_1 and F_2 are antiderivatives of f_1 and f_2 respectively, then

$$D_x(F_1 + F_2) = D_xF_1 + D_xF_2 = f_1 + f_2 \tag{4.36}$$

that is, $F_1 + F_2$ is an antiderivative of $f_1 + f_2$. This result can be extended to any finite sum of functions. The previous fact may be stated as follows: "*An antiderivative of a sum is the sum of the antiderivatives.*" As usual, when working with several functions we assume that the domain is restricted to the intersection of the domains of the individual functions. A similar result is true for differences.

Example 2 Find the most general antiderivative $F(x)$ of $f(x) = 3x^4 - x + 4 + (5/x^3)$.

Solution Writing the last term of $f(x)$ as $5x^{-3}$ and applying the Power Rule for Antidifferentiation (4.35) to each term gives us

$$F(x) = \frac{3}{5}x^5 - \frac{1}{2}x^2 + 4x - \frac{5}{2}x^{-2} + C.$$

It is unnecessary to introduce an arbitrary constant for each of the four antiderivatives, since they could be added together to produce the one constant C.

Equations which involve derivatives of an unknown function f are very common in mathematical applications. Such equations are called **differential equations.** The function f is called a **solution** of the differential equation. To **solve** a differential equation means to find all solutions. Sometimes, in addition to the differential equation we may know certain values of f, called **boundary values**, as illustrated in the next example.

Example 3 Solve the differential equation $f'(x) = 6x^2 + x - 5$ with boundary value $f(0) = 2$.

Solution From our discussion of antiderivatives.

$$f(x) = 2x^3 + \frac{1}{2}x^2 - 5x + C$$

for some number C. Letting $x = 0$ and using the given boundary value, we obtain

$$f(0) = 0 + 0 - 0 + C = 2$$

and hence $C = 2$. Consequently, the solution f of the differential equation with the given boundary value is

$$f(x) = 2x^3 + \frac{1}{2}x^2 - 5x + 2.$$

If a point P is moving rectilinearly, then its position function s is an antiderivative of its velocity function, that is, $s'(t) = v(t)$. Similarly, since $v'(t) = a(t)$, the velocity function is an antiderivative of the acceleration function. If the velocity or acceleration function is known, then given sufficient boundary conditions it is possible to determine the position function. The particular boundary conditions corresponding to $t = 0$ are sometimes called the **initial conditions**. The next example illustrates these remarks.

Example 4 A point moves rectilinearly such that $a(t) = 12t - 4$. If the initial conditions are $v(0) = 8$ and $s(0) = 15$, find $s(t)$.

Solution From $v'(t) = 12t - 4$ we obtain, by antidifferentiation,

$$v(t) = 6t^2 - 4t + C$$

for some number C. Substitution of 0 for t and use of the fact that $v(0) = 8$ gives us $8 = 0 - 0 + C = C$. Thus

$$v(t) = 6t^2 - 4t + 8$$

or, equivalently,

$$s'(t) = 6t^2 - 4t + 8.$$

The most general antiderivative of $s'(t)$ is

$$s(t) = 2t^3 - 2t^2 + 8t + D$$

where D is some number. Substitution of 0 for t and use of the fact that $s(0) = 15$ leads to $15 = 0 - 0 + 0 + D = D$ and consequently the desired position function is given by

$$s(t) = 2t^3 - 2t^2 + 8t + 15.$$

An object on or near the surface of the Earth is acted upon by a force called **gravity**, which produces a constant acceleration denoted by g. The approximation to g which is employed for most problems is 32 ft/sec^2 or 980 cm/sec^2. The use of this important physical constant is illustrated in the following example.

Example 5 A stone is thrown vertically upward from a position 144 feet above the ground with a velocity of 96 ft/sec. If air resistance is neglected, find its distance above the ground after t seconds. For what length of time does the stone rise? When, and with what velocity, does it strike the ground?

Solution The motion of the stone may be represented by a point moving rectilinearly on a vertical line with origin at ground level and positive direction upward. The distance above the ground at time t is $s(t)$ and the initial conditions are $s(0) = 144$ and $v(0) = 96$. Since the velocity is decreasing, $v'(t) < 0$; that is, the acceleration is negative. Hence, by the remarks preceding this example,

$$a(t) = -32.$$

Since v is an antiderivative of a,

$$v(t) = -32t + C,$$

for some number C. Substituting 0 for t and using the fact that $v(0) = 96$ gives us $96 = 0 + C = C$ and consequently

$$v(t) = -32t + 96.$$

Since $s'(t) = v(t)$ we obtain, by antidifferentiation,

$$s(t) = -16t^2 + 96t + D$$

for some number D. Letting $t = 0$ and using $s(0) = 144$ leads to $144 = 0 + 0 + D = D$. It follows that the distance from the ground to the stone at time t is given by

$$s(t) = -16t^2 + 96t + 144.$$

The stone will rise until $v(t) = 0$, that is until $-32t + 96 = 0$, or $t = 3$. The stone will strike the ground when $s(t) = 0$, that is when $-16t^2 + 96t + 144 = 0$ or, equivalently, $t^2 - 6t - 9 = 0$. Applying the quadratic formula, $t = 3 \pm 3\sqrt{2}$. The solution $3 - 3\sqrt{2}$ is extraneous (Why?) and hence the stone strikes the ground after $3 + 3\sqrt{2}$ seconds. The velocity at that time is

$$\begin{aligned} v(3 + 3\sqrt{2}) &= -32(3 + 3\sqrt{2}) + 96 \\ &= -96\sqrt{2} \approx -135.8 \text{ ft/sec.} \end{aligned}$$

EXERCISES 4.9

Find the most general antiderivatives of the functions defined in Exercises 1–20.

1 $f(x) = 9x^2 - 4x + 3$

2 $f(x) = 4x^2 - 8x + 1$

3 $f(x) = 2x^3 - x^2 + 3x - 7$

4 $f(x) = 10x^4 - 6x^3 + 5$

5 $f(x) = \dfrac{1}{x^3} - \dfrac{3}{x^2}$

6 $f(x) = \dfrac{4}{x^7} - \dfrac{7}{x^4} + x$

7 $f(x) = 3\sqrt{x} + (1/\sqrt{x})$

8 $f(x) = \sqrt{x^3} - \frac{1}{2}x^{-2} + 5$

9 $f(x) = (6/\sqrt[3]{x}) - (\sqrt[3]{x}/6) + 7$

10 $f(x) = 3x^5 - \sqrt[3]{x^5}$

11 $f(x) = 2x^{5/4} + 6x^{1/4} + 3x^{-4}$

12 $f(x) = \left(x - \dfrac{1}{x}\right)^2$

13 $f(x) = (3x - 1)^2$

14 $f(x) = (2x - 5)(3x + 1)$

15 $f(x) = \dfrac{8x - 5}{\sqrt[3]{x}}$

16 $f(x) = \dfrac{2x^2 - x + 3}{\sqrt{x}}$

17 $f(x) = \sqrt[5]{32x^4}$

18 $f(x) = \sqrt[3]{64x^5}$

19 $f(x) = \dfrac{x^3 - 1}{x - 1}$

20 $f(x) = \dfrac{x^3 + 3x^2 - 9x - 2}{x - 2}$

Solve the differential equations in Exercises 21–24 subject to the given boundary conditions.

21 $f'(x) = 12x^2 - 6x + 1,\ f(1) = 5$

22 $f'(x) = 9x^2 + x - 8,\ f(-1) = 1$

23 $f''(x) = 4x - 1,\ f'(2) = -2,\ f(1) = 3$

24 $f'''(x) = 6x,\ f''(0) = 2,\ f'(0) = -1,\ f(0) = 4$

25 A point moves rectilinearly such that $a(t) = 2 - 6t$. If the initial conditions are $v(0) = -5$ and $s(0) = 4$, find $s(t)$.

26 Work Exercise 25 if $a(t) = 3t^2$, $v(0) = 20$, and $s(0) = 5$.

27 A projectile is fired vertically upward from ground level with a velocity of 1600 ft/sec. If air resistance is neglected, find its distance $s(t)$ above ground at time t. What is its maximum height?

28 An object is dropped from a height of 1000 feet. Neglecting air resistance, find the distance it falls in t seconds. What is its velocity at the end of 3 seconds? When will it strike the ground?

29 A stone is thrown directly downward from a height of 96 ft with a velocity of 16 ft/sec. Find (a) its distance above the ground after t seconds, (b) when it will strike the ground, and (c) the velocity at which it strikes the ground.

30 Work Exercise 29 if the stone is thrown directly upward with a velocity of 16 ft/sec.

31 If a projectile is fired vertically upward from a height of s_0 feet above the ground with a velocity of v_0 ft/sec, prove that if air resistance is neglected, its distance $s(t)$ above the ground after t seconds is given by $s(t) = -\frac{1}{2}gt^2 + v_0t + s_0$, where g is the gravitational constant.

32 A ball rolls down an inclined plane with an acceleration of 2 ft/sec^2. If the ball is given no initial velocity how far will it roll in t seconds? What initial velocity must be given in order for the ball to roll 100 feet in 5 seconds?

33 If an automobile starts from rest, what constant acceleration will enable it to travel 500 feet in 10 seconds?

34 If a car is traveling at a speed of 60 miles per hour, what constant (negative) acceleration will enable it to stop in 9 seconds?

35 If C and F denote Celsius and Fahrenheit temperature readings, then the rate of change of F with respect to C is given by $dF/dC = 9/5$. If $F = 32$ when $C = 0$, use antidifferentiation to obtain a general formula for F in terms of C.

36 The rate of change of the temperature T of a solution is given by $dT/dt = (1/4)t + 10$, where t is the time in minutes and T is measured in degrees Celsius. If the temperature is 5°C at $t = 0$, find a formula which gives T at time t.

37 The volume V of a balloon is changing with respect to time t at a rate given by $dV/dt = 3\sqrt{t} + (t/4)$ ft^3/sec. If, at $t = 4$, the volume is 20 ft^3, express V as a function of t.

38 Suppose the slope of the tangent line at any point P on the graph of an equation equals the square of the abscissa of P. Find the equation if the graph contains (a) the origin; (b) the point $(3, 6)$; (c) the point $(-1, 1)$. Sketch the graph in each case.

39 Suppose F and G are antiderivatives of f and g, respectively. Prove or disprove each of the following.

(a) FG is an antiderivative of fg.

(b) F/G is an antiderivative of f/g.

40 Suppose F is an antiderivative of f. Prove that

(a) if F is an even function, then f is odd.

(b) if F is an odd function, then f is even.

4.10 APPLICATIONS TO ECONOMICS

Calculus has become an important tool for problems which arise in economics. As will be seen in this section, if a function f is used to describe some economic entity, then the adjective *marginal* is employed to specify the derivative f'.

If the cost of producing x units of a certain commodity is denoted by $C(x)$, then C is called a **cost function**. The **average cost** $c(x)$ of one unit is defined by $c(x) = C(x)/x$. In order to use the techniques of calculus, x is regarded as a real number, even though this variable may take on only integer values. We always assume that $x \geq 0$, since the production of a negative number of units has no practical significance. The derivative C' of the cost function is called the **marginal cost function**. If we interpret the derivative as a rate of change (see Section 7), then $C'(x)$ is the rate at which the cost changes with respect to the number of units produced. The number $C'(x)$ is referred to as the **marginal cost** associated with the production of x units. Evidently $C'(x) > 0$, since the cost should increase as more units are produced.

If C is a cost function and n is a positive integer, then by Theorem (3.7)

$$C'(n) = \lim_{h \to 0} \frac{C(n+h) - C(n)}{h}.$$

Hence, if h is small, then

$$C'(n) \approx \frac{C(n+h) - C(n)}{h}.$$

If the number n of units produced is large, economists often let $h = 1$ in the last formula to approximate the marginal cost, obtaining

$$C'(n) \approx C(n+1) - C(n).$$

In this context, the marginal cost associated with the production of n units is (approximately) the cost of producing one more unit.

Some companies find that the cost $C(x)$ of producing x units of a certain commodity is given by a formula such as

$$C(x) = a + bx + dx^2 + kx^3.$$

The constant a represents a fixed overhead charge for items such as rent, heat, light, etc., which are independent of the number of units produced. If the cost of producing one unit were b dollars and no other factors were involved, then the second term bx in the formula would represent the cost of producing x units. If x becomes very large, then the terms dx^2 and kx^3 may significantly affect production costs.

Example 1 A company estimates that the cost (in dollars) of producing x units of a certain commodity is given by

$$C(x) = 200 + 0.05x + 0.0001x^2.$$

Find the cost, the average cost, and the marginal cost of producing (a) 500 units; (b) 1,000 units; (c) 5,000 units.

Solution The average cost of producing x units is given by

$$c(x) = \frac{C(x)}{x} = \frac{200}{x} + 0.05 + 0.0001x.$$

The marginal cost is

$$C'(x) = 0.05 + 0.0002x.$$

We leave it to the reader to verify the entries in the following table, where numbers in the last three columns represent dollars, rounded off to the nearest cent.

x	$C(x)$	$c(x) = \dfrac{C(x)}{x}$	$C'(x)$
500	250.00	0.50	0.15
1,000	350.00	0.35	0.25
5,000	2950.00	0.59	1.05

The derivative $c'(x)$ of the average cost $c(x)$ is called the **marginal average cost**. Applying the quotient rule to $c(x) = C(x)/x$ we obtain

$$c'(x) = \frac{xC'(x) - C(x)}{x^2}$$

and consequently the average cost will have an extremum if

$$xC'(x) - C(x) = 0$$

that is, if

$$C'(x) = \frac{C(x)}{x} = c(x).$$

Thus, a minimum average cost can occur only when the marginal and average costs are equal.

Example 2 In Example 1, find (a) the number of units which will minimize the average cost, and (b) the minimum average cost.

Solution (a) From the preceding discussion we must have $C'(x) = c(x)$, that is,

$$0.05 + 0.0002x = \frac{200}{x} + 0.05 + 0.0001x.$$

This equation reduces to

$$0.0001x = \frac{200}{x}$$

or

$$x^2 = \frac{200}{0.0001} = 2{,}000{,}000.$$

Consequently

$$x = \sqrt{2{,}000{,}000} \approx 1414.$$

The First or Second Derivative Test can be used to verify that the average cost is minimal for this number of units.

(b) Using the result obtained in part (a), the minimum average cost is

$$c(1414) = \frac{200}{1414} + 0.05 + 0.0001(1414) \approx 0.33.$$

Example 3 A company determines that the cost $C(x)$ of manufacturing x units of a commodity may be approximated by

$$C(x) = 100 + \frac{10}{x} + \frac{x^2}{200}.$$

How many units should be produced in order to minimize the cost?

Solution Since the marginal cost is

$$C'(x) = -\frac{10}{x^2} + \frac{x}{100}$$

the cost function will have an extremum if

$$-\frac{10}{x^2} + \frac{x}{100} = 0$$

or equivalently

$$\frac{-1000 + x^3}{100x^2} = 0.$$

Solving for x gives us $x = 10$. It is left to the reader to show that C has an absolute minimum at $x = 10$.

A company must consider many factors in order to determine a selling price. In addition to the cost of production and the profit desired, the seller should be aware of the manner in which consumer demand will vary if the price increases. For some commodities there is a constant demand, and changes in price have little effect on sales. For items which are not necessities of life, a price increase will probably lead to a decrease in the number of units sold. Suppose a company knows from past experience that it can sell x units when the price per unit is given by $p(x)$, where p is some type of function. We sometimes say that $p(x)$ is the price per unit when there is a **demand** for x units, and we refer to p as the **demand function** for the commodity. The total income, or total revenue, is the number of units sold times the price per unit, that is, $x \cdot p(x)$. For this reason the function R, defined by

$$R(x) = xp(x)$$

is called the **total revenue function**. The derivatives p' and R' are called the **marginal demand** and **marginal revenue** functions, respectively. They are used to find the rates of change of the demand and total revenue functions with respect to the number of units sold.

If we let $S = p(x)$, then S is the selling price per unit associated with a demand of x units. Since a decrease in S would ordinarily be associated with an increase in x, a demand function p is usually decreasing, that is $p'(x) < 0$ for all x. Demand functions are sometimes defined implicitly by an equation involving S and x, as in the next example.

Example 4 The demand for x units of a certain commodity is related to a selling price of S dollars per unit by means of the equation $2x + S^2 - 12000 = 0$. Find the demand function, the marginal demand function, the total revenue function, and the marginal revenue function. Find the number of units and the price per unit which yield the maximum revenue. What is the maximum revenue?

Solution Since $S^2 = 12000 - 2x$ and S is positive, we see that the demand function p is given by

$$S = p(x) = \sqrt{12000 - 2x}.$$

The domain of p consists of all x such that $12000 - 2x > 0$, or equivalently, $2x < 12000$. Thus $0 \leq x < 6000$. The graph of p is sketched in Figure 4.45. In theory, there are no sales if the selling price is $\sqrt{12000}$, or approximately \$109.54, and when the selling price is close to \$0 the demand is close to 6000.

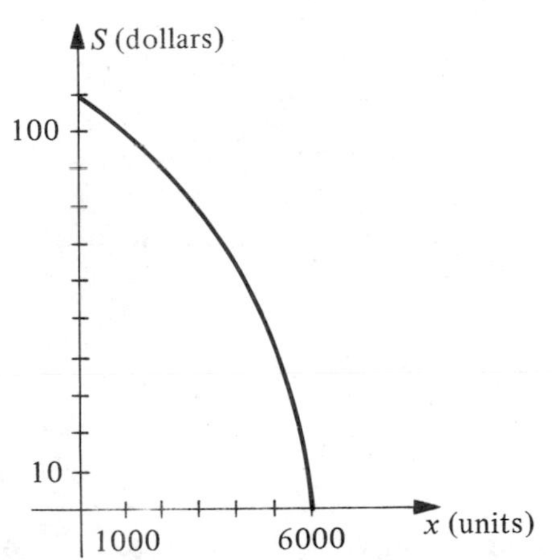

Figure 4.45. Graph of demand function

The marginal demand function p' is given by

$$p'(x) = \frac{-1}{\sqrt{12000 - 2x}}.$$

The negative sign indicates that a decrease in price is associated with an increase in demand.

The total revenue function R is given by

$$R(x) = xp(x) = x\sqrt{12000 - 2x}.$$

Differentiating and simplifying gives us the marginal revenue function R', where

$$R'(x) = \frac{12000 - 3x}{\sqrt{12000 - 2x}}.$$

Thus, $x = 12000/3 = 4000$ is a critical number for the total revenue function R. Since $R'(x)$ is positive if $0 \leq x < 4000$ and negative if $4000 < x < 6000$, there is a maximum total revenue when 4000 units are produced. This corresponds to a selling price of

$$p(4000) = \sqrt{12000 - 2(4000)} \approx \$63.25.$$

The maximum revenue, obtained from selling 4000 units at this price, is

$$4000(63.25) = \$253,000.$$

If x units of a commodity are sold at a price of $p(x)$ per unit, then the **profit** $P(x)$ is given by

$$P(x) = R(x) - C(x)$$

where R and C are the total revenue and cost functions, respectively. We call P the **profit function** and P' the **marginal profit function**. The critical numbers of P are solutions of the equation

$$P'(x) = R'(x) - C'(x) = 0.$$

This shows that there may be a maximum (or minimum) profit when $R'(x) = C'(x)$, or equivalently when $R'(x) = c(x)$. Thus if P has an extremum then the marginal revenue, the marginal cost, and the average cost are all equal.

Example 5 A company estimates that the weekly cost (in dollars) of manufacturing x units of a certain commodity is given by

$$C(x) = x^3 - 3x^2 - 80x + 500.$$

Each unit produced is sold for \$2,800. What weekly production rate will maximize the profit? What is the largest possible profit per week?

Solution Since the income obtained from selling x units is $2800x$, the total revenue function R is given by $R(x) = 2800x$. From the discussion preceding this example, the maximum profit occurs if $C'(x) = R'(x)$, that is, if

$$3x^2 - 6x - 80 = 2800.$$

This equation reduces to

$$x^2 - 2x - 960 = 0, \quad \text{or} \quad (x - 32)(x + 30) = 0$$

and hence either $x = 32$ or $x = -30$. Since the negative solution is extraneous, it suffices to check $x = 32$.

The second derivative of the profit function $P = R - C$ is given by

$$P''(x) = R''(x) - C''(x) = 0 - (6x - 6).$$

Consequently,

$$P''(32) = -6(32) + 6 < 0$$

which means that a maximum profit occurs if 32 units per week are manufactured. The maximum weekly profit is

$$\begin{aligned} P(32) &= R(32) - C(32) \\ &= 2800(32) - [(32)^3 - 3(32)^2 - 80(32) + 500] \\ &= \$61{,}964. \end{aligned}$$

As a final remark on economic applications, if a marginal function is known, then it is sometimes possible to use antidifferentiation to find the function, as illustrated in the next example.

Example 6 A company finds that the marginal cost (in dollars) associated with the production of x units of a certain commodity is given by $30 - 0.02x$. If the cost of producing one unit is \$35, find the cost function and the cost of producing 100 units.

Solution If C is the cost function, then the marginal cost is the rate of change of C with respect to x, that is

$$C'(x) = 30 - 0.02x.$$

Antidifferentiation gives us

$$C(x) = 30x - 0.01x^2 + K$$

for some real number K. Letting $x = 1$ and using $C(1) = 35$ we obtain

$$35 = 30 - 0.01 + K$$

and hence $K = 5.01$. Consequently

$$C(x) = 30x - 0.01x^2 + 5.01.$$

In particular, the cost of producing 100 units is

$$C(100) = 3000 - 100 + 5.01 = \$2{,}905.01.$$

EXERCISES 4.10

In each of Exercises 1–4, C is the cost function for a certain commodity.
(a) Find the cost of producing 100 units.
(b) Find the average and marginal cost functions, and their values at $x = 100$.
(c) Find the minimum average cost.
(d) If possible, find the value of x that will minimize the cost.

1 $C(x) = 800 + 0.04x + 0.0002x^2$

2 $C(x) = 6400 + 6.5x + 0.003x^2$

3 $C(x) = 250 + 100x + 0.001x^3$

4 $C(x) = 200 + (100/\sqrt{x}) + (\sqrt{x}/1000)$

5 A manufacturer of small motors estimates that the cost (in dollars) of producing x motors per day is given by $C(x) = 25 + 2x + 256/\sqrt{x}$. What daily production will minimize the average cost?

6 A company finds that the cost of producing x liters of a certain chemical is given by $C(x) = 3 + x + (10/x)$. Compare the marginal cost of producing ten liters with the cost of producing the eleventh liter.

In each of Exercises 7–10 the given equation relates the demand for x units of a certain commodity to a selling price of S dollars per unit. Find the demand function, the marginal demand function, the total revenue function, and the marginal revenue function. If possible, find the number of units and the price per unit which will yield the maximum revenue.

7 $3S + 4x - 800 = 0$

8 $x^2 + S + xS - 400 = 0$

9 $S^2 - 8S - x + 16 = 0$

10 $Sx^2 - 1000x + 144S - 5000 = 0$

In Exercises 11–14, R denotes the total revenue function when there is a demand for x units of a certain commodity. Find (a) the marginal revenue function, (b) the maximum total revenue, and (c) the demand function.

11 $R(x) = 2x\sqrt{400 - x}$

12 $R(x) = 70x - x^2$

13 $R(x) = x(300 - x^2)$

14 $R(x) = 300x - 2x^{3/2}$

In Exercises 15 and 16 the demand and cost functions of a commodity are denoted by p and C, respectively. Find (a) the marginal demand function, (b) the total revenue function, (c) the profit function, (d) the marginal profit function, (e) the maximum profit, and (f) the marginal cost when the demand is 10 units.

15 $p(x) = 50 - (x/10)$; $C(x) = 10 + 2x$

16 $p(x) = 80 - \sqrt{x - 1}$; $C(x) = 75x + 2\sqrt{x - 1}$

17 A company estimates that in order to sell x units of a certain commodity the selling price per unit should be $1800 - 2x$ dollars, where $1 \leq x \leq 100$. If the manufacturing cost for x units is $1000 + x + 0.01x^2$ dollars, find (a) the total revenue function, (b) the profit function, (c) the number of units that will maximize the profit, and (d) the maximum profit.

18 A manufacturer determines that x units of a product will be sold if the selling price is $400 - 0.05x$ dollars for each unit. If the production cost for x units is $500 + 10x$, find (a) the total revenue function, (b) the profit function, (c) the number of units that maximize the profit, and (d) the price per unit when the marginal revenue is 300.

19 If the cost of manufacturing x units per day of a certain commodity is $500 + 0.02x + 0.001x^2$ dollars, and if each unit is sold for \$8.00, what rate of production will maximize the profit? What is the maximum daily profit?

20 A company that conducts bus tours found that when the price was \$9.00 per person, it averaged 1000 customers per week. When it reduced the price to \$7.00 per person, the average number of customers increased to 1500 per week. Assuming that the demand function is linear, what price should be charged to obtain the greatest weekly revenue?

21 When a company sold a certain commodity at \$50 per unit, there was a demand for 1000 units per week. After the price rose to \$70, the demand dropped to 800 units per week. Assuming that the demand function p is linear, find p and the total revenue function.

22 If a demand function p is defined by $p(x) = ax^2 + b$ where $a < 0$ and $b > 0$, what value of x will maximize the total revenue?

23 Prove that if the demand function of a commodity is linear and decreasing for all $x > 0$, then so is the marginal revenue function.

24 Show that the marginal revenue of a commodity can be found by adding to the selling price the product of the number of units sold and the marginal demand.

25 If the marginal cost of producing x units of a certain commodity is given in dollars by $20 - 0.015x$, and if the cost of producing one unit is \$25, find the cost function and the cost of producing 50 units.

26 If the marginal cost function is given by $2/(x + 6)^{1/3}$, and if the cost of producing two units is \$20, find the cost function and the cost of producing 120 units.

27 If, in a certain company, the marginal revenue is given by $x^2 - 6x + 15$, find the total revenue function, and the marginal demand function.

28 Work Exercise 27 if the marginal revenue is given by $4/(x + 2)^{3/2}$.

4.11 REVIEW

Concepts

Define or discuss each of the following.

1. Local maximum or local minimum of a function
2. Absolute maximum or absolute minimum of a function
3. Critical numbers of a function
4. Rolle's Theorem
5. The Mean Value Theorem
6. The First Derivative Test
7. End-point extrema of a function
8. Upward or downward concavity
9. Tests for concavity
10. Point of inflection
11. The Second Derivative Test

12 $\lim_{x\to\infty} f(x) = L$; $\lim_{x\to-\infty} f(x) = L$

13 $\lim_{x\to a} f(x) = \infty$; $\lim_{x\to a} f(x) = -\infty$

14 Vertical and horizontal asymptotes

15 The derivative as a rate of change

16 Velocity and acceleration in rectilinear motion

17 Related rates

18 Antiderivative of a function

19 The power rule for antidifferentiation

20 Differential equation

Exercises

1 Find equations of the tangent and normal lines to the graph of the equation $6x^2 - 2xy + y^3 = 9$ at the point $P(2, -3)$.

2 Find equations of the tangent lines to the graph of the equation $x^2 + y^2 = 1$ which contain the point $P(2, 1)$.

In Exercises 3 and 4, find the abscissas of points on the graph of the given equation where the tangent line is either horizontal or vertical.

3 $y = (x^2 + 3x + 2)^{1/3}$

4 $x^2 - 2xy + y^2 - 4 = 0$

5 If $f(x) = x^3 + x^2 + x + 1$, find a number c which satisfies the conclusion of the Mean Value Theorem (4.12) on the interval $[0, 4]$.

In Exercises 6–8 find the local extrema of f by means of the First Derivative Test. Describe the intervals in which f is increasing or decreasing and sketch the graph of f.

6 $f(x) = -x^3 + 4x^2 - 3x$

7 $f(x) = 1/(1 + x^2)$

8 $f(x) = (4 - x^2)x^{1/3}$

In Exercises 9–11 find the local extrema of f by means of the Second Derivative Test. Discuss concavity, find abscissas of points of inflection, and sketch the graph of f.

9 $f(x) = -x^3 + 4x^2 - 3x$

10 $f(x) = 1/(1 + x^2)$

11 $f(x) = 40x^3 - x^6$

12 The position function of a point moving rectilinearly is given by $s(t) = (t^2 + 3t + 1)/(t^2 + 1)$. Find the velocity and acceleration at time t and describe the motion of the point during the time interval $[-2, 2]$.

13 A stone is thrown directly downward from a height of 900 feet with a velocity of 30 ft/sec. Find its distance above ground after t seconds. What is its velocity after 5 sec? When will it strike the ground?

In Exercises 14–18 find the most general antiderivative of f.

14 $f(x) = \dfrac{8x^2 - 4x + 5}{x^4}$

15 $f(x) = 3x^5 + 2x^3 - x$

16 $f(x) = 100$

17 $f(x) = x^{3/5}(2x - \sqrt{x})$

18 $f(x) = (2x + 1)^3$

19 Solve the differential equation $f''(x) = x^{1/3} - 5$ if $f'(1) = 2$ and $f(1) = -8$.

20 A man wishes to put a fence around a rectangular field and then subdivide this field into three smaller rectangular plots by placing two fences parallel to one of the sides. If he can afford only 1000 yds of fencing, what dimensions will give him the maximum area?

21 Find the altitude of the right circular cylinder of maximum curved surface area that can be inscribed in a sphere of radius a.

22 A wire 5 ft long is to be cut into two pieces. One of the pieces is to be bent into the shape of a circle and the other into the shape of a square. Where should the wire be cut in order that the sum of the areas of the circle and square is (a) a maximum; (b) a minimum?

23 The interior of a half-mile race track consists of a rectangle with semicircles at two opposite ends. Find the dimensions which will maximize the area of the rectangle.

24 A rectangle has its vertices on the x-axis, the y-axis, the origin, and the graph of $y = 4 - x^2$. Of all such rectangles, find the dimensions of the one with maximum area.

25 A water tank has the shape of a right circular cone of altitude 12 ft and base radius 4 ft, with vertex at the bottom of the tank. If water is being taken out of the tank at a rate of $10\ \text{ft}^3/\text{min}$, how fast is the water level falling when the depth is 5 ft?

26 The ends of a horizontal trough 10 ft long are isosceles trapezoids with lower base 3 ft, upper base 5 ft, and altitude 2 ft. If the water level is rising at a rate of 1/4 in./min when the depth is 1 ft, how fast is water entering the trough?

27 Two cars are approaching the same intersection along roads which run at right angles to each other. Car A is traveling at 20 mi/hr and car B is traveling at 40 mi/hr. If, at a certain instant, A is 1/4 mile from the intersection and B is 1/2 mile from the intersection, find the rate at which they are approaching one another at that instant.

28 A point $P(x, y)$ moves on the graph of $y^2 = 2x^3$ such that $dy/dt = x$, where t is time. Find dx/dt at the point $(2, 4)$.

29 Boyle's Law states that $pv = c$, where p is pressure, v is volume, and c is a constant. Find a formula for the rate of change of p with respect to v.

30 A railroad bridge is 20 ft above, and at right angles to, a river. A man in a train traveling 60 mi/hr passes over the center of the bridge at the same instant that a man in a motor boat traveling 20 mi/hr passes under the center of the bridge. How fast are the two men separating 10 sec later?

Find each of the limits in Exercises 31–38, if it exists.

31 $\displaystyle\lim_{x \to -\infty} \frac{(2x - 5)(3x + 1)}{(x + 7)(4x - 9)}$

32 $\displaystyle\lim_{x \to \infty} \frac{2x + 11}{\sqrt{x + 1}}$

33 $\displaystyle\lim_{x \to -\infty} \frac{6 - 7x}{(3 + 2x)^4}$

34 $\displaystyle\lim_{x \to -3} \sqrt[3]{\frac{x + 3}{x^3 + 27}}$

35 $\displaystyle\lim_{x \to 2/3^+} \frac{x^2}{4 - 9x^2}$

36 $\displaystyle\lim_{x \to 3/5^-} \frac{1}{5x - 3}$

37 $\displaystyle\lim_{x \to 0^+} \left(\sqrt{x} - \frac{1}{\sqrt{x}}\right)$

38 $\displaystyle\lim_{x \to 1} (x - 1)/\sqrt{(x - 1)^2}$

Find horizontal and vertical asymptotes and sketch the graph of f in Exercises 39 and 40.

39 $f(x) = \dfrac{3x^2}{9x^2 - 25}$

40 $f(x) = \dfrac{x^2}{(x - 1)^2}$

The Definite Integral

Calculus consists of two main parts, differential calculus *and* integral calculus. *Differential calculus is based upon the derivative. In this chapter we define the concept which is the basis for integral calculus: the* definite integral. *One of the most important results we shall discuss is the* Fundamental Theorem of Calculus. *This theorem demonstrates that differential and integral calculus are very closely related.*

5.1 AREA

In our development of the definite integral we shall employ sums of many numbers. To express such sums compactly, it is convenient to use **summation notation**. To illustrate, given a collection of numbers $\{a_1, a_2, \ldots, a_n\}$, the symbol $\Sigma_{i=1}^{n} a_i$ represents their sum, that is,

$$\sum_{i=1}^{n} a_i = a_1 + a_2 + a_3 + \cdots + a_n.$$

The Greek capital letter Σ (sigma) indicates a sum and the symbol a_i represents the ith term. The letter i is called the **index of summation** or the **summation variable**, and the numbers 1 and n indicate the extreme values of the summation variable.

Example 1 Find $\sum_{i=1}^{4} i^2(i-3)$.

Solution In this case, $a_i = i^2(i-3)$. To find the indicated sum we merely substitute, in succession, the integers 1, 2, 3, and 4 for i and add the resulting terms. Thus,

$$\begin{aligned}\sum_{i=1}^{4} i^2(i-3) &= 1^2(1-3) + 2^2(2-3) + 3^2(3-3) + 4^2(4-3)\\ &= (-2) + (-4) + 0 + 16 = 10.\end{aligned}$$

The letter used for the summation variable is arbitrary. The sum in Example 1 can be written as

$$\sum_{k=1}^{4} k^2(k-3), \quad \sum_{j=1}^{4} j^2(j-3)$$

or in many other ways.

If a_i is the same for each i, say $a_i = c$, then for example,

$$\sum_{i=1}^{3} a_i = a_1 + a_2 + a_3 = c + c + c = 3c.$$

In general,

(5.1)
$$\sum_{i=1}^{n} c = nc$$

for every real number c.

The domain of the summation variable does not have to begin at 1. For example, the following is self-explanatory:

$$\sum_{i=4}^{8} a_i = a_4 + a_5 + a_6 + a_7 + a_8.$$

Example 2 Find $\sum_{i=0}^{3} \frac{2^i}{(i+1)}$.

Solution

$$\sum_{i=0}^{3} \frac{2^i}{(i+1)} = \frac{2^0}{(0+1)} + \frac{2^1}{(1+1)} + \frac{2^2}{(2+1)} + \frac{2^3}{(3+1)}$$
$$= 1 + 1 + \frac{4}{3} + 2 = \frac{16}{3}.$$

The next theorem is important for calculations with sums.

(5.2) **Theorem**

If n is any positive integer and $\{a_1, a_2, \ldots, a_n\}$ and $\{b_1, b_2, \ldots, b_n\}$ are sets of numbers, then

(i) $\sum_{i=1}^{n} (a_i + b_i) = \sum_{i=1}^{n} a_i + \sum_{i=1}^{n} b_i$;

(ii) $\sum_{i=1}^{n} ca_i = c\left(\sum_{i=1}^{n} a_i\right)$, for any number c;

(iii) $\sum_{i=1}^{n} (a_i - b_i) = \sum_{i=1}^{n} a_i - \sum_{i=1}^{n} b_i$.

Proof. To prove formula (i) we begin by writing

$$\sum_{i=1}^{n} (a_i + b_i) = (a_1 + b_1) + (a_2 + b_2) + (a_3 + b_3) + \cdots + (a_n + b_n).$$

The terms on the right may be rearranged to produce

$$\sum_{i=1}^{n} (a_i + b_i) = (a_1 + a_2 + a_3 + \cdots + a_n) + (b_1 + b_2 + b_3 + \cdots + b_n).$$

Expressing the right side in summation notation gives us formula (i).

For formula (ii) we have

$$\sum_{i=1}^{n} (ca_i) = ca_1 + ca_2 + ca_3 + \cdots + ca_n$$

$$= c(a_1 + a_2 + a_3 + \cdots + a_n) = c\left(\sum_{i=1}^{n} a_i\right).$$

The proof of (iii) is left as an exercise.

Before the definition of the definite integral is stated, it will be instructive to consider the area of a certain region in a plane. Physical examples could be used; however, we prefer to postpone them until the next chapter. It is important to remember that the discussion of area in this section is *not* to be considered as the definition of the definite integral. It is included only to help motivate the work in Section 2 in the same way that slopes of tangent lines and velocities were used to motivate the definition of derivative.

It is easy to calculate the area of a plane region bounded by straight lines. For example, the area of a rectangle is the product of its length and width. The area of a triangle is one-half the product of the altitude and base. The area of any polygon can be found by subdividing it into triangles. In order to find areas of more complicated regions, whose boundaries involve graphs of functions, it is necessary to introduce a limiting process and use methods of calculus. In particular, let us consider a region S in a coordinate plane, bounded by vertical lines with x-intercepts a and b, by the x-axis, and by the graph of a function f which is continuous and nonnegative on the closed interval $[a, b]$. A region of this type is illustrated in Figure 5.1. Since $f(x) \geq 0$ for every x in $[a, b]$, no part of the graph lies below the x-axis. For convenience we shall refer to S as the region **under the graph of f from a to b**. Our objective is to define the area of S.

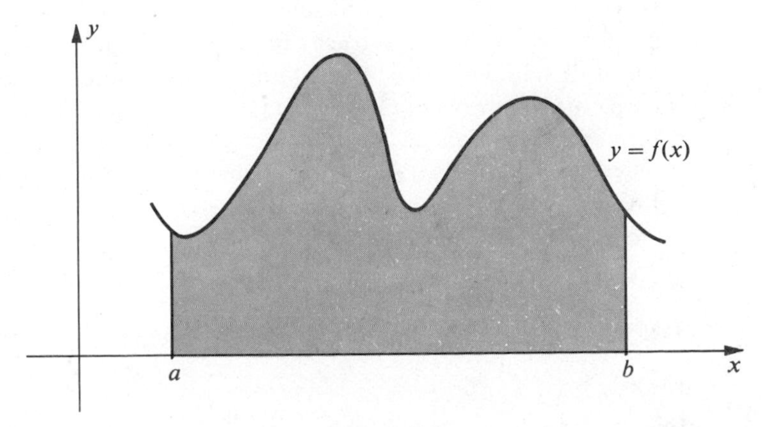

Figure 5.1

If n is any positive integer, let us begin by dividing the interval $[a, b]$ into n subintervals, all having the same length $(b - a)/n$. This can be accomplished by choosing numbers $x_0, x_1, x_2, \ldots, x_n$ where $a = x_0, b = x_n$, and

$$x_i - x_{i-1} = \frac{b - a}{n}$$

for $i = 1, 2, \ldots, n$. If the length $(b - a)/n$ of each subinterval is denoted by Δx, then for each i we have

$$\Delta x = x_i - x_{i-1}, \quad \text{and} \quad x_i = x_{i-1} + \Delta x$$

as illustrated in Figure 5.2.

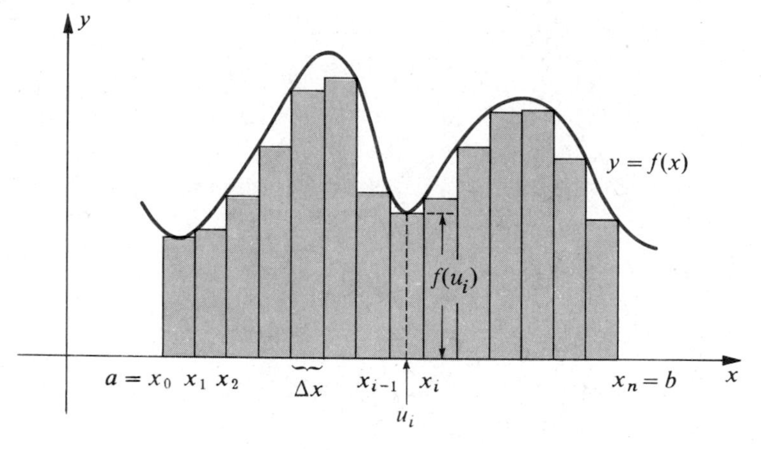

Figure 5.2

Referring to the figure we see that

$$\textbf{(5.3)} \quad \begin{aligned} &x_0 = a, \quad x_1 = a + \Delta x, \quad x_2 = a + 2\,\Delta x, \quad \ldots, \\ &x_i = a + i\,\Delta x, \quad \ldots, \quad x_n = a + n\,\Delta x = b. \end{aligned}$$

Since f is continuous on each subinterval $[x_{i-1}, x_i]$, it follows from (4.4) that f takes on a minimum value at some number u_i in $[x_{i-1}, x_i]$. As shown in Figure 5.2, for each i we can construct a rectangle with one side of length $x_i - x_{i-1}$ and the other side of length equal to the minimum distance $f(u_i)$ from the x-axis to the graph of f. The area of this ith rectangle is $f(u_i)\,\Delta x$. The boundary of the region formed by the totality of these rectangles is called the **inscribed rectangular polygon** associated with the subdivision of $[a, b]$ into n subintervals. The area of this inscribed polygon is the sum of the areas of the n rectangles, that is,

$$f(u_1)\,\Delta x + f(u_2)\,\Delta x + \cdots + f(u_n)\,\Delta x.$$

If we use summation notation, then this sum can be written

$$\textbf{(5.4)} \quad \sum_{i=1}^{n} f(u_i)\,\Delta x.$$

Referring to Figure 5.2, we see that if n is very large or, equivalently, if Δx is very small, then the sum of the rectangular areas appears to be close to what we wish to consider as the area of S. Indeed, reasoning intuitively, if there exists a number A which has the property that the sum (5.4) gets closer and closer to A as Δx gets closer and closer to 0 (but $\Delta x \neq 0$), we shall call A the **area** of S and write

(5.5) $$A = \lim_{\Delta x \to 0} \sum_{i=1}^{n} f(u_i)\,\Delta x.$$

The meaning of this "limit of a sum" is not the same as that for limit of a function introduced in Chapter 2. In order to eliminate the hazy phrase "closer and closer" and arrive at a satisfactory definition of A, let us take a slightly different point of view. If A denotes the area of S, then the difference

(5.6) $$A - \sum_{i=1}^{n} f(u_i)\Delta x$$

is the area of the unshaded portion in Figure 5.2 which lies under the graph of f and over the inscribed rectangular polygon. This number may be thought of as the error involved in using the area of the inscribed rectangular polygon to approximate A. If we have the proper notion of area, then we should be able to make the difference (5.6) arbitrarily small by choosing the width Δx of the rectangles sufficiently small. This is the motivation for the following definition, where the notation is the same as that used in the preceding discussion.

(5.7) Definition

Let A be a real number. The statement

$$A = \lim_{\Delta x \to 0} \sum_{i=1}^{n} f(u_i)\,\Delta x$$

means that for every $\varepsilon > 0$ there corresponds a $\delta > 0$ such that if $0 < \Delta x < \delta$, then

$$A - \sum_{i=1}^{n} f(u_i)\,\Delta x < \varepsilon.$$

If A is the indicated limit and we let $\varepsilon = 10^{-9}$, then (5.7) states that by using sufficiently thin rectangles, the difference between A and the area of the inscribed polygon is less than one-billionth of a square unit! Similarly, if $\varepsilon = 10^{-12}$ we can make this difference less than one-trillionth of a square unit. In general, the difference can be made less than *any* preassigned ε.

If f is continuous on $[a, b]$, then it is shown in more advanced texts that a number A satisfying (5.7) actually exists. We shall call A **the area under the graph of f from a to b.**

The area A may also be obtained by means of **circumscribed rectangular polygons** of the type illustrated in Figure 5.3. In this case we select the number v_i in each interval $[x_{i-1}, x_i]$, such that $f(v_i)$ is the maximum value of f on $[x_{i-1}, x_i]$. The area of this circumscribed polygon is then given by

(5.8)
$$\sum_{i=1}^{n} f(v_i)\,\Delta x.$$

The limit of this sum as $\Delta x \to 0$ is defined as in (5.7), where the only change is that we use

$$\sum_{i=1}^{n} f(v_i)\,\Delta x - A < \varepsilon$$

in the definition since we want this difference to be nonnegative. It can be proved that the same number A is obtained using either inscribed or circumscribed rectangles.

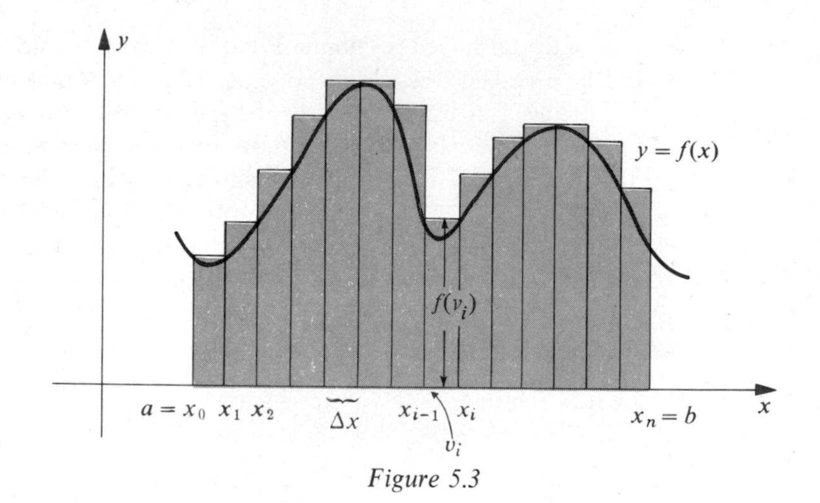

Figure 5.3

The following formulas will be useful in some illustrations of Definition (5.7).

(5.9)
$$\sum_{i=1}^{n} i = 1 + 2 + \cdots + n = \frac{n(n+1)}{2}$$

(5.10)
$$\sum_{i=1}^{n} i^2 = 1^2 + 2^2 + \cdots + n^2 = \frac{n(n+1)(2n+1)}{6}$$

(5.11)
$$\sum_{i=1}^{n} i^3 = 1^3 + 2^3 + \cdots + n^3 = \left[\frac{n(n+1)}{2}\right]^2.$$

These may be proved by means of mathematical induction (see Appendix I).

The next two examples provide concrete illustrations of how summation properties may be used in conjunction with Definition (5.7) to find the areas of certain regions in a coordinate plane.

Example 3 If $f(x) = 16 - x^2$, find the area of the region under the graph of f from 0 to 3.

Solution The region is illustrated in Figure 5.4 where for clarity we have used different scales on the x- and y-axes. If the interval $[0, 3]$ is divided into n equal subintervals, then the length Δx of a typical subinterval is $3/n$. Employing the notation in (5.3) with $a = 0$ and $b = 3$,

$$x_0 = 0,\ x_1 = \Delta x,\ x_2 = 2(\Delta x),\ \ldots,\ x_i = i(\Delta x),\ \ldots,\ x_n = n(\Delta x) = 3.$$

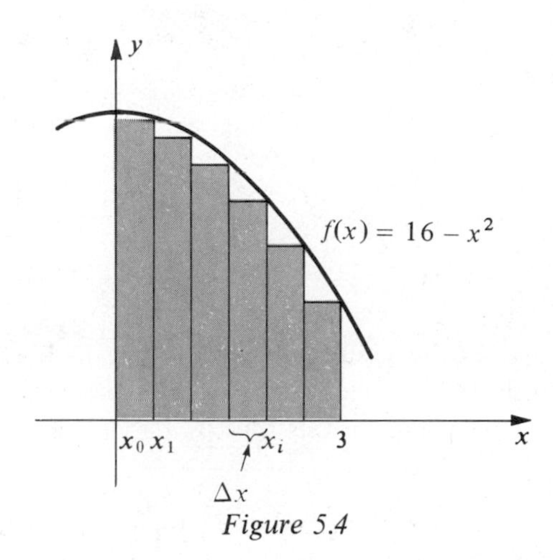

Figure 5.4

Using the fact that $\Delta x = 3/n$ we may write

$$x_i = i(\Delta x) = i\left(\frac{3}{n}\right) = \frac{3i}{n}.$$

Since f is decreasing on $[0, 3]$, the number u_i in $[x_{i-1}, x_i]$ at which f takes on its minimum value is always the right-hand end-point x_i, that is, $u_i = x_i = 3i/n$. Since

$$f(u_i) = f\left(\frac{3i}{n}\right) = 16 - \left(\frac{3i}{n}\right)^2 = 16 - \frac{9i^2}{n^2}$$

the summation (5.4) may be written

$$\begin{aligned}\sum_{i=1}^{n} f(u_i)\,\Delta x &= \sum_{i=1}^{n}\left(16 - \frac{9i^2}{n^2}\right)\left(\frac{3}{n}\right)\\ &= \sum_{i=1}^{n}\left(\frac{48}{n} - \frac{27i^2}{n^3}\right).\end{aligned}$$

Using Theorem (5.2) and (5.1), the last sum may be simplified as follows:

$$\sum_{i=1}^{n}\frac{48}{n} - \sum_{i=1}^{n}\frac{27i^2}{n^3} = \left(\frac{48}{n}\right)n - \frac{27}{n^3}\sum_{i=1}^{n} i^2.$$

Next, applying (5.10) we obtain

$$\begin{aligned}\sum_{i=1}^{n} f(u_i)\,\Delta x &= 48 - \frac{27}{n^3}\left[\frac{n(n+1)(2n+1)}{6}\right]\\ &= 48 - \frac{9}{2n^3}\,[2n^3 + 3n^2 + n].\end{aligned}$$

In order to find the area we must now let Δx approach 0. Since $\Delta x = (b - a)/n$, this can be accomplished by letting n increase without bound. Although our discussion on limits involving infinity in Section 5 of Chapter 4 was concerned with a real variable x, a similar discussion can be given for the integer variable n. Assuming that this is true, and that we can replace $\Delta x \to 0$ by $n \to \infty$, we have

$$\begin{aligned}\lim_{\Delta x \to 0} \sum_{i=1}^{n} f(u_i)\,\Delta x &= \lim_{n\to\infty} \left\{48 - \frac{9}{2n^3}[2n^3 + 3n^2 + n]\right\} \\ &= \lim_{n\to\infty} 48 - \frac{9}{2}\lim_{n\to\infty}\left[\frac{2n^3 + 3n^2 + n}{n^3}\right] \\ &= 48 - \frac{9}{2}\lim_{n\to\infty}\left[2 + \frac{3}{n} + \frac{1}{n^2}\right] \\ &= 48 - \frac{9}{2}[2 + 0 + 0] = 48 - 9 = 39.\end{aligned}$$

Because of the assumptions we have made, our solution is not completely rigorous. Indeed, one reason for introducing the definite integral is to enable us to solve problems of this type in a simple and precise manner.

The area in the preceding example may also be found by using circumscribed rectangular polygons. In this case we select, in each subinterval $[x_{i-1}, x_i]$, the number $v_i = (i - 1)(3/n)$ at which f takes on its maximum value.

The next example illustrates the use of circumscribed rectangles in finding an area.

Example 4 If $f(x) = x^3$, find the area under the graph of f from 0 to b, where $b > 0$.

Solution If we subdivide the interval $[0, b]$ into n equal parts, then Figure 5.5 illustrates a typical circumscribed rectangular polygon where, as in Example 2,

$$\Delta x = \frac{b}{n} \quad \text{and} \quad x_i = i(\Delta x).$$

For clarity different scales have been used on the x- and y-axes.

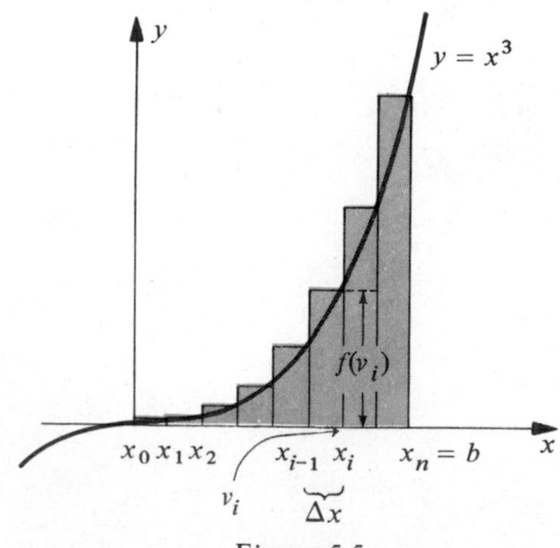

Figure 5.5

Since f is an increasing function, the maximum value $f(v_i)$ in the interval $[x_{i-1}, x_i]$ occurs at the right-hand end-point, that is,

$$v_i = x_i = i(\Delta x) = i\left(\frac{b}{n}\right) = \frac{bi}{n}.$$

As in (5.8), the sum of the areas of the circumscribed rectangles is

$$\begin{aligned}\sum_{i=1}^{n} f(v_i)\,\Delta x &= \sum_{i=1}^{n} \left(\frac{bi}{n}\right)^3 \left(\frac{b}{n}\right)\\ &= \sum_{i=1}^{n} \frac{b^4}{n^4} i^3\\ &= \frac{b^4}{n^4} \sum_{i=1}^{n} i^3\\ &= \frac{b^4}{n^4}\left[\frac{n(n+1)}{2}\right]^2\end{aligned}$$

where the final step follows from (5.11). The reader may now verify that

$$\sum_{i=1}^{n} f(v_i)\,\Delta x = \frac{b^4}{4}\left(\frac{n^4 + 2n^3 + n^2}{n^4}\right) = \frac{b^4}{4}\left(1 + \frac{2}{n} + \frac{1}{n^2}\right).$$

If we let Δx approach 0, n increases without bound and the expression in parentheses approaches 1. It follows that the area under the graph is

$$\lim_{\Delta x \to 0} \sum_{i=1}^{n} f(v_i)\,\Delta x = \frac{b^4}{4}.$$

EXERCISES 5.1

Find the numbers in Exercises 1–10.

1 $\sum_{i=1}^{5} (3i - 10)$

2 $\sum_{i=1}^{6} (9 - 2i)$

3 $\sum_{j=1}^{4} (j^2 + 1)$

4 $\sum_{n=1}^{10} [1 + (-1)^n]$

5 $\sum_{k=0}^{5} k(k - 1)$

6 $\sum_{k=0}^{4} (k - 2)(k - 3)$

7 $\sum_{i=1}^{8} 2^i$

8 $\sum_{s=1}^{6} \frac{5}{s}$

9 $\sum_{i=1}^{50} 10$

10 $\sum_{k=1}^{1000} 2$

11 Prove (iii) of Theorem (5.2).

12 Extend (i) of Theorem (5.2) to $\Sigma_{i=1}^{n} (a_i + b_i + c_i)$.

13 Find $\Sigma_{i=1}^{n} (i^2 + 3i + 5)$. (*Hint:* Write the sum as $\Sigma_{i=1}^{n} i^2 + 3\,\Sigma_{i=1}^{n} i + \Sigma_{i=1}^{n} 5$ and employ (5.10), (5.9), and (5.1).)

14 Find $\sum_{i=1}^{n} (3i^2 - 2i + 1)$.

15 Find $\sum_{k=1}^{n} (2k - 3)^2$.

16 Find $\sum_{k=1}^{n} (k^3 + 2k^2 - k + 4)$.

In each of Exercises 17–26 find the area under the graph of f from a to b using (a) inscribed rectangles; (b) circumscribed rectangles. In each case sketch the graph and typical rectangles, labeling the drawing as in Figures 5.4 and 5.5.

17 $f(x) = 2x + 3$; $a = 0, b = 4$

18 $f(x) = 8 - 3x$; $a = 0, b = 2$

19 $f(x) = x^2$; $a = 0, b = 5$

20 $f(x) = x^2 + 2$; $a = 1, b = 3$

21 $f(x) = 3x^2 + 5$; $a = 1, b = 4$ (*Hint:* $x_i = 1 + (3i/n)$.)

22 $f(x) = 7$; $a = -2$, $b = 6$

23 $f(x) = 9 - x^2$; $a = 0, b = 3$

24 $f(x) = 4x^2 + 3x + 2$; $a = 1, b = 5$

25 $f(x) = x^3 + 1$; $a = 1, b = 2$

26 $f(x) = 4x + x^3$; $a = 0, b = 2$

27 Use Definition (5.7) to prove that the area of a right triangle of altitude h and base b is $\frac{1}{2}bh$. (*Hint:* Consider the area under the graph of $f(x) = (h/b)x$ from 0 to b.)

28 If $f(x) = px^2 + qx + r$ and $f(x) \geq 0$ for all x, prove that the area under the graph of f from 0 to b is

$$p\left(\frac{b^3}{3}\right) + q\left(\frac{b^2}{2}\right) + rb.$$

29 Use mathematical induction to prove (5.9)–(5.11).

30 Prove (5.9) by writing

$$\begin{aligned} S &= 1 + \quad 2 \quad + \cdots + n \\ S &= n + (n-1) + \cdots + 1 \end{aligned}$$

and then adding corresponding sides of these equations.

5.2 DEFINITION OF DEFINITE INTEGRAL

Limiting processes similar to that used in the preceding section for areas arise frequently in mathematics and its applications. The situations which occur often lead to limits of the form

$$\lim_{\Delta x \to 0} \sum_{i=1}^{n} f(w_i)\Delta x. \tag{5.12}$$

A special instance of this limit appeared in Definition (5.7); however, in the general case, there are several major differences from our work with areas. In our discussion of area in the previous section we made the following assumptions.

1. The function f is continuous on a closed interval $[a, b]$.
2. $f(x)$ is nonnegative for all x in $[a, b]$.
3. All the subintervals $[x_{i-1}, x_i]$ determined by the subdivision of $[a, b]$ have the same length Δx.

4. The numbers w_i are chosen such that $f(w_i)$ is always the minimum (or maximum) value of f on $[x_{i-1}, x_i]$.

These four conditions are not always present in problems which require methods similar to those used for area. For this reason it is necessary to generalize the limit in (5.12) to the following cases.

1′. The function f may be discontinuous at some numbers in $[a, b]$.

2′. $f(x)$ may be negative for some x in $[a, b]$.

3′. The lengths of the subintervals $[x_{i-1}, x_i]$ may be different.

4′. The number w_i is *any* number in $[x_{i-1}, x_i]$.

We shall begin by introducing some new terminology and notation. A **partition** P of a closed interval $[a, b]$ is any decomposition of $[a, b]$ into subintervals of the form

$$[x_0, x_1], [x_1, x_2], [x_2, x_3], \ldots, [x_{n-1}, x_n]$$

where n is a positive integer and the x_i are numbers such that

$$a = x_0 < x_1 < x_2 < x_3 < \cdots < x_{n-1} < x_n = b.$$

The length of the ith subinterval $[x_{i-1}, x_i]$ will be denoted by Δx_i, that is,

$$\Delta x_i = x_i - x_{i-1}.$$

A typical partition of $[a, b]$ is illustrated in Figure 5.6. The largest of the numbers $\Delta x_1, \Delta x_2, \ldots, \Delta x_n$ is called the **norm** of the partition P and is denoted by $\|P\|$.

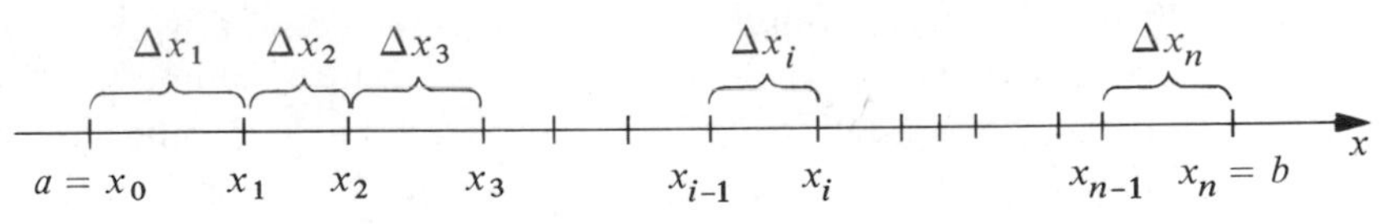

Figure 5.6. A partition of $[a, b]$

The following concept, named after the mathematician G. Riemann (1826–1866), is fundamental for the definition of the definite integral.

(5.13) Definition

Let f be a function that is defined on a closed interval $[a, b]$ and let P be a partition of $[a, b]$. A **Riemann sum** of f for P is any expression R_P of the form

$$R_P = \sum_{i=1}^{n} f(w_i)\Delta x_i$$

where w_i is some number in $[x_{i-1}, x_i]$ for $i = 1, 2, \ldots, n$.

The sums in (5.4) and (5.8), which represent sums of areas of certain rectangles, are special types of Riemann sums. However, in Definition (5.13) $f(w_i)$

is not necessarily a maximum or minimum value of f on $[x_{i-1}, x_i]$, and hence if we construct a rectangle of length $f(w_i)$ and width Δx_i as illustrated in Figure 5.7, the rectangle may be neither inscribed nor circumscribed.

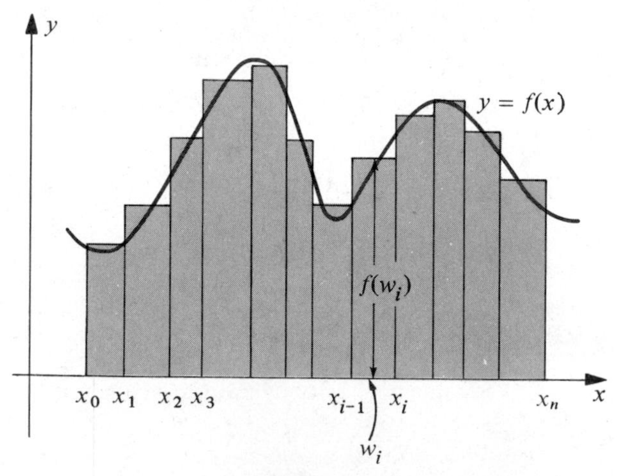

Figure 5.7

In general, since $f(x)$ may be negative for some x in $[a, b]$, some terms of R_P in (5.13) may be negative. Consequently, a Riemann sum does not always represent a sum of areas of rectangles. It is possible, however, to interpret the sums geometrically as follows. If R_P is defined as in (5.13), then for each subinterval $[x_{i-1}, x_i]$ let us construct a horizontal line segment through the point $(w_i, f(w_i))$, thereby obtaining a set of rectangles. If $f(w_i)$ is positive, the rectangle lies above the x-axis as illustrated by the shaded rectangles in Figure 5.8, and the product $f(w_i)\Delta x_i$ is the area of this rectangle. If $f(w_i)$ is negative, then the rectangle lies below the x-axis as illustrated by the unshaded rectangles in Figure 5.8. In this case the product $f(w_i)\Delta x_i$ is the *negative* of the area of a rectangle. It follows that R_P is the sum of the areas of the rectangles which lie above the x-axis and the *negatives* of the areas of the rectangles which lie below the x-axis.

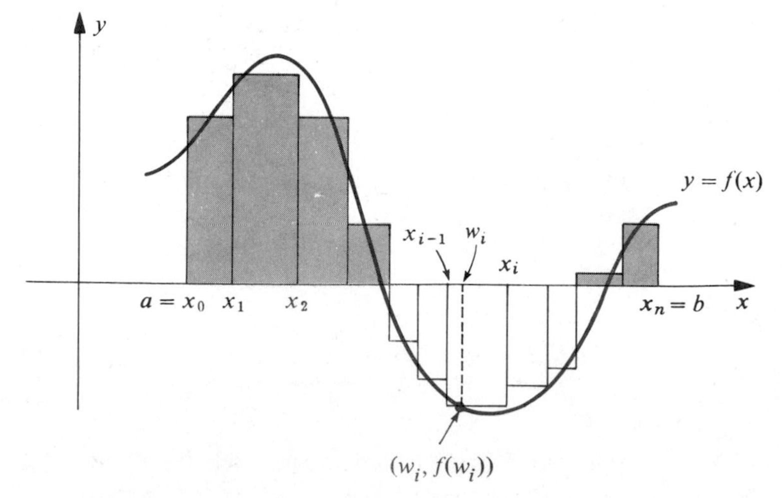

Figure 5.8

Example Suppose $f(x) = 8 - (x^2/2)$ and P is the partition of $[0, 6]$ into the five subintervals determined by $x_0 = 0$, $x_1 = 1.5$, $x_2 = 2.5$, $x_3 = 4.5$, $x_4 = 5$, and $x_5 = 6$. Find (a) the norm of the partition and (b) the Riemann sum R_P if $w_1 = 1$, $w_2 = 2$, $w_3 = 3.5$, $w_4 = 5$, and $w_5 = 5.5$.

Solution The graph of f is sketched in Figure 5.9. Also shown in the figure are the points on the x-axis which correspond to x_i and the rectangles of lengths $|f(w_i)|$ for $i = 1, 2, 3, 4$, and 5. Thus,

$$\Delta x_1 = 1.5,\ \Delta x_2 = 1,\ \Delta x_3 = 2,\ \Delta x_4 = 0.5,\ \Delta x_5 = 1,$$

and hence the norm $\|P\|$ of the partition is Δx_3, or 2. By Definition (5.13)

$$\begin{aligned} R_P &= f(w_1)\Delta x_1 + f(w_2)\Delta x_2 + f(w_3)\Delta x_3 + f(w_4)\Delta x_4 + f(w_5)\Delta x_5 \\ &= f(1)(1.5) + f(2)(1) + f(3.5)(2) + f(5)(0.5) + f(5.5)(1) \\ &= (7.5)(1.5) + (6)(1) + (1.875)(2) + (-4.5)(0.5) + (-7.125)(1) \end{aligned}$$

which reduces to $R_P = 11.625$.

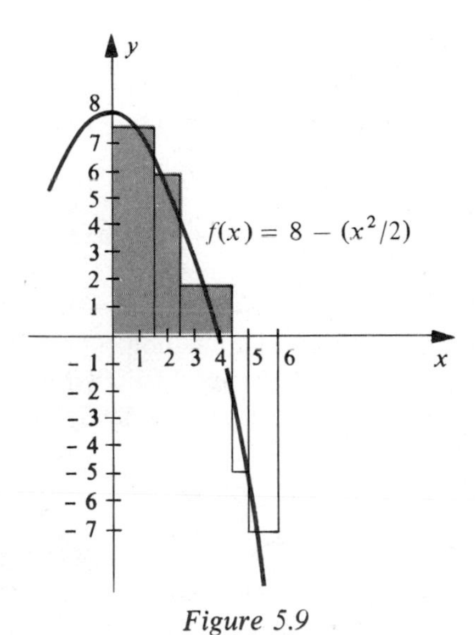

Figure 5.9

In the future the number n of subintervals in a partition P will not always be specified. In this event the Riemann sum (5.13) will be written

(5.14)
$$R_P = \sum_i f(w_i)\Delta x_i,$$

where it is understood that terms of the form $f(w_i)\Delta x_i$ are to be summed over all subintervals $[x_{i-1}, x_i]$ of the partition P.

In a manner similar to that used in formulating Definition (5.7), we shall next define what is meant by

$$\lim_{\|P\| \to 0} \sum_i f(w_i)\Delta x_i = I$$

where I is a real number. Intuitively, it will mean that if the norm $\|P\|$ of the partition P is close to 0, then every Riemann sum for P is close to I.

(5.15) Definition

Let f be a function that is defined on a closed interval $[a, b]$, and let I be a real number. The statement

$$\lim_{\|P\| \to 0} \sum_i f(w_i)\,\Delta x_i = I$$

means that for every $\varepsilon > 0$ there exists a $\delta > 0$, such that if P is a partition of $[a, b]$ with $\|P\| < \delta$, then

$$\left| \sum_i f(w_i)\,\Delta x_i - I \right| < \varepsilon$$

for any choice of numbers w_i in the subintervals $[x_{i-1}, x_i]$ of P. The number I is called a **limit of a sum**.

Note that for every $\delta > 0$ there are infinitely many partitions P such that $\|P\| < \delta$. Moreover, for each such partition P there are infinitely many ways of choosing the numbers w_i in $[x_{i-1}, x_i]$. Consequently, there may be an infinite number of different Riemann sums associated with *each* partition P. However, if the limit I exists, then for any ε, each of these Riemann sums is within ε units of I, provided a small enough norm is chosen. Although Definition (5.15) differs from the definition of limit of a function, a proof similar to that given in Appendix II may be used to show that if the limit I exists, then it is unique.

We next define the definite integral as a limit of a sum.

(5.16) Definition

Let f be a function that is defined on a closed interval $[a, b]$. The **definite integral of f from a to b**, denoted by $\int_a^b f(x)\,dx$, is given by

$$\int_a^b f(x)\,dx = \lim_{\|P\| \to 0} \sum_i f(w_i)\,\Delta x_i,$$

provided the limit exists.

If the definite integral of f from a to b exists, then f is said to be **integrable** on the closed interval $[a, b]$, or we say that the integral $\int_a^b f(x)\,dx$ **exists**. The process of finding the number represented by the above limit is called **evaluating the integral**.

The symbol $\int$ in Definition (5.16) is called an **integral sign**. It may be thought of as an elongated letter S (the first letter of the word *sum*) and is used to indicate the connection between definite integrals and Riemann sums. The numbers a and b are referred to as the **limits of integration**, where a is called the **lower limit** and b the **upper limit**. The word *limit* involved in this terminology is used to denote the smallest and largest numbers in the interval $[a, b]$ and has no connection with any definitions of limits given earlier. The expression $f(x)$ which appears to the right of the integral sign (we sometimes say "*behind* the integral sign") is called the **integrand**. Finally, the symbol dx which follows $f(x)$ should not be confused with the differential of x defined in Chapter 3. At this stage of our work it is merely used to indicate the variable. Later in the text the use of the differential symbol will have certain practical advantages.

Letters other than x may be used in the notation for the definite integral. This follows from the fact that when we describe a function, the symbol used for the independent variable is immaterial. Thus, if f is integrable on $[a, b]$, then

$$\int_a^b f(x)\,dx = \int_a^b f(s)\,ds = \int_a^b f(t)\,dt$$

and so on. For this reason the letter x in Definition (5.16) is sometimes referred to as a **dummy variable**.

Whenever an interval $[a, b]$ is employed it is assumed that $a < b$. Consequently Definition (5.16) does not take into account the cases in which the lower limit of integration is greater than or equal to the upper limit. The definition may be extended to include the case where the lower limit is greater than the upper limit, as follows.

(5.17) Definition

If $c > d$, then

$$\int_c^d f(x)\,dx = -\int_d^c f(x)\,dx$$

provided the latter integral exists.

In words, Definition (5.17) may be phrased "interchanging the limits of integration changes the sign of the integral." One reason for the form of Definition (5.17) will become apparent later, after we have considered the Fundamental Theorem of Calculus.

The case in which the lower and upper limits of integration are equal is covered by the next definition.

(5.18) Definition

If $f(a)$ exists, then $\int_a^a f(x)\,dx = 0$.

Not every function f is integrable. For example, if $f(x)$ becomes positively or negatively infinite at some number in $[a, b]$, then the definite integral does not exist. To illustrate, if f is defined on $[a, b]$ and $\lim_{x \to a^+} f(x) = \infty$, then in the first subinterval $[x_0, x_1]$ of any partition P of $[a, b]$ a number w_1 can be found such that $f(w_1)\Delta x_1$ is larger than any given number M. It follows that for any partition P, we can find a Riemann sum $\Sigma_i f(w_i)\Delta x_i$ which is arbitrarily large. Hence if I is any real number and P any partition of $[a, b]$, then there exist Riemann sums R_P such that $R_P - I$ is arbitrarily large. This implies that f is not integrable. A similar argument can be given if f becomes infinite at any other number in $[a, b]$. Consequently, *if a function f is integrable on $[a, b]$, then it is bounded on $[a, b]$*; that is, there is a real number M such that $|f(x)| \le M$ for all x in $[a, b]$.

The reader might be tempted to conjecture that if a function is discontinuous somewhere in $[a, b]$ then it is not integrable. This conjecture is false. Definite integrals of discontinuous functions may or may not exist, depending on the nature of the discontinuities. However, according to the next theorem, continuous functions are *always* integrable.

(5.19) Theorem

If f is continuous on $[a, b]$, then f is integrable on $[a, b]$.

This result provides a large class of functions which are integrable. A proof can be found in most texts on advanced calculus.

If f is integrable, then the limit in Definition (5.16) exists for all choices of w_i in $[x_{i-1}, x_i]$. This fact allows us to specialize the choice of w_i if we wish to do so. For example, we could always take w_i as the smallest number x_{i-1} in the subinterval, or as the largest number x_i, or as the midpoint of the subinterval, or as the number which always produces the minimum or maximum value in $[x_{i-1}, x_i]$, and so on. In addition, since the limit is independent of the partitions P of $[a, b]$ (as long as $||P||$ is sufficiently small) we may specialize the partitions to the case in which all the subintervals $[x_{i-1}, x_i]$ have the same length Δx. Such a partition is called a **regular partition**. We will see in the next chapter that specializations of the types we have described are often used in applications. As an immediate illustration, if f is continuous and $f(x) \ge 0$ for all x in $[a, b]$ and if A is the area of the region under the graph of f from a to b, then

$$A = \lim_{\Delta x \to 0} \sum_i f(u_i)\Delta x = \int_a^b f(x)\,dx \tag{5.20}$$

where $f(u_i)$ is the minimum value of f on $[x_{i-1}, x_i]$.

EXERCISES 5.2

In each of Exercises 1–4, the given numbers $\{x_0, x_1, x_2, \ldots, x_n\}$ determine a partition P of the indicated interval $[a, b]$. Find $\Delta x_1, \Delta x_2, \ldots, \Delta x_n$ and the norm $||P||$ of the partition.

1 $[0, 5]$; $\{0, 1.1, 2.6, 3.7, 4.1, 5\}$

2 $[2, 6]$; $\{2, 3, 3.7, 4, 5.2, 6\}$

3 $[-3, 1]$; $\{-3, -2.7, -1, 0.4, 0.9, 1\}$

4 $[1, 4]$; $\{1, 1.6, 2, 3.5, 4\}$

In Exercises 5 and 6 find the Riemann sum R_P of f, where P is the regular partition of $[1, 5]$ into the four equal subintervals determined by $x_0 = 1, x_1 = 2, x_2 = 3, x_3 = 4, x_4 = 5$ and
(a) w_i is the right-hand end-point x_i of $[x_{i-1}, x_i]$.
(b) w_i is the left-hand end-point x_{i-1} of $[x_{i-1}, x_i]$.
(c) w_i is the midpoint of $[x_{i-1}, x_i]$.

5 $f(x) = 2x + 3$

6 $f(x) = 3 - 4x$

7 If $f(x) = 8 - (x^2/2)$, find the Riemann sum R_P of f where P is the regular partition of $[0, 6]$ into the six equal subintervals determined by $x_0 = 0$, $x_1 = 1$, $x_2 = 2$, $x_3 = 3$, $x_4 = 4$, $x_5 = 5$, $x_6 = 6$, and w_i is the midpoint of the interval $[x_{i-1}, x_i]$.

8 If $f(x) = 8 - (x^2/2)$, find the Riemann sum R_P of f where P is the partition of $[0, 6]$ into the four subintervals determined by $x_0 = 0$, $x_1 = 1.5$, $x_2 = 3$, $x_3 = 4.5$, $x_4 = 6$, and $w_1 = 1$, $w_2 = 2$, $w_3 = 4$, and $w_4 = 5$.

9 Suppose $f(x) = x^3$ and P is the partition of $[-2, 4]$ into the four subintervals determined by $x_0 = -2$, $x_1 = 0$, $x_2 = 1$, $x_3 = 3$, and $x_4 = 4$. Find the Riemann sum R_P if $w_1 = -1$, $w_2 = 1$, $w_3 = 2$, and $w_4 = 4$.

10 Suppose $f(x) = \sqrt{x}$ and P is the partition of $[1, 16]$ into the five subintervals determined by $x_0 = 1, x_1 = 3, x_2 = 5, x_3 = 7, x_4 = 9$, and $x_5 = 16$. Find the Riemann sum R_P if $w_1 = 1$, $w_2 = 4$, $w_3 = 5$, $w_4 = 9$, and $w_5 = 9$.

In each of Exercises 11–14, use Definition (5.16) to express each limit as a definite integral on the indicated closed interval $[a, b]$.

11 $\lim_{||P|| \to 0} \sum_{i=1}^{n} (3w_i^2 - 2w_i + 5)\Delta x_i$; $[-1, 2]$

12 $\lim_{||P|| \to 0} \sum_{i=1}^{n} \pi(w_i^2 - 4)\Delta x_i$; $[2, 3]$

13 $\lim_{||P|| \to 0} \sum_{i=1}^{n} 2\pi w_i(1 + w_i^3)\Delta x_i$; $[0, 4]$

14 $\lim_{||P|| \to 0} \sum_{i=1}^{n} (\sqrt[3]{w_i} + 4w_i)\Delta x_i$; $[-4, -3]$

15 If $\int_1^4 \sqrt{x}\, dx = 14/3$, find $\int_4^1 \sqrt{x}\, dx$.

16 Find $\int_3^3 x^2\, dx$.

17 If $\int_1^2 (5x^4 - 1)\, dx = 30$, find $\int_1^2 (5r^4 - 1)\, dr$.

18 If $\int_{-1}^8 \sqrt[3]{s}\, ds = 45/4$, find $\int_8^{-1} \sqrt[3]{t}\, dt$.

In each of Exercises 19–22 find the value of the definite integral by interpreting it as the area under the graph of a function f.

19 $\int_{-3}^2 (2x + 6)\, dx$

20 $\int_{-1}^2 (7 - 3x)\, dx$

21 $\int_0^3 \sqrt{9 - x^2}\, dx$

22 $\int_{-a}^a \sqrt{a^2 - x^2}\, dx, a > 0$

23 Find $\int_0^5 x^3\, dx$. (*Hint:* See Example 4 of Section 1.)

24 Let c be an arbitrary real number and suppose $f(x) = c$ for all x. If P is any partition of $[a, b]$, show that every Riemann sum R_P of f equals $c(b - a)$. Use this fact to prove that $\int_a^b c\, dx = c(b - a)$. Interpret this geometrically if $c > 0$.

25 Give an example of a function which is continuous on the interval (0, 1) such that $\int_0^1 f(x)\,dx$ does not exist. Why doesn't this contradict Theorem (5.19)?

26 Give an example of a function which is not continuous on [0, 1] such that $\int_0^1 f(x)\,dx$ exists.

5.3 PROPERTIES OF THE DEFINITE INTEGRAL

This section contains some fundamental properties of the definite integral. Most of the proofs are rather technical and have been placed in Appendix II, where the reader may study them whenever time permits.

If f is the constant function defined by $f(x) = k$ for all x in $[a, b]$ and P is a partition of $[a, b]$, then for every Riemann sum of f,

$$\sum_i f(w_i)\,\Delta x_i = \sum_i k\,\Delta x_i = k\sum_i \Delta x_i = k(b - a),$$

since the sum $\Sigma_i\,\Delta x_i$ is the length of the interval $[a, b]$. Consequently

$$\left|\sum_i f(w_i)\,\Delta x_i - k(b - a)\right| = |k(b - a) - k(b - a)| = 0,$$

which is less than any positive number ε *regardless* of the size of $\|P\|$. Therefore,

$$\lim_{\|P\|\to 0} \sum_i f(w_i)\,\Delta x_i = \lim_{\|P\|\to 0} \sum_i k\,\Delta x_i = k(b - a);$$

that is,

(5.21) $$\int_a^b k\,dx = k(b - a).$$

This equality is in agreement with the discussion of area in Section 1, for if $k > 0$, then the graph of f is a horizontal line k units above the x-axis, and the region under the graph from a to b is a rectangle with sides of length k and $b - a$. Hence the area of the rectangle is $k(b - a)$.

Example 1 Evaluate $\int_{-2}^{3} 7\,dx$.

Solution Using (5.21),

$$\int_{-2}^{3} 7dx = 7[3 - (-2)] = 7(5) = 35.$$

For the special case of (5.21) with $k = 1$ we shall abbreviate the integrand by writing

$$\int_a^b dx = b - a.$$

If a function f is integrable on $[a, b]$ and k is a real number, then by Theorem (5.2) a Riemann sum of the function kf may be written

$$\sum_i kf(w_i)\,\Delta x_i = k\sum_i f(w_i)\,\Delta x_i.$$

It is proved in Appendix II that the limit of the sum on the left is equal to k times the limit of the sum on the right. Restating this fact in terms of definite integrals gives us the next theorem.

(5.22) Theorem

> If f is integrable on $[a, b]$ and k is any real number, then kf is integrable on $[a, b]$ and
>
> $$\int_a^b kf(x)\,dx = k\int_a^b f(x)\,dx.$$

The conclusion of Theorem (5.22) is sometimes stated "*a constant factor in the integrand may be taken outside the integral sign.*" It is *not* permissible to take expressions involving variables outside the integral sign in this manner.

If two functions f and g are defined on $[a, b]$, then by Theorem (5.2) a Riemann sum of $f + g$ may be written

$$\sum_i [f(w_i) + g(w_i)]\,\Delta x_i = \sum_i f(w_i)\,\Delta x_i + \sum_i g(w_i)\,\Delta x_i.$$

It can be shown that if f and g are integrable, then the limit of the sum on the left may be found by adding the limits of the two sums on the right. This fact is stated in integral form as follows.

(5.23) Theorem

> If f and g are integrable on $[a, b]$, then $f + g$ is integrable on $[a, b]$ and
>
> $$\int_a^b [f(x) + g(x)]\,dx = \int_a^b f(x)\,dx + \int_a^b g(x)\,dx.$$

A proof is given in Appendix II. The analogue for differences may also be established; that is

(5.24)
$$\int_a^b [f(x) - g(x)]\,dx = \int_a^b f(x)\,dx - \int_a^b g(x)\,dx$$

provided f and g are integrable on $[a, b]$.

Theorem (5.23) may be extended to any finite number of functions. Specifically, if $f_1, f_2, \ldots, f_n$ are integrable on $[a, b]$, then so is their sum and

$$\int_a^b [f_1(x) + f_2(x) + \cdots + f_n(x)]\,dx$$

$$= \int_a^b f_1(x)\,dx + \int_a^b f_2(x)\,dx + \cdots + \int_a^b f_n(x)\,dx.$$

Example 2 Given $\int_0^2 x^3\,dx = 4$ and $\int_0^2 x\,dx = 2$, evaluate $\int_0^2 (5x^3 - 3x + 6)\,dx$.

Solution We may proceed as follows:

$$\int_0^2 (5x^3 - 3x + 6)\,dx = \int_0^2 5x^3\,dx - \int_0^2 3x\,dx + \int_0^2 6\,dx$$

$$= 5\int_0^2 x^3\,dx - 3\int_0^2 x\,dx + 6(2 - 0)$$

$$= 5(4) - 3(2) + 12 = 26.$$

If f is continuous on $[a, b]$ and $f(x) \geq 0$ for all x in $[a, b]$, then by the discussion in Section 1, the integral $\int_a^b f(x)\,dx$ is the area under the graph of f from a to b. In like manner, if $a < c < b$, then the integrals $\int_a^c f(x)\,dx$ and $\int_c^b f(x)\,dx$ are the areas under the graph of f from a to c and from c to b, respectively. It follows that

$$\int_a^b f(x)\,dx = \int_a^c f(x)\,dx + \int_c^b f(x)\,dx.$$

The next theorem shows that this equality is also true under a more general hypothesis. The proof is given in Appendix II.

(5.25) Theorem

If $a < c < b$ and f is integrable on both $[a, c]$ and $[c, d]$, then f is integrable on $[a, b]$ and

$$\int_a^b f(x)\,dx = \int_a^c f(x)\,dx + \int_c^b f(x)\,dx.$$

The following result shows that Theorem (5.25) can be generalized to the case where c is not necessarily between a and b.

(5.26) Theorem

If f is integrable on a closed interval and if a, b, and c are any three numbers in the interval, then

$$\int_a^b f(x)\,dx = \int_a^c f(x)\,dx + \int_c^b f(x)\,dx.$$

Proof. If a, b, and c are all different, then there are six possible ways of ordering these three numbers. The theorem should be verified for each of these cases as well as for the cases in which two, or all three, of the numbers are equal. We shall verify one case and leave the remaining parts as exercises. Thus, suppose the numbers are ordered such that $c < a < b$. Using Theorem (5.25),

$$\int_c^b f(x)\,dx = \int_c^a f(x)\,dx + \int_a^b f(x)\,dx$$

which, in turn, may be written

$$\int_a^b f(x)\,dx = -\int_c^a f(x)\,dx + \int_c^b f(x)\,dx.$$

The desired conclusion now follows, since interchanging the limits of integration changes the sign of the integral (see Definition (5.17)).

If f and g are continuous on $[a, b]$ and $f(x) \geq g(x) \geq 0$ for all x in $[a, b]$, then the area under the graph of f from a to b is greater than or equal to the area under the graph of g from a to b. The Corollary to the next theorem is a generalization of this fact to arbitrary integrable functions. The proof of the theorem is given in Appendix II.

(5.27) Theorem

If f is integrable on $[a, b]$ and if $f(x) \geq 0$ for all x in $[a, b]$, then

$$\int_a^b f(x)\,dx \geq 0.$$

(5.28) Corollary

If f and g are integrable on $[a, b]$ and $f(x) \geq g(x)$ for all x in $[a, b]$, then

$$\int_a^b f(x)\,dx \geq \int_a^b g(x)\,dx.$$

Proof. Since $f - g$ is integrable and $f(x) - g(x) \geq 0$ for all x in $[a, b]$, then by Theorem (5.27),

$$\int_a^b [f(x) - g(x)]\,dx \geq 0.$$

Applying (5.24) leads to the desired conclusion.

It can be shown that (5.27) and (5.28) are true if $\geq$ is replaced by $>$ (see Exercise 28).

EXERCISES 5.3

Evaluate the definite integrals in Exercises 1–6.

1 $\int_{-2}^{4} 5\,dx$ **2** $\int_{1}^{10} \sqrt{2}\,dx$ **3** $\int_{6}^{2} 3\,dx$

4 $\int_{4}^{-3} dx$ **5** $\int_{-1}^{1} dx$ **6** $\int_{2}^{2} 100\,dx$

It will follow from our work in Section 5 that

$$\int_1^4 x^2\,dx = 21, \quad \int_1^4 x\,dx = 15/2, \quad \text{and} \quad \int_1^4 \sqrt{x}\,dx = 14/3.$$

Use these facts to evaluate the integrals in Exercises 7–14.

7 $\int_1^4 (3x^2 + 5)\,dx$ **8** $\int_1^4 (6x - 1)\,dx$

9 $\int_1^4 (2 - 9x - 4x^2)\,dx$ **10** $\int_1^4 (3x + 2)^2\,dx$

11 $\int_4^1 \sqrt{5x}\,dx$ **12** $\int_1^4 2x(x + 1)\,dx$

13 $\int_1^4 (\sqrt{x} - 5)^2\,dx$ **14** $\int_4^1 (3\sqrt{x} + 1)(\sqrt{x} - 2)\,dx$

Verify the inequalities in Exercises 15 and 16 without evaluating the integrals.

15 $\int_1^2 (3x^2 + 4)\,dx \geq \int_1^2 (2x^2 + 5)\,dx$ **16** $\int_2^4 (5x^2 - 4\sqrt{x} + 2)\,dx > 0$

In Exercises 17 and 18 assume that f is integrable on $[a, b]$.

17 If $f(x) \leq M$ for all x in $[a, b]$, prove that $\int_a^b f(x)\,dx \leq M(b - a)$. Illustrate this result graphically.

18 If $m \leq f(x)$ for all x in $[a, b]$, prove that $m(b-a) \leq \int_a^b f(x)\,dx$. Illustrate this result graphically.

In Exercises 19–22 express each sum or difference as a single integral of the form $\int_a^b f(x)\,dx$.

19 $\int_5^1 f(x)\,dx + \int_{-3}^5 f(x)\,dx$

20 $\int_4^1 f(x)\,dx + \int_6^4 f(x)\,dx$

21 $\int_c^{c+h} f(x)\,dx - \int_c^h f(x)\,dx$

22 $\int_{-2}^6 f(x)\,dx - \int_{-2}^2 f(x)\,dx$

23 Prove (5.24). (*Hint:* Let $f(x) - g(x) = f(x) + [-g(x)]$ and use Theorems (5.23) and (5.22).)

24 If f and g are integrable functions on $[a, b]$ and if p and q are any real numbers, prove that

$$\int_a^b [pf(x) + qg(x)]\,dx = p\int_a^b f(x)\,dx + q\int_a^b g(x)\,dx.$$

(*Hint:* Use Theorems (5.23) and (5.22).)

25 Complete the proof of Theorem (5.26) by considering all other orderings of the numbers a, b, and c.

26 Use Theorem (5.26) to prove that if f is integrable on a closed interval and if $c_1, c_2, \ldots, c_n$ are any numbers in the interval, then

$$\int_{c_1}^{c_n} f(x)\,dx = \sum_{i=1}^{n-1} \int_{c_i}^{c_{i+1}} f(x)\,dx.$$

27 If f is integrable on $[a, b]$ and if $f(x) \leq 0$ for all x in $[a, b]$, prove that $\int_a^b f(x)\,dx \leq 0$. (*Hint:* If $f(x) \leq 0$, then $-f(x) \geq 0$.)

28 If f is continuous on $[a, b]$ and $f(x) > 0$ for all x in $[a, b]$, prove that $\int_a^b f(x)\,dx > 0$. (*Hint:* Let $f(u)$ be the minimum value of f on $[a, b]$ and prove that $R_P \geq f(u)(b-a)$, where P is any partition of $[a, b]$ and R_P is any Riemann sum of f for P.)

29 If f is continuous on $[a, b]$, prove that

$$\left|\int_a^b f(x)\,dx\right| \leq \int_a^b |f(x)|\,dx.$$

30 Refer to Exercise 29. Suppose $f_1, f_2, \ldots, f_n$ are continuous functions on $[a, b]$, where n is any positive integer. Use mathematical induction to prove that

$$\left|\sum_{i=1}^{n} \int_a^b f_i(x)\,dx\right| \leq \sum_{i=1}^{n} \int_a^b |f_i(x)|\,dx.$$

5.4 THE MEAN VALUE THEOREM FOR DEFINITE INTEGRALS

Suppose f is continuous and $f(x) \geq 0$ for all x in a closed interval $[a, b]$. If $f(c) > 0$ for some c in $[a, b]$, then $\lim_{x \to c} f(x) > 0$ and, by an argument similar to that used in the proof of Theorem (2.10), there is an interval $[a', b']$ contained in $[a, b]$ throughout which $f(x)$ is positive. Let $f(u)$ be the minimum value of f on $[a', b']$, as illustrated in Figure 5.10. It follows that the area under the graph of f from a to b is at least as large as the area $f(u)(b' - a')$ of the pictured rectangle. Consequently, $\int_a^b f(x)\,dx > 0$. This result can also be proved directly from the definition of the definite integral.

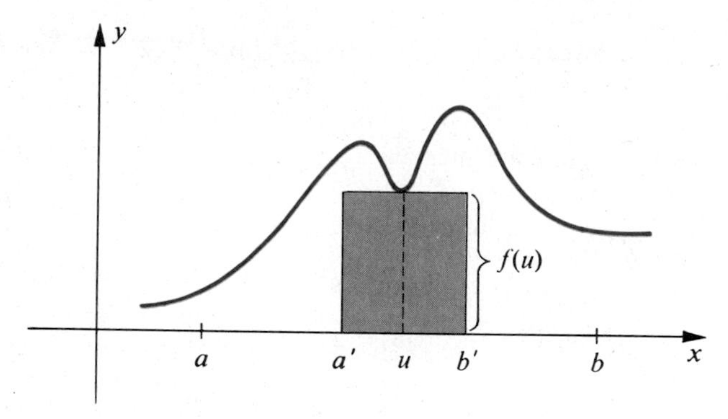

Figure 5.10

If f and g are continuous on $[a, b]$ and if $f(x) \geq g(x)$ for all x in $[a, b]$, but $f \neq g$, then $f(x) - g(x) > 0$ for some x and, by the previous discussion, $\int_a^b [f(x) - g(x)]\,dx > 0$. Consequently $\int_a^b f(x)\,dx > \int_a^b g(x)\,dx$. This fact will be used in the proof of the next theorem.

(5.29) The Mean Value Theorem for Definite Integrals

If f is continuous on a closed interval $[a, b]$, then there is some number z in the open interval (a, b) such that

$$\int_a^b f(x)\,dx = f(z)(b - a).$$

Proof. If f is a constant function, then the result follows trivially from (5.21) where z is *any* number in (a, b). Next assume that f is not a constant function and suppose m and M are the minimum and maximum values of f, respectively, on $[a, b]$. Let $f(u) = m$ and $f(v) = M$ where u and v are in $[a, b]$. Since f is not a constant function, $m < f(x) < M$ for some x in $[a, b]$ and hence by the remark immediately preceding this theorem,

$$\int_a^b m\,dx < \int_a^b f(x)\,dx < \int_a^b M\,dx.$$

Employing (5.21),

$$m(b - a) < \int_a^b f(x)\,dx < M(b - a).$$

Dividing by $b - a$ and replacing m and M by $f(u)$ and $f(v)$, respectively, we obtain

$$f(u) < \frac{1}{b - a}\int_a^b f(x)\,dx < f(v).$$

Since $[1/(b - a)]\int_a^b f(x)\,dx$ is a number between $f(u)$ and $f(v)$, it follows from the Intermediate Value Theorem (2.37) that there is a number z, strictly between u and v, such that

$$f(z) = \frac{1}{b - a}\int_a^b f(x)\,dx.$$

Multiplying both sides by $b - a$ gives us the conclusion of the theorem.

The number z of Theorem (5.29) is not necessarily unique. Indeed, as pointed out in the proof, if f is a constant function then *any* number z can be used. The theorem guarantees that at *least* one number z will produce the desired result.

The Mean Value Theorem has an interesting geometric interpretation if $f(x) \geq 0$ on $[a, b]$. In this case $\int_a^b f(x)\,dx$ is the area under the graph of f from a to b and the number $f(z)$ in Theorem (5.29) is the ordinate of the point P on the graph of f having abscissa z (see Figure 5.11). If a horizontal line is drawn through P, then the area of the rectangular region bounded by this line, the x-axis, and the lines $x = a$ and $x = b$ is $f(z)(b - a)$ which, according to Theorem (5.29), is the same as the area under the graph of f from a to b.

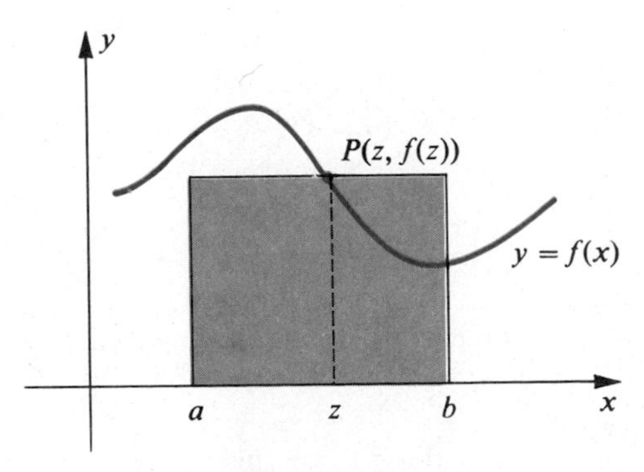

Figure 5.11

Example It can be proved that $\int_0^3 [4 - (x^2/4)]\,dx = 39/4$. Find a number which satisfies the conclusion of the Mean Value Theorem for this integral.

Solution According to the Mean Value Theorem for Definite Integrals, there is a number z between 0 and 3 such that

$$\int_0^3 \left(4 - \frac{x^2}{4}\right) dx = \left(4 - \frac{z^2}{4}\right)(3 - 0)$$

or, equivalently,

$$\frac{39}{4} = \left(\frac{16 - z^2}{4}\right)(3).$$

Multiplying both sides of the last equation by 4/3 leads to $13 = 16 - z^2$ and, therefore, $z^2 = 3$. Consequently $\sqrt{3}$ satisfies the conclusion of Theorem (5.29).

The Mean Value Theorem for Definite Integrals can be used to help prove a number of important theorems. One of the most important is the *Fundamental Theorem of Calculus* given in the next section.

EXERCISES 5.4

The definite integrals given in Exercises 1–10 may be verified by later methods. In each exercise find numbers which satisfy the conclusion of the Mean Value Theorem for Integrals (5.29).

1 $\int_0^3 3x^2\, dx = 27$

2 $\int_{-1}^3 (3x^2 - 2x + 3)\, dx = 32$

3 $\int_{-1}^8 3\sqrt{x+1}\, dx = 54$

4 $\int_{-2}^{-1} 8x^{-3}\, dx = -3$

5 $\int_1^2 (4x^3 - 1)\, dx = 14$

6 $\int_1^4 (2 + 3\sqrt{x})\, dx = 20$

7 $\int_2^7 \frac{1}{(x+3)^2}\, dx = \frac{1}{10}$

8 $\int_1^8 4\sqrt[3]{x}\, dx = 45$

9 $\int_0^a \sqrt{a^2 - x^2}\, dx = \frac{\pi}{4}a^2, a > 0$

10 $\int_1^3 \left(x^2 + \frac{1}{x^2}\right) dx = \frac{28}{3}$

11 If $f(x) = k$ for all x in $[a, b]$, prove that every number z in $[a, b]$ satisfies the conclusion of Theorem (5.29). Interpret this fact geometrically.

12 If $f(x) = x$ and $0 < a < b$, find (without integrating) a number z in (a, b) such that

$$\int_a^b f(x)\, dx = f(z)(b - a).$$

5.5 THE FUNDAMENTAL THEOREM OF CALCULUS

The task of evaluating a definite integral by means of Definition (5.16) is quite difficult even in the simplest cases. This section contains a theorem which can be used to find the definite integral without using limits of sums. Due to its importance in evaluating definite integrals, and because it exhibits the connection between differentiation and integration, the theorem is aptly called *The Fundamental Theorem of Calculus.* This theorem was discovered independently by Sir Isaac Newton (1642–1727) in England and by Gottfried Leibniz (1646–1716) in Germany. It is primarily because of this discovery that these outstanding mathematicians are credited with the invention of calculus.

To avoid confusion, in the following discussion we shall use the variable t and denote the definite integral of f from a to b by $\int_a^b f(t)\,dt$. If f is continuous on $[a, b]$ and if x is in $[a, b]$, then f is continuous on $[a, x]$ and hence by Theorem (5.19) f is integrable on $[a, x]$ whenever $a \leq x \leq b$. Consequently, the equation

(5.30)
$$G(x) = \int_a^x f(t)\,dt$$

defines a function G with domain $[a, b]$, since for each x in $[a, b]$ there corresponds a unique number $G(x)$ given by (5.30). The next theorem brings out the remarkable fact that G is an antiderivative of f. In addition, it shows how any antiderivative may be used to find a definite integral of f.

(5.31) The Fundamental Theorem of Calculus

Suppose f is continuous on a closed interval $[a, b]$.

PART I. If the function G is defined by

$$G(x) = \int_a^x f(t)\,dt$$

for all x in $[a, b]$, then G is an antiderivative of f on $[a, b]$.

PART II. If F is any antiderivative of f, then

$$\int_a^b f(x)\,dx = F(b) - F(a).$$

Proof. To establish Part I we must show that if x is in $[a, b]$, then $G'(x) = f(x)$, that is,

$$\lim_{h \to 0} \frac{G(x + h) - G(x)}{h} = f(x).$$

Before giving a formal proof, it is instructive to consider some geometric aspects of this formula. If $f(x) \geq 0$ throughout $[a, b]$, then $G(x)$ is the area under the graph of f from a to x, as illustrated in Figure 5.12. If $h > 0$, then the difference $G(x + h) - G(x)$ is the area under the graph of f from x to $x + h$, the number h is the length of the interval $[x, x + h]$, and $f(x)$ is the ordinate of the point with abscissa x on the graph of f. We will show below that $[G(x + h) - G(x)]/h = f(z)$, where z is between x and $x + h$. Reasoning intuitively, it appears that if $h \to 0$, then $z \to x$ and $f(z) \to f(x)$, which is what we wish to prove.

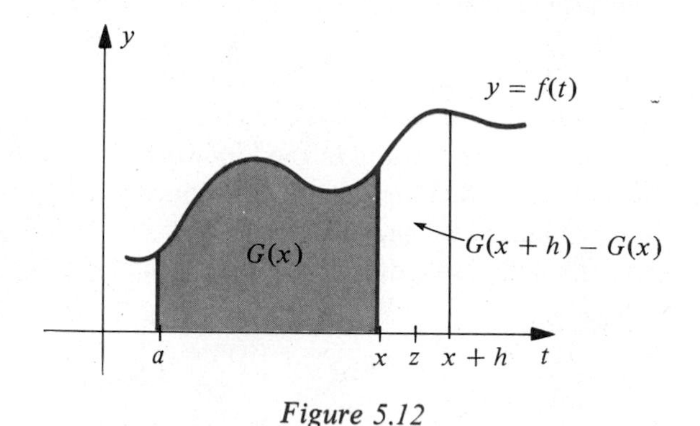

Figure 5.12

Let us now give a rigorous proof that $G'(x) = f(x)$ if f is continuous on $[a, b]$. If x and $x + h$ are in $[a, b]$, then using the definition of G, together with (5.17) and (5.25),

$$\begin{aligned} G(x + h) - G(x) &= \int_a^{x+h} f(t)\,dt - \int_a^x f(t)\,dt \\ &= \int_a^{x+h} f(t)\,dt + \int_x^a f(t)\,dt \\ &= \int_x^{x+h} f(t)\,dt. \end{aligned}$$

Consequently, if $h \neq 0$,

$$\frac{G(x + h) - G(x)}{h} = \frac{1}{h}\int_x^{x+h} f(t)\,dt.$$

If $h > 0$, then by the Mean Value Theorem for Integrals (5.29), there is a number z (depending on h) in the open interval $(x, x + h)$ such that

$$\int_x^{x+h} f(t)\,dt = f(z)h$$

and, therefore,

$$\frac{G(x + h) - G(x)}{h} = f(z). \tag{5.32}$$

Since $x < z < x + h$ it follows that

$$\lim_{h \to 0^+} f(z) = \lim_{z \to x^+} f(z) = f(x)$$

and hence from (5.32),

$$\lim_{h\to 0^+} \frac{G(x+h)-G(x)}{h} = f(x).$$

If $h < 0$, then we may prove in similar fashion that

$$\lim_{h\to 0^-} \frac{G(x+h)-G(x)}{h} = f(x).$$

The last two one-sided limits imply that

$$G'(x) = \lim_{h\to 0} \frac{G(x+h)-G(x)}{h} = f(x),$$

which is what we wished to prove.

To prove Part II, let F be any antiderivative of f and let G be the special antiderivative defined by (5.30). It follows from Theorem (4.33) that F and G differ by a constant; that is, there is a number C such that $G(x) - F(x) = C$ for all x in $[a, b]$. Hence, from the definition of G,

$$\int_a^x f(t)\,dt - F(x) = C$$

for all x in $[a, b]$. If we let $x = a$ and use Definition (5.18), then $0 - F(a) = C$. Consequently

$$\int_a^x f(t)\,dt - F(x) = -F(a).$$

Since this is an identity for all x in $[a, b]$ we may substitute b for x, obtaining

$$\int_a^b f(t)\,dt - F(b) = -F(a).$$

Adding $F(b)$ to both sides of this equation and replacing the variable t by x gives us the desired conclusion.

In terms of the differential operator D_x, the first part of the Fundamental Theorem (5.31) implies that $D_x G(x) = G'(x) = f(x)$. This gives us the following formula:

(5.33) $$D_x \int_a^x f(t)\,dt = f(x).$$

It is customary to denote the difference $F(b) - F(a)$ either by the symbol $F(x)]_a^b$ or by $[F(x)]_a^b$. We may then write

(5.34) $$\int_a^b f(x)\,dx = F(x)\Big]_a^b = F(b) - F(a)$$

where $F'(x) = f(x)$. The preceding formula is also valid if $a \geq b$, for if $a > b$, then by Definition (5.17)

$$\begin{aligned} \int_a^b f(x)\,dx &= -\int_b^a f(x)\,dx \\ &= -[F(a) - F(b)] \\ &= F(b) - F(a). \end{aligned}$$

If $a = b$, then by Definition (5.18)

$$\int_a^a f(x)\,dx = 0 = F(a) - F(a).$$

Example 1 Evaluate $\int_{-2}^{3} (6x^2 - 5)\,dx$

Solution An antiderivative of $6x^2 - 5$ is given by $F(x) = 2x^3 - 5x$. Applying (5.34),

$$\begin{aligned} \int_{-2}^{3} (6x^2 - 5)\,dx = \Big[2x^3 - 5x\Big]_{-2}^{3} &= [2(3)^3 - 5(3)] - [2(-2)^3 - 5(-2)] \\ &= [54 - 15] - [-16 + 10] = 45. \end{aligned}$$

Note that if $F(x) + C$ is used in place of $F(x)$ in (5.34) the same result is obtained, since

$$\begin{aligned} \Big[F(x) + C\Big]_a^b &= \{F(b) + C\} - \{F(a) + C\} \\ &= F(b) - F(a) = \Big[F(x)\Big]_a^b. \end{aligned}$$

This is in keeping with the statement that *any* antiderivative F may be employed in the Fundamental Theorem. Also note that for any number k,

$$\Big[kF(x)\Big]_a^b = kF(b) - kF(a) = k[F(b) - F(a)] = k\Big[F(x)\Big]_a^b;$$

that is, a constant factor can be "taken out" of the bracket when using this notation. This result is analogous to Theorem (5.22) on definite integrals.

If f is defined by $f(x) = kx^r$ where k is a real number and r is a rational number different from -1, then the function F defined by $F(x) = (k/(r + 1))x^{r+1}$ is an antiderivative of f, whenever f is defined. Applying (5.34) gives us the following formula for definite integrals:

(5.35)
$$\int_a^b kx^r\,dx = \left[\left(\frac{k}{r+1}\right)x^{r+1}\right]_a^b = \left(\frac{k}{r+1}\right)(b^{r+1} - a^{r+1})$$

provided $r \neq -1$. If the integrand is a sum of terms of the form kx^r where $r \neq -1$, then (5.35) may be applied to each term.

Example 2 Evaluate $\int_{-1}^{2} (x^3 + 1)^2\, dx$.

Solution Squaring the integrand and then applying (5.35) to each term gives us

$$\int_{-1}^{2} (x^3 + 1)^2\, dx = \int_{-1}^{2} (x^6 + 2x^3 + 1)\, dx$$

$$= \left[\frac{1}{7}x^7 + \frac{2}{4}x^4 + x\right]_{-1}^{2}$$

$$= \left[\frac{1}{7}(2)^7 + \frac{1}{2}(2)^4 + 2\right] - \left[\frac{1}{7}(-1)^7 + \frac{1}{2}(-1)^4 + (-1)\right] = \frac{405}{14}.$$

It is very important to note that

$$\int_{-1}^{2} (x^3 + 1)^2\, dx \neq \left.\frac{(x^3 + 1)^3}{3}\right]_{-1}^{2}.$$

Example 3 Evaluate $\int_{1}^{4} \left(5x - 2\sqrt{x} + \frac{32}{x^3}\right) dx$.

Solution We begin by changing the form of the integrand so that (5.35) may be applied to each term. Thus

$$\int_{1}^{4} (5x - 2x^{1/2} + 32x^{-3})\, dx = \left[\frac{5}{2}x^2 - \frac{2}{(3/2)}x^{3/2} + \frac{32}{-2}x^{-2}\right]_{1}^{4}$$

$$= \left[\frac{5}{2}x^2 - \frac{4}{3}x^{3/2} - \frac{16}{x^2}\right]_{1}^{4}$$

$$= \left[\frac{5}{2}(4)^2 - \frac{4}{3}(4)^{3/2} - \frac{16}{4^2}\right] - \left[\frac{5}{2} - \frac{4}{3} - 16\right]$$

$$= 259/6.$$

Example 4 Evaluate $\int_{-2}^{3} |x|\, dx$.

Solution By Definition (1.2), $|x| = -x$ if $x < 0$ and $|x| = x$ if $x \geq 0$. This suggests that we use Theorem (5.25) to express the given integral as a sum of two definite integrals as follows:

$$\begin{aligned}\int_{-2}^{3} |x|\,dx &= \int_{-2}^{0} |x|\,dx + \int_{0}^{3} |x|\,dx\\ &= \int_{-2}^{0} (-x)\,dx + \int_{0}^{3} x\,dx\\ &= -\left[\frac{x^2}{2}\right]_{-2}^{0} + \left[\frac{x^2}{2}\right]_{0}^{3}\\ &= -\left[0 - \frac{4}{2}\right] + \left[\frac{9}{2} - 0\right]\\ &= 2 + \frac{9}{2} = \frac{13}{2}.\end{aligned}$$

It should be observed that any formula about derivatives can be used to obtain a formula for definite integrals. For example,

$$D_x(\sqrt{x^2+1}) = \frac{x}{\sqrt{x^2+1}} \quad \text{implies that} \quad \int_a^b \frac{x}{\sqrt{x^2+1}}\,dx = \sqrt{x^2+1}\,\Big]_a^b$$

$$D_x(x-5)^{10} = 10(x-5)^9 \quad \text{implies that} \quad \int_a^b 10(x-5)^9\,dx = (x-5)^{10}\Big]_a^b.$$

These formulas should not be memorized. In the next section a method will be discussed which will enable us to evaluate integrals of the above types.

EXERCISES 5.5

Evaluate the definite integrals in Exercises 1–26.

1 $\int_1^4 (x^2 - 4x - 3)\,dx$

2 $\int_{-2}^{3} (5 + x - 6x^2)\,dx$

3 $\int_{-2}^{3} (8z^3 + 3z - 1)\,dz$

4 $\int_0^2 (w^4 - 2w^3)\,dw$

5 $\int_7^{12} dx$

6 $\int_{-6}^{-1} 8\,dx$

7 $\int_1^2 \frac{5}{8x^6}\,dx$

8 $\int_1^4 \sqrt{16x^5}\,dx$

9 $\int_4^9 \frac{t-3}{\sqrt{t}}\,dt$

10 $\int_{-1}^{-2} \frac{2s-7}{s^3}\,ds$

11 $\int_{-8}^{8} (\sqrt[3]{s^2} + 2)\,ds$

12 $\int_1^0 t^2(\sqrt[3]{t} - \sqrt{t})\,dt$

13 $\int_0^1 (2x-3)(5x+1)\,dx$

14 $\int_{-1}^{1} (x^2+1)^2\,dx$

15 $\int_{-1}^{0} (2w + 3)^2 \, dw$

16 $\int_{5}^{5} \sqrt[3]{x^2 + \sqrt{x^5 + 1}} \, dx$

17 $\int_{3}^{2} \frac{x^2 - 1}{x - 1} \, dx$

18 $\int_{0}^{-1} \frac{x^3 + 8}{x + 2} \, dx$

19 $\int_{1}^{1} (4x^2 - 5)^{100} \, dx$

20 $\int_{1}^{2} (4u^{-5} + 6u^{-4}) \, du$

21 $\int_{1}^{3} \frac{2x^3 - 4x^2 + 5}{x^2} \, dx$

22 $\int_{-2}^{-1} \left(r - \frac{1}{r}\right)^2 dr$

23 $\int_{-3}^{6} |x - 4| \, dx$

24 $\int_{-1}^{1} (x + 1)(x + 2)(x + 3) \, dx$

25 $\int_{0}^{4} \sqrt{3t}(\sqrt{t} + \sqrt{3}) \, dt$

26 $\int_{-1}^{5} |2x - 3| \, dx$

Verify the identities in Exercises 27 and 28 by first using the Fundamental Theorem of Calculus (5.31) and then differentiating.

27 $D_x \int_{0}^{x} (t^3 - 4\sqrt{t} + 5) \, dt = x^3 - 4\sqrt{x} + 5$ if $x \geq 0$

28 $D_x \int_{0}^{x} (5t + 3)^2 \, dt = (5x + 3)^2$

29 Find $D_x \int_{1}^{x} \frac{1}{t} \, dt$, if $x > 0$.

30 Find $D_x \int_{0}^{x} \frac{1}{\sqrt{1 - t^2}} \, dt$, if $|x| < 1$.

31 If $f(x) = x^2 + 1$, find the area of the region under the graph of f from -1 to 2.

32 If $f(x) = x^3$, find the area of the region under the graph of f from 1 to 3.

Verify the formulas in Exercises 33–36.

33 $\int_{a}^{b} \frac{x}{\sqrt{(x^2 + 1)^3}} \, dx = \left.\frac{-1}{\sqrt{x^2 + 1}}\right]_{a}^{b}$

34 $\int_{a}^{b} \frac{x^3}{\sqrt{x^2 + 1}} \, dx = \left[\frac{1}{3}\sqrt{(x^2 + 1)^3} - \sqrt{x^2 + 1}\right]_{a}^{b}$

35 $\int_{a}^{b} \frac{x}{(x^2 + c^2)^{n+1}} \, dx = \left.\frac{-1}{2n(x^2 + c^2)^n}\right]_{a}^{b}, \quad n \neq 0$

36 $\int_{a}^{b} \frac{x^2}{(x^3 + c^3)^{n+1}} \, dx = \left.\frac{-1}{3n(x^3 + c^3)^n}\right]_{a}^{b}, \quad n \neq 0$

Find numbers which satisfy the conclusion of the Mean Value Theorem for Integrals (5.29) for the definite integrals in Exercises 37–40.

37 $\int_{0}^{4} (\sqrt{x} + 1) \, dx$

38 $\int_{-1}^{1} (2x + 1)^2 \, dx$

39 $\int_{-1}^{2} (3x^3 + 2) \, dx$

40 $\int_{1}^{9} \frac{3}{x^2} \, dx$

41 If f is continuous on $[a, b]$, then the **average value of f on $[a, b]$** is defined as the number

$$\frac{1}{b - a} \int_{a}^{b} f(x) \, dx.$$

(a) If $f(x) = x^2 + 3x - 1$, find the average value of f on $[-1, 2]$.

(b) Suppose a point P which moves on a coordinate line has a continuous velocity function v. Show that the average velocity during the time interval $[a, b]$ equals the average value of v on $[a, b]$.

42 Refer to Exercise 41. If a function f has a continuous derivative on $[a, b]$, show that the average rate of change of $f(x)$ with respect to x on $[a, b]$ (see Section 4.7) equals the average value of f' on $[a, b]$.

5.6 INDEFINITE INTEGRALS AND CHANGE OF VARIABLES

The connection between antiderivatives and definite integrals provided by the Fundamental Theorem of Calculus has made it customary to use integral signs to denote antiderivatives. In order to distinguish the latter from definite integrals, no limits of integration are attached to the integral sign. Specifically, we write

(5.36) $$\int f(x)\,dx = F(x) + C \quad \text{if and only if} \quad F'(x) = f(x)$$

where C is an arbitrary constant. The expression on the left in (5.36) is called an **indefinite integral** of f (or of $f(x)$) and is merely another way of specifying the most general antiderivative of f. Instead of using the terminology *antidifferentiation* for the process of finding F when f is given, we now use the phrase **indefinite integration**. The arbitrary constant C is called the **constant of integration**, $f(x)$ is called the **integrand**, and x is called the **variable of integration**. We often refer to the process of finding $F(x) + C$ in (5.36) as **evaluating the indefinite integral**. The domain of F will not usually be stated explicitly. It is always assumed that a suitable interval over which f is integrable has been chosen. In particular, a closed interval on which f is continuous could be used. As with definite integrals, the symbol employed for the variable of integration is insignificant since, for example, $\int f(t)\,dt$, $\int f(u)\,du$, and so on, give rise to the same function F as $\int f(x)\,dx$.

As illustrations of this new notation for antiderivatives we have

$$\int x^2\,dx = \tfrac{1}{3}x^3 + C, \quad \text{and}$$

$$\int \left(8t^3 - 5 + \frac{1}{t^2}\right) dt = 2t^4 - 5t - \frac{1}{t} + C.$$

In order to check equalities of this type, we may differentiate the expression on the right-hand side and see whether or not the integrand is obtained.

The following formulas are immediate consequences of the definition of indefinite integral:

(5.37) $$D_x \int f(x)\,dx = f(x)$$

(5.38) $$\int D_x[f(x)]\,dx = f(x) + C.$$

Thus from (5.36),

$$D_x \int f(x)\,dx = D_x[F(x) + C] = F'(x) = f(x).$$

The second formula follows from the fact that $f(x) + C$ is an antiderivative of $D_x[f(x)]$.

Since the indefinite integral of f is an antiderivative, the Fundamental Theorem of Calculus gives us the following relationship between definite and indefinite integrals:

(5.39)
$$\int_a^b f(x)\,dx = \left[\int f(x)\,dx\right]_a^b.$$

Thus, if the indefinite integral of a function f is known, then definite integrals of f can be evaluated. In future chapters, methods for finding indefinite integrals of various functions will be developed. At the present time it is possible to state several useful rules. For example, the **power rule for (indefinite) integration** is

(5.40)
$$\int x^r\,dx = \left(\frac{1}{r+1}\right)x^{r+1} + C,$$

where the exponent r is a rational number and $r \neq -1$. In Chapter 8 this rule will be extended to real exponents.

Formula (4.36), which states that an antiderivative of a sum is the sum of the antiderivatives, may be translated into the notation of indefinite integrals as follows:

(5.41)
$$\int [f(x) + g(x)]\,dx = \int f(x)\,dx + \int g(x)\,dx.$$

Another integration formula is

(5.42)
$$\int c\,f(x)\,dx = c\int f(x)\,dx$$

where c is any real number. To prove this rule we merely differentiate the right-hand side, obtaining

$$D_x\left[c\int f(x)\,dx\right] = c \cdot D_x \int f(x)\,dx = c\,f(x)$$

which is the integrand on the left side of (5.42). By combining the last three formulas it is easy to find the indefinite integral of any polynomial.

As is the case with definite integrals, every rule for differentiation can be transformed into a corresponding rule for indefinite integration. For example,

$$D_x\sqrt{x^2+5} = \frac{x}{\sqrt{x^2+5}} \quad \text{implies that} \quad \int \frac{x}{\sqrt{x^2+5}}\,dx = \sqrt{x^2+5} + C;$$

$$D_x(x^3+2x)^{10} = 10(x^3+2x)^9(3x^2+2) \quad \text{implies that}$$

$$\int 10(x^3+2x)^9(3x^2+2)\,dx = (x^3+2x)^{10} + C.$$

A similar technique may be used in conjunction with the Chain Rule to obtain a general formula for indefinite integration. Thus, suppose F is an antiderivative of a function f, so that $F' = f$. In addition, suppose g is another differentiable function such that $g(x)$ is in the domain of F for every x in some closed interval $[a, b]$. We may then consider the composite function defined by $F(g(x))$ for all x in $[a, b]$. Applying the Chain Rule (3.36) and the fact that $F' = f$, we obtain

$$D_x F(g(x)) = F'(g(x))g'(x) = f(g(x))g'(x).$$

This, in turn, gives us the integration formula

$$\int f(g(x))g'(x)\,dx = F(g(x)) + C, \quad \text{where } F' = f.$$

There is a simple way to remember this formula. If we let $u = g(x)$ and formally replace $g'(x)\,dx$ by the differential du, we obtain

(5.43)
$$\int f(g(x))g'(x)\,dx = \int f(u)\,du = F(u) + C = F(g(x)) + C.$$

This memorization device indicates that $g'(x)\,dx$ may be regarded as the product of $g'(x)$ and dx. Indeed, formula (5.43) is one of the main reasons for using the symbol dx in the integral notation.

The technique of finding indefinite integrals by means of (5.43) will be referred to as a **change of variable** or as **the method of substitution.**

As a first application, we may extend the Power Rule (5.40) to the following formula involving powers of functions:

(5.44)
$$\int u^r\,du = \frac{1}{r+1}u^{r+1} + C$$

where $u = g(x)$, $du = g'(x)\,dx$, and $r \neq -1$. This is the special case of (5.43) with $u = g(x)$ and $f(x) = x^r$. The solutions to most of the exercises in this section will make use of formula (5.44). Later we shall apply the method of substitution to other types of integrals.

Example 1 Find $\int (2x^3 + 1)^7 x^2\,dx$.

Solution If we let $u = 2x^3 + 1$, then $du = 6x^2\,dx$. In order to obtain the form (5.43), it is necessary to introduce the factor 6 in the integrand. Doing this, and compensating by multiplying the integral by 1/6 (this is legitimate by (5.42)), we obtain

$$\int (2x^3 + 1)^7 x^2\,dx = \frac{1}{6}\int (2x^3 + 1)^7 6x^2\,dx.$$

Making the indicated substitution and integrating,

$$\int (2x^3 + 1)^7 x^2\,dx = \frac{1}{6}\int u^7\,du = \frac{1}{6}\left[\frac{1}{8}u^8 + K\right]$$

where K is a constant. It is now necessary to return to the original variable x. Since $u = 2x^3 + 1$, the last equality gives us

$$\int (2x^3 + 1)^7 x^2\, dx = \frac{1}{6}\left[\frac{1}{8}(2x^3 + 1)^8 + K\right]$$
$$= \frac{1}{48}(2x^3 + 1)^8 + \frac{1}{6}K.$$

Instead of employing constants of integration such as $\frac{1}{6}K$ it is customary to write this result as

$$\int (2x^3 + 1)^7 x^2\, dx = \frac{1}{48}(2x^3 + 1)^8 + C.$$

The relationship between K and C is $C = \frac{1}{6}K$; however, there is no practical advantage in remembering this fact. In the future we shall manipulate constants of integration in various ways without making explicit mention of relationships which exist. Moreover, instead of proceeding as above, it should be clear that it is permissible to integrate as follows:

$$\frac{1}{6}\int u^7\, du = \frac{1}{6}\left[\frac{1}{8}u^8\right] + C.$$

Formal substitutions in indefinite integrals can be made in different ways. To illustrate, another method for solving Example 1 is to let

$$u = 2x^3 + 1, \quad du = 6x^2\, dx, \quad \frac{1}{6}du = x^2\, dx.$$

We then substitute directly for $x^2\, dx$ instead of introducing the number 6 in the integrand as follows:

$$\int (2x^3 + 1)^7 x^2\, dx = \int u^7 \frac{1}{6}\, du = \frac{1}{6}\int u^7\, du$$
$$= \frac{1}{48}u^8 + C = \frac{1}{48}(2x^3 + 1)^8 + C.$$

The change of variable technique for indefinite integrals is a powerful tool, provided the student can recognize that the integrand is of the form $f(g(x))g'(x)$, or $f(g(x))kg'(x)$ where k is a real number. The ability to recognize this form is directly proportional to the number of exercises worked!

Example 2 Find $\int x\sqrt[3]{7 - 6x^2}\, dx$.

Solution Note that the integrand contains the term $x\, dx$. If the factor x were missing, the change of variable technique could not be applied. Letting

$$u = 7 - 6x^2, \quad du = -12x\, dx,$$

introducing the factor -12 in the integrand, and then compensating by multiplying the integral by $-1/12$, gives us

$$\begin{aligned}\int x\sqrt[3]{7-6x^2}\,dx &= -\frac{1}{12}\int \sqrt[3]{7-6x^2}(-12)x\,dx \\ &= -\frac{1}{12}\int \sqrt[3]{u}\,du \\ &= -\frac{1}{12}\int u^{1/3}\,du \\ &= -\frac{1}{12}\frac{u^{4/3}}{(4/3)} + C \\ &= -\frac{1}{16}(7-6x^2)^{4/3} + C.\end{aligned}$$

As in the second solution to Example 1, we could begin with $du = -12x\,dx$ and write $x\,dx = (-1/12)\,du$. The change of variables then takes on the form

$$\int \sqrt[3]{7-6x^2}\,x\,dx = \int \sqrt[3]{u}\left(-\frac{1}{12}\right)du = -\frac{1}{12}\int \sqrt[3]{u}\,du.$$

The remainder of the solution now proceeds exactly as before.

The method of substitution may also be used to evaluate definite integrals. We simply use (5.43) to find an indefinite integral (that is, an antiderivative) and then apply the Fundamental Theorem of Calculus. Another method, which is sometimes shorter, is to change the limits of integration. Using (5.43) together with the Fundamental Theorem gives us the following formula, where $F' = f$:

$$\int_a^b f(g(x))g'(x)\,dx = F(g(x))\Big]_a^b.$$

The number on the right of this equality may be written

$$F(g(b)) - F(g(a)) = F(u)\Big]_{g(a)}^{g(b)} = \int_{g(a)}^{g(b)} f(u)\,du.$$

This gives us the following important result, where for brevity we have not restated the restrictions on f and g.

(5.45) The Change of Variable Theorem

$$\boxed{\int_a^b f(g(x))g'(x)\,dx = \int_{g(a)}^{g(b)} f(u)\,du, \quad \text{where } u = g(x).}$$

Theorem (5.45) states that after making the substitutions $u = g(x)$ and $du = g'(x)\,dx$, we may use the values of g which correspond to $x = a$ and $x = b$, respectively, as the limits of the integral involving u. It is then unnecessary to return to the variable x after integrating. This technique is illustrated in the next example.

Example 3 Evaluate $\int_2^{10} \frac{3}{\sqrt{5x-1}}\,dx$.

Solution Let us begin by writing the integral as

$$3\int_2^{10} \frac{1}{\sqrt{5x-1}}\,dx.$$

The form of the integrand suggests the substitution $u = 5x - 1$. Of course,

$$\text{if} \quad u = 5x - 1, \quad \text{then} \quad du = 5\,dx.$$

To change the limits of integration we note that

$$\text{if} \quad x = 2, \quad \text{then} \quad u = 9, \quad \text{and}$$
$$\text{if} \quad x = 10, \quad \text{then} \quad u = 49.$$

Applying Theorem (5.45) leads to the following evaluation:

$$\begin{aligned} 3\int_2^{10} \frac{1}{\sqrt{5x-1}}\,dx &= \frac{3}{5}\int_2^{10} \frac{1}{\sqrt{5x-1}}5\,dx \\ &= \frac{3}{5}\int_9^{49} \frac{1}{\sqrt{u}}\,du \\ &= \frac{3}{5}\int_9^{49} u^{-1/2}\,du \\ &= \left(\frac{3}{5}\right)2u^{1/2}\Big]_9^{49} \\ &= \frac{6}{5}\left[49^{1/2} - 9^{1/2}\right] = \frac{24}{5}. \end{aligned}$$

EXERCISES 5.6

Evaluate the integrals in Exercises 1–22.

1 $\int (3x+1)^4\,dx$

2 $\int (2x^2-3)^5 x\,dx$

3 $\int \sqrt{t^3-1}\,t^2\,dt$

4 $\int \sqrt{9-z^2}\,z\,dz$

5 $\int \frac{x-2}{(x^2-4x+3)^3}\,dx$

6 $\int \frac{x^2+x}{(4-3x^2-2x^3)^4}\,dx$

7 $\int \frac{s}{\sqrt[3]{1-2s^2}}\,ds$

8 $\int \sqrt[5]{t^4-t^2}(10t^3-5t)\,dt$

9 $\displaystyle\int \frac{(\sqrt{u}+3)^4}{\sqrt{u}}\,du$

10 $\displaystyle\int \left(1+\frac{1}{u}\right)^{-3}\left(\frac{1}{u^2}\right)du$

11 $\displaystyle\int_1^4 \sqrt{5-x}\,dx$

12 $\displaystyle\int_1^5 \sqrt[3]{2x-1}\,dx$

13 $\displaystyle\int_{-1}^1 (t^2-1)^3 t\,dt$

14 $\displaystyle\int_{-2}^0 \frac{v^2}{(v^3-2)^2}\,dv$

15 $\displaystyle\int_0^1 \frac{1}{(3-2v)^2}\,dv$

16 $\displaystyle\int_0^4 \frac{x}{\sqrt{x^2+9}}\,dx$

17 $\displaystyle\int (x^2+1)^3\,dx$

18 $\displaystyle\int (3-x^3)^2 x\,dx$

19 $\displaystyle\int 5\sqrt{8x+5}\,dx$

20 $\displaystyle\int \frac{6}{\sqrt{4-5t}}\,dt$

21 $\displaystyle\int_1^4 \frac{1}{\sqrt{x}(\sqrt{x}+1)^3}\,dx$

22 $\displaystyle\int (3-x^4)^3 x^3\,dx$

Evaluate the integrals in Exercises 23–26 by (a) the method of substitution and (b) expanding the integrand. In what way do the constants of integration differ?

23 $\displaystyle\int (x+4)^2\,dx$

24 $\displaystyle\int (x^2+4)^2 x\,dx$

25 $\displaystyle\int \frac{(\sqrt{x}+3)^2}{\sqrt{x}}\,dx$

26 $\displaystyle\int \left(1+\frac{1}{x}\right)^2 \frac{1}{x^2}\,dx$

Verify the identities in Exercises 27 and 28 by first integrating and then differentiating.

27 $\displaystyle D_x \int x^3\sqrt{x^4+5}\,dx = x^3\sqrt{x^4+5}$

28 $\displaystyle D_x \int (3x+2)^7\,dx = (3x+2)^7$

29 Find $\displaystyle D_x \int \frac{1}{\sqrt{x^3+x+5}}\,dx$.

30 Find $\displaystyle \int D_x \frac{x}{\sqrt[3]{x+1}}\,dx$.

31 Find $\displaystyle \int_0^3 D_x\sqrt{x^2+16}\,dx$.

32 Find $\displaystyle D_x \int_0^1 \sqrt{x^2+4}\,x\,dx$.

33 If $f(x) = \sqrt{x+1}$, find the area of the region under the graph of f from 0 to 3.

34 If $f(x) = x/(x^2+1)^2$, find the area of the region under the graph of f from 1 to 2.

In Exercises 35–38 find numbers which satisfy the conclusion of the Mean Value Theorem for Integrals.

35 $\displaystyle\int_0^4 \frac{x}{\sqrt{x^2+9}}\,dx$

36 $\displaystyle\int_{-2}^0 \sqrt[3]{x+1}\,dx$

37 $\displaystyle\int_0^5 \sqrt{x+4}\,dx$

38 $\displaystyle\int_{-3}^2 \sqrt{6-x}\,dx$

Verify the formulas in Exercises 39 and 40.

39 $\displaystyle\int \frac{\sqrt{x^2-a^2}}{x^4}\,dx = \frac{\sqrt{(x^2-a^2)^3}}{3a^2x^3} + C$

40 $\displaystyle\int \frac{1}{x^2\sqrt{x^2-a^2}}\,dx = \frac{\sqrt{x^2-a^2}}{a^2x} + C$

41 Let f be continuous on $[-a, a]$. If f is an even function show that

$$\int_{-a}^{a} f(x)\,dx = 2\int_{0}^{a} f(x)\,dx$$

and interpret this result geometrically.

42 Let f be continuous on $[-a, a]$. If f is an odd function show that $\int_{-a}^{a} f(x)\,dx = 0$ and interpret this result geometrically.

5.7 NUMERICAL INTEGRATION

In order to evaluate a definite integral $\int_a^b f(x)\,dx$ by means of the Fundamental Theorem of Calculus, it is necessary to find an antiderivative of f. If an antiderivative cannot be found, then numerical methods may be used to approximate the integral to any degree of accuracy. For example, if the norm of a partition of $[a, b]$ is small, then by Definition (5.16) the definite integral can be approximated by any Riemann sum of f. In particular, if we use a regular partition with $\Delta x = (b - a)/n$, then

(5.46)
$$\int_a^b f(x)\,dx \approx \sum_{i=1}^{n} f(w_i)\,\Delta x$$

where w_i is any number in the ith subinterval $[x_{i-1}, x_i]$ of the partition. Of course, the accuracy of the approximation depends upon the nature of f and the magnitude of Δx. It may be necessary to make Δx very small in order to obtain the desired degree of accuracy. This, in turn, means that n is large, and hence the sum in (5.46) contains many terms. Figure 5.2 illustrates the case in which $f(w_i)$ is the minimum value of f in $[x_{i-1}, x_i]$. In this case the error involved in the approximation is numerically the same as the area of the unshaded region which lies under the graph of f and over the inscribed rectangles.

If we take $w_i = x_{i-1}$, that is, if f is evaluated at the left-hand end-point of each subinterval $[x_{i-1}, x_i]$, then (5.46) becomes

$$\int_a^b f(x)\,dx \approx \sum_{i=1}^{n} f(x_{i-1})\,\Delta x.$$

If we let $w_i = x_i$, that is, if f is evaluated at the right-hand end-point of $[x_{i-1}, x_i]$, then

$$\int_a^b f(x)\,dx \approx \sum_{i=1}^{n} f(x_i)\,\Delta x.$$

Another, and usually more accurate approximation, can be obtained by using the average of the last two approximations, that is,

$$\frac{1}{2}\left[\sum_{i=1}^{n} f(x_{i-1})\,\Delta x + \sum_{i=1}^{n} f(x_i)\,\Delta x\right].$$

With the exception of $f(x_0)$ and $f(x_n)$, each $f(x_i)$ appears twice, and hence we may write this expression as

$$\frac{\Delta x}{2}\left[f(x_0) + \sum_{i=1}^{n-1} 2f(x_i) + f(x_n)\right].$$

Since $\Delta x = (b - a)/n$, this gives us the following rule.

(5.47) Trapezoidal Rule

If f is continuous on $[a, b]$ and if a regular partition of $[a, b]$ is determined by the numbers $a = x_0, x_1, \ldots, x_n = b$, then

$$\int_a^b f(x)\,dx \approx \frac{b-a}{2n}[f(x_0) + 2f(x_1) + 2f(x_2) + \cdots + 2f(x_{n-1}) + f(x_n)].$$

The term *trapezoidal* arises from the case in which $f(x)$ is nonnegative on $[a, b]$. As illustrated in Figure 5.13, if P_k is the point with abscissa x_k on the graph of $y = f(x)$, then for each $i = 1, 2, \ldots, n$, the points on the x-axis with abscissas x_{i-1} and x_i, together with P_{i-1} and P_i, are vertices of a trapezoid having area

$$\frac{\Delta x}{2}[f(x_{i-1}) + f(x_i)].$$

The sum of the areas of these trapezoids is the same as the sum in Rule (5.47). Hence, in geometric terms, the Trapezoidal Rule gives us an approximation to the area under the graph of f from a to b by means of trapezoids instead of the rectangles associated with Riemann sums.

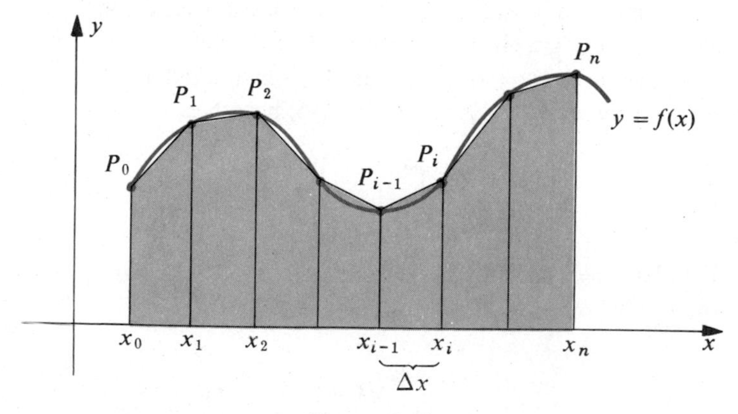

Figure 5.13

The next result provides information about the maximum error which can occur if the Trapezoidal Rule is used to approximate a definite integral. The proof requires advanced methods and is, therefore, omitted.

(5.48) Error Estimate for the Trapezoidal Rule

> If M is a positive real number such that $|f''(x)| \le M$ for all x in $[a, b]$, then the error involved in using the Trapezoidal Rule (5.47) is not greater than $M(b - a)^3/12n^2$.

Example 1 Approximate $\int_1^2 (1/x)\,dx$ by using the Trapezoidal Rule with $n = 10$. Estimate the maximum error in the approximation.

Solution Note that at this stage of our work it is impossible to find a function F such that $F'(x) = 1/x$ and hence we cannot use the Fundamental Theorem of Calculus to evaluate the integral. It is convenient to arrange our work as follows, where each $f(x_i)$ was obtained with a hand-held calculator and is accurate to four decimal places.

i	x_i	$f(x_i)$	m	$mf(x_i)$
0	1.0	1.0000	1	1.0000
1	1.1	0.9091	2	1.8182
2	1.2	0.8333	2	1.6666
3	1.3	0.7692	2	1.5384
4	1.4	0.7143	2	1.4286
5	1.5	0.6667	2	1.3334
6	1.6	0.6250	2	1.2500
7	1.7	0.5882	2	1.1764
8	1.8	0.5556	2	1.1112
9	1.9	0.5263	2	1.0526
10	2.0	0.5000	1	0.5000

The sum of the numbers in the last column is 13.8754. Since

$$(b - a)/2n = (2 - 1)/20 = 0.05,$$

it follows from (5.47) that

$$\int_1^2 \frac{1}{x}\,dx \approx (0.05)(13.8754) \approx 0.6938.$$

The error in the approximation may be estimated by means of (5.48). Since $f(x) = 1/x$ we have $f'(x) = -1/x^2$ and $f''(x) = 2/x^3$. The maximum value of $f''(x)$ on the interval $[1, 2]$ occurs at $x = 1$ and hence

$$|f''(x)| \le 2/(1)^3 = 2.$$

Applying (5.48) with $M = 2$ we see that the maximum error is not greater than

$$\frac{2(2-1)^3}{12(10)^2} = \frac{1}{600} < 2 \times 10^{-3}.$$

It will be shown in Chapter 8 that the integral in Example 1 equals the natural logarithm of 2, denoted by ln 2. Using a hand-held calculator it is possible to verify that to five decimal places, ln 2 is approximated by 0.69315. In order to obtain this approximation by means of the Trapezoidal Rule it is necessary to use a very large value of n.

The following Rule is usually more accurate than the Trapezoidal Rule.

(5.49) Simpson's Rule

Suppose f is continuous on $[a, b]$ and n is an even integer. If a regular partition is determined by $a = x_0, x_1, \ldots, x_n = b$, then

$$\int_a^b f(x)\,dx \approx \frac{b-a}{3n}[f(x_0) + 4f(x_1) + 2f(x_2) + 4f(x_3) + \cdots + 2f(x_{n-2}) + 4f(x_{n-1}) + f(x_n)].$$

The idea behind the proof of Simpson's Rule is that instead of using trapezoids to approximate the graph of f, we use portions of graphs of equations of the form $y = cx^2 + dx + e$. If $c \neq 0$, then the graph of such an equation is a parabola. If $P_0(x_0, y_0)$, $P_1(x_1, y_1)$, and $P_2(x_2, y_2)$ are points, where $x_0 < x_1 < x_2$, then substituting the coordinates of P_0, P_1, and P_2, respectively, in this equation produces three equations which may be solved for c, d, and e. As a special case, suppose h, y_0, y_1, and y_2 are positive, and consider the points $P_0(-h, y_0)$, $P_1(0, y_1)$, and $P_2(h, y_2)$, as illustrated in (i) of Figure 5.14. The area A under the graph of the equation from $-h$ to h is

$$A = \int_{-h}^{h} (cx^2 + dx + e)\,dx = \left[\frac{cx^3}{3} + \frac{dx^2}{2} + ex\right]_{-h}^{h} = \frac{h}{3}(2ch^2 + 6e).$$

Since the coordinates of P_0, P_1, and P_2 satisfy the equation $y = cx^2 + dx + e$, we have

$$\begin{aligned} y_0 &= ch^2 - dh + e \\ y_1 &= e \\ y_2 &= ch^2 + dh + e \end{aligned}$$

from which it follows that

$$y_0 + 4y_1 + y_2 = 2ch^2 + 6e.$$

Consequently,

$$A = \frac{h}{3}(y_0 + 4y_1 + y_2). \tag{5.50}$$

If the points P_0, P_1, and P_2 are translated horizontally, as illustrated in (ii) of Figure 5.14, then the area under the graph remains the same. Consequently (5.50) is true for *any* points P_0, P_1, and P_2, provided $x_1 - x_0 = x_2 - x_1$.

If $f(x) \geq 0$ on $[a, b]$, then Simpson's Rule is derived by regarding the definite integral as the area under the graph of f from a to b. Thus, suppose n is an even integer and $h = (b - a)/n$. We divide $[a, b]$ into n subintervals, each of length h, by choosing numbers $a = x_0, x_1, \ldots, x_n = b$. Let $P_k(x_k, y_k)$ be the point on the graph

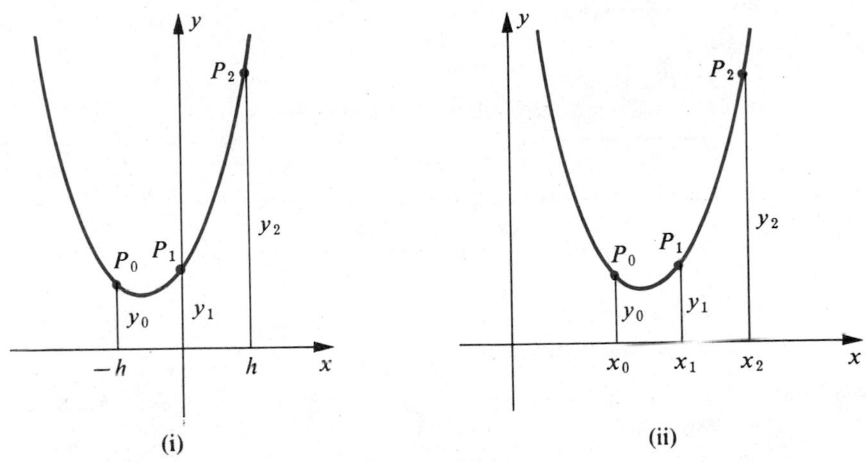

Figure 5.14

of f with abscissa x_k, as illustrated in Figure 5.15. If the arc through P_0, P_1, and P_2 is approximated by the graph of an equation $y = cx^2 + dx + e$, then the area under the graph of f from x_0 to x_2 is approximated by (5.50). If the arc through P_2, P_3 and P_4 is approximated in similar fashion, then the area under the graph of f from x_2 to x_4 is approximately

$$\frac{h}{3}(y_2 + 4y_3 + y_4).$$

We continue in this manner until we reach the last triple of points P_{n-2}, P_{n-1}, P_n and the corresponding approximation to the area under the graph, namely,

$$\frac{h}{3}(y_{n-2} + 4y_{n-1} + y_n).$$

Summing these approximations gives us

$$\int_a^b f(x)\,dx \approx \frac{h}{3}(y_0 + 4y_1 + 2y_2 + 4y_3 + \cdots + 2y_{n-2} + 4y_{n-1} + y_n),$$

which is the same as the sum in (5.49). If f is negative in $[a, b]$, then negatives of areas may be used to establish Simpson's Rule.

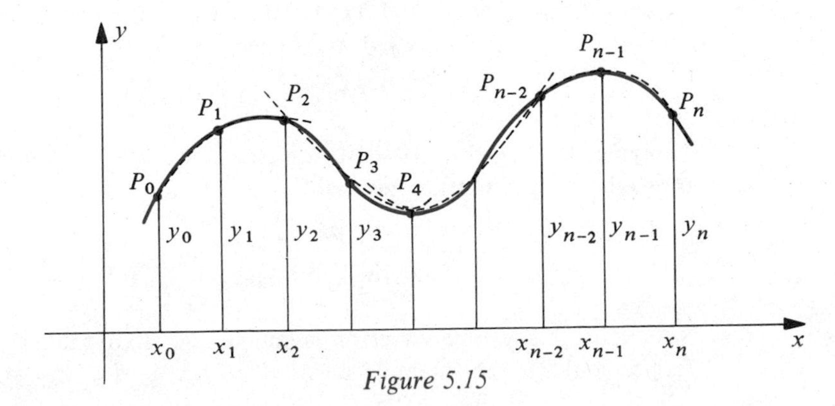

Figure 5.15

The following result may be established using advanced methods.

(5.51) Error Estimate for Simpson's Rule

If M is a positive real number such that $|f^{(4)}(x)| \leq M$ for all x in $[a, b]$, then the error involved in using Simpson's Rule (5.49) is not greater than $M(b-a)^5/180n^4$.

Example 2 Approximate $\int_1^2 (1/x)\,dx$ by using Simpson's Rule with $n = 10$. Estimate the error in the approximation.

Solution This is the same integral considered in Example 1. We arrange our work as follows:

i	x_i	$f(x_i)$	m	$mf(x_i)$
0	1.0	1.0000	1	1.0000
1	1.1	0.9091	4	3.6364
2	1.2	0.8333	2	1.6666
3	1.3	0.7692	4	3.0768
4	1.4	0.7143	2	1.4286
5	1.5	0.6667	4	2.6668
6	1.6	0.6250	2	1.2500
7	1.7	0.5882	4	2.3528
8	1.8	0.5556	2	1.1112
9	1.9	0.5263	4	2.1052
10	2.0	0.5000	1	0.5000

The sum of the numbers in the last column is 20.7944. Since

$$(b-a)/3n = (2-1)/30,$$

it follows from (5.49) that

$$\int_1^2 \frac{1}{x}\,dx \approx (1/30)(20.7944) \approx 0.6932.$$

We shall use (5.51) to estimate the error in the approximation. If $f(x) = 1/x$, the reader may verify that $f^{(4)}(x) = 24/x^5$. The maximum value of $f^{(4)}(x)$ on the interval $[1, 2]$ occurs at $x = 1$ and hence

$$|f^{(4)}(x)| \leq 24/(1)^3 = 24.$$

Applying (5.51) with $M = 24$, we see that the maximum error in the approximation is not greater than

$$\frac{24(2-1)^3}{180(10)^4} = \frac{2}{150000} < 1.4 \times 10^{-5}.$$

Note that this estimated error is much less than that obtained using the Trapezoidal Rule in Example 1.

An important aspect of numerical integration is that it can be used to approximate the integral of a function which is described in tabular form. To illustrate, suppose it is found experimentally that two physical variables x and y are related as shown in the following table.

x	1.0	1.5	2.0	2.5	3.0	3.5	4.0
y	3.1	4.0	4.2	3.8	2.9	2.8	2.7

If we regard y as a function of x, say $y = f(x)$, where f is continuous, then the definite integral $\int_1^4 f(x)\,dx$ may have an important physical meaning. In the present illustration, this integral may be approximated without knowing the explicit formula for $f(x)$. In particular, if the Trapezoidal Rule (5.47) is used, with $n = 6$ and $(b - a)/2n = (4 - 1)/12 = 0.25$, then

$$\int_1^4 f(x)\,dx \approx 0.25[3.1 + 2(4.0) + 2(4.2) + 2(3.8) + 2(2.9) + 2(2.8) + 2.7]$$

$$\approx 10.3.$$

Since the number of subdivisions is even, we could also approximate the integral by means of Simpson's Rule.

EXERCISES 5.7

In Exercises 1–6 use (a) the Trapezoidal Rule (5.47) and (b) Simpson's Rule (5.49) to approximate the definite integral for the stated value of n. Use approximations to four decimal places for $f(x_i)$ and round off answers to two decimal places.

1 $\int_1^4 \frac{1}{x}\,dx,\ n = 6$

2 $\int_0^3 \frac{1}{1 + x}\,dx,\ n = 8$

3 $\int_0^1 \frac{1}{\sqrt{1 + x^2}}\,dx,\ n = 4$

4 $\int_2^3 \sqrt{1 + x^3}\,dx,\ n = 4$

5 $\int_0^2 \frac{1}{4 + x^2}\,dx,\ n = 10$

6 $\int_0^{0.6} \frac{1}{\sqrt{4 - x^2}}\,dx,\ n = 6$

7 Use the Trapezoidal Rule (5.47) with $(b - a)/n = 0.1$ to show that

$$\int_1^{2.7} \frac{1}{x}\,dx < 1 < \int_1^{2.8} \frac{1}{x}\,dx.$$

8 Find upper bounds for the errors in parts (a) and (b) of Exercise 1.

Suppose the tables in Exercises 9 and 10 indicate the relationship between the two physical variables x and y. Assuming that $y = f(x)$ where f is continuous, approximate $\int_2^4 f(x)\,dx$ by means of (a) the Trapezoidal Rule and (b) Simpson's Rule.

9

x	2.00	2.25	2.50	2.75	3.00	3.25	3.50	3.75	4.00
y	4.12	3.76	3.21	3.58	3.94	4.15	4.69	5.44	7.52

10

x	2.0	2.2	2.4	2.6	2.8	3.0	3.2	3.4	3.6	3.8	4.0
y	12.1	11.4	9.7	8.4	6.3	6.2	5.8	5.4	5.1	5.9	5.6

5.8 REVIEW

Concepts

Define or discuss each of the following.

1 The area under the graph of a nonnegative continuous function f from a to b

2 Partition of $[a, b]$

3 Norm of a partition

4 Riemann sum

5 $\lim_{\|P\| \to 0} \sum_i f(w_i)\,\Delta x_i = I$

6 The definite integral of f from a to b

7 Upper and lower limits of integration

8 Integrand

9 Properties of the definite integral

10 The Mean Value Theorem for Integrals

11 The Fundamental Theorem of Calculus

12 Indefinite integral

13 Constant of integration

14 Power rule for indefinite integration

15 The method of substitution

16 Numerical integration

Exercises

Find the numbers in Exercises 1–3.

1 $\sum_{k=1}^{5} (k^2 + 3)$

2 $\sum_{k=1}^{50} 8$

3 $\sum_{k=0}^{4} (-1/2)^{k-2}$

4 Suppose f is defined on the interval $[-2, 3]$ by $f(x) = 1 - x^2$ and let P be the partition of $[-2, 3]$ into five equal subintervals. Find the Riemann sum R_P if f is evaluated at the midpoint of each subinterval.

5 Work Exercise 4 if f is evaluated at the right-hand end-point of each subinterval.

6 Given $\int_1^4 (x^2 + 2x - 5)\,dx$, find numbers which satisfy the conclusion of the Mean Value Theorem for Integrals (5.29)

Evaluate the integrals in Exercises 7–27.

7 $\int_0^1 \sqrt[3]{8x^7}\,dx$

8 $\int \sqrt[3]{5t + 1}\,dt$

9 $\int_0^1 \frac{z^2}{(1 + z^3)^2}\,dz$

10 $\int (x^2 + 4)^2\,dx$

11 $\int (1 - 2x^2)^3 x\,dx$

12 $\int \frac{(1 + \sqrt{x})^2}{\sqrt[3]{x}}\,dx$

13 $\int_1^2 \frac{w + 1}{\sqrt{w^2 + 2w}}\,dw$

14 $\int_1^2 \frac{s^2 + 2}{s^2}\,ds$

15 $\int \frac{1}{\sqrt{x}(1 + \sqrt{x})^2}\,dx$

16 $\int_1^2 \frac{x^2 - x - 6}{x + 2}\,dx$

17 $\int (3 - 2x - 5x^3)\,dx$

18 $\int (y + y^{-1})^2\,dy$

19 $\int_0^2 x^2\sqrt{x^3 + 1}\,dx$

20 $\int_1^1 3x^2\sqrt{x^3 + x}\,dx$

21 $\int (4t + 1)(4t^2 + 2t - 7)^2\,dt$

22 $\int \frac{\sqrt[4]{1 - v^{-1}}}{v^2}\,dv$

23 $\int (2x^{-3} - 3x^{-2})\,dx$

24 $\int_1^9 \sqrt{2r + 7}\,dr$

25 $\int D_y \sqrt[5]{y^4 + 2y^2 + 1}\,dy$

26 $D_z \int_0^z (x^2 + 1)^{10}\,dx$

27 $D_x \int_0^1 (x^3 + x^2 - 7)^5\,dx$

28 Evaluate $\int_0^{10} \sqrt{1 + x^4}\,dx$ by using (a) the Trapezoidal Rule (5.47) with $n = 5$ and (b) Simpson's Rule (5.49) with $n = 8$. Use approximations to four decimal places for $f(x_i)$ and round off answers to two decimal places.

29 If $f(x) = x^4\sqrt{x^5 + 4}$, find the area of the region under the graph of f from 0 to 2.

30 Use a definite integral to prove that the area of a right triangle of altitude a and base b is $ab/2$. (*Hint:* Take the vertices at the points $(0, 0)$, $(b, 0)$, and (b, a).)

Applications of the Definite Integral

The definite integral is useful for solving a large variety of applied problems. In this chapter we shall discuss area, volume, work, liquid force, lengths of curves, and some problems from economics and biology. Other applications will be considered later in the text.

6.1 AREA

If a function f is continuous on a closed interval $[a, b]$, and $f(x) \geq 0$ for all x in $[a, b]$, then by (5.20) the area under the graph of f from a to b equals $\int_a^b f(x)\,dx$. If g is another continuous nonnegative valued function on $[a, b]$, and if $f(x) \geq g(x)$ for all x in $[a, b]$, then the area A of the region bounded by the graphs of f, g, $x = a$, and $x = b$ (see Figure 6.1) can be found by subtracting the area under the graph of g from the area under the graph of f, that is,

$$A = \int_a^b f(x)\,dx - \int_a^b g(x)\,dx = \int_a^b [f(x) - g(x)]\,dx.$$

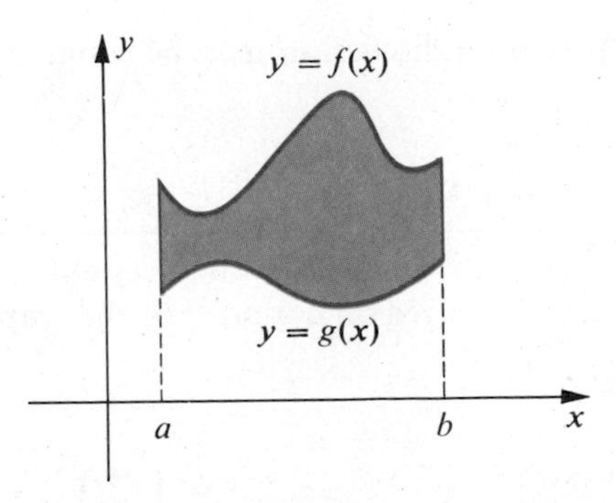

Figure 6.1

This formula for A can be extended to the case in which f or g is negative for some x in $[a, b]$, as illustrated in (i) of Figure 6.2. To find the area of the indicated region, let us choose a negative number d less than the minimum value of g on $[a, b]$ and consider the two functions f_1 and g_1 defined by

$$f_1(x) = f(x) - d, \quad g_1(x) = g(x) - d$$

for all x in $[a, b]$. Since d is negative, values of f_1 and g_1 are found by adding the positive number $|d|$ to corresponding values of f and g, respectively. Geometrically, this amounts to raising the graphs of f and g a distance $|d|$, giving us a region having the same shape as the original region, but lying entirely above the x-axis.

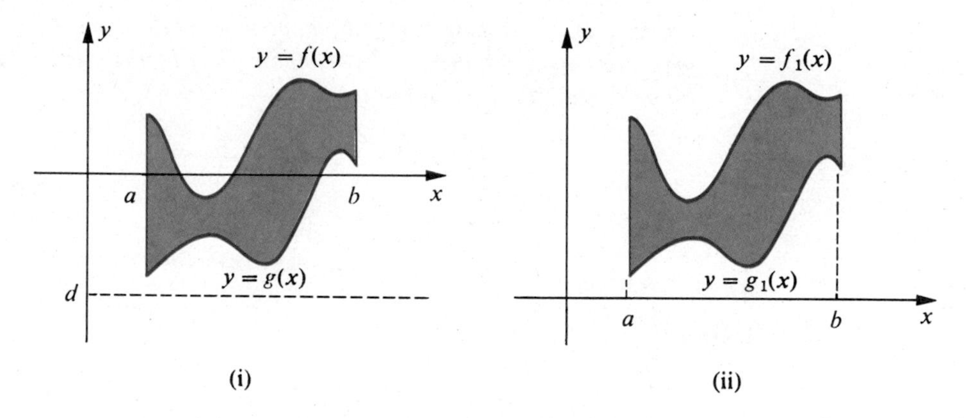

Figure 6.2

If A is the area of the region in (ii) of Figure 6.2, then

$$\begin{aligned} A &= \int_a^b [f_1(x) - g_1(x)]\,dx \\ &= \int_a^b \{[f(x) - d] - [g(x) - d]\}\,dx \\ &= \int_a^b [f(x) - g(x)]\,dx. \end{aligned}$$

The preceding discussion may be summarized as follows.

(6.1) Area between two graphs

> If f and g are continuous on $[a, b]$ and $f(x) \geq g(x)$ for all x in $[a, b]$, then the **area** A of the region bounded by the graphs of f, g, $x = a$, and $x = b$ is given by
>
> $$A = \int_a^b [f(x) - g(x)]\,dx.$$

If $g(x)$ is zero for all x in $[a, b]$, then the graph of g is the x-axis and (6.1) reduces to (5.20). If $f(x)$ is zero throughout $[a, b]$, then the graph of f is the x-axis and the situation illustrated in (i) of Figure 6.3 occurs. In this case (6.1) takes on the form

(6.2)
$$A = \int_a^b [0 - g(x)]\,dx = -\int_a^b g(x)\,dx.$$

This formula can also be derived by observing that the graph of the equation $y = -g(x)$ is the reflection of the graph of $y = g(x)$ through the x-axis (see (ii) of Figure 6.3). Consequently, the area of the region bounded by the x-axis and the graphs of g, $x = a$, and $x = b$ is the same as the area of the region *under* the graph of $-g$ from a to b. The latter area is given by (6.2).

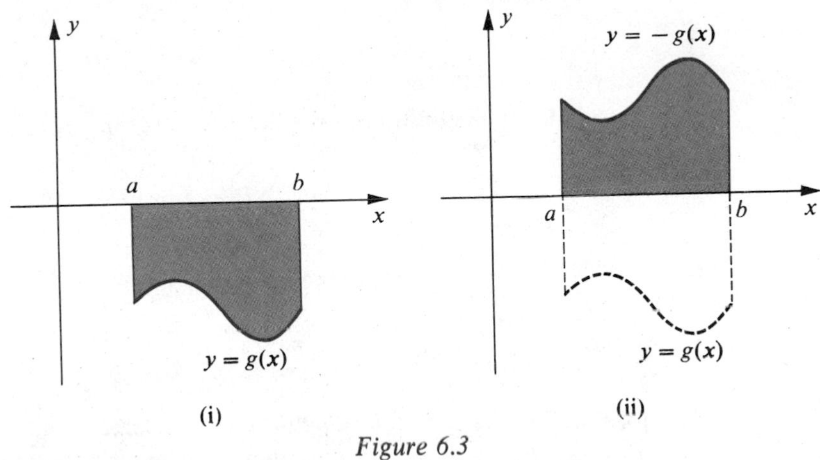

Figure 6.3

The formula for A in (6.1) may be interpreted as a limit of a sum. If we define the function h by $h(x) = f(x) - g(x)$, for all x in $[a, b]$, then $h(x)$ is the distance between the graphs of f and g, that is, the distance from the point on the graph of g with abscissa x to the point on the graph of f with abscissa x. As in the discussion of Riemann sums in Chapter 5, let P denote a partition of $[a, b]$ determined by the numbers $a = x_0, x_1, \ldots, x_n = b$. For each i, let $\Delta x_i = x_i - x_{i-1}$, and let w_i be an arbitrary number in the ith subinterval $[x_{i-1}, x_i]$ of P. By the definition of h,

$$h(w_i)\,\Delta x_i = [f(w_i) - g(w_i)]\,\Delta x_i$$

which, in geometric terms, is the area of a rectangle of length $f(w_i) - g(w_i)$ and width Δx_i (see Figure 6.4).

The Riemann sum

$$\sum_i h(w_i)\,\Delta x_i = \sum_i [f(w_i) - g(w_i)]\,\Delta x_i$$

is the sum of the areas of all rectangles pictured in Figure 6.4, and may be thought of as an approximation to the area of the region between the graphs of f and g

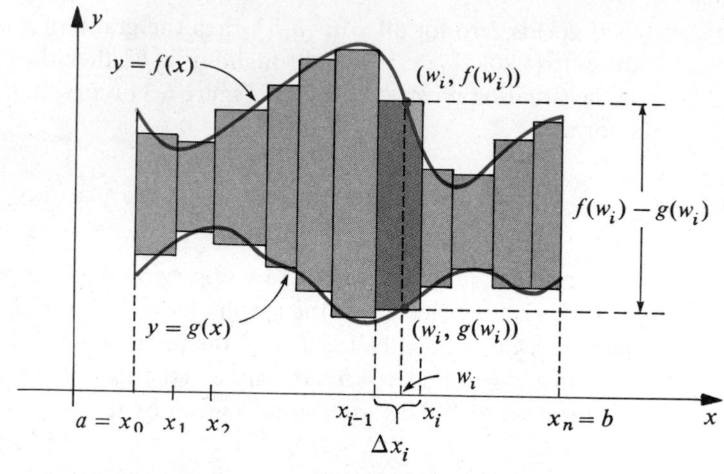

Figure 6.4

from a to b. By the definition of the definite integral,

$$\lim_{||P|| \to 0} \sum_i h(w_i)\,\Delta x_i = \int_a^b h(x)\,dx$$

or, equivalently,

(6.3) $$\lim_{||P|| \to 0} \sum_i [f(w_i) - g(w_i)]\,\Delta x_i = \int_a^b [f(x) - g(x)]\,dx = A.$$

This formula enables us to take a point of view different from that given by (6.1). When we use (6.3) we initially think of approximating the region by means of rectangles of the type shown in Figure 6.4. After writing a formula for the area of a typical rectangle, we sum all such rectangles and, as in (6.3), take the limit of this sum to obtain the area of the region.

Example 1 Find the area of the region bounded by the graphs of the equations $y = x^2$ and $y = \sqrt{x}$.

Solution We shall employ the Riemann sum approach discussed above. The region and a typical rectangle are sketched in Figure 6.5. As indicated in the figure, the length of the rectangle is $\sqrt{w_i} - w_i^2$ and its area is $(\sqrt{w_i} - w_i^2)\,\Delta x_i$. Using (6.3) we obtain

$$A = \lim_{||P|| \to 0} \sum_i (\sqrt{w_i} - w_i^2)\,\Delta x_i$$

$$= \int_0^1 (\sqrt{x} - x^2)\,dx$$

$$= \left[\frac{2}{3}x^{3/2} - \frac{1}{3}x^3\right]_0^1 = \frac{2}{3} - \frac{1}{3} = \frac{1}{3}.$$

The area can also be found by direct substitution in (6.1), with $f(x) = \sqrt{x}$ and $g(x) = x^2$.

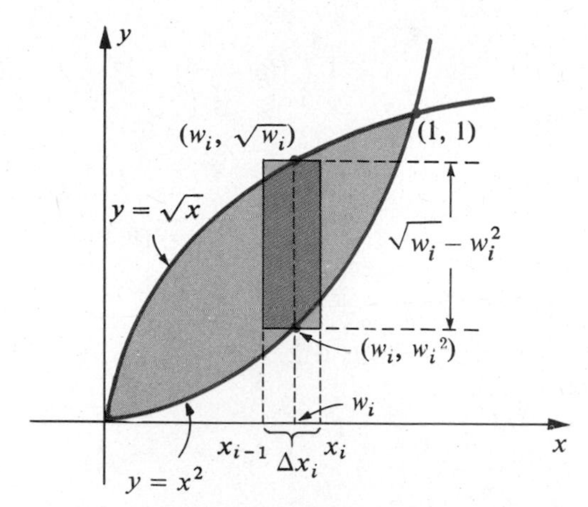

Figure 6.5

As illustrated in Example 1, to find areas by using limits of sums, we always begin by sketching the region together with at least one typical rectangle, labeling the drawing appropriately. The reason for stressing the summation technique is that similar limiting processes will be employed later for calculating many other mathematical and physical quantities. Treating areas as limits of sums will make it easier to understand those future applications. At the same time it will help solidify the meaning of the definite integral.

Example 2 Find the area of the region bounded by the graphs of $y + x^2 = 6$ and $y + 2x - 3 = 0$.

Solution The region and a typical rectangle are sketched in Figure 6.6. The points of intersection $(-1, 5)$ and $(3, -3)$ of the two graphs may be found by solving the two given equations simultaneously. In order to use (6.1) or (6.3) it is necessary to solve each equation for y in terms of x, obtaining

$$y = 6 - x^2 \quad \text{and} \quad y = 3 - 2x.$$

The functions f and g are then given by

$$f(x) = 6 - x^2 \quad \text{and} \quad g(x) = 3 - 2x.$$

As shown in Figure 6.6, the length of a typical rectangle is

$$(6 - w_i^2) - (3 - 2w_i)$$

where w_i is some number in the ith subinterval of a partition P of $[-1, 3]$. The area of this rectangle is

$$[(6 - w_i^2) - (3 - 2w_i)]\,\Delta x_i.$$

Applying (6.3) we obtain

$$\begin{aligned}
A &= \lim_{||P|| \to 0} \sum_i [(6 - w_i^2) - (3 - 2w_i)] \Delta x_i \\
&= \int_{-1}^{3} [(6 - x^2) - (3 - 2x)]\, dx \\
&= \int_{-1}^{3} (3 - x^2 + 2x)\, dx \\
&= \left[3x - \frac{x^3}{3} + x^2\right]_{-1}^{3} \\
&= \left[9 - \frac{27}{3} + 9\right] - \left[-3 - \left(-\frac{1}{3}\right) + 1\right] = \frac{32}{3}.
\end{aligned}$$

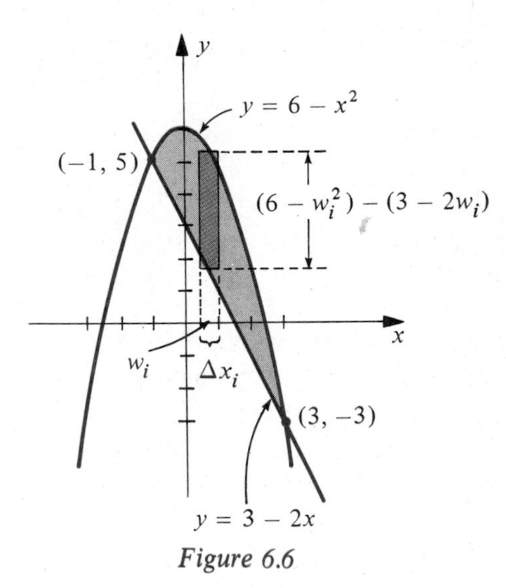

Figure 6.6

After students thoroughly understand the limit of a sum technique illustrated in Examples 1 and 2, it is customary to bypass the subscript part of the solution and proceed immediately to "setting up" the integral. To illustrate, in Example 2 we could regard dx as the width of a typical rectangle. The length of the rectangle is represented by the distance

$$(6 - x^2) - (3 - 2x)$$

between the upper and lower boundaries of the region, where x is an arbitrary number in $[-1, 3]$. Thus the area of a rectangle may be represented, in nonsubscript notation, by

$$[(6 - x^2) - (3 - 2x)]\, dx.$$

Placing the integral sign $\int_{-1}^{3}$ in front of this expression may be regarded as summing all such terms and simultaneously letting the widths of the rectangles approach 0.

Since one of our objectives is to emphasize the definition of the definite integral (5.16), we will continue to use limits of sums in many of the remaining examples in this chapter. The symbol w_i employed in solutions always denotes a number in the ith subinterval of a partition. Those who wish to use the nonsubscript approach may (at the discretion of the instructor) proceed immediately to the definite integral which follows a stated limit of a sum.

The following example illustrates that it is sometimes necessary to subdivide a region and use formulas (6.1) or (6.3) more than once in order to find the area.

Example 3 Find the area of the region R bounded by the graphs of $y - x = 6$, $y - x^3 = 0$, and $2y + x = 0$.

Solution The graphs and region are sketched in (i) of Figure 6.7 where, as indicated, each equation has been solved for y in terms of x so that appropriate functions may be introduced. Typical rectangles are shown extending from the lower boundary to the upper boundary of R. Since the lower boundary consists of portions of two different graphs, the area cannot be found by using only one definite integral. However, if R is divided into two subregions R_1 and R_2, as shown in (ii) of Figure 6.7, then we can determine the area of each and add them together.

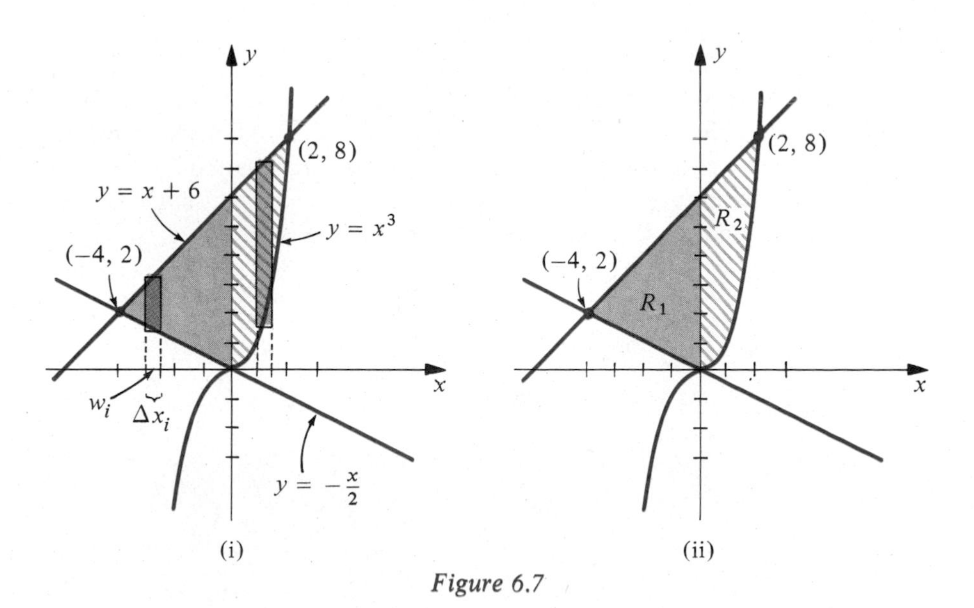

Figure 6.7

For region R_1 the upper and lower boundaries are the lines $y = x + 6$ and $y = -x/2$, respectively, and hence the length of a typical rectangle is

$$(w_i + 6) - (-w_i/2).$$

Applying (6.3), the area A_1 of R_1 is

$$\begin{aligned} A_1 &= \lim_{\|P\|\to 0} \sum_i [(w_i + 6) - (-w_i/2)]\Delta x_i \\ &= \int_{-4}^{0} [(x+6) - (-x/2)]\,dx \\ &= \int_{-4}^{0} \left[\frac{3}{2}x + 6\right] dx \\ &= \left[\frac{3}{4}x^2 + 6x\right]_{-4}^{0} = 0 - (12 - 24) = 12. \end{aligned}$$

Region R_2 has the same upper boundary $y = x + 6$ as R_1; however, the lower boundary is given by $y = x^3$. In this case the length of a typical rectangle is

$$(w_i + 6) - w_i^3$$

and the area A_2 of R_2 is

$$\begin{aligned} A_2 &= \lim_{\|P\|\to 0} \sum_i [(w_i + 6) - w_i^3]\Delta x_i \\ &= \int_{0}^{2} [(x+6) - x^3]\,dx \\ &= \left[\frac{x^2}{2} + 6x - \frac{x^4}{4}\right]_0^2 \\ &= (2 + 12 - 4) - 0 = 10. \end{aligned}$$

The area of the entire region R is, therefore, $A_1 + A_2 = 12 + 10 = 22$.

Sometimes it is necessary to find the area A of the region bounded by the graphs of $y = c$ and $y = d$, and of two equations of the form $x = f(y)$ and $x = g(y)$, where f and g are continuous functions and $f(y) \geq g(y)$ for all y in $[c, d]$ (see Figure 6.8). In a manner similar to our earlier discussion, but with the roles of x and y interchanged, we obtain the formula

$$A = \int_{c}^{d} [f(y) - g(y)]\,dy$$

where we now regard y as the independent variable.

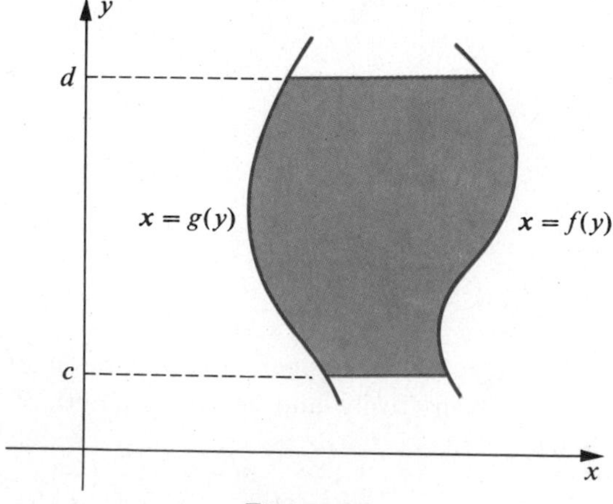

Figure 6.8

Summation techniques can also be applied to regions of the type shown in Figure 6.8. In this case we select points on the y-axis with ordinates $y_0, y_1, \ldots, y_n$, where $y_0 = c$ and $y_n = d$, thereby obtaining a partition of the interval $[c, d]$ into subintervals of width $\Delta y_i = y_i - y_{i-1}$. For each i we choose a number w_i in $[y_{i-1}, y_i]$ and consider rectangles of areas $[f(w_i) - g(w_i)]\Delta y_i$, as illustrated in Figure 6.9. This leads to

(6.4) $$A = \lim_{||P|| \to 0} \sum_i [f(w_i) - g(w_i)]\Delta y_i = \int_c^d [f(y) - g(y)]\,dy,$$

where the last equality follows from the definition of the definite integral.

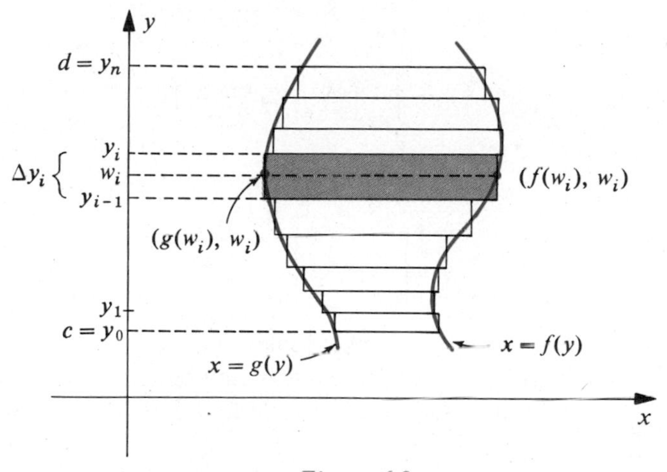

Figure 6.9

Example 4 Find the area of the region bounded by the graphs of the equations $2y^2 = x + 4$ and $x = y^2$.

Solution Two sketches of the region are shown in Figure 6.10, where (i) illustrates the situation that occurs if we use vertical rectangles (integration with respect to x), and (ii) is the case if we use horizontal rectangles (integration with respect to y).

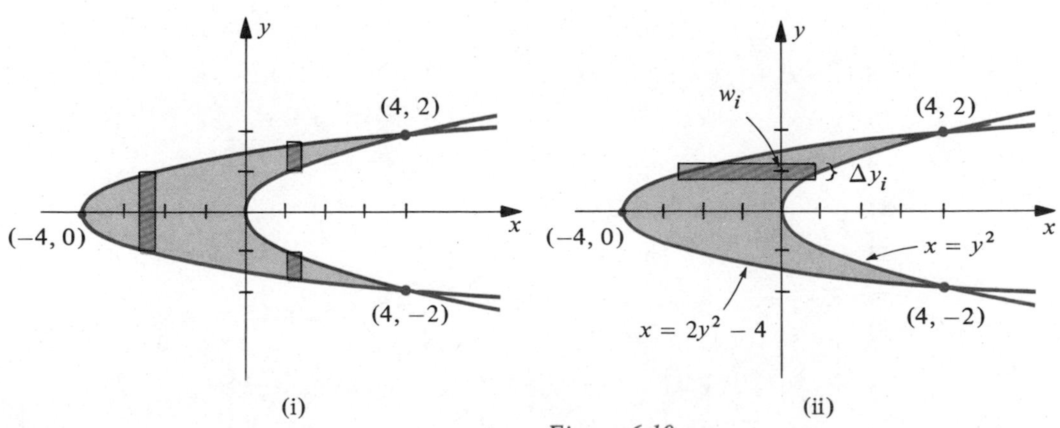

Figure 6.10

Referring to (i) of the figure we see that several definite integrals are required to find the area. (Why?) However, in (ii) we can use (6.4) to find the area with only one integration. Letting $f(y) = y^2$, $g(y) = 2y^2 - 4$, and referring to (ii) of Figure 6.10, the length $f(w_i) - g(w_i)$ of a *horizontal* rectangle is

$$w_i^2 - (2w_i^2 - 4).$$

Since the width is Δy_i, the area of the rectangle is

$$[w_i^2 - (2w_i^2 - 4)]\,\Delta y_i.$$

Using (6.4), the area of R is

$$\begin{aligned} A &= \lim_{\|P\|\to 0} \sum_i [w_i^2 - (2w_i^2 - 4)]\,\Delta y_i \\ &= \int_{-2}^{2} [y^2 - (2y^2 - 4)]\,dy \\ &= \int_{-2}^{2} (4 - y^2)\,dy \\ &= \left[4y - \frac{y^3}{3}\right]_{-2}^{2} = \left[8 - \frac{8}{3}\right] - \left[-8 - \left(-\frac{8}{3}\right)\right] = \frac{32}{3}. \end{aligned}$$

Actually, the integration could have been further simplified. For example, since the x-axis bisects the region, it is sufficient to find the area of that part of the region which lies above the x-axis and double it, obtaining

$$A = 2\int_{0}^{2} [y^2 - (2y^2 - 4)]\,dy.$$

EXERCISES 6.1

In each of Exercises 1–20 sketch the region bounded by the graphs of the given equations, show a typical vertical or horizontal rectangle, and find the area of the region.

1 $y = 1/x^2,\ y = -x^2,\ x = 1,\ x = 2$

2 $y = \sqrt{x},\ y = -x,\ x = 1,\ x = 4$

3 $y^2 = -x,\ x - y = 4,\ y = -1,\ y = 2$

4 $x = y^2,\ y - x = 2,\ y = -2,\ y = 3$

5 $y = x^2 + 1,\ y = 5$

6 $y = 4 - x^2,\ y = -4$

7 $y = x^2,\ y = 4x$

8 $y = x^3,\ y = x^2$

9 $y = 1 - x^2,\ y = x - 1$

10 $x + y = 3,\ y + x^2 = 3$

11 $y^2 = 4 + x, y^2 + x = 2$

12 $x = y^2,\ x - y - 2 = 0$

13 $y = x,\ y = 3x,\ x + y = 4$

14 $x - y + 1 = 0,\ 7x - y - 17 = 0,\ 2x + y + 2 = 0$

15 $y = x^3 - x, y = 0$

16 $y = x^3 - x^2 - 6x, y = 0$

17 $x = 4y - y^3, x = 0$

18 $x = y^{2/3}, x = y^2$

19 $y = x\sqrt{4 - x^2}, y = 0$

20 $y = x\sqrt{x^2 - 9}, y = 0, x = 5$

21 If R is the region bounded by the graph of $(x - 4)^2 + y^2 = 9$, express the area A of R as a limit of a sum. Find A without integrating.

22 If R is the region bounded by the graphs of $2x + 3y = 6$, $x = 0$, and $y = 0$, express the area A of R as a limit of a sum. Find A without integrating.

Each of Exercises 23–30 represents a limit of a sum for a function f on the indicated interval. Interpret each limit as an area of a region R and find its value. Sketch the region R.

23 $\lim_{||P|| \to 0} \sum_i (4w_i + 1)\Delta x_i$; $[0, 1]$

24 $\lim_{||P|| \to 0} \sum_i (w_i - w_i^3)\Delta x_i$; $[0, 1]$

25 $\lim_{||P|| \to 0} \sum_i (4 - w_i^2)\Delta y_i$; $[0, 1]$

26 $\lim_{||P|| \to 0} \sum_i \sqrt{3w_i + 1}\,\Delta y_i$; $[0, 1]$

27 $\lim_{||P|| \to 0} \sum_i [w_i/(w_i^2 + 1)^2]\Delta x_i$; $[2, 5]$

28 $\lim_{||P|| \to 0} \sum_i (5w_i/\sqrt{9 + w_i^2})\Delta x_i$; $[1, 4]$

29 $\lim_{||P|| \to 0} \sum_i [(5 + \sqrt{w_i})/\sqrt{w_i}]\Delta y_i$; $[1, 4]$

30 $\lim_{||P|| \to 0} \sum_i w_i^{-2}\Delta y_i$; $[-5, -2]$

6.2 SOLIDS OF REVOLUTION

If a region in a plane is revolved about a line in the plane, the resulting solid is called a **solid of revolution** and the solid is said to be **generated** by the region. The line about which the revolution takes place is called an **axis of revolution**. If the region bounded by the graph of a continuous nonnegative valued function f, the x-axis, and the graphs of $x = a$ and $x = b$ (see (i) of Figure 6.11) is revolved about the x-axis, a solid of the type shown in (ii) of Figure 6.11 is generated. For example, if f is a constant function, then the region is rectangular and the solid generated is a right circular cylinder. If the graph of f is a semicircle with end-points of a diameter at the points $(a, 0)$ and $(b, 0)$ where $b > a$, then the solid of

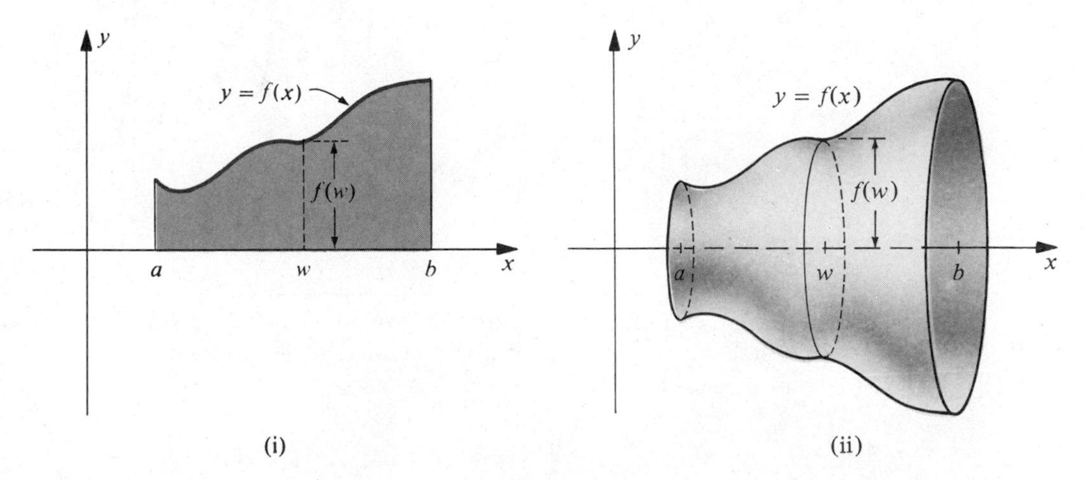

Figure 6.11

revolution is a sphere with diameter $b - a$. If the given region is a right triangle with base on the x-axis and two vertices at the points $(a, 0)$ and $(b, 0)$ with the right angle at one of these points, then a right circular cone is generated.

If a plane perpendicular to the x-axis intersects the solid shown in (ii) of Figure 6.11, a circular cross section is obtained. If, as indicated in the figure, the plane passes through the point on the x-axis with abscissa w, then the radius of the circle is $f(w)$ and hence its area is $\pi[f(w)]^2$. We shall arrive at a definition for the volume of such a solid of revolution by using Riemann sums in a manner similar to that used for areas in the previous section.

Suppose f is continuous and $f(x) \geq 0$ for all x in $[a, b]$. Consider a Riemann sum $\Sigma_i f(w_i)\Delta x_i$, where w_i is any number in the ith subinterval $[x_{i-1}, x_i]$ of a partition P of $[a, b]$. Geometrically, this gives us a sum of areas of rectangles of the type shown in (i) of Figure 6.12. The solid generated by the polygon formed by these rectangles has the appearance shown in (ii) of the figure. Observe that the ith rectangle generates a circular disc (that is, a "flat" right circular cylinder) of base radius $f(w_i)$ and altitude, or "thickness," $\Delta x_i = x_i - x_{i-1}$. The volume of this disc is the area of the base times the altitude, that is, $\pi[f(w_i)]^2\,\Delta x_i$. The sum of the volumes of all such discs is the volume of the solid shown in (ii) of Figure 6.12 and is given by

$$\sum_i \pi[f(w_i)]^2\,\Delta x_i.$$

This may be regarded as a Riemann sum for the function h defined by $h(x) = \pi[f(x)]^2$. Intuitively, it appears that if $\|P\|$ is close to zero, then the sum is close to the volume of the solid. It is natural, therefore, to define the volume of revolution as the limit of this sum.

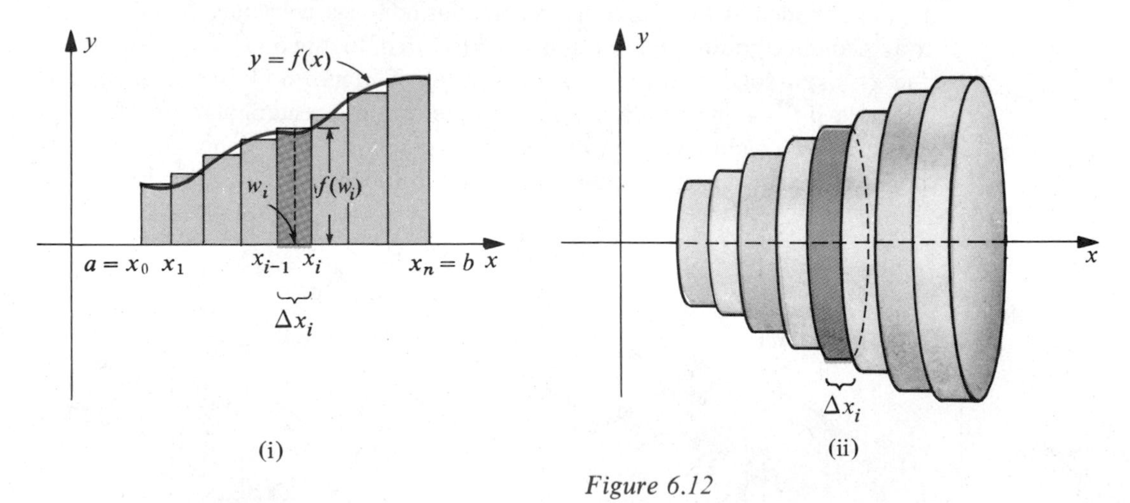

Figure 6.12

(6.5) Definition

> Let f be continuous on $[a, b]$. The **volume** V of the solid of revolution generated by revolving the region bounded by the graphs of f, $x = a$, $x = b$, and the x-axis about the x-axis is given by
>
> $$V = \lim_{\|P\| \to 0} \sum_i \pi[f(w_i)]^2\,\Delta x_i = \int_a^b \pi[f(x)]^2\,dx.$$

The fact that the limit of the sum in Definition (6.5) equals $\int_a^b \pi [f(x)]^2\, dx$ follows from the definition of the definite integral. Hereafter, when considering limits of Riemann sums, the meanings of all symbols will not be explicitly pointed out. Instead, it will be assumed that the reader is aware of the significance of symbols such as $\|P\|$, w_i, and Δx_i.

The requirement that $f(x) \geq 0$ for all x in $[a, b]$ was omitted in Definition (6.5). If f is negative for some x, as illustrated in (i) of Figure 6.13, and if the region bounded by the graphs of f, $x = a$, $x = b$, and the x-axis is revolved about the x-axis, a solid of the type shown in (ii) of the figure is obtained. This solid is the same as that generated by revolving the region under the graph of $y = |f(x)|$ from a to b about the x-axis. Since $|f(x)|^2 = [f(x)]^2$, the limit in Definition (6.5) is the desired volume.

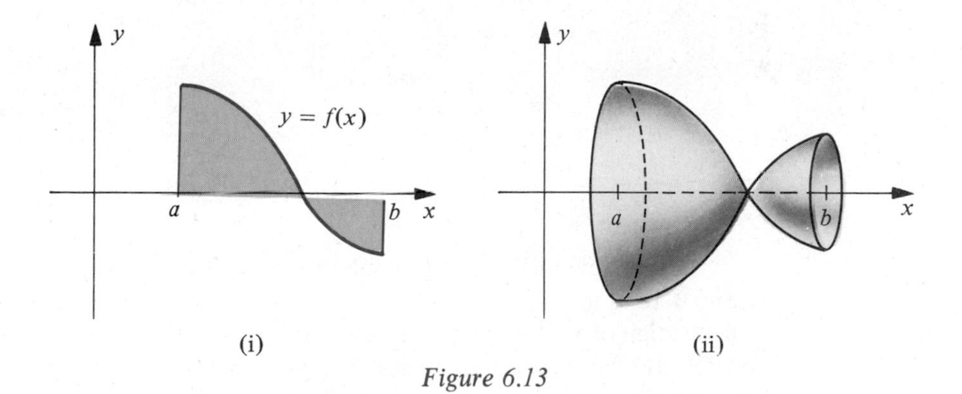

Figure 6.13

Example 1 If $f(x) = x^2 + 1$, find the volume of the solid generated by revolving the region under the graph of f from -1 to 1 about the x-axis.

Solution The solid is illustrated in Figure 6.14. Included in the sketch is a typical rectangle and the disc it generates. Since the radius of the disc is $w_i^2 + 1$, its volume is

$$\pi(w_i^2 + 1)^2 \Delta x_i$$

and as in Definition (6.5),

$$\begin{aligned} V &= \lim_{\|P\| \to 0} \sum_i \pi(w_i^2 + 1)^2 \Delta x_i \\ &= \int_{-1}^{1} \pi(x^2 + 1)^2\, dx \\ &= \pi \int_{-1}^{1} (x^4 + 2x^2 + 1)\, dx \\ &= \pi\left[\frac{1}{5}x^5 + \frac{2}{3}x^3 + x\right]_{-1}^{1} \\ &= \pi\left[\left(\frac{1}{5} + \frac{2}{3} + 1\right) - \left(-\frac{1}{5} - \frac{2}{3} - 1\right)\right] = \frac{56}{15}\pi. \end{aligned}$$

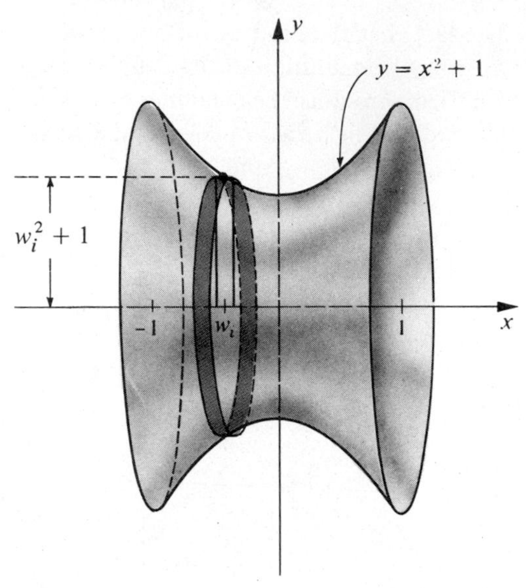

Figure 6.14

The volume in Example 1 could also have been calculated by integrating from 0 to 1 and doubling the result. (Why?) Another solution consists of substitution of $x^2 + 1$ for $f(x)$ in the formula $V = \int_a^b \pi[f(x)]^2\,dx$. We shall not ordinarily specify the units of measure. If the unit of linear measurement is inches, the volume is expressed in cubic inches. If x is in cm, then V is in cm^3, etc.

Let us next consider a region bounded by horizontal lines with y-intercepts c and d, by the y-axis, and by the graph of $x = g(y)$, where the function g is continuous and $g(y) \geq 0$ for all y in $[c, d]$. If this region is revolved about the y-axis, the volume V of the resulting solid may be found by interchanging the roles of x and y in Definition (6.5), taking y as the independent variable. This gives us

$$V = \int_c^d \pi[g(y)]^2\,dy.$$

The preceding formula can be interpreted as a limit of a sum as follows. Suppose P is the partition of the interval $[c, d]$ determined by the numbers $c = y_0, y_1, \ldots, y_n = d$. Let $\Delta y_i = y_i - y_{i-1}$, let w_i be any number in the ith subinterval, and consider the rectangles of length $g(w_i)$ and width Δy_i as illustrated in (i) of Figure 6.15. The solid generated by revolving these rectangles about the y-axis is illustrated in (ii) of the figure. The volume of the disc generated by the ith rectangle is $\pi[g(w_i)]^2\,\Delta y_i$. Summing and taking the limit gives us

(6.6)
$$V = \lim_{\|P\|\to 0} \sum_i \pi[g(w_i)]^2\,\Delta y_i = \int_c^d \pi[g(y)]^2\,dy.$$

Example 2 The region bounded by the y-axis and the graphs of $y = x^3$, $y = 1$, and $y = 8$ is revolved about the y-axis. Find the volume of the resulting solid.

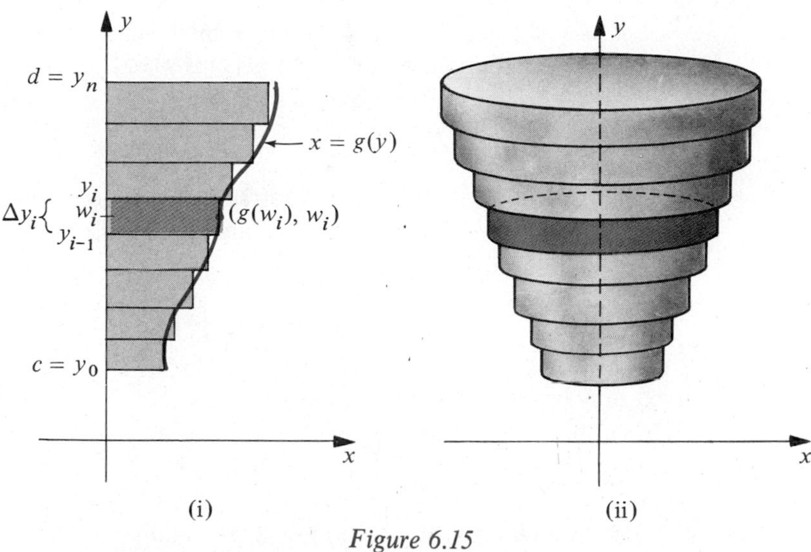

Figure 6.15

Solution The solid is sketched in Figure 6.16 together with a disc generated by a typical rectangle. Since we wish to integrate with respect to y it is necessary to solve the equation $y = x^3$ for x in terms of y, obtaining $x = y^{1/3}$. If we let $x = g(y) = y^{1/3}$, then as shown in Figure 6.16, the radius of a typical disc is $g(w_i) = w_i^{1/3}$ and its volume is $\pi(w_i^{1/3})^2\,\Delta y_i$. Applying (6.6) with $g(y) = y^{1/3}$ gives us

$$
\begin{aligned}
V &= \lim_{\|P\|\to 0} \sum_i \pi(w_i^{1/3})^2\,\Delta y_i \\
&= \int_1^8 \pi(y^{1/3})^2\,dy = \pi \int_1^8 y^{2/3}\,dy \\
&= \pi\left(\frac{3}{5}\right)\left[y^{5/3}\right]_1^8 = \frac{3}{5}\pi\left[8^{5/3} - 1\right] = \frac{93}{5}\pi.
\end{aligned}
$$

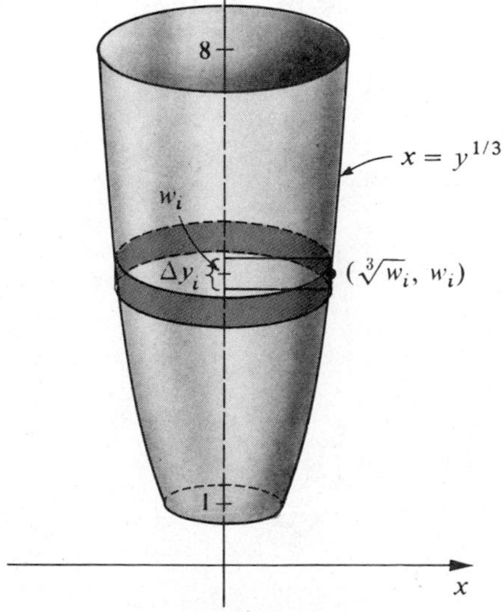

Figure 6.16

Let us next consider a region bounded by the graphs of $x = a$, $x = b$, and of two continuous functions f and g where $f(x) \geq g(x) \geq 0$ for all x in $[a, b]$. If this region is revolved about the x-axis, a solid of the type illustrated in (i) of Figure 6.17 may be obtained. Note that if $g(x) > 0$ for all x in $[a, b]$, then there is a hole through the solid.

The volume V may be found by subtracting the volume of the solid generated by the smaller region from the volume of the solid generated by the larger region. Using Definition (6.5) gives us

$$V = \int_a^b \pi[f(x)]^2\,dx - \int_a^b \pi[g(x)]^2\,dx$$

which may be written in the form

(6.7)
$$V = \int_a^b \pi\{[f(x)]^2 - [g(x)]^2\}\,dx.$$

The last formula has an interesting interpretation as a limit of a sum. As illustrated in (ii) of Figure 6.17, a rectangle extending from the graph of g to the graph of f, through the points with abscissa w_i, generates a washer-shaped solid whose volume is

$$\pi[f(w_i)]^2\,\Delta x_i - \pi[g(w_i)]^2\,\Delta x_i$$

or equivalently

$$\pi\{[f(w_i)]^2 - [g(w_i)]^2\}\,\Delta x_i.$$

Summing the volumes of all such washers and taking the limit gives us (6.7). Thus, formula (6.7) may be thought of as a limit of a sum of volumes of washers. When working problems of this type it is often convenient to use the following general formula:

(6.8) Volume of a washer $= \pi[(\text{outer radius})^2 - (\text{inner radius})^2] \cdot (\text{thickness})$.

In integration problems the thickness will be given by either Δx_i or Δy_i.

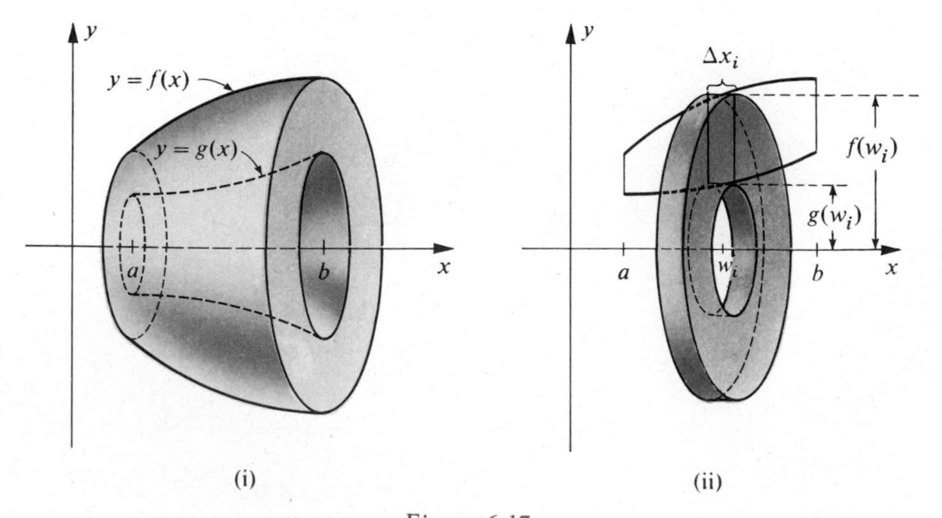

Figure 6.17

Example 3 The region bounded by the graphs of $x^2 = y - 2$, $2y - x - 2 = 0$, $x = 0$, and $x = 1$ is revolved about the x-axis. Find the volume of the resulting solid.

Solution The region and a typical rectangle are sketched in (i) of Figure 6.18. Since we wish to integrate with respect to x we solve the first two equations for y in terms of x, obtaining $y = x^2 + 2$ and $y = \frac{1}{2}x + 1$. The washer generated by the rectangle in (i) is illustrated in (ii) of Figure 6.18. Since the outer radius is $w_i^2 + 2$ and the inner radius is $\frac{1}{2}w_i + 1$, its volume (see (6.8)) is

$$\pi[(w_i^2 + 2)^2 - (\tfrac{1}{2}w_i + 1)^2]\Delta x_i.$$

Taking the limit of a sum of such volumes gives us

$$\begin{aligned} V &= \int_0^1 \pi\left[\left(x^2 + 2\right)^2 - \left(\frac{1}{2}x + 1\right)^2\right] dx \\ &= \pi \int_0^1 \left(x^4 + \frac{15}{4}x^2 - x + 3\right) dx \\ &= \pi\left[\frac{1}{5}x^5 + \frac{5}{4}x^3 - \frac{1}{2}x^2 + 3x\right]_0^1 = \frac{79\pi}{20}. \end{aligned}$$

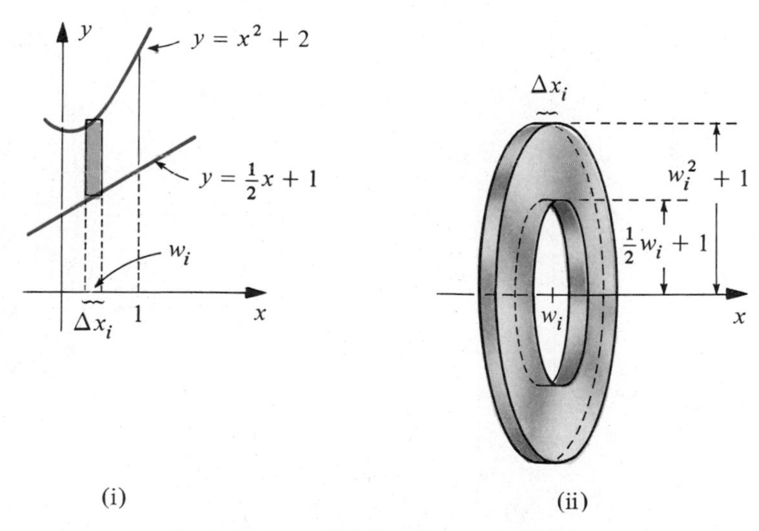

Figure 6.18

Example 4 Find the volume of the solid generated by revolving the region described in Example 3 about the line $y = 3$.

Solution The region and a typical rectangle are resketched in (i) of Figure 6.19, together with the axis of revolution $y = 3$. The washer generated by the rectangle is illustrated in (ii) of Figure 6.19. Note that the radii of this washer are as follows:

$$\text{inner radius} = 3 - (w_i^2 + 2) = 1 - w_i^2$$
$$\text{outer radius} = 3 - (\tfrac{1}{2}w_i + 1) = 2 - \tfrac{1}{2}w_i.$$

Using (6.8), the volume of the washer is

$$\pi\left[\left(2 - \frac{1}{2}w_i\right)^2 - (1 - w_i^2)^2\right]\Delta x_i.$$

Taking a limit of a sum of such terms gives us

$$\begin{aligned}
V &= \int_0^1 \pi\left[\left(2 - \frac{1}{2}x\right)^2 - (1 - x^2)^2\right] dx \\
&= \pi \int_0^1 \left[\left(4 - 2x + \frac{1}{4}x^2\right) - (1 - 2x^2 + x^4)\right] dx \\
&= \pi \int_0^1 \left(3 - 2x + \frac{9}{4}x^2 - x^4\right) dx \\
&= \pi\left[3x - x^2 + \frac{3}{4}x^3 - \frac{1}{5}x^5\right]_0^1 \\
&= \pi\left[3 - 1 + \frac{3}{4} - \frac{1}{5}\right] = \frac{51}{20}\pi \approx 8.01.
\end{aligned}$$

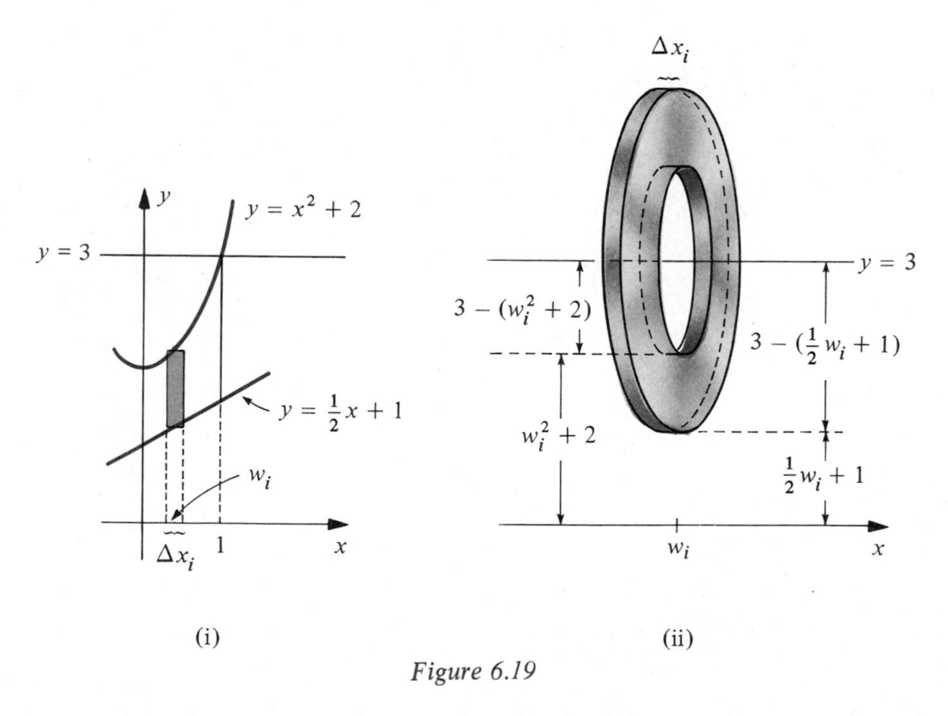

Figure 6.19

By interchanging the roles of x and y we can apply the techniques discussed in this section to solids generated by revolving regions about the y-axis, or a line parallel to the y-axis, as in the next example.

Example 5 The region in the first quadrant bounded by the graphs of $y = \frac{1}{8}x^3$ and $y = 2x$ is revolved about the y-axis. Find the volume of the resulting solid.

Solution The region and a typical rectangle are shown in (i) of Figure 6.20. Since we wish to integrate with respect to y, we solve the given equations for x in terms of y, obtaining

$$x = \tfrac{1}{2}y \quad \text{and} \quad x = 2y^{1/3}.$$

As shown in (ii) of Figure 6.20, the inner and outer radii of the washer generated by the rectangle are $\frac{1}{2}w_i$ and $2w_i^{1/3}$, respectively. Since the thickness is Δy_i it follows from (6.8) that the volume of the washer is

$$\pi[(2w_i^{1/3})^2 - (\tfrac{1}{2}w_i)^2]\,\Delta y_i$$

or

$$\pi[4w_i^{2/3} - \tfrac{1}{4}w_i^2]\,\Delta y_i.$$

Taking a limit of a sum of such terms gives us

$$\begin{aligned} V &= \int_0^8 \pi\left[4y^{2/3} - \frac{1}{4}y^2\right]dy \\ &= \pi\left[\frac{12}{5}y^{5/3} - \frac{1}{12}y^3\right]_0^8 \\ &= \pi\left[\frac{12}{5}(8^{5/3}) - \frac{1}{12}(8^3)\right] \\ &= \frac{512}{15}\pi \approx 107.2. \end{aligned}$$

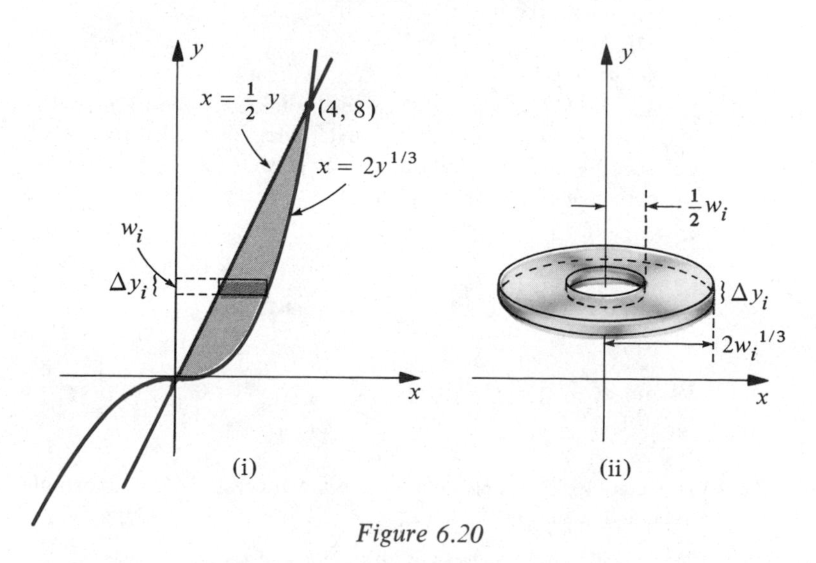

Figure 6.20

EXERCISES 6.2

In each of Exercises 1–12 sketch the region R bounded by the graphs of the given equations and find the volume of the solid generated by revolving R about the indicated axis. In each case show a typical rectangle, together with the disc or washer it generates.

1 $y = 1/x$, $x = 1$, $x = 3$, $y = 0$; about the x-axis.

2 $y = \sqrt{x}$, $y = 0$, $x = 4$; about the x-axis.

3 $y = x^2$, $y = 2$; about the y-axis.

4 $y = 1/x$, $x = 0$, $y = 1$, $y = 3$; about the y-axis.

5 $y = x^2 - 4x$, $y = 0$; about the x-axis.

6 $y = x^3$, $x = -2$, $y = 0$; about the x-axis.

7 $y^2 = x$, $2y = x$; about the y-axis.

8 $y = 2x$, $y = 4x^2$; about the y-axis.

9 $y = x^2$, $y = 4 - x^2$; about the x-axis.

10 $x = y^3$, $x^2 + y = 0$; about the x-axis.

11 $x = y^2$, $y - x + 2 = 0$; about the y-axis.

12 $x + y = 1$, $y = x + 1$, $x = 2$; about the y-axis.

13 Find the volume of the solid generated by revolving the region bounded by the graphs of $y = x^2$ and $y = 4$ as follows.
(a) About the line $y = 4$
(b) About the line $y = 5$
(c) About the line $x = 2$

14 Find the volume of the solid generated by revolving the region bounded by the graphs of $y = \sqrt{x}$, $y = 0$, and $x = 4$ as follows.
(a) About the line $x = 4$
(b) About the line $x = 6$
(c) About the line $y = 2$

In Exercises 15–20 sketch the region R bounded by the graphs of the equations given in (a) and then set up (but do not evaluate) integrals needed to find the volume of the solid obtained by revolving R about the line given in (b). As usual, show typical rectangles and the corresponding discs or washers.

15 (a) $y = x^3$, $y = 4x$
(b) $y = 8$

16 (a) $y = x^3$, $y = 4x$
(b) $x = 4$

17 (a) $x + y = 3$, $y + x^2 = 3$
(b) $x = 2$

18 (a) $y = 1 - x^2$, $x - y = 1$
(b) $y = 3$

19 (a) $x^2 + y^2 = 1$
(b) $x = 5$

20 (a) $y = x^{2/3}$, $y = x^2$
(b) $y = -1$

In each of Exercises 21–24 use a definite integral to derive a formula for the volume of the indicated solid.

21 A right circular cone of altitude h and radius of base r.

22 A sphere of radius r.

23 A frustum of a right circular cone of altitude h, lower base radius r_1, and upper base radius r_2.

24 The volume of a spherical segment of height h and radius of sphere r.

Exercises 25 and 26 each represent a limit of a sum for a function f on the interval $[0, 1]$. Interpret each limit as a volume and find its value.

25 $\lim_{||P|| \to 0} \sum_i \pi(w_i^4 - w_i^6)\Delta x_i$

26 $\lim_{||P|| \to 0} \sum_i \pi(w_i - w_i^8)\Delta y_i$

6.3 VOLUMES USING CYLINDRICAL SHELLS

There is another method for finding volumes of solids of revolution which, in certain cases, is simpler to apply than those discussed in Section 2. The method employs hollow circular cylinders, that is, thin cylindrical shells of the type illustrated in Figure 6.21.

The volume of a shell having outer radius r_2, inner radius r_1, and altitude h is $\pi r_2^2 h - \pi r_1^2 h$. This expression may also be written

$$\begin{aligned}\pi(r_2^2 - r_1^2)h &= \pi(r_2 + r_1)(r_2 - r_1)h \\ &= 2\pi\left(\frac{r_2 + r_1}{2}\right)h(r_2 - r_1).\end{aligned}$$

If we let $r = (r_2 + r_1)/2$ (the **average radius** of the shell) and $\Delta r = r_2 - r_1$ (the **thickness** of the shell), then the volume of the shell is given by $2\pi rh\,\Delta r$, or equivalently, by

(6.9) Volume of a shell = 2π(average radius)(altitude)(thickness).

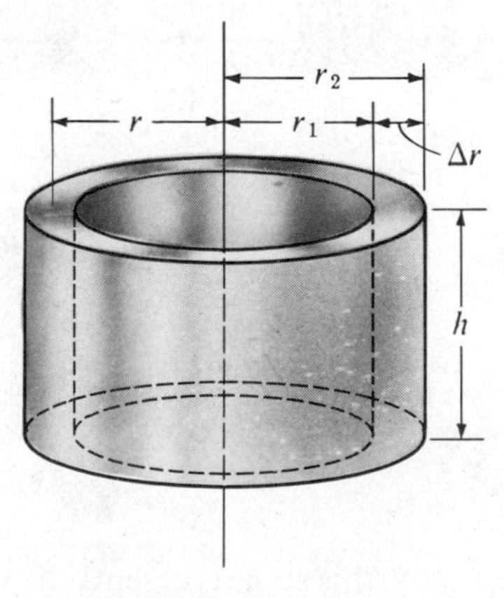

Figure 6.21

Let f be continuous on $[a, b]$ and $f(x) \geq 0$ for all x in $[a, b]$. Let R be the region bounded by the graph of f, the x-axis, and by the graphs of $x = a$ and $x = b$ where $b > a \geq 0$. The solid generated by revolving R about the y-axis is illustrated in (i) Figure 6.22. Note that if $a > 0$, there is a hole through the solid. Let P be a partition of $[a, b]$ and consider a rectangle with base corresponding to the interval $[x_{i-1}, x_i]$ and altitude $f(w_i)$, where w_i is the midpoint of $[x_{i-1}, x_i]$. If this rectangle is revolved about the y-axis, then as illustrated in (ii) of Figure 6.22 there results a cylindrical shell with average radius w_i, altitude $f(w_i)$, and thickness $\Delta x_i = x_i - x_{i-1}$. By (6.9) we may express its volume as

$$2\pi w_i f(w_i)\,\Delta x_i.$$

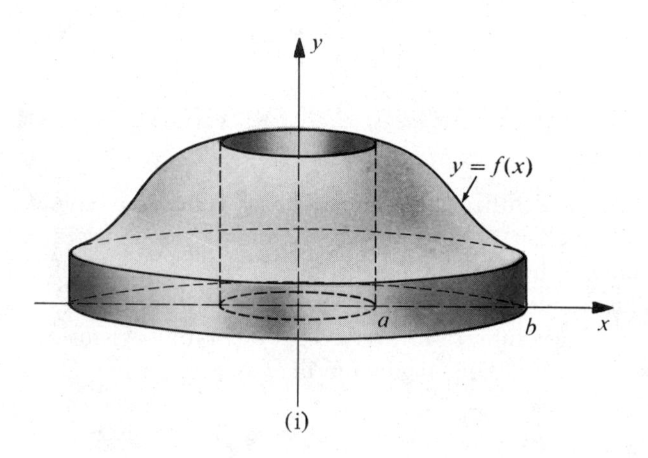

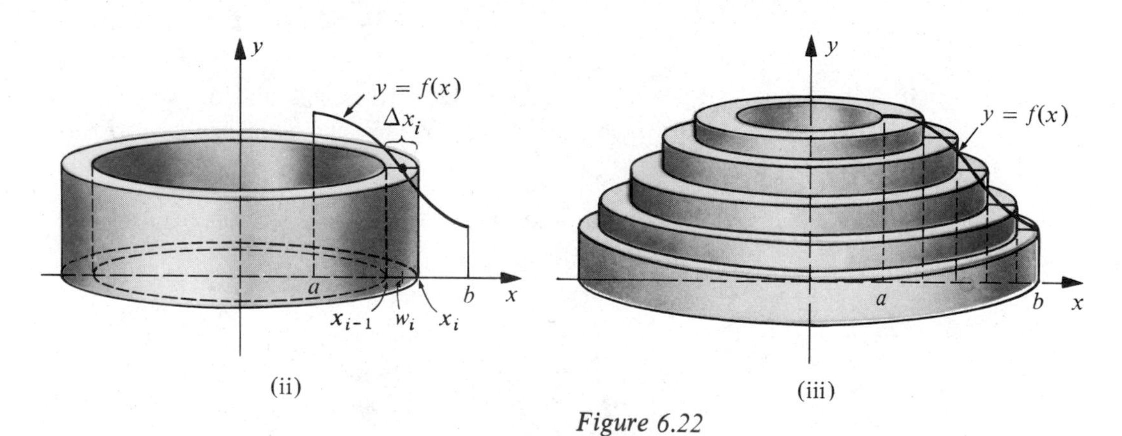

Figure 6.22

Doing this for each subinterval in the partition and adding gives

$$\sum_i 2\pi w_i f(w_i)\,\Delta x_i.$$

Geometrically, this sum represents the volume of a solid of the type illustrated in (iii) of Figure 6.22. Evidently, the smaller the norm $\|P\|$ of the partition, the better the sum approximates the volume V of the solid generated by R. Indeed, it appears that the volume of the solid illustrated in (i) of the figure is given by

(6.10) $$V = \lim_{\|P\| \to 0} \sum_i 2\pi w_i f(w_i)\,\Delta x_i = \int_a^b 2\pi x f(x)\,dx$$

where the last equality follows from the definition of the definite integral.

The above argument is incomplete since it should be proved that if the methods of Section 2 are also applicable, then both methods lead to the same answer.

Example 1 The region bounded by the graph of $y = 2x - x^2$ and the x-axis is revolved about the y-axis. Find the volume of the resulting solid.

Solution The region and a shell generated by a typical rectangle are sketched in Figure 6.23, where w_i is the midpoint of $[x_{i-1}, x_i]$. Since the average radius of the shell is w_i, the altitude is $2w_i - w_i^2$, and the thickness is Δx_i, it follows from (6.9) that the volume of the shell is

$$2\pi w_i(2w_i - w_i^2)\,\Delta x_i.$$

Consequently, the volume V of the solid is given by

$$\begin{aligned} V &= \lim_{\|P\|\to 0} \sum_i 2\pi w_i(2w_i - w_i^2)\,\Delta x_i \\ &= \int_0^2 2\pi x(2x - x^2)\,dx \\ &= 2\pi \int_0^2 (2x^2 - x^3)\,dx \\ &= 2\pi\left[\frac{2}{3}x^3 - \frac{1}{4}x^4\right]_0^2 = 8\pi/3. \end{aligned}$$

The volume V can also be found using washers; however, the calculations would be more involved since the given equation would have to be solved for x in terms of y.

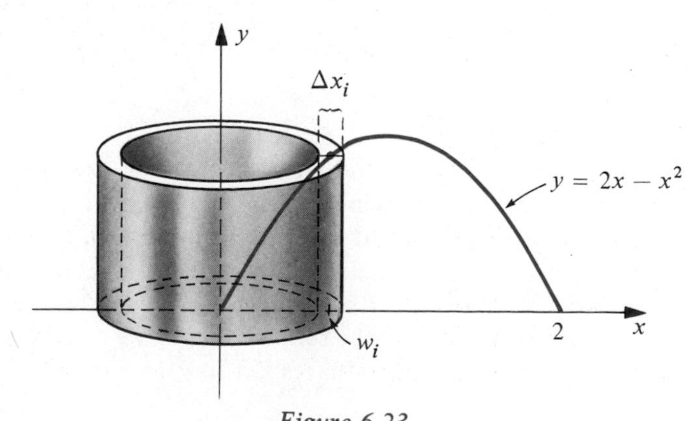

Figure 6.23

Example 2 The region bounded by the graphs of $y = x^2$ and $y = x + 2$ is revolved about the line $x - 3$. Express the volume of the resulting solid as a definite integral.

Solution The region and a typical rectangle are shown in (i) of Figure 6.24, where w_i represents the midpoint of the ith subinterval $[x_{i-1}, x_i]$. The cylindrical shell generated by the rectangle is illustrated in (ii) of Figure 6.24. We see that this shell has the following dimensions:

$$\text{altitude} = (w_i + 2) - w_i^2$$
$$\text{average radius} = 3 - w_i$$
$$\text{thickness} = \Delta x_i.$$

Hence, by (6.9) its volume is

$$2\pi(3 - w_i)[(w_i + 2) - w_i^2]\,\Delta x_i.$$

Taking the limit of a sum of such terms gives us

$$V = \int_{-1}^{2} 2\pi(3 - x)(x + 2 - x^2)\,dx.$$

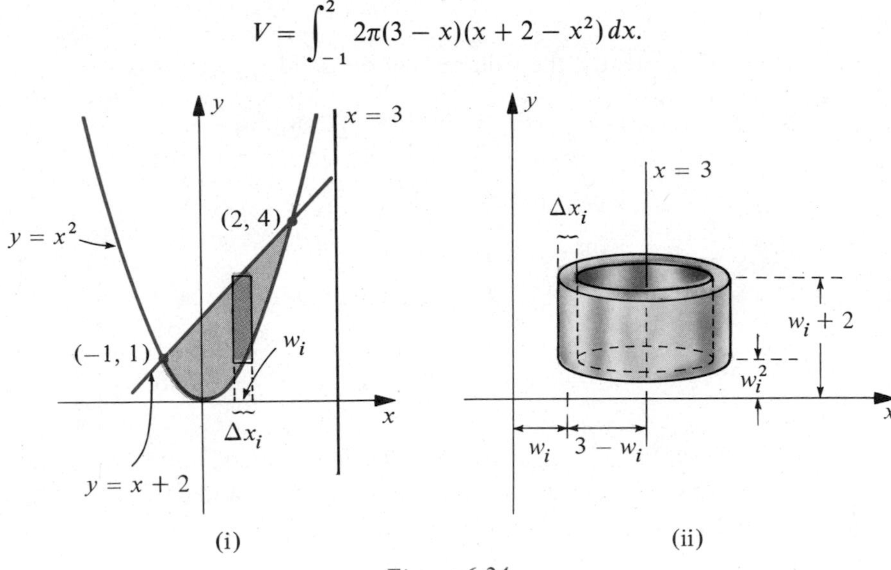

Figure 6.24

The definite integral in the preceding example could be evaluated by multiplying the factors in the integrand and then integrating each term. Since we have worked a sufficient number of problems of this type, it would not be very instructive to carry out all of these details. For convenience we shall refer to the process of expressing V in terms of an integral as *setting up the integral for V.*

By interchanging the roles of x and y we can find certain volumes by using shells and integrating with respect to y, as illustrated in the next example.

Example 3 The region in the first quadrant bounded by the graph of $x = 2y^3 - y^4$ and the y-axis is revolved about the x-axis. Set up the integral for the volume of the resulting solid.

Solution The region is sketched in (i) of Figure 6.25 together with a typical (horizontal) rectangle. The cylindrical shell generated by the rectangle is illustrated in (ii) of

Figure 6.25. This shell has the following dimensions:

$$\text{altitude} = 2w_i^3 - w_i^4$$
$$\text{average radius} = w_i$$
$$\text{thickness} = \Delta y_i.$$

Hence, by (6.9) its volume is

$$2\pi w_i(2w_i^3 - w_i^4)\,\Delta y_i.$$

Taking a limit of a sum of such terms and remembering that y is the independent variable gives us

$$V = \int_0^2 2\pi y(2y^3 - y^4)\,dy.$$

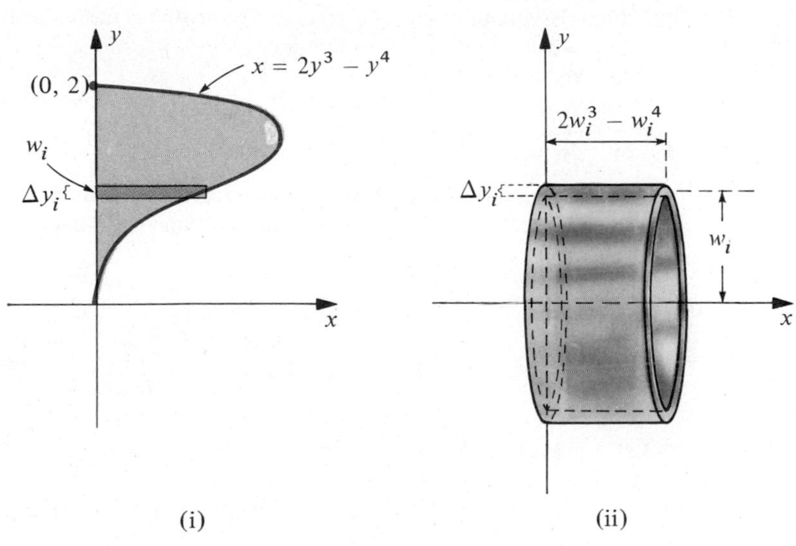

Figure 6.25

It is worth noting that in the preceding example we were forced to use shells and to integrate with respect to y, since use of discs and integration with respect to x would require that the equation $x = 2y^3 - y^4$ be solved for y in terms of x—a formidable task, to put it mildly!

EXERCISES 6.3

In each of Exercises 1–10 sketch the region R bounded by the graphs of the given equations and use the methods of this section to find the volume of the solid generated by revolving R about the indicated axis. In each case show a typical rectangle together with the cylindrical shell it generates.

1 $y = \sqrt{x}$, $x = 4$, $y = 0$; about the y-axis.

2 $y = 1/x$, $x = 1$, $x = 2$, $y = 0$; about the y-axis.

3 $y = x^2$, $y^2 = 8x$; about the y-axis.

4 $y = x^2 - 5x$, $y = 0$; about the y-axis.

5 $2x - y - 12 = 0$, $x - 2y - 3 = 0$, $x = 4$; about the y-axis.

6 $y = x^3 + 1$, $x + 2y = 2$, $x = 1$; about the y-axis.

7 $x^2 = 4y$, $y = 4$; about the x-axis.

8 $y^3 = x$, $y = 3$, $x = 0$; about the x-axis.

9 $y = 2x$, $y = 6$, $x = 0$; about the x-axis.

10 $2y = x$, $y = 4$, $x = 1$; about the x-axis.

11 Find the volume of the solid generated by revolving the region bounded by the graphs of $y = x^2 + 1$, $x = 0$, $x = 2$, and $y = 0$ as follows.
(a) About the line $x = 3$
(b) About the line $x = -1$

12 Find the volume of the solid generated by revolving the region bounded by the graphs of $y = 4 - x^2$ and $y = 0$ as follows.
(a) About the line $x = 2$
(b) About the line $x = -3$

13–24 Use the methods of this section to solve Exercises 13–24 of Section 2.

The expressions in Exercises 25 and 26 each represent a limit of a sum for a function f on the interval $[0, 1]$. Interpret each limit as a volume and find its value.

25 $\lim_{||P|| \to 0} \sum_i 2\pi(w_i^2 - w_i^3)\Delta x_i$

26 $\lim_{||P|| \to 0} \sum_i 2\pi(w_i^5 + w_i^{3/2})\Delta x_i$

6.4 VOLUMES BY SLICING

If a plane intersects a solid, then the region common to the plane and the solid is called a **cross section** of the solid. In Section 2 we encountered circular and washer-shaped cross sections. We shall now consider solids which have the property that for every x in a closed interval $[a, b]$ on a coordinate line l, the plane perpendicular to l at the point with coordinate x intersects the solid in a cross section whose area is given by $A(x)$, where A is a continuous function on $[a, b]$. Figures 6.26 and 6.27 illustrate solids of the type we wish to discuss. The solid is called a *cylinder* if, as illustrated in Figure 6.26, all cross sections are the same. If

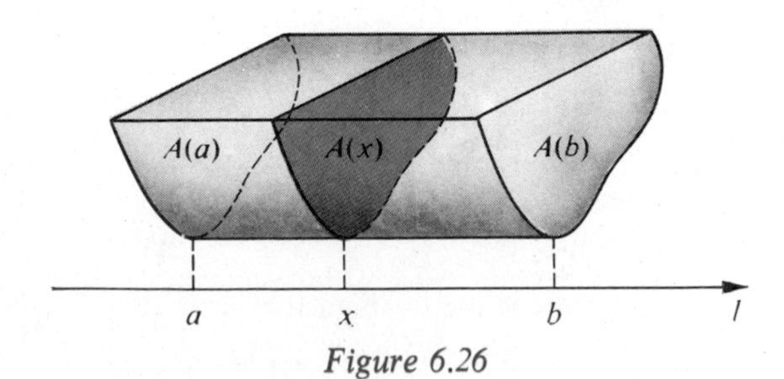

Figure 6.26

we are only interested in that part of the graph bounded by planes through the points with coordinates a and b, then the cross sections determined by these planes are called the **bases** of the cylinder and the distance between the bases is called the **altitude**. By definition, the volume of such a cylinder is the area of a base multiplied by the altitude. As a special case, a right circular cylinder of base radius r and altitude h has volume $\pi r^2 h$.

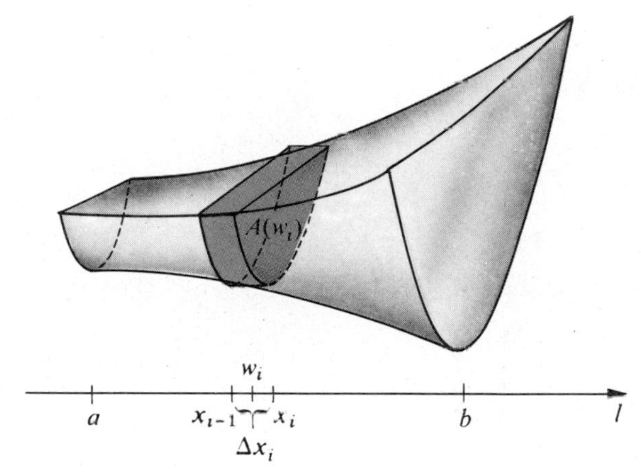

Figure 6.27

The solid illustrated in Figure 6.27 is not a cylinder since cross sections by planes perpendicular to l are not all the same. To find the volume we begin with a partition P of $[a, b]$ by choosing $a = x_0, x_1, x_2, \ldots, x_n = b$. The planes perpendicular to l at the points with these coordinates slice the solid into smaller pieces. The ith such slice is shown in Figure 6.27. As usual, let $\Delta x_i = x_i - x_{i-1}$ and choose any number w_i in $[x_{i-1}, x_i]$. It appears that if Δx_i is small, then the volume of the slice can be approximated by the volume of the cylinder of base area $A(w_i)$ and altitude Δx_i, that is, by $A(w_i)\,\Delta x_i$. Consequently, the total volume of the solid is approximated by the Riemann sum $\Sigma_{i=1}^{n} A(w_i)\,\Delta x_i$. Since the approximation improves as $\|P\|$ gets smaller, we define the volume V of the solid as the limit of this sum. Since A is continuous on $[a, b]$,

(6.11)
$$V = \lim_{\|P\| \to 0} \sum_i A(w_i)\,\Delta x_i = \int_a^b A(x)\,dx.$$

Example 1 Find the volume of a right pyramid which has altitude h and square base of side a.

Solution If, as shown in Figure 6.28, we introduce a coordinate line l along the axis of the pyramid, with origin O at the vertex, then cross sections by planes perpendicular to l are squares. If $A(x)$ is the cross-sectional area determined by the plane which intersects the axis x units from O, then

$$A(x) = (2y)^2 = 4y^2$$

where y is the distance indicated in the figure. By similar triangles

$$\frac{y}{x} = \frac{a/2}{h}, \quad \text{or} \quad y = \frac{ax}{2h}$$

and hence

$$A(x) = 4y^2 = \frac{4a^2x^2}{4h^2} = \frac{a^2}{h^2}x^2.$$

Applying (6.11),

$$V = \int_0^h \left(\frac{a^2}{h^2}\right) x^2\,dx$$
$$= \left(\frac{a^2}{h^2}\right)\left[\frac{x^3}{3}\right]_0^h = \frac{a^2h}{3}.$$

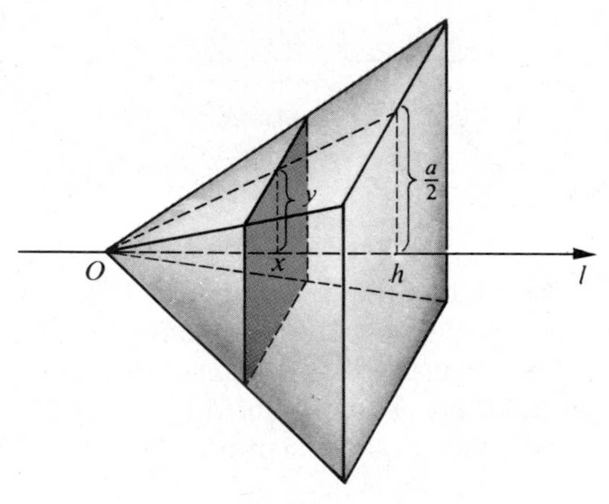

Figure 6.28

Example 2 A solid has, as its base, the circular region in the xy-plane bounded by the graph of $x^2 + y^2 = a^2$, where $a > 0$. Find the volume of the solid if every cross section by a plane perpendicular to the x-axis is an equilateral triangle with one side in the base.

Solution A typical cross section by a plane x units from the origin is illustrated in Figure 6.29. If the point $P(x, y)$ is on the circle, then the length of a side of the triangle is $2y$ and the altitude is $\sqrt{3}y$. Hence the area $A(x)$ of the pictured triangle is

$$A(x) = \tfrac{1}{2}(2y)(\sqrt{3}y) = \sqrt{3}y^2 = \sqrt{3}(a^2 - x^2).$$

Applying (6.11) gives us

$$V = \int_{-a}^{a} \sqrt{3}(a^2 - x^2)\,dx = \sqrt{3}\left[a^2x - \frac{x^3}{3}\right]_{-a}^{a} = \frac{4\sqrt{3}a^3}{3}.$$

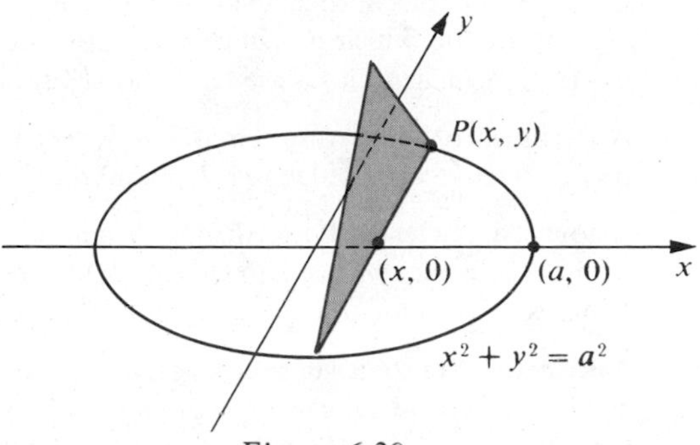

Figure 6.29

EXERCISES 6.4

1 The base of a solid is the circular region in the xy-plane bounded by the graph of $x^2 + y^2 = a^2$, where $a > 0$. Find the volume of the solid if every cross section by a plane perpendicular to the x-axis is a square.

2 Work Exercise 1 if every cross section is an isosceles triangle with base on the xy-plane and altitude equal to the length of the base.

3 A solid has as its base the region in the xy-plane bounded by the graphs of $y = 4$ and $y = x^2$. Find the volume of the solid if every cross section by a plane perpendicular to the x-axis is an isosceles right triangle with hypotenuse on the xy-plane.

4 Work Exercise 3 if each cross section is a square.

5 Find the volume of a pyramid of the type illustrated in Figure 6.28 if the altitude is h and the base is a rectangle of dimensions a and $2a$.

6 A solid has as its base the region in the xy-plane bounded by the graphs of $y = x$ and $y^2 = x$. Find the volume of the solid if every cross section by a plane perpendicular to the x-axis is a semicircle with diameter in the xy-plane.

7 A solid has as its base the region in the xy-plane bounded by the graphs of $y^2 = 4x$ and $x = 4$. If every cross section by a plane perpendicular to the y-axis is a semicircle, find the volume of the solid.

8 A solid has as its base the region in the xy-plane bounded by the graphs of $x^2 = 16y$ and $y = 2$. Every cross section by a plane perpendicular to the y-axis is a rectangle whose height is twice that of the side in the xy-plane. Find the volume of the solid.

9 A wedge-shaped solid is obtained by intersecting a right circular cylinder of radius a by two planes. One of the planes is perpendicular to the axis of the cylinder and the second plane intersects the first at an angle of 45° along a diameter of the cross section. Find the volume of the solid.

10 The axes of two right circular cylinders of radius a intersect at right angles. Find the volume of the solid bounded by the cylinders.

11 The base of a solid is the circular region in the xy-plane bounded by the graph of $x^2 + y^2 = a^2$, where $a > 0$. Find the volume of the solid if every cross section by a plane perpendicular to the x-axis is an isosceles triangle of constant altitude b. (*Hint:* Interpret $\int_{-a}^{a} \sqrt{a^2 - x^2}\, dx$ as an area.)

12 Cross sections of a horn-shaped solid by planes perpendicular to its axis are circles. If a cross section which is s inches from the smaller end of the solid has diameter $6 + (s^2/36)$ inches, and if the length of the solid is 2 feet, find its volume.

13 A tetrahedron has three mutually perpendicular faces and three mutually perpendicular edges of lengths 2, 3, and 4 cm, respectively. Find its volume.

14 A hole of diameter d is bored through a spherical solid of radius r such that the axis of the hole coincides with a diameter of the sphere. Find the volume of the solid which remains.

15 A **prismatoid** is a solid whose cross-sectional areas by planes parallel to, and a distance x from, a fixed plane can be expressed as $ax^2 + bx + c$ where a, b, and c are real numbers. Prove that the volume V of a prismatoid is given by

$$V = \frac{h}{6}(B_1 + B_2 + 4B)$$

where B_1 and B_2 are the areas of the bases, B is the cross-sectional area parallel to and halfway between the bases, and h is the distance between the bases. Show that the frustum of a right circular cone of base radii r_1 and r_2 and altitude h is a prismatoid and find its volume.

16 **Cavalieri's Theorem** states that if two solids have equal altitudes and if all cross sections by planes parallel to their bases and at the same distances from their bases have equal areas, then the solids have the same volume. Prove this theorem.

6.5 WORK

The concept of **force** may be thought of intuitively as the physical entity which is used to describe a push or pull on an object. For example, a force is needed to push or pull an object along a horizontal plane, to lift an object off the ground, or to move a charged particle through an electromagnetic field.

Forces are often measured in pounds. If an object weighs ten pounds, then by definition the force required to lift it (or hold it off the ground) is ten pounds. A force of this type is a **constant force**, since its magnitude does not change while it is applied to the given object.

If a constant force F is applied to an object, moving it a distance d in the direction of the force, then by definition the **work** W done on the object is given by

(6.12) $$W = Fd.$$

If F is measured in pounds and d in feet, then the units for W are foot-pounds (ft-lb). In the metric system a **dyne** is defined as the force which, when applied to a mass of 1 gm, induces an acceleration of 1 cm/sec^2. If F is expressed in dynes and d in centimeters, then the units for W are dyne-centimeters, or **ergs**.

Example 1 Find the work done in pushing an automobile along a level road from a point A to another point B, 20 feet from A, while exerting a constant force of 300 pounds.

Solution The problem is illustrated in Figure 6.30, where we have pictured the road as part of a line l. Since the constant force is $F = 300$ lb and the distance the automobile moves is $d = 20$ ft, it follows from (6.12) that the work done is

$$W = (300)(20) = 6{,}000 \text{ ft-lb.}$$

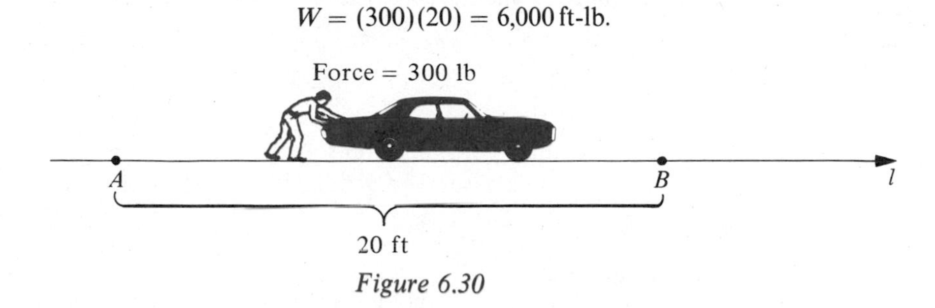

Figure 6.30

Anyone who has pushed an automobile (or some other object) is aware of the fact that the force applied usually varies from one point to another. Thus, if an automobile is stalled at point A in Figure 6.30, then it may require a larger force to get it moving than that which is needed after it is in motion. In addition, the force applied at points between A and B could vary considerably because of friction since, for example, part of the road may be smooth concrete and another part rough gravel. Similarly, someone inside the automobile could change the force required to move it by applying the brakes from time to time. Forces which are not constant are sometimes referred to as *variable forces.*

If a variable force is applied to an object, moving it a certain distance in the direction of the force, then methods of calculus are needed to find the work done. For the present we shall assume that the object moves along a straight line l. The case of motion along a nonlinear path will be considered in Chapter 18. Let us introduce a coordinate system on l and assume that the object moves from the point A with coordinate a to the point B with coordinate b where $b > a$. In order to attack the problem, it is essential to know the force at the point on l with abscissa x, for every x in the interval $[a, b]$. This force will be denoted by $f(x)$. For simplicity, it will be assumed that the function f obtained in this manner is continuous on $[a, b]$. Let P denote the partition of $[a, b]$ determined by the numbers $a = x_0, x_1, \ldots, x_n = b$ and let $\Delta x_i = x_i - x_{i-1}$ (see Figure 6.31). If w_i is a number in $[x_{i-1}, x_i]$, then the force at the point Q with coordinate w_i is $f(w_i)$. If Δx_i is small, then since f is continuous, the values of f change very little in the interval $[x_{i-1}, x_i]$. Roughly speaking, the function f is almost constant on $[x_{i-1}, x_i]$. It appears, therefore, that the work W_i done as the object moves through the ith subinterval may be approximated by means of (6.12); that is,

$$W_i \approx f(w_i)\,\Delta x_i.$$

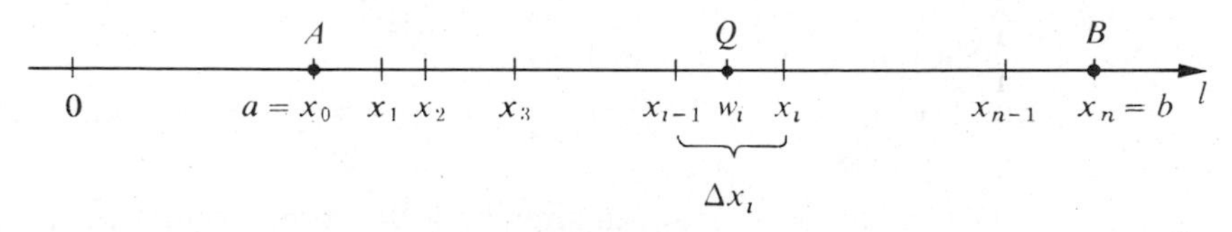

Figure 6.31

It seems evident that the smaller we choose Δx_i, the better $f(w_i)\,\Delta x_i$ approximates the work done in the interval $[x_{i-1}, x_i]$. If it is also assumed that work is additive, in the sense that the work W done as the object moves from A to B can be found by adding the work done over each subinterval, then

$$W \approx \sum_{i=1}^{n} f(w_i)\,\Delta x_i.$$

Since we expect this approximation to improve as the norm $\|P\|$ of the partition becomes smaller, it is natural to define W as the limit of the preceding sum. This limit leads to a definite integral.

(6.13) Definition

Let the force at the point with coordinate x on a coordinate line l be $f(x)$, where f is continuous on $[a, b]$. The work W done in moving an object from the point with coordinate a to the point with coordinate b is given by

$$W = \lim_{\|P\| \to 0} \sum_i f(w_i)\,\Delta x_i = \int_a^b f(x)\,dx.$$

The formula in Definition (6.13) can be used to find the work done in stretching or compressing a spring. In order to solve problems of this type it is necessary to use the following law from physics.

(6.14) Hooke's Law

The force $f(x)$ required to stretch a spring x units beyond its natural length is given by

$$f(x) = kx$$

where k is a constant called the **spring constant**.

The same formula is used to find the work done in compressing a spring x units from its natural length.

Example 2 A force of 9 pounds is required to stretch a spring from its natural length of 6 inches to a length of 8 inches.

(a) Find the work done in stretching the spring from its natural length to a length of 10 inches.

(b) Find the work done in stretching the spring from a length of 7 inches to a length of 9 inches.

Solution (a) Let us introduce a coordinate line l as shown in Figure 6.32, where one end of the spring is attached to some point to the left of the origin and the end to be pulled is located at the origin. According to Hooke's Law (6.14), the force $f(x)$ required to stretch a spring x units beyond its natural length is given by

$$f(x) = kx$$

for some constant k. Using the given data, $f(2) = 9$. Substituting in $f(x) = kx$ we obtain $9 = k \cdot 2$, and hence the spring constant is $k = 9/2$. Consequently, for this spring, Hooke's Law has the form

$$f(x) = \frac{9}{2}x.$$

By Definition (6.13) the work done in stretching the spring 4 inches is given by

$$W = \int_0^4 \frac{9}{2}x\,dx = \frac{9}{4}x^2\Big]_0^4 = 36 \text{ in.-lb.}$$

(b) We use the same function f but change the interval to $[1, 3]$, obtaining

$$W = \int_1^3 \frac{9}{2}x\,dx = \frac{9}{4}x^2\Big]_1^3 = \frac{81}{4} - \frac{9}{4} = 18 \text{ in.-lb.}$$

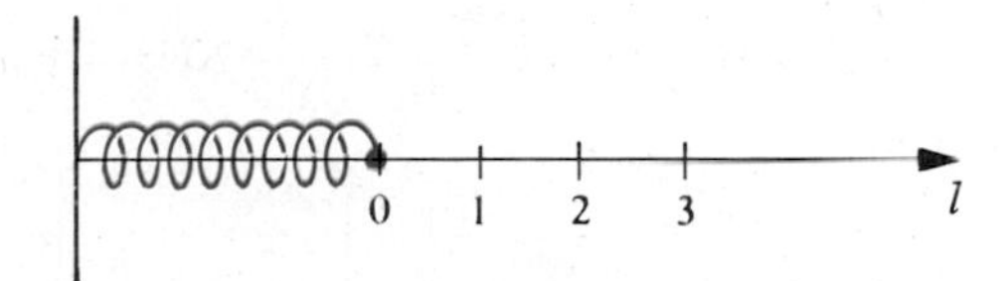

Natural length

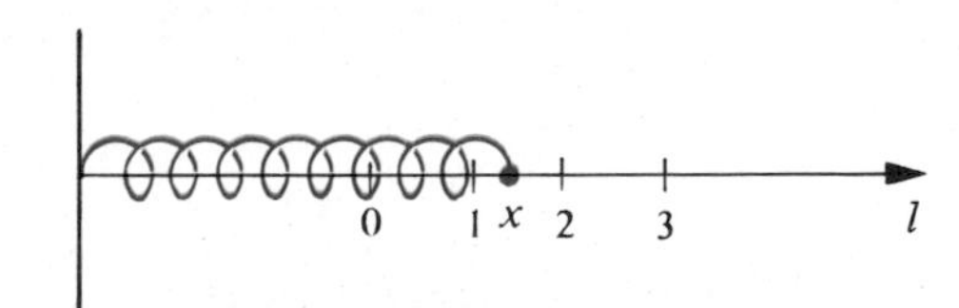

Stretched x units

Figure 6.32

Example 3 A right circular conical tank of altitude 20 feet and radius of base 5 feet has its vertex at ground level and axis vertical. If the tank is full of water, find the work done in pumping the water over the top of the tank.

Solution We begin by introducing a coordinate system as shown in Figure 6.33. The cone intersects the xy-plane along the line of slope 4 through the origin. An equation for this line is $y = 4x$.

Let P denote the partition of the interval $[0, 20]$ determined by $0 = y_0, y_1, y_2, \ldots, y_n = 20$, let $\Delta y_i = y_i - y_{i-1}$, and let x_i be the abscissa of the point on $y = 4x$ with ordinate y_i. If the cone is subdivided by means of planes perpendicular to the y-axis at each y_i, then we may think of the water as being sliced into n parts. As illustrated in Figure 6.33, the volume of the ith slice may be approximated by the volume $\pi x_i^2\,\Delta y_i$ of a circular disc or, since $x_i = y_i/4$, by $\pi(y_i/4)^2\,\Delta y_i$. This leads to the approximation

$$\text{Volume of } i\text{th slice} \approx \pi\left(\frac{y_i^2}{16}\right)\Delta y_i.$$

Assuming that water weighs 62.5 lb/ft^3, the weight of the disc in Figure 6.33 is approximately $62.5\pi(y_i^2/16)\,\Delta y_i$. By (6.12), the work done in lifting the disc to the top of the tank is the product of the distance $20 - y_i$ and the weight, that is,

$$\text{Work done in lifting } i\text{th slice} \approx (20 - y_i)62.5\pi\left(\frac{y_i^2}{16}\right)\Delta y_i.$$

Since the last number is an approximation to the work done in lifting the ith slice to the top, the work done in emptying the entire tank is approximately

$$\sum_{i=1}^{n}(20 - y_i)62.5\pi\left(\frac{y_i^2}{16}\right)\Delta y_i.$$

The actual work W is obtained by taking the limit of this sum as the norm $\|P\|$ approaches zero. This gives us

$$\begin{aligned} W &= \int_0^{20}(20 - y)62.5\pi\left(\frac{y^2}{16}\right)dy \\ &= \frac{62.5\pi}{16}\int_0^{20}(20y^2 - y^3)\,dy \\ &= \frac{62.5\pi}{16}\left[\frac{20y^3}{3} - \frac{y^4}{4}\right]_0^{20} \\ &= \frac{62.5\pi}{16}\left(\frac{40{,}000}{3}\right) \approx 163{,}625 \text{ ft-lb}. \end{aligned}$$

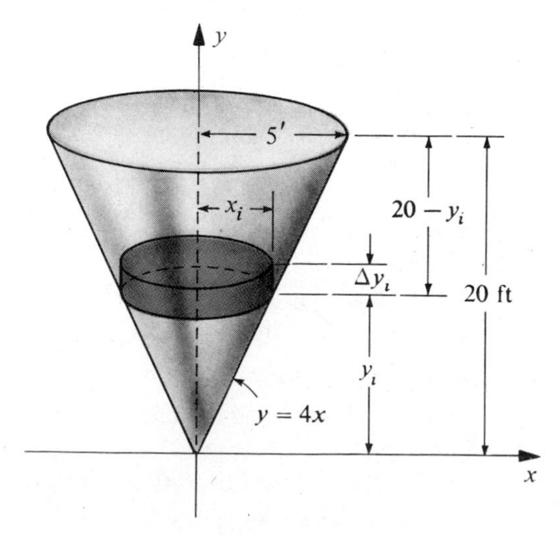

Figure 6.33

Example 4 The pressure p (lb/in.²) and volume v (in.³) of an enclosed expanding gas are related by the formula $pv^k = c$, where k and c are constants. If the gas expands from $v = a$ to $v = b$, show that the work done (in.-lb) is given by

$$W = \int_a^b p\,dv.$$

Solution Since the work done is independent of the shape of the container, we may assume that the gas is enclosed in a right circular cylinder of radius r and that the expansion takes place against a piston head as illustrated in Figure 6.34. Let P denote a partition of $[a, b]$ determined by $a = v_0, v_1, \ldots, v_n = b$, and let $\Delta v_i = v_i - v_{i-1}$. We shall regard the expansion as taking place in volume increments $\Delta v_1, \Delta v_2, \ldots, \Delta v_n$, and let $d_1, d_2, \ldots, d_n$ be the corresponding distances that the piston head moves (see Figure 6.34). It follows that for each i,

$$\Delta v_i = \pi r^2 d_i$$

and hence

$$d_i = \left(\frac{1}{\pi r^2}\right) \Delta v_i.$$

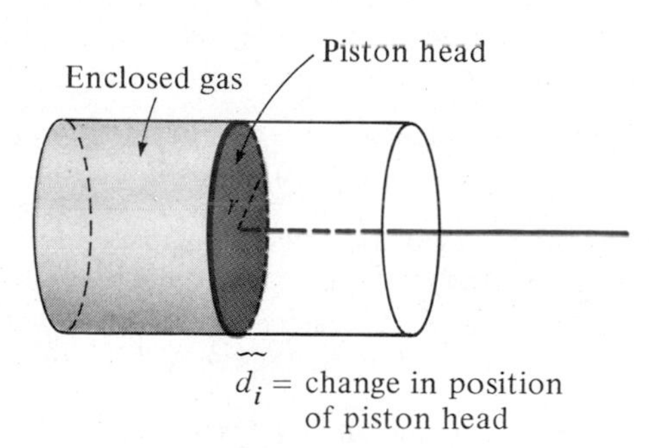

Figure 6.34

If p_i represents a value of the pressure p corresponding to the ith increment, then the force against the piston head is the product of p_i and the area of the piston head, that is, $p_i \pi r^2$. Hence the work done in this increment is

$$p_i \pi r^2 (d_i) = p_i \pi r^2 \left(\frac{1}{\pi r^2}\right) \Delta v_i = p_i \Delta v_i$$

and, therefore,

$$W \approx \sum_i p_i \Delta v_i.$$

Since this approximation improves as the Δv_i approach zero, we conclude that

$$W = \lim_{\|P\| \to 0} \sum_i p_i \Delta v_i = \int_a^b p\,dv.$$

EXERCISES 6.5

1 A spring of natural length 10 inches stretches 1.5 inches under a weight of 8 pounds.
(a) Find the work done in stretching the spring from its natural length to a length of 14 inches.
(b) Find the work done in stretching the spring from a length of 11 inches to a length of 13 inches.

2 A force of 400 dynes is required to compress a spring of natural length 12 cm to a length of 10 cm. Find the work done in compressing the spring from its natural length to a length of 8 cm.

3 If a spring is 12 cm long, compare the work done in stretching it from 12 cm to 13 cm with that done in stretching it from 13 cm to 14 cm.

4 It requires 60 dyne-cm of work to stretch a certain spring from a length of 6 cm to 7 cm, and another 120 dyne-cm of work to stretch it from 7 cm to 8 cm. Find the spring constant and the natural length of the spring.

5 A fishtank has a rectangular base of width 2 ft and length 4 ft, and rectangular sides of height 3 ft. If the tank is filled with water, what work is required to pump the water out over the top of the tank?

6 Generalize Example 3 of this section to the case of a conical tank of altitude h and radius of base a which is filled with a liquid of density (weight per unit volume) ρ.

7 A freight elevator weighing 3000 lb is supported by a 12-foot long cable weighing 15 lb per linear foot. Find the work required to lift the elevator 10 ft by winding the cable onto a winch.

8 A 170-pound man climbs a vertical telephone pole 15 feet high. What work is done if he reaches the top in (a) 10 seconds? (b) 5 seconds?

9 A vertical cylindrical tank of diameter 3 feet and height 6 feet is full of water.
(a) Find the work required to pump the water out over the top of the tank.
(b) Find the work required to pump the water out through a pipe which rises to a height 4 feet above the top of the tank.

10 Answer parts (a) and (b) of Exercise 9 if the tank is only half full of water.

11 The ends of a water trough 8 feet long are equilateral triangles having sides of width 2 feet. If the trough is full of water, find the work required to pump it out over the top.

12 A cistern has the shape of a hemisphere of radius 5 feet, with the circular part at the top. If the cistern is full of water, find the work required to pump all the water to a point 4 feet above the top of the cistern.

13 The force (in dynes) with which two electrons repel one another is inversely proportional to the square of the distance (in centimeters) between them.
(a) If one electron is held fixed at the point $(5, 0)$, find the work done in moving a second electron along the x-axis from the origin to the point $(3, 0)$.
(b) If two electrons are held fixed at the points $(5, 0)$ and $(-5, 0)$, respectively, find the work done in moving a third electron from the origin to $(3, 0)$.

14 A uniform cable 40 feet long and weighing 60 pounds hangs vertically from the top of a building. If a 500-pound weight is attached to the end of the cable, what work is required to pull it to the top?

15 The volume and pressure of a certain gas vary in accordance with the law $pv^{1.2} = 115$, where the units of measurement are inches and pounds. Find the work done if the gas expands from 32 to 40 cubic inches. (*Hint:* See Example 4.)

16 The pressure and volume of a quantity of enclosed steam are related by the formula $pv^{1.14} = c$, where c is a constant. If the initial pressure and volume are p_0 and v_0, respectively, find a formula for the work done if the steam expands to twice its volume. (*Hint:* See Example 4.)

17 Newton's Law of Gravitation states that the force F of attraction between two particles having masses m_1 and m_2 is given by $F = gm_1m_2/s^2$, where g is a constant and s is the distance between the particles. If the mass m_1 of the Earth is regarded as concentrated at the center of the Earth, and a rocket of mass m_2 is on the surface (a distance of 4,000 miles from the center), find a general formula for the work done in firing the rocket vertically upward to an altitude h.

18 In the study of electricity the formula $F = q/kr^2$, where k is a constant, is used to find the force (in dynes) with which a positive charge Q of strength q units repels a unit positive charge located r cm from Q. Find the work done in moving a unit charge from a point d cm from Q to a point $d/2$ cm from Q.

Suppose the tables in Exercises 19 and 20 were obtained experimentally, where $f(x)$ denotes the force acting at the point with coordinate x on a coordinate line l. Use the Trapezoidal Rule (5.47) to approximate the work done on the interval $[a, b]$, where a and b are the smallest and largest values of x, respectively.

19

x ft	0	0.5	1.0	1.5	2.0	2.5	3.0	3.5	4.0	4.5	5.0
$f(x)$ lb	7.4	8.1	8.4	7.8	6.3	7.1	5.9	6.8	7.0	8.0	9.2

20

x cm	1	2	3	4	5	6	7	8	9
$f(x)$ dynes	125	120	130	146	165	157	150	143	140

6.6 FORCE EXERTED BY A LIQUID

In physics the **pressure** p at a depth h in a liquid is defined as the weight of the liquid contained in a column having a cross-sectional area of one unit and altitude h. Pressure may also be regarded as the force per unit area exerted by the liquid. If the **density** (the weight per unit volume) of the liquid is denoted by ρ, then at a depth h,

(6.15)
$$p = \rho h$$

where the units of measurement for ρ and h must be consistent. For example, if the density of water is 62.5 lb/ft³, then at a depth of 4 feet the pressure is

$$p = \rho h = (62.5)(4) = 250 \text{ lb/ft}^2.$$

Note that this is the weight of a column of water having cross-sectional area 1 ft² and altitude 4 ft. It can be verified experimentally that the pressure at any depth is exerted equally in all directions.

If a rectangular tank, such as an ordinary fish aquarium, is filled with water, then the total force exerted by the water on the (horizontal) base can be found by multiplying the pressure at the bottom of the tank by the area of the base. For example, if the depth of water is 4 ft and the area of the base is 10 ft^2, then the pressure at the bottom is 250 lb/ft^2 and the total force acting on the base is

$$\left(250\frac{\text{lb}}{\text{ft}^2}\right)\cdot(10\,\text{ft}^2) = 2500\,\text{lb}.$$

This corresponds to 10 columns of water, each having cross-sectional area 1 ft^2 and each weighing 250 lb.

It is more complicated to find the force exerted on one of the sides of the tank since the pressure is not constant there, but increases as the depth increases. Instead of investigating this particular problem, we shall consider a more general situation.

Consider a flat plate which is submerged in a liquid of density ρ such that the face of the plate is perpendicular to the surface of the liquid. Suppose that the shape of the plate is the same as that of the region in an xy-plane bounded by the graphs of the equations $y = c$, $y = d$, $x = f(y)$, and $x = g(y)$, where f and g are continuous functions on $[c, d]$ and $f(y) \geq g(y)$ for all y in $[c, d]$. Furthermore, suppose that the surface of the water contains the line $y = k$. A region of this type is illustrated in Figure 6.35, although in general it is not required that the region lie entirely above the x-axis as shown in the figure. Let P denote the partition of $[c, d]$ determined by $c = y_0, y_1, \ldots, y_n = d$ and let $\Delta y_i = y_i - y_{i-1}$. Select a number w_i in each subinterval $[y_{i-1}, y_i]$ and consider the rectangle of area $[f(w_i) - g(w_i)]\,\Delta y_i$. If Δy_i is small, then all points in the rectangular region are roughly the same distance, $k - w_i$, from the surface of the liquid and hence by (6.15) the pressure at any point within the rectangle may be approximated by $\rho(k - w_i)$. It appears, therefore, that the force on the ith rectangle is approximately equal to this pressure times the area of the rectangle, that is,

$$\rho(k - w_i)\,[f(w_i) - g(w_i)]\,\Delta y_i.$$

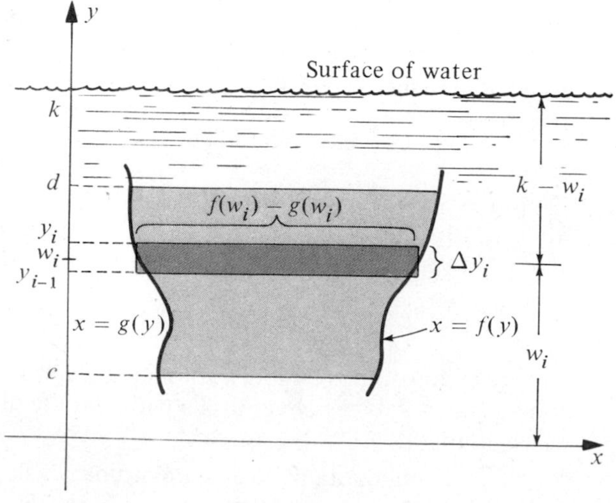

Figure 6.35

The total force on the region shown in Figure 6.35 may be approximated by the Riemann sum

$$\sum_i \rho(k - w_i)[f(w_i) - g(w_i)]\,\Delta y_i.$$

Since we expect this approximation to improve as the norm of the partition decreases we arrive at the following definition.

(6.16) Definition

The **force *F* exerted by a liquid** of constant density ρ on a region of the type illustrated in Figure 6.35, where the functions f and g are continuous on $[c, d]$, is

$$F = \lim_{\|P\| \to 0} \sum_i \rho(k - w_i)[f(w_i) - g(w_i)]\,\Delta y_i$$

$$= \int_c^d \rho(k - y)[f(y) - g(y)]\,dy.$$

If a more complicated region is divided into subregions of the type used above, we apply Definition (6.16) to each subregion and add the resulting numbers. The coordinate system may be introduced in various ways. In Example 2 we shall choose the x-axis along the surface of the water and the positive direction of the y-axis downward. In this case, Definition (6.16) must be changed accordingly.

It is often convenient to use the following formula to specify the force on a typical horizontal rectangle:

(6.17) (Force on one rectangle) = (density) · (depth) · (area of rectangle)

and then take a limit of a sum to obtain the total force on the region.

Example 1 The ends of a water trough 8 feet long have the shape of isosceles trapezoids of lower base 4 feet, upper base 6 feet, and altitude 4 feet. Find the total force on one end if the trough is full of water.

Solution Figure 6.36 illustrates one of the ends of the trough superimposed on a rectangular coordinate system and appropriately labeled. The equation of the line through the points $(2, 0)$ and $(3, 4)$ is $y = 4x - 8$ or, equivalently, $x = (1/4)(y + 8)$. It follows that the indicated rectangle has area

$$2x_i\,\Delta y_i = 2\left(\frac{1}{4}\right)(w_i + 8)\,\Delta y_i = \left(\frac{1}{2}\right)(w_i + 8)\,\Delta y_i.$$

Using (6.17), we find that the force exerted by the liquid on this rectangle is approximately

$$(62.5)(4 - w_i)\left(\frac{1}{2}\right)(w_i + 8)\,\Delta y_i.$$

Summing and taking the limit gives us the total force. Thus,

$$F = \int_0^4 (62.5)(4 - y)\left(\frac{1}{2}\right)(y + 8)\,dy$$

$$= 31.25 \int_0^4 (32 - 4y - y^2)\,dy$$

$$= 31.25\left[32y - 2y^2 - \frac{y^3}{3}\right]_0^4$$

$$= 31.25\left[128 - 32 - \frac{64}{3}\right] = \frac{7{,}000}{3}\text{ pounds.}$$

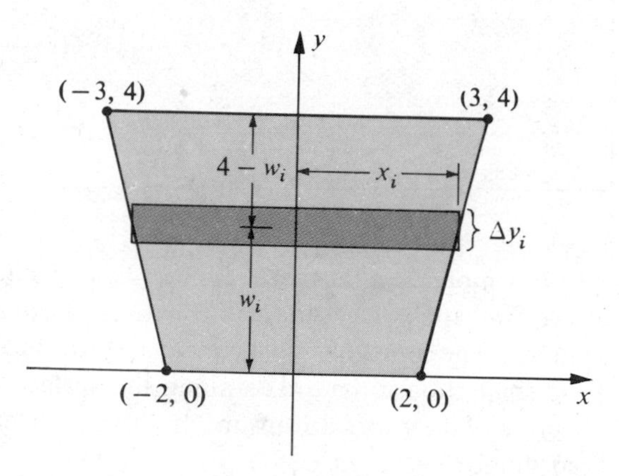

Figure 6.36

Example 2 A cylindrical tank 6 feet in diameter and 10 feet long is lying on its side. If the tank is half full of oil weighing 58 lb/ft^3, find the force exerted by the oil on one side of the tank.

Solution Let us introduce a coordinate system so that the end of the tank is a circle of radius 3 feet with center at the origin. If we choose the positive direction of the y-axis *downward*, then as shown in Figure 6.37, w_i represents the depth at a point in a typical horizontal rectangle.

From the equation of the circle, the length of the rectangle is $2\sqrt{9 - w_i^2}$. Using (6.17), the force acting on the pictured rectangle is

$$58w_i(2\sqrt{9 - w_i^2})\,\Delta y_i.$$

As usual, we take a limit of a sum of such terms to obtain the total force F on the end of the tank. This gives us

$$F = \int_0^3 58y(2\sqrt{9 - y^2})\,dy.$$

The integral may be evaluated by letting

$$u = 9 - y^2 \quad \text{and} \quad du = -2y\,dy.$$

Note that if $y = 0$, then $u = 9$, whereas if $y = 3$, then $u = 0$. We may, therefore, transform the integral as follows:

$$F = -58 \int_0^3 \sqrt{9 - y^2}(-2y\,dy)$$

$$= -58 \int_9^0 u^{1/2}\,du$$

$$= -58 \frac{u^{3/2}}{(3/2)}\Big]_9^0$$

$$= -\frac{116}{3}\Big[u^{3/2}\Big]_9^0$$

$$= -\frac{116}{3}[0 - 9^{3/2}] = 1044 \text{ lb.}$$

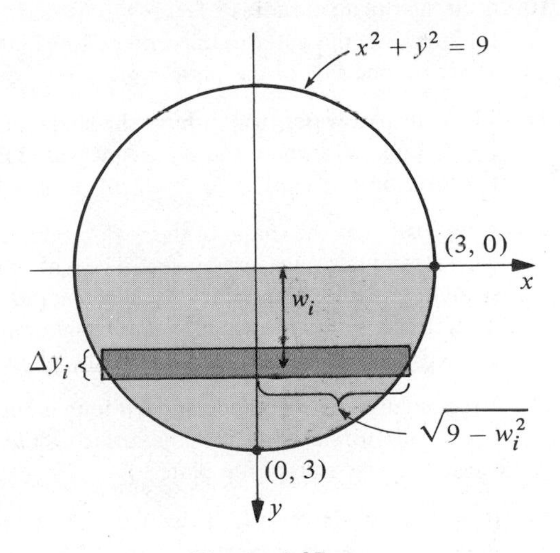

Figure 6.37

EXERCISES 6.6

1 A glass tank to be used as an aquarium is 3 feet long and has square ends of width 1 foot. If the tank is filled with water, find the force exerted by the water on (a) one end and (b) one side.

2 If one of the square ends of the tank in Exercise 1 is divided into two parts by means of a diagonal, find the force exerted on each part.

3 The ends of a water trough 6 feet long have the shape of isosceles triangles with equal sides of length 2 feet and the third side of length $2\sqrt{3}$ feet at the top of the trough.
(a) Find the force exerted by the water on one end of the trough if the trough is full of water.
(b) Find the force if the trough is half full of water.

4 Answer (a) and (b) of Exercise 3 if the altitude is h feet, where $0 < h < 2$.

5 A cylindrical tank 4 feet in diameter and 5 feet long is lying on its side. If the tank is half full of oil weighing 60 lb/ft^3, find the force exerted by the oil on one end of the tank.

6 A rectangular gate in a dam is 5 feet long and 3 feet high. If the gate is vertical, with the top of the gate parallel to the surface of the water and 6 feet below it, find the force of the water against the gate.

7 A rectangular swimming pool is 20 feet wide and 40 feet long. The depth of the water in the pool varies uniformly from 3 feet at one end to 9 feet at the other end. Find the total force exerted by the water on the bottom of the pool.

8 Find the force exerted by the water on the side of the swimming pool described in Exercise 7.

9 A plate having the shape of an isosceles trapezoid with upper base 4 feet long and lower base 8 feet long is submerged vertically in water such that the bases are parallel to the surface. If the distances from the surface of the water to the lower and upper bases are 10 feet and 6 feet, respectively, find the force exerted by the water on one side of the plate.

10 A circular plate of radius 2 feet is submerged vertically in water. If the distance from the surface of the water to the center of the plate is 6 feet, find the force exerted by the water on one side of the plate.

11 The ends of a water trough have the shape of the region bounded by the graphs of $y = x^2$ and $y = 4$ where x and y are measured in feet. If the trough is full of water, find the force on one end.

12 A flat plate has the shape of the region bounded by the graphs of $y = x^4$ and $y = 1$ where x and y are measured in feet. The plate is submerged vertically in water with the straight part of its boundary parallel to (and closest to) the surface. If the distance from the surface of the water to the straight part of the boundary is 4 ft, find the force exerted by the water on one side of the plate.

13 A rectangular plate 3 ft wide and 6 ft long is submerged vertically in oil with its short side parallel to, and 2 ft below, the surface. If the oil weighs 50 lb/ft^3, find the total force exerted on one side of the plate.

14 If the plate in Exercise 13 is divided into two parts by means of a diagonal, find the force exerted on each part.

6.7 ARC LENGTH

A function f is said to be **smooth** on an interval if it has a derivative f' which is continuous throughout the interval. Roughly speaking, this means that a small change in x produces a small change in the slope $f'(x)$ of the tangent line to the graph of f. Thus, there are no sharp corners on the graph of a smooth function. In this section we shall define what is meant by the **length of arc** between the two points A and B on a smooth curve.

If f is smooth on a closed interval $[a, b]$, the points $A(a, f(a))$ and $B(b, f(b))$ will be called the **end-points** of the graph of f. Let us consider the partition P of

$[a, b]$ determined by $a = x_0, x_1, x_2, \ldots, x_n = b$, and let Q_i denote the point with coordinates $(x_i, f(x_i))$. This gives us $n + 1$ points $Q_0, Q_1, Q_2, \ldots, Q_n$ on the graph of f. If, as illustrated in Figure 6.38, we connect each Q_{i-1} to Q_i by a straight line segment of length $d(Q_{i-1}, Q_i)$, then the length L_P of the resulting broken line is

$$L_P = \sum_{i=1}^{n} d(Q_{i-1}, Q_i).$$

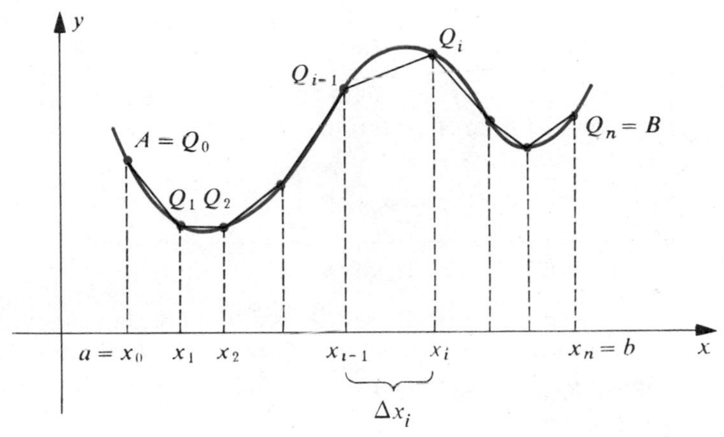

Figure 6.38

If the norm $\|P\|$ is small, then Q_{i-1} is close to Q_i for each i, and we expect L_P to be an approximation to the length of arc between A and B. This gives us a clue to a suitable definition of arc length. Specifically, we shall consider the limit of the sum L_P as $\|P\| \to 0$. In order to formulate this concept precisely, and at the same time arrive at a formula for calculating arc length, let us proceed as follows. By the Distance Formula (1.7),

$$d(Q_{i-1}, Q_i) = \sqrt{(x_i - x_{i-1})^2 + [f(x_i) - f(x_{i-1})]^2}.$$

Applying the Mean Value Theorem (4.12),

$$f(x_i) - f(x_{i-1}) = f'(w_i)(x_i - x_{i-1})$$

where w_i is in the open interval (x_{i-1}, x_i). Substituting this in the preceding formula and letting $\Delta x_i = x_i - x_{i-1}$, we obtain

$$\begin{aligned} d(Q_{i-1}, Q_i) &= \sqrt{(\Delta x_i)^2 + [f'(w_i)\,\Delta x_i]^2} \\ &= \sqrt{1 + [f'(w_i)]^2}\,\Delta x_i. \end{aligned}$$

Consequently,

$$L_p = \sum_{i=1}^{n} \sqrt{1 + [f'(w_i)]^2}\,\Delta x_i.$$

Observe that L_P is a Riemann sum for the function g defined by $g(x) = \sqrt{1 + [f'(x)]^2}$. In addition, g is continuous on $[a, b]$ since f' is continuous. As have mentioned, if the norm $\|P\|$ is small, then the length L_P of the broken line should approximate the length of the graph of f from A to B. Moreover, this approximation should improve as $\|P\|$ decreases. It is natural, therefore, to define the *length* (also called the *arc length*) of the graph of f from A to B as the limit of the sum L_P. Since $g = \sqrt{1 + (f')^2}$ is a continuous function the limit exists and equals the definite integral $\int_a^b \sqrt{1 + [f'(x)]^2}\,dx$. This arc length will be denoted by the symbol L_a^b.

(6.18) Definition

> Let the function f be smooth on a closed interval $[a, b]$. The **arc length of the graph** of f from $A(a, f(a))$ to $B(b, f(b))$ is given by
>
> $$L_a^b = \int_a^b \sqrt{1 + [f'(x)]^2}\,dx.$$

Definition (6.18) will be extended to more general graphs in Chapter 13. A graph which has arc length is said to be **rectifiable**. If a function f is defined implicitly by an equation in x and y, then we shall also refer to the *arc length of the graph of the equation.*

Example 1 If $f(x) = 3x^{2/3} - 10$, find the arc length of the graph of f from the point $A(8, 2)$ to $B(27, 17)$.

Solution The graph of f is sketched in Figure 6.39.

Since $f'(x) = 2x^{-1/3} = 2/(x^{1/3})$, we have

$$L_8^{27} = \int_8^{27} \sqrt{1 + \left(\frac{2}{x^{1/3}}\right)^2}\,dx = \int_8^{27} \sqrt{1 + \frac{4}{x^{2/3}}}\,dx$$

$$= \int_8^{27} \frac{\sqrt{x^{2/3} + 4}}{x^{1/3}}\,dx.$$

In order to evaluate this integral let

$$u = x^{2/3} + 4 \quad \text{and} \quad du = \frac{2}{3}x^{-1/3}\,dx = \frac{2}{3}\frac{1}{x^{1/3}}\,dx.$$

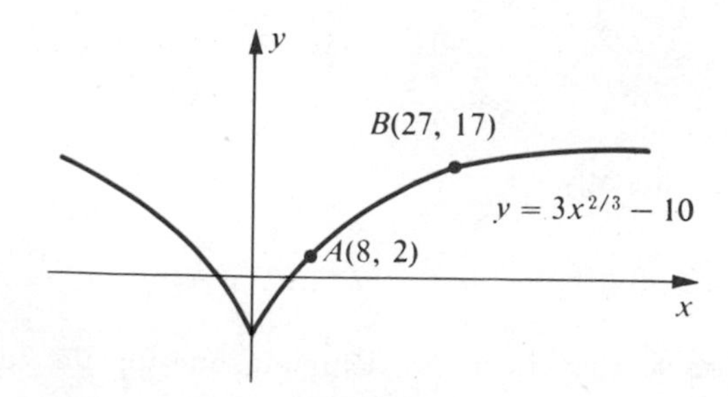

Figure 6.39

The integral can be expressed in a suitable form for integration by introducing the factor 2/3 in the integrand and compensating for this by multiplying the integral by 3/2. Thus

$$L_8^{27} = \frac{3}{2}\int_8^{27} \sqrt{x^{2/3}+4}\left(\frac{2}{3}\frac{1}{x^{1/3}}\right) dx.$$

If $x = 8$, then $u = (8)^{2/3} + 4 = 8$, whereas if $x = 27$, then $u = (27)^{2/3} + 4 = 13$. Making the substitution and changing the limits of integration,

$$L_8^{27} = \frac{3}{2}\int_8^{13} \sqrt{u}\, du = u^{3/2}\Big]_8^{13} = 13^{3/2} - 8^{3/2} \approx 24.2.$$

If a function g is defined by $x = g(y)$, where g is continuous on an interval $[c, d]$, then an argument similar to that given earlier, but with y regarded as the independent variable, leads to the following formula for the length L_c^d of the graph of g from $(g(c), c)$ to $(g(d), d)$:

(6.19)
$$L_c^d = \int_c^d \sqrt{1 + [g'(y)]^2}\, dy.$$

Example 2 Set up an integral for finding the arc length of the graph of $y = y^3 - x$ from $A(0, -1)$ to $B(6, 2)$. Approximate the integral by using Simpson's Rule (5.49) with $n = 6$.

Solution The equation is not of the form $y = f(x)$ and hence Definition (6.18) cannot be applied. However, if we write $x = y^3 - y$, then we can employ (6.19) with $g(y) = y^3 - y$. The graph of the equation is sketched in Figure 6.40.

Using (6.19) with $c = -1$ and $d = 2$ gives us

$$L_{-1}^2 = \int_{-1}^{2} \sqrt{1 + (3y^2 - 1)^2}\, dy$$

$$= \int_{-1}^{2} \sqrt{9y^4 - 6y^2 + 2}\, dy.$$

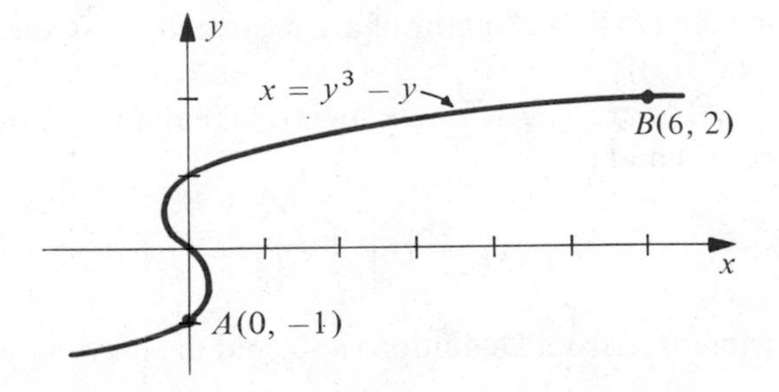

Figure 6.40

To use Simpson's Rule, we let $f(y) = \sqrt{9y^4 - 6y^2 + 2}$ and arrange our work as we did in Section 5.7. The reader may verify the entries in the following table.

i	y_i	$f(y_i)$	m	$mf(y_i)$
0	−1.0	2.2361	1	2.2361
1	−0.5	1.0308	4	4.1232
2	0.0	1.4142	2	2.8284
3	0.5	1.0308	4	4.1232
4	1.0	2.2361	2	4.4722
5	1.5	5.8363	4	23.3452
6	2.0	11.0454	1	11.0454

The sum of the numbers in the last column is 52.1737. Hence by Simpson's Rule (5.49) with $a = -1$, $b = 2$, and $n = 6$,

$$\int_{-1}^{2} \sqrt{9y^4 - 6y^2 + 2} \approx \frac{1}{6}[52.1737] \approx 8.7.$$

If a graph can be decomposed into a finite number of parts, each of which is the graph of a smooth function, then the arc length of the graph is defined as the sum of the arc lengths of the individual graphs. A function of this type is said to be **piecewise smooth** on its domain.

In order to avoid any misunderstanding in the following discussion, the variable of integration will be denoted by t. In this case the arc length formula given in Definition (6.18) becomes

$$L_a^b = \int_a^b \sqrt{1 + [f'(t)]^2}\, dt.$$

If f is smooth on $[a, b]$, then f is smooth on $[a, x]$ for every x in $[a, b]$, and the length of the graph from the point $A(a, f(a))$ to the point $Q(x, f(x))$ is given by

$$L_a^x = \int_a^x \sqrt{1 + [f'(t)]^2}\, dt.$$

If we change the notation and use the symbol $s(x)$ in place of L_a^x, then s may be regarded as a function with domain $[a, b]$, since to each x in $[a, b]$ there corresponds a unique number $s(x)$. We shall call s the **arc length function** for the graph of f (relative to the point A). The values $s(x)$ of s are represented geometrically as lengths of arc measured along the graph of f from $A(a, f(a))$ to $Q(x, f(x))$.

The arc length function s is differentiable on $[a, b]$. Indeed, by the definition of $s(x)$ and (5.33),

$$D_x[s(x)] = D_x\left[\int_a^x \sqrt{1 + [f'(t)]^2}\, dt\right] = \sqrt{1 + [f'(x)]^2}.$$

Moreover, from Definition (3.30) and the previous equality, the differential of s is

(6.20) $$ds = s'(x)\, dx = \sqrt{1 + [f'(x)]^2}\, dx$$

or, bringing the differential dx underneath the radical,

$$ds = \sqrt{(dx)^2 + [f'(x)\,dx]^2}.$$

Letting $y = f(x)$ and using Definition (3.30), this may be written

$$ds = \sqrt{(dx)^2 + (dy)^2}.$$

There is an interesting geometric interpretation for the last formula. First, squaring both sides gives us

(6.21) $$(ds)^2 = (dx)^2 + (dy)^2.$$

Next, consider $y = f(x)$ and give x an increment Δx. Let Δy denote the change in y and Δs the change in arc length corresponding to Δx. These increments are illustrated in Figure 6.41 where dy is the amount that the tangent line rises or falls if the independent variable changes from x to $x + \Delta x$ (see also Figure 3.9). It follows from (6.21) that ds may be thought of as the length of the hypotenuse of a right triangle with sides $|dx|$ and $|dy|$ as illustrated in Figure 6.41. This figure also illustrates that if Δx is small, then ds may be used to approximate the increment Δs of arc length.

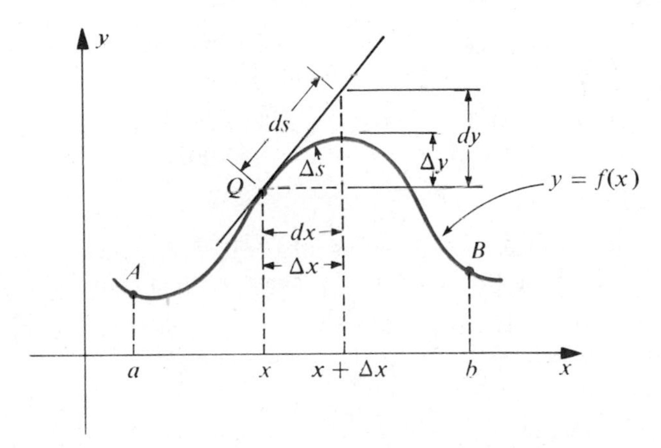

Figure 6.41

Example 3 Use differentials to approximate the arc length of the graph of $y = x^3 + 2x$ from $A(1, 3)$ to $B(1.2, 4.128)$.

Solution If we let $f(x) = x^3 + 2x$, then by (6.20),

$$ds = \sqrt{1 + (3x^2 + 2)^2}\,dx.$$

An approximation may be obtained by letting $x = 1$ and $dx = 0.2$. Thus

$$ds = \sqrt{1 + 5^2}\,(0.2) = \sqrt{26}(0.2) \approx 1.02.$$

EXERCISES 6.7

In Exercises 1–4 use Definition (6.18) to find the arc length of the graph of the given equation from A to B.

1 $8x^2 = 27y^3$; $A(1, 2/3)$, $B(8, 8/3)$

2 $(y + 1)^2 = (x - 4)^3$; $A(5, 0)$, $B(8, 7)$

3 $y = 5 - \sqrt{x^3}$; $A(1, 4)$, $B(4, -3)$

4 $y = 6\sqrt[3]{x^2} + 1$; $A(-1, 7)$, $B(-8, 25)$

5 Solve Exercise 1 by means of (6.19).

6 Solve Exercise 2 by means of (6.19).

In Exercises 7–10 find the arc length from A to B of the graph of the given equation.

7 $y = (x^3/12) + (1/x)$; $A(1, 13/12)$, $B(2, 7/6)$

8 $y + 1/(4x) + (x^3/3) = 0$; $A(2, 67/24)$, $B(3, 109/12)$

9 $30xy^3 - y^8 = 15$; $A(8/15, 1)$, $B(271/240, 2)$

10 $x = (y^4/16) + (1/2y^2)$; $A(9/8, -2)$, $B(9/16, -1)$

In Exercises 11 and 12 set up (but do not evaluate) the integral for finding the arc length of the graph of the given equation from A to B.

11 $2y^3 - 7y + 2x - 8 = 0$; $A(3, 2)$, $B(4, 0)$

12 $11x - 4x^3 - 7y + 7 = 0$; $A(1, 2)$, $B(0, 1)$

13 Find the arc length of the graph of $x^{2/3} + y^{2/3} = 1$.

14 Find the arc length of the graph of $y = (3x^8 + 5)/30x^3$ from the point with abscissa 1 to the point with abscissa 2.

15 If f is defined by $f(x) = \sqrt[3]{x^2}$, what is the arc length function s relative to the point $(1, 1)$? If x increases from 1 to 1.1, find Δs and ds.

16 Work Exercise 15 if $f(x) = \sqrt{x^3}$.

17 Use differentials to approximate the arc length of the graph of $y = x^2$ from $A(2, 4)$ to $B(2.1, 4.41)$. Illustrate this approximation graphically and compare it with $d(A, B)$.

18 Use differentials to approximate the arc length of the graph of $y + x^3 = 0$ from $A(1, -1)$ to $B(1.1, -1.331)$. Illustrate this approximation graphically and compare it with $d(A, B)$.

In each of Exercises 19 and 20 use Simpson's Rule (5.49) with $n = 4$ to approximate the arc length of the graph of the given equation from A to B.

19 $y = x^3$; $A(0, 0)$, $B(2, 8)$

20 $y = x^2 + x + 3$; $A(-2, 5)$, $B(2, 9)$

6.8 OTHER APPLICATIONS

In this section we shall consider several miscellaneous applications in order to give the reader some idea of the versatility of the definite integral. The interested person can find many other illustrations. The first three applications have to do with the study of economics and the fourth occurs in biology.

It is important for people in business to plan for depreciation of equipment. The most elementary technique employed is the **straight-line method**, in which the rate of depreciation is considered constant. For example, if the depreciation each year is \$200, then the total depreciation $f(t)$ at the end of t years is given by $f(t) = 200t$. Note that the rate of depreciation is $f'(t)$. In many instances the rate of depreciation is not constant. To illustrate, an automobile depreciates very rapidly during the first few years after it is purchased and then more slowly as it gets older. For other items the rate of depreciation may increase from year to year. In general, suppose that the rate of depreciation of a certain piece of equipment over the time interval $[0, t]$ may be approximated by $g(t)$, where g is a continuous function. If the total depreciation in $[0, t]$ is $f(t)$, then $f'(t) = g(t)$ and we may write

$$f(t) = \int_0^t f'(x)\,dx = \int_0^t g(x)\,dx.$$

Suppose further that an additional fixed cost C is required to overhaul the equipment after time t. Thus, in the time interval $[0, t]$, the expense $h(t)$ connected with the equipment is given by

$$h(t) = C + \int_0^t g(x)\,dx. \tag{6.22}$$

The average expense, that is, the expense per year, is

$$k(t) = \frac{h(t)}{t}. \tag{6.23}$$

If no other factors are involved, then the best time to overhaul the equipment corresponds to the value of t for which the function k has a relative minimum. This occurs when

$$k'(t) = \frac{th'(t) - h(t)}{t^2} = 0$$

or equivalently, when

$$h'(t) = \frac{h(t)}{t} = k(t).$$

Since $h'(t) = g(t)$, the best time to overhaul occurs when $g(t) = k(t)$, that is, when the rate of depreciation is the same as the average expense.

Example 1 Using the notation of the preceding discussion, suppose that for a given piece of equipment $g(t) = 300\sqrt{t}$ and $C = 500$, where t is in years. When should the equipment be overhauled in order to minimize the average expense?

Solution From (6.22) and (6.23), the average expense is given by

$$k(t) = \left(500 + \int_0^t 300\sqrt{x}\,dx\right) \div t$$
$$= (500 + 200t^{3/2})/t.$$

From the previous discussion, k will have a minimum value if

$$(500 + 200t^{3/2})/t = 300t^{1/2}.$$

Solving for t we obtain $t = 5^{2/3} \approx 2.924$ or, to the nearest month, 2 years and 11 months.

In economics, the process which a corporation uses to increase its accumulated wealth is called **capital formation.** If the amount K of capital at time t can be approximated by $K = f(t)$, where f is a differentiable function, then the rate of change of K with respect to t is called the **net investment flow**. Hence, if I denotes the investment flow, then

$$I = \frac{dK}{dt} = f'(t).$$

Conversely, if I is given by $g(t)$, where the function g is continuous on an interval $[a, b]$, then the increase in capital over this time interval is

(6.24)
$$\int_a^b g(t)\,dt = f(b) - f(a).$$

Example 2 Suppose a corporation wishes to have its net investment flow approximated by $g(t) = t^{1/3}$, where t is in years and $g(t)$ is in millions of dollars per year. If $t = 0$ corresponds to the present time, estimate the amount of capital formation over the next eight years.

Solution Applying (6.24) we obtain

$$\int_0^8 t^{1/3}\,dt = \frac{3}{4}t^{4/3}\bigg]_0^8 = 12.$$

Consequently, the amount of capital formation is \$12,000,000.

In many types of employment, a worker must perform the same assignment repeatedly. For example, a boy hired by a bicycle shop may be asked to assemble new bicycles. As he assembles one bicycle after another, the boy should become more proficient and, up to a certain point, should assemble each one in less time than the preceding one. Another example of this process of learning by repetition is that of a key-punch operator who must translate information from written forms to punched cards. The time required to punch each card should decrease as the number of cards increases. As a final illustration, the time required for a person to trace a path through a maze should improve with practice.

Let us consider a general situation in which a certain task is to be repeated many times. Suppose experience has shown that the time required to perform the task for the ith time can be approximated by $f(i)$, where f is a continuous decreasing function on a suitable interval. The total time required to perform the task k times is given by the sum

$$\sum_{i=1}^{k} f(i) = f(1) + f(2) + \cdots + f(k).$$

If we consider the graph of f, then as illustrated in Figure 6.42 the previous sum equals the area of the pictured inscribed rectangular polygon and, therefore, may be approximated by the definite integral $\int_0^k f(x)\,dx$. Evidently, the approximation will be close to the actual sum if f decreases slowly on $[0, k]$. If f changes rapidly per unit change in x, then an integral should not be used as an approximation.

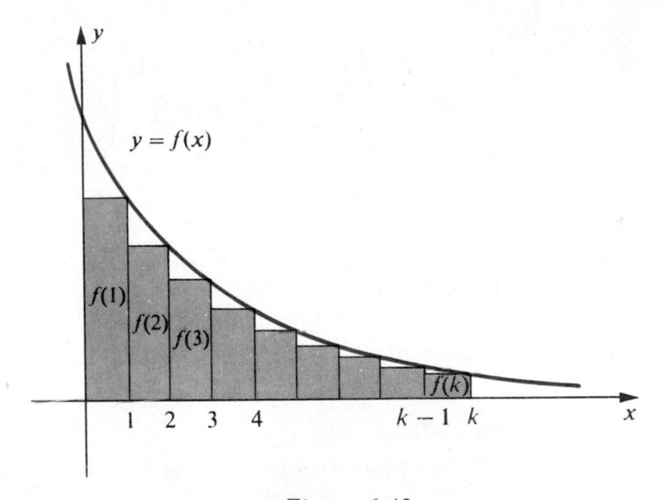

Figure 6.42

Example 3 A company which conducts polls by means of telephone interviews finds that the time required by an employee to complete one interview depends on the number of interviews that the employee has completed previously. Suppose it is estimated that for a certain survey, the number of minutes required to complete the ith interview is given by $f(i) = 6(1 + i)^{-1/5}$. Use a definite integral to approximate the time required for an employee to complete (a) 100 interviews; (b) 200 interviews. If an interviewer receives \$3.60 per hour, estimate how much more expensive it is to have two employees each conduct 100 interviews than it is to have one employee conduct 200 interviews.

Solution As in the preceding discussion, the time required for 100 interviews is approximately

$$\int_0^{100} 6(1 + x)^{-1/5}\,dx = 6 \cdot \frac{5}{4}(1 + x)^{4/5}\Big]_0^{100}$$

$$= \frac{15}{2}[(101)^{4/5} - 1]$$

$$\approx 293.5 \text{ minutes.}$$

The time required for 200 interviews is approximately

$$\int_0^{200} 6(1 + x)^{-1/5}\,dx = \frac{15}{2}[(201)^{4/5} - 1]$$

$$\approx 514.4 \text{ minutes.}$$

Since an interviewer receives \$0.06 per minute, the cost for one employee to conduct 200 interviews is roughly (\$0.06)(514.4) or \$30.86. If two employees each conduct 100 interviews the cost is about 2(\$0.06)(293.5) or \$35.22, which is \$4.36 more than the cost of one employee. Note, however, that the time saved in using two people is approximately 220 minutes.

A computing machine may be used to show that

$$\sum_{i=1}^{100} 6(1 + i)^{-1/5} \approx 291.75$$

and

$$\sum_{i=1}^{200} 6(1 + i)^{-1/5} \approx 512.57.$$

Hence the result obtained by integration (the area under the graph of f) is roughly 2 units more than the value of the corresponding sum (the area of the inscribed rectangular polygon).

Our next application of the definite integral is taken from the field of biology. If a liquid flows through a cylindrical tube and if the velocity is a constant v_0, then the amount of liquid passing a fixed point per unit time is given by v_0A, where A is the area of a cross section of the tube (see Figure 6.43). A more complicated formula is required to study the flow of blood in an arteriole. For the latter case the flow is in layers, as illustrated in Figure 6.44. In the layer closest to the wall of the arteriole, the blood tends to stick to the wall, and its velocity may be considered zero. The velocity increases as the layers approach the center of the arteriole. For computational purposes, we may regard the blood flow as consisting of thin cylindrical shells which slide along, with the outer shell fixed and the velocity of the shells increasing as the radii of the shells decrease (see Figure 6.44). If the velocity in each shell is considered constant, then from the theory of liquids in motion, the velocity $v(r)$ in a shell having average radius r is given by

$$v(r) = \frac{P}{4\nu l}(R^2 - r^2),$$

where R is the radius of the arteriole (in cm), l is the length of the arteriole (in cm), P is the pressure difference between the two ends of the arteriole (in dyne/cm^2) and ν is the viscosity of the blood (in dyne-sec/cm^2). Note that the formula gives zero velocity if $r = R$ and maximum velocity $PR^2/4\nu l$ as r approaches 0. If the radius of the kth shell is r_k and the thickness of the shell is Δr_k, then by (6.9) the volume of blood in this shell is

$$2\pi r_k v(r_k)\,\Delta r_k = \frac{2\pi r_k P}{4vl}(R^2 - r_k^2)\,\Delta r_k.$$

If there are n shells, then the total flow in the arteriole per unit time may be approximated by

$$\sum_{k=1}^{n} \frac{2\pi r_k P}{4vl}(R^2 - r_k^2)\,\Delta r_k.$$

In order to estimate the total flow F, that is, the volume of blood per unit time, we take the limit of this sum as n increases without bound. This leads to the following definite integral:

$$F = \int_0^R \frac{2\pi r P}{4vl}(R^2 - r^2)\,dr$$

$$= \frac{2\pi P}{4vl}\int_0^R (R^2 r - r^3)\,dr$$

$$= \frac{\pi P}{2vl}\left[\frac{1}{2}R^2 r^2 - \frac{1}{4}r^4\right]_0^R.$$

Substituting the limits of integration gives us

$$F = \frac{\pi P R^4}{8vl}\,\text{cm}^3.$$

This formula for F is not exact since the thickness of the shells cannot be made arbitrarily small. Indeed, the lower limit is the width of a red blood cell, or approximately 2×10^{-4} cm. However, we may assume that the formula gives a reasonable estimate. It is interesting to observe that a small change in the radius of an arteriole produces a large change in the flow, since F is directly proportional to the fourth power of R. A small change in the pressure difference has a lesser effect, since P appears to only the first power.

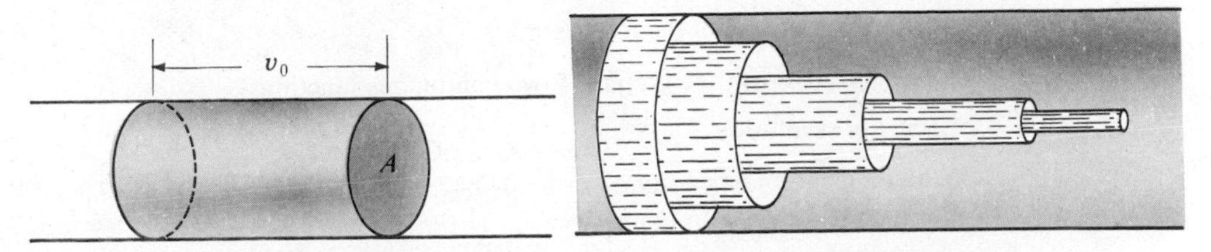

Figure 6.43

Figure 6.44

EXERCISES 6.8

1 The rate of depreciation of a certain piece of equipment over the time interval $[0, 3]$ may be approximated by $g(t) = 1 - t^2/9$ where t is in years and $g(t)$ is in hundreds of dollars. Find the total depreciation at the end of the following periods.
(a) 6 months (b) 1 year (c) 18 months (d) 2 years

2 The rate of depreciation of a certain piece of equipment over the time interval $[0, 6]$ may be approximated by $g(t) = \sqrt{10 - t}$, where t is in years and $g(t)$ is in thousands of dollars. If, after 6 years, an additional cost of \$420 is needed to overhaul the equipment, what is the average expense for the six years?

3 If, in Example 1, $g(t) = 100t^{2/3}$ and $C = 400$, find the best time to overhaul the equipment.

4 If, in Example 2, the rate of investment is approximated by $g(t) = 2t(3t + 1)$, where $g(t)$ is in thousands of dollars, find the amount of capital formation over the intervals $[0, 5]$ and $[5, 10]$.

5 A key-punch operator is required to transfer registration data of college students from written sheets to punched cards. It is estimated that the number of minutes required to punch the ith card is approximately $f(i) = 6(1 + i)^{-1/3}$. Use a definite integral to estimate the time required for (a) one operator to punch 600 cards; (b) two operators to punch 300 cards each.

6 It is estimated that the number of minutes needed for a person to trace a path through a certain maze without error is given by $f(i) = 5i^{-1/2}$, where i is the number of trials previously completed. How much time is required to complete 10 trials?

7 A small parts manufacturer estimates that the time required for a worker to assemble a certain item depends on the number of items the worker has previously assembled. If the time (in minutes) required to assemble the nth item is given by $f(n) = 20\,(n+1)^{-.04} + 3$, approximate the time required to assemble the following quantities.

(a) 1 item (b) 4 items (c) 8 items (d) 16 items

8 Use a definite integral to approximate the sum $\sum_{k=1}^{100} k(k^2 + 1)^{-1/4}$.

6.9 REVIEW

Concepts

Define or discuss each of the following.

1 The area between the graphs of two continuous functions

2 Solid of revolution

3 Methods of finding volumes of solids of revolution

4 Volumes by slicing

5 Work

6 Force exerted by a liquid

7 Smooth function

8 Piecewise smooth function

9 Arc length of a graph

10 The arc length function

Exercises

In Exercises 1 and 2, sketch the region bounded by the graphs of the given equations and find the area in two ways, first by integrating with respect to x and second by integrating with respect to y.

1 $y = -x^2$, $y = x^2 - 8$

2 $y^2 = 4 - x$, $x + 2y - 1 = 0$

In Exercises 3 and 4 find the area of the region bounded by the graphs of the given equations.

3 $x = y^2$, $x + y = 1$

4 $y + x^3 = 0$, $y = \sqrt{x}$, $3y + 7x - 10 = 0$

In Exercises 5–8 sketch the region R bounded by the graphs of the given equations and find the volume of the solid generated by revolving R about the indicated axis.

5 $y = \sqrt{4x + 1}$, $y = 0$, $x = 0$, $x = 2$; about the x-axis

6 $y = x^4$, $y = 0$, $x = 1$; about the y-axis

7 $y = x^3 + 1$, $x = 0$, $y = 2$; about the y-axis

8 $y = \sqrt[3]{x}$, $y = \sqrt{x}$; about the x-axis

9 Find the volume of the solid generated by revolving the region bounded by the graphs of $y = 4x^2$ and $4x + y - 8 = 0$ as follows.
(a) About the x-axis
(b) About the line $x = 1$
(c) About the line $y = 16$

10 Find the volume of the solid generated by revolving the region bounded by the graphs of $y = x^3$, $x = 2$, and $y = 0$ as follows.
(a) About the x-axis
(b) About the y-axis
(c) About the line $x = 2$
(d) About the line $x = 3$
(e) About the line $y = 8$
(f) About the line $y = -1$

11 Find the arc length of the graph of the equation $(x + 3)^2 = 8(y - 1)^3$ from the point $A(-2, 3/2)$ to the point $B(5, 3)$.

12 A solid has, for its base, the region in the xy-plane bounded by the graphs of $y^2 = 4x$ and $x = 4$. Find the volume of the solid if every cross section by a plane perpendicular to the x-axis is an isosceles right triangle with one of the equal sides on the base of the solid.

13 An above-ground swimming pool has the shape of a right circular cylinder of diameter 12 feet and height 5 feet. If the depth of the water in the pool is 4 feet, find the work required to empty the pool by pumping the water out over the top.

14 As a bucket is raised a distance of 30 feet from the bottom of a well, water leaks out at a uniform rate. Find the work done if the bucket originally contains 24 pounds of water and 1/3 leaks out. Assume that the weight of the empty bucket is 4 pounds and neglect the weight of the rope.

15 A square plate of side 4 feet is submerged vertically in water such that one of the diagonals is parallel to the surface. If the distance from the surface to the center of the plate is 6 feet, find the force exerted by the water on one side of the plate.

16 Use differentials to approximate the arc length of the graph of $y = 3x^2 - 4x + 2$ from $A(1, 1)$ to $B(1.1, 1.23)$.

If $\lim_{\|P\| \to 0} \sum_i \pi w_i^4 \Delta x_i$ represents the limit of a sum for a function f on the interval $[0, 1]$, solve Exercises 17–20.

17 Find the value of the limit.

18 Interpret the limit as the area of a region in the xy-plane.

19 Interpret the limit as the volume of a solid of revolution.

20 Interpret the limit as the work done by a force.

CHAPTER 7

Topics in Analytic Geometry

Plane geometry includes the study of figures such as lines, circles, and triangles which lie in a plane. Theorems are proved by reasoning deductively from certain postulates. In analytic *geometry, plane geometric figures are investigated by introducing a coordinate system and then using equations and formulas of various types. If the study of analytic geometry were to be summarized by means of one statement, perhaps the following would be appropriate: "Given an equation, find its graph and, conversely, given a graph, find its equation." In this chapter we shall apply coordinate methods to several basic plane figures.*

7.1 CONIC SECTIONS

Each of the geometric figures to be discussed in this chapter can be obtained by intersecting a double-napped right circular cone with a plane. For this reason they are called **conic sections** or simply **conics**. If, as in (i) of Figure 7.1, the plane cuts entirely across one nappe of the cone and is not perpendicular to the axis, then the curve of intersection is called an **ellipse**. If the plane is perpendicular to the axis of the cone, a **circle** results. If the plane does not cut across one entire nappe and does not intersect both nappes, as illustrated in (ii) of Figure 7.1, then the curve of intersection is a **parabola**. If the plane cuts through both nappes of the cone, as in (iii) of Figure 7.1, then the resulting figure is called a **hyperbola**.

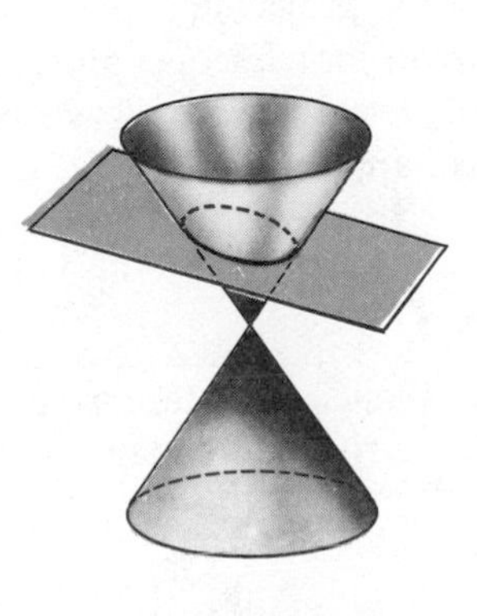

(i) Ellipse

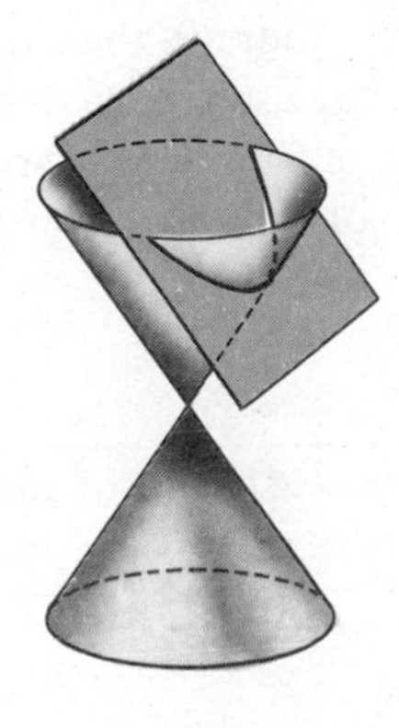

(ii) Parabola

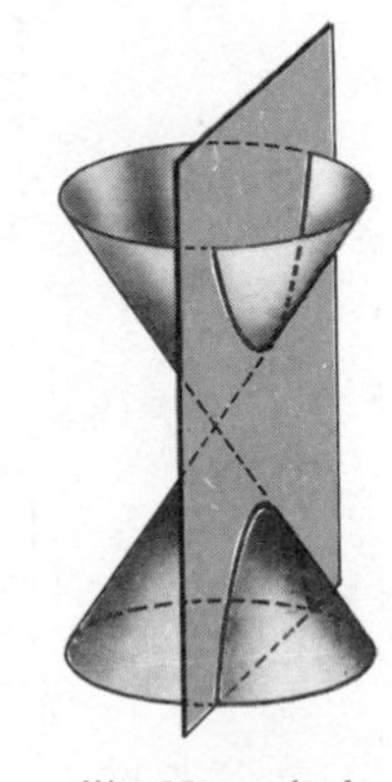

(iii) Hyperbola

Figure 7.1

By changing the position of the plane and the shape of the cone, conics can be made to vary considerably. For certain positions of the plane there result what are called **degenerate conics**. For example, if the plane intersects the cone only at the vertex, then the conic consists of one point. If the axis of the cone lies on the plane, then a pair of intersecting lines is obtained. Finally, if we begin with the parabolic case, as in (ii) of Figure 7.1, and move the plane parallel to its initial position until it coincides with one of the generators of the cone, a line results.

The conic sections were studied extensively by the early Greek mathematicians, who used the methods of Euclidean geometry. They discovered the properties which enable us to define conics in terms of points (foci) and lines (directrices) in the plane of the conic. Reconciliation of the latter definitions with the previous discussion requires proofs which we shall not go into here.

A remarkable fact about conic sections is that although they were studied thousands of years ago, they are far from obsolete. Indeed, they are important tools for present-day investigations in outer space and for the study of the behavior of atomic particles. It is shown in physics that if a particle moves under the influence of what is called an *inverse square force field*, then its path may be described by means of a conic section. Examples of inverse square fields are gravitational and electromagnetic fields. Planetary orbits are elliptical. If the ellipse is very "flat," the curve resembles the path of a comet. The hyperbola is useful for describing the path of an alpha particle in the electric field of the nucleus of an atom. The interested person can find many other applications of conic sections.

7.2 PARABOLAS

Parabolas are very useful in applications of mathematics to the physical world. For example, it can be shown that if a projectile is fired and it is assumed that it is acted upon only by the force of gravity (that is, air resistance and other outside factors are ignored), then the path of the projectile is parabolic. Properties of parabolas are used in the design of mirrors for telescopes and searchlights, and in the design of field microphones used in television broadcasts of football games. These are only a few of many physical applications.

(7.1) Definition

A **parabola** is the set of all points in a plane equidistant from a fixed point F (the **focus**) and a fixed line l (the **directrix**) in the plane.

We shall assume that F is not on l, for otherwise the parabola degenerates into a line. If P is any point in the plane and P' is the point on l determined by a line through P which is perpendicular to l, then according to Definition (7.1), P is on the parabola if and only if $d(P, F) = d(P, P')$. A typical situation is illustrated in Figure 7.2, where the dashes indicate possible positions of P. The line through F, perpendicular to the directrix, is called the **axis** of the parabola. The point V on the axis, half-way from F to l, is called the **vertex** of the parabola.

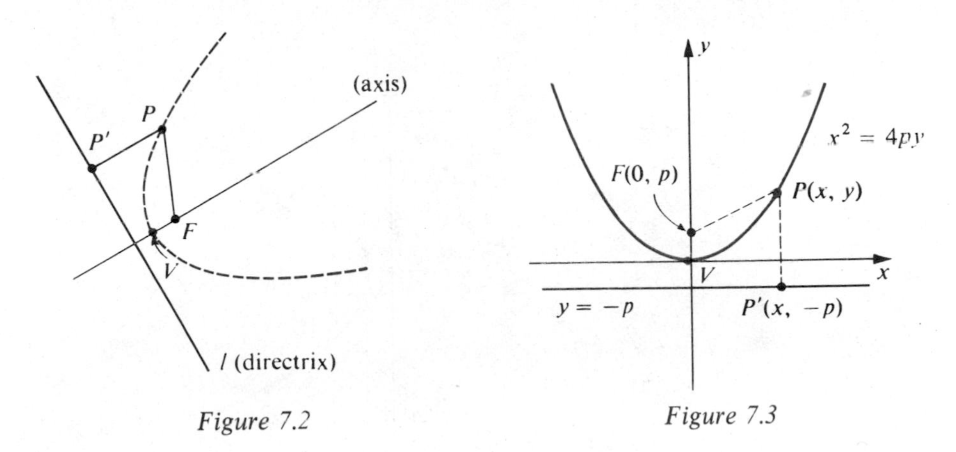

Figure 7.2

Figure 7.3

In order to obtain a simple equation for a parabola, let us choose the y-axis along the axis of the parabola, with the origin at the vertex V, as illustrated in Figure 7.3. In this case, the focus F has coordinates $(0, p)$ for some real number $p \neq 0$, and the equation of the directrix is $y = -p$. By the Distance Formula, a point $P(x, y)$ is on the parabola if and only if

$$\sqrt{(x-0)^2 + (y-p)^2} = \sqrt{(x-x)^2 + (y+p)^2}.$$

Squaring both sides gives us

$$(x-0)^2 + (y-p)^2 = (y+p)^2$$

or

$$x^2 + y^2 - 2py + p^2 = y^2 + 2py + p^2$$

which simplifies to

$$x^2 = 4py. \qquad (7.2)$$

We have shown that the coordinates of every point (x, y) on the parabola satisfy $x^2 = 4py$. Conversely, if (x, y) is a solution of this equation, then by reversing the previous steps we see that the point (x, y) is on the parabola. We shall call (7.2) the **standard form** for the equation of a parabola with focus at $F(0, p)$ and directrix $y = -p$. If $p > 0$, the parabola **opens upward**, as in Figure 7.3, whereas if $p < 0$, the parabola **opens downward**.

An analogous situation exists if the axis of the parabola is taken along the x-axis. If the vertex is $V(0, 0)$, the focus $F(p, 0)$, and the directrix has equation

$x = -p$ (see Figure 7.4), then using the same type of argument we obtain the **standard form**

(7.3)
$$y^2 = 4px.$$

If $p > 0$, the parabola opens to the right, as in Figure 7.4, whereas if $p < 0$, it opens to the left.

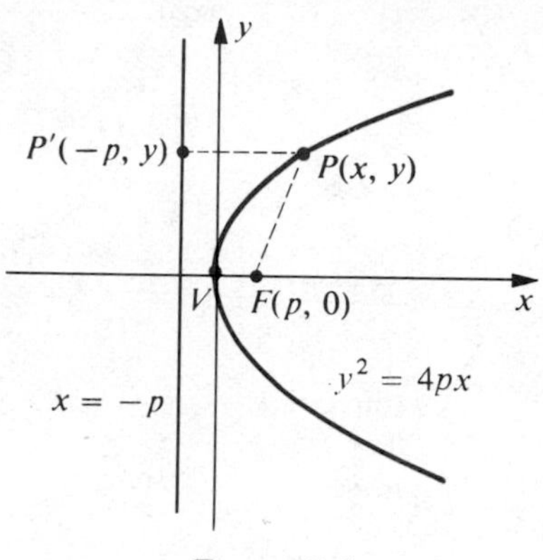

Figure 7.4

Example 1 Find the focus and directrix of the parabola having equation $y^2 = -6x$, and sketch the graph.

Solution The equation has the form (7.3) with $4p = -6$, and hence $p = -3/2$. From the previous discussion the focus is $F(p, 0)$, that is, $F(-3/2, 0)$. The equation of the directrix is $x = -p$, or $x = 3/2$. The graph is sketched in Figure 7.5.

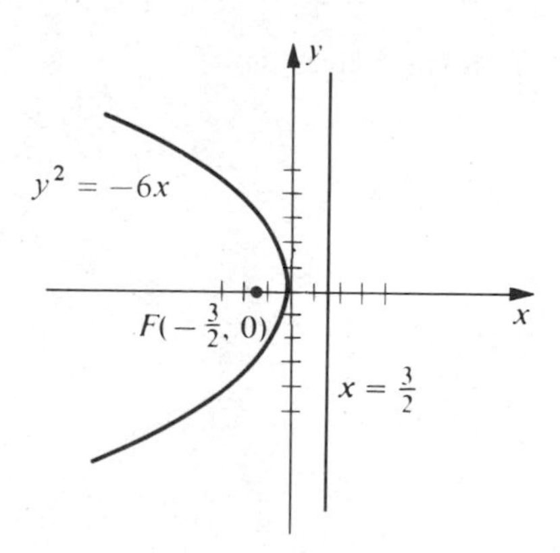

Figure 7.5

Example 2 Find an equation of the parabola that has its vertex at the origin, opens upward, and passes through the point $P(-3, 7)$.

Solution The general form of the equation is given by $x^2 = 4py$ (see (7.2)). If P is on the parabola, then $(-3, 7)$ is a solution of the equation. Hence we must have $(-3)^2 = 4p(7)$, or $p = 9/28$. Substituting for p in (7.2) leads to the desired equation $x^2 = (9/7)y$, or $7x^2 = 9y$.

If the coordinate plane in Figure 7.3 is folded along the y-axis, then the graph which lies in the left half of the plane coincides with that in the right half. We say that **the graph is symmetric with respect to the y-axis.** As in (i) of Figure 7.6, a graph is symmetric with respect to the y-axis provided that the point $(-x, y)$ is on the graph whenever (x, y) is on the graph. Similarly, as in (ii) of Figure 7.6, **a graph is symmetric with respect to the x-axis** if whenever a point (x, y) is on the graph, then $(x, -y)$ is also on the graph. In the latter case if we fold the coordinate plane along the x-axis, that part of the graph which lies above the x-axis will coincide with the part which lies below. The previous remarks give us the results stated in (7.4).

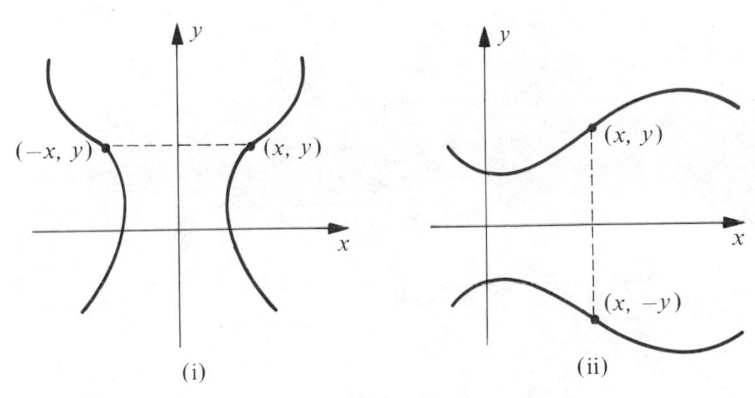

Figure 7.6

(7.4) Tests for Symmetry with Respect to an Axis

(i) The graph of an equation is symmetric with respect to the y-axis if substitution of $-x$ for x does not change the solutions of the equation.

(ii) The graph of an equation is symmetric with respect to the x-axis if substitution of $-y$ for y does not change the solutions of the equation.

As an illustration of (7.4), replacing x by $-x$ in (7.2) gives us $(-x)^2 = 4py$, which is the same as $x^2 = 4py$, and hence there is no change in the solutions. Thus the parabola with equation $x^2 = 4py$ is symmetric with respect to the y-axis. This is also evident from the graph in Figure 7.3. Similarly, the graph of $y^2 = 4px$ in Figure 7.4 is symmetric with respect to the x-axis, since the equation is unchanged if y is replaced by $-y$.

If symmetry with respect to an axis exists, then it is sufficient to determine the graph in half of the coordinate plane, since the remainder of the graph is a mirror image, or reflection, of that half.

It is not difficult to extend our work to the case in which the axis of a parabola is parallel to one of the coordinate axes. In Figure 7.7 we have taken the vertex at the point $V(h, k)$, the focus at $F(h, k + p)$, and the directrix as $y = k - p$, where $p > 0$. As before, the point $P(x, y)$ is on the parabola if and only if $d(P, F) = d(P, P')$, that is, if and only if

$$\sqrt{(x - h)^2 + (y - k - p)^2} = \sqrt{(x - x)^2 + (y - k + p)^2}.$$

We leave it to the reader to show that the equation simplifies to

$$(x - h)^2 = 4p(y - k) \tag{7.5}$$

which is called the **standard form** for the equation of a parabola with vertex at (h, k) and axis parallel to the y-axis. As a special case, if $(h, k) = (0, 0)$, then (7.5) reduces to (7.2).

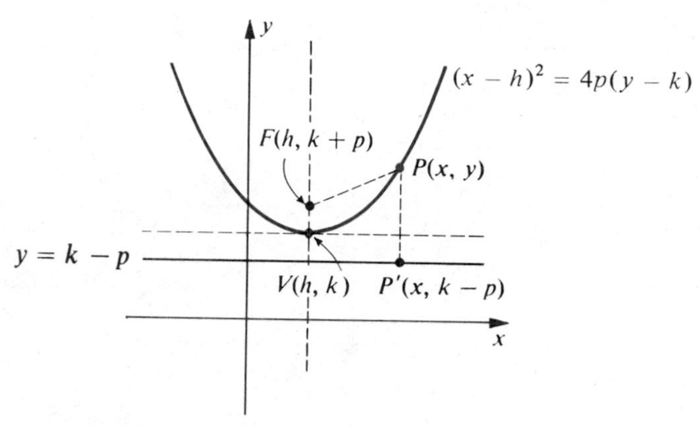

Figure 7.7

Squaring the left side of (7.5) and simplifying leads to an equation of the form

$$y = ax^2 + bx + c$$

where a, b, and c are real numbers. Conversely, given such an equation, we may complete the square in x to arrive at the standard form (7.5). This technique will be illustrated in Example 3. Consequently, if $a \neq 0$, then the graph of $y = ax^2 + bx + c$ is a parabola with a vertical axis.

Example 3 Discuss and sketch the graph of $y = 2x^2 - 6x + 4$.

Solution The graph is a parabola with vertical axis. To obtain form (7.5) we begin by writing the equation as

$$y - 4 = 2x^2 - 6x = 2(x^2 - 3x).$$

Next we complete the square for the expression $x^2 - 3x$. Recall that to complete the square for *any* expression of the form $x^2 + qx$ we add the square of half the coefficient of x, that is, $(q/2)^2$. Thus, for $x^2 - 3x$ we must add $(-3/2)^2$, or $9/4$. However, if we add $9/4$ to $x^2 - 3x$ in the equation $y - 4 = 2(x^2 - 3x)$, then because there is a factor 2 outside the parentheses, this amounts to adding $9/2$ to the right side of the equation, and hence we must compensate by adding $9/2$ to the left side. This gives us

$$y - 4 + \frac{9}{2} = 2\left(x^2 - 3x + \frac{9}{4}\right)$$

$$y + \frac{1}{2} = 2\left(x - \frac{3}{2}\right)^2$$

or equivalently

$$\left(x - \frac{3}{2}\right)^2 = \frac{1}{2}\left(y + \frac{1}{2}\right).$$

The last equation is in form (7.5) with $h = 3/2$, $k = -1/2$, and $4p = 1/2$, or $p = 1/8$. Hence the vertex of the parabola is $V(3/2, -1/2)$. The focus is $F(h, k + p)$, or $F(3/2, -3/8)$, and the directrix is

$$y = k - p = (-1/2) - (1/8), \quad \text{or} \quad y = -5/8.$$

As an aid to sketching the graph we note that the y-intercept is 4. To find the x-intercepts we solve $2x^2 - 6x + 4 = 0$ or the equivalent equation $(2x - 2)(x - 2) = 0$, obtaining $x = 2$ and $x = 1$. The graph is sketched in Figure 7.8.

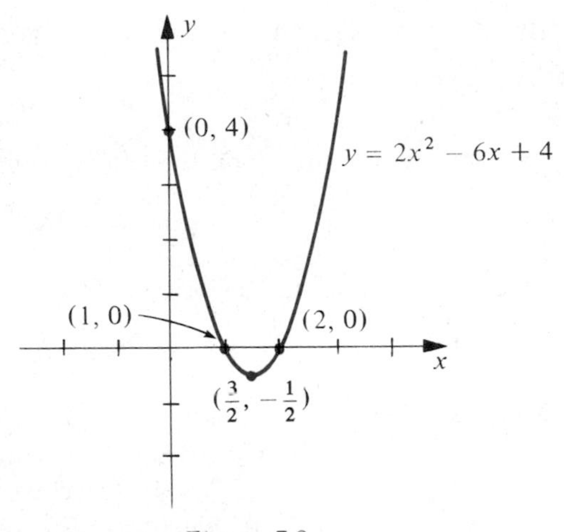

Figure 7.8

To obtain an equation for an arbitrary parabola with a *horizontal* axis we choose vertex $V(h, k)$, focus $F(h + p, k)$, and directrix $x = h - p$, as illustrated in Figure 7.9. It is left as an exercise to show that $d(P, F) = d(P, P')$ if and only if

(7.6) $$(y - k)^2 = 4p(x - h).$$

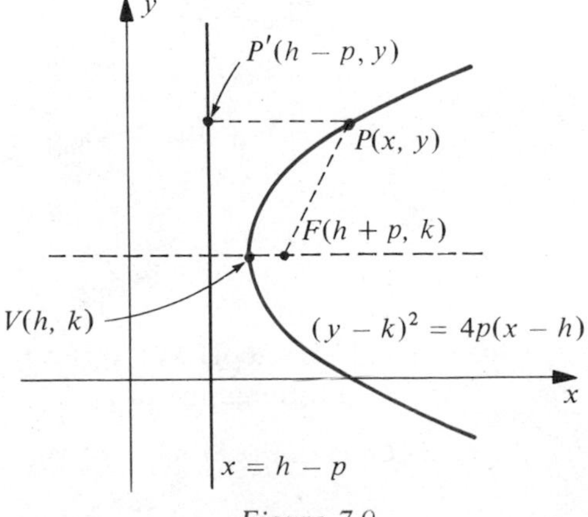

Figure 7.9

The equation in (7.6) may be written as

$$x = ay^2 + by + c$$

where a, b, and c are real numbers. Conversely, the latter equation can be expressed in form (7.6) by completing the square in y as illustrated next, in Example 4. Hence, if $a \neq 0$, the graph of $x = ay^2 + by + c$ is a parabola with a horizontal axis.

Example 4 Discuss and sketch the graph of the equation

$$2x = y^2 + 8y + 22.$$

Solution By our previous remarks, the graph is a parabola with a horizontal axis. Writing

$$y^2 + 8y = 2x - 22$$

we complete the square on the left by adding 16 to both sides. This gives us

$$y^2 + 8y + 16 = 2y - 6.$$

The last equation may be written

$$(y + 4)^2 = 2(x - 3)$$

which is in the form (7.6) with $h = 3$, $k = -4$, and $4p = 2$, or $p = 1/2$. Hence the vertex is $V(3, -4)$. Since $p = 1/2 > 0$, the parabola opens to the right with focus at $F(h + p, k)$, that is, $F(7/2, -4)$. The equation of the directrix is $x = h - p$, or $x = 5/2$. The parabola is sketched in Figure 7.10.

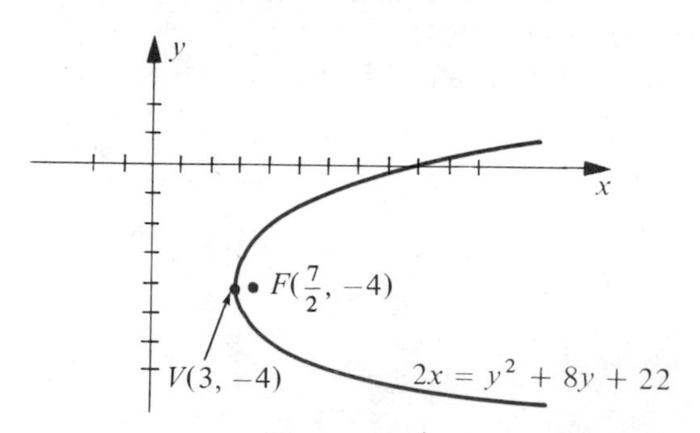

Figure 7.10

Example 5 Find an equation of the parabola with vertex $(4, -1)$, with axis parallel to the y-axis, and which passes through the origin.

Solution By (7.5) the equation is of the form

$$(x - 4)^2 = 4p(y + 1).$$

If the origin is on the parabola, then (0, 0) is a solution of this equation, and hence $(0-4)^2 = 4p(0+1)$. Consequently $16 = 4p$ and $p = 4$. The desired equation is, therefore,

$$(x-4)^2 = 16(y+1).$$

We shall conclude this section by pointing out an important property of parabolas. Suppose l is the tangent line at a point $P(x_1, y_1)$ on the graph of $y^2 = 4px$, and let F be the focus. As in Figure 7.11, let α denote the angle between l and the line segment FP, and let β denote the angle between l and the indicated horizontal half-line with end-point P. In Exercise 34, the reader is asked to prove that $\alpha = \beta$. This property has many practical applications. For example, the shape of the mirror in a searchlight is obtained by revolving a parabola about its axis. If a light source is placed at F, then by a law of physics a beam of light will be reflected along a line parallel to the axis. The same principle is employed in the construction of mirrors for telescopes or solar ovens, where a beam of light coming toward the parabolic mirror, and parallel to the axis, will be reflected into the focus. Antennas for radar systems and field microphones used at football games also make use of this property.

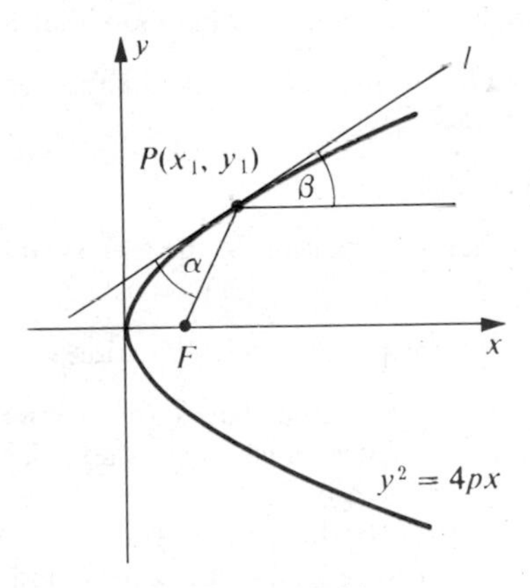

Figure 7.11

EXERCISES 7.2

In each of Exercises 1–16 find the vertex, focus, and directrix of the parabola with the given equation and sketch the graph.

1 $x^2 = -12y$

2 $y^2 = (1/2)x$

3 $2y^2 = -3x$

4 $x^2 = -3y$

5 $8x^2 = y$

6 $y^2 = -100x$

7 $y^2 - 12 = 12x$

8 $y = 40x - 97 - 4x^2$

9 $y = x^2 - 4x + 2$

10 $y = 8x^2 + 16x + 10$

11 $y^2 - 4y - 2x - 4 = 0$

12 $y^2 + 14y + 4x + 45 = 0$

13 $4x^2 + 40x + y + 106 = 0$

14 $y^2 - 20y + 100 = 6x$

15 $x^2 + 20y = 10$

16 $4x^2 + 4x + 4y + 1 = 0$

17 Describe a method for using a derivative to locate the vertex of a parabola which has the equation $y = ax^2 + bx + c$. Use this technique to find the vertices in Exercises 8–10. Describe a similar method for equations of the form $x = ay^2 + by + c$.

18 Describe how a second derivative may be used to determine whether the parabola with equation $y = ax^2 + bx + c$ opens upward or downward. Illustrate this technique with Exercises 8–10.

In each of Exercises 19–24 find an equation for the parabola that satisfies the given conditions.

19 Focus $(2, 0)$, directrix $x = -2$

20 Focus $(0, -4)$, directrix $y = 4$

21 Focus $(6, 4)$, directrix $y = -2$

22 Focus $(-3, -2)$, directrix $y = 1$

23 Vertex at the origin, symmetric to the y-axis, and passing through the point $A(2, -3)$

24 Vertex $V(-3, 5)$, axis parallel to the x-axis, and passing through $A(5, 9)$

25 A searchlight reflector is designed so that a cross section through its axis is a parabola and the light source is at the focus. Find the focus if the reflector is 3 feet across at the opening and 1 foot deep.

26 Prove that the point on a parabola that is closest to the focus is the vertex.

27 Find an equation of the parabola with a vertical axis which passes through the points $A(2, 3)$, $B(-1, 6)$, and $C(1, 0)$.

28 Derive (7.6).

29 Prove that the equation of the tangent line to the parabola $x^2 = 4py$ at the point $P(x_1, y_1)$ is $x_1x - 2py - 2py_1 = 0$.

30 Prove that the parabola with equation $y = ax^2 + bx + c$ has no points of inflection.

31 Let R denote the region in the xy-plane bounded by the parabola $x^2 = 4y$ and the line l through the focus which is perpendicular to the axis of the parabola.
(a) Find the area of R.
(b) If R is revolved about the y-axis, find the volume of the resulting solid.
(c) If R is revolved about the x-axis, find the volume of the resulting solid.

32 Let R denote the region in the xy-plane bounded by the graphs of $y^2 = 2x - 6$ and $x = 5$.
(a) Find the area of R.
(b) If R is revolved about the y-axis, find the volume of the resulting solid.
(c) If R is revolved about the x-axis, find the volume of the resulting solid.

33 A vertical gate in a dam has the shape of the parabolic region of Exercise 31. If the line l lies along the surface of the water, find the force exerted on the gate by the water.

34 In Figure 7.11, prove that $\alpha = \beta$.

35 A line segment which passes through the focus of a parabola and has its end-points on the parabola is called a **focal chord**. If AB is a focal chord, prove that the tangent lines at A and B are perpendicular.

36 If AB is a focal chord of a parabola (see Exercise 35), prove that the tangent lines at A and B intersect on the directrix.

7.3 ELLIPSES

An ellipse may be defined as follows:

(7.7) Definition

> An **ellipse** is the set of all points in a plane, the sum of whose distances from two fixed points in the plane (the **foci**) is constant.

It is known that the orbits of planets in the solar system are elliptical, with the sun at one of the foci. This is only one of many important applications of ellipses.

There is an easy way to construct an ellipse on paper. We may begin by inserting two thumbtacks in the paper at points labeled F and F' and fastening the ends of a piece of string to the thumbtacks. If the string is now looped around a pencil and drawn taut at point P, as in Figure 7.12, then moving the pencil and at the same time keeping the string taut, the sum of the distances $d(F, P)$ and $d(F', P)$ is the length of the string, and hence is constant. The pencil will, therefore, trace out a figure which resembles an ellipse with foci at F and F'. By varying the positions of F and F' but keeping the length of string fixed, the shape of the ellipse can be made to change considerably. If F and F' are far apart, in the sense that $d(F, F')$ is almost the same as the length of the string, then the ellipse is quite flat. On the other hand, if $d(F, F')$ is close to zero, the ellipse is almost circular. Indeed, if $F = F'$, a circle is obtained.

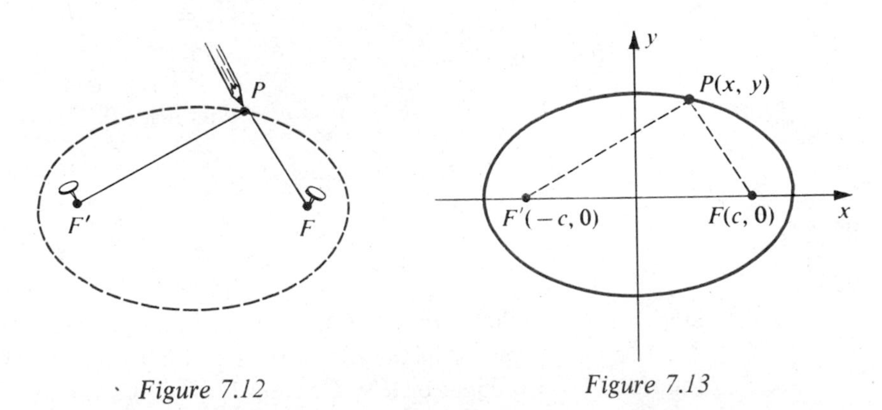

Figure 7.12 Figure 7.13

By introducing suitable coordinate systems we may derive simple equations for ellipses. Let us choose the x-axis as the line through the two foci F and F', with the origin at the midpoint of the segment $F'F$. This point is called the **center** of the ellipse. If F has coordinates $(c, 0)$, where $c > 0$, then, as shown in Figure 7.13, F' has coordinates $(-c, 0)$ and hence the distance between F and F' is $2c$. Let the constant sum of the distances of P from F and F' be denoted by $2a$, where in order to get points that are not on the x-axis we must have $2a > 2c$, that is, $a > c$. (Why?) By Definition (7.7), $P(x, y)$ is on the ellipse if and only if

$$d(P, F) + d(P, F') = 2a$$

or, by the Distance Formula,

$$\sqrt{(x-c)^2+(y-0)^2}+\sqrt{(x+c)^2+(y-0)^2}=2a.$$

Writing the above equation as

$$\sqrt{(x-c)^2+y^2}=2a-\sqrt{(x+c)^2+y^2}$$

and squaring both sides, we obtain

$$x^2-2cx+c^2+y^2=4a^2-4a\sqrt{(x+c)^2+y^2}+x^2+2cx+c^2+y^2$$

which simplifies to

$$a\sqrt{(x+c)^2+y^2}=a^2+cx.$$

Squaring both sides gives us

$$a^2(x^2+2cx+c^2+y^2)=a^4+2a^2cx+c^2x^2$$

which may be written in the form

$$x^2(a^2-c^2)+a^2y^2=a^2(a^2-c^2).$$

Dividing both sides by $a^2(a^2-c^2)$ leads to

$$\frac{x^2}{a^2}+\frac{y^2}{a^2-c^2}=1.$$

For convenience, we let

(7.8) $$b^2=a^2-c^2 \quad \text{where } b>0$$

in the preceding equation, obtaining

(7.9) $$\frac{x^2}{a^2}+\frac{y^2}{b^2}=1.$$

Since $c>0$, it follows from (7.8) that $a^2>b^2$ and hence $a>b$.

We have shown that the coordinates of every point (x, y) on the ellipse in Figure 7.13 satisfy equation (7.9). Conversely, if (x, y) is a solution of this equation, then by reversing the preceding steps we see that the point (x, y) is on the ellipse. Equation (7.9) is called the **standard form** for the equation of an ellipse with foci on the x-axis and center at the origin.

The x-intercepts may be found by setting $y=0$. Doing so gives us $x^2/a^2=1$, or $x^2=a^2$, and consequently the x-intercepts are a and $-a$. The corresponding points $V(a, 0)$ and $V'(-a, 0)$ on the graph are called the **vertices** of the ellipse, and the line segment $V'V$ is referred to as the **major axis**. Setting $x=0$ in (7.9), we obtain the y-intercepts b and $-b$. The segment from $M'(0, -b)$ to $M(0, b)$ is called the **minor axis** of the ellipse. Note that the major axis is longer than the minor axis, since $a>b$.

By the Tests for Symmetry (7.4) we see that the ellipse is symmetric to both the x-axis and the y-axis. It is also **symmetric with respect to the origin**, in the sense that if the point $P(x, y)$ is on the graph, then the point $P'(-x, -y)$ is also on the graph. Note that P' can be found by extending the line segment from P to the origin O *through* O a distance equal to $d(O, P)$. The following result is useful for investigating symmetry with respect to the origin.

(7.10) Test for Symmetry with Respect to the Origin

> The graph of an equation is symmetric with respect to the origin if the simultaneous substitution of $-x$ for x and $-y$ for y does not change the solutions of the equation.

Example 1 Discuss and sketch the graph of the equation

$$4x^2 + 18y^2 = 36.$$

Solution To obtain the standard form, we divide both sides of the given equation by 36 and simplify. This leads to

$$\frac{x^2}{9} + \frac{y^2}{2} = 1$$

which is in the form (7.9) with $a^2 = 9$ and $b^2 = 2$. Thus $a = 3$, $b = \sqrt{2}$, and hence the end-points of the major axis are $(\pm 3, 0)$ and the end-points of the minor axis are $(0, \pm\sqrt{2})$. From (7.8) we obtain

$$c^2 = a^2 - b^2 = 9 - 2 = 7, \quad \text{or} \quad c = \sqrt{7}.$$

Consequently, the foci are $(\pm\sqrt{7}, 0)$. The graph is sketched in Figure 7.14.

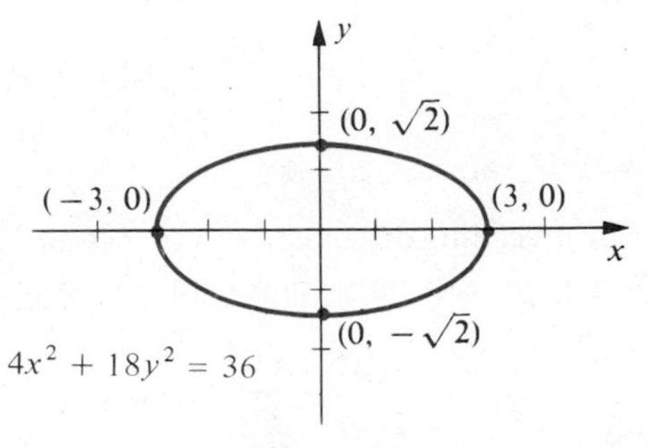

Figure 7.14

Example 2 Find an equation of the ellipse with vertices $(\pm 4, 0)$ and foci $(\pm 2, 0)$.

Solution Substituting $a = 4$ and $c = 2$ in (7.8), we have $b^2 = 16 - 4 = 12$. Employing (7.9) gives us the equation

$$\frac{x^2}{16} + \frac{y^2}{12} = 1.$$

Multiplying both sides by 48 leads to $3x^2 + 4y^2 = 48$.

It is sometimes convenient to choose the major axis of the ellipse along the y-axis. If the foci are $(0, \pm c)$, then by the same type of argument used to derive (7.9), we obtain the equation

$$\frac{x^2}{b^2} + \frac{y^2}{a^2} = 1 \qquad \textbf{(7.11)}$$

where $a > b$. As before, the connection between a, b, and c is given by $b^2 = a^2 - c^2$ or, equivalently, by $c^2 = a^2 - b^2$. In this case the vertices are $V(0, a)$ and $V'(0, -a)$. The end-points of the minor axis are $M(b, 0)$ and $M'(-b, 0)$. A typical graph is sketched in Figure 7.15.

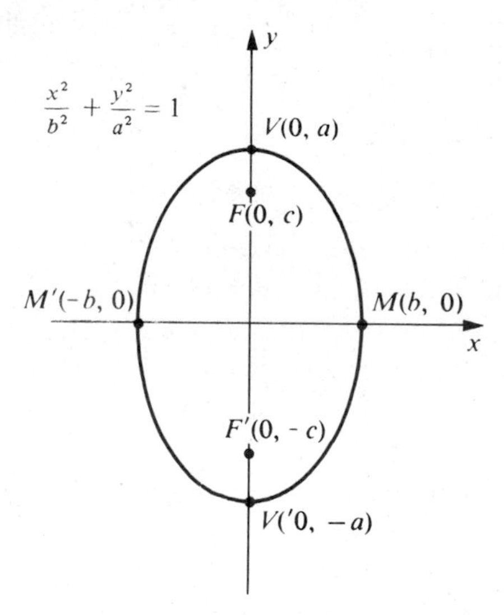

Figure 7.15

The preceding discussion shows that an equation of an ellipse with center at the origin and foci on a coordinate axis can always be written in the form

$$\frac{x^2}{p} + \frac{y^2}{q} = 1 \quad \text{or} \quad qx^2 + py^2 = pq$$

where p and q are positive. If $p > q$, then the major axis lies on the x-axis, whereas if $q > p$, then the major axis is on the y-axis. It is unnecessary to memorize these facts, since in any given problem the major axis can be determined by examining the x- and y-intercepts.

Example 3 Sketch the graph of the equation $9x^2 + 4y^2 = 25$.

Solution The graph is an ellipse with center at the origin and foci on one of the coordinate axes. To find the x-intercepts, we let $y = 0$, obtaining $9x^2 = 25$, or $x = \pm 5/3$. Similarly, to find the y-intercepts, we let $x = 0$, obtaining $4y^2 = 25$, or $y = \pm 5/2$. This enables us to sketch the ellipse (see Figure 7.16). Since $5/3 < 5/2$, the major axis is on the y-axis.

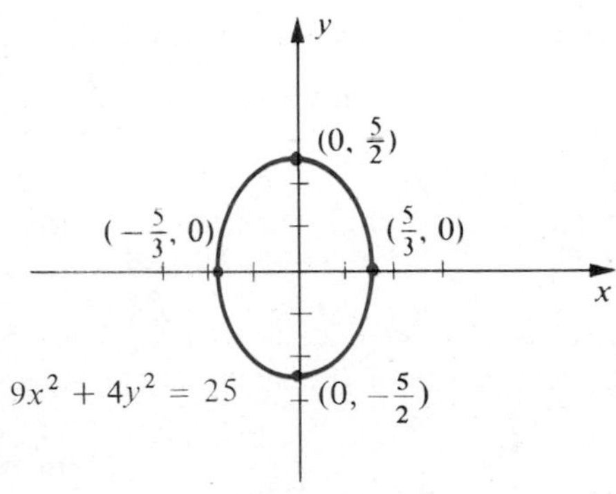

Figure 7.16

Example 4 Find the area of the region bounded by an ellipse whose major and minor axes have lengths $2a$ and $2b$, respectively.

Solution An equation for the ellipse is given by (7.9). Solving for y gives us

$$y = (\pm b/a)\sqrt{a^2 - x^2}.$$

The graph of the ellipse has the general shape shown in Figure 7.14 and hence, by symmetry, it is sufficient to find the area of the region in the first quadrant and multiply the result by 4. Using (5.20),

$$A = \frac{4b}{a}\int_0^a \sqrt{a^2 - x^2}\, dx.$$

Since the graph of $y = \sqrt{a^2 - x^2}$ from $x = 0$ to $x = a$ is one-fourth of a circle of radius a with center at the origin, the integral equals $\pi a^2/4$. Consequently,

$$A = \left(\frac{4b}{a}\right)\left(\frac{\pi a^2}{4}\right) = \pi ab.$$

As a special case, if $b = a$ the ellipse is a circle and $A = \pi a^2$.

EXERCISES 7.3

In each of Exercises 1–8 sketch the graph of the equation and give coordinates of the vertices and foci.

1 $\dfrac{x^2}{9} + \dfrac{y^2}{4} = 1$

2 $\dfrac{x^2}{25} + \dfrac{y^2}{16} = 1$

3 $4x^2 + y^2 = 16$

4 $y^2 + 9x^2 = 9$

5 $5x^2 + 2y^2 = 10$

6 $(1/2)x^2 + 2y^2 = 8$

7 $4x^2 + 25y^2 = 1$

8 $10y^2 + x^2 = 5$

In each of Exercises 9–14 find an equation for the ellipse satisfying the given conditions.

9 Vertices $V(\pm 8, 0)$, foci $F(\pm 5, 0)$

10 Vertices $V(0, \pm 7)$, foci $F(0, \pm 2)$

11 Vertices $V(0, \pm 5)$, length of minor axis 3

12 Foci $F(\pm 3, 0)$, length of minor axis 2

13 Vertices $V(0, \pm 6)$, passing through $(3, 2)$

14 Center at the origin, symmetric with respect to both axes, passing through $A(2, 3)$ and $B(6, 1)$

In Exercises 15 and 16 find the points of intersection of the graphs of the given equations. Sketch both graphs on the same coordinate axes, showing points of intersection.

15 $\begin{cases} x^2 + 4y^2 = 20 \\ x + 2y = 6 \end{cases}$

16 $\begin{cases} x^2 + 4y^2 = 36 \\ x^2 + y^2 = 12 \end{cases}$

17 An arch of a bridge is semi-elliptical with major axis horizontal. The base of the arch is 30 feet across and the highest part of the arch is 10 feet above the horizontal roadway. Find the height of the arch 6 feet from the center of the base.

18 The **eccentricity** of an ellipse is defined as the ratio $(\sqrt{a^2 - b^2})/a$. If a is fixed and b varies, describe the general shape of the ellipse when the eccentricity is close to 1 and when it is close to zero.

19 Find an equation of the tangent line to the ellipse $5x^2 + 4y^2 = 56$ at the point $P(-2, 3)$.

20 Prove that an equation of the tangent line to the ellipse $x^2/a^2 + y^2/b^2 = 1$ at the point $P(x_1, y_1)$ is $xx_1/a^2 + yy_1/b^2 = 1$.

21 If tangent lines to the ellipse $9x^2 + 4y^2 = 36$ intersect the y-axis at the point $(0, 6)$, find the points of tangency.

22 If tangent lines to the ellipse $b^2x^2 + a^2y^2 = a^2b^2$ intersect the y-axis at the point $(0, d)$, where $d > b$, find the points of tangency.

23 Find the volume of the solid obtained by revolving the region bounded by the ellipse $b^2x^2 + a^2y^2 = a^2b^2$ about the x-axis.

24 Find the volume of the solid obtained by revolving the region bounded by the ellipse $b^2x^2 + a^2y^2 = a^2b^2$ about the y-axis.

25 The base of a solid is a plane region bounded by an ellipse with major and minor axes of lengths 16 and 9, respectively. Find the volume of the solid if every cross section by a plane perpendicular to the major axis is (a) a square; (b) an equilateral triangle.

26 The base of a right elliptic cone has major and minor axes of lengths $2a$ and $2b$, respectively. Find the volume if the altitude is h. (*Hint:* see Example 4.)

27 Find the dimensions of the rectangle of maximum area that can be inscribed in an ellipse of semi-axes a and b, if two sides of the rectangle are parallel to the major axis.

28 A cylindrical tank whose cross sections are elliptical with axes of lengths 6 feet and 4 feet, respectively, is lying on its side. If the tank is half full of water, find the force exerted by the water on one end of the tank.

29 Let l denote the tangent line at the point P on the ellipse illustrated in Figure 7.13. If α is the angle between $F'P$ and l, and if β is the angle between FP and l, prove that $\alpha = \beta$. (This is analogous to the reflective property of the parabola illustrated in Figure 7.11.)

30 Derive (7.11).

7.4 HYPERBOLAS

The definition of a hyperbola is similar to that of an ellipse. The only change is that instead of using the *sum* of distances from two fixed points we use the *difference.*

(7.12) Definition

> A **hyperbola** is the set of all points in a plane, the difference of whose distances from two fixed points in the plane (the **foci**) is a positive constant.

To find a simple equation for a hyperbola, we choose a coordinate system with foci at $F(c, 0)$ and $F'(-c, 0)$, and denote the (constant) distance by $2a$. Referring to Figure 7.17, we see that a point $P(x, y)$ is on the hyperbola if and only if either one of the following is true:

(7.13)
$$d(P, F') - d(P, F) = 2a$$
$$d(P, F) - d(P, F') = 2a.$$

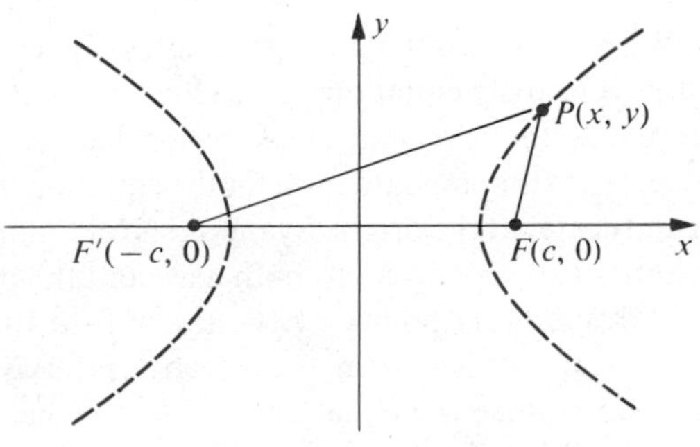

Figure 7.17

For hyperbolas (unlike ellipses) we need $a < c$ in order to obtain points on the hyperbola which are not on the x-axis, for if P is such a point, then from Figure 7.17 we see that

$$d(P, F) < d(F', F) + d(P, F')$$

since the length of one side of a triangle is always less than the sum of the lengths of the other two sides. Similarly,

$$d(P, F') < d(F', F) + d(P, F).$$

Equivalent forms for the previous two inequalities are

$$d(P, F) - d(P, F') < d(F', F)$$
$$d(P, F') - d(P, F) < d(F', F).$$

From (7.13) and the fact that $d(F', F) = 2c$, the last two inequalities imply that $2a < 2c$, or $a < c$.

Equations (7.13) may be replaced by the single equation

$$|d(P, F) - d(P, F')| = 2a.$$

It then follows from the Distance Formula that an equation of the hyperbola is given by

$$|\sqrt{(x - c)^2 + (y - 0)^2} - \sqrt{(x + c)^2 + (y - 0)^2}| = 2a.$$

Employing the type of simplification procedure used to derive an equation for an ellipse, we arrive at the equivalent equation

$$\frac{x^2}{a^2} - \frac{y^2}{c^2 - a^2} = 1.$$

For convenience, let

(7.14) $$b^2 = c^2 - a^2 \quad \text{where } b > 0$$

in the preceding equation, obtaining

(7.15) $$\frac{x^2}{a^2} - \frac{y^2}{b^2} = 1.$$

We have shown that the coordinates of every point (x, y) on the hyperbola in Figure 7.17 satisfy equation (7.15). Conversely, if (x, y) is a solution of (7.15), then by reversing the preceding steps we see that the point (x, y) is on the hyperbola. We call (7.15) the **standard form** for the equation of a hyperbola with foci on the x-axis and center at the origin. By the Tests for Symmetry we see that this hyperbola is symmetric with respect to both axes and the origin. The x-intercepts are $\pm a$. The corresponding points $V(a, 0)$ and $V'(-a, 0)$ are called the **vertices**, and the line segment $V'V$ is known as the **transverse axis** of the hyperbola. There are no y-intercepts, since the equation $-y^2/b^2 = 1$ has no solutions.

If (7.15) is solved for y, we obtain

(7.16) $$y = \pm\frac{b}{a}\sqrt{x^2 - a^2}.$$

Hence there are no points (x, y) on the graph if $x^2 - a^2 < 0$, that is, if $-a < x < a$. However, there *are* points $P(x, y)$ on the graph if $x \geq a$ or $x \leq -a$.

The line $y = (b/a)x$ is an asymptote for the hyperbola (7.15) in the sense that the distance $d(x)$ between the point $P(x, y)$ on the hyperbola and the corresponding point $P'(x, y_1)$ on the line approaches zero as x increases without bound. To prove this we note from (7.16) that if $x > 0$, then

$$d(x) = \frac{b}{a}x - \frac{b}{a}\sqrt{x^2 - a^2} = \frac{b}{a}(x - \sqrt{x^2 - a^2}).$$

Since

$$x - \sqrt{x^2 - a^2} = \frac{a^2}{x + \sqrt{x^2 - a^2}}$$

we have

$$\lim_{x\to\infty} d(x) = \lim_{x\to\infty} \frac{b}{a} \frac{a^2}{x + \sqrt{x^2 - a^2}} = 0.$$

Similarly, $d(x)$ has the limit zero as x becomes negatively infinite. It can be shown, in like manner, that the line $y = (-b/a)x$ is an asymptote for the hyperbola (7.15).

The asymptotes serve as excellent guides for sketching the graph. A convenient way to sketch the asymptotes is first to plot the vertices $V(a, 0)$, $V'(-a, 0)$ and the points $W(0, b)$, $W'(0, -b)$ (see Figure 7.18). The line segment $W'W$ of length $2b$ is called the **conjugate axis** of the hyperbola. If horizontal and vertical lines are drawn through the end-points of the conjugate and transverse axes respectively, then the diagonals of the resulting rectangle have slopes b/a and $-b/a$. Hence, by extending these diagonals we obtain lines with equations $y = (\pm b/a)x$. The hyperbola is then sketched as in Figure 7.18, using the asymptotes as guides. The two curves which make up the hyperbola are called the **branches** of the hyperbola.

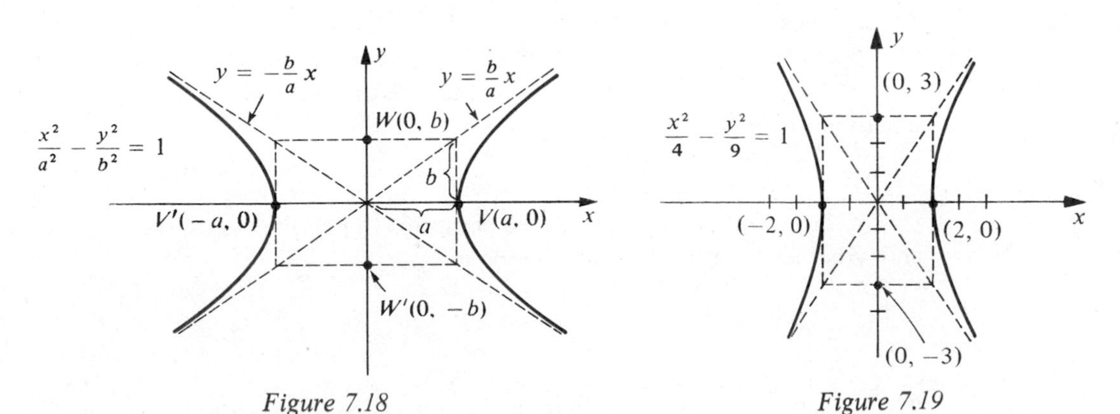

Figure 7.18 *Figure 7.19*

Example 1 Discuss and sketch the graph of the equation $9x^2 - 4y^2 = 36$.

Solution Dividing both sides by 36, we have

$$\frac{x^2}{4} - \frac{y^2}{9} = 1,$$

which is in the standard form (7.15) with $a^2 = 4$ and $b^2 = 9$. Hence $a = 2$ and $b = 3$. The vertices $(\pm 2, 0)$ and the end-points $(0, \pm 3)$ of the conjugate axis determine a rectangle whose diagonals (extended) give us the asymptotes. The graph of the equation is sketched in Figure 7.19. The equations of the asymptotes, $y = \pm\frac{3}{2}x$, can be found by referring to the graph or to the equations $y = (\pm b/a)x$. From (7.14) we have $c^2 = a^2 + b^2 = 4 + 9 = 13$, and consequently the foci are $(\pm\sqrt{13}, 0)$.

The preceding example indicates that for hyperbolas it is not always true that $a < b$, as was the case for ellipses. Indeed, we may have $a < b$, $a > b$, or $a = b$.

Example 2 Find an equation, the foci, and the asymptotes of a hyperbola which has vertices $(\pm 3, 0)$ and passes through the point $P(5, 2)$.

Solution Substituting $a = 3$ in (7.15), we obtain the equation

$$\frac{x^2}{9} - \frac{y^2}{b^2} = 1.$$

If $(5, 2)$ is a solution of this equation, then

$$\frac{25}{9} - \frac{4}{b^2} = 1.$$

This gives us $b^2 = 9/4$ and hence the desired equation is

$$\frac{x^2}{9} - \frac{4y^2}{9} = 1,$$

or equivalently, $x^2 - 4y^2 = 9$.

From (7.14), $c^2 = a^2 + b^2 = 9 + (9/4) = 45/4$ and, therefore, the foci are $(\pm\frac{3}{2}\sqrt{5}, 0)$. Substituting for b and a in $y = \pm(b/a)x$ and simplifying, we obtain equations $y = \pm\frac{1}{2}x$ for the asymptotes.

If the foci of a hyperbola are the points $(0, \pm c)$ on the y-axis, then we may show that an equation for the hyperbola is

$$\frac{y^2}{a^2} - \frac{x^2}{b^2} = 1 \tag{7.17}$$

where again $b^2 = c^2 - a^2$. Equation (7.17) is called the **standard form** for the equation of a hyperbola with foci on the y-axis and center at the origin. The points $V(0, a)$ and $V'(0, -a)$ are the vertices of the hyperbola, and the end-points of the conjugate axis are now $W(b, 0)$ and $W'(-b, 0)$. The asymptotes are found, as before, by using the diagonals of the rectangle determined by these points and lines parallel to the coordinate axes. The graph is sketched in Figure 7.20. The equations of the asymptotes are $y = (\pm a/b)x$. Note the difference between these equations and the equations $y = (\pm b/a)x$ for the asymptotes of the hyperbola given by (7.15).

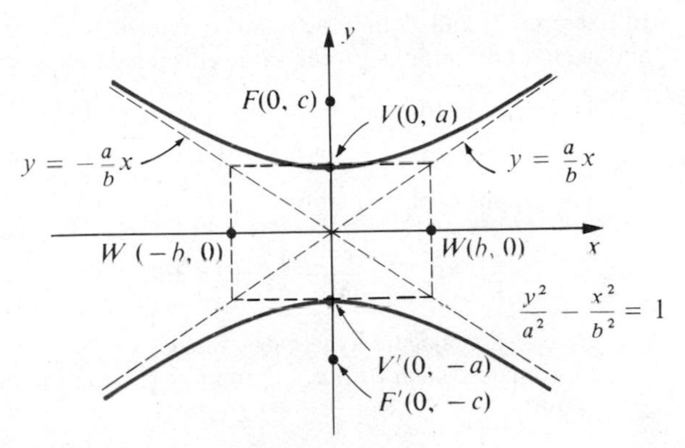

Figure 7.20

Example 3 Discuss and sketch the graph of the equation $4y^2 - 2x^2 = 1$.

Solution The standard form is

$$\frac{y^2}{1/4} - \frac{x^2}{1/2} = 1.$$

Thus $a^2 = 1/4$, $b^2 = 1/2$, and $c^2 = 1/4 + 1/2 = 3/4$. Consequently $a = 1/2$, $b = \sqrt{2}/2$, and $c = \sqrt{3}/2$. The vertices are $(0, \pm 1/2)$ and the foci are $(0, \pm\sqrt{3}/2)$. The graph has the general appearance of the graph shown in Figure 7.20.

EXERCISES 7.4

In Exercises 1–12 sketch the graph of the equation, find the coordinates of the vertices and foci, and write equations for the asymptotes.

1 $\frac{x^2}{9} - \frac{y^2}{4} = 1$ **2** $\frac{y^2}{49} - \frac{x^2}{16} = 1$ **3** $\frac{y^2}{9} - \frac{x^2}{4} = 1$

4 $\frac{x^2}{49} - \frac{y^2}{16} = 1$ **5** $y^2 - 4x^2 = 16$ **6** $x^2 - 2y^2 = 8$

7 $x^2 - y^2 = 1$ **8** $y^2 - 16x^2 = 1$ **9** $x^2 - 5y^2 = 25$

10 $4y^2 - 4x^2 = 1$ **11** $3x^2 - y^2 = -3$ **12** $16x^2 - 36y^2 = 1$

In Exercises 13–20 find an equation for the hyperbola satisfying the given conditions.

13 Foci $F(0, \pm 4)$, vertices $V(0, \pm 1)$

14 Foci $F(\pm 8, 0)$, vertices $V(\pm 5, 0)$

15 Foci $F(\pm 5, 0)$, vertices $V(\pm 3, 0)$

16 Foci $F(0, \pm 3)$, vertices $V(0, \pm 2)$

17 Foci $F(0, \pm 5)$, length of conjugate axis 4

18 Vertices $V(\pm 4, 0)$, passing through $P(8, 2)$

19 Vertices $V(\pm 3, 0)$, equations of asymptotes $y = \pm 2x$

20 Foci $F(0, \pm 10)$, equations of asymptotes $y = \pm(1/3)x$

In Exercises 21 and 22 find the points of intersection of the graphs of the given equations and sketch both graphs on the same coordinate axes, showing points of intersection.

21 $\begin{cases} y^2 - 4x^2 = 16 \\ y - x = 4 \end{cases}$

22 $\begin{cases} x^2 - y^2 = 4 \\ y^2 - 3x = 0 \end{cases}$

23 The graphs of the equations

$$\frac{x^2}{a^2} - \frac{y^2}{b^2} = 1 \quad \text{and} \quad \frac{x^2}{a^2} - \frac{y^2}{b^2} = -1$$

are called **conjugate hyperbolas**. Sketch the graphs of both equations on the same coordinate system with $a = 2$ and $b = 5$. Describe the relationship between the two graphs.

24 Derive (7.15) and prove that the line $y = (-b/a)x$ is an asymptote for this hyperbola.

25 Find an equation of the tangent line to the hyperbola $2x^2 - 5y^2 = 3$ at the point $P(-2, 1)$.

26 Prove that an equation of the tangent line to the graph of the hyperbola $x^2/a^2 - y^2/b^2 = 1$ at the point $P(x_1, y_1)$ is

$$\frac{x_1 x}{a^2} - \frac{y_1 y}{b^2} = 1.$$

27 If tangent lines to the hyperbola $9x^2 - y^2 = 36$ intersect the y-axis at the point $(0, 6)$, find the points of tangency.

28 If tangent lines to the hyperbola $b^2x^2 - a^2y^2 = a^2b^2$ intersect the y-axis at the point $(0, d)$, find the points of tangency.

29 Find an equation of a line through $P(2, -1)$ which is tangent to the hyperbola $x^2 - 4y^2 = 16$.

30 Find equations of tangent lines to the hyperbola $8x^2 - 3y^2 = 48$ which are parallel to the line $2x - y = 10$.

31 The region bounded by the hyperbola $b^2x^2 - a^2y^2 = a^2b^2$ and a vertical line through a focus is revolved about the x-axis. Find the volume of the resulting solid.

32 If the region described in Exercise 31 is revolved about the y-axis, find the volume of the resulting solid.

7.5 TRANSLATION OF AXES

If a and b are the coordinates of two points A and B, respectively, on a coordinate line l, the distance between A and B was defined in (1.6) by $d(A, B) = |b - a|$. If we wish to take into account the direction of l, then we use the **directed distance** $\overline{AB}$ from A to B where, by definition,

$$\overline{AB} = b - a.$$

Since $\overline{BA} = a - b$, we have $\overline{AB} = -\overline{BA}$. If the positive direction on l is to the right, then B is to the right of A if and only if $\overline{AB} > 0$, and is to the left of A if and only if $\overline{AB} < 0$. If C is any other point on l with coordinate c, it follows that

(7.18) $$\overline{AC} = \overline{AB} + \overline{BC}$$

since $(c - a) = (b - a) + (c - b)$. We shall use (7.18) in the following discussion to develop formulas for **translation of axes** in two dimensions.

Suppose that $C(h, k)$ is an arbitrary point in an xy-coordinate plane. Let us introduce a new $x'y'$-coordinate system with origin O' at C such that the x'- and y'-axes are parallel to, and have the same unit lengths and positive directions as, the x- and y-axes, respectively. A typical situation of this type is illustrated in Figure 7.21 where, for simplicity, we have placed C in the first quadrant. We shall use primes on letters to denote coordinates of points in the $x'y'$-coordinate system in order to distinguish them from coordinates with respect to the xy-coordinate system. Thus the point $P(x, y)$ in the xy-system will be denoted by $P(x', y')$ in the $x'y'$-system. If we label projections of P on the various axes as indicated in Figure 7.21, and let A and B denote projections of C on the x- and y-axes, respectively, then using (7.18) we have

$$x = \overline{OQ} = \overline{OA} + \overline{AQ} = \overline{OA} + \overline{O'Q'} = h + x'$$

$$y = \overline{OR} = \overline{OB} + \overline{BR} = \overline{OB} + \overline{O'R'} = k + y'.$$

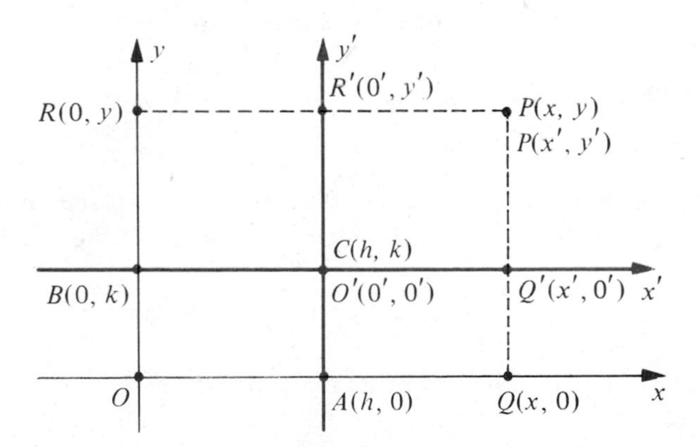

Figure 7.21

To summarize, if (x, y) are the coordinates of a point P relative to the xy-coordinate system, and if (x', y') are the coordinates of P relative to an $x'y'$-coordinate system with origin at the point $C(h, k)$ of the xy-system, then

(7.19) $$x = x' + h, \quad y = y' + k$$

or equivalently,

(7.20) $$x' = x - h, \quad y' = y - k.$$

The above formulas enable us to go from either coordinate system to the other. Their major use is to change the form of equations of graphs. To be specific, if, in the xy-plane, a certain collection of points is the graph of an equation in x and y, then to find an equation in x' and y' which has the same graph in the $x'y'$-plane, we may substitute $x' + h$ for x and $y' + k$ for y in the given equation. Conversely, if a set of points in the $x'y'$-plane is the graph of an equation in x' and y', then to find the corresponding equation in x and y we substitute $x - h$ for x' and $y - k$ for y'.

As a simple illustration of the preceding remarks, the equation

$$(x')^2 + (y')^2 = r^2$$

has, for its graph in the $x'y'$-plane, a circle of radius r with center at the origin O'. Using (7.20), an equation for this circle in the xy-plane is

$$(x - h)^2 + (y - k)^2 = r^2,$$

which is in agreement with the formula for a circle of radius r with center at $C(h, k)$ in the xy-plane. As another illustration, we know that

$$(x')^2 = 4py'$$

is an equation of a parabola with vertex at the origin O' of the $x'y'$-plane. Using (7.20), we see that

$$(x - h)^2 = 4p(y - k)$$

is an equation of the same parabola in the xy-plane. This agrees with formula (7.5), which was derived for a parabola with vertex at the point $V(h, k)$ in the xy-plane. It should now be evident how this technique can be applied to all the conics. For example, by (7.9), the graph of

$$\frac{(x')^2}{a^2} + \frac{(y')^2}{b^2} = 1$$

is an ellipse with center at O' in the $x'y'$-plane, as illustrated in Figure 7.22. According to (7.20), its equation relative to the xy-coordinate system is

(7.21)
$$\frac{(x - h)^2}{a^2} + \frac{(y - k)^2}{b^2} = 1.$$

A similar situation exists for hyperbolas.

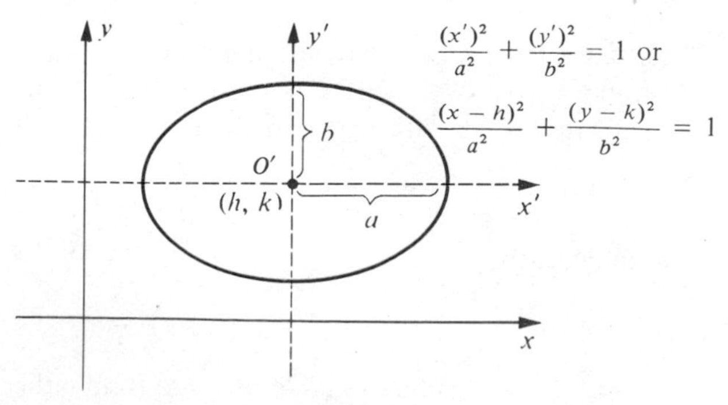

Figure 7.22

In certain cases, given an equation in x and y we may, by a proper translation of axes, obtain a simpler equation in x' and y' which has the same graph. In particular, this is true for an equation in x and y of the form

(7.22) $$Ax^2 + Cy^2 + Dx + Ey + F = 0$$

where the coefficients are real numbers. The graph of (7.22) is a conic, except for the degenerate cases in which points, lines, or no graphs are obtained. We shall not give a general proof of that fact but will, instead, illustrate the procedure by means of examples.

Example 1 Discuss and sketch the graph of the equation

$$16x^2 + 9y^2 + 64x - 18y - 71 = 0.$$

Solution In order to determine the origin of a new $x'y'$-coordinate system which will enable us to simplify the given equation, we begin by writing the given equation in the form

$$16(x^2 + 4x) + 9(y^2 - 2y) = 71.$$

Next, we complete the squares for the expressions within parentheses, obtaining

$$16(x^2 + 4x + 4) + 9(y^2 - 2y + 1) = 71 + 64 + 9.$$

Note that by adding 4 to the expression within the first parentheses we have added 64 to the left side of the equation and hence must compensate by adding 64 to the right side. Similarly, by adding 1 to the expression within the second parentheses, 9 is added to the left side and consequently 9 must also be added to the right side. The last equation may be written

$$16(x + 2)^2 + 9(y - 1)^2 = 144.$$

Dividing by 144 we obtain

$$\frac{(x + 2)^2}{9} + \frac{(y - 1)^2}{16} = 1$$

which is of the form

(7.23) $$\frac{(x')^2}{9} + \frac{(y')^2}{16} = 1$$

where $x' = x + 2$ and $y' = y - 1$. This shows that if we let $h = -2$ and $k = 1$ in (7.20), then (7.23) reduces to the given equation. Since the graph of (7.23) is an ellipse with center at the origin O' in the $x'y'$-plane it follows that the given equation is an ellipse with center $C(-2, 1)$ in the xy-plane and with axes parallel to the coordinate axes. The graph is sketched in Figure 7.23.

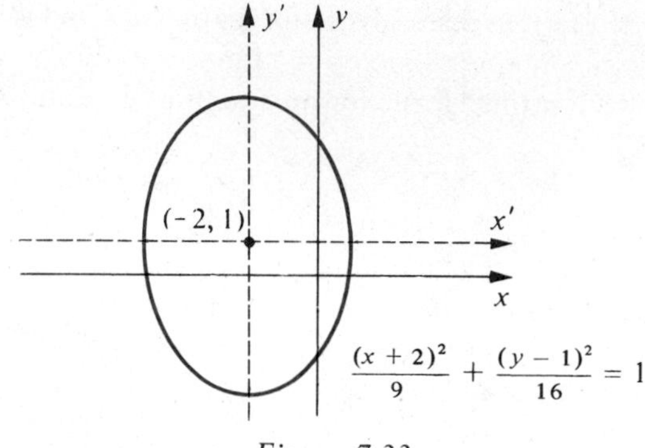

Figure 7.23

Example 2 Discuss and sketch the graph of the equation

$$9x^2 - 4y^2 - 54x - 16y + 29 = 0.$$

Solution As in Example 1, we arrange our work as follows:

$$9(x^2 - 6x) - 4(y^2 + 4y) = -29$$
$$9(x^2 - 6x + 9) - 4(y^2 + 4y + 4) = -29 + 81 - 16$$
$$9(x-3)^2 - 4(y+2)^2 = 36$$
$$\frac{(x-3)^2}{4} - \frac{(y+2)^2}{9} = 1.$$

If we substitute $h = 3$ and $k = -2$ in (7.20), then the given equation reduces to the standard form (7.15) for the equation of a hyperbola, namely

$$\frac{(x')^2}{4} - \frac{(y')^2}{9} = 1.$$

By translating the x- and y-axes to the new origin $C(3, -2)$ we obtain the sketch shown in Figure 7.24.

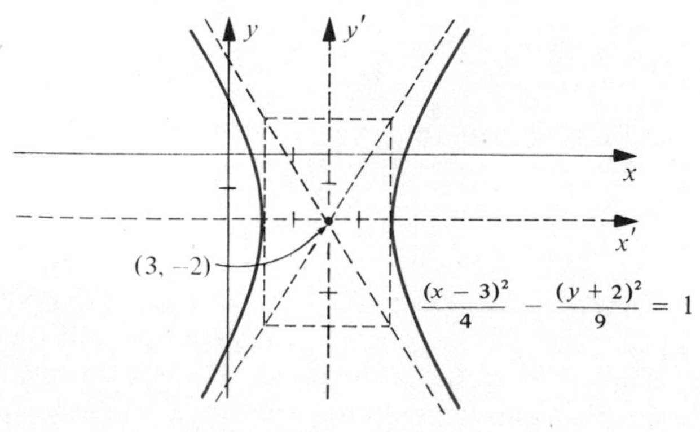

Figure 7.24

Example 3 Discuss and sketch the graph of the equation

$$2x = y^2 + 8y + 22.$$

Solution This example is the same as Example 4 of Section 2, where we completed the square in y and obtained the equation $(y + 4)^2 = 2(x - 3)$. By the methods of the present section, if we let $h = 3$ and $k = -4$ in (7.20), then the given equation reduces to the standard form $(y')^2 = 2x'$ of a parabola with vertex at O'. Consequently, a translation of axes to the new origin $C(3, -4)$ leads to the sketch shown in Figure 7.10.

Although we have only considered special examples of (7.22), our methods are perfectly general. If A and C are equal and not zero, then the graph of (7.22), when it exists, is a circle or, in exceptional cases, a point. If A and C are unequal but have the same sign, then by completing squares and properly translating axes we obtain an equation whose graph, when it exists, is an ellipse (or a point). If A and C have opposite signs, an equation of a hyperbola is obtained, or possibly, in the degenerate case, two intersecting straight lines. Finally, if either A or C (but not both) is zero, the graph is a parabola or, in certain cases, a pair of parallel straight lines.

EXERCISES 7.5

Discuss and sketch the graphs of the equations in Exercises 1–26 after making suitable translations of axes.

1 $(y - 5)^2 = 8(x + 1)$

2 $(x + 3)^2 = -5(y - 4)$

3 $\dfrac{(x - 4)^2}{9} + \dfrac{(y - 2)^2}{4} = 1$

4 $4(x + 1)^2 + 8(y - 5)^2 = 32$

5 $\dfrac{(x + 5)^2}{16} - \dfrac{(y - 1)^2}{25} = 1$

6 $\dfrac{(y - 6)^2}{4} - (x + 2)^2 = 1$

7 $4(x + 5)^2 + (y - 3)^2 = 1$

8 $(x - 3)^2 + 4y^2 = 16$

9 $100y^2 - 16(x - 5)^2 = 1600$

10 $(x - 5)^2 - (y + 1)^2 = 1$

11 $9x^2 + 16y^2 + 54x - 32y - 47 = 0$

12 $4x^2 + 9y^2 + 24x + 18y + 9 = 0$

13 $25x^2 + 4y^2 - 250x - 16y + 541 = 0$

14 $4x^2 + y^2 = 2y$

15 $4x^2 - 32y + 4x - 49 = 0$

16 $4y^2 - x - 16y + 13 = 0$

17 $4y^2 - x^2 + 40y - 4x + 60 = 0$

18 $25x^2 - 9y^2 - 100x - 54y + 10 = 0$

19 $9y^2 - x^2 - 36y + 12x - 36 = 0$

20 $4x^2 - y^2 + 32x - 8y + 49 = 0$

21 $y = |x - 5| - 4$

22 $x + 2 = \sqrt{(y - 1)^2}$

23 $x = 3 + (y - 6)^3$

24 $(x - 7)^3 - y - 2 = 0$

25 $2y + 10 - (x + 2)^4 = 0$

26 $(y + 7)(x - 5) = 1$

27 Find an equation of the hyperbola with foci $(h \pm c, k)$ and vertices $(h \pm a, k)$, where $0 < a < c$ and $c^2 = a^2 + b^2$.

28 Find an equation of the hyperbola with vertices $(h, k \pm a)$ and asymptotes $y - k = \pm(a/b)(x - h)$, where $b > 0$.

29 Find an equation of the tangent line to the ellipse $4x^2 + y^2 - 8x + 2y - 12 = 0$ at the point $P(3, -2)$.

30 Find an equation of the normal line to the hyperbola $5x^2 - 2y^2 + 8x + 7y + 7 = 0$ at the point $P(-1, 4)$.

7.6 ROTATION OF AXES

The $x'y'$-coordinate system discussed in Section 5 may be thought of as having been obtained by moving the origin O of the xy-system to a new position $C(h, k)$ while, at the same time, not changing the positive directions of the axes or the units of length. We shall now introduce a new coordinate system by keeping the origin O fixed and rotating the x- and y-axes about O to another position denoted by x' and y'. A transformation of this type will be referred to as a **rotation of axes.**

Let us consider a rotation of axes and, as shown in Figure 7.25, let ϕ denote the angle through which the positive x-axis must be rotated in order to coincide with the positive x'-axis. If (x, y) are the coordinates of a point P relative to the xy-plane, then as before (x', y') will denote its coordinates relative to the new $x'y'$-coordinate system. Let the projections of P on the various axes be denoted as in Figure 7.25 and let θ denote angle POQ'. If $p = d(O, P)$, then

(7.24) $$x' = p\cos\theta, \qquad y' = p\sin\theta,$$

(7.25) $$x = p\cos(\theta + \phi), \qquad y = p\sin(\theta + \phi).$$

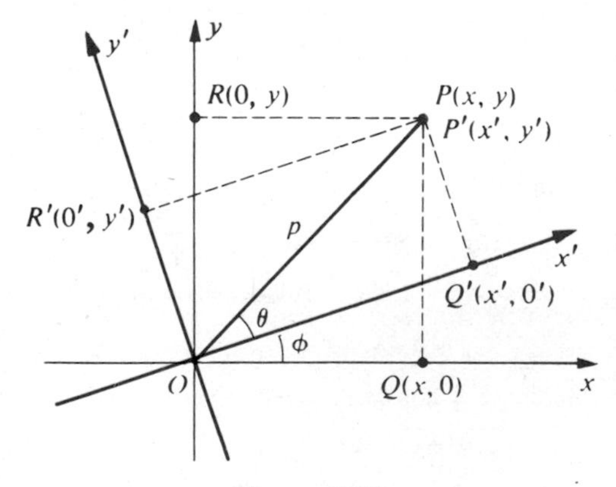

Figure 7.25

Applying the addition formulas to (7.25),

$$x = p\cos\theta\cos\phi - p\sin\theta\sin\phi$$
$$y = p\sin\theta\cos\phi + p\cos\theta\sin\phi.$$

By (7.24) we may replace $p\cos\theta$ and $p\sin\theta$ by x' and y', respectively. This gives us

(7.26)
$$x = x'\cos\phi - y'\sin\phi$$
$$y = x'\sin\phi + y'\cos\phi.$$

If the equations in (7.26) are solved for x' and y', we obtain

(7.27)
$$x' = x\cos\phi + y\sin\phi$$
$$y' = -x\sin\phi + y\cos\phi.$$

The formulas in (7.26) and (7.27) can be used to change from either coordinate system to the other. We shall call them the **formulas for rotation of axes through the angle ϕ.**

Example 1 The graph of the equation $xy = 1$, or equivalently $y = 1/x$, is sketched in Figure 1.30. If the coordinate axes are rotated through an angle of 45°, find the equation of the graph relative to the new $x'y'$-coordinate system.

Solution Letting $\phi = 45°$ in (7.26), we obtain

$$x = x'\left(\frac{\sqrt{2}}{2}\right) - y'\left(\frac{\sqrt{2}}{2}\right) = \left(\frac{\sqrt{2}}{2}\right)(x' - y')$$

$$y = x'\left(\frac{\sqrt{2}}{2}\right) + y'\left(\frac{\sqrt{2}}{2}\right) = \left(\frac{\sqrt{2}}{2}\right)(x' + y').$$

Substituting for x and y in the equation $xy = 1$ gives us

$$\left(\frac{\sqrt{2}}{2}\right)(x' - y')\left(\frac{\sqrt{2}}{2}\right)(x' + y') = 1.$$

This reduces to

$$\frac{(x')^2}{2} - \frac{(y')^2}{2} = 1$$

which is the standard equation of a hyperbola with vertices $(\pm\sqrt{2}, 0)$ on the x'-axis. Figure 7.26 shows the graph, together with the new coordinate axes. Note that the asymptotes for the hyperbola have equations $y' = \pm x'$ in the new system. These correspond to the original x- and y-axes.

Example 1 illustrates a method for eliminating a term of an equation which contains the product xy. This method can be used to transform any equation of the form

(7.28)
$$Ax^2 + Bxy + Cy^2 + Dx + Ey + F = 0$$

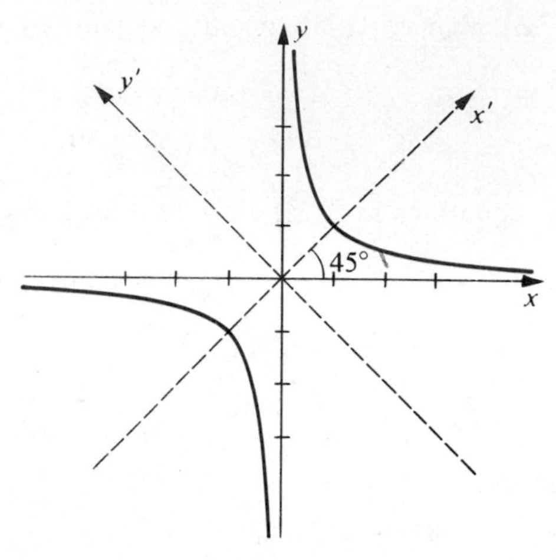

Figure 7.26

where $B \neq 0$, into an equation in x' and y' which contains no $x'y'$ term. Let us prove that this may always be done. If we rotate the axes through an angle ϕ, then substituting the expressions of (7.26) for x and y in (7.28) gives us

$$\begin{aligned}&A(x'\cos\phi - y'\sin\phi)^2\\&\quad + B(x'\cos\phi - y'\sin\phi)(x'\sin\phi + y'\cos\phi)\\&\quad + C(x'\sin\phi + y'\cos\phi)^2 + D(x'\cos\phi - y'\sin\phi)\\&\quad + E(x'\sin\phi + y'\cos\phi) + F = 0.\end{aligned}$$

The last equation may be written in the form

$$A'(x')^2 + B'x'y' + C'(y')^2 + D'x' + E'y' + F = 0 \tag{7.29}$$

where the coefficient B' of $x'y'$ is given by

$$B' = 2(C - A)\sin\phi\cos\phi - B(\cos^2\phi - \sin^2\phi).$$

In order to eliminate the $x'y'$ term in (7.29) we must select ϕ so that

$$2(C - A)\sin\phi\cos\phi + B(\cos^2\phi - \sin^2\phi) = 0.$$

Using the double-angle formulas, the last equation may be written

$$(C - A)\sin 2\phi + B\cos 2\phi = 0$$

which is equivalent to

$$\cot 2\phi = \frac{A - C}{B}, \quad B \neq 0. \tag{7.30}$$

Thus, to eliminate the xy term in (7.28), we may choose ϕ such that (7.30) is true and then employ the rotation of axes formulas (7.26). The resulting equation will contain no $x'y'$ term and therefore can be analyzed by previous methods. This proves that if the graph of (7.28) exists, it is a conic (except for degenerate cases).

Example 2 Discuss and sketch the graph of the equation

$$41x^2 - 24xy + 34y^2 - 25 = 0.$$

Solution Using the notation of (7.28) we have $A = 41$, $B = -24$, and $C = 34$. Applying (7.30),

$$\cot 2\phi = \frac{41 - 34}{-24} = -\frac{7}{24}.$$

Since $\cot 2\phi$ is negative we may choose 2ϕ such that $90° < 2\phi < 180°$, and consequently $\cos 2\phi = -7/25$. (Why?) We now use the half-angle formulas to obtain

$$\sin \phi = \sqrt{\frac{1 - \cos 2\phi}{2}} = \sqrt{\frac{1 - (-7/25)}{2}} = \frac{4}{5}$$

$$\cos \phi = \sqrt{\frac{1 + \cos 2\phi}{2}} = \sqrt{\frac{1 + (-7/25)}{2}} = \frac{3}{5}.$$

It follows that the desired rotation formulas (7.26) are

$$x = \tfrac{3}{5}x' - \tfrac{4}{5}y', \quad y = \tfrac{4}{5}x' + \tfrac{3}{5}y'.$$

We leave it to the reader to show that after substituting for x and y in the given equation and simplifying, we obtain the equation

$$(x')^2 + 2(y')^2 = 1.$$

Thus the graph is an ellipse with vertices at $(\pm 1, 0)$ on the x'-axis. Since $\tan \phi = \sin \phi/\cos \phi = (4/5)/(3/5) = 4/3$, we obtain $\phi = \tan^{-1}(4/3)$. To the nearest minute, $\phi \approx 53° 8'$. The graph is sketched in Figure 7.27.

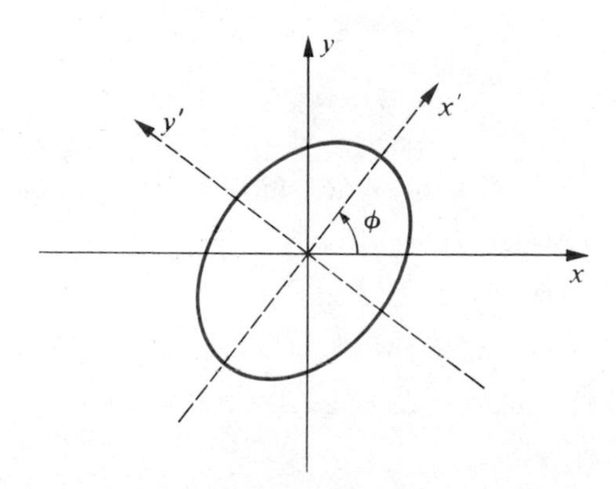

Figure 7.27

EXERCISES 7.6

After a suitable rotation of axes, describe and sketch the graph of the equation in each of Exercises 1–14.

1 $32x^2 - 72xy + 53y^2 = 80$

2 $7x^2 - 48xy - 7y^2 = 225$

3 $11x^2 + 10\sqrt{3}xy + y^2 = 4$

4 $x^2 - xy + y^2 = 3$

5 $5x^2 - 8xy + 5y^2 = 9$

6 $11x^2 - 10\sqrt{3}xy + y^2 = 20$

7 $16x^2 - 24xy + 9y^2 - 60x - 80y + 100 = 0$

8 $x^2 + 2\sqrt{3}xy + 3y^2 + 8\sqrt{3}x - 8y + 32 = 0$

9 $5x^2 + 6\sqrt{3}xy - y^2 + 8x - 8\sqrt{3}y - 12 = 0$

10 $18x^2 - 48xy + 82y^2 + 6\sqrt{10}x + 2\sqrt{10}y - 80 = 0$

11 $x^2 + 4xy + 4y^2 + 6\sqrt{5}x - 18\sqrt{5}y + 45 = 0$

12 $15x^2 + 20xy - 4\sqrt{5}x + 8\sqrt{5}y - 100 = 0$

13 $40x^2 - 36xy + 25y^2 - 8\sqrt{13}x - 12\sqrt{13}y = 0$

14 $64x^2 - 240xy + 225y^2 + 1020x - 544y = 0$

15 Prove that, except for degenerate cases, the graph of equation (7.28) is
(a) a parabola if $B^2 - 4AC = 0$.
(b) an ellipse if $B^2 - 4AC < 0$.
(c) a hyperbola if $B^2 - 4AC > 0$.

16 Use the results of Exercise 15 to determine the nature of the graphs in Exercises 1–14.

7.7 REVIEW

Concepts

Define or discuss each of the following:

1 Conic sections

2 Parabola

3 Focus, directrix, vertex, and axis of a parabola

4 Tests for symmetry

5 Ellipse

6 Major and minor axes of an ellipse

7 Foci and vertices of an ellipse

8 Hyperbola

9 Transverse and conjugate axes of a hyperbola

10 Foci and vertices of a hyperbola

11 Asymptotes of a hyperbola

12 Translation of axes

13 Rotation of axes

Exercises

In each of Exercises 1–10, find the foci and vertices and sketch the graph of the conic which has the given equation.

1 $y^2 = 64x$

2 $y - 1 = 8(x + 2)^2$

3 $9y^2 = 144 - 16x^2$

4 $9y^2 = 144 + 16x^2$

5 $x^2 - y^2 - 4 = 0$

6 $25x^2 + 36y^2 = 1$

7 $25y = 100 - x^2$

8 $3(x - 3)^2 + 4(y + 1)^2 = 12$

9 $(x + 4)^2 - 9(y - 5)^2 = 1$

10 $x = 2y^2 + 8y + 3$

Find equations for the conics in Exercises 11–18.

11 The hyperbola with vertices $V(0, \pm 7)$ and end-points of conjugate axes $(\pm 3, 0)$

12 The parabola with focus $F(-4, 0)$ and directrix $x = 4$

13 The parabola with focus $(0, -10)$ and directrix $y = 10$

14 The parabola with vertex at the origin, symmetric to the x-axis, and passing through the point $(5, -1)$

15 The ellipse with vertices $V(0, \pm 10)$ and foci $F(0, \pm 5)$

16 The hyperbola with foci $F(\pm 10, 0)$ and vertices $V(\pm 5, 0)$

17 The hyperbola with vertices $V(0, \pm 6)$ and asymptotes which have equations $y = \pm 9x$

18 The ellipse with foci $F(\pm 2, 0)$ and passing through the point $(2, \sqrt{2})$

Discuss and sketch the graph of each of the equations in Exercises 19–23 after making a suitable translation of axes.

19 $4x^2 + 9y^2 + 24x - 36y + 36 = 0$

20 $4x^2 - y^2 - 40x - 8y + 88 = 0$

21 $y^2 - 8x + 8y + 32 = 0$

22 $4x^2 + y^2 - 24x + 4y + 36 = 0$

23 $x^2 - 9y^2 + 8x + 7 = 0$

24 Find equations of the tangent and normal lines to the hyperbola $4x^2 - 9y^2 - 8x + 6y - 36 = 0$ at the point $P(-3, 2)$.

25 Tangent lines to the parabola $y = 2x^2 + 3x + 1$ pass through the point $P(2, -1)$. Find the abscissas of the points of tangency.

26 Let R be the region bounded by the parabola $9y = 5x^2 - 39x + 90$ and an asymptote of the hyperbola $9y^2 - 4x^2 = 36$.
(a) Find the area of R.
(b) Find the volume of the solid obtained by revolving R about the y-axis.

27 An ellipse having axes of lengths 8 and 4 is revolved about its major axis. Find the volume of the resulting solid.

28 A solid has, for its base, the region in the xy-plane bounded by the graph of $x^2 + y^2 = r^2$. Find the volume of the solid if every cross section by a plane perpendicular to the x-axis is half of an ellipse with one axis always of length c.

After a suitable rotation of axes, describe and sketch the graphs of the equations in Exercises 29 and 30.

29 $x^2 - 8xy + 16y^2 - 12\sqrt{17}x - 3\sqrt{17}y = 0$

30 $8x^2 + 12xy + 17y^2 - 16\sqrt{5}x - 12\sqrt{5}y = 0$

Exponential and Logarithmic Functions

Most of the functions considered in preceding chapters were algebraic functions. Functions which are not algebraic are called transcendental. *In this chapter we shall define two important transcendental functions and investigate some of their properties.*

8.1 THE NATURAL LOGARITHMIC FUNCTION

If a is a positive real number it is easy to define a^n for every integral or rational exponent n, and then prove the following **Laws of Exponents**.

(8.1)
$$a^m a^n = a^{m+n} \qquad (a^m)^n = a^{mn} \qquad (ab)^n = a^n b^n$$
$$\frac{a^m}{a^n} = a^{m-n} \qquad \left(\frac{a}{b}\right)^n = \frac{a^n}{b^n}$$

However, the extension to irrational exponents such as $a^{\sqrt{2}}$ or a^{π} requires deeper concepts than those discussed in elementary mathematics. Indeed, in precalculus courses it is always *assumed* that if $a > 0$, then a^u can be defined for every real number u such that the laws in (8.1) are true for all *real* exponents. Having made this assumption, logarithms with base a are introduced by requiring that

(8.2)
$$u = \log_a v \quad \text{if and only if} \quad a^u = v.$$

The following **Laws of Logarithms** can then be established.

(8.3)
$$\log_a pq = \log_a p + \log_a q$$
$$\log_a \left(\frac{p}{q}\right) = \log_a p - \log_a q$$
$$\log_a p^r = r \log_a p$$

where p and q are positive real numbers and r is any real number.

In this chapter we shall use concepts developed in calculus to state definitions of $\log_a x$ for every positive real number x and of a^x for every real number x. Our approach to these definitions will be first to use a definite integral to introduce the **natural logarithmic function**, denoted by ln. Later, the natural logarithmic function will be used to define the **natural exponential function**. Finally, we will give meaning to the expressions a^x and $\log_a x$. The reason for proceeding in this fashion is that it provides a precise method of defining general logarithmic and exponential functions without invoking specialized limiting processes. Our technique will also allow us to establish results on continuity, derivatives, and integrals in a very simple manner. Moreover, as will be seen, the end result will be the familiar rule (8.2) that students encounter before taking calculus.

If a function f is continuous on a closed interval $[a, b]$, then as in (5.30) we can define a function F by

$$F(x) = \int_a^x f(t)\,dt$$

where x is any number in $[a, b]$. If $f(t) \geq 0$ throughout $[a, b]$, then $F(x)$ equals the area under the graph of f from a to x. Let us consider the special case $f(t) = t^n$, where n is any integer different from -1. By the Fundamental Theorem of Calculus (5.31),

$$\begin{aligned} F(x) &= \int_a^x t^n\,dt = \frac{1}{n+1}t^{n+1}\Big]_a^x \\ &= \frac{1}{n+1}(x^{n+1} - a^{n+1}) \end{aligned}$$

provided t^n is defined throughout $[a, x]$. It is necessary to exclude $f(t) = 1/t$ in the last formula since $1/(n + 1)$ is undefined if $n = -1$. Indeed, up to this stage of our work it has been impossible to find an antiderivative for $f(x) = 1/x$. We shall now remedy this situation by introducing a function whose derivative is $1/x$.

(8.4) Definition

The **natural logarithmic function**, denoted by **ln**, is defined by

$$\ln x = \int_1^x \frac{1}{t}\,dt$$

for all $x > 0$.

The expression $\ln x$ is called *the natural logarithm of* x. The restriction $x > 0$ is necessary since $\int_1^x (1/t)\,dt$ does not exist if $x \leq 0$. (Why?)

Using Part I of the Fundamental Theorem of Calculus (5.31) with $f(t) = 1/t$ and $a = 1$, we see that for all $x > 0$,

$$D_x \int_1^x \frac{1}{t}\,dt = \frac{1}{x}$$

or equivalently,

(8.5) $$D_x(\ln x) = \frac{1}{x}.$$

This proves that **ln** $\boldsymbol{x}$ **is an antiderivative of** $\boldsymbol{1/x}$. Moreover, since $\ln x$ is differentiable, and its derivative $1/x$ is positive for all $x > 0$, it follows from (3.8) and Theorem (4.13) that *the natural logarithmic function is continuous and increasing throughout its domain.*

Let us sketch the graph of $y = \ln x$. The ordinate y of the point with abscissa x is given by the integral in Definition (8.4). In particular, if $x = 1$, then

$$\ln 1 = \int_1^1 \frac{1}{t}\,dt = 0.$$

Thus the graph of $y = \ln x$ has x-intercept 1. Since ln is an increasing function, it follows that if $x > 1$, then the point (x, y) on the graph lies above the x-axis, whereas if $0 < x < 1$, then (x, y) lies below the x-axis. To estimate y if $x \neq 1$ we may apply either the Trapezoidal Rule or Simpson's Rule. If $x = 2$, then by Example 2 in Section 5.7,

$$\ln 2 = \int_1^2 \frac{1}{t}\,dt \approx 0.6932 \approx 0.69.$$

It will be shown in Theorem (8.7) that if $a > 0$, then $\ln a^r = r \ln a$ for every rational number r. Using this result we obtain

$$\begin{aligned}
\ln 4 &= \ln 2^2 = 2\ln 2 \approx 2(0.69) = 1.38\\
\ln 8 &= \ln 2^3 = 3\ln 2 \approx 2.07\\
\ln(1/2) &= \ln 2^{-1} = -\ln 2 \approx -0.69\\
\ln(1/4) &= \ln 2^{-2} = -2\ln 2 \approx -1.38\\
\ln(1/8) &= \ln 2^{-3} = -3\ln 2 \approx -2.07.
\end{aligned}$$

Table III provides a list of natural logarithms of many other numbers correct to three decimal places. Hand-held calculators may also be used to estimate values of ln.

Plotting the points which correspond to the coordinates we have found and using the fact that ln is continuous and increasing gives us the sketch in Figure 8.1. At the end of this section it will be proved that

$$\lim_{x \to \infty} \ln x = \infty$$

that is, the values of ln can be made arbitrarily large by choosing x sufficiently large. It will also be shown that

$$\lim_{x \to 0^+} \ln x = -\infty$$

which means that the y-axis is a vertical asymptote for the graph.

Finally, to investigate concavity we note that

$$D_x^2 (\ln x) = D_x\left(\frac{1}{x}\right) = -\frac{1}{x^2}$$

which is negative for every $x > 0$. Hence by (4.16), the graph of the natural logarithmic function is concave downward at every point $P(c, \ln c)$ on the graph.

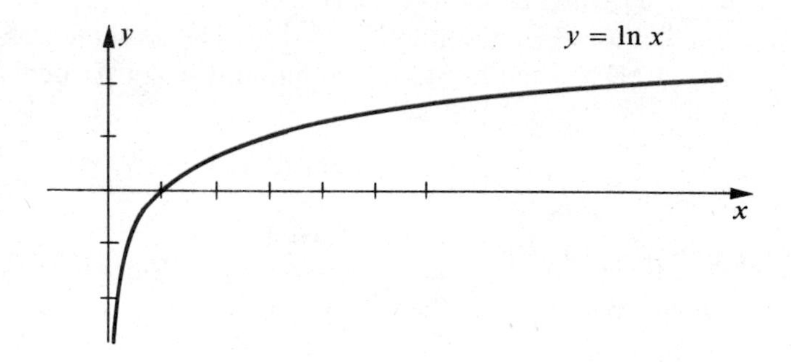

Figure 8.1

The Chain Rule (3.36) can be used to generalize the differentiation formula (8.5). Specifically, if we let $y = \ln u$ and $u = g(x)$, where g is differentiable and $g(x) > 0$, then

$$D_x y = (D_u y)(D_x u) = \frac{1}{u} D_x u.$$

Thus

(8.6) $$D_x \ln u = \frac{1}{u} D_x u, \quad \text{where } u = g(x) > 0.$$

Note that if $u = x$, then (8.6) reduces to (8.5).

Example 1 (a) Find $f'(x)$ if $f(x) = \ln(x^2 + 6)$.
(b) Find y' if $y = \ln\sqrt{x + 1}$, where $x > -1$.

Solution (a) Letting $u = x^2 + 6$ in (8.6) we obtain

$$f'(x) = D_x \ln(x^2 + 6) = \frac{1}{x^2 + 6} D_x(x^2 + 6) = \frac{2x}{x^2 + 6}.$$

(b) Letting $u = \sqrt{x+1}$ in (8.6) we obtain

$$y' = \frac{1}{\sqrt{x+1}} D_x \sqrt{x+1} = \frac{1}{\sqrt{x+1}} \cdot \frac{1}{2}(x+1)^{-1/2}$$

$$= \frac{1}{\sqrt{x+1}} \cdot \frac{1}{2} \frac{1}{\sqrt{x+1}} = \frac{1}{2(x+1)}.$$

The next theorem shows that natural logarithms obey the same laws (8.3) studied in precalculus mathematics courses.

(8.7) Theorem

> If $p > 0$ and $q > 0$, then
> (i) $\ln pq = \ln p + \ln q$.
> (ii) $\ln \left(\frac{p}{q}\right) = \ln p - \ln q$.
> (iii) $\ln p^r = r \ln p$, where r is any rational number.

Proof. (i) If $p > 0$, then using (8.6) with $u = px$ gives us

$$D_x \ln (px) = \frac{1}{px} D_x(px) = \frac{1}{px}(p) = \frac{1}{x}.$$

Since $\ln px$ has the same derivative as $\ln x$ for all $x > 0$, it follows from Theorem (4.33) that these expressions differ by a constant; that is,

$$\ln px = \ln x + C$$

for some real number C. Substituting 1 for x we obtain

$$\ln p = \ln 1 + C.$$

Since $\ln 1 = 0$ we see that $C = \ln p$, and hence

$$\ln px = \ln x + \ln p.$$

Substituting q for x in the last equation gives us

$$\ln pq = \ln q + \ln p$$

which is what we wished to prove.

(ii) Using the formula established in part (i), with $p = 1/q$, we see that

$$\ln \frac{1}{q} + \ln q = \ln \left(\frac{1}{q} \cdot q\right) = \ln 1 = 0$$

and hence

$$\ln \frac{1}{q} = -\ln q.$$

Consequently

$$\ln\frac{p}{q} = \ln\left(p\cdot\frac{1}{q}\right) = \ln p + \ln\frac{1}{q} = \ln p - \ln q.$$

(iii) If r is any rational number and $x > 0$, then by (8.6) with $u = x^r$,

$$D_x(\ln x^r) = \frac{1}{x^r}D_x(x^r) = \frac{1}{x^r}rx^{r-1} = r\left(\frac{1}{x}\right).$$

We may also write

$$D_x(r\ln x) = r\,D_x(\ln x) = r\left(\frac{1}{x}\right).$$

Consequently

$$D_x(\ln x^r) = D_x(r\ln x).$$

Since $\ln x^r$ and $r\ln x$ have the same derivative for all $x > 0$ it follows from Theorem (4.33) that

$$\ln x^r = r\ln x + C$$

for some constant C. If we let $x = 1$ in the last formula we obtain

$$\ln 1 = r\ln 1 + C.$$

Since $\ln 1 = 0$ this implies that $C = 0$ and, therefore,

$$\ln x^r = r\ln x.$$

In Section 4 we shall extend this law to irrational exponents (see (8.24)).

The next example illustrates the fact that it is sometimes convenient to use Theorem (8.7) before differentiating.

Example 2 Find $f'(x)$ if $f(x) = \ln[\sqrt{6x-1}(4x+5)^3]$, where $x > 1/6$.

Solution We first write $\sqrt{6x-1} = (6x-1)^{1/2}$ and then use (i) and (iii) of Theorem (8.7), obtaining

$$\begin{aligned} f(x) &= \ln[(6x-1)^{1/2}(4x+5)^3] \\ &= \ln(6x-1)^{1/2} + \ln(4x+5)^3 \\ &= \frac{1}{2}\ln(6x-1) + 3\ln(4x+5). \end{aligned}$$

We may now differentiate with ease, using (8.6). Thus

$$\begin{aligned} f'(x) &= \frac{1}{2}\cdot\frac{1}{6x-1}(6) + 3\cdot\frac{1}{4x+5}(4) \\ &= \frac{3}{6x-1} + \frac{12}{4x+5} \\ &= \frac{84x+3}{(6x-1)(4x+5)}. \end{aligned}$$

In future examples and exercises, if a function is defined in terms of the natural logarithmic function its domain will not usually be stated explicitly. Instead it will be assumed that x is restricted to values for which the given expression has meaning.

Example 3 Find y' if $y = \ln \sqrt[3]{\dfrac{x^2-1}{x^2+1}}$.

Solution We first use Theorem (8.7) to change the form of y as follows:

$$\begin{aligned} y &= \ln\left(\frac{x^2-1}{x^2+1}\right)^{1/3} = \frac{1}{3}\ln\left(\frac{x^2-1}{x^2+1}\right) \\ &= \frac{1}{3}[\ln(x^2-1) - \ln(x^2+1)]. \end{aligned}$$

Next we apply (8.6), obtaining

$$\begin{aligned} y' &= \frac{1}{3}\left[\frac{1}{x^2-1}(2x) - \frac{1}{x^2+1}(2x)\right] \\ &= \frac{2x}{3}\left[\frac{1}{x^2-1} - \frac{1}{x^2+1}\right] \\ &= \frac{2x}{3}\left[\frac{2}{(x^2-1)(x^2+1)}\right] = \frac{4x}{3(x^4-1)}. \end{aligned}$$

We shall conclude this section by investigating the behavior of $\ln x$ as $x \to \infty$ and as $x \to 0^+$.

If $x > 1$ and t is in the interval $[1, x]$, then $1/t > 0$ and hence we may interpret the integral $\int_1^x (1/t)\,dt = \ln x$ as the area of the region in a ty-coordinate system bounded by the graphs of $y = 1/t$, $t = 1$, $t = x$, and the t-axis. A region of this type is illustrated in Figure 8.2. Next, observe that the sum of the areas of the three rectangles shown in Figure 8.3 is

$$\frac{1}{2} + \frac{1}{3} + \frac{1}{4} = \frac{13}{12}.$$

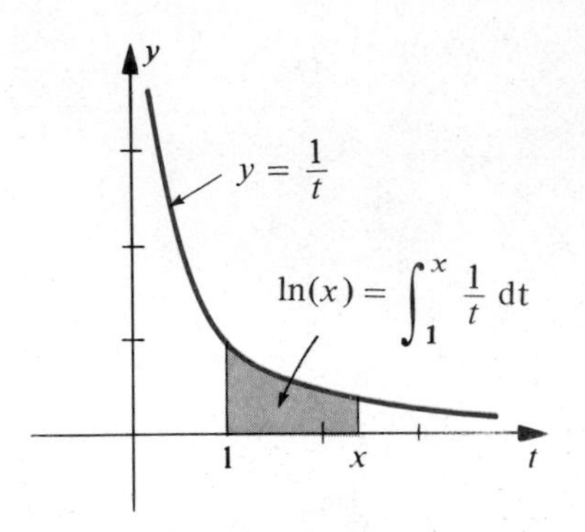

Figure 8.2

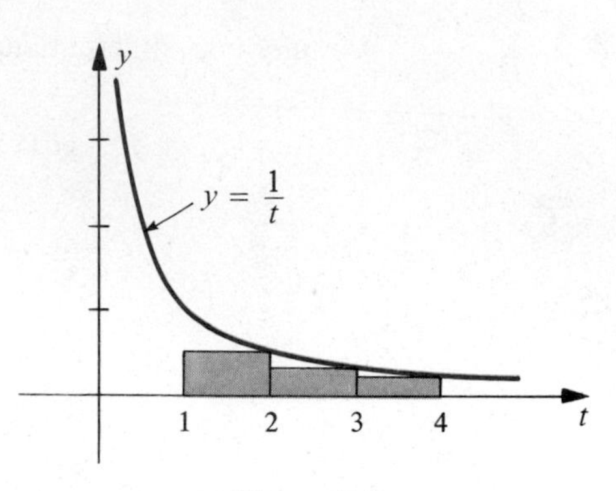

Figure 8.3

Since the area under the graph of $y = 1/t$ from $t = 1$ to $t = 4$ is $\ln 4$ we may write

$$\ln 4 > \frac{13}{12} > 1.$$

It follows that if M is any positive rational number, then

$$M \ln 4 > M, \quad \text{or} \quad \ln 4^M > M.$$

If $x > 4^M$, then since ln is an increasing function,

$$\ln x > \ln 4^M > M.$$

This proves that $\ln x$ can be made as large as desired by choosing x sufficiently large, that is

$$\lim_{x \to \infty} \ln x = \infty.$$

Next we note that

$$-\ln \frac{1}{x} = -(\ln 1 - \ln x) = \ln x$$

and hence

$$\lim_{x \to 0^+} \ln x = \lim_{x \to 0^+} \left(-\ln \frac{1}{x} \right).$$

As x approaches zero through positive values, $1/x$ becomes positively infinite and, therefore, so does $\ln(1/x)$. Consequently, $-\ln(1/x)$ becomes negatively infinite, that is,

$$\lim_{x \to 0^+} \ln x = -\infty.$$

EXERCISES 8.1

In Exercises 1–24 find $f'(x)$ if $f(x)$ equals the given expression.

1 $\ln(9x+4)$

2 $\ln(x^4+1)$

3 $\ln(2-3x)^5$

4 $\ln(5x^2+1)^3$

5 $\ln\sqrt{7-2x^3}$

6 $\ln\sqrt[3]{6x+7}$

7 $\ln(3x^2-2x+1)$

8 $\ln(4x^3-x^2+2)$

9 $\ln\sqrt[3]{4x^2+7x}$

10 $\ln\sqrt{1-x^2}$

11 $x\ln x$

12 $\dfrac{x^2}{\ln x}$

13 $\ln\sqrt{x}+\sqrt{\ln x}$

14 $\ln x^3+(\ln x)^3$

15 $\dfrac{1}{\ln x}+\ln\left(\dfrac{1}{x}\right)$

16 $\ln\sqrt{\dfrac{4+x^2}{4-x^2}}$

17 $\ln(5x-7)^4(2x+3)^3$

18 $\ln\sqrt[3]{4x-5}(3x+8)^2$

19 $\ln\dfrac{\sqrt{x^2+1}}{(9x-4)^2}$

20 $\ln\dfrac{x^2(2x-1)^3}{(x+5)^2}$

21 $\ln\sqrt[3]{\dfrac{x^2-1}{x^2+1}}$

22 $\ln(\ln x)$

23 $\ln(x+\sqrt{x^2-1})$

24 $\ln(x+\sqrt{x^2+1})$

In Exercises 25–28 use implicit differentiation to find y'.

25 $3y-x^2+\ln xy=2$

26 $y^2+\ln(x/y)-4x+3=0$

27 $x\ln y-y\ln x=1$

28 $y^3+x^2\ln y=5x+3$

29 Find an equation of the tangent line to the graph of $y=x^2+\ln(2x-5)$ at the point on the graph with abscissa 3.

30 Find an equation of the tangent line to the graph of $y=x+\ln x$ which is perpendicular to the line $2x+6y=5$.

31 What is the difference between the graphs of $y=\ln(x^2)$ and $y=2\ln x$?

32 If $0<a<b$, show that the natural logarithmic function satisfies the hypotheses of the Mean Value Theorem (4.12) on $[a,b]$ and find a general formula for the number c in the conclusion of (4.12).

33 Find the points on the graph of $y=x^2+4\ln x$ at which the tangent line is parallel to the line $y-6x+3=0$.

34 Find an equation of the tangent line to the graph of $x^3-x\ln y+y^3=2x+5$ at the point $(2,1)$.

35 If $\ln 2.00\approx 0.6932$ use differentials to approximate $\ln 2.01$.

36 If $f(x)=\ln x$, find a formula for the nth derivative $f^{(n)}(x)$, where n is any positive integer.

37 The position function of a point moving on a coordinate line is given by $s(t)=t^2-4\ln(t+1)$, where $0\le t\le 4$. Find the velocity and acceleration at time t and describe the motion of the point during the time interval $[0,4]$.

38 The equation

$$t = \frac{1}{c(a-b)} \ln \frac{b(a-x)}{a(b-x)}$$

occurs in the study of certain chemical reactions, where x is the concentration of a substance at time t and a, b, c are constants. Prove that $dx/dt = c(a-x)(b-x)$.

39 Use Table III to sketch the graph of $y = \ln x$. Find the slope of the tangent line to the graph at the points with abscissas 1, 5, 10, 100, and 1000. What is true as the abscissa a of the point of tangency increases without bound? What is true if a approaches 0?

40 Sketch the graphs of the following equations.

(a) $y = \ln|x|$ (b) $y = |\ln x|$

Verify the formulas in Exercises 41–44.

41 $\int \ln x \, dx = x \ln x - x + C$

42 $\int (\ln x)^2 \, dx = x(\ln x)^2 - 2x \ln x + 2x + C$

43 $\int \frac{1}{\sqrt{x^2 + a^2}} \, dx = \ln (x + \sqrt{x^2 + a^2}) + C$

44 $\int \frac{1}{a^2 - x^2} \, dx = \frac{1}{2a} \ln \frac{a+x}{a-x} + C, a^2 > x^2$

8.2 THE NATURAL EXPONENTIAL FUNCTION

In Section 1 we proved that

$$\lim_{x \to \infty} \ln x = \infty \quad \text{and} \quad \lim_{x \to 0^+} \ln x = -\infty.$$

These facts are used in the proof of the following important result.

(8.8) Theorem

> To every real number x there corresponds a unique positive real number y such that $\ln y = x$.

Proof. First note that if $x = 0$, then $\ln 1 = 0$. Moreover, 1 is the only value of y such that $\ln y = 0$. (Why?)

If x is positive, then we may choose a number b such that

$$\ln 1 < x < \ln b.$$

Since ln is continuous, the Intermediate Value Theorem (2.37) guarantees the existence of a number y between 1 and b such that $\ln y = x$. Moreover, since ln is an increasing function, there is only one such number.

Finally, if x is negative, then there is a number $b > 0$ such that

$$\ln b < x < \ln 1$$

and as before, there is precisely one number y between b and 1 such that $\ln y = x$. This completes the proof.

Theorem (8.8) enables us to define a function from the set of real numbers to the set of positive real numbers, since to each x there corresponds a unique $y > 0$. The formal definition of this function is as follows.

(8.9) Definition

> The **natural exponential function**, denoted by **exp**, is defined by
>
> $$\exp x = y \quad \text{if and only if} \quad \ln y = x$$
>
> for all x, where $y > 0$.

The next theorem specifies the relationship between ln and exp.

(8.10) Theorem

> The natural logarithmic and natural exponential functions are inverse functions of one another.

Proof. According to Definition (1.23) we must prove

$$\text{(i)}\quad \ln(\exp x) = x \text{ for every } x, \quad \text{and}$$

$$\text{(ii)}\quad \exp(\ln y) = y \text{ for every } y > 0.$$

These statements follow immediately from Definition (8.9), for if $\exp x = y$, then $\ln y = x$, and substitution of $\exp x$ for y gives us (i). Similarly, if $\ln y = x$, then $\exp x = y$, and substitution of $\ln y$ for x gives us (ii).

As pointed out in Chapter 1, if two functions are inverses of one another, then their graphs are reflections through the line $y = x$. Hence the graph of $y = \exp x$ can be obtained by reflecting the graph of $y = \ln x$ through this line, as illustrated in Figure 8.4. Evidently,

$$\lim_{x \to \infty} \exp x = \infty \quad \text{and} \quad \lim_{x \to -\infty} \exp x = 0.$$

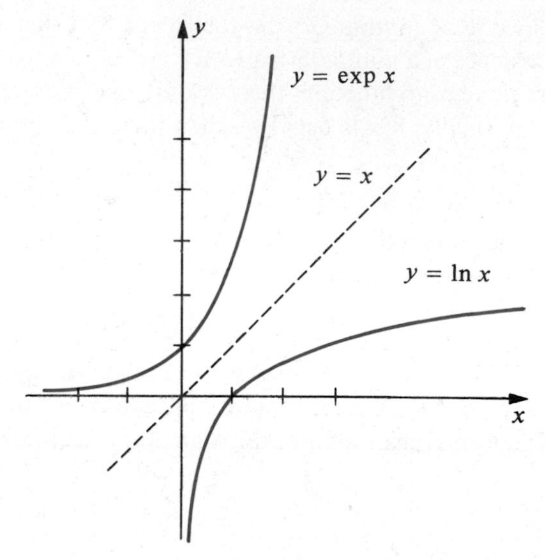

Figure 8.4

By Theorem (8.8) there is a unique positive real number whose natural logarithm is 1. This number is denoted by e, to commemorate the Swiss mathematician Leonard Euler (1707–1783) who was among the first to study its properties extensively.

(8.11) Definition

> The letter e denotes the unique positive real number such that $\ln e = 1$.

Several values of ln were calculated in Section 1. It can be shown, by means of the Trapezoidal Rule (5.47), that

$$\int_1^{2.7} \frac{1}{t}\,dt < 1 < \int_1^{2.8} \frac{1}{t}\,dt$$

(see Exercise 7 in Section 5.7). Consequently, by Definition (8.4),

$$\ln 2.7 < \ln e < \ln 2.8$$

which implies that

$$2.7 < e < 2.8.$$

Later, in Theorem (8.32), it will be shown that e may be expressed as the following limit:

$$e = \lim_{h \to 0} (1 + h)^{1/h}.$$

This formula can be used to approximate e to any degree of accuracy. In particular, to fifteen decimal places,

$$e \approx 2.718281828459045.$$

It has been proved that e is an irrational number.

If r is any rational number, then using (iii) of Theorem (8.7),

$$\ln e^r = r \ln e = r(1) = r.$$

This provides the motivation for the following definition of e^x.

(8.12) Definition

> If x is any real number, then e^x is the (unique) real number y such that $\ln y = x$.

Comparing (8.12) with Definition (8.9), we see that

$$e^x = \exp x$$

for every real number x. This is the reason for calling exp an *exponential* function. Indeed, it is often called the **exponential function with base** e. Perhaps the best way to remember Definition (8.12) is by means of the following statement:

$$e^x = y \quad \text{if and only if} \quad \ln y = x. \tag{8.13}$$

The fact that $\ln(\exp x) = x$ and $\exp(\ln y) = y$ for every x and every $y > 0$ (see Theorem (8.10)) may now be written in the form

$$\ln e^x = x \quad \text{and} \quad e^{\ln y} = y. \tag{8.14}$$

Since e^x was not defined as a power of e, it is not obvious that the usual laws of exponents hold. However, these laws may be established easily, as indicated in the next theorem.

(8.15) Theorem

> If p and q are real numbers and r is a rational number, then
>
> (i) $e^p e^q = e^{p+q}$.
>
> (ii) $\dfrac{e^p}{e^q} = e^{p-q}$.
>
> (iii) $(e^p)^r = e^{pr}$

Proof. Using Theorem (8.7) and (8.14),

$$\ln e^p e^q = \ln e^p + \ln e^q = p + q = \ln e^{p+q}.$$

Since the natural logarithmic function is one-to-one it follows that

$$e^p e^q = e^{p+q}.$$

This proves (i). The proofs for (ii) and (iii) are similar and are left as an exercise. It will be shown in Section 4 that (iii) is also true if r is irrational.

It is not difficult to show that if x is rational, then e^x has the same meaning as e raised to the power x. For example, we may use (8.14) as follows:

$$e^0 = e^{\ln 1} = 1$$
$$e^1 = e^{\ln e} = e.$$

Next, from Theorem (8.15),

$$e^2 = e^{1+1} = e^1 e^1 = ee$$
$$e^3 = e^{2+1} = e^2 e^1 = (ee)e$$

and in general, if n is a positive integer, e^n is a product of n factors, all equal to e. Negative exponents also have the usual properties, that is,

$$e^{-1} = e^{0-1} = \frac{e^0}{e^1} = \frac{1}{e}$$

and, in general,

$$e^{-n} = \frac{1}{e^n}$$

if n is a positive integer. Rational powers of e may also be interpreted as they are in elementary algebra.

The graph of $y = e^x$ is the same as that of $y = \exp x$ illustrated in Figure 8.4. Hereafter we shall use e^x instead of $\exp x$ to denote values of the natural exponential function.

In precalculus mathematics, graphs of equations such as $y = 2^x$ and $y = 3^x$ are sketched by *assuming* that these exponential expressions are defined for all real x and increase as x increases. Using this intuitive point of view, a rough sketch of the graph of $y = e^x$ can be obtained by sketching $y = (2.7)^x$.

It will be shown in Section 6 that the inverse function of a differentiable function is differentiable. Anticipating this result, let us find the derivative of the natural exponential function implicitly. Thus, if

$$y = e^x, \quad \text{then} \quad \ln y = x.$$

Differentiating implicitly gives us

$$D_x(\ln y) = D_x(x), \quad \text{or} \quad \frac{1}{y} D_x y = 1.$$

Multiplying both sides of the last equation by y we obtain

$$D_x y = y.$$

This shows that for every real number x,

(8.16) $$D_x e^x = e^x$$

that is, the natural exponential function is its own derivative!

Example 1 Find $f'(x)$ if $f(x) = x^2 e^x$.

Solution By the Product Rule (3.20),

$$\begin{aligned} f'(x) &= x^2 D_x e^x + e^x D_x x^2 \\ &= x^2 e^x + e^x(2x) \\ &= (x + 2)xe^x. \end{aligned}$$

We can employ the Chain Rule (3.36) in the usual way to generalize (8.16). Specifically, if we let $y = e^u$ and $u = g(x)$, where g is differentiable, then

$$D_x y = (D_u y)(D_x u) = e^u D_x u$$

that is,

(8.17) $$D_x e^u = e^u D_x u, \quad \text{where } u = g(x).$$

If $u = x$ then (8.17) reduces to (8.16).

Example 2 Find y' if $y = e^{\sqrt{x^2+1}}$.

Solution By (8.17),

$$\begin{aligned} D_x e^{\sqrt{x^2+1}} &= e^{\sqrt{x^2+1}} D_x \sqrt{x^2+1} = e^{\sqrt{x^2+1}} D_x (x^2+1)^{1/2} \\ &= e^{\sqrt{x^2+1}} \left(\tfrac{1}{2}\right)(x^2+1)^{-1/2}(2x) = e^{\sqrt{x^2+1}} \cdot \frac{x}{\sqrt{x^2+1}} \\ &= \frac{xe^{\sqrt{x^2+1}}}{\sqrt{x^2+1}}. \end{aligned}$$

Functions of the type considered in the next example arise in the branch of mathematics called *probability*.

Example 3 If $f(x) = e^{-x^2/2}$, find the local extrema of f. Discuss concavity, find the points of inflection, and sketch the graph of f.

Solution By (8.17),

$$f'(x) = e^{-x^2/2} D_x\left(-\frac{x^2}{2}\right) = e^{-x^2/2}(-2x/2) = -xe^{-x^2/2}.$$

Since $e^{-x^2/2}$ is always positive, the only critical number of f is 0. If $x < 0$, then $f'(x) > 0$ whereas if $x > 0$, then $f'(x) < 0$. It follows from the First Derivative Test that f has a local maximum at 0. The maximum value is $f(0) = e^{-0} = 1$.

Applying the Product Rule to $f'(x)$,

$$\begin{aligned} f''(x) &= -xe^{-x^2/2}(-x) - e^{-x^2/2} \\ &= e^{-x^2/2}(x^2 - 1), \end{aligned}$$

and hence the second derivative is zero at -1 and 1. If $-1 < x < 1$, then $f''(x) < 0$ and, by (4.16), the graph of f is concave downward in the open interval $(-1, 1)$. If $x < -1$ or $x > 1$, then $f''(x) > 0$ and, therefore, the graph is concave upward throughout the infinite intervals $(-\infty, -1)$ and $(1, \infty)$. Consequently $P(-1, e^{-1/2})$ and $Q(1, e^{-1/2})$ are points of inflection. Writing

$$f(x) = \frac{1}{e^{x^2/2}}$$

it is evident that as x increases numerically, $f(x)$ approaches 0. It is left as an exercise to prove that $\lim_{x \to \infty} f(x) = 0$ and $\lim_{x \to -\infty} f(x) = 0$; that is, the x-axis is a horizontal asymptote. The graph of f is sketched in Figure 8.5.

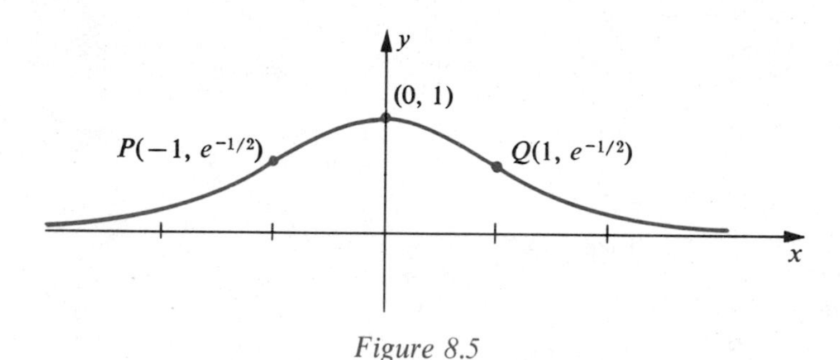

Figure 8.5

EXERCISES 8.2

In Exercises 1–26 find $f'(x)$ if $f(x)$ equals the given expression.

1 e^{-5x}

2 e^{3x}

3 e^{3x^2}

4 e^{1-x^3}

5 $\sqrt{1 + e^{2x}}$

6 $1/(e^x + 1)$

7 $e^{\sqrt{x+1}}$

8 xe^{-x}

9 x^2e^{-2x}

10 $\sqrt{e^{2x} + 2x}$

11 $e^x/(x^2 + 1)$

12 x/e^{x^2}

13 $(e^{4x} - 5)^3$

14 $(e^{3x} - e^{-3x})^4$

15 $e^{1/x} + 1/e^x$

16 $e^{\sqrt{x}} + \sqrt{e^x}$

17 $\dfrac{e^x - e^{-x}}{e^x + e^{-x}}$

18 e^{e^x}

19 $e^{-2x}\ln x$

20 $e^{x \ln x}$

21 $\ln\dfrac{e^x+1}{e^x-1}$

22 $\dfrac{\ln(e^x+1)}{\ln(e^x-1)}$

23 $\sqrt{\ln(e^{2x}+e^{-2x})}$

24 $\ln\sqrt{e^{2x}+e^{-2x}}$

25 $e^{\ln x}$

26 $\ln e^x$

In Exercises 27–30 use implicit differentiation to find y'.

27 $e^{xy} - x^3 + 3y^2 = 11$

28 $xe^y + 2x - \ln(y+1) = 3$

29 $y^3 + xe^y = 3x^2 - 10$

30 $xe^y - ye^x = 2$

31 Find an equation of the tangent line to the graph of $y = (x-1)e^x + 3\ln x + 2$ at the point $P(1, 2)$.

32 Find an equation of the tangent line to the graph of $y = x - e^{-x}$ which is parallel to the line $6x - 2y = 7$.

33 Show that the natural exponential function satisfies the hypotheses of the Mean Value Theorem (4.12) on every closed interval $[a, b]$, and find a general formula for the number c in the conclusion of (4.12).

34 Compare the graphs of $y = \sqrt{e^x}$ and $y = e^{\sqrt{x}}$.

35 Find a point on the graph of $y = e^{2x}$ at which the tangent line passes through the origin.

36 Find an equation of the normal line to the graph of $y = 4xe^{x^2-1}$ at the point $P(1, 4)$.

In Exercises 37–40 find the local extrema of f. Determine where f is increasing or decreasing, discuss concavity, find the points of inflection, and sketch the graph of f.

37 $f(x) = xe^x$

38 $f(x) = x^2e^{-2x}$

39 $f(x) = e^{-2x}$

40 $f(x) = xe^{-x}$

41 Use differentials to approximate the change in $f(x) = xe^{x^2}$ if x changes from 1.00 to 1.01. What is the approximate value of $f(1.01)$?

42 If $f(x) = e^{2x}$, find a formula for the nth derivative $f^{(n)}(x)$.

In each of Exercises 43–46 find values of c such that $y = e^{cx}$ is a solution of the given differential equation.

43 $y'' - 3y' + 2y = 0$

44 $y'' - 9y = 0$

45 $y''' - y'' - 4y' + 4y = 0$

46 $y''' - y'' - 6y' = 0$

47 Prove the following.

(a) $\lim_{x\to\infty} e^{-x^2/2} = 0$

(b) $\lim_{x\to-\infty} e^{-x^2/2} = 0$

48 Prove the following.

(a) $\lim_{x\to-\infty} e^x = 0$

(b) $\lim_{x\to\infty} e^x = \infty$

49 Prove (ii) and (iii) of Theorem (8.15).

50 Verify the following.

(a) $\int e^{ax} \ln x \, dx = \frac{1}{a} e^{ax} \ln x - \frac{1}{a} \int \frac{1}{x} e^{ax} \, dx$

(b) $\int x^n e^{ax} \, dx = \frac{1}{a} x^n e^{ax} - \frac{n}{a} \int x^{n-1} e^{ax} \, dx$

8.3 DIFFERENTIATION AND INTEGRATION

The formula for the derivative of $\ln u$ established in Section 1 can be extended as indicated in the next theorem.

(8.18) Theorem

If $u = g(x)$, where g is differentiable and $g(x) \neq 0$, then

$$D_x \ln |u| = \frac{1}{u} D_x u.$$

Proof. If $x < 0$, then $\ln |x| = \ln(-x)$ and by (8.6),

$$D_x \ln(-x) = \frac{1}{(-x)} D_x(-x) = \frac{1}{(-x)}(-1) = \frac{1}{x}.$$

Consequently

$$D_x \ln |x| = \frac{1}{x} \quad \text{for all } x \neq 0.$$

The Chain Rule may now be used to complete the proof.

Observe that if $u = g(x) > 0$, then Theorem (8.18) is the same as (8.6).

Example 1 Find $f'(x)$ if $f(x) = \ln |4 + 5x - 2x^3|$.

Solution Using (8.18) with $u = 4 + 5x - 2x^3$,

$$f'(x) = \frac{1}{4 + 5x - 2x^3}(5 - 6x^2) = \frac{5 - 6x^2}{4 + 5x - 2x^3}.$$

We may use differentiation formulas for ln to obtain rules for integration. In particular, since $D_x \ln |x| = 1/x$ it follows that

$$\int \frac{1}{x} \, dx = \ln |x| + C. \tag{8.19}$$

More generally, since

$$D_x \ln |g(x)| = \frac{1}{g(x)} g'(x)$$

the following integration formula holds:

(8.20)
$$\int \frac{1}{u}\, du = \ln |u| + C, \quad \text{where } u = g(x) \neq 0.$$

Of course, if $u > 0$, then the absolute value sign may be deleted.

When evaluating an indefinite integral by means of (8.20) we often use the method of substitution, letting $u = g(x)$ and $du = g'(x)\, dx$ for an appropriate $g(x)$.

Example 2 Find $\int \frac{x}{3x^2 - 5}\, dx$.

Solution If we let $u = 3x^2 - 5$, then $du = 6x\, dx$. We may arrive at the form (8.20) by introducing a factor 6 in the integrand as follows:

$$\begin{aligned} \int \frac{x}{3x^2 - 5}\, dx &= \frac{1}{6} \int \frac{1}{3x^2 - 5} 6x\, dx \\ &= \frac{1}{6} \int \frac{1}{u}\, du \\ &= \frac{1}{6} \ln |u| + C \\ &= \frac{1}{6} \ln |3x^2 - 5| + C. \end{aligned}$$

Another technique would be to replace the expression $x\, dx$ in the integral by $\frac{1}{6} du$ and then integrate.

Example 3 Evaluate $\int_2^4 \frac{1}{9 - 2x}\, dx$.

Solution Since $1/(9 - 2x)$ is continuous on $[2, 4]$, the definite integral exists. One method of evaluation consists of using an indefinite integral to find an antiderivative of $1/(9 - 2x)$. If we let $u = 9 - 2x$, then $du = -2\, dx$, and we may proceed as follows:

$$\begin{aligned} \int \frac{1}{9 - 2x}\, dx &= -\frac{1}{2} \int \frac{1}{9 - 2x} (-2)\, dx \\ &= -\frac{1}{2} \int \frac{1}{u}\, du \\ &= -\frac{1}{2} \ln |u| + C \\ &= -\frac{1}{2} \ln |9 - 2x| + C. \end{aligned}$$

Applying the Fundamental Theorem of Calculus (5.31),

$$\int_2^4 \frac{1}{9-2x}\,dx = -\frac{1}{2}\Big[\ln|9-2x|\Big]_2^4$$
$$= -\frac{1}{2}[\ln 1 - \ln 5]$$
$$= \frac{1}{2}\ln 5.$$

Another method is to use the same substitution in the *definite* integral and change the limits of integration. Referring to $u = 9 - 2x$ we see that if $x = 2$ then $u = 5$, and if $x = 4$ then $u = 1$. Consequently

$$\int_2^4 \frac{1}{9-2x}\,dx = -\frac{1}{2}\int_2^4 \frac{1}{9-2x}(-2)\,dx$$
$$= -\frac{1}{2}\int_5^1 \frac{1}{u}\,du$$
$$= -\frac{1}{2}\Big[\ln|u|\Big]_5^1$$
$$= -\frac{1}{2}[\ln 1 - \ln 5] = \frac{1}{2}\ln 5.$$

Example 4 Find $\int \frac{\sqrt{\ln x}}{x}\,dx.$

Solution If we let $u = \ln x$, then $du = (1/x)\,dx$ and

$$\int \frac{\sqrt{\ln x}}{x}\,dx = \int \sqrt{\ln x}\cdot\frac{1}{x}\,dx$$
$$= \int u^{1/2}\,du$$
$$= \frac{2}{3}u^{3/2} + C$$
$$= \frac{2}{3}(\ln x)^{3/2} + C.$$

It is easy to find integration formulas for the natural exponential function. Since $D_x e^x = e^x$ we may write

(8.21) $$\int e^x\,dx = e^x + C.$$

If we use the Chain Rule, then this can be extended to

(8.22)
$$\int e^u \, du = e^u + C$$

where $u = g(x)$ is differentiable.

Example 5 Find $\int \frac{e^{3/x}}{x^2} \, dx$.

Solution If we let $u = 3/x$, then $du = (-3/x^2) \, dx$ and the integrand may be put in the form (8.22) by introducing the factor -3. Doing this and compensating by multiplying the integral by $-1/3$,

$$\begin{aligned} \int \frac{e^{3/x}}{x^2} \, dx &= (-\tfrac{1}{3}) \int e^{3/x}\left(-\frac{3}{x^2}\right) dx \\ &= (-\tfrac{1}{3}) \int e^u \, du \\ &= (-\tfrac{1}{3})e^u + C \\ &= (-\tfrac{1}{3})e^{3/x} + C. \end{aligned}$$

Example 6 Find $\int_1^2 \frac{e^{3/x}}{x^2} \, dx$.

Solution An antiderivative was found in Example 5. Applying the Fundamental Theorem of Calculus (5.31),

$$\begin{aligned} \int_1^2 \frac{e^{3/x}}{x^2} \, dx &= (-\tfrac{1}{3})\Big[e^{3/x}\Big]_1^2 \\ &= (-\tfrac{1}{3})(e^{3/2} - e^3). \end{aligned}$$

This example can also be solved by using the method of substitution. If, as in Example 5 we let $u = 3/x$, then $du = -3/x^2$. Next we note that if $x = 1$, then $u = 3$, and if $x = 2$, then $u = 3/2$. Consequently,

$$\begin{aligned} \int_1^2 \frac{e^{3/x}}{x^2} \, dx &= (-\tfrac{1}{3}) \int_3^{3/2} e^u \, du \\ &= (-\tfrac{1}{3})e^u \Big]_3^{3/2} \\ &= (-\tfrac{1}{3})(e^{3/2} - e^3). \end{aligned}$$

Example 7 Find the area of the region bounded by the graphs of the equations $y = e^x$, $y = \sqrt{x}$, $x = 0$, and $x = 1$.

Solution The region and a typical rectangle of the type considered in Chapter 6 are shown in Figure 8.6. The area A is given by

$$\begin{aligned} A &= \lim_{\|P\| \to 0} \sum_i (e^{w_i} - \sqrt{w_i})\Delta x_i \\ &= \int_0^1 (e^x - \sqrt{x})\,dx \\ &= \left[e^x - \frac{2}{3}x^{3/2}\right]_0^1 \\ &= \left(e - \frac{2}{3}\right) - (e^0 - 0) = e - \frac{5}{3}. \end{aligned}$$

If an approximation is desired, then to the nearest thousandth

$$A \approx 2.718 - 1.666 = 1.052.$$

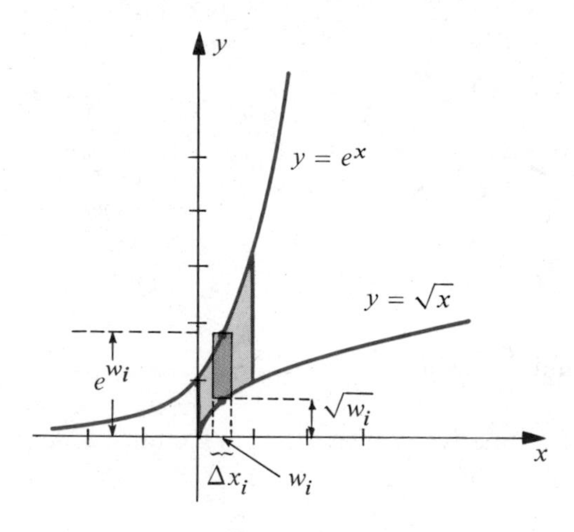

Figure 8.6

Given $y = f(x)$ it is sometimes convenient to find $D_x y$ by a process called **logarithmic differentiation**. This method is especially useful if $f(x)$ involves complicated products, quotients, or powers. Later in the text we shall also apply it to expressions such as $y = (x^2 + 1)^x$. The process may be outlined as follows where, at the outset, it is assumed that $f(x) > 0$.

Steps in Logarithmic Differentiation

1.	$y = f(x)$	(Given)
2.	$\ln y = \ln f(x)$	(Take natural logarithms and simplify)
3.	$D_x[\ln y] = D_x[\ln f(x)]$	(Differentiate implicitly)
4.	$\frac{1}{y}D_x y = D_x[\ln f(x)]$	(By (8.6))
5.	$D_x y = f(x)D_x[\ln f(x)]$	(Multiply by $y = f(x)$)

Of course, to complete the solution it is necessary to differentiate $\ln f(x)$ at some stage after Step 3. If $f(x) < 0$ for some x, then Step 2 is invalid, since $\ln f(x)$ is undefined. However, in this event we could replace Step 1 by $|y| = |f(x)|$ and take natural logarithms, obtaining $\ln|y| = \ln|f(x)|$. If we now differentiate implicitly and use (8.18) we again arrive at Step 4. Thus negative values of $f(x)$ do not change the outcome and it is unnecessary to be concerned whether $f(x)$ is positive or negative. The method should not be used to find $f'(a)$ if $f(a) = 0$, since $\ln 0$ is undefined.

Example 8 Use logarithmic differentiation to find $D_x y$ if $y = \dfrac{(5x-4)^3}{\sqrt{2x+1}}$.

Solution As in the outline we begin by taking the natural logarithm of each side and simplifying, obtaining

$$\ln y = 3\ln(5x-4) - \left(\frac{1}{2}\right)\ln(2x+1).$$

Differentiating both sides with respect to x we obtain

$$\begin{aligned}\frac{1}{y}D_x y &= 3\frac{1}{5x-4}(5) - \left(\frac{1}{2}\right)\frac{1}{2x+1}(2)\\ &= \frac{25x+19}{(5x-4)(2x+1)}.\end{aligned}$$

Finally, multiplying both sides of the last equation by y (that is, by $(5x-4)^3/\sqrt{2x+1}$),

$$\begin{aligned}D_x y &= \frac{25x+19}{(5x-4)(2x+1)} \cdot \frac{(5x-4)^3}{\sqrt{2x+1}}\\ &= \frac{(25x+19)(5x-4)^2}{(2x+1)^{3/2}}.\end{aligned}$$

This result may be checked by applying the quotient rule to y.

EXERCISES 8.3

Evaluate the integrals in Exercises 1–28.

1 $\displaystyle\int \frac{x}{x^2+1}\,dx$

2 $\displaystyle\int \frac{1}{8x+3}\,dx$

3 $\displaystyle\int \frac{1}{7-5x}\,dx$

4 $\displaystyle\int \frac{x^3}{x^4-5}\,dx$

5 $\displaystyle\int \frac{x-2}{x^2-4x+9}\,dx$

6 $\displaystyle\int \frac{(2+\ln x)^3}{x}\,dx$

7 $\displaystyle\int \frac{x^2}{x^3+1}\,dx$

8 $\displaystyle\int_1^2 \frac{3x}{x^2+4}\,dx$

9 $\int_{-2}^{1} \frac{1}{2x+7}\,dx$

10 $\int_{-1}^{0} \frac{1}{4-5x}\,dx$

11 $\int_{1}^{4} \frac{2}{\sqrt{x}(\sqrt{x}+4)}\,dx$

12 $\int \frac{x-1}{3x^2-6x+2}\,dx$

13 $\int (x+e^{5x})\,dx$

14 $\int (1+e^{-3x})\,dx$

15 $\int \frac{\ln x}{x}\,dx$

16 $\int \frac{1}{x(\ln x)^2}\,dx$

17 $\int_{1}^{3} e^{-4x}\,dx$

18 $\int_{0}^{1} e^{2x+3}\,dx$

19 $\int \frac{e^{\sqrt{x}}}{\sqrt{x}}\,dx$

20 $\int xe^{x^2}\,dx$

21 $\int \frac{(e^x+1)^2}{e^x}\,dx$

22 $\int \frac{e^x}{(e^x+1)^2}\,dx$

23 $\int \frac{e^x-e^{-x}}{e^x+e^{-x}}\,dx$

24 $\int \frac{e^x}{e^x+1}\,dx$

25 $\int \frac{1}{x^2+2x+1}\,dx$

26 $\int \frac{(x^2-4)^2}{2x}\,dx$

27 $\int \frac{2x^2-5x-7}{x-3}\,dx$

28 $\int \frac{x^2+3x+1}{x}\,dx$

In each of Exercises 29–32 find the area of the region bounded by the graphs of the given equations.

29 $y = e^{2x},\ y = 0,\ x = 0,\ x = \ln 3$

30 $xy = 1,\ y = 0,\ x = 1,\ x = e$

31 $y = e^{-x},\ xy = 1,\ x = 1,\ x = 2$

32 $y = e^{-2x},\ y = -e^x,\ x = 0,\ x = 2$

33 The region bounded by the graphs of $y = e^{-x^2}$, $y = 0$, $x = 0$, and $x = 1$ is revolved about the y-axis. Find the volume of the resulting solid.

34 The region bounded by the graphs of $y = 1/\sqrt{x}$, $y = 0$, $x = 1$, and $x = 4$ is revolved about the x-axis. Find the volume of the resulting solid.

35 A particle moves on a straight line such that its velocity at time t is e^{-3t}. If the particle is at the origin at $t = 0$, how far does it travel during the time interval $[0, 2]$?

36 A solid has, for its base, the region in the xy-plane bounded by the graphs of $y = e^{2x}$, $y = e^x$, and $x = 1$. Find the volume of the solid if every cross section by a plane perpendicular to the x-axis is a square.

In Exercises 37 and 38 do the following.
(a) Set up an integral for finding the arc length L of the graph of the given equation from A to B.
(b) Write out the formula for approximating L by means of the Trapezoidal Rule (5.47) with $n = 5$.
(c) If a calculator is available, find the value of the approximation given by the formula in part (b).

37 $y = e^x;\ A(0, 1),\ B(1, e)$

38 $y = e^{-x^2};\ A(0, 1),\ B(1, e^{-1})$

In Exercises 39–46 find $D_x y$ by logarithmic differentiation.

39 $y = (5x + 2)^3(6x + 1)^2$

40 $y = \sqrt{4x + 7}(x - 5)^3$

41 $y = \dfrac{(x^2 + 3)^5}{\sqrt{x + 1}}$

42 $y = (x + 1)^2(x + 2)^3(x + 3)^4$

43 $y = \sqrt[3]{2x + 1}(4x - 1)^2(3x + 5)^4$

44 $y = \dfrac{(2x - 3)^2}{\sqrt{x + 1}(7x + 2)^3}$

45 $y = \sqrt{(3x^2 + 2)\sqrt{6x - 7}}$

46 $y = \dfrac{(x^2 + 3)^{2/3}(3x - 4)^4}{\sqrt{x}}$

In Exercises 47–50 find $f'(x)$.

47 $f(x) = \ln|3 - 2x|$

48 $f(x) = \ln|4 - 5x^3|^2$

49 $f(x) = \ln|1 - e^{-2x}|^3$

50 $f(x) = \dfrac{1}{\ln|2x - 15|}$

8.4 GENERAL EXPONENTIAL AND LOGARITHMIC FUNCTIONS

We shall now give meaning to the expression a^x, where $a > 0$ and x is any real number. If the exponent is a *rational* number r, then applying (8.14) and (iii) of Theorem (8.7) we see that

$$a^r = e^{\ln a^r} = e^{r \ln a}.$$

We shall use this formula as our motivation for the following definition of a^x.

(8.23) **Definition**

If x is any real number and a is any positive real number, then

$$a^x = e^{x \ln a}.$$

The function f defined by $f(x) = a^x$ is called the **exponential function with base** $\boldsymbol{a}$. Since e^x is positive for all x, so is a^x. In order to approximate values of a^x we may refer to tables of logarithmic and exponential functions, or use a calculator.

It is now possible to prove that the Law of Logarithms stated in (iii) of Theorem (8.7) is also true for irrational exponents. Thus, if u is any *real* number, then by Definition (8.23) and (8.14),

(8.24) $$\ln a^u = \ln e^{u \ln a} = u \ln a.$$

We may also prove that the Laws of Exponents stated in (8.1) are true for all real exponents. For example, to show that $a^u a^v = a^{u+v}$ if u and v are real we use Definition (8.23) and (i) of Theorem (8.15) as follows:

$$\begin{aligned} a^u a^v &= e^{u \ln a} e^{v \ln a} \\ &= e^{u \ln a + v \ln a} \\ &= e^{(u+v) \ln a} \\ &= a^{u+v}. \end{aligned}$$

In order to prove that $(a^u)^v = a^{uv}$ we first use Definition (8.23) with a^u in place of a and $v = x$ to write

$$(a^u)^v = e^{v \ln a^u}.$$

Applying (8.24) and Definition (8.23) we obtain

$$(a^u)^v = e^{vu \ln a} = a^{vu} = a^{uv}.$$

The proofs of the remaining laws in (8.1) are left as exercises.

It is easy to find the derivative of a^x by applying (8.23) and (8.17) as follows:

$$\begin{aligned} D_x(a^x) = D_x(e^{x \ln a}) &= e^{x \ln a} D_x(x \ln a) \\ &= e^{x \ln a}(\ln a). \end{aligned}$$

Again using Definition (8.23) we obtain

(8.25) $$D_x(a^x) = a^x \ln a.$$

It should be noted that if $a = e$, then (8.25) reduces to (8.16). If $a > 1$, then $\ln a > 0$ and, therefore, $D_x(a^x) > 0$. Hence a^x is increasing on the interval $(-\infty, \infty)$ if $a > 1$. If $0 < a < 1$, then $\ln a < 0$ and $D_x(a^x) < 0$; that is, a^x is decreasing for all x.

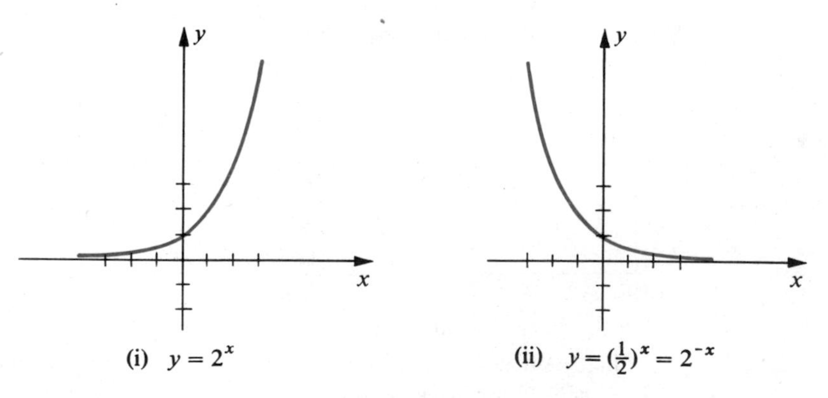

Figure 8.7

The graphs of $y = 2^x$ and $y = (1/2)^x = 2^{-x}$ illustrated in Figure 8.7 may be sketched by plotting some representative points. The graph of the equation $y = a^x$ has the general shape illustrated in (i) or (ii) of Figure 8.7 depending on whether $a > 1$ or $0 < a < 1$, respectively.

The Chain Rule (3.36) may be used to extend (8.25) as follows:

(8.26) $$D_x(a^u) = (a^u \ln a)D_x u, \quad \text{where } u = g(x).$$

Example 1 Find y' if $y = 3^{\sqrt{x}}$.

Solution Using (8.26) with $a = 3$ and $u = \sqrt{x}$,

$$\begin{aligned} y' &= (3^{\sqrt{x}} \ln 3)\left(\frac{1}{2}x^{-1/2}\right) \\ &= \frac{3^{\sqrt{x}} \ln 3}{2\sqrt{x}}. \end{aligned}$$

If $u = g(x)$ it is very important to distinguish between expressions of the form a^u and u^a where a is a real number. In order to differentiate a^u we use (8.26), whereas for u^a the Power Rule must be employed, as illustrated in the next example.

Example 2 Find y' if $y = (x^2 + 1)^{10} + 10^{x^2+1}$.

Solution Using the Power Rule for Functions (3.38) and (8.26), we obtain

$$\begin{aligned} y' &= 10(x^2 + 1)^9(2x) + 10^{x^2+1} \ln 10(2x) \\ &= 20x\,[(x^2 + 1)^9 + 10^{x^2} \ln 10]. \end{aligned}$$

The integration formula

(8.27) $$\int a^x \, dx = \left(\frac{1}{\ln a}\right)a^x + C$$

may be verified by showing that the integrand is the derivative of the expression on the right side of the equation.

It follows from (8.26) that

(8.28) $$\int a^u \, du = \left(\frac{1}{\ln a}\right)a^u + C, \quad \text{where } u = g(x).$$

Example 3 Evaluate $\int 3^{x^2} x \, dx$.

Solution If we let $u = x^2$, then $du = 2x\,dx$ and we may proceed as follows:

$$\begin{aligned}\int 3^{x^2} x\,dx &= \frac{1}{2}\int 3^{x^2}(2x)\,dx \\ &= \frac{1}{2}\int 3^u\,du \\ &= \left(\frac{1}{2}\right)\left(\frac{1}{\ln 3}\right)3^u + C \\ &= \left(\frac{1}{2\ln 3}\right)3^{x^2} + C.\end{aligned}$$

If $a \neq 1$ and $f(x) = a^x$, then f is a one-to-one function. Its inverse function is denoted by $\log_a$ and is called the *logarithmic function with base a*. Thus

(8.29) $$y = \log_a x \quad \text{if and only if} \quad x = a^y.$$

This is in agreement with formula (8.2) used in precalculus mathematics. The expression $\log_a x$ is called *the logarithm of x with base a*. Using this terminology, natural logarithms are logarithms with base e, that is,

$$\ln x = \log_e x.$$

The fact that the Laws of Logarithms (8.3) are true is left as an exercise.

In order to obtain the relationship between $\log_a$ and ln, consider $y = \log_a x$ or, equivalently, $x = a^y$. Taking the natural logarithm of both sides of the last equation gives us $\ln x = y \ln a$, or $y = \ln x/\ln a$. This proves that

$$\log_a x = \frac{\ln x}{\ln a}.$$

Differentiating both sides of the last equation gives us

(8.30) $$D_x(\log_a x) = \frac{1}{x \ln a}.$$

Using the Chain Rule (3.36) and generalizing to absolute values as in (8.18), we obtain

(8.31) $$D_x \log_a |u| = \frac{1}{u \ln a} D_x u$$

where $u = g(x)$ is differentiable and $g(x) \neq 0$ for all x.

Example 4 Find $f'(x)$ if $f(x) = \log_{10}\sqrt[3]{(2x + 5)^2}$.

Solution We begin by using the analogue of (iii) of Theorem (8.7) for logarithms with any base together with the fact that $(2x + 5)^2 = |2x + 5|^2$. Thus

$$f(x) = \frac{1}{3}\log_{10}|2x + 5|^2$$
$$= \frac{2}{3}\log_{10}|2x + 5|.$$

Employing (8.31),

$$f'(x) = \left(\frac{2}{3}\right)\frac{1}{(2x + 5)\ln 10}(2)$$
$$= \frac{4}{(6x + 15)\ln 10}.$$

Since irrational exponents have been given meaning, we may now consider the **general power function** f defined by $f(x) = x^c$ where c is any real number. If c is irrational, then by definition, the domain of f is the set of positive real numbers. Using (8.23), (8.17), and (8.5),

$$D_x x^c = D_x(e^{c \ln x})$$
$$= e^{c \ln x} D_x(c \ln x)$$
$$= e^{c \ln x}\left(\frac{c}{x}\right)$$
$$= x^c\left(\frac{c}{x}\right)$$
$$= cx^{c-1}.$$

This proves that the Power Rule (3.12) is true for irrational as well as rational exponents. The Power Rule for Functions (3.38) may also be extended to irrational exponents.

Example 5 Find y' if (a) $y = x^{\sqrt{2}}$; (b) $y = (1 + e^{2x})^{\pi}$.

Solution

(a) $$y' = \sqrt{2}x^{\sqrt{2}-1}$$

(b) $$y' = \pi(1 + e^{2x})^{\pi-1}D_x(1 + e^{2x})$$
$$= \pi(1 + e^{2x})^{\pi-1}(2e^{2x})$$
$$= 2\pi e^{2x}(1 + e^{2x})^{\pi-1}$$

Example 6 Find $D_x y$ if $y = x^x$, where $x > 0$.

Solution Since the exponent in x^x is a variable, the Power Rule (3.12) may not be used. Similarly, (8.25) is not applicable since the base a is not a fixed real number.

However, by Definition (8.23), $x^x = e^{x \ln x}$ for all $x > 0$, and hence

$$\begin{aligned} D_x(x^x) &= D_x(e^{x \ln x}) \\ &= e^{x \ln x} D_x(x \ln x) \\ &= e^{x \ln x}\left[x\left(\frac{1}{x}\right) + (1) \ln x\right] \\ &= x^x(1 + \ln x). \end{aligned}$$

Another way of attacking this problem is to use the method of logarithmic differentiation introduced in the previous section. In this case we take the natural logarithm of both sides of the equation $y = x^x$ and then differentiate implicitly as follows:

$$\begin{aligned} \ln y &= \ln x^x = x \ln x \\ D_x(\ln y) &= D_x(x \ln x) \\ \frac{1}{y} D_x y &= 1 + \ln x \\ D_x y &= y(1 + \ln x) = x^x(1 + \ln x). \end{aligned}$$

We shall conclude this section by expressing the number e as a limit. This was not done in Section 2 because the proof of the next theorem makes use of (8.24) and a general exponential function.

(8.32) Theorem (i) $\lim_{h \to 0} (1 + h)^{1/h} = e$ (ii) $\lim_{n \to \infty} \left(1 + \frac{1}{n}\right)^n = e$

Proof. Applying the derivative formula (3.1) to $f(x) = \ln x$ and using Laws of Logarithms gives us

$$\begin{aligned} f'(x) &= \lim_{h \to 0} \frac{\ln (x + h) - \ln x}{h} \\ &= \lim_{h \to 0} \frac{1}{h} \ln \frac{x + h}{x} \\ &= \lim_{h \to 0} \ln \left(1 + \frac{h}{x}\right)^{1/h}. \end{aligned}$$

Since $f'(x) = 1/x$ we have, for $x = 1$,

$$1 = \lim_{h \to 0} \ln (1 + h)^{1/h}.$$

We next observe, from (8.12), that

$$(1 + h)^{1/h} = e^{\ln (1 + h)^{1/h}}.$$

Since the natural exponential function is continuous at 1, it follows from Theorem (2.31) that

$$\begin{aligned}\lim_{h\to 0}(1+h)^{1/h} &= \lim_{h\to 0}\,[e^{\ln(1+h)^{1/h}}]\\ &= e^{[\lim_{h\to 0}\ln(1+h)^{1/h}]}\\ &= e^1 = e.\end{aligned}$$

This establishes part (i) of the theorem. The limit in part (ii) may be obtained by introducing the change of variable $n = 1/h$ (see Exercise 43).

The formulas in Theorem (8.32) are very important in mathematics. Indeed, they are often used to *define* the number *e*. If a calculator is available the student will find it instructive to calculate $(1+h)^{1/h}$ for numerically small values of h. Some approximate values are given in the following table.

h	$(1+h)^{1/h}$	h	$(1+h)^{1/h}$
0.01	2.704814	−0.01	2.731999
0.001	2.716924	−0.001	2.719642
0.0001	2.718146	−0.0001	2.718418
0.00001	2.718268	−0.00001	2.718295
0.000001	2.718280	−0.000001	2.718283

To five decimal places, $e \approx 2.71828$.

EXERCISES 8.4

In Exercises 1–22 find $f'(x)$ if $f(x)$ is defined by the given expression.

1 7^x

2 5^{-x}

3 8^{x^2+1}

4 $9^{\sqrt{x}}$

5 $\log_{10}(x^4 + 3x^2 + 1)$

6 $\log_3|6x - 7|$

7 5^{3x-4}

8 3^{2-x^2}

9 $(x^2 + 1)10^{1/x}$

10 $(10^x + 10^{-x})^{10}$

11 $7^{\sqrt{x^4+9}}$

12 $x/(6^x + x^6)$

13 $\log_{10}(3x^2 + 2)^5$

14 $\log_{10}\sqrt{x^2 + 1}$

15 $\log_5\left|\dfrac{6x+4}{2x-3}\right|$

16 $\log_{10}\left|\dfrac{1-x^2}{2-5x^3}\right|$

17 $\log_{10}\ln x$

18 $\ln\log_{10} x$

19 $x^e + e^x$

20 $x^\pi \pi^x$

21 $(x + 1)^x$

22 x^{x^2+4}

Evaluate the integrals in Exercises 23–34.

23 $\int 10^{3x}\,dx$

24 $\int 5^{-2x}\,dx$

25 $\int x(3^{-x^2})\,dx$

26 $\int \frac{(2^x + 1)^2}{2^x}\,dx$

27 $\int \frac{2^x}{2^x + 1}\,dx$

28 $\int \frac{3^x}{\sqrt{3^x + 4}}\,dx$

29 $\int_1^2 5^{-2x}\,dx$

30 $\int_{-1}^1 2^{3x-1}\,dx$

31 $\int x^2 2^{x^3}\,dx$

32 $\int \frac{10^{\sqrt{x}}}{\sqrt{x}}\,dx$

33 $\int \frac{1}{x\log_{10} x}\,dx$

34 $\int \frac{10^x + 10^{-x}}{10^x - 10^{-x}}\,dx$

35 Find the area of the region bounded by the graphs of $y = 2^x$, $x + y = 1$, and $x = 1$.

36 The region under the graph of $y = 3^{-x}$ from $x = 1$ to $x = 2$ is revolved about the x-axis. Find the volume of the resulting solid.

37 Find equations of the tangent and normal lines to the graph of the equation $y = x3^x$ at the point $P(1, 3)$.

38 Find the abscissa of the point on the graph of $y = 2^x$ at which the tangent line is parallel to the line with equation $2y - 8x + 3 = 0$.

39 Prove that for all real numbers u and v, and any $a > 0$, $b > 0$:

$$(ab)^u = a^u b^u; \quad \frac{a^u}{a^v} = a^{u-v}$$

40 Prove the Laws of Logarithms (8.3).

In Exercises 41 and 42 sketch the graphs of the two given equations on the same coordinate plane and discuss the relationship between the graphs.

41 $y = 10^x$, $y = \log_{10} x$

42 $y = 2^x$, $y = \log_2 x$

43 Establish (ii) of Theorem (8.32) by using the limit in part (i) and the change of variable $n = 1/h$.

8.5 LAWS OF GROWTH AND DECAY

Suppose a physical quantity varies with time and that the magnitude of the quantity at time t may be approximated by $q(t)$, where q is a differentiable function. The derivative $q'(t)$ may then be used to measure the rate of change of $q(t)$ with respect to time. In many applications, the rate of change at a given instant is directly proportional to the magnitude of the quantity at that instant. The number of bacteria in certain cultures behaves in this way. If the number of bacteria is small, then the rate of increase is small; however, as the number of bacteria increases, the rate of increase becomes larger. In many cases, this change in the rate of change is very great and we may regard the result as a microscopic view of a population explosion. The decay of a radioactive substance obeys a similar law, since as the amount of matter decreases, the rate of decay—that is, the amount of radiation—also decreases. As a final illustration, suppose an electrical condenser is allowed to discharge. If there is a large charge on the condenser at the outset, then the rate of discharge is also large, but as thc charge becomes weaker, the condenser discharges less rapidly. There are many other examples of this phenomenon. All of them are special cases of the general situation described below.

Suppose q is a differentiable function of t and the rate of change of q with respect to t is directly proportional to $q(t)$, where $q(t) > 0$ for all t under consideration. This fact can be expressed by means of the differential equation

$$q'(t) = cq(t) \tag{8.33}$$

where c is a real number. An equivalent equation is

$$\frac{q'(t)}{q(t)} = c.$$

If the interval $[0, x]$ is in the domain of q'/q, then

$$\int_0^x \frac{q'(t)}{q(t)}\,dt = \int_0^x c\,dt.$$

Since $\ln q(t)$ is an antiderivative of $q'(t)/q(t)$ it follows from the Fundamental Theorem of Calculus (5.31) that

$$\ln q(t)\Big]_0^x = ct\Big]_0^x$$

which can also be written in any of the following equivalent forms:

$$\ln q(x) - \ln q(0) = cx$$

$$\ln \frac{q(x)}{q(0)} = cx$$

$$\frac{q(x)}{q(0)} = e^{cx}$$

$$q(x) = q(0)e^{cx}.$$

If the variable t is used instead of x, then the last equation takes on the form

$$q(t) = q(0)e^{ct}. \tag{8.34}$$

We have proved that if the rate of change of $q(t)$ is directly proportional to $q(t)$, then q may be expressed in terms of an exponential function. If q increases with t, then (8.34) is called the **law of growth** of q, whereas if q decreases, it is referred to as the **law of decay** of q. The number $q(0)$ is sometimes called the **initial value** of q.

In applied problems the preceding discussion is often phrased in terms of a differential equation in which the expression $q(t)$ is suppressed. Specifically, if we let $y = q(t)$, then (8.33) may be written

$$\frac{dy}{dt} = cy$$

or, in terms of differentials, as

$$dy = (cy)\,dt.$$

Dividing both sides of this equation by y we obtain

$$\frac{1}{y}\,dy = c\,dt. \tag{8.35}$$

Since it is possible to separate the variables y and t in the sense that they can be placed on opposite sides of the equals sign, as in (8.35), the differential equation $dy/dt = cy$ is called a **separable differential equation**. Such equations will be studied in more detail in Chapter 19, where it is shown that solutions can be found by integrating both sides of the "separated" equation. Thus, from (8.35),

$$\int \frac{1}{y}\,dy = \int c\,dt$$

and assuming $y > 0$,

$$\ln y = ct + d$$

where the two constants of integration have been combined into the one constant d. It follows that

$$y = e^{ct+d} = e^d e^{ct}. \tag{8.36}$$

If y_0 denotes the initial value of y (that is, the value corresponding to $t = 0$), then letting $t = 0$ in the last equation gives us

$$y_0 = e^d e^0 = e^d$$

and hence the solution (8.36) may be written as

$$y = y_0 e^{ct}$$

which is an alternate form of (8.34).

Example 1 The number of bacteria in a certain culture increases from 600 to 1800 in 2 hours. Assuming that the exponential law of growth holds, find a formula for the number of bacteria in the culture at any time t. What is the number of bacteria at the end of 4 hours?

Solution Let $t = 0$ correspond to the time at which 600 bacteria are present in the culture and let $q(t)$ denote the number of bacteria after t hours. Thus $q(0) = 600$ and $q(2) = 1800$. Since the growth function is exponential, $q(t)$ has the form (8.34) with $q(0) = 600$; that is,

$$q(t) = 600e^{ct}.$$

Letting $t = 2$ we obtain

$$1800 = 600e^{2c}, \quad \text{or} \quad e^{2c} = 3.$$

Consequently

$$2c = \ln 3, \quad \text{or} \quad c = \tfrac{1}{2}\ln 3.$$

Thus the formula for $q(t)$ is

$$q(t) = 600e^{(1/2 \ln 3)t}.$$

If we use the fact that $e^{\ln 3} = 3$ (see (8.12)), then this law of growth can be expressed in terms of an exponential function with base 3 as follows:

$$q(t) = 600(e^{\ln 3})^{t/2} = 600(3)^{t/2}.$$

In particular, at the end of 4 hours,

$$q(4) = 600(3)^{4/2} = 600(9) = 5400.$$

Example 2 Radium decays exponentially and has a half-life of approximately 1600 years; that is, given any quantity, one-half of it will disintegrate in 1600 years. Find a formula for the amount $q(t)$ remaining from 50 milligrams of pure radium after t years. When will there be 20 mg left?

Solution By hypothesis, $q(0) = 50$ and $q(1600) = 25$. Applying (8.34),

$$q(t) = 50e^{ct}.$$

Letting $t = 1600$ we obtain

$$25 = 50e^{1600c}, \quad \text{or} \quad e^{1600c} = \frac{1}{2}.$$

Hence

$$1600c = \ln(1/2) = \ln 1 - \ln 2 = -\ln 2$$

and

$$c = -\frac{\ln 2}{1600}.$$

Consequently

$$q(t) = 50e^{-(\ln 2/1600)t}.$$

As in Example 1 we may write this formula in terms of a different base as follows:

$$q(t) = 50(e^{\ln 2})^{-t/1600}$$

or

$$q(t) = 50(2)^{-t/1600}.$$

To find the value of t at which $q(t) = 20$ it is necessary to solve the equation

$$20 = 50(2)^{-t/1600} \quad \text{or} \quad 2^{t/1600} = 5/2.$$

Taking the natural logarithm of both sides,

$$\frac{t}{1600}\ln 2 = \ln\frac{5}{2}, \quad \text{or} \quad t = \frac{1600\ln(5/2)}{\ln 2}.$$

If an approximation is desired, then from Table III,

$$t \approx 1600(0.916)/(0.693) \approx 2{,}115 \text{ years.}$$

Example 3 According to Newton's Law of Cooling, the rate at which an object cools is directly proportional to the difference in temperature between the object and the surrounding medium. If a certain object cools from 125° to 100° in half an hour when surrounded by air at a temperature of 75°, find its temperature at the end of another half hour.

Solution If y denotes the temperature of the object after t hours of cooling, then by Newton's Law,

$$\frac{dy}{dt} = c(y - 75)$$

where c is a constant. Separating variables we obtain

$$\frac{1}{y - 75}\,dy = c\,dt.$$

Integrating both sides leads to

$$\ln(y - 75) = ct + b$$

where b is a constant or, equivalently,

$$y - 75 = e^{ct+b} = e^b e^{ct}.$$

If we let $k = e^b$, then the last formula may be written

$$y = ke^{ct} + 75.$$

Since $y = 125$ when $t = 0$ we see that

$$125 = ke^0 + 75 = k + 75 \quad \text{or} \quad k = 50.$$

Hence

$$y = 50e^{ct} + 75.$$

Since $y = 100$ when $t = 1/2$ hour,

$$100 = 50e^{c/2} + 75 \quad \text{or} \quad e^{c/2} = 25/50 = 1/2.$$

This implies that

$$c = 2\ln(1/2) = \ln(1/4).$$

Consequently, the temperature y after t hours is given by

$$y = 50e^{t\ln(1/4)} + 75.$$

In particular, if $t = 1$,

$$\begin{aligned} y &= 50e^{\ln(1/4)} + 75 \\ &= 50(1/4) + 75 = 87.5°. \end{aligned}$$

In many cases, growth circumstances are more stable than that given by (8.34). Typical situations which occur involve populations, sales of products, and values of assets. In biology, a function G is sometimes used as follows to estimate the size of a quantity at time t:

(8.37) $$G(t) = ke^{(-Ae^{-Bt})}$$

where k, A, and B are positive constants. The function G is always positive and increasing, but has a limit as t increases without bound. The graph of G is called a **Gompertz growth curve.**

Example 4 Discuss and sketch the graph of the function G given by (8.37).

Solution We first observe that the y-intercept is $G(0) = ke^{-A}$ and that $G(t) > 0$ for all t. Differentiating we obtain

$$\begin{aligned} G'(t) &= ke^{(-Ae^{-Bt})}D_t(-Ae^{-Bt}) \\ &= ABke^{(-Bt-Ae^{-Bt})} \end{aligned}$$

and

$$\begin{aligned} G''(t) &= ABke^{(-Bt-Ae^{-Bt})}D_t(-Bt - Ae^{-Bt}) \\ &= ABk(-B + ABe^{-Bt})e^{-Bt-Ae^{-Bt}}. \end{aligned}$$

Since $G'(t) > 0$ for all t, the function G is increasing on $[0, \infty)$. The second derivative $G''(t)$ is zero if

$$-B + ABe^{-Bt} = 0, \quad \text{or} \quad e^{Bt} = A.$$

Solving the latter equation for t gives us $t = (1/B)\ln A$, which is a critical number for the function G'. It is left as an exercise to show that at this time the rate of growth G' has a maximum value Bk/e. It is also left to the reader to show that

$$\lim_{t\to\infty} G'(t) = 0 \quad \text{and} \quad \lim_{t\to\infty} G(t) = k.$$

Hence, as t increases without bound, the rate of growth approaches 0 and the graph of G has a horizontal asymptote $y = k$. A typical graph is sketched in Figure 8.8.

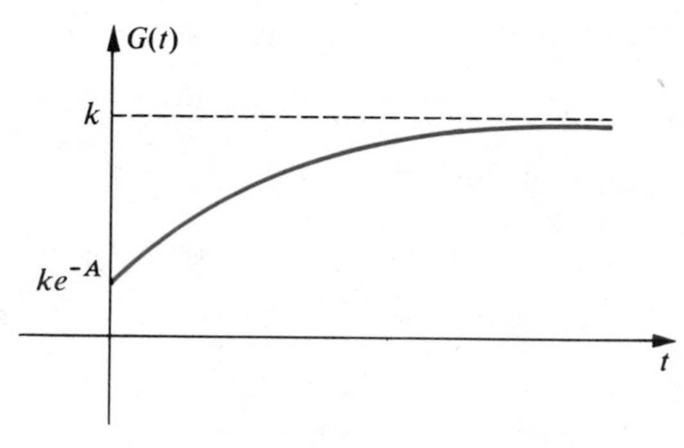

Figure 8.8

EXERCISES 8.5

1 The number of bacteria in a certain culture increases from 5000 to 15,000 in 10 hours. Assuming that the rate of increase is proportional to the number of bacteria present, find a formula for the number of bacteria in the culture at any time t. Estimate the number at the end of 20 hours. When will the number be 50,000?

2 The polonium isotope ^{210}Po has a half-life of approximately 140 days. If a sample weighs 20 mg initially, how much remains after t days? Approximately how much will be left after two weeks?

3 If the temperature is constant, then the rate of change of atmospheric pressure p with respect to altitude h is proportional to p. If $p = 30$ at sea level and $p = 29$ when $h = 1000$ ft, find the pressure at an altitude of 5000 ft.

4 The population of a certain city is increasing at the rate of 5 per cent per year. If the present population is 500,000, what will the population be in 10 years?

5 It is usually assumed that 1/4 acre of land is required to provide food for one person. It is also estimated that there are 10 billion acres of tillable land in the world, and hence a maximum population of 40 billion people can be sustained if no other food source is available. The world population at the beginning of 1970 was 3.6 billion. Assuming that the population increases at a rate of 2 per cent per year, when will the maximum population be reached?

6 An object cools from 180° to 150° in 20 minutes when surrounded by air at a temperature of 60°. Use Newton's Law of Cooling (see Example 3) to approximate its temperature at the end of one hour of cooling. When will the temperature be 100°?

7 An outdoor thermometer registers a temperature of 40°. Five minutes after it is brought into a room where the temperature is 70°, the thermometer registers 60°. When will it register 65°? (See Example 3.)

8 The rate at which salt dissolves in water is directly proportional to the amount which remains undissolved. If 10 pounds of salt is placed in a container of water and 4 pounds dissolves in 20 minutes, how long will it take 2 more pounds to dissolve?

9 The equation

$$E = Ri + L\frac{di}{dt}$$

occurs in the theory of electrical circuits, where the constants E, R, and L denote the electromotive force, the resistance, and the inductance, respectively, and i denotes the current at time t. If the electromagnetic force is removed at time $t = 0$ and if the current is I at the instant of removal, prove that

$$i = Ie^{-Rt/L}.$$

10 The rate at which sugar decomposes in solution is directly proportional to the amount of sugar remaining. If a solution contains k pounds of sugar initially and if y is the amount which decomposes in time t, express y in terms of t.

11 If a sum of money P is invested at an interest rate of $100r$ per cent per year, compounded m times per year, then the principal at the end of t years is given by

$$P\left(1 + \frac{r}{m}\right)^{mt}.$$

If we regard m as a real number and let m increase without bound, then the interest is said to be *compounded continuously*. Use Theorem (8.32) to show that in this case the principal after t years is Pe^{rt}.

12 Use the results of Exercise 11 to find the principal after one year if \$1000 is invested at a rate of 10 per cent, compounded as follows.
(a) Monthly (b) Daily (c) Hourly (d) Continuously

13 In Example 4, verify the following.
(a) Bk/e is a maximum value for G'. (b) $\lim_{t\to\infty} G'(t) = 0$ (c) $\lim_{t\to\infty} G(t) = k$

14 Sketch the graph of (8.37) if $k = 10$, $A = 1/2$, and $B = 1$.

15 In the **Law of Logistic Growth**, it is assumed that at time t, the rate of growth $f'(t)$ of a quantity $f(t)$ is given by

$$f'(t) = Af(t)[B - f(t)]$$

where A and B are constants. If $f(0) = C$, show that

$$f(t) = \frac{BC}{C + (B - C)e^{-ABt}}.$$

16 Discuss and sketch the graph of the growth function f defined by

$$f(x) = a + b(1 - e^{-cx}),$$

where a, b, and c are positive constants.

17 If a radioactive substance decays according to the law $q(t) = q_0e^{-ct}$, where $q_0 = q(0)$, find its half-life if (a) $c = 100$; (b) $c = 1000$; (c) c is any positive constant.

18 If one-fourth of a certain radioactive substance disintegrates in 10 days, what is its half-life?

19 A machine which was purchased for \$20,000 had a value of \$16,000 after two years of use. Assuming that the value decreases exponentially, find the value of the machine at the end of another year.

20 The cost of real estate in a certain city has appreciated at a rate of 10% per year since 1975. If this rate continues, what will a house which was purchased for \$60,000 in 1975 be worth in 1985?

8.6 DERIVATIVES OF INVERSE FUNCTIONS

Inverse functions were defined in Section 6 of Chapter 1. At that time we introduced techniques for finding $f^{-1}(x)$ if $f(x)$ is stated in terms of a simple algebraic expression. It was also pointed out that the graphs of f and f^{-1} are reflections of one another through the line $y = x$. Earlier in this chapter we proved that the natural logarithmic and exponential functions are inverse functions of one another. Inverses of trigonometric functions will be discussed in Chapter 9. In this section we shall prove several general theorems which hold for all inverse functions, and obtain a formula for finding their derivatives.

(8.38) Theorem

> If a function f is continuous and increasing on an interval $[a, b]$, then f has an inverse function f^{-1} which is continuous and increasing on the interval $[f(a), f(b)]$.

Proof. The theorem is intuitively evident if we regard the graph of f^{-1} as a reflection, through the line $y = x$, of the graph of f. However, since a graphical illustration does not constitute a proof, we shall proceed as follows.

In order to establish the existence of f^{-1} it is sufficient to show that f is a one-to-one function with range $[f(a), f(b)]$. Consider numbers x_1 and x_2 in $[a, b]$. If $x_1 < x_2$, then since f is increasing, $f(x_1) < f(x_2)$. Similarly if $x_2 < x_1$, then $f(x_2) < f(x_1)$. Consequently, if $x_1 \neq x_2$, then $f(x_1) \neq f(x_2)$, that is, f is one-to-one. By the Intermediate Value Theorem (2.37), f takes on every value between $f(a)$ and $f(b)$ and, therefore, the range of f is $[f(a), f(b)]$. Thus the inverse function f^{-1} exists, with domain $[f(a), f(b)]$ and range $[a, b]$.

To prove that f^{-1} is increasing, it must be shown that if $w_1 < w_2$ in $[f(a), f(b)]$, then $f^{-1}(w_1) < f^{-1}(w_2)$ in $[a, b]$. We shall give an indirect proof of this fact. Thus, *suppose* $f^{-1}(w_2) \le f^{-1}(w_1)$. Since f is increasing, $f(f^{-1}(w_2)) \le f(f^{-1}(w_1))$ and hence, by (1.24), $w_2 \le w_1$, a contradiction. Consequently $f^{-1}(w_1) < f^{-1}(w_2)$.

It remains to prove that f^{-1} is continuous on $[f(a), f(b)]$. Since f^{-1} is the inverse of f, it follows that $y = f(x)$ if and only if $x = f^{-1}(y)$. In particular, if y_0 is in the open interval $(f(a), f(b))$, let x_0 denote the number in the interval (a, b) such that $y_0 = f(x_0)$ or, equivalently, $x_0 = f^{-1}(y_0)$. We wish to show that

$$\lim_{y \to y_0} f^{-1}(y) = f^{-1}(y_0) = x_0. \tag{8.39}$$

It is enlightening to use a geometric representation of a function and its inverse as described in Chapter 1. In this event, the domain $[a, b]$ of f is represented by points on an x-axis and the domain $[f(a), f(b)]$ by points on a y-axis. Arrows are drawn from one axis to the other to represent functional values. To prove (8.39), consider any interval $(x_0 - \varepsilon, x_0 + \varepsilon)$ where $\varepsilon > 0$. It is sufficient to find an interval $(y_0 - \delta, y_0 + \delta)$, of the type sketched in (i) of Figure 8.9, such that whenever y is in $(y_0 - \delta, y_0 + \delta)$, then $f^{-1}(y)$ is in $(x_0 - \varepsilon, x_0 + \varepsilon)$. We may assume, without loss of generality, that $x_0 - \varepsilon$ and $x_0 + \varepsilon$ are in $[a, b]$. As illustrated in (ii) of Figure 8.9, let $\delta_1 = y_0 - f(x_0 - \varepsilon)$ and $\delta_2 = f(x_0 + \varepsilon) - y_0$. Since the function f determines a one-to-one correspondence between the numbers in $(x_0 - \varepsilon, x_0 + \varepsilon)$ and $(y_0 - \delta_1, y_0 + \delta_2)$, then f^{-1} maps the numbers in $(y_0 - \delta_1, y_0 + \delta_2)$ onto the numbers in $(x_0 - \varepsilon, x_0 + \varepsilon)$. Consequently, if δ denotes the smaller of δ_1 and δ_2, then whenever y is in $(y_0 - \delta, y_0 + \delta)$, $f^{-1}(y)$ is in $(x_0 - \varepsilon, x_0 + \varepsilon)$. This proves that f^{-1} is continuous on the open interval $(f(a), f(b))$. The continuity at the end-points $f(a)$ and $f(b)$ is proved in a similar manner using one-sided limits.

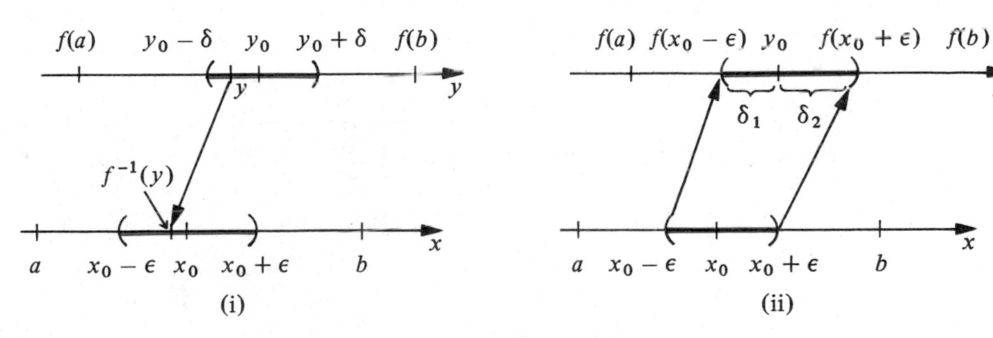

Figure 8.9

The proof of the next result is omitted since it is similar to that of Theorem (8.38).

(8.40) Theorem

> If a function f is continuous and decreasing on an interval $[a, b]$, then f has an inverse function which is continuous and decreasing on the interval $[f(b), f(a)]$.

Example 1 Verify Theorem (8.38) if $f(x) = x^2 - 3$ on the interval $[0, 6]$.

Solution Since f is a polynomial function, it is continuous on $[0, 6]$. Moreover, since $f'(x) = 2x$, $f'(x) > 0$ for all x in $(0, 6]$ and, therefore, f is increasing on $[0, 6]$. It was shown in Example 2 of Section 1.6, that f^{-1} is given by

$$f^{-1}(x) = \sqrt{x+3}, \quad \text{where } x \geq -3.$$

In the present example the domain of f^{-1} is $[f(0), f(6)]$, that is, $[-3, 33]$. Since $\lim_{x \to a} \sqrt{x+3} = \sqrt{a+3}$ for every a in $(-3, 33]$, f^{-1} is continuous on this interval. It is also continuous at -3. Finally, since the derivative

$$D_x f^{-1}(x) = \frac{1}{2\sqrt{x+3}}$$

is positive on $(-3, 33]$, it follows that f^{-1} is increasing on $[-3, 33]$.

The next theorem provides a method for finding the derivative of an inverse function.

(8.41) Theorem

> If a differentiable function f has an inverse function g, and if $f'(g(c)) \neq 0$, then g is differentiable at c and
>
> $$g'(c) = \frac{1}{f'(g(c))}.$$

Proof. Using the formula in Theorem (3.7),

$$g'(c) = \lim_{x \to c} \frac{g(x) - g(c)}{x - c}.$$

We shall now introduce a new variable z such that $z = g(x)$ and let $a = g(c)$. Since f and g are inverse functions of one another, it follows that

$$g(x) = z \quad \text{if and only if} \quad f(z) = x,$$
$$g(c) = a \quad \text{if and only if} \quad f(a) = c.$$

The statement $x \to c$ in the above limit may be replaced by $z \to a$, since if $x \to c$, then $g(x) \to g(c)$ and if $z \to a$, then $f(z) \to f(a)$. Thus we may write

$$\begin{aligned} g'(c) &= \lim_{z \to a} \frac{z - a}{f(z) - f(a)} \\ &= \lim_{z \to a} \frac{1}{\dfrac{f(z) - f(a)}{z - a}} \\ &= \frac{1}{f'(a)} = \frac{1}{f'(g(c))}. \end{aligned}$$

By Theorem (8.41), if g is the inverse function of a differentiable function f, then

(8.42)

$$g'(x) = \frac{1}{f'(g(x))}, \quad \text{if } f'(g(x)) \neq 0.$$

Example 2 Use (8.42) and the fact that the natural exponential function is the inverse of the natural logarithmic function to prove that $D_x e^x = e^x$.

Solution If $f(x) = \ln x$ and $g(x) = e^x$, then $f'(x) = 1/x$ and $f'(g(x)) = f'(e^x) = 1/e^x$. Hence, from (8.42),

$$g'(x) = \frac{1}{(1/e^x)} = e^x.$$

Example 3 If $f(x) = x^3 + 2x - 1$, prove that f has an inverse function g and find the slope of the tangent line to the graph of g at the point $P(2, 1)$.

Solution Since f is continuous and $f'(x) = 3x^2 + 2 > 0$ for all x, we may conclude from Theorem (8.38) that f has an inverse function g. Since $f(1) = 2$ it follows that $g(2) = 1$ and consequently the point $P(2, 1)$ is on the graph of g. Using (8.42), the slope of the desired tangent line is

$$g'(2) = \frac{1}{f'(g(2))} = \frac{1}{f'(1)} = \frac{1}{5}.$$

There is an easy way to remember (8.42) if f is defined by an equation of the form $y = f(x)$. If g is the inverse function of f, then $g(y) = g(f(x)) = x$. From (8.42), we obtain

(8.43)

$$g'(y) = \frac{1}{f'(g(y))} = \frac{1}{f'(x)}$$

or, using differential notation,

(8.44)

$$\frac{dx}{dy} = \frac{1}{\left(\dfrac{dy}{dx}\right)}.$$

This shows that in a sense, the derivative of the inverse function g is the reciprocal of the derivative of f. A disadvantage of using (8.43) or (8.44) is that these formulas are not stated in terms of the independent variable for the inverse function. To illustrate, suppose in Example 3 we let $y = x^3 + 2x - 1$ and $x = g(y)$. Then by (8.43),

$$\frac{dx}{dy} = \frac{1}{3x^2 + 2},$$

that is,

$$g'(y) = \frac{1}{3x^2 + 2} = \frac{1}{3(g(y))^2 + 2}.$$

This may also be written in the form

$$g'(x) = \frac{1}{3(g(x))^2 + 2}.$$

Consequently, in order to find $g'(x)$ it is necessary to know $g(x)$, just as in (8.42).

EXERCISES 8.6

In Exercises 1–10 prove that the function f, defined on the given interval, has an inverse function f^{-1} and state its domain. Find $f^{-1}(x)$. Sketch the graphs of f and f^{-1} on the same coordinate plane. Find $D_x f^{-1}(x)$ directly and also by means of (8.42).

1 $f(x) = \sqrt{2x + 3}$, $[1, 11]$

2 $f(x) = \sqrt[3]{5x + 2}$, $[0, 5]$

3 $f(x) = 4 - x^2$, $[0, 7]$

4 $f(x) = x^2 - 4x + 5$, $[-1, 1]$

5 $f(x) = 1/x$, $(0, \infty)$

6 $f(x) = \sqrt{9 - x^2}$, $[0, 3]$

7 $f(x) = e^{-x^2}$, $[0, \infty)$

8 $f(x) = \ln(3 - 2x)$, $(-\infty, 3/2)$

9 $f(x) = e^x - e^{-x}$, $(-\infty, \infty)$

10 $f(x) = e^x + e^{-x}$, $[0, \infty)$

In each of Exercises 11–14 prove that f has an inverse function f^{-1} on every closed interval $[a, b]$ and find the slope of the tangent line to the graph of f^{-1} at the indicated point P.

11 $f(x) = x^5 + 3x^3 + 2x - 1$, $P(5, 1)$

12 $f(x) = 2 - x - x^3$, $P(-8, 2)$

13 $f(x) = (e^{2x} - 1)/(e^{2x} + 1)$, $P(0, 0)$

14 $f(x) = e^{2-3x}$, $P(1/e, 1)$

In each of Exercises 15–18, prove that if the domain of f is $\mathbb{R}$, then f has no inverse function. Also prove that if the domain is suitably restricted, then f^{-1} exists.

15 $f(x) = x^4 + 3x^2 + 7$

16 $f(x) = |x - 2|$

17 $f(x) = 10^{-x^2}$

18 $f(x) = \ln(x^2 + 1)$

19 Complete the proof of (8.38) by showing that f^{-1} is continuous at $f(a)$ and $f(b)$.

20 Prove Theorem (8.40).

8.7 REVIEW

Concepts

Define or discuss each of the following.

1 The natural logarithmic function

2 The natural exponential function

3 Laws of logarithms

4 Derivative formulas for $\ln u$ and e^u

5 Logarithmic differentiation

6 a^x, where $a > 0$

7 The number e

8 $\log_a$

9 Derivative formulas for $\log_a u$ and a^u

10 General power function

11 Laws of growth and decay

12 Derivatives of inverse functions

Exercises

Find $f'(x)$ if $f(x)$ is defined as in Exercises 1–20.

1 $(1 - 2x)\ln|1 - 2x|$

2 $\sqrt{\ln\sqrt{x}}$

3 $\ln\dfrac{(3x + 2)^4\sqrt{6x - 5}}{(8x - 7)}$

4 $\log_{10}\left|\dfrac{2 - 9x}{1 - x^2}\right|$

5 $\dfrac{1}{\ln(2x^2 + 3)}$

6 $\dfrac{\ln x}{e^{2x} + 1}$

7 $e^{\ln(x^2 + 1)}$

8 $\ln(e^{4x} + 9)$

9 $10^x\log_{10} x$

10 $\ln\sqrt[4]{\dfrac{x}{3x + 5}}$

11 $x^{\ln x}$

12 $4^{\sqrt{2x + 3}}$

13 $x^2 e^{1 - x^2}$

14 $\sqrt{e^{3x} + e^{-3x}}$

15 $2^{-1/x}/(x^3 + 4)$

16 $5^{3x} + (3x)^5$

17 $(1 + \sqrt{x})^e$

18 $\ln e^{\sqrt{x}}$

19 $10^{\ln x}$

20 $(x^2 + 1)^{2x}$

Find y' in Exercises 21 and 22.

21 $1 + xy = e^{xy}$

22 $\ln(x + y) + x^2 - 2y^3 = 1$

23 Show, by substitution, that $y = c_1e^{ax} + c_2xe^{ax}$ is a solution of the differential equation $y'' - 2ay' + a^2y = 0$, where c_1, c_2, and a are any constants.

24 The position of a point on a coordinate line during the time interval $[-5, 5]$ is given by $f(t) = (t^2 + 2t)e^{-t}$.
(a) When is the velocity 0?
(b) When is the acceleration 0?
(c) When does the point move in the positive direction?
(d) When does the point move in the negative direction?

25 A particle moves on a straight line with an acceleration at time t of $e^{t/2}$ cm/sec^2. At $t = 0$ the particle is at the origin and its velocity is 6 cm/sec. How far does it travel during the time interval $[0, 4]$?

Evaluate the integrals in Exercises 26–41.

26 $\int_0^1 e^{-3x+2}\, dx$

27 $\int \frac{(1 + e^x)^2}{e^{2x}}\, dx$

28 $\int \frac{1}{x \ln x}\, dx$

29 $\int_1^4 \frac{1}{\sqrt{x}e^{\sqrt{x}}}\, dx$

30 $\int_1^2 \frac{x^2 + 1}{x^3 + 3x}\, dx$

31 $\int \frac{x^2}{3x + 2}\, dx$

32 $\int \frac{e^{1/x}}{x^2}\, dx$

33 $\int_0^1 x4^{x^2}\, dx$

34 $\int \frac{(e^{2x} + e^{3x})^2}{e^{5x}}\, dx$

35 $\int \frac{1}{x\sqrt{\log_{10} x}}\, dx$

36 $\int \frac{x}{x^4 + 2x^2 + 1}\, dx$

37 $\int \frac{x^2 + 1}{x + 1}\, dx$

38 $\int \frac{2e^x}{1 + e^x}\, dx$

39 $\int 5^x e^x\, dx$

40 $\int x10^{x^2}\, dx$

41 $\int x^e\, dx$

42 Find the local extrema for $f(x) = x^2 \ln x$, $x > 0$. Discuss concavity, find the points of inflection, and sketch the graph of f.

43 Find an equation of the tangent line to the graph of $y = xe^{1/x^3} + \ln|2 - x^2|$ at the point $P(1, e)$.

44 Find the area of the region bounded by the graphs of the equations $y = e^{2x}$, $y = x/(x^2 + 1)$, $x = 0$, and $x = 1$.

45 The region bounded by the graphs of $y = e^{4x}$, $x = -2$, $x = -3$, and $y = 0$ is revolved about the x-axis. Find the volume of the resulting solid.

46 If $f(x) = 2x^3 - 8x + 5$ on $[-1, 1]$, prove that f has an inverse function g and find $g'(5)$.

47 Suppose $f(x) = e^{2x} + 2e^x + 1$, where $x \geq 0$.
(a) Prove that f has an inverse function f^{-1} and state its domain.
(b) Find $f^{-1}(x)$ and $D_x f^{-1}(x)$.
(c) Find the slope of the tangent line to the graph of f at the point $(0, 4)$ and the slope of the tangent line to the graph of f^{-1} at $(4, 0)$.

48 A certain radioactive substance has a half-life of 5 days. How long will it take for an amount A to disintegrate to the time at which only 1 per cent of A remains?

49 The rate at which sugar decomposes in water is proportional to the amount which remains undecomposed. If 10 pounds of sugar is placed in a container of water at 1:00 P.M., and if half is dissolved at 4:00 P.M., answer these questions.
(a) How long will it take 2 more pounds to dissolve?
(b) How much of the 10 pounds will be dissolved at 8:00 P.M.?

50 According to Newton's Law of Cooling, the rate at which an object cools is directly proportional to the difference in temperature between the object and its surrounding medium. If $f(t)$ denotes the temperature at time t, show that

$$f(t) = T + [f(0) - T]e^{-kt}$$

where T is the temperature of the surrounding medium and k is a positive constant.

Other Transcendental Functions

In this chapter we shall develop formulas for limits, derivatives, and integrals of trigonometric and inverse trigonometric functions. The last two sections contain a discussion of the hyperbolic and inverse hyperbolic functions. A review of the trigonometric functions and trigonometric identities is provided in Appendix III. Students who need a refresher on these topics should read it carefully before proceeding to Section 1.

9.1 LIMITS OF TRIGONOMETRIC FUNCTIONS

The results of this section are important in the development of differentiation formulas for the trigonometric functions. When we discuss limits, derivatives, and integrals of trigonometric expressions involving $\sin t$, $\cos x$, $\tan \theta$, etc., we shall assume that each variable represents a real number or the radian measure of an angle. Since $\sin 0 = 0$, the first theorem tells us that the sine function is continuous at 0.

(9.1) Theorem

$$\boxed{\lim_{t \to 0} \sin t = 0}$$

Proof. Let us first prove that $\lim_{t \to 0^+} \sin t = 0$. Since we are only interested in positive values of t near zero, there is no loss of generality in assuming that $0 < t < \pi/2$. Let U be the circle of radius 1 with center at the origin of a rectangular coordinate system, and let A be the point $(1, 0)$. If, as illustrated in Figure 9.1, $P(x, y)$ is the point on U such that $\widehat{AP} = t$, then the radian measure of angle AOP is t. Referring to the figure we see that

$$0 < y < \widehat{AP} = t$$

or, since $y = \sin t$,

$$0 < \sin t < t.$$

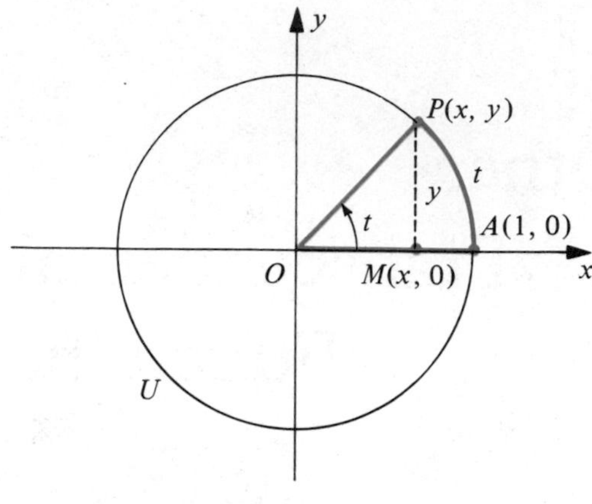

Figure 9.1

Since $\lim_{t \to 0^+} t = 0$, it follows from the Sandwich Theorem (2.23) that $\lim_{t \to 0^+} \sin t = 0$.

To complete the proof it is sufficient to show that $\lim_{t \to 0^-} \sin t = 0$. If $-\pi/2 < t < 0$, then $0 < -t < \pi/2$ and hence, from the first part of the proof,

$$0 < \sin(-t) < -t.$$

Multiplying the last inequality by -1 and using the fact that $\sin(-t) = -\sin t$ gives us

$$t < \sin t < 0.$$

Since $\lim_{t \to 0^-} t = 0$, it follows from the Sandwich Theorem (2.23) that $\lim_{t \to 0^-} \sin t = 0$.

(9.2) Corollary

$$\boxed{\lim_{t \to 0} \cos t = 1}$$

Proof. If $-\pi/2 < t < \pi/2$, then $\cos t = \sqrt{1 - \sin^2 t}$.

Consequently

$$\lim_{t \to 0} \cos t = \lim_{t \to 0} \sqrt{1 - \sin^2 t} = \sqrt{\lim_{t \to 0} (1 - \sin^2 t)}$$
$$= \sqrt{1 - 0} = 1.$$

Since $\cos 0 = 1$, Corollary (9.2) states that the cosine function is continuous at 0.

We shall see in Section 2 that to find the derivative of the sine function it is essential to know the limits of $(\sin t)/t$ and $(1 - \cos t)/t$ as t approaches 0. The next two theorems provide these important results.

(9.3) Theorem

$$\lim_{t \to 0} \frac{\sin t}{t} = 1$$

Proof. If $0 < t < \pi/2$ we have the situation illustrated in Figure 9.2 where U is a unit circle.

If A_1 is the area of triangle AOP, A_2 the area of circular sector AOP, and A_3 the area of triangle AOQ, then

$$A_1 < A_2 < A_3.$$

Since

$$A_1 = \tfrac{1}{2}(1)d(M, P) = \tfrac{1}{2}y = \tfrac{1}{2}\sin t$$
$$A_2 = \tfrac{1}{2}(1)^2 t = \tfrac{1}{2}t$$
$$A_3 = \tfrac{1}{2}(1)d(A, Q) = \tfrac{1}{2}\tan t$$

we have

$$\tfrac{1}{2}\sin t < \tfrac{1}{2}t < \tfrac{1}{2}\tan t.$$

Dividing by $\frac{1}{2}\sin t$ and using the fact that $\tan t = \sin t/\cos t$ gives us

$$1 < \frac{t}{\sin t} < \frac{1}{\cos t}$$

or, equivalently,

$$1 > \frac{\sin t}{t} > \cos t. \tag{9.4}$$

If $-\pi/2 < t < 0$, then $0 < -t < \pi/2$, and from the result just established,

$$1 > \frac{\sin(-t)}{-t} > \cos(-t).$$

Since $\sin(-t) = -\sin t$ and $\cos(-t) = \cos t$, this inequality reduces to (9.4). This shows that (9.4) is also true if $-\pi/2 < t < 0$. Thus (9.4) holds for all t in the open interval $(-\pi/2, \pi/2)$ except $t = 0$. Since $\lim_{t \to 0} \cos t = 1$, and $(\sin t)/t$ is always between $\cos t$ and 1, it follows that

$$\lim_{t \to 0} \frac{\sin t}{t} = 1.$$

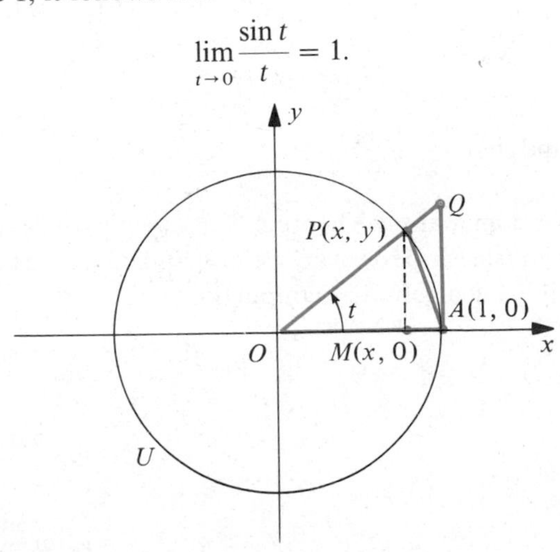

Figure 9.2

Roughly speaking, Theorem (9.3) implies that if t is close to 0, then $(\sin t)/t$ is close to 1. Another way of stating this is to write $\sin t \approx t$ for small values of t. It is important to remember that if t denotes an angle, then *radian measure must be used in Theorem (9.3)* and in the approximation formula $\sin t \approx t$. To illustrate, trigonometric tables show that to five decimal places,

$$\sin(0.06) \approx 0.05996$$
$$\sin(0.05) \approx 0.04998$$
$$\sin(0.04) \approx 0.03999$$
$$\sin(0.03) \approx 0.03000.$$

(9.5) **Theorem**

$$\lim_{t\to 0}\frac{1-\cos t}{t} = 0$$

Proof. We may change the form of $(1 - \cos t)/t$ as follows:

$$\begin{aligned}\frac{1-\cos t}{t} &= \frac{1-\cos t}{t}\cdot\frac{1+\cos t}{1+\cos t}\\ &= \frac{\sin^2 t}{t(1+\cos t)}\\ &= \frac{\sin t}{t}\cdot\frac{\sin t}{1+\cos t}.\end{aligned}$$

Consequently

$$\begin{aligned}\lim_{t\to 0}\frac{1-\cos t}{t} &= \lim_{t\to 0}\left(\frac{\sin t}{t}\cdot\frac{\sin t}{1+\cos t}\right)\\ &= \left(\lim_{t\to 0}\frac{\sin t}{t}\right)\left(\lim_{t\to 0}\frac{\sin t}{1+\cos t}\right)\\ &= 1\cdot\left(\frac{0}{1+1}\right) = 1\cdot 0 = 0.\end{aligned}$$

Example 1 Find $\lim_{x\to 0}\dfrac{\sin 5x}{2x}$.

Solution We cannot apply Theorem (9.3) directly, since the given expression is not in the form $(\sin t)/t$. However, we may introduce this form (with $t = 5x$) by using the following algebraic manipulation:

$$\begin{aligned}\lim_{x\to 0}\frac{\sin 5x}{2x} &= \lim_{x\to 0}\frac{1}{2}\frac{\sin 5x}{x}\\ &= \lim_{x\to 0}\frac{5}{2}\frac{\sin 5x}{5x}\\ &= \frac{5}{2}\lim_{x\to 0}\frac{\sin 5x}{5x}.\end{aligned}$$

It follows from the definition of limit that $x \to 0$ may be replaced by $5x \to 0$. Hence, by Theorem (9.3), with $t = 5x$, we see that

$$\lim_{x \to 0} \frac{\sin 5x}{2x} = \frac{5}{2}(1) = \frac{5}{2}.$$

Warning: When working problems of this type, remember that $\sin 5x \neq 5 \sin x$.

Example 2 Find $\lim_{t \to 0} \frac{\tan t}{2t}$.

Solution Using the fact that $\tan t = \sin t / \cos t$,

$$\lim_{t \to 0} \frac{\tan t}{2t} = \lim_{t \to 0} \left(\frac{1}{2} \cdot \frac{\sin t}{t} \cdot \frac{1}{\cos t} \right)$$
$$= \tfrac{1}{2} \cdot 1 \cdot 1 = \tfrac{1}{2}.$$

EXERCISES 9.1

Find the limits in Exercises 1–20.

1 $\lim_{t \to 0} \frac{\cos t}{1 - \sin t}$

2 $\lim_{t \to 0} \frac{\sin t}{1 + \cos t}$

3 $\lim_{t \to 0} \frac{1 - \cos t}{\sin t}$

4 $\lim_{x \to 0} \frac{\sin (x/2)}{x}$

5 $\lim_{x \to 0} \frac{x + \tan x}{\sin x}$

6 $\lim_{t \to 0} \frac{\sin^2 2t}{t^2}$

7 $\lim_{x \to 0} x \cot x$

8 $\lim_{x \to 0} \frac{\csc 2x}{\cot x}$

9 $\lim_{x \to 0} \frac{\sin^2 (x/2)}{\sin x}$

10 $\lim_{x \to \pi} \frac{\sin x}{x - \pi}$

11 $\lim_{w \to 2} \frac{\sin (3w - 6)}{7w - 14}$

12 $\lim_{v \to 0} \frac{\cos (v + \pi/2)}{v}$

13 $\lim_{x \to 3} \frac{x^2 - 6x + 9}{\sin^2 (x - 3)}$

14 $\lim_{x \to -1} \frac{\sin (x + 1)}{x^2 + 2x + 1}$

15 $\lim_{\theta \to 0} \frac{\cos \theta}{\theta}$

16 $\lim_{\alpha \to 0} \alpha^2 \csc^2 \alpha$

17 $\lim_{x \to 0^+} \sqrt{x} \csc x$

18 $\lim_{x \to 0} \frac{\sin 3x}{\sin 5x}$

19 $\lim_{x \to \infty} x \cos \frac{1}{x}$

20 $\lim_{x \to \infty} e^x \cos x$

Establish the limits in Exercises 21–24, where a and b are any nonzero real numbers.

21 $\lim_{x \to 0} \dfrac{\sin ax}{bx} = \dfrac{a}{b}$

22 $\lim_{x \to 0} \dfrac{1 - \cos ax}{bx} = 0$

23 $\lim_{x \to 0} \dfrac{\sin ax}{\sin bx} = \dfrac{a}{b}$

24 $\lim_{x \to 0} \dfrac{\cos ax}{\cos bx} = 1$

9.2 DERIVATIVES OF TRIGONOMETRIC FUNCTIONS

To find the derivative of the sine function, we let $f(x) = \sin x$. If $f'(x)$ exists, then

$$\begin{aligned} f'(x) &= \lim_{h \to 0} \frac{f(x+h) - f(x)}{h} \\ &= \lim_{h \to 0} \frac{\sin(x+h) - \sin x}{h} \\ &= \lim_{h \to 0} \frac{\sin x \cos h + \cos x \sin h - \sin x}{h} \\ &= \lim_{h \to 0} \frac{\sin x(\cos h - 1) + \cos x \sin h}{h} \\ &= \lim_{h \to 0} \left[\sin x \left(\frac{\cos h - 1}{h} \right) + \cos x \left(\frac{\sin h}{h} \right) \right]. \end{aligned}$$

Employing Theorems (9.5) and (9.3),

$$f'(x) = (\sin x)(0) + (\cos x)(1) = \cos x,$$

that is,

(9.6) $$D_x \sin x = \cos x.$$

We have shown that the derivative of the sine function is the cosine function.

If $y = \sin u$, where $u = g(x)$ is differentiable, then applying the Chain Rule (3.36) and formula (9.6) with u in place of x,

$$D_x y = (D_u y)(D_x u) = \cos u \, D_x u$$

that is,

(9.7) $$D_x \sin u = \cos u \, D_x u.$$

Of course, if $u = x$, then (9.7) reduces to (9.6).

Example 1 (a) Find $D_x \sin(4x^3)$.
(b) Find $D_x \sin^3 4x$.

Solution (a) From (9.7),

$$D_x \sin(4x^3) = \cos(4x^3)\, D_x(4x^3)$$
$$= 12x^2 \cos(4x^3).$$

(b) Since $\sin^3 4x = (\sin 4x)^3$, we first use the Power Rule for Functions (3.38), obtaining

$$D_x \sin^3 4x = 3(\sin 4x)^2\, D_x \sin 4x$$
$$= (3 \sin^2 4x)\, D_x \sin 4x.$$

Next applying (9.7),

$$D_x \sin^3 4x = (3 \sin^2 4x) \cos 4x\, D_x 4x$$
$$= (3 \sin^2 4x)(\cos 4x)(4)$$
$$= 12 \sin^2 4x \cos 4x.$$

The derivatives of the other trigonometric functions may be obtained from (9.7) with the aid of trigonometric identities. If we use the addition formulas for sines and cosines we see that $\cos x = \sin(\pi/2 - x)$ and $\sin x = \cos(\pi/2 - x)$. Using these facts together with (9.7),

$$D_x \cos x = D_x \sin(\pi/2 - x)$$
$$= \cos(\pi/2 - x)\, D_x(\pi/2 - x)$$
$$= (\sin x)(-1)$$
$$= -\sin x.$$

As before, if $u = g(x)$ where g is differentiable, then we may write

(9.8) $$D_x \cos u = -\sin u\, D_x u.$$

Applying the Quotient Rule to $\tan x = \sin x/\cos x$,

$$D_x \tan x = \frac{\cos x\, D_x \sin x - \sin x\, D_x \cos x}{\cos^2 x}$$
$$= \frac{\cos^2 x + \sin^2 x}{\cos^2 x}$$
$$= \frac{1}{\cos^2 x} = \sec^2 x.$$

More generally, if $u = g(x)$ and g is differentiable, then

(9.9) $$D_x \tan u = \sec^2 u\, D_x u.$$

Similarly,

$$\begin{aligned} D_x \sec x = D_x \frac{1}{\cos x} &= \frac{(\cos x) D_x 1 - (1) D_x \cos x}{\cos^2 x} \\ &= \frac{0 + \sin x}{\cos^2 x} \\ &= \frac{1}{\cos x} \frac{\sin x}{\cos x} \\ &= \sec x \tan x. \end{aligned}$$

If $u = g(x)$ and g is differentiable, then

(9.10) $$D_x \sec u = \sec u \tan u \, D_x u.$$

It is left as an exercise to prove each of the following:

(9.11) $$D_x \cot u = -\csc^2 u \, D_x u.$$

(9.12) $$D_x \csc u = -\csc u \cot u \, D_x u.$$

Example 2 Find $f'(x)$ if (a) $f(x) = e^{-2x} \tan 4x$; (b) $f(x) = \sec^2 (3x - 1)$.

Solution (a) Using the Product Rule (3.20),

$$\begin{aligned} f'(x) &= e^{-2x} D_x \tan 4x + \tan 4x \, D_x e^{-2x} \\ &= e^{-2x} (\sec^2 4x)(4) + \tan 4x (e^{-2x})(-2) \\ &= 2e^{-2x} (2 \sec^2 4x - \tan 4x). \end{aligned}$$

(b) By the Power Rule (3.12) and (9.10),

$$\begin{aligned} f'(x) &= 2 \sec (3x - 1) D_x \sec (3x - 1) \\ &= 2 \sec (3x - 1) \sec (3x - 1) \tan (3x - 1)(3) \\ &= 6 \sec^2 (3x - 1) \tan (3x - 1). \end{aligned}$$

Portions of the graphs of the equations $y = \sin x$, $y = \cos x$, and $y = \tan x$ are sketched in Figure 9.3. The dashed lines for the graph of $y = \tan x$ indicate vertical asymptotes. Since the sine function is differentiable, it is continuous at every real number a. Moreover, since $D_x \sin x = \cos x$, the critical numbers of the sine function are solutions of the equation $\cos x = 0$. These solutions are $x = \pi/2 + n\pi$, where n is any integer. It can be shown, by means of the Second Derivative Test, that local maxima occur if n is even and local minima occur if n is odd. This gives us the high and low points on the graph of $y = \sin x$. Since $D_x^2 \sin x = D_x \cos x = -\sin x$, it follows that the graph of the sine function is concave downward whenever $\sin x$ is positive and concave upward whenever $\sin x$ is negative. Consequently, points of inflection have abscissas $n\pi$ for every integer n. A similar analysis can be made for the graph of each of the remaining trigonometric functions.

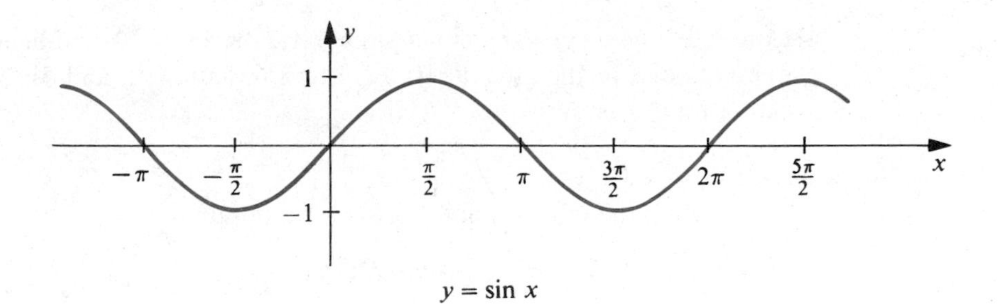

$y = \sin x$

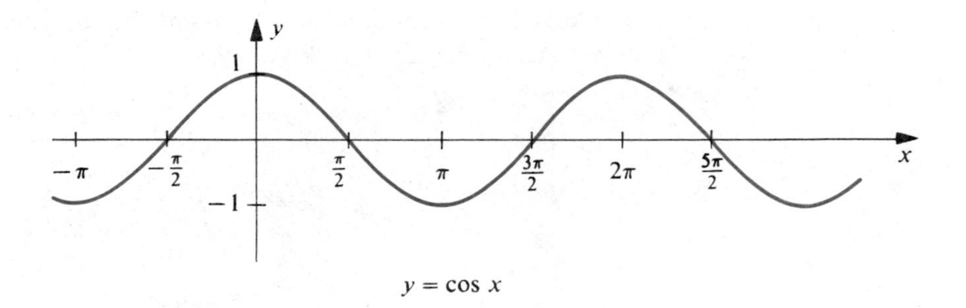

$y = \cos x$

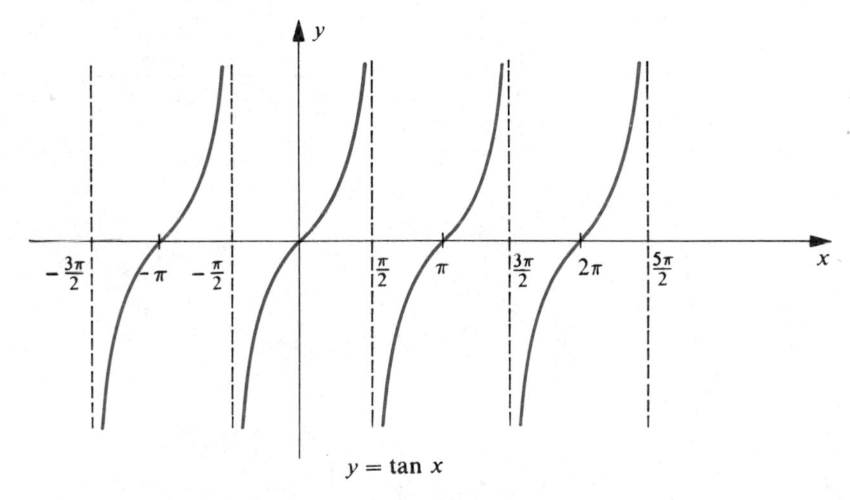

$y = \tan x$

Figure 9.3

Example 3 If $f(x) = 2 \sin x + \cos 2x$, find the local extrema and sketch the graph of f on the interval $[0, 2\pi]$.

Solution Differentiating we obtain

$$f'(x) = 2 \cos x - 2 \sin 2x.$$

Since $\sin 2x = 2 \sin x \cos x$, this may be rewritten

$$\begin{aligned} f'(x) &= 2 \cos x - 4 \sin x \cos x \\ &= 2 \cos x(1 - 2 \sin x). \end{aligned}$$

Consequently, $f'(x) = 0$ if either $\sin x = 1/2$ or $\cos x = 0$, and hence the critical numbers of f in the interval $[0, 2\pi]$ are $\pi/6$, $5\pi/6$, $\pi/2$, and $3\pi/2$. The second derivative of f is

$$f''(x) = -2\sin x - 4\cos 2x.$$

Substituting the critical numbers for x we obtain

$$f''(\pi/6) = -3, \quad f''(5\pi/6) = -3, \quad f''(\pi/2) = 2, \quad f''(3\pi/2) = 6.$$

Applying the Second Derivative Test we see that there are local maxima at $\pi/6$ and $5\pi/6$, and local minima at $\pi/2$ and $3\pi/2$. This information, together with the following table, leads to the sketch in Figure 9.4.

x	0	$\pi/6$	$\pi/2$	$5\pi/6$	π	$3\pi/2$	2π
$f(x)$	1	3/2	1	3/2	1	-3	1

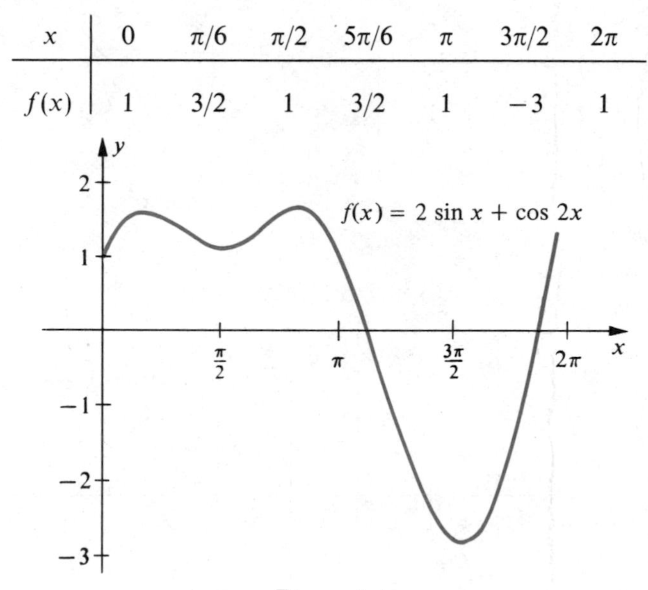

Figure 9.4

Example 4 Suppose

$$f(x) = \begin{cases} x\sin\dfrac{1}{x} & \text{if} \quad x \neq 0 \\ 0 & \text{if} \quad x = 0. \end{cases}$$

Prove that f is continuous at every real number and sketch the graph of f.

Solution Let us show that $\lim_{x\to a} f(x) = f(a)$ for every real number a. If $a \neq 0$, then

$$\begin{aligned}\lim_{x\to a}\left(x\sin\frac{1}{x}\right) &= \left(\lim_{x\to a} x\right)\left(\lim_{x\to a}\sin\frac{1}{x}\right)\\ &= a\sin\frac{1}{a} = f(a).\end{aligned}$$

From Example 8 in Section 2.3,

$$\lim_{x\to 0} x\sin\frac{1}{x} = 0 = f(0).$$

Hence by Definition (2.27), f is continuous at every real number.

The graph of $y = f(x)$ is symmetric with respect to the y-axis, since substitution of $-x$ for x gives us

$$y = -x \sin \frac{1}{-x} = -x\left(-\sin \frac{1}{x}\right) = x \sin \frac{1}{x}.$$

Thus we may restrict our discussion to points (x, y) with $x \geq 0$, since the remainder of the graph can be determined by a reflection through the y-axis.

To find the x-intercepts we solve the equation $x \sin(1/x) = 0$, obtaining $x = 0$ and $1/x = n\pi$, where n is any integer. In particular, in the interval $[0, 1]$ the graph has an infinite number of x-intercepts of the form $1/n\pi$. Several of these are

$$\frac{1}{\pi}, \quad \frac{1}{2\pi}, \quad \frac{1}{3\pi}, \quad \frac{1}{4\pi}, \quad \text{and} \quad \frac{1}{5\pi}.$$

It is of interest to note that the function f is continuous on the interval $[0, 1]$, but its graph cannot be sketched without lifting the pencil from the paper since there is no "first" x-intercept greater than 0.

It is not difficult to show that

$$f(x) < 0 \quad \text{if} \quad \frac{1}{2\pi} < x < \frac{1}{\pi} \quad \text{and}$$

$$f(x) > 0 \quad \text{if} \quad \frac{1}{3\pi} < x < \frac{1}{2\pi}.$$

Similar inequalities can be obtained if $1/((n+1)\pi) < x < 1/n\pi$, where n is any positive integer.

If $x \neq 0$, then $|\sin(1/x)| \leq 1$ and hence

$$|f(x)| = \left|x \sin \frac{1}{x}\right| = |x| \left|\sin \frac{1}{x}\right| \leq |x|.$$

It follows that the graph of f lies between (and on) the lines $y = x$ and $y = -x$. The graph intersects these lines if $|f(x)| = |x|$, that is, if $|\sin(1/x)| = 1$. Observe that $|\sin(1/x)| = 1$ if

$$\frac{1}{x} = \frac{\pi}{2} + n\pi = \frac{\pi(1+2n)}{2}, \quad \text{or} \quad x = \frac{2}{\pi(1+2n)}$$

where n is any integer. The abscissas of several such points of intersection are

$$\frac{2}{\pi}, \quad \frac{2}{3\pi}, \quad \frac{2}{5\pi}, \quad \frac{2}{7\pi}, \quad \text{and} \quad \frac{2}{9\pi}.$$

In Exercise 60 the reader is asked to show that the graph of f is tangent to $y = x$ or $y = -x$ at each of these points, and that the graph has a horizontal tangent line at $(x, f(x))$ if $\tan(1/x) = 1/x$. The latter points correspond to relative extrema of f.

Using the information we have obtained leads to the graph in Figure 9.5.

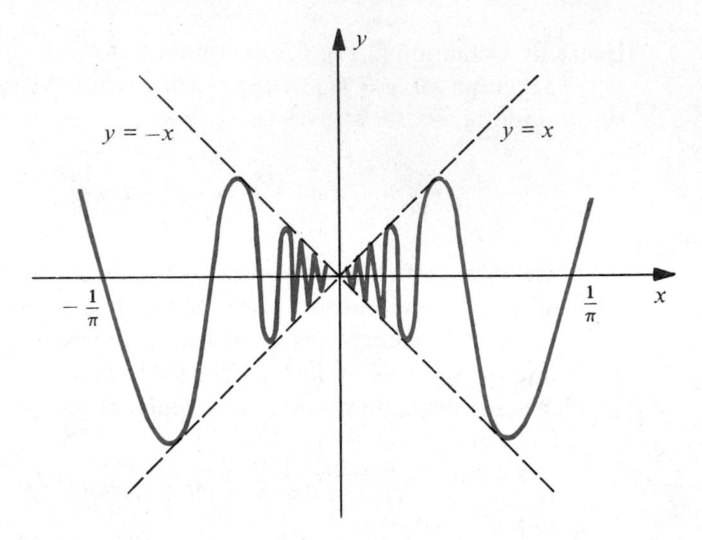

Figure 9.5

Example 5 A revolving light located 200 feet from the nearest point P on a straight shoreline makes one revolution every 15 seconds. Find the rate at which a ray from the light moves along the shore at a point 400 feet from P.

Solution The problem is diagrammed in Figure 9.6, where the horizontal line represents the straight shoreline, L denotes the position of the light, and ϕ is the angle between LP and a ray to a point S on the shore x units from P.

Since the light revolves four times per minute, the angle ϕ changes at a rate of 8π radians per minute, that is, $d\phi/dt = 8\pi$. Using triangle PLS we see that $\tan\phi = x/200$, or $x = 200\tan\phi$. Consequently

$$\frac{dx}{dt} = 200\sec^2\phi\frac{d\phi}{dt} = (200\sec^2\phi)(8\pi) = 1600\pi\sec^2\phi.$$

If $x = 400$, then

$$\tan\phi = 400/200 = 2, \quad \text{and} \quad \sec\phi = \sqrt{1+\tan^2\phi} = \sqrt{1+4} = \sqrt{5}.$$

Hence

$$\frac{dx}{dt} = 1600\pi(\sqrt{5})^2 = 8000\pi \text{ ft/min}.$$

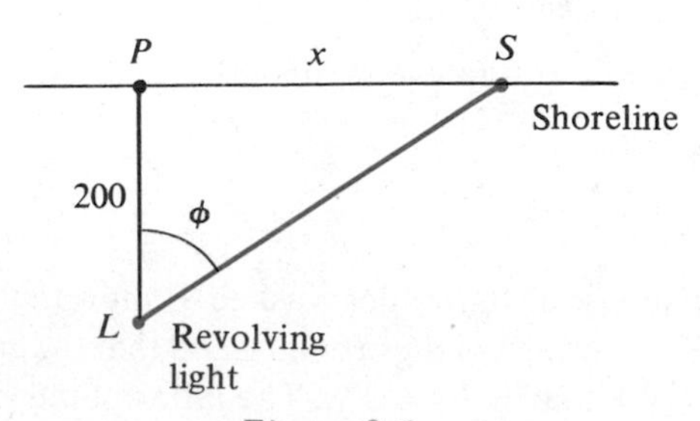

Figure 9.6

EXERCISES 9.2

1 Prove (9.11) and (9.12).

2 Sketch the graphs of the cosecant, secant, and cotangent functions. In each case discuss continuity, determine where the function increases or decreases, and find the local extrema. Discuss concavity and find the points of inflection.

In each of Exercises 3–34 find $f'(x)$ if $f(x)$ equals the given expression.

3 $\sin(8x + 3)$

4 $\cos(x/2)$

5 $\sec\sqrt{x - 1}$

6 $\csc(x^2 + 4)$

7 $\cot(x^3 - 2x)$

8 $\tan\sqrt[3]{5 - 6x}$

9 $\cos 3x^2$

10 $\tan^3 6x$

11 $\cos^2 3x$

12 $\sin e^{-2x}$

13 $x^2 \csc 5x$

14 $x \csc(1/x)$

15 $\tan^2 x \sec^3 x$

16 $x^2 \sec^3 4x$

17 $(\sin 5x - \cos 5x)^5$

18 $\sin\sqrt{x} + \sqrt{\sin x}$

19 $\cot^3(3x + 1)$

20 $e^{\cos 2x}$

21 $\dfrac{\cos 4x}{1 - \sin 4x}$

22 $\dfrac{\sec 2x}{\tan 2x + 1}$

23 $\ln|\csc x + \cot x|$

24 $\sin(2x + 3)^4$

25 $e^{-3x}\tan\sqrt{x}$

26 $\ln\cos^2 3x$

27 $\ln\ln\sec 2x$

28 $\csc(\cot 4x)$

29 $\tan^3 2x - \sec^3 2x$

30 $(\tan 2x - \sec 2x)^3$

31 $\dfrac{\csc 3x}{x^3 + 1}$

32 $4x^3 - x^2\cot^3(1/x)$

33 $x^{\sin x}$

34 $(\tan x)^{3x}$

In Exercises 35–40 find dy/dx and d^2y/dx^2.

35 $y = \sec^2 3x$

36 $y = \cot^3 5x$

37 $y = \sin x - x\cos x$

38 $y = \ln\tan x$

39 $y = \sqrt{\tan x}$

40 $y = \dfrac{\cos x - 1}{\cos x + 1}$

In Exercises 41–44 use implicit differentiation to find y'.

41 $y = x\sin y$

42 $xy = \tan xy$

43 $e^x\cos y = xe^y$

44 $\cos(x - y) = y\cos x$

In Exercises 45–48 find the local extrema and sketch the graph of f for $0 \le x \le 4\pi$.

45 $f(x) = 2\cos x + \sin 2x$

46 $f(x) = \cos x - \sin x$

47 $f(x) = e^{-x}\sin x$

48 $f(x) = (x/2) - \sin x$

In Exercises 49 and 50 find equations of the tangent and normal lines to the graph of the given equation at the point P.

49 $y = 8\sin^3 x$; $P(\pi/6, 1)$

50 $y = \tan x - \sqrt{2}\sin x$; $P(3\pi/4, -2)$

51 At a point 20 feet from the base of a flagpole, the angle of elevation of the top of the pole is measured as 60°, with a possible error of 15′. Use differentials to approximate the error in the calculated height of the pole.

52 Use differentials to approximate the change in cot x if x changes from 45° to 46°.

53 An airplane at an altitude of 10,000 feet is flying at a constant speed on a line which will take it directly over an observer on the ground. If, at a given instant, the observer notes that the angle of elevation of the airplane is 60° and is increasing at a rate of 1° per second, find the speed of the airplane.

54 An isosceles triangle has equal sides 6 inches long. If the angle θ between the sides is changing at a rate of 2° per minute, how fast is the area of the triangle changing when $\theta = 30°$?

If a point moves on a coordinate line such that its distance $s(t)$ from the origin is given by either $s(t) = a\cos\omega t$ or $s(t) = a\sin\omega t$, where a and ω are constants, then the motion of the point is termed **simple harmonic**. In Exercises 55 and 56 find the velocity and acceleration, and discuss the motion of a point for the given $s(t)$.

55 $s(t) = 3\cos 2t$

56 $s(t) = 2\sin \pi t$

57 A water cup in the shape of a right circular cone is to be constructed by removing a circular sector from a circular sheet of paper and then joining the two straight edges of the remaining paper. Find the central angle of the sector which will maximize the volume of the cup.

58 Two corridors 3 feet and 4 feet wide, respectively, meet at right angles. Find the length of the longest thin straight rod that will pass horizontally around the corner, as illustrated in (i) of Figure 9.7.

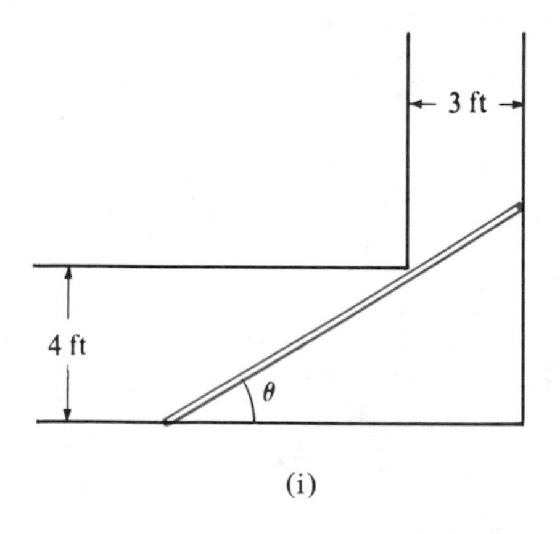

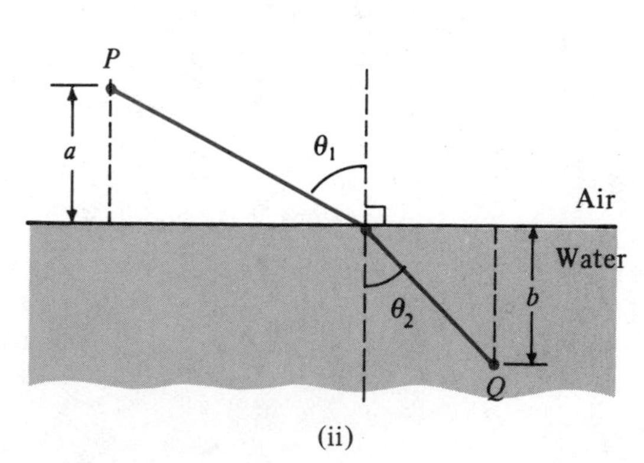

Figure 9.7

59 When light travels from one point to another it takes the path which requires the least amount of time. Suppose that light has velocity v_1 in air and v_2 in water, where $v_1 > v_2$. If light travels from a point P in air to a point Q in water, as illustrated in (ii) of Figure 9.7, show that the path requiring the least amount of time occurs if

$$\frac{\sin\theta_1}{\sin\theta_2} = \frac{v_1}{v_2}.$$

(This is an example of **Snell's Law of Refraction**.)

60 Let f be the function defined in Example 4.
(a) Prove that the graph of f has a horizontal tangent line at $(x, f(x))$ if $\tan(1/x) = 1/x$. Show that there are an infinite number of such points by sketching the graphs of $y = \tan t$ and $y = t$ on the same coordinate axes.
(b) Show that the lines $y = x$ and $y = -x$ are tangent to the graph of f at an infinite number of points.

9.3 INTEGRALS OF TRIGONOMETRIC FUNCTIONS

The rules for differentiation developed in Section 2 can be used to obtain formulas for integrals. For example, since $D_x \sin x = \cos x$, it follows from the definition of indefinite integral that

(9.13)
$$\int \cos x\, dx = \sin x + C.$$

More generally, from (9.7), if g is a differentiable function, then $D_x \sin g(x) = \cos g(x) \cdot g'(x)$ and hence

$$\int \cos g(x) \cdot g'(x)\, dx = \sin g(x) + C.$$

The preceding formula can be remembered easily by using the change of variable technique introduced in (5.44). Specifically, if we let $u = g(x)$, then $du = g'(x)\, dx$, and formal substitution for $g(x)$ and $g'(x)\, dx$ gives us

(9.14)
$$\int \cos u\, du = \sin u + C.$$

This has the same form as (9.13); however, in (9.14) u may be regarded as an expression in x and du as the differential of u.

Example 1 Find $\int x \cos x^2\, dx$.

Solution If $u = x^2$, then $du = 2x\, dx$ and form (9.14) may be obtained by introducing the factor 2 in the integrand. This leads to

$$\begin{aligned}\int x \cos x^2\, dx &= \tfrac{1}{2} \int \cos x^2 (2x)\, dx \\ &= \tfrac{1}{2} \int \cos u\, du \\ &= \tfrac{1}{2} \sin u + C = \tfrac{1}{2} \sin x^2 + C.\end{aligned}$$

In a manner similar to the derivation of (9.14), we may use (9.8)–(9.12) to obtain the following, where $u = g(x)$ and $du = g'(x)\, dx$.

(9.15) $$\int \sin u \, du = -\cos u + C$$

(9.16) $$\int \sec^2 u \, du = \tan u + C$$

(9.17) $$\int \sec u \tan u \, du = \sec u + C$$

(9.18) $$\int \csc^2 u \, du = -\cot u + C$$

(9.19) $$\int \csc u \cot u \, du = -\csc u + C$$

Example 2 Evaluate $\int \sin 5x \, dx$.

Solution We let $u = 5x$ so that $du = 5\, dx$, and proceed as follows:

$$\begin{aligned}\int \sin 5x \, dx &= \tfrac{1}{5}\int (\sin 5x)5\, dx \\ &= \tfrac{1}{5}\int \sin u \, du \\ &= -\tfrac{1}{5}\cos u + C \\ &= -\tfrac{1}{5}\cos 5x + C.\end{aligned}$$

Example 3 Find $\int \sec x(\sec x + \tan x)\, dx$.

Solution We may proceed as follows:

$$\begin{aligned}\int \sec x(\sec x + \tan x)\, dx &= \int (\sec^2 x + \sec x \tan x)\, dx \\ &= \int \sec^2 dx + \int \sec x \tan x \, dx \\ &= \tan x + \sec x + C.\end{aligned}$$

Example 4 Find $\int \left(\frac{1}{\sqrt{x}}\right) \csc^2 \sqrt{x}\, dx$.

Solution If

$$u = \sqrt{x}, \quad \text{then} \quad du = \left(\frac{1}{2\sqrt{x}}\right) dx$$

and the integrand can be written in the form (9.18) by introducing the factor 1/2.

Thus

$$\int \left(\frac{1}{\sqrt{x}}\right) \csc^2 \sqrt{x}\, dx = 2\int \csc^2 \sqrt{x}\left(\frac{1}{2\sqrt{x}}\right) dx$$

$$= 2\int \csc^2 u\, du$$

$$= -2 \cot u + C$$

$$= -2 \cot \sqrt{x} + C.$$

An integration formula for $\tan x$ may be obtained by first writing

$$\int \tan x\, dx = \int \frac{\sin x}{\cos x}\, dx.$$

If we now let $v = \cos x$, then $dv = -\sin x\, dx$ and hence

$$\int \tan x\, dx = -\int \frac{1}{\cos x}(-\sin x)\, dx = -\int \frac{1}{v}\, dv$$

$$= -\ln |v| + C.$$

Consequently, if $\cos x \neq 0$, then

(9.20) $$\int \tan x\, dx = -\ln |\cos x| + C.$$

Since $-\ln |\cos x| = \ln (1/|\cos x|) = \ln |\sec x|$, we may also write

(9.21) $$\int \tan x\, dx = \ln |\sec x| + C.$$

If $u = g(x)$ where g is a differentiable function, then (9.21) and (9.20) may be generalized to

(9.22) $$\int \tan u\, du = \ln |\sec u| + C = -\ln |\cos u| + C$$

where $du = g'(x)\, dx$.

Similarly, if we write $\cot u = \cos u/\sin u$ it follows that

(9.23) $$\int \cot u\, du = \ln |\sin u| + C$$

provided $\sin u \neq 0$.

Example 5 Evaluate $\int_0^{\pi/2} \tan \frac{x}{2}\, dx$.

Solution If $u = x/2$, then $du = (1/2)\, dx$. Making this change of variable and noting that

$u = 0$ if $x = 0$, and $u = \pi/4$ if $x = \pi/2$, we obtain

$$\begin{aligned}\int_0^{\pi/2} \tan\frac{x}{2}\,dx &= 2\int_0^{\pi/2} \tan\frac{x}{2}\cdot\frac{1}{2}\,dx\\ &= 2\int_0^{\pi/4} \tan u\,du\\ &= 2\ln\sec u\Big]_0^{\pi/4}\end{aligned}$$

where it is unnecessary to employ the absolute value sign as in (9.22) since $\sec u$ is positive if u is between 0 and $\pi/4$. Since $\ln\sec\pi/4 = \ln\sqrt{2} = (1/2)\ln 2$, and $\ln\sec 0 = \ln 1 = 0$, it follows that

$$\int_0^{\pi/2} \tan\frac{x}{2}\,dx = 2\cdot\frac{1}{2}\ln 2 = \ln 2.$$

An integral formula for $\sec x$ may be found by using the following technique:

$$\begin{aligned}\int \sec x\,dx &= \int \sec x\,\frac{\sec x + \tan x}{\sec x + \tan x}\,dx\\ &= \int \frac{\sec^2 x + \sec x\tan x}{\sec x + \tan x}\,dx.\end{aligned}$$

If $v = \tan x + \sec x$, then $dv = (\sec^2 x + \sec x\tan x)\,dx$ and formal substitution gives us

$$\begin{aligned}\int \sec x\,dx &= \int \frac{1}{v}\,dv\\ &= \ln|v| + C\\ &= \ln|\sec x + \tan x| + C.\end{aligned}$$

As usual, if $u = g(x)$ and g is differentiable, then

(9.24) $$\int \sec u\,du = \ln|\sec u + \tan u| + C.$$

It can be shown in similar fashion that

(9.25) $$\int \csc u\,du = \ln|\csc u - \cot u| + C.$$

Example 6 Evaluate $\int e^{2x}\sec e^{2x}\,dx$.

Solution If we let $u = e^{2x}$, then $du = 2e^{2x}\,dx$, and

$$\begin{aligned}\int e^{2x} \sec e^{2x}\, dx &= \tfrac{1}{2} \int \sec e^{2x}(2e^{2x})\, dx \\ &= \tfrac{1}{2} \int \sec u\, du \\ &= \tfrac{1}{2} \ln |\sec u + \tan u| + C \\ &= \tfrac{1}{2} \ln |\sec e^{2x} + \tan e^{2x}| + C.\end{aligned}$$

Example 7 Evaluate $\int (\csc x - 1)^2\, dx$.

Solution

$$\begin{aligned}\int (\csc x - 1)^2\, dx &= \int (\csc^2 x - 2 \csc x + 1)\, dx \\ &= \int \csc^2 x\, dx - 2 \int \csc x\, dx + \int dx \\ &= -\cot x - 2 \ln |\csc x - \cot x| + x + C.\end{aligned}$$

Additional methods for integrating trigonometric expressions will be discussed in Chapter 10.

EXERCISES 9.3

Evaluate the integrals in Exercises 1–28.

1 $\int \sin 4x\, dx$

2 $\int \sec^2 5x\, dx$

3 $\int \tan 3x \sec 3x\, dx$

4 $\int x^2 \cot x^3 \csc x^3\, dx$

5 $\int (\tan 3x + \sec 3x)\, dx$

6 $\int \frac{1}{\sec 2x}\, dx$

7 $\int \frac{1}{\cos 2x}\, dx$

8 $\int \frac{\cot \sqrt[3]{x}}{\sqrt[3]{x^2}}\, dx$

9 $\int x \csc^2 (x^2 + 1)\, dx$

10 $\int (x + \csc 8x)\, dx$

11 $\int \cot 6x \sin 6x\, dx$

12 $\int \sin 2x \tan 2x\, dx$

13 $\int_0^{\pi/4} \tan x \sec^2 x\, dx$

14 $\int \csc^2 x \cot x\, dx$

15 $\int \frac{\tan^2 2x}{\sec 2x}\, dx$

16 $\int_{\pi/6}^{\pi/2} \frac{\cos^2 x}{\sin x}\, dx$

17 $\int_{\pi/6}^{\pi} \sin x \cos x\, dx$

18 $\int \frac{\cos^2 x}{\csc x}\, dx$

19 $\displaystyle\int \frac{1 - \sin x}{x + \cos x}\,dx$

20 $\displaystyle\int \frac{e^x}{\cos e^x}\,dx$

21 $\displaystyle\int_{\pi/4}^{\pi/3} \frac{1 + \sin x}{\cos^2 x}\,dx$

22 $\displaystyle\int_0^{\pi/4} (1 + \sec x)^2\,dx$

23 $\displaystyle\int e^x(1 + \cos e^x)\,dx$

24 $\displaystyle\int (\csc^2 x)2^{\cot x}\,dx$

25 $\displaystyle\int \frac{e^{\cos x}}{\csc x}\,dx$

26 $\displaystyle\int \frac{\tan e^{-3x}}{e^{3x}}\,dx$

27 $\displaystyle\int \frac{\sec^2 x}{2\tan x + 1}\,dx$

28 $\displaystyle\int \frac{2\sin x}{1 + 3\cos x}\,dx$

29 Find the area of the region under the graph of $y = \sin(x/2)$ from $x = 0$ to $x = \pi$.

30 Find the area of the region under the graph of $y = 2\tan x$ from $x = 0$ to $x = \pi/4$.

31 Find the area of the region bounded by the graphs of $y = \sec x$, $y = x$, $x = -\pi/4$, and $x = \pi/4$.

32 Find the area of the region bounded by the graphs of $y = \sin x$, $y = \cos x$, $x = -\pi/2$, and $x = \pi/6$.

33 The region bounded by the graphs of $y = \sec x$, $x = -\pi/3$, $x = \pi/3$, and $y = 0$ is revolved about the x-axis. Find the volume of the resulting solid.

34 The region under the graph of $y = \sin x$ from $x = 0$ to $x = \pi$ is revolved about the x-axis. Find the volume of the resulting solid. (*Hint:* $\sin^2 x = (1 - \cos 2x)/2$.)

35 Derive (9.25).

36 Derive (9.23).

37 Show, by evaluating in two different ways, that

$$\int \tan x \sec^2 x\,dx = \tfrac{1}{2}\tan^2 x + C = \tfrac{1}{2}\sec^2 x + D.$$

How can the answers be reconciled?

38 Show, by evaluating in three different ways, that

$$\int \sin x \cos x\,dx = \tfrac{1}{2}\sin^2 x + C = -\tfrac{1}{2}\cos^2 x + D = -\tfrac{1}{4}\cos 2x + E.$$

How can the answers be reconciled?

In Exercises 39 and 40, (a) set up an integral for finding the arc length L, from A to B, of the graph of the given equation. (b) Write out the formula for approximating L by means of Simpson's Rule (5.49) with $n = 4$. (c) If a calculator is available, find the approximation given by the formula in part (b).

39 $y = \sin x$; $A(0, 0)$, $B(\pi/2, 1)$

40 $y = \tan x$; $A(0, 0)$, $B(\pi/4, 1)$

9.4 INVERSE TRIGONOMETRIC FUNCTIONS

As in Section 1.6, if f is a one-to-one function with domain Y and range X, then its inverse function f^{-1} has domain X and range Y. Moreover,

(9.26) $$f^{-1}(x) = y \quad \text{if and only if} \quad f(y) = x$$

for every x in X and every y in Y. Since the trigonometric functions are not one-to-one they do not have inverses. However, by restricting the domains it is possible to obtain functions that behave in the same way as the trigonometric functions (over the smaller domains) and which do possess inverse functions.

To begin, consider the sine function with domain $\mathbb{R}$ and range the set of real numbers in the interval $[-1, 1]$. This function is not one-to-one since, for example, numbers such as $\pi/6$, $5\pi/6$, and $-7\pi/6$ lead to the same functional value, $1/2$. It is easy to find a subset S of $\mathbb{R}$ with the property that as x ranges through S, $\sin x$ takes on each value between -1 and 1 once and only once. It is convenient to choose the interval $[-\pi/2, \pi/2]$ for S. The new function obtained by restricting the domain of the sine function to $[-\pi/2, \pi/2]$ is continuous and increasing and hence, by Theorem (8.38), has an inverse function. Applying (9.26) leads to the following definition.

(9.27) Definition

The **inverse sine function**, denoted $\sin^{-1}$, is defined by

$$\sin^{-1} x = y \quad \text{if and only if} \quad \sin y = x$$

where $-1 \le x \le 1$ and $-\pi/2 \le y \le \pi/2$.

It is also customary to refer to this function as the **arcsine function** and use $\arcsin x$ in place of $\sin^{-1} x$. We shall employ both notations in our work. The -1 in $\sin^{-1}$ is not to be regarded as an exponent, but rather as a means of denoting this inverse function. Observe that by (9.27),

$$-\pi/2 \le \sin^{-1} x \le \pi/2 \quad \text{or} \quad -\pi/2 \le \arcsin x \le \pi/2.$$

It follows from Definition (9.27) that the graph of $y = \sin^{-1} x$ (or $y = \arcsin x$) may be found by sketching the graph of $x = \sin y$, where $-\pi/2 \le y \le \pi/2$ (see Figure 9.8).

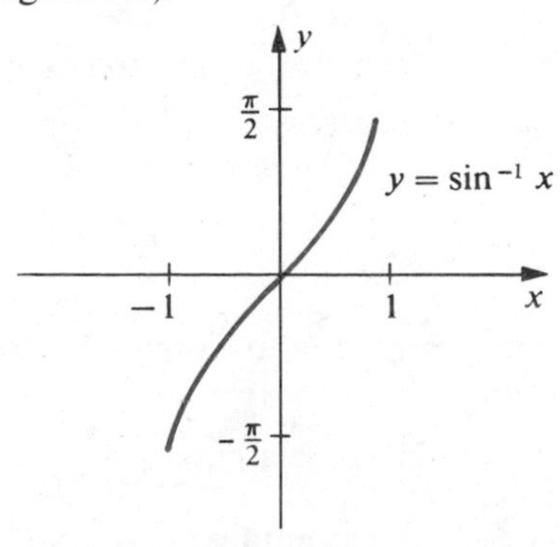

Figure 9.8

As in (1.24) we may write

(9.28)
$$\sin^{-1}(\sin x) = x \quad \text{if} \quad -\pi/2 \le x \le \pi/2$$
$$\sin(\sin^{-1} x) = x \quad \text{if} \quad -1 \le x \le 1.$$

Of course, the formulas in (9.28) may also be written

$$\arcsin(\sin x) = x \quad \text{and} \quad \sin(\arcsin x) = x.$$

Example 1 Find $\sin^{-1}(\sqrt{2}/2)$ and $\arcsin(-1/2)$.

Solution If $y = \sin^{-1}(\sqrt{2}/2)$, then by (9.27), $\sin y = \sqrt{2}/2$ and consequently $y = \pi/4$. Note that it is essential to choose y in the interval $[-\pi/2, \pi/2]$. A number such as $3\pi/4$ is incorrect, even though $\sin(3\pi/4) = \sqrt{2}/2$.

In like manner, if $y = \arcsin(-1/2)$, then $\sin y = -1/2$ and hence $y = -\pi/6$.

Similar discussions may be given for the other trigonometric functions. The procedure is first to determine a convenient subset of the domain such that a one-to-one function is obtained and then to apply (9.26).

If the domain of the cosine function is restricted to the interval $[0, \pi]$ there results a continuous, decreasing function which has a continuous decreasing inverse function by Theorem (8.40). This leads to the next definition.

(9.29) Definition

The **inverse cosine function**, denoted by $\cos^{-1}$, is defined by

$$\cos^{-1} x = y \quad \text{if and only if} \quad \cos y = x$$

where $-1 \le x \le 1$ and $0 \le y \le \pi$.

The inverse cosine function is also referred to as the **arccosine function** and the notation $\arccos x$ is used interchangeably with $\cos^{-1} x$.

As in (9.28),

$$\cos^{-1}(\cos x) = \arccos(\cos x) = x \quad \text{if} \quad 0 \le x \le \pi$$
$$\cos(\cos^{-1} x) = \cos(\arccos x) = x \quad \text{if} \quad -1 \le x \le 1.$$

(9.30) Definition

The **inverse tangent** or **arctangent function**, denoted by $\tan^{-1}$ or arctan, is defined by

$$\tan^{-1} x = \arctan x = y \quad \text{if and only if} \quad \tan y = x$$

where x is any real number and $-\pi/2 < y < \pi/2$.

The graphs of the inverse cosine and inverse tangent functions are sketched in Figure 9.9.

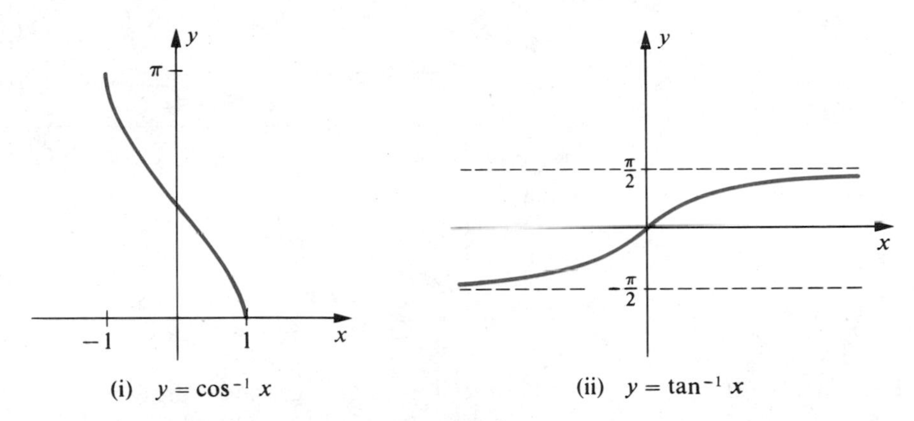

Figure 9.9

Example 2 If $-1 \le x \le 1$, rewrite $\cos(\sin^{-1} x)$ as an algebraic expression in x.

Solution Let $y = \sin^{-1} x$, so that $\sin y = x$. We wish to express $\cos y$ in terms of x. Since $-\pi/2 \le y \le \pi/2$, it follows that $\cos y \ge 0$, and hence

$$\cos y = \sqrt{1 - \sin^2 y} = \sqrt{1 - x^2}.$$

Consequently,

$$\cos(\sin^{-1} x) = \sqrt{1 - x^2}.$$

If y varies through the two intervals $[0, \pi/2)$ and $[\pi, 3\pi/2)$, then $\sec y$ takes on each of its values once and only once and an inverse function may be defined.

(9.31) Definition

The **inverse secant** or **arcsecant function**, denoted by $\sec^{-1}$ or arcsec, is defined by

$$\sec^{-1} x = \operatorname{arcsec} x = y \quad \text{if and only if} \quad \sec y = x$$

where $|x| \ge 1$ and y is in $[0, \pi/2)$ or in $[\pi, 3\pi/2)$.

The graph of $y = \sec^{-1} x$ is sketched in Figure 9.10. The main reason for choosing y as in (9.31) instead of the more natural intervals $[0, \pi/2)$ and $(\pi/2, \pi]$ is that the differentiation formula for the inverse secant is simpler.

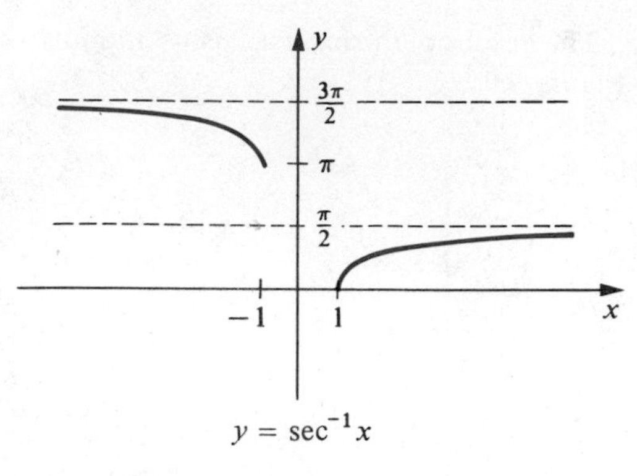

$y = \sec^{-1} x$

Figure 9.10

Example 3 If $|x| \geq 1$, rewrite $\tan(\sec^{-1} x)$ as an algebraic expression in x.

Solution Let $y = \sec^{-1} x$, so that $\sec y = x$. Our objective is to express $\tan y$ in terms of x. From Definition (9.31), either $0 \leq y < \pi/2$ or $\pi \leq y < 3\pi/2$. Consequently, $\tan y$ is always nonnegative and we may write

$$\tan y = \sqrt{\sec^2 y - 1} = \sqrt{x^2 - 1}$$

that is

$$\tan(\sec^{-1} x) = \sqrt{x^2 - 1}.$$

EXERCISES 9.4

Find the exact values in Exercises 1–18.

1 (a) $\sin^{-1} \sqrt{3}/2$ (b) $\sin^{-1}(-\sqrt{3}/2)$

2 (a) $\sin^{-1} 0$ (b) $\arccos 0$

3 (a) $\cos^{-1} \sqrt{2}/2$ (b) $\cos^{-1}(-\sqrt{2}/2)$

4 (a) $\arcsin(-1)$ (b) $\cos^{-1}(-1)$

5 (a) $\tan^{-1} \sqrt{3}$ (b) $\arctan(-\sqrt{3})$

6 (a) $\tan^{-1}(-1)$ (b) $\arccos(1/2)$

7 $\sin[\cos^{-1} \sqrt{3}/2]$

8 $\cos[\sin^{-1} 0]$

9 $\sin[\arccos(3/5)]$

10 $\tan[\tan^{-1} 10]$

11 $\arcsin(\sin \sqrt{5})$

12 $\tan^{-1}(\cos 0)$

13 $\cos[\sin^{-1} \frac{3}{5} + \tan^{-1} \frac{4}{3}]$

14 $\sin[\arcsin \frac{1}{2} + \arccos 0]$

15 $\tan[\arctan \frac{3}{4} + \arccos \frac{3}{5}]$

16 $\cos[2 \sin^{-1} \frac{8}{17}]$

17 $\sin[2 \arccos(-\frac{4}{5})]$

18 $\sin(\arctan \frac{1}{2} - \arccos \frac{4}{5})$

Rewrite the expressions in Exercises 19–22 as algebraic expressions in x.

19 $\sin(\tan^{-1} x)$

20 $\tan(\arccos x)$

21 $\cos((1/2)\arccos x)$

22 $\cos(2\tan^{-1} x)$

Verify the identities in Exercises 23–30.

23 $\sin^{-1} x + \cos^{-1} x = \pi/2$

24 $\arctan x + \arctan(1/x) = \pi/2, x > 0$

25 $\arcsin \dfrac{2x}{1+x^2} = 2\arctan x, |x| \leq 1$

26 $2\cos^{-1} x = \cos^{-1}(2x^2 - 1), 0 \leq x \leq 1$

27 $\sin^{-1}(-x) = -\sin^{-1} x$

28 $\arccos(-x) = \pi - \arccos x$

29 $\sin^{-1} x = \tan^{-1} \dfrac{x}{\sqrt{1-x^2}}$

30 $\tan^{-1} x + \tan^{-1} y = \tan^{-1} \dfrac{x+y}{1-xy}$

31 Define $\cot^{-1}$ by restricting the domain of cot to the interval $(0, \pi)$.

32 Define $\csc^{-1}$ by restricting the domain of csc to $(0, \pi/2]$ and $(\pi, 3\pi/2]$.

Sketch the graphs of the equations in Exercises 33–38.

33 $y = \sin^{-1} 2x$

34 $y = \frac{1}{2}\tan^{-1} x$

35 $y = 2\sin^{-1} x$

36 $y = \tan^{-1} \frac{1}{2}x$

37 $y - 2 = \sin^{-1}(x + 4)$

38 $y = -1 + \tan^{-1}(x - 2)$

Show that the equations in Exercises 39 and 40 are *not* identities.

39 $\tan^{-1} x = \dfrac{1}{\tan x}$

40 $(\arcsin x)^2 + (\arccos x)^2 = 1$

9.5 DERIVATIVES AND INTEGRALS OF INVERSE TRIGONOMETRIC FUNCTIONS

If we let $f(x) = \sin x$ and $g(x) = \sin^{-1} x$ in Theorem (8.41), then it follows that the inverse sine function g is differentiable if $|x| < 1$. We shall use implcit differentiation to find $g'(x)$. First note that by Definition (9.27)

$$y = \sin^{-1} x \quad \text{if and only if} \quad \sin y = x$$

where $-1 < x < 1$, and $-\pi/2 < y < \pi/2$. Differentiating the last equation implicitly we obtain

$$\cos y\, D_x y = 1$$

and hence

$$D_x y = D_x \sin^{-1} x = \frac{1}{\cos y}.$$

Since $-\pi/2 < y < \pi/2$, $\cos y$ is positive and, therefore,

$$\cos y = \sqrt{1 - \sin^2 y} = \sqrt{1 - x^2}.$$

Substitution in the previous formula for $D_x y$ gives us

$$D_x \sin^{-1} x = \frac{1}{\sqrt{1 - x^2}} \tag{9.32}$$

provided $|x| < 1$. The inverse sine function is not differentiable at ± 1. This fact is geometrically evident from Figure 9.8, since vertical tangent lines occur at the end-points of the graph.

If $u = k(x)$, where k is differentiable and $|k(x)| < 1$, then an application of the Chain Rule gives us

$$D_x \sin^{-1} u = \frac{1}{\sqrt{1 - u^2}} D_x u. \tag{9.33}$$

In similar fashion, to find the derivative of the inverse cosine function we begin with the equivalent equations

$$y = \cos^{-1} x \quad \text{and} \quad \cos y = x$$

where $|x| < 1$ and $0 < y < \pi$. Differentiating the second equation implicitly gives us

$$-\sin y \, D_x y = 1$$

and, therefore,

$$D_x y = D_x \cos^{-1} x = -\frac{1}{\sin y}.$$

Since $0 < y < \pi$, $\sin y = \sqrt{1 - \cos^2 y} = \sqrt{1 - x^2}$. Consequently, if $|x| < 1$, then

$$D_x \cos^{-1} x = -\frac{1}{\sqrt{1 - x^2}}.$$

More generally, if $u = k(x)$ where k is differentiable and $|k(x)| < 1$, then

$$D_x \cos^{-1} u = -\frac{1}{\sqrt{1 - u^2}} D_x u. \tag{9.34}$$

Example 1 Find dy/dx if $y = \sin^{-1} 3x - \cos^{-1} 3x$.

Solution Applying (9.33) and (9.34) with $u = 3x$,

$$\frac{dy}{dx} = \frac{3}{\sqrt{1 - 9x^2}} + \frac{3}{\sqrt{1 - 9x^2}} = \frac{6}{\sqrt{1 - 9x^2}}.$$

It follows from Theorem (8.10) that the inverse tangent function is differentiable at every real number. If we consider the equivalent equations

$$y = \tan^{-1} x \quad \text{and} \quad \tan y = x$$

where $-\pi/2 < y < \pi/2$, and differentiate the second equation implicitly, there results

$$\sec^2 y\, D_x y = 1.$$

Consequently,

$$D_x y = D_x \tan^{-1} x = \frac{1}{\sec^2 y}.$$

Using the fact that $\sec^2 y = 1 + \tan^2 y = 1 + x^2$ gives us

(9.35)
$$D_x \tan^{-1} x = \frac{1}{1 + x^2}.$$

Using the Chain Rule, the preceding formula can be generalized to

(9.36)
$$D_x \tan^{-1} u = \frac{1}{1 + u^2} D_x u$$

where $u = k(x)$ and k is differentiable.

Example 2 Find $f'(x)$ if $f(x) = \tan^{-1} e^{2x}$.

Solution Using (9.36) with $u = e^{2x}$,

$$f'(x) = \frac{1}{1 + (e^{2x})^2} D_x e^{2x} = \frac{2e^{2x}}{1 + e^{4x}}.$$

Finally, consider the equivalent equations

$$y = \sec^{-1} x \quad \text{and} \quad \sec y = x$$

where y is chosen in either $[0, \pi/2)$ or $[\pi, 3\pi/2)$. Differentiating the second equation implicitly gives us

$$\sec y \tan y\, D_x y = 1$$

or, if $\tan y \neq 0$

$$D_x y = D_x \sec^{-1} x = \frac{1}{\sec y \tan y}.$$

Using the fact that $\tan y = \sqrt{\sec^2 y - 1} = \sqrt{x^2 - 1}$, we have

(9.37)
$$D_x \sec^{-1} x = \frac{1}{x\sqrt{x^2 - 1}}$$

provided $|x| > 1$. The inverse secant function is not differentiable at $x = \pm 1$.

Indeed, the graph has vertical tangent lines at the points with these abscissas (see Figure 9.10).

By the Chain Rule, if $u = k(x)$ where k is differentiable and $|k(x)| > 1$, then

(9.38)
$$D_x \sec^{-1} u = \frac{1}{u\sqrt{u^2 - 1}} D_x u.$$

Example 3 Find y' if $y = \sec^{-1}(x^2)$.

Solution Applying (9.38) with $u = x^2$,

$$y' = \frac{1}{x^2\sqrt{x^4 - 1}}(2x) = \frac{2}{x\sqrt{x^4 - 1}}.$$

Following the same pattern employed many times before, we may use the differentiation formulas for the inverse trigonometric functions to obtain integration formulas. The proof of each of the following is left to the reader:

(9.39)
$$\int \frac{1}{\sqrt{1 - u^2}}\, du = \sin^{-1} u + C$$

(9.40)
$$\int \frac{1}{1 + u^2}\, du = \tan^{-1} u + C$$

(9.41)
$$\int \frac{1}{u\sqrt{u^2 - 1}}\, du = \sec^{-1} u + C$$

where it is assumed that u is suitably restricted.

Example 4 Evaluate $\int \frac{e^{2x}}{\sqrt{1 - e^{4x}}}\, dx$.

Solution If we let $u = e^{2x}$, then $du = 2e^{2x}\, dx$ and the integral may be written in the form (9.39) as follows:

$$\begin{aligned}
\int \frac{e^{2x}}{\sqrt{1 - e^{4x}}}\, dx &= \frac{1}{2}\int \frac{2e^{2x}}{\sqrt{1 - (e^{2x})^2}}\, dx \\
&= \frac{1}{2}\int \frac{1}{\sqrt{1 - u^2}}\, du \\
&= \frac{1}{2}\sin^{-1} u + C \\
&= \frac{1}{2}\sin^{-1} e^{2x} + C.
\end{aligned}$$

Formulas (9.39)–(9.41) can be extended as follows:

(9.42)
$$\int \frac{1}{\sqrt{a^2 - u^2}}\,du = \sin^{-1}\frac{u}{a} + C$$

(9.43)
$$\int \frac{1}{a^2 + u^2}\,du = \frac{1}{a}\tan^{-1}\frac{u}{a} + C$$

(9.44)
$$\int \frac{1}{u\sqrt{u^2 - a^2}}\,du = \frac{1}{a}\sec^{-1}\frac{u}{a} + C.$$

Let us prove (9.43). If $u = g(x)$ and g is differentiable, then by (9.35),

$$\begin{aligned} D_x\left[\frac{1}{a}\tan^{-1}\frac{g(x)}{a}\right] &= \frac{1}{a}\,\frac{1}{1 + [g(x)/a]^2}\,D_x\left[\frac{1}{a}g(x)\right] \\ &= \frac{1}{a}\,\frac{a^2}{a^2 + [g(x)]^2}\,\frac{1}{a}g'(x) \\ &= \frac{1}{a^2 + [g(x)]^2}g'(x). \end{aligned}$$

Consequently,

$$\int \frac{1}{a^2 + [g(x)]^2}g'(x)\,dx = \frac{1}{a}\tan^{-1}\frac{g(x)}{a} + C$$

which is the same as (9.43). The verifications of (9.42) and (9.44) are left as exercises.

Example 5 Evaluate $\int \frac{x^2}{5 + x^6}\,dx.$

Solution If we let $u = x^3$, then $du = 3x^2\,dx$ and the given integral may be written in form (9.43) as follows:

$$\begin{aligned} \int \frac{x^2}{5 + x^6}\,dx &= \frac{1}{3}\int \frac{3x^2}{5 + (x^3)^2}\,dx \\ &= \frac{1}{3}\int \frac{1}{(\sqrt{5})^2 + u^2}\,du \\ &= \frac{1}{3}\cdot\frac{1}{\sqrt{5}}\tan^{-1}\frac{u}{\sqrt{5}} + C \\ &= \frac{\sqrt{5}}{15}\tan^{-1}\frac{x^3}{\sqrt{5}} + C. \end{aligned}$$

Example 6 Evaluate $\int \frac{1}{x\sqrt{x^4 - 9}}\,dx.$

Solution If we let $u = x^2$, then $du = 2x\,dx$ and the integral may be transformed into form (9.44) as follows:

$$\begin{aligned}\int \frac{1}{x\sqrt{x^4 - 9}}\,dx &= \frac{1}{2}\int \frac{1}{x^2\sqrt{(x^2)^2 - 9}}\,2x\,dx \\ &= \frac{1}{2}\int \frac{1}{u\sqrt{u^2 - 9}}\,du \\ &= \frac{1}{2}\cdot\frac{1}{3}\sec^{-1}\frac{u}{3} + C \\ &= \frac{1}{6}\sec^{-1}\frac{x^2}{3} + C.\end{aligned}$$

EXERCISES 9.5

In Exercises 1–24 find $f'(x)$ if $f(x)$ equals the given expression.

1 $\tan^{-1}(3x - 5)$

2 $\sin^{-1}(x/3)$

3 $\sin^{-1}\sqrt{x}$

4 $\tan^{-1} x^2$

5 $e^{-x}\operatorname{arcsec} e^{-x}$

6 $\sqrt{\operatorname{arcsec} 3x}$

7 $x^2 \arctan x^2$

8 $\tan^{-1}\sin 2x$

9 $(1 + \cos^{-1} 3x)^3$

10 $x^2\sec^{-1} 5x$

11 $\ln\arctan x^2$

12 $\arcsin\ln x$

13 $1/\sin^{-1} x$

14 $\arctan\dfrac{x+1}{x-1}$

15 $\sec^{-1}\sqrt{x^2 - 1}$

16 $\left(\dfrac{1}{x} - \arcsin\dfrac{1}{x}\right)^4$

17 $(\arctan x)/(x^2 + 1)$

18 $\cos^{-1}\cos e^x$

19 $\sqrt{x}\sec^{-1}\sqrt{x}$

20 $\dfrac{e^{2x}}{\sin^{-1} 5x}$

21 $3^{\arcsin x^3}$

22 $x\arccos\sqrt{4x + 1}$

23 $(\tan x)^{\arctan x}$

24 $(\tan^{-1} 4x)e^{\tan^{-1} 4x}$

Find y' in Exercises 25 and 26.

25 $x^2 + x\sin^{-1} y = ye^x$

26 $\ln(x + y) = \tan^{-1} xy$

Evaluate the integrals in Exercises 27–42.

27 $\displaystyle\int_0^4 \frac{1}{x^2 + 16}\,dx$

28 $\displaystyle\int_0^1 \frac{e^x}{1 + e^{2x}}\,dx$

29 $\displaystyle\int_0^{\sqrt{2}/2} \frac{x}{\sqrt{1 - x^4}}\,dx$

30 $\displaystyle\int_{2/\sqrt{3}}^{2} \frac{1}{x\sqrt{x^2 - 1}}\,dx$

31 $\int \frac{\sin x}{\cos^2 x + 1}\,dx$

32 $\int \frac{\cos x}{\sqrt{9 - \sin^2 x}}\,dx$

33 $\int \frac{1}{\sqrt{x}(1 + x)}\,dx$

34 $\int \frac{1}{e^x\sqrt{1 - e^{-2x}}}\,dx$

35 $\int \frac{e^x}{\sqrt{16 - e^{2x}}}\,dx$

36 $\int \frac{\sec x \tan x}{1 + \sec^2 x}\,dx$

37 $\int \frac{1}{x\sqrt{x^6 - 4}}\,dx$

38 $\int \frac{x}{\sqrt{36 - x^2}}\,dx$

39 $\int \frac{x}{x^2 + 9}\,dx$

40 $\int \frac{1}{x\sqrt{x - 1}}\,dx$

41 $\int \frac{1}{\sqrt{e^{2x} - 25}}\,dx$

42 $\int \frac{e^x}{\sqrt{4 - e^x}}\,dx$

43 Find the area of the region bounded by the graphs of the equations $y = 4/\sqrt{16 - x^2}$, $x = -2$, $x = 2$, and $y = 0$.

44 If $f(x) = x^2/(1 + x^6)$, find the area of the region under the graph of f from $x = 0$ to $x = 1$.

45 The sides opposite and adjacent to an angle θ of a right triangle are measured as 10 feet and 7 feet, respectively, with a possible error of 1/2 inch in the 10-foot measurement. Use the differential of an inverse trigonometric function to approximate the error in the calculated value of θ.

46 Use differentials to approximate the change in $\arcsin x$ if x changes from 0.25 to 0.26.

47 An airplane at an altitude of 5 miles and a speed of 500 miles per hour is flying directly away from an observer on the ground. Use inverse trigonometric functions to find the rate at which the angle of elevation is changing when the airplane is over a point 2 miles from the observer.

48 A searchlight located 1/8 of a mile from the nearest point P on a straight road is trained on an automobile traveling on the road at a rate of 50 miles per hour. Use inverse trigonometric functions to find the rate at which the searchlight is rotating when the car is 1/4 of a mile from P.

49 A billboard 20 feet high is located on top of a building, with its lower edge 60 feet above the level of a viewer's eye. Use inverse trigonometric functions to find how far from a point directly below the sign a viewer should stand in order to maximize the angle between the lines of sight of the top and bottom of the billboard.

50 Given points $A(3, 1)$ and $B(6, 4)$ on a rectangular coordinate system, find the abscissa of the point P on the x-axis such that angle APB has its largest value.

51 Prove (9.42).

52 Prove (9.44).

53 Find equations for the tangent and normal lines to the graph of $y = \sin^{-1}(x - 1)$ at the point $(3/2, \pi/6)$.

54 Find the points on the graph of $y = \tan^{-1} 2x$ at which the tangent line is parallel to the line $13y - 2x + 5 = 0$.

55 Find the intervals in which the graph of $y = \tan^{-1} x$ is (a) concave upward; (b) concave downward.

56 The velocity, at time t, of a point moving on a straight line is $(1 + t^2)^{-1}$ ft/sec. If the point is at the origin at $t = 0$, find its position at the instant that the acceleration and velocity have the same absolute value.

57 The region bounded by the graphs of $y = e^x$, $y = 1/\sqrt{x^2 + 1}$, and $x = 1$ is revolved about the x-axis. Find the volume of the resulting solid.

58 Use differentials to approximate the arc length of the graph of $y = \tan^{-1} x$ from the point $(0, 0)$ to the point $(0.1, \tan^{-1} 0.1)$.

59 A missile is fired vertically from a point which is 5 miles from a tracking station and at the same elevation. For the first 20 seconds of flight its angle of elevation changes at a constant rate of 2°/sec. Find the velocity of the missile when the angle of elevation is 30°.

60 Prove that $\int_0^1 \frac{4}{1 + x^2}\,dx = \pi$ and then apply Simpson's Rule with $n = 10$ to approximate π.

9.6 HYPERBOLIC FUNCTIONS

The exponential expressions

$$\frac{e^x - e^{-x}}{2} \quad \text{and} \quad \frac{e^x + e^{-x}}{2}$$

occur frequently in applied mathematics and engineering. Their behavior is, in many ways, similar to that of $\sin x$ and $\cos x$. Moreover, as we shall see, they are related to a hyperbola in much the same way that $\sin x$ and $\cos x$ are related to a unit circle. Because of these facts, the expressions are called the *hyperbolic sine* and the *hyperbolic cosine* of x and are used to define the following **hyperbolic functions**.

(9.45) Definition

The **hyperbolic sine function**, denoted by **sinh**, and the **hyperbolic cosine function**, denoted by **cosh**, are given by

$$\sinh x = \frac{e^x - e^{-x}}{2} \quad \text{and} \quad \cosh x = \frac{e^x + e^{-x}}{2}$$

where x is any real number.

The graph of $y = \cosh x$ may be found by the method called **addition of ordinates**. To use this technique, the graphs of $y = (1/2)e^x$ and $y = (1/2)e^{-x}$ are sketched as indicated by the dashes in Figure 9.11. The ordinates of points on the graph of $y = \cosh x$ may then be obtained by adding ordinates of points on the other two graphs. This results in the sketch shown in the figure. It is evident from the graph that the range of cosh is $[1, \infty)$. This can also be proved directly.

The graph of $y = \sinh x$ may be obtained by adding ordinates of the graphs of $y = (1/2)e^x$ and $y = -(1/2)e^{-x}$ as illustrated in Figure 9.12.

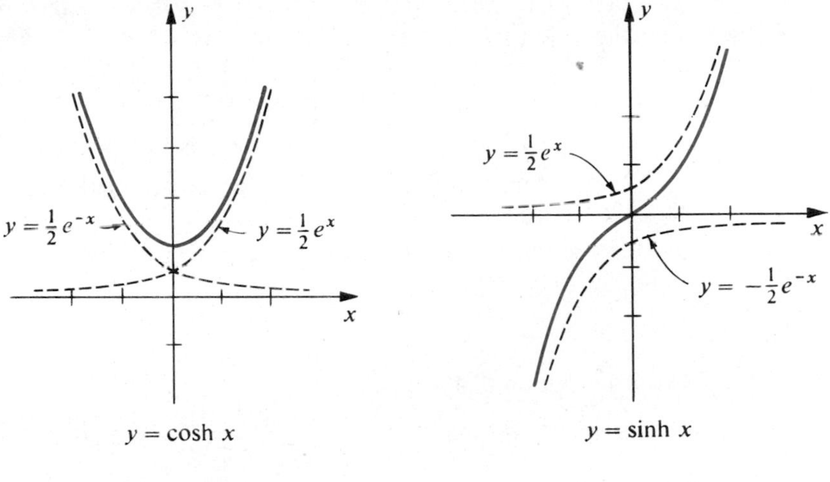

Figure 9.11 *Figure 9.12*

The hyperbolic cosine function can be used to describe the shape of a uniform flexible cable, or chain, whose ends are supported from the same height. This is often the case for telephone or power lines, as illustrated in Figure 9.13. At first glance the shape of the cable has the general appearance of a parabola, but this is not the case. If we introduce a coordinate system as indicated in the figure, then it can be shown that an equation which corresponds to the shape of the cable is $y = a\cosh(x/a)$, where a is a real number. The graph is called a **catenary**, after the Latin word for *chain*.

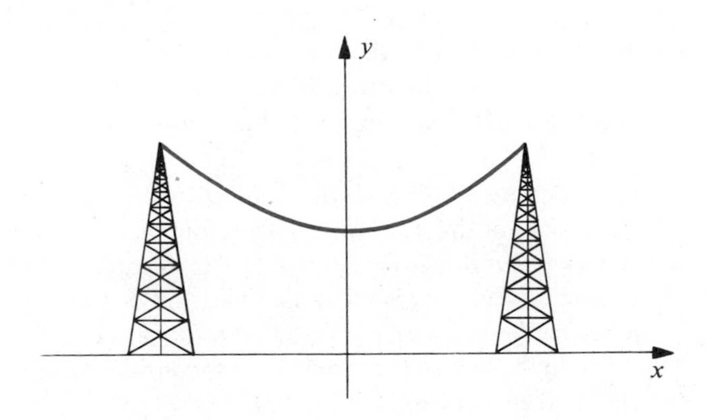

Figure 9.13. Catenary; $y = a\cosh x/a$

Another application of the hyperbolic cosine function occurs in the analysis of motion in a resisting medium. If an object is dropped from a given height and if air resistance is neglected, then the distance y that it falls in t seconds is given by

$y = (1/2)gt^2$, where g is a gravitational constant. However, air resistance cannot always be neglected. Indeed, as the velocity of the object increases, air resistance may significantly affect its motion. For example, if the air resistance is directly proportional to the square of the velocity, then it can be shown that the distance y that the object falls in t seconds is given by

$$y = A \ln(\cosh Bt)$$

where A and B are constants.

Many identities similar to those for trigonometric functions hold for the hyperbolic sine and cosine functions. For example, if $\cosh^2 x$ and $\sinh^2 x$ denote $(\cosh x)^2$ and $(\sinh x)^2$, respectively, let us verify that

(9.46) $$\cosh^2 x - \sinh^2 x = 1.$$

From Definition (9.45),

$$\begin{aligned}\cosh^2 x - \sinh^2 x &= \left(\frac{e^x + e^{-x}}{2}\right)^2 - \left(\frac{e^x - e^{-x}}{2}\right)^2 \\ &= \frac{e^{2x} + 2 + e^{-2x}}{4} - \frac{e^{2x} - 2 + e^{-2x}}{4} \\ &= \frac{e^{2x} + 2 + e^{-2x} - e^{2x} + 2 - e^{-2x}}{4} \\ &= \frac{4}{4} = 1.\end{aligned}$$

Identity (9.46) is analogous to the trigonometric identity $\cos^2 x + \sin^2 x = 1$. Many other identities may be established. Some will be found in the exercises. In each case it is sufficient to express the hyperbolic functions in terms of exponential functions and show that one side of the equation can be transformed into the other. The hyperbolic identities are similar to (but not always identical to) certain trigonometric identities. Any differences which occur usually involve signs of terms.

Identity (9.46) can be used to justify the adjective "hyperbolic" in the definitions of sinh and cosh. The trigonometric functions are sometimes referred to as *circular* functions, because if P is the point on the unit circle shown in (i) of Figure 9.14, and if t is the radian measure of angle POB, then P has coordinates $(\cos t, \sin t)$. As t varies from 0 to 2π, the point P traverses the circle once in the counterclockwise direction. If we next consider a point P with coordinates $(\cosh t, \sinh t)$, then by (9.46), P is on the hyperbola $x^2 - y^2 = 1$ (see (ii) of Figure 9.14). If we let t vary through the set of all real numbers, the point P traces the right-hand branch of the hyperbola. In this case the variable t does not represent an angle, as was true for circular functions. We can, however, find an interesting relationship. First note that the area A of the circular sector POB in (i) of Figure 9.14 is

$$A = \tfrac{1}{2}(1)^2 t = \tfrac{1}{2}t$$

and hence

$$t = 2A.$$

It can be shown that in (ii) of Figure 9.14, $t = 2A$ where A is the area of the indicated *hyperbolic* sector.

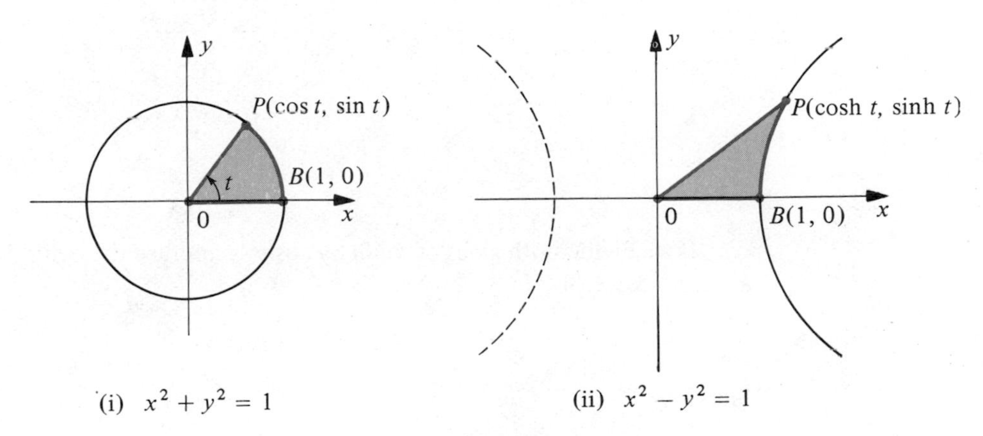

Figure 9.14

The striking analogy between the trigonometric and hyperbolic sine and cosine makes it natural to introduce functions which correspond to the four remaining trigonometric functions. Specifically, the hyperbolic tangent, cotangent, secant, and cosecant functions, denoted by tanh, coth, sech, and csch, respectively, are defined by

(9.47)

$$\tanh x = \frac{\sinh x}{\cosh x} = \frac{e^x - e^{-x}}{e^x + e^{-x}}$$

$$\coth x = \frac{\cosh x}{\sinh x} = \frac{e^x + e^{-x}}{e^x - e^{-x}}, \quad x \neq 0$$

$$\operatorname{csch} x = \frac{1}{\sinh x} = \frac{2}{e^x - e^{-x}}, \quad x \neq 0$$

$$\operatorname{sech} x = \frac{1}{\cosh x} = \frac{2}{e^x + e^{-x}}.$$

The graph of $y = \tanh x$ is sketched in Figure 9.15. Note that the lines $y = 1$ and $y = -1$ are horizontal asymptotes. We leave the verification of this, and the sketching of the graphs of the remaining hyperbolic functions, as exercises.

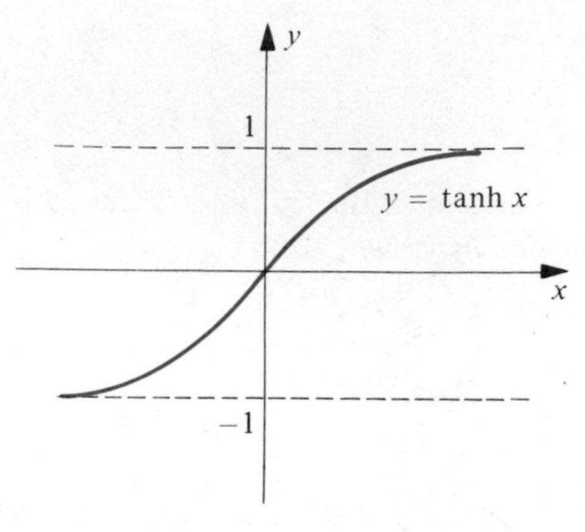

Figure 9.15

If we divide both sides of (9.46) by $\cosh^2 x$ and use the definitions of $\tanh x$ and $\operatorname{sech} x$ we obtain

(9.48)
$$1 - \tanh^2 x = \operatorname{sech}^2 x.$$

Similarly

$$\coth^2 x - 1 = \operatorname{csch}^2 x.$$

Once again the reader should carefully note the differences between these and the analogous identities studied in trigonometry. Other identities involving (9.47) appear in the exercises.

It is easy to find formulas for the derivatives of the hyperbolic functions. For example, since $D_x e^x = e^x$ and $D_x e^{-x} = -e^{-x}$,

$$D_x \sinh x = D_x\left(\frac{e^x - e^{-x}}{2}\right) = \frac{e^x + e^{-x}}{2} = \cosh x$$

$$D_x \cosh x = D_x\left(\frac{e^x + e^{-x}}{2}\right) = \frac{e^x - e^{-x}}{2} = \sinh x.$$

To differentiate $\tanh x$ we apply the Quotient Rule (3.22) as follows.

$$D_x \tanh x = D_x \frac{\sinh x}{\cosh x}$$

$$= \frac{\cosh x\, D_x \sinh x - \sinh x\, D_x \cosh x}{\cosh^2 x}$$

$$= \frac{\cosh^2 x - \sinh^2 x}{\cosh^2 x}$$

$$= \frac{1}{\cosh^2 x} = \operatorname{sech}^2 x.$$

The derivative formulas for the remaining three functions listed in (9.49) are left to the reader. As usual we may employ the Chain Rule to generalize each differentiation formula to the case $u = g(x)$ where g is differentiable; that is

(9.49)
$$\begin{aligned} D_x \sinh u &= \cosh u \, D_x u \\ D_x \cosh u &= \sinh u \, D_x u \\ D_x \tanh u &= \operatorname{sech}^2 u \, D_x u \\ D_x \coth u &= -\operatorname{csch}^2 u \, D_x u \\ D_x \operatorname{sech} u &= -\operatorname{sech} u \tanh u \, D_x u \\ D_x \operatorname{csch} u &= -\operatorname{csch} u \coth u \, D_x u. \end{aligned}$$

Example 1 Find $f'(x)$ if $f(x) = \cosh(x^2 + 1)$.

Solution Applying the second formula in (9.48), with $u = x^2 + 1$,

$$\begin{aligned} f'(x) &= \sinh(x^2 + 1) \cdot D_x(x^2 + 1) \\ &= 2x \sinh(x^2 + 1). \end{aligned}$$

The integration formulas corresponding to (9.49) are

(9.50)
$$\begin{aligned} \int \sinh u \, du &= \cosh u + C \\ \int \cosh u \, du &= \sinh u + C \\ \int \operatorname{sech}^2 u \, du &= \tanh u + C \\ \int \operatorname{csch}^2 u \, du &= -\coth u + C \\ \int \operatorname{sech} u \tanh u \, du &= -\operatorname{sech} u + C \\ \int \operatorname{csch} u \coth u \, du &= -\operatorname{csch} u + C. \end{aligned}$$

Example 2 Evaluate $\int x^2 \sinh(x^3) \, dx$.

Solution If we let $u = x^3$, then $du = 3x^2\,dx$ and

$$\begin{aligned}\int x^2 \sinh(x^3)\,dx &= \tfrac{1}{3}\int \sinh(x^3)(3x^2)\,dx \\ &= \tfrac{1}{3}\int \sinh u\,du \\ &= \tfrac{1}{3}\cosh u + C \\ &= \tfrac{1}{3}\cosh x^3 + C.\end{aligned}$$

EXERCISES 9.6

Verify the identities in Exercises 1–14.

1 $\cosh x + \sinh x = e^x$

2 $\cosh x - \sinh x = e^{-x}$

3 $\sinh(-x) = -\sinh x$

4 $\cosh(-x) = \cosh x$

5 $\sinh(x + y) = \sinh x \cosh y + \cosh x \sinh y$

6 $\cosh(x + y) = \cosh x \cosh y + \sinh x \sinh y$

7 $\sinh 2x = 2 \sinh x \cosh x$

8 $\cosh 2x = \cosh^2 x + \sinh^2 x$

9 $\tanh(x + y) = \dfrac{\tanh x + \tanh y}{1 + \tanh x \tanh y}$

10 $\tanh 2x = \dfrac{2 \tanh x}{1 + \tanh^2 x}$

11 $\cosh \dfrac{x}{2} = \sqrt{\dfrac{1 + \cosh x}{2}}$

12 $\tanh \dfrac{x}{2} = \dfrac{\sinh x}{1 + \cosh x}$

13 $\sinh x + \sinh y = 2 \sinh \frac{1}{2}(x + y) \cosh \frac{1}{2}(x - y)$

14 $(\cosh x + \sinh x)^n = \cosh nx + \sinh nx$ for every positive integer n (*Hint:* Use Exercise 1.)

In Exercises 15–30 find $f'(x)$ if $f(x)$ equals the given expression.

15 $\sinh 5x$

16 $\cosh \sqrt{4x^2 + 3}$

17 $\sqrt{x} \tanh \sqrt{x}$

18 $x \operatorname{csch} e^{4x}$

19 $\dfrac{\operatorname{sech} x^2}{x^2 + 1}$

20 $\dfrac{\coth x}{\cot x}$

21 $\cosh x^3$

22 $\sinh(x^2 + 1)$

23 $\cosh^3 x$

24 $\sinh^2 3x$

25 $\ln \sinh 2x$

26 $\arctan \tanh x$

27 $e^{3x} \operatorname{sech} x$

28 $\sqrt{\operatorname{sech} 5x}$

29 $\dfrac{1}{\tanh x + 1}$

30 $\dfrac{1 + \cosh x}{1 - \cosh x}$

31 Find y' if $\sinh xy = ye^x$.

32 Find y' if $x^2 \tanh y = \ln y$.

Evaluate the integrals in Exercises 33–44.

33 $\int \frac{\sinh \sqrt{x}}{\sqrt{x}} \, dx$

34 $\int \frac{\cosh \ln x}{x} \, dx$

35 $\int \frac{\cosh x}{\sinh x} \, dx$

36 $\int \frac{1}{\cosh^2 3x} \, dx$

37 $\int \sinh x \cosh x \, dx$

38 $\int \operatorname{sech}^2 x \tanh x \, dx$

39 $\int \tanh 3x \operatorname{sech} 3x \, dx$

40 $\int \sinh x \sqrt{\cosh x} \, dx$

41 $\int \tanh^2 3x \operatorname{sech}^2 3x \, dx$

42 $\int \tanh x \, dx$

43 $\int \frac{\operatorname{sech}^2 x}{1 - 2 \tanh x} \, dx$

44 $\int \frac{e^{\sinh x}}{\operatorname{sech} x} \, dx$

45 Find the area of the region bounded by the graphs of $y = \sinh 3x$, $y = 0$, and $x = 1$.

46 Find the arc length of the graph of $y = \cosh x$ from $x = 0$ to $x = 1$.

47 Find the points on the graph of $y = \sinh x$ at which the tangent line has slope 2.

48 The region bounded by the graphs of $y = \cosh x$, $x = -1$, $x = 1$, and $y = 0$ is revolved about the x-axis. Find the volume of the resulting solid.

49 Verify the graph of $y = \tanh x$ in Figure 9.15.

Sketch the graphs of the equations in Exercises 50–52.

50 $y = \coth x$

51 $y = \operatorname{sech} x$

52 $y = \operatorname{csch} x$

9.7 INVERSE HYPERBOLIC FUNCTIONS

The hyperbolic sine function is continuous and increasing for all x and hence, by Theorem (8.38), has a continuous, increasing inverse function, denoted by $\sinh^{-1}$. Moreover,

$$y = \sinh^{-1} x \quad \text{if and only if} \quad x = \sinh y.$$

The last equation can be used to find an explicit form for $\sinh^{-1} x$. Thus, if

$$x = \sinh y = \frac{e^y - e^{-y}}{2}$$

then

$$e^y - 2x - e^{-y} = 0.$$

Multiplying both sides by e^y yields

$$e^{2y} - 2xe^y - 1 = 0$$

and, applying the quadratic formula,

$$e^y = \frac{2x \pm \sqrt{4x^2+4}}{2} \quad \text{or} \quad e^y = x \pm \sqrt{x^2+1}.$$

Since e^y is never negative, the minus sign must be discarded. Doing this and then taking the natural logarithm of both sides of the equation, we obtain

$$y = \ln(x + \sqrt{x^2+1})$$

that is,

(9.51) $$\sinh^{-1} x = \ln(x + \sqrt{x^2+1}).$$

It follows from (9.51) that

$$\begin{aligned} D_x \sinh^{-1} x &= \frac{1}{x+\sqrt{x^2+1}}\left(1 + \frac{x}{\sqrt{x^2+1}}\right) \\ &= \frac{\sqrt{x^2+1}+x}{(x+\sqrt{x^2+1})\sqrt{x^2+1}} \\ &= \frac{1}{\sqrt{x^2+1}}. \end{aligned}$$

This formula can be extended in the usual way. Specifically, if $u = g(x)$, where g is a differentiable function, then

(9.52) $$D_x \sinh^{-1} u = \frac{1}{\sqrt{u^2+1}} D_x u.$$

It is left as an exercise to show that

(9.53) $$\int \frac{1}{\sqrt{u^2+a^2}}\, du = \sinh^{-1}\frac{u}{a} + C, \quad a > 0.$$

The remaining inverse hyperbolic functions are defined in similar fashion. As was the case with the trigonometric functions, it is sometimes necessary to restrict the domain in order for an inverse function to exist. For example, if the domain of cosh is restricted to the set of nonnegative real numbers, then the resulting function is continuous and increasing and its inverse function $\cosh^{-1}$ is defined by

$$y = \cosh^{-1} x \quad \text{if and only if} \quad \cosh y = x, \quad y \geq 0.$$

Employing the techniques used to analyze $\sinh^{-1} x$ we obtain

(9.54) $$\cosh^{-1} x = \ln(x + \sqrt{x^2-1}), \quad x \geq 1$$

(9.55) $$D_x \cosh^{-1} u = \frac{1}{\sqrt{u^2-1}} D_x u, \quad u > 1$$

(9.56)
$$\int \frac{1}{\sqrt{u^2 - a^2}}\,du = \cosh^{-1}\frac{u}{a} + C, \quad u > a > 0.$$

Similarly, the inverse hyperbolic tangent function, denoted by $\tanh^{-1}$, is defined by

$$y = \tanh^{-1} x \quad \text{if and only if} \quad \tanh y = x,$$

from which it follows that

(9.57)
$$\tanh^{-1} x = \frac{1}{2}\ln\frac{1+x}{1-x}, \quad |x| < 1$$

(9.58)
$$D_x \tanh^{-1} u = \frac{1}{1-u^2} D_x u, \quad |u| < 1$$

(9.59)
$$\int \frac{1}{a^2 - u^2}\,du = \frac{1}{a}\tanh^{-1}\frac{u}{a} + C, \quad a > 0, \quad |u| < a.$$

If the domain of sech is restricted to nonnegative numbers, there results a one-to-one function and we define

$$y = \operatorname{sech}^{-1} x \quad \text{if and only if} \quad \operatorname{sech} y = x, \quad y \geq 0.$$

It can be proved that

(9.60)
$$\operatorname{sech}^{-1} x = \ln\left(\frac{1 + \sqrt{1 - x^2}}{x}\right), \quad 0 < x \leq 1$$

(9.61)
$$D_x \operatorname{sech}^{-1} u = \frac{-1}{u\sqrt{1-u^2}} D_x u, \quad 0 < u < 1$$

(9.62)
$$\int \frac{1}{u\sqrt{a^2 - u^2}}\,du = -\frac{1}{a}\operatorname{sech}^{-1}\frac{|u|}{a} + C, \quad a > 0, \quad 0 < |u| < a.$$

Similar discussions can be given for $\coth^{-1}$ and csch^{-1}.

Example 1 Evaluate $\displaystyle\int \frac{1}{\sqrt{9x^2 + 25}}\,dx.$

Solution The integral may be expressed in form (9.53) by letting $u = 3x$ and $du = 3\,dx$. Thus

$$\int \frac{1}{\sqrt{9x^2+25}}\,dx = \frac{1}{3}\int \frac{1}{\sqrt{(3x)^2 + (5)^2}}\,3\,dx$$
$$= \frac{1}{3}\sinh^{-1}\frac{3x}{5} + C.$$

Example 2 Evaluate $\displaystyle\int \frac{e^x}{16 - e^{2x}}\,dx.$

Solution Letting $u = e^x$, $du = e^x\,dx$, and applying (9.59) we obtain

$$\begin{aligned}\int \frac{e^x}{16 - e^{2x}}\,dx &= \int \frac{1}{4^2 - (e^x)^2} e^x\,dx \\ &= \int \frac{1}{4^2 - u^2}\,du \\ &= \frac{1}{4}\tanh^{-1}\frac{u}{4} + C \\ &= \frac{1}{4}\tanh^{-1}\frac{e^x}{4} + C\end{aligned}$$

provided $e^x < 4$.

EXERCISES 9.7

Derive the formulas listed in Exercises 1–6.

1 (9.54) **2** (9.57) **3** (9.55)

4 (9.58) **5** (9.56) **6** (9.59)

Sketch the graphs of the equations in Exercises 7–10.

7 $y = \sinh^{-1} x$ **8** $y = \cosh^{-1} x$

9 $y = \tanh^{-1} x$ **10** $y = \operatorname{sech}^{-1} x$

In Exercises 11–20 find $f'(x)$ if $f(x)$ equals the given expression.

11 $\sinh^{-1} 5x$ **12** $\sinh^{-1} e^x$

13 $\cosh^{-1} \sqrt{x}$ **14** $\sqrt{\cosh^{-1} x}$

15 $\tanh^{-1} (x^2 - 1)$ **16** $\tanh^{-1} \sin 3x$

17 $x \sinh^{-1} (1/x)$ **18** $1/\sinh^{-1} x^2$

19 $\ln \cosh^{-1} 4x$ **20** $\cosh^{-1} \ln 4x$

Evaluate the integrals in Exercises 21–28.

21 $\displaystyle\int \frac{1}{\sqrt{9x^2 + 25}}\,dx$ **22** $\displaystyle\int \frac{1}{\sqrt{16x^2 - 9}}\,dx$

23 $\displaystyle\int \frac{1}{49 - 4x^2}\,dx$ **24** $\displaystyle\int \frac{\sin x}{\sqrt{1 + \cos^2 x}}\,dx$

25 $\displaystyle\int \frac{e^x}{\sqrt{e^{2x} - 16}}\,dx$ **26** $\displaystyle\int \frac{2}{5 - 3x^2}\,dx$

27 $\displaystyle\int \frac{1}{x\sqrt{9 - x^4}}\,dx$ **28** $\displaystyle\int \frac{1}{\sqrt{5 - e^{2x}}}\,dx$

29 Use (9.53) and (9.51) to prove that if $a > 0$, then

$$\int \frac{1}{\sqrt{u^2 + a^2}}\,du = \ln(u + \sqrt{u^2 + a^2}) + C.$$

30 Use (9.56) and (9.54) to prove that if $u > a > 0$, then

$$\int \frac{1}{\sqrt{u^2 - a^2}}\,du = \ln(u + \sqrt{u^2 - a^2}) + C.$$

9.8 REVIEW

Concepts

Define or discuss each of the following.

1 Differentiation formulas for trigonometric functions

2 Integration formulas for trigonometric functions

3 The inverse trigonometric functions

4 Differentiation and integration formulas for inverse trigonometric functions

5 Hyperbolic functions

6 Inverse hyperbolic functions

Exercises

Find $f'(x)$ if $f(x)$ is defined as in Exercises 1–44.

1 $\cos\sqrt{3x^2 + x}$

2 $x^2 \cot 2x$

3 $(\sec x + \tan x)^5$

4 $\sqrt[3]{x^3 + \csc 6x}$

5 $x^2 \operatorname{arcsec} x^2$

6 $\tan^{-1}(\ln 3x)$

7 $\dfrac{(3x + 7)^4}{\sin^{-1} 5x}$

8 $\dfrac{\sin 8x}{4x^2 - x}$

9 $(\cos x)^{x+1}$

10 $7^{\sin 3x}$

11 $\sqrt{2x^2 + \operatorname{sech} 4x}$

12 $x \sec^{-1} 4x$

13 $\ln(\csc^3 2x)$

14 $\log_{10}(\sin 2x)$

15 $\dfrac{1}{2x + \sec^2 x}$

16 $\cot\dfrac{1}{x} + \dfrac{1}{\cot x}$

17 $e^{\cos x} + (\cos x)^e$

18 $\dfrac{\ln \sinh x}{x}$

19 $\cosh e^{-5x}$

20 $(\cos x)^{\cot x}$

21 $\tanh^{-1}(\tanh \sqrt[3]{x})$

22 $\sec(\sec x)$

23 $2^{\arctan 2x}$

24 $(1 + \operatorname{arcsec} 2x)^{\sqrt{2}}$

25 $\sin^3 e^{-2x}$

26 $\ln \cos^2 3x$

27 $e^{-x^2} \cot x^2$

28 $\sec 5x \tan 5x$

29 $\dfrac{\csc x + 1}{\cot x + 1}$

30 $\dfrac{1 - x^2}{\arccos x}$

31 $\sin^{-1} \sqrt{1 - x^2}$

32 $\sqrt{\sin^{-1}(1 - x^2)}$

33 $\tan(\sin 3x)$

34 $\sqrt{\sin \sqrt{x}}$

35 $(\tan x + \tan^{-1} x)^4$

36 $e^{4x} \sec^{-1} e^{4x}$

37 $\tan^{-1}(\tan^{-1} x)$

38 $e^{x \cosh x}$

39 $e^{-x} \sinh e^{-x}$

40 $\ln \tanh(5x + 1)$

41 $\dfrac{\sinh x}{\cosh x - \sinh x}$

42 $\dfrac{1}{x} \tanh \dfrac{1}{x}$

43 $\sinh^{-1} x^2$

44 $\cosh^{-1} \tan x$

Evaluate the integrals in Exercises 45–76.

45 $\int \sin(3 - 5x)\, dx$

46 $\int \csc(x/2) \cot(x/2)\, dx$

47 $\int \dfrac{\sec^2 \sqrt{x}}{\sqrt{x}}\, dx$

48 $\int \dfrac{\sec(1/x)}{x^2}\, dx$

49 $\int (\cot 9x + \csc 9x)\, dx$

50 $\int x \csc^2(3x^2 + 4)\, dx$

51 $\int e^x \tan e^x\, dx$

52 $\int \cot 2x \csc 2x\, dx$

53 $\int (\csc 3x + 1)^2\, dx$

54 $\int \dfrac{\sin x + 1}{\cos x}\, dx$

55 $\int \dfrac{\sin 4x}{\tan 4x}\, dx$

56 $\int x \cot x^2\, dx$

57 $\int \dfrac{x}{4 + 9x^2}\, dx$

58 $\int \dfrac{1}{4 + 9x^2}\, dx$

59 $\int \dfrac{e^{2x}}{\sqrt{1 - e^{2x}}}\, dx$

60 $\int \dfrac{e^x}{\sqrt{1 - e^{2x}}}\, dx$

61 $\int \dfrac{x}{\operatorname{sech} x^2}\, dx$

62 $\int \dfrac{1}{x\sqrt{x^4 - 1}}\, dx$

63 $\int_{-1/2}^{1/2} \dfrac{1}{\sqrt{1 - x^2}}\, dx$

64 $\int_0^{\pi/2} \dfrac{\cos x}{1 + \sin^2 x}\, dx$

65 $\int \sec^2 x (1 + \tan x)^2\, dx$

66 $\int (1 + \cos^2 x) \sin x\, dx$

67 $\int \dfrac{\csc^2 x}{2 + \cot x}\, dx$

68 $\int (\sin x) e^{\cos x + 1}\, dx$

69 $\int \frac{\sinh(\ln x)}{x}\,dx$

70 $\int \operatorname{sech}^2(1-2x)\,dx$

71 $\int \frac{1}{\sqrt{9-4x^2}}\,dx$

72 $\int \frac{x}{\sqrt{9-4x^2}}\,dx$

73 $\int \frac{1}{x\sqrt{9-4x^2}}\,dx$

74 $\int \frac{1}{x\sqrt{4x^2-9}}\,dx$

75 $\int \frac{x}{\sqrt{25x^2+36}}\,dx$

76 $\int \frac{1}{\sqrt{25x^2+36}}\,dx$

77 If $f(x) = 2\sin x - \cos 2x$ find the local extrema and sketch the graph of f for $0 \le x \le 4\pi$. Find the equations of the tangent and normal lines to the graph of f at the point $(\pi/6, 1/2)$.

78 Find the area of the region bounded by the graphs of $y = x/(1+x^4)$, $x = 1$, and $y = 0$.

79 Find the points on the graph of $y = \sin^{-1} 3x$ at which the tangent line is parallel to the line through $A(2, -3)$ and $B(4, 7)$.

80 Find the equation of the tangent line to the graph of

$$y = \pi \tan \frac{y}{x} + x^2 - 16$$

at the point $(4, \pi)$.

81 The region between the graphs of $y = \cos(x^2)$ and the x-axis, from $x = 0$ to $x = \sqrt{\pi/2}$, is revolved about the y-axis. Find the volume of the resulting solid.

82 The position of a moving point on a coordinate line is given by

$$s(t) = a \sin kt + b \cos kt$$

where a, b, and k are constants. Prove that the magnitude of the acceleration is directly proportional to the distance from the origin.

83 Use differentials to approximate $(0.98)^2 \tan^{-1}(0.98)$.

84 Prove that $y = ae^{-x}\cos 2x + be^{-x}\sin 2x$, where a and b are constants, is a solution of the differential equation $y'' + 2y' + 5y = 0$.

85 Find the points of inflection and discuss the concavity of the graph of $y = x\sin^{-1} x$.

86 Two sides of a triangle are measured as 10 inches and 12 inches. If the included angle is measured as 30° with a possible error of 0.1°, use differentials to approximate the error in the calculated area of the triangle.

87 A ladder 12 feet long leans against a house. If the lower end slides along level ground at a rate of 3 feet per second, how fast is the angle between the ladder and the ground changing when the lower end is 6 feet from the house?

88 A V-shaped water gutter is to be constructed from two rectangular sheets of metal 10 inches wide. Find the angle between the sheets which will maximize the carrying capacity of the gutter.

Additional Techniques and Applications of Integration

In this chapter we shall discuss some techniques which can be used to evaluate many types of integrals. We also consider applications of the definite integral to the problems of finding moments and centers of mass of certain solids.

10.1 INTEGRATION BY PARTS

If f and g are differentiable functions, then by the Product Rule (3.20)

$$D_x[f(x)g(x)] = f(x)g'(x) + g(x)f'(x)$$

or, equivalently,

$$f(x)g'(x) = D_x[f(x)g(x)] - g(x)f'(x).$$

Integrating both sides of the previous equation gives us

$$\int f(x)g'(x)\,dx = \int D_x[f(x)g(x)]\,dx - \int g(x)f'(x)\,dx.$$

By (5.37) the first integral on the right side equals $f(x)g(x) + C$. Since another constant of integration results from the second integral, it is unnecessary to include C in the formula; that is,

$$\int f(x)g'(x)\,dx = f(x)g(x) - \int g(x)f'(x)\,dx.$$

If we let $u = f(x)$ and $v = g(x)$, so that $du = f'(x)\,dx$ and $dv = g'(x)\,dx$, then the preceding formula may be written

(10.1)
$$\int u\,dv = uv - \int v\,du.$$

In order to apply (10.1) to a given integral we let u represent part of the integrand and dv the remaining part (including dx). For this reason the technique of using formula (10.1) is called **integration by parts**. The following examples illustrate this important method of integration.

Example 1 Find $\int xe^{2x}\,dx$.

Solution There are four possible choices for dv, namely dx, $x\,dx$, $e^{2x}\,dx$, or $xe^{2x}\,dx$. If we let $dv = e^{2x}\,dx$, then the remaining part of the integrand is u; that is, $u = x$. To find v we integrate dv, obtaining $v = \frac{1}{2}e^{2x}$. Note that a constant of integration is not added at this stage of the solution. Since $u = x$ we see that $du = dx$. For ease of reference it is convenient to display these expressions as follows:

$$\begin{aligned} u &= x & dv &= e^{2x}\,dx \\ du &= dx & v &= \tfrac{1}{2}e^{2x}. \end{aligned}$$

Substituting the above expressions in formula (10.1), that is, *integrating by parts*, we obtain

$$\int xe^{2x}\,dx = x(\tfrac{1}{2}e^{2x}) - \int \tfrac{1}{2}e^{2x}\,dx.$$

The integral on the right side may be found by means of (8.22). This gives us

$$\int xe^{2x}\,dx = \tfrac{1}{2}xe^{2x} - \tfrac{1}{4}e^{2x} + C.$$

It takes considerable practice to become proficient in making a suitable choice for dv. To illustrate, if we had chosen $dv = x\,dx$ in Example 1, then it would have been necessary to let $u = e^{2x}$, giving us

$$\begin{aligned} u &= e^{2x} & dv &= x\,dx \\ du &= 2e^{2x}\,dx & v &= \tfrac{1}{2}x^2. \end{aligned}$$

Integrating by parts we obtain

$$\int xe^{2x}\,dx = \tfrac{1}{2}x^2e^{2x} - \int x^2e^{2x}\,dx.$$

Since the exponent associated with x has increased, the integral on the right is more complicated than the given integral. This indicates an incorrect choice for dv.

Example 2 Evaluate $\int x\sec^2 x\,dx$.

Solution Since $\sec^2 x$ can be integrated readily by means of (9.16), we let $dv = \sec^2 x\,dx$. Thus

$$\begin{aligned} u &= x & dv &= \sec^2 x\,dx \\ du &= dx & v &= \tan x \end{aligned}$$

and integration by parts gives us

$$\int x \sec^2 x \, dx = x \tan x - \int \tan x \, dx$$

$$= x \tan x - \ln |\sec x| + C.$$

We did not state a formula for $\int \ln x \, dx$ in Chapter 8. The reason for not doing so is that integration by parts is needed to find an antiderivative of the natural logarithmic function, as shown in the next example.

Example 3 Find $\displaystyle\int \ln x \, dx$.

Solution Let

$$u = \ln x \qquad dv = dx$$

$$du = \frac{1}{x} dx \qquad v = x.$$

Integrating by parts we obtain

$$\int \ln x \, dx = x \ln x - \int x \left(\frac{1}{x}\right) dx$$

$$= x \ln x - \int dx$$

$$= x \ln x - x + C.$$

Sometimes it is necessary to use integration by parts more than once in the same problem. This is illustrated in the next example.

Example 4 Find $\displaystyle\int x^2 e^{2x} \, dx$.

Solution Let

$$u = x^2 \qquad dv = e^{2x} \, dx$$

$$du = 2x \, dx \qquad v = \tfrac{1}{2} e^{2x} \, dx.$$

Integrating by parts we obtain

$$\int x^2 e^{2x} \, dx = x^2 (\tfrac{1}{2} e^{2x}) - \int (\tfrac{1}{2} e^{2x}) 2x \, dx$$

$$= \tfrac{1}{2} x^2 e^{2x} - \int x e^{2x} \, dx.$$

To evaluate the integral on the right side of the last equation we must again integrate by parts. Proceeding exactly as in Example 1 leads to

$$\int x^2 e^{2x}\,dx = \tfrac{1}{2}x^2 e^{2x} - \tfrac{1}{2}xe^{2x} + \tfrac{1}{4}e^{2x} + C.$$

The following example illustrates another device for finding certain integrals by means of two applications of the integration by parts formula (10.1).

Example 5 Find $\int e^x \cos x\,dx$.

Solution Let

$$u = e^x \qquad dv = \cos x\,dx$$
$$du = e^x\,dx \qquad v = \sin x.$$

Integrating by parts,

$$\text{(a)} \qquad \int e^x \cos x\,dx = e^x \sin x - \int e^x \sin x\,dx.$$

We next apply integration by parts to the integral on the right side of equation (a). Letting

$$u = e^x \qquad dv = \sin x\,dx$$
$$du = e^x\,dx \qquad v = -\cos x$$

and integrating by parts leads to

$$\text{(b)} \qquad \int e^x \sin x\,dx = -e^x \cos x + \int e^x \cos x\,dx.$$

If we now use equation (b) to substitute on the right side of equation (a) we obtain

$$\int e^x \cos x\,dx = e^x \sin x - \left[-e^x \cos x + \int e^x \cos x\,dx\right]$$

or

$$\int e^x \cos x\,dx = e^x \sin x + e^x \cos x - \int e^x \cos x\,dx.$$

Adding $\int e^x \cos x\,dx$ to both sides gives us

$$2\int e^x \cos x\,dx = e^x(\sin x + \cos x).$$

Finally, dividing both sides by 2 and adding the constant of integration, we have

$$\int e^x \cos x\,dx = \tfrac{1}{2}e^x(\sin x + \cos x) + C.$$

Some care must be taken when evaluating integrals of the type given in Example 5. To illustrate, suppose in the evaluation of the integral on the right in equation (a) of the solution we had used

$$u = \sin x \qquad dv = e^x\,dx$$
$$du = \cos x\,dx \qquad v = e^x.$$

In this event integration by parts leads to

$$\int e^x \sin x\,dx = e^x \sin x - \int e^x \cos x\,dx.$$

If we now substitute in (a) there results

$$\int e^x \cos x\,dx = e^x \sin x - \left[e^x \sin x - \int e^x \cos x\,dx \right]$$

which reduces to

$$\int e^x \cos x\,dx = \int e^x \cos x\,dx.$$

Although this is a true statement, it is not a solution to the problem! Incidentally, the integral in Example 5 *can* be evaluated by using $dv = e^x\,dx$ for *both* the first and second applications of the integration by parts formula (10.1).

Integration by parts may sometimes be employed to obtain **reduction formulas** for integrals. Such formulas can be used to write an integral involving powers of an expression in terms of integrals which involve lower powers of the expression.

Example 6 Find a reduction formula for $\int \sin^n x\,dx$.

Solution Let

$$u = \sin^{n-1} x \qquad dv = \sin x\,dx$$
$$du = (n-1)\sin^{n-2} x \cos x\,dx \qquad v = -\cos x.$$

Integrating by parts,

$$\int \sin^n x\,dx = -\cos x \sin^{n-1} x + (n-1)\int \sin^{n-2} x \cos^2 x\,dx.$$

Since $\cos^2 x = 1 - \sin^2 x$, we may write

$$\int \sin^n x\,dx = -\cos x \sin^{n-1} x + (n-1)\int \sin^{n-2} x\,dx - (n-1)\int \sin^n x\,dx.$$

Consequently,

$$\int \sin^n x\,dx + (n-1)\int \sin^n x\,dx = -\cos x \sin^{n-1} x + (n-1)\int \sin^{n-2} x\,dx.$$

The left side of the last equation reduces to $n \int \sin^n x\,dx$. Dividing both sides by n, we obtain

(10.2)
$$\int \sin^n x\,dx = -\frac{1}{n}\cos x \sin^{n-1} x + \frac{(n-1)}{n}\int \sin^{n-2} x\,dx.$$

As an illustration of the use of (10.2) let us consider the case $n = 4$. Thus

$$\int \sin^4 x\,dx = -\tfrac{1}{4}\cos x \sin^3 x + \tfrac{3}{4}\int \sin^2 x\,dx.$$

Applying (10.2) to the integral on the right,

$$\begin{aligned}\int \sin^2 x\,dx &= -\tfrac{1}{2}\cos x \sin x + \tfrac{1}{2}\int dx \\ &= -\tfrac{1}{2}\cos x \sin x + \tfrac{1}{2}x + C_1.\end{aligned}$$

Consequently

$$\int \sin^4 x\,dx = -\tfrac{1}{4}\cos x \sin^3 x - \tfrac{3}{8}\cos x \sin x + \tfrac{3}{8}x + C.$$

It should be evident that by repeated applications of (10.2) we can find $\int \sin^n x\,dx$ for any positive integer n because these reductions either end with $\int \sin x\,dx$ or $\int \sin^2 x\,dx$, and each of these can be evaluated easily.

EXERCISES 10.1

Evaluate the integrals in Exercises 1–38.

1 $\int xe^{-x}\,dx$

2 $\int x \sin x\,dx$

3 $\int x^2 e^{3x}\,dx$

4 $\int x^2 \sin 4x\,dx$

5 $\int x \cos 5x\,dx$

6 $\int xe^{-2x}\,dx$

7 $\int x \sec x \tan x\,dx$

8 $\int x \csc^2 3x\,dx$

9 $\int x^2 \cos x\,dx$

10 $\int x^3 e^{-x}\,dx$

11 $\int \tan^{-1} x\,dx$

12 $\int \sin^{-1} x\,dx$

13 $\int \sqrt{x} \ln x \, dx$

14 $\int x^2 \ln x \, dx$

15 $\int x \csc^2 x \, dx$

16 $\int x \tan^{-1} x \, dx$

17 $\int e^{-x} \sin x \, dx$

18 $\int e^{3x} \cos 2x \, dx$

19 $\int \sin x \ln \cos x \, dx$

20 $\int_0^1 x^3 e^{-x^2} \, dx$

21 $\int \sec^3 x \, dx$

22 $\int \csc^5 x \, dx$

23 $\int_0^1 \frac{x^3}{\sqrt{x^2 + 1}} \, dx$

24 $\int \sin \ln x \, dx$

25 $\int_0^{\pi/2} x \sin 2x \, dx$

26 $\int_{\pi/6}^{\pi/4} x \sec^2 x \, dx$

27 $\int x(2x + 3)^{99} \, dx$

28 $\int \frac{x^5}{\sqrt{1 - x^3}} \, dx$

29 $\int e^{4x} \sin 5x \, dx$

30 $\int x^3 \cos (x^2) \, dx$

31 $\int (\ln x)^2 \, dx$

32 $\int x \, 2^x \, dx$

33 $\int x^3 \sinh x \, dx$

34 $\int (x + 4) \cosh 4x \, dx$

35 $\int \cos \sqrt{x} \, dx$

36 $\int \cot^{-1} 3x \, dx$

37 $\int \cos^{-1} x \, dx$

38 $\int (x + 1)^{10}(x + 2) \, dx$

Use integration by parts to derive the reduction formulas in Exercises 39–42.

39 $\int x^m e^x \, dx = x^m e^x - m \int x^{m-1} e^x \, dx$

40 $\int x^m \sin x \, dx = -x^m \cos x + m \int x^{m-1} \cos x \, dx$

41 $\int (\ln x)^m \, dx = x(\ln x)^m - m \int (\ln x)^{m-1} \, dx$

42 $\int \sec^m x \, dx = \frac{\sec^{m-2} x \tan x}{m - 1} + \frac{m - 2}{m - 1} \int \sec^{m-2} x \, dx, \; m \neq 1$

43. Use Exercise 39 to evaluate $\int x^5 e^x \, dx$.

44 Use Exercise 41 to evaluate $\int (\ln x)^4 \, dx$.

45 If $f(x) = \sin \sqrt{x}$, find the area of the region under the graph of f from 0 to π^2.

46 The region between the graph of $y = x\sqrt{\sin x}$ and the x-axis from $x = 0$ to $x = \pi/2$ is revolved about the x-axis. Find the volume of the resulting solid.

47 The region bounded by the graphs of $y = \ln x$, $y = 0$, and $x = e$ is revolved about the y-axis. Find the volume of the resulting solid.

48 Suppose the force $f(x)$ acting at the point with coordinate x on a coordinate line l is given by $f(x) = x^5\sqrt{x^3 + 1}$. Find the work done in moving an object from $x = 0$ to $x = 1$.

49 The ends of a water trough have the shape of the region between the graph of $y = \sin x$ and the x-axis from $x = \pi$ to $x = 2\pi$. If the trough is full of water, find the total force on one end.

50 The velocity (at time t) of a point moving along a coordinate line is t/e^{2t} ft/sec. If the point is at the origin at $t = 0$, find its position at time t.

10.2 TRIGONOMETRIC INTEGRALS

In Example 6 of Section 10.1 a reduction formula was obtained for $\int \sin^n x\,dx$. Integrals of this type may also be found without resorting to integration by parts. If n is an odd positive integer, we begin by writing

$$\int \sin^n x\,dx = \int \sin^{n-1} x \sin x\,dx.$$

Since the integer $n - 1$ is even, we may then use the fact that $\sin^2 x = 1 - \cos^2 x$ to obtain a form which is easy to integrate.

Example 1 Evaluate $\int \sin^5 x\,dx$.

Solution As in the preceding discussion we have

$$\begin{aligned}\int \sin^5 x\,dx &= \int \sin^4 x \sin x\,dx \\ &= \int (\sin^2 x)^2 \sin x\,dx \\ &= \int (1 - \cos^2 x)^2 \sin x\,dx \\ &= \int (1 - 2\cos^2 x + \cos^4 x) \sin x\,dx.\end{aligned}$$

We next employ the method of substitution, letting

$$u = \cos x \quad \text{and} \quad du = -\sin x\,dx.$$

Thus

$$\begin{aligned}\int \sin^5 x\,dx &= -\int (1 - 2\cos^2 x + \cos^4 x)(-\sin x)\,dx \\ &= -\int (1 - 2u^2 + u^4)\,du \\ &= -u + \tfrac{2}{3}u^3 - \tfrac{1}{5}u^5 + C \\ &= -\cos x + \tfrac{2}{3}\cos^3 x - \tfrac{1}{5}\cos^5 x + C.\end{aligned}$$

A similar technique can be employed for odd powers of $\cos x$. Specifically, we write

$$\int \cos^n x\,dx = \int \cos^{n-1} x \cos x\,dx$$

and use the fact that $\cos^2 x = 1 - \sin^2 x$ in order to obtain an integrable form.

If the integrand is $\sin^n x$ or $\cos^n x$ and n is *even*, then the half-angle formulas

$$\sin^2 x = \frac{1 - \cos 2x}{2} \quad \text{or} \quad \cos^2 x = \frac{1 + \cos 2x}{2}$$

may be used to simplify the integrand.

Example 2 Evaluate $\int \cos^2 x\,dx$.

Solution

$$\begin{aligned}\int \cos^2 x\,dx &= \tfrac{1}{2}\int (1 + \cos 2x)\,dx \\ &= \tfrac{1}{2}x + \tfrac{1}{4}\sin 2x + C.\end{aligned}$$

Example 3 Evaluate $\int \sin^4 x\,dx$.

Solution

$$\begin{aligned}\int \sin^4 x\,dx &= \int (\sin^2 x)^2\,dx \\ &= \int \left(\frac{1 - \cos 2x}{2}\right)^2 dx \\ &= \tfrac{1}{4}\int (1 - 2\cos 2x + \cos^2 2x)\,dx.\end{aligned}$$

We apply a half-angle formula again and write

$$\cos^2 2x = \tfrac{1}{2}(1 + \cos 4x) = \tfrac{1}{2} + \tfrac{1}{2}\cos 4x.$$

Substituting in the last integral and simplifying gives us

$$\begin{aligned}\int \sin^4 x\,dx &= \tfrac{1}{4}\int (\tfrac{3}{2} - 2\cos 2x + \tfrac{1}{2}\cos 4x)\,dx \\ &= \tfrac{3}{8}x - \tfrac{1}{4}\sin 2x + \tfrac{1}{32}\sin 4x + C.\end{aligned}$$

Integrals of the form $\int \sin^m x \cos^n x\, dx$ where m and n are positive integers may be found by using variations of the previous techniques. If m and n are both even, then half-angle formulas should be employed first. If n is odd, we can write

$$\int \sin^m x \cos^n x\, dx = \int \sin^m x \cos^{n-1} x \cos x\, dx$$

and express $\cos^{n-1} x$ in terms of $\sin x$ by using the identity $\cos^2 x = 1 - \sin^2 x$. The substitution $u = \sin x$ then leads to an integrand which can be handled easily. A similar technique may be used if m is odd.

Example 4 Evaluate $\displaystyle\int \cos^3 x \sin^4 x\, dx$.

Solution We proceed as follows:

$$\begin{aligned}\int \cos^3 x \sin^4 x\, dx &= \int \cos^2 x \sin^4 x \cos x\, dx \\ &= \int (1 - \sin^2 x) \sin^4 x \cos x\, dx.\end{aligned}$$

If we let $u = \sin x$, then $du = \cos x\, dx$ and the integral may be written

$$\begin{aligned}\int \cos^3 x \sin^4 x\, dx &= \int (1 - u^2)u^4\, du \\ &= \int (u^4 - u^6)\, du \\ &= \tfrac{1}{5}u^5 - \tfrac{1}{7}u^7 + C \\ &= \tfrac{1}{5}\sin^5 x - \tfrac{1}{7}\sin^7 x + C.\end{aligned}$$

For integrals of the form $\int \tan^m x \sec^n x\, dx$ we may proceed as follows:

1. If n is even, write the integral as

$$\int \tan^m x \sec^{n-2} x \sec^2 x\, dx$$

   and express $\sec^{n-2} x$ in terms of $\tan x$ by using the fact that $\sec^2 x = 1 + \tan^2 x$. The substitution $u = \tan x$ leads to a simple integral.
2. If m is odd, write the integral as

$$\int \tan^{m-1} x \sec^{n-1} x \sec x \tan x\, dx.$$

   Since $m - 1$ is even, $\tan^{m-1} x$ may be expressed in terms of $\sec x$ by means of the identity $\tan^2 x = \sec^2 x - 1$. The substitution $u = \sec x$ then leads to a form which is readily integrable.
3. If n is odd and m is even, then another method such as integration by parts should be used.

Example 5 Evaluate $\int \tan^2 x \sec^4 x \, dx$.

Solution Using method 1 above,

$$\int \tan^2 x \sec^4 x \, dx = \int \tan^2 x \sec^2 x \sec^2 x \, dx$$
$$= \int \tan^2 x (\tan^2 x + 1) \sec^2 x \, dx.$$

If we let $u = \tan x$, then $du = \sec^2 x \, dx$ and

$$\int \tan^2 x \sec^4 x \, dx = \int u^2(u^2 + 1) \, du$$
$$= \int (u^4 + u^2) \, du$$
$$= \tfrac{1}{5}u^5 + \tfrac{1}{3}u^3 + C$$
$$= \tfrac{1}{5}\tan^5 x + \tfrac{1}{3}\tan^3 x + C.$$

Example 6 Evaluate $\int \tan^3 x \sec^5 x \, dx$.

Solution Using method 2 above,

$$\int \tan^3 x \sec^5 x \, dx = \int \tan^2 x \sec^4 x (\sec x \tan x) \, dx$$
$$= \int (\sec^2 x - 1) \sec^4 x (\sec x \tan x) \, dx.$$

Substituting $u = \sec x$ and $du = \sec x \tan x \, dx$, we obtain

$$\int \tan^3 x \sec^5 x \, dx = \int (u^2 - 1)u^4 \, du$$
$$= \int (u^6 - u^4) \, du$$
$$= \tfrac{1}{7}u^7 - \tfrac{1}{5}u^5 + C$$
$$= \tfrac{1}{7}\sec^7 x - \tfrac{1}{5}\sec^5 x + C.$$

Integrals of the form $\int \cot^m x \csc^n x \, dx$ may be evaluated in similar fashion.

Finally, integrals of the form $\int \sin mx \cos nx \, dx$ may be evaluated by means of the product formulas (see Appendix III), as illustrated in the next example.

Example 7 Evaluate $\int \cos 5x \cos 3x \, dx$.

Solution We may write

$$\cos 5x \cos 3x = \tfrac{1}{2}(\cos 8x + \cos 2x).$$

Consequently,

$$\int \cos 5x \cos 3x \, dx = \tfrac{1}{2} \int (\cos 8x + \cos 2x) \, dx$$
$$= \tfrac{1}{16} \sin 8x + \tfrac{1}{4} \sin 2x + C.$$

EXERCISES 10.2

Evaluate the integrals in Exercises 1–30.

1 $\int \cos^3 x \, dx$

2 $\int \sin^2 2x \, dx$

3 $\int \sin^2 x \cos^2 x \, dx$

4 $\int \cos^7 x \, dx$

5 $\int \sin^3 x \cos^2 x \, dx$

6 $\int \sin^5 x \cos^3 x \, dx$

7 $\int \sin^6 x \, dx$

8 $\int \sin^4 x \cos^2 x \, dx$

9 $\int \tan^3 x \sec^4 x \, dx$

10 $\int \sec^6 x \, dx$

11 $\int \tan^3 x \sec^3 x \, dx$

12 $\int \tan^5 x \sec x \, dx$

13 $\int \tan^6 x \, dx$

14 $\int \cot^4 x \, dx$

15 $\int \sqrt{\sin x} \cos^3 x \, dx$

16 $\int \frac{\cos^3 x}{\sqrt{\sin x}} \, dx$

17 $\int (\tan x + \cot x)^2 \, dx$

18 $\int \cot^3 x \csc^3 x \, dx$

19 $\int_0^{\pi/4} \sin^3 x \, dx$

20 $\int_0^1 \tan^2 (\pi x/4) \, dx$

21 $\int \sin 5x \sin 3x \, dx$

22 $\int_0^{\pi/4} \cos x \cos 5x \, dx$

23 $\int_0^{\pi/2} \sin 3x \cos 2x \, dx$

24 $\int \sin 4x \cos 3x \, dx$

25 $\int \csc^4 x \cot^4 x \, dx$

26 $\int (1 + \sqrt{\cos x})^2 \sin x \, dx$

27 $\int \frac{\cos x}{2 - \sin x} \, dx$

28 $\int \frac{\tan^2 x - 1}{\sec^2 x} \, dx$

29 $\int \frac{\sec^2 x}{(1 + \tan x)^2} \, dx$

30 $\int \frac{\sec x}{\cot^5 x} \, dx$

31 The region bounded by the x-axis and the arc of $y = \sin x$ from $x = 0$ to $x = \pi$ is revolved about the x-axis. Find the volume of the resulting solid.

32 The region between the graphs of $y = \tan^2 x$ and $y = 0$ from $x = 0$ to $x = \pi/4$ is revolved about the x-axis. Find the volume of the resulting solid.

33 Suppose the velocity (at time t) of a point moving rectilinearly is $\cos^2 \pi t$ ft/sec. How far does the point travel in 5 seconds?

34 The acceleration (at time t) of a point moving along a coordinate line is $\sin^2 t \cos t$ ft/sec^2. At $t = 0$ the point is at the origin and its velocity is 10 ft/sec. Find its position at time t.

10.3 TRIGONOMETRIC SUBSTITUTIONS

If an integrand contains the expression $\sqrt{a^2 - x^2}$ where $a > 0$, then the **trigonometric substitution** $x = a \sin \theta$ leads to

$$\begin{aligned}\sqrt{a^2 - x^2} &= \sqrt{a^2 - a^2 \sin^2 \theta} \\ &= \sqrt{a^2(1 - \sin^2 \theta)} \\ &= \sqrt{a^2 \cos^2 \theta} \\ &= a\,|\cos \theta|.\end{aligned}$$

When making this substitution or the other trigonometric substitutions discussed in this section, we shall assume that θ is in the range of the corresponding inverse trigonometric function. Thus, for the sine substitution above $-\pi/2 \le \theta \le \pi/2$. Consequently $\cos \theta \ge 0$ and $\sqrt{a^2 - x^2} = a \cos \theta$. Of course, if $\sqrt{a^2 - x^2}$ occurs in a denominator we make the further restriction $-\pi/2 < \theta < \pi/2$.

Example 1 Evaluate $\displaystyle\int \frac{1}{x^2\sqrt{a^2 - x^2}}\,dx$, where $a > 0$.

Solution Let $x = a \sin \theta$, where $-\pi/2 < \theta < \pi/2$. It follows that

$$\sqrt{a^2 - x^2} = \sqrt{a^2 - a^2 \sin^2 \theta} = a\sqrt{1 - \sin^2 \theta} = a \cos \theta.$$

Since $x = a \sin \theta$, we have $dx = a \cos \theta \, d\theta$. Substituting in the given integral,

$$\begin{aligned}\int \frac{1}{x^2\sqrt{a^2 - x^2}}\,dx &= \int \frac{1}{(a^2 \sin^2 \theta) a \cos \theta} a \cos \theta \, d\theta \\ &= \frac{1}{a^2} \int \frac{1}{\sin^2 \theta}\,d\theta \\ &= \frac{1}{a^2} \int \csc^2 \theta \, d\theta \\ &= -\frac{1}{a^2} \cot \theta + C.\end{aligned}$$

It is now necessary to return to the original variable of integration x. A simple method for doing so is to use the following geometric device. If $0 < \theta < \pi/2$, then since $\sin\theta = x/a$, we may interpret θ as an acute angle of a right triangle having opposite side and hypotenuse of lengths x and a, respectively (see Figure 10.1). The length $\sqrt{a^2 - x^2}$ of the adjacent side is calculated by means of the Pythagorean Theorem. Referring to the triangle we see that

$$\cot\theta = \frac{\sqrt{a^2 - x^2}}{x}.$$

It can be shown that the last formula is also true if $-\pi/2 < \theta < 0$. Thus, Figure 10.1 may be used whether θ is positive or negative.

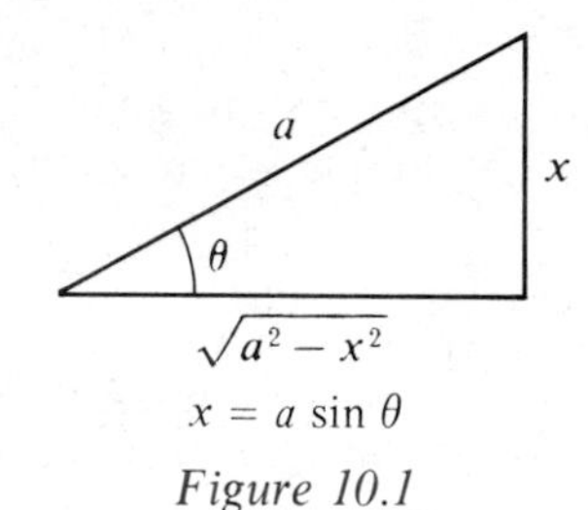

$x = a \sin\theta$

Figure 10.1

Substituting $\sqrt{a^2 - x^2}/x$ for $\cot\theta$ in our integral evaluation gives us

$$\begin{aligned}\int \frac{1}{x^2\sqrt{a^2 - x^2}}\,dx &= -\frac{1}{a^2}\cdot\frac{\sqrt{a^2 - x^2}}{x} + C \\ &= -\frac{\sqrt{a^2 - x^2}}{a^2 x} + C.\end{aligned}$$

If an integrand contains $\sqrt{a^2 + x^2}$, where $a > 0$, then the substitution $x = a\tan\theta$ will eliminate the radical sign. When using this substitution it will be assumed that θ is in the range of the inverse tangent function; that is, $-\pi/2 < \theta < \pi/2$. In this event

$$\begin{aligned}\sqrt{a^2 + x^2} &= \sqrt{a^2 + a^2\tan^2\theta} \\ &= \sqrt{a^2(1 + \tan^2\theta)} \\ &= \sqrt{a^2\sec^2\theta} \\ &= a\sec\theta.\end{aligned}$$

After making this substitution and evaluating the resulting trigonometric integral, it is necessary to return to the variable x. The preceding formulas show that

$$\tan\theta = \frac{x}{a} \quad \text{and} \quad \sec\theta = \frac{\sqrt{a^2 + x^2}}{a}.$$

As in the solution of Example 1, the trigonometric functions of θ can be found by referring to the triangle in (i) of Figure 10.2, whether θ is positive or negative.

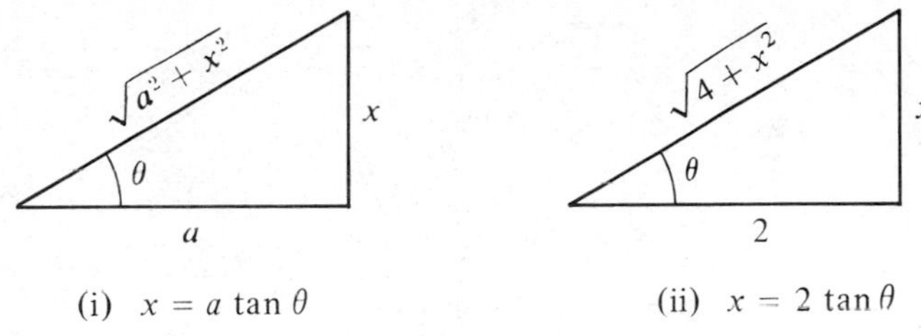

Figure 10.2

Example 2 Evaluate $\int \frac{1}{\sqrt{4 + x^2}} dx.$

Solution Let us substitute as follows:

$$x = 2 \tan \theta, \qquad dx = 2 \sec^2 \theta \, d\theta.$$

Consequently

$$\sqrt{4 + x^2} = \sqrt{4 + 4 \tan^2 \theta} = 2\sqrt{1 + \tan^2 \theta} = 2 \sec \theta$$

and

$$\begin{aligned} \int \frac{1}{\sqrt{4 + x^2}} dx &= \int \frac{1}{2 \sec \theta} 2 \sec^2 \theta \, d\theta \\ &= \int \sec \theta \, d\theta \\ &= \ln |\sec \theta + \tan \theta| + C. \end{aligned}$$

Since $\tan \theta = x/2$ we see from the triangle in (ii) of Figure 10.2 that

$$\sec \theta = \frac{\sqrt{4 + x^2}}{2}$$

and hence

$$\int \frac{1}{\sqrt{4 + x^2}} dx = \ln \left| \frac{\sqrt{4 + x^2}}{2} + \frac{x}{2} \right| + C.$$

The expression on the right may be written

$$\ln \left| \frac{\sqrt{4 + x^2} + x}{2} \right| + C = \ln |\sqrt{4 + x^2} + x| - \ln 2 + C.$$

Since $\sqrt{4 + x^2} + x > 0$ for all x, the absolute value sign is unnecessary. If we also let $D = -\ln 2 + C$, then

$$\int \frac{1}{\sqrt{4 + x^2}} dx = \ln (\sqrt{4 + x^2} + x) + D.$$

For integrands containing $\sqrt{x^2 - a^2}$ we substitute $x = a \sec \theta$, where θ is chosen in the range of the inverse secant function; that is, either $0 \le \theta < \pi/2$ or $\pi \le \theta < 3\pi/2$. In this case

$$\begin{aligned} \sqrt{x^2 - a^2} &= \sqrt{a^2 \sec^2 \theta - a^2} \\ &= \sqrt{a^2(\sec^2 \theta - 1)} \\ &= \sqrt{a^2 \tan^2 \theta} \\ &= a \tan \theta. \end{aligned}$$

Since

$$\sec \theta = \frac{x}{a} \quad \text{and} \quad \tan \theta = \frac{\sqrt{x^2 - a^2}}{a}$$

it follows that we may refer to the triangle in (i) of Figure 10.3 when changing from the variable θ to the variable x.

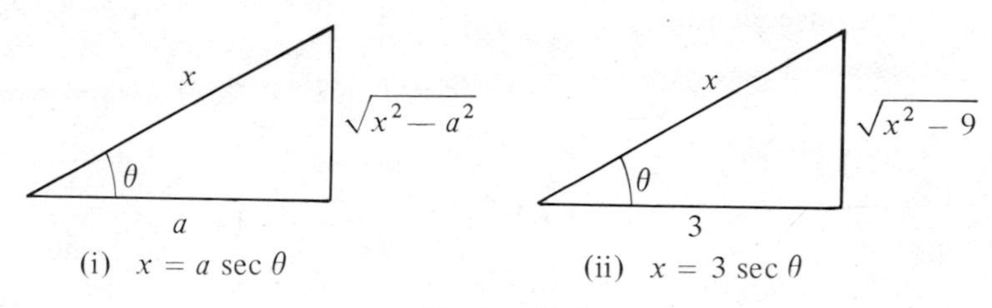

Figure 10.3

Example 3 Evaluate $\displaystyle\int \frac{\sqrt{x^2 - 9}}{x}\,dx.$

Solution Let us substitute as follows:

$$x = 3 \sec \theta, \qquad dx = 3 \sec \theta \tan \theta \, d\theta.$$

Consequently

$$\sqrt{x^2 - 9} = \sqrt{9 \sec^2 \theta - 9} = 3\sqrt{\sec^2 \theta - 1} = 3 \tan \theta$$

and, therefore,

$$\begin{aligned} \int \frac{\sqrt{x^2 - 9}}{x}\,dx &= \int \frac{3 \tan \theta}{3 \sec \theta} 3 \sec \theta \tan \theta \, d\theta \\ &= 3 \int \tan^2 \theta \, d\theta \\ &= 3 \int (\sec^2 \theta - 1) \, d\theta \\ &= 3 (\tan \theta - \theta) + C. \end{aligned}$$

Since $\sec\theta = x/3$ we may refer to the triangle in (ii) of Figure 10.3 and write

$$\int \frac{\sqrt{x^2-9}}{x}\,dx = 3\left[\frac{\sqrt{x^2-9}}{3} - \sec^{-1}\left(\frac{x}{3}\right)\right] + C$$

$$= \sqrt{x^2-9} - 3\sec^{-1}\left(\frac{x}{3}\right) + C.$$

For reference let us summarize the trigonometric substitutions discussed in this section as follows.

Given expression	Trigonometric substitution
$\sqrt{a^2 - x^2}$	$x = a\sin\theta$
$\sqrt{a^2 + x^2}$	$x = a\tan\theta$
$\sqrt{x^2 - a^2}$	$x = a\sec\theta$

Hyperbolic functions may also be used to simplify certain integrations. For example, $\cosh^2 u = 1 + \sinh^2 u$ and hence, if an integrand contains the expression $\sqrt{a^2 + x^2}$, the substitution $x = a\sinh u$ leads to

$$\sqrt{a^2+x^2} = \sqrt{a^2 + a^2\sinh^2 u}$$
$$= \sqrt{a^2(1+\sinh^2 u)}$$
$$= \sqrt{a^2\cosh^2 u}$$
$$= a\cosh u.$$

Example 4 Evaluate $\int \frac{1}{\sqrt{4+x^2}}\,dx$ by using hyperbolic functions.

Solution The given integral is the same as that considered in Example 2. Let us substitute as follows:

$$x = 2\sinh u, \quad dx = 2\cosh u\,du.$$

It follows that

$$\sqrt{4+x^2} = \sqrt{4+4\sinh^2 u} = \sqrt{4\cosh^2 u} = 2\cosh u$$

and hence

$$\int \frac{1}{\sqrt{4+x^2}}\,dx = \int \frac{1}{2\cosh u}2\cosh u\,du$$

$$= \int du = u + C$$

$$= \sinh^{-1}\left(\frac{x}{2}\right) + C.$$

The fact that this is equivalent to our first solution follows from (9.51).

Although additional integration techniques are now available, the student should also keep earlier methods in mind. For example, the integral $\int (x/\sqrt{9+x^2})\,dx$ could be evaluated by means of the trigonometric substitution $x = 3\tan\theta$. However, it is simpler to use the algebraic substitution $u = 9 + x^2$ and $du = 2x\,dx$, for in this event the integral takes on the form $(1/2)\int u^{-1/2}\,du$, which is readily integrated by means of the Power Rule. In the following exercises we shall include integrals which can be evaluated using simpler techniques than trigonometric substitutions.

EXERCISES 10.3

Evaluate the integrals in Exercises 1–22.

1 $\displaystyle\int \frac{x^2}{\sqrt{4-x^2}}\,dx$

2 $\displaystyle\int \frac{\sqrt{4-x^2}}{x^2}\,dx$

3 $\displaystyle\int \frac{1}{x\sqrt{9+x^2}}\,dx$

4 $\displaystyle\int \frac{1}{x^2\sqrt{x^2+9}}\,dx$

5 $\displaystyle\int \frac{1}{x^2\sqrt{x^2-25}}\,dx$

6 $\displaystyle\int \frac{1}{x^3\sqrt{x^2-25}}\,dx$

7 $\displaystyle\int \frac{x}{\sqrt{4-x^2}}\,dx$

8 $\displaystyle\int \frac{x}{x^2+9}\,dx$

9 $\displaystyle\int \frac{1}{(x^2-1)^{3/2}}\,dx$

10 $\displaystyle\int \frac{1}{\sqrt{4x^2-25}}\,dx$

11 $\displaystyle\int \frac{1}{(36+x^2)^2}\,dx$

12 $\displaystyle\int \frac{1}{(16-x^2)^{5/2}}\,dx$

13 $\displaystyle\int \sqrt{9-4x^2}\,dx$

14 $\displaystyle\int \frac{\sqrt{x^2+1}}{x}\,dx$

15 $\displaystyle\int \frac{x}{(16-x^2)^2}\,dx$

16 $\displaystyle\int x\sqrt{x^2-9}\,dx$

17 $\displaystyle\int \frac{x^3}{\sqrt{9x^2+49}}\,dx$

18 $\displaystyle\int \frac{1}{x\sqrt{25x^2+16}}\,dx$

19 $\displaystyle\int \frac{1}{x^4\sqrt{x^2-3}}\,dx$

20 $\displaystyle\int \frac{x^2}{(1-9x^2)^{3/2}}\,dx$

21 $\displaystyle\int \frac{(4+x^2)^2}{x^3}\,dx$

22 $\displaystyle\int \frac{3x-5}{\sqrt{1-x^2}}\,dx$

23–28 Use trigonometric substitutions to obtain formulas (9.39)–(9.44).

29 The region bounded by the graphs of $y = x(x^2+25)^{-1/2}$, $y = 0$, and $x = 5$ is revolved about the y-axis. Find the volume of the resulting solid.

30 Find the area of the region bounded by the graph of $y = x^3(10-x^2)^{-1/2}$, the x-axis, and the line $x = 1$.

31 Find the arc length of the graph of $y = x^2/2$ from $A(0, 0)$ to $B(2, 2)$.

32 The region bounded by the graphs of $y = 1/(x^2 + 1)$, $y = 0$, $x = 0$, and $x = 2$ is revolved about the x-axis. Find the volume of the resulting solid.

33 Use a trigonometric substitution to find the area of the region bounded by the ellipse $b^2x^2 + a^2y^2 = a^2b^2$.

34 Find the area of the region bounded by the hyperbola $b^2x^2 - a^2y^2 = a^2b^2$ and a line through a focus perpendicular to the transverse axis.

35 Suppose $y = f(x)$ and $x\,dy - \sqrt{x^2 - 16}\,dx = 0$. If $f(4) = 0$, find $f(x)$.

36 Suppose two variables x and y are related such that $\sqrt{1 - x^2}\,dy = x^3\,dx$. If $y = 0$ when $x = 0$, express y as a function of x.

Evaluate the integrals in Exercises 37–40 by means of a hyperbolic substitution.

37 $\displaystyle\int \frac{1}{x^2\sqrt{25 + x^2}}\,dx$

38 $\displaystyle\int \frac{x^2}{(x^2 + 9)^{3/2}}\,dx$

39 $\displaystyle\int \frac{1}{x^2\sqrt{1 - x^2}}\,dx$

40 $\displaystyle\int \frac{1}{16 - x^2}\,dx$

(*Hint*: Let $x = \tanh u$ and use (9.48).)

41 Use trigonometric substitutions to establish the following formulas in the Table of Integrals: 21, 27, 31, 36, 41, 44.

10.4 PARTIAL FRACTIONS

It is easy to verify that

$$\frac{2}{x^2 - 1} = \frac{1}{x - 1} + \frac{-1}{x + 1}.$$

The expression on the right side of this equation is called *the partial fraction decomposition* of $2/(x^2 - 1)$. This decomposition may be used to find the indefinite integral of $2/(x^2 - 1)$. We merely integrate each of the fractions which make up the decomposition independently, obtaining

$$\begin{aligned}\int \frac{2}{x^2 - 1}\,dx &= \int \frac{1}{x - 1}\,dx + \int \frac{-1}{x + 1}\,dx \\ &= \ln|x - 1| - \ln|x + 1| + C \\ &= \ln\left|\frac{x - 1}{x + 1}\right| + C.\end{aligned}$$

It is theoretically possible to write *any* rational expression $f(x)/g(x)$ as a sum of rational expressions whose denominators involve powers of polynomials of degree not greater than two. More specifically, if $f(x)$ and $g(x)$ are polynomials *and the degree of* $f(x)$ *is less than the degree of* $g(x)$, then it follows from a theorem in algebra that

$$\frac{f(x)}{g(x)} = F_1 + F_2 + \cdots + F_k \tag{10.3}$$

where each F_i has one of the forms

$$\frac{A}{(px+q)^m} \quad \text{or} \quad \frac{Cx+D}{(ax^2+bx+c)^n}$$

for some nonnegative integers m and n, and where $ax^2 + bx + c$ is **irreducible**, in the sense that this quadratic expression has no real zeros, that is, $b^2 - 4ac < 0$. The sum on the right in (10.3) is called the **partial fraction decomposition** of $f(x)/g(x)$ and each F_i is called a **partial fraction**. We shall not prove this algebraic result but will, instead, give rules for obtaining the decomposition.

To find the partial fraction decomposition of a rational expression $f(x)/g(x)$ *it is essential* that $f(x)$ have lower degree than $g(x)$. If this is not the case, then long division should be employed to arrive at such an expression. For example, given

$$\frac{x^3 - 6x^2 + 5x - 3}{x^2 - 1}$$

we obtain, by long division,

$$\frac{x^3 - 6x^2 + 5x - 3}{x^2 - 1} = x - 6 + \frac{6x - 9}{x^2 - 1}.$$

The partial fraction decomposition is then found for $(6x - 9)/(x^2 - 1)$.

In order to obtain the decomposition (10.3), we begin by expressing the denominator $g(x)$ as a product of factors $px + q$ or irreducible quadratic factors $ax^2 + bx + c$. Repeated factors are then collected so that $g(x)$ is a product of *different* factors of the form $(px + q)^m$ or $(ax^2 + bx + c)^n$, where m and n are nonnegative integers and $ax^2 + bx + c$ is irreducible. We then apply the following rules.

Rule 1. For each factor of the form $(px + q)^m$ where $m \geq 1$, the decomposition (10.3) contains a sum of m partial fractions of the form

$$\frac{A_1}{px+q} + \frac{A_2}{(px+q)^2} + \cdots + \frac{A_m}{(px+q)^m}$$

where each A_i is a real number.

Rule 2. For each factor of the form $(ax^2 + bx + c)^n$ where $n \geq 1$ and $b^2 - 4ac < 0$, the decomposition (10.3) contains a sum of n partial fractions of the form

$$\frac{A_1x + B_1}{ax^2+bx+c} + \frac{A_2x + B_2}{(ax^2+bx+c)^2} + \cdots + \frac{A_nx + B_n}{(ax^2+bx+c)^n}$$

where, for each i, A_i and B_i are real numbers.

Example 1 Evaluate $\int \frac{4x^2 + 13x - 9}{x^3 + 2x^2 - 3x}\,dx.$

Solution The denominator of the integrand has the factored form $x(x + 3)(x - 1)$. Each of the linear factors is handled under Rule 1, with $m = 1$. Thus, for the factor x there corresponds a partial fraction of the form A/x. Similarly, for the factors $x+3$ and $x - 1$ there correspond partial fractions $B/(x + 3)$ and $C/(x - 1)$, respectively. The decomposition (10.3) then has the form

$$\frac{4x^2 + 13x - 9}{x(x + 3)(x - 1)} = \frac{A}{x} + \frac{B}{x + 3} + \frac{C}{x - 1}.$$

Multiplying by the lowest common denominator gives us

$$4x^2 + 13x - 9 = A(x + 3)(x - 1) + Bx(x - 1) + Cx(x + 3). \tag{10.4}$$

In a case such as this, in which the factors are all linear and nonrepeated, the values for A, B, and C can be found by substituting values for x which make the various factors zero. If we let $x = 0$ in (10.4), then

$$-9 = -3A \quad \text{or} \quad A = 3.$$

Letting $x = 1$ in (10.4) gives us

$$8 = 4C \quad \text{or} \quad C = 2.$$

Finally, if $x = -3$, then

$$-12 = 12B \quad \text{or} \quad B = -1.$$

The partial fraction decomposition is, therefore,

$$\frac{4x^2 + 13x - 9}{x(x + 3)(x - 1)} = \frac{3}{x} + \frac{-1}{x + 3} + \frac{2}{x - 1}.$$

Integrating,

$$\begin{aligned}\int \frac{4x^2 + 13x - 9}{x(x + 3)(x - 1)}\,dx &= \int \frac{3}{x}\,dx + \int \frac{-1}{x + 3}\,dx + \int \frac{2}{x - 1}\,dx \\ &= 3\ln|x| - \ln|x + 3| + 2\ln|x - 1| + D \\ &= \ln|x^3| - \ln|x + 3| + \ln|x - 1|^2 + D \\ &= \ln\left|\frac{x^3(x - 1)^2}{x + 3}\right| + D.\end{aligned}$$

Another technique for finding A, B, and C is to compare coefficients of x. If the right-hand side of (10.4) is expanded and like powers of x are collected, then

$$4x^2 + 13x - 9 = (A + B + C)x^2 + (2A - B + 3C)x - 3A.$$

We now use the fact that if two polynomials are equal, then coefficients of like powers are the same. Thus

$$\begin{cases} A + B + C = 4 \\ 2A - B + 3C = 13 \\ -3A = -9. \end{cases}$$

It is left to the reader to show that the solution of this system of equations is $A = 3$, $B = -1$, and $C = 2$.

Example 2 Evaluate $\int \frac{3x^3 - 18x^2 + 29x - 4}{(x+1)(x-2)^3}\,dx$.

Solution By Rule 1, there is a partial fraction of the form $A/(x+1)$ corresponding to the factor $x + 1$ in the denominator of the integrand. For the factor $(x-2)^3$ we apply Rule 1 (with $m = 3$), obtaining a sum of three partial fractions $B/(x-2)$, $C/(x-2)^2$, and $D/(x-2)^3$. Consequently, the decomposition (10.3) has the form

$$\frac{3x^3 - 18x^2 + 29x - 4}{(x+1)(x-2)^3} = \frac{A}{x+1} + \frac{B}{x-2} + \frac{C}{(x-2)^2} + \frac{D}{(x-2)^3}.$$

Multiplying both sides by $(x+1)(x-2)^3$ gives us

$$\begin{aligned} &3x^3 - 18x^2 + 29x - 4 \\ &\quad = A(x-2)^3 + B(x+1)(x-2)^2 + C(x+1)(x-2) + D(x+1). \end{aligned} \tag{10.5}$$

Two of the unknown constants may be determined easily. If we let $x = 2$ in (10.5), then

$$24 - 72 + 58 - 4 = 3D, \quad 6 = 3D, \quad \text{and} \quad D = 2.$$

Similarly, letting $x = -1$ in (10.5),

$$-3 - 18 - 29 - 4 = -27A, \quad -54 = -27A, \quad \text{and} \quad A = 2.$$

The remaining constants may be found by comparing coefficients. If the right side of (10.5) is expanded and like powers of x collected, we see that the coefficient of x^3 is $A + B$. This must equal the coefficient of x^3 on the left, that is,

$$A + B = 3.$$

Since $A = 2$, it follows that $B = 3 - A = 3 - 2 = 1$. Finally, we compare the constant terms in (10.5) by letting $x = 0$. This gives us

$$-4 = -8A + 4B - 2C + D.$$

Substituting the values we have found for A, B, and D leads to

$$-4 = -16 + 4 - 2C + 2$$

which has the solution $C = -3$. The partial fraction decomposition is, therefore,

$$\frac{3x^3 - 18x^2 + 29x - 4}{(x+1)(x-2)^3} = \frac{2}{x+1} + \frac{1}{x-2} + \frac{-3}{(x-2)^2} + \frac{2}{(x-2)^3}.$$

To find the given integral we integrate each of the partial fractions on the right side of the last equation. This gives us

$$2\ln|x+1| + \ln|x-2| + \frac{3}{x-2} - \frac{1}{(x-2)^2} + E$$

which may be written in the more compact form

$$\ln(x+1)^2|x-2| + \frac{3x-7}{(x-2)^2} + E.$$

Example 3 Evaluate $\displaystyle\int \frac{x^2 - x - 21}{2x^3 - x^2 + 8x - 4}\,dx.$

Solution The denominator may be factored by grouping as follows:

$$2x^3 - x^2 + 8x - 4 = x^2(2x-1) + 4(2x-1) = (x^2+4)(2x-1).$$

Applying Rule 2 to the irreducible quadratic factor $x^2 + 4$ we see that one of the partial fractions has the form $(Ax + B)/(x^2 + 4)$. By Rule 1, there is also a partial fraction $C/(2x - 1)$ corresponding to the factor $2x - 1$. Consequently

$$\frac{x^2 - x - 21}{2x^3 - x^2 + 8x - 4} = \frac{Ax+B}{x^2+4} + \frac{C}{2x-1}.$$

As in previous examples, this leads to

$$x^2 - x - 21 = (Ax + B)(2x - 1) + C(x^2 + 4). \tag{10.6}$$

Substituting $x = 1/2$ we obtain $(1/4) - (1/2) - 21 = (17/4)C$, which has the solution $C = -5$. The remaining constants may be found by comparing coefficients. Rearranging the right side of (10.6) gives us

$$x^2 - x - 21 = (2A + C)x^2 + (-A + 2B)x - B + 4C.$$

Comparing the coefficients of x^2 we see that $2A + C = 1$. Since $C = -5$ it follows that $2A = 6$ or $A = 3$. Similarly, comparing the constant terms, $-B + 4C = -21$ and hence $-B - 20 = -21$ or $B = 1$. Thus the partial fraction decomposition of the integrand is

$$\begin{aligned}\frac{x^2 - x - 21}{2x^3 - x^2 + 8x - 4} &= \frac{3x+1}{x^2+4} + \frac{-5}{2x-1}\\ &= \frac{3x}{x^2+4} + \frac{1}{x^2+4} - \frac{5}{2x-1}.\end{aligned}$$

The given integral may now be found by integrating the right side of the last equation. This gives us

$$\frac{3}{2}\ln(x^2+4)+\frac{1}{2}\tan^{-1}\frac{x}{2}-\frac{5}{2}\ln|2x-1|+D.$$

Example 4 Evaluate $\int \frac{5x^3-3x^2+7x-3}{(x^2+1)^2}\,dx$.

Solution Applying Rule 2, with $n = 2$,

$$\frac{5x^3-3x^2+7x-3}{(x^2+1)^2}=\frac{Ax+B}{x^2+1}+\frac{Cx+D}{(x^2+1)^2}$$

and, therefore,

$$5x^3-3x^2+7x-3=(Ax+B)(x^2+1)+Cx+D$$

or

$$5x^3-3x^2+7x-3=Ax^3+Bx^2+(A+C)x+(B+D).$$

Comparing the coefficients of x^3 and x^2 we obtain $A = 5$ and $B = -3$. From the coefficients of x we see that $A + C = 7$ or $C = 7 - A = 7 - 5 = 2$. Finally, the constant terms give us $B + D = -3$ or $D = -3 - B = -3 - (-3) = 0$. Therefore,

$$\begin{aligned}\frac{5x^3-3x^2+7x-3}{(x^2+1)^2}&=\frac{5x-3}{x^2+1}+\frac{2x}{(x^2+1)^2}\\&=\frac{5x}{x^2+1}-\frac{3}{x^2+1}+\frac{2x}{(x^2+1)^2}.\end{aligned}$$

Integrating, we obtain

$$\int\frac{5x^3-3x^2+7x-3}{(x^2+1)^2}\,dx=\tfrac{5}{2}\ln(x^2+1)-3\tan^{-1}x-\frac{1}{x^2+1}+E.$$

EXERCISES 10.4

Evaluate the integrals in Exercises 1–32.

1 $\int \frac{5x-12}{x(x-4)}\,dx$

2 $\int \frac{x+34}{(x-6)(x+2)}\,dx$

3 $\int \frac{37-11x}{(x+1)(x-2)(x-3)}\,dx$

4 $\int \frac{4x^2+54x+134}{(x-1)(x+5)(x+3)}\,dx$

5 $\int \frac{6x-11}{(x-1)^2}\,dx$

6 $\int \frac{-19x^2+50x-25}{x^2(3x-5)}\,dx$

7 $\int \frac{x + 16}{x^2 + 2x - 8}\,dx$

8 $\int \frac{11x + 2}{2x^2 - 5x - 3}\,dx$

9 $\int \frac{5x^2 - 10x - 8}{x^3 - 4x}\,dx$

10 $\int \frac{4x^2 - 5x - 15}{x^3 - 4x^2 - 5x}\,dx$

11 $\int \frac{2x^2 - 25x - 33}{(x + 1)^2(x - 5)}\,dx$

12 $\int \frac{2x^2 - 12x + 4}{x^3 - 4x^2}\,dx$

13 $\int \frac{9x^4 + 17x^3 + 3x^2 - 8x + 3}{x^5 + 3x^4}\,dx$

14 $\int \frac{5x^2 + 30x + 43}{(x + 3)^3}\,dx$

15 $\int \frac{x^3 + 3x^2 + 3x + 63}{(x^2 - 9)^2}\,dx$

16 $\int \frac{1}{(x - 7)^5}\,dx$

17 $\int \frac{5x^2 + 11x + 17}{x^3 + 5x^2 + 4x + 20}\,dx$

18 $\int \frac{4x^3 - 3x^2 + 6x - 27}{x^4 + 9x^2}\,dx$

19 $\int \frac{x^2 + 3x + 1}{x^4 + 5x^2 + 4}\,dx$

20 $\int \frac{4x}{(x^2 + 1)^3}\,dx$

21 $\int \frac{2x^3 + 10x}{(x^2 + 1)^2}\,dx$

22 $\int \frac{x^4 + 2x^2 + 4x + 1}{(x^2 + 1)^3}\,dx$

23 $\int \frac{x^3 + 3x - 2}{x^2 - x}\,dx$

24 $\int \frac{x^4 + 2x^2 + 3}{x^3 - 4x}\,dx$

25 $\int \frac{x^6 - x^3 + 1}{x^4 + 9x^2}\,dx$

26 $\int \frac{x^5}{(x^2 + 4)^2}\,dx$

27 $\int \frac{2x^3 - 5x^2 + 46x + 98}{(x^2 + x - 12)^2}\,dx$

28 $\int \frac{-2x^4 - 3x^3 - 3x^2 + 3x + 1}{x^2(x + 1)^3}\,dx$

29 $\int \frac{4x^3 + 2x^2 - 5x - 18}{(x - 4)(x + 1)^3}\,dx$

30 $\int \frac{10x^2 + 9x + 1}{2x^3 + 3x^2 + x}\,dx$

31 $\int \frac{2x^4 - 2x^3 + 6x^2 - 5x + 1}{x^3 - x^2 + x - 1}\,dx$

32 $\int \frac{x^5 - x^4 - 2x^3 + 4x^2 - 15x + 5}{(x^2 + 1)^2(x^2 + 4)}\,dx$

Use partial fractions to evaluate the integrals in Exercises 33–36 (see formulas 19, 49, 50, and 52 in the Table of Integrals).

33 $\int \frac{1}{a^2 - u^2}\,du$

34 $\int \frac{1}{u(a + bu)}\,du$

35 $\int \frac{1}{u^2(a + bu)}\,du$

36 $\int \frac{1}{u(a + bu)^2}\,du$

37 If $f(x) = x/(x^2 - 2x - 3)$, find the area of the region under the graph of f from $x = 0$ to $x = 2$.

38 The region bounded by the graphs of $y = 1/(x - 1)(4 - x)$, $y = 0$, $x = 2$, and $x = 3$ is revolved about the y-axis. Find the volume of the resulting solid.

39 If the region described in Exercise 38 is revolved about the x-axis, find the volume of the resulting solid.

40 Suppose the velocity of a point moving along a coordinate line is $(t + 3)/(t^3 + t)$ ft/sec, where t is the time in seconds. How far does the point travel during the time interval $[1, 2]$?

10.5 QUADRATIC EXPRESSIONS

Partial fraction decompositions may lead to integrands containing an irreducible quadratic expression $ax^2 + bx + c$. If $b \neq 0$ it is often necessary to complete the square as follows:

$$ax^2 + bx + c = a\left(x^2 + \frac{b}{a}x\right) + c$$
$$= a\left(x + \frac{b}{2a}\right)^2 + c - \frac{b^2}{4a}.$$

The substitution $u = x + b/2a$ may then lead to an integrable form.

Example 1 Evaluate $\int \frac{2x - 1}{x^2 - 6x + 13} dx.$

Solution Note that the quadratic expression $x^2 - 6x + 13$ is irreducible, since $b^2 - 4ac = 36 - 52 = -16 < 0$. We complete the square as follows:

$$x^2 - 6x + 13 = (x^2 - 6x \qquad) + 13$$
$$= (x^2 - 6x + 9) + 13 - 9 = (x - 3)^2 + 4.$$

If we let $u = x - 3$, then $x = u + 3$, $dx = du$, and hence

$$\int \frac{2x - 1}{(x - 3)^2 + 4} dx = \int \frac{2(u + 3) - 1}{u^2 + 4} du$$
$$= \int \frac{2u + 5}{u^2 + 4} du$$
$$= \int \frac{2u}{u^2 + 4} du + 5 \int \frac{1}{u^2 + 4} du$$
$$= \ln(u^2 + 4) + \tfrac{5}{2} \tan^{-1} \frac{u}{2} + C$$
$$= \ln(x^2 - 6x + 13) + \tfrac{5}{2} \tan^{-1} \frac{x - 3}{2} + C.$$

The technique of completing the square may also be employed if quadratic expressions appear under a radical sign.

Example 2 Evaluate $\int \frac{1}{\sqrt{8 + 2x - x^2}} dx.$

Solution We may complete the square for the quadratic expression $8 + 2x - x^2$ as follows:

$$8 + 2x - x^2 = 8 - (x^2 - 2x) = 8 + 1 - (x^2 - 2x + 1)$$
$$= 9 - (x - 1)^2.$$

Next, letting $u = x - 1$ we have $du = dx$, and hence

$$\int \frac{1}{\sqrt{8 + 2x - x^2}}\,dx = \int \frac{1}{\sqrt{9 - u^2}}\,du$$

$$= \sin^{-1}\frac{u}{3} + C$$

$$= \sin^{-1}\frac{x-1}{3} + C.$$

In the next example it is necessary to make a trigonometric substitution after completing the square.

Example 3 Evaluate $\int \frac{1}{\sqrt{x^2 + 8x + 25}}\,dx.$

Solution We complete the square for the quadratic expression as follows:

$$\begin{aligned} x^2 + 8x + 25 &= (x^2 + 8x \qquad) + 25 \\ &= (x^2 + 8x + 16) + 25 - 16 \\ &= (x + 4)^2 + 9. \end{aligned}$$

Hence

$$\int \frac{1}{\sqrt{x^2 + 8x + 25}}\,dx = \int \frac{1}{\sqrt{(x + 4)^2 + 9}}\,dx.$$

If we next make the trigonometric substitution

$$x + 4 = 3\tan\theta$$

then

$$dx = 3\sec^2\theta\,d\theta$$

$$\sqrt{(x + 4)^2 + 9} = \sqrt{9\tan^2\theta + 9} = 3\sqrt{\tan^2\theta + 1} = 3\sec\theta$$

and hence

$$\int \frac{1}{\sqrt{x^2 + 8x + 25}}\,dx = \int \frac{1}{3\sec\theta}3\sec^2\theta\,d\theta$$

$$= \int \sec\theta\,d\theta$$

$$= \ln|\sec\theta + \tan\theta| + C.$$

In order to return to the variable x we use the triangle in Figure 10.4. This gives us

$$\int \frac{1}{\sqrt{x^2 + 8x + 25}}\,dx = \ln\left|\frac{\sqrt{x^2 + 8x + 25}}{3} + \frac{x + 4}{3}\right| + C$$

$$= \ln\left|\sqrt{x^2 + 8x + 25} + x + 4\right| + D$$

where $D = C - \ln 3$.

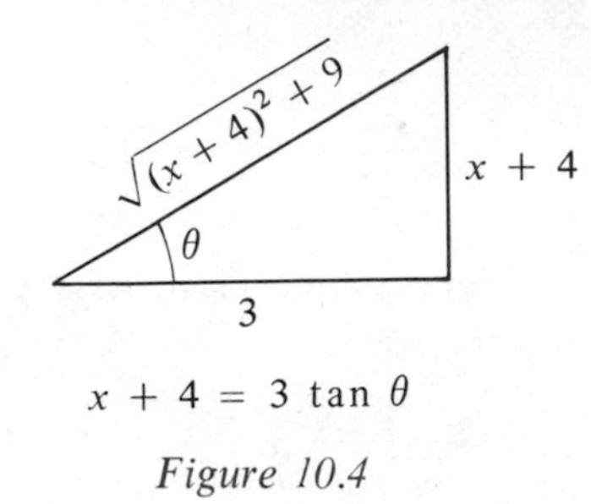

$x + 4 = 3 \tan \theta$

Figure 10.4

EXERCISES 10.5

Evaluate the integrals in Exercises 1–14.

1 $\displaystyle\int \frac{1}{x^2 - 4x + 8}\,dx$

2 $\displaystyle\int \frac{1}{\sqrt{7 + 6x - x^2}}\,dx$

3 $\displaystyle\int \frac{1}{\sqrt{4x - x^2}}\,dx$

4 $\displaystyle\int \frac{1}{x^2 - 2x + 2}\,dx$

5 $\displaystyle\int \frac{2x + 3}{\sqrt{9 - 8x - x^2}}\,dx$

6 $\displaystyle\int \frac{x + 5}{9x^2 + 6x + 17}\,dx$

7 $\displaystyle\int \frac{1}{x^3 - 1}\,dx$

8 $\displaystyle\int \frac{x^3}{x^3 - 1}\,dx$

9 $\displaystyle\int \frac{1}{(x^2 + 4x + 5)^2}\,dx$

10 $\displaystyle\int \frac{1}{x^4 - 4x^3 + 13x^2}\,dx$

11 $\displaystyle\int \frac{1}{(x^2 + 6x + 13)^{3/2}}\,dx$

12 $\displaystyle\int \sqrt{x(6 - x)}\,dx$

13 $\displaystyle\int \frac{1}{2x^2 - 3x + 9}\,dx$

14 $\displaystyle\int \frac{2x}{(x^2 + 2x + 5)^2}\,dx$

15 Find the area of the region bounded by the graphs of $y = (x^3 + 1)^{-1}$, $y = 0$, $x = 0$, and $x = 1$.

16 The region bounded by the graph of $y = 1/(x^2 + 2x + 10)$, the coordinate axes, and the line $x = 2$ is revolved about the x-axis. Find the volume of the resulting solid.

17 If the region described in Exercise 16 is revolved about the y-axis, find the volume of the resulting solid.

18 The velocity (at time t) of a point moving along a coordinate line is $(75 + 10t - t^2)^{-1/2}$ ft/sec. How far does the point travel during the time interval $[0, 5]$?

10.6 MISCELLANEOUS SUBSTITUTIONS

We have often used a change of variables to aid in the evaluation of a definite or indefinite integral. In this section we shall consider additional substitutions which are sometimes useful. The first example indicates that if an integral contains an expression of the form $\sqrt[n]{f(x)}$, then one of the substitutions $u = \sqrt[n]{f(x)}$ or $u = f(x)$ may simplify the evaluation.

Example 1 Evaluate $\displaystyle\int \frac{x^3}{\sqrt[3]{x^2+4}}\,dx$.

Solution 1 The substitution $u = \sqrt[3]{x^2+4}$ leads to the following equivalent equations:

$$u = \sqrt[3]{x^2+4}, \quad u^3 = x^2 + 4, \quad x^2 = u^3 - 4.$$

Taking the differential of each side of the last equation we obtain

$$2x\,dx = 3u^2\,du, \quad \text{or} \quad x\,dx = \tfrac{3}{2}u^2\,du.$$

We now substitute in the given integral as follows:

$$\begin{aligned}
\int \frac{x^3}{\sqrt[3]{x^2+4}}\,dx &= \int \frac{x^2}{\sqrt[3]{x^2+4}} \cdot x\,dx \\
&= \int \frac{u^3-4}{u} \cdot \frac{3}{2}u^2\,du \\
&= \frac{3}{2}\int (u^4 - 4u)\,du \\
&= \frac{3}{2}\left(\frac{1}{5}u^5 - 2u^2\right) + C \\
&= \frac{3}{10}u^2(u^3 - 10) + C \\
&= \frac{3}{10}(x^2+4)^{2/3}(x^2 - 6) + C.
\end{aligned}$$

Solution 2 If we substitute u for the expression *underneath* the radical, then

$$u = x^2 + 4 \qquad x^2 = u - 4$$

$$2x\,dx = du \qquad x\,dx = \tfrac{1}{2}\,du.$$

In this case we may write

$$\begin{aligned}
\int \frac{x^3}{\sqrt[3]{x^2+4}}\,dx &= \int \frac{x^2}{\sqrt[3]{x^2+4}} \cdot x\,dx \\
&= \int \frac{u-4}{u^{1/3}} \cdot \frac{1}{2}\,du \\
&= \frac{1}{2}\int (u^{2/3} - 4u^{-1/3})\,du \\
&= \frac{1}{2}\left[\frac{3}{5}u^{5/3} - 6u^{2/3}\right] + C \\
&= \frac{3}{10}u^{2/3}[u - 10] + C \\
&= \frac{3}{10}(x^2+4)^{2/3}(x^2-6) + C.
\end{aligned}$$

Example 2 Evaluate $\int \frac{1}{\sqrt{x} + \sqrt[3]{x}}\,dx$.

Solution If we let $z = \sqrt[6]{x}$, then

$$x = z^6, \quad \sqrt{x} = z^3, \quad \sqrt[3]{x} = z^2, \quad dx = 6z^5\,dz$$

and

$$\int \frac{1}{\sqrt{x} + \sqrt[3]{x}}\,dx = \int \frac{1}{z^3 + z^2} 6z^5\,dz = 6\int \frac{z^3}{z+1}\,dz.$$

By long division,

$$\frac{z^3}{z+1} = z^2 - z + 1 - \frac{1}{z+1}.$$

Consequently

$$\begin{aligned}
\int \frac{1}{\sqrt{x} + \sqrt[3]{x}}\,dx &= 6\int \left[z^2 - z + 1 - \frac{1}{z+1}\right] dz \\
&= 6\left(\tfrac{1}{3}z^3 - \tfrac{1}{2}z^2 + z - \ln|z+1|\right) + C \\
&= 2\sqrt{x} - 3\sqrt[3]{x} + 6\sqrt[6]{x} - 6\ln(\sqrt[6]{x} + 1) + C.
\end{aligned}$$

If an integrand is a rational expression in $\sin x$ and $\cos x$, then the substitution $z = \tan(x/2)$ where $-\pi < x < \pi$ will transform it into a rational (algebraic) expression in z. To prove this, first note that

$$\cos(x/2) = \frac{1}{\sec(x/2)} = \frac{1}{\sqrt{1 + \tan^2(x/2)}} = \frac{1}{\sqrt{1+z^2}},$$

$$\sin(x/2) = \tan(x/2)\cos(x/2) = z\frac{1}{\sqrt{1+z^2}}.$$

Consequently

$$\sin x = 2\sin(x/2)\cos(x/2) = \frac{2z}{1+z^2},$$

$$\cos x = 1 - 2\sin^2(x/2) = 1 - \frac{2z^2}{1+z^2} = \frac{1-z^2}{1+z^2}.$$

Moreover, since $x/2 = \tan^{-1} z$, we have $x = 2\tan^{-1} z$ and, therefore,

$$dx = \frac{2}{1+z^2}\,dz.$$

It follows that replacement of $\sin x$, $\cos x$, and dx by

(10.7) $$\sin x = \frac{2z}{1+z^2}, \quad \cos x = \frac{1-z^2}{1+z^2}, \quad dx = \frac{2}{1+z^2}\,dz$$

will produce an integral which is a rational expression in z.

Example 3 Evaluate $\displaystyle\int \frac{1}{4\sin x - 3\cos x}\,dx$.

Solution Applying (10.7) and simplifying the integrand,

$$\begin{aligned}\int \frac{1}{4\sin x - 3\cos x}\,dx &= \int \frac{1}{4\left(\dfrac{2z}{1+z^2}\right) - 3\left(\dfrac{1-z^2}{1+z^2}\right)} \cdot \frac{2}{1+z^2}\,dz \\ &= \int \frac{2}{8z - 3(1-z^2)}\,dz \\ &= 2\int \frac{1}{3z^2 + 8z - 3}\,dz.\end{aligned}$$

Using partial fractions,

$$\frac{1}{3z^2+8z-3} = \frac{1}{10}\left(\frac{3}{3z-1} - \frac{1}{z+3}\right)$$

and hence

$$\begin{aligned}\int \frac{1}{4\sin x - 3\cos x}\,dx &= \frac{1}{5}\int\left(\frac{3}{3z-1} - \frac{1}{z+3}\right)dz \\ &= \frac{1}{5}(\ln|3z-1| - \ln|z+3|) + C \\ &= \frac{1}{5}\ln\left|\frac{3z-1}{z+3}\right| + C \\ &= \frac{1}{5}\ln\left|\frac{3\tan(x/2) - 1}{\tan(x/2) + 3}\right| + C.\end{aligned}$$

Other substitutions are sometimes useful; however, it is impossible to state rules which apply to all situations. Whether or not one can express an integrand in a suitable form is often a matter of individual ingenuity.

EXERCISES 10.6

Evaluate the integrals in Exercises 1–26.

1 $\int x\sqrt[3]{x+9}\,dx$

2 $\int x^2\sqrt{2x+1}\,dx$

3 $\int \frac{x}{\sqrt[5]{3x+2}}\,dx$

4 $\int \frac{5x}{(x+3)^{2/3}}\,dx$

5 $\int \frac{1}{\sqrt{x}+4}\,dx$

6 $\int \frac{1}{\sqrt{4+\sqrt{x}}}\,dx$

7 $\int \frac{\sqrt{x}}{1+\sqrt[3]{x}}\,dx$

8 $\int \frac{1}{\sqrt[4]{x}+\sqrt[3]{x}}\,dx$

9 $\int \frac{1}{(x+1)\sqrt{x-2}}\,dx$

10 $\int \frac{2x+3}{\sqrt{1+2x}}\,dx$

11 $\int \frac{x+1}{(x+4)^{1/3}}\,dx$

12 $\int \frac{x^{1/3}+1}{x^{1/3}-1}\,dx$

13 $\int e^{3x}\sqrt{1+e^x}\,dx$

14 $\int \frac{e^{2x}}{\sqrt[3]{1+e^x}}\,dx$

15 $\int \frac{e^{2x}}{e^x+4}\,dx$

16 $\int \frac{\sin 2x}{\sqrt{1+\sin x}}\,dx$

17 $\int \sin\sqrt{x+4}\,dx$

18 $\int \sqrt{x}e^{\sqrt{x}}\,dx$

19 $\int \frac{x}{(x-1)^6}\,dx$ (*Hint:* Let $u = x - 1$.)

20 $\int \frac{x^2}{(3x+4)^{10}}\,dx$ (*Hint:* Let $u = 3x + 4$.)

21 $\int \frac{1}{2+\sin x}\,dx$

22 $\int \frac{1}{3+2\cos x}\,dx$

23 $\int \frac{1}{1+\sin x+\cos x}\,dx$

24 $\int \frac{1}{\tan x+\sin x}\,dx$

25 $\int \frac{\sec x}{4-3\tan x}\,dx$

26 $\int \frac{1}{\sin x-\sqrt{3}\cos x}\,dx$

10.7 TABLES OF INTEGRALS

Mathematicians and scientists who use integrals in their work often refer to tables of integrals. Many of the formulas contained in these tables may be obtained by methods we have studied. In general, tables of integrals should not be used until the student has had sufficient experience with the standard methods of integration. Indeed, for complicated integrals it is often necessary to make substitutions, or to use partial fractions, integration by parts, or other techniques in order to obtain integrands for which the table is applicable.

The following examples illustrate the use of several formulas stated in the brief Table of Integrals printed on the inside covers of this text.

Example 1 Evaluate $\int \frac{1}{x^2\sqrt{3+5x^2}}\,dx$, where $x > 0$.

Solution The integrand suggests that we use that part of the table dealing with the form $\sqrt{a^2+u^2}$. Specifically, Formula 28 states that

$$\int \frac{du}{u^2\sqrt{a^2+u^2}} = -\frac{\sqrt{a^2+u^2}}{a^2u} + C$$

where for compactness the differential du is placed in the numerator instead of to the right of the integrand. In order to use this formula we must adjust the given integral so that it matches *exactly* with the formula. If we let

$$a^2 = 3 \quad \text{and} \quad u^2 = 5x^2$$

then the expression underneath the radical is taken care of; however, we also need

(i) u^2 to the left of the radical.

(ii) du in the numerator.

We can achieve (i) by writing the integral as

$$5\int \frac{1}{5x^2\sqrt{3+5x^2}}\,dx.$$

To accomplish (ii) we note that

$$u = \sqrt{5}x \quad \text{and} \quad du = \sqrt{5}\,dx$$

and write the preceding integral as

$$5\cdot\frac{1}{\sqrt{5}}\int \frac{\sqrt{5}\,dx}{5x^2\sqrt{3+5x^2}}.$$

The integral now matches exactly with that in Formula 28 and hence

$$\int \frac{1}{x^2\sqrt{3+5x^2}}\,dx = \sqrt{5}\left[-\frac{\sqrt{3+5x^2}}{3(\sqrt{5}x)}\right] + C$$
$$= -\frac{\sqrt{3+5x^2}}{3x} + C.$$

In order to guard against algebraic errors when using tables of integrals, answers should be checked by differentiating.

Example 2 Evaluate $\int x^3 \cos x\,dx$.

Solution We first use reduction Formula 85 in the Table of Integrals with $n = 3$ and $u = x$, obtaining

$$\int x^3 \cos x\,dx = x^3 \sin x - 3\int x^2 \sin x\,dx.$$

Next we apply Formula 84 with $n = 2$, and then Formula 83, obtaining

$$\int x^2 \sin x\,dx = -x^2 \cos x + 2\int x \cos x\,dx$$
$$= -x^2 \cos x + 2\,[\cos x + x \sin x] + C.$$

Substitution in the first expression gives us

$$\int x^3 \cos x\,dx = x^3 \sin x + 3x^2 \cos x - 6 \cos x - 6x \sin x + C.$$

This answer may be verified by differentiation.

EXERCISES 10.7

Use the Table of Integrals printed on the inside covers of this text to evaluate the integrals in Exercises 1–20.

1 $\int \frac{\sqrt{4+9x^2}}{x}\,dx$

2 $\int \frac{1}{x\sqrt{2+3x^2}}\,dx$

3 $\int (16 - x^2)^{3/2}\,dx$

4 $\int x^2\sqrt{4x^2 - 16}\,dx$

5 $\int x\sqrt{2 - 3x}\,dx$

6 $\int x^2\sqrt{5 + 2x}\,dx$

7 $\int \sin^6 3x\,dx$

8 $\int x \cos^5 (x^2)\,dx$

9 $\int \csc^4 x\,dx$

10 $\int \sin 5x \cos 3x\,dx$

11 $\int x \sin^{-1} x \, dx$

12 $\int x^2 \tan^{-1} x \, dx$

13 $\int e^{-3x} \sin 2x \, dx$

14 $\int x^5 \ln x \, dx$

15 $\int \frac{\sqrt{5x - 9x^2}}{x} dx$

16 $\int \frac{1}{x\sqrt{3x - 2x^2}} dx$

17 $\int \frac{x}{5x^4 - 3} dx$

18 $\int \cos x \sqrt{\sin^2 x - 4} \, dx$

19 $\int e^{2x} \cos^{-1} e^x \, dx$

20 $\int \sin^2 x \cos^3 x \, dx$

10.8 MOMENTS AND CENTROIDS OF PLANE REGIONS

Some applications of the definite integral were considered in Chapter 6. In this section and the next, several other applications are discussed.

Let l be a coordinate line and P a point on l with coordinate x. If a particle of mass m is located at P, then the **moment** (more precisely, the **first moment**) of the particle with respect to the origin O is defined as the product mx of the mass and the directed distance from the origin to P. Suppose l is horizontal with positive direction to the right and let us imagine that l is free to rotate about O as if a fulcrum were positioned as shown in Figure 10.5. If an object has its mass m_1 concentrated at a point with positive coordinate x_1, then the moment m_1x_1 is positive, and the weight of the object would make l rotate in a clockwise direction. If an object of mass m_2 is at a point with negative coordinate x_2, then its moment m_2x_2 is negative, and l would rotate in a counterclockwise direction. The system consisting of both objects is said to be in **equilibrium** if $m_1x_1 = m_2|x_2|$. Since $x_2 < 0$, this is equivalent to $m_1x_1 = -m_2x_2$, or

$$m_1x_1 + m_2x_2 = 0;$$

that is, *the sum of the moments with respect to the origin is zero.* This situation is similar to a seesaw which balances at the point O if two persons of weights m_1 and m_2 are located as indicated in Figure 10.5.

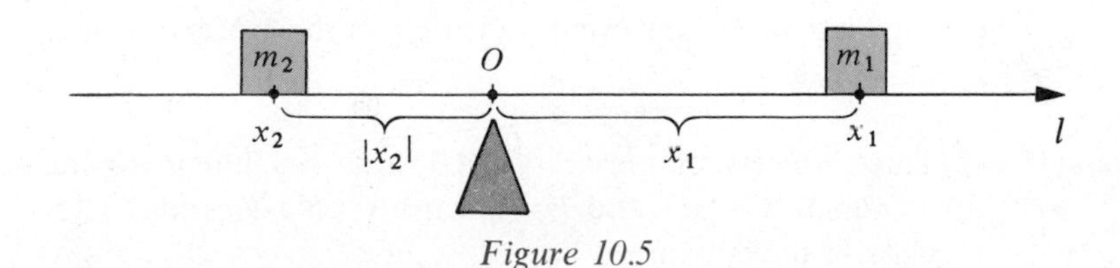

Figure 10.5

If the system is not in equilibrium, then as illustrated in Figure 10.6 there is a "balance" point P with coordinate $\bar{x}$, in the sense that

$$m_1(x_1 - \bar{x}) = m_2|x_2 - \bar{x}| = -m_2(x_2 - \bar{x})$$

where $|x_2 - \bar{x}| = -(x_2 - \bar{x})$ since $x_2 < \bar{x}$.

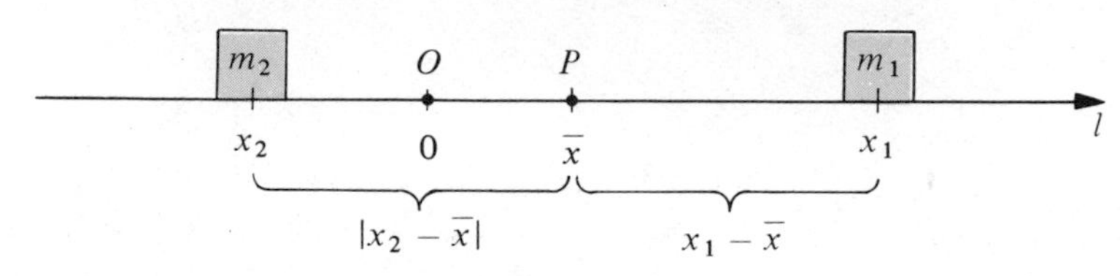

Figure 10.6

To locate P we may solve for $\bar{x}$ as follows:

$$m_1(x_1 - \bar{x}) + m_2(x_2 - \bar{x}) = 0$$
$$m_1x_1 + m_2x_2 - (m_1 + m_2)\bar{x} = 0$$
$$(m_1 + m_2)\bar{x} = m_1x_1 + m_2x_2$$

(10.8)
$$\bar{x} = \frac{m_1x_1 + m_2x_2}{m_1 + m_2}.$$

Thus, to find $\bar{x}$ we divide the sum of the moments with respect to the origin by the total mass $m = m_1 + m_2$. The point $\bar{x}$ is called the **center of mass** (or **center of gravity**) of the system.

More generally, if a system of n particles of masses $m_1, m_2, \ldots, m_n$ are located at points on l with coordinates $x_1, x_2, \ldots, x_n$, respectively, then the sum of the moments $\Sigma_{i=1}^n m_ix_i$ is called the **moment of the system with respect to the origin**. If $m = \Sigma_{i=1}^n m_i$ denotes the total mass of the system then in analogy with (10.8) we define $\bar{x}$ by

(10.9)
$$\bar{x} = \frac{\sum_{i=1}^{n} m_ix_i}{m} \quad \text{or} \quad m\bar{x} = \sum_{i=1}^{n} m_ix_i.$$

The number $m\bar{x}$ may be regarded as the moment with respect to the origin of a particle of mass m located at the point with coordinate $\bar{x}$. The second formula in (10.9) then states that $\bar{x}$ gives the position at which the total mass m could be concentrated without changing the moment of the system with respect to the origin. The point with coordinate $\bar{x}$ is the balance point of the system in the sense of our seesaw illustration and is called the **center of mass** (or **center of gravity**) of the system. If $\bar{x} = 0$, then by (10.9) $\Sigma_{i=1}^n m_ix_i = 0$ and the system is said to be in equilibrium. In this event the origin is the center of mass.

Example 1 Three objects of masses 40, 60, and 100 units are located at points with coordinates -2, 3, and 7, respectively, on a coordinate line l. Find the center of mass of the system.

Solution If we denote the three masses by m_1, m_2, and m_3, then we have the situation illustrated in Figure 10.7, where $x_1 = -2$, $x_2 = 3$, and $x_3 = 7$. Applying (10.9), the coordinate $\bar{x}$ of the center of mass is given by

$$\begin{aligned}\bar{x} &= \frac{40(-2) + 60(3) + 100(7)}{40 + 60 + 100} \\ &= \frac{-80 + 180 + 700}{200} = \frac{800}{200} = 4.\end{aligned}$$

$m_1 = 40$ $m_2 = 60$ $m_3 = 100$

-2 0 3 7 l

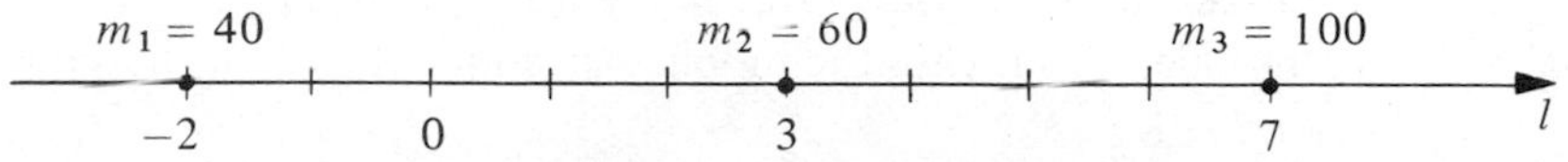

Figure 10.7

The preceding concepts may be extended to two dimensions as follows. Let n particles of masses $m_1, m_2, \ldots, m_n$ be located at points $P_1(x_1, y_1)$, $P_2(x_2, y_2), \ldots,$ $P_n(x_n, y_n)$, respectively, in a coordinate plane as illustrated in Figure 10.8. The **moment of the *i*th particle with respect to the *x*-axis** is defined as $m_i y_i$, that is, the product of the mass and the (directed) distance from the x-axis. Similarly, the **moment of the *i*th particle with respect to the *y*-axis** is defined as $m_i x_i$. We then define **the moments M_x and M_y of the system with respect to the *x*-axis and *y*-axis**, respectively, by

(10.10)
$$M_x = \sum_{i=1}^{n} m_i y_i \quad \text{and} \quad M_y = \sum_{i=1}^{n} m_i x_i.$$

In words, to find M_x, multiply the mass of each particle by its directed distance from the x-axis and add. For M_y we use directed distances from the y-axis.

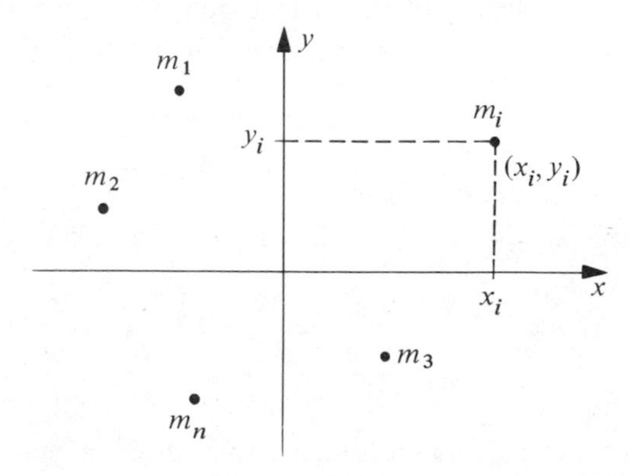

Figure 10.8

In analogy with (10.9), if $m = \sum_{i=1}^{n} m_i$ denotes the total mass of the system, then the **center of mass** (or **center of gravity**) of the system is the point $P(\bar{x}, \bar{y})$ given by the equations

(10.11)
$$m\bar{x} = M_y \quad \text{and} \quad m\bar{y} = M_x.$$

The number $m\bar{x}$ may be regarded as the moment with respect to the y-axis of a particle of mass m located at the point $P(\bar{x}, \bar{y})$, and $m\bar{y}$ may be thought of as its moment with respect to the x-axis. Consequently, equations (10.11) imply that the center of mass of the system of particles is the point at which the total mass could be concentrated without changing the moments of the system with respect to the coordinate axes. To get an intuitive picture of this situation, we may think of the n particles as fastened to the center of mass P by weightless rods, in a manner similar to the way spokes of a wheel are attached to the center of the wheel. The system would then balance when supported from the ceiling of a room by a cord attached to P, as illustrated in Figure 10.9. The appearance would be similar to a mobile; in our case the mobile has all its objects in the same horizontal plane.

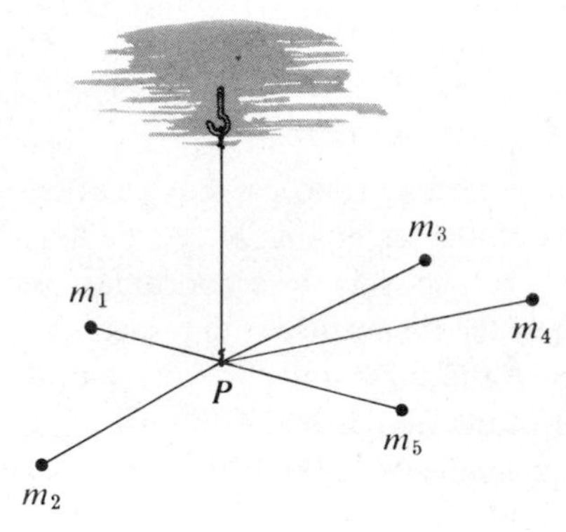

Figure 10.9

Example 2 Particles of masses 4, 8, 3, and 2 units are located at the points $P_1(-2, 3)$, $P_2(2, -6)$, $P_3(7, -3)$, and $P_4(5, 1)$, respectively. Find the moments M_x and M_y and the coordinates of the center of mass of the system.

Solution The system is illustrated in Figure 10.10, where we have also anticipated the position of $(\bar{x}, \bar{y})$. Applying (10.10),

$$M_x = (4)(3) + (8)(-6) + (3)(-3) + (2)(1) = -43$$
$$M_y = (4)(-2) + (8)(2) + (3)(7) + (2)(5) = 39.$$

Since $m = 4 + 8 + 3 + 2 = 17$, it follows from (10.11) that

$$\bar{x} = \frac{M_y}{m} = \frac{39}{17} \approx 2.3, \quad \bar{y} = \frac{M_x}{m} = -\frac{43}{17} \approx -2.5.$$

Let us now consider a thin sheet of material (called a **lamina**) which is **homogeneous**, that is, has a constant density. We wish to define the center of mass P so that it is the balance point, in a sense analogous to that for systems of particles. If a lamina has the shape of a rectangle, then it is evident that the center of mass is the intersection of the diagonals. If the rectangle lies in an xy-plane, then we assume that the mass of the rectangle (more precisely, of the rectangular lamina) can be concentrated at the center of mass without changing its moments with respect to the x- or y-axes.

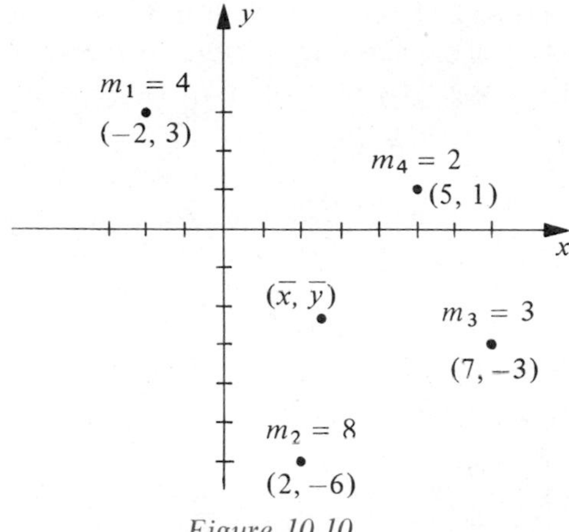

Figure 10.10

Let us first consider a lamina which has the shape of a region of the type illustrated in Figure 10.11, where f is continuous on the closed interval $[a, b]$. We partition $[a, b]$ by choosing numbers $a = x_0 < x_1 < \cdots < x_n = b$ and, for each i, choose w_i as the midpoint of the subinterval $[x_{i-1}, x_i]$, that is, $w_i = (x_{i-1} + x_i)/2$. The Riemann sum $\Sigma_{i=1}^{n} f(w_i)\Delta x_i$ may be thought of as the sum of areas of rectangles of the type shown in Figure 10.11. If the density is ρ, then the mass of that part of the lamina corresponding to the ith rectangle is $\rho f(w_i)\Delta x_i$, and consequently the mass of the rectangular polygon associated with the partition is $\Sigma_i \rho f(w_i)\Delta x_i$. As the norm $\|P\|$ of the partition approaches zero, the area of the rectangular polygon approaches the area of the face of the lamina, and the latter sum should approach the mass of the lamina. It is natural, therefore, to define the mass m of the lamina by

(10.12)
$$m = \lim_{\|P\| \to 0} \sum_i \rho f(w_i)\Delta x_i = \rho \int_a^b f(x)\,dx.$$

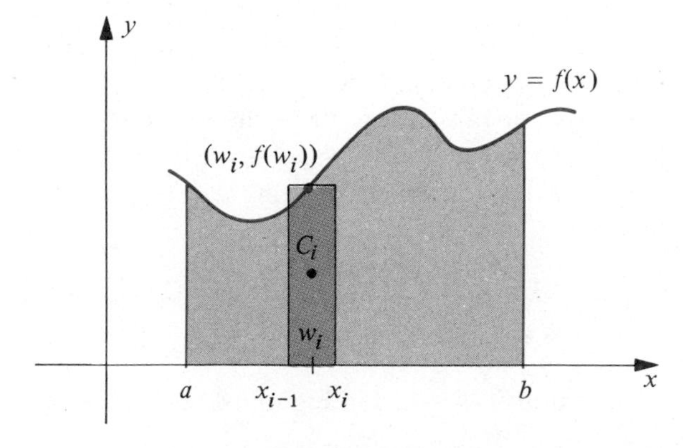

Figure 10.11

The center of mass of the ith rectangular lamina illustrated in Figure 10.11 is located at the point $C_i(w_i, \frac{1}{2}f(w_i))$. If we assume that the mass is concentrated at C_i, then its moment with respect to the x-axis can be found by multiplying the distance $\frac{1}{2}f(w_i)$ from the x-axis to C_i by the mass $\rho f(w_i)\,\Delta x_i$. Using the additive property of moments, the moment of the rectangular polygon associated with the partition is $\Sigma_i \frac{1}{2}f(w_i) \cdot \rho f(w_i)\,\Delta x_i$. The moment M_x of the lamina is defined as the limit of this sum, that is,

(10.13)
$$M_x = \lim_{\|P\| \to 0} \sum_i \tfrac{1}{2}f(w_i) \cdot \rho f(w_i)\,\Delta x_i = \rho \int_a^b \tfrac{1}{2}f(x)^2\,dx.$$

In like manner, using the distance w_i from the y-axis to the center of mass of the ith rectangle, we arrive at the definition of the moment M_y of the lamina with respect to the y-axis. Specifically,

(10.14)
$$M_y = \lim_{\|P\| \to 0} \sum_i w_i \cdot \rho f(w_i)\,\Delta x_i = \rho \int_a^b x f(x)\,dx.$$

Finally, as with particles (see (10.11)), the coordinates $\bar{x}$ and $\bar{y}$ of the center of mass of the lamina are defined by

(10.15)
$$m\bar{x} = M_y \quad \text{and} \quad m\bar{y} = M_x.$$

Substituting the integral forms in these equations and solving for $\bar{x}$ and $\bar{y}$ gives us

(10.16)
$$\bar{x} = \frac{M_y}{m} = \frac{\rho \int_a^b x f(x)\,dx}{\rho \int_a^b f(x)\,dx}$$
$$\bar{y} = \frac{M_x}{m} = \frac{\rho \int_a^b \frac{1}{2}f(x)^2\,dx}{\rho \int_a^b f(x)\,dx}.$$

Since the constant ρ in these formulas may be canceled, we see that the coordinates of the center of mass of a homogeneous lamina are independent of the density ρ; that is, they depend only on the shape of the lamina and not on the density. For this reason the point $(\bar{x}, \bar{y})$ is sometimes referred to as the center of mass of a *region* in the plane, or as the **centroid** of the region. We can obtain formulas for centroids by letting $\rho = 1$ in the preceding formulas.

Example 3 Find the coordinates of the centroid of the region bounded by the graphs of $y = e^x$, $x = 0$, $x = 1$, and $y = 0$.

Solution The region is sketched in Figure 10.12. Using (10.12) and (10.13), with $\rho = 1$, we have

$$m = \int_0^1 e^x\,dx = e^x\Big]_0^1 = e - 1$$
$$M_x = \int_0^1 \tfrac{1}{2}e^{2x}\,dx = \tfrac{1}{4}e^{2x}\Big]_0^1 = \tfrac{1}{4}(e^2 - 1).$$

Consequently, by (10.16),

$$\bar{y} = \frac{\frac{1}{4}(e^2 - 1)}{e - 1} = \frac{e + 1}{4} \approx 0.93.$$

Next, by (10.14),

$$M_y = \int_0^1 xe^x \, dx.$$

Integrating by parts (with $u = x$ and $dv = e^x \, dx$) we obtain

$$\begin{aligned} M_y &= \Big[xe^x - e^x \Big]_0^1 \\ &= (e - e) - (0 - 1) = 1. \end{aligned}$$

Using (10.16),

$$\bar{x} = \frac{1}{e - 1} \approx 0.58.$$

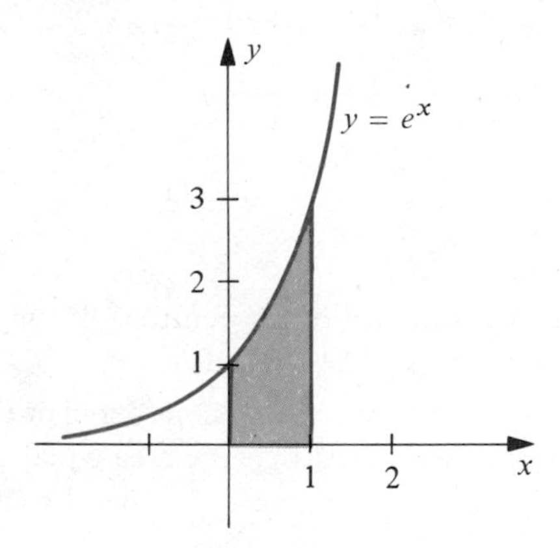

Figure 10.12

Formulas similar to those in (10.16) may be obtained for more complicated regions. Thus, consider a lamina of constant density ρ which has the shape illustrated in Figure 10.13, where f and g are continuous functions and $f(x) \geq g(x)$ for all x in $[a, b]$. Partitioning $[a, b]$, choosing w_i as before, and then applying the Midpoint Formula (1.8) we see that the center of mass of the ith rectangular lamina pictured in Figure 10.13 is the point $C_i(w_i, \frac{1}{2}[f(w_i) + g(w_i)])$. Using an argument similar to that given previously, the moment of the ith rectangle with respect to the x-axis is the distance from the x-axis to C_i multiplied by the mass; that is,

$$\tfrac{1}{2}[f(w_i) + g(w_i)]\rho[f(w_i) - g(w_i)] \, \Delta x_i.$$

Summing and taking the limit as the norm of the partition approaches zero gives us

(10.17)

$$M_x = \rho \int_a^b \tfrac{1}{2}[f(x) + g(x)]\,[f(x) - g(x)]\,dx, \quad \text{or}$$

$$M_x = \rho \int_a^b \tfrac{1}{2}\{[f(x)]^2 - [g(x)]^2\}\,dx.$$

In like manner,

(10.18)

$$M_y = \rho \int_a^b x[f(x) - g(x)]\,dx.$$

The formulas in (10.15) may then be used to find $\bar{x}$ and $\bar{y}$.

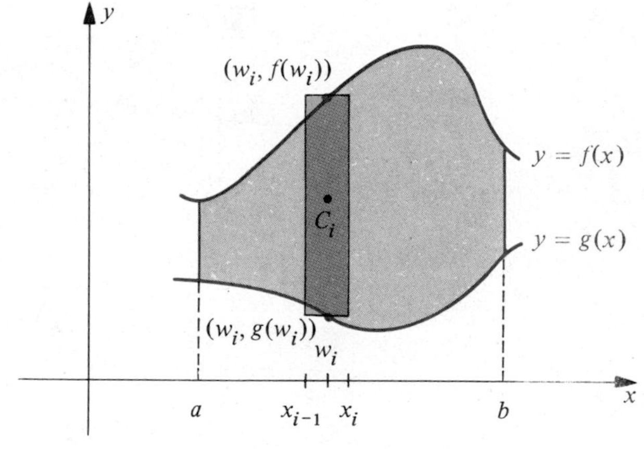

Figure 10.13

Example 4 Find the coordinates of the centroid of the region bounded by the graphs of $y + x^2 = 6$ and $y + 2x - 3 = 0$.

Solution The region is the same as that considered in Example 2 of Section 6.1 (see Figure 6.6) where it was found that the area equals 32/3. If, as in that example, we let $f(x) = 6 - x^2$ and $g(x) = 3 - 2x$, then by (10.17), with $\rho = 1$,

$$M_x = \tfrac{1}{2}\int_{-1}^{3} [(6 - x^2)^2 - (3 - 2x)^2]\,dx$$

$$= \tfrac{1}{2}\int_{-1}^{3} (x^4 - 16x^2 + 12x + 27)\,dx.$$

It is left to the reader to show that $M_x = 416/15$. Hence, by (10.16)

$$\bar{y} = \frac{M_x}{m} = \frac{416/15}{32/3} = \frac{13}{5}.$$

Next applying (10.18), with $\rho = 1$,

$$M_y = \int_{-1}^{3} x(3 - x^2 + 2x)\,dx$$

$$= \int_{-1}^{3} (3x - x^3 + 2x^2)\,dx$$

from which it follows that $M_y = 32/3$. Consequently

$$\bar{x} = \frac{M_y}{m} = \frac{32/3}{32/3} = 1.$$

Formulas for moments may also be obtained for regions of the type illustrated in Figure 6.8. This is left as an exercise for the student.

EXERCISES 10.8

1 Particles of masses 2, 7, and 5 are located at the points $A(4, -1)$, $B(-2, 0)$, and $C(-8, -5)$, respectively. Find the moments M_x and M_y and the coordinates of the center of mass of this system.

2 Particles of masses 10, 3, 4, 1, and 8 are located at the points $A(-5, -2)$, $B(3, 7)$, $C(0, -3)$, $D(-8, -3)$, and $O(0, 0)$. Find the moments M_x and M_y and the coordinates of the center of mass of this system.

In each of Exercises 3–12, sketch the region bounded by the graphs of the given equations and find the centroid of the region.

3 $y = x^3, y = 0, x = 1$

4 $y = \sqrt{x}, y = 0, x = 9$

5 $y = \sin x, y = 0, x = 0, x = \pi$

6 $y = \sec^2 x, y = 0, x = -\pi/4, x = \pi/4$

7 $y = 1 - x^2, y = x - 1$

8 $x = y^2, x - y - 2 = 0$

9 $y = 1/\sqrt{16 + x^2}, y = 0, x = 0, x = 3$

10 $xy = 1, y = 0, x = 1, x = e$

11 $y = e^{2x}, y = 0, x = -1, x = 0$

12 $y = \cosh x, y = 0, x = -1, x = 1$

13 Prove that the centroid of a triangle coincides with the intersection of the medians. (*Hint:* Take the vertices at the points $(0, 0)$, (a, b), and $(0, c)$, where a, b, and c are positive.)

14 Find the centroid of the region in the first quadrant bounded by the circle $x^2 + y^2 = a^2$ and the coordinate axes.

15 Find the centroid of a semicircular region of radius a.

16 Find the centroid of the region bounded by the parabola $x^2 = 4py$ and a line through the focus perpendicular to the y-axis.

17 A plane region has the shape of a square of side $2a$ surmounted by a semicircle of radius a. Find the centroid. (*Hint:* Use Exercise 15 and the fact that moments are additive.)

18 A region has the shape of a square of side a surmounted by an equilateral triangle of side a. Find the centroid. (*Hint:* Use Exercise 13 and the fact that moments are additive.)

19 Find the centroid of the semi-elliptical region bounded by the graph of $y = (b/a)\sqrt{a^2 - x^2}$ and the x-axis.

20 Find M_y for the region bounded by the hyperbola $b^2x^2 - a^2y^2 = a^2b^2$ and a line through a focus perpendicular to the transverse axis.

10.9 CENTROIDS OF SOLIDS OF REVOLUTION

The center of mass of a homogeneous solid of revolution is a certain point on the axis of revolution. In the special case of a right circular cylinder of finite altitude, the center of mass is the point on the axis which lies halfway between the two bases. In order to develop formulas for locating the center of mass of a solid we begin with an xy-plane and introduce a third coordinate axis, called the ***z*-axis**, perpendicular to both the x- and y-axes at the origin, as illustrated in Figure 10.14. The y- and z-axes determine a coordinate plane called the ***yz*-plane**. Similarly, the plane determined by the x- and z-axes is referred to as the ***xz*-plane**.

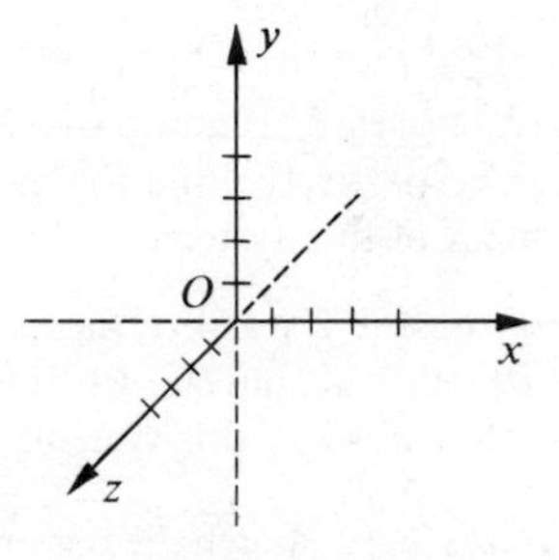

Figure 10.14

Let us consider a solid of revolution of the type shown in Figure 10.15; that is, a solid obtained by revolving about the x-axis the region bounded by the graph of a continuous function f and the lines $x = a$, $x = b$, and $y = 0$. It will be assumed that the resulting solid has constant density (weight per unit volume) ρ and that its mass m is the product of ρ and the volume of the solid. Thus, according to Definition (6.5),

(10.19) $$m = \rho \int_a^b \pi [f(x)]^2 \, dx.$$

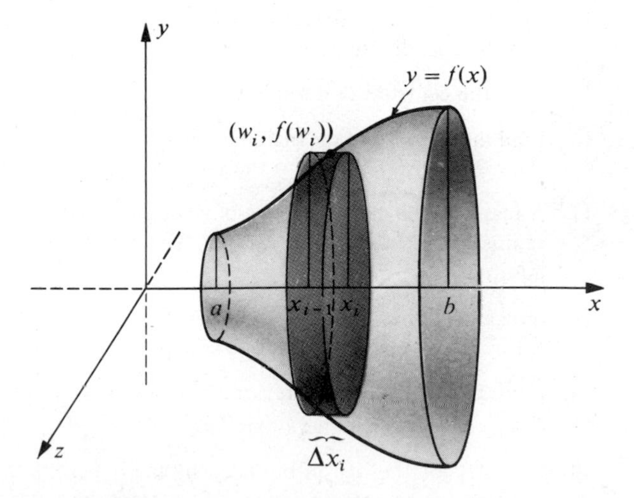

Figure 10.15

Since the center of mass C of the solid in Figure 10.15 is on the x-axis, it is sufficient to define the directed distance $\bar{x}$ from the origin to C. Let us partition $[a, b]$, in the usual way, choose w_i as the midpoint of the ith subinterval $[x_{i-1}, x_i]$, and consider the rectangle with its base on $[x_{i-1}, x_i]$ and with altitude $f(w_i)$. The mass of the disc generated by revolving this rectangle about the x-axis is $\rho\pi[f(w_i)]^2 \Delta x_i$. The center of mass of this disc is the point on the x-axis with coordinate w_i. The **moment with respect to the yz-plane of the ith disc** is defined as the product of the distance w_i from the yz-plane to the center of mass of the ith disc and the mass of the disc, that is,

$$w_i \cdot \rho\pi[f(w_i)]^2 \Delta x_i.$$

Assuming that moments are additive, the moment with respect to the yz-plane of the solid which consists of the discs determined by all such rectangles is a sum of terms of the above form. The **moment M_{yz} of the solid of revolution with respect to the yz-plane** is defined as the limit of this sum, that is,

(10.20) $$M_{yz} = \lim_{\|P\| \to 0} \sum w_i \cdot \rho\pi[f(w_i)]^2 \Delta x_i = \rho\pi \int_a^b x[f(x)]^2\, dx.$$

The x-coordinate $\bar{x}$ of the center of mass is defined by

(10.21) $$m\bar{x} = M_{yz}, \quad \text{or} \quad \bar{x} = \frac{M_{yz}}{m}.$$

If the integral forms (10.19) and (10.20) are substituted in (10.21), the density factor ρ cancels. Consequently the center of mass of a homogeneous solid of revolution is independent of the density; that is, it depends only on the shape of the solid. For this reason we shall often refer to the **centroid of a geometric solid** instead of to center of mass. For convenience we may let $\rho = 1$ in (10.19) and (10.20) to find centroids of geometric solids of revolution.

Example 1 Find the center of mass of a homogeneous right circular cone of altitude h and radius of base r.

Solution If the triangular region with vertices at the points $(0, 0)$, $(h, 0)$, and (h, r) is revolved about the x-axis, a cone of the desired type is generated (see Figure 10.16). An equation of the line through $(0, 0)$ and (h, r) is $y = (r/h)x$, and hence $f(x) = (r/h)x$ defines a function f of the type used in the preceding discussion.

Applying (10.20),

$$\begin{aligned} M_{yz} &= \rho\pi \int_0^h x\left(\frac{r}{h}x\right)^2 dx \\ &= \rho\pi\left(\frac{r^2}{h^2}\right) \int_0^h x^3\, dx \\ &= \rho\pi\left(\frac{r^2}{h^2}\right)\left[\frac{x^4}{4}\right]_0^h \\ &= \rho \cdot \frac{1}{4}\pi r^2 h^2. \end{aligned}$$

The mass of the cone could be found from (10.19); however, since the volume of a right circular cone of altitude h and base radius r is $\frac{1}{3}\pi r^2 h$, the mass m is $\rho \cdot \frac{1}{3}\pi r^2 h$. Finally, using (10.21)

$$\bar{x} = \frac{M_{yz}}{m} = \frac{\rho \cdot \frac{1}{4}\pi r^2 h^2}{\rho \cdot \frac{1}{3}\pi r^2 h} = \frac{3}{4}h.$$

Thus, the center of mass is on the axis of the cone, 3/4 of the way from the vertex to the base.

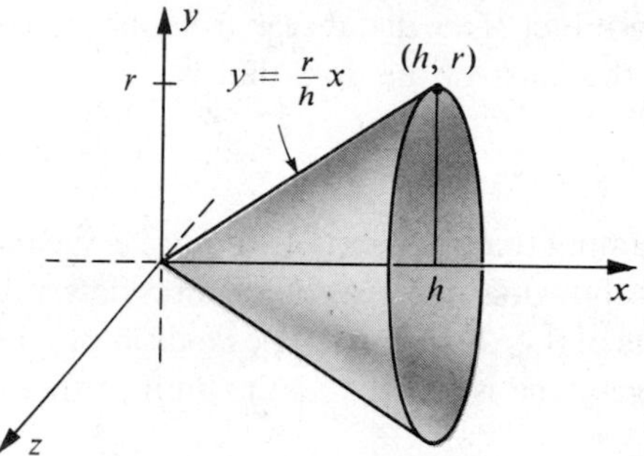

Figure 10.16

A formula analogous to (10.20) can be obtained for solids which are generated by revolving regions about the y-axis. The derivation merely involves interchanging the roles of x and y in the previous discussion. Referring to Figure 10.17, we see that if w_i is the midpoint of $[y_{i-1}, y_i]$, then

(10.22) $$M_{xz} = \lim_{\|P\| \to 0} \sum_i \rho \cdot w_i \pi [g(w_i)]^2 \, \Delta y_i = \rho\pi \int_c^d y[g(y)]^2 \, dy.$$

The center of mass is on the y-axis and its coordinate $\bar{y}$ is given by

(10.23) $$m\bar{y} = M_{xz} \quad \text{or} \quad \bar{y} = \frac{M_{xz}}{m}.$$

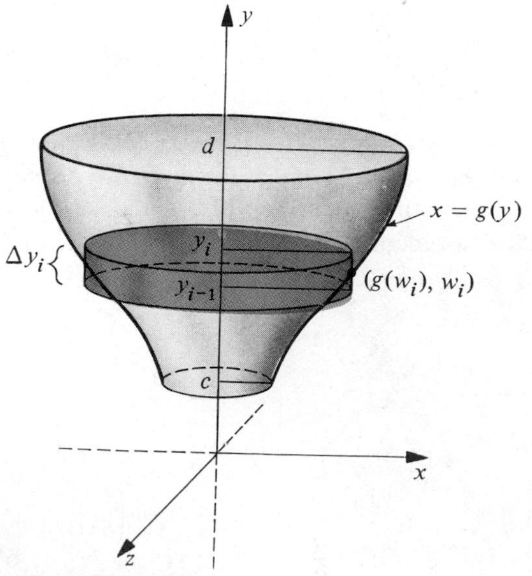

Figure 10.17

Example 2 Find the centroid of a hemisphere of radius a.

Solution If the region in the first quadrant under the graph of the equation $x^2 + y^2 = a^2$ is revolved about the y-axis, there results a hemisphere of radius a (see Figure 10.18).

Solving the equation of the circle for x in terms of y gives us $x = \sqrt{a^2 - y^2}$, and hence $g(y) = \sqrt{a^2 - y^2}$ defines a function g of the type discussed above. Using (10.22) with $\rho = 1$,

$$\begin{aligned} M_{xz} &= \pi \int_0^a y(\sqrt{a^2 - y^2})^2\, dy \\ &= \pi \int_0^a (a^2 y - y^3)\, dy \\ &= \pi \left[\frac{1}{2}a^2 y^2 - \frac{1}{4}y^4\right]_0^a \\ &= \pi \left[\frac{1}{2}a^4 - \frac{1}{4}a^4\right] = \frac{1}{4}\pi a^4. \end{aligned}$$

Since the volume of a sphere of radius a is $\frac{4}{3}\pi a^3$, the hemisphere has volume $\frac{2}{3}\pi a^3$. From (10.23),

$$\bar{y} = \frac{\frac{1}{4}\pi a^4}{\frac{2}{3}\pi a^3} = \frac{3a}{8}.$$

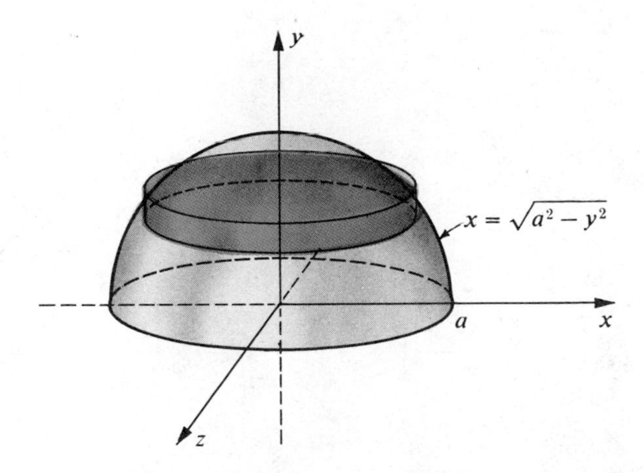

Figure 10.18

Centroids may also be found by using washer-shaped elements of volume or cylindrical shells. The next example illustrates the shell method.

Example 3 The region bounded by the graph of $y = 2x - x^2$ and the x-axis is revolved about the y-axis. Find the centroid of the resulting solid.

Solution The region is the same as that considered in Example 1 of Section 6.3, and is resketched in Figure 10.19 along with the shell generated by a typical rectangle.

As in the solution of that example, the volume of the shell is $2\pi w_i(2w_i - w_i^2)\,\Delta x_i$. The centroid of the shell is on the y-axis and its distance from the x-axis is one-half the altitude of the shell, or $\frac{1}{2}(2w_i - w_i^2)$. Consequently, the moment of the shell with respect to the xz-plane is

$$\tfrac{1}{2}(2w_i - w_i^2) \cdot 2\pi w_i(2w_i - w_i^2)\,\Delta x_i.$$

Summing and taking the limit as $\|P\|$ approaches zero gives us the moment M_{xz} for the solid, namely

$$\begin{aligned} M_{xz} &= \int_0^2 \tfrac{1}{2}(2x - x^2) \cdot 2\pi x(2x - x^2)\,dx \\ &= \pi \int_0^2 (4x^3 - 4x^4 + x^5)\,dx \\ &= \pi\left[x^4 - \frac{4}{5}x^5 + \frac{1}{6}x^6\right]_0^2 = \frac{16\pi}{15}. \end{aligned}$$

From Example 1 in Section 6.3, $m = 1V = 8\pi/3$ and therefore

$$\bar{y} = \frac{M_{xz}}{m} = \frac{16\pi/15}{8\pi/3} = \frac{2}{5}.$$

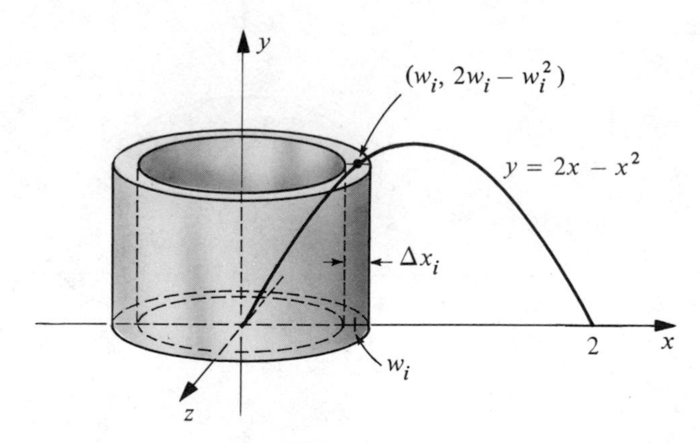

Figure 10.19

We shall conclude this chapter by stating a useful theorem about solids of revolution. To illustrate a special case of the theorem, let f and g be continuous functions such that $f(x) \geq g(x)$ for all x in an interval $[a, b]$, where $a \geq 0$. Let R denote the region bounded by the graphs of f, g, and the vertical lines $x = a$ and $x = b$. A typical region of this type is sketched in Figure 10.13. If R is revolved about the y-axis, then the volume of the resulting solid is given by

$$V = \int_a^b 2\pi x[f(x) - g(x)]\,dx.$$

Suppose the region R has centroid $(\bar{x}, \bar{y})$ and area A. If we use (10.15) and (10.18) with $\rho = 1$ (in which case $m = \rho A = A$) we obtain

$$\bar{x} = \frac{M_y}{A} = \frac{\int_a^b x[f(x) - g(x)]\,dx}{A} = \frac{V/2\pi}{A} = \frac{V}{2\pi A}$$

and hence $V = 2\pi\bar{x}A$. Since $\bar{x}$ is the distance from the y-axis to the centroid of R, the last formula states that the volume V of the solid of revolution may be found by multiplying the area A of R by the distance $2\pi\bar{x}$ which the centroid travels when R is revolved once about the y-axis. A similar statement is true if $g(x) \geq 0$ and R is revolved about the x-axis. In general, it is possible to prove the following theorem, which is named after the mathematician Pappus of Alexandria (*ca.* 300 A.D.).

(10.24) Theorem of Pappus

> Let R be a region in a plane which lies entirely on one side of a line l in the plane. If R is revolved once about l, then the volume of the resulting solid is the product of the area of R and the distance traveled by the centroid of R.

Example 4 A circle of radius a is revolved about a line in the plane of the circle which is a distance b from the center of the circle, where $b > a$. Find the volume V of the resulting solid. (This doughnut-shaped solid is called a **torus**.)

Solution The region bounded by the circle has area πa^2 and the distance traveled by the centroid is $2\pi b$. Hence by the Theorem of Pappus,

$$V = (2\pi b)(\pi a^2) = 2\pi^2 a^2 b.$$

EXERCISES 10.9

In each of Exercises 1–6, find the centroid of the solid generated by revolving the region bounded by the graphs of the given equations about the x-axis.

1 $y = 1/x$, $y = 0$, $x = 1$, $x = 2$

2 $y = (\sin x)^{1/2}$, $y = 0$, $x = 0$, $x = \pi$

3 $y = e^x$, $y = 0$, $x = 0$, $x = 1$

4 $y = 1/\sqrt{16 + x^2}$, $y = 0$, $x = 0$, $x = 4$

5 $y = x^2$, $x = y^3$

6 $x^2 = y$, $y = 2x$

In Exercises 7–12, find the centroid of the solid generated by revolving the region bounded by the graphs of the given equations about the y-axis.

7 $y^2 - x^2 = 4$, $y = 4$

8 $x = \sqrt{y - 1}$, $x = 0$, $y = 1$, $y = 5$

9 $y = e^{2x}$, $y = 0$, $x = 0$, $x = 2$

10 $y = e^{-x}$, $y = 0$, $x = 0$, $x = -1$

11 $y = 1/(x^2 + 25)$, $y = 0$, $x = 0$, $x = 5$

12 $y = 1/(9 - x^2)$, $y = 0$, $x = 0$, $x = 2$

13 The region bounded by the upper half of the ellipse $b^2x^2 + a^2y^2 = a^2b^2$ and the x-axis is revolved about the y-axis. Find the centroid of the resulting solid.

14 The region bounded by the hyperbola $b^2y^2 - a^2x^2 = a^2b^2$ and the line $y = c$, where $c > a$, is revolved about the y-axis. Find the centroid of the resulting solid.

15 The region between the graph of $y = \cos x$ and the x-axis, from $x = 0$ to $x = \pi/2$, is revolved about the y-axis. Find the centroid of the resulting solid.

16 If the region described in Exercise 15 is revolved about the x-axis, find the centroid of the resulting solid.

In Exercises 17 and 18 use the Theorem of Pappus to find the volume generated by revolving the quadrilateral with vertices A, B, C, D about (a) the y-axis; (b) the x-axis.

17 $A(1,0)$; $B(3,6)$; $C(11,6)$; $D(9,0)$

18 $A(2,2)$; $B(1,3)$; $C(4,6)$; $D(5,5)$

19 Use the Theorem of Pappus to find the centroid of the region in the first quadrant bounded by the graph of $y = \sqrt{a^2 - x^2}$ and the coordinate axes.

20 Use the Theorem of Pappus to find the centroid of the triangle with vertices $O(0,0)$, $A(0,a)$, $B(b,0)$.

10.10 REVIEW

Concepts

Discuss each of the following.

1 Integration by parts

2 Trigonometric substitutions

3 Partial fractions

4 Moments and centroids of plane areas

5 Moments and centroids of solids of revolution

Exercises

Evaluate the integrals in Exercises 1–100.

1 $\int x \sin^{-1} x\, dx$

2 $\int \csc^3 x\, dx$

3 $\int_0^1 \ln(1 + x)\, dx$

4 $\int_0^1 e^{\sqrt{x}}\, dx$

5 $\int \cos^3 2x \sin^2 2x\, dx$

6 $\int \cos^4 x\, dx$

7 $\int \tan x \sec^5 x\, dx$

8 $\int \tan x \sec^6 x\, dx$

9 $\int \frac{1}{(x^2 + 25)^{3/2}}\, dx$

10 $\int \frac{1}{x^2\sqrt{16 - x^2}}\, dx$

11 $\int \frac{\sqrt{4-x^2}}{x}\,dx$

12 $\int \frac{x}{(x^2+1)^2}\,dx$

13 $\int \frac{x^3+1}{x(x-1)^3}\,dx$

14 $\int \frac{1}{x+x^3}\,dx$

15 $\int \frac{x^3-20x^2-63x-198}{x^4-81}\,dx$

16 $\int \frac{x-1}{(x+2)^5}\,dx$

17 $\int \frac{x}{\sqrt{4+4x-x^2}}\,dx$

18 $\int \frac{x}{x^2+6x+13}\,dx$

19 $\int \frac{\sqrt[3]{x+8}}{x}\,dx$

20 $\int \frac{\sin x}{2\cos x+3}\,dx$

21 $\int e^{2x}\sin 3x\,dx$

22 $\int \cos(\ln x)\,dx$

23 $\int \sin^3 x\cos^3 x\,dx$

24 $\int \cot^2 3x\,dx$

25 $\int \frac{x}{\sqrt{4-x^2}}\,dx$

26 $\int \frac{1}{x\sqrt{9x^2+4}}\,dx$

27 $\int \frac{x^5-x^3+1}{x^3+2x^2}\,dx$

28 $\int \frac{x^3}{x^3-3x^2+9x-27}\,dx$

29 $\int \frac{1}{x^{3/2}+x^{1/2}}\,dx$

30 $\int \frac{2x+1}{(x+5)^{100}}\,dx$

31 $\int e^x\sec e^x\,dx$

32 $\int x\tan x^2\,dx$

33 $\int x^2\sin 5x\,dx$

34 $\int \sin 2x\cos x\,dx$

35 $\int \sin^3 x\cos^{1/2} x\,dx$

36 $\int \sin 3x\cot 3x\,dx$

37 $\int e^x\sqrt{1+e^x}\,dx$

38 $\int x(4x^2+25)^{-1/2}\,dx$

39 $\int \frac{x^2}{\sqrt{4x^2+25}}\,dx$

40 $\int \frac{3x+2}{x^2+8x+25}\,dx$

41 $\int \sec^2 x\tan^2 x\,dx$

42 $\int \sin^2 x\cos^5 x\,dx$

43 $\int x\cot x\csc x\,dx$

44 $\int (1+\csc 2x)^2\,dx$

45 $\int x^2(8-x^3)^{1/3}\,dx$

46 $\int x(\ln x)^2\,dx$

47 $\int \cos\sqrt{x}\,dx$

48 $\int x\sqrt{5-3x}\,dx$

49 $\int \frac{e^{3x}}{1+e^x}\,dx$

50 $\int \frac{e^{2x}}{4+e^{4x}}\,dx$

51 $\int \frac{x^2 - 4x + 3}{\sqrt{x}}\,dx$

52 $\int \frac{\cos^3 x}{\sqrt{1 + \sin x}}\,dx$

53 $\int \frac{x^3}{\sqrt{16 - x^2}}\,dx$

54 $\int \frac{x}{25 - 9x^2}\,dx$

55 $\int \frac{1 - 2x}{x^2 + 12x + 35}\,dx$

56 $\int \frac{7}{x^2 - 6x + 18}\,dx$

57 $\int \tan^{-1} 5x\,dx$

58 $\int \sin^4 3x\,dx$

59 $\int \frac{e^{\tan x}}{\cos^2 x}\,dx$

60 $\int \frac{x}{\csc 5x^2}\,dx$

61 $\int \frac{1}{\sqrt{7 + 5x^2}}\,dx$

62 $\int \frac{2x + 3}{x^2 + 4}\,dx$

63 $\int \cot^6 x\,dx$

64 $\int \cot^5 x \csc x\,dx$

65 $\int x^3\sqrt{x^2 - 25}\,dx$

66 $\int (\sin x)10^{\cos x}\,dx$

67 $\int (x^2 - \operatorname{sech}^2 4x)\,dx$

68 $\int x \cosh x\,dx$

69 $\int x^2 e^{-4x}\,dx$

70 $\int x^5\sqrt{x^3 + 1}\,dx$

71 $\int \frac{3}{\sqrt{11 - 10x - x^2}}\,dx$

72 $\int \frac{12x^3 + 7x}{x^4}\,dx$

73 $\int \tan 7x \cos 7x\,dx$

74 $\int e^{1 + \ln 5x}\,dx$

75 $\int \frac{4x^2 - 12x - 10}{(x - 2)(x^2 - 4x + 3)}\,dx$

76 $\int \frac{1}{x^4\sqrt{16 - x^2}}\,dx$

77 $\int (x^3 + 1)\cos x\,dx$

78 $\int (x - 3)^2(x + 1)\,dx$

79 $\int \frac{\sqrt{9 - 4x^2}}{x^2}\,dx$

80 $\int \frac{4x^3 - 15x^2 - 6x + 81}{x^4 - 18x^2 + 81}\,dx$

81 $\int (x - \cot 3x)^2\,dx$

82 $\int x(x^2 + 5)^{3/4}\,dx$

83 $\int \frac{1}{x(\sqrt{x} + \sqrt[4]{x})}\,dx$

84 $\int \frac{x}{\cos^2 4x}\,dx$

85 $\int \frac{\sin x}{\sqrt{1 + \cos x}}\,dx$

86 $\int \frac{4x^2 - 6x + 4}{(x^2 + 4)(x - 2)}\,dx$

87 $\int \frac{x^2}{(25 + x^2)^2}\,dx$

88 $\int \sin^4 x \cos^3 x\,dx$

89 $\int \tan^3 x \sec x\,dx$

90 $\int \frac{x}{\sqrt{4 + 9x^2}}\,dx$

91 $\int \frac{2x^3 + 4x^2 + 10x + 13}{x^4 + 9x^2 + 20}\,dx$

92 $\int \frac{\sin x}{(1 + \cos x)^3}\,dx$

93 $\int \frac{(x^2 - 2)^2}{x}\,dx$

94 $\int \cot^2 x \csc x\,dx$

95 $\int x^{3/2} \ln x\,dx$

96 $\int \frac{x}{\sqrt[3]{x} - 1}\,dx$

97 $\int \frac{x^2}{\sqrt[3]{2x + 3}}\,dx$

98 $\int \frac{1 - \sin x}{\cot x}\,dx$

99 $\int x^3 e^{x^2}\,dx$

100 $\int (x + 2)^2(x + 1)^{10}\,dx$

101 The region between the graph of $y = \sin x$ and the x-axis from $x = 0$ to $x = \pi$ is revolved about the y-axis. Find the volume of the resulting solid.

102 The region bounded by the graphs of $y = \tan x$, $y = 0$, $x = \pi/6$, and $x = \pi/4$ is revolved about the x-axis. Find the volume of the resulting solid.

103 Find the arc length of the graph of $y = \ln \sec x$ from $A(0, 0)$ to $B(\pi/3, \ln 2)$.

104 Find the area of the region bounded by the coordinate axes and the graphs of $y = (9 + 4x^2)^{-1/2}$ and $x = 2$.

In Exercises 105 and 106 sketch the region bounded by the graphs of the given equations and find the centroid.

105 $y = x^3, y = x^2$

106 $y = \cos x, y = 0, x = 0, x = \pi/2$

In Exercises 107 and 108 find the centroid of the solid generated by revolving the region bounded by the graphs of the given equations about the x-axis.

107 $y = \sqrt{x}$, $y = 0$, $x = 4$

108 $y = \sec x$, $y = 0$, $x = 0$, $x = \pi/4$

109 The region bounded by the graphs of $y = e^{-3x}$, $y = 0$, $x = 0$, and $x = 1$ is revolved about the y-axis. Find the centroid of the resulting solid.

110 The region bounded by the graphs of $y = \sin(x^2)$, $y = 0$, $x = 0$, and $x = \sqrt{\pi}$ is revolved about the y-axis. Find the centroid of the resulting solid.

Indeterminate Forms, Improper Integrals, and Taylor's Formula

In this chapter we introduce a technique called L'Hôpital's Rule, *which is very useful in the investigation of certain limits. We shall also enlarge the class of functions for which definite integrals can be found. Section 5 contains a method for approximating functions by means of polynomials. These topics have many mathematical and physical applications. Our most important use for them will occur in the next chapter, when* infinite series *are studied. The final section is concerned with Newton's method of approximating roots.*

11.1 THE INDETERMINATE FORMS 0/0 AND ∞/∞

In our early work with limits we encountered expressions of the form $\lim_{x \to c} f(x)/g(x)$, where both f and g have the limit 0 as x approaches c. In this event, $f(x)/g(x)$ is said to have the **indeterminate form 0/0** at $x = c$. The word *indeterminate* is used because a further analysis is necessary in order to conclude whether or not the limit exists. Perhaps the most important examples of the indeterminate form 0/0 occur in the use of the derivative formula

$$f'(c) = \lim_{x \to c} \frac{f(x) - f(c)}{x - c}.$$

If f and g become positively or negatively infinite as x approaches c, we say that $f(x)/g(x)$ has the **indeterminate form** ∞/∞ at $x = c$.

Indeterminate forms can sometimes be investigated by employing algebraic manipulations of some type. To illustrate, in Chapter 2 we considered

$$\lim_{x \to 2} \frac{2x^2 - 5x + 2}{5x^2 - 7x - 6}.$$

The indicated quotient has the indeterminate form 0/0; however, the limit may be found as follows:

$$\lim_{x \to 2} \frac{(x - 2)(2x - 1)}{(x - 2)(5x + 3)} = \lim_{x \to 2} \frac{2x - 1}{5x + 3} = \frac{3}{13}.$$

Other indeterminate forms require more complicated techniques. For example, in Chapter 9 a geometric argument was used to show that

$$\lim_{x \to 0} \frac{\sin x}{x} = 1.$$

In this section we shall establish **L'Hôpital's Rule** and illustrate how it can be used to investigate many indeterminate forms. The proof makes use of the following formula, which bears the name of the famous French mathematician A. Cauchy (1789–1857).

(11.1) Cauchy's Formula

If the functions f and g are continuous on a closed interval $[a, b]$, differentiable on the open interval (a, b), and if $g'(x) \neq 0$ for all x in (a, b), then there is a number w in (a, b) such that

$$\frac{f(b) - f(a)}{g(b) - g(a)} = \frac{f'(w)}{g'(w)}.$$

Proof. We first note that $g(b) - g(a) \neq 0$, for otherwise $g(a) = g(b)$, and by Rolle's Theorem (4.9) there is a number c in (a, b) such that $g'(c) = 0$, contrary to hypothesis.

It is convenient to introduce a new function h as follows:

$$h(x) = [f(b) - f(a)]g(x) - [g(b) - g(a)]f(x)$$

for all x in $[a, b]$. It follows that h is continuous on $[a, b]$, differentiable on (a, b), and that $h(a) = h(b)$. By Rolle's Theorem, there is a number w in (a, b) such that $h'(w) = 0$, that is,

$$[f(b) - f(a)]g'(w) - [g(b) - g(a)]f'(w) = 0.$$

The preceding equation may be written in the form stated in the conclusion of the theorem.

As a special case, if we let $g(x) = x$ in (11.1), then the conclusion has the form

$$\frac{f(b) - f(a)}{b - a} = \frac{f'(w)}{1}$$

which is equivalent to

$$f(b) - f(a) = f'(w)(b - a).$$

This shows that Cauchy's Formula is a generalization of the Mean Value Theorem (4.12).

The next result is the main theorem on indeterminate forms. It is named after the French mathematician G. L'Hôpital (1661–1704).

(11.2) L'Hôpital's Rule

Suppose the functions f and g are differentiable at every number except possibly at c in an interval (a, b). If $g'(x) \neq 0$ for $x \neq c$, and if $f(x)/g(x)$ has the indeterminate form 0/0 or ∞/∞ at $x = c$, then

$$\lim_{x \to c} \frac{f(x)}{g(x)} = \lim_{x \to c} \frac{f'(x)}{g'(x)},$$

provided $f'(x)/g'(x)$ has a limit or becomes infinite as x approaches c.

Proof. Suppose $f(x)/g(x)$ has the indeterminate form 0/0 at $x = c$ and $\lim_{x \to c} f'(x)/g'(x) = L$ for some number L. We wish to prove that $\lim_{x \to c} f(x)/g(x) = L$. It is convenient to introduce functions F and G where

$$F(x) = f(x) \quad \text{if } x \neq c \text{ and } F(c) = 0,$$
$$G(x) = g(x) \quad \text{if } x \neq c \text{ and } G(c) = 0.$$

Since

$$\lim_{x \to c} F(x) = \lim_{x \to c} f(x) = 0 = F(c),$$

the function F is continuous at c and hence is continuous *throughout* the interval (a, b). Similarly, G is continuous on (a, b). Moreover, at every $x \neq c$ we have $F'(x) = f'(x)$ and $G'(x) = g'(x)$. It follows from Cauchy's Formula, applied either to the interval $[c, x]$ or to $[x, c]$, that there is a number w between c and x such that

$$\frac{F(x) - F(c)}{G(x) - G(c)} = \frac{F'(w)}{G'(w)} = \frac{f'(w)}{g'(w)}.$$

Using the fact that $F(x) = f(x)$, $G(x) = g(x)$, and $F(c) = G(c) = 0$ gives us

$$\frac{f(x)}{g(x)} = \frac{f'(w)}{g'(w)}.$$

Since w is always between c and x it follows that

$$\lim_{x \to c} \frac{f(x)}{g(x)} = \lim_{x \to c} \frac{f'(w)}{g'(w)} = \lim_{w \to c} \frac{f'(w)}{g'(w)} = L$$

which is what we wished to prove. A similar argument may be given if $f'(x)/g'(x)$ becomes infinite as x approaches c. The proof for the indeterminate form ∞/∞ is more difficult and may be found in more advanced texts.

Beginning students sometimes use L'Hôpital's Rule incorrectly by applying the quotient rule to $f(x)/g(x)$. Note that (11.2) states that the derivatives of $f(x)$ and $g(x)$ are taken *separately*, after which the limit of the quotient $f'(x)/g'(x)$ is investigated.

Example 1 Find $\lim_{x\to 0} \dfrac{\cos x + 2x - 1}{3x}$.

Solution The quotient has the indeterminate form 0/0 at $x = 0$. By L'Hôpital's Rule (11.2),

$$\lim_{x\to 0} \frac{\cos x + 2x - 1}{3x} = \lim_{x\to 0} \frac{-\sin x + 2}{3} = \frac{2}{3}.$$

To be completely rigorous in Example 1 we should have determined whether or not $\lim_{x\to 0}(-\sin x + 2)/3$ existed *before* equating it to the given expression; however, to simplify solutions it is customary to proceed as indicated.

Sometimes it is necessary to employ L'Hôpital's Rule several times in the same problem, as illustrated in the next example.

Example 2 Find $\lim_{x\to 0} \dfrac{e^x + e^{-x} - 2}{1 - \cos 2x}$.

Solution The quotient has the indeterminate form 0/0. By L'Hôpital's Rule,

$$\lim_{x\to 0} \frac{e^x + e^{-x} - 2}{1 - \cos 2x} = \lim_{x\to 0} \frac{e^x - e^{-x}}{2\sin 2x}$$

provided the second limit exists. Since the latter quotient has the indeterminate form 0/0, we apply L'Hôpital's Rule a second time, obtaining

$$\lim_{x\to 0} \frac{e^x - e^{-x}}{2\sin 2x} = \lim_{x\to 0} \frac{e^x + e^{-x}}{4\cos 2x} = \frac{1}{2}.$$

It follows that the given limit exists and equals 1/2.

L'Hôpital's Rule is also valid for one-sided limits, as illustrated in the following example.

Example 3 Find $\lim_{x\to \frac{1}{2}\pi^-} \dfrac{4\tan x}{1 + \sec x}$.

Solution The indeterminate form is ∞/∞. By L'Hôpital's Rule,

$$\lim_{x\to \frac{1}{2}\pi^-} \frac{4\tan x}{1 + \sec x} = \lim_{x\to \frac{1}{2}\pi^-} \frac{4\sec^2 x}{\sec x \tan x} = \lim_{x\to \frac{1}{2}\pi^-} \frac{4\sec x}{\tan x}.$$

The last quotient again has the indeterminate form ∞/∞ at $x = \pi/2$; however, any additional applications of L'Hôpital's Rule always produce the form ∞/∞. In this case the limit may be found by using trigonometric identities to change the last quotient as follows:

$$\frac{4\sec x}{\tan x} = \frac{4/\cos x}{\sin x/\cos x} = \frac{4}{\sin x}.$$

Consequently,

$$\lim_{x\to\frac{1}{2}\pi^-} \frac{4\tan x}{1+\sec x} = \lim_{x\to\frac{1}{2}\pi^-} \frac{4}{\sin x} = 4.$$

It can be shown that the statement of L'Hôpital's Rule remains true if the symbol $x \to c$ is replaced by $x \to \infty$ or $x \to -\infty$. Let us give a partial proof of this fact. Suppose

$$\lim_{x\to\infty} f(x) = \lim_{x\to\infty} g(x) = 0.$$

If we let $u = 1/x$ and apply L'Hôpital's Rule,

$$\lim_{x\to\infty} \frac{f(x)}{g(x)} = \lim_{u\to 0^+} \frac{f(1/u)}{g(1/u)} = \lim_{u\to 0^+} \frac{D_u f(1/u)}{D_u g(1/u)}. \tag{11.3}$$

By the Chain Rule (3.36),

$$D_u f(1/u) = f'(1/u)(-1/u^2) \quad \text{and} \quad D_u g(1/u) = g'(1/u)(-1/u^2).$$

Substituting in (11.3) and simplifying we obtain

$$\lim_{x\to\infty} \frac{f(x)}{g(x)} = \lim_{u\to 0^+} \frac{f'(1/u)}{g'(1/u)} = \lim_{x\to\infty} \frac{f'(x)}{g'(x)}.$$

We shall also refer to this as L'Hôpital's Rule. The next two examples illustrate the application of the rule to the form ∞/∞.

Example 4 Find $\displaystyle\lim_{x\to\infty} \frac{\ln x}{\sqrt{x}}$.

Solution The indeterminate form is ∞/∞. By L'Hôpital's Rule,

$$\lim_{x\to\infty} \frac{\ln x}{\sqrt{x}} = \lim_{x\to\infty} \frac{1/x}{1/(2\sqrt{x})} = \lim_{x\to\infty} \frac{2\sqrt{x}}{x} = \lim_{x\to\infty} \frac{2}{\sqrt{x}} = 0.$$

Example 5 Find $\displaystyle\lim_{x\to\infty} \frac{e^{3x}}{x^2}$, if it exists.

Solution The indeterminate form is ∞/∞. In this case we must apply L'Hôpital's Rule twice, as follows.

$$\lim_{x\to\infty} \frac{e^{3x}}{x^2} = \lim_{x\to\infty} \frac{3e^{3x}}{2x} = \lim_{x\to\infty} \frac{9e^{3x}}{2} = \infty$$

Thus the given quotient increases without bound as x becomes infinite.

It is important to verify that a given quotient has the indeterminate form 0/0 or ∞/∞ before using L'Hôpital's Rule. Indeed, if the rule is applied to a nonindeterminate form, an incorrect conclusion may be obtained, as illustrated in the next example.

Example 6 Find $\lim_{x \to 0} \dfrac{e^x + e^{-x}}{x^2}$, if it exists.

Solution Suppose we overlook the fact that the quotient does *not* have either of the indeterminate forms 0/0 or ∞/∞ at $x = 0$. If we (incorrectly) apply L'Hôpital's Rule we obtain

$$\lim_{x \to 0} \frac{e^x + e^{-x}}{x^2} = \lim_{x \to 0} \frac{e^x - e^{-x}}{2x}.$$

Since the last quotient has the indeterminate form 0/0 we may apply L'Hôpital's Rule, obtaining

$$\lim_{x \to 0} \frac{e^x - e^{-x}}{2x} = \lim_{x \to 0} \frac{e^x + e^{-x}}{2} = \frac{1 + 1}{2} = 1.$$

This would lead us to the (wrong) conclusion that the given limit exists and equals 1.

A correct method for investigating the limit is to observe that

$$\lim_{x \to 0} \frac{e^x + e^{-x}}{x^2} = \lim_{x \to 0} (e^x + e^{-x})\left(\frac{1}{x^2}\right).$$

Since

$$\lim_{x \to 0} (e^x + e^{-x}) = 2 \quad \text{and} \quad \lim_{x \to 0} \frac{1}{x^2} = \infty$$

it follows from (ii) of Theorem 4.27 that

$$\lim_{x \to 0} \frac{e^x + e^{-x}}{x^2} = \infty.$$

EXERCISES 11.1

Find the limits in Exercises 1–50, if they exist.

1 $\lim_{x \to 0} \dfrac{\sin x}{2x}$

2 $\lim_{x \to 0} \dfrac{5x}{\tan x}$

3 $\lim_{x \to 5} \dfrac{\sqrt{x - 1} - 2}{x^2 - 25}$

4 $\lim_{x \to 4} \dfrac{x - 4}{\sqrt[3]{x + 4} - 2}$

5 $\lim_{x \to 2} \dfrac{2x^2 - 5x + 2}{5x^2 - 7x - 6}$

6 $\lim_{x \to -3} \dfrac{x^2 + 2x - 3}{2x^2 + 3x - 9}$

7 $\lim_{x\to 1} \frac{x^3 - 3x + 2}{x^2 - 2x - 1}$

8 $\lim_{x\to 2} \frac{x^2 - 5x + 6}{2x^2 - x - 7}$

9 $\lim_{x\to 0} \frac{\sin x - x}{\tan x - x}$

10 $\lim_{x\to 0} \frac{\sin x}{x - \tan x}$

11 $\lim_{x\to 0} \frac{x + 1 - e^x}{x^2}$

12 $\lim_{x\to 0^+} \frac{x + 1 - e^x}{x^3}$

13 $\lim_{x\to 0} \frac{x - \sin x}{x^3}$

14 $\lim_{x\to \pi/2} \frac{1 - \sin x}{\cos x}$

15 $\lim_{x\to \pi/2} \frac{1 + \sin x}{\cos^2 x}$

16 $\lim_{x\to 0^+} \frac{\cos x}{x}$

17 $\lim_{x\to \frac{1}{2}\pi^-} \frac{2 + \sec x}{3\tan x}$

18 $\lim_{x\to 0^+} \frac{\ln x}{\cot x}$

19 $\lim_{x\to \infty} \frac{x^2}{\ln x}$

20 $\lim_{x\to \infty} \frac{\ln x}{x^2}$

21 $\lim_{x\to 0^+} \frac{\ln \sin x}{\ln \sin 2x}$

22 $\lim_{x\to 0} \frac{2x}{\tan^{-1} x}$

23 $\lim_{x\to 0} \frac{e^x - e^{-x} - 2\sin x}{x \sin x}$

24 $\lim_{x\to 2} \frac{\ln(x - 1)}{x - 2}$

25 $\lim_{x\to 0} \frac{x\cos x + e^{-x}}{x^2}$

26 $\lim_{x\to 0} \frac{2e^x - 3x - e^{-x}}{x^2}$

27 $\lim_{x\to \infty} \frac{2x^2 + 3x + 1}{5x^2 + x + 4}$

28 $\lim_{x\to \infty} \frac{x^3 + x + 1}{3x^3 + 4}$

29 $\lim_{x\to \infty} \frac{x\ln x}{x + \ln x}$

30 $\lim_{x\to \infty} \frac{e^{3x}}{\ln x}$

31 $\lim_{x\to \infty} \frac{x^n}{e^x}, n > 0$

32 $\lim_{x\to \infty} \frac{e^x}{x^n}, n > 0$

33 $\lim_{x\to 2^+} \frac{\ln(x - 1)}{(x - 2)^2}$

34 $\lim_{x\to 0} \frac{\sin^2 x + 2\cos x - 2}{\cos^2 x - x\sin x - 1}$

35 $\lim_{x\to 0} \frac{\sin^{-1} 2x}{\sin^{-1} x}$

36 $\lim_{x\to \infty} \frac{\ln(\ln x)}{\ln x}$

37 $\lim_{x\to 0} \frac{\tan x - \sin x}{x^3 \tan x}$

38 $\lim_{x\to 1} \frac{2x^3 - 5x^2 + 6x - 3}{x^3 - 2x^2 + x - 1}$

39 $\lim_{x\to -\infty} \frac{3 - 3^x}{5 - 5^x}$

40 $\lim_{x\to 0} \frac{2 - e^x - e^{-x}}{1 - \cos^2 x}$

41 $\lim_{x\to 1} \frac{x^4 + x^3 - 3x^2 - x + 2}{x^4 - 5x^3 + 9x^2 - 7x + 2}$

42 $\lim_{x\to 1} \frac{x^4 - x^3 - 3x^2 + 5x - 2}{x^4 - 5x^3 + 9x^2 - 7x + 2}$

43 $\lim_{x\to 0} \frac{x\sin^{-1} x}{x - \sin x}$

44 $\lim_{x\to \infty} \frac{e^{-x}}{1 + e^{-x}}$

45 $\lim_{x\to \infty} \frac{x^{3/2} + 5x - 4}{x\ln x}$

46 $\lim_{x\to 0} \frac{x - \tan^{-1} x}{x\sin x}$

47 $\lim_{x\to\infty} \dfrac{\sqrt{x^2+1}}{\tan^{-1} x}$

48 $\lim_{x\to\infty} \dfrac{3^x+2x}{x^3+1}$

49 $\lim_{x\to\infty} \dfrac{2e^{3x}+\ln x}{e^{3x}+x^2}$

50 $\lim_{x\to\pi/2} \dfrac{\tan x}{\cot 2x}$

11.2 OTHER INDETERMINATE FORMS

If $\lim_{x\to c} f(x) = 0$ and $\lim_{x\to c} |g(x)| = \infty$, then $f(x)g(x)$ is said to have the **indeterminate form** $0 \cdot \infty$ at $x = c$. The same terminology is used for one-sided limits or if x becomes positively or negatively infinite. This form may be changed to one of the indeterminate forms $0/0$ or ∞/∞ by writing

$$f(x)g(x) = \frac{f(x)}{1/g(x)} \quad \text{or} \quad f(x)g(x) = \frac{g(x)}{1/f(x)}.$$

Example 1 Find $\lim_{x\to 0^+} x^2 \ln x$.

Solution The indeterminate form is $0 \cdot \infty$. We first write

$$x^2 \ln x = \frac{\ln x}{(1/x^2)}$$

and then apply L'Hôpital's Rule to the resulting indeterminate form ∞/∞. Thus

$$\lim_{x\to 0^+} x^2 \ln x = \lim_{x\to 0^+} \frac{\ln x}{(1/x^2)} = \lim_{x\to 0^+} \frac{(1/x)}{(-2/x^3)}.$$

The last quotient has the indeterminate form ∞/∞; however, further applications of L'Hôpital's Rule would again lead to ∞/∞. In this case we simplify the quotient algebraically and find the limit as follows:

$$\lim_{x\to 0^+} \frac{(1/x)}{(-2/x^3)} = \lim_{x\to 0^+} \frac{x^3}{-2x} = \lim_{x\to 0^+} \frac{x^2}{-2} = 0.$$

Example 2 Find $\lim_{x\to\pi/2} (2x - \pi)\sec x$.

Solution The indeterminate form is $0 \cdot \infty$. Hence we begin by writing

$$(2x - \pi)\sec x = \frac{2x - \pi}{1/\sec x} = \frac{2x - \pi}{\cos x}.$$

Since the last expression has the indeterminate form $0/0$ at $x = \pi/2$, L'Hôpital's

Rule may be applied as follows:

$$\lim_{x \to \pi/2} \frac{2x - \pi}{\cos x} = \lim_{x \to \pi/2} \frac{2}{-\sin x} = -2.$$

Indeterminate forms denoted by 0^0, ∞^0, and 1^∞ arise from expressions such as $f(x)^{g(x)}$. One method for dealing with these forms is to write

$$y = f(x)^{g(x)}$$

and take the natural logarithm of both sides, obtaining

$$\ln y = \ln f(x)^{g(x)} = g(x) \ln f(x).$$

Note that if the indeterminate form for y is 0^0 or ∞^0, then the indeterminate form for $\ln y$ is $0 \cdot \infty$, which may be handled using previous methods. Similarly, if the form for y is 1^∞, then the indeterminate form for $\ln y$ is $\infty \cdot 0$. It follows that

$$\text{if} \quad \lim_{x \to c} \ln y = L, \quad \text{then} \quad \lim_{x \to c} y = \lim_{x \to c} e^{\ln y} = e^L,$$

that is,

$$\lim_{x \to c} f(x)^{g(x)} = e^L.$$

This procedure may be outlined as follows.

Outline for investigating $\lim_{x \to c} f(x)^{g(x)}$ if the indeterminate form is $0^0, 1^\infty$, or ∞^0

1. Let $y = f(x)^{g(x)}$.
2. Take logarithms: $\ln y = \ln f(x)^{g(x)} = g(x) \ln f(x)$.
3. Find $\lim_{x \to c} \ln y$, if it exists.
4. If $\lim_{x \to c} \ln y = L$, then $\lim_{x \to c} y = e^L$.

A common error is to stop after showing $\lim_{x \to c} \ln y = L$ and conclude that the given expression has the limit L. Remember that *we wish to find the limit of* y, and if $\ln y$ has the limit L, then y has the limit e^L. The outline may also be used if $x \to \infty$, or $x \to -\infty$, or for one-sided limits.

Example 3 Find $\lim_{x \to 0} (1 + 3x)^{1/2x}$.

Solution The indeterminate form is 1^∞. Following the preceding outline, we begin by writing

1. $$y = (1 + 3x)^{1/2x}.$$

2. $$\ln y = \frac{1}{2x} \ln (1 + 3x) = \frac{\ln (1 + 3x)}{2x}.$$

The last expression has the indeterminate form 0/0 at $x = 0$. By L'Hôpital's Rule,

3. $$\lim_{x \to 0} \ln y = \lim_{x \to 0} \frac{\ln(1 + 3x)}{2x} = \lim_{x \to 0} \frac{3/(1 + 3x)}{2} = \frac{3}{2}.$$

Consequently,

4. $$\lim_{x \to 0} (1 + 3x)^{1/2x} = \lim_{x \to 0} y = e^{3/2}.$$

If $\lim_{x \to c} f(x) = \lim_{x \to c} g(x) = \infty$, then $f(x) - g(x)$ has the indeterminate form $\infty - \infty$ at $x = c$. In this case the expression should be changed so that one of the forms we have discussed is obtained.

Example 4 Find $\lim_{x \to 0} \left(\frac{1}{e^x - 1} - \frac{1}{x} \right)$.

Solution The form is $\infty - \infty$; however, if the difference is written as a single fraction, then

$$\lim_{x \to 0} \left(\frac{1}{e^x - 1} - \frac{1}{x} \right) = \lim_{x \to 0} \frac{x - e^x + 1}{xe^x - x}.$$

This gives us the indeterminate form 0/0. It is necessary to apply L'Hôpital's Rule twice, since the first application leads to the indeterminate form 0/0. Thus

$$\lim_{x \to 0} \frac{x - e^x + 1}{xe^x - x} = \lim_{x \to 0} \frac{1 - e^x}{xe^x + e^x - 1}$$
$$= \lim_{x \to 0} \frac{-e^x}{xe^x + 2e^x} = -\frac{1}{2}.$$

EXERCISES 11.2

Find the limits in Exercises 1–40, if they exist.

1 $\lim_{x \to 0^+} x \ln x$

2 $\lim_{x \to \frac{1}{2}\pi^-} \tan x \ln \sin x$

3 $\lim_{x \to \infty} (x^2 - 1)e^{-x^2}$

4 $\lim_{x \to \infty} x(e^{1/x} - 1)$

5 $\lim_{x \to 0} e^{-x} \sin x$

6 $\lim_{x \to -\infty} x \tan^{-1} x$

7 $\lim_{x \to 0^+} \sin x \ln \sin x$

8 $\lim_{x \to \infty} x\left(\frac{\pi}{2} - \tan^{-1} x\right)$

9 $\lim_{x \to \infty} x \sin \frac{1}{x}$

10 $\lim_{x \to \infty} e^{-x} \ln x$

11 $\lim_{x \to 0} x \sec^2 x$

12 $\lim_{x \to 0} (\cos x)^{x+1}$

13 $\lim_{x \to \infty} \left(1 + \frac{1}{x}\right)^{5x}$

14 $\lim_{x \to 0^+} (e^x + 3x)^{1/x}$

15 $\lim_{x\to 0^+} (e^x - 1)^x$

16 $\lim_{x\to 0^+} x^x$

17 $\lim_{x\to\infty} x^{1/x}$

18 $\lim_{x\to \frac{1}{2}\pi^-} (\tan x)^{\cos x}$

19 $\lim_{x\to\pi/2^-} (\tan x)^x$

20 $\lim_{x\to 2^+} (x - 2)^x$

21 $\lim_{x\to 0^+} (2x + 1)^{\cot x}$

22 $\lim_{x\to 0^+} (1 + 3x)^{\csc x}$

23 $\lim_{x\to\infty} \left(\dfrac{x^2}{x-1} - \dfrac{x^2}{x+1}\right)$

24 $\lim_{x\to 1} \left(\dfrac{1}{x-1} - \dfrac{1}{\ln x}\right)$

25 $\lim_{x\to 0} \left(\dfrac{1}{x} - \dfrac{1}{\sin x}\right)$

26 $\lim_{x\to \frac{1}{2}\pi^-} (\sec x - \tan x)$

27 $\lim_{x\to 1^-} (1 - x)^{\ln x}$

28 $\lim_{x\to\infty} (1 + e^x)^{e^{-x}}$

29 $\lim_{x\to 0} \left(\dfrac{1}{\sqrt{x^2+1}} - \dfrac{1}{x}\right)$

30 $\lim_{x\to 0} (\cot^2 x - \csc^2 x)$

31 $\lim_{x\to 0} \cot 2x \tan^{-1} x$

32 $\lim_{x\to\infty} x^3 2^{-x}$

33 $\lim_{x\to 0} (\cot^2 x - e^{-x})$

34 $\lim_{x\to\infty} (\sqrt{x^2+4} - \tan^{-1} x)$

35 $\lim_{x\to\pi/2^-} (1 + \cos x)^{\tan x}$

36 $\lim_{x\to 0} (1 + ax)^{b/x}$

37 $\lim_{x\to -3} \left(\dfrac{x}{x^2+2x-3} - \dfrac{4}{x+3}\right)$

38 $\lim_{x\to\infty} (\sqrt{x^4 + 5x^2 + 3} - x^2)$

39 $\lim_{x\to 0} (x + \cos 2x)^{\csc 3x}$

40 $\lim_{x\to\pi/2} \sec x \cos 3x$

11.3 INTEGRALS WITH INFINITE LIMITS OF INTEGRATION

Suppose a function f is continuous and nonnegative on an infinite interval $[a, \infty)$ and $\lim_{x\to\infty} f(x) = 0$. If $t > a$, then the area $A(t)$ under the graph of f from a to t (see Figure 11.1) is given by

(11.4)
$$A(t) = \int_a^t f(x)\,dx.$$

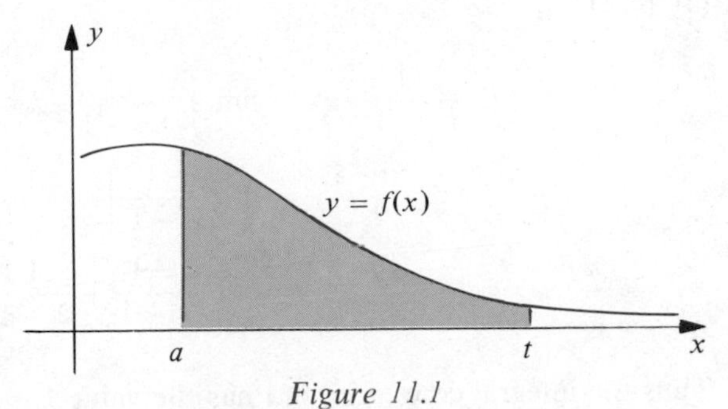

Figure 11.1

If, in (11.4), $\lim_{t\to\infty} A(t)$ exists, then the limit may be interpreted as the area of the *unbounded* region which lies under the graph of f, over the x-axis, and to the right of $x = a$. The symbol $\int_a^\infty f(x)\,dx$ is used to denote this number. More generally, if f is *any* function which is continuous on $[a, \infty)$, then by definition,

(11.5)
$$\int_a^\infty f(x)\,dx = \lim_{t\to\infty} \int_a^t f(x)\,dx,$$

provided the limit exists.

In similar fashion, if f is continuous on $(-\infty, a]$, then we define

(11.6)
$$\int_{-\infty}^a f(x)\,dx = \lim_{t\to-\infty} \int_t^a f(x)\,dx,$$

provided the limit exists. If $f(x) \geq 0$ for all x, then (11.6) may be regarded as the area under the graph of f, over the x-axis, and to the *left* of $x = a$.

The expressions in (11.5) and (11.6) are called **improper integrals**. They differ from definite integrals since one of the limits of integration is not a real number. These integrals are said to **converge** if, as $|t|$ increases without bound, the limits on the right side of the equations exist. Otherwise the integrals **diverge**. Improper integrals also occur with *two* infinite limits of integration. Specifically, if f is continuous for all x and a is any real number, then by definition,

(11.7)
$$\int_{-\infty}^\infty f(x)\,dx = \int_{-\infty}^a f(x)\,dx + \int_a^\infty f(x)\,dx,$$

provided *both* of the latter integrals converge. If one of the integrals diverges, then $\int_{-\infty}^\infty f(x)\,dx$ is said to diverge. It can be shown that (11.7) is independent of the real number a (see Exercise 32). We may also show that $\int_{-\infty}^\infty f(x)\,dx$ is not necessarily the same as $\lim_{t\to\infty} \int_{-t}^t f(x)\,dx$ (see Exercise 31).

Example 1 Determine whether the following improper integrals converge or diverge.

(a) $\displaystyle\int_2^\infty \frac{1}{(x-1)^2}\,dx$ (b) $\displaystyle\int_2^\infty \frac{1}{x-1}\,dx$

Solution (a) By (11.5),

$$\begin{aligned}\int_2^\infty \frac{1}{(x-1)^2}\,dx &= \lim_{t\to\infty} \int_2^t \frac{1}{(x-1)^2}\,dx \\ &= \lim_{t\to\infty} \frac{-1}{x-1}\Big]_2^t \\ &= \lim_{t\to\infty} \left(\frac{-1}{t-1} + \frac{1}{2-1}\right) = 0 + 1 = 1.\end{aligned}$$

Thus the integral converges and has the value 1.

(b) By (11.5),

$$\begin{aligned}\int_2^{\infty} \frac{1}{x-1}\,dx &= \lim_{t\to\infty} \int_2^t \frac{1}{x-1}\,dx \\ &= \lim_{t\to\infty} \ln(x-1)\Big]_2^t \\ &= \lim_{t\to\infty} [\ln(t-1) - \ln(2-1)] \\ &= \lim_{t\to\infty} \ln(t-1) = \infty.\end{aligned}$$

Consequently this improper integral diverges.

The graphs of the two functions defined by the integrands in parts (a) and (b) of Example 1 are sketched in Figure 11.2. Note that although the graphs have the same general shape for $x \geq 2$, we may assign an area to the region under the graph shown in (i) of the figure, whereas this is not true for the graph in (ii).

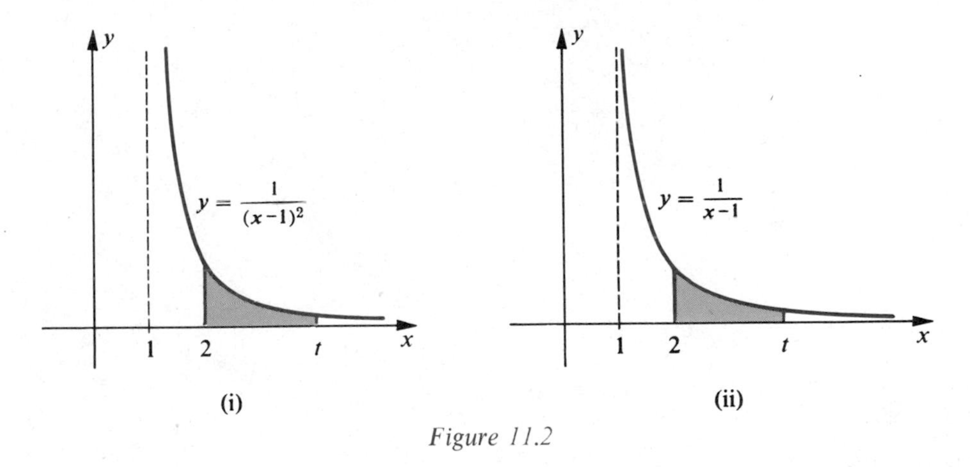

Figure 11.2

There is an interesting sidelight to the graph in (ii) of Figure 11.2. If the region under the graph of $y = 1/(x-1)$ from 2 to t is revolved about the x-axis, then by Definition (6.5) the volume of the resulting solid is

$$\pi \int_2^t \frac{1}{(x-1)^2}\,dx.$$

The improper integral

$$\pi \int_2^{\infty} \frac{1}{(x-1)^2}\,dx$$

may be regarded as the volume of the *unbounded* solid obtained by revolving, about the x-axis, the region under the graph of $y = 1/(x-1)$ for $x \geq 2$. By (a) of Example 1, the value of this improper integral is $\pi \cdot 1$ or π. This gives us the rather curious fact that although the area of the region is infinite, the volume of the solid of revolution it generates is finite.

Example 2 Assign an area to the region which lies under the graph of $y = e^x$, over the x-axis, and to the left of $x = 1$.

Solution The region bounded by the graphs of $y = e^x$, $y = 0$, $x = 1$, and $x = t$, where $t < 1$, is sketched in Figure 11.3.

By our previous remarks, the desired area is given by

$$\begin{aligned}\int_{-\infty}^{1} e^x\,dx &= \lim_{t\to-\infty}\int_t^1 e^x\,dx\\ &= \lim_{t\to-\infty} e^x\Big]_t^1\\ &= \lim_{t\to-\infty}(e - e^t) = e - 0 = e.\end{aligned}$$

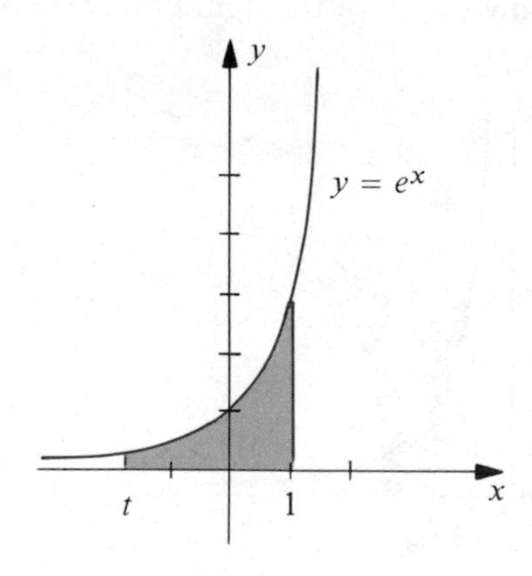

Figure 11.3

Example 3 Evaluate $\int_{-\infty}^{\infty}\frac{1}{1+x^2}\,dx$. Sketch the graph of $f(x) = 1/(1 + x^2)$ and interpret the integral as an area.

Solution Using (11.7) with $a = 0$,

$$\int_{-\infty}^{\infty}\frac{1}{1+x^2}\,dx = \int_{-\infty}^{0}\frac{1}{1+x^2}\,dx + \int_{0}^{\infty}\frac{1}{1+x^2}\,dx.$$

Next, applying (11.5),

$$\begin{aligned}\int_0^{\infty}\frac{1}{1+x^2}\,dx &= \lim_{t\to\infty}\int_0^t\frac{1}{1+x^2}\,dx\\ &= \lim_{t\to\infty}\arctan x\Big]_0^t\\ &= \lim_{t\to\infty}(\arctan t - \arctan 0) = \pi/2 - 0 = \pi/2.\end{aligned}$$

Similarly, we may show by means of (11.6) that

$$\int_{-\infty}^{0} \frac{1}{1 + x^2}\, dx = \pi/2.$$

Consequently, the given improper integral converges and has the value $\pi/2 + \pi/2 = \pi$.

The graph of $y = 1/(1 + x^2)$ is sketched in Figure 11.4. As in our previous discussion, the unbounded region which lies under the graph and above the x-axis may be assigned an area of π square units.

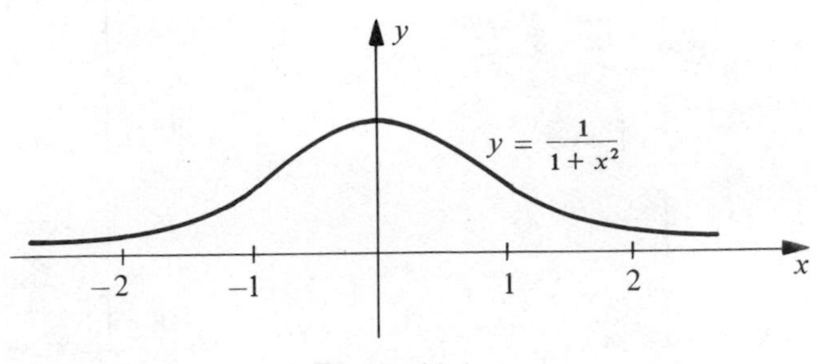

Figure 11.4

Improper integrals with infinite limits of integration have many applications in the physical world. To illustrate, suppose a and b are the coordinates of two points A and B on a coordinate line l, as shown in Figure 11.5. If $F(x)$ is the force acting at the point P with coordinate x, then by Definition (6.13), the work done as P moves from A to B is given by

$$W = \int_{a}^{b} F(x)\, dx.$$

In similar fashion, the improper integral $\int_a^\infty F(x)\, dx$ is used to define the work done as P moves indefinitely to the right (in applications, the terminology "to infinity" is sometimes used). For example, if $F(x)$ is the force of attraction between a particle fixed at point A and a (movable) particle at P, and if $c > a$, then $\int_c^\infty F(x)\, dx$ represents the work required to move P from the point with coordinate c to infinity (see Exercises 33 and 34).

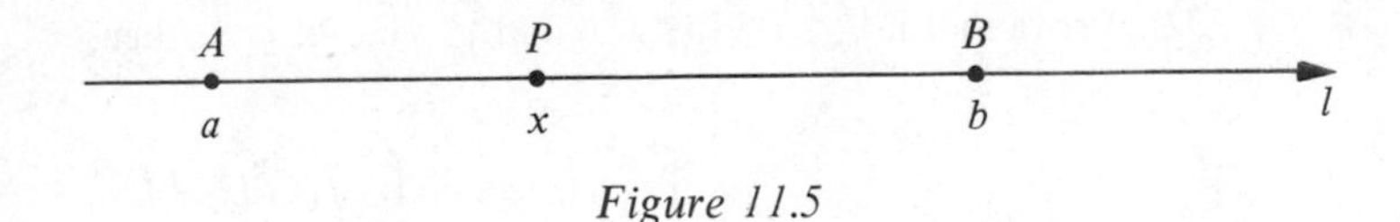

Figure 11.5

EXERCISES 11.3

Determine whether the integrals in Exercises 1–24 converge or diverge and evaluate those which converge.

1 $\displaystyle\int_{1}^{\infty} \frac{1}{x^{4/3}}\, dx$

2 $\displaystyle\int_{-\infty}^{0} \frac{1}{(x - 1)^3}\, dx$

3 $\int_1^{\infty} \frac{1}{x^{3/4}}\,dx$

4 $\int_0^{\infty} \frac{x}{1+x^2}\,dx$

5 $\int_{-\infty}^{2} \frac{1}{5-2x}\,dx$

6 $\int_{-\infty}^{\infty} \frac{x}{x^4+9}\,dx$

7 $\int_0^{\infty} e^{-2x}\,dx$

8 $\int_{-\infty}^{0} e^{x}\,dx$

9 $\int_{-\infty}^{-1} \frac{1}{x^3}\,dx$

10 $\int_0^{\infty} \frac{1}{\sqrt[3]{x+1}}\,dx$

11 $\int_{-\infty}^{0} \frac{1}{(x-8)^{2/3}}\,dx$

12 $\int_1^{\infty} \frac{x}{(1+x^2)^2}\,dx$

13 $\int_1^{\infty} \frac{x}{1+x^2}\,dx$

14 $\int_{-\infty}^{2} \frac{1}{x^2+4}\,dx$

15 $\int_{-\infty}^{\infty} xe^{-x^2}\,dx$

16 $\int_{-\infty}^{\infty} \cos^2 x\,dx$

17 $\int_1^{\infty} \frac{\ln x}{x}\,dx$

18 $\int_3^{\infty} \frac{1}{x^2-1}\,dx$

19 $\int_0^{\infty} \cos x\,dx$

20 $\int_{-\infty}^{\pi/2} \sin 2x\,dx$

21 $\int_{-\infty}^{\infty} \operatorname{sech} x\,dx$

22 $\int_0^{\infty} xe^{-x}\,dx$

23 $\int_{-\infty}^{0} \frac{1}{x^2-3x+2}\,dx$

24 $\int_4^{\infty} \frac{x+18}{x^2+x-12}\,dx$

In Exercises 25–28 assign, if possible, a value to (a) the area of the region R, and (b) the volume of the solid obtained by revolving R about the x-axis.

25 $R = \{(x, y): x \geq 1,\ 0 \leq y \leq 1/x\}$

26 $R = \{(x, y): x \geq 1,\ 0 \leq y \leq 1/\sqrt{x}\}$

27 $R = \{(x, y): x \geq 4,\ 0 \leq y \leq x^{-3/2}\}$

28 $R = \{(x, y): x \geq 8,\ 0 \leq y \leq x^{-2/3}\}$

29 Find all values of n for which the integral $\int_1^{\infty} x^n\,dx$ (a) converges; (b) diverges.

30 Find all integer values of n for which the integral $\int_{-\infty}^{-1} x^n\,dx$ (a) converges; (b) diverges.

31 Find a function f such that $\lim_{t\to\infty} \int_{-t}^{t} f(x)\,dx$ exists and $\int_{-\infty}^{\infty} f(x)\,dx$ diverges.

32 Prove that if $\int_{-\infty}^{a} f(x)\,dx = L$ and $\int_a^{\infty} f(x)\,dx = K$, then

$$\int_{-\infty}^{b} f(x)\,dx + \int_b^{\infty} f(x)\,dx = L + K$$

for every real number b.

33 In Figure 11.5 take $a = 0$ and let A represent the center of the earth. If P represents a point above the surface, then the force exerted by gravity at P is given by $F(x) = k/x^2$, where k is a constant. If the radius of the earth is 4000 miles, find the work required to project an object weighing 100 pounds along l, from the surface to infinity. (*Hint:* $F(4000) = 100$.)

34 The force (in dynes) with which two electrons repel one another is inversely proportional to the square of the distance (in centimeters) between them. If, in Figure 11.5, one electron is fixed at A, find the work done if another electron is repelled along l from a point B, which is one cm from A, to infinity.

In the theory of differential equations, if f is a function, then the **Laplace Transform** L of $f(x)$ is defined by

$$L[f(x)] = \int_0^{\infty} e^{-sx} f(x)\, dx$$

for every real number s for which the improper integral converges. In each of Exercises 35-40 find $L[f(x)]$ if $f(x)$ is the indicated expression.

35 1 **36** x

37 $\cos x$ **38** $\sin x$

39 e^{ax} **40** $\sin ax$

11.4 INTEGRALS WITH INFINITE INTEGRANDS

If a function f is continuous on a closed interval $[a, b]$, then by Theorem (5.19) the definite integral $\int_a^b f(x)\, dx$ exists. If f has an infinite discontinuity at some number in the interval it may still be possible to assign a value to the integral. Suppose, for example, that f is continuous and nonnegative on the half-open interval $[a, b)$ and $\lim_{x \to b^-} f(x) = \infty$. If $a < t < b$, then the area $A(t)$ under the graph of f from a to t (see Figure 11.6) is given by

$$A(t) = \int_a^t f(x)\, dx.$$

If $\lim_{t \to b^-} A(t)$ exists, then the limit may be interpreted as the area of the unbounded region which lies under the graph of f, over the x-axis, and between $x = a$ and $x = b$. It is natural to denote this number by $\int_a^b f(x)\, dx$.

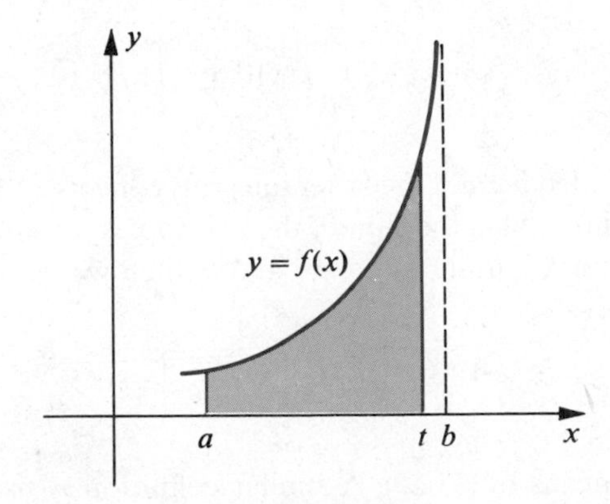

Figure 11.6

In general, if f is any function which is continuous on the half-open interval $[a,b)$ and becomes infinite at b, then by definition,

$$\int_a^b f(x)\,dx = \lim_{t\to b^-} \int_a^t f(x)\,dx \tag{11.8}$$

provided the limit exists.

Similarly, if f is continuous on $(a,b]$ and becomes infinite at a, then by definition,

$$\int_a^b f(x)\,dx = \lim_{t\to a^+} \int_t^b f(x)\,dx \tag{11.9}$$

provided the limit exists. A situation of this type is illustrated in Figure 11.7.

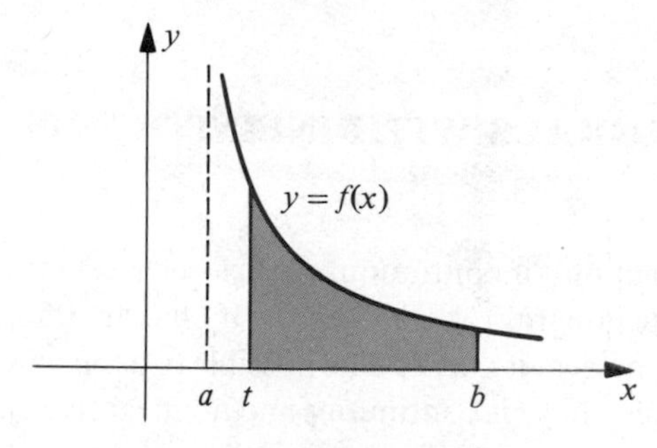

Figure 11.7

As in the preceding section, (11.8) and (11.9) are referred to as **improper integrals** and they are said to **converge** if the indicated limits exist. Otherwise they **diverge**.

If f has an infinite discontinuity at a number c in (a,b) but is continuous elsewhere in $[a,b]$ (as, for example, in Figure 11.8) then we define

$$\int_a^b f(x)\,dx = \int_a^c f(x)\,dx + \int_c^b f(x)\,dx \tag{11.10}$$

provided *both* of the latter integrals converge. If both converge, then the value of the integral is the sum of the two values. Finally, if f is continuous on (a,b) but becomes infinite at *both* a and b, then we again define

$$\int_a^b f(x)\,dx$$

by means of (11.10). A similar definition is used if f has any finite number of discontinuities in $[a,b]$.

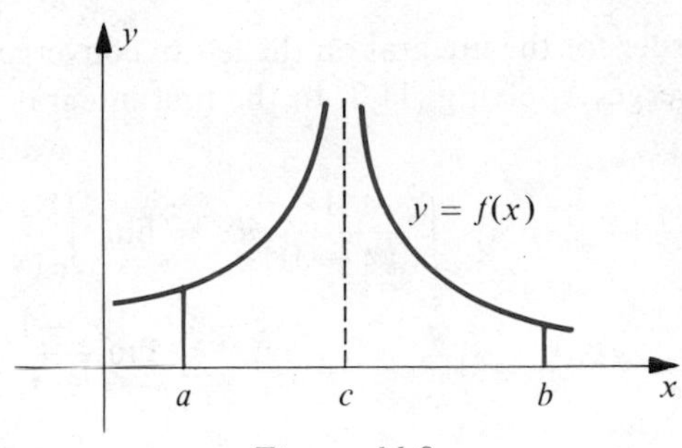

Figure 11.8

Example 1 Evaluate $\int_0^3 \frac{1}{\sqrt{3-x}}\,dx$.

Solution Since the integrand has an infinite discontinuity at $x = 3$ we apply (11.8) as follows:

$$\begin{aligned}\int_0^3 \frac{1}{\sqrt{3-x}}\,dx &= \lim_{t\to 3^-} \int_0^t \frac{1}{\sqrt{3-x}}\,dx \\ &= \lim_{t\to 3^-} \Big[-2\sqrt{3-x}\Big]_0^t \\ &= \lim_{t\to 3^-} \left[-2\sqrt{3-t} + 2\sqrt{3}\right] \\ &= 0 + 2\sqrt{3} = 2\sqrt{3}.\end{aligned}$$

Example 2 Determine whether $\int_0^1 \frac{1}{x}\,dx$ converges or diverges.

Solution The integrand is undefined at $x = 0$. Applying (11.9),

$$\begin{aligned}\int_0^1 \frac{1}{x}\,dx &= \lim_{t\to 0^+} \int_t^1 \frac{1}{x}\,dx \\ &= \lim_{t\to 0^+} \Big[\ln x\Big]_t^1 \\ &= \lim_{t\to 0^+} [0 - \ln t] = \infty.\end{aligned}$$

Since the limit does not exist, the improper integral diverges.

Example 3 Determine whether $\int_0^4 \frac{1}{(x-3)^2}\,dx$ converges or diverges.

Solution The integrand is undefined at $x = 3$. Since this number is in the interior of the interval $[0, 4]$ we use (11.10) with $c = 3$, obtaining

$$\int_0^4 \frac{1}{(x-3)^2}\,dx = \int_0^3 \frac{1}{(x-3)^2}\,dx + \int_3^4 \frac{1}{(x-3)^2}\,dx.$$

In order for the integral on the left to converge, *both* integrals on the right must converge. Applying (11.8) to the first integral,

$$\begin{aligned}\int_0^3 \frac{1}{(x-3)^2}\,dx &= \lim_{t\to 3^-}\int_0^t \frac{1}{(x-3)^2}\,dx\\ &= \lim_{t\to 3^-} \frac{-1}{x-3}\Big]_0^t\\ &= \lim_{t\to 3^-}\left(\frac{-1}{t-3}-\frac{1}{3}\right) = \infty.\end{aligned}$$

It follows that the given improper integral diverges.

It is important to note that the Fundamental Theorem of Calculus (5.31) cannot be applied to the integral in Example 3 since the function given by the integrand is not continuous on $[0,4]$. Indeed, if we had applied the Fundamental Theorem we would have obtained

$$\begin{aligned}\int_0^4 \frac{1}{(x-3)^2}\,dx &= \frac{-1}{(x-3)}\Big]_0^4\\ &= -1-\frac{1}{3} = -\frac{4}{3}.\end{aligned}$$

This result is obviously incorrect since the integrand is never negative.

Example 4 Evaluate $\displaystyle\int_{-2}^7 \frac{1}{(x+1)^{2/3}}\,dx$.

Solution The integrand is undefined at $x=-1$, a number between -2 and 7. Consequently we apply (11.10) with $c=-1$, as follows:

$$\int_{-2}^7 \frac{1}{(x+1)^{2/3}}\,dx = \int_{-2}^{-1} \frac{1}{(x+1)^{2/3}}\,dx + \int_{-1}^7 \frac{1}{(x+1)^{2/3}}\,dx.$$

We next investigate each integral on the right. Using (11.8) with $b=-1$ gives us

$$\begin{aligned}\int_{-2}^{-1} \frac{1}{(x+1)^{2/3}}\,dx &= \lim_{t\to -1^-}\int_{-2}^t \frac{1}{(x+1)^{2/3}}\,dx\\ &= \lim_{t\to -1^-}\left[3(x+1)^{1/3}\right]_{-2}^t\\ &= \lim_{t\to -1^-}\left[3(t+1)^{1/3}-3(-1)^{1/3}\right]\\ &= 0+3=3.\end{aligned}$$

In similar fashion, using (11.9) with $a = -1$,

$$\int_{-1}^{7} \frac{1}{(x+1)^{2/3}}\,dx = \lim_{t \to -1^+} \int_{t}^{7} \frac{1}{(x+1)^{2/3}}\,dx$$

$$= \lim_{t \to -1^+} \Big[3(x+1)^{1/3} \Big]_t^7$$

$$= \lim_{t \to -1^+} \left[3(8)^{1/3} - 3(t+1)^{1/3} \right]$$

$$= 6 - 0 = 6.$$

Since both integrals converge, the given integral converges and has the value $3 + 6 = 9$.

Improper integrals of the types considered in this section arise in certain physical applications. Figure 11.9 is a schematic drawing of a spring with an attached weight which is oscillating between points with coordinates $-d$ and d on a coordinate line y (the y-axis has been positioned at the right for clarity).

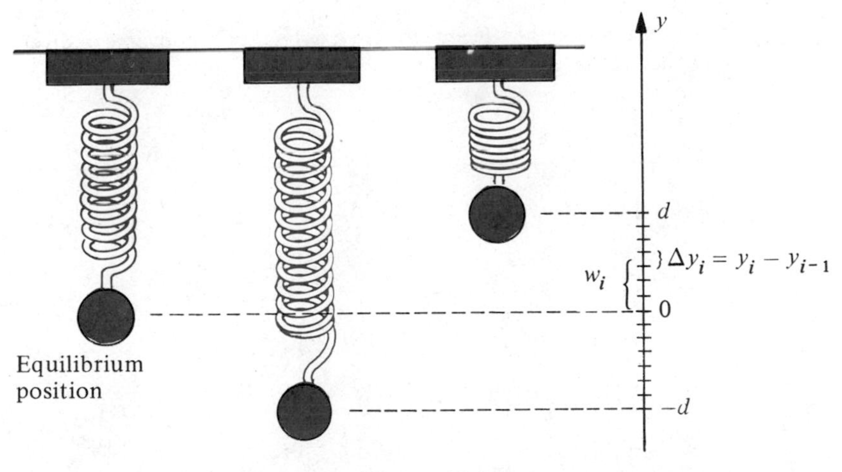

Figure 11.9

The **period** T is the time required for one complete oscillation, that is, *twice* the time required for the weight to cover the interval $[-d, d]$. Let $v(y)$ denote the velocity of the weight when it is at the point with coordinate y in $[-d, d]$. We partition $[-d, d]$ in the usual way and let $\Delta y_i = y_i - y_{i-1}$ denote the distance the weight travels during the time interval Δt_i. If w_i is any number in $[y_{i-1}, y_i]$, then $v(w_i)$ is the velocity of the weight when it is at the point with coordinate w_i. If the norm of the partition is small and we assume v is a continuous function, then the distance Δy_i may be approximated by the product $v(w_i)\Delta t_i$, that is,

$$\Delta y_i \approx v(w_i)\,\Delta t_i.$$

Hence the time required for the weight to cover the distance Δy_i may be approximated by

$$\Delta t_i \approx \frac{1}{v(w_i)} \Delta y_i$$

and, therefore,

$$T = 2\sum_i \Delta t_i \approx 2\sum_i \frac{1}{v(w_i)} \Delta y_i.$$

Taking the limit of the sum on the right, and using the definition of the definite integral,

$$T = 2\int_{-d}^{d} \frac{1}{v(y)} dy.$$

Since $v(d) = 0$ and $v(-d) = 0$, the integral is improper (but necessarily convergent).

EXERCISES 11.4

Determine whether the integrals in Exercises 1–26 converge or diverge and evaluate those which converge.

1 $\displaystyle\int_0^8 \frac{1}{\sqrt[3]{x}} dx$

2 $\displaystyle\int_0^9 \frac{1}{\sqrt{x}} dx$

3 $\displaystyle\int_{-3}^1 \frac{1}{x^2} dx$

4 $\displaystyle\int_{-2}^{-1} \frac{1}{(x+2)^{5/4}} dx$

5 $\displaystyle\int_0^{\pi/2} \sec^2 x \, dx$

6 $\displaystyle\int_0^1 \frac{e^{\sqrt{x}}}{\sqrt{x}} dx$

7 $\displaystyle\int_0^4 \frac{1}{(4-x)^{3/2}} dx$

8 $\displaystyle\int_0^{-1} \frac{1}{\sqrt[3]{x+1}} dx$

9 $\displaystyle\int_0^4 \frac{1}{(4-x)^{2/3}} dx$

10 $\displaystyle\int_1^2 \frac{x}{x^2-1} dx$

11 $\displaystyle\int_{-2}^2 \frac{1}{(x+1)^3} dx$

12 $\displaystyle\int_{-1}^1 x^{-4/3} dx$

13 $\displaystyle\int_{-2}^0 \frac{1}{\sqrt{4-x^2}} dx$

14 $\displaystyle\int_{-2}^0 \frac{x}{\sqrt{4-x^2}} dx$

15 $\displaystyle\int_{-1}^2 \frac{1}{x} dx$

16 $\displaystyle\int_0^4 \frac{1}{x^2-x-2} dx$

17 $\displaystyle\int_0^1 x \ln x \, dx$

18 $\displaystyle\int_0^{\pi/2} \tan^2 x \, dx$

19 $\displaystyle\int_0^{\pi/2} \tan x \, dx$

20 $\displaystyle\int_0^{\pi/2} \frac{1}{1-\cos x} dx$

21 $\displaystyle\int_{2}^{4} \frac{x-2}{x^2-5x+4}\,dx$

22 $\displaystyle\int_{1/e}^{e} \frac{1}{x(\ln x)^2}\,dx$

23 $\displaystyle\int_{-1}^{2} \frac{1}{x^2}\cos\frac{1}{x}\,dx$

24 $\displaystyle\int_{0}^{\pi} \sec x\,dx$

25 $\displaystyle\int_{0}^{\pi} \frac{\cos x}{\sqrt{1-\sin x}}\,dx$

26 $\displaystyle\int_{0}^{9} \frac{x}{\sqrt[3]{x-1}}\,dx$

In Exercises 27–30 assign, if possible, a value to (a) the area of the region R, and (b) the volume of the solid obtained by revolving R about the x-axis.

27 $R = \{(x, y)\colon 0 \le x \le 1, 0 \le y \le 1/\sqrt{x}\}$

28 $R = \{(x, y)\colon 0 \le x \le 1, 0 \le y \le 1/\sqrt[3]{x}\}$

29 $R = \{(x, y)\colon -4 \le x \le 4, 0 \le y \le 1/(x+4)\}$

30 $R = \{(x, y)\colon 1 \le x \le 2, 0 \le y \le 1/(x-1)\}$

Find all values of n for which the integrals in Exercises 31 and 32 converge.

31 $\displaystyle\int_{0}^{1} x^n\,dx$

32 $\displaystyle\int_{0}^{1} x^n \ln x\,dx$

11.5 TAYLOR'S FORMULA

Recall that f is a **polynomial function of degree n** if

$$f(x) = a_0 + a_1x + a_2x^2 + \cdots + a_nx^n$$

for all x, where each a_i is a real number and the exponents are nonnegative integers. Polynomial functions are the simplest functions to use for calculations in the sense that their values can be found by employing only additions and multiplications of real numbers. More complicated operations are needed to calculate values of logarithmic, exponential, trigonometric, or other transcendental functions. However, sometimes it is possible to *approximate* values by using polynomials. For example, since $\lim_{x\to 0} (\sin x)/x = 1$, it follows that if x is small, then $\sin x \approx x$; that is, the value of the sine function is almost the same as the value of the polynomial x. We say that $\sin x$ *may be approximated by the polynomial* x (provided x is close to 0).

As a second illustration, let f be the natural exponential function, that is, $f(x) = e^x$ for every x. Suppose we are interested in calculating values of f when x is close to 0. Since $f'(x) = e^x$, the slope of the tangent line at the point $(0, 1)$ on the graph of f is $f'(0) = e^0 = 1$. Hence an equation of the tangent line is

$$y - 1 = 1(x - 0), \quad \text{or} \quad y = 1 + x.$$

Referring to Figure 11.10, it is evident that if x is very close to 0, then the point $(x, 1 + x)$ on the tangent line is close to the point (x, e^x) on the graph of f and hence we may write

$$e^x \approx 1 + x.$$

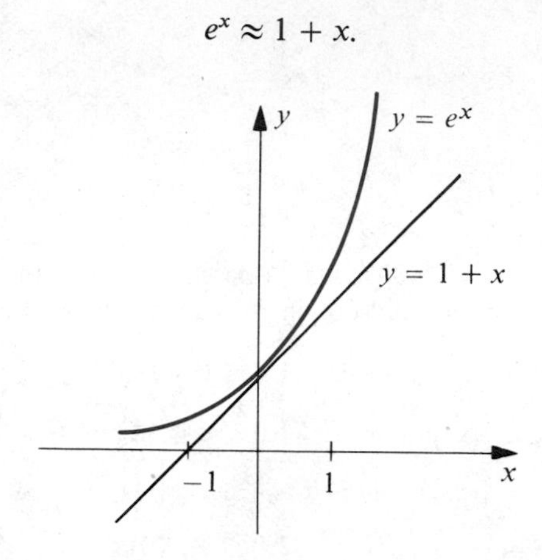

Figure 11.10

The previous formula allows us to approximate e^x by means of a polynomial of degree 1. Since the approximation is obviously poor unless x is very close to 0, let us seek a second-degree polynomial

$$g(x) = a + bx + cx^2$$

such that $e^x \approx g(x)$ when x is numerically small. The first and second derivatives of $g(x)$ are

$$g'(x) = b + 2cx$$
$$g''(x) = 2c.$$

If we want the graph of g (a parabola) to have the same y-intercept, as well as the same tangent line and concavity, as the graph of f at the point (0, 1), then we must have

$$g(0) = f(0), \quad g'(0) = f'(0), \quad g''(0) = f''(0).$$

Since all derivatives of e^x equal e^x, and $e^0 = 1$ these three equations imply that

$$a = 1, \quad b = 1, \quad \text{and} \quad 2c = 1$$

and hence

$$e^x \approx g(x) = 1 + x + \tfrac{1}{2}x^2.$$

The graphs of f and g are sketched in Figure 11.11. Comparing with Figure 11.10 it appears that if x is close to 0, then $1 + x + \frac{1}{2}x^2$ is closer to e^x than $1 + x$.

If we wish to approximate $f(x) = e^x$ by means of a *third*-degree polynomial $h(x)$, it is natural to require that $h(0)$, $h'(0)$, $h''(0)$, and $h'''(0)$ be the same as $f(0)$, $f'(0)$, $f''(0)$, and $f'''(0)$, respectively. The graphs of f and h then have the same tangent line and concavity at (0, 1) and, in addition, their *rates of change of concavity* with respect to x (that is, the third derivatives) are equal. Using the same technique employed previously would give us

$$e^x \approx h(x) = 1 + x + \frac{1}{2}x^2 + \frac{1}{3!}x^3.$$

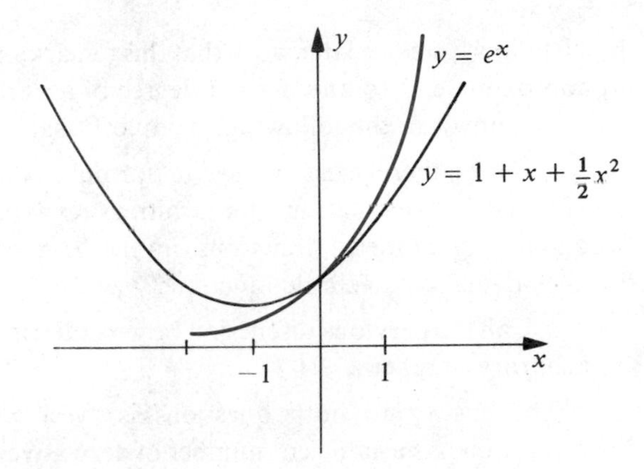

Figure 11.11

In order to get some idea of the accuracy of this approximation, several values of e^x and $h(x)$, approximated to the nearest hundredth, are displayed in the following table.

x	−1.5	−1.0	−0.5	0	0.5	1.0	1.5
e^x	0.22	0.37	0.61	1	1.65	2.72	4.48
$h(x)$	0.06	0.33	0.60	1	1.65	2.67	4.19

Observe that the error in the approximation increases as x increases numerically. The graphs of f and h are sketched in Figure 11.12. Notice the improvement in the approximation near $x = 0$.

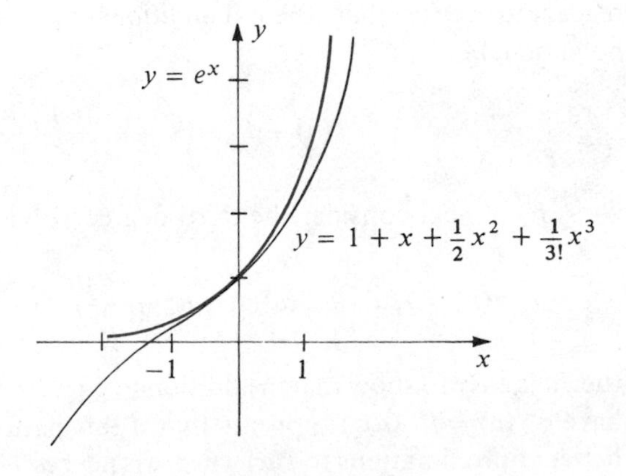

Figure 11.12

If we continued in this manner and determined a polynomial of degree n whose first n derivatives coincide with those of f at $x = 0$, we would arrive at

$$e^x \approx 1 + x + \frac{1}{2}x^2 + \frac{1}{3!}x^3 + \cdots + \frac{1}{n!}x^n.$$

It will follow from our later work that this remarkably simple formula can be used to approximate e^x to any desired degree of accuracy.

We now ask the following two questions:

1. Does there exist a *general* formula which may be used to obtain polynomial approximations for arbitrary exponential functions, logarithmic functions, trigonometric functions, inverse trigonometric functions, and other transcendental or algebraic functions?

2. Can the previous discussion be generalized to the case where x is close to an arbitrary number $a \neq 0$?

The answer to both questions is "yes" provided the function under consideration has a sufficient number of derivatives. Specifically, formula (11.12) of this section may be used to obtain polynomial approximations of a wide variety of functions. In order to see why (11.12) is a reasonable formula to expect, suppose f is a function which has many derivatives, and consider a number a in the domain of f. Let us proceed as we did for the special case of e^x discussed previously, but with a in place of 0. First we note that the only polynomial of degree 0 which coincides with f at a is the constant $f(a)$. To approximate $f(x)$ near a by a polynomial of degree 1, we choose the polynomial whose graph is the tangent line to the graph of f at $(a, f(a))$. Since the equation of this tangent line is

$$y - f(a) = f'(a)(x - a), \quad \text{or} \quad y = f(a) + f'(a)(x - a)$$

the desired first-degree polynomial is

$$f(a) + f'(a)(x - a).$$

A better approximation should be obtained by using a polynomial whose graph has the same *concavity* and tangent line as that of f at $(a, f(a))$. It is left to the reader to verify that these conditions are fulfilled by the second-degree polynomial

$$f(a) + f'(a)(x - a) + \frac{f''(a)}{2}(x - a)^2.$$

If we next consider the third-degree polynomial

$$g(x) = f(a) + f'(a)(x - a) + \frac{f''(a)}{2!}(x - a)^2 + \frac{f'''(a)}{3!}(x - a)^3$$

then it is easy to show that in addition to $g'(a) = f'(a)$ and $g''(a) = f''(a)$, we also have $g'''(a) = f'''(a)$. It appears that if this pattern is continued, we should get better approximations to $f(x)$ when x is near a. This leads to the first $n + 1$ terms on the right side of (11.12). The form of the last term is a consequence of the next result, which bears the name of the English mathematician Brook Taylor (1685–1731).

(11.11) Taylor's Formula

Let f be a function and n a positive integer such that the derivative $f^{(n+1)}(x)$ exists for every x in an interval I. If a and b are distinct numbers in I, then there is a number z between a and b such that

$$f(b) = f(a) + \frac{f'(a)}{1!}(b - a) + \frac{f''(a)}{2!}(b - a)^2 + \cdots + \frac{f^{(n)}(a)}{n!}(b - a)^n + \frac{f^{(n+1)}(z)}{(n+1)!}(b - a)^{n+1}.$$

Proof. There exists a number R_n (depending on a, b, and n) such that

$$f(b) = f(a) + \frac{f'(a)}{1!}(b - a) + \frac{f''(a)}{2!}(b - a)^2 + \cdots + \frac{f^{(n)}(a)}{n!}(b - a)^n + R_n.$$

Indeed, to find R_n we merely subtract the sum of the first $n + 1$ terms on the right side of this equation from $f(b)$. We wish to show that R_n is the same as the last term of the formula given in the statement of the theorem.

Let g be the function defined by

$$g(x) = f(b) - f(x) - \frac{f'(x)}{1!}(b - x) - \frac{f''(x)}{2!}(b - x)^2 - \cdots - \frac{f^{(n)}(x)}{n!}(b - x)^n - R_n\frac{(b - x)^{n+1}}{(b - a)^{n+1}}$$

for all x in I. If we differentiate each side of this equation, then many terms on the right cancel. As a matter of fact, it can be shown (see Exercise 41) that

$$g'(x) = -\frac{f^{(n+1)}(x)}{n!}(b - x)^n + R_n(n + 1)\frac{(b - x)^n}{(b - a)^{n+1}}.$$

It is easy to see that $g(b) = 0$. Moreover, substituting a for x in the expression for $g(x)$ and making use of the first equation of the proof gives us $g(a) = 0$. According to Rolle's Theorem (4.9), there is a number z between a and b such that $g'(z) = 0$. Evaluating $g'(x)$ at z and solving for R_n we see that

$$R_n = \frac{f^{(n+1)}(z)}{(n + 1)!}(b - a)^{n+1}$$

which is what we wished to prove.

If we replace b by x in (11.11) we obtain the following formula.

(11.12) Taylor's Formula with the Remainder

$$f(x) = f(a) + \frac{f'(a)}{1!}(x-a) + \frac{f''(a)}{2!}(x-a)^2 + \cdots + \frac{f^{(n)}(a)}{n!}(x-a)^n + \frac{f^{(n+1)}(z)}{(n+1)!}(x-a)^{n+1}$$

where z is a number between a and x.

The sum of the first $n + 1$ terms on the right side of (11.12) is often denoted by $P_n(x)$ and called the **nth-degree Taylor polynomial of *f* at *a***, that is,

$$P_n(x) = f(a) + \frac{f'(a)}{1!}(x-a) + \frac{f''(a)}{2!}(x-a)^2 + \cdots + \frac{f^{(n)}(a)}{n!}(x-a)^n. \tag{11.13}$$

The following expression is called the **remainder after *n* + 1 terms**:

$$R_n(x) = \frac{f^{(n+1)}(z)}{(n+1)!}(x-a)^{n+1}. \tag{11.14}$$

Thus we may write

$$f(x) = P_n(x) + R_n(x). \tag{11.15}$$

It follows that *if the numerical value of $R_n(x)$ is small, then $f(x)$ may be approximated by the nth-degree Taylor polynomial of f at a, that is,*

$$f(x) \approx P_n(x) \quad \text{if} \quad x \approx a. \tag{11.16}$$

Since $|f(x) - P_n(x)| = |R_n(x)|$, the error involved in using this polynomial approximation equals the absolute value of $|R_n(x)|$.

Example 1 If $f(x) = \ln x$, find Taylor's Formula with the Remainder for $n = 3$ and $a = 1$.

Solution If $n = 3$ in (11.12), then we need the first four derivatives of f. It is convenient to arrange our work as follows.

$$\begin{array}{ll} f(x) = \ln x & f(1) = 0 \\ f'(x) = x^{-1} & f'(1) = 1 \\ f''(x) = -x^{-2} & f''(1) = -1 \\ f'''(x) = 2x^{-3} & f'''(1) = 2 \\ f^{(4)}(x) = -3!x^{-4} & f^{(4)}(z) = -6z^{-4} \end{array}$$

By Taylor's Formula (11.12),

$$\ln x = 0 + \frac{1}{1!}(x-1) - \frac{1}{2!}(x-1)^2 + \frac{2}{3!}(x-1)^3 - \frac{6z^{-4}}{4!}(x-1)^4,$$

where z is between 1 and x. The last formula simplifies to

$$\ln x = (x - 1) - \frac{1}{2}(x - 1)^2 + \frac{1}{3}(x - 1)^3 - \frac{1}{4z^4}(x - 1)^4.$$

Example 2 Use the formula obtained in Example 1 to approximate $\ln(1.1)$, and estimate the accuracy of this approximation.

Solution Substituting 1.1 for x in the formula of Example 1 gives us

$$\ln(1.1) = 0.1 - \frac{1}{2}(0.1)^2 + \frac{1}{3}(0.1)^3 - \frac{1}{4z^4}(0.1)^4$$

where $1 < z < 1.1$. Summing the first three terms we obtain $\ln(1.1) \approx 0.0953$. Since $z > 1$, $1/z < 1$ and, therefore, $1/z^4 < 1$. Consequently

$$|R_3(1.1)| = \left|\frac{(0.1)^4}{4z^4}\right| < \left|\frac{0.0001}{4}\right| = 0.000025.$$

Consequently, the approximation $\ln(1.1) \approx 0.0953$ is accurate to four decimal places.

If we wish to approximate a functional value $f(x)$ for some x, it is desirable to choose the number a in (11.12) such that the remainder $R_n(x)$ is very close to 0 when n is relatively small (say $n = 3$ or $n = 4$). This will be true if we choose a close to x. In addition, a should be chosen in such a way that the values of the first $n + 1$ derivatives of f at a are easy to calculate. This was done in Example 2, where to approximate $\ln x$ for $x = 1.1$ we selected $a = 1$ (see Example 1). The next example provides another illustration of a suitable choice for a.

Example 3 Use a Taylor Polynomial to approximate $\cos 61°$, and estimate the accuracy of the approximation.

Solution We wish to approximate $f(x) = \cos x$ if $x = 61°$. Let us begin by observing that $61°$ is close to $60°$, or $\pi/3$ radians, and that it is easy to calculate values of trigonometric functions at $\pi/3$. This suggests that we choose $a = \pi/3$ in (11.12). The choice of n will depend on the accuracy we wish to attain. Let us try $n = 2$. In this event the first three derivatives of f are required and we arrange our work as follows:

$$\begin{aligned} f(x) &= \cos x & f(\pi/3) &= 1/2 \\ f'(x) &= -\sin x & f'(\pi/3) &= -\sqrt{3}/2 \\ f''(x) &= -\cos x & f''(\pi/3) &= -1/2 \\ f'''(x) &= \sin x & f'''(z) &= \sin z \end{aligned}$$

By (11.13), the second-degree Taylor polynomial of f at $\pi/3$ is

$$P_2(x) = \frac{1}{2} - \frac{(\sqrt{3}/2)}{1!}\left(x - \frac{\pi}{3}\right) - \frac{(1/2)}{2!}\left(x - \frac{\pi}{3}\right)^2.$$

Since x represents a real number, 61° must be converted to radian measure before substitution on the right side. Writing

$$61° = 60° + 1° = \frac{\pi}{3} + \frac{\pi}{180}$$

and substituting in $P_2(x)$ we obtain

$$P_2\left(\frac{\pi}{3} + \frac{\pi}{180}\right) = \frac{1}{2} - \left(\frac{\sqrt{3}}{2}\right)\left(\frac{\pi}{180}\right) - \frac{1}{4}\left(\frac{\pi}{180}\right)^2 \approx 0.48481.$$

Thus by (11.16),

$$\cos 61° \approx 0.48481.$$

To estimate the accuracy of this approximation, we see from (11.14) that

$$|R_2(x)| = \left|\frac{\sin z}{3!}\left(x - \frac{\pi}{3}\right)^3\right|.$$

Substituting $x = \pi/3 + \pi/180$ and using the fact that $|\sin z| \leq 1$,

$$\left|R_2\left(\frac{\pi}{3} + \frac{\pi}{180}\right)\right| = \left|\frac{\sin z}{3!}\left(\frac{\pi}{180}\right)^3\right| \leq \left|\frac{1}{3!}\left(\frac{\pi}{180}\right)^3\right| \leq 0.000001.$$

Consequently, the approximation $\cos 61° \approx 0.48481$ is accurate to five decimal places. If more accuracy is desired, then it is necessary to find a value of n such that the maximum value of $|R_n(\pi/3 + \pi/180)|$ is within the desired range.

In Examples 2 and 3 we stated that the approximations were accurate to four and five decimal places, respectively. Throughout the remainder of the text an approximation will be considered accurate to k decimal places if the error E is less than $5 \times 10^{-(k+1)}$ in absolute value. Using this convention we have, for example, accuracy to

1 decimal place if $|E| < 5 \times 10^{-2} = 0.05$
2 decimal places if $|E| < 5 \times 10^{-3} = 0.005$
3 decimal places if $|E| < 5 \times 10^{-4} = 0.0005$.

If 0 is used for a in (11.12), then we obtain the formula

(11.17) $$f(x) = f(0) + \frac{f'(0)}{1!}x + \frac{f''(0)}{2!}x^2 + \cdots + \frac{f^{(n)}(0)}{n!}x^n + \frac{f^{(n+1)}(z)}{(n+1)!}x^{n+1},$$

where z is between 0 and x. This important special case of (11.12) is called **Maclaurin's Formula**, named after the Scottish mathematician Colin Maclaurin (1698–1746).

Example 4 Find Maclaurin's Formula (11.17) for $f(x) = e^x$ if n is any positive integer.

Solution For every positive integer k, $f^{(k)}(x) = e^x$ and hence $f^{(k)}(0) = e^0 = 1$. Substituting in (11.17) gives us

$$e^x = 1 + x + \frac{x^2}{2!} + \cdots + \frac{x^n}{n!} + \frac{e^z}{(n+1)!}x^{n+1}$$

where z is between 0 and x. Note that the first $n + 1$ terms are the same as those obtained in our discussion at the beginning of this section.

The formula derived in Example 4 may be used to approximate values of the natural exponential function. Another important application will be discussed in the next chapter in conjunction with representations of functions by means of infinite series.

Example 5 Find Maclaurin's Formula for $f(x) = \sin x$ and $n = 8$.

Solution We need the first nine derivatives of $f(x)$. Let us begin as follows:

$$\begin{aligned} f(x) &= \sin x & f(0) &= 0 \\ f'(x) &= \cos x & f'(0) &= 1 \\ f''(x) &= -\sin x & f''(0) &= 0 \\ f'''(x) &= -\cos x & f'''(0) &= -1 \end{aligned}$$

Since $f^{(4)}(x) = \sin x$, the remaining derivatives follow the same pattern, and we arrive at

$$f^{(9)}(x) = \cos x \qquad f^{(9)}(z) = \cos z.$$

Substituting in (11.17) and noting that the constant term and the coefficients of x^2, x^4, x^6, and x^8 are 0, we obtain

$$\sin x = \frac{1}{1!}x + \frac{(-1)}{3!}x^3 + \frac{1}{5!}x^5 + \frac{(-1)}{7!}x^7 + \frac{\cos z}{9!}x^9$$

which may be written

$$\sin x = x - \frac{x^3}{3!} + \frac{x^5}{5!} - \frac{x^7}{7!} + \frac{\cos z}{9!}x^9.$$

Incidentally, if we used the first four nonzero terms of the formula found in Example 5 to approximate $\sin(0.1)$, then the error would be $|R_8(0.1)|$.

Since $|\cos z| \le 1$,

$$|R_8(0.1)| = \left|\frac{\cos z}{9!}(0.1)^9\right| \le \frac{(0.1)^9}{9!} < 2.7 \times 10^{-15}.$$

According to our convention concerning accuracy, this means that the approximation would be correct to 14 decimal places!

EXERCISES 11.5

In Exercises 1–12, find Taylor's Formula with the Remainder for the given values of a and n.

1 $f(x) = \sin x, a = \pi/2, n = 3$

2 $f(x) = \cos x, a = \pi/4, n = 3$

3 $f(x) = \sqrt{x}, a = 4, n = 3$

4 $f(x) = e^{-x}, a = 1, n = 3$

5 $f(x) = \tan x, a = \pi/4, n = 4$

6 $f(x) = 1/(x-1)^2, a = 2, n = 5$

7 $f(x) = 1/x, a = -2, n = 5$

8 $f(x) = \sqrt[3]{x}, a = -8, n = 3$

9 $f(x) = \tan^{-1} x, a = 1, n = 2$

10 $f(x) = \ln \sin x, a = \pi/6, n = 3$

11 $f(x) = xe^x, a = -1, n = 4$

12 $f(x) = \log_{10} x, a = 10, n = 2$

In Exercises 13–24, find Maclaurin's Formula for the given values of n.

13 $f(x) = \ln(x + 1), n = 4$

14 $f(x) = \sin x, n = 7$

15 $f(x) = \cos x, n = 8$

16 $f(x) = \tan^{-1} x, n = 3$

17 $f(x) = e^{2x}, n = 5$

18 $f(x) = \sec x, n = 3$

19 $f(x) = \dfrac{1}{(x-1)^2}, n = 5$

20 $f(x) = \sqrt{4 - x}, n = 3$

21 $f(x) = \arcsin x, n = 2$

22 $f(x) = e^{-x^2}, n = 3$

23 $f(x) = 2x^4 - 5x^3 + x^2 - 3x + 7, n = 4$ and $n = 5$

24 $f(x) = \cosh x, n = 4$ and $n = 5$

Approximate the numbers in Exercises 25–28 to four decimal places by using the indicated exercise. In each case prove that your answer is correct by showing that $|R_n(x)| < 5 \times 10^{-5}$.

25 $\sin 89°$ (Use Exercise 1 and $\pi/180 \approx 0.0175$.)

26 $\cos 47°$ (Use Exercise 2 and $\pi/180 \approx 0.0175$.)

27 $\sqrt{4.03}$ (Use Exercise 3.)

28 $e^{-1.02}$ (Use Exercise 4.)

In each of Exercises 29–34 approximate the number by using the indicated exercise, and estimate the error in the approximation by means of $R_n(x)$.

29 $-1/2.2$; Exercise 7

30 $\sqrt[3]{-8.5}$; Exercise 8

31 $\ln(1.25)$; Exercise 13

32 $\sin 0.1$; Exercise 14

33 $\cos 30°$; Exercise 15

34 $\log_{10} 10.01$; Exercise 12

Use Maclaurin's Formula to establish the approximation formulas in Exercises 35–40 and state, in terms of decimal places, the accuracy of the approximation if $|x| \leq 0.1$.

35 $\cos x \approx 1 - \dfrac{x^2}{2}$

36 $\sqrt[3]{1 + x} \approx 1 + \dfrac{1}{3}x$

37 $e^x \approx 1 + x + \dfrac{x^2}{2}$

38 $\sin x \approx x - \dfrac{x^3}{6}$

39 $\ln (1 + x) \approx x - \dfrac{x^2}{2} + \dfrac{x^3}{3}$

40 $\cosh x \approx 1 + \dfrac{x^2}{2}$

41 In the proof of Taylor's Formula (11.11) verify the formula for $g'(x)$.

42 If $f(x)$ is a polynomial of degree n and a is any real number, prove that $f(x) = P_n(x)$, where $P_n(x)$ is given by (11.13).

11.6 NEWTON'S METHOD

In the preceding section Taylor's Formula with the Remainder (11.12) was used to find polynomial approximations of functions. As a special case, the first-degree polynomial

$$f(a) + f'(a)(x - a)$$

can be used to approximate $f(x)$ if x is near a. We shall now describe a technique for using this polynomial to approximate a real root of $f(x)$, that is, a real number r such that $f(r) = 0$. We begin by making a first approximation x_1 to the root r. Since r is an x-intercept of the graph of f, a suitable number x_1 can usually be found by referring to a rough sketch of the graph. If we consider the tangent line l to the graph of f at the point $(x_1, f(x_1))$ and if x_1 is sufficiently close to r, then as illustrated in Figure 11.13, the x-intercept x_2 of l should be a better approximation to r.

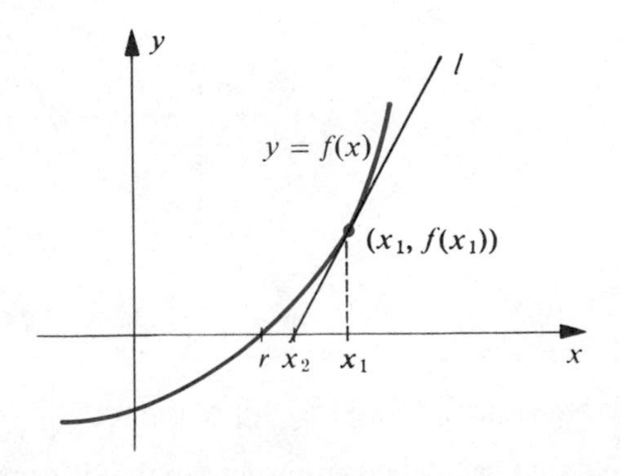

Figure 11.13

Since an equation of the tangent line l is

$$y = f(x_1) + f'(x_1)(x - x_1)$$

the x-intercept x_2 is given by

$$0 = f(x_1) + f'(x_1)(x_2 - x_1).$$

If $f'(x_1) \neq 0$, the preceding equation is equivalent to

$$x_2 = x_1 - \frac{f(x_1)}{f'(x_1)}.$$

If we take x_2 as a second approximation to r, then the process may be repeated by using the tangent line at $(x_2, f(x_2))$. If $f'(x_2) \neq 0$, this leads to a third approximation x_3, given by

$$x_3 = x_2 - \frac{f(x_2)}{f'(x_2)}.$$

The process is continued until the desired degree of accuracy is obtained. In general, if x_n is any approximation, then the next approximation x_{n+1} is given by

(11.18) $$x_{n+1} = x_n - \frac{f(x_n)}{f'(x_n)}$$

provided $f'(x_n) \neq 0$. This technique of successive approximations of real roots is referred to as **Newton's Method**. Some care must be exercised in choosing the first approximation x_1. Indeed, if x_1 is not sufficiently close to r, it is possible for the second approximation x_2 to be worse than x_1, as illustrated in Figure 11.14. It is evident from the denominator in (11.18) that we should not choose a number x_n such that $f'(x_n)$ is close to 0. In particular, values which are near critical numbers of f should not be used.

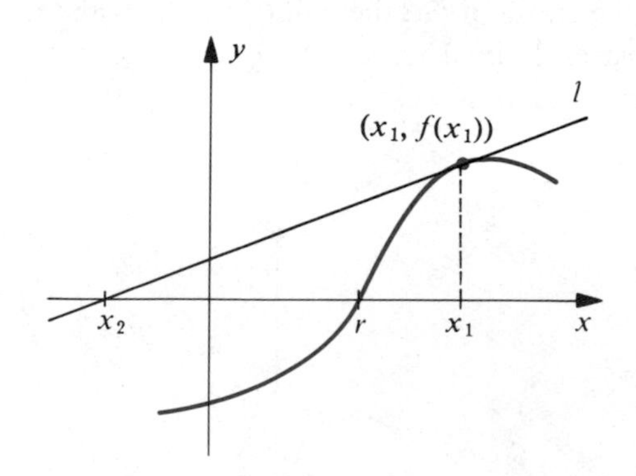

Figure 11.14

Example 1 Use Newton's Method to approximate $\sqrt{7}$.

Solution The stated problem is equivalent to that of approximating the positive real zero r of $f(x) = x^2 - 7$. Since $f(2) = -3$ and $f(3) = 2$, it follows from the continuity of f

that $2 < r < 3$. Moreover, since f is increasing there can be only one zero in the interval $(2, 3)$. If x_n is any approximation to r, then by (11.18) the next approximation x_{n+1} is given by

$$x_{n+1} = x_n - \frac{f(x_n)}{f'(x_n)} = x_n - \frac{x_n^2 - 7}{2x_n}$$

which simplifies to

$$x_{n+1} = \frac{x_n^2 + 7}{2x_n}. \tag{11.19}$$

If we choose $x_1 = 3$ as a first approximation, then by (11.19) with $n = 1$,

$$x_2 = \frac{(3)^2 + 7}{2(3)} = \frac{16}{6} = \frac{8}{3}.$$

Using (11.19) again, with $n = 2$, the next approximation is

$$x_3 = \frac{(8/3)^2 + 7}{2(8/3)} = \frac{127}{48} \approx 2.6458.$$

It can be shown that to six decimal places, $\sqrt{7} \approx 2.645751$. Hence, two applications of Newton's Method resulted in accuracy to three decimal places.

Example 2 Approximate the real root of the equation $2x - e^{-x} = 0$.

Solution We wish to find a value of x such that $e^{-x} = 2x$. This coincides with the abscissa of the point of intersection of the graphs of $y = e^{-x}$ and $y = 2x$. It appears from Figure 11.15 that $x_1 = 0.3$ is a reasonable first approximation.

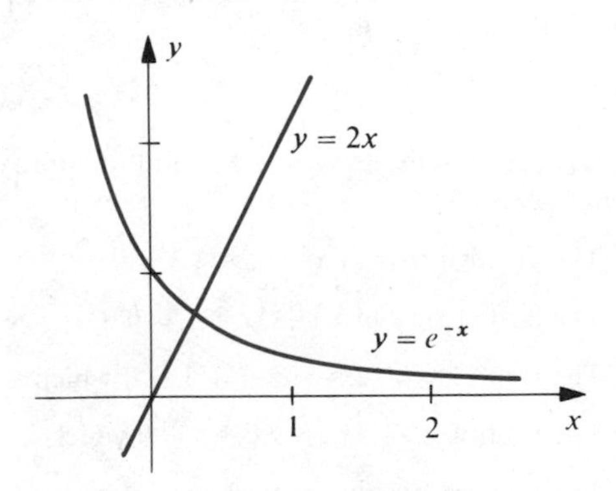

Figure 11.15

If we let $f(x) = 2x - e^{-x}$, then by (11.18)

$$x_2 = x_1 - \frac{f(x_1)}{f'(x_1)} = x_1 - \frac{2x_1 - e^{-x_1}}{2 + e^{-x_1}}$$

or

$$x_2 = 0.3 - \frac{0.6 - e^{-0.3}}{2 + e^{-0.3}}.$$

Using Table II,

$$x_2 \approx 0.3 - \frac{0.6 - (0.7408)}{2 + (0.7408)} \approx 0.351.$$

If we use $x_2 = 0.35$ in (11.18), then

$$\begin{aligned} x_3 &= 0.35 - \frac{0.7 - e^{-0.35}}{2 + e^{-0.35}} \\ &\approx 0.35 - \frac{0.7 - 0.7047}{2 + 0.7047} \\ &\approx 0.352. \end{aligned}$$

If a computer is used, it can be shown that the first five decimal places of r are 0.35173.

The iterative process used in Newton's Method is continued until repetitions of digits occur from one approximation to the next. For example, if we desire accuracy to the nearest hundredth, we ordinarily continue the process until the first three decimal places remain fixed and then round off to two decimal places.

EXERCISES 11.6

In Exercises 1 and 2 use Newton's Method to approximate the given number to four decimal places.

1 $\sqrt[3]{2}$ **2** $\sqrt[5]{3}$

In Exercises 3–10 use Newton's Method to approximate the indicated real root to four decimal places.

3 The positive root of $x^3 + 5x - 3 = 0$

4 The largest root of $x^3 - 3x + 1 = 0$

5 The root of $x^4 + 2x^3 - 5x^2 + 1 = 0$ which is between 1 and 2

6 The root of $x^4 - 5x^2 + 2x - 5 = 0$ which is between 2 and 3

7 The smallest positive root of $\cos x - e^{-x} = 0$

8 The root of $\cos x + x = 2$ **9** The root of $x + \ln x = 2$

10 The largest root of $e^x - \sin x = 0$

In each of Exercises 11–18 use Newton's Method to find all real roots of the given equation to two decimal places.

11 $x^4 = 125$ **12** $10x^2 - 1 = 0$

13 $x^4 - x - 2 = 0$

14 $x^5 - 2x^2 + 4 = 0$

15 $2x^3 + e^x = 0$

16 $2e^x + x - 1 = 0$

17 $2x - 5 - \sin x = 0$

18 $x^2 - 2 - e^{-2x} = 0$

11.7 REVIEW

Concepts

Define or discuss each of the following.

1 Indeterminate forms

2 L'Hôpital's Rule

3 Cauchy's Formula

4 Improper integrals

5 Taylor's Formula

6 Maclaurin's Formula

7 Newton's Method

Exercises

Find the limits in Exercises 1–14, if they exist.

1 $\lim_{x \to 0} \dfrac{\ln(2 - x)}{1 + e^{2x}}$

2 $\lim_{x \to 0} \dfrac{\sin 2x - \tan 2x}{x^2}$

3 $\lim_{x \to \infty} \dfrac{x^2 + 2x + 3}{\ln(x + 1)}$

4 $\lim_{x \to 0} \dfrac{\tan^{-1} x}{\sin^{-1} x}$

5 $\lim_{x \to 0} \dfrac{e^{2x} - e^{-2x} - 4x}{x^3}$

6 $\lim_{x \to \frac{1}{2}\pi^-} \dfrac{\tan x}{\sec x}$

7 $\lim_{x \to \infty} \dfrac{x^e}{e^x}$

8 $\lim_{x \to \frac{1}{2}\pi^-} \cos x \ln \cos x$

9 $\lim_{x \to \infty} (1 - 2e^{1/x})x$

10 $\lim_{x \to 0} \tan^{-1} x \csc x$

11 $\lim_{x \to 0} (1 + 8x^2)^{1/x^2}$

12 $\lim_{x \to 1} (\ln x)^{x-1}$

13 $\lim_{x \to \infty} (e^x + 1)^{1/x}$

14 $\lim_{x \to 0} \left(\dfrac{1}{\tan x} - \dfrac{1}{x}\right)$

Determine whether the integrals in Exercises 15–26 converge or diverge, and evaluate those which converge.

15 $\int_4^\infty \dfrac{1}{\sqrt{x}}\,dx$

16 $\int_4^\infty \dfrac{1}{x\sqrt{x}}\,dx$

17 $\int_{-\infty}^{0} \frac{1}{x+2}\,dx$

18 $\int_{0}^{\infty} \sin x\,dx$

19 $\int_{-8}^{1} \frac{1}{\sqrt[3]{x}}\,dx$

20 $\int_{-4}^{0} \frac{1}{x+4}\,dx$

21 $\int_{0}^{2} \frac{x}{(x^2-1)^2}\,dx$

22 $\int_{1}^{2} \frac{1}{x\sqrt{x^2-1}}\,dx$

23 $\int_{-\infty}^{\infty} \frac{1}{e^x+e^{-x}}\,dx$

24 $\int_{-\infty}^{0} xe^x\,dx$

25 $\int_{0}^{1} \frac{\ln x}{x}\,dx$

26 $\int_{0}^{\pi/2} \csc x\,dx$

27 Find Taylor's Formula with the Remainder for the following.

(a) $f(x) = \ln\cos x, a = \pi/6, n = 3$

(b) $f(x) = \sqrt{x-1}, a = 2, n = 4$

28 Find Maclaurin's Formula for the following.

(a) $f(x) = e^{-x^3}, n = 3$

(b) $f(x) = 1/(1-x), n = 6$

29 Use Taylor's Formula to approximate $\sin^2 43°$ to four decimal places. (*Hint:* $\sin^2 x = (1 - \cos 2x)/2$.)

30 Establish the approximation formula $\sin x \approx x - (x^3/6)$, and state the accuracy of the approximation if $|x| \leq 0.2$.

31 Use Newton's Method to approximate the largest root of $2x^3 - 4x^2 - 3x + 1 = 0$ to three decimal places.

32 Use Newton's Method to approximate $\sqrt[4]{5}$ to three decimal places.

Infinite Series

Infinite series are useful in advanced courses in mathematics, physics, and engineering because they may be employed to represent functions in a new way. In this chapter we shall discuss some of the fundamental results associated with this important mathematical concept.

12.1 INFINITE SEQUENCES

A function f from a set X to a set Y is a correspondence that associates with each element x of X a unique element $f(x)$ of Y (see Definition (1.18)). Up to now the domain X has usually been an interval of real numbers. In this section we shall consider a different class of functions.

(12.1) Definition

An **infinite sequence** is a function whose domain is the set of positive integers.

For convenience we sometimes refer to infinite sequences merely as *sequences*. In this book the range of an infinite sequence will be a set of real numbers.

If f is an infinite sequence, then to each positive integer n there corresponds a real number $f(n)$. These numbers in the range of f may be listed by writing

$$f(1), f(2), f(3), \ldots, f(n), \ldots$$

where the dots at the end indicate that the sequence does not terminate. The number $f(1)$ is called the **first term** of the sequence, $f(2)$ the **second term** and, in general, $f(n)$ the ***n*th term** of the sequence. It is customary to use a subscript notation instead of the functional notation and write these numbers as

$$a_1, a_2, a_3, \ldots, a_n, \ldots \qquad (12.2)$$

where it is understood that for each positive integer n, the symbol a_n denotes the real number $f(n)$. In this way we obtain an infinite collection of real numbers which is *ordered* in the sense that there is a first number, a second number, a forty-fifth number, and so on. Although sequences are functions, a collection such as

(12.2) will also be referred to as an infinite sequence. If we wish to convert (12.2) to a function f we let $f(n) = a_n$ for all positive integers n.

From the definition of equality of functions we see that a sequence

$$a_1, a_2, a_3, \ldots, a_n, \ldots$$

is **equal** to a sequence

$$b_1, b_2, b_3, \ldots, b_n, \ldots$$

if and only if $a_i = b_i$ for every positive integer i. Infinite sequences are often defined by stating a formula for the nth term, as in the following example.

Example 1 List the first four terms and the tenth term of the sequence whose nth term a_n is as follows.

(a) $a_n = \dfrac{n}{n+1}$ (b) $a_n = 2 + (0.1)^n$

(c) $a_n = (-1)^{n+1}\dfrac{n^2}{3n-1}$ (d) $a_n = 4$

Solution To find the first four terms we substitute, successively, $n = 1, 2, 3$, and 4 in the formula for a_n. The tenth term is found by substituting 10 for n. Doing this and simplifying gives us the following:

	First four terms	*Tenth term*
(a)	$\frac{1}{2}, \frac{2}{3}, \frac{3}{4}, \frac{4}{5}$	$\frac{10}{11}$
(b)	2.1, 2.01, 2.001, 2.0001	2.0000000001
(c)	$\frac{1}{2}, -\frac{4}{5}, \frac{9}{8}, -\frac{16}{11}$	$-\frac{100}{29}$
(d)	4, 4, 4, 4	4

We sometimes denote the sequence (12.2) by $\{a_n\}$. For example, the sequence $\{2^n\}$ has nth term $a_n = 2^n$. According to Definition (12.1), the sequence $\{2^n\}$ is the function f such that $f(n) = 2^n$ for every positive integer n.

Some infinite sequences $\{a_n\}$ have the property that as n increases a_n gets very close to some real number L. Another way of stating this is to say that the numerical difference $|a_n - L|$ is almost zero when n is large. As an illustration, consider the sequence $\{a_n\}$ where

$$a_n = 2 + \left(-\frac{1}{2}\right)^n.$$

The first few terms are

$$2 - \frac{1}{2}, 2 + \frac{1}{4}, 2 - \frac{1}{8}, 2 + \frac{1}{16}, 2 - \frac{1}{32}, \ldots$$

and it appears that the terms get closer to 2 as n increases. As a matter of fact, for every positive integer n,

$$|a_n - 2| = \left|2 + \left(-\frac{1}{2}\right)^n - 2\right| = \left|\left(-\frac{1}{2}\right)^n\right| = \left(\frac{1}{2}\right)^n = \frac{1}{2^n}$$

and the number $1/2^n$, and hence $|a_n - 2|$, *can be made arbitrarily close to zero by choosing n sufficiently large.* According to Definition (12.3) below, the given sequence *has the limit* 2, and we write

$$\lim_{n\to\infty}\left[2 + \left(-\frac{1}{2}\right)^n\right] = 2.$$

The situation here is almost identical to that in Chapter 4, where $\lim_{x\to\infty} f(x) = L$ was defined. The only difference is that if $f(n) = a_n$, then the domain of f is the set of positive integers and not an infinite interval of real numbers. As in Definition (4.19), but using a_n instead of $f(x)$, we state the following definition.

(12.3) **Definition**

A sequence $\{a_n\}$ has the limit L, written

$$\lim_{n\to\infty} a_n = L$$

if for every $\varepsilon > 0$, there exists a positive number N such that

$$\text{if} \quad n > N, \quad \text{then} \quad |a_n - L| < \varepsilon.$$

If $\lim_{n\to\infty} a_n$ does not exist in the sense of Definition (12.3), then the sequence $\{a_n\}$ has no limit.

A geometric interpretation similar to that shown in Figure 4.25 can be given for the limit of a sequence. The only difference is that the coordinate x of the point shown on the x-axis is always a positive integer. In Figure 12.1 we have illustrated the behavior of the points (k, a_k) for a special case in which $\lim_{n\to\infty} a_n = L$. Note that for any $\varepsilon > 0$, the points (n, a_n) lie between the lines $y = L \pm \varepsilon$, provided n is sufficiently large. Of course, the approach to L may vary from that illustrated in the figure. Another geometric interpretation of Definition (12.3) may be obtained by plotting the points corresponding to a_k on a coordinate line, as shown in Figure 12.2. In this case if we consider any open interval $(L - \varepsilon, L + \varepsilon)$, then a_n is in the interval whenever n is sufficiently large.

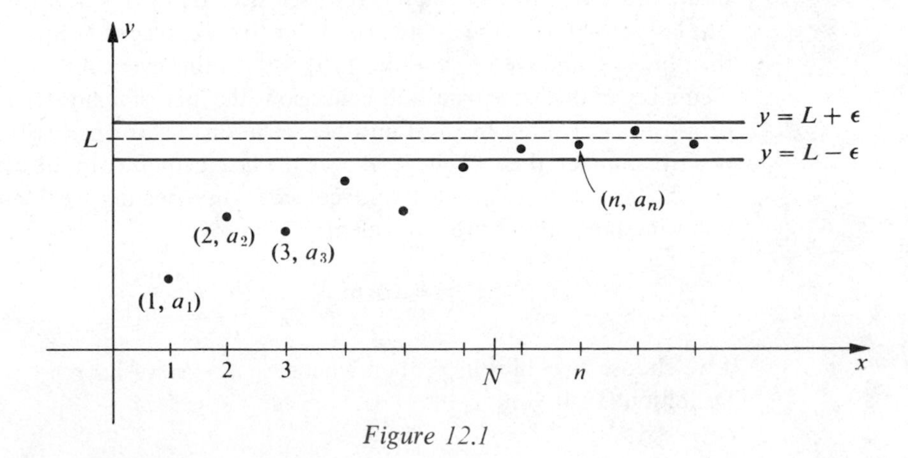

Figure 12.1

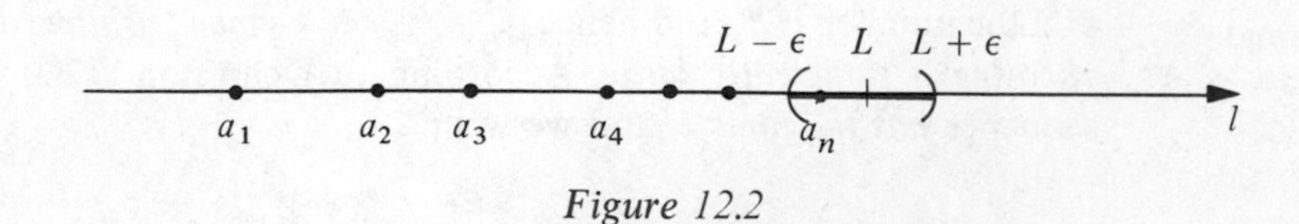

Figure 12.2

If a_n can be made as large as desired by choosing n sufficiently large, then the sequence $\{a_n\}$ has no limit, but we still write $\lim_{n\to\infty} a_n = \infty$. A more precise way of specifying this behavior is as follows:

(12.4) Definition

The statement $\lim_{n\to\infty} a_n = \infty$ means that for every positive real number P, there exists a number N such that if $n > N$, then $a_n > P$.

The proof of the next theorem illustrates the use of the preceding definitions.

(12.5) Theorem

(i) $\lim_{n\to\infty} r^n = 0 \quad \text{if } |r| < 1$

(ii) $\lim_{n\to\infty} |r^n| = \infty \quad \text{if } |r| > 1$

Proof. If $r = 0$ it follows trivially that the limit is 0. Let us assume that $0 < |r| < 1$. To prove (i) by means of Definition (12.3) we must show that for every $\varepsilon > 0$ there exists a positive number N such that

$$\text{if} \quad n > N, \quad \text{then} \quad |r^n - 0| < \varepsilon.$$

The inequality $|r^n - 0| < \varepsilon$ is equivalent to each inequality in the following list:

$$|r|^n < \varepsilon, \quad n \ln |r| < \ln \varepsilon, \quad n > \frac{\ln \varepsilon}{\ln |r|}$$

where the final inequality sign is reversed because $\ln |r|$ is negative if $0 < |r| < 1$. The last inequality in the list gives us a clue to the choice of N. Specifically, if $\varepsilon < 1$, then $\ln \varepsilon < 0$ and we let $N = \ln \varepsilon / \ln |r| > 0$. In this event if $n > N$, then the last inequality in the list is true, and hence so is the first, which is what we wished to prove. If $\varepsilon \geq 1$, then $\ln \varepsilon \geq 0$ and hence $\ln \varepsilon / \ln |r| \leq 0$. In this case if N is *any* positive number, then whenever $n > N$ the last inequality in the list is again true.

To prove (ii) let $|r| > 1$ and consider any positive real number P. The following inequalities are equivalent:

$$|r|^n > P, \quad n \ln |r| > \ln P, \quad n > \frac{\ln P}{\ln |r|}.$$

If we choose $N = \ln P / \ln |r|$, then whenever $n > N$ we have $|r|^n > P$. Hence by Definition (12.4), $\lim_{n\to\infty} |r|^n = \infty$.

Example 2 List the first four terms and, if possible, find the limits of each of the following sequences.

(a) $\{(-2/3)^n\}$ (b) $\{(1.01)^n\}$

Solution (a) The first four terms are

$$-\frac{2}{3}, \frac{4}{9}, -\frac{8}{27}, \frac{16}{81}.$$

According to Theorem (12.5),

$$\lim_{n\to\infty}\left(-\frac{2}{3}\right)^n = 0.$$

(b) The first four terms are

$$1.01, \quad 1.0201, \quad 1.030301, \quad 1.04060401.$$

According to Theorem (12.5),

$$\lim_{n\to\infty}(1.01)^n = \infty.$$

Theorems which are analogous to some of the limit theorems in Chapter 2 may be established for infinite sequences. In particular, given two infinite sequences $\{a_n\}$ and $\{b_n\}$, corresponding terms may be added to form the infinite sequence $\{a_n + b_n\}$, which is called the **sum** of the given sequences. Similarly we could form the **difference**, **product**, or **quotient** by subtracting, multiplying, or dividing corresponding terms (provided no zero denominators occur in the division process).

If $\lim_{n\to\infty} a_n = L$ and $\lim_{n\to\infty} b_n = M$ where L and M are real numbers, then it can be shown that

$$\lim_{n\to\infty}(a_n + b_n) = L + M$$

$$\lim_{n\to\infty}(a_n - b_n) = L - M$$

$$\lim_{n\to\infty} a_n b_n = LM$$

$$\lim_{n\to\infty}\frac{a_n}{b_n} = \frac{L}{M}.$$

In the last formula we must have $M \neq 0$ and $b_n \neq 0$ for all n.

If $a_n = c$ for all n, so that the infinite sequence is $c, c, \ldots, c, \ldots$, then

$$\lim_{n\to\infty} c = c.$$

Similarly, if c is a real number and k is a positive rational number, then, as in the proof of Theorem (4.21),

$$\lim_{n\to\infty} \frac{c}{n^k} = 0.$$

Example 3 Find the limit of the sequence $\left\{\dfrac{2n}{5n-3}\right\}$.

Solution We wish to find $\lim_{n\to\infty} a_n$, where $a_n = 2n/(5n-3)$. Dividing numerator and denominator of a_n by n and applying limit theorems we obtain

$$\begin{aligned}\lim_{n\to\infty} \frac{2n}{5n-3} &= \lim_{n\to\infty} \frac{2}{5-(3/n)} \\ &= \frac{\lim_{n\to\infty} 2}{\lim_{n\to\infty} [5-(3/n)]} \\ &= \frac{2}{\lim_{n\to\infty} 5 - \lim_{n\to\infty} (3/n)} \\ &= \frac{2}{5-0} = \frac{2}{5}.\end{aligned}$$

Hence the sequence has the limit 2/5.

Let $\{a_n\}$ be a sequence and consider the function f, where $f(n) = a_n$. It is often true that $f(x)$ is defined for every *real* number $x \geq 1$. In this case it follows from Definitions (12.3) and (4.19) that

(12.6) $$\text{if} \quad \lim_{x\to\infty} f(x) = L, \quad \text{then} \quad \lim_{n\to\infty} f(n) = L.$$

This fact enables us to apply theorems about limits of functions (as $x \to \infty$) to limits of sequences. Of special importance is L'Hôpital's Rule (11.2), as illustrated in the next example.

Example 4 Find the limit of the sequence $\{5n/e^{2n}\}$.

Solution As indicated in the preceding discussion, we consider $f(x) = 5x/e^{2x}$ where x is real and note that f takes on the indeterminate form ∞/∞ as $x \to \infty$. Using L'Hôpital's Rule,

$$\lim_{x\to\infty} \frac{5x}{e^{2x}} = \lim_{x\to\infty} \frac{5}{2e^{2x}} = 0.$$

Hence by (12.6),

$$\lim_{n\to\infty} \frac{5n}{e^{2n}} = 0.$$

The next theorem is analogous to Theorem (2.23). It states that if the terms of an infinite sequence are always sandwiched between corresponding terms of two sequences which have the same limit L, then the given sequence also has the limit L.

(12.7) The Sandwich Theorem for Infinite Sequences

If $\{a_n\}$, $\{b_n\}$, and $\{c_n\}$ are infinite sequences such that $a_n \le b_n \le c_n$ for all n, and if

$$\lim_{n\to\infty} a_n = L = \lim_{n\to\infty} c_n$$

then $\lim_{n\to\infty} b_n = L$.

Proof. For every $\varepsilon > 0$ there corresponds a number M such that if $n > M$, then both $|a_n - L| < \varepsilon$ and $|c_n - L| < \varepsilon$; that is,

$$L - \varepsilon < a_n < L + \varepsilon \quad \text{and} \quad L - \varepsilon < c_n < L + \varepsilon.$$

Consequently if $n > M$, then

$$L - \varepsilon < a_n \quad \text{and} \quad c_n < L + \varepsilon.$$

Since $a_n \le b_n \le c_n$, it follows that if $n > M$, then $L - \varepsilon < b_n < L + \varepsilon$ or, equivalently, $|b_n - L| < \varepsilon$. This completes the proof.

Example 5 Find the limit of the sequence $\left\{\dfrac{\cos^2 n}{3^n}\right\}$.

Solution Since $0 < \cos^2 n < 1$ for every positive integer n, we may write

$$0 < \frac{\cos^2 n}{3^n} < \frac{1}{3^n}.$$

Applying Theorem (12.5) with $r = 1/3$,

$$\lim_{n\to\infty} \frac{1}{3^n} = \lim_{n\to\infty} \left(\frac{1}{3}\right)^n = 0.$$

Moreover $\lim_{n\to\infty} 0 = 0$. It follows from the Sandwich Theorem (12.7) with $a_n = 0$, $b_n = (\cos^2 n)/3^n$, and $c_n = 1/3^n$, that

$$\lim_{n\to\infty} \frac{\cos^2 n}{3^n} = 0.$$

Hence the limit of the sequence is 0.

The proof of the next theorem is left as an exercise.

(12.8) Theorem

Let $\{a_n\}$ be a sequence. If $\lim_{n\to\infty} |a_n| = 0$, then $\lim_{n\to\infty} a_n = 0$.

Example 6 Suppose the nth term of a sequence is $a_n = (-1)^{n+1}\left(\frac{1}{n}\right)$. Prove that $\lim_{n\to\infty} a_n = 0$.

Solution The terms of the sequence are alternately positive and negative. For example, the first five terms are

$$1, \quad -\frac{1}{2}, \ \frac{1}{3}, \quad -\frac{1}{4}, \ \frac{1}{5}.$$

Since

$$\lim_{n\to\infty} |a_n| = \lim_{n\to\infty} \frac{1}{n} = 0,$$

it follows from (12.8) that $\lim_{n\to\infty} a_n = 0$.

A sequence is said to be **monotonic** if successive terms are nondecreasing in the sense that

$$a_1 \leq a_2 \leq \cdots \leq a_n \leq \cdots$$

or nonincreasing in the sense that

$$a_1 \geq a_2 \geq \cdots \geq a_n \geq \cdots.$$

A sequence is **bounded** if there is a positive real number M such that $|a_k| \leq M$ for all k. The next theorem is fundamental for later developments.

(12.9) Theorem

A bounded, monotonic, infinite sequence has a limit.

In order to prove (12.9), it is necessary to use an important property of real numbers. Let us first state several definitions. If S is a nonempty set of real numbers, then a real number u is called an **upper bound** of S if $x \leq u$ for every x in S. A number v is a **least upper bound** of S if v is an upper bound and no number less than v is an upper bound of S. The least upper bound is, therefore, the smallest real number that is greater than or equal to every number in S. To illustrate, if S is the open interval (a, b), then any number greater than b is an upper bound of S; however, the least upper bound of S is unique, and equals b (see Exercise 40).

The following statement is an axiom for the real number system.

(12.10) The Completeness Property

If a nonempty set S of real numbers has an upper bound, then S has a least upper bound.

We shall now use the Completeness Property to prove Theorem (12.9). If $\{a_n\}$ is a bounded monotonic sequence with nondecreasing terms, then there is a number M such that $a_k \le M$ for every positive integer k. Thus the set

$$S = \{a_k : k = 1, 2, \ldots\}$$

has an upper bound M and, by (12.10), S has a least upper bound L, where $L \le M$. If $\varepsilon > 0$, then $L - \varepsilon$ is not an upper bound of S and hence at least one term of $\{a_n\}$ is greater than $L - \varepsilon$; that is,

$$L - \varepsilon < a_N \quad \text{for some positive integer } N.$$

Since the terms of $\{a_n\}$ are nondecreasing, we have

$$a_N \le a_{N+1} \le a_{N+2} \le \cdots$$

and, therefore,

$$L - \varepsilon < a_n \quad \text{for every } n > N.$$

It follows that if $n > N$, then

$$0 \le L - a_n < \varepsilon, \quad \text{or} \quad |L - a_n| < \varepsilon.$$

By Definition (12.3) this implies that

$$\lim_{n\to\infty} a_n = L \le M;$$

that is, $\{a_n\}$ has a limit. The proof for a sequence $\{a_n\}$ of nonincreasing terms may be obtained by a similar proof or by considering the sequence $\{-a_n\}$.

EXERCISES 12.1

Given the nth term a_n of an infinite sequence as in Exercises 1–16, find the first four terms and $\lim_{n\to\infty} a_n$, if it exists.

1 $a_n = \dfrac{n}{3n + 2}$

2 $a_n = \dfrac{6n - 5}{5n + 1}$

3 $a_n = \dfrac{7 - 4n^2}{3 + 2n^2}$

4 $a_n = \dfrac{4}{8 - 7n}$

5 $a_n = -5$

6 $a_n = \sqrt{2}$

7 $a_n = \dfrac{(2n - 1)(3n + 1)}{n^3 + 1}$

8 $a_n = 8n + 1$

9 $a_n = \dfrac{2}{\sqrt{n^2 + 9}}$

10 $a_n = \dfrac{100n}{n^{3/2} + 4}$

11 $a_n = (-1)^{n+1}\dfrac{3n}{n^2 + 4n + 5}$

12 $a_n = (-1)^{n+1}\dfrac{\sqrt{n}}{n + 1}$

13 $a_n = 1 + (0.1)^n$

14 $a_n = 1 - (1/2^n)$

15 $a_n = 1 + (-1)^{n+1}$

16 $a_n = (n + 1)/\sqrt{n}$

In each of Exercises 17–34, find the limit of the sequence, if it exists.

17 $\{6(-5/6)^n\}$

18 $\{8 - (7/8)^n\}$

19 $\{\arctan n\}$

20 $\left\{\dfrac{\tan^{-1} n}{n}\right\}$

21 $\{1000 - n\}$

22 $\{(1.0001)^n/1000\}$

23 $\left\{(-1)^n \dfrac{\ln n}{n}\right\}$

24 $\left\{\dfrac{n^2}{\ln(n+1)}\right\}$

25 $\left\{\dfrac{4n^4 + 1}{2n^2 - 1}\right\}$

26 $\left\{\dfrac{\cos n}{n}\right\}$

27 $\left\{\dfrac{e^n}{n^4}\right\}$

28 $\{e^{-n} \ln n\}$

29 $\left\{\left(1 + \dfrac{1}{n}\right)^n\right\}$

30 $\{(-1)^n n^3 3^{-n}\}$

31 $\{2^{-n} \sin n\}$

32 $\left\{\dfrac{4n^3 + 5n + 1}{2n^3 - n^2 + 5}\right\}$

33 $\left\{\dfrac{n^2}{2n - 1} - \dfrac{n^2}{2n + 1}\right\}$

34 $\left\{n \sin \dfrac{1}{n}\right\}$

35 A test question lists the first four terms of a sequence as 2, 4, 6, and 8 and asks the person being tested to list the fifth term. Show that the fifth term can be any real number a by finding the nth term of a sequence which has for its first five terms 2, 4, 6, 8, and a.

36 Prove, directly from Definition (12.3), that if $\lim_{n\to\infty} a_n = L$, then $\lim_{n\to\infty} |a_n| = |L|$.

37 If k is an integer greater than 1, prove that $\lim_{n\to\infty} 1/k^n = 0$. (*Hint*: Use Theorem (12.7).)

38 Prove Theorem (12.8).

39 Prove Theorem (12.9) if the terms of $\{a_n\}$ are nonincreasing.

40 Prove that b is the least upper bound of the open interval (a, b).

12.2 CONVERGENT OR DIVERGENT INFINITE SERIES

If $\{a_n\}$ is an infinite sequence, then an expression of the form

$$a_1 + a_2 + \cdots + a_n + \cdots$$

is called an **infinite series**, or simply a **series**. In summation notation, this series will be denoted by either

$$\sum_{n=1}^{\infty} a_n \quad \text{or} \quad \sum a_n$$

where it is understood that the summation variable in the last sum is n. Each number a_i is called a **term** of the series and a_n is called the ***n*th term**. Since only finite sums may be added algebraically, it is necessary to *define* what is meant by an infinite sum. To arrive at a definition we begin by considering, for each positive integer n, the **nth partial sum S_n** of the series, where

$$S_n = a_1 + a_2 + \cdots + a_n.$$

Thus

$$S_1 = a_1$$
$$S_2 = a_1 + a_2$$
$$S_3 = a_1 + a_2 + a_3$$
$$S_4 = a_1 + a_2 + a_3 + a_4$$

and so on. The infinite sequence

$$S_1, S_2, \ldots, S_n, \ldots$$

is called the **sequence of partial sums** associated with the infinite series $\sum a_n$. To illustrate, consider the series

$$0.6 + 0.06 + 0.006 + \cdots + \frac{6}{10^n} + \cdots.$$

The first few terms of the sequence of partial sums are

$$0.6, \quad 0.66, \quad 0.666, \quad 0.6666, \quad 0.66666, \quad \ldots.$$

Later in this section we shall show that this sequence has the limit 2/3. Since the partial sums S_n of the series approach 2/3 as n increases, it is natural to refer to 2/3 as the *sum* of the infinite series and write

$$\frac{2}{3} = 0.6 + 0.06 + 0.006 + \cdots + \frac{6}{10^n} + \cdots.$$

Indeed, this is the basis for expressing the rational number 2/3 as the infinite repeating decimal 0.6666.... The following definition extends these remarks to any infinite series.

(12.11) Definition

An infinite series

$$a_1 + a_2 + \cdots + a_n + \cdots$$

with sequence of partial sums $S_1, S_2, \ldots, S_n, \ldots$ **is convergent** (or **converges**) if $\lim_{n\to\infty} S_n = S$ for some real number S. The series **is divergent** (or **diverges**) if this limit does not exist.

If $a_1 + a_2 + \cdots + a_n + \cdots$ is a convergent infinite series and $\lim_{n\to\infty} S_n = S$, then S is called the **sum of the series** and we write

$$S = a_1 + a_2 + \cdots + a_n + \cdots.$$

If a series diverges, it has no sum.

Example 1 Prove that the infinite series

$$\frac{1}{1\cdot 2}+\frac{1}{2\cdot 3}+\frac{1}{3\cdot 4}+\cdots+\frac{1}{n(n+1)}+\cdots$$

converges and find its sum.

Solution The partial fraction decomposition of a_n is

$$a_n=\frac{1}{n(n+1)}=\frac{1}{n}-\frac{1}{n+1}.$$

Consequently, the nth partial sum of the series may be written

$$\begin{aligned}S_n &= a_1+a_2+a_3+\cdots+a_n\\ &= \left(1-\frac{1}{2}\right)+\left(\frac{1}{2}-\frac{1}{3}\right)+\left(\frac{1}{3}-\frac{1}{4}\right)+\cdots+\left(\frac{1}{n}-\frac{1}{n+1}\right)\\ &= 1-\frac{1}{n+1}=\frac{n}{n+1}.\end{aligned}$$

Since

$$\lim_{n\to\infty} S_n=\lim_{n\to\infty}\frac{n}{n+1}=1$$

the series converges and has the sum 1.

Certain types of infinite series arise frequently in applications. One example is a **geometric series**

$$a+ar+ar^2+\cdots+ar^{n-1}+\cdots$$

where a and r are real numbers and $a\neq 0$. The next result is extremely important for work later in this chapter.

(12.12) Theorem

> The geometric series
>
> $$a+ar+ar^2+\cdots+ar^{n-1}+\cdots$$
>
> with $a\neq 0$
>
> (i) *converges* and has the sum $\dfrac{a}{1-r}$ if $|r|<1$.
>
> (ii) *diverges* if $|r|\geq 1$.

Proof. If $r=1$, then $S_n=a+a+\cdots+a=na$ and the series diverges since $\lim_{n\to\infty} S_n$ does not exist.

If $r \neq 1$ consider

$$S_n = a + ar + ar^2 + \cdots + ar^{n-1}$$

and

$$rS_n = ar + ar^2 + ar^3 + \cdots + ar^n.$$

Subtracting corresponding sides of these equations we obtain

$$(1 - r)S_n = a - ar^n.$$

Dividing both sides by $1 - r$ gives us

$$S_n = \frac{a}{1 - r} - \frac{ar^n}{1 - r}.$$

Consequently

$$\begin{aligned}\lim_{n \to \infty} S_n &= \lim_{n \to \infty} \left(\frac{a}{1 - r} - \frac{ar^n}{1 - r} \right) \\ &= \lim_{n \to \infty} \frac{a}{1 - r} - \lim_{n \to \infty} \frac{ar^n}{1 - r} \\ &= \frac{a}{1 - r} - \frac{a}{1 - r} \lim_{n \to \infty} r^n.\end{aligned}$$

If $|r| < 1$, then $\lim_{n \to \infty} r^n = 0$ (see Theorem (12.5)) and hence

$$\lim_{n \to \infty} S_n = \frac{a}{1 - r}.$$

If $|r| > 1$, then $\lim_{n \to \infty} r^n$ does not exist (see Theorem (12.5)) and hence $\lim_{n \to \infty} S_n$ does not exist. In this case the series diverges.

The series

$$0.6 + 0.06 + 0.006 + \cdots + \frac{6}{10^n} + \cdots$$

considered earlier is geometric with $a = 0.6$ and $r = 0.1$. By Theorem (12.12) it converges and has the sum

$$S = \frac{0.6}{1 - 0.1} = \frac{0.6}{0.9} = \frac{2}{3}.$$

Example 2 Prove that the infinite series

$$2 + \frac{2}{3} + \frac{2}{3^2} + \cdots + \frac{2}{3^{n-1}} + \cdots$$

converges and find its sum.

Solution The series converges since it is geometric with $r = 1/3 < 1$. By (i) of Theorem (12.12), the sum is

$$S = \frac{2}{1 - (1/3)} = \frac{2}{(2/3)} = 3.$$

The nth term a_n of an infinite series can be expressed as

$$a_n = S_n - S_{n-1}.$$

If $\lim_{n\to\infty} S_n = S$, then also $\lim_{n\to\infty} S_{n-1} = S$ and

$$\lim_{n\to\infty} a_n = \lim_{n\to\infty} (S_n - S_{n-1}) = \lim_{n\to\infty} S_n - \lim_{n\to\infty} S_{n-1} = 0.$$

This gives us the following theorem.

(12.13) Theorem

> If an infinite series $\Sigma\, a_n$ is convergent, then $\lim_{n\to\infty} a_n = 0$.

The preceding theorem states that *if* a series converges, *then* its nth term has the limit 0 as $n \to \infty$. The converse is false, that is, if $\lim_{n\to\infty} a_n = 0$ *it does not necessarily follow* that the series $\Sigma\, a_n$ is convergent. (See Example 4 of this section.)

The next result is an immediate corollary of Theorem (12.13).

(12.14) Test for Divergence

> If $\lim_{n\to\infty} a_n \neq 0$, then the infinite series $\Sigma\, a_n$ is divergent.

Example 3 Determine whether the series

$$\frac{1}{3} + \frac{2}{5} + \frac{3}{7} + \cdots + \frac{n}{2n+1} + \cdots$$

converges or diverges.

Solution Since

$$\lim_{n\to\infty} a_n = \lim_{n\to\infty} \frac{n}{2n+1} = \frac{1}{2} \neq 0,$$

the series diverges by (12.14).

(12.15) Theorem

> If an infinite series $\Sigma\, a_n$ is convergent, then for every $\varepsilon > 0$ there exists an integer N such that $|S_k - S_l| < \varepsilon$ whenever $k, l > N$.

Proof. If the sum of the series is S, then

$$|S_k - S_l| = |(S_k - S) + (S - S_l)| \leq |S_k - S| + |S - S_l|.$$

Since $\lim_{k \to \infty} S_k = S = \lim_{l \to \infty} S_l$, for every $\varepsilon > 0$ there is an integer N such that

$$|S_k - S| < \frac{\varepsilon}{2} \quad \text{and} \quad |S - S_l| < \frac{\varepsilon}{2}$$

whenever $k, l > N$. The conclusion of the theorem is now immediate.

Example 4 Prove that the series $1 + \frac{1}{2} + \frac{1}{3} + \cdots + \frac{1}{n} + \cdots$ is divergent.

Solution If $n > 1$, then

$$\begin{aligned} S_{2n} - S_n &= \frac{1}{n+1} + \frac{1}{n+2} + \cdots + \frac{1}{2n} \\ &> \frac{1}{2n} + \frac{1}{2n} + \cdots + \frac{1}{2n} = \frac{1}{2}. \end{aligned}$$

If the given series is convergent, then by Theorem (12.15) with $\varepsilon = 1/2$, $k = 2n$, and $l = n$, $|S_{2n} - S_n| < 1/2$ provided n is sufficiently large. Since the latter inequality is *never* true, it follows that the given series diverges.

The infinite series in Example 4 is called the **harmonic series** and is an illustration of a divergent series $\Sigma\, a_n$ for which $\lim_{n \to \infty} a_n = 0$. This shows that the converse of (12.13) is false. Consequently, to establish convergence of an infinite series *it is not enough* to prove that $\lim_{n \to \infty} a_n = 0$, since that may be true for divergent as well as convergent series.

The next theorem states that if corresponding terms of two infinite series are identical after a certain term, then both series converge or both series diverge.

(12.16) Theorem

> If $\Sigma\, a_n$ and $\Sigma\, b_n$ are infinite series such that $a_i = b_i$ for all $i > k$, where k is a positive integer, then both series converge or both series diverge.

Proof. By hypothesis we may write

$$\sum a_n = a_1 + a_2 + \cdots + a_k + a_{k+1} + \cdots + a_n + \cdots$$

and

$$\sum b_n = b_1 + b_2 + \cdots + b_k + a_{k+1} + \cdots + a_n + \cdots.$$

Let S_n and T_n denote the nth partial sums of $\Sigma\, a_n$ and $\Sigma\, b_n$, respectively. It follows that if $n \geq k$, then

$$S_n - S_k = T_n - T_k$$

or

$$S_n = T_n + (S_k - T_k).$$

Consequently

$$\lim_{n \to \infty} S_n = \lim_{n \to \infty} T_n + (S_k - T_k)$$

and hence either both of the limits exist or both do not exist. This gives us the desired conclusion. Evidently, if both series converge, then their sums differ by $S_k - T_k$.

Theorem (12.16) implies that changing a finite number of terms of an infinite series has no effect on its convergence or divergence (although it does change the sum of a convergent series). In particular, if we replace the first k terms of $\Sigma\, a_n$ by 0, convergence is unaffected. It follows that the series

$$a_{k+1} + a_{k+2} + \cdots + a_n + \cdots$$

converges or diverges according as $\Sigma\, a_n$ converges or diverges. The series $a_{k+1} + a_{k+2} + \cdots$ is said to have been obtained from $\Sigma\, a_n$ by **deleting the first k terms**.

Example 5 Show that the series

$$\frac{1}{3 \cdot 4} + \frac{1}{4 \cdot 5} + \cdots + \frac{1}{(n+2)(n+3)} + \cdots$$

converges.

Solution The series converges since it can be obtained by deleting the first two terms of the convergent series in Example 1.

The proof of the next theorem follows directly from Definition (12.11) and is left as an exercise.

(12.17) Theorem

If $\Sigma\, a_n$ and $\Sigma\, b_n$ are convergent series with sums A and B respectively, then

(i) $\Sigma\,(a_n + b_n)$ converges and has sum $A + B$.

(ii) if c is a real number, $\Sigma\, ca_n$ converges and has sum cA.

(iii) $\Sigma\,(a_n - b_n)$ converges and has sum $A - B$.

It is also easy to show that if $\Sigma\, a_n$ diverges, then so does $\Sigma\, ca_n$ for every $c \neq 0$.

Example 6 Prove that the series

$$\sum_{n=1}^{\infty} \left[\frac{7}{n(n+1)} + \frac{2}{3^{n-1}}\right]$$

converges and find its sum.

Solution The series $\Sigma\, 1/[n(n+1)]$ was considered in Example 1, where it was shown that it converges and has the sum 1. Using (ii) of Theorem (12.17) with $c = 7$ and $a_n = 1/[n(n+1)]$, we see that $\Sigma\, 7/[n(n+1)]$ converges and has the sum $7(1) = 7$.

The geometric series $\Sigma\, 2/3^{n-1}$ converges and has the sum 3 (see Example 2). Hence by (i) of (12.17), the given series converges and has the sum $7 + 3$, or 10.

(12.18) Theorem

> If $\Sigma\, a_n$ is a convergent series and $\Sigma\, b_n$ is divergent, then $\Sigma(a_n + b_n)$ is divergent.

Proof. We shall give an indirect proof. Thus suppose $\Sigma\,(a_n + b_n)$ is convergent. Applying (iii) of Theorem (12.17), $\Sigma\,[(a_n + b_n) - a_n] = \Sigma\, b_n$ is convergent, a contradiction. Hence our supposition is false, that is, $\Sigma\,(a_n + b_n)$ is divergent.

Example 7 Determine the convergence or divergence of

$$\sum_{n=1}^{\infty}\left(\frac{1}{5^n} + \frac{1}{n}\right).$$

Solution Since $\Sigma\,(1/5^n)$ is a convergent geometric series and $\Sigma\,(1/n)$ is the divergent harmonic series, the given series diverges by Theorem (12.18).

EXERCISES 12.2

In each of Exercises 1–10 determine whether the geometric series converges or diverges, and if it converges find its sum.

1 $3 + \dfrac{3}{4} + \cdots + \dfrac{3}{4^{n-1}} + \cdots$

2 $3 + \dfrac{3}{(-4)} + \cdots + \dfrac{3}{(-4)^{n-1}} + \cdots$

3 $1 + \dfrac{(-1)}{\sqrt{5}} + \cdots + \left(\dfrac{-1}{\sqrt{5}}\right)^{n-1} + \cdots$

4 $1 + \left(\dfrac{e}{3}\right) + \cdots + \left(\dfrac{e}{3}\right)^{n-1} + \cdots$

5 $0.37 + 0.0037 + \cdots + \dfrac{37}{(100)^n} + \cdots$

6 $0.628 + 0.000628 + \cdots + \dfrac{628}{(1000)^n} + \cdots$

7 $\displaystyle\sum_{n=1}^{\infty} 2^{-n}3^{n-1}$

8 $\displaystyle\sum_{n=1}^{\infty} (-5)^{n-1}4^{-n}$

9 $\displaystyle\sum_{n=1}^{\infty} (-1)^{n-1}$

10 $\Sigma(\sqrt{2})^{n-1}$

Use the examples and theorems discussed in this section to determine whether the series in Exercises 11–28 converge or diverge.

11 $\dfrac{1}{4\cdot 5} + \dfrac{1}{5\cdot 6} + \cdots + \dfrac{1}{(n+3)(n+4)} + \cdots$

12 $\dfrac{-1}{1\cdot 2} + \dfrac{-1}{2\cdot 3} + \cdots + \dfrac{-1}{n(n+1)} + \cdots$

13 $\dfrac{5}{1\cdot 2}+\dfrac{5}{2\cdot 3}+\cdots+\dfrac{5}{n(n+1)}+\cdots$

14 $\dfrac{1}{4}+\dfrac{1}{5}+\cdots+\dfrac{1}{n+3}+\cdots$

15 $3+\dfrac{3}{2}+\cdots+\dfrac{3}{n}+\cdots$

16 $\dfrac{1}{2}+\dfrac{2}{3}+\cdots+\dfrac{n}{n+1}+\cdots$

17 $\sum_{n=1}^{\infty}\dfrac{3n}{5n-1}$

18 $\sum_{n=1}^{\infty}\dfrac{1}{1+(0.3)^n}$

19 $\sum_{n=1}^{\infty}\dfrac{1}{\sqrt[n]{e}}$

20 $\sum_{n=1}^{\infty}\dfrac{n}{\ln(n+1)}$

21 $\sum_{n=1}^{\infty}\left(\dfrac{1}{8^n}+\dfrac{1}{n(n+1)}\right)$

22 $\sum\left(\dfrac{1}{3^n}-\dfrac{1}{4^n}\right)$

23 $\sum_{n=1}^{\infty}\left(\dfrac{5}{n+2}-\dfrac{5}{n+3}\right)$

24 $\sum_{n=1}^{\infty}\ln\left(\dfrac{2n}{7n-5}\right)$

25 $\sum_{n=1}^{\infty}\left[\left(\dfrac{3}{2}\right)^n+\left(\dfrac{2}{3}\right)^n\right]$

26 $\sum_{n=1}^{\infty}\left[\dfrac{1}{n(n+1)}-\dfrac{4}{n}\right]$

27 $\sum_{n=1}^{\infty} n\sin\dfrac{1}{n}$

28 $\sum_{n=1}^{\infty}(2^{-n}-2^{-3n})$

Use the method illustrated in Example 1 to prove that the series in Exercises 29 and 30 converge and find their sums.

29 $\sum_{n=1}^{\infty}\dfrac{1}{4n^2-1}$

30 $\sum\dfrac{-1}{9n^2+3n-2}$

31 Prove that if $\Sigma\, a_n$ diverges, then so does $\Sigma\, ca_n$ for every $c \neq 0$.

32 Prove Theorem (12.17).

33 Prove or disprove: If $\Sigma\, a_n$ and $\Sigma\, b_n$ both diverge, then $\Sigma\,(a_n + b_n)$ diverges.

34 What is wrong with the following "proof" that the divergent geometric series $\Sigma_{n=1}^{\infty}(-1)^{n+1}$ has the sum 0?

$$\begin{aligned}\sum_{n=1}^{\infty}(-1)^{n+1} &= [1+(-1)]+[1+(-1)]+[1+(-1)]+\cdots\\ &= 0+0+0+\cdots = 0\end{aligned}$$

In Exercises 35–38, the bar indicates that the digits underneath it repeat indefinitely. Express each repeating decimal as an infinite series and use Theorem (12.12) to find the rational number it represents.

35 $0.\overline{23}$

36 $5.\overline{146}$

37 $3.2\overline{394}$

38 $2.7\overline{1828}$

39 A rubber ball is dropped from a height of 10 meters. If it rebounds approximately one-half the distance after each fall, use an infinite geometric series to approximate the total distance the ball travels before coming to rest.

40 The bob of a pendulum swings through an arc 24 cm long on its first swing. If each successive swing is approximately five-sixths the length of the preceding swing, use an infinite geometric series to approximate the total distance it travels before coming to rest.

12.3 POSITIVE TERM SERIES

It is difficult to apply the definition of convergence or divergence (12.11) to an infinite series $\Sigma\, a_n$ since in most cases a simple formula for S_n cannot be found. We can, however, develop techniques for using the nth term a_n to test a series for convergence or divergence. When applying these tests we shall be concerned not with the *sum* of the series but with whether the series converges or diverges.

In this section we shall consider only **positive term series**, that is, infinite series for which every term is positive. If $\{S_n\}$ is the sequence of partial sums of a positive term series, then

$$S_1 < S_2 < \cdots < S_n < \cdots$$

and therefore $\{S_n\}$ is monotonic. If there exists a number M such that $S_n < M$ for every n, then as in the proof of Theorem (12.9),

$$\lim_{n\to\infty} S_n = S \leq M$$

for some S, and hence the series converges. If no such M exists, then $\lim_{n\to\infty} S_n = \infty$ and the series diverges. This proves the next theorem.

(12.19) Theorem

If $\Sigma\, a_n$ is a positive term series and if there exists a number M such that $S_n < M$ for every n, then the series converges and has a sum $S \leq M$. If no such M exists the series diverges.

Suppose a function f is defined for every real number $x \geq 1$. We may then consider the infinite series

$$\sum_{n=1}^{\infty} f(n) = f(1) + f(2) + \cdots + f(n) + \cdots.$$

For example, if $f(x) = 1/x^2$, then

$$\sum_{n=1}^{\infty} f(n) = \sum_{n=1}^{\infty} \frac{1}{n^2} = \frac{1}{1^2} + \frac{1}{2^2} + \cdots + \frac{1}{n^2} + \cdots.$$

A series of this type may be tested for convergence or divergence by means of an improper integral, as indicated in the following result.

(12.20) The Integral Test

If a function f is positive valued, continuous, and decreasing for $x \geq 1$, then the infinite series

$$f(1) + f(2) + \cdots + f(n) + \cdots$$

(i) converges if $\int_1^{\infty} f(x)\,dx$ converges.

(ii) diverges if $\int_1^{\infty} f(x)\,dx$ diverges.

Proof. If n is a positive integer greater than 1, then the area of the inscribed rectangular polygon illustrated in (i) of Figure 12.3 is given by

$$\sum_{k=2}^{n} f(k) = f(2) + f(3) + \cdots + f(n).$$

Similarly, the area of the circumscribed rectangular polygon illustrated in (ii) of the figure is

$$\sum_{k=1}^{n-1} f(k) = f(1) + f(2) + \cdots + f(n-1).$$

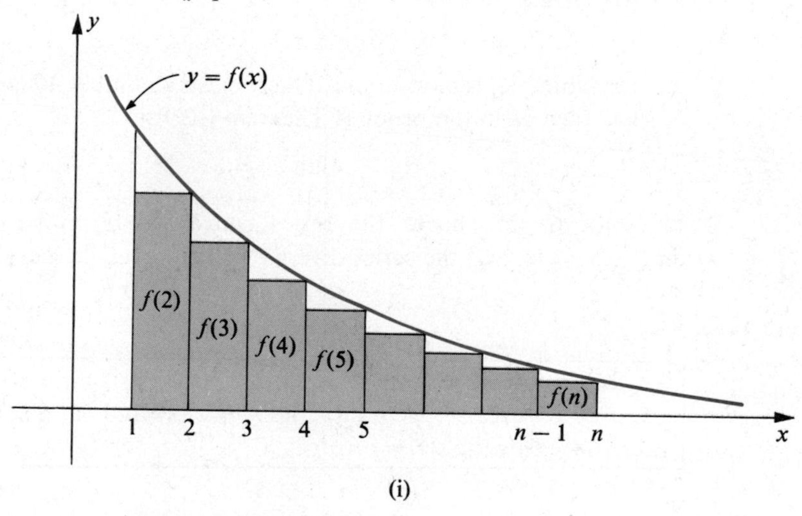

(i)

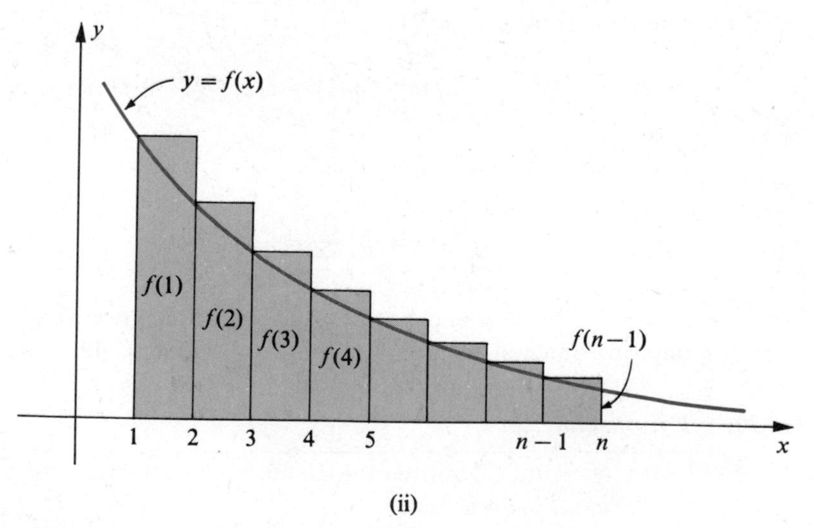

(ii)

Figure 12.3

Since $\int_1^n f(x)\,dx$ is the area under the graph of f from 1 to n,

$$\sum_{k=2}^{n} f(k) \leq \int_1^n f(x)\,dx \leq \sum_{k=1}^{n-1} f(k).$$

If S_n denotes the nth partial sum of the series $f(1) + f(2) + \cdots + f(n) + \cdots$, then this inequality may be written

$$S_n - f(1) \leq \int_1^n f(x)\,dx \leq S_{n-1}.$$

The preceding inequality implies that if the integral $\int_1^\infty f(x)\,dx$ converges and equals $K > 0$, then

$$S_n - f(1) \leq K, \quad \text{or} \quad S_n \leq K + f(1)$$

for every positive integer n. Hence, by Theorem (12.19), the series $\Sigma f(n)$ converges. If the improper integral diverges, then

$$\lim_{n \to \infty} \int_1^n f(x)\,dx = \infty$$

and since $\int_1^n f(x)\,dx \leq S_{n-1}$ we also have $\lim_{n\to\infty} S_{n-1} = \infty$; that is, the series $\Sigma f(n)$ diverges.

Example 1 Use the Integral Test to prove that the harmonic series

$$1 + \frac{1}{2} + \frac{1}{3} + \cdots + \frac{1}{n} + \cdots$$

diverges.

Solution If we let $f(x) = 1/x$, then f is positive-valued, continuous, and decreasing for $x \geq 1$ and, therefore, the Integral Test may be applied. Since

$$\begin{aligned} \int_1^\infty \frac{1}{x}\,dx &= \lim_{t\to\infty} \int_1^t \frac{1}{x}\,dx \\ &= \lim_{t\to\infty} \Big[\ln x\Big]_1^t \\ &= \lim_{t\to\infty} [\ln t - \ln 1] = \infty \end{aligned}$$

the series diverges by (ii) of (12.20).

Example 2 Determine whether the infinite series $\Sigma\, ne^{-n^2}$ converges or diverges.

Solution If we let $f(x) = xe^{-x^2}$, then the given series is the same as $\Sigma\, f(n)$. If $x \geq 1$, f is positive valued and continuous. The first derivative may be used to determine whether f is decreasing. Since

$$f'(x) = e^{-x^2} - 2x^2e^{-x^2} = e^{-x^2}(1 - 2x^2) < 0$$

f is decreasing on $[1, \infty)$. We may, therefore, apply the Integral Test as follows.

$$\int_1^{\infty} xe^{-x^2}\,dx = \lim_{t\to\infty}\int_1^t xe^{-x^2}\,dx$$
$$= \lim_{t\to\infty}\left(-\frac{1}{2}\right)e^{-x^2}\Big]_1^t$$
$$= \left(-\frac{1}{2}\right)\lim_{t\to\infty}\left[\frac{1}{e^{t^2}} - \frac{1}{e}\right] = \frac{1}{2e}$$

Hence the series converges by (i) of (12.20).

An integral test may also be used if the function f satisfies the conditions of (12.20) for all $x \geq m$, where m is any positive integer. In this case we merely replace the integral in (12.20) by $\int_m^{\infty} f(x)\,dx$. This corresponds to deleting the first $m - 1$ terms of the series.

If $f(x) = 1/x^p$ where $p > 0$, then the series $\Sigma f(n)$ has the form

$$1 + \frac{1}{2^p} + \frac{1}{3^p} + \cdots + \frac{1}{n^p} + \cdots$$

and is called the ***p*-series**. This series will be very useful when we apply comparison tests later in this section. The following theorem provides information about convergence or divergence.

(12.21) Theorem

> The p-series $\sum_{n=1}^{\infty} \frac{1}{n^p}$
>
> (i) converges if $p > 1$.
>
> (ii) diverges if $p \leq 1$.

Proof. First we note that the special case $p = 1$ is the divergent harmonic series. Next suppose that p is any positive real number different from 1. If we let $f(x) = 1/x^p = x^{-p}$, then f is positive-valued and continuous for $x \geq 1$. Moreover, for these values of x we see that $f'(x) = -px^{-p-1} < 0$, and hence f is decreasing. Thus f satisfies the conditions stated in the Integral Test (12.20) and we consider

$$\int_1^{\infty} \frac{1}{x^p}\,dx = \lim_{t\to\infty}\int_1^t x^{-p}\,dx$$
$$= \lim_{t\to\infty} \frac{x^{1-p}}{1-p}\Big]_1^t$$
$$= \frac{1}{1-p}\lim_{t\to\infty}\left[t^{1-p} - 1\right].$$

If $p > 1$, then $p - 1 > 0$ and the last expression may be written

$$\frac{1}{1-p}\lim_{t\to\infty}\left[\frac{1}{t^{p-1}} - 1\right] = \frac{1}{1-p}(0 - 1) = \frac{1}{p-1}.$$

Thus by (i) of (12.20), the p-series converges if $p > 1$.

If $0 < p < 1$, then $1 - p > 0$ and

$$\frac{1}{1-p} \lim_{t \to \infty} [t^{1-p} - 1] = \infty.$$

Hence by (ii) of (12.20), the p-series diverges.

If $p \leq 0$, then $\lim_{n \to \infty} (1/n^p) \neq 0$, and by (12.14) the series diverges.

Example 3 Determine whether the following series converge or diverge.

(a) $1 + \dfrac{1}{2^2} + \dfrac{1}{3^2} + \cdots + \dfrac{1}{n^2} + \cdots$

(b) $5 + \dfrac{5}{\sqrt{2}} + \dfrac{5}{\sqrt{3}} + \cdots + \dfrac{5}{\sqrt{n}} + \cdots$

Solution (a) The series $\Sigma\, 1/n^2$ converges since it is the p-series with $p = 2 > 1$.

(b) The series $\Sigma\, 1/\sqrt{n}$ diverges since it is the p-series with $p = 1/2 < 1$. Consequently $\Sigma\, 5/\sqrt{n}$ diverges. (See Exercise 31 of Section 2.)

The next theorem allows us to use known convergent (divergent) series to establish the convergence (divergence) of other series.

(12.22) Comparison Test

> Suppose $\Sigma\, a_n$ and $\Sigma\, b_n$ are positive term series.
>
> (i) If $\Sigma\, b_n$ converges and $a_n \leq b_n$ for every positive integer n, then $\Sigma\, a_n$ converges.
>
> (ii) If $\Sigma\, b_n$ diverges and $a_n \geq b_n$ for every positive integer n, then $\Sigma\, a_n$ diverges.

Proof. Let S_n and T_n denote the nth partial sums of $\Sigma\, a_n$ and $\Sigma\, b_n$, respectively. Suppose $\Sigma\, b_n$ converges and has the sum T. If $a_n \leq b_n$ for every n, then $S_n \leq T_n < T$ and hence by Theorem (12.19) $\Sigma\, a_n$ converges. This proves part (i).

To prove (ii), suppose $\Sigma\, b_n$ diverges and $a_n \geq b_n$ for every n. Then $S_n \geq T_n$ and, since T_n increases without bound as n becomes infinite, so does S_n. Consequently $\Sigma\, a_n$ diverges.

Since convergence or divergence of a series is not affected by deleting a finite number of terms, the conditions $a_n \geq b_n$ or $a_n \leq b_n$ of (12.22) are only required from the kth term on, where k is some fixed positive integer.

A series $\Sigma\, d_n$ is said to **dominate** a series $\Sigma\, c_n$ if $0 < c_n \leq d_n$ for every positive integer n. Using this terminology, part (i) of (12.22) states that a positive term series which is dominated by a convergent series is also convergent. Part (ii) states that a series which dominates a divergent series also diverges.

Example 4 Determine whether the following series converge or diverge.

(a) $\sum_{n=1}^{\infty} \frac{1}{2 + 5^n}$ (b) $\sum_{n=2}^{\infty} \frac{3}{\sqrt{n} - 1}$

Solution (a) For every $n \geq 1$,

$$\frac{1}{2 + 5^n} < \frac{1}{5^n} = \left(\frac{1}{5}\right)^n.$$

Since $\Sigma(1/5)^n$ is a convergent geometric series, the given series converges, by (i) of (12.22).

(b) The p-series $\Sigma\, 1/\sqrt{n}$ diverges and hence so does the series obtained by discarding the first term $1/\sqrt{1}$. Since

$$\frac{3}{\sqrt{n} - 1} > \frac{1}{\sqrt{n}} \quad \text{if } n \geq 2,$$

it follows from (ii) of (12.22) that the given series diverges.

(12.23) Limit Comparison Test

> If $\Sigma\, a_n$ and $\Sigma\, b_n$ are positive term series and if
>
> $$\lim_{n \to \infty} \frac{a_n}{b_n} = k > 0$$
>
> then either both series converge or both diverge.

Proof. If $\lim_{n \to \infty} (a_n/b_n) = k > 0$, then there exists a number N such that

$$\frac{k}{2} < \frac{a_n}{b_n} < \frac{3k}{2} \quad \text{whenever } n > N.$$

This is equivalent to

$$\frac{k}{2} b_n < a_n < \frac{3k}{2} b_n \quad \text{whenever } n > N.$$

If $\Sigma\, a_n$ converges, then $\Sigma\, (k/2)b_n$ also converges, since it is dominated by $\Sigma\, a_n$. Applying (ii) of (12.17),

$$\sum b_n = \sum \left(\frac{2}{k}\right)\left(\frac{k}{2}\right) b_n$$

converges. Conversely, if $\Sigma\, b_n$ converges, then so does $\Sigma\, a_n$, since it is dominated by the convergent series $\Sigma\,(3k/2)b_n$. We have proved that $\Sigma\, a_n$ converges if and only if $\Sigma\, b_n$ converges. Consequently $\Sigma\, a_n$ diverges if and only if $\Sigma\, b_n$ diverges.

Two other comparison tests are stated in Exercises 49 and 50.

Example 5 Determine whether the following series converge or diverge.

(a) $\displaystyle\sum_{n=1}^{\infty} \frac{1}{\sqrt[3]{n^2+1}}$ (b) $\displaystyle\sum_{n=1}^{\infty} \frac{3n^2+5n}{2^n(n^2+1)}$

Solution (a) We shall apply (12.23) with

$$a_n = \frac{1}{\sqrt[3]{n^2+1}} \quad \text{and} \quad b_n = \frac{1}{\sqrt[3]{n^2}} = \frac{1}{n^{2/3}}.$$

The series $\Sigma\, b_n$ is the p-series with $p = 2/3 < 1$ and, therefore, diverges. Since

$$\lim_{n\to\infty} \frac{a_n}{b_n} = \lim_{n\to\infty} \frac{\sqrt[3]{n^2}}{\sqrt[3]{n^2+1}} = \lim_{n\to\infty} \sqrt[3]{\frac{n^2}{n^2+1}} = 1 > 0,$$

$\Sigma\, a_n$ also diverges.

(b) If we let

$$a_n = \frac{3n^2+5n}{2^n(n^2+1)} \quad \text{and} \quad b_n = \frac{1}{2^n}$$

then

$$\lim_{n\to\infty} \frac{a_n}{b_n} = \lim_{n\to\infty} \frac{3n^2+5n}{n^2+1} = 3 > 0.$$

Since $\Sigma\, b_n$ is a convergent geometric series, the given series $\Sigma\, a_n$ is also convergent by (12.23).

When seeking a suitable series $\Sigma\, b_n$ to be used for comparison with $\Sigma\, a_n$, where a_n is a fraction, a good procedure is to discard all terms in the numerator and denominator except those which have the most effect on the magnitude. For example, if

$$a_n = \frac{3n^2+\sqrt{n}}{5+n^{-1}+n^{7/2}}$$

and we discard all but the highest powers of n in the numerator and denominator, there results

$$\frac{3n^2}{n^{7/2}} = \frac{3}{n^{3/2}}.$$

This suggests that we choose $b_n = 1/n^{3/2}$. In this event

$$\lim_{n\to\infty} \frac{a_n}{b_n} = \lim_{n\to\infty} \frac{3n^{7/2} + n^2}{5 + n^{-1} + n^{7/2}} = 3.$$

Since $\Sigma\, b_n$ is a convergent p-series with $p = 3/2 > 1$, it follows from (12.23) that $\Sigma\, a_n$ is also convergent.

If we employ this technique in part (b) of Example 5, where

$$a_n = \frac{3n^2 + 5n}{2^n n^2 + 2^n}$$

we are led to

$$\frac{3n^2}{2^n n^2} = \frac{3}{2^n}.$$

This suggests taking $b_n = 1/2^n$, as was done in the solution.

We shall conclude this section with several general remarks about positive term series. Suppose $\Sigma\, a_n$ is a positive term series and the terms are grouped in some manner. For example, we could have

$$(a_1 + a_2) + a_3 + (a_4 + a_5 + a_6 + a_7) + \cdots.$$

If we denote the last series by $\Sigma\, b_n$, so that

$$b_1 = a_1 + a_2, \quad b_2 = a_3, \quad b_3 = a_4 + a_5 + a_6 + a_7, \quad \ldots$$

then any partial sum of $\Sigma\, b_n$ is also a partial sum of $\Sigma\, a_n$. It follows that if $\Sigma\, a_n$ converges, then $\Sigma\, b_n$ converges and has the same sum. A similar argument may be used for any grouping of the terms of $\Sigma\, a_n$. Thus, *if a positive term series converges, then the series obtained by grouping the terms in any manner also converges and has the same sum.*

This result is not necessarily true for a series which contains both positive and negative terms. Also, we cannot make a similar statement about arbitrary divergent series. For example, the terms of the divergent series $\Sigma\,(-1)^n$ may be grouped to produce a convergent series (see Exercise 34 of Section 12.2).

Next, suppose that a convergent positive term series $\Sigma\, a_n$ has the sum S and that a new series $\Sigma\, b_n$ is formed by rearranging the terms in some way. For example, $\Sigma\, b_n$ could be the series

$$a_2 + a_8 + a_1 + a_5 + a_7 + a_3 + \cdots.$$

If T_n is the nth partial sum of $\Sigma\, b_n$, then it is a sum of terms of $\Sigma\, a_n$. If m is the largest subscript associated with the terms a_i in T_n, then $T_n \le S_m < S$. Consequently $T_n < S$ for every n. Applying Theorem (12.19), $\Sigma\, b_n$ converges and has a sum $T \le S$. The preceding proof is independent of the particular rearrangement of terms. We may also regard the series $\Sigma\, a_n$ as having been obtained by rearranging the terms of $\Sigma\, b_n$ and hence, by the same argument, $S \le T$. We have proved that *if the terms of a convergent positive term series $\Sigma\, a_n$ are rearranged in any manner, then the resulting series converges and has the same sum.*

EXERCISES 12.3

Use the Integral Test to determine whether the series in Exercises 1–16 converge or diverge.

1 $\sum_{n=1}^{\infty} \frac{1}{(3+2n)^2}$

2 $\sum_{n=1}^{\infty} \frac{1}{(4+n)^{3/2}}$

3 $\sum_{n=1}^{\infty} \frac{1}{4n+7}$

4 $\sum_{n=2}^{\infty} \frac{1}{n(\ln n)^2}$

5 $\sum_{n=1}^{\infty} \frac{\ln n}{n}$

6 $\sum_{n=1}^{\infty} \frac{n}{n^2+1}$

7 $\sum_{n=1}^{\infty} \frac{1}{\sqrt[3]{2n+1}}$

8 $\sum_{n=1}^{\infty} \frac{1}{1+16n^2}$

9 $\sum_{n=1}^{\infty} \frac{\arctan n}{1+n^2}$

10 $\sum_{n=1}^{\infty} ne^{-n}$

11 $\sum_{n=3}^{\infty} \frac{1}{n(2n-5)}$

12 $\sum_{n=1}^{\infty} \frac{1}{n(n+1)(n+2)}$

13 $\sum_{n=1}^{\infty} n2^{-n^2}$

14 $\sum_{n=1}^{\infty} \frac{1}{\sqrt{n+9}}$

15 $\sum_{n=2}^{\infty} \frac{1}{n\sqrt[3]{\ln n}}$

16 $\sum_{n=2}^{\infty} \frac{1}{n\sqrt{n^2-1}}$

Use comparison tests to determine whether the series in Exercises 17–38 converge or diverge.

17 $\sum_{n=1}^{\infty} \frac{1}{n^4+n^2+1}$

18 $\sum_{n=1}^{\infty} \frac{\sqrt{n}}{n^2+1}$

19 $\sum_{n=1}^{\infty} \frac{1}{n3^n}$

20 $\sum_{n=1}^{\infty} \frac{n^2}{n^3+1}$

21 $\sum_{n=1}^{\infty} \frac{2n+n^2}{n^3+1}$

22 $\sum_{n=1}^{\infty} \frac{2}{3+\sqrt{n}}$

23 $\sum_{n=1}^{\infty} \frac{\sqrt{n}}{n+4}$

24 $\sum_{n=1}^{\infty} \frac{n^5+4n^3+1}{2n^8+n^4+2}$

25 $\sum_{n=2}^{\infty} \frac{1}{\sqrt{4n^3-5n}}$

26 $\sum_{n=4}^{\infty} \frac{3n}{2n^2-7}$

27 $\sum_{n=1}^{\infty} \frac{8n^2-7}{e^n(n+1)^2}$

28 $\sum_{n=1}^{\infty} \frac{\sin^2 n}{2^n}$

29 $\sum_{n=1}^{\infty} \frac{1+2^n}{1+3^n}$

30 $\sum_{n=1}^{\infty} \frac{\ln n}{n^4}$ (*Hint:* $\ln n < n$)

31 $\sum_{n=1}^{\infty} \frac{2+\cos n}{n^2}$

32 $\sum_{n=1}^{\infty} \frac{\arctan n}{n^2}$

33 $\sum_{n=1}^{\infty} \frac{(2n+1)^3}{(n^3+1)^2}$

34 $\sum_{n=1}^{\infty} \frac{1}{\sqrt{n(n+1)(n+2)}}$

35 $\sum_{n=1}^{\infty} \frac{1}{\sqrt[3]{5n^2+1}}$

36 $\sum_{n=1}^{\infty} \frac{3n+5}{n \cdot 2^n}$

37 $\sum_{n=1}^{\infty} \frac{1}{n^n}$

38 $\sum_{n=1}^{\infty} \frac{1}{n!}$

Determine whether the series in Exercises 39–46 converge or diverge.

39 $\sum_{n=1}^{\infty} \frac{n+\ln n}{n^3+n+1}$

40 $\sum_{n=1}^{\infty} \frac{n+\ln n}{n^2+1}$

41 $\sum_{n=1}^{\infty} \sin \frac{1}{n^2}$

42 $\sum_{n=1}^{\infty} \tan \frac{1}{n}$

43 $\sum_{n=1}^{\infty} \frac{\ln n}{n^3}$

44 $\sum_{n=1}^{\infty} \ln\left(1+\frac{1}{2^n}\right)$

45 $\sum_{n=1}^{\infty} \frac{n^2+2^n}{n+3^n}$

46 $\sum_{n=1}^{\infty} \frac{\sin n + 2^n}{n+5^n}$

In Exercises 47 and 48 find all real numbers k for which the series converge.

47 $\sum_{n=2}^{\infty} \frac{1}{n^k \ln n}$

48 $\sum_{n=2}^{\infty} \frac{1}{n(\ln n)^k}$

49 Suppose $\Sigma\, a_n$ and $\Sigma\, b_n$ are positive term series. Prove that if $\lim_{n\to\infty} (a_n/b_n) = 0$ and $\Sigma\, b_n$ converges, then $\Sigma\, a_n$ converges. (This is not necessarily true for series which contain negative terms.)

50 Prove that if $\lim_{n\to\infty} (a_n/b_n) = \infty$ and $\Sigma\, b_n$ diverges, then $\Sigma\, a_n$ diverges.

12.4 ALTERNATING SERIES

An infinite series whose terms are alternately positive and negative is called an **alternating series**. It is customary to express an alternating series in one of the forms

$$a_1 - a_2 + a_3 - a_4 + \cdots + (-1)^{n-1}a_n + \cdots$$

or

$$-a_1 + a_2 - a_3 + a_4 - \cdots + (-1)^n a_n + \cdots$$

where each $a_i > 0$. The next theorem provides the main test for convergence of these series.

(12.24) **Alternating Series Test**

If $a_k \geq a_{k+1} > 0$ for every positive integer k and $\lim_{n\to\infty} a_n = 0$, then the alternating series $\Sigma\,(-1)^{n-1}a_n$ is convergent.

Proof. Let us first consider the partial sums

$$S_2, S_4, S_6, \ldots, S_{2n}, \ldots$$

which contain an even number of terms of the series. Since

$$S_{2n} = (a_1 - a_2) + (a_3 - a_4) + \cdots + (a_{2n-1} - a_{2n})$$

and $a_k - a_{k+1} \geq 0$ for every k, we see that

$$0 \leq S_2 \leq S_4 \leq \cdots \leq S_{2n} \leq \cdots,$$

that is, $\{S_{2n}\}$ is a monotonic sequence. The formula for S_{2n} can also be written

$$S_{2n} = a_1 - (a_2 - a_3) - (a_4 - a_5) - \cdots - (a_{2n-2} - a_{2n-1}) - a_{2n}$$

and hence $S_{2n} \leq a_1$ for every positive integer n. As in the proof of Theorem (12.9),

$$\lim_{n\to\infty} S_{2n} = S \leq a_1$$

for some number S. If we consider a partial sum S_{2n+1} having an *odd* number of terms of the series, then $S_{2n+1} = S_{2n} + a_{2n+1}$ and, since $\lim_{n\to\infty} a_{2n+1} = 0$,

$$\lim_{n\to\infty} S_{2n+1} = \lim_{n\to\infty} S_{2n} = S.$$

It follows that

$$\lim_{n\to\infty} S_n = S \leq a_1$$

that is, the series converges.

Example 1 Determine whether the following alternating series converge or diverge.

(a) $\displaystyle\sum_{n=1}^{\infty} (-1)^{n-1} \frac{2n}{4n^2 - 3}$ (b) $\displaystyle\sum_{n=1}^{\infty} (-1)^{n-1} \frac{2n}{4n - 3}$

Solution (a) Let

$$a_n = f(n) = \frac{2n}{4n^2 - 3}.$$

To apply the Alternating Series Test we must show that:

(i) $a_k \geq a_{k+1}$ for every positive integer k.

(ii) $\lim_{n\to\infty} a_n = 0$.

One method of proving (i) is to show that $f(x) = 2x/(4x^2 - 3)$ is decreasing for $x \geq 1$. By the Quotient Rule,

$$f'(x) = \frac{(4x^2 - 3)(2) - (2x)(8x)}{(4x^2 - 3)^2}$$

$$= \frac{-8x^2 - 6}{(4x^2 - 3)^2} < 0.$$

It follows that f is decreasing and, therefore, $a_k \geq a_{k+1}$ for every positive integer k.

To prove (ii) we see that

$$\lim_{n\to\infty} a_n = \lim_{n\to\infty} \frac{2n}{4n^2 - 3} = 0.$$

Thus the alternating series converges.

(b) It can be shown that $a_k \geq a_{k+1}$ for every k; however,

$$\lim_{n\to\infty} \frac{2n}{4n - 3} = \frac{1}{2} \neq 0$$

and hence the series diverges by (12.14).

In part (a) of Example 1 we used a derivative to show that $a_k \geq a_{k+1}$ for every k. This can also be done directly by proving that $a_k - a_{k+1} \geq 0$. Specifically, if $a_n = 2n/(4n^2 - 3)$,

$$a_k - a_{k+1} = \frac{2k}{4k^2 - 3} - \frac{2(k+1)}{4(k+1)^2 - 3}$$

$$= \frac{8k^2 + 8k + 6}{(4k^2 - 3)(4k^2 + 8k + 1)} \geq 0$$

for every positive integer k. Another technique for proving that $a_k \geq a_{k+1}$ is to show that $a_{k+1}/a_k \leq 1$.

If an infinite series converges, then the nth partial sum S_n can be used to approximate the sum S of the series. In most cases it is difficult to determine the accuracy of the approximation. However, for an *alternating* series, the next theorem provides a simple way of estimating the error which is involved.

(12.25) **Theorem**

> If $\Sigma(-1)^{n-1}a_n$ is an alternating series such that $a_k > a_{k+1} > 0$ for every positive integer k and if $\lim_{n\to\infty} a_n = 0$, then the error involved in approximating the sum S of the series by the nth partial sum S_n is numerically less than a_{n+1}.

Proof. Note that the alternating series $\Sigma(-1)^{n-1}a_n$ satisfies the conditions of the Alternating Series Test and hence has a sum S. The series obtained by deleting the first n terms, namely

$$_{n+1} + (-1)^{n+1}a_{n+2} + (-1)^{n+2}a_{n+3} + \cdots$$

also satisfies the conditions of (12.24) and, therefore, has a sum R_n. Thus

$$S - S_n = R_n = (-1)^n(a_{n+1} - a_{n+2} + a_{n+3} - \cdots)$$

and

$$|R_n| = a_{n+1} - a_{n+2} + a_{n+3} - \cdots.$$

Employing the same argument used in the proof of the Alternating Series Test (but replacing $\leq$ by $<$) we see that $|R_n| < a_{n+1}$. Consequently

$$|S - S_n| = |R_n| < a_{n+1}$$

which is what we wished to prove.

Example 2 Prove that the series

$$1 - \frac{1}{3!} + \frac{1}{5!} - \cdots + (-1)^{n-1}\frac{1}{(2n-1)!} + \cdots$$

is convergent and approximate its sum S to five decimal places.

Solution Evidently $a_n = 1/(2n-1)!$ has the limit 0 as $n \to \infty$, and $a_k > a_{k+1}$ for every positive integer k. Hence the series converges by the Alternating Series Test. If we use S_4 to approximate the sum S of the series, then

$$S \approx 1 - \frac{1}{3!} + \frac{1}{5!} - \frac{1}{7!}$$

$$= 1 - \frac{1}{6} + \frac{1}{120} - \frac{1}{5040} \approx 0.84147.$$

By (12.25) the error involved in the approximation is less than

$$a_5 = \frac{1}{9!} < 0.000003.$$

Thus the approximation 0.84147 is accurate to five decimal places. It will follow from (12.43) that the sum of the given series equals $\sin 1$ and hence $\sin 1 \approx 0.84147$. This illustrates one technique for constructing trigonometric tables.

EXERCISES 12.4

Determine whether the series in Exercises 1–12 converge or diverge.

1 $\sum_{n=1}^{\infty} (-1)^{n-1}\frac{1}{\sqrt{2n+1}}$

2 $\sum_{n=1}^{\infty} (-1)^{n-1}\frac{1}{n^{2/3}}$

3 $\sum_{n=1}^{\infty} (-1)^{n-1}\frac{1}{\ln(n+1)}$

4 $\sum_{n=1}^{\infty} (-1)^{n-1}\frac{n}{n^2+4}$

5 $\sum_{n=2}^{\infty} (-1)^{n}\frac{n}{\ln n}$

6 $\sum_{n=1}^{\infty} (-1)^{n}\frac{\ln n}{n}$

7 $\sum_{n=1}^{\infty} (-1)^{n-1} \frac{n^2+1}{n^3+1}$

8 $\sum_{n=1}^{\infty} (-1)^{n-1} \frac{3n+4}{5n+7}$

9 $\sum_{n=1}^{\infty} (-1)^n \frac{e^n}{n^4}$

10 $\sum_{n=0}^{\infty} (-1)^n \frac{\sqrt{n+1}}{8n+5}$

11 $\sum_{n=1}^{\infty} (-1)^n \frac{\sqrt[3]{n}}{2n+5}$

12 $\sum_{n=0}^{\infty} (-1)^n \frac{1+4^n}{1+3^n}$

In Exercises 13–16, approximate the sum of each series to three decimal places.

13 $\sum_{n=0}^{\infty} (-1)^n \frac{1}{n!}$

14 $\sum_{n=0}^{\infty} (-1)^{n+1} \frac{1}{(2n)!}$

15 $\sum_{n=1}^{\infty} (-1)^{n-1} \frac{n+1}{5^n}$

16 $\sum_{n=1}^{\infty} (-1)^{n-1} \frac{1}{n^5}$

In each of Exercises 17–20, use (12.25) to find a positive integer n such that S_n approximates the sum of the given series to four decimal places.

17 $\sum_{n=1}^{\infty} (-1)^n \frac{1}{n^2}$

18 $\sum_{n=1}^{\infty} (-1)^n \frac{1}{\sqrt{n}}$

19 $\sum_{n=1}^{\infty} (-1)^n \frac{1}{n^n}$

20 $\sum_{n=1}^{\infty} (-1)^n \frac{1}{n^3+1}$

12.5 ABSOLUTE CONVERGENCE

The following concept plays an important role in work with infinite series.

(12.26) Definition

An infinite series Σa_n is **absolutely convergent** if the series

$$\sum |a_n| = |a_1| + |a_2| + \cdots + |a_n| + \cdots$$

obtained by taking the absolute value of each term is convergent.

Note that if Σa_n is a positive term series, then $|a_n| = a_n$, and in this case absolute convergence is the same as convergence.

Example 1 Prove that the alternating series

$$1 - \frac{1}{2^2} + \frac{1}{3^2} - \frac{1}{4^2} + \cdots + (-1)^n \frac{1}{n^2} + \cdots$$

is absolutely convergent.

Solution Taking the absolute value of each term gives us

$$1 + \frac{1}{2^2} + \frac{1}{3^2} + \frac{1}{4^2} + \cdots + \frac{1}{n^2} + \cdots$$

which is a convergent p-series. Hence by Definition (12.26), the given alternating series is absolutely convergent.

The next theorem tells us that absolute convergence implies convergence.

(12.27) **Theorem**

If an infinite series $\Sigma\, a_n$ is absolutely convergent, then $\Sigma\, a_n$ is convergent.

Proof. If we let $b_n = a_n + |a_n|$ and use the property $-|a_n| \leq a_n \leq |a_n|$, then

$$0 \leq a_n + |a_n| \leq 2|a_n|, \quad \text{or} \quad 0 \leq b_n \leq 2|a_n|.$$

If $\Sigma\, a_n$ is absolutely convergent, then $\Sigma\, |a_n|$ is convergent, and hence by (ii) of Theorem (12.17), $\Sigma\, 2|a_n|$ is convergent. Applying the Comparison Test (12.22) it follows that $\Sigma\, b_n$ is convergent. Using (iii) of (12.17), $\Sigma\, (b_n - |a_n|)$ is convergent. Since $b_n - |a_n| = a_n$, this completes the proof.

Example 2 Determine whether the series

$$\sin 1 + \frac{\sin 2}{2^2} + \frac{\sin 3}{3^2} + \cdots + \frac{\sin n}{n^2} + \cdots$$

is convergent or divergent.

Solution The series contains both positive and negative terms, but it is not an alternating series since, for example, the first three terms are positive and the next three are negative. The series of absolute values is

$$\sum_{n=1}^{\infty} \left|\frac{\sin n}{n^2}\right| = \sum_{n=1}^{\infty} \frac{|\sin n|}{n^2}.$$

Since

$$\frac{|\sin n|}{n^2} \leq \frac{1}{n^2}$$

the series of absolute values $\Sigma\, |(\sin n)/n^2|$ is dominated by the convergent p-series $\Sigma\, 1/n^2$ and hence is convergent. Thus the given series is absolutely convergent and, therefore, is convergent by (12.27).

Series which are convergent, but are not absolutely convergent, are given a special name.

(12.28) Definition

An infinite series $\Sigma\, a_n$ is **conditionally convergent** if $\Sigma\, a_n$ is convergent and $\Sigma\, |a_n|$ is divergent.

Example 3 Show that the alternating series

$$1 - \frac{1}{2} + \frac{1}{3} - \frac{1}{4} + \cdots + (-1)^{n-1}\frac{1}{n} + \cdots$$

is conditionally convergent.

Solution The given series is convergent by the Alternating Series Test. Taking the absolute value of each term we obtain

$$1 + \frac{1}{2} + \frac{1}{3} + \frac{1}{4} + \cdots + \frac{1}{n} + \cdots$$

which is the divergent harmonic series. Hence by Definition (12.28), the given series is conditionally convergent.

We see from the preceding discussion that an arbitrary infinite series may be classified in exactly *one* of the following ways: (i) absolutely convergent, (ii) conditionally convergent, (iii) divergent. Of course, for positive term series we need only determine convergence or divergence.

One of the most important tests for absolute convergence is the following.

(12.29) The Ratio Test

Let $\Sigma\, a_n$ be an infinite series of nonzero terms.

(i) If $\lim_{n\to\infty} \left|\frac{a_{n+1}}{a_n}\right| = L < 1$, the series is absolutely convergent.

(ii) If $\lim_{n\to\infty} \left|\frac{a_{n+1}}{a_n}\right| = L > 1$, or $\lim_{n\to\infty} \left|\frac{a_{n+1}}{a_n}\right| = \infty$, the series is divergent.

(iii) If $\lim_{n\to\infty} \left|\frac{a_{n+1}}{a_n}\right| = 1$, the series may be absolutely convergent, conditionally convergent, or divergent.

Proof. (i) Suppose $\lim_{n\to\infty} |a_{n+1}/a_n| = L < 1$. If r is any number such that $0 \le L < r < 1$, then there exists an integer N such that

$$\left|\frac{a_{n+1}}{a_n}\right| < r \quad \text{whenever } n \ge N.$$

This implies that

$$|a_{N+1}| < |a_N|r$$
$$|a_{N+2}| < |a_{N+1}|r < |a_N|r^2$$
$$|a_{N+3}| < |a_{N+2}|r < |a_N|r^3$$

and, in general,

$$|a_{N+m}| < |a_N|r^m \quad \text{whenever } m > 0.$$

It follows from the Comparison Test (12.22) that the series

$$|a_{N+1}| + |a_{N+2}| + \cdots + |a_{N+m}| + \cdots$$

converges, since its terms are less than the corresponding terms of the convergent geometric series

$$|a_N|r + |a_N|r^2 + \cdots + |a_N|r^n + \cdots.$$

Since convergence or divergence is unaffected by discarding a finite number of terms, the series $\Sigma_{n=1}^{\infty} |a_n|$ also converges, that is, $\Sigma\, a_n$ is absolutely convergent.

(ii) Suppose next that $\lim_{n\to\infty} |a_{n+1}/a_n| = L > 1$. If r is a number such that $1 < r < L$, then there exists an integer N such that

$$\left|\frac{a_{n+1}}{a_n}\right| > r > 1 \quad \text{whenever } n \geq N.$$

Consequently $|a_{n+1}| > |a_n|$ if $n \geq N$ and therefore $\lim_{n\to\infty} |a_n| \neq 0$. Thus $\lim_{n\to\infty} a_n \neq 0$ and, by (12.14), the series $\Sigma\, a_n$ diverges. The proof for $\lim_{n\to\infty} |a_{n+1}/a_n| = \infty$ is similar and is left as an exercise.

(iii) The Ratio Test provides no useful information if $\lim_{n\to\infty} |a_{n+1}/a_n| = 1$. Indeed, it is easy to verify that the limit is 1 for the absolutely convergent series $\Sigma\,(-1)^n/n^2$, for the conditionally convergent series $\Sigma\,(-1)^n/n$ and for the divergent series $\Sigma\, 1/n$. Consequently, *if the limit is* 1, *then a different test should be employed.*

Example 4 Determine whether the following series are absolutely convergent, conditionally convergent, or divergent.

(a) $\displaystyle\sum_{n=1}^{\infty} (-1)^n \frac{3^n}{n!}$ (b) $\displaystyle\sum_{n=1}^{\infty} \frac{3^n}{n^2}$

Solution (a) Since

$$\lim_{n\to\infty}\left|\frac{a_{n+1}}{a_n}\right| = \lim_{n\to\infty}\left|\frac{3^{n+1}}{(n+1)!}\cdot\frac{n!}{3^n}\right|$$
$$= \lim_{n\to\infty}\frac{3}{n+1} = 0 < 1$$

the series is absolutely convergent by the Ratio Test.

(b) Since the series contains only positive terms the absolute value signs in Theorem (12.29) may be deleted. Thus

$$\lim_{n\to\infty} \frac{a_{n+1}}{a_n} = \lim_{n\to\infty} \frac{3^{n+1}}{(n+1)^2} \cdot \frac{n^2}{3^n}$$

$$= \lim_{n\to\infty} \frac{3n^2}{n^2+2n+1} = 3 > 1$$

and hence the series diverges by the Ratio Test.

Example 5 Determine the convergence or divergence of $\sum_{n=1}^{\infty} \frac{n^n}{n!}$.

Solution Since all terms of the series are positive, we may drop the absolute value signs in the Ratio Test (12.29). Thus

$$\lim_{n\to\infty} \frac{a_{n+1}}{a_n} = \lim_{n\to\infty} \frac{(n+1)^{n+1}}{(n+1)!} \cdot \frac{n!}{n^n} = \lim_{n\to\infty} \frac{(n+1)^{n+1}}{(n+1)} \cdot \frac{1}{n^n}$$

$$= \lim_{n\to\infty} \frac{(n+1)^n}{n^n} = \lim_{n\to\infty} \left(\frac{n+1}{n}\right)^n$$

$$= \lim_{n\to\infty} \left(1 + \frac{1}{n}\right)^n = e$$

where the last equality is a consequence of Theorem (8.32). Since $e > 1$, the given series diverges by the Ratio Test.

The following test is very useful if a_n contains only powers of n. It is not as versatile as the Ratio Test because it cannot be applied if a_n contains factorials.

(12.30) The Root Test

> Let $\Sigma\, a_n$ be an infinite series.
>
> (i) If $\lim_{n\to\infty} \sqrt[n]{|a_n|} = L < 1$, the series is absolutely convergent.
>
> (ii) If $\lim_{n\to\infty} \sqrt[n]{|a_n|} = L > 1$, or $\lim_{n\to\infty} \sqrt[n]{|a_n|} = \infty$, the series is divergent.
>
> (iii) If $\lim_{n\to\infty} \sqrt[n]{|a_n|} = 1$, the series may be absolutely convergent, conditionally convergent, or divergent.

Proof. The proof is similar to that used for the Ratio Test. If $L < 1$ as in (i), let us consider any number r such that $L < r < 1$. By the definition of limit, there exists a positive integer N such that if $n \ge N$,

$$\sqrt[n]{|a_n|} < r, \quad \text{or} \quad |a_n| < r^n.$$

Since $0 < r < 1$, $\Sigma_{n=N}^{\infty} r^n$ is a convergent geometric series, and hence by the Comparison Test (12.22), $\Sigma_{n=N}^{\infty} |a_n|$ converges. Consequently $\Sigma_{n=1}^{\infty} |a_n|$ converges; that is, $\Sigma\, a_n$ is absolutely convergent. This proves (i). The remainder of the proof is left as an exercise.

Example 6 Determine the convergence or divergence of $\displaystyle\sum_{n=1}^{\infty} \frac{2^{3n+1}}{n^n}$.

Solution Since all terms are positive we may delete the absolute value signs in the Root Test (12.30). Thus

$$\lim_{n\to\infty} \sqrt[n]{\frac{2^{3n+1}}{n^n}} = \lim_{n\to\infty} \left(\frac{2^{3n+1}}{n^n}\right)^{1/n}$$
$$= \lim_{n\to\infty} \frac{2^{3+(1/n)}}{n} = 0 < 1.$$

Hence the series converges by the Root Test.

As a final remark, it can be proved that if a series $\Sigma\, a_n$ is absolutely convergent and if the terms are rearranged in any manner, then the resulting series converges and has the same sum as the given series. This is not true for conditionally convergent series. Indeed, if $\Sigma\, a_n$ is conditionally convergent, then by suitably rearranging terms one can obtain either a divergent series, or a series which converges and has any desired sum S.*

EXERCISES 12.5

In Exercises 1–38, determine whether (a) a series which contains both positive and negative terms is absolutely convergent, conditionally convergent, or divergent; or (b) a positive term series is convergent or divergent.

1 $\displaystyle\sum_{n=1}^{\infty} (-1)^{n+1} \frac{1}{\sqrt{n}}$

2 $\displaystyle\sum_{n=1}^{\infty} (-1)^n \frac{n}{n^2+1}$

3 $\displaystyle\sum_{n=2}^{\infty} (-1)^n \frac{5}{n^4-1}$

4 $\displaystyle\sum_{n=1}^{\infty} (-1)^n e^{-n}$

5 $\displaystyle\sum_{n=1}^{\infty} \frac{1000-n}{n!}$

6 $\displaystyle\sum_{n=1}^{\infty} \frac{n!}{e^n}$

7 $\displaystyle\sum_{n=1}^{\infty} (-1)^{n-1} \frac{3n+1}{2^n}$

8 $\displaystyle\sum_{n=1}^{\infty} (-1)^{n-1} \frac{3^n}{n^2+4}$

9 $\displaystyle\sum_{n=1}^{\infty} \frac{5^n}{n(3^{n+1})}$

10 $\displaystyle\sum_{n=1}^{\infty} \frac{2^{n-1}}{5^n(n+1)}$

* See, for example, R. C. Buck, *Advanced Calculus*, Third Edition (New York: McGraw-Hill, 1978), pp. 238–239.

11 $\sum_{n=1}^{\infty} \frac{(-100)^n}{n!}$

12 $\sum_{n=1}^{\infty} \frac{10 - n^2}{n!}$

13 $\sum_{n=1}^{\infty} (-1)^{n-1} \frac{\sqrt{n}}{n^2 + 1}$

14 $\sum_{n=1}^{\infty} (-1)^{n-1} \frac{\sqrt{n}}{n + 1}$

15 $\sum_{n=1}^{\infty} (-1)^n \frac{n^2 + 1}{n^3 + 1}$

16 $\sum_{n=1}^{\infty} (-1)^n \frac{2n + 1}{n^2 + n^3}$

17 $\sum_{n=2}^{\infty} (-1)^n \frac{1}{n(\ln n)^5}$

18 $\sum_{n=1}^{\infty} (-1)^n \frac{n!}{(n + 1)^5}$

19 $\sum_{n=1}^{\infty} (-1)^{n-1} \frac{2}{n^3 + e^n}$

20 $\sum_{n=1}^{\infty} (-1)^{n-1} \frac{n3^{2n}}{5^{n-1}}$

21 $\sum_{n=1}^{\infty} (-1)^n \frac{\arctan n}{n^2}$

22 $\sum_{n=1}^{\infty} \frac{\cos n - 1}{n^{3/2}}$

23 $\sum_{n=1}^{\infty} \frac{n^n}{10^n}$

24 $\sum_{n=1}^{\infty} \frac{10 - 2^n}{n!}$

25 $\sum_{n=1}^{\infty} \frac{n!}{(-5)^n}$

26 $\sum_{n=1}^{\infty} \frac{\sec^{-1} n}{\tan^{-1} n}$

27 $\sum_{n=1}^{\infty} \frac{(n!)^2}{(2n)!}$

28 $\sum_{n=1}^{\infty} (-1)^n \frac{n^2 + 3}{(2n - 5)^2}$

29 $\sum_{n=1}^{\infty} \frac{\sin \sqrt{n}}{\sqrt{n^3 + 1}}$

30 $\sum_{n=1}^{\infty} \frac{n!}{n^n}$

31 $1 + \frac{1 \cdot 3}{2!} + \frac{1 \cdot 3 \cdot 5}{3!} + \cdots + \frac{1 \cdot 3 \cdot 5 \cdots (2n - 1)}{n!} + \cdots$

32 $\frac{1}{2} + \frac{1 \cdot 4}{2 \cdot 4} + \frac{1 \cdot 4 \cdot 7}{2 \cdot 4 \cdot 6} + \cdots + \frac{1 \cdot 4 \cdot 7 \cdots (3n - 2)}{2 \cdot 4 \cdot 6 \cdots (2n)} + \cdots$

33 $\sum_{n=2}^{\infty} \frac{1}{(\ln n)^n}$

34 $\sum_{n=1}^{\infty} \frac{(2n)^n}{(5n + 3n^{-1})^n}$

35 $\sum_{n=1}^{\infty} (-1)^n \frac{\ln n}{(1.01)^n}$

36 $\sum_{n=1}^{\infty} (-1)^n 3^{1/n}$

37 $\sum_{n=1}^{\infty} (-1)^n n \tan \frac{1}{n}$

38 $\sum_{n=1}^{\infty} \frac{\cos (n\pi/6)}{n^2}$

39 Complete the proof of the Ratio Test (12.29) by showing that if $\lim_{n \to \infty} |a_{n+1}/a_n| = \infty$, then $\Sigma\, a_n$ is divergent.

40 Complete the proof of the Root Test (12.30).

12.6 POWER SERIES

In the previous sections we concentrated on infinite series with constant terms. Of major importance in applications are series whose terms contain variables. In particular, if x is a variable, then a series of the form

$$\sum_{n=0}^{\infty} a_n x^n = a_0 + a_1 x + a_2 x^2 + \cdots + a_n x^n + \cdots$$

is called a **power series in x**. If a number is substituted for x we obtain a series of constant terms which may be tested for convergence or divergence. In order to simplify the general term of the power series it is assumed that $x^0 = 1$ even if $x = 0$. The main objective of this section is to determine all values of x for which a power series converges. Evidently every power series in x converges if $x = 0$. To find other numbers which produce a convergent series we often employ the Ratio Test, as illustrated in the following examples.

Example 1 Find all values of x for which the power series

$$1 + \frac{1}{5}x + \frac{2}{5^2}x^2 + \cdots + \frac{n}{5^n}x^n + \cdots$$

is absolutely convergent.

Solution If we let $u_n = (n/5^n)x^n = nx^n/5^n$, then

$$\begin{aligned}\lim_{n\to\infty}\left|\frac{u_{n+1}}{u_n}\right| &= \lim_{n\to\infty}\left|\frac{(n+1)x^{n+1}}{5^{n+1}}\cdot\frac{5^n}{nx^n}\right| \\ &= \lim_{n\to\infty}\left|\frac{(n+1)x}{5n}\right| = \lim_{n\to\infty}\left(\frac{n+1}{5n}\right)|x| = \frac{1}{5}|x|.\end{aligned}$$

By the Ratio Test, the given series is absolutely convergent if the following equivalent inequalities are true:

$$\tfrac{1}{5}|x| < 1, \quad |x| < 5, \quad -5 < x < 5.$$

The series diverges if $\frac{1}{5}|x| > 1$, that is if $x > 5$ or $x < -5$. If $\frac{1}{5}|x| = 1$, the Ratio Test gives no information and hence the numbers 5 and -5 require special consideration. Substituting 5 for x in the given series we obtain

$$\sum_{n=0}^{\infty}\frac{n}{5^n}5^n = \sum_{n=0}^{\infty} n = 0 + 1 + 2 + 3 + \cdots$$

which is divergent. If we let $x = -5$ we obtain

$$\sum_{n=0}^{\infty}\frac{n}{5^n}(-5)^n = \sum_{n=1}^{\infty} n(-1)^n = 0 - 1 + 2 - 3 + \cdots$$

which is also divergent. Consequently, the given series $\Sigma\,(n/5^n)x^n$ is absolutely convergent for every x in the open interval $(-5, 5)$ and diverges elsewhere.

Example 2 Find all values of x for which the power series

$$1 + \frac{1}{1!}x + \frac{1}{2!}x^2 + \cdots + \frac{1}{n!}x^n + \cdots$$

is absolutely convergent.

Solution We shall employ the same technique used in Example 1. If we let

$$u_n = \frac{1}{n!}x^n = \frac{x^n}{n!}$$

then

$$\begin{aligned}\lim_{n\to\infty}\left|\frac{u_{n+1}}{u_n}\right| &= \lim_{n\to\infty}\left|\frac{x^{n+1}}{(n+1)!}\cdot\frac{n!}{x^n}\right| \\ &= \lim_{n\to\infty}\left|\frac{x}{n+1}\right| \\ &= \lim_{n\to\infty}\frac{1}{n+1}|x| = 0.\end{aligned}$$

Since the limit 0 is less than 1 for every value of x, it follows from the Ratio Test that the given series is absolutely convergent for all real numbers.

Example 3 Find all values of x for which $\Sigma\, n!\, x^n$ is convergent.

Solution Let $u_n = n!\, x^n$. If $x \neq 0$, then

$$\begin{aligned}\lim_{n\to\infty}\left|\frac{u_{n+1}}{u_n}\right| &= \lim_{n\to\infty}\left|\frac{(n+1)!\, x^{n+1}}{n!\, x^n}\right| \\ &= \lim_{n\to\infty}|(n+1)x| \\ &= \lim_{n\to\infty}(n+1)|x| = \infty\end{aligned}$$

and, by the Ratio Test, the series diverges. Hence the series is convergent only if $x = 0$.

It will be shown in Theorem (12.32) that the solutions of the preceding examples are typical, in the sense that if a power series converges for nonzero values of x then either it is absolutely convergent for all real numbers or it is absolutely convergent throughout some open interval $(-r, r)$ and diverges outside of the closed interval $[-r, r]$. The proof of this fact depends on the next theorem.

(12.31) Theorem

(i) If a power series $\Sigma\, a_n x^n$ converges for a nonzero number c, then it is absolutely convergent whenever $|x| < |c|$.

(ii) If a power series $\Sigma\, a_n x^n$ diverges for a nonzero number d, then it diverges whenever $|x| > |d|$.

Proof. If $\Sigma\, a_n c^n$ converges, where $c \neq 0$, then by Theorem (12.13), $\lim_{n\to\infty} a_n c^n = 0$. Employing Definition (12.3) with $\varepsilon = 1$, there is a positive integer N such that

$$|a_n c^n| < 1 \quad \text{whenever } n \geq N$$

and, therefore,

$$|a_n x^n| = \left|\frac{a_n c^n x^n}{c^n}\right| = |a_n c^n| \left|\frac{x}{c}\right|^n < \left|\frac{x}{c}\right|^n$$

provided $n \geq N$. If $|x| < |c|$, then $|x/c| < 1$ and $\Sigma\, |x/c|^n$ is a convergent geometric series. Hence, by the Comparion Test (12.22), the series obtained by deleting the first N terms of $\Sigma\, |a_n x^n|$ is convergent. It follows that the series $\Sigma\, |a_n x^n|$ is also convergent, which proves (i).

To prove (ii), suppose the series diverges for $x = d \neq 0$. If the series converges for some c_1, where $|c_1| > |d|$, then by (i) it converges whenever $|x| < |c_1|$. In particular it converges for $x = d$, contrary to our supposition. Hence the series diverges whenever $|x| > |d|$.

(12.32) Theorem

> If $\Sigma\, a_n x^n$ is a power series, then precisely one of the following is true.
>
> (i) The series converges only if $x = 0$.
>
> (ii) The series is absolutely convergent for all x.
>
> (iii) There is a positive number r such that the series is absolutely convergent if $|x| < r$ and divergent if $|x| > r$.

Proof. If neither (i) nor (ii) is true, then there exist nonzero numbers c and d such that the series converges if $x = c$ and diverges if $x = d$. Let S denote the set of all real numbers for which the series is absolutely convergent. From Theorem (12.31), the series diverges if $|x| > |d|$ and hence every number in S is less than $|d|$. By the Completeness Property (12.10), S has a least upper bound r. It follows that the series is absolutely convergent if $|x| < r$ and diverges if $|x| > r$.

If (iii) of Theorem (12.32) is true, then the power series $\Sigma\, a_n x^n$ is absolutely convergent throughout the open interval $(-r, r)$ and diverges outside of the closed interval $[-r, r]$. The number r is called the **radius of convergence** of the series. Either convergence or divergence may occur at $-r$ or r, depending on the nature of the series. The totality of numbers for which a power series converges is called its **interval of convergence**. If the radius of convergence r is positive, then the interval of convergence is one of the following:

$$(-r, r), \quad (-r, r], \quad [-r, r), \quad \text{or} \quad [-r, r].$$

If (i) or (ii) of (12.32) is true, then the radius of convergence is denoted by 0 or ∞, respectively. In Example 1 of this section, the interval of convergence is $(-5, 5)$ and the radius of convergence is 5. In Example 2 the interval of convergence is

$(-\infty, \infty)$ and we write $r = \infty$. In Example 3, $r = 0$. The next example illustrates the case of a half-open interval of convergence.

Example 4 Find the interval of convergence of the power series $\sum_{n=1}^{\infty} \frac{1}{\sqrt{n}} x^n$.

Solution Note that in this example the coefficient of x^0 is 0 and the summation begins with $n = 1$. We let $u_n = x^n/\sqrt{n}$ and consider

$$\begin{aligned}\lim_{n \to \infty} \left| \frac{u_{n+1}}{u_n} \right| &= \lim_{n \to \infty} \left| \frac{x^{n+1}}{\sqrt{n+1}} \cdot \frac{\sqrt{n}}{x^n} \right| \\ &= \lim_{n \to \infty} \left| \frac{\sqrt{n}}{\sqrt{n+1}} x \right| \\ &= \lim_{n \to \infty} \sqrt{\frac{n}{n+1}} |x| \\ &= (1)|x| = |x|.\end{aligned}$$

It follows from the Ratio Test that the power series is absolutely convergent if $|x| < 1$, that is, if $-1 < x < 1$. The series diverges if $|x| > 1$, that is, if $x > 1$ or $x < -1$. The numbers 1 and -1 must be checked by direct substitution in the given series. If we let $x = 1$ we obtain

$$\sum_{n=1}^{\infty} \frac{1}{\sqrt{n}} (1)^n = 1 + \frac{1}{\sqrt{2}} + \frac{1}{\sqrt{3}} + \cdots + \frac{1}{\sqrt{n}} + \cdots$$

which is a divergent p-series. If we substitute $x = -1$ the result is

$$\sum_{n=1}^{\infty} \frac{1}{\sqrt{n}} (-1)^n = -1 + \frac{1}{\sqrt{2}} - \frac{1}{\sqrt{3}} + \cdots + \frac{(-1)^n}{\sqrt{n}} + \cdots$$

which converges by the Alternating Series Test. Hence the power series converges if $-1 \le x < 1$, and the interval of convergence is $[-1, 1)$.

If c is a real number, then

$$\sum_{n=0}^{\infty} a_n (x - c)^n = a_0 + a_1(x - c) + a_2(x - c)^2 + \cdots + a_n(x - c)^n + \cdots$$

is called a **power series in $x - c$**. In order to simplify the nth term, it is assumed that $(x - c)^0 = 1$ even if $x = c$. If we employ the same reasoning used to prove Theorem (12.32), and replace x by $x - c$, then it can be shown that precisely one of the following is true.

(i) The series converges only if $x - c = 0$, that is, if $x = c$.

(ii) The series is absolutely convergent for all x.

(iii) There is a positive number r such that the series is absolutely convergent if $|x - c| < r$ and divergent if $|x - c| > r$.

Whenever (iii) occurs, then the series $\Sigma\, a_n(x-c)^n$ is absolutely convergent if

$$-r < x - c < r, \quad \text{or} \quad c - r < x < c + r$$

that is, if x is the interval $(c - r, c + r)$. The end-points of the interval must be checked separately. As before, the totality of numbers for which the series converges is called the *interval of convergence*, and r is called the *radius of convergence*.

Example 5 Find the interval of convergence of

$$1 - \frac{1}{2}(x-3) + \frac{1}{3}(x-3)^2 + \cdots + (-1)^n \frac{1}{n+1}(x-3)^n + \cdots.$$

Solution If we let $u_n = (-1)^n(x-3)^n/(n+1)$, then

$$\begin{aligned}
\lim_{n\to\infty}\left|\frac{u_{n+1}}{u_n}\right| &= \lim_{n\to\infty}\left|\frac{(x-3)^{n+1}}{n+2}\cdot\frac{n+1}{(x-3)^n}\right| \\
&= \lim_{n\to\infty}\left|\frac{n+1}{n+2}(x-3)\right| \\
&= \lim_{n\to\infty}\left(\frac{n+1}{n+2}\right)|x-3| \\
&= (1)|x-3| = |x-3|.
\end{aligned}$$

By the Ratio Test the series is absolutely convergent if $|x-3| < 1$, that is, if

$$-1 < x - 3 < 1 \quad \text{or} \quad 2 < x < 4.$$

The series diverges if $x < 2$ or $x > 4$. The numbers 2 and 4 must be checked separately. If $x = 4$ the resulting series is

$$1 - \frac{1}{2} + \frac{1}{3} - \cdots + (-1)^n \frac{1}{n+1} + \cdots$$

which converges by the Alternating Series Test. For $x = 2$ the series becomes

$$1 + \frac{1}{2} + \frac{1}{3} + \cdots + \frac{1}{n+1} + \cdots$$

which is the divergent harmonic series. Hence the interval of convergence is $(2, 4]$.

EXERCISES 12.6

Find the interval of convergence of the power series in Exercises 1–26.

1 $\displaystyle\sum_{n=0}^{\infty} \frac{1}{n+4}x^n$

2 $\displaystyle\sum_{n=0}^{\infty} \frac{1}{n^2+4}x^n$

3 $\sum_{n=0}^{\infty} \frac{n^2}{2^n} x^n$

4 $\sum_{n=1}^{\infty} \frac{(-3)^n}{n} x^{n+1}$

5 $\sum_{n=1}^{\infty} (-1)^{n-1} \frac{1}{\sqrt{n}} x^n$

6 $\sum_{n=1}^{\infty} \frac{1}{\ln(n+1)} x^n$

7 $\sum_{n=2}^{\infty} \frac{n}{n^2+1} x^n$

8 $\sum_{n=1}^{\infty} \frac{1}{4^n \sqrt{n}} x^n$

9 $\sum_{n=2}^{\infty} \frac{\ln n}{n^3} x^n$

10 $\sum_{n=0}^{\infty} \frac{10^{n+1}}{3^{2n}} x^n$

11 $\sum_{n=0}^{\infty} \frac{n+1}{10^n} (x-4)^n$

12 $\sum_{n=1}^{\infty} \frac{1}{n(n+1)} (x-2)^n$

13 $\sum_{n=0}^{\infty} \frac{n!}{100^n} x^n$

14 $\sum_{n=0}^{\infty} \frac{(3n)!}{(2n)!} x^n$

15 $\sum_{n=0}^{\infty} \frac{1}{(-4)^n} x^{2n+1}$

16 $\sum_{n=1}^{\infty} (-1)^{n-1} \frac{1}{\sqrt[3]{n} 3^n} x^n$

17 $\sum_{n=0}^{\infty} \frac{2^n}{(2n)!} x^{2n}$

18 $\sum_{n=0}^{\infty} \frac{10^n}{n!} x^n$

19 $\sum_{n=0}^{\infty} \frac{3^{2n}}{n+1} (x-2)^n$

20 $\sum_{n=1}^{\infty} \frac{1}{n5^n} (x-5)^n$

21 $\sum_{n=0}^{\infty} \frac{n^2}{2^{3n}} (x+4)^n$

22 $\sum_{n=0}^{\infty} \frac{1}{2n+1} (x+3)^n$

23 $\sum_{n=1}^{\infty} \frac{\ln n}{e^n} (x-e)^n$

24 $\sum_{n=0}^{\infty} \frac{n}{3^{2n-1}} (x-1)^{2n}$

25 $\sum_{n=1}^{\infty} (-1)^n \frac{1}{n6^n} (2x-1)^n$

26 $\sum_{n=0}^{\infty} \frac{1}{\sqrt{3n+4}} (3x+4)^n$

Find the radius of convergence of each power series in Exercises 27–30.

27 $\sum_{n=1}^{\infty} (-1)^n \frac{1 \cdot 3 \cdot 5 \cdot \cdots \cdot (2n-1)}{3 \cdot 6 \cdot 9 \cdot \cdots \cdot (3n)} x^n$

28 $\sum_{n=1}^{\infty} \frac{2 \cdot 4 \cdot 6 \cdot \cdots \cdot (2n)}{4 \cdot 7 \cdot 10 \cdot \cdots \cdot (3n+1)} x^n$

29 $\sum_{n=1}^{\infty} \frac{n^n}{n!} x^n$

30 $\sum_{n=0}^{\infty} \frac{(n+1)!}{10^n} (x-5)^n$

12.7 POWER SERIES REPRESENTATIONS OF FUNCTIONS

A power series $\Sigma a_n x^n$ may be used to define a function f whose domain is the interval of convergence of the series. Specifically, for each x in this interval we let $f(x)$ equal the sum of the series, that is,

(12.33) $$f(x) = a_0 + a_1 x + a_2 x^2 + \cdots + a_n x^n + \cdots.$$

If a function f is defined in this way we say that $\Sigma a_n x^n$ is a **power series**

representation for $f(x)$ (or *of* $f(x)$). We also use the phrase f *is represented by the power series.*

A power series representation as in (12.33) enables us to find functional values in a new way. Specifically, if c is in the interval of convergence, then $f(c)$ can be found (or approximated) by finding (or approximating) the sum of the series

$$a_0 + a_1 c + a_2 c^2 + \cdots + a_n c^n + \cdots.$$

Power series representations will also enable us to solve problems involving differentiation and integration by using techniques different from those considered earlier in the text.

To illustrate the previous remarks, if $|x| < 1$, then by Theorem (12.12), the geometric series

$$1 - x + x^2 - x^3 + \cdots + (-1)^n x^n + \cdots$$

converges and has the sum $1/[1 - (-x)] = 1/(1 + x)$, and hence we may write

(12.34) $$\frac{1}{1+x} = 1 - x + x^2 - x^3 + \cdots + (-1)^n x^n + \cdots.$$

This gives us a power series representation for $f(x) = 1/(1 + x)$, if $|x| < 1$.

Most of the examples and exercises in this section are concerned with geometric series, since the sum can be determined by means of a specific formula. In the next section we shall consider a more difficult problem, namely, given a function f, find a power series representation for f.

A function f defined by a power series as in (12.33) has properties similar to a polynomial. In particular, it can be shown that f has a derivative whose power series representation may be found by differentiating each term of the given series. Similarly, definite integrals of f may be obtained by integrating each term of the series. The following theorem, which is stated without proof, summarizes these remarks.

(12.35) Theorem

Suppose a power series $\Sigma\, a_n x^n$ has a nonzero radius of convergence r and let the function f be defined by

$$f(x) = \sum_{n=0}^{\infty} a_n x^n = a_0 + a_1 x + a_2 x^2 + \cdots + a_n x^n + \cdots$$

for every x in the interval of convergence. If $-r < x < r$, then

(i) $$f'(x) = \sum_{n=0}^{\infty} D_x(a_n x^n) = \sum_{n=1}^{\infty} n a_n x^{n-1}$$
$$= a_1 + 2a_2 x + 3a_3 x^2 + \cdots + n a_n x^{n-1} + \cdots.$$

(ii) $$\int_0^x f(t)\,dt = \sum_{n=0}^{\infty} \int_0^x (a_n t^n)\,dt = \sum_{n=0}^{\infty} \frac{a_n}{n+1} x^{n+1}$$
$$= a_0 x + \frac{1}{2} a_1 x^2 + \frac{1}{3} a_2 x^3 + \cdots + \frac{1}{n+1} a_n x^{n+1} + \cdots.$$

It can be shown that the series in (i) and (ii) of Theorem (12.35) have the same radius of convergence as $\Sigma\, a_n x^n$. As a corollary of (i), *a function which is represented by a power series in an interval* $(-r, r)$ *is continuous throughout* $(-r, r)$. (See Theorem (3.8).) Similar results are true for representations of functions by means of power series of the form $\Sigma\, a_n(x - c)^n$.

Example 1 Use (12.34) to find a power series representation for $1/(1 + x)^2$.

Solution If $f(x) = 1/(1 + x)$, then $f'(x) = -1/(1 + x)^2$. Hence if we differentiate each term of the series in (12.34), then by (i) of Theorem (12.35),

$$-\frac{1}{(1 + x)^2} = -1 + 2x - 3x^2 - \cdots + (-1)^n n x^{n-1} + \cdots$$

provided $|x| < 1$. Consequently,

$$\frac{1}{(1 + x)^2} = 1 - 2x + 3x^2 + \cdots + (-1)^{n+1} n x^{n-1} + \cdots.$$

Example 2 Find a power series representation for $\ln(1 + x)$ if $|x| < 1$.

Solution If $|x| < 1$, then

$$\begin{aligned} \ln(1 + x) &= \int_0^x \frac{1}{1 + t}\, dt \\ &= \int_0^x [1 - t + t^2 - \cdots + (-1)^n t^n + \cdots]\, dt \end{aligned}$$

where the last equality follows from (12.34). As in (ii) of Theorem (12.35), we may integrate each term of the series to obtain

$$\begin{aligned} \ln(1 + x) &= \int_0^x 1\, dt - \int_0^x t\, dt + \int_0^x t^2\, dt - \cdots + (-1)^n \int_0^x t^n\, dt + \cdots \\ &= t\Big]_0^x - \frac{t^2}{2}\Big]_0^x + \frac{t^3}{3}\Big]_0^x - \cdots + (-1)^n \frac{t^{n+1}}{n + 1}\Big]_0^x + \cdots. \end{aligned}$$

Hence

$$\ln(1 + x) = x - \frac{x^2}{2} + \frac{x^3}{3} - \cdots + (-1)^n \frac{x^{n+1}}{n + 1} + \cdots$$

if $|x| < 1$.

Example 3 Calculate $\ln(1.1)$ to five decimal places.

Solution In Example 2 we found a series representation for $\ln(1 + x)$ if $|x| < 1$. Substituting 0.1 for x in that series gives us

$$\begin{aligned} \ln(1.1) &= 0.1 - \frac{(0.1)^2}{2} + \frac{(0.1)^3}{3} - \frac{(0.1)^4}{4} + \frac{(0.1)^5}{5} - \cdots \\ &= 0.1 - 0.005 + 0.000333 - 0.000025 + 0.000002 - \cdots. \end{aligned}$$

If we sum the first four terms on the right and round off to five decimal places, then

$$\ln(1.1) \approx 0.09531.$$

By (12.25), the error is less than the absolute value 0.000002 of the fifth term and, therefore, the number 0.09531 is accurate to five decimal places. The reader should compare this solution to that of Example 2 in Section 11.5.

Example 4 Find a power series representation for $\arctan x$.

Solution We first observe that

$$\arctan x = \int_0^x \frac{1}{1+t^2}\,dt.$$

Next we note that if $|t| < 1$, then by (12.12), with $a = 1$ and $r = -t^2$,

$$\frac{1}{1+t^2} = 1 - t^2 + t^4 - \cdots + (-1)^n t^{2n} + \cdots.$$

By (ii) of Theorem (12.35) we may integrate each term of the series to obtain

$$\arctan x = x - \frac{x^3}{3} + \frac{x^5}{5} - \cdots + (-1)^n \frac{x^{2n+1}}{2n+1} + \cdots$$

provided $|x| < 1$.

Example 5 Prove that e^x has the power series representation

$$e^x = 1 + x + \frac{x^2}{2!} + \frac{x^3}{3!} + \cdots + \frac{x^n}{n!} + \cdots.$$

Solution The indicated power series was considered in Example 2 of the preceding section, where it was shown to be absolutely convergent for every real number x. If we let f denote the function represented by the series, then

$$f(x) = \sum_{n=0}^{\infty} \frac{x^n}{n!}.$$

Applying (i) of Theorem (12.35),

$$f'(x) = \sum_{n=1}^{\infty} \frac{nx^{n-1}}{n!} = \sum_{n=1}^{\infty} \frac{x^{n-1}}{(n-1)!}$$

$$= 1 + x + \frac{x^2}{2!} + \frac{x^3}{3!} + \cdots + \frac{x^n}{n!} + \cdots$$

that is,

$$f'(x) = f(x) \quad \text{for every } x.$$

If, in (8.33), we let $q = f$, $t = x$, and $c = 1$, then as in the proof of (8.34) we obtain

$$f(x) = f(0)e^x.$$

However,

$$f(0) = 1 + 0 + \frac{0^2}{2!} + \cdots + \frac{0^n}{n!} + \cdots = 1$$

and hence

$$f(x) = e^x$$

which is what we wished to prove.

Note that Example 5 allows us to express the number e as the sum of a convergent positive term series, namely,

$$e = 1 + 1 + \frac{1}{2!} + \frac{1}{3!} + \cdots + \frac{1}{n!} + \cdots.$$

Example 6 Approximate $\int_0^{0.1} e^{-x^2}\,dx$.

Solution Letting $x = -t^2$ in the series of Example 5 gives us

$$e^{-t^2} = 1 - t^2 + \frac{t^4}{2!} - \cdots + \frac{(-1)^n t^{2n}}{n!} + \cdots$$

which is true for all t. Applying (ii) of Theorem (12.35),

$$\begin{aligned}\int_0^{0.1} e^{-x^2}\,dx &= \int_0^{0.1} e^{-t^2}\,dt \\ &= t\Big]_0^{0.1} - \frac{t^3}{3}\Big]_0^{0.1} + \frac{t^5}{10}\Big]_0^{0.1} - \cdots \\ &= 0.1 - \frac{(0.1)^3}{3} + \frac{(0.1)^5}{10} - \cdots.\end{aligned}$$

If we use the first two terms to approximate the sum of this convergent alternating series, then by Theorem (12.25), the error is less than the third term $(0.1)^5/10 = 0.000001$. Hence

$$\int_0^{0.1} e^{-x^2}\,dx \approx 0.1 - \frac{0.001}{3} \approx 0.99667$$

which is accurate to five decimal places.

Thus far, the techniques used to obtain power series representations of functions were *indirect*, in the sense that we started with known series and then differentiated or integrated, and perhaps used other knowledge about functional behavior, to obtain series representations. In the next section we shall discuss a *direct* method which can be used to find power series representations for a large variety of functions.

EXERCISES 12.7

In each of Exercises 1–10, find a power series representation for $f(x)$ and specify the radius of convergence.

1 $f(x) = 1/(1 - x)$

2 $f(x) = 1/(1 + x^2)$

3 $f(x) = 1/(1 - x)^2$

4 $f(x) = 1/(1 - 4x)$

5 $f(x) = x^2/(1 - x^2)$

6 $f(x) = x/(1 - x^4)$

7 $f(x) = x/(2 - 3x)$

8 $f(x) = x^3/(4 - x^3)$

9 $f(x) = (x^2 + 1)/(x - 1)$

10 $f(x) = 3/(2x + 5)$

11 Prove that

$$\ln(1 - x) = \sum_{n=1}^{\infty} \frac{-x^n}{n} \quad \text{if } |x| < 1,$$

and use this fact to approximate $\ln(1.2)$ to three decimal places.

12 Use Example 4 to prove that the sum of the series $\Sigma_{n=0}^{\infty}(-1/3)^n/\sqrt{3}(2n + 1)$ is $\pi/6$.

In each of Exercises 13–18 use the result obtained in Example 5 to find a power series representation for $f(x)$.

13 $f(x) = e^{-x}$

14 $f(x) = e^{2x}$

15 $f(x) = \cosh x$

16 $f(x) = \sinh x$

17 $f(x) = xe^{3x}$

18 $f(x) = x^2 e^{x^2}$

Use infinite series to approximate each of the integrals in Exercises 19–24 to four decimal places.

19 $\displaystyle\int_0^{1/3} \frac{1}{1 + x^6}\, dx$

20 $\displaystyle\int_0^{1/2} \arctan x^2\, dx$

21 $\displaystyle\int_0^{0.1} \frac{\arctan x}{x}\, dx$

22 $\displaystyle\int_0^{0.2} \frac{x^3}{1 + x^5}\, dx$

23 $\displaystyle\int_0^{1} e^{-x^2/10}\, dx$

24 $\displaystyle\int_0^{0.5} e^{-x^3}\, dx$

25 Use the power series representation for $(1 - x^2)^{-1}$ to find a power series representation for $2x(1 - x^2)^{-2}$.

26 Use the method of Example 2 to find a power series representation for $\ln(3 + 2x)$.

In each of Exercises 27–30 find a power series representation for $f(x)$.

27 $f(x) = \displaystyle\int_0^{x} \frac{e^{-t^2} - 1}{t}\, dt$

28 $f(x) = \displaystyle\int_0^{x} \frac{e^{t} - 1}{t}\, dt$

29 $f(x) = \displaystyle\int_0^{x} \frac{\ln(1 - t)}{t}\, dt$

30 $f(x) = \displaystyle\int_0^{x} \ln(1 + t^2)\, dt$

12.8 TAYLOR AND MACLAURIN SERIES

Suppose a function f is represented by a power series in $x - c$, such that

(12.36)
$$f(x) = \sum_{n=0}^{\infty} a_n(x - c)^n$$
$$= a_0 + a_1(x - c) + a_2(x - c)^2 + a_3(x - c)^3 + a_4(x - c)^4 + \cdots$$

where the domain of f is an open interval containing c. As in the preceding section, power series representations may be found for $f'(x)$, $f''(x), \ldots,$ by differentiating the terms of the series in (12.36). Thus

$$f'(x) = \sum_{n=1}^{\infty} na_n(x - c)^{n-1}$$
$$= a_1 + 2a_2(x - c) + 3a_3(x - c)^2 + 4a_4(x - c)^3 + \cdots$$
$$f''(x) = \sum_{n=2}^{\infty} n(n - 1)a_n(x - c)^{n-2}$$
$$= 2a_2 + (3 \cdot 2)a_3(x - c) + (4 \cdot 3)a_4(x - c)^2 + \cdots$$
$$f'''(x) = \sum_{n=3}^{\infty} n(n - 1)(n - 2)a_n(x - c)^{n-3}$$
$$= (3 \cdot 2)a_3 + (4 \cdot 3 \cdot 2)a_4(x - c) + \cdots$$

and, for every positive integer k,

$$f^{(k)}(x) = \sum_{n=k}^{\infty} n(n - 1)\cdots(n - k + 1)a_n(x - c)^{n-k}.$$

Moreover, each series obtained by differentiation has the same radius of convergence as the original series. Substituting c for x in each of these series representations, we obtain

$$f(c) = a_0, \quad f'(c) = a_1, \quad f''(c) = 2a_2, \quad f'''(c) = (3 \cdot 2)a_3$$

and, for every positive integer n,

$$f^{(n)}(c) = n!\,a_n, \quad \text{or} \quad a_n = \frac{f^{(n)}(c)}{n!}.$$

We have proved the following result.

(12.37) Theorem

> If f is a function such that
>
> $$f(x) = \sum_{n=0}^{\infty} a_n(x - c)^n$$
>
> for all x in an open interval containing c, then
>
> $$f(x) = f(c) + f'(c)(x - c) + \frac{f''(c)}{2!}(x - c)^2 + \cdots + \frac{f^{(n)}(c)}{n!}(x - c)^n + \cdots.$$

The series which appears in the conclusion of the last theorem is called the **Taylor series for $f(x)$ at c.** The special case $c = 0$ is very important and hence is stated separately as a corollary.

(12.38) Corollary

If f is a function such that $f(x) = \Sigma\, a_n x^n$ for all x in an open interval $(-r, r)$, then

$$f(x) = f(0) + f'(0)x + \frac{f''(0)}{2!}x^2 + \cdots + \frac{f^{(n)}(0)}{n!}x^n + \cdots.$$

The series appearing in this Corollary is called the **Maclaurin series for $f(x)$**. Each example in the preceding section involves a Maclaurin series.

Theorem (12.37) states that *if* a function f is represented by a power series in $x - c$, then the power series *must* be the Taylor series. However, the theorem does not state conditions which guarantee that a power series representation actually exists. We shall now obtain such conditions. Let us begin by noting that the $(n + 1)$st partial sum of the Taylor series stated in (12.37) is the nth-degree Taylor Polynomial $P_n(x)$ of f at c (see (11.13)). Moreover, by (11.15),

$$P_n(x) = f(x) - R_n(x) \tag{12.39}$$

where

$$R_n(x) = \frac{f^{(n+1)}(z)}{(n+1)!}(x - c)^{n+1} \tag{12.40}$$

for some number z between c and x. In the next theorem we use $R_n(x)$ to specify sufficient conditions for the existence of a power series representation of $f(x)$.

(12.41) Theorem

If a function f has derivatives of all orders throughout an interval containing c, and if

$$\lim_{n\to\infty} R_n(x) = 0$$

for every x in that interval, then $f(x)$ is represented by the Taylor series for $f(x)$ at c.

Proof. The polynomial $P_n(x)$ is a general term for the sequence of partial sums of the Taylor series for $f(x)$ at c. Moreover, from (12.39),

$$\lim_{n\to\infty} P_n(x) = f(x) - \lim_{n\to\infty} R_n(x) = f(x).$$

Hence the sequence of partial sums converges to $f(x)$. This proves the theorem.

In Example 2 of Section 6 we proved that the power series $\Sigma\, x^n/n!$ is absolutely convergent for all real numbers. Consequently, by Theorem (12.13),

(12.42)
$$\lim_{n\to\infty}\left|\frac{x^n}{n!}\right| = 0$$

for every real number x. We shall use this fact in the solution of the following example.

Example 1 Find the Maclaurin series for $\sin x$ and prove that it represents $\sin x$ for every real number x.

Solution Let us arrange our work as follows:

$$\begin{aligned} f(x) &= \sin x & f(0) &= 0 \\ f'(x) &= \cos x & f'(0) &= 1 \\ f''(x) &= -\sin x & f''(0) &= 0 \\ f'''(x) &= -\cos x & f'''(0) &= -1 \end{aligned}$$

Successive derivatives follow this same pattern. Substitution in (12.38) yields the Maclaurin series

(12.43)
$$\sin x = x - \frac{x^3}{3!} + \frac{x^5}{5!} - \frac{x^7}{7!} + \cdots + (-1)^n \frac{x^{2n+1}}{(2n+1)!} + \cdots.$$

At this stage all we know is that *if* $\sin x$ is represented by a power series in x, then it is given by (12.43). In order to prove that the representation is actually true for every number x, let us employ Theorem (12.41). If n is a positive integer, then either

$$|f^{(n+1)}(x)| = |\cos x| \quad \text{or} \quad |f^{(n+1)}(x)| = |\sin x|.$$

Hence $|f^{(n+1)}(z)| \leq 1$ for every number z and, using (12.40) with $c = 0$,

$$|R_n(x)| = \frac{|f^{(n+1)}(z)|}{(n+1)!}|x|^{n+1} \leq \frac{|x|^{n+1}}{(n+1)!}.$$

It follows from (12.42) and the Sandwich Theorem (12.7) that $\lim_{n\to\infty} |R_n(x)| = 0$. Consequently $\lim_{n\to\infty} R_n(x) = 0$, and the Maclaurin series representation (12.43) for $\sin x$ is true for every x.

Example 2 Find the Maclaurin series for $\cos x$.

Solution We could proceed directly as in Example 1; however, let us obtain the desired series by differentiating (12.43). Thus,

(12.44)
$$\cos x = 1 - \frac{x^2}{2!} + \frac{x^4}{4!} - \frac{x^6}{6!} + \cdots + (-1)^n \frac{x^{2n}}{(2n)!} + \cdots.$$

This series can also be obtained by integrating the terms of (12.43).

The Maclaurin series for e^x was obtained in Example 5 of the preceding section by using an indirect technique. We shall next give a direct derivation of this extremely important formula.

Example 3 Find a Maclaurin series which represents e^x for every real number x.

Solution If $f(x) = e^x$, then $f^{(n)}(x) = e^x$ for every positive integer n. Hence $f^{(n)}(0) = 1$ and substitution in (12.38) gives us

(12.45)
$$e^x = 1 + x + \frac{x^2}{2!} + \frac{x^3}{3!} + \cdots + \frac{x^n}{n!} + \cdots.$$

As in the solution of Example 1, we now use Theorem (12.41) to prove that this power series representation of e^x is true for every real number x. From (12.40), with $c = 0$,

$$R_n(x) = \frac{e^z}{(n+1)!}x^{n+1}$$

where z is a number between 0 and x. If $0 < x$, then $e^z < e^x$ since the natural exponential function is increasing, and hence, for every positive integer n,

$$0 < R_n(x) < \frac{e^x}{(n+1)!}x^{n+1}.$$

Using (12.42),

$$\lim_{n\to\infty} \frac{e^x}{(n+1)!}x^{n+1} = e^x \lim_{n\to\infty} \frac{x^{n+1}}{(n+1)!} = 0$$

and, by the Sandwich Theorem (12.7),

$$\lim_{n\to\infty} R_n(x) = 0.$$

If $x < 0$, then also $z < 0$ and hence $e^z < e^0 = 1$. Consequently

$$0 < |R_n(x)| < \left|\frac{x^{n+1}}{(n+1)!}\right|$$

and we again see that $R_n(x)$ has the limit 0 as n increases without bound. It follows from Theorem (12.41) that the power series representation (12.45) for e^x is valid for all nonzero x. Finally, note that if $x = 0$, then (12.45) reduces to $e^0 = 1$.

Example 4 Find the Taylor series for $\sin x$ in powers of $x - \pi/6$.

Solution The derivatives of $f(x) = \sin x$ are listed in Example 1. If we evaluate them at $c = \pi/6$, there results

$$f(\pi/6) = 1/2, \quad f'(\pi/6) = \sqrt{3}/2, \quad f''(\pi/6) = -1/2, \quad f'''(\pi/6) = -\sqrt{3}/2,$$

and this pattern of four numbers repeats itself indefinitely. Substitution in (12.37) gives us

$$\sin x = \frac{1}{2} + \frac{\sqrt{3}}{2}\left(x - \frac{\pi}{6}\right) - \frac{1}{2(2!)}\left(x - \frac{\pi}{6}\right)^2 - \frac{\sqrt{3}}{2(3!)}\left(x - \frac{\pi}{6}\right)^3 + \cdots.$$

The general term u_n of this series is given by

$$u_n = \begin{cases} (-1)^{n/2}\dfrac{1}{2n!}\left(x - \dfrac{\pi}{6}\right)^n & \text{if } n = 0, 2, 4, 6, \ldots \\ (-1)^{(n-1)/2}\dfrac{\sqrt{3}}{2n!}\left(x - \dfrac{\pi}{6}\right)^n & \text{if } n = 1, 3, 5, 7, \ldots. \end{cases}$$

The proof that the series represents $\sin x$ for all x is similar to that given in Example 1 and is therefore omitted.

The series obtained in this section can be used to obtain power series representations for other functions. To illustrate, since (12.45) is true for every x, a power series representation for e^{-x^2} can be found by substituting $-x^2$ for x. This gives us

(12.46) $$e^{-x^2} = 1 - x^2 + \frac{x^4}{2!} - \cdots + (-1)^n\frac{x^{2n}}{n!} + \cdots.$$

Similarly, replacing x by $-x$ in (12.45) leads to

(12.47) $$e^{-x} = 1 - x + \frac{x^2}{2!} - \cdots + (-1)^n\frac{x^n}{n!} + \cdots.$$

Since $\cosh x = \frac{1}{2}(e^x + e^{-x})$, a power series for $\cosh x$ can be found by adding corresponding terms of (12.45) and (12.47) and then multiplying by 1/2. Thus

(12.48) $$\cosh x = 1 + \frac{x^2}{2!} + \frac{x^4}{4!} + \cdots + \frac{x^{2n}}{(2n)!} + \cdots.$$

It is interesting to compare this with the series (12.44) for $\cos x$. Similar examples are left as exercises.

Taylor and Maclaurin series may be used to approximate values of functions in a manner similar to that used in the preceding section. For example, to find $\sin(0.1)$ we could use (12.43) to write

$$\sin(0.1) = 0.1 - \frac{0.001}{6} + \frac{0.00001}{120} - \cdots.$$

By Theorem (12.25), the error involved in using the sum of the first two terms as an approximation is less than 0.00001/120. More generally, we have the following polynomial approximation formula:

$$\sin x \approx x - \frac{x^3}{6},$$

where the error is less than $|x^5|/5!$.

The next example illustrates the use of infinite series in approximating values of definite integrals.

Example 5 Approximate $\int_0^1 \sin x^2\, dx$ to four decimal places.

Solution From (12.43),

$$\sin x^2 = x^2 - \frac{x^6}{3!} + \frac{x^{10}}{5!} - \frac{x^{14}}{7!} + \cdots.$$

Integrating each term of the series we obtain

$$\int_0^1 \sin x^2\, dx = \tfrac{1}{3} - \tfrac{1}{42} + \tfrac{1}{1320} - \tfrac{1}{75600} + \cdots.$$

Summing the first three terms,

$$\int_0^1 \sin x^2\, dx \approx 0.3103$$

where the error is less than $1/75600 \approx 0.000013$.

Note that in the preceding example we achieved accuracy to four decimal places by summing only *three* terms of the integrated series for $\sin x^2$. In order to obtain this degree of accuracy by means of the Trapezoidal Rule or Simpson's Rule, it would be necessary to use an extremely fine partition of the interval $[0, 1]$. This brings out an important point. For numerical applications, in addition to analyzing a given problem we should also strive to find the most efficient method for computing the answer.

In order to obtain a Taylor or Maclaurin series representation for a function f by means of Theorem (12.37) or Corollary (12.38), respectively, it is necessary to find a general formula for $f^{(n)}(x)$ and, in addition, investigate $\lim_{n\to\infty} R_n(x)$. Because of this, our examples were restricted to expressions such as $\sin x$, $\cos x$, and e^x. The method cannot be used if, for example, $f(x)$ equals $\tan x$ or $\sin^{-1} x$, since $f^{(n)}(x)$ becomes very complicated as n increases. Most of the exercises which follow are based on functions whose nth derivatives can be determined easily, or on series representations which we have already established. In more complicated cases we shall restrict our attention to only the first few terms of a Taylor or Maclaurin series representation.

EXERCISES 12.8

In each of Exercises 1–4 use (12.38) to find the Maclaurin series for $f(x)$ and prove that the series is valid for all x by means of (12.41).

1 $f(x) = \cos x$

2 $f(x) = e^{-x}$

3 $f(x) = e^{2x}$

4 $f(x) = \cosh x$

In Exercises 5–12 use established series to find a Maclaurin series for $f(x)$ and state the radius of convergence.

5 $f(x) = x^2 e^x$

6 $f(x) = xe^{-2x}$

7 $f(x) = \sinh x$

8 $f(x) = x^2 \sin x$

9 $f(x) = x \sin 3x$

10 $f(x) = \cos x^2$

11 $f(x) = \cos^2 x$ (*Hint:* Use $\cos^2 x = (1 + \cos 2x)/2$.)

12 $f(x) = \sin^2 x$

In Exercises 13–18 find the Taylor Series for $f(x)$ at the indicated number c. (Do not verify that $\lim_{n \to \infty} R_n(x) = 0$.)

13. $f(x) = \sin x;\ c = \pi/4$

14 $f(x) = \cos x;\ c = \pi/3$

15 $f(x) = 1/x;\ c = 2$

16 $f(x) = e^x;\ c = -3$

17 $f(x) = 10^x;\ c = 0$

18 $f(x) = \ln(1 + x);\ c = 0$

19 Find a series representation for e^{2x} in powers of $x + 1$.

20 Find a series representation for $\ln x$ in powers of $x - 1$.

In Exercises 21–26, find the first four terms of the Taylor series for $f(x)$ at the indicated value of c.

21 $f(x) = \sec x;\ c = \pi/3$

22 $f(x) = \tan x;\ c = \pi/4$

23 $f(x) = \sin^{-1} x;\ c = 1/2$

24 $f(x) = \csc x;\ c = 2\pi/3$

25 $f(x) = xe^x;\ c = -1$

26 $f(x) = \operatorname{sech} x;\ c = 0$

In each of Exercises 27–36, use an infinite series to approximate the given number to four decimal places.

27 $\sqrt{e}$

28 $\sin 1°$

29 $\int_0^1 e^{-x^2}\, dx$

30 $\int_0^{1/2} x \cos x^3\, dx$

31 $\int_0^1 \frac{1 - \cos x}{x^2}\, dx$

32 $\int_0^1 \frac{\sin x}{x}\, dx$

33 $\int_0^{1/2} \frac{\ln(1 + x)}{x}\, dx$

34 $\int_0^{0.5} \cos x^2\, dx$

35 $\int_{-1}^0 \frac{e^{3x} - 1}{10x}\, dx$

36 $\int_0^1 \frac{1 - e^{-x}}{x}\, dx$

12.9 THE BINOMIAL SERIES

The Binomial Theorem states that if k is a positive integer, then for all numbers a and b,

$$(a+b)^k = a^k + ka^{k-1}b + \frac{k(k-1)}{2!}a^{k-2}b^2 + \cdots + \frac{k(k-1)\cdots(k-n+1)}{n!}a^{k-n}b^n + \cdots + b^k.$$

If we let $a = 1$ and $b = x$, then

$$(1+x)^k = 1 + kx + \frac{k(k-1)}{2!}x^2 + \cdots + \frac{k(k-1)\cdots(k-n+1)}{n!}x^n + \cdots + x^k. \tag{12.49}$$

If k is not a positive integer (or 0), it is useful to study the power series $\Sigma\, a_n x^n$, where $a_0 = 1$ and $a_n = k(k-1)\cdots(k-n+1)/n!$ for $n \geq 1$. This series has the form

$$1 + kx + \frac{k(k-1)}{2!}x^2 + \cdots + \frac{k(k-1)\cdots(k-n+1)}{n!}x^n + \cdots \tag{12.50}$$

and is called the **binomial series**. If k is a nonnegative integer, then (12.50) reduces to the finite sum (12.49). Otherwise the binomial series does not terminate. It is left as an exercise to show that

$$\lim_{n\to\infty}\left|\frac{a_{n+1}x^{n+1}}{a_n x^n}\right| = \lim_{n\to\infty}\left|\frac{k-n}{n+1}\right||x| = |x|. \tag{12.51}$$

Hence, by the Ratio Test, (12.50) is absolutely convergent if $|x| < 1$ and diverges if $|x| > 1$. Thus the binomial series represents a functon f, where

$$f(x) = \sum_{n=0}^{\infty} a_n x^n \quad \text{if } |x| < 1.$$

It has already been noted that if k is a nonnegative integer, then $f(x) = (1+x)^k$. We shall now prove that the same is true for *every* real number k. Since $f(x)$ is given by (12.50),

$$f'(x) = k + k(k-1)x + \cdots + \frac{nk(k-1)\cdots(k-n+1)}{n!}x^{n-1} + \cdots$$

and, therefore,

$$xf'(x) = kx + k(k-1)x^2 + \cdots + \frac{nk(k-1)\cdots(k-n+1)}{n!}x^n + \cdots.$$

If we add corresponding terms of the last two power series, then the coefficient of x^n is

$$\frac{(n+1)k(k-1)\cdots(k-n)}{(n+1)!} + \frac{nk(k-1)\cdots(k-n+1)}{n!}$$

which simplifies to

$$[(k-n)+n]\frac{k(k-1)\cdots(k-n+1)}{n!} = ka_n.$$

Consequently

$$f'(x) + xf'(x) = \sum_{n=0}^{\infty} ka_n x^n = kf(x)$$

and, therefore,

$$f'(x)(1+x) - kf(x) = 0.$$

If we define the function g by $g(x) = f(x)/(1+x)^k$, then

$$g'(x) = \frac{(1+x)^k f'(x) - f(x)k(1+x)^{k-1}}{(1+x)^{2k}}$$

$$= \frac{(1+x)f'(x) - kf(x)}{(1+x)^{k+1}} = 0.$$

It follows from Theorem (4.34) that $g(x) = c$ for some constant c, that is,

$$\frac{f(x)}{(1+x)^k} = c.$$

Since $f(0) = 1$, we see that $c = 1$ and hence $f(x) = (1+x)^k$, which is what we wished to prove. To summarize, if $|x| < 1$ then

(12.52) $$(1+x)^k = 1 + kx + \frac{k(k-1)}{2!}x^2 + \cdots + \frac{k(k-1)\cdots(k-n+1)}{n!}x^n + \cdots$$

for every real number k.

Example 1 Find a power series representation for $\sqrt[3]{1+x}$.

Solution Using (12.52) with $a = 1/3$,

$$\sqrt[3]{1+x} = 1 + \frac{1}{3}x + \frac{\frac{1}{3}(\frac{1}{3}-1)}{2!}x^2 + \frac{\frac{1}{3}(\frac{1}{3}-1)(\frac{1}{3}-2)}{3!}x^3 + \cdots$$

$$+ \frac{\frac{1}{3}(\frac{1}{3}-1)\cdots(\frac{1}{3}-n+1)}{n!}x^n + \cdots$$

which may be written

$$\sqrt[3]{1+x} = 1 + \frac{1}{3}x - \frac{2}{3^2 \cdot 2!}x^2 + \frac{1 \cdot 2 \cdot 5}{3^3 \cdot 3!}x^3 + \cdots$$
$$+ (-1)^{n+1}\frac{1 \cdot 2 \cdots (3n-4)}{3^n \cdot n!}x^n + \cdots$$

where $|x| < 1$. The formula for the nth term of this series is valid provided $n \geq 2$.

Example 2 Find a power series representation for $\sqrt[3]{1+x^4}$.

Solution The desired series can be obtained by substituting x^4 for x in the series of Example 1. Hence, if $|x| < 1$, then

$$\sqrt[3]{1+x^4} = 1 + \frac{1}{3}x^4 - \frac{2}{3^2 \cdot 2!}x^8 + \cdots$$
$$+ (-1)^{n+1}\frac{1 \cdot 2 \cdots (3n-4)}{3^n \cdot n!}x^{4n} + \cdots.$$

Example 3 Approximate $\int_0^{0.3} \sqrt[3]{1+x^4}\,dx$.

Solution Integrating the terms of the series obtained in Example 2 gives us

$$\int_0^{0.3} \sqrt[3]{1+x^4}\,dx = 0.3 + 0.000162 - 0.000000243 + \cdots.$$

Consequently, the integral may be approximated by 0.300162, which is accurate to six decimal places since the error is less than 0.000000243. (Why?)

The binomial series can be used to obtain polynomial approximation formulas for $(1+x)^k$. To illustrate, if $|x| < 1$, then from Example 1

$$\sqrt[3]{1+x} \approx 1 + \tfrac{1}{3}x,$$

where the error involved in this approximation is less than the next term $x^2/9$.

EXERCISES 12.9

In each of Exercises 1–6 find a power series representation for $f(x)$ and state the radius of convergence.

1 (a) $f(x) = \sqrt{1+x}$ (b) $f(x) = \sqrt{1-x^3}$

2 (a) $f(x) = \dfrac{1}{\sqrt[3]{1+x}}$ (b) $f(x) = \dfrac{1}{\sqrt[3]{1-x^2}}$

3 $f(x) = (1+x)^{-3}$

4 $f(x) = x(1+2x)^{-2}$

5 $f(x) = \sqrt[3]{8+x}$

6 $f(x) = (4+x)^{3/2}$

7 Obtain a power series representation for $\sin^{-1} x$ by using

$$\sin^{-1} x = \int_0^x \frac{1}{\sqrt{1-t^2}}\,dt.$$

What is the radius of convergence?

8 Obtain a power series representation for $\sinh^{-1} x$ by using

$$\sinh^{-1} x = \int_0^x \frac{1}{\sqrt{1+t^2}}\,dt.$$

What is the radius of convergence?

Approximate the integrals in Exercises 9 and 10 to three decimal places.

9 $\displaystyle\int_0^{1/2} \sqrt{1+x^3}\,dx$ (see Exercise 1) **10** $\displaystyle\int_0^{1/2} \frac{1}{\sqrt[3]{1+x^2}}\,dx$ (see Exercise 2)

12.10 REVIEW

Concepts

Define or discuss each of the following.

1 Infinite sequence

2 Limit of an infinite sequence

3 Theorems on limits of sequences

4 Monotonic sequence

5 Sequence of partial sums of an infinite series

6 Convergent infinite series

7 Divergent infinite series

8 Geometric series

9 The harmonic series

10 The p-series

11 The Integral Test

12 Comparison tests

13 The Alternating Series Test

14 Absolute convergence

15 Conditional convergence

16 The Ratio Test

17 The Root Test

18 Power series

19 Radius of convergence

20 Interval of convergence

21 Differentiation and integration of power series

22 Taylor series

23 Maclaurin series

24 Binomial series

25 Approximations by series

Exercises

In each of Exercises 1–6 find the limit of the sequence, if it exists.

1 $\left\{\dfrac{\ln(n^2+1)}{n}\right\}$

2 $\{100(0.99)^n\}$

3 $\{10^n/n^{10}\}$

4 $\left\{\dfrac{1}{n}+(-2)^n\right\}$

5 $\left\{\dfrac{n}{\sqrt{n}+4}-\dfrac{n}{\sqrt{n}+9}\right\}$

6 $\left\{\left(1+\dfrac{2}{n}\right)^{2n}\right\}$

In Exercises 7–32, determine whether (a) a positive term series is convergent or divergent, or (b) a series which contains negative terms is absolutely convergent, conditionally convergent, or divergent.

7 $\displaystyle\sum_{n=1}^{\infty}\frac{1}{\sqrt[3]{n(n+1)(n+2)}}$

8 $\displaystyle\sum_{n=0}^{\infty}\frac{(2n+3)^2}{(n+1)^3}$

9 $\displaystyle\sum_{n=1}^{\infty}(-2/3)^{n-1}$

10 $\displaystyle\sum_{n=0}^{\infty}\frac{1}{2+(1/2)^n}$

11 $\displaystyle\sum_{n=1}^{\infty}\frac{3^{2n+1}}{n5^{n-1}}$

12 $\displaystyle\sum_{n=1}^{\infty}\frac{1}{3^n+2}$

13 $\displaystyle\sum_{n=1}^{\infty}\frac{n!}{\ln(n+1)}$

14 $\displaystyle\sum_{n=1}^{\infty}\frac{n^2-1}{n^2+1}$

15 $\displaystyle\sum_{n=1}^{\infty}(n^2+9)(-2)^{1-n}$

16 $\displaystyle\sum_{n=1}^{\infty}\frac{n+\cos n}{n^3+1}$

17 $\displaystyle\sum_{n=1}^{\infty}\frac{e^n}{n^e}$

18 $\displaystyle\sum_{n=1}^{\infty}(-1)^{n-1}\frac{n}{n^2+1}$

19 $\displaystyle\sum_{n=1}^{\infty}(-1)^n\frac{1}{\sqrt[n]{n}}$

20 $\displaystyle\sum_{n=2}^{\infty}(-1)^n\frac{(0.9)^n}{\ln n}$

21 $\displaystyle\sum_{n=1}^{\infty}\frac{\sin(n5\pi/3)}{n^{5\pi/3}}$

22 $\displaystyle\sum_{n=2}^{\infty}(-1)^n\frac{\sqrt[3]{n}-1}{n^2-1}$

23 $\displaystyle\sum_{n=1}^{\infty}(-1)^{n-1}\frac{\sqrt{n}}{n+1}$

24 $\displaystyle\sum_{n=1}^{\infty}(-1)^n\frac{2n+3}{n!}$

25 $\sum_{n=1}^{\infty} \frac{1 - \cos n}{n^2}$

26 $\frac{2}{1!} - \frac{2 \cdot 4}{2!} + \cdots + (-1)^{n-1}\frac{2 \cdot 4 \cdots (2n)}{n!} + \cdots$

27 $\sum_{n=1}^{\infty} \frac{(2n)^n}{n^{2n}}$

28 $\sum_{n=1}^{\infty} \frac{3^{n-1}}{n^2 + 9}$

29 $\sum_{n=1}^{\infty} \frac{e^{2n}}{(2n-1)!}$

30 $\sum_{n=1}^{\infty} \left(\frac{1}{3^n} - \frac{5}{\sqrt{n}}\right)$

31 $\sum_{n=2}^{\infty} (-1)^n \frac{\sqrt{\ln n}}{n}$

32 $\sum_{n=1}^{\infty} \frac{\tan^{-1} n}{\sqrt{1 + n^2}}$

Use the integral test to determine the convergence or divergence of the series in Exercises 33–38.

33 $\sum_{n=1}^{\infty} \frac{1}{(3n+2)^3}$

34 $\sum_{n=2}^{\infty} \frac{n}{\sqrt{n^2 - 1}}$

35 $\sum_{n=1}^{\infty} n^{-2} e^{1/n}$

36 $\sum_{n=2}^{\infty} \frac{1}{n(\ln n)^3}$

37 $\sum_{n=1}^{\infty} \frac{10}{\sqrt[3]{n + 8}}$

38 $\sum_{n=5}^{\infty} \frac{1}{n^2 - 4n}$

Approximate the sums of the series in Exercises 39 and 40 to three decimal places.

39 $\sum_{n=1}^{\infty} (-1)^{n-1} \frac{1}{(2n+1)!}$

40 $\sum_{n=1}^{\infty} (-1)^{n-1} \frac{1}{n^2(n^2+1)}$

Find the interval of convergence of each of the series in Exercises 41–44.

41 $\sum_{n=0}^{\infty} \frac{n+1}{(-3)^n} x^n$

42 $\sum_{n=0}^{\infty} (-1)^n \frac{4^{2n}}{\sqrt{n+1}} x^n$

43 $\sum_{n=1}^{\infty} \frac{1}{n \cdot 2^n} (x + 10)^n$

44 $\sum_{n=2}^{\infty} \frac{1}{n(\ln n)^2} (x - 1)^n$

Find the radius of convergence of the series in Exercises 45 and 46.

45 $\sum_{n=0}^{\infty} \frac{(2n)!}{(n!)^2} x^n$

46 $\sum_{n=0}^{\infty} \frac{1}{(n+5)!} (x + 5)^n$

In each of Exercises 47–52 find the Maclaurin series for $f(x)$ and state the radius of convergence.

47 $f(x) = \frac{1 - \cos x}{x}$ if $x \neq 0$, $f(0) = 0$

48 $f(x) = xe^{-2x}$

49 $f(x) = \sin x \cos x$

50 $f(x) = \ln(2 + x)$

51 $f(x) = (1 + x)^{2/3}$

52 $f(x) = \frac{1}{\sqrt{1 - x^2}}$

53 Find a series representation for e^{-x} in powers of $x + 2$.

54 Find a series representation for $\cos x$ in powers of $x - \pi/2$.

55 Find a series representation for $\sqrt{x}$ in powers of $x - 4$.

Use infinite series to approximate the numbers in Exercises 56–60 to three decimal places.

56 $1/\sqrt[3]{3e}$

57 $\int_0^1 x^2 e^{-x^2}\,dx$

58 $\int_0^1 \frac{\sin x}{\sqrt{x}}\,dx$

59 $\sqrt[5]{1.01}$

60 $e^{-0.25}$

Plane Curves and Polar Coordinates

In this chapter parametric and polar equations of curves are discussed. Applications include tangent lines, areas, arc length, and surfaces of revolution.

13.1 PLANE CURVES AND PARAMETRIC EQUATIONS

The graph of an equation $y = f(x)$, where f is a function, is often called a *plane curve*. However, to use this as a definition is unnecessarily restrictive, since it rules out most of the conic sections and many other useful graphs. The following statement is satisfactory for most applications.

(13.1) Definition

A **plane curve** is a set C of ordered pairs of the form

$$(f(t), g(t))$$

where the functions f and g are continuous on an interval I.

For simplicity, we shall often refer to a plane curve as a **curve**. The **graph** of the curve C in (13.1) is the set of all points $P(t) = (f(t), g(t))$ in a rectangular coordinate system that correspond to the ordered pairs. Each $P(t)$ is referred to as a *point* on the curve. We shall use the term *curve* interchangeably with *graph of a curve*. Sometimes it is convenient to think that the point $P(t)$ traces the curve C as t varies through the interval I. This is especially true in applications for which t represents time and $P(t)$ is the position of a moving particle at time t.

The graphs of several curves are sketched in Figure 13.1 for the case where I is a closed interval $[a, b]$. If, as in (i) of the figure, $P(a) \neq P(b)$, then $P(a)$ and $P(b)$ are called the **end-points** of C. Note that the curve illustrated in (i) intersects itself in the sense that two different values of t give rise to the same point. If $P(a) = P(b)$, as illustrated in (ii) of Figure 13.1, then C is called a **closed curve**. If $P(a) = P(b)$ and C does not intersect itself at any other point, as illustrated in (iii) of the figure, then C is called a **simple closed curve**.

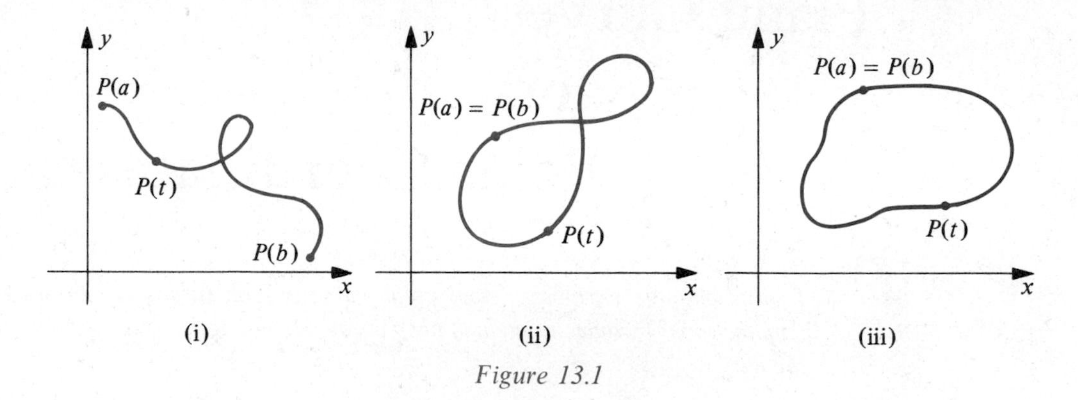

Figure 13.1

If C is the curve in (13.1), then the equations

(13.2)
$$x = f(t), \quad y = g(t)$$

where t is in I are called **parametric equations** for C and t is called a **parameter**. As t varies through I, the point $P(x, y)$ traces the curve. Sometimes it is possible to eliminate the parameter in (13.2) and obtain an equation for C which involves the variables x and y.

The continuity of f and g in Definition (13.1) implies that a small change in t produces a small change in the position of the point $(f(t), g(t))$ on C. This fact may be used to obtain a rough sketch of the graph by plotting many points and connecting them in the order of increasing t, as illustrated in Example 1.

Example 1 Describe and sketch the graph of the curve $C = \{(2t, t^2 - 1): -1 \leq t \leq 2\}$.

Solution In this example $f(t) = 2t$, $g(t) = t^2 - 1$, and parametric equations for C are

$$x = 2t, \quad y = t^2 - 1, \quad \text{where } -1 \leq t \leq 2.$$

These equations can be used to tabulate coordinates for points $P(x, y)$ on C as in the following table.

t	-1	$-\frac{1}{2}$	0	$\frac{1}{2}$	1	$\frac{3}{2}$	2
x	-2	-1	0	1	2	3	4
y	0	$-\frac{3}{4}$	-1	$-\frac{3}{4}$	0	$\frac{5}{4}$	3

Plotting points and using the continuity of f and g gives us the sketch in Figure 13.2.

A precise description of the graph may be obtained by eliminating the parameter. To illustrate, solving the first parametric equation for t we obtain

$t = x/2$. If we next substitute for t in the second equation the result is

$$y = \left(\frac{x}{2}\right)^2 - 1, \quad \text{or} \quad y + 1 = \frac{1}{4}x^2.$$

The graph of the last equation is a parabola with vertical axis and vertex at the point $(0, -1)$. The curve C is that part of the parabola shown in Figure 13.2.

Parametric equations of curves are not unique. The curve C of this example is also given by

$$C = \{(t, \tfrac{1}{4}t^2 - 1): -2 \leq t \leq 4\}$$
$$C = \{(t^3, \tfrac{1}{4}t^6 - 1): \sqrt[3]{-2} \leq t \leq \sqrt[3]{4}\}$$

or by many other expressions.

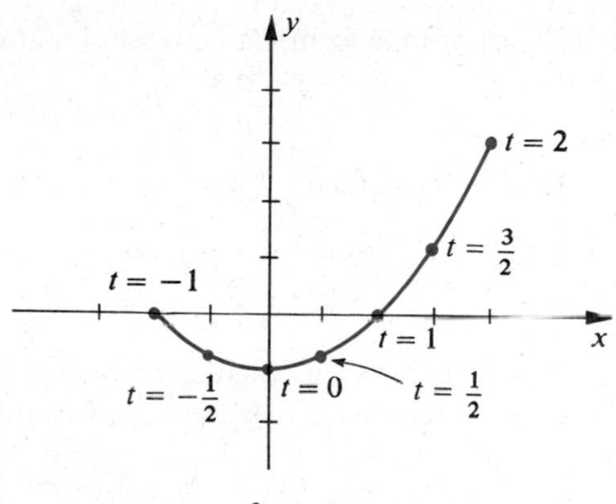

$x = 2t, y = t^2 - 1; -1 \leqslant t \leqslant 2$

Figure 13.2

Example 2 Describe the graph of the curve C having parametric equations

$$x = \cos t, \quad y = \sin t, \quad 0 \leq t \leq 2\pi.$$

Solution Eliminating the parameter gives us $x^2 + y^2 = 1$ and hence points on C are on the unit circle with center at the origin. As t increases from 0 to 2π, $P(t)$ starts at the point $A(1, 0)$ and traverses the circle once in the counterclockwise direction. In this example the parameter may be interpreted geometrically as the length of arc from A to P, as illustrated in Figure 9.1.

If a curve C is described by means of an equation $y = f(x)$, where f is a continuous function, then an easy way to obtain parametric equations is to let

$$x = t, \quad y = f(t)$$

where t is in the domain of f. For example, if $y = x^3$, then parametric equations are

$$x = t, \quad y = t^3, \quad \text{where } t \text{ is in } \mathbb{R}.$$

We can use many different substitutions for x, provided that as t varies through some interval, x takes on all values in the domain of f. Thus the graph of $y = x^3$ is also given by

$$x = t^{1/3}, \quad y = t, \quad \text{where } t \text{ is in } \mathbb{R}.$$

Note, however, that the parametric equations

$$x = \sin t, \quad y = \sin^3 t, \quad \text{where } t \text{ is in } \mathbb{R}$$

give only that part of the graph of $y = x^3$ which lies between $(-1, -1)$ and $(1, 1)$.

Example 3 Find parametric equations for the line of slope m through the point (x_1, y_1).

Solution By the point-slope form, an equation for the line is

$$y - y_1 = m(x - x_1).$$

If we let $x = t$, then $y - y_1 = m(t - x_1)$ and parametric equations for the line are

$$x = t, \quad y = y_1 + m(t - x_1) \quad \text{where } t \text{ is in } \mathbb{R}.$$

Simpler parametric equations can be found by letting $x - x_1 = t$. In this case $y - y_1 = mt$ and hence the line is given parametrically by

$$x = x_1 + t, \quad y = y_1 + mt, \quad \text{where } t \text{ is in } \mathbb{R}.$$

As a third illustration, if we let $x - x_1 = \tan t$ we obtain

$$x = x_1 + \tan t, \quad y = y_1 + m \tan t, \quad \text{where } -\frac{\pi}{2} < t < \frac{\pi}{2}.$$

There are, of course, many other ways to represent the line parametrically.

Suppose a curve C has parametric equations $x = f(t)$, $y = g(t)$, where t is in an interval I. We call C a **smooth curve** if the derivatives f' and g' are continuous on I and are not simultaneously zero, except possibly at any end-points of I. The curve C is **piecewise smooth** if the interval I can be partitioned into subintervals such that C is smooth on each subinterval. The graph of a smooth curve has no corners. The curves given in Examples 1–3 are smooth. The curve described in the next example is piecewise smooth.

Example 4 The curve traced by a fixed point P on the circumference of a circle as the circle rolls along a straight line in a plane is called a **cycloid**. Find parametric equations for a cycloid. Determine whether the cycloid is smooth or piecewise smooth.

Solution Suppose the circle has radius a and that it rolls along (and above) the x-axis in the positive direction. If one position of P is the origin, then Figure 13.3 displays part of the curve and a possible position of the circle.

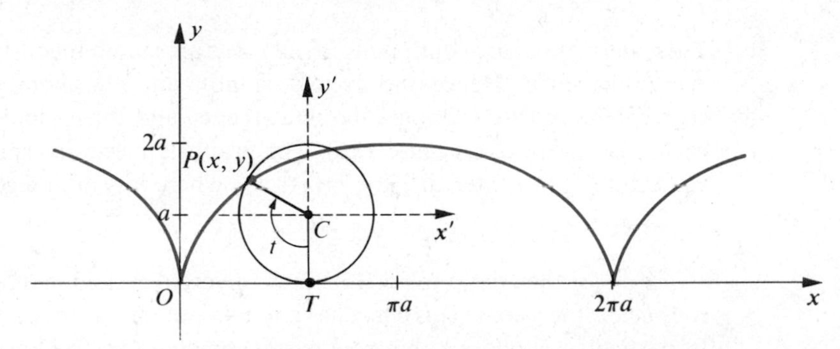

Figure 13.3

Let C denote the center of the circle and T the point of tangency with the x-axis. We introduce a parameter t as the radian measure of angle TCP. Since $\overline{OT}$ is the distance the circle has rolled, $\overline{OT} = at$. Consequently, the coordinates of C are (at, a). If we consider an $x'y'$-coordinate system with origin at $C(at, a)$ and if $P(x', y')$ denotes the point P relative to this system, then by (7.19), with $h = at$ and $k = a$,

$$x = at + x', \quad y = a + y'.$$

If, as in Figure 13.4, θ denotes an angle in standard position on the $x'y'$-system, then $\theta = 3\pi/2 - t$. Hence

$$\begin{aligned} x' &= a\cos\theta = a\cos(3\pi/2 - t) = -a\sin t \\ y' &= a\sin\theta = a\sin(3\pi/2 - t) = -a\cos t, \end{aligned}$$

and substitution in $x = at + x', y = a + y'$ gives us parametric equations for the cycloid, namely

(13.3) $$x = a(t - \sin t), \quad y = a(1 - \cos t)$$

where t is in $\mathbb{R}$.

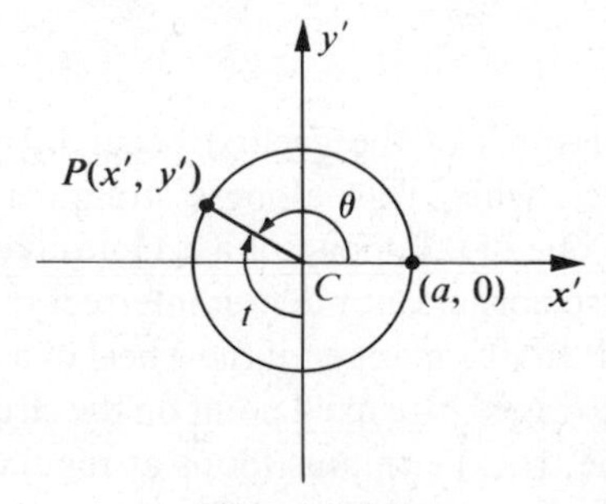

Figure 13.4

To examine whether the cycloid is smooth we consider

$$\frac{dx}{dt} = a(1 - \cos t), \quad \frac{dy}{dt} = a \sin t.$$

These derivatives are continuous for all t, but are simultaneously 0 if $t = 2\pi n$ for every integer n. Hence the cycloid is not smooth. Note that the points corresponding to $t = 2\pi n$ are the x-intercepts, and the cycloid has a corner at each such point (see Figure 13.3). The graph is piecewise smooth, since it is smooth in each t-interval $[2\pi n, 2\pi(n+1)]$ where n is an integer.

If $a < 0$, then the graph of (13.3) is the inverted cycloid that results if the circle rolls *below* the x-axis. This curve has a number of important physical properties. In particular, suppose a thin wire passes through two fixed points A and B as illustrated in Figure 13.5, and that the shape of the wire can be changed by bending it in any manner. Suppose further, that a bead is allowed to slide along the wire and the only force acting on the bead is gravity. We now ask which of all the possible paths will allow the bead to slide from A to B in the least amount of time. It is natural to conjecture that the desired path is the straight line segment from A to B; however, this is not the correct answer. The path which requires the least time coincides with the graph of (13.3) where $a < 0$ and A is the origin. The proof of this result is difficult and may be found in more advanced texts.

To cite another interesting property of this curve, suppose that A is the origin and B is the point with abscissa $\pi|a|$, that is, the lowest point on the cycloid occurring in the first arc to the right of A. It can be shown that if the bead is released at *any* point between A and B, the time required for it to reach B is always the same!

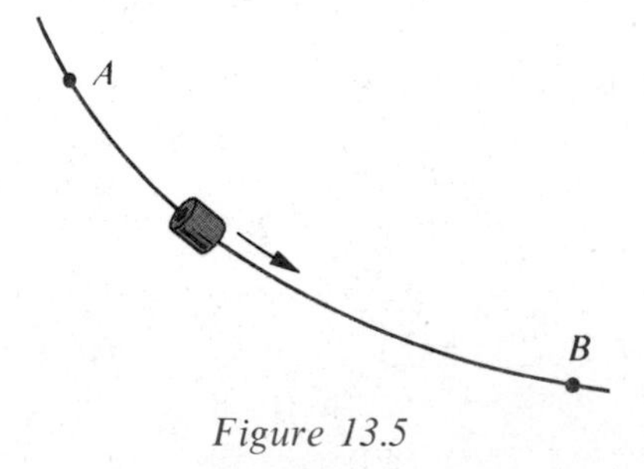

Figure 13.5

Variations of the cycloid occur in practical problems. For example, if a motorcycle wheel rolls along a straight road, then the curve traced by a fixed point on one of the spokes is a cycloid-like curve. In this case, the curve does not have sharp corners, nor does it intersect the road (the x-axis) as does the graph of a cycloid. In like manner, if the wheel of a train rolls along a railroad track, then the curve traced by a fixed point on the circumference of the wheel (which extends below the track) contains loops at regular intervals. Several other cycloids are defined in Exercises 33 and 34.

EXERCISES 13.1

In each of Exercises 1–20, (a) sketch the graph of the curve C having the indicated parametric equations, and (b) find a rectangular equation of a graph which contains the points on C.

1 $x = t - 2, y = 2t + 3; 0 \leq t \leq 5$

2 $x = 1 - 2t, y = 1 + t; -1 \leq t \leq 4$

3 $x = t^2 + 1, y = t^2 - 1; -2 \leq t \leq 2$

4 $x = t^3 + 1, y = t^3 - 1; -2 \leq t \leq 2$

5 $x = 4t^2 - 5, y = 2t + 3; t$ in $\mathbb{R}$

6 $x = t^3, y = t^2; t$ in $\mathbb{R}$

7 $x = e^t, y = e^{-2t}; t$ in $\mathbb{R}$

8 $x = \sqrt{t}, y = 3t + 4; t \geq 0$

9 $x = 2 \sin t, y = 3 \cos t; 0 \leq t \leq 2\pi$

10 $x = \cos t - 2, y = \sin t + 3; 0 \leq t \leq 2\pi$

11 $x = \sec t, y = \tan t; -\pi/2 < t < \pi/2$

12 $x = \cos 2t, y = \sin t; -\pi \leq t \leq \pi$

13 $x = t^2, y = 2 \ln t; t > 0$

14 $x = \cos^3 t, y = \sin^3 t; 0 \leq t \leq 2\pi$

15 $x = \sin t, y = \csc t; 0 < t \leq \pi/2$

16 $x = e^t, y = e^{-t}; t$ in $\mathbb{R}$

17 $x = \cosh t, y = \sinh t; t$ in $\mathbb{R}$

18 $x = 3 \cosh t, y = 2 \sinh t; t$ in $\mathbb{R}$

19 $x = t, y = \sqrt{t^2 - 1}; |t| \geq 1$

20 $x = -2\sqrt{1 - t^2}, y = t; |t| \leq 1$

In Exercises 21–24, sketch the graph of the curve C having the indicated parametric equations and determine if C is smooth or piecewise smooth.

21 $x = t, y = \sqrt{t^2 - 2t + 1}; 0 \leq t \leq 4$

22 $x = 2t, y = 8t^3; -1 \leq t \leq 1$

23 $x = (t + 1)^3, y = (t + 2)^2; 0 \leq t \leq 2$

24 $x = \tan t, y = 1; -\pi/2 < t < \pi/2$

In Exercises 25 and 26 curves C_1, C_2, C_3, and C_4 are given parametrically, where t is in $\mathbb{R}$. Sketch the graphs of C_1, C_2, C_3, C_4, and discuss their similarities and differences.

25 $C_1: x = t^2, y = t$
$C_2: x = t^4, y = t^2$
$C_3: x = \sin^2 t, y = \sin t$
$C_4: x = e^{2t}, y = -e^t$

26 $C_1: x = t, y = 1 - t$
$C_2: x = 1 - t^2, y = t^2$
$C_3: x = \cos^2 t, y = \sin^2 t$
$C_4: x = \ln t - t, y = 1 + t - \ln t$

The parametric equations in each of Exercises 27 and 28 give the position of a point $P(x, y)$, where t represents time. Describe the motion of the point during the indicated time interval.

27 (a) $x = \cos t, y = \sin t; 0 \leq t \leq \pi$
(b) $x = \sin t, y = \cos t; 0 \leq t \leq \pi$
(c) $x = t, y = \sqrt{1 - t^2}; -1 \leq t \leq 1$

28 (a) $x = t^2, y = 1 - t^2; 0 \leq t \leq 1$
(b) $x = 1 - \ln t, y = \ln t; 1 \leq t \leq e$
(c) $x = \cos^2 t, y = \sin^2 t; 0 \leq t \leq 2\pi$

29 If $P_1(x_1, y_1)$ and $P_2(x_2, y_2)$ are distinct points, show that

$$x = (x_2 - x_1)t + x_1, \quad y = (y_2 - y_1)t + y_1,$$

where t is in $\mathbb{R}$, are parametric equations of the line l through P_1 and P_2. Find three other pairs of parametric equations for the line l. Show that there are an infinite number of different pairs of parametric equations for l.

30 What is the difference between the graph of the hyperbola $b^2x^2 - a^2y^2 = a^2b^2$ and the graph of $x = a \cosh t, y = b \sinh t$, where t is in $\mathbb{R}$?

31 Show that

$$x = a\cos t + h, \quad y = b\sin t + k,$$

where $0 \le t \le 2\pi$ are parametric equations of an ellipse with center at the point (h, k) and semi-axes of lengths a and b.

32 Find parametric equations for the parabola with (a) vertex $V(0, 0)$ and focus $F(0, p)$; (b) vertex $V(h, k)$ and focus $F(h, k + p)$.

33 A circle C of radius b rolls on the inside of a second circle having equation $x^2 + y^2 = a^2$, where $b < a$. Let P be a fixed point on C and let the initial position of P be $A(a, 0)$. If the parameter t is the angle from the positive x-axis to the line segment from O to the center of C, show that parametric equations for the curve traced by P (called a **hypocycloid**) are

$$x = (a - b)\cos t + b\cos\frac{a - b}{b}t, \qquad y = (a - b)\sin t - b\sin\frac{a - b}{b}t$$

where $0 \le t \le 2\pi$. If $b = a/4$ show that

$$x = a\cos^3 t, \quad y = a\sin^3 t$$

and sketch the graph of the curve.

34 If the circle C of Exercise 33 rolls on the outside of the second circle, find parametric equations for the curve traced by P. (This curve is called an **epicycloid**.)

13.2 TANGENT LINES TO CURVES

In Example 1 of the preceding section we saw that the curve C having parametric equations

$$x = 2t, \quad y = t^2 - 1, \quad \text{where } -1 \le t \le 2$$

can also be described by using a rectangular equation of the form $y = k(x)$, where k is a function defined on a suitable interval. As a matter of fact we may take

$$y = k(x) = \tfrac{1}{4}x^2 - 1, \quad \text{where } -2 \le x \le 4.$$

It follows that the slope of the tangent line at any point $P(x, y)$ on C is

$$k'(x) = \tfrac{1}{2}x \quad \text{or} \quad k'(x) = \tfrac{1}{2}(2t) = t.$$

We shall now introduce a technique for finding the slope directly from the parametric equations.

Let $x = f(t)$, $y = g(t)$ be parametric equations of a curve C and suppose C can also be represented by a rectangular equation of the form $y = k(x)$, where k is a function. Let us also assume that f, g, and k are differentiable throughout their domains. If $P(f(t), g(t))$ is a point on C, then the coordinates satisfy the equation $y = k(x)$, that is,

$$g(t) = k(f(t)).$$

Applying the Chain Rule (3.36),

$$g'(t) = k'(f(t))f'(t) = k'(x)f'(t)$$

and hence the slope of the tangent line at P is

(13.4)
$$k'(x) = \frac{g'(t)}{f'(t)}, \quad \text{provided } f'(t) \neq 0.$$

If $f'(t) = 0$ and $g'(t) \neq 0$ for some t, then C has a vertical tangent line at the point corresponding to t. An easy way to remember formula (13.4) is to use the differential notation:

(13.5)
$$\frac{dy}{dx} = \frac{dy/dt}{dx/dt}, \quad \text{provided } \frac{dx}{dt} \neq 0.$$

The conditions imposed in the preceding discussion are fulfilled if C is a smooth curve and if $f'(t) \neq 0$ for all t in a closed interval $[a, b]$. In this case f' is continuous and hence either $f'(t) > 0$ or $f'(t) < 0$ throughout $[a, b]$. (Why?) It follows from Theorem (8.38) or Theorem (8.40) that f has an inverse function f^{-1}, and since $x = f(t)$ we may write $t = f^{-1}(x)$, where x is in the interval $[f(a), f(b)]$. Hence

$$y = g(t) = g(f^{-1}(x)) = k(x)$$

where k is the composite function $g \circ f^{-1}$.

Example 1 Find the slopes of the tangent and normal lines at the point $P(t)$ on the curve $C = \{(2t, t^2 - 1): -1 \leq t \leq 2\}$.

Solution The curve C is the same as that considered in Example 1 of the preceding section (see Figure 13.2). Using (13.5) with $x = 2t$ and $y = t^2 - 1$, the slope of the tangent line at $P(t)$ is

$$\frac{dy}{dx} = \frac{dy/dt}{dx/dt} = \frac{2t}{2} = t.$$

Note that this agrees with the discussion at the beginning of this section. The slope of the normal line is $-1/t$, provided $t \neq 0$.

Example 2 If a curve C has parametric equations $x = t^3 - 3t$, $y = t^2 - 5t - 1$, where t is in $\mathbb{R}$, find an equation of the tangent line at the point corresponding to $t = 2$. For what values of t is the tangent line horizontal or vertical?

Solution Applying (13.5),

$$\frac{dy}{dx} = \frac{dy/dt}{dx/dt} = \frac{2t - 5}{3t^2 - 3}.$$

Thus at the point P corresponding to $t = 2$, the slope m of the tangent line is

$$m = \frac{2(2) - 5}{3(2^2) - 3} = -\frac{1}{9}.$$

Substituting $t = 2$ in the parametric equations gives us the coordinates $(2, -7)$ of P. Hence an equation of the tangent line is

$$y + 7 = -\tfrac{1}{9}(x - 2)$$

or equivalently

$$x + 9y + 61 = 0.$$

The tangent line is horizontal if $dy/dt = 0$, that is, if $2t - 5 = 0$ or $t = 5/2$. The tangent line is vertical if $3t^2 - 3 = 0$. (Why?) Thus there are vertical tangent lines at the points corresponding to $t = 1$ and $t = -1$.

If the graph of a curve C coincides with the graph of an equation $y = k(x)$, then

$$y' = \frac{dy}{dx} = \frac{dy/dt}{dx/dt}.$$

If y' is a differentiable function of t, then we can find d^2y/dx^2 by applying (13.4) to y' as follows:

$$\frac{d^2y}{dx^2} = \frac{d}{dx}(y') = \frac{dy'/dt}{dx/dt}.$$

Example 3 Suppose C has parametric equations

$$x = e^{-t}, \quad y = e^{2t}, \quad \text{where } t \text{ is in } \mathbb{R}.$$

(a) Find d^2y/dx^2 directly from the parametric equations.

(b) Define a function k which has the same graph as C and check the answer to part (a).

(c) Discuss the concavity of C.

Solution (a) First we find

$$y' = \frac{dy}{dx} = \frac{dy/dt}{dx/dt} = \frac{2e^{2t}}{-e^{-t}} = -2e^{3t}.$$

Second,

$$\frac{d^2y}{dx^2} = \frac{dy'}{dx} = \frac{dy'/dt}{dx/dt} = \frac{-6e^{3t}}{-e^{-t}} = 6e^{4t}.$$

(b) Using the fact that $x = e^{-t} = 1/e^t$, we obtain $e^t = 1/x = x^{-1}$. Since $y = e^{2t} = (e^t)^2$, it follows that C is the graph of the equation

$$y = (x^{-1})^2 = x^{-2}, \quad \text{where } x > 0.$$

The restriction $x > 0$ is necessary because $e^{-t} > 0$ for all t. Thus a function k whose graph coincides with the graph of C may be defined by

$$k(x) = x^{-2}, \quad \text{where } x > 0.$$

Since $k'(x) = -2x^{-3}$,

$$k''(x) = 6x^{-4} = 6(e^{-t})^{-4} = 6e^{4t}$$

which is in agreement with part (a).

(c) Since $d^2y/dx^2 = 6e^{4t} > 0$ for all t, the curve C is concave upward at every point.

EXERCISES 13.2

Find the slopes of the tangent lines at the point corresponding to $t = 1$ on each of the curves defined in Exercises 1–10. Compare with Exercises 1–10 of Section 1.

1 $x = t - 2, y = 2t + 3; 0 \leq t \leq 5$

2 $x = 1 - 2t, y = 1 + t; -1 \leq t \leq 4$

3 $x = t^2 + 1, y = t^2 - 1; -2 \leq t \leq 2$

4 $x = t^3 + 1, y = t^3 - 1; -2 \leq t \leq 2$

5 $x = 4t^2 - 5, y = 2t + 3; t$ in $\mathbb{R}$

6 $x = t^3, y = t^2; t$ in $\mathbb{R}$

7 $x = e^t, y = e^{-2t}; t$ in $\mathbb{R}$

8 $x = \sqrt{t}, y = 3t + 4; t \geq 0$

9 $x = 2 \sin t, y = 3 \cos t; 0 \leq t \leq 2\pi$

10 $x = \cos t - 2, y = \sin t + 3; 0 \leq t \leq 2\pi$

11 Let l be the line with parametric equations $x = t/2, y = 2t - 5$. If a curve C is given by $x = t^4 - 7, y = 8t + 3$, where t is in $\mathbb{R}$, find a rectangular equation of the normal line to C which is parallel to l.

12 If a curve C has parametric equations $x = 6t + 1, y = t^3 - 2t$, where t is in $\mathbb{R}$, find the points on C at which the tangent line is perpendicular to the line $3x + 5y - 8 = 0$.

13 If C is given parametrically by $x = 4t^2, y = 2t - 5$, where t is in $\mathbb{R}$, find a rectangular equation of the line through the point $(4, 1)$ which is tangent to C.

14 Find a rectangular equation of the line through the origin which is normal to the curve having parametric equations $x = 3t^2 - 2$, $y = 2t^3$, where t is in $\mathbb{R}$.

In each of Exercises 15–18 find the points on the indicated curve at which the tangent line is either horizontal or vertical. Sketch the graph, showing the horizontal and vertical tangent lines. Find d^2y/dx^2 in each case.

15 $x = 4t^2, y = t^3 - 12t; t$ in $\mathbb{R}$

16 $x = t^3 + 1, y = t^2 - 2t; t$ in $\mathbb{R}$

17 $x = 3t^2 - 6t, y = \sqrt{t}; t \geq 0$

18 $x = \sqrt[3]{t}, y = \sqrt[3]{t} - t; t$ in $\mathbb{R}$

19 What is the slope of the tangent line at the point $P(t)$ on the cycloid

$$x = a(t - \sin t), \quad y = a(1 - \cos t),$$

where t is in $\mathbb{R}$? At what points is the tangent line horizontal or vertical? Where does the slope of the tangent line equal 1?

20 Answer the questions of Exercise 19 for the hypocycloid $x = a\cos^3 t$, $y = a\sin^3 t$ where $0 \leq t \leq 2\pi$.

13.3 POLAR COORDINATE SYSTEMS

We have previously specified points in a plane in terms of rectangular coordinates, using the ordered pair (a, b) to denote the point whose directed distances from the x- and y-axes are b and a, respectively. Another important method for representing points is by means of **polar coordinates**. In order to introduce a system of polar coordinates in a plane we begin with a fixed point O (called the **origin**, or **pole**) and a directed half-line (called the **polar axis**) with end-point O. Next we consider any point P in the plane different from O. If, as illustrated in Figure 13.6, $r = d(O, P)$ and θ denotes the measure of any angle determined by the polar axis and OP, then r and θ are called **polar coordinates** of P and the symbols (r, θ) or $P(r, \theta)$ are used to denote P. As usual θ is considered positive if the angle is generated by a counterclockwise rotation of the polar axis and negative if the rotation is clockwise. Either radians or degrees may be used for the measure of θ.

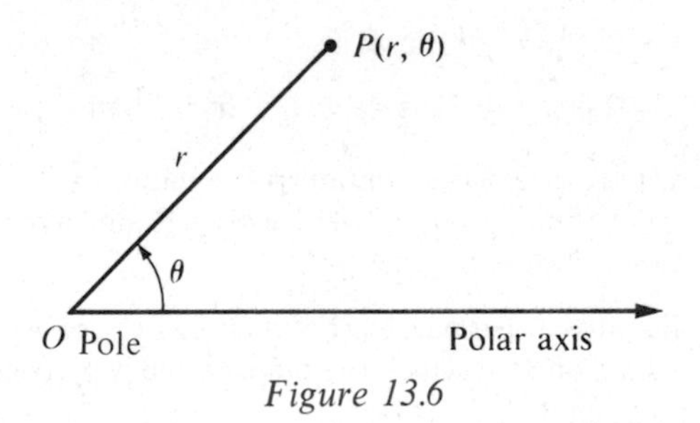

Figure 13.6

If the polar axis is the initial side, then there are many angles with the same terminal side. Hence the polar coordinates of a point are not unique. For example, $(3, \pi/4)$, $(3, 9\pi/4)$, and $(3, -7\pi/4)$ all represent the same point (see Figure 13.7). We shall also allow r to be negative. In this event, instead of measuring $|r|$ units along the terminal side of the angle θ, we measure along the half-line with end-point O which has direction opposite to that of the terminal side. The points corresponding to the pairs $(-3, 5\pi/4)$ and $(-3, -3\pi/4)$ are also plotted in Figure 13.7.

Finally, we agree that the pole O has polar coordinates $(0, \theta)$ for *any* θ. An assignment of ordered pairs of the form (r, θ) to points in a plane will be referred to as a **polar coordinate system** and the plane will be called an $\boldsymbol{r\theta}$**-plane**.

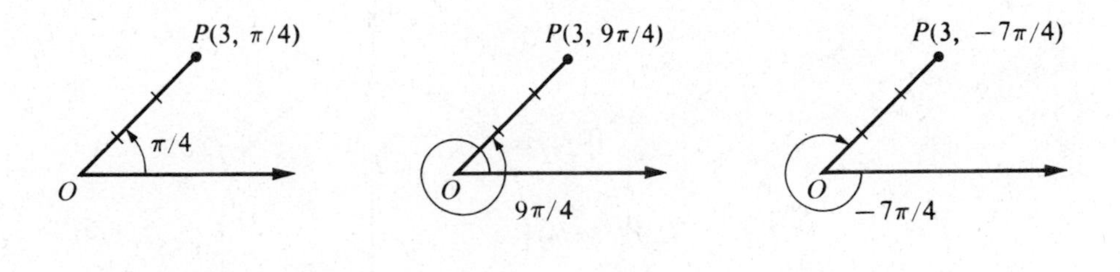

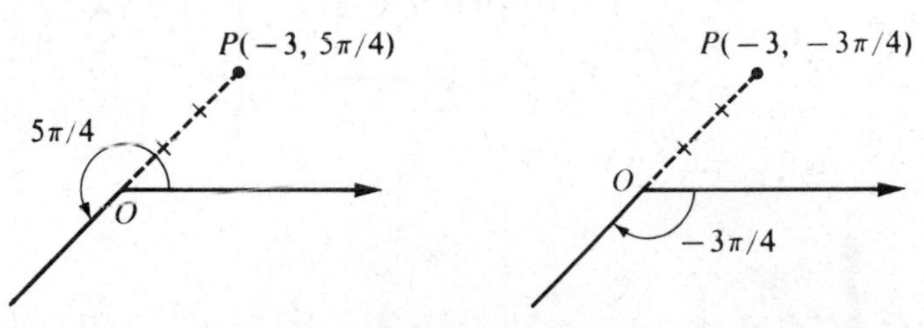

Figure 13.7

A **polar equation** is an equation in r and θ. A **solution** of a polar equation is an ordered pair (a, b) which leads to equality if a is substituted for r and b for θ. The **graph** of a polar equation is the set of all points (in an $r\theta$-plane) which correspond to the solutions.

Example 1 Sketch the graph of the polar equation $r = 4\sin\theta$.

Solution The following table contains some solutions of the equation.

θ	0	$\pi/6$	$\pi/4$	$\pi/3$	$\pi/2$	$2\pi/3$	$3\pi/4$	$5\pi/6$	π
r	0	2	$2\sqrt{2}$	$2\sqrt{3}$	4	$2\sqrt{3}$	$2\sqrt{2}$	2	0

In rectangular coordinates, the graph of the given equation consists of sine waves of amplitude 4 and period 2π. However, if polar coordinates are used, then the points which correspond to the pairs in the table lie on a circle of radius 2 and we draw the graph accordingly (see Figure 13.8). As an aid to plotting points, we have extended the polar axis in the negative direction and introduced a vertical line through the pole. The proof that the graph is a circle will be given in Example 4. Additional points obtained by letting θ vary from π to 2π lie on the same circle. For example, the solution $(-2, 7\pi/6)$ gives us the same point as $(2, \pi/6)$; the point corresponding to $(-2\sqrt{2}, 5\pi/4)$ is the same as that obtained from $(2\sqrt{2}, \pi/4)$; and so on. If we let θ increase through all real numbers we obtain the same points over and over because of the periodicity of the sine function.

Example 2 Sketch the graph of the equation $r = 2 + 2\cos\theta$.

Solution Since the cosine function decreases from 1 to -1 as θ varies from 0 to π, it follows that r decreases from 4 to 0 in this θ-interval. The following table exhibits some solutions of the given equation.

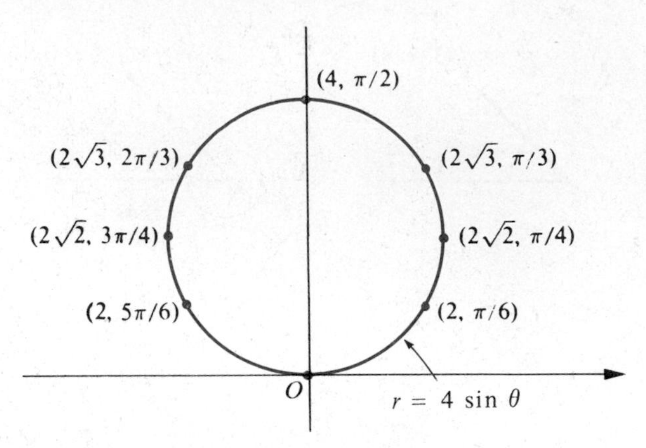

Figure 13.8

θ	0	$\pi/6$	$\pi/4$	$\pi/3$	$\pi/2$	$2\pi/3$	$3\pi/4$	$5\pi/6$	π
r	4	$2+\sqrt{3}$	$2+\sqrt{2}$	3	2	1	$2-\sqrt{2}$	$2-\sqrt{3}$	0

If θ increases from π to 2π, then $\cos\theta$ increases from -1 to 1, and consequently r increases from 0 to 4. Plotting points and connecting them with a smooth curve leads to the sketch shown in Figure 13.9, where we have used polar coordinate graph paper which displays lines through O at various angles and circles with centers at the pole. The same graph may be obtained by taking other intervals for θ.

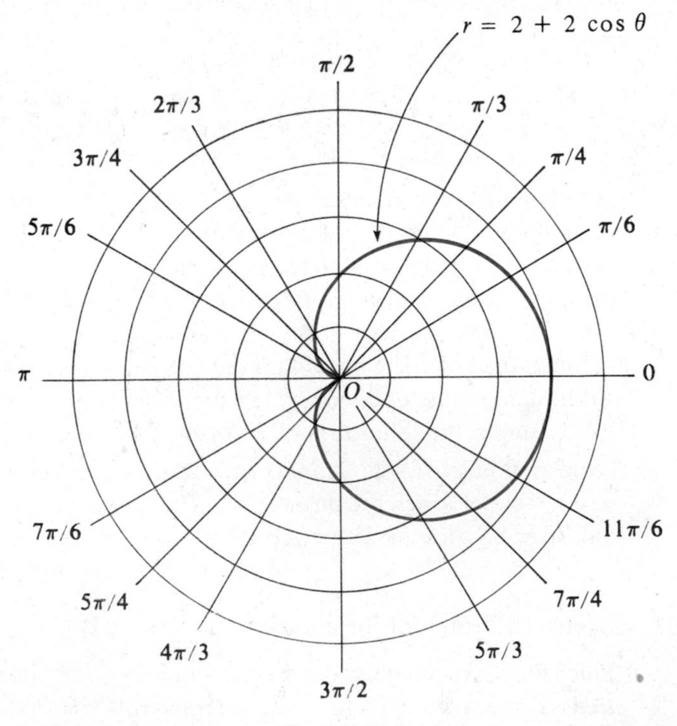

Figure 13.9

The heart-shaped graph in Example 2 is called a **cardioid**. In general, the graph of any polar equation of the form

$$r = a(1 + \cos\theta), \qquad r = a(1 + \sin\theta)$$
$$r = a(1 - \cos\theta), \qquad r = a(1 - \sin\theta)$$

where a is a real number, is a cardioid.

Example 3 Sketch the graph of the equation $r = a\sin 2\theta$, where $a > 0$.

Solution Instead of tabulating solutions, let us reason as follows. If θ increases from 0 to $\pi/4$, then 2θ varies from 0 to $\pi/2$ and hence $\sin 2\theta$ increases from 0 to 1. It follows that r increases from 0 to a in the θ-interval $[0, \pi/4]$. If we next let θ increase from $\pi/4$ to $\pi/2$, then 2θ changes from $\pi/2$ to π. Consequently r decreases from a to 0 in the θ-interval $[\pi/4, \pi/2]$. (Why?) The corresponding points on the graph constitute a loop as illustrated in Figure 13.10. We shall leave it to the reader to show that as θ increases from $\pi/2$ to π, a similar loop is obtained directly *below* the first loop. Note that for this range of θ we have $\pi < 2\theta < 2\pi$ and hence $\sin 2\theta$ is negative. Similar loops are obtained for the θ-intervals $[\pi, 3\pi/2]$ and $[3\pi/2, 2\pi]$. In Figure 13.10 we have plotted only those points on the graph which correspond to the largest numerical values of r.

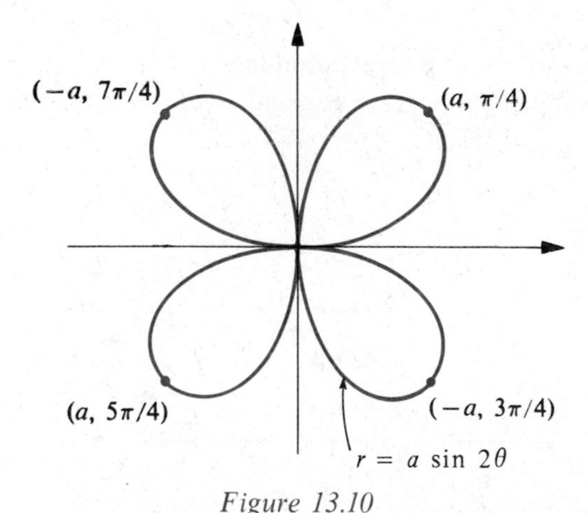

Figure 13.10

The graph in Example 3 is called a **four-leaved rose**. In general, any equation of the form

$$r = a\sin n\theta \quad \text{or} \quad r = a\cos n\theta$$

where n is a positive integer greater than 1 and a is a real number, has a graph which consists of a number of loops attached to the origin. If n is even there are $2n$ loops, whereas if n is odd there are n loops (see, for example, Exercises 9, 10, and 22).

Many other interesting graphs result from polar equations. Some are included in the exercises at the end of this section. Polar coordinates are very useful in applications involving circles with centers at the origin or lines that pass through the origin, since equations which have these graphs may be written in the simple forms $r = k$ or $\theta = k$ for some fixed number k. (Verify this fact!)

Let us now superimpose an xy-plane on an $r\theta$-plane in such a way that the positive x-axis coincides with the polar axis. Any point P in the plane may then be assigned rectangular coordinates (x, y) or polar coordinates (r, θ). It is not difficult to obtain formulas which specify the relationship between the two coordinate systems. If $r > 0$ we have a situation similar to that illustrated in (i) of Figure 13.11. If $r < 0$ we have that shown in (ii) of the figure where, for later purposes, we have also plotted the point P' having polar coordinates $(|r|, \theta)$ and rectangular coordinates $(-x, -y)$. Although we have pictured θ as an acute angle, the discussion which follows is valid for all angles. On the one hand, if $r > 0$ as in (i) of Figure 13.11, then

(13.6) $$x = r\cos\theta, \quad y = r\sin\theta.$$

On the other hand, if $r < 0$ then $|r| = -r$, and from (ii) of Figure 13.11 we see that

$$\cos\theta = \frac{-x}{|r|} = \frac{-x}{-r} = \frac{x}{r},$$

$$\sin\theta = \frac{-y}{|r|} = \frac{-y}{-r} = \frac{y}{r}.$$

Multiplication by r produces (13.6) and therefore the latter formulas hold whether r is positive or negative. If $r = 0$, then the point is the pole and we again see that (13.6) is true.

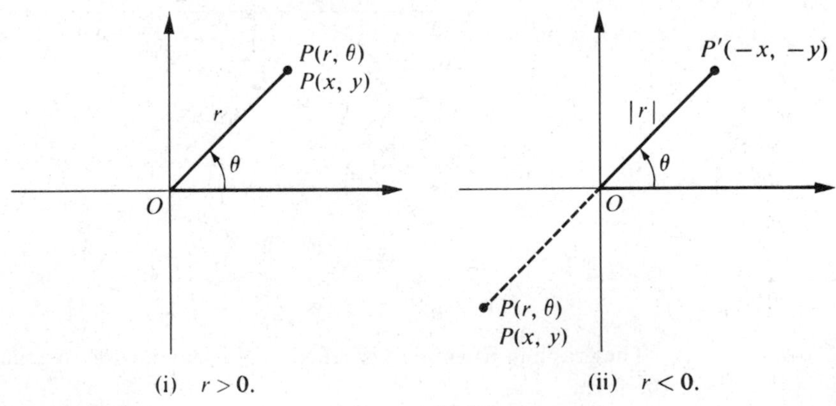

Figure 13.11

The following formulas are further consequences of our discussion:

(13.7) $$\tan\theta = \frac{y}{x}, \quad r^2 = x^2 + y^2.$$

We may use (13.6) and (13.7) to change from one system of coordinates to the other. A more important use is for transforming a polar equation to an equation in x and y and vice versa. This is illustrated in the next two examples.

Example 4 Find an equation in x and y which has the same graph as $r = 4\sin\theta$.

Solution The given equation was considered in Example 1. It is convenient to multiply both sides by r, obtaining $r^2 = 4r\sin\theta$. Applying (13.6) and (13.7) gives us $x^2 + y^2 = 4y$. The latter equation is equivalent to $x^2 + (y - 2)^2 = 4$ whose graph is a circle of radius 2 with center at $(0, 2)$ in the xy-plane.

Example 5 Find a polar equation of an arbitrary line.

Solution We know that every straight line in an xy-coordinate system is the graph of a linear equation $ax + by + c = 0$. Substituting for x and y from (13.6) leads to the polar equation

$$r(a\cos\theta + b\sin\theta) + c = 0.$$

If we superimpose an xy-plane on an $r\theta$-plane, then the graph of a polar equation may be symmetric with respect to the x-axis, the y-axis, or the origin. It is left to the reader to show that if a substitution listed in the following table does not change the solutions of a polar equation, then the graph has the indicated symmetry.

Substitution	*Symmetry*
$-\theta$ for θ	x-axis
$-r$ for r	origin
$\pi - \theta$ for θ	y-axis

To illustrate, since $\cos(-\theta) = \cos\theta$, the graph of the equation in Example 2 (see Figure 13.9) is symmetric with respect to the x-axis. Since $\sin(\pi - \theta) = \sin\theta$, the graph in Example 1 is symmetric with respect to the y-axis. The graph in Example 3 is symmetric to both axes and the origin. Other tests for symmetry may be stated; however, those listed above are among the easiest to apply.

EXERCISES 13.3

Sketch the graph of each of the polar equations in Exercises 1–24.

1 $r = 5$

2 $\theta = \pi/4$

3 $\theta = -\pi/6$

4 $r = -2$

5 $r = 4\cos\theta$

6 $r = -2\sin\theta$

7 $r = 4(1 - \sin\theta)$ (cardioid)

8 $r = 1 + 2\cos\theta$ (limaçon)

9 $r = 8\cos 3\theta$ (three-leaved rose)

10 $r = 2\sin 4\theta$ (eight-leaved rose)

11 $r^2 = 4\cos 2\theta$ (lemniscate)

12 $r = 6\sin^2(\frac{1}{2}\theta)$ (cardioid)

13 $r = 4 \csc \theta$

14 $r = -3 \sec \theta$

15 $r = 2 - \cos \theta$ (limaçon)

16 $r = 2 + 2 \sec \theta$ (conchoid)

17 $r = 2^{\theta}, \theta \geq 0$ (spiral)

18 $r\theta = 1, \theta > 0$ (spiral)

19 $r = -6(1 + \cos \theta)$ (cardioid)

20 $r = e^{2\theta}$ (logarithmic spiral)

21 $r = 2 + 4 \cos \theta$ (limaçon)

22 $r = 8 \cos 5\theta$ (five-leaved rose)

23 $r^2 = -16 \sin 2\theta$ (lemniscate)

24 $4r = \theta$ (spiral)

In exercises 25–32 find a polar equation which has the same graph as the given equation.

25 $x = -3$

26 $y = 2$

27 $x^2 + y^2 = 16$

28 $x^2 = 8y$

29 $y = 6$

30 $y = 6x$

31 $x^2 - y^2 = 16$

32 $9x^2 + 4y^2 = 36$

In exercises 33–48 find an equation in x and y which has the same graph as the given polar equation and use it as an aid in sketching the graph in a polar coordinate system.

33 $r \cos \theta = 5$

34 $r \sin \theta = -2$

35 $r - 6 \sin \theta = 0$

36 $r = 6 \cot \theta$

37 $r = 2$

38 $\theta = \pi/4$

39 $r = \tan \theta$

40 $r = 4 \sec \theta$

41 $r^2(4 \sin^2 \theta - 9 \cos^2 \theta) = 36$

42 $r^2(\cos^2 \theta + 4 \sin^2 \theta) = 16$

43 $r^2 \cos 2\theta = 1$

44 $r^2 \sin 2\theta = 4$

45 $r(\sin \theta - 2 \cos \theta) = 6$

46 $r(\sin \theta + r \cos^2 \theta) = 1$

47 $r = \dfrac{1}{1 + \cos \theta}$

48 $r = \dfrac{4}{2 + \sin \theta}$

49 If $P_1(r_1, \theta_1)$ and $P_2(r_2, \theta_2)$ are points in an $r\theta$-plane, use the Law of Cosines to prove that $[d(P_1, P_2)]^2 = r_1^2 + r_2^2 - 2r_1r_2 \cos(\theta_2 - \theta_1)$.

50 Prove that the graphs of the following polar equations are circles and find the center and radius in each case.

(a) $r = a \sin \theta, a \neq 0$
(b) $r = b \cos \theta, b \neq 0$
(c) $r = a \sin \theta + b \cos \theta, ab \neq 0$

13.4 POLAR EQUATIONS OF CONICS

The following theorem provides another method for describing the conic sections.

(13.8) Theorem

> Let F be a fixed point and l a fixed line in a plane. The set of all points P in the plane such that the ratio $d(P, F)/d(P, Q)$ is a positive constant e, where $d(P, Q)$ is the distance from P to l, is a conic section. Moreover, the conic is a parabola if $e = 1$, an ellipse if $0 < e < 1$, and a hyperbola if $e > 1$.

The constant e is called the **eccentricity** of the conic and should not be confused with the base of the natural logarithms. It will be seen that the point F is a focus of the conic. The line l is called a **directrix**. We shall prove (13.8) if $e \leq 1$ and leave the case $e > 1$ to the reader.

If $e = 1$ in Theorem (13.8), then $d(P, F) = d(P, Q)$ and, according to Definition (7.1), a parabola with focus F and directrix l is obtained.

Suppose next that $0 < e < 1$. It is convenient to introduce a polar coordinate system in the plane with F as the pole and with l perpendicular to the polar axis at the point $D(d, 0)$, where $d > 0$. If $P(r, \theta)$ is a point in the plane such that $d(P, F)/d(P, Q) = e < 1$, then from Figure 13.12 we see that P lies to the left of l.

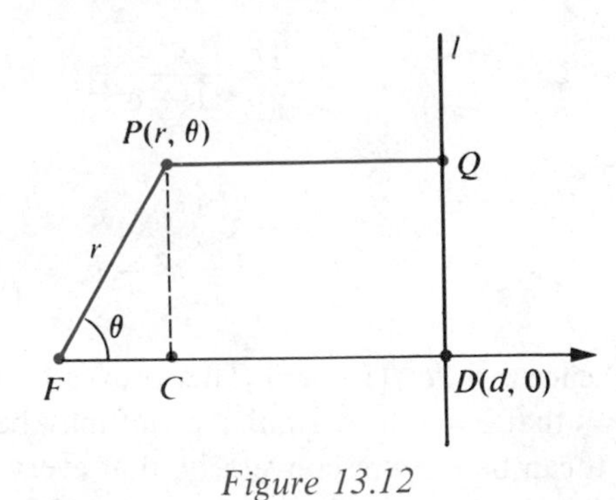

Figure 13.12

Let C be the projection of P on the polar axis. Since

$$d(P, F) = r \quad \text{and} \quad d(P, Q) = \overline{FD} - \overline{FC} = d - r\cos\theta$$

it follows that P satisfies the condition in (13.8) if and only if

$$\frac{r}{d - r\cos\theta} = e$$

or, equivalently,

$$r = de - er\cos\theta. \tag{13.9}$$

Solving for r gives us

(13.10)
$$r = \frac{de}{1 + e\cos\theta}$$

which is a polar equation of the graph. Actually, the same equation is obtained if $e = 1$; however, in this event there is no point (r, θ) on the graph if $1 + \cos\theta = 0$.

The rectangular equation corresponding to (13.9) is

$$\sqrt{x^2 + y^2} = de - ex.$$

Squaring both sides and rearranging terms leads to

$$(1 - e^2)x^2 + 2de^2x + y^2 = d^2e^2.$$

Completing the square in the previous equation and simplifying we obtain

$$\left(x + \frac{de^2}{1 - e^2}\right)^2 + \frac{y^2}{1 - e^2} = \frac{d^2e^2}{(1 - e^2)^2}.$$

Finally, dividing both sides by $d^2e^2/(1 - e^2)^2$ gives us the form

$$\frac{(x - h)^2}{a^2} + \frac{(y - k)^2}{b^2} = 1$$

(see (7.21)). Consequently the graph is an ellipse with center at the point $(-de^2/(1 - e^2), 0)$ on the x-axis and where

$$a^2 = \frac{d^2e^2}{(1 - e^2)^2}, \quad b^2 = \frac{d^2e^2}{1 - e^2}.$$

By (7.8),

$$c^2 = a^2 - b^2 = \frac{d^2e^4}{(1 - e^2)^2}$$

and hence $c = de^2/(1 - e^2)$. This proves that F is a focus of the ellipse. It also follows that $e = c/a$. A similar proof may be given for the case $e > 1$.

It can be shown, conversely, that every conic which is not a circle may be described by means of the statement in (13.8). This gives us a formulation of conic sections which is equivalent to the approach used in Chapter 7. Since Theorem (13.8) includes all three types of conics, it is sometimes regarded as a definition for the conic sections.

If we had chosen the focus F to the *right* of the directrix, as illustrated in Figure 13.13 (where $d > 0$), then an equation similar to (13.10) would have resulted, but with a minus sign in place of the plus sign. Other sign changes occur if d is allowed to be negative.

If l is taken *parallel* to the polar axis through one of the points $(d, \pi/2)$ or $(d, 3\pi/2)$, then the corresponding equations would contain $\sin\theta$ instead of $\cos\theta$. The proofs of these facts are left to the reader.

The following theorem summarizes our discussion.

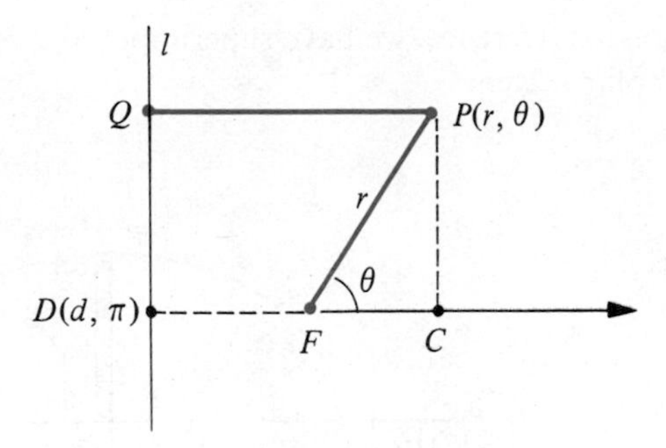

Figure 13.13

(13.11) **Theorem**

A polar equation having one of the forms

$$r = \frac{de}{1 \pm e\cos\theta}, \quad r = \frac{de}{1 \pm e\sin\theta}$$

is a conic section. Moreover, the conic is a parabola if $e = 1$, an ellipse if $0 < e < 1$, or a hyperbola if $e > 1$.

Example 1 Describe and sketch the graph of the equation

$$r = \frac{10}{3 + 2\cos\theta}.$$

Solution Dividing numerator and denominator of the given fraction by 3 gives us

$$r = \frac{10/3}{1 + \frac{2}{3}\cos\theta}$$

which has one of the forms in Theorem (13.11) with $e = 2/3$. Thus the graph is an ellipse with focus F at the pole and major axis along the polar axis. The endpoints of the major axis may be found by setting θ equal to 0 and π. This gives us $V(2, 0)$ and $V'(10, \pi)$. Hence

$$2a = d(V', V) = 12, \quad \text{or} \quad a = 6.$$

The center of the ellipse is the midpoint of the segment $V'V$, namely $(4, \pi)$. Using the fact that $e = c/a$ we obtain

$$c = ae = 6(2/3) = 4.$$

Hence

$$b^2 = a^2 - c^2 = 36 - 16 = 20$$

that is, the semiminor axis has length $\sqrt{20}$. The graph is sketched in Figure 13.14

where, for reference, we have superimposed a rectangular coordinate system on the polar system.

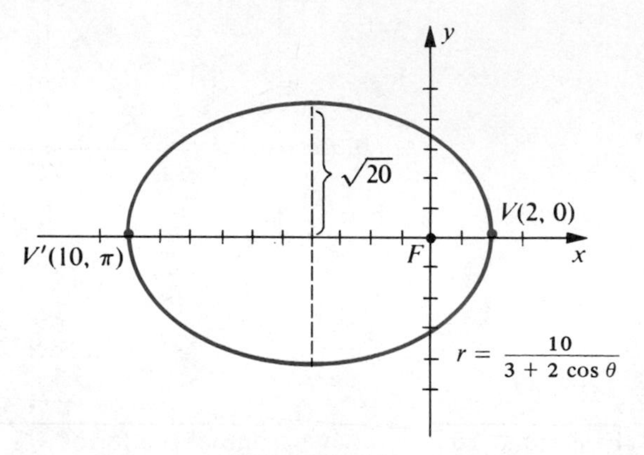

Figure 13.14

Example 2 Describe and sketch the graph of the equation

$$r = \frac{10}{2 + 3\sin\theta}.$$

Solution To express the equation in one of the forms in Theorem (13.11) we divide numerator and denominator of the given fraction by 2, obtaining

$$r = \frac{5}{1 + \frac{3}{2}\sin\theta}.$$

Thus $e = 3/2$ and the graph is a hyperbola with a focus at the pole. The expression $\sin\theta$ tells us that the transverse axis of the hyperbola is perpendicular to the polar axis. To find the vertices we let θ equal $\pi/2$ and $3\pi/2$ in the given equation. This gives us the points $V(2, \pi/2)$, $V'(-10, 3\pi/2)$, and hence

$$2a = d(V, V') = 8, \quad \text{or} \quad a = 4.$$

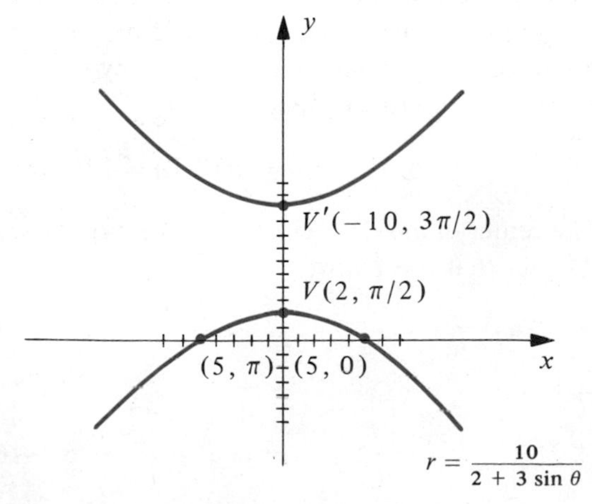

Figure 13.15

The points $(5, 0)$ and $(5, \pi)$ on the graph can be used to get a rough estimate of the lower branch of the hyperbola. The upper branch is obtained by symmetry, as illustrated in Figure 13.15. If more accuracy or additional information is desired, we may calculate

$$c = ae = 4(\tfrac{3}{2}) = 6 \quad \text{and}$$
$$b^2 = c^2 - a^2 = 36 - 16 = 20.$$

Asymptotes may then be constructed in the usual way.

Example 3 Sketch the graph of the equation

$$r = \frac{15}{4 - 4\cos\theta}.$$

Solution To obtain one of the forms in Theorem (13.11) we divide numerator and denominator of the given fraction by 4, obtaining

$$r = \frac{(15/4)}{1 - \cos\theta}.$$

Consequently $e = 1$ and the graph is a parabola with focus at the pole. A rough sketch can be found by plotting the points which correspond to the x- and y-intercepts. These are indicated in the following table.

θ	$\pi/2$	π	$3\pi/2$
r	15/4	15/8	15/4

Note that $\theta = 0$ is excluded from consideration. (Why?) Plotting these three points and using the fact that the graph is a parabola with focus at the pole gives us the sketch in Figure 13.16.

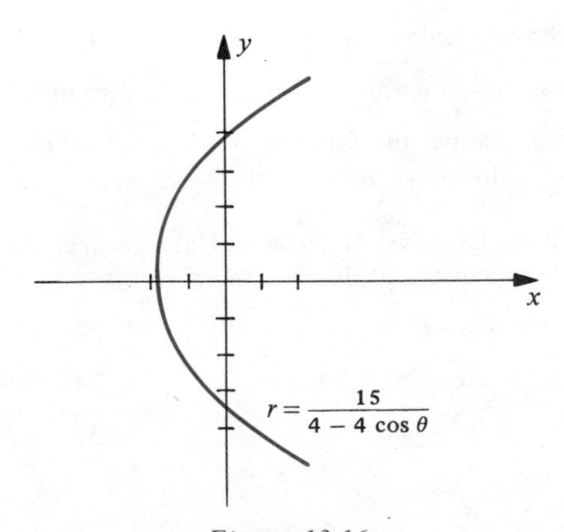

Figure 13.16

If only a rough sketch of a conic is desired, then the technique employed in Example 3 is recommended. To use this method we plot (if possible) points corresponding to $\theta = 0, \pi/2, \pi$, and $3\pi/2$. These points, together with the type of conic (obtained from the value of e), readily lead to the sketch.

EXERCISES 13.4

In each of Exercises 1–10, identify and sketch the graph of the given equation.

1 $r = \dfrac{12}{6 + 2\sin\theta}$

2 $r = \dfrac{12}{6 - 2\sin\theta}$

3 $r = \dfrac{12}{2 - 6\cos\theta}$

4 $r = \dfrac{12}{2 + 6\cos\theta}$

5 $r = \dfrac{3}{2 + 2\cos\theta}$

6 $r = \dfrac{3}{2 - 2\sin\theta}$

7 $r = \dfrac{4}{\cos\theta - 2}$

8 $r = \dfrac{4\sec\theta}{2\sec\theta - 1}$

9 $r = \dfrac{6\csc\theta}{2\csc\theta + 3}$

10 $r = \csc\theta(\csc\theta - \cot\theta)$

11–20 Find rectangular equations for the graphs in Exercises 1–10.

In each of Exercises 21–26 use Theorem (13.8) to find a polar equation of the conic with focus at the pole and the given eccentricity and equation of directrix.

21 $e = 1/3, r = 2\sec\theta$

22 $e = 2/5, r = 4\csc\theta$

23 $e = 4, r = -3\csc\theta$

24 $e = 3, r = -4\sec\theta$

25 $e = 1, r\cos\theta = 5$

26 $e = 1, r\sin\theta = -2$

27 Find a polar equation of the parabola with focus at the pole and vertex $(4, \pi/2)$.

28 Find a polar equation of an ellipse with eccentricity 2/3, a vertex at $(1, 3\pi/2)$, and a focus at the pole.

29 Prove Theorem (13.8) for the case $e > 1$.

30 (a) Use Figure 13.13 to derive the formula $r = de/(1 - e\cos\theta)$.

(b) Derive the formulas $r = de/(1 \pm e\sin\theta)$ in Theorem (13.11) by taking the directrix parallel to the polar axis.

In each of Exercises 31–36, express the given equation in one of the forms in Theorem (13.11) and then find the eccentricity and an equation for the directrix.

31 $y^2 = 4 - 4x$

32 $x^2 = 1 - 2y$

33 $3y^2 - 16y - x^2 + 16 = 0$

34 $5x^2 + 9y^2 = 32x + 64$

35 $8x^2 + 9y^2 + 4x = 4$

36 $4x^2 - 5y^2 + 36y - 36 = 0$

13.5 AREAS IN POLAR COORDINATES

Areas of certain regions bounded by graphs of polar equations can be found by employing limits of sums of the areas of circular sectors. Thus suppose R is a region in an $r\theta$-plane bounded by the lines through O with equations $\theta = a$ and $\theta = b$, where $0 \le a < b \le 2\pi$, and by the graph of $r = f(\theta)$, where f is continuous and $f(\theta) \ge 0$ on $[a, b]$. An illustration of such a region appears in (i) of Figure 13.17. Let P denote a partition of $[a, b]$ determined by

$$a = \theta_0 < \theta_1 < \theta_2 < \cdots < \theta_n = b$$

and let $\Delta\theta_i = \theta_i - \theta_{i-1}$ for $i = 1, 2, \ldots, n$. The lines with equations $\theta = \theta_i$ divide R into wedge-shaped subregions. If $f(u_i)$ is the minimum value and $f(v_i)$ is the maximum value of f on $[\theta_{i-1}, \theta_i]$, then as illustrated in (ii) of Figure 13.17 the area ΔA_i of the ith subregion is between the areas of the inscribed and circumscribed circular sectors having central angle $\Delta\theta_i$ and radii $f(u_i)$ and $f(v_i)$, respectively. Hence,

$$\tfrac{1}{2}[f(u_i)]^2\,\Delta\theta_i \le \Delta A_i \le \tfrac{1}{2}[f(v_i)]^2\,\Delta\theta_i.$$

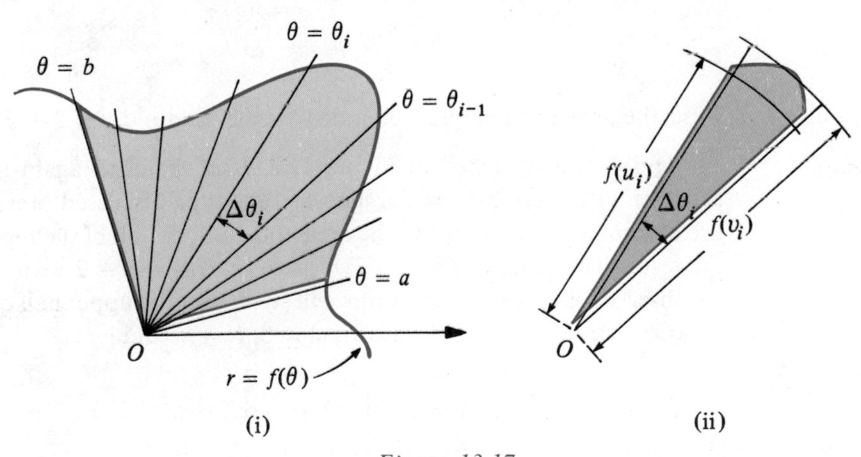

Figure 13.17

Summing from $i = 1$ to $i = n$, and using the fact that the sum of the ΔA_i is the area A of R, we obtain

$$\sum_{i=1}^{n} \tfrac{1}{2}[f(u_i)]^2\,\Delta\theta_i \le A \le \sum_{i=1}^{n} \tfrac{1}{2}[f(v_i)]^2\,\Delta\theta_i.$$

The limit of each sum, as the norm $\|P\|$ of the subdivision approaches zero, equals the integral $\int_a^b \frac{1}{2}[f(\theta)]^2\,d\theta$. Consequently, the **area A of the region bounded by the graphs of $r = f(\theta)$, $\theta = a$, and $\theta = b$** is

(13.12) $$A = \tfrac{1}{2}\int_a^b r^2\,d\theta.$$

This integral can also be interpreted as a limit of a sum by writing

(13.13)
$$A = \lim_{\|P\| \to 0} \sum_{i=1}^{n} \tfrac{1}{2}[f(w_i)]^2 \, \Delta\theta_i = \tfrac{1}{2} \int_a^b [f(\theta)]^2 \, d\theta$$

where w_i is *any* number in the interval $[\theta_{i-1}, \theta_i]$. The sketch in Figure 13.18 provides a geometric illustration of a typical Riemann sum. It is convenient to regard (13.13) as the process of starting with the area of a typical circular sector of the type shown in Figure 13.18 and then sweeping out the region R by letting θ vary from a to b while at the same time each $\Delta\theta_i$ approaches zero.

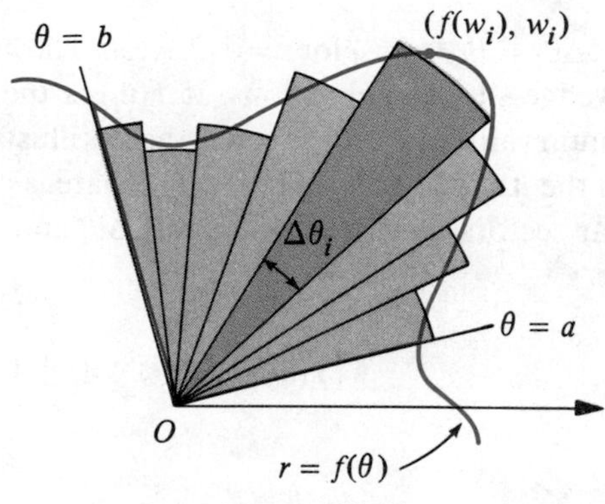

Figure 13.18

Example 1 Find the area of the region bounded by the cardioid $r = 2 + 2\cos\theta$.

Solution The graph was sketched in Figure 13.9. It is sketched again in Figure 13.19 together with a typical circular sector of the type discussed previously. Making use of symmetry we shall find the area of the upper half of the region and double the result. The function f of (13.13) is given by $f(\theta) = 2 + 2\cos\theta$. To find a and b we observe that the circular sectors will sweep out the upper half of the cardioid if θ varies from 0 to π. Consequently $a = 0$, $b = \pi$, and

$$\begin{aligned} A &= 2 \cdot \tfrac{1}{2} \int_0^{\pi} (2 + 2\cos\theta)^2 \, d\theta \\ &= \int_0^{\pi} (4 + 8\cos\theta + 4\cos^2\theta) \, d\theta. \end{aligned}$$

Using the fact that $\cos^2\theta = \tfrac{1}{2}(1 + \cos 2\theta)$,

$$\begin{aligned} A &= \int_0^{\pi} (6 + 8\cos\theta + 2\cos 2\theta) \, d\theta \\ &= \Big[6\theta + 8\sin\theta + \sin 2\theta \Big]_0^{\pi} = 6\pi. \end{aligned}$$

The area could also have been found by using (13.13) with $a = 0$ and $b = 2\pi$.

Let us next consider two continuous functions f and g such that $f(\theta) \geq g(\theta) \geq 0$ for every θ in an interval $[a, b]$. Let R denote the region bounded

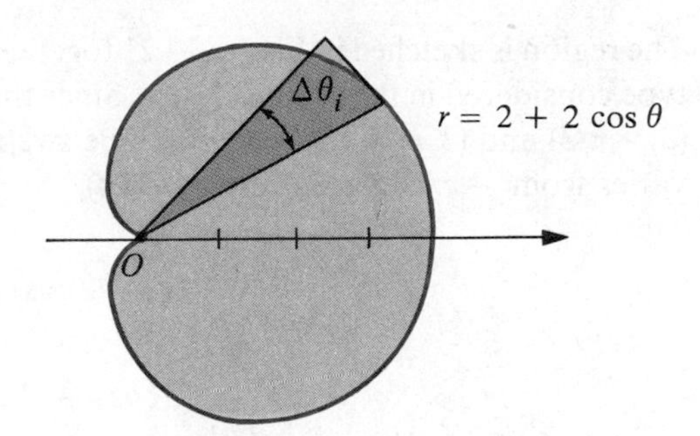

Figure 13.19

by the graphs of $r = f(\theta)$, $r = g(\theta)$, $\theta = a$, and $\theta = b$ as illustrated in Figure 13.20. Evidently the area A of R can be found by subtracting the areas of two regions of the type considered previously. Thus

$$A = \frac{1}{2}\int_a^b [f(\theta)]^2\,d\theta - \frac{1}{2}\int_a^b [g(\theta)]^2\,d\theta$$

or

(13.14)
$$A = \frac{1}{2}\int_a^b ([f(\theta)]^2 - [g(\theta)]^2)\,d\theta.$$

The last integral can be expressed as a limit of a sum as follows:

(13.15)
$$A = \lim_{\|P\|\to 0} \sum_i \tfrac{1}{2}([f(w_i)]^2 - [g(w_i)]^2)\,\Delta\theta_i.$$

The sum in (13.15) may be thought of as the process of starting with the area of a region of the type illustrated by the darker shaded portion of Figure 13.20 and then sweeping out R by letting θ vary from a to b.

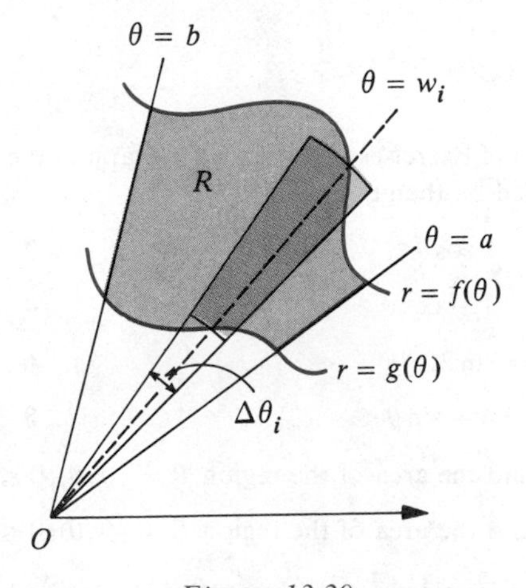

Figure 13.20

Example 2 Find the area of the region which is inside the cardioid $r = 2 + 2\cos\theta$ and outside the circle $r = 3$.

Solution The region is sketched in Figure 13.21 together with a polar element of area of the type considered in the sum in (13.15). Since the two graphs intersect at the points $(3, -\pi/3)$ and $(3, \pi/3)$, the region will be swept out by the indicated element if θ varies from $-\pi/3$ to $\pi/3$. Using (13.14),

$$\begin{aligned} A &= \tfrac{1}{2}\int_{-\pi/3}^{\pi/3} [(2 + 2\cos\theta)^2 - 3^2]\, d\theta \\ &= \tfrac{1}{2}\int_{-\pi/3}^{\pi/3} (8\cos\theta + 4\cos^2\theta - 5)\, d\theta. \end{aligned}$$

As in Example 1, the integral may be evaluated by letting $\cos^2\theta = \frac{1}{2}(1 + \cos 2\theta)$. It is left to the reader to show that $A = (9\sqrt{3}/2) - \pi$.

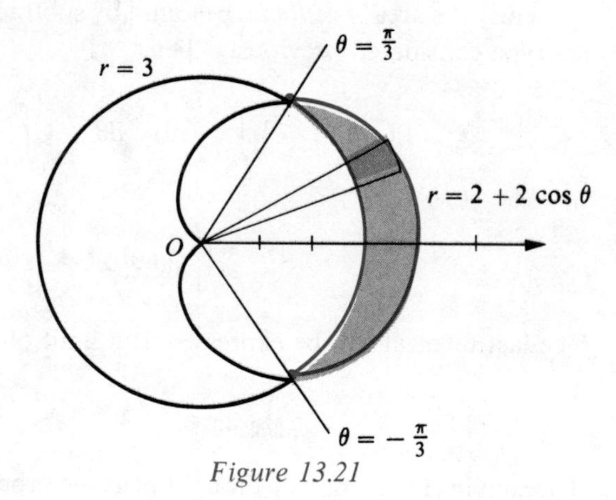

Figure 13.21

EXERCISES 13.5

In each of Exercises 1–8 sketch the graph of the equation and find the area of the region bounded by the graph.

1 $r = 2\cos\theta$

2 $r = 5\sin\theta$

3 $r = 1 - \cos\theta$

4 $r = 6 - 6\sin\theta$

5 $r = \sin 2\theta$

6 $r^2 = 9\cos 2\theta$

7 $r = 4 + \sin\theta$

8 $r = 3 + 2\cos\theta$

9 Find the area of the region $R = \{(r, \theta): 0 \le \theta \le \pi/2, 0 \le r \le e^{\theta}\}$.

10 Find the area of the region $R = \{(r, \theta): 0 \le \theta \le \pi, 0 \le r \le 2\theta\}$.

In Exercises 11–14, find the area of the region bounded by one loop of the graph of the given equation.

11 $r^2 = 4\cos 2\theta$

12 $r = 2\cos 3\theta$

13 $r = 3\cos 5\theta$

14 $r = \sin 6\theta$

In each of Exercises 15–18 find the area of the region which is outside the graph of the first equation and inside the graph of the second equation.

15 $r = 2 + 2\cos\theta, r = 3$

16 $r = 2, r = 4\cos\theta$

17 $r = 2, r^2 = 8\sin 2\theta$

18 $r = 1 - \sin\theta, r = 3\sin\theta$

In each of Exercises 19–22, find the area of the region which is inside the graphs of *both* of the given equations.

19 $r = \sin\theta, r = \sqrt{3}\cos\theta$

20 $r = 2(1 + \sin\theta), r = 1$

21 $r = 1 + \sin\theta, r = 5\sin\theta$

22 $r^2 = 4\cos 2\theta, r = 1$

In each of Exercises 23–26, find the area of the region bounded by the graphs of the equations.

23 $r = 2\sec\theta, \theta = \pi/6, \theta = \pi/3$

24 $r = \csc\theta\cot\theta, \theta = \pi/6, \theta = \pi/4$

25 $r = 4/(1 - \cos\theta), \theta = \pi/4$

26 $r(1 + \sin\theta) = 2, \theta = \pi/3$

13.6 LENGTHS OF CURVES

If a curve is the graph of an equation $y = f(x)$, where f is smooth on an interval $[a, b]$, then its arc length may be found by means of Definition (6.18). If the curve is given parametrically or in terms of polar coordinates, then (6.18) is not necessarily applicable. In this section we shall develop a formula which can be used for more general curves.

Suppose a curve C is given parametrically by

$$x = f(t), y = g(t)$$

where f and g are differentiable on an interval $[a, b]$. Furthermore, suppose C does not intersect itself, that is, different values of t between a and b determine different points on C. Consider a partition P of $[a, b]$ given by $a = t_0 < t_1 < t_2 < \cdots < t_n = b$. Let $\Delta t_i = t_i - t_{i-1}$ and let $P_i = (f(t_i), g(t_i))$ be the point on C determined by t_i. If $d(P_{i-1}, P_i)$ is the length of the line segment $P_{i-1}P_i$, then the length L_P of the broken line shown in Figure 13.22 is

$$L_P = \sum_{i=1}^{n} d(P_{i-1}, P_i).$$

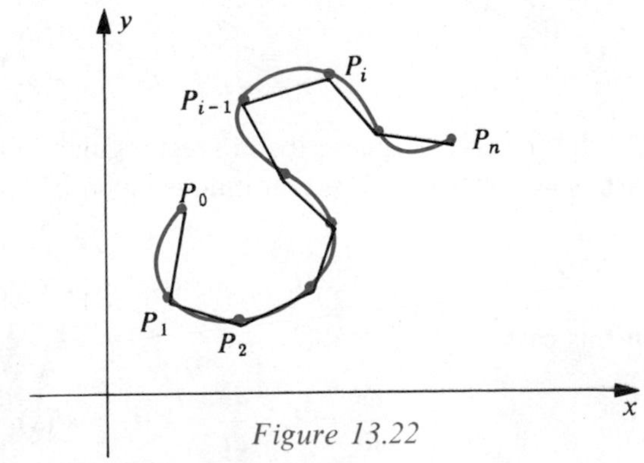

Figure 13.22

In a manner similar to Definition (5.15) we write

$$L = \lim_{\|P\| \to 0} L_P$$

and call L the **length of C from P_0 to P_n** if, for every $\varepsilon > 0$, there exists a $\delta > 0$ such that $|L_P - L| < \varepsilon$ for all partitions P with $\|P\| < \delta$.

By the Distance Formula,

$$d(P_{i-1}, P_i) = \sqrt{[f(t_i) - f(t_{i-1})]^2 + [g(t_i) - g(t_{i-1})]^2}.$$

Applying the Mean Value Theorem (4.12), there exist numbers w_i and z_i in the open interval (t_{i-1}, t_i) such that

$$f(t_i) - f(t_{i-1}) = f'(w_i)\,\Delta t_i,$$
$$g(t_i) - g(t_{i-1}) = g'(z_i)\,\Delta t_i.$$

Substituting in the formula for $d(P_{i-1}, P_i)$ and removing the common factor $(\Delta t_i)^2$ from the radicand gives us

(13.16) $$d(P_{i-1}, P_i) = \sqrt{[f'(w_i)]^2 + [g'(z_i)]^2}\,\Delta t_i.$$

Consequently,

$$L = \lim_{\|P\| \to 0} L_P = \lim_{\|P\| \to 0} \sum_{i=1}^{n} \sqrt{[f'(w_i)]^2 + [g'(z_i)]^2}\,\Delta t_i,$$

provided the limit exists. If $w_i = z_i$ for all i, then this sum is a Riemann sum for the function k defined by $k(t) = \sqrt{[f'(t)]^2 + [g'(t)]^2}$. If C is smooth, then the limit of the sum exists and the length of C from P_0 to P_n is given by

(13.17) $$L = \int_a^b \sqrt{[f'(t)]^2 + [g'(t)]^2}\,dt.$$

It can be shown that the limit exists even if $w_i \neq z_i$; however, the proof requires advanced methods and hence is omitted. To summarize, (13.17) gives the length of a parametrically defined curve C, provided f and g are smooth functions on $[a, b]$ and C does not intersect itself, except possibly for the case $P(a) = P(b)$. Formula (13.17) may be written in terms of the differential notation as

(13.18) $$L = \int_a^b \sqrt{\left(\frac{dx}{dt}\right)^2 + \left(\frac{dy}{dt}\right)^2}\,dt.$$

If the curve C is described in rectangular coordinates by an equation of the form $y = k(x)$, where k' is continuous on $[a, b]$, then parametric equations for C are

$$x = t, \quad y = k(t), \quad \text{where } a \le t \le b.$$

In this case

$$\frac{dx}{dt} = 1, \quad \frac{dy}{dt} = k'(t) = k'(x), \quad dt = dx$$

and (13.18) may be written

$$L = \int_a^b \sqrt{1 + [k'(x)]^2}\, dx.$$

This is in agreement with the arc length formula given in Definition (6.18).

Example 1 Find the length of one arch of the cycloid having parametric equations $x = t - \sin t$, $y = 1 - \cos t$.

Solution The graph has the shape illustrated in Figure 13.3, where the radius a of the circle is 1. One arch is obtained if t varies from 0 to 2π. Applying (13.18)

$$L = \int_0^{2\pi} \sqrt{(1 - \cos t)^2 + (\sin t)^2}\, dt.$$

Since $\cos^2 t + \sin^2 t = 1$, the integrand reduces to

$$\sqrt{2 - 2\cos t} = \sqrt{2}\sqrt{1 - \cos t}.$$

Thus

$$L = \int_0^{2\pi} \sqrt{2}\sqrt{1 - \cos t}\, dt.$$

By a half-angle formula, $\sin^2 (t/2) = (1 - \cos t)/2$, or equivalently,

$$1 - \cos t = 2\sin^2 (t/2).$$

Hence

$$\sqrt{1 - \cos t} = \sqrt{2\sin^2 (t/2)} = \sqrt{2}\,|\sin (t/2)|.$$

The absolute value sign may be deleted since if $0 \le t \le 2\pi$, then $0 \le t/2 \le \pi$ and hence $\sin (t/2) \ge 0$. Consequently

$$\begin{aligned} L &= \int_0^{2\pi} \sqrt{2}\sqrt{2}\sin (t/2)\, dt \\ &= 2\int_0^{2\pi} \sin (t/2)\, dt \\ &= \Big[-4\cos (t/2) \Big]_0^{2\pi} \\ &= (-4)(-1) - (-4)(1) = 8. \end{aligned}$$

If a curve C is the graph of a polar equation $r = f(\theta)$ where f is smooth on $[a, b]$, then by (13.6), parametric equations for C are $x = f(\theta)\cos\theta$, $y = f(\theta)\sin\theta$

and hence

$$\frac{dx}{d\theta} = -f(\theta)\sin\theta + f'(\theta)\cos\theta$$

$$\frac{dy}{d\theta} = f(\theta)\cos\theta + f'(\theta)\sin\theta.$$

It is left as an exercise to show that

(13.19) $$\left(\frac{dx}{d\theta}\right)^2 + \left(\frac{dy}{d\theta}\right)^2 = [f(\theta)]^2 + [f'(\theta)]^2.$$

It follows from (13.18), with $\theta = t$, that the length L of C from $\theta = a$ to $\theta = b$ is

(13.20) $$L = \int_a^b \sqrt{[f(\theta)]^2 + [f'(\theta)]^2}\, d\theta.$$

Example 2 Find the length of the spiral having polar equation $r = e^{\theta/2}$, from $\theta = 1$ to $\theta = 2$.

Solution A portion of the spiral is sketched in Figure 13.23, where polar coordinates are shown for the points corresponding to $\theta = 1$ and $\theta = 2$. Using (13.20) with $f(\theta) = e^{\theta/2}$,

$$\begin{aligned} L &= \int_1^2 \sqrt{(e^{\theta/2})^2 + (\tfrac{1}{2}e^{\theta/2})^2}\, d\theta \\ &= \int_1^2 \sqrt{\tfrac{5}{4}e^{\theta}}\, d\theta \\ &= \frac{\sqrt{5}}{2}\int_1^2 e^{\theta/2}\, d\theta \\ &= \sqrt{5}e^{\theta/2}\Big]_1^2 = \sqrt{5}(e - \sqrt{e}). \end{aligned}$$

If an approximation is desired, $L \approx 2.4$.

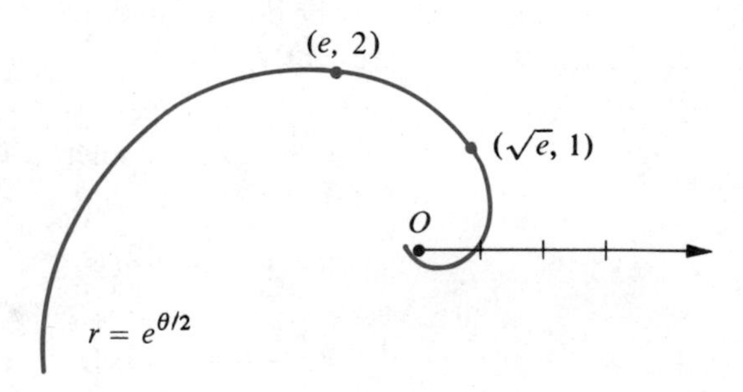

Figure 13.23

EXERCISES 13.6

Find the lengths of the curves defined in Exercises 1–7.

1 $x = 5t^2, y = 2t^3; 0 \le t \le 1$

2 $x = 3t, y = 2t^{3/2}; 0 \le t \le 4$

3 $x = e^t \cos t, y = e^t \sin t; 0 \le t \le \pi/2$

4 $x = 3t^2 - 5, y = 2t^2 + 1; \; 0 \le t \le 2$

5 $x = \ln \cos t, y = t; 0 \le t \le \pi/3$

6 $x = \cos^3 t;\ y = \sin^3 t; 0 \le t \le \pi/2$

7 $x = 2t, y = t^4 + (1/8)t^{-2}; 1 \le t \le 2$

8 Find the length of the curve having polar equation $r = \sin^3(\theta/3)$.

9 Find the length of the spiral $r = e^{-\theta}$ from $\theta = 0$ to $\theta = 2\pi$.

10 Find the length of the spiral $r = \theta$ from $\theta = 0$ to $\theta = 4\pi$.

11 Find the length of the curve $r = \cos^2(\theta/2)$ from $\theta = 0$ to $\theta = \pi$.

12 Find the length of the spiral $r = 2^\theta$ from $\theta = 0$ to $\theta = \pi$.

The parametric equations in each of Exercises 13–16 give the position (x, y) of a particle at time t. Find the distance the particle travels during the indicated time interval.

13 $x = 4t + 3, y = 2t^2; 0 \le t \le 5$

14 $x = \cos 2t, y = \sin^2 t; 0 \le t \le \pi$

15 $x = t \cos t - \sin t, y = t \sin t + \cos t; 0 \le t \le \pi/2$

16 $x = t^2, y = 2t; 0 \le t \le 4$

17 Verify (13.19).

18 In a manner analogous to the work with plane areas, define the *centroid* $(\bar{x}, \bar{y})$ of a curve and state integral formulas for finding $(\bar{x}, \bar{y})$ when the curve is given parametrically. (See (10.13)-(10.15).)

19 Find the length of the curve defined by $x = 1/t, y = \ln t; 1 \le t \le 2$.

20 Find the length of the cardioid $r = 1 + \cos\theta$.

13.7 SURFACES OF REVOLUTION

If a plane curve C is revolved about a line in the plane, a **surface of revolution** is generated. For example, if a circle is revolved about a diameter, the surface of a sphere is obtained. Formulas for surface area may be developed provided C is sufficiently well behaved. In particular, suppose C is given parametrically by $x = f(t), y = g(t)$, where f' and g' are continuous on an interval $[a, b]$ and $g(t) \ge 0$ for all t. The last condition implies that the graph of C is above the x-axis. We shall use the notation of the previous section, letting P denote a partition of $[a, b]$ and $P_i = (f(t_i), g(t_i))$ denote the point on C which corresponds to t_i, as illustrated in (i) of Figure 13.24. Reasoning intuitively, we observe that if the norm $\|P\|$ is close to zero, then the broken line C' obtained by connecting P_{i-1} to P_i for each i is an approximation to C, and hence the area of the surface generated by revolving C' about the x-axis should approximate the area of the surface generated by C. As illustrated in (ii) of Figure 13.24, each segment $P_{i-1}P_i$ generates the lateral surface of a frustum of a right circular cone having base radii $g(t_{i-1})$ and $g(t_i)$, and slant height $d(P_{i-1}, P_i)$. We know from geometry that the lateral area of a frustum of a right circular cone of base radii r_1 and r_2 and slant height s is given by $\pi(r_1 + r_2)s$. Hence the area of the surface generated by $P_{i-1}P_i$ is

$$\pi[g(t_{i-1}) + g(t_i)]d(P_{i-1}, P_i).$$

Summing terms of this form from $i = 1$ to $i = n$ gives us the area S_P of the surface generated by the broken line C'.

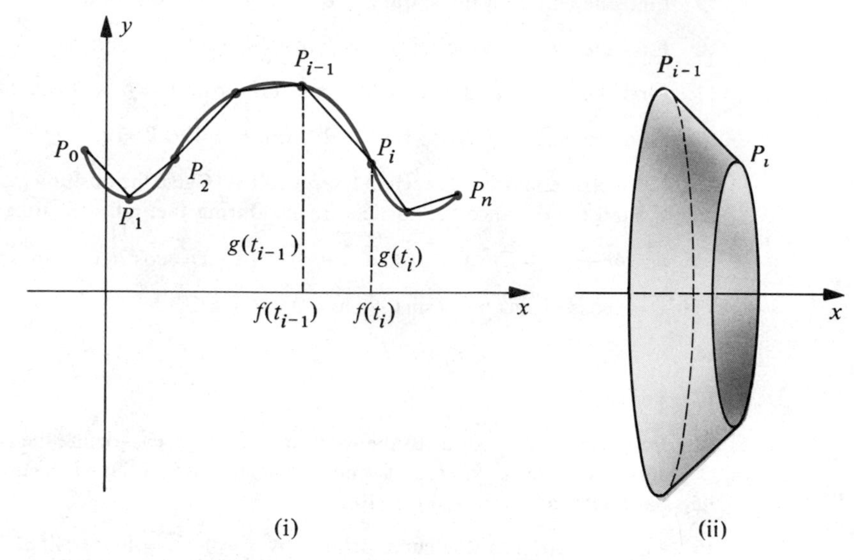

Figure 13.24

If we substitute the expression in (13.16) for $d(P_{i-1}, P_i)$, then

(13.21)
$$S_P = \sum_{i=1}^{n} \pi[g(t_i) + g(t_{i-1})]\sqrt{[f'(w_i)]^2 + [g'(z_i)]^2}\,\Delta t_i$$

where w_i and z_i are in the interval (t_{i-1}, t_i). As in the previous section, the length of C is the limit of the length L_P of C' as $\|P\| \to 0$, and hence it is natural to define the area S of the surface generated by C as

$$S = \lim_{\|P\| \to 0} S_P.$$

It is reasonable to expect, from (13.21), that this limit is given by

(13.22)
$$S = \int_a^b 2\pi g(t)\sqrt{[f'(t)]^2 + [g'(t)]^2}\,dt$$

or, in terms of the differential notation,

(13.23)
$$S = \int_a^b 2\pi y\sqrt{\left(\frac{dx}{dt}\right)^2 + \left(\frac{dy}{dt}\right)^2}\,dt.$$

The proof of (13.22) is beyond the scope of this book.

There is an easy way to remember (13.23). If we regard $\sqrt{(dx/dt)^2 + (dy/dt)^2}\,dt$ as the slant height of a typical frustum, and if (x, y)

represents a point in the corresponding subarc, then *the integrand in* (13.23) *may be regarded as the product of the slant height and the circumference* $2\pi y$ *of the circle traced by* (x, y). It follows in similar fashion that if the curve C is revolved about the y-axis, then S is given by

(13.24)
$$S = \int_a^b 2\pi x \sqrt{\left(\frac{dx}{dt}\right)^2 + \left(\frac{dy}{dt}\right)^2}\, dt,$$

where again we may regard $2\pi x$ as the circumference of a circle traced by a point (x, y) on C.

Example 1 Verify that the surface area of a sphere of radius a is $4\pi a^2$.

Solution If C is the upper half of the circle $x^2 + y^2 = a^2$, then the desired surface may be obtained by revolving C about the x-axis. Parametric equations for C are

$$x = a\cos t, \quad y = a\sin t$$

where $0 \le t \le \pi$. Applying (13.23) and using the identity $\sin^2 t + \cos^2 t = 1$,

$$\begin{aligned} S &= \int_0^\pi 2\pi a \sin t \sqrt{a^2\sin^2 t + a^2\cos^2 t}\, dt \\ &= 2\pi a^2 \int_0^\pi \sin t\, dt \\ &= -2\pi a^2 [\cos t]_0^\pi \\ &= -2\pi a^2[-1 - 1] = 4\pi a^2. \end{aligned}$$

If C is the graph of an equation $y = f(x)$, where f is smooth on an interval $[a, b]$, then parametric equations for C are

$$x = t, \qquad y = f(t)$$

where $a \le t \le b$. Using (13.23) and the fact that $t = x$, the area of the surface generated by revolving C around the x-axis is given by

(13.25)
$$S = \int_a^b 2\pi f(x)\sqrt{1 + [f'(x)]^2}\, dx.$$

If C is revolved about the y-axis, then as in (13.24)

(13.26)
$$S = \int_a^b 2\pi x\sqrt{1 + [f'(x)]^2}\, dx.$$

Analogous formulas may be stated if C is the graph of an equation $x = g(y)$ where $c \le y \le d$.

Example 2 The arc of the parabola $y^2 = x$ from $(1, 1)$ to $(4, 2)$ is revolved about the x-axis. Find the area of the resulting surface.

Solution The surface is illustrated in Figure 13.25. Using (13.25) with $y = f(x) = x^{1/2}$ we obtain

$$\begin{aligned} S &= \int_1^4 2\pi x^{1/2} \sqrt{1 + \left(\frac{1}{2x^{1/2}}\right)^2}\, dx \\ &= \int_1^4 2\pi x^{1/2} \sqrt{\frac{4x+1}{4x}}\, dx \\ &= \pi \int_1^4 \sqrt{4x+1}\, dx \\ &= \frac{\pi}{6}(4x+1)^{3/2}\Big]_1^4 = \frac{\pi}{6}(17^{3/2} - 5^{3/2}). \end{aligned}$$

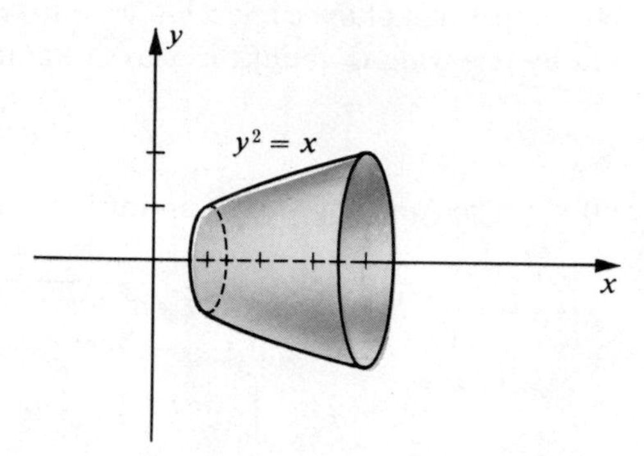

Figure 13.25

Formulas for S may also be stated if C is the graph of a polar equation $r = f(\theta)$, where $a \leq \theta \leq b$. In this case parametric equations for C are

$$x = f(\theta)\cos\theta, \quad y = f(\theta)\sin\theta$$

where $a \leq \theta \leq b$. If C is revolved about the polar axis, then substituting in (13.23) and using (13.19) gives us

(13.27) $$S = \int_a^b 2\pi f(\theta)\sin\theta\sqrt{[f(\theta)]^2 + [f'(\theta)]^2}\, d\theta.$$

Example 3 The part of the spiral $r = e^{\theta/2}$ from $\theta = 0$ to $\theta = \pi$ is revolved about the x-axis. Find the area of the resulting surface.

Solution The surface is illustrated in Figure 13.26. Using (13.27) with $f(\theta) = e^{\theta/2}$,

$$\begin{aligned} S &= \int_0^\pi 2\pi e^{\theta/2}\sin\theta\sqrt{(e^{\theta/2})^2 + (\tfrac{1}{2}e^{\theta/2})^2}\, d\theta \\ &= \int_0^\pi 2\pi e^{\theta/2}\sin\theta\sqrt{\tfrac{5}{4}e^\theta}\, d\theta \\ &= \sqrt{5}\pi\int_0^\pi e^\theta\sin\theta\, d\theta. \end{aligned}$$

Integration by parts leads to

$$S = \frac{\sqrt{5}\pi}{2} e^{\theta}(\sin\theta - \cos\theta)\Big]_0^{\pi} = \frac{\sqrt{5}\pi}{2}(e^{\pi} + 1).$$

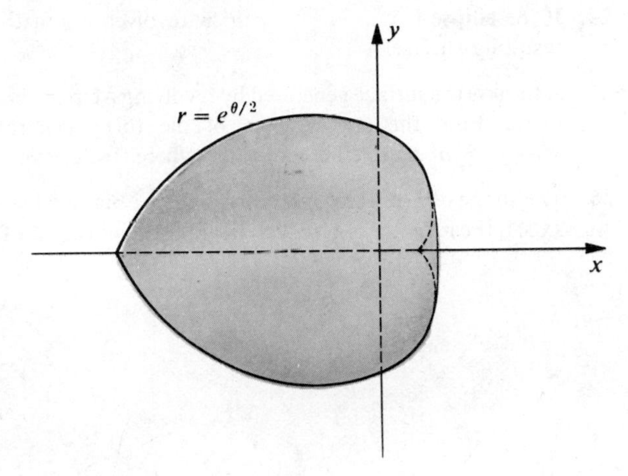

Figure 13.26

EXERCISES 13.7

In each of Exercises 1–8 find the area of the surface generated by revolving the given curve about the x-axis.

1 $x = t^2, y = 2t; 0 \le t \le 4$

2 $x = 4t, y = t^3; 1 \le t \le 2$

3 $x = t^2, y = t - \frac{1}{3}t^3; 0 \le t \le 1$

4 $x = 4t^2 + 1, y = 3 - 2t; -2 \le t \le 0$

5 $y = x^3$ from $x = 1$ to $x = 2$

6 $y = e^{-x}$ from $x = 0$ to $x = 1$

7 $x = a(t - \sin t),\ y = a(1 - \cos t);\ 0 \le t \le 2\pi$

8 $12y = 4x^3 + (3/x)$ from $x = 1$ to $x = 2$

In each of Exercises 9–16 find the area of the surface generated by revolving the given arc about the y-axis.

9 $x = 4t^{1/2}, y = \frac{1}{2}t^2 + t^{-1}; 1 \le t \le 4$

10 $x = 3t, y = t + 1; 0 \le t \le 5$

11 $x = e^t \sin t, y = e^t \cos t; 0 \le t \le \pi/2$

12 $x = 3t^2, y = 2t^3; 0 \le t \le 1$

13 $x^2 = 16y$ from $(0, 0)$ to $(8, 4)$

14 $8x = y^3$ from $(1, 2)$ to $(8, 4)$

15 $y = \ln x$ from $x = 1$ to $x = 2$

16 $y = \cosh x$ from $x = 1$ to $x = 2$

In each of Exercises 17–20 find the area of the surface generated by revolving the given curve about the polar axis.

17 $r = 2 + 2\cos\theta$

18 $r^2 = 4\cos 2\theta$

19 $r = 2a\sin\theta$

20 $r = 2a\cos\theta$

21 The smaller arc of the circle $x^2 + y^2 = 25$ between the points $(-3, 4)$ and $(3, 4)$ is revolved about the y-axis. Find the area of the resulting surface.

22 If the arc described in Exercise 21 is revolved about the x-axis, find the area of the resulting surface.

23 Prove that the surface area of a right circular cone of altitude a and base radius b is $\pi b\sqrt{a^2 + b^2}$.

24 If the ellipse $b^2x^2 + a^2y^2 = a^2b^2$ is revolved about the y-axis, find the area of the resulting surface.

25 A **torus** is the surface generated by revolving a circle about a nonintersecting line in its plane. Find the surface area of the torus generated by revolving the circle $x^2 + y^2 = a^2$ about the line $x = b$, where $0 < a < b$.

26 The shape of a reflector in a searchlight is obtained by revolving a parabola about its axis. If the reflector is 4 feet across at the opening and 1 foot deep, find its surface area.

13.8 REVIEW

Concepts

Define or discuss each of the following.

1 Plane curves

2 Closed curves

3 Parametric equations of a curve

4 Polar coordinate systems

5 Graphs of polar equations

6 Areas in polar coordinates

7 Length of a curve

8 Surfaces of revolution

Exercises

In Exercises 1–3 sketch the graph of the curve, find a rectangular equation of a graph which contains the points on the curve, and find the slope of the tangent line at the point corresponding to $t = 1$.

1 $x = 1/t + 1,\ y = 2/t - t;\ 0 < t \leq 4$

2 $x = \cos^2 t - 2,\ y = \sin t + 1;\ 0 \leq t \leq 2\pi$

3 $x = \sqrt{t},\ y = e^{-t};\ t \geq 0$

4 Let C be the curve with parametric equations $x = t^2, y = 2t^3 + 4t - 1$, where t is in $\mathbb{R}$. Find the abscissas of the points on C at which the tangent line passes through the origin.

5 Compare the graphs of the following curves, where t is in $\mathbb{R}$.
(a) $x = t^2$, $y = t^3$ (b) $x = t^4$, $y = t^6$
(c) $x = e^{2t}$, $y = e^{3t}$ (d) $x = 1 - \sin^2 t$, $y = \cos^3 t$

In each of Exercises 6–19, sketch the graph of the equation and find an equation in x and y which has the same graph.

6 $r = 3 + 2\cos\theta$

7 $r = 6\cos 2\theta$

8 $r = 4 - 4\cos\theta$

9 $r^2 = -4\sin 2\theta$

10 $r = 2\sin 3\theta$

11 $r(3\cos\theta - 2\sin\theta) = 6$

12 $r = e^{-\theta}$, $\theta \geq 0$

13 $r^2 = \sec 2\theta$

14 $r = 8\sec\theta$

15 $r = 4\cos^2(\theta/2)$

16 $r = 6 - r\cos\theta$

17 $r - 1 = 0$

18 $r = \dfrac{8}{1 - 3\sin\theta}$

19 $r = \dfrac{8}{3 + \cos\theta}$

Find polar equations for the graphs in Exercises 20 and 21.

20 $x^2 + y^2 = 2xy$

21 $y^2 = x^2 - 2x$

22 Find a polar equation of the hyperbola which has focus at the pole, eccentricity 2, and equation of directrix $r = 6\sec\theta$.

23 Find the area of the region bounded by one loop of the graph of $r^2 = 4\sin 2\theta$.

24 Find the area of the region which is inside the graph of $r = 3 + 2\sin\theta$ and outside the graph of $r = 4$.

25 The position (x, y) of a particle at time t is given by $x = 2\sin t$, $y = \sin^2 t$. Find the distance the particle travels from $t = 0$ to $t = \pi/2$.

26 Find the length of the spiral $r = 1/\theta$ from $(1, 1)$ to $(1/2, 2)$.

27 Find the area of the surface generated by revolving the graph of $y = \cosh x$ from $x = 0$ to $x = 1$ about the x-axis.

28 The curve $x = 2t^2 + 1$, $y = 4t - 3$ where $0 \leq t \leq 1$ is revolved about the y-axis. Find the area of the resulting surface.

29 The arc of the spiral $r = e^{\theta}$ from $(1, 0)$ to $(e, 1)$ is revolved about the line $\theta = \pi/2$. Find the area of the resulting surface.

30 Find the area of the surface generated by revolving the lemniscate $r^2 = a^2\cos 2\theta$ about the polar axis.

Vectors and Solid Analytic Geometry

Vectors have two different natures, one geometric and the other algebraic. For applications it is necessary to have an understanding of both aspects. Because of this, our development will oscillate between the two approaches, sometimes emphasizing geometry and other times algebra. For simplicity we shall begin with vectors in a plane. The extension to three dimensions is then easily accomplished by simply taking a third component into account. We shall also discuss some basic topics from solid analytic geometry including lines, planes, cylinders, quadric surfaces, and space curves.

14.1 VECTORS IN TWO DIMENSIONS

A physical entity which may be completely characterized by a specific real number is often referred to as a **scalar quantity** or simply a **scalar**. Examples of scalar quantities are temperature, mass, or density. **Vector quantities** possess both magnitude and direction. Some examples are the velocity of a particle in motion, a force acting at a point, or displacement of an object along a straight path. A vector quantity may be represented geometrically by means of a directed line segment, where the length and direction of the segment indicate the magnitude and direction, respectively, of the quantity. Such directed line segments are called **vectors**.

If a vector extends from a point A (called the **initial point**) to a point B (called the **terminal point**) it is customary to place an arrowhead at B and use $\overrightarrow{AB}$ to denote the directed line segment (see Figure 14.1). We shall also use boldface letters such as **a** or **b** to denote vectors. For written work, where it is difficult to represent boldface type, a notation such as $\vec{a}$ or $\vec{b}$ may be employed.

If $\overrightarrow{AB}$ is a vector, then the length of the line segment is called the **magnitude** of the vector and is denoted by $|\overrightarrow{AB}|$. The vectors $\overrightarrow{AB}$ and $\overrightarrow{CD}$ are said to be **equal**, and we write $\overrightarrow{AB} = \overrightarrow{CD}$, if they have the same magnitude and direction, as illustrated in Figure 14.1. Consequently, vectors may be translated from one position to another, provided neither the magnitude nor direction is changed. Vectors of this type are often referred to as **free vectors**.

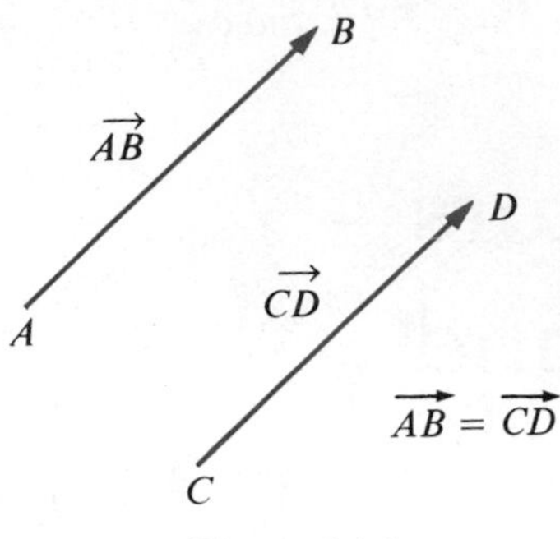

Figure 14.1

Suppose we use $\overrightarrow{AB}$ to represent a displacement of a point along the line segment from A to B, that is, the path taken by a point as it moves from A to B. As illustrated in Figure 14.2, a displacement $\overrightarrow{AB}$ followed by a displacement $\overrightarrow{BC}$ will lead to the same point as the single displacement $\overrightarrow{AC}$. By definition, the vector $\overrightarrow{AC}$ is called the **sum** of $\overrightarrow{AB}$ and $\overrightarrow{BC}$, and we write

$$\overrightarrow{AC} = \overrightarrow{AB} + \overrightarrow{BC}.$$

Since we are working with free vectors, *any* two vectors can be added by placing the initial point of one on the terminal point of the other and then proceeding as in (i) of Figure 14.2.

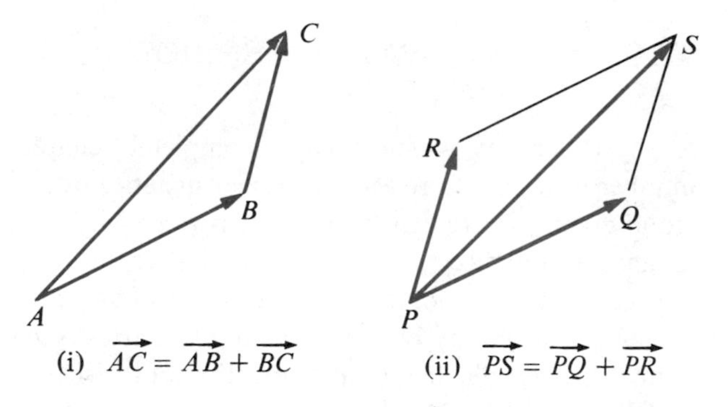

Figure 14.2

Another way to find the sum of two vectors is to consider vectors which are equal to the given ones and have the same initial point, as illustrated by $\overrightarrow{PQ}$ and $\overrightarrow{PR}$ in (ii) of Figure 14.2. If we construct the parallelogram $RPQS$ with adjacent sides $\overrightarrow{PR}$ and $\overrightarrow{PQ}$, then since $\overrightarrow{PR} = \overrightarrow{QS}$, it follows that

$$\overrightarrow{PS} = \overrightarrow{PQ} + \overrightarrow{PR}.$$

If $\overrightarrow{PQ}$ and $\overrightarrow{PR}$ are two forces acting at P, then $\overrightarrow{PS}$ is the **resultant force**, that is, the single force that produces the same effect as the two combined forces.

If c is a real number and $\overrightarrow{AB}$ is a vector, then the product $c\overrightarrow{AB}$ is defined as a vector whose magnitude is $|c|$ times the magnitude of $\overrightarrow{AB}$ and whose direction is the same as $\overrightarrow{AB}$ if $c > 0$, and opposite that of $\overrightarrow{AB}$ if $c < 0$. Geometric illustrations are given in Figure 14.3.

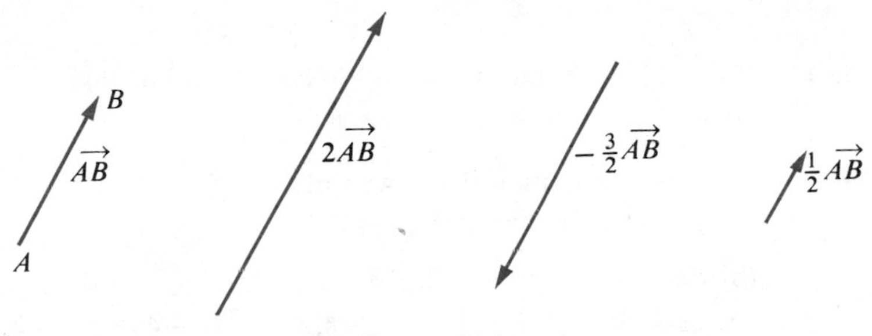

Figure 14.3

If all vectors under discussion lie in a fixed coordinate plane, then initial and terminal points can be assigned coordinates, that is, ordered pairs of real numbers, and statements about vectors can be translated into the language of algebra. We may then dispose of visual interpretations and discuss vectors from an algebraic point of view. In this way vectors take on a dual role, with the algebraic viewpoint emerging from the previous work involving geometry. It is possible to reverse the order of this development and *begin* with an algebraic definition of vectors and *then* obtain directed line segments as an alternate interpretation. There are several advantages to this approach. One is that proofs of properties of vectors become simple exercises in algebraic manipulations instead of problems in geometry. Also, it is then possible to obtain geometric facts by using algebraic techniques. Another advantage is that the algebraic approach to vectors is easily generalized to vectors in three or more dimensions. With these remarks in mind we shall next state an algebraic definition for vectors. Geometric representations which agree with our previous discussion will be introduced after the definition.

The algebraic formulation of vectors in two dimensions involves ordered pairs of real numbers. To avoid confusion with the notation used for open intervals or points in a plane, we shall use $\langle x, y\rangle$ for an ordered pair which refers to a vector. Such ordered pairs will also be denoted by symbols such as $\mathbf{a}$ or $\vec{a}$. Real numbers will be called **scalars**.

(14.1) Definition

The **two-dimensional vector space V_2** is the set of all ordered pairs $\langle x, y\rangle$ of real numbers, called **vectors**, subject to the following axioms.

(i) **Addition of vectors**. If $\mathbf{a} = \langle a_1, a_2\rangle$ and $\mathbf{b} = \langle b_1, b_2\rangle$ are vectors, then

$$\mathbf{a} + \mathbf{b} = \langle a_1 + b_1, a_2 + b_2\rangle.$$

(ii) **Multiplication of vectors by scalars**. If $\mathbf{a} = \langle a_1, a_2\rangle$ and c is a scalar, then

$$c\mathbf{a} = \langle ca_1, ca_2\rangle.$$

The numbers a_1 and a_2 in $\langle a_1, a_2 \rangle$ are called the **components** of the vector. Thus, to add two vectors we add corresponding components. The vector $c\mathbf{a}$ obtained by multiplying each component of $\mathbf{a}$ by c is called a **scalar multiple** of $\mathbf{a}$.

Example 1 If $\mathbf{a} = \langle 3, -2 \rangle$ and $\mathbf{b} = \langle -6, 7 \rangle$, find the following.
(a) $\mathbf{a} + \mathbf{b}$ (b) $4\mathbf{a}$ (c) $2\mathbf{a} + 3\mathbf{b}$

Solution Applying Definition (14.1), we obtain

(a) $\mathbf{a} + \mathbf{b} = \langle 3, -2 \rangle + \langle -6, 7 \rangle = \langle -3, 5 \rangle$

(b) $4\mathbf{a} = 4\langle 3, -2 \rangle = \langle 12, -8 \rangle$

(c) $2\mathbf{a} + 3\mathbf{b} = 2\langle 3, -2 \rangle + 3\langle -6, 7 \rangle = \langle 6, -4 \rangle + \langle -18, 21 \rangle = \langle -12, 17 \rangle$

The **zero vector 0** and the **negative** $-\mathbf{a}$ of a vector $\mathbf{a} = \langle a_1, a_2 \rangle$ are defined as follows:

$$\mathbf{0} = \langle 0, 0 \rangle \qquad -\mathbf{a} = -\langle a_1, a_2 \rangle = \langle -a_1, -a_2 \rangle.$$

A nonzero vector $\mathbf{a} = \langle a_1, a_2 \rangle$ will be represented in a coordinate plane by a directed line segment $\overrightarrow{PQ}$ with any initial point $P(x, y)$ and terminal point $Q(x + a_1, y + a_2)$. Several representations are shown in Figure 14.4, where the symbol $\mathbf{a}$ is placed next to any directed line segment which represents the vector $\mathbf{a}$ of V_2. The zero vector $\mathbf{0} = \langle 0, 0 \rangle$ will be represented by any point in the plane. Strictly speaking, a directed line segment $\overrightarrow{PQ}$ *represents* a vector; however, we shall also refer to $\overrightarrow{PQ}$ as a *vector*. It should always be clear from the discussion whether the term "vector" refers to an ordered pair or a directed line segment.

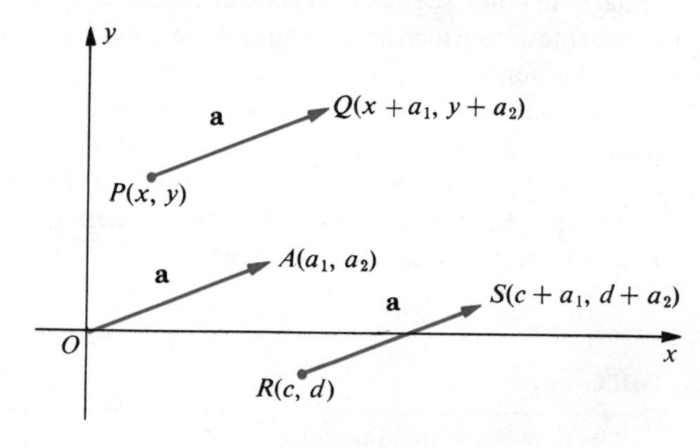

Figure 14.4

(14.2) Definition

> The **magnitude** $|\mathbf{a}|$ of a vector $\mathbf{a} = \langle a_1, a_2 \rangle$ is given by
>
> $$|\mathbf{a}| = \sqrt{a_1^2 + a_2^2}.$$

It follows from the distance formula that $|\mathbf{a}|$ is the length of any directed line segment used to represent $\mathbf{a}$. Note also that $|\mathbf{a}| \geq 0$, where $|\mathbf{a}| = 0$, if and only if $\mathbf{a} = \mathbf{0}$. The symbol for the magnitude of a vector should not be confused with the symbol $|c|$ for the absolute value of a real number c. As an illustration, if $\mathbf{a} = \langle 3, -2 \rangle$, then

$$|\mathbf{a}| = |\langle 3, -2 \rangle| = \sqrt{9 + 4} = \sqrt{13}.$$

If P is any point in a coordinate plane and O is the origin, then $\overrightarrow{OP}$ is called the **position vector** of P. If $\mathbf{a} = \langle a_1, a_2 \rangle$, then the position vector $\overrightarrow{OA}$ of the point $A(a_1, a_2)$ (see Figure 14.4) is called the **position vector corresponding to a**. Although $\mathbf{a}$ has many geometric representations, there is precisely one position vector corresponding to $\mathbf{a}$.

(14.3) **Theorem**

If $P_1(x_1, y_1)$ and $P_2(x_2, y_2)$ are points, then the vector $\mathbf{a}$ in V_2 which has geometric representation $\overrightarrow{P_1P_2}$ is

$$\mathbf{a} = \langle x_2 - x_1, y_2 - y_1 \rangle.$$

Proof. If $P_1(x_1, y_1)$ is the initial point of a vector corresponding to $\mathbf{a} = \langle a_1, a_2 \rangle$, then the terminal point is $Q(x_1 + a_1, y_1 + a_2)$. If Q is to coincide with $P_2(x_2, y_2)$, then we must have $x_2 = x_1 + a_1$ and $y_2 = y_1 + a_2$. Consequently, $\mathbf{a} = \langle x_2 - x_1, y_2 - y_1 \rangle$. One illustration of $\overrightarrow{P_1P_2}$ is shown in Figure 14.5, together with the position vector corresponding to $\mathbf{a}$.

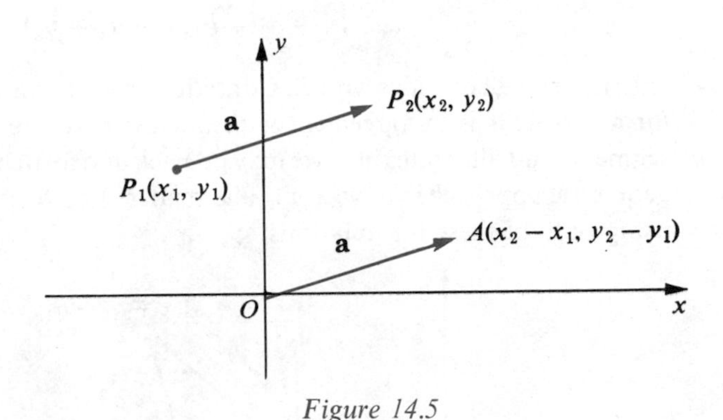

Figure 14.5

Example 2 Given the points $P(-2, 3)$ and $Q(4, 5)$, find vectors $\mathbf{a}$ and $\mathbf{b}$ which have geometric representations $\overrightarrow{PQ}$ and $\overrightarrow{QP}$, respectively. Sketch $\overrightarrow{PQ}$, $\overrightarrow{QP}$, and the position vectors corresponding to $\mathbf{a}$ and $\mathbf{b}$.

Solution Applying Theorem (14.3), the vector corresponding to $\overrightarrow{PQ}$ is

$$\mathbf{a} = \langle 4 - (-2), 5 - 3 \rangle = \langle 6, 2 \rangle$$

and the vector corresponding to $\overrightarrow{QP}$ is

$$\mathbf{b} = \langle -2 - 4, 3 - 5 \rangle = \langle -6, -2 \rangle.$$

The sketches are shown in Figure 14.6. Observe that $\mathbf{b} = -\mathbf{a}$

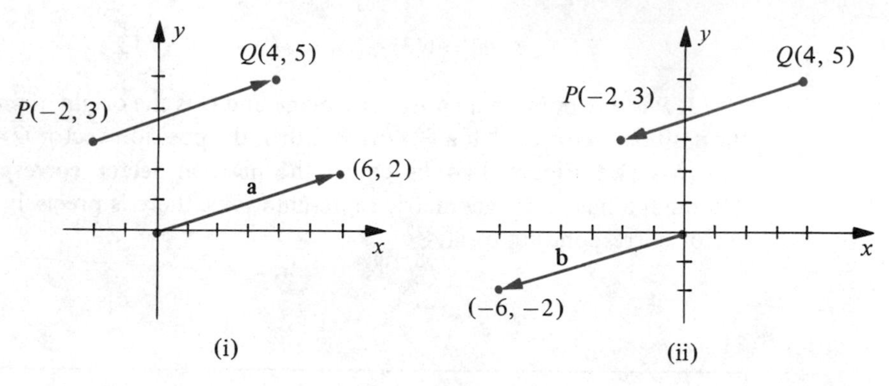

Figure 14.6

It is easy to obtain a geometric representation for the sum of two vectors $\mathbf{a} = \langle a_1, a_2 \rangle$ and $\mathbf{b} = \langle b_1, b_2 \rangle$. First we represent $\mathbf{a}$ by a vector $\overrightarrow{PQ}$ with initial point $P(x, y)$ and terminal point $Q(x + a_1, y + a_2)$. Next, as illustrated in Figure 14.7, $\mathbf{b}$ is represented by $\overrightarrow{QR}$ where R is the point

$$R(x + a_1 + b_1, y + a_2 + b_2).$$

If we represent $\mathbf{a} + \mathbf{b} = \langle a_1 + b_1, a_2 + b_2 \rangle$ by a vector with initial point P, then the terminal point has coordinates

$$(x + a_1 + b_1, y + a_2 + b_2)$$

and, therefore, coincides with R. Consequently, $\overrightarrow{PR}$ is a geometric representation for $\mathbf{a} + \mathbf{b}$. This is in agreement with our definition of the sum of directed line segments and illustrates how we may go back and forth between the algebraic and geometric approaches to vectors. The reader should strive to become proficient using both of these formulations.

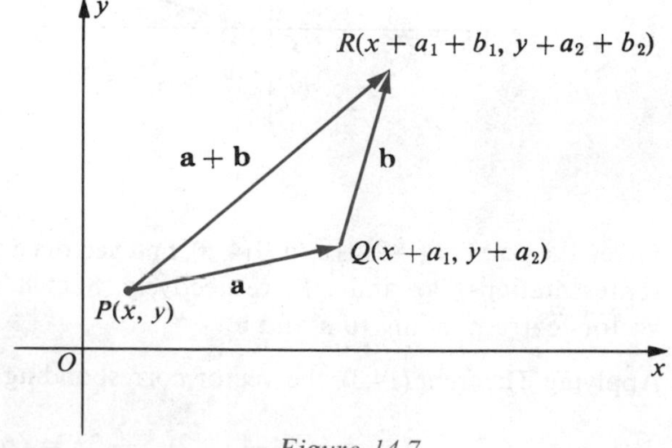

Figure 14.7

Four important properties of vectors are stated in the next theorem.

(14.4) Theorem

If **a**, **b**, **c** are any vectors in V_2, then
(i) $\mathbf{a} + \mathbf{b} = \mathbf{b} + \mathbf{a}$
(ii) $\mathbf{a} + (\mathbf{b} + \mathbf{c}) = (\mathbf{a} + \mathbf{b}) + \mathbf{c}$
(iii) $\mathbf{a} + \mathbf{0} = \mathbf{a}$
(iv) $\mathbf{a} + (-\mathbf{a}) = \mathbf{0}$

The proof of Theorem (14.4) follows readily from the definition of vector addition and properties of real numbers. For example, for part (i) we let $\mathbf{a} = \langle a_1, a_2 \rangle$ and $\mathbf{b} = \langle b_1, b_2 \rangle$. Since $a_1 + b_1 = b_1 + a_1$ and $a_2 + b_2 = b_2 + a_2$ we may write

$$\mathbf{a} + \mathbf{b} = \langle a_1 + b_1, a_2 + b_2 \rangle = \langle b_1 + a_1, b_2 + a_2 \rangle = \mathbf{b} + \mathbf{a}.$$

The remainder of the proof is left as an exercise. The reader should also give geometric interpretations for the conclusions of (14.4).

The operation of **subtraction** of vectors, denoted by $-$, is defined as follows.

(14.5) Definition

Let **a** and **b** be vectors in V_2. The **difference** $\mathbf{a} - \mathbf{b}$ is given by

$$\mathbf{a} - \mathbf{b} = \mathbf{a} + (-\mathbf{b}).$$

If we let $\mathbf{a} = \langle a_1, a_2 \rangle$ and $\mathbf{b} = \langle b_1, b_2 \rangle$, then $-\mathbf{b} = \langle -b_1, -b_2 \rangle$ and it follows from Definitions (14.5) and (14.1) that

$$\mathbf{a} - \mathbf{b} = \langle a_1, a_2 \rangle + \langle -b_1, -b_2 \rangle = \langle a_1 - b_1, a_2 - b_2 \rangle.$$

Thus, to find $\mathbf{a} - \mathbf{b}$ we merely subtract the components of **b** from the corresponding components of **a**.

Example 3 If $\mathbf{a} = \langle 5, -4 \rangle$ and $\mathbf{b} = \langle -3, 2 \rangle$, find (a) $\mathbf{a} - \mathbf{b}$; (b) $2\mathbf{a} - 3\mathbf{b}$.

Solution

(a) $\mathbf{a} - \mathbf{b} = \langle 5, -4 \rangle - \langle -3, 2 \rangle$
$= \langle 5 - (-3), -4 - 2 \rangle = \langle 8, -6 \rangle$

(b) $2\mathbf{a} - 3\mathbf{b} = 2\langle 5, -4 \rangle - 3\langle -3, 2 \rangle$
$= \langle 10, -8 \rangle - \langle -9, 6 \rangle = \langle 19, -14 \rangle$

It is easy to show that if **a** and **b** are arbitrary vectors, then

$$\mathbf{b} + (\mathbf{a} - \mathbf{b}) = \mathbf{a}$$

that is, $\mathbf{a} - \mathbf{b}$ is the vector which, when added to $\mathbf{b}$, gives us $\mathbf{a}$. This fact leads to a simple geometric interpretation for $\mathbf{a} - \mathbf{b}$. Specifically, if we represent $\mathbf{a}$ and $\mathbf{b}$ by vectors $\overrightarrow{PQ}$ and $\overrightarrow{PR}$ with the same initial point, as illustrated in Figure 14.8, then

$$\overrightarrow{PR} + \overrightarrow{RQ} = \overrightarrow{PQ}$$

and hence $\overrightarrow{RQ}$ represents $\mathbf{a} - \mathbf{b}$.

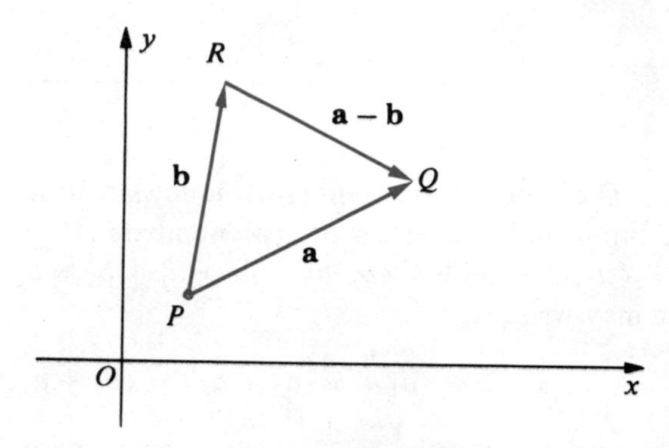

Figure 14.8

At the beginning of this section the symbol $c\overrightarrow{AB}$ was used to denote a vector having the same direction as $\overrightarrow{AB}$ if $c > 0$ or the opposite direction if $c < 0$. The next definition is the analogue of this concept for vectors in V_2.

(14.6) Definition

Nonzero vectors $\mathbf{a}$ and $\mathbf{b}$ of V_2 are said to have

(i) the **same direction** if $\mathbf{b} = c\mathbf{a}$ for some scalar $c > 0$.

(ii) **opposite directions** if $\mathbf{b} = c\mathbf{a}$ for some scalar $c < 0$.

It is not difficult to prove that if $\mathbf{b} = c\mathbf{a}$, then position vectors corresponding to $\mathbf{a}$ and $\mathbf{b}$ lie on the same line through the origin. Moreover, if $c > 0$, then the terminal points A and B of the position vectors lie in the same quadrant (or on the same positive or negative coordinate axis). If $c < 0$, and A is not on a coordinate axis, then A and B lie in diagonally opposite quadrants.

It is convenient to assume that the zero vector $\mathbf{0}$ has *all* directions.

We say that two vectors $\mathbf{a}$ and $\mathbf{b}$ are **parallel** if and only if $\mathbf{b} = c\mathbf{a}$ for some scalar c. In particular, nonzero vectors $\mathbf{a}$ and $\mathbf{b}$ are parallel if they have the same, or opposite, directions. The next theorem brings out the relationship between the magnitudes of $\mathbf{a}$ and $c\mathbf{a}$.

(14.7) Theorem If $\mathbf{a}$ is a vector and c is a scalar, then $|c\mathbf{a}| = |c|\,|\mathbf{a}|$.

Proof. If $\mathbf{a} = \langle a_1, a_2 \rangle$, then $c\mathbf{a} = \langle ca_1, ca_2 \rangle$. Using Definition (14.2) and properties of real numbers,

$$\begin{aligned} |c\mathbf{a}| &= \sqrt{(ca_1)^2 + (ca_2)^2} = \sqrt{c^2(a_1^2 + a_2^2)} \\ &= \sqrt{c^2}\sqrt{a_1^2 + a_2^2} \\ &= |c|\,|\mathbf{a}|. \end{aligned}$$

Theorem (14.7) implies that the length of a directed line segment which represents $c\mathbf{a}$ is $|c|$ times the length of a directed line segment which represents $\mathbf{a}$. This agrees with the geometric interpretation discussed at the beginning of this section and illustrated in Figure 14.3.

Some properties of scalar multiples of vectors are stated in the following theorem.

(14.8) Theorem If $\mathbf{a}$, $\mathbf{b}$ are vectors in V_2 and c, d are scalars, then

(i) $c(\mathbf{a} + \mathbf{b}) = c\mathbf{a} + c\mathbf{b}$.

(ii) $(c + d)\mathbf{a} = c\mathbf{a} + d\mathbf{a}$.

(iii) $(cd)\mathbf{a} = c(d\mathbf{a}) = d(c\mathbf{a})$.

(iv) $1\mathbf{a} = \mathbf{a}$.

(v) $0\mathbf{a} = \mathbf{0} = c\mathbf{0}$.

Proof. We shall prove (i) and leave proofs of the remaining properties as exercises. Letting $\mathbf{a} = \langle a_1, a_2 \rangle$ and $\mathbf{b} = \langle b_1, b_2 \rangle$ we have

$$\begin{aligned} c(\mathbf{a} + \mathbf{b}) &= c\langle a_1 + b_1, a_2 + b_2 \rangle \\ &= \langle ca_1 + cb_1, ca_2 + cb_2 \rangle \\ &= \langle ca_1, ca_2 \rangle + \langle cb_1, cb_2 \rangle \\ &= c\mathbf{a} + c\mathbf{b}. \end{aligned}$$

There should be no difficulty in remembering the properties in Theorem (14.8) since they resemble familiar properties of real numbers.

The special vectors $\mathbf{i}$ and $\mathbf{j}$ are defined by

$$\mathbf{i} = \langle 1, 0 \rangle, \quad \mathbf{j} = \langle 0, 1 \rangle.$$

These vectors can be used to obtain an alternate way of denoting vectors in two dimensions. Specifically, if $\mathbf{a} = \langle a_1, a_2 \rangle$ we may write

$$\mathbf{a} = \langle a_1, 0 \rangle + \langle 0, a_2 \rangle = a_1\langle 1, 0 \rangle + a_2\langle 0, 1 \rangle$$

that is,

$$\mathbf{a} = \langle a_1, a_2 \rangle = a_1\mathbf{i} + a_2\mathbf{j}. \tag{14.9}$$

The vector sum on the right in (14.9) is called a **linear combination** of **i** and **j**. If this notation is employed, then previous rules for addition, subtraction, and multiplication by a scalar may be written as follows, where $\mathbf{b} = \langle b_1, b_2 \rangle = b_1\mathbf{i} + b_2\mathbf{j}$:

$$(a_1\mathbf{i} + a_2\mathbf{j}) + (b_1\mathbf{i} + b_2\mathbf{j}) = (a_1 + b_1)\mathbf{i} + (a_2 + b_2)\mathbf{j}$$
$$(a_1\mathbf{i} + a_2\mathbf{j}) - (b_1\mathbf{i} + b_2\mathbf{j}) = (a_1 - b_1)\mathbf{i} + (a_2 - b_2)\mathbf{j}$$
$$c(a_1\mathbf{i} + a_2\mathbf{j}) = (ca_1)\mathbf{i} + (ca_2)\mathbf{j}.$$

These formulas show that linear combinations of **i** and **j** may be regarded as ordinary algebraic sums.

Example 4 If $\mathbf{a} = 5\mathbf{i} + \mathbf{j}$ and $\mathbf{b} = 4\mathbf{i} - 7\mathbf{j}$, express $3\mathbf{a} - 2\mathbf{b}$ as a linear combination of **i** and **j**.

Solution

$$\begin{aligned} 3\mathbf{a} - 2\mathbf{b} &= 3(5\mathbf{i} + \mathbf{j}) - 2(4\mathbf{i} - 7\mathbf{j}) \\ &= (15\mathbf{i} + 3\mathbf{j}) - (8\mathbf{i} - 14\mathbf{j}) \\ &= 7\mathbf{i} + 17\mathbf{j} \end{aligned}$$

A **unit vector** is a vector of magnitude 1. The vectors **i** and **j** are unit vectors.

(14.10) **Theorem**

If **a** is a nonzero vector, then a unit vector **u** having the same direction as **a** is given by

$$\mathbf{u} = \frac{1}{|\mathbf{a}|}\mathbf{a}.$$

Proof. If we let $c = 1/|\mathbf{a}|$, then $c > 0$ and it follows from (14.6) that the vector $\mathbf{u} = c\mathbf{a}$ has the same direction as **a**. Moreover, by Theorem (14.7),

$$|\mathbf{u}| = |c\mathbf{a}| = |c|\,|\mathbf{a}| = \frac{1}{|\mathbf{a}|}|\mathbf{a}| = 1.$$

This completes the proof.

Example 5 Find a unit vector **u** having the same direction as $3\mathbf{i} - 4\mathbf{j}$.

Solution Since $|\mathbf{a}| = \sqrt{9 + 16} = 5$ we have, from Theorem (14.10),

$$\mathbf{u} = \tfrac{1}{5}(3\mathbf{i} - 4\mathbf{j}) = \tfrac{3}{5}\mathbf{i} - \tfrac{4}{5}\mathbf{j}.$$

The formula $\mathbf{a} = a_1\mathbf{i} + a_2\mathbf{j}$ for the vector $\mathbf{a} = \langle a_1, a_2 \rangle$ has an interesting geometric interpretation. Position vectors corresponding to **i**, **j**, and **a** are illustrated in (i) of Figure 14.9. Since **i** and **j** are unit vectors, $a_1\mathbf{i}$ and $a_2\mathbf{j}$ may be represented by horizontal and vertical vectors of magnitudes $|a_1|$ and $|a_2|$, respectively, as illustrated in (ii) of Figure 14.9. The position vector for **a** may be regarded as the sum of these vectors. For this reason a_1 is called the **horizontal component** and a_2 the **vertical component** of the vector **a**.

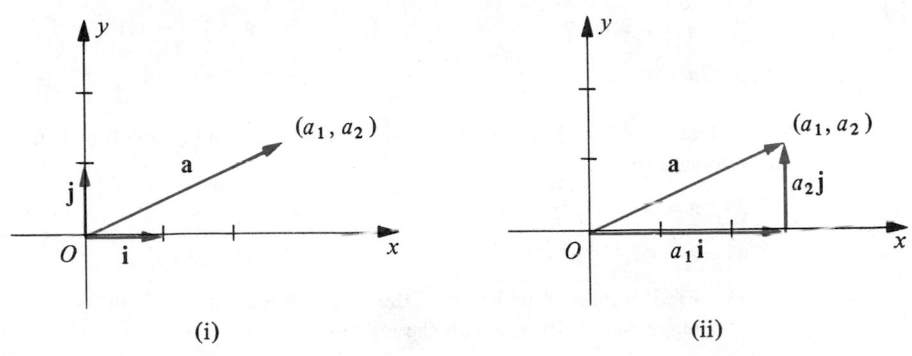

Figure 14.9

EXERCISES 14.1

1 Prove (ii)–(iv) of Theorem (14.4). Also give geometric arguments which justify the conclusions.

2 Prove (ii)–(v) of Theorem (14.8).

Prove each of the properties in Exercises 3–10, where **a**, **b** are vectors and c is a scalar.

3 $(-1)\mathbf{a} = -\mathbf{a}$

4 $(-c)\mathbf{a} = -c\mathbf{a}$

5 $-(\mathbf{a} + \mathbf{b}) = -\mathbf{a} - \mathbf{b}$

6 $c(\mathbf{a} - \mathbf{b}) = c\mathbf{a} - c\mathbf{b}$

7 If $\mathbf{a} + \mathbf{b} = \mathbf{0}$, then $\mathbf{b} = -\mathbf{a}$.

8 If $\mathbf{a} + \mathbf{b} = \mathbf{a}$, then $\mathbf{b} = \mathbf{0}$.

9 If $c\mathbf{a} = \mathbf{0}$ and $c \neq 0$, then $\mathbf{a} = \mathbf{0}$.

10 If $c\mathbf{a} = \mathbf{0}$ and $\mathbf{a} \neq \mathbf{0}$, then $c = 0$.

In each of Exercises 11–20 find $\mathbf{a} + \mathbf{b}$, $\mathbf{a} - \mathbf{b}$, $4\mathbf{a} + 5\mathbf{b}$, and $4\mathbf{a} - 5\mathbf{b}$.

11 $\mathbf{a} = \langle 2, -3\rangle, \mathbf{b} = \langle 1, 4\rangle$

12 $\mathbf{a} = \langle -2, 6\rangle, \mathbf{b} = \langle 2, 3\rangle$

13 $\mathbf{a} = -\langle 7, -2\rangle, \mathbf{b} = 4\langle -2, 1\rangle$

14 $\mathbf{a} = 2\langle 5, -4\rangle, \mathbf{b} = -\langle 6, 0\rangle$

15 $\mathbf{a} = \mathbf{i} + 2\mathbf{j}, \mathbf{b} = 3\mathbf{i} - 5\mathbf{j}$

16 $\mathbf{a} = -3\mathbf{i} + \mathbf{j}, \mathbf{b} = -3\mathbf{i} + \mathbf{j}$

17 $\mathbf{a} = -(4\mathbf{i} - \mathbf{j}), \mathbf{b} = 2(\mathbf{i} - 3\mathbf{j})$

18 $\mathbf{a} = 8\mathbf{j}, \mathbf{b} = (-3)(-2\mathbf{i} + \mathbf{j})$

19 $\mathbf{a} = 2\mathbf{j}, \mathbf{b} = -3\mathbf{i}$

20 $\mathbf{a} = \mathbf{0}, \mathbf{b} = \mathbf{i} + \mathbf{j}$

In Exercises 21–26 find a vector **a** which has geometric representation $\overrightarrow{PQ}$. Sketch $\overrightarrow{PQ}$ and a position vector corresponding to **a**.

21 $P(1, -4), Q(5, 3)$

22 $P(7, -3), Q(-2, 4)$

23 $P(2, 5), Q(-4, 5)$

24 $P(-4, 6), Q(-4, -2)$

25 $P(-3, -1), Q(6, -4)$

26 $P(2, 3), Q(-6, 0)$

In Exercises 27–30 sketch vectors corresponding to **a**, **b**, $\mathbf{a} + \mathbf{b}$, $\mathbf{a} - \mathbf{b}$, $2\mathbf{a}$, and $-3\mathbf{b}$.

27 $\mathbf{a} = 3\mathbf{i} + 2\mathbf{j}, \mathbf{b} = -\mathbf{i} + 5\mathbf{j}$

28 $\mathbf{a} = -5\mathbf{i} + 2\mathbf{j}, \mathbf{b} = \mathbf{i} - 3\mathbf{j}$

29 $\mathbf{a} = \langle -4, 6\rangle, \mathbf{b} = \langle -2, 3\rangle$

30 $\mathbf{a} = \langle 2, 0\rangle, \mathbf{b} = \langle -2, 0\rangle$

In Exercises 31–38 find the magnitude of **a**.

31 $\mathbf{a} = \langle 3, -3\rangle$

32 $\mathbf{a} = \langle -2, -2\sqrt{3}\rangle$

33 $\mathbf{a} = \langle -5, 0\rangle$

34 $\mathbf{a} = \langle 0, 10\rangle$

35 $\mathbf{a} = -4\mathbf{i} + 5\mathbf{j}$

36 $\mathbf{a} = 10\mathbf{i} - 10\mathbf{j}$

37 $\mathbf{a} = -18\mathbf{j}$

38 $\mathbf{a} = 0\mathbf{i} + 0\mathbf{j}$

In Exercises 39–42 find a unit vector having (a) the same direction as **a**; (b) the direction opposite to **a**.

39 $\mathbf{a} = -8\mathbf{i} + 15\mathbf{j}$

40 $\mathbf{a} = 5\mathbf{i} - 3\mathbf{j}$

41 $\mathbf{a} = \langle 2, -5\rangle$

42 $\mathbf{a} = \langle 0, 6\rangle$

43 Find a vector which has the same direction as $\langle -6, 3\rangle$ and (a) twice the magnitude; (b) one-half the magnitude.

44 Find a vector which has the opposite direction of $8\mathbf{i} - 5\mathbf{j}$ and (a) three times the magnitude; (b) one-third the magnitude.

45 Find a vector of magnitude 6 which has the same direction as $\mathbf{a} = 4\mathbf{i} - 7\mathbf{j}$.

46 Find a vector of magnitude 4 whose direction is opposite that of $\mathbf{a} = \langle 2, -5\rangle$.

47 Demonstrate graphically that $|\mathbf{a} + \mathbf{b}| \le |\mathbf{a}| + |\mathbf{b}|$. Under what conditions is $|\mathbf{a} + \mathbf{b}| = |\mathbf{a}| + |\mathbf{b}|$?

48 (a) Let $\mathbf{a} = \langle a_1, a_2\rangle$ be any nonzero vector and let $\overrightarrow{OA}$ be the position vector corresponding to **a**. If θ is the smallest nonnegative angle from the positive x-axis to $\overrightarrow{OA}$, show that

$$\mathbf{a} = |\mathbf{a}|(\cos\theta\mathbf{i} + \sin\theta\mathbf{j}).$$

(b) Show that every unit vector in V_2 can be expressed in the form $\cos\theta\mathbf{i} + \sin\theta\mathbf{j}$ for some θ.

49 If $P_1, P_2, \ldots, P_n$ are arbitrary points in a coordinate plane, show that

$$\sum_{i=1}^{n-1} \overrightarrow{P_iP_{i+1}} + \overrightarrow{P_nP_1}$$

is the zero vector. Illustrate this fact graphically if $n = 5$.

50 If $P_1, P_2, \ldots, P_n$ are vertices of a regular polygon and if O is the center of the polygon, prove that $\Sigma_{i=1}^{n} \overrightarrow{OP_i}$ is the zero vector.

14.2 RECTANGULAR COORDINATE SYSTEMS IN THREE DIMENSIONS

In order to discuss geometric aspects of vectors in space, it is necessary to introduce a three-dimensional coordinate system. This is accomplished by means of ordered triples of real numbers. An **ordered triple** (a, b, c) is a set $\{a, b, c\}$ of three numbers in which a is considered as the first number of the set, b as the second number, and c as the third number. The totality of all such ordered triples is often denoted by $\mathbb{R} \times \mathbb{R} \times \mathbb{R}$. Two ordered triples (a_1, a_2, a_3) and (b_1, b_2, b_3) are said to be **equal**, and we write $(a_1, a_2, a_3) = (b_1, b_2, b_3)$, if and only if $a_1 = b_1, a_2 = b_2$, and $a_3 = b_3$.

In order to specify points in space we choose a fixed point O (called the **origin**) and consider three mutually perpendicular coordinate lines (called the x-, y-, and z-axes) with common origin O, as illustrated in (i) of Figure 14.10. To visualize this configuration, we may imagine that the y- and z-axes lie in the plane of the paper and that the x-axis projects out from the paper. A coordinate system of this type is called *right-handed.* If the x- and y-axes are interchanged, the resulting coordinate system is said to be *left-handed.* The coordinate plane determined by the x- and y-axes is called the *xy-plane.* Similarly, the coordinate plane determined by the y- and z-axes is called the *yz-plane* and that determined by the x- and z-axes is called the *xz-plane.* The terminology *right-handed* may be justified as follows. If, as in (ii) of Figure 14.10, the fingers of the right hand are curled in the direction of a 90° rotation of axes in the xy-plane (so that the positive x-axis is transformed into the positive y-axis), then the extended thumb points in the direction of the positive z-axis.

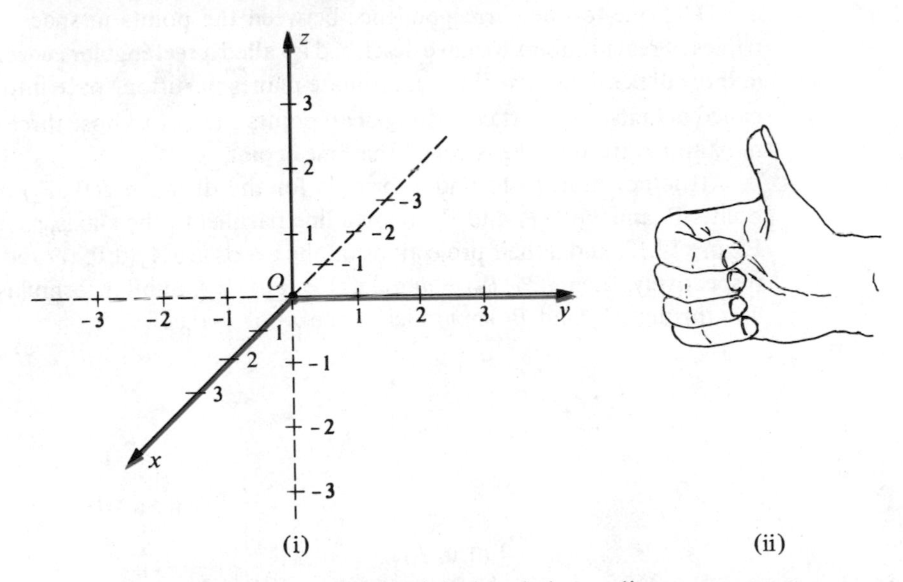

Figure 14.10. Right-handed coordinate system

If P is a point, then the projection of P on the x-axis has some coordinate a, called the **x-coordinate** of P. The x-coordinate of P may also be thought of as a directed distance from the yz-plane to P. Similarly, the coordinates b and c of the projections of P on the y- and z-axes, respectively, are called the **y-coordinate** and **z-coordinate** of P. We shall use the ordered triple (a, b, c) to denote the coordinates of P, as well as P itself. The notation $P(a, b, c)$ will also be used for the point P with coordinates a, b, and c. If P is not on a coordinate plane, then the three planes through P which are parallel to the coordinate planes, together with the coordinate planes, determine a rectangular parallelepiped. This is illustrated in (i) of Figure 14.11, where the eight vertices are labeled in the manner just described.

Conversely, to each ordered triple (a, b, c) of real numbers, there corresponds a point P having coordinates a, b, and c. The concept of plotting points is similar to that used in two dimensions. As an aid to plotting, it is often convenient to construct a parallelepiped as shown in (i) of Figure 14.11. The points $A(-1, -3, 1)$ and $B(3, 4, -2)$ are plotted in (ii) of the figure.

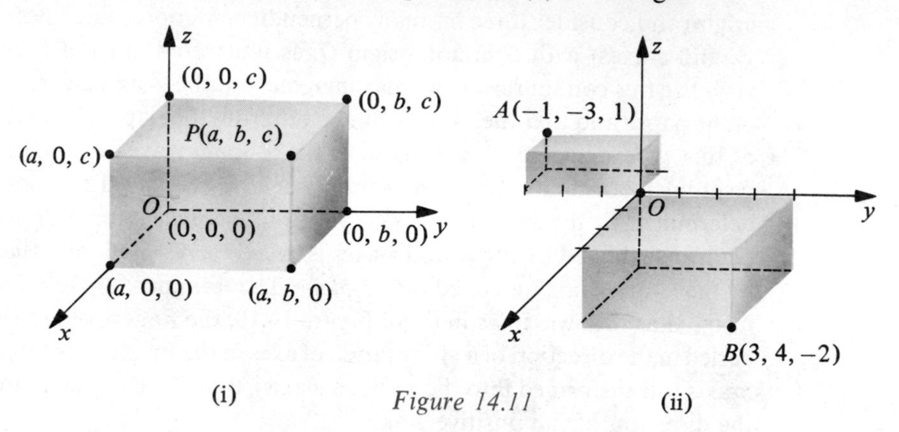

Figure 14.11

The one-to-one correspondence between the points in space and ordered triples of real numbers we have described is called a **rectangular coordinate system in three dimensions**. The three coordinate planes partition space into eight parts, called **octants**. The part consisting of all points $P(a, b, c)$ whose three coordinates a, b, and c are positive is called the **first octant**.

It is not difficult to find a formula for the distance $d(P_1, P_2)$ between two points P_1 and P_2. If P_1 and P_2 are on a line parallel to the z-axis, as illustrated in Figure 14.12, and if their projections on the z-axis are $A_1(0, 0, z_1)$ and $A_2(0, 0, z_2)$, respectively, then $d(P_1, P_2) = d(A_1, A_2) = |z_2 - z_1|$. Similar formulas hold if the line through P_1 and P_2 is parallel to the x- or y-axis.

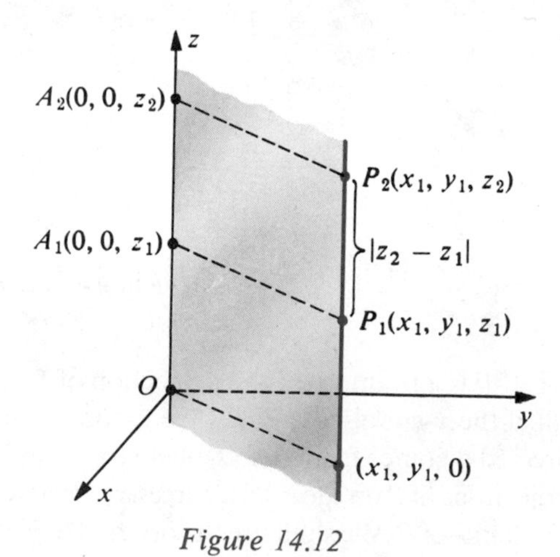

Figure 14.12

If we have a situation of the type illustrated in Figure 14.13, then triangle P_1AP_2 is a right triangle, and hence by the Pythagorean Theorem,

$$d(P_1, P_2) = \sqrt{[d(P_1, A)]^2 + [d(A, P_2)]^2}.$$

Since P_1 and A are in a plane parallel to the xy-plane, it follows from the Distance Formula in two dimensions that $[d(P_1, A)]^2 = (x_2 - x_1)^2 + (y_2 - y_1)^2$, whereas from our previous remarks $[d(A, P_2)]^2 = (z_2 - z_1)^2$. Substituting in the above formula for $d(P_1, P_2)$ gives us the following **Distance Formula in three dimensions**:

(14.11) $$d(P_1, P_2) = \sqrt{(x_2 - x_1)^2 + (y_2 - y_1)^2 + (z_2 - z_1)^2}.$$

Note that if P_1 and P_2 are on the xy-plane, so that $z_1 = z_2 = 0$, then (14.11) reduces to the two-dimensional Distance Formula.

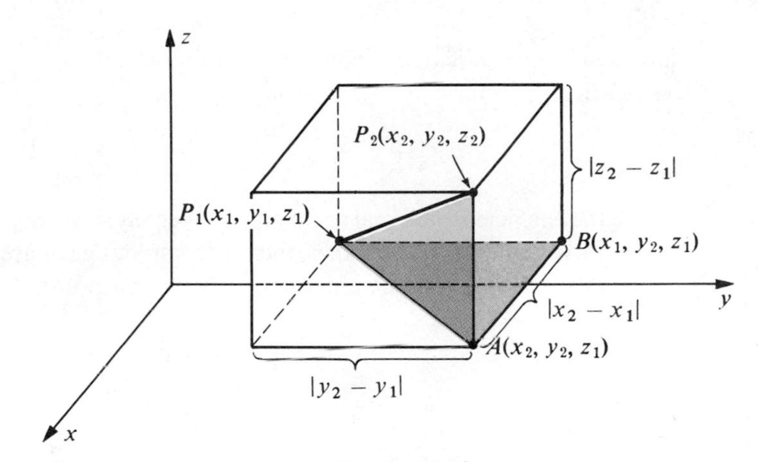

Figure 14.13

Example 1 Find the distance between $A(-1, -3, 1)$ and $B(3, 4, -2)$.

Solution Points A and B are plotted in Figure 14.11. Using the Distance Formula (14.11),

$$d(A, B) = \sqrt{(3 + 1)^2 + (4 + 3)^2 + (-2 - 1)^2}$$
$$= \sqrt{16 + 49 + 9} = \sqrt{74}.$$

By referring to Figure 14.13 and using similar triangles the following result can be proved.

(14.12) **Midpoint Formula**

The midpoint of the line segment from $P_1(x_1, y_1, z_1)$ to $P_2(x_2, y_2, z_2)$ is

$$\left(\frac{x_1 + x_2}{2}, \frac{y_1 + y_2}{2}, \frac{z_1 + z_2}{2}\right).$$

As an illustration of (14.12), if A and B are the points in Example 1, then the midpoint of segment AB has coordinates $(1, 1/2, -1/2)$.

The graph of an equation in three variables x, y, and z is defined as the set of all points $P(a, b, c)$ in a rectangular coordinate system such that the ordered triple

(a, b, c) is a solution of the equation; that is, such that equality is obtained when a, b, and c are substituted for x, y, and z, respectively. The graph of such an equation is a **surface** of some type. It is easy to derive an equation which has as its graph a sphere of radius r with center at the point $C(h, k, l)$. As illustrated in Figure 14.14 a point $P(x, y, z)$ is on the sphere if and only if $[d(C, P)]^2 = r^2$. Applying the Distance Formula (14.11) gives us the following **standard equation of a sphere of radius r with center $C(h, k, l)$**:

(14.13)
$$(x - h)^2 + (y - k)^2 + (z - l)^2 = r^2.$$

If we square the indicated expressions and simplify, then equation (14.13) may be written in the form

$$x^2 + y^2 + z^2 + ax + by + cz + d = 0$$

where the coefficients are real numbers. Conversely, if we begin with an equation of this form and if the graph exists, then by completing squares we can obtain the form (14.13), and hence the graph is a sphere or a point.

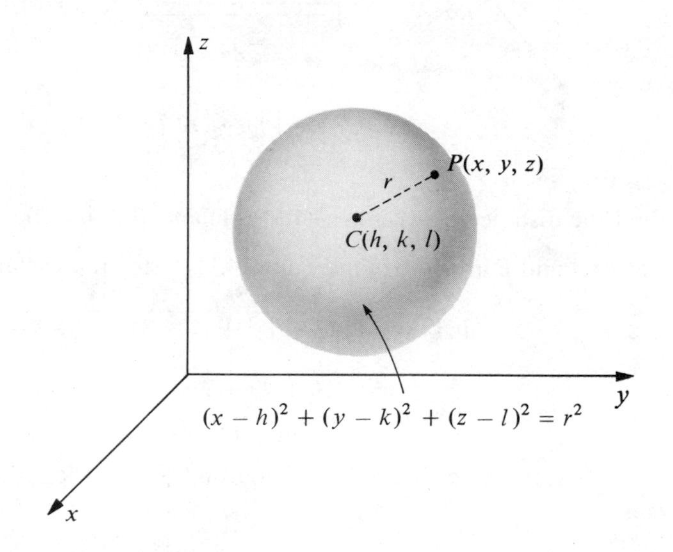

Figure 14.14

Example 2 Discuss the graph of the equation $x^2 + y^2 + z^2 - 6x + 8y + 4z + 4 = 0$.

Solution We complete squares as follows:

$$(x^2 - 6x) + (y^2 + 8y) + (z^2 + 4z) = -4$$
$$(x^2 - 6x + 9) + (y^2 + 8y + 16) + (z^2 + 4z + 4) = -4 + 9 + 16 + 4$$
$$(x - 3)^2 + (y + 4)^2 + (z + 2)^2 = 25.$$

Comparing the last equation with (14.13) it follows that the graph is a sphere of radius 5 with center $C(3, -4, -2)$.

EXERCISES 14.2

In Exercises 1–6 plot the points A and B and find (a) $d(A, B)$; (b) the midpoint of AB.

1 $A(2, 4, -5), B(4, -2, 3)$

2 $A(1, -2, 7), B(2, 4, -1)$

3 $A(-4, 0, 1), B(3, -2, 1)$

4 $A(0, 5, -4), B(1, 1, 0)$

5 $A(1, 0, 0), B(0, 1, 1)$

6 $A(0, 0, 0), B(-8, -1, 4)$

In Exercises 7 and 8, if A and B are opposite vertices of a parallelepiped having its faces parallel to the coordinate planes, find the coordinates of the other vertices.

7 $A(2, 5, -3), B(-4, 2, 1)$

8 $A(-3, 2, 6), B(1, 5, -1)$

In Exercises 9 and 10 prove that A, B, and C are vertices of a right triangle and find its area.

9 $A(2, 0, 1), B(3, 1, 2), C(1, 2, 0)$

10 $A(4, -3, 2), B(6, -2, 1), C(7, -6, 5)$

In Exercises 11–14 find an equation of the sphere with center C and radius r.

11 $C(3, -1, 2), r = 3$

12 $C(4, -5, 1), r = 5$

13 $C(-5, 0, 1), r = 1/2$

14 $C(0, -3, -6), r = \sqrt{3}$

15 Find an equation of the sphere with center $(-2, 4, -6)$ which is as follows.
(a) Tangent to the yz-plane
(b) Tangent to the xz-plane
(c) Tangent to the xy-plane

16 Find an equation of the sphere having end-points of a diameter at $A(1, 4, -2)$ and $B(-7, 1, 2)$.

In Exercises 17–22 find the center and radius of the sphere having the given equation.

17 $x^2 + y^2 + z^2 + 4x - 2y + 2z + 2 = 0$

18 $x^2 + y^2 + z^2 - 6x - 10y + 6z + 34 = 0$

19 $x^2 + y^2 + z^2 - 8x + 8z + 16 = 0$

20 $4x^2 + 4y^2 + 4z^2 - 4x + 8y - 3 = 0$

21 $x^2 + y^2 + z^2 + 4y = 0$

22 $x^2 + y^2 + z^2 - z = 0$

23 Find an equation for the set of all points equidistant from $A(2, -1, 3)$ and $B(-1, 5, 1)$. Describe the graph of the equation.

24 Work Exercise 23 using $A(5, 0, -4)$ and $B(2, -1, 7)$.

25 Describe the graph of each of the following equations.
(a) $z = 5$
(b) $y = 2$
(c) $x = 0$

26 Describe the graph of the equation $xyz = 0$.

In each of Exercises 27–34, describe the given region R in a three-dimensional coordinate system.

27 $R = \{(x, y, z): x^2 + y^2 + z^2 \leq 1\}$

28 $R = \{(x, y, z): x^2 + y^2 + z^2 > 1\}$

29 $R = \{(x, y, z): |x| \leq 1, |y| \leq 2, |z| \leq 3\}$

30 $R = \{(x, y, z): x^2 + y^2 = 1\}$

31 $R = \{(x, y, z): 0 < x^2 + y^2 \leq 1\}$ **32** $R = \{(x, y, z): 4 < x^2 + y^2 + z^2 < 9\}$

33 $R = \{(x, y, z): x^2 + y^2 + z^2 - 4x - 2y + 1 < 0\}$

34 $R = \{(x, y, z): y = x\}$

14.3 VECTORS IN THREE DIMENSIONS

The system V_3 of vectors in three dimensions is defined as the collection of all ordered triples $\langle a_1, a_2, a_3 \rangle$ of real numbers, called **vectors**, subject to definitions of addition and multiplication by scalars which are similar to Definition (14.1). The numbers a_1, a_2, and a_3 are called the **components** of the vector $\langle a_1, a_2, a_3 \rangle$.

If $\mathbf{a} = \langle a_1, a_2, a_3 \rangle$ and $\mathbf{b} = \langle b_1, b_2, b_3 \rangle$ are vectors and c is a scalar, then addition and scalar multiplication are defined by

$$\mathbf{a} + \mathbf{b} = \langle a_1 + b_1, a_2 + b_2, a_3 + b_3 \rangle$$
$$c\mathbf{a} = \langle ca_1, ca_2, ca_3 \rangle.$$

The **zero vector** $\mathbf{0}$ and the **negative** $-\mathbf{a}$ of $\mathbf{a}$ are

$$\mathbf{0} = \langle 0, 0, 0 \rangle \quad \text{and} \quad -\mathbf{a} = -\langle a_1, a_2, a_3 \rangle = \langle -a_1, -a_2, -a_3 \rangle.$$

Subtraction is defined by

$$\mathbf{a} - \mathbf{b} = \mathbf{a} + (-\mathbf{b}) = \langle a_1 - b_1, a_2 - b_2, a_3 - b_3 \rangle.$$

Properties of vectors in two dimensions may be extended without difficulty to V_3 by simply taking the third component into account. In particular, the properties listed in Theorem (14.4) and Theorem (14.8) are readily proved.

A vector $\mathbf{a} = \langle a_1, a_2, a_3 \rangle$ in V_3 may be represented in a rectangular coordinate system by a directed line segment $\overrightarrow{PQ}$ with arbitrary initial point $P(x, y, z)$ and terminal point $Q(x + a_1, y + a_2, z + a_3)$, as illustrated in Figure 14.15. As before, directed line segments are also referred to as *vectors*. The vector $\overrightarrow{OP}$ from the origin to a point P is called the **position vector of *P***. If P has coordinates (a_1, a_2, a_3), then $\overrightarrow{OP}$ (see Figure 14.15) is called the **position vector corresponding to** $\mathbf{a} = \langle a_1, a_2, a_3 \rangle$. The geometric interpretation of vector addition in three dimensions is exactly the same as that in two dimensions.

By definition, the **magnitude** $|\mathbf{a}|$ of a vector $\mathbf{a} = \langle a_1, a_2, a_3 \rangle$ is the length of any of its geometric representations. Applying the Distance Formula (14.11) gives us

$$|\mathbf{a}| = \sqrt{a_1^2 + a_2^2 + a_3^2}.$$

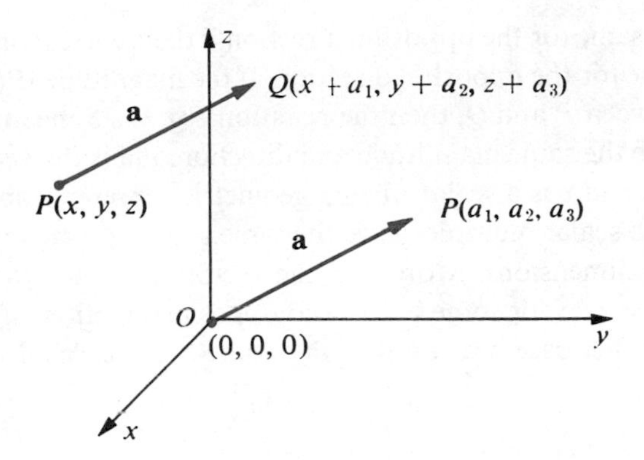

Figure 14.15

Observe that $|\mathbf{a}| \geq 0$ and that $|\mathbf{a}| = 0$ if and only if $\mathbf{a} = \mathbf{0}$. As in two dimensions, it can be proved that $|c\mathbf{a}| = |c|\,|\mathbf{a}|$.

Example 1 If $\mathbf{a} = \langle 2, 5, -3 \rangle$ and $\mathbf{b} = \langle -4, 1, 7 \rangle$ find $\mathbf{a} + \mathbf{b}$, $2\mathbf{a} - 3\mathbf{b}$, and $|\mathbf{a}|$.

Solution

$$\begin{aligned} \mathbf{a} + \mathbf{b} &= \langle 2, 5, -3 \rangle + \langle -4, 1, 7 \rangle = \langle -2, 6, 4 \rangle \\ 2\mathbf{a} - 3\mathbf{b} &= 2\langle 2, 5, -3 \rangle - 3\langle -4, 1, 7 \rangle = \langle 4, 10, -6 \rangle - \langle -12, 3, 21 \rangle \\ &= \langle 16, 7, -27 \rangle \\ |\mathbf{a}| &= \sqrt{(2)^2 + (5)^2 + (-3)^2} = \sqrt{38} \end{aligned}$$

Exactly as we did for vectors in two dimensions, we say that two nonzero vectors $\mathbf{a}$ and $\mathbf{b}$ of V_3 have the **same direction** if $\mathbf{b} = c\mathbf{a}$ for some scalar $c > 0$, or **opposite directions** if $\mathbf{b} = c\mathbf{a}$ for some $c < 0$. To illustrate, given the vectors

$$\mathbf{a} = \langle 15, -6, 24 \rangle, \quad \mathbf{b} = \langle 5, -2, 8 \rangle, \quad \mathbf{c} = \langle -\tfrac{15}{2}, 3, -12 \rangle$$

the vectors $\mathbf{a}$ and $\mathbf{b}$ have the same direction since $\mathbf{a} = 3\mathbf{b}$ or, equivalently, $\mathbf{b} = (1/3)\mathbf{a}$; whereas $\mathbf{c}$ and $\mathbf{a}$ have opposite directions since $\mathbf{c} = (-1/2)\mathbf{a}$, or $\mathbf{a} = -2\mathbf{c}$.

As in two dimensions, we say that the vectors $\mathbf{a}$ and $\mathbf{b}$ are **parallel** if $\mathbf{b} = c\mathbf{a}$ for some scalar c.

The proof of the next result is similar to the proof of Theorem (14.3).

(14.14) Theorem

> If $P_1(x_1, y_1, z_1)$ and $P_2(x_2, y_2, z_2)$ are points, then the vector $\mathbf{a}$ in V_3 which has geometric representation $\overrightarrow{P_1P_2}$ is
>
> $$\mathbf{a} = \langle x_2 - x_1, y_2 - y_1, z_2 - z_1 \rangle.$$

An illustration of Theorem (14.14) is given in Figure 14.16, where $\overrightarrow{OA}$ is the position vector corresponding to $\mathbf{a}$. Two directed line segments are said to have the same (or the opposite) direction if their corresponding vectors in V_3 have the same (or the opposite) direction. If the magnitude $|\overrightarrow{PQ}|$ is defined as the distance between P and Q, then the notation $\overrightarrow{PQ} = \overrightarrow{RS}$ means that the indicated vectors have the same magnitude and direction. If **a** is the vector in V_3 corresponding to $\overrightarrow{PQ}$, and c is a scalar, then a geometric representation of $c\mathbf{a}$ is denoted by $c\overrightarrow{PQ}$. Such scalar multiples have the same geometric meaning as their counterparts in two dimensions. Moreover, the vectors $\overrightarrow{PQ}$ and $\overrightarrow{RS}$ have the same direction if $\overrightarrow{PQ} = c\overrightarrow{RS}$ for some $c > 0$, and opposite directions if $\overrightarrow{PQ} = c\overrightarrow{RS}$ for some $c < 0$. In either case, we say that $\overrightarrow{PQ}$ and $\overrightarrow{RS}$ are **parallel**.

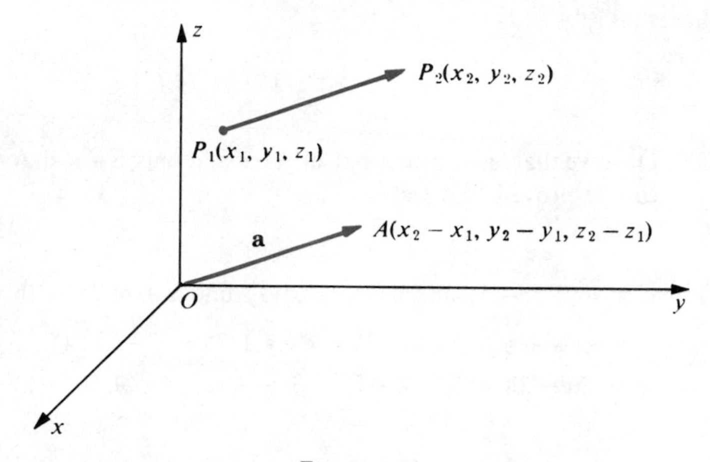

Figure 14.16

Example 2 Given the points $P_1(5, 6, -2)$ and $P_2(-3, 8, 7)$,

(a) find the position vector corresponding to $\overrightarrow{P_1P_2}$.

(b) find $|\overrightarrow{P_1P_2}|$.

Solution (a) By Theorem (14.14), the vector $\mathbf{a} = \langle -3 - 5, 8 - 6, 7 + 2 \rangle = \langle -8, 2, 9 \rangle$ has geometric representation $\overrightarrow{P_1P_2}$. Thus $A(-8, 2, 9)$ is the terminal point of the position vector $\overrightarrow{OA}$ corresponding to $\overrightarrow{P_1P_2}$.

(b) $|\overrightarrow{P_1P_2}| = |\mathbf{a}| = \sqrt{64 + 4 + 81} = \sqrt{149}$

A vector **a** is a **unit vector** if $|\mathbf{a}| = 1$. The special unit vectors

$$\mathbf{i} = \langle 1, 0, 0 \rangle, \quad \mathbf{j} = \langle 0, 1, 0 \rangle, \quad \mathbf{k} = \langle 0, 0, 1 \rangle$$

are important since any vector $\mathbf{a} = \langle a_1, a_2, a_3 \rangle$ can be expressed as a linear combination of **i**, **j**, and **k**. Specifically,

$$\mathbf{a} = \langle a_1, a_2, a_3 \rangle = a_1\mathbf{i} + a_2\mathbf{j} + a_3\mathbf{k}.$$

Geometric representations of **i**, **j**, and **k** are sketched in (i) of Figure 14.17. Part (ii) of the figure illustrates how the position vector for $\mathbf{a} = \langle a_1, a_2, a_3 \rangle$ may be regarded as the sum of three vectors corresponding to $a_1\mathbf{i}$, $a_2\mathbf{j}$, and $a_3\mathbf{k}$.

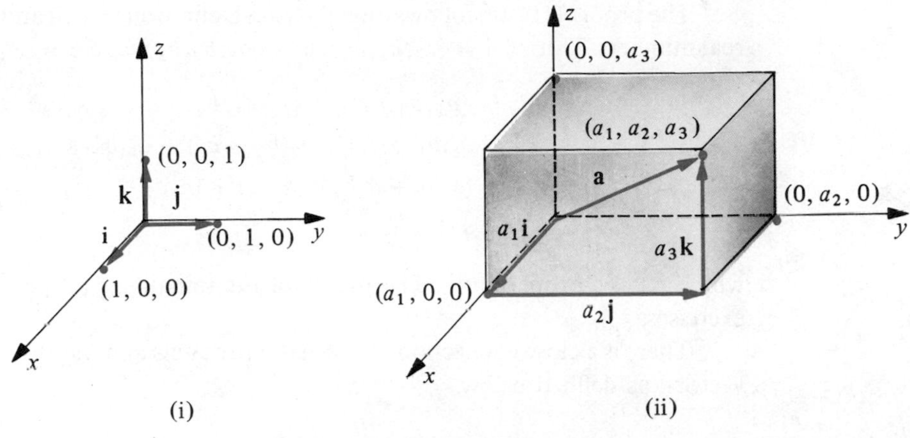

Figure 14.17

As in Section 1, rules for addition, subtraction, and multiplication by scalars may be easily translated into the **i**, **j**, **k** notation. It is often convenient to regard V_2 as a subset of V_3 by identifying the vector $\langle a_1, a_2\rangle$ with $\langle a_1, a_2, 0\rangle$. If this is done, then there is no essential difference between the vectors $\mathbf{i} = \langle 1, 0\rangle$, $\mathbf{j} = \langle 0, 1\rangle$, and the vectors **i**, **j** defined above.

The next concept to be introduced has many mathematical and physical applications. In the manner of our previous work we shall begin with an algebraic definition and give geometric and physical interpretations later in the section.

(14.15) Definition

If $\mathbf{a} = \langle a_1, a_2, a_3\rangle$ and $\mathbf{b} = \langle b_1, b_2, b_3\rangle$ then the **dot product** $\mathbf{a}\cdot\mathbf{b}$ of **a** and **b** is given by

$$\mathbf{a}\cdot\mathbf{b} = a_1b_1 + a_2b_2 + a_3b_3.$$

The dot product is also referred to as the **scalar product** or **inner product**. It is important to note that $\mathbf{a}\cdot\mathbf{b}$ is a scalar, not a vector. For example,

$$\langle 2, 4, -3\rangle\cdot\langle -1, 5, 2\rangle = (2)(-1) + (4)(5) + (-3)(2) = 12$$
$$(3\mathbf{i} - 2\mathbf{j} + \mathbf{k})\cdot(4\mathbf{i} + 5\mathbf{j} - 2\mathbf{k}) = (3)(4) + (-2)(5) + (1)(-2) = 0.$$

(14.16) Properties of the Dot Product

If **a**, **b**, **c** are vectors and c is a scalar, then

(i) $\mathbf{a}\cdot\mathbf{a} = |\mathbf{a}|^2$.

(ii) $\mathbf{a}\cdot\mathbf{b} = \mathbf{b}\cdot\mathbf{a}$.

(iii) $\mathbf{a}\cdot(\mathbf{b} + \mathbf{c}) = \mathbf{a}\cdot\mathbf{b} + \mathbf{a}\cdot\mathbf{c}$.

(iv) $(c\mathbf{a})\cdot\mathbf{b} = c(\mathbf{a}\cdot\mathbf{b}) = \mathbf{a}\cdot(c\mathbf{b})$.

(v) $\mathbf{0}\cdot\mathbf{a} = 0$.

The proof of (14.16) follows directly from Definition (14.15) and properties of real numbers. Thus, if $\mathbf{a} = \langle a_1, a_2, a_3 \rangle$, $\mathbf{b} = \langle b_1, b_2, b_3 \rangle$, and $\mathbf{c} = \langle c_1, c_2, c_3 \rangle$, then

$$\begin{aligned}\mathbf{a} \cdot (\mathbf{b} + \mathbf{c}) &= \langle a_1, a_2, a_3 \rangle \cdot \langle b_1 + c_1, b_2 + c_2, b_3 + c_3 \rangle \\ &= a_1(b_1 + c_1) + a_2(b_2 + c_2) + a_3(b_3 + c_3) \\ &= (a_1b_1 + a_2b_2 + a_3b_3) + (a_1c_1 + a_2c_2 + a_3c_3) \\ &= \mathbf{a} \cdot \mathbf{b} + \mathbf{a} \cdot \mathbf{c}\end{aligned}$$

which proves property (iii). The proofs of the remaining properties are left as exercises.

There is a close connection between dot products and the angle between two vectors, as defined below.

(14.17) Definition

Let **a** and **b** be nonzero vectors. If **b** is not a scalar multiple of **a**, and if $\overrightarrow{OA}$ and $\overrightarrow{OB}$ are the position vectors corresponding to **a** and **b**, respectively, then **the angle θ between a and b** (or between $\overrightarrow{OA}$ and $\overrightarrow{OB}$) is defined as angle AOB of the triangle determined by points A, O, and B (see Figure 14.18). If $\mathbf{b} = c\mathbf{a}$ for some scalar c, then $\theta = 0$ or $\theta = \pi$ according as $c > 0$ or $c < 0$.

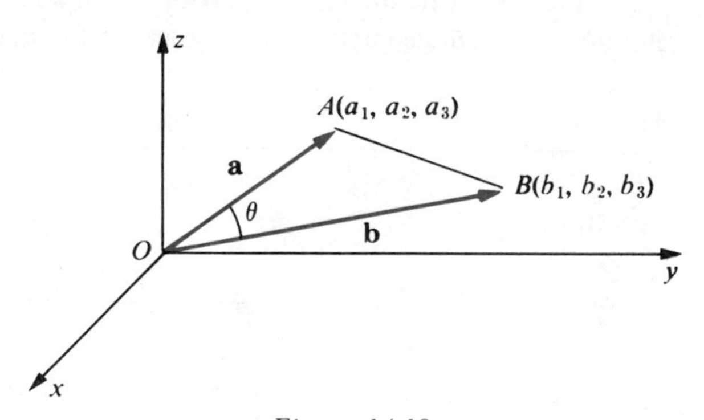

Figure 14.18

Note that if **a** and **b** are parallel then, by (14.17), $\theta = 0$ or $\theta = \pi$. The vectors **a** and **b** are said to be **orthogonal** if $\theta = \pi/2$. For convenience we shall assume that the zero vector **0** is parallel and orthogonal to every vector **a**.

(14.18) Theorem

If θ is the angle between nonzero vectors **a** and **b**, then

$$\mathbf{a} \cdot \mathbf{b} = |\mathbf{a}|\,|\mathbf{b}| \cos \theta.$$

Proof. If $\mathbf{b} \neq c\mathbf{a}$ we have a situation similar to that illustrated in Figure 14.18. Applying the Law of Cosines to triangle AOB gives us

$$|\overrightarrow{AB}|^2 = |\mathbf{a}|^2 + |\mathbf{b}|^2 - 2|\mathbf{a}|\,|\mathbf{b}| \cos \theta.$$

Consequently,

$$(b_1 - a_1)^2 + (b_2 - a_2)^2 + (b_3 - a_3)^2$$
$$= a_1^2 + a_2^2 + a_3^2 + b_1^2 + b_2^2 + b_3^2 - 2|\mathbf{a}|\,|\mathbf{b}|\cos\theta$$

which reduces to

$$-2a_1b_1 - 2a_2b_2 - 2a_3b_3 = -2|\mathbf{a}|\,|\mathbf{b}|\cos\theta.$$

Dividing both sides of the last equation by -2 gives us the desired conclusion.

If $\mathbf{b} = c\mathbf{a}$, then by properties (iv) and (i) of (14.16),

$$\mathbf{a}\cdot\mathbf{b} = \mathbf{a}\cdot(c\mathbf{a}) = c(\mathbf{a}\cdot\mathbf{a}) = c|\mathbf{a}|^2.$$

Also,

$$|\mathbf{a}|\,|\mathbf{b}|\cos\theta = |\mathbf{a}|\,|c\mathbf{a}|\cos\theta = |c|\,|\mathbf{a}|^2\cos\theta.$$

If $c > 0$, then $|c| = c$, $\theta = 0$, and $|c|\,|\mathbf{a}|^2\cos\theta$ reduces to $c|\mathbf{a}|^2$. Consequently $|\mathbf{a}|\,|\mathbf{b}|\cos\theta = \mathbf{a}\cdot\mathbf{b}$. If $c < 0$, then $|c| = -c$, $\theta = \pi$, and again $|c|\,|\mathbf{a}|^2\cos\theta$ reduces to $c|\mathbf{a}|^2$. This completes the proof of the theorem.

(14.19) **Corollary**

> If θ is the angle between nonzero vectors $\mathbf{a}$ and $\mathbf{b}$, then
>
> $$\cos\theta = \frac{\mathbf{a}\cdot\mathbf{b}}{|\mathbf{a}|\,|\mathbf{b}|}.$$

Example 3 Find the cosine of the angle between $\mathbf{a} = \langle 4, -3, 1\rangle$ and $\mathbf{b} = \langle 1, 2, -2\rangle$.

Solution Applying Corollary (14.19),

$$\cos\theta = \frac{\mathbf{a}\cdot\mathbf{b}}{|\mathbf{a}|\,|\mathbf{b}|} = \frac{(4)(1) + (-3)(2) + (1)(-2)}{\sqrt{16+9+1}\sqrt{1+4+4}} = \frac{-4}{3\sqrt{26}} = -\frac{4\sqrt{26}}{78}.$$

The following theorem is an immediate consequence of Theorem (14.18).

(14.20) **Theorem**

> Two vectors $\mathbf{a}$ and $\mathbf{b}$ are orthogonal if and only if $\mathbf{a}\cdot\mathbf{b} = 0$.

Example 4 Prove that the following pairs of vectors are orthogonal.

(a) $\mathbf{i}, \mathbf{j}$ (b) $3\mathbf{i} - 7\mathbf{j} + 2\mathbf{k}, 10\mathbf{i} + 4\mathbf{j} - \mathbf{k}$

Solution The proof follows from Theorem (14.20). Thus

(a) $\mathbf{i} \cdot \mathbf{j} = \langle 1, 0, 0 \rangle \cdot \langle 0, 1, 0 \rangle = (1)(0) + (0)(1) + (0)(0) = 0$

(b) $(3\mathbf{i} - 7\mathbf{j} + 2\mathbf{k}) \cdot (10\mathbf{i} + 4\mathbf{j} - \mathbf{k}) = 30 - 28 - 2 = 0$

The next result has important applications in the study of vectors.

(14.21) **The Cauchy–Schwarz Inequality**

> If $\mathbf{a}$ and $\mathbf{b}$ are any vectors in V_3, then
>
> $$|\mathbf{a} \cdot \mathbf{b}| \leq |\mathbf{a}|\,|\mathbf{b}|.$$

Proof. The result is trivial if either $\mathbf{a}$ or $\mathbf{b}$ is $\mathbf{0}$. If $\mathbf{a}$ and $\mathbf{b}$ are nonzero vectors, then by Theorem (14.18), $|\mathbf{a} \cdot \mathbf{b}| = |\mathbf{a}|\,|\mathbf{b}|\,|\cos \theta|$, where θ is the angle between $\mathbf{a}$ and $\mathbf{b}$. Since $|\cos \theta| \leq 1$, it follows that $|\mathbf{a} \cdot \mathbf{b}| \leq |\mathbf{a}|\,|\mathbf{b}|$.

(14.22) **The Triangle Inequality**

> If $\mathbf{a}$ and $\mathbf{b}$ are vectors, then
>
> $$|\mathbf{a} + \mathbf{b}| \leq |\mathbf{a}| + |\mathbf{b}|.$$

Proof. Using properties of the dot product we may write

$$\begin{aligned} |\mathbf{a} + \mathbf{b}|^2 &= (\mathbf{a} + \mathbf{b}) \cdot (\mathbf{a} + \mathbf{b}) = \mathbf{a} \cdot \mathbf{a} + 2\mathbf{a} \cdot \mathbf{b} + \mathbf{b} \cdot \mathbf{b} \\ &= |\mathbf{a}|^2 + 2\mathbf{a} \cdot \mathbf{b} + |\mathbf{b}|^2. \end{aligned}$$

Since $\mathbf{a} \cdot \mathbf{b} \leq |\mathbf{a} \cdot \mathbf{b}| \leq |\mathbf{a}|\,|\mathbf{b}|$ we have

$$|\mathbf{a} + \mathbf{b}|^2 \leq |\mathbf{a}|^2 + 2|\mathbf{a}|\,|\mathbf{b}| + |\mathbf{b}|^2 = (|\mathbf{a}| + |\mathbf{b}|)^2.$$

Taking square roots gives us

$$|\mathbf{a} + \mathbf{b}| \leq |\mathbf{a}| + |\mathbf{b}|.$$

The reason for the name "Triangle Inequality" is apparent if we represent $\mathbf{a}$ and $\mathbf{b}$ geometrically as in Figure 14.19. In this case (14.22) states that the length of a side of a triangle is not greater than the sum of the lengths of the other two sides.

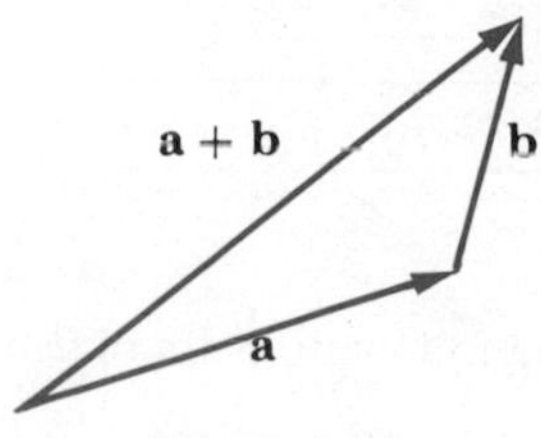

Figure 14.19

The angle between two vectors $\overrightarrow{PQ}$ and $\overrightarrow{TR}$ is defined as the angle θ between their corresponding vectors $\mathbf{a}$ and $\mathbf{b}$ in V_3. If $\theta = \pi/2$, then $\overrightarrow{PQ}$ and $\overrightarrow{TR}$ are **orthogonal**, or **perpendicular**. A typical angle is illustrated in Figure 14.20, where $\overrightarrow{OA}$ and $\overrightarrow{OB}$ are position vectors corresponding to $\mathbf{a}$ and $\mathbf{b}$, respectively. The **dot product** of $\overrightarrow{PQ}$ and $\overrightarrow{TR}$ is then defined by

$$\overrightarrow{PQ} \cdot \overrightarrow{TR} = \mathbf{a} \cdot \mathbf{b} = |\mathbf{a}|\,|\mathbf{b}| \cos\theta.$$

Since $|\overrightarrow{PQ}| = |\mathbf{a}|$ and $|\overrightarrow{TR}| = |\mathbf{b}|$, this may be written

(14.23) $$\overrightarrow{PQ} \cdot \overrightarrow{TR} = |\overrightarrow{PQ}|\,|\overrightarrow{TR}| \cos\theta.$$

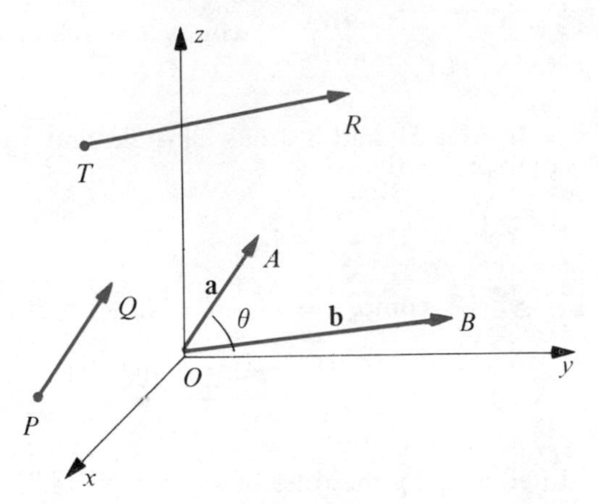

Figure 14.20

If $\overrightarrow{PQ}$ and $\overrightarrow{PR}$ have the same initial point, and if S is the projection of Q on the line through P and R (see Figure 14.21), then the scalar $|\overrightarrow{PQ}| \cos\theta$ is called the **component of $\overrightarrow{PQ}$ along $\overrightarrow{PR}$** abbreviated **comp$_{\overrightarrow{PR}}$ $\overrightarrow{PQ}$**. Note that $|\overrightarrow{PQ}| \cos\theta$ is positive or negative according as $0 \le \theta < \pi/2$ or $\pi/2 < \theta \le \pi$, respectively. If $\theta = \pi/2$, the component is 0. It follows that

(14.24) $$\text{comp}_{\overrightarrow{PR}}\, \overrightarrow{PQ} = |\overrightarrow{PQ}| \cos\theta = \overrightarrow{PQ} \cdot \frac{1}{|\overrightarrow{PR}|}\overrightarrow{PR}.$$

This shows that *to find the component of $\overrightarrow{PQ}$ along $\overrightarrow{PR}$, we may dot $\overrightarrow{PQ}$ with a unit vector having the same direction as $\overrightarrow{PR}$.*

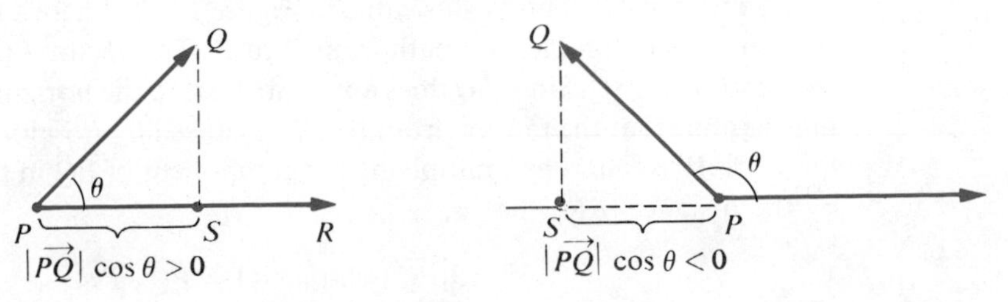

Figure 14.21. Comp$_{\overrightarrow{PR}}$ $\overrightarrow{PQ}$

The preceding concept may also be applied to vectors in V_3. We merely represent **a** and **b** geometrically by $\overrightarrow{PQ}$ and $\overrightarrow{PR}$, respectively, as in Figure 14.21. The **component of a along b**, denoted by $\text{comp}_{\mathbf{b}}\,\mathbf{a}$, is then defined as in (14.24); that is,

(14.25)
$$\text{comp}_{\mathbf{b}}\,\mathbf{a} = \mathbf{a}\cdot\frac{1}{|\mathbf{b}|}\mathbf{b}.$$

In particular, if $\mathbf{a} = a_1\mathbf{i} + a_2\mathbf{j} + a_3\mathbf{k}$, then by (14.25)

$$\text{comp}_{\mathbf{i}}\,\mathbf{a} = \mathbf{a}\cdot\mathbf{i} = a_1$$
$$\text{comp}_{\mathbf{j}}\,\mathbf{a} = \mathbf{a}\cdot\mathbf{j} = a_2$$
$$\text{comp}_{\mathbf{k}}\,\mathbf{a} = \mathbf{a}\cdot\mathbf{k} = a_3.$$

Example 5 If $\mathbf{a} = 4\mathbf{i} - \mathbf{j} + 5\mathbf{k}$ and $\mathbf{b} = 6\mathbf{i} + 3\mathbf{j} - 2\mathbf{k}$, find
(a) $\text{comp}_{\mathbf{b}}\,\mathbf{a}$. (b) $\text{comp}_{\mathbf{a}}\,\mathbf{b}$.

Solution (a) Using (14.25),

$$\text{comp}_{\mathbf{b}}\,\mathbf{a} = \mathbf{a}\cdot\frac{1}{|\mathbf{b}|}\mathbf{b} = (4\mathbf{i} - \mathbf{j} + 5\mathbf{k})\cdot\frac{1}{7}(6\mathbf{i} + 3\mathbf{j} - 2\mathbf{k})$$
$$= \frac{24 - 3 - 10}{7} = \frac{11}{7}.$$

(b) Interchanging the roles of **a** and **b** in (14.25),

$$\text{comp}_{\mathbf{a}}\,\mathbf{b} = \mathbf{b}\cdot\frac{1}{|\mathbf{a}|}\mathbf{a}$$
$$= (6\mathbf{i} + 3\mathbf{j} - 2\mathbf{k})\cdot\frac{1}{\sqrt{42}}(4\mathbf{i} - \mathbf{j} + 5\mathbf{k})$$
$$= \frac{24 - 3 - 10}{\sqrt{42}} = \frac{11}{\sqrt{42}}.$$

We shall conclude this section with a physical interpretation for the dot product. Recall from (6.12) that the work done if a constant force F is exerted through a distance d is given by $W = Fd$. This formula is very restrictive since it can only be used if the force is applied along the line of motion. More generally, suppose a vector $\overrightarrow{PQ}$ represents a force, and that its point of application moves along a vector $\overrightarrow{PR}$. This is illustrated in Figure 14.22, where a force $\overrightarrow{PQ}$ is used to pull an object along a level path from P to R. The vector $\overrightarrow{PQ}$ is the sum of the vectors $\overrightarrow{PS}$ and $\overrightarrow{SQ}$. Since $\overrightarrow{SQ}$ does not contribute to the horizontal movement, we may assume that the motion from P to R is caused by $\overrightarrow{PS}$ alone. Applying (6.12), the work W is found by multiplying the component of $\overrightarrow{PQ}$ in the direction of $\overrightarrow{PR}$ by the distance $|\overrightarrow{PR}|$, that is,

$$W = (|\overrightarrow{PQ}|\cos\theta)|\overrightarrow{PR}| = \overrightarrow{PQ}\cdot\overrightarrow{PR}.$$

This leads to the following definition.

(14.26) **Definition**

> The **work done by a constant force** $\overrightarrow{PQ}$ as its point of application moves along the vector $\overrightarrow{PR}$ is given by $\overrightarrow{PQ} \cdot \overrightarrow{PR}$.

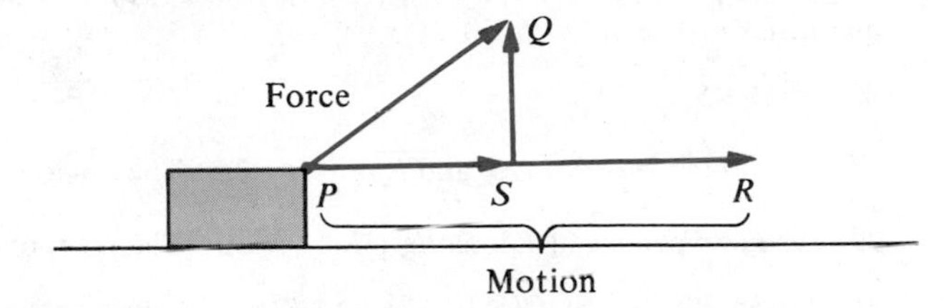

Figure 14.22

Example 6 The magnitude and direction of a constant force are given by $\mathbf{a} = 5\mathbf{i} + 2\mathbf{j} + 6\mathbf{k}$. Find the work done if the point of application of the force moves from $P(1, -1, 2)$ to $R(4, 3, -1)$.

Solution By Theorem (14.14) the vector in V_3 which corresponds to $\overrightarrow{PR}$ is $\mathbf{b} = \langle 3, 4, -3\rangle$. If $\overrightarrow{PQ}$ is a geometric representation for $\mathbf{a}$, then by (14.26) the work done is given by

$$\overrightarrow{PQ} \cdot \overrightarrow{PR} = \mathbf{a} \cdot \mathbf{b} = 15 + 8 - 18 = 5.$$

If, for example, the unit of length is feet and the magnitude of the force is measured in pounds, then the work done is 5 ft-lb.

EXERCISES 14.3

1 Complete the proof of (14.16).

2 Prove Theorems (14.4) and (14.8) for V_3.

In Exercises 3–8 find $\mathbf{a} + \mathbf{b}$, $5\mathbf{a} - 4\mathbf{b}$, $\mathbf{a} \cdot \mathbf{b}$, $|\mathbf{a}|$, $|3\mathbf{a}|$, $|-3\mathbf{a}|$, $\mathbf{a} - \mathbf{b}$, and $|\mathbf{a} - \mathbf{b}|$.

3 $\mathbf{a} = \langle -2, 6, 1\rangle, \mathbf{b} = \langle 3, -3, -1\rangle$

4 $\mathbf{a} = \langle 1, 2, -3\rangle, \mathbf{b} = \langle -4, 0, 1\rangle$

5 $\mathbf{a} = 3\mathbf{i} - 4\mathbf{j} + 2\mathbf{k}, \mathbf{b} = \mathbf{i} + 2\mathbf{j} - 5\mathbf{k}$

6 $\mathbf{a} = 2\mathbf{i} - \mathbf{j} + 4\mathbf{k}, \mathbf{b} = \mathbf{i} - \mathbf{k}$

7 $\mathbf{a} = \mathbf{i} + \mathbf{j}, \mathbf{b} = -\mathbf{j} + \mathbf{k}$

8 $\mathbf{a} = 2\mathbf{i}, \mathbf{b} = 3\mathbf{k}$

In Exercises 9 and 10, sketch geometric representations for $\mathbf{a}$, $\mathbf{b}$, $2\mathbf{a}$, $-3\mathbf{b}$, $\mathbf{a} + \mathbf{b}$, and $\mathbf{a} - \mathbf{b}$.

9 $\mathbf{a} = \langle 2, 3, 4\rangle, \mathbf{b} = \langle 1, -2, 2\rangle$

10 $\mathbf{a} = -\mathbf{i} + 2\mathbf{j} + 3\mathbf{k}, \mathbf{b} = -2\mathbf{j} + \mathbf{k}$

If $\mathbf{a} = \langle -2, 3, 1\rangle$, $\mathbf{b} = \langle 7, 4, 5\rangle$, and $\mathbf{c} = \langle 1, -5, 2\rangle$ find the numbers in Exercises 11–20.

11 $\mathbf{a} \cdot \mathbf{b}$

12 $\mathbf{b} \cdot \mathbf{c}$

13 $\mathbf{a} \cdot (\mathbf{b} + \mathbf{c})$

14 $\mathbf{b} \cdot (\mathbf{a} - \mathbf{c})$

15 $(2\mathbf{a} + \mathbf{b}) \cdot 3\mathbf{c}$

16 $(\mathbf{a} - \mathbf{b}) \cdot (\mathbf{b} + \mathbf{c})$

17 $\text{comp}_{\mathbf{c}}\, \mathbf{b}$

18 $\text{comp}_{\mathbf{a}}\, \mathbf{c}$

19 $\text{comp}_{\mathbf{b}}\, (\mathbf{a} + \mathbf{c})$

20 $\text{comp}_{\mathbf{c}}\mathbf{c}$

In each of Exercises 21–24, find the cosine of the angle between **a** and **b**.

21 $\mathbf{a} = -4\mathbf{i} + 8\mathbf{j} - 3\mathbf{k}, \mathbf{b} = 2\mathbf{i} + \mathbf{j} + \mathbf{k}$

22 $\mathbf{a} = \mathbf{i} - 7\mathbf{j} + 4\mathbf{k}, \mathbf{b} = 5\mathbf{i} - \mathbf{k}$

23 $\mathbf{a} = -2\mathbf{i} - 3\mathbf{j}, \mathbf{b} = -6\mathbf{i} + 4\mathbf{k}$

24 $\mathbf{a} = \langle 3, -5, -1 \rangle, \mathbf{b} = \langle 2, 1, -3 \rangle$

Given the points $P(3, -2, -1)$, $Q(1, 5, 4)$, $R(2, 0, -6)$, and $S(-4, 1, 5)$, find each of the quantities in Exercises 25–30.

25 $\overrightarrow{PQ} \cdot \overrightarrow{RS}$

26 $\overrightarrow{QS} \cdot \overrightarrow{RP}$

27 The angle between $\overrightarrow{PQ}$ and $\overrightarrow{RS}$

28 The angle between $\overrightarrow{QS}$ and $\overrightarrow{RP}$

29 The component of $\overrightarrow{PS}$ along $\overrightarrow{QR}$

30 The component of $\overrightarrow{QR}$ along $\overrightarrow{PS}$

31 Given $P(8, -3, 5)$, $Q(6, 1, -7)$, and $R(x, y, z)$, state an equation in x, y, and z which guarantees that $\overrightarrow{PR}$ is orthogonal to $\overrightarrow{PQ}$. Give a geometric description of all such points $R(x, y, z)$.

32 Prove, by means of vectors, that $P(2, -3, 1)$, $Q(-5, 1, 7)$, and $R(6, 1, 3)$ are vertices of a right triangle and find its area.

In Exercises 33 and 34, the magnitude and direction of a force are given by the vector **a**. Find the work done if the point of application moves from P to Q.

33 $\mathbf{a} = -\mathbf{i} + 5\mathbf{j} - 3\mathbf{k}$; $P(4, 0, -7)$, $Q(2, 4, 0)$

34 $\mathbf{a} = \langle 8, 0, -4 \rangle$; $P(-1, 2, 5)$, $Q(4, 1, 0)$

35 A constant force of magnitude 4 has the same direction as the vector $\mathbf{a} = \mathbf{i} + \mathbf{j} + \mathbf{k}$. Find the work done if the point of application moves along the y-axis from $(0, 2, 0)$ to $(0, -1, 0)$.

36 A constant force of magnitude 5 has the same direction as the positive z-axis. Find the work done if the point of application moves along a straight line from the origin to the point $P(1, 2, 3)$.

37 Determine c such that the vectors $3\mathbf{i} - \mathbf{j} + c\mathbf{k}$ and $2c\mathbf{i} + 3\mathbf{j} + 4\mathbf{k}$ are orthogonal.

38 Determine all values of c such that the vectors $\langle 3c, 1, -4c \rangle$ and $\langle c, 4, 2 \rangle$ are orthogonal.

39 Find a unit vector orthogonal to both $\langle 7, -10, 1 \rangle$ and $\langle -2, 2, 3 \rangle$.

40 Find two unit vectors in V_2 orthogonal to $\mathbf{a} = \langle -2, -1 \rangle$.

41 If $\mathbf{a} = 14\mathbf{i} - 15\mathbf{j} + 6\mathbf{k}$,
(a) find a vector having the same direction and twice the magnitude.
(b) find a vector having the opposite direction and 1/3 the magnitude.

42 Work Exercise 41 if $\mathbf{a} = \langle -6, -3, 6 \rangle$.

43 Under what conditions are the following true?
(a) $|\mathbf{a} \cdot \mathbf{b}| = |\mathbf{a}|\,|\mathbf{b}|$
(b) $|\mathbf{a} + \mathbf{b}| = |\mathbf{a}| + |\mathbf{b}|$

44 Prove that $|\mathbf{a} - \mathbf{b}| \geq |\mathbf{a}| - |\mathbf{b}|$ for all $\mathbf{a}, \mathbf{b}$ in V_3. (*Hint:* Let $\mathbf{a} = \mathbf{b} + (\mathbf{a} - \mathbf{b})$ and use the Triangle Inequality.)

45 Prove that $(\mathbf{a} + \mathbf{b}) \cdot (\mathbf{a} - \mathbf{b}) = |\mathbf{a}|^2 - |\mathbf{b}|^2$ for all $\mathbf{a}$, $\mathbf{b}$ in V_3.

46 Extend (14.22) to any finite sum of vectors.

47 If A, B, C are any three points in space and M is the midpoint of the line segment BC, prove that $\overrightarrow{AM} = \frac{1}{2}(\overrightarrow{AB} + \overrightarrow{AC})$.

48 Use vectors to prove that the line segments which join the midpoints of consecutive side of a quadrilateral form a parallelogram. (*Hint*: See Exercise 47.)

49 Prove that a quadrilateral is a parallelogram if and only if the diagonals bisect one another.

50 Use the dot product to prove that the diagonals of a rhombus are perpendicular.

14.4 DIRECTION ANGLES AND DIRECTION COSINES

In this section we shall introduce concepts which are useful for the study of lines and planes in space.

(14.27) Definition

The **direction angles** of a nonzero vector $\mathbf{a} = \langle a_1, a_2, a_3 \rangle$ are the angles α, β, and γ between the vectors **i**, **j**, and **k**, respectively, and the vector **a**. The **direction cosines** of **a** are $\cos \alpha$, $\cos \beta$, and $\cos \gamma$.

The direction angles of a vector will always be stated in the specific order α, β, γ. If $\overrightarrow{OA}$ is the position vector corresponding to **a**, then it is convenient to regard these angles as being measured *from* each coordinate axis *to* $\overrightarrow{OA}$, as indicated by the curved arrows in Figure 14.23. Note that each direction angle is in the interval $[0, \pi]$. Of course, the direction angles and cosines of a vector $\overrightarrow{PQ}$ are defined as those of the vector **a** in V_3 which corresponds to $\overrightarrow{PQ}$.

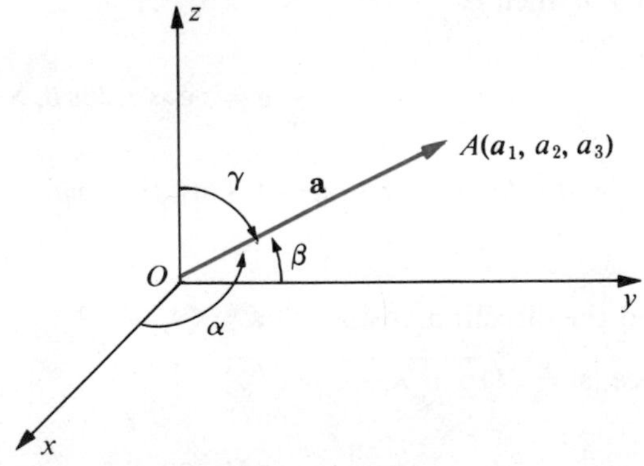

Figure 14.23

(14.28) Theorem

If α, β, and γ are the direction angles of a nonzero vector $\mathbf{a} = \langle a_1, a_2, a_3 \rangle$, then

(i) $\cos \alpha = \dfrac{a_1}{|\mathbf{a}|}, \cos \beta = \dfrac{a_2}{|\mathbf{a}|}, \cos \gamma = \dfrac{a_3}{|\mathbf{a}|}.$

(ii) $\cos^2 \alpha + \cos^2 \beta + \cos^2 \gamma = 1.$

(iii) the direction angles of $-\mathbf{a}$ are $\pi - \alpha$, $\pi - \beta$, and $\pi - \gamma$.

Proof. Since α is the angle between $\mathbf{a}$ and $\mathbf{i}$ we have, by Theorem (14.18),

$$\mathbf{a} \cdot \mathbf{i} = |\mathbf{a}|\,|\mathbf{i}| \cos \alpha.$$

Hence

$$a_1 = |\mathbf{a}| \cos \alpha \quad \text{or} \quad \cos \alpha = \frac{a_1}{|\mathbf{a}|}.$$

The formulas for $\cos \beta$ and $\cos \gamma$ can be found in similar fashion by considering $\mathbf{a} \cdot \mathbf{j}$ and $\mathbf{a} \cdot \mathbf{k}$, or by using Corollary (14.19).

To prove part (ii), we square the expressions for the direction cosines obtained in (i) and add, obtaining

$$\frac{a_1^2}{|\mathbf{a}|^2} + \frac{a_2^2}{|\mathbf{a}|^2} + \frac{a_3^2}{|\mathbf{a}|^2} = \frac{|\mathbf{a}|^2}{|\mathbf{a}|^2} = 1.$$

Part (iii) follows from the fact that the position vectors for $\mathbf{a}$ and $-\mathbf{a}$ are on the same straight line.

We see from (i) of Theorem (14.28) that any vector $\mathbf{a} = \langle a_1, a_2, a_3 \rangle$ can be expressed in the form

$$\mathbf{a} = \langle |\mathbf{a}| \cos \alpha, |\mathbf{a}| \cos \beta, |\mathbf{a}| \cos \gamma \rangle$$

or equivalently,

$$\mathbf{a} = |\mathbf{a}| \langle \cos \alpha, \cos \beta, \cos \gamma \rangle.$$

If $\mathbf{a} \neq \mathbf{0}$, then

$$\frac{1}{|\mathbf{a}|}\mathbf{a} = \langle \cos \alpha, \cos \beta, \cos \gamma \rangle$$

that is, *the direction cosines of* $\mathbf{a}$ *are the components of the unit vector* $(1/|\mathbf{a}|)\,\mathbf{a}$.

Example 1 Find the direction cosines of $\mathbf{a} = \langle 4, -3, 2 \rangle$.

Solution Since $|\mathbf{a}| = \sqrt{16 + 9 + 4} = \sqrt{29}$,

$$\frac{1}{|\mathbf{a}|}\mathbf{a} = \langle 4/\sqrt{29}, -3/\sqrt{29}, 2/\sqrt{29} \rangle.$$

Hence by the remark preceding this example,

$$\cos \alpha = \frac{4}{\sqrt{29}}, \quad \cos \beta = \frac{-3}{\sqrt{29}}, \quad \cos \gamma = \frac{2}{\sqrt{29}}.$$

Example 2 If all components of a vector $\mathbf{a}$ are positive, and if two of its direction angles are $\alpha = 60°$ and $\beta = 45°$, find γ.

Solution Since $\cos\alpha = 1/2$ and $\cos\beta = \sqrt{2}/2$, substitution in (ii) of Theorem (14.28) gives us

$$\left(\frac{1}{2}\right)^2 + \left(\frac{\sqrt{2}}{2}\right)^2 + \cos^2\gamma = 1$$

or equivalently,

$$\cos^2\gamma = 1 - \frac{3}{4} = \frac{1}{4}.$$

Since the components of **a** are positive,

$$\cos\gamma = 1/2 \quad \text{and} \quad \gamma = 60°.$$

EXERCISES 14.4

In each of Exercises 1–4 find the direction cosines of **a**, 3**a**, and −2**a**.

1 $\mathbf{a} = \langle -2, 1, 5\rangle$

2 $\mathbf{a} = \langle 2, -1, -2\rangle$

3 $\mathbf{a} = 4\mathbf{j} - 3\mathbf{k}$

4 $\mathbf{a} = 5\mathbf{i} + \mathbf{k}$

5 Find the direction angles and direction cosines of **i**, **j**, and **k**.

6 Find two unit vectors each of whose direction cosines are all the same.

In Exercises 7–10 find the direction cosines of $\overrightarrow{PQ}$.

7 $P(7, -2, 4), Q(3, 2, -1)$

8 $P(-1, 5, 0), Q(2, -3, 3)$

9 $P(-5, 1, 3), Q(-4, 1, -2)$

10 $P(4, -3, 2), Q(4, -7, 2)$

11 Find a vector which has positive components, magnitude 2, and whose direction angles are all the same.

12 Show that it is impossible to find a vector **a** having direction angles $\alpha = 30°$ and $\beta = 30°$.

13 If all components of **a** are positive, and if $\cos\alpha = 1/3$ and $\cos\beta = 1/3$, find $\cos\gamma$.

14 If two direction cosines of **a** are $\beta = 120°$ and $\gamma = 60°$, what can be said about α?

15 Three nonzero numbers l, m, and n are called **direction numbers** of a nonzero vector **a** (or $\overrightarrow{PQ}$) if they are proportional to the direction cosines, that is, if there exists a positive number k such that

$$l = k\cos\alpha, \quad m = k\cos\beta, \quad n = k\cos\gamma.$$

(a) If l, m, n are direction numbers of **a** and $d = (l^2 + m^2 + n^2)^{1/2}$, prove that $\cos\alpha = l/d$, $\cos\beta = m/d$, $\cos\gamma = n/d$.

(b) Given points $P_1(x_1, y_1, z_1)$ and $P_2(x_2, y_2, z_2)$, prove that $x_2 - x_1, y_2 - y_1$, and $z_2 - z_1$ are direction numbers of $\overrightarrow{P_1P_2}$.

16 Refer to Exercise 15. If l_1, m_1, n_1 and l_2, m_2, n_2 are direction numbers of **a** and **b**, respectively, prove the following.

(a) **a** and **b** are orthogonal if and only if $l_1l_2 + m_1m_2 + n_1n_2 = 0$.

(b) **a** and **b** are parallel if and only if there is a number k such that $l_1 = kl_2, m_1 = km_2$ and $n_1 = kn_2$.

17 If $\mathbf{a} = \langle a_1, a_2, a_3 \rangle$ and $\mathbf{b} = \langle b_1, b_2, b_3 \rangle$ are nonzero vectors in V_3 prove each of the following.

(i) **a** and **b** have the same direction if and only if their direction angles are the same.

(ii) **a** and **b** have opposite directions if and only if their direction angles are supplements of one another.

14.5 LINES IN SPACE

If $P_1(x_1, y_1, z_1)$ and $P_2(x_2, y_2, z_2)$ are distinct points, then by Theorem (14.14), the vector $\mathbf{a} = \langle a_1, a_2, a_3 \rangle$ in V_3 with geometric representation $\overrightarrow{P_1P_2}$ is given by $\mathbf{a} = \langle x_2 - x_1, y_2 - y_1, z_2 - z_1 \rangle$. To find equations for the line l through P_1 and P_2 we observe that a point $P(x, y, z)$ is on l if and only if $\overrightarrow{P_1P}$ and $\overrightarrow{P_1P_2}$ are parallel. This is illustrated in Figure 14.24, where the position vector $\overrightarrow{OA}$ corresponding to **a** is also shown. From our work in Section 3, an equivalent statement is

$$\overrightarrow{P_1P} = t\overrightarrow{P_1P_2}, \quad \text{for some scalar } t.$$

This condition may be stated algebraically as follows:

$$\langle x - x_1, y - y_1, z - z_1 \rangle = t\langle a_1, a_2, a_3 \rangle = \langle a_1t, a_2t, a_3t \rangle.$$

Equating coefficients and solving for x, y, and z gives us

$$x = x_1 + a_1t, \quad y = y_1 + a_2t, \quad z = z_1 + a_3t \tag{14.29}$$

where t is a real number. The equations in (14.29) are called **parametric equations for the line** l, and the variable t is called a **parameter**. The points $P(x, y, z)$ on l are obtained by letting t take on all real values. For example, P_1 corresponds to $t = 0$, the midpoint of P_1P_2 to $t = 1/2$, and P_2 to $t = 1$.

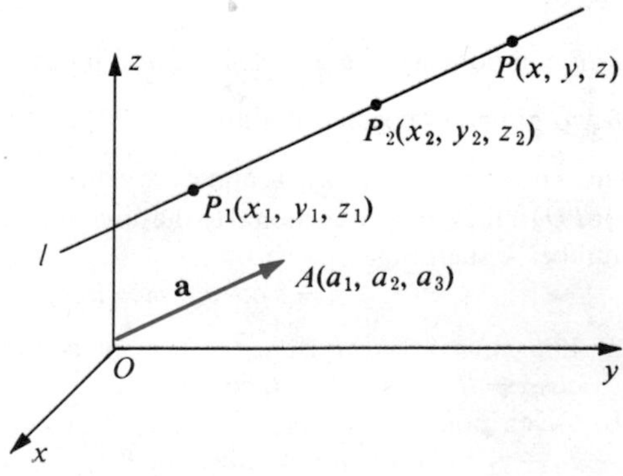

Figure 14.24

If, in the preceding discussion, we let

$$\mathbf{r} = x\mathbf{i} + y\mathbf{j} + z\mathbf{k} \quad \text{and} \quad \mathbf{r}_1 = x_1\mathbf{i} + y_1\mathbf{j} + z_1\mathbf{k}$$

then $\overrightarrow{OP}$ and $\overrightarrow{OP_1}$ are position vectors of $\mathbf{r}$ and $\mathbf{r}_1$, respectively, and $\overrightarrow{P_1P}$ represents the vector $\mathbf{r} - \mathbf{r}_1$. Thus we may write

$$\mathbf{r} - \mathbf{r}_1 = t\mathbf{a}.$$

This leads to the following **vector equation for a line**:

$$\mathbf{r} = t\mathbf{a} + \mathbf{r}_1.$$

Example 1 Find parametric equations for the line l through $P_1(3, 1, -2)$ and $P_2(-2, 7, -4)$. At what point does l intersect the xy-plane?

Solution The vector $\mathbf{a}$ in V_3 corresponding to $\overrightarrow{P_1P_2}$ is

$$\mathbf{a} = \langle -2 - 3, 7 - 1, -4 + 2\rangle = \langle -5, 6, -2\rangle.$$

Applying (14.29), parametric equations for l are

$$x = 3 - 5t, \quad y = 1 + 6t, \quad z = -2 - 2t.$$

The line intersects the xy-plane at the point $R(x, y, z)$ such that $z = -2 - 2t = 0$, that is, $t = -1$. Thus R is the point with coordinates $(8, -5, 0)$.

If $\mathbf{a} = \langle a_1, a_2, a_3\rangle$ is a nonzero vector in V_3 and $P_1(x_1, y_1, z_1)$ is any point, then there is a unique line l through P_1 parallel to the position vector $\overrightarrow{OA}$ corresponding to $\mathbf{a}$. Indeed, if P_2 is the point $P_2(x_1 + a_1, y_1 + a_2, z_1 + a_3)$, then l is precisely the line determined by $\overrightarrow{P_1P_2}$, as illustrated in Figure 14.24. In this context we shall refer to l as **the line through P_1 parallel to a**. It follows that (14.29) are parametric equations for l. If $\mathbf{b}$ is any nonzero vector which is parallel to $\mathbf{a}$, then the same line is determined, for in this case

$$\mathbf{b} = c\mathbf{a} = \langle ca_1, ca_2, ca_3\rangle$$

and parametric equations for the line through P_1 determined by $\mathbf{b}$ are

$$x = x_1 + (ca_1)v, \quad y = y_1 + (ca_2)v, \quad z = z_1 + (ca_3)v$$

where the parameter v ranges through all real numbers. These equations determine the same line as (14.29), since the point given by t can be obtained by letting $v = t/c$.

Example 2 Find parametric equations for the line l through $P(5, -6, 2)$ parallel to $\mathbf{a} = \langle 1/2, 2, -4/3\rangle$.

Solution To avoid fractions we shall use the vector $\mathbf{b} = 6\mathbf{a} = \langle 3, 12, -8\rangle$ instead of $\mathbf{a}$. Applying (14.29), parametric equations for l are

$$x = 5 + 3t, \quad y = -6 + 12t, \quad z = 2 - 8t.$$

If l_1 and l_2 are lines parallel to vectors $\mathbf{a} = \langle a_1, a_2, a_3 \rangle$ and $\mathbf{b} = \langle b_1, b_2, b_3 \rangle$, respectively, then the angle between l_1 and l_2 is defined as the angle θ between **a** and **b**, whether or not the lines intersect. The lines are **orthogonal** if $\mathbf{a} \cdot \mathbf{b} = 0$ or, equivalently, if $a_1b_1 + a_2b_2 + a_3b_3 = 0$. The lines are **parallel** if $\mathbf{b} = c\mathbf{a}$ for some scalar c, that is, if $b_1 = ca_1$, $b_2 = ca_2$, $b_3 = ca_3$.

In the next section we shall discuss representations of lines as intersections of planes.

EXERCISES 14.5

In Exercises 1–4 find parametric equations for the line through P_1 and P_2. Determine (if possible) the points at which the line intersects each of the coordinate planes.

1 $P_1(5, -2, 4), P_2(2, 6, 1)$

2 $P_1(-3, 1, -1), P_2(7, 11, -8)$

3 $P_1(2, 0, 5), P_2(-6, 0, 3)$

4 $P_1(2, -2, 4), P_2(2, -2, -3)$

In Exercises 5–8 find parametric equations for the line through P parallel to **a**.

5 $P(4, 2, -3), \mathbf{a} = \langle 1/3, 2, 1/2 \rangle$

6 $P(5, 0, -2), \mathbf{a} = \langle -1, -4, 1 \rangle$

7 $P(0, 0, 0), \mathbf{a} = \langle 0, 1, 0 \rangle$

8 $P(1, 2, 3), \mathbf{a} = \langle 1, 2, 3 \rangle$

9 If a line l has parametric equations $x = 5 - 3t, y = -2 + t, z = 1 + 9t$, find parametric equations for a line through $P(-6, 4, -3)$ which is parallel to l.

10 If l_1 is the line through $P(5, -2, 4)$ and $Q(2, 6, 1)$, and if l_2 is the line through $R(-3, 1, -1)$ and $S(7, 11, -8)$, find the angle between l_1 and l_2.

11 Find the angle between the two lines having parametric equations $x = 7 - 2t$, $y = 4 + 3t$, $z = 5t$ and $x = -1 + t$, $y = 3 + 4t$, $z = 1 + t$, respectively.

12 Find parametric equations for the line through $P(4, -1, 0)$ which is parallel to the line through $P_1(-3, 9, -2)$ and $P_2(5, 7, -3)$.

In each of Exercises 13–16, determine whether the two lines intersect and, if so, find the point of intersection.

13 $x = 1 + 2t, y = 1 - 4t, z = 5 - t; x = 4 - v, y = -1 + 6v, z = 4 + v$

14 $x = 1 - 6t, y = 3 + 2t, z = 1 - 2t; x = 2 + 2v, y = 6 + v, z = 2 + v$

15 $x = 3 + t, y = 2 - 4t, z = t; x = 4 - v, y = 3 + v, z = -2 + 3v$

16 $x = 2 - 5t, y = 6 + 2t, z = -3 - 2t; x = 4 - 3v, y = 7 + 5v, z = 1 + 4v$

In each of Exercises 17 and 18, show that the lines intersect orthogonally.

17 $x = 2 + 3t, y = -4 - 2t, z = -1 + 4t; x = 6 + 4v, y = -2 + 2v, z = -3 - 2v$

18 $x = 4 - t, y = -1 - t, z = 4 + 3t; x = 3 + 2v, y = -1 + v, z = v$

19 Use the dot product to find the distance d from the point $A(2, -6, 1)$ to the line l through $B(3, 4, -2)$ and $C(7, -1, 5)$. (*Hint:* if D is the projection on l, then $d = |\overrightarrow{AD}|$.)

20 Work Exercise 19 for the points $A(1, 5, 0)$, $B(-2, 1, -4)$, and $C(0, -3, 2)$.

14.6 PLANES

If P_1 and P_2 are distinct points, then all points P such that $\overrightarrow{P_1P}$ is orthogonal to $\overrightarrow{P_1P_2}$ lie on a plane Γ through P_1, as illustrated in Figure 14.25. We shall call $\overrightarrow{P_1P_2}$, or the vector $\mathbf{a} = \langle a_1, a_2, a_3 \rangle$ with geometric representation $\overrightarrow{P_1P_2}$, a **normal vector** to Γ. It follows from Theorem (14.20) that P is on Γ if and only if $\overrightarrow{P_1P_2} \cdot \overrightarrow{P_1P} = 0$ or, equivalently,

$$\langle a_1, a_2, a_3 \rangle \cdot \langle x - x_1, y - y_1, z - z_1 \rangle = 0.$$

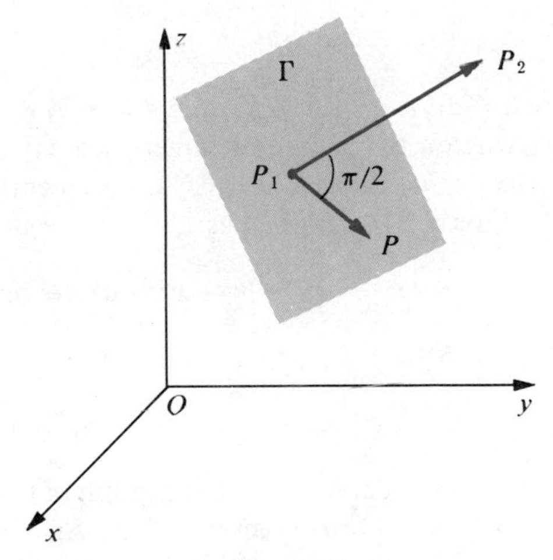

Figure 14.25

Applying the definition of scalar product gives us the next result.

(14.30) Theorem

An equation of the plane through $P_1(x_1, y_1, z_1)$ with normal vector $\mathbf{a} = \langle a_1, a_2, a_3 \rangle$ is

$$a_1(x - x_1) + a_2(y - y_1) + a_3(z - z_1) = 0.$$

If we let

$$\mathbf{r} = x\mathbf{i} + y\mathbf{j} + z\mathbf{k} \quad \text{and} \quad \mathbf{r_1} = x_1\mathbf{i} + y_1\mathbf{j} + z_1\mathbf{k}$$

then $\overrightarrow{OP}$ and $\overrightarrow{OP_1}$ are position vectors of $\mathbf{r}$ and $\mathbf{r_1}$, respectively, and $\overrightarrow{P_1P}$ represents the vector $\mathbf{r} - \mathbf{r_1}$. This leads to the following **vector equation for a plane**:

$$\mathbf{a} \cdot (\mathbf{r} - \mathbf{r_1}) = 0.$$

Example 1 Find an equation of the plane through the point $(5, -2, 4)$ with normal vector $\mathbf{a} = \langle 1, 2, 3 \rangle$.

Solution Applying Theorem (14.30) we obtain

$$1(x - 5) + 2(y + 2) + 3(z - 4) = 0$$

which reduces to

$$x + 2y + 3z - 13 = 0.$$

The equation of the plane in Theorem (14.30) may be written in the form

$$ax + by + cz + d = 0 \tag{14.31}$$

where $a = a_1$, $b = b_1$, $c = c_1$, and $d = -a_1x_1 - a_2y_1 - a_3z_1$. Conversely, given (14.31) where a, b, and c are not all zero, we may choose numbers x_1, y_1, and z_1 such that $ax_1 + by_1 + cz_1 + d = 0$. Consequently $d = -ax_1 - by_1 - cz_1$, and substitution in (14.31) gives us

$$ax + by + cz - ax_1 - by_1 - cz_1 = 0$$

or, equivalently,

$$a(x - x_1) + b(y - y_1) + c(z - z_1) = 0.$$

According to Theorem (14.30), the graph of the last equation is a plane through $P(x_1, y_1, z_1)$ with normal vector $\langle a, b, c\rangle$. An equation of the form (14.31), where a, b, and c are not all zero, is called a **linear equation in three variables** x, y, and z. We have established the following theorem.

(14.32) Theorem

The graph of every linear equation $ax + by + cz + d = 0$ is a plane with normal vector $\langle a, b, c\rangle$.

For simplicity we often use the phrase "the plane $ax + by + cz + d = 0$" instead of the more accurate statement "the plane having equation $ax + by + cz + d = 0$."

In order to sketch the graph of a linear equation we often find, if possible, the **trace** of the graph in each coordinate plane, that is, the line in which the graph intersects the coordinate plane. To find the trace in the xy-plane we substitute 0 for z in (14.31), since this will lead to all points of the graph which lie on the xy-plane. Similarly, to find the trace in the yz-plane or the xz-plane we let $x = 0$ or $y = 0$, respectively, in (14.31).

Example 2 Sketch the graph of the equation $2x + 3y + 4z = 12$.

Solution There are three points on the plane which are easily found, namely the points of intersection of the plane with the coordinate axes. Substituting 0 for both y and z in the equation, we obtain $2x = 12$ or $x = 6$. Thus the point $(6, 0, 0)$ is on the graph. As in two dimensions, 6 is called the ***x*-intercept** of the graph. Similarly, substitution of 0 for x and z gives us the ***y*-intercept** 4, and hence the point $(0, 4, 0)$ is on the graph. The point $(0, 0, 3)$ (or ***z*-intercept** 3) is obtained in like manner. The trace in the xy-plane is found by substituting 0 for z in the given equation. This leads to $2x + 3y = 12$ which has as its graph in the xy-plane a straight line with x-intercept 6 and y-intercept 4. This trace and the traces of the graph in the xz- and yz-planes are illustrated in Figure 14.26.

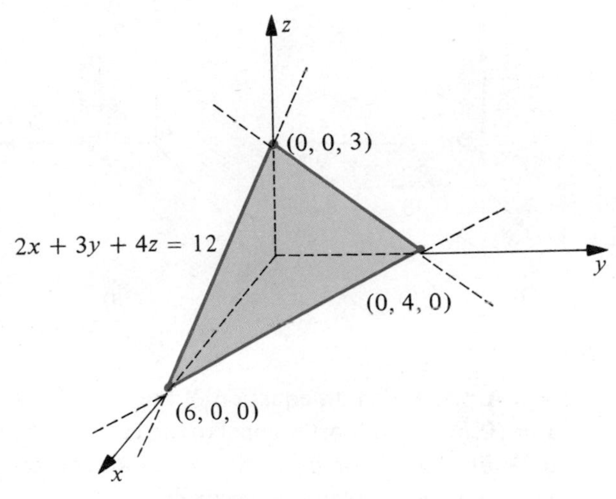

Figure 14.26

(14.33) Definition

> Two planes with normal vectors $\mathbf{a}$ and $\mathbf{b}$ are
> (i) **parallel** if $\mathbf{a}$ and $\mathbf{b}$ are parallel.
> (ii) **orthogonal** if $\mathbf{a}$ and $\mathbf{b}$ are orthogonal.

Example 3 Prove that the graphs of $2x - 3y - z - 5 = 0$ and $-6x + 9y + 3z + 2 = 0$ are parallel planes.

Solution By Theorem (14.32), the graphs are planes with normal vectors $\mathbf{a} = \langle 2, -3, -1 \rangle$ and $\mathbf{b} = \langle -6, 9, 3 \rangle$. Since $\mathbf{b} = -3\mathbf{a}$, the vectors $\mathbf{a}$ and $\mathbf{b}$ are parallel and hence, by Definition (14.33), so are the planes.

Example 4 Find an equation of the plane Γ through $P(5, -2, 4)$ which is parallel to the plane $3x + y - 6z + 8 = 0$.

Solution The vector $\mathbf{a} = \langle 3, 1, -6 \rangle$ may be used as a normal vector for Γ and hence the desired equation may be written in the form $3x + y - 6z + d = 0$ for some number d. If $P(5, -2, 4)$ is on Γ, then its coordinates must satisfy this equation, that is, $3(5) + (-2) - 6(4) + d = 0$ or $d = 11$. Hence an equation for Γ is $3x + y - 6z + 11 = 0$.

The vector $\mathbf{i} = \langle 1, 0, 0 \rangle$ is a normal vector for the yz-plane. A plane which has an equation of the form $x - x_1 = 0$ (or $x = x_1$) also has normal vector $\mathbf{i}$ and hence is parallel to the yz-plane (and orthogonal to both the xy- and xz-planes). A portion of the graph of $x = a$ is sketched in (i) of Figure 14.27. Similarly, the graph of $y = b$ is a plane parallel to the xz-plane with y-intercept b, whereas the graph of $z = c$ is a plane parallel to the xy-plane with z-intercept c (see (ii) and (iii) of Figure 14.27).

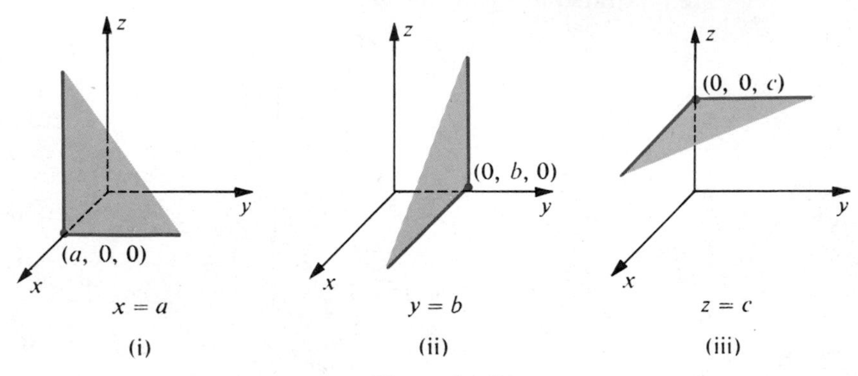

Figure 14.27

A plane with an equation of the form $by + cz + d = 0$ has normal vector $\mathbf{a} = \langle 0, b, c \rangle$ and is orthogonal to the yz-plane since $\mathbf{a} \cdot \mathbf{i} = 0$. Similarly, graphs of $ax + by + d = 0$ or $ax + cz + d = 0$ are planes which are orthogonal to the xy-plane and xz-plane, respectively.

Example 5 Sketch the graph of the equation $3x + 5z = 10$.

Solution The graph is a plane orthogonal to the xz-plane with x-intercept 10/3 and z-intercept 2. Note that the trace in the yz-plane has equation $5z = 10$ and hence is a line parallel to the y-axis with z-intercept 2. Similarly, the trace in the xy-plane has equation $3x = 10$ and is a line parallel to the y-axis with x-intercept 10/3. A portion of the graph showing traces in the three coordinate planes is sketched in Figure 14.28.

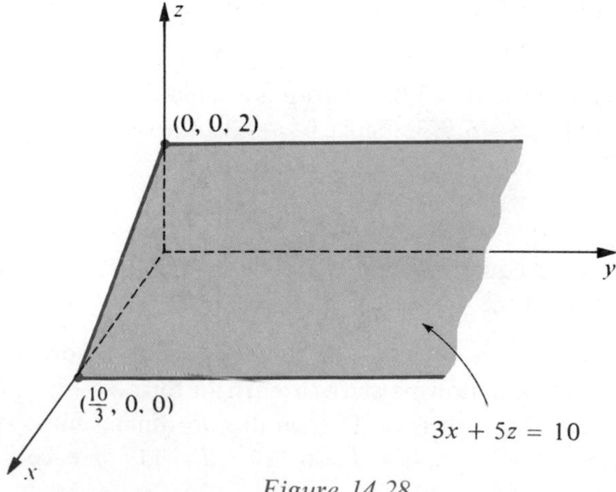

Figure 14.28

Lines may be described as intersections of planes. If a line l is given parametrically by (14.29) and if a_1, a_2, a_3 are different from zero, we may solve each equation for t, obtaining

$$t = \frac{x - x_1}{a_1}, \quad t = \frac{y - y_1}{a_2}, \quad t = \frac{z - z_1}{a_3}.$$

It follows that a point $P(x, y, z)$ is on l if and only if

(14.34)
$$\frac{x - x_1}{a_1} = \frac{y - y_1}{a_2} = \frac{z - z_1}{a_3}.$$

The equations in (14.34) are called a **symmetric form** for the line. By taking the indicated expressions in pairs, say

$$\frac{x - x_1}{a_1} = \frac{y - y_1}{a_2} \quad \text{and} \quad \frac{x - x_1}{a_1} = \frac{z - z_1}{a_3},$$

we obtain a description of l as an intersection of two planes, the first orthogonal to the xy-plane and the second orthogonal to the xz-plane. If one of the numbers a_1, a_2, or a_3 is zero, we cannot solve each equation in (14.29) for t. For example, if $a_3 = 0$ and $a_1 a_2 \neq 0$, then the third equation in (14.29) reduces to $z = z_1$, and a symmetric form may be written as

$$\frac{x - x_1}{a_1} = \frac{y - y_1}{a_2}, \quad z = z_1$$

which again expresses l as an intersection of two planes. A similar situation exists if $a_1 = 0$ or $a_2 = 0$.

Example 6 Find a symmetric form for the line through $P_1(3, 1, -2)$ and $P_2(-2, 7, -4)$.

Solution As in Example 1 of the preceding section, a vector **a** corresponding to $\overrightarrow{P_1P_2}$ is

$$\mathbf{a} = \langle -2 - 3, 7 - 1, -4 + 2 \rangle = \langle -5, 6, -2 \rangle$$

and by (14.29) a parametric representation for the line is

$$x = 3 - 5t, \quad y = 1 + 6t, \quad z = -2 - 2t.$$

Solving each equation for t and equating the results we obtain the symmetric form

$$\frac{x - 3}{-5} = \frac{y - 1}{6} = \frac{z + 2}{-2}.$$

A symmetric form is not unique since in (14.34) we may use three numbers b_1, b_2, b_3 which are proportional to a_1, a_2, a_3, or any point on l other than (x_1, y_1, z_1).

Example 7 Find a formula for the distance h from a point $P(x_0, y_0, z_0)$ to the plane $ax + by + cz + d = 0$.

Solution Let $R(x_1, y_1, z_1)$ be any point on the plane and let **n** be a normal vector to the plane. The vector $\mathbf{p} = \langle x_0 - x_1, y_0 - y_1, z_0 - z_1 \rangle$ has a geometric representation $\overrightarrow{RP}$. As illustrated in Figure 14.29, the distance h is given by

$$h = |\text{comp}_{\mathbf{n}}\, \mathbf{p}| = \left| \mathbf{p} \cdot \frac{1}{|\mathbf{n}|}\mathbf{n} \right|$$

(see (14.25)). Since $\langle a, b, c \rangle$ is a normal vector to the plane we may let

$$\frac{1}{|\mathbf{n}|}\mathbf{n} = \frac{1}{\sqrt{a^2 + b^2 + c^2}}\langle a, b, c \rangle.$$

Consequently,

$$h = \left| \mathbf{p} \cdot \frac{1}{|\mathbf{n}|}\mathbf{n} \right| = \left| \frac{a(x_0 - x_1) + b(y_0 - y_1) + c(z_0 - z_1)}{\sqrt{a^2 + b^2 + c^2}} \right|$$

$$= \frac{|(ax_0 + by_0 + cz_0) + (-ax_1 - by_1 - cz_1)|}{\sqrt{a^2 + b^2 + c^2}}.$$

Since R is on the plane, $ax_1 + by_1 + cz_1 + d = 0$, so $d = -ax_1 - by_1 - cz_1$, and hence the preceding formula may be written

$$h = \frac{|ax_0 + by_0 + cz_0 + d|}{\sqrt{a^2 + b^2 + c^2}}.$$

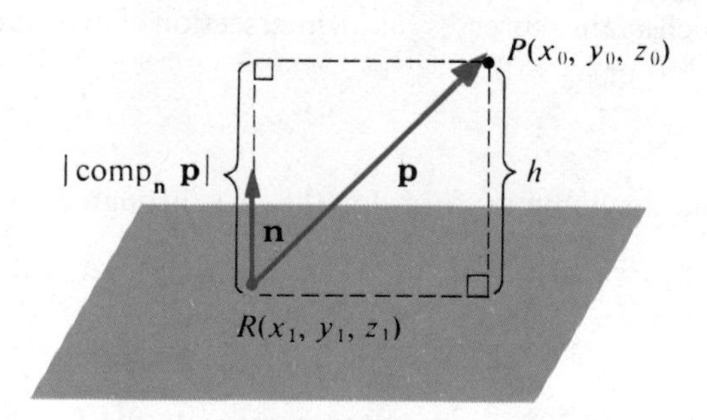

Figure 14.29

Example 8 Find the distance from the point $P(-6, 2, 3)$ to the plane $4x - 5y + 8z - 7 = 0$.

Solution Using the formula obtained in Example 7 we have

$$h = \frac{|4(-6) - 5(2) + 8(3) - 7|}{\sqrt{16 + 25 + 64}} = \frac{17}{\sqrt{105}}.$$

EXERCISES 14.6

Sketch the graphs of the equations in Exercises 1–8.

1 (a) $x = 3$ (b) $y = -2$ (c) $z = 5$

2 (a) $x = -4$ (b) $y = 0$ (c) $z = -2/3$

3 $2x + y - 6 = 0$

4 $3x - 2z - 24 = 0$

5 $4y - 2z - 15 = 0$

6 $5x + y - 4z + 20 = 0$

7 $2x - y + 5z + 10 = 0$

8 $x + y + z = 0$

In each of Exercises 9–18, find an equation of the plane which satisfies the stated conditions.

9 Through $P(6, -7, 4)$, parallel to (a) the xy-plane; (b) the yz-plane; (c) the xz-plane

10 Through $P(-2, 5, -8)$ with normal vector (a) $\mathbf{i}$; (b) $\mathbf{j}$; (c) $\mathbf{k}$

11 Through $P(-11, 4, -2)$ with normal vector $\mathbf{a} = \langle 6, -5, -1 \rangle$

12 Through $P(4, 2, -9)$ with normal vector $\overrightarrow{OP}$

13 Through $P(2, 5, -6)$ and parallel to the plane $3x - y + 2z = 10$

14 Through the origin and parallel to the plane $x - 6y + 4z = 7$

15 Through the origin and the points $P(0, 2, 5)$ and $Q(1, 4, 0)$

16 Through the points $P(3, 2, 1)$, $Q(-1, 1, -2)$, and $R(3, -4, 1)$

17 Through $P(5, 2, -3)$ and orthogonal to the planes $2x - y + 4z = 7$ and $3y - z = 8$

18 Through $P(-4, 1, 6)$ and having the same trace in the xz-plane as $x + 4y - 5z = 8$

In Exercises 19–22, find a symmetric form for the line through P_1 and P_2.

19 $P_1(5, -2, 4)$, $P_2(2, 6, 1)$

20 $P_1(-3, 1, -1)$, $P_2(7, 11, -8)$

21 $P_1(4, 2, -3)$, $P_2(-3, 2, 5)$

22 $P_1(5, -7, 4)$, $P_2(-2, -1, 4)$

In each of Exercises 23–26, determine whether the given lines (a) are parallel; (b) are orthogonal; (c) intersect.

23 $(x + 1)/4 = (y + 7)/(-2) = (z - 5)/3$; $(x - 3)/3 = (y + 2)/3 = z/(-2)$

24 $(2x - 3)/6 = (y - 4)/(-1) = (z + 1)/2$; $(x + 3)/(-6) = (y + 1)/2 = (z - 5)/(-4)$

25 $(x + 1)/3 = (y - 3)/(-1) = (z + 5)/2$; $x/(-4) = (y - 4)/2 = (z - 9)/4$

26 $(x - 3)/5 = (y + 6)/(-2) = (z - 2)$; $x/7 = (y - 4)/3 = (1 - z)/5$.

In each of Exercises 27–30, find the distance from the point P to the given plane.

27 $3x - 7y + z - 5 = 0$, $P(1, -1, 2)$

28 $2x + 4y - 5z + 1 = 0$, $P(3, 1, -2)$

29 $4x - 3z = 2$, $P(5, -8, 1)$

30 $6y - z = 10$, $P(4, 1, -3)$

31 If a line l has parametric equations $x = 3t + 1$, $y = -2t + 4$, $z = t - 3$, find the equation of the plane which contains l and the point $P(5, 0, 2)$.

32 Find parametric equations for the line of intersection of the planes $2x + y + 4z = 8$ and $x + 3y - z = -1$.

33 Find an equation of the plane through the point $A(1, 2, -3)$ and the line of intersection of the planes $x + 3y - 5z = 10$ and $6x - y + z = 0$.

34 Let $\mathbf{a}, \mathbf{b}, \mathbf{c}$ be vectors in V_3. Prove that their position vectors $\overrightarrow{OA}, \overrightarrow{OB}, \overrightarrow{OC}$ are coplanar if and only if $\mathbf{c} = m\mathbf{a} + n\mathbf{b}$ for some scalars m and n.

14.7 THE VECTOR PRODUCT

In this section we introduce the *vector product* (or *cross product*) $\mathbf{a} \times \mathbf{b}$ of two vectors **a** and **b**. Unlike the dot product, which is a scalar, this new operation produces another vector. The vector product was first used as a tool for physical problems involving moments of forces. It is possible to define $\mathbf{a} \times \mathbf{b}$ geometrically and then obtain an algebraic form by introducing a rectangular coordinate system. We shall reverse this process and begin with an algebraic definition. This approach disguises the geometric nature of $\mathbf{a} \times \mathbf{b}$; however, it leads to simpler proofs of properties.

It is convenient to use *determinants* when working with vector products. A **determinant of order 2** is defined by

(14.35)
$$\begin{vmatrix} a_1 & a_2 \\ b_1 & b_2 \end{vmatrix} = a_1 b_2 - a_2 b_1$$

where all letters represent real numbers. For example,

$$\begin{vmatrix} 2 & -3 \\ 4 & 5 \end{vmatrix} = (2)(5) - (-3)(4) = 10 + 12 = 22.$$

A **determinant of order 3** is given by

(14.36)
$$\begin{vmatrix} c_1 & c_2 & c_3 \\ a_1 & a_2 & a_3 \\ b_1 & b_2 & b_3 \end{vmatrix} = \begin{vmatrix} a_2 & a_3 \\ b_2 & b_3 \end{vmatrix} c_1 - \begin{vmatrix} a_1 & a_3 \\ b_1 & b_3 \end{vmatrix} c_2 + \begin{vmatrix} a_1 & a_2 \\ b_1 & b_2 \end{vmatrix} c_3.$$

This is sometimes called the *expansion of the determinant by the first row*. The numerical value can be found by applying (14.35) to the second-order determinants on the right side of the equation.

Example 1 Find the value of $\begin{vmatrix} 2 & -1 & 3 \\ -2 & 5 & 1 \\ 1 & 2 & -4 \end{vmatrix}$.

Solution Using (14.36), and (14.35),

$$\begin{vmatrix} 2 & -1 & 3 \\ -2 & 5 & 1 \\ 1 & 2 & -4 \end{vmatrix} = \begin{vmatrix} 5 & 1 \\ 2 & -4 \end{vmatrix}(2) - \begin{vmatrix} -2 & 1 \\ 1 & -4 \end{vmatrix}(-1) + \begin{vmatrix} -2 & 5 \\ 1 & 2 \end{vmatrix}(3)$$
$$= (-20 - 2)(2) - (8 - 1)(-1) + (-4 - 5)(3)$$
$$= -44 + 7 - 27 = -64.$$

The **vector product** (or **cross product**) $\mathbf{a} \times \mathbf{b}$ of the vectors $\mathbf{a} = \langle a_1, a_2, a_3 \rangle$ and $\mathbf{b} = \langle b_1, b_2, b_3 \rangle$ is obtained by replacing c_1, c_2, c_3 on the right side of (14.36) by the unit vectors $\mathbf{i}, \mathbf{j}, \mathbf{k}$. Thus, by definition

(14.37) $$\mathbf{a} \times \mathbf{b} = \begin{vmatrix} a_2 & a_3 \\ b_2 & b_3 \end{vmatrix} \mathbf{i} - \begin{vmatrix} a_1 & a_3 \\ b_1 & b_3 \end{vmatrix} \mathbf{j} + \begin{vmatrix} a_1 & a_2 \\ b_1 & b_2 \end{vmatrix} \mathbf{k}.$$

A convenient way to remember this formula is to use the notation

(14.38) $$\mathbf{a} \times \mathbf{b} = \begin{vmatrix} \mathbf{i} & \mathbf{j} & \mathbf{k} \\ a_1 & a_2 & a_3 \\ b_1 & b_2 & b_3 \end{vmatrix}.$$

The symbol on the right side of this equation is not a determinant, since the first row contains vectors instead of scalars. Consequently, properties of determinants may not be applied to (14.38). The determinant notation is used as a mnemonic device for remembering the more cumbersome formula (14.37). With this warning in mind we shall use (14.38) to find vector products, as in the following example.

Example 2 Find $\mathbf{a} \times \mathbf{b}$ if $\mathbf{a} = \langle 2, -1, 6 \rangle$ and $\mathbf{b} = \langle -3, 5, 1 \rangle$.

Solution Writing

$$\mathbf{a} \times \mathbf{b} = \begin{vmatrix} \mathbf{i} & \mathbf{j} & \mathbf{k} \\ 2 & -1 & 6 \\ -3 & 5 & 1 \end{vmatrix}$$

we obtain

$$\begin{aligned} \mathbf{a} \times \mathbf{b} &= \begin{vmatrix} -1 & 6 \\ 5 & 1 \end{vmatrix} \mathbf{i} - \begin{vmatrix} 2 & 6 \\ -3 & 1 \end{vmatrix} \mathbf{j} + \begin{vmatrix} 2 & -1 \\ -3 & 5 \end{vmatrix} \mathbf{k} \\ &= (-1 - 30)\mathbf{i} - (2 + 18)\mathbf{j} + (10 - 3)\mathbf{k} \\ &= -31\mathbf{i} - 20\mathbf{j} + 7\mathbf{k}. \end{aligned}$$

If $\mathbf{a}$ is any vector in V_3, then

$$\mathbf{a} \times \mathbf{0} = \mathbf{0} = \mathbf{0} \times \mathbf{a}$$

for if one of the vectors in (14.37) is $\mathbf{0}$, then each determinant has a row of zeros. Also, $\mathbf{a} \times \mathbf{a} = \mathbf{0}$ for every $\mathbf{a}$, since in this case each determinant in (14.37) has equal rows.

The next theorem brings out an important property of vector products.

(14.39) Theorem

The vector $\mathbf{a} \times \mathbf{b}$ is orthogonal to both $\mathbf{a}$ and $\mathbf{b}$.

Proof. By Theorem (14.20) it is sufficient to show that

$$(\mathbf{a} \times \mathbf{b}) \cdot \mathbf{a} = 0 \quad \text{and} \quad (\mathbf{a} \times \mathbf{b}) \cdot \mathbf{b} = 0$$

If we apply the definition of dot product to (14.37) and if $\mathbf{a} = \langle a_1, a_2, a_3 \rangle$, then

$$\begin{aligned}(\mathbf{a} \times \mathbf{b}) \cdot \mathbf{a} &= \begin{vmatrix} a_2 & a_3 \\ b_2 & b_3 \end{vmatrix} a_1 - \begin{vmatrix} a_1 & a_3 \\ b_1 & b_3 \end{vmatrix} a_2 + \begin{vmatrix} a_1 & a_2 \\ b_1 & b_2 \end{vmatrix} a_3 \\ &= (a_2 b_3 - a_3 b_2) a_1 - (a_1 b_3 - a_3 b_1) a_2 + (a_1 b_2 - a_2 b_1) a_3 \\ &= a_2 b_3 a_1 - a_3 b_2 a_1 - a_1 b_3 a_2 + a_3 b_1 a_2 + a_1 b_2 a_3 - a_2 b_1 a_3 \\ &= 0.\end{aligned}$$

Hence $\mathbf{a} \times \mathbf{b}$ is orthogonal to $\mathbf{a}$. The verification that $(\mathbf{a} \times \mathbf{b}) \cdot \mathbf{b} = 0$ is left to the reader.

In geometric terms, Theorem (14.39) implies that if nonzero vectors **a** and **b** are represented by vectors $\overrightarrow{PQ}$ and $\overrightarrow{PR}$ with the same initial point P, then $\mathbf{a} \times \mathbf{b}$ may be represented by a vector $\overrightarrow{PS}$ which is normal to the plane determined by P, Q, and R, as illustrated in (i) of Figure 14.30. We shall write

$$\overrightarrow{PS} = \overrightarrow{PQ} \times \overrightarrow{PR}.$$

It can be shown that the direction of $\overrightarrow{PS}$ may be obtained using the right-hand rule illustrated in (ii) of Figure 14.30. Specifically, if θ denotes the angle between $\overrightarrow{PQ}$ and $\overrightarrow{PR}$, and if the fingers of the right hand are curled such that a rotation through θ will transform $\overrightarrow{PQ}$ into a vector having the same direction as $\overrightarrow{PR}$, then the extended thumb points in the direction of $\overrightarrow{PQ} \times \overrightarrow{PR}$.

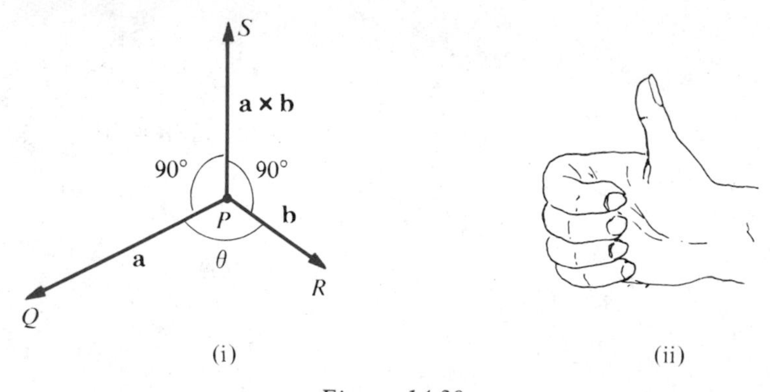

Figure 14.30

Example 3 Find an equation of the plane determined by the points $P(4, -3, 1)$, $Q(6, -4, 7)$, and $R(1, 2, 2)$.

Solution Applying Theorem (14.14), vectors **a** and **b** corresponding to $\overrightarrow{PQ}$ and $\overrightarrow{PR}$ are

$$\mathbf{a} = \langle 2, -1, 6 \rangle \quad \text{and} \quad \mathbf{b} = \langle -3, 5, 1 \rangle.$$

By the preceding discussion, the vector $\mathbf{a} \times \mathbf{b}$ is normal to the plane determined by P, Q, and R. From Example 2,

$$\mathbf{a} \times \mathbf{b} = -31\mathbf{i} - 20\mathbf{j} + 7\mathbf{k}.$$

Using Theorem (14.30) with $P_1 = P$ gives us the equation

$$-31(x - 4) - 20(y + 3) + 7(z - 1) = 0$$

which reduces to

$$-31x - 20y + 7z + 57 = 0.$$

The next result provides information about the magnitude of $\mathbf{a} \times \mathbf{b}$.

(14.40) Theorem

If θ is the angle between nonzero vectors $\mathbf{a}$ and $\mathbf{b}$, then

$$|\mathbf{a} \times \mathbf{b}| = |\mathbf{a}|\,|\mathbf{b}| \sin \theta.$$

Proof. Applying Definitions (14.2) and (14.37),

$$\begin{aligned}
|\mathbf{a} \times \mathbf{b}|^2 &= \begin{vmatrix} a_2 & a_3 \\ b_2 & b_3 \end{vmatrix}^2 + \begin{vmatrix} a_1 & a_3 \\ b_1 & b_3 \end{vmatrix}^2 + \begin{vmatrix} a_1 & a_2 \\ b_1 & b_2 \end{vmatrix}^2 \\
&= (a_2b_3 - a_3b_2)^2 + (a_1b_3 - a_3b_1)^2 + (a_1b_2 - a_2b_1)^2 \\
&= a_2^2b_3^2 - 2a_2a_3b_2b_3 + a_3^2b_2^2 + a_1^2b_3^2 - 2a_1a_3b_1b_3 \\
&\quad + a_3^2b_1^2 + a_1^2b_2^2 - 2a_1a_2b_1b_2 + a_2^2b_1^2 \\
&= (a_1^2 + a_2^2 + a_3^2)(b_1^2 + b_2^2 + b_3^2) - (a_1b_1 + a_2b_2 + a_3b_3)^2.
\end{aligned}$$

The last equality may be verified by multiplying the indicated expressions. The vector form of this identity is

$$|\mathbf{a} \times \mathbf{b}|^2 = (|\mathbf{a}|\,|\mathbf{b}|)^2 - (\mathbf{a} \cdot \mathbf{b})^2$$

or, since $\mathbf{a} \cdot \mathbf{b} = |\mathbf{a}|\,|\mathbf{b}| \cos \theta$,

$$\begin{aligned}
|\mathbf{a} \times \mathbf{b}|^2 &= (|\mathbf{a}|\,|\mathbf{b}|)^2 - (|\mathbf{a}|\,|\mathbf{b}|)^2 \cos^2 \theta \\
&= (|\mathbf{a}|\,|\mathbf{b}|)^2(1 - \cos^2 \theta) \\
&= (|\mathbf{a}|\,|\mathbf{b}|)^2 \sin^2 \theta.
\end{aligned}$$

Finally, taking square roots, we obtain

$$|\mathbf{a} \times \mathbf{b}| = |\mathbf{a}|\,|\mathbf{b}| \sin \theta$$

which is what we wished to prove.

(14.41) Corollary

Two vectors $\mathbf{a}$ and $\mathbf{b}$ are parallel if and only if $\mathbf{a} \times \mathbf{b} = \mathbf{0}$.

Proof. Suppose **a** and **b** are nonzero vectors. If θ is the angle between **a** and **b**, then the vectors are parallel if and only if $\theta = 0$ or $\theta = \pi$, or equivalently $\sin\theta = 0$. By Theorem (14.40) the last statement is equivalent to $\mathbf{a} \times \mathbf{b} = \mathbf{0}$. If either **a** or **b** is the zero vector the proof is trivial.

To interpret $|\mathbf{a} \times \mathbf{b}|$ geometrically let us represent **a** and **b** by vectors $\overrightarrow{PQ}$ and $\overrightarrow{PR}$ having the same initial point P. Let S be the point such that segments PQ and PR are adjacent sides of a parallelogram with vertices P, Q, R, and S, as illustrated in Figure 14.31. An altitude of the parallelogram is $|\mathbf{b}|\sin\theta$ and hence its area is $|\mathbf{a}|\,|\mathbf{b}|\sin\theta$. Comparing this with (14.40) we see that *the magnitude of the vector product* $\mathbf{a} \times \mathbf{b}$ *equals the area of the parallelogram determined by* **a** *and* **b**. Physical interpretations of the vector product will be given in the next chapter.

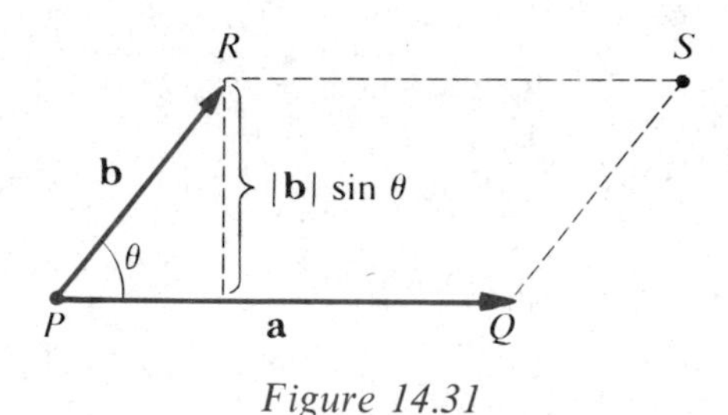

Figure 14.31

Example 4 Find the area of the triangle determined by $P(4, -3, 1)$, $Q(6, -4, 7)$, and $R(1, 2, 2)$.

Solution As in the solution of Example 3, the vectors corresponding to $\overrightarrow{PQ}$ and $\overrightarrow{PR}$ are $\mathbf{a} = \langle 2, -1, 6\rangle$ and $\mathbf{b} = \langle -3, 5, 1\rangle$, respectively, and $\mathbf{a} \times \mathbf{b} = -31\mathbf{i} - 20\mathbf{j} + 7\mathbf{k}$. Hence the area of the parallelogram with adjacent sides PQ and PR is

$$|\mathbf{a} \times \mathbf{b}| = \sqrt{961 + 400 + 49} = \sqrt{1410}.$$

It follows that the area of the triangle is $\frac{1}{2}\sqrt{1410}$.

The vector products of the special unit vectors **i**, **j**, and **k** are of interest. For example, using (14.37) with $\mathbf{a} = \mathbf{i} = \langle 1, 0, 0\rangle$ and $\mathbf{b} = \mathbf{j} = \langle 0, 1, 0\rangle$,

$$\mathbf{i} \times \mathbf{j} = \begin{vmatrix} 0 & 0 \\ 1 & 0 \end{vmatrix}\mathbf{i} - \begin{vmatrix} 1 & 0 \\ 0 & 0 \end{vmatrix}\mathbf{j} + \begin{vmatrix} 1 & 0 \\ 0 & 1 \end{vmatrix}\mathbf{k} = \mathbf{k}.$$

In general, the following are true:

(14.42)
$$\begin{array}{lll} \mathbf{i} \times \mathbf{j} = \mathbf{k}, & \mathbf{j} \times \mathbf{k} = \mathbf{i}, & \mathbf{k} \times \mathbf{i} = \mathbf{j} \\ \mathbf{j} \times \mathbf{i} = -\mathbf{k}, & \mathbf{k} \times \mathbf{j} = -\mathbf{i}, & \mathbf{i} \times \mathbf{k} = -\mathbf{j} \\ \mathbf{i} \times \mathbf{i} = \mathbf{j} \times \mathbf{j} = \mathbf{k} \times \mathbf{k} = \mathbf{0}. & & \end{array}$$

The proof is left as an exercise.

The fact that $\mathbf{i} \times \mathbf{j} \neq \mathbf{j} \times \mathbf{i}$ shows that the vector product is not commutative. The associative law does not hold either since, for example,

$$\mathbf{i} \times (\mathbf{j} \times \mathbf{j}) = \mathbf{i} \times \mathbf{0} = \mathbf{0}$$

whereas

$$(\mathbf{i} \times \mathbf{j}) \times \mathbf{j} = \mathbf{k} \times \mathbf{j} = -\mathbf{i}.$$

The next theorem lists some important properties of the vector product.

(14.43) Theorem

If **a**, **b**, and **c** are vectors and m is a scalar, then

(i) $\mathbf{a} \times \mathbf{b} = -\mathbf{b} \times \mathbf{a}$.

(ii) $(m\mathbf{a}) \times \mathbf{b} = m(\mathbf{a} \times \mathbf{b}) = \mathbf{a} \times (m\mathbf{b})$.

(iii) $\mathbf{a} \times (\mathbf{b} + \mathbf{c}) = \mathbf{a} \times \mathbf{b} + \mathbf{a} \times \mathbf{c}$.

(iv) $(\mathbf{a} + \mathbf{b}) \times \mathbf{c} = \mathbf{a} \times \mathbf{c} + \mathbf{b} \times \mathbf{c}$.

(v) $(\mathbf{a} \times \mathbf{b}) \cdot \mathbf{c} = \mathbf{a} \cdot (\mathbf{b} \times \mathbf{c})$.

(vi) $\mathbf{a} \times (\mathbf{b} \times \mathbf{c}) = (\mathbf{a} \cdot \mathbf{c})\mathbf{b} - (\mathbf{a} \cdot \mathbf{b})\mathbf{c}$.

All of the properties in Theorem (14.43) may be established by straightforward (but sometimes lengthy) applications of Definition (14.37). For example, if $\mathbf{a} = \langle a_1, a_2, a_3 \rangle$ and $\mathbf{b} = \langle b_1, b_2, b_3 \rangle$, then

$$\mathbf{b} \times \mathbf{a} = \begin{vmatrix} b_2 & b_3 \\ a_2 & a_3 \end{vmatrix} \mathbf{i} - \begin{vmatrix} b_1 & b_3 \\ a_1 & a_3 \end{vmatrix} \mathbf{j} + \begin{vmatrix} b_1 & b_2 \\ a_1 & a_2 \end{vmatrix} \mathbf{k}.$$

Since interchanging two rows of a determinant changes its sign, we have

$$\begin{aligned} \mathbf{b} \times \mathbf{a} &= -\begin{vmatrix} a_2 & a_3 \\ b_2 & b_3 \end{vmatrix} \mathbf{i} + \begin{vmatrix} a_1 & a_3 \\ b_1 & b_3 \end{vmatrix} \mathbf{j} - \begin{vmatrix} a_1 & a_2 \\ b_1 & b_2 \end{vmatrix} \mathbf{k} \\ &= -\mathbf{a} \times \mathbf{b}. \end{aligned}$$

This proves property (i).

If $\mathbf{c} = \langle c_1, c_2, c_3 \rangle$, then the **i** component of $\mathbf{a} \times (\mathbf{b} + \mathbf{c})$ is

$$\begin{aligned} \begin{vmatrix} a_2 & a_3 \\ b_2 + c_2 & b_3 + c_3 \end{vmatrix} &= a_2(b_3 + c_3) - a_3(b_2 + c_2) \\ &= (a_2b_3 - a_3b_2) + (a_2c_3 - a_3c_2) \\ &= \begin{vmatrix} a_2 & a_3 \\ b_2 & b_3 \end{vmatrix} + \begin{vmatrix} a_2 & a_3 \\ c_2 & c_3 \end{vmatrix} \end{aligned}$$

which is equal to the **i** component of $\mathbf{a} \times \mathbf{b} + \mathbf{a} \times \mathbf{c}$. A similar calculation can be used to prove that the **j** and **k** components of $\mathbf{a} \times (\mathbf{b} + \mathbf{c})$ are the same as those of $\mathbf{a} \times \mathbf{b} + \mathbf{a} \times \mathbf{c}$. This establishes (iii) of (14.43). The proofs of the remaining properties are left to the reader.

We shall conclude this section with several geometric applications of the vector product.

Example 5 Find a formula for the distance d from a point R to a line l.

Solution As illustrated in Figure 14.32, let P and Q be points on l, and let θ be the angle between $\overrightarrow{PQ}$ and $\overrightarrow{PR}$. Since

$$d = |\overrightarrow{PR}| \sin \theta$$

and

$$|\overrightarrow{PQ} \times \overrightarrow{PR}| = |\overrightarrow{PQ}||\overrightarrow{PR}| \sin \theta$$

we see that

$$d = \frac{1}{|\overrightarrow{PQ}|}|\overrightarrow{PQ} \times \overrightarrow{PR}|.$$

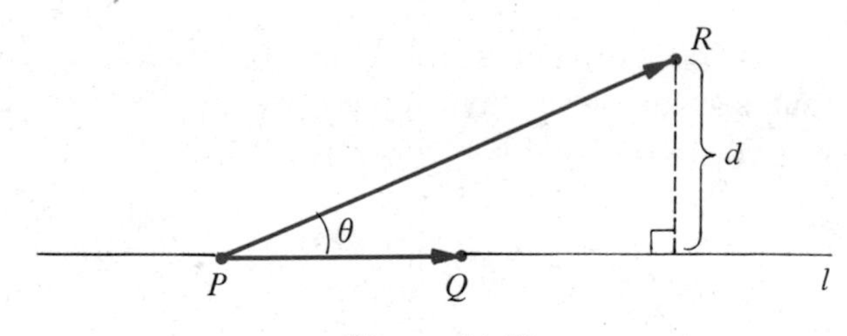

Figure 14.32

Example 6 Suppose $\overrightarrow{PQ}$, $\overrightarrow{PR}$, and $\overrightarrow{PS}$ represent adjacent sides of a rectangular parallelepiped. If $\mathbf{a}$, $\mathbf{b}$, and $\mathbf{c}$ are the corresponding vectors in V_3, show that $|(\mathbf{a} \times \mathbf{b}) \cdot \mathbf{c}|$ is the volume of the parallelepiped.

Solution One illustration of the parallelepiped is given in Figure 14.33. The area of the base is $|\mathbf{a} \times \mathbf{b}|$. Let θ be the angle between $\mathbf{c}$ and $\mathbf{a} \times \mathbf{b}$. Since the vector corresponding to $\mathbf{a} \times \mathbf{b}$ is perpendicular to the base, it follows that the altitude h of the parallelepiped is given by

$$h = |\mathbf{c}|\,|\cos \theta|.$$

It is necessary to use the absolute value $|\cos \theta|$ since θ may be an obtuse angle. Hence the volume V of the parallelepiped is given by

$$\begin{aligned} V &= (\text{area of base})(\text{altitude}) \\ &= |\mathbf{a} \times \mathbf{b}|\,|\mathbf{c}|\,|\cos \theta| \\ &= |(\mathbf{a} \times \mathbf{b}) \cdot \mathbf{c}|. \end{aligned}$$

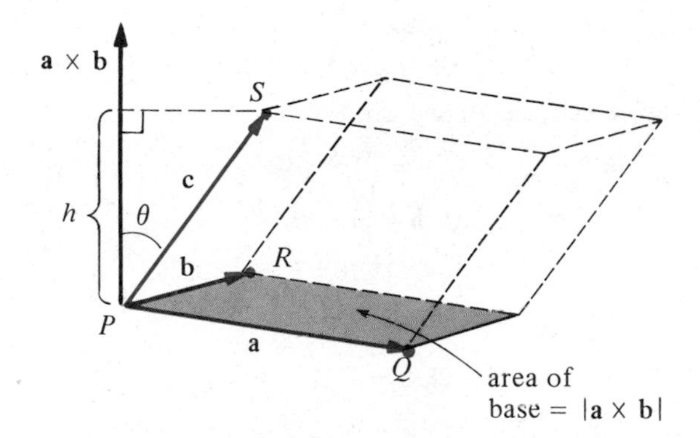

Figure 14.33

Example 7 Find a formula for the shortest distance d between two skew lines l_1 and l_2.

Solution Two lines are skew if they are not parallel and do not intersect. Choose points P_1, Q_1 on l_1 and P_2, Q_2 on l_2, as illustrated in Figure 14.34. It follows that the vector

$$\mathbf{n} = \frac{1}{|\overrightarrow{P_1Q_1} \times \overrightarrow{P_2Q_2}|}(\overrightarrow{P_1Q_1} \times \overrightarrow{P_2Q_2})$$

is a unit vector orthogonal to both $\overrightarrow{P_1Q_1}$ and $\overrightarrow{P_2Q_2}$. (Why?) Let us consider planes Γ_1 and Γ_2 through P_1 and P_2, respectively, each having normal vector $\mathbf{n}$. Thus Γ_1 and Γ_2 are parallel planes containing l_1 and l_2, respectively. The distance d between the planes is measured along a line parallel to the common normal $\mathbf{n}$, as shown in the figure. It follows that d is the shortest distance between l_1 and l_2. Moreover,

$$\begin{aligned} d &= |\text{comp}_{\mathbf{n}} \overrightarrow{P_1P_2}| = |\mathbf{n} \cdot \overrightarrow{P_1P_2}| \\ &= \frac{1}{|\overrightarrow{P_1Q_1} \times \overrightarrow{P_2Q_2}|}|(\overrightarrow{P_1Q_1} \times \overrightarrow{P_2Q_2}) \cdot \overrightarrow{P_1P_2}|. \end{aligned}$$

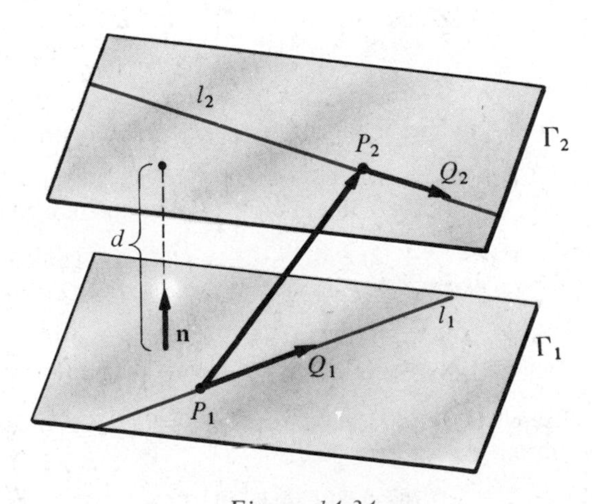

Figure 14.34

EXERCISES 14.7

In Exercises 1–10 find $\mathbf{a} \times \mathbf{b}$.

1 $\mathbf{a} = \langle 1, -2, 3\rangle, \mathbf{b} = \langle 2, 1, -4\rangle$

2 $\mathbf{a} = \langle -5, 1, -1\rangle, \mathbf{b} = \langle 3, 6, -2\rangle$

3 $\mathbf{a} = 5\mathbf{i} - 6\mathbf{j} - \mathbf{k}, \mathbf{b} = 3\mathbf{i} + \mathbf{k}$

4 $\mathbf{a} = 2\mathbf{i} + \mathbf{j}, \mathbf{b} = -5\mathbf{j} + 2\mathbf{k}$

5 $\mathbf{a} = \langle 0, 1, 2\rangle, \mathbf{b} = \langle 1, 2, 0\rangle$

6 $\mathbf{a} = -3\mathbf{i} + \mathbf{j} + 2\mathbf{k}, \mathbf{b} = 9\mathbf{i} - 3\mathbf{j} - 6\mathbf{k}$

7 $\mathbf{a} = 3\mathbf{i} - \mathbf{j} + 8\mathbf{k}, \mathbf{b} = 5\mathbf{j}$

8 $\mathbf{a} = \langle 0, 0, 4\rangle, \mathbf{b} = \langle -7, 1, 0\rangle$

9 $\mathbf{a} = 4\mathbf{i} - 6\mathbf{j} + 2\mathbf{k}, \mathbf{b} = -2\mathbf{i} + 3\mathbf{j} - \mathbf{k}$

10 $\mathbf{a} = 3\mathbf{i}, \mathbf{b} = 4\mathbf{k}$

11 If $\mathbf{a} = \langle 2, 0, -1\rangle, \mathbf{b} = \langle -3, 1, 0\rangle$, and $\mathbf{c} = \langle 1, -2, 4\rangle$, find $\mathbf{a} \times (\mathbf{b} \times \mathbf{c})$ and $(\mathbf{a} \times \mathbf{b}) \times \mathbf{c}$.

12 If $\mathbf{a}$, $\mathbf{b}$, and $\mathbf{c}$ are the vectors of Exercise 11, find $\mathbf{a} \times (\mathbf{b} - \mathbf{c})$ and $\mathbf{a} \times \mathbf{b} - \mathbf{a} \times \mathbf{c}$.

13 Prove that $(\mathbf{a} \times \mathbf{b}) \cdot \mathbf{b} = 0$ for all vectors $\mathbf{a}$ and $\mathbf{b}$.

14 Prove (14.42).

15 Complete the proof of Theorem (14.43).

16 Prove that $(\mathbf{a} + \mathbf{b}) \times (\mathbf{a} - \mathbf{b}) = 2\mathbf{b} \times \mathbf{a}$ for all $\mathbf{a}$ and $\mathbf{b}$.

17 If $\mathbf{a} \times \mathbf{b} = \mathbf{a} \times \mathbf{c}$ and $\mathbf{a} \neq \mathbf{0}$, does it follow that $\mathbf{b} = \mathbf{c}$? Explain.

18 Prove that for all vectors $\mathbf{a}$, $\mathbf{b}$, and $\mathbf{c}$,

$$\mathbf{a} \times (\mathbf{b} \times \mathbf{c}) + \mathbf{b} \times (\mathbf{c} \times \mathbf{a}) + \mathbf{c} \times (\mathbf{a} \times \mathbf{b}) = \mathbf{0}.$$

(*Hint:* Use (vi) of Theorem (14.43).)

In each of Exercises 19–22 find (a) a vector orthogonal to the plane determined by the points P, Q, and R; (b) an equation of the plane through P, Q, R; (c) the area of the triangle determined by P, Q, and R.

19 $P(1, -1, 2), Q(0, 3, -1), R(3, -4, 1)$

20 $P(-3, 0, 5), Q(2, -1, -3), R(4, 1, -1)$

21 $P(4, 0, 0), Q(0, 5, 0), R(0, 0, 2)$

22 $P(-1, 2, 0), Q(0, 2, -3), R(5, 0, 1)$

23 If $\mathbf{a} = \langle a_1, a_2, a_3\rangle, \mathbf{b} = \langle b_1, b_2, b_3\rangle$, and $\mathbf{c} = \langle c_1, c_2, c_3\rangle$, prove that

$$\mathbf{a} \cdot (\mathbf{b} \times \mathbf{c}) = (\mathbf{a} \times \mathbf{b}) \cdot \mathbf{c} = \begin{vmatrix} a_1 & a_2 & a_3 \\ b_1 & b_2 & b_3 \\ c_1 & c_2 & c_3 \end{vmatrix}$$

(This number is called the **triple scalar product** of $\mathbf{a}$, $\mathbf{b}$, and $\mathbf{c}$.)

24 If $\mathbf{a}$, $\mathbf{b}$, and $\mathbf{c}$ are represented by vectors with a common initial point, show that $\mathbf{a} \cdot (\mathbf{b} \times \mathbf{c}) = 0$ if and only if the vectors are coplanar.

25 Given $P(1, -1, 2), Q(0, 3, -1)$, and $R(3, -4, 1)$, use Exercise 24 to find the volume of the parallelepiped having adjacent sides OP, OQ, and OR.

26 Given $A(2, 1, -1)$, $B(3, 0, 2)$, $C(4, -2, 1)$, and $D(5, -3, 0)$, use Exercise 24 to find the volume of the parallelepiped having adjacent sides AB, AC, and AD.

In Exercises 27 and 28 let l_1 be the line through A, B and let l_2 be the line through C, D. Find the distance between l_1 and l_2.

27 $A(1, -2, 3), B(2, 0, 5), C(4, 1, -1), D(-2, 3, 4)$

28 $A(1, 3, 0), B(0, 4, 5), C(-2, -1, 2), D(5, 1, 0)$

29 Find the distance from the point $C(2, 1, -2)$ to the line through the points $A(3, -4, 1)$ and $B(-1, 2, 5)$.

30 A line l has parametric equations $x = 2t + 1$, $y = -t + 3$, $z = 5t$. Find the distance from $A(3, 1, -1)$ to l.

Verify the identities in Exercises 31 and 32, where $\mathbf{a}, \mathbf{b}, \mathbf{c}, \mathbf{d}$ are arbitrary vectors.

31 $(\mathbf{a} \times \mathbf{b}) \times \mathbf{c} = (\mathbf{a} \cdot \mathbf{c})\,\mathbf{b} - (\mathbf{b} \cdot \mathbf{c})\,\mathbf{a}$

32 $(\mathbf{a} \times \mathbf{b}) \cdot (\mathbf{c} \times \mathbf{d}) = \begin{vmatrix} \mathbf{a} \cdot \mathbf{c} & \mathbf{b} \cdot \mathbf{c} \\ \mathbf{a} \cdot \mathbf{d} & \mathbf{b} \cdot \mathbf{d} \end{vmatrix}$

14.8 CYLINDERS AND SURFACES OF REVOLUTION

In this section and the next we shall consider equations in x, y, and z whose graphs are fundamental in the study of analytic geometry. We previously defined the concept of *trace* in a coordinate plane. More generally, the trace of a surface in *any* plane is the intersection of the surface and the plane. To find the shape of a surface from its equation, we often make considerable use of traces in planes which are parallel to coordinate planes.

Example 1 Find traces, in various planes, of the surface having equation $z = x^2 + y^2$, and sketch the graph of the equation.

Solution Substitution of 0 for x in the equation gives us $z = y^2$, and hence the trace of the surface in the yz-plane is a parabola with vertex at the origin and opening upward, as shown in Figure 14.35. Similarly, substituting 0 for y leads to $z = x^2$, and hence

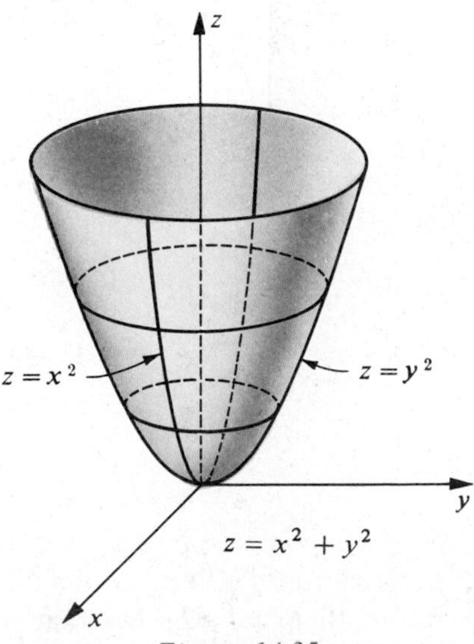

Figure 14.35

the trace in the xz-plane is also a parabola. It is instructive to find traces in planes parallel to the xy-plane, that is, planes with equations of the form $z = c$. Substituting c for z in the given equation produces $x^2 + y^2 = c$. Thus, if $c > 0$, then the trace in the plane $z = c$ is a circle of radius $\sqrt{c}$. Three such circles are sketched in Figure 14.35. If $c < 0$, then $x^2 + y^2 = c$ has no graph and consequently none of the points beneath the xy-plane is on the surface. The trace in the xy-plane has equation $x^2 + y^2 = 0$ and, therefore, consists of only one point, the origin. Although traces in planes parallel to the xz- or yz-planes could be determined, those we have obtained are sufficient for an accurate description of the graph. It will follow from the discussion at the end of this section that the surface in this example may be regarded as having been generated by revolving the graph of the parabola $z = y^2$ in the yz-plane about the z-axis. This surface is called a **circular paraboloid** or **paraboloid of revolution**.

(14.44) Definition

If C is a curve in a plane and l is a line not in the plane, then the set of points on all lines which intersect C and are parallel to l is called a **cylinder**.

The curve C in (14.44) is called a **directrix** for the cylinder, and each line through C parallel to l is a **ruling** of the cylinder. The most familiar type of cylinder is a **right circular cylinder**, obtained by letting C be a circle in a plane and l a line perpendicular to the plane, as illustrated in (i) of Figure 14.36. Although we have cut off the cylinder in the figure, it is to be understood that the rulings extend indefinitely. It is not required that the directrix C in Definition (14.44) be a closed curve. This is illustrated in (ii) of Figure 14.36, where C is a parabola.

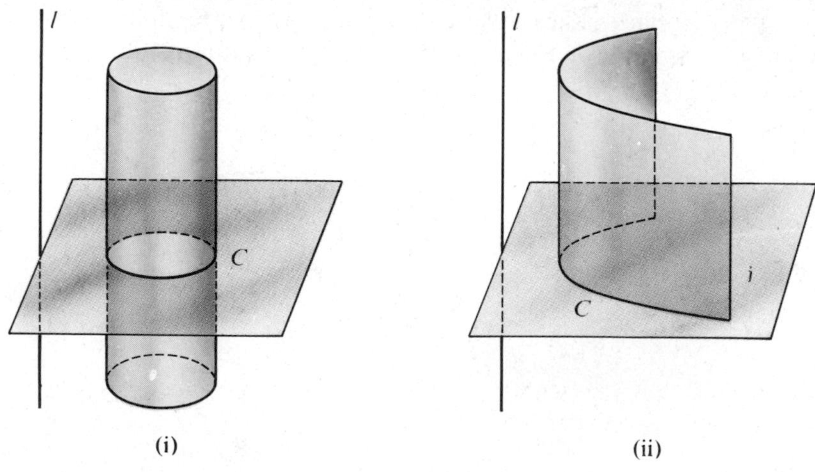

Figure 14.36

In the discussion to follow we shall only consider the case in which the directrix C is on a coordinate plane and the line l is parallel to the coordinate axis which is not on the plane. Suppose that C is on the xy-plane and has equation

$y = f(x)$, where f is a function, and that the rulings are parallel to the z-axis. As illustrated in Figure 14.37, a point $P(x, y, z)$ is on the cylinder if and only if $Q(x, y, 0)$ is on C; that is, if and only if the first two coordinates x and y of P satisfy the equation $y = f(x)$. It follows that the equation of the cylinder is $y = f(x)$, and hence is the same as the equation of the directrix in the xy-plane.

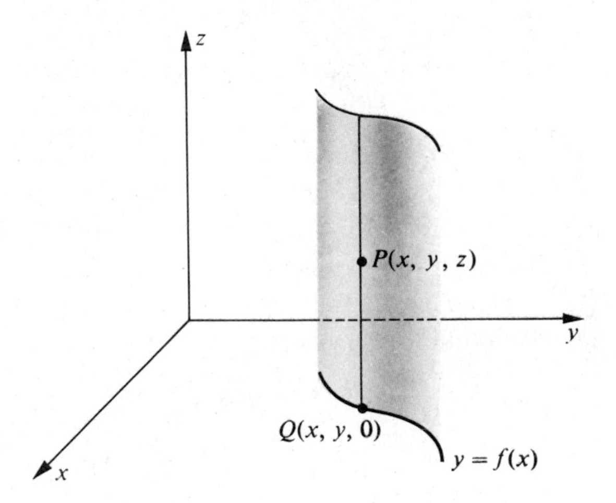

Figure 14.37

Example 2 Sketch the graph of $\dfrac{x^2}{4} + \dfrac{y^2}{9} = 1$.

Solution From our previous remarks, the graph is a cylinder with rulings parallel to the z-axis. We begin by sketching the graph of $x^2/4 + y^2/9 = 1$ in the xy-plane. This ellipse is a directrix for the cylinder. All traces in planes parallel to the xy-plane are ellipses congruent to this directrix. A portion of the graph is shown in Figure 14.38. This surface is called an **elliptic cylinder**.

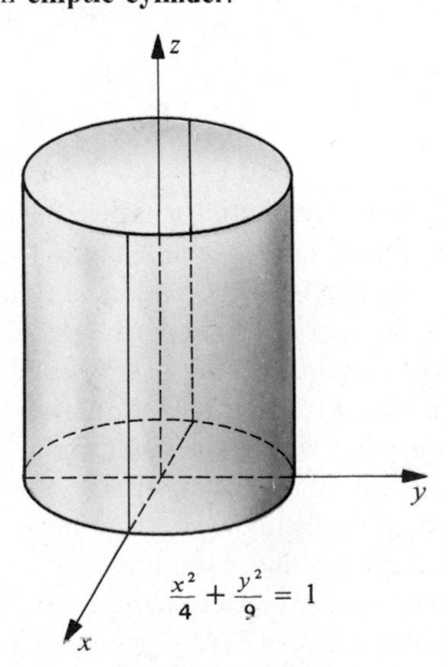

Figure 14.38

It can be shown that the graph of an equation which contains only the variables y and z is a cylinder with rulings parallel to the x-axis and whose trace (directrix) in the yz-plane is the graph of the given equation. Similarly, the graph of an equation which does not contain the variable y is a cylinder with rulings parallel to the y-axis and whose directrix is the graph of the given equation in the xz-plane.

Example 3 Sketch the graphs of the following equations.
(a) $y^2 = 9 - z$ (b) $z = \sin x$

Solution (a) The graph is a cylinder with rulings parallel to the x-axis. A directrix in the yz-plane is the graph of $y^2 = 9 - z$. Part of the graph is sketched in (i) of Figure 14.39. This surface is called a **parabolic cylinder**.

(b) The graph is a cylinder with rulings parallel to the y-axis and whose directrix in the xz-plane is the graph of the equation $z = \sin x$. A portion of the graph is sketched in (ii) of Figure 14.39.

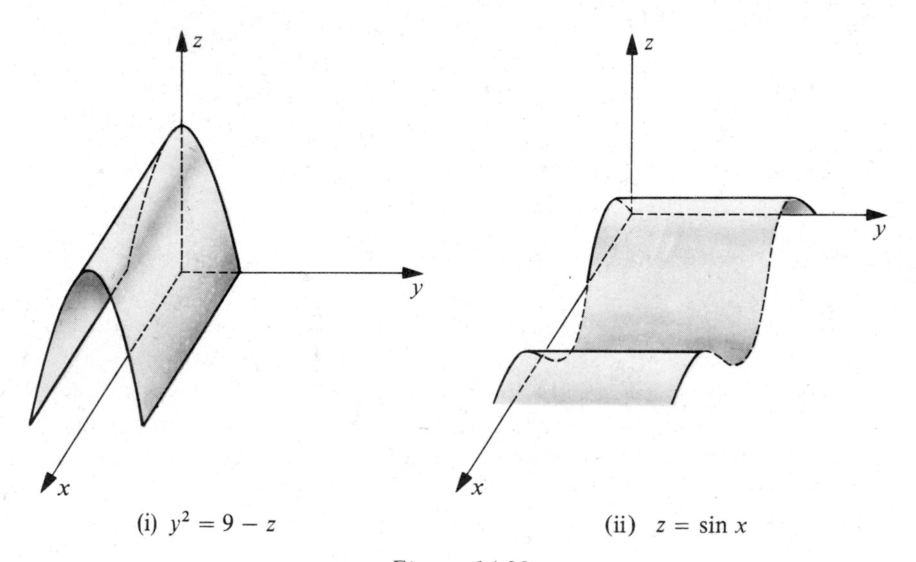

(i) $y^2 = 9 - z$ (ii) $z = \sin x$

Figure 14.39

A surface of revolution was defined in Chapter 13 as the surface generated by revolving a plane curve C about a line in the plane, called the axis of revolution. In the discussion to follow, C will always lie in a coordinate plane and the axis of revolution will be one of the coordinate axes. It will be convenient to use the symbol $f(x, y)$ to denote an algebraic or transcendental expression in the variables x and y. In this case $f(a, b)$ will denote the number obtained by substituting a for x and b for y. This notation will be discussed more thoroughly in Chapter 16.

The graph of the equation $f(x, y) = 0$ in the xy-plane is a curve C. (We are interested here only in the graph in the xy-plane and not in the three-dimensional graph, which is a cylinder.) Suppose, for simplicity, that x and y are nonnegative for all points (x, y) on C and let S denote the surface obtained by revolving C about the y-axis. As illustrated in Figure 14.40, a point $P(x, y, z)$ is on S if and only if $Q(x_1, y, 0)$ is on C, where $x_1 = \sqrt{x^2 + z^2}$. Consequently $P(x, y, z)$ is on S if and only if

$f(\sqrt{x^2 + z^2}, y) = 0$. Thus, to find an equation for S we replace the variable x in the equation for C by $\sqrt{x^2 + z^2}$. Similarly, if the graph of $f(x, y) = 0$ is revolved about the x-axis, then an equation for the resulting surface may be found by replacing y by $\sqrt{y^2 + z^2}$. For some curves which contain points (x, y) where x or y is negative, the preceding discussion can be extended by substituting $\pm\sqrt{x^2 + z^2}$ for x or $\pm\sqrt{y^2 + z^2}$ for y. If, as in the next example, x or y appear only to even powers, then there is no need to make this distinction since the radical disappears when the equation is simplified.

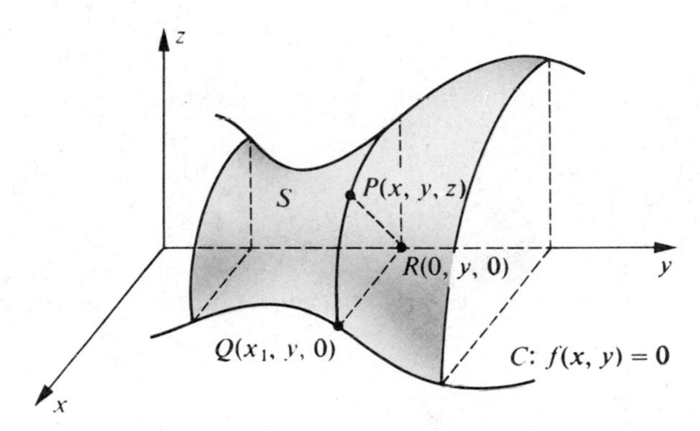

Figure 14.40

Example 4 The graph of $9x^2 + 4y^2 = 36$ is revolved about the y-axis. Find an equation for the resulting surface.

Solution The desired equation may be found by substituting $x^2 + z^2$ for x^2. This gives us

$$9(x^2 + z^2) + 4y^2 = 36.$$

The surface is called an **ellipsoid of revolution**.

A similar discussion can be given for curves which lie in the yz-plane or the xz-plane. For example, if a suitable curve C in the xz-plane is revolved about the z-axis, then an equation for the resulting surface may be found by replacing x by $\sqrt{x^2 + y^2}$. If C is revolved about the x-axis we replace z by $\sqrt{y^2 + z^2}$.

Finally, note that equations for surfaces of revolution are characterized by the fact that two of the variables appear in combinations such as $x^2 + y^2$, $y^2 + z^2$, or $x^2 + z^2$.

Example 5 Describe the graph of the equation $y^2 + z^2 - 4x^2 = 16$ as a surface of revolution.

Solution The given equation may be obtained from the equation $y^2 - 4x^2 = 16$ by substituting $\sqrt{y^2 + z^2}$ for y. Hence the graph is a surface generated by revolving the hyperbola in the xy-plane having equation $y^2 - 4x^2 = 16$ about the x-axis. We could also describe the surface as that generated by revolving the hyperbola in the xz-plane with equation $z^2 - 4x^2 = 16$ about the x-axis.

EXERCISES 14.8

In Exercises 1–10 sketch the graph of the equation in three dimensions.

1 $x^2 + y^2 = 9$

2 $y^2 + z^2 = 16$

3 $4y^2 + 9z^2 = 36$

4 $x^2 + 5z^2 = 25$

5 $x^2 = 9z$

6 $x^2 - 4y = 0$

7 $y^2 - x^2 = 16$

8 $xz = 1$

9 $z = e^y$

10 $z = \log_{10} x$

In Exercises 11–16, find an equation of the surface obtained by revolving the graph of the given equation about the indicated axis. Sketch the graph of the surface.

11 $x^2 + 4y^2 = 16$; y-axis

12 $y^2 = 4x$; x-axis

13 $z = 4 - y^2$; z-axis

14 $z = e^{-y^2}$; y-axis

15 $z^2 - x^2 = 1$; x-axis

16 $xz = 1$; z-axis

In Exercises 17–20, sketch the graph of the given equation.

17 $x^2 + 4y^2 + z^2 = 16$

18 $x^2 - 4y^2 + z^2 = 16$

19 $y - 4z^2 = 4x^2$

20 $x^2 - y^2 + z^2 = 1$

14.9 QUADRIC SURFACES

In Chapter 7 it was shown that in two dimensions the graph of a second-degree equation in x and y is a conic section. In three-dimensional analytic geometry, the graph of a second-degree equation in x, y, and z is referred to as a **quadric surface**. In the present section we shall investigate standard equations for such surfaces.

The graph of

$$\frac{x^2}{a^2} + \frac{y^2}{b^2} + \frac{z^2}{c^2} = 1 \qquad (14.45)$$

where a, b, and c are positive real numbers, is called an **ellipsoid**. Traces of this surface in planes parallel to coordinate planes are ellipses. For example, the trace in the xy-plane is the ellipse with equation $x^2/a^2 + y^2/b^2 = 1$. Similarly, the traces in the yz- and xz-planes are ellipses, as indicated in Figure 14.41. Let us find the trace in an arbitrary plane parallel to the xy-plane, that is, in a plane whose equation is of the form $z = k$. Substituting k for z in (14.45) leads to the equation

$$\frac{x^2}{a^2} + \frac{y^2}{b^2} = 1 - \frac{k^2}{c^2}.$$

If $|k| > c$, then $1 - k^2/c^2 < 0$ and there is no graph. Thus the graph of (14.45) lies between the planes $z = -c$ and $z = c$. If $|k| < c$, then $1 - k^2/c^2 > 0$ and hence the trace in the plane $z = k$ is an ellipse. Similarly, traces in planes parallel to the other two coordinate planes are ellipses, provided they do not intersect the x-axis outside of the open interval $(-a, a)$ or the y-axis outside of $(-b, b)$. Note that if $a = b = c$ then the graph of (14.45) is a sphere of radius a with center at the origin.

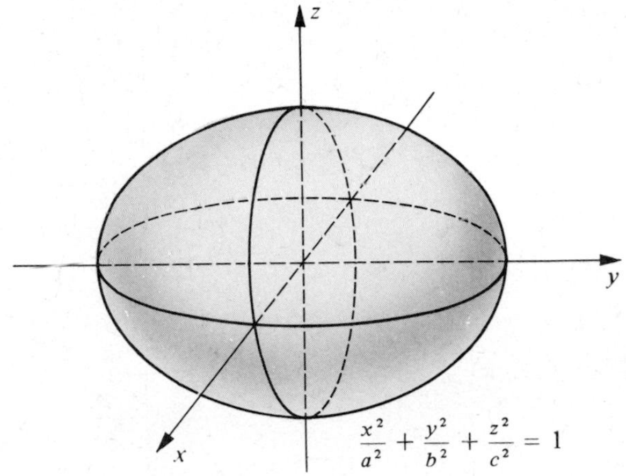

Figure 14.41

The graph of

(14.46)
$$\frac{x^2}{a^2} + \frac{y^2}{b^2} - \frac{z^2}{c^2} = 1$$

is called a **hyperboloid of one sheet**. The traces in the xz- and yz-planes are hyperbolas with equations

$$\frac{x^2}{a^2} - \frac{z^2}{c^2} = 1 \quad \text{and} \quad \frac{y^2}{b^2} - \frac{z^2}{c^2} = 1,$$

respectively. Traces on planes parallel to the xy-plane have equations of the form

$$\frac{x^2}{a^2} + \frac{y^2}{b^2} = 1 + \frac{k^2}{c^2},$$

where k is a real number and, therefore, are ellipses. The graph of (14.46) is sketched in Figure 14.42. The z-axis is called the *axis* of the hyperboloid.

The graph of

$$\frac{x^2}{a^2} - \frac{y^2}{b^2} + \frac{z^2}{c^2} = 1$$

is also a hyperboloid of one sheet; however, in this case the axis of the hyperboloid is the y-axis. If the term involving x^2 is negative and the other terms are positive, then the axis of the hyperboloid coincides with the x-axis.

The graph of

(14.47)
$$\frac{x^2}{a^2} - \frac{y^2}{b^2} - \frac{z^2}{c^2} = 1$$

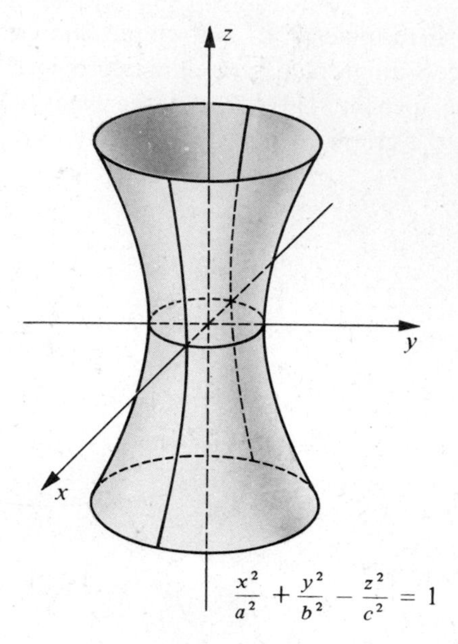

Figure 14.42

is called a **hyperboloid of two sheets**. Traces in the xy- and xz-planes are hyperbolas, whereas traces in planes with equations of the form $x = k$ where $|k| > a$ are ellipses. We leave it to the reader to show that the graph of (14.47) has the appearance shown in Figure 14.43. The x-axis is called the axis of the hyperboloid. By using minus signs on different terms, we can obtain a hyperboloid whose axis is the y- or the z-axis.

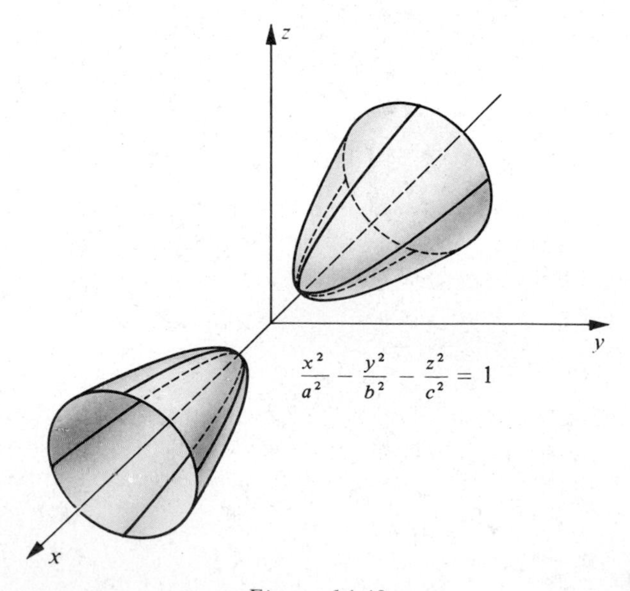

Figure 14.43

The graph of

$$\frac{x^2}{a^2} + \frac{y^2}{b^2} - \frac{z^2}{c^2} = 0 \tag{14.48}$$

is a **cone** which has the z-axis as its axis. The trace in the yz-plane has equation $y^2/b^2 - z^2/c^2 = 0$. Solving for y we obtain $y = \pm(b/c)z$, which gives us the equations of two straight lines through the origin. Similarly, the trace in the xz-plane is a pair of straight lines which intersect at the origin. Traces in planes parallel to the xy-plane are ellipses. (Why?) The graph is sketched in Figure 14.44. By changing signs of the terms in (14.48) we obtain a cone whose axis is either the x- or the y-axis.

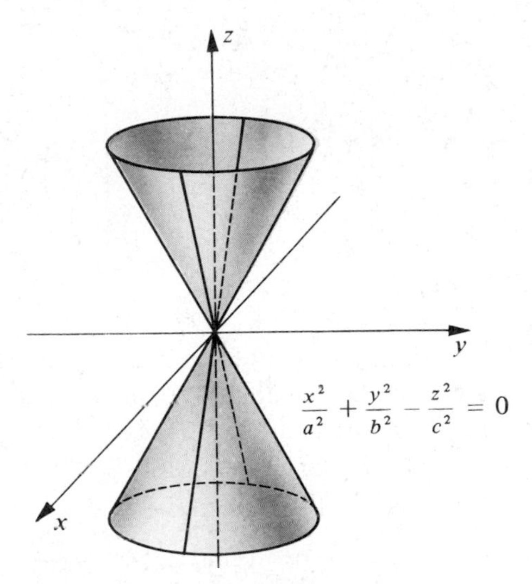

Figure 14.44

The graph of an equation of the form

$$\frac{x^2}{a^2} + \frac{y^2}{b^2} = cz \tag{14.49}$$

is called a **paraboloid**. Example 1 in the previous section is the special case of (14.49) in which $a = b = c = 1$. If $c > 0$, then the graph of (14.49) is similar to that shown in Figure 14.35, except that if $a \neq b$, then traces in planes parallel to the xy-plane are ellipses instead of circles. If $c < 0$, then the paraboloid opens downward. The z-axis is called the *axis* of the paraboloid. The graphs of the equations

$$\frac{x^2}{a^2} + \frac{z^2}{b^2} = cy \quad \text{and} \quad \frac{y^2}{a^2} + \frac{z^2}{b^2} = cx$$

are paraboloids whose axes are the y-axis and x-axis, respectively.

Finally, the graph of the equation

$$\frac{y^2}{a^2} - \frac{x^2}{b^2} = cz \tag{14.50}$$

is called a **hyperbolic paraboloid**. A typical sketch of this saddle-shaped surface for the case $c > 0$ is shown in Figure 14.45. Variations are obtained by interchanging x, y, and z in (14.50).

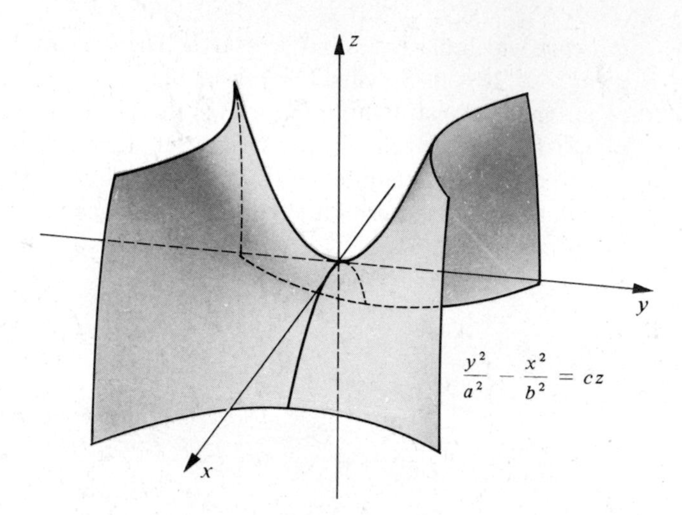

Figure 14.45

It is possible to obtain formulas for translation or rotation of axes in three dimensions which are analogous to those in two dimensions. These can be used to show that the graph of an equation of degree two in x, y, and z, is one of the surfaces discussed in this chapter, except for degenerate cases.

Example 1 Sketch the graph of $16x^2 - 9y^2 + 36z^2 = 144$ and identify the surface.

Solution Dividing both sides of the equation by 144 leads to

$$\frac{x^2}{9} - \frac{y^2}{16} + \frac{z^2}{4} = 1.$$

The trace in the xy-plane is the hyperbola $x^2/9 - y^2/16 = 1$, with vertices at $(\pm 3, 0, 0)$. The trace in the yz-plane is another hyperbola $z^2/4 - y^2/16 = 1$, with vertices at $(0, 0, \pm 2)$. The trace in the xz-plane is the ellipse $x^2/9 + z^2/4 = 1$. These traces are illustrated in Figure 14.46. To find traces on planes which are parallel to the xz-plane we substitute $y = k$ in the given equation, obtaining

$$\frac{x^2}{9} + \frac{z^2}{4} = 1 + \frac{k^2}{16}.$$

This shows that every cross section in a plane parallel to the xz-plane is an ellipse. For example, if $k = \pm 8$ we obtain $x^2/9 + z^2/4 = 5$ or, equivalently, $x^2/45 + z^2/20 = 1$. These traces are illustrated in Figure 14.46. Traces on planes parallel to the xy-plane or the yz-plane may be found by substituting $z = k$ or $x = k$, respectively, in the given equation. We leave it to the reader to verify that in either case the trace is a hyperbola. The surface is a hyperboloid of one sheet with the y-axis as its axis.

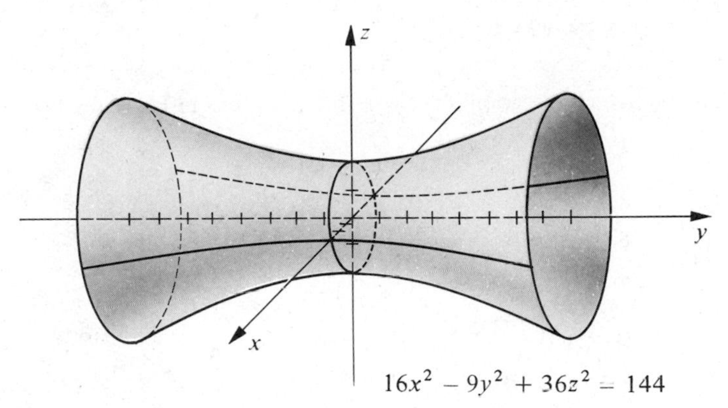

Figure 14.46

Example 2 Sketch the graph of $y^2 + 4z^2 = x$ and identify the surface.

Solution The trace in the xy-plane is the parabola $y^2 = x$. The trace in the xz-plane is the parabola $4z^2 = x$. These traces are illustrated in Figure 14.47. The trace in the yz-plane is the graph of $y^2 + 4z^2 = 0$ and hence consists of only one point, the origin. To find traces in planes parallel to the yz-plane we let $x = k$ in the given equation, obtaining

$$y^2 + 4z^2 = k.$$

If $k < 0$ there is no graph. (Why?) If $k > 0$ the trace is an ellipse. For example, letting $k = 9$ we obtain

$$\frac{y^2}{9} + \frac{z^2}{(9/4)} = 1$$

which is an ellipse with semi-axes of lengths 3 and 3/2, as illustrated in Figure 14.47. The reader should verify that traces on planes parallel to either the xz- or xy-planes are parabolas. The surface is a paraboloid having the x-axis as its axis.

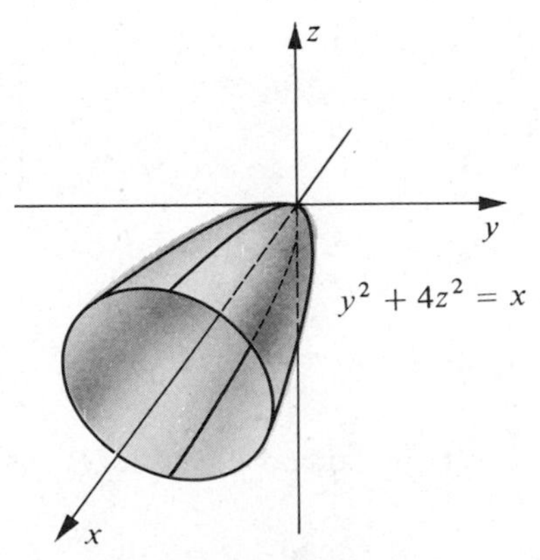

Figure 14.47

EXERCISES 14.9

Sketch the graph of each of the equations in Exercises 1–20 and identify the surface.

1 $4x^2 + 9y^2 = 36z$

2 $8x^2 + 4y^2 + z^2 = 16$

3 $16x^2 + 100y^2 - 25z^2 = 400$

4 $25x^2 - 225y^2 + 9z^2 = 225$

5 $3x^2 - 4y^2 - z^2 = 12$

6 $4x^2 + y^2 = 9z^2$

7 $x^2 - 16y^2 = 4z^2$

8 $4y^2 - 25z^2 = 100x$

9 $9x^2 + 4y^2 + z^2 = 36$

10 $16x^2 - 25y^2 + 100z^2 = 200$

11 $16x^2 - 4y^2 - z^2 + 1 = 0$

12 $36x = 9y^2 + z^2$

13 $y^2 - 9x^2 - z^2 - 9 = 0$

14 $z^2 - x^2 - y^2 = 1$

15 $4y^2 + 25z^2 + 100x = 0$

16 $16y = x^2 + 4z^2$

17 $36x^2 - 16y^2 + 9z^2 = 0$

18 $4y^2 + 9z^2 = 9x^2$

19 $4y = x^2 - z^2$

20 $4x^2 + 16y = z^2$

14.10 CYLINDRICAL AND SPHERICAL COORDINATE SYSTEMS

The system of polar coordinates can be extended to three dimensions in several ways. The simplest technique is to represent a point P by an ordered triple (r, θ, z) where z is the usual (third) rectangular coordinate of P and (r, θ) are polar coordinates for the projection P' of P onto the xy-plane (see Figure 14.48).

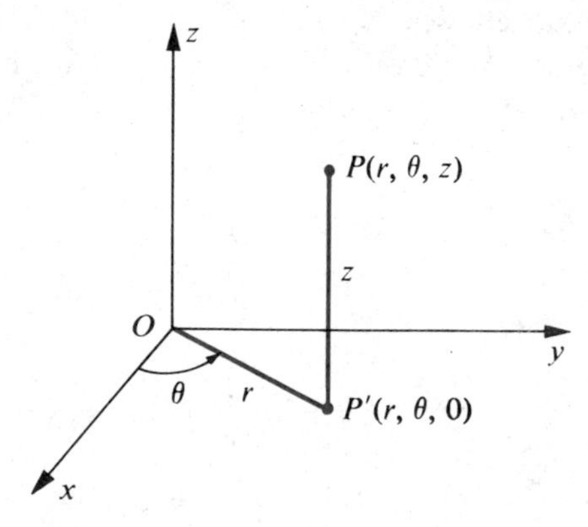

Figure 14.48

If the rectangular coordinates of P are (x, y, z), then we may use the formulas

(14.51)
$$x = r\cos\theta, \quad y = r\sin\theta, \quad \tan\theta = \frac{y}{x}, \quad r^2 = x^2 + y^2$$

to change from one system of coordinates to the other (see (13.6) and (13.7)). If $k > 0$, then the graph of the equation $r = k$, or equivalently of $x^2 + y^2 = k^2$, is a circular cylinder of radius k with axis along the z-axis. It is for this reason that the system of coordinates we have described is called the **cylindrical coordinate system.** Note that the graph of $\theta = k$ is a plane containing the z-axis, whereas the graph of $z = k$ is a plane perpendicular to the z-axis.

Example 1 Find a rectangular equation for, and describe the graph of, each of the following equations.
(a) $z = r^2$ (b) $r = 4\sin\theta$

Solution (a) By (13.7), a rectangular equation for the graph of $z = r^2$ is $z = x^2 + y^2$. The graph is the paraboloid sketched in Figure 14.35.
(b) As in Example 4 of Section 13.3, a rectangular equation for the graph of $r = 4\sin\theta$ is $x^2 + y^2 = 4y$. The graph is a cylinder with rulings parallel to the z-axis. The directrix of the cylinder is a circle of radius 2 in the xy-plane whose center in rectangular coordinates is $(0, 2, 0)$.

Example 2 Find an equation in cylindrical coordinates for, and describe the graph of, each of the following equations.
(a) $z^2 = x^2 - y^2$ (b) $z^2 = x^2 + y^2$

Solution (a) Applying the first two equations in (14.51),

$$\begin{aligned} z^2 &= r^2\cos^2\theta - r^2\sin^2\theta \\ &= r^2(\cos^2\theta - \sin^2\theta) \end{aligned}$$

or

$$z^2 = r^2\cos 2\theta.$$

The graph is a circular cone with axis along the x-axis.
(b) By the last equation in (14.51),

$$z^2 = r^2 \quad \text{or} \quad z = r.$$

The graph is a circular cone with axis along the z-axis.

A system called **spherical coordinates** may also be introduced in three dimensions. In this system a point P different from the origin is represented by an ordered triple (ρ, θ, ϕ), where $\rho = |\overrightarrow{OP}|$, θ is a polar angle associated with the projection P' of P onto the xy-plane, and ϕ is the angle between the positive z-axis and $\overrightarrow{OP}$ (see (i) of Figure 14.49). The origin is represented by any triple of the form $(0, \theta, \phi)$. The term *spherical* arises from the fact that the graph of the equation $\rho = k$, where $k > 0$, is a sphere with center at O. As in cylindrical coordinates, the graph of $\theta = k$ is a half-plane containing the z-axis. The graph of $\phi = k$ is usually a half-cone with vertex at O.

The relationship between spherical coordinates (ρ, θ, ϕ) and rectangular coordinates (x, y, z) of a point P can be found by referring to (ii) of Figure 14.49. Using the fact that

$$x = |\overrightarrow{OP}'|\cos\theta \quad \text{and} \quad y = |\overrightarrow{OP}'|\sin\theta$$

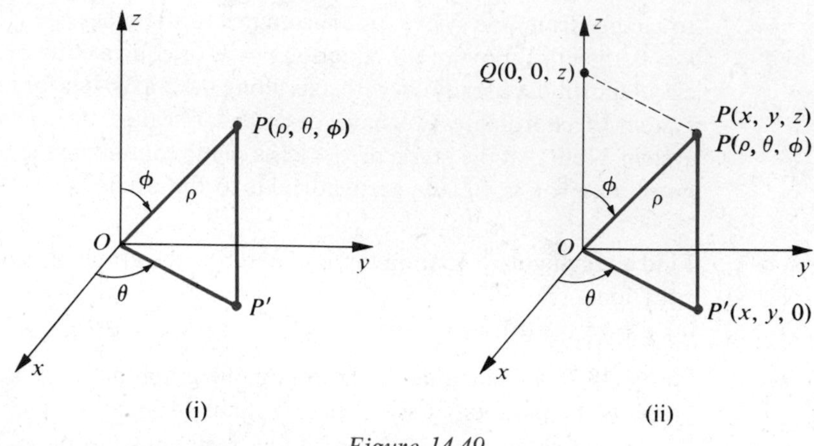

Figure 14.49

together with

$$|\overrightarrow{OP}'| = |\overrightarrow{QP}| = \rho \sin \phi \quad \text{and} \quad |\overrightarrow{OQ}| = \rho \cos \phi$$

we obtain

(14.52) $$x = \rho \sin \phi \cos \theta, \quad y = \rho \sin \phi \sin \theta, \quad z = \rho \cos \phi.$$

It is also evident, from the Distance Formula, that

(14.53) $$\rho^2 = x^2 + y^2 + z^2.$$

Example 3 Find a rectangular equation for $\rho = 2 \sin \phi \cos \theta$ and describe the graph of this equation.

Solution Multiplying both sides of the given equation by ρ we obtain $\rho^2 = 2\rho \sin \phi \cos \theta$. Applying (14.53) and (14.52) gives us $x^2 + y^2 + z^2 = 2x$ or, equivalently, $(x - 1)^2 + y^2 + z^2 = 1$. By (14.13), the graph is a sphere of radius 1 whose center in rectangular coordinate is (1, 0, 0).

Example 4 If a point P has spherical coordinates $(4, \pi/3, \pi/6)$, find rectangular and cylindrical coordinates for P.

Solution Using (14.52) with $\rho = 4$, $\theta = \pi/3$, and $\phi = \pi/6$ we obtain the rectangular coordinates.

$$x = 4 \sin \frac{\pi}{6} \cos \frac{\pi}{3} = 4\left(\frac{1}{2}\right)\left(\frac{1}{2}\right) = 1$$

$$y = 4 \sin \frac{\pi}{6} \sin \frac{\pi}{3} = 4\left(\frac{1}{2}\right)\left(\frac{\sqrt{3}}{2}\right) = \sqrt{3}$$

$$z = 4 \cos \frac{\pi}{6} = 4\left(\frac{\sqrt{3}}{2}\right) = 2\sqrt{3}.$$

To find the cylindrical coordinates we note that $r^2 = x^2 + y^2 = 1 + 3 = 4$. Hence cylindrical coordinates for P are (r, θ, z) or $(2, \pi/3, 2\sqrt{3})$.

EXERCISES 14.10

1 Change the following from cylindrical coordinates to rectangular coordinates.
(a) $(5, \pi/2, 3)$ (b) $(6, \pi/3, -5)$

2 Change the following from spherical coordinates to rectangular coordinates.
(a) $(4, \pi/2, \pi/6)$ (b) $(1, 2\pi/3, 3\pi/4)$

3 Change the following from rectangular coordinates to spherical coordinates.
(a) $(1, 1, -2\sqrt{2})$ (b) $(1, \sqrt{3}, 0)$

4 Change the following from rectangular coordinates to cylindrical coordinates.
(a) $(2\sqrt{3}, 2, -2)$ (b) $(\sqrt{2}, -\sqrt{2}, 1)$

5 Change the following from spherical coordinates to cylindrical coordinates.
(a) $(4, \pi/3, \pi/3)$ (b) $(2, \pi/4, 5\pi/6)$

6 Change the following from cylindrical coordinates to spherical coordinates.
(a) $(\sqrt{2}, \pi/4, 1)$ (b) $(3, \pi/3, 1)$

In each of Exercises 7–20 describe the graph of the equation.

7 $r = 4$

8 $\theta = \pi/4$

9 $\phi = \pi/6$

10 $\rho = 3$

11 $r = 4\cos\theta$

12 $r = \sin 2\theta$

13 $\rho = 4\cos\phi$

14 $\rho = \sec\phi$

15 $\phi = 0$

16 $\phi = \pi/2$

17 $r^2 = \cos 2\theta$

18 $z = 1 - r^2$

19 $\rho^2 - 3\rho + 2 = 0$

20 $\phi^2 - \pi\phi + (3\pi^2/16) = 0$

Find equations in cylindrical coordinates and in spherical coordinates for the graphs of the equations in Exercises 21–26.

21 $x^2 + y^2 + z^2 = 4$

22 $x^2 + y^2 = 4z$

23 $3x + y - 4z = 12$

24 $x^2 - y^2 - z^2 = 1$

25 $y^2 + z^2 = 9$

26 $x^2 + z^2 = 9$

14.11 REVIEW

Concepts

Define or discuss each of the following.

1 Scalar

2 Vectors in V_2 and V_3

3 Components of a vector

4 Addition of vectors

5 Scalar multiples of vectors

6 Zero vector

7 Position vector

8 Magnitude of a vector

9 Unit vector

10 Direction angles and direction cosines

11 The dot product and its properties

12 The vector product and its properties

13 The angle between two vectors

14 Parallel vectors

15 Orthogonal vectors

16 Cauchy-Schwarz Inequality

17 Triangle inequality

18 Work done by a force

19 Equations of a line

20 Equations of a plane

21 Cylinders

22 Quadric surfaces

23 Spherical coordinates

24 Cylindrical coordinates

Exercises

Find the vectors or scalars indicated in Exercises 1–10 if $\mathbf{a} = 2\mathbf{i} + 5\mathbf{j}$ and $\mathbf{b} = 4\mathbf{i} - \mathbf{j}$.

1 $4\mathbf{a} + \mathbf{b}$

2 $2\mathbf{a} - 3\mathbf{b}$

3 $(\mathbf{a} + \mathbf{b}) \cdot (\mathbf{a} - 2\mathbf{b})$

4 $3\mathbf{a} \cdot (5\mathbf{b} + \mathbf{i})$

5 $|\mathbf{a}| - |\mathbf{b}|$

6 $|\mathbf{a} - \mathbf{b}|$

7 The angle between **a** and **i**

8 A unit vector having the same direction as **b**

9 A unit vector orthogonal to **a**

10 The cosine of the angle between **a** and **b**

11 Given the points $A(5, -3, 2)$ and $B(-1, -4, 3)$, find the following.
(a) $d(A, B)$
(b) The coordinates of the midpoint of the line segment AB
(c) An equation of the sphere with center B and tangent to the xz-plane
(d) An equation of the plane through B parallel to the xz-plane
(e) Parametric equations for the line through A and B
(f) An equation of the plane through A with normal vector $\overrightarrow{AB}$

12 Find an equation for the plane through $A(0, 4, 9)$ and $B(0, -3, 7)$ which is perpendicular to the yz-plane.

13 Find an equation for the plane with x-intercept 5, y-intercept -2, and z-intercept 6.

14 Find an equation for the cylinder which is perpendicular to the xy-plane and has, for its directrix, the circle in the xy-plane with center $C(4, -3, 0)$ and radius 5.

15 Find an equation for an ellipsoid with center O which has x-intercept 8, y-intercept 3, and z-intercept 1.

16 Find an equation for the surface obtained by revolving the graph of the equation $z = x$ about the z-axis.

In each of Exercises 17–27, sketch the graph of the given equation.

17 $x^2 + y^2 + z^2 - 14x + 6y - 8z + 10 = 0$

18 $4y - 3z - 15 = 0$

19 $3x - 5y + 2z = 10$

20 $y = z^2 + 1$

21 $9x^2 + 4z^2 = 36$

22 $x^2 + 4y + 9z^2 = 0$

23 $z^2 - 4x^2 = 9 - 4y^2$

24 $2x^2 + 4z^2 - y^2 = 0$

25 $z^2 - 4x^2 - y^2 = 4$

26 $x^2 + 2y^2 + 4z^2 = 16$

27 $x^2 - 4y^2 = 4z$

If $\mathbf{a} = 3\mathbf{i} - \mathbf{j} - 4\mathbf{k}$, $\mathbf{b} = 2\mathbf{i} + 5\mathbf{j} - 2\mathbf{k}$, and $\mathbf{c} = -\mathbf{i} + 6\mathbf{k}$, find the vectors or scalars in Exercises 28–44.

28 $3\mathbf{a} - 2\mathbf{b}$

29 $\mathbf{a} \cdot (\mathbf{b} - \mathbf{c})$

30 $|\mathbf{b} + \mathbf{c}|$

31 $|\mathbf{b}| + |\mathbf{c}|$

32 A unit vector having the same direction as **a**

33 The direction cosines of **a**

34 The cosine of the angle between **a** and **c**

35 $\mathbf{a} \times \mathbf{b}$

36 $\mathbf{a} \cdot (\mathbf{b} \times \mathbf{c})$

37 $\text{comp}_{\mathbf{b}}\, \mathbf{a}$

38 $\text{comp}_{\mathbf{a}}\, (\mathbf{b} \times \mathbf{c})$

39 $\mathbf{a} \cdot \mathbf{a}$

40 $\mathbf{a} \times \mathbf{a}$

41 $(\mathbf{a} \times \mathbf{c}) + (\mathbf{c} \times \mathbf{a})$

42 $\mathbf{a} \times (\mathbf{b} + \mathbf{c})$

43 A vector having the opposite direction of **b** and twice the magnitude of **b**

44 Two unit vectors orthogonal to both **b** and **c**

45 Given the points $P(2, -1, 1)$, $Q(-3, 2, 0)$, and $R(4, -5, 3)$ find the following.
(a) The direction cosines of $\overrightarrow{PR}$
(b) A unit normal vector orthogonal to the plane determined by P, Q, and R
(c) An equation for the plane determined by P, Q, and R
(d) Parametric equations for a line through P which is parallel to the line through Q and R
(e) $\overrightarrow{QP} \cdot \overrightarrow{QR}$
(f) The angle between $\overrightarrow{QP}$ and $\overrightarrow{QR}$
(g) The area of the triangle with vertices P, Q, and R

46 Find the angle between the two lines $(x - 3)/2 = (y + 1)/(-4) = (z - 5)/8$ and $(x + 1)/7 = (6 - y)/2 = (2z + 7)/(-4)$.

47 Find parametric equations for each of the lines in Exercise 46.

48 Determine whether the following lines intersect and, if so, find the point of intersection:

$$x = 2 + t, y = 1 + t, z = 4 + 7t; x = -4 + 5t, y = 2 - 2t, z = 1 - 4t.$$

49 Find the angle between the lines in Exercise 48.

50 The position of a particle at time t is $(t, t \sin t, t \cos t)$. Find the distance the particle moves during the time interval $[0, 5]$.

51 If the rectangular coordinates of a point P are $(2, -2, 1)$, find cylindrical and spherical coordinates for P.

52 If spherical coordinates of a point P are $(12, 3\pi/4, \pi/6)$, find cylindrical and rectangular coordinates for P.

Find a rectangular equation and describe the graph of each equation in Exercises 53–56.

53 $\phi = 3\pi/4$

54 $r = \cos 2\theta$

55 $\rho \sin \phi \cos \theta = 1$

56 $\rho^2 - 3\rho = 0$

Find equations in cylindrical coordinates and in spherical coordinates for the graphs of the equations in each of Exercises 57–60.

57 $x^2 + y^2 = 1$

58 $z = x^2 - y^2$

59 $x^2 + y^2 + z^2 - 2z = 0$

60 $2x + y - 3z = 4$

Vector-Valued Functions

In this chapter we shall extend the theory of limits, derivatives, and integrals to functions whose ranges consist of vectors instead of real numbers. Our principal application of these concepts will be to the study of motion in two or three dimensions. Further applications will be given in later chapters.

15.1 DEFINITIONS

A function was defined in (1.18) as a correspondence that associates with each element of a set X a unique element of a set Y. If X is a set of real numbers and Y is a set of vectors, then the function will be called a **vector-valued function** and denoted by symbols such as $\mathbf{r}$ or $\vec{r}$. Thus, if $\mathbf{r}$ is a vector-valued function, then for each real number t in the domain X of $\mathbf{r}$ there exists a unique vector $\mathbf{r}(t) = \langle x, y, z \rangle$. If, with each t, we associate the first component of $\mathbf{r}(t)$, we obtain a function f from X to $\mathbb{R}$, where $x = f(t)$. Similarly, there are functions g and h from X to $\mathbb{R}$ such that $y = g(t)$ and $z = h(t)$, respectively. Hence we may write

$$\mathbf{r}(t) = \langle f(t), g(t), h(t) \rangle = f(t)\mathbf{i} + g(t)\mathbf{j}, + h(t)\mathbf{k} \tag{15.1}$$

for every number t in X.

Conversely, if f, g, and h are functions from X to $\mathbb{R}$, then we may define a vector-valued function $\mathbf{r}$ by means of (15.1). Consequently, $\mathbf{r}$ *is a vector-valued function if and only if* $\mathbf{r}(t)$ *is expressible in the form* (15.1). If it is not stated explicitly, the domain of $\mathbf{r}$ is assumed to be the intersection of the domains of f, g, and h.

Example 1 If $\mathbf{r}(t) = 3t^2\mathbf{i} + 2t^3\mathbf{j} + (t - 2)\mathbf{k}$, find $\mathbf{r}(1)$, $\mathbf{r}(2)$, and $\mathbf{r}(0)$.

Solution The specified vectors may be found by substituting the appropriate numbers for t. Thus

$$\mathbf{r}(1) = 3\mathbf{i} + 2\mathbf{j} - \mathbf{k}$$
$$\mathbf{r}(2) = 12\mathbf{i} + 16\mathbf{j} + 0\mathbf{k} = 12\mathbf{i} + 16\mathbf{j}$$
$$\mathbf{r}(0) = 0\mathbf{i} + 0\mathbf{j} - 2\mathbf{k} = -2\mathbf{k}.$$

There is a close connection between vector-valued functions and curves in space. As in Definition (13.1) for plane curves, we define a **space curve**, or a **curve in**

three dimensions, as a set C of ordered triples of the form

$$(f(t), g(t), h(t))$$

where the functions f, g, and h are continuous on an interval I. The graph of C in a rectangular coordinate system is the set of all points $P(f(t), g(t), h(t))$ which correspond to the ordered triples in C. The equations

$$x = f(t), \quad y = g(t), \quad z = h(t)$$

where t is in I, are called **parametric equations** for C. The notions of **end-points** of a curve, **closed curve**, **simple closed curve**, and **smooth curve** have the same meanings as in the two-dimensional case. If $I = [a, b]$ and C does not intersect itself, except possibly for the case $P(a) = P(b)$, then it can be shown that the **length** L of C from $P(a)$ to $P(b)$ is given by

$$\text{(15.2)} \qquad L = \int_a^b \sqrt{[f'(t)]^2 + [g'(t)]^2 + [h'(t)]^2}\, dt = \int_a^b \sqrt{\left(\frac{dx}{dt}\right)^2 + \left(\frac{dy}{dt}\right)^2 + \left(\frac{dz}{dt}\right)^2}\, dt.$$

The reader should compare these formulas with (13.17) and (13.18).

Let us now consider a vector function $\mathbf{r}$, where $\mathbf{r}(t) = \langle f(t), g(t), h(t)\rangle$. As illustrated in Figure 15.1, if $\overrightarrow{OP}$ is the position vector corresponding to $\mathbf{r}(t)$, then as t varies, the terminal point P traces the curve with parametric equations $x = f(t)$, $y = g(t)$, $z = h(t)$. This geometric representation of $\mathbf{r}(t)$ is especially useful if t is time and P is the position of a particle in motion.

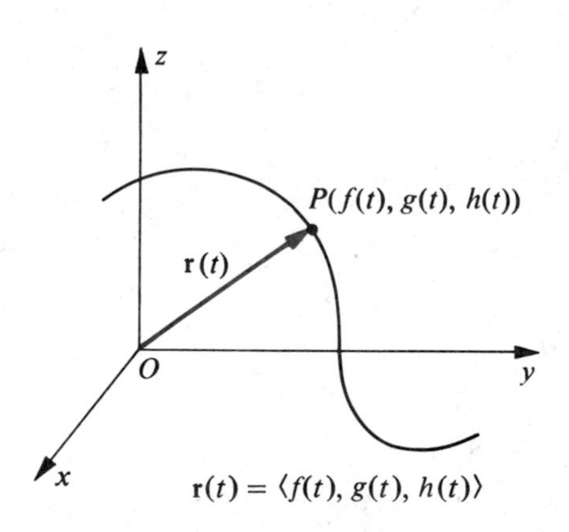

Figure 15.1

Example 2 Suppose $\mathbf{r}(t) = a\cos t\mathbf{i} + a\sin t\mathbf{j} + bt\mathbf{k}$, where t is in $\mathbb{R}$, and let $\overrightarrow{OP}$ be the position vector of $\mathbf{r}(t)$.

(a) Sketch the graph of the curve C traced by P as t varies.

(b) Find the length of C between the points corresponding to $t = 0$ and $t = 2\pi$.

Solution (a) Parametric equations for C are $x = a\cos t$, $y = a\sin t$, $z = bt$. If the point $P(x, y, z)$ is on C, then P is on the circular cylinder having equation $x^2 + y^2 = a^2$. (Why?) As t varies from 0 to 2π, the point P starts at $(a, 0, 0)$ and winds around this cylinder one time. Other intervals of length 2π lead to similar loops (see Figure 15.2). This curve is called a **circular helix**.

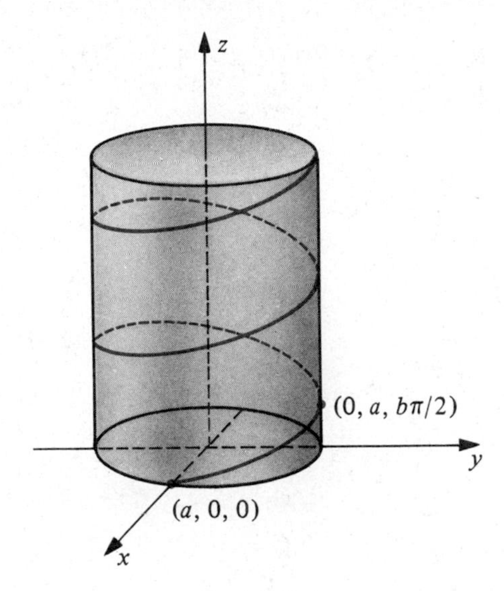

Figure 15.2

(b) The length of C between the points corresponding to $t = 0$ and $t = 2\pi$ can be found by means of (15.2). Thus

$$L = \int_0^{2\pi} \sqrt{a^2 \sin^2 t + a^2 \cos^2 t + b^2}\, dt$$

$$= \int_0^{2\pi} \sqrt{a^2 + b^2}\, dt = \sqrt{a^2 + b^2}\, t \Big]_0^{2\pi} = \sqrt{a^2 + b^2}\, 2\pi.$$

Example 3 If $\mathbf{r}(t) = (2t + 1)\mathbf{i} + (3t - 4)\mathbf{j} - 5t\mathbf{k}$, and if $\overrightarrow{OP}$ is the position vector of $\mathbf{r}(t)$, describe the graph of the curve C traced by P as t varies.

Solution Parametric equations for C are

$$x = 2t + 1, \quad y = 3t - 4, \quad z = -5t.$$

By (14.29) the graph is a line through the point $(1, -4, 0)$ which is parallel to the vector $\langle 2, 3, -5 \rangle$.

As in the preceding chapter, we may regard V_2 as the subset of V_3 consisting of all vectors whose third component is 0. In this case (15.1) takes on the form

$$\mathbf{r}(t) = f(t)\mathbf{i} + g(t)\mathbf{j}$$

and as t varies the curve traced by the terminal point of the position vector $\overrightarrow{OP}$ lies in the xy-plane. In the next section we shall consider the case in which t is time and P is the position of a moving point.

EXERCISES 15.1

In each of Exercises 1–12 sketch the graph of the curve C traced by the end-point of the position vector of $\mathbf{r}(t)$ if t varies as indicated.

1 $\mathbf{r}(t) = \langle t, 4\cos t, 9\sin t\rangle;\ t \geq 0$

2 $\mathbf{r}(t) = \langle \tan t, \sec t, 2\rangle;\ -\pi/2 < t < \pi/2$

3 $\mathbf{r}(t) = \langle t, t^2, t^3\rangle;\ t$ in $\mathbb{R}$

4 $\mathbf{r}(t) = \langle t^3, t^2, t\rangle;\ 0 \leq t \leq 4$

5 $\mathbf{r}(t) = (t^2 + 1)\mathbf{i} + t\mathbf{j} + 3\mathbf{k};\ t$ in $\mathbb{R}$

6 $\mathbf{r}(t) = 6\sin t\mathbf{i} + 4\mathbf{j} + 25\cos t\mathbf{k};\ -2\pi \leq t \leq 2\pi$

7 $\mathbf{r}(t) = t\mathbf{i} + t\mathbf{j} + \sin t\mathbf{k};\ t$ in $\mathbb{R}$

8 $\mathbf{r}(t) = t\mathbf{i} + 2t\mathbf{j} + e^t\mathbf{k};\ t$ in $\mathbb{R}$

9 $\mathbf{r}(t) = \langle 3t, 1 - 9t^2\rangle;\ t$ in $\mathbb{R}$

10 $\mathbf{r}(t) = \langle 1 - t^3, t\rangle;\ t \geq 0$

11 $\mathbf{r}(t) = e^t\cos t\mathbf{i} + e^t\sin t\mathbf{j};\ 0 \leq t \leq \pi$

12 $\mathbf{r}(t) = 2\cosh t\mathbf{i} + 3\sinh t\mathbf{j};\ t$ in $\mathbb{R}$

In each of Exercises 13–18 find the length of the given curve.

13 $C = \{(5t, 4t^2, 3t^2): 0 \leq t \leq 2\}$

14 $C = \{(t^2, t\sin t, t\cos t): 0 \leq t \leq 1\}$

15 $x = e^t\cos t,\ y = e^t,\ z = e^t\sin t;\ 0 \leq t \leq 2\pi$

16 $x = 2t,\ y = 4\sin 3t,\ z = 4\cos 3t;\ 0 \leq t \leq 2\pi$

17 $x = 3t^2,\ y = t^3,\ z = 6t;\ 0 \leq t \leq 1$

18 $x = 1 - 2t^2,\ y = 4t,\ z = 3 + 2t^2;\ 0 \leq t \leq 2$

15.2 DERIVATIVES AND INTEGRALS OF VECTOR-VALUED FUNCTIONS

Limits of vector-valued functions may be defined as follows.

(15.3) Definition

If $\mathbf{r}(t) = \langle f(t), g(t), h(t)\rangle$ then

$$\lim_{t\to a}\mathbf{r}(t) = \langle \lim_{t\to a} f(t), \lim_{t\to a} g(t), \lim_{t\to a} h(t)\rangle$$

provided f, g, and h have limits as t approaches a.

If $\mathbf{r}(t) = f(t)\mathbf{i} + g(t)\mathbf{j} + h(t)\mathbf{k}$, then we write

$$\lim_{t\to a}\mathbf{r}(t) = \left[\lim_{t\to a} f(t)\right]\mathbf{i} + \left[\lim_{t\to a} g(t)\right]\mathbf{j} + \left[\lim_{t\to a} h(t)\right]\mathbf{k}.$$

The preceding definition may be extended to one-sided limits as was done in Chapter 2. If, in (15.3),

$$\lim_{t \to a} f(t) = a_1, \quad \lim_{t \to a} g(t) = a_2, \quad \text{and} \quad \lim_{t \to a} h(t) = a_3$$

then

$$\lim_{t \to a} \mathbf{r}(t) = \langle a_1, a_2, a_3 \rangle = \mathbf{a}.$$

Intuitively, we have a situation of the type illustrated in Figure 15.3, where if $\overrightarrow{OP}$ and $\overrightarrow{OA}$ are position vectors corresponding to $\mathbf{r}(t)$ and $\mathbf{a}$, respectively, then as t approaches a, $\overrightarrow{OP}$ approaches $\overrightarrow{OA}$, in the sense that P gets closer and closer to A. Theorems on limits of vector-valued functions similar to those in Chapter 2 can be established. These are left as exercises. Limits of vector-valued functions may also be formulated using an ε-δ approach similar to that in Definition (2.6) (see Exercise 35).

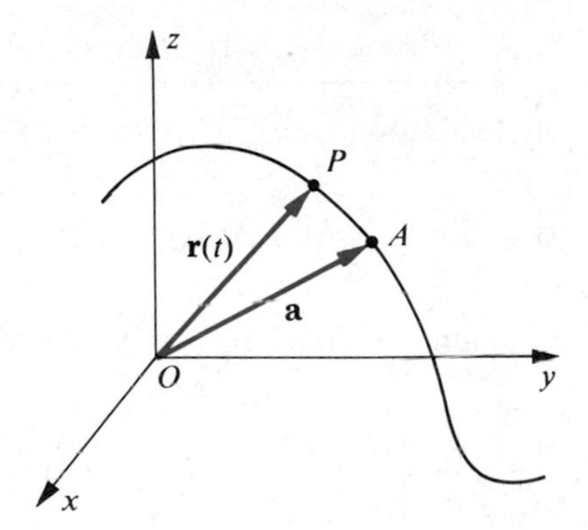

Figure 15.3

(15.4) Definition

> A vector-valued function $\mathbf{r}$ is **continuous** at a if
>
> $$\lim_{t \to a} \mathbf{r}(t) = \mathbf{r}(a).$$

It follows that if $\mathbf{r}(t) = \langle f(t), g(t), h(t) \rangle$, then $\mathbf{r}$ is continuous at a if and only if f, g, and h are continuous at a. Continuity in an interval is defined in the usual way.

(15.5) Definition

> If $\mathbf{r}$ is a vector-valued function, then the **derivative** of $\mathbf{r}$ is the vector-valued function $\mathbf{r}'$ defined by
>
> $$\mathbf{r}'(t) = \lim_{\Delta t \to 0} \frac{1}{\Delta t}[\mathbf{r}(t + \Delta t) - \mathbf{r}(t)]$$
>
> for all t such that the limit exists.

If we adopt the convention that

$$\frac{1}{\Delta t}[\mathbf{r}(t + \Delta t) - \mathbf{r}(t)] = \frac{\mathbf{r}(t + \Delta t) - \mathbf{r}(t)}{\Delta t}$$

then Definition (15.5) takes on the familiar form for derivatives introduced in Chapter 3.

The next result shows that to find $\mathbf{r}'(t)$ we may differentiate each component of $\mathbf{r}(t)$.

(15.6) Theorem

> If
>
> $$\mathbf{r}(t) = \langle f(t), g(t), h(t) \rangle = f(t)\mathbf{i} + g(t)\mathbf{j} + h(t)\mathbf{k}$$
>
> where f, g, and h are differentiable, then
>
> $$\mathbf{r}'(t) = \langle f'(t), g'(t), h'(t) \rangle = f'(t)\mathbf{i} + g'(t)\mathbf{j} + h'(t)\mathbf{k}.$$

Proof. By Definition (15.5),

$$\begin{aligned}
\mathbf{r}'(t) &= \lim_{\Delta t \to 0} \frac{1}{\Delta t}[\langle f(t + \Delta t), g(t + \Delta t), h(t + \Delta t)\rangle - \langle f(t), g(t), h(t)\rangle] \\
&= \lim_{\Delta t \to 0} \frac{1}{\Delta t}\langle f(t + \Delta t) - f(t), g(t + \Delta t) - g(t), h(t + \Delta t) - h(t)\rangle \\
&= \lim_{\Delta t \to 0} \left\langle \frac{f(t + \Delta t) - f(t)}{\Delta t}, \frac{g(t + \Delta t) - g(t)}{\Delta t}, \frac{h(t + \Delta t) - h(t)}{\Delta t} \right\rangle.
\end{aligned}$$

Taking the limit of each component gives us the desired result.

If $\mathbf{r}'(t)$ exists we say that the function $\mathbf{r}$ is **differentiable** at t. We also denote derivatives by

$$\mathbf{r}'(t) = D_t\mathbf{r}(t) = \frac{d}{dt}\mathbf{r}(t).$$

Higher derivatives may be obtained in like manner. For example,

$$\mathbf{r}''(t) = \langle f''(t), g''(t), h''(t) \rangle = f''(t)\mathbf{i} + g''(t)\mathbf{j} + h''(t)\mathbf{k}. \tag{15.7}$$

Example 1 If $\mathbf{r}(t) = (\ln t)\mathbf{i} + e^{-3t}\mathbf{j} + t^2\mathbf{k}$,

(a) find the domain of $\mathbf{r}$ and determine where $\mathbf{r}$ is continuous.

(b) find $\mathbf{r}'(t)$ and $\mathbf{r}''(t)$.

Solution (a) Since $\ln t$ is undefined if $t \le 0$, the domain of $\mathbf{r}$ is the set of positive real numbers. Moreover, $\mathbf{r}$ is continuous throughout its domain since each component determines a continuous function.

(b) By (15.6) and (15.7),

$$\mathbf{r}'(t) = (1/t)\mathbf{i} - 3e^{-3t}\mathbf{j} + 2t\mathbf{k}$$
$$\mathbf{r}''(t) = (-1/t^2)\mathbf{i} + 9e^{-3t}\mathbf{j} + 2\mathbf{k}.$$

Since the definition of $\mathbf{r}'(t)$ in (15.5) has the same form as that given for $f'(x)$ in Chapter 3, it is reasonable to expect that differentiation formulas for vector-valued functions are similar to those for real-valued functions. The next theorem brings out this fact. Other differentiation formulas may be found in the Exercises.

(15.8) Theorem

If $\mathbf{u}$ and $\mathbf{v}$ are differentiable vector-valued functions and c is a scalar, then

(i) $D_t[\mathbf{u}(t) + \mathbf{v}(t)] = \mathbf{u}'(t) + \mathbf{v}'(t)$.

(ii) $D_t[c\mathbf{u}(t)] = c\mathbf{u}'(t)$.

(iii) $D_t[\mathbf{u}(t) \cdot \mathbf{v}(t)] = \mathbf{u}(t) \cdot \mathbf{v}'(t) + \mathbf{u}'(t) \cdot \mathbf{v}(t)$.

(iv) $D_t[\mathbf{u}(t) \times \mathbf{v}(t)] = \mathbf{u}(t) \times \mathbf{v}'(t) + \mathbf{u}'(t) \times \mathbf{v}(t)$.

Proof. We shall prove (iii) and leave the other parts as exercises. If we write

$$\mathbf{u}(t) = f_1(t)\mathbf{i} + f_2(t)\mathbf{j} + f_3(t)\mathbf{k}$$
$$\mathbf{v}(t) = g_1(t)\mathbf{i} + g_2(t)\mathbf{j} + g_3(t)\mathbf{k}$$

where each f_i and g_i is a differentiable function of t, then

$$\mathbf{u}(t) \cdot \mathbf{v}(t) = \sum_{i=1}^{3} f_i(t)g_i(t).$$

Consequently,

$$\begin{aligned} D_t[\mathbf{u}(t) \cdot \mathbf{v}(t)] &= D_t \sum_{i=1}^{3} f_i(t)g_i(t) = \sum_{i=1}^{3} D_t[f_i(t)g_i(t)] \\ &= \sum_{i=1}^{3} [f_i(t)g_i'(t) + f_i'(t)g_i(t)] \\ &= \sum_{i=1}^{3} f_i(t)g_i'(t) + \sum_{i=1}^{3} f_i'(t)g_i(t) \\ &= \mathbf{u}(t) \cdot \mathbf{v}'(t) + \mathbf{u}'(t) \cdot \mathbf{v}(t). \end{aligned}$$

There is an interesting geometric interpretation for the limit in Definition (15.5). Given $\mathbf{r}(t) = \langle f(t), g(t), h(t) \rangle$ where f, g, and h are differentiable, let C denote the curve with parametric equations $x = f(t)$, $y = g(t)$, $z = h(t)$. As in (i) of Figure 15.4, if $\overrightarrow{OP}$ and $\overrightarrow{OQ}$ are the position vectors corresponding to $\mathbf{r}(t)$ and $\mathbf{r}(t + \Delta t)$, respectively, then

$$\overrightarrow{PQ} = \overrightarrow{OQ} - \overrightarrow{OP}$$

is a geometric vector corresponding to $\mathbf{r}(t + \Delta t) - \mathbf{r}(t)$. It follows that $(1/\Delta t)\overrightarrow{PQ}$ is a vector corresponding to $(1/\Delta t)[\mathbf{r}(t + \Delta t) - \mathbf{r}(t)]$. If $\Delta t > 0$, then $(1/\Delta t)\overrightarrow{PQ}$ is a vector $\overrightarrow{PR}$ having the same direction as $\overrightarrow{PQ}$. Moreover, if $0 < \Delta t < 1$, then $(1/\Delta t) > 1$ and $|\overrightarrow{PR}| > |\overrightarrow{PQ}|$ as illustrated in (ii) of Figure 15.4. If we let $\Delta t \to 0^+$ then Q approaches P and it appears that the vector $\overrightarrow{PR}$ approaches a vector which lies on the tangent line to C at P. A similar discussion may be given if $\Delta t < 0$. For this reason we refer to $\mathbf{r}'(t)$, or any of its geometric representations, as a **tangent vector** to C at P. By definition the **tangent line** to C at P is the line through P parallel to $\mathbf{r}'(t)$.

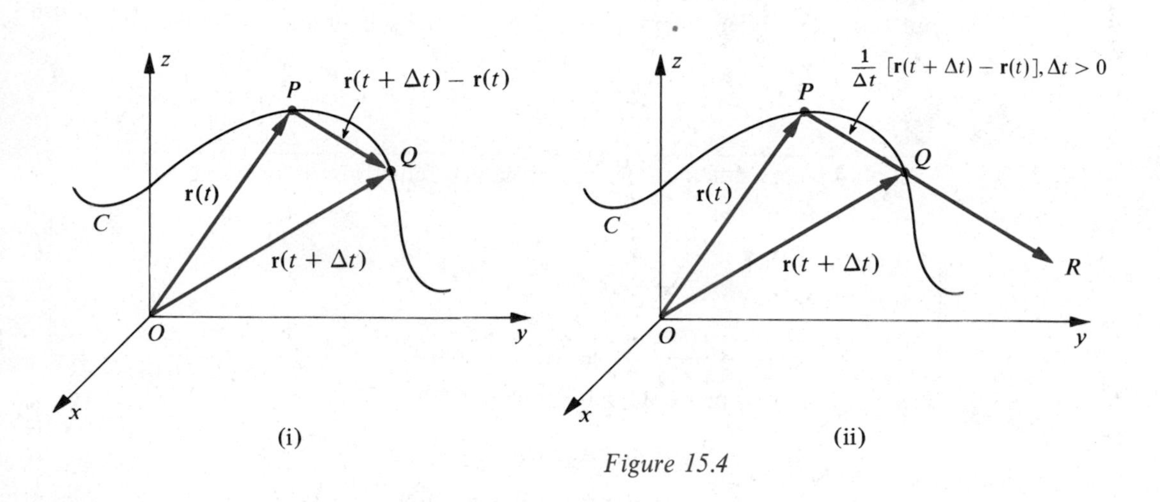

Figure 15.4

For the two-dimensional case $\mathbf{r}(t) = f(t)\mathbf{i} + g(t)\mathbf{j}$ we have

$$\mathbf{r}'(t) = f'(t)\mathbf{i} + g'(t)\mathbf{j}.$$

If $f'(t) \neq 0$, then the slope of the line parallel to $\mathbf{r}'(t)$ is $g'(t)/f'(t)$. This is in agreement with formula (13.4) obtained for tangent lines to plane curves. These remarks are illustrated in the next example.

Example 2 If $\mathbf{r}(t) = 2t\mathbf{i} + (4 - t^2)\mathbf{j}$, where $-2 \leq t \leq 2$, find $\mathbf{r}'(t)$. Sketch the curve C with parametric equations $x = 2t$, $y = 4 - t^2$, where $-2 \leq t \leq 2$. Illustrate $\mathbf{r}(1)$ and $\mathbf{r}'(1)$ geometrically.

Solution Since we are dealing with V_2, it is sufficient to use an xy-plane to represent vectors. Eliminating the parameter in $x = 2t$, $y = 4 - t^2$, we obtain $y = 4 - (x/2)^2$, which is an equation of a parabola. Coordinates of points on C which correspond to integer values of t are listed in the following table.

t	-2	-1	0	1	2
x	-4	-2	0	2	4
y	0	3	4	3	0

The curve C is that part of the parabola illustrated in Figure 15.5. Since $\mathbf{r}(1) = 2\mathbf{i} + 3\mathbf{j}$, the position vector corresponding to $\mathbf{r}(1)$ is $\overrightarrow{OP}$, where P is the point on C with coordinates $(2, 3)$. By differentiation we obtain $\mathbf{r}'(t) = 2\mathbf{i} - 2t\mathbf{j}$. In the figure we have represented $\mathbf{r}'(1) = 2\mathbf{i} - 2\mathbf{j}$ by a vector with initial point P in order to illustrate that it is a tangent vector to C.

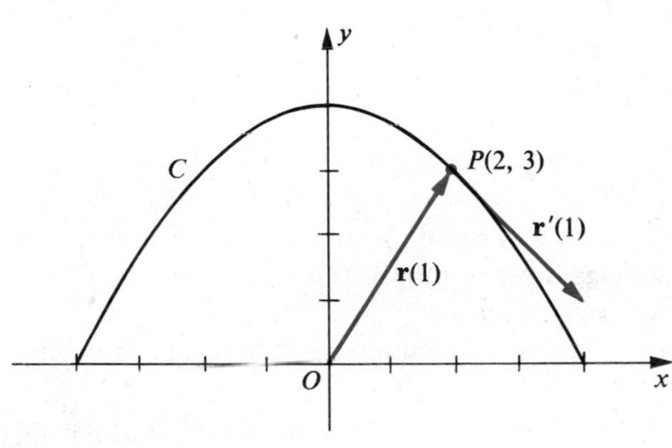

Figure 15.5

Example 3 Let C be the curve with parametric equations

$$x = t^2 - 5, \quad y = t^3, \quad z = 3t + 1$$

where t is in $\mathbb{R}$. Find parametric equations for the tangent line to C at the point corresponding to $t = 2$.

Solution The point $P(t)$ on C corresponding to t is $(t^2 - 5, t^3, 3t + 1)$. If we let

$$\mathbf{r}(t) = \langle t^2 - 5, t^3, 3t + 1\rangle = (t^2 - 5)\mathbf{i} + t^3\mathbf{j} + (3t + 1)\mathbf{k}$$

then, from the previous discussion, a tangent vector to C at $P(t)$ is

$$\mathbf{r}'(t) = \langle 2t, 3t^2, 3\rangle = 2t\mathbf{i} + 3t^2\mathbf{j} + 3\mathbf{k}.$$

In particular, the point corresponding to $t = 2$ is $P(2) = (-1, 8, 7)$ and a tangent vector is

$$\mathbf{r}'(2) = \langle 4, 12, 3\rangle = 4\mathbf{i} + 12\mathbf{j} + 3\mathbf{k}.$$

Applying (14.29), parametric equations of the tangent line are

$$x = 4t - 1, \quad y = 12t + 8, \quad z = 3t + 7.$$

Example 4 Prove that if $|\mathbf{r}(t)|$ is constant, then $\mathbf{r}'(t)$ is orthogonal to $\mathbf{r}(t)$ for every t.

Solution By hypothesis

$$\mathbf{r}(t) \cdot \mathbf{r}(t) = |\mathbf{r}(t)|^2 = c$$

for some scalar c. Writing $\mathbf{r}(t) = \langle f(t), g(t), h(t) \rangle$, the last equation may be written

$$[f(t)]^2 + [g(t)]^2 + [h(t)]^2 = c.$$

Differentiating implicitly,

$$2f(t)f'(t) + 2g(t)g'(t) + 2h(t)h'(t) = 0$$

that is

$$2\mathbf{r}(t) \cdot \mathbf{r}'(t) = 0.$$

The desired conclusion now follows from Theorem (14.20).

The result stated in Example 4 is also geometrically evident, for if $|\mathbf{r}(t)|$ is constant, then as t varies the end point of the position vector corresponding to $\mathbf{r}(t)$ moves on a curve C which lies on the surface of a sphere with center at O. The **tangent vector $\mathbf{r}'(t)$ to C is always orthogonal to the position vector and hence to $\mathbf{r}(t)$.**

Definite integrals of vector-valued functions may also be considered. Specifically, if

$$\mathbf{r}(t) = f(t)\mathbf{i} + g(t)\mathbf{j} + h(t)\mathbf{k}$$

where f, g, and h are integrable on $[a, b]$, then by definition

(15.9)
$$\int_a^b \mathbf{r}(t)\,dt = \left(\int_a^b f(t)\,dt\right)\mathbf{i} + \left(\int_a^b g(t)\,dt\right)\mathbf{j} + \left(\int_a^b h(t)\,dt\right)\mathbf{k}.$$

In this case we say that $\mathbf{r}$ is **integrable** on $[a, b]$. If $\mathbf{R}(t)$ is an antiderivative of $\mathbf{r}(t)$ in the sense that $\mathbf{R}'(t) = \mathbf{r}(t)$ for all t in $[a, b]$, then, as in the Fundamental Theorem of Calculus (5.31),

(15.10)
$$\int_a^b \mathbf{r}(t)\,dt = \mathbf{R}(t)\Big]_a^b = \mathbf{R}(b) - \mathbf{R}(a).$$

The proof is left as an exercise.

Example 5 Find $\int_0^2 \mathbf{r}(t)\,dt$ if $\mathbf{r}(t) = 12t^3\mathbf{i} + 4e^{2t}\mathbf{j} + (t+1)^{-1}\mathbf{k}$.

Solution By finding an antiderivative of each component of $\mathbf{r}(t)$ we obtain

$$\mathbf{R}(t) = 3t^4\mathbf{i} + 2e^{2t}\mathbf{j} + \ln(t+1)\mathbf{k}.$$

Since $\mathbf{R}'(t) = \mathbf{r}(t)$ it follows from (15.10) that

$$\begin{aligned}\int_0^2 \mathbf{r}(t)\,dt &= \mathbf{R}(2) - \mathbf{R}(0)\\ &= (48\mathbf{i} + 2e^4\mathbf{j} + \ln 3\mathbf{k}) - (0\mathbf{i} + 2\mathbf{j} + 0\mathbf{k})\\ &= 48\mathbf{i} + 2(e^4 - 1)\mathbf{j} + \ln 3\mathbf{k}.\end{aligned}$$

Indefinite integrals of vector-valued functions have a theory similar to that developed for real-valued functions. The proofs of theorems require only minor modifications of those given earlier and are omitted. If $\mathbf{R}(t)$ is an antiderivative of $\mathbf{r}(t)$, then every antiderivative has the form $\mathbf{R}(t) + \mathbf{c}$ for some (constant) vector $\mathbf{c}$ and we write

(15.11)
$$\int \mathbf{r}(t)\,dt = \mathbf{R}(t) + \mathbf{c}, \quad \text{where } \mathbf{R}'(t) = \mathbf{r}(t).$$

If boundary conditions are known, then $\mathbf{c}$ may be determined (see Exercises 27–30).

EXERCISES 15.2

In Exercises 1–4, find (a) the domain of $\mathbf{r}$ and determine where $\mathbf{r}$ is continuous; (b) $\mathbf{r}'(t)$ and $\mathbf{r}''(t)$.

1 $\mathbf{r}(t) = \sqrt{t-1}\,\mathbf{i} + \sqrt{2-t}\,\mathbf{j}$

2 $\mathbf{r}(t) = (1/t)\mathbf{i} + \sin 3t\mathbf{j}$

3 $\mathbf{r}(t) = \langle \tan t, t^2 + 8t \rangle$

4 $\mathbf{r}(t) = \langle e^{t^2}, \sin^{-1} t \rangle$

In Exercises 5–12, (a) sketch the curve in the xy-plane determined by the components of $\mathbf{r}(t)$; (b) find $\mathbf{r}'(t)$ and sketch geometric vectors corresponding to $\mathbf{r}(t)$ and $\mathbf{r}'(t)$ at the indicated value of t.

5 $\mathbf{r}(t) = \langle -t^4/4, t^2 \rangle, t = 2$

6 $\mathbf{r}(t) = \langle e^{2t}, e^{-4t} \rangle, t = 0$

7 $\mathbf{r}(t) = 4\cos t\mathbf{i} + 2\sin t\mathbf{j}, t = 3\pi/4$

8 $\mathbf{r}(t) = 2\sec t\mathbf{i} + 3\tan t\mathbf{j}, t = \pi/4$

9 $\mathbf{r}(t) = t^3\mathbf{i} + t^{-3}\mathbf{j}, t = 1$

10 $\mathbf{r}(t) = t^2\mathbf{i} + t^3\mathbf{j}, t = -1$

11 $\mathbf{r}(t) = \langle 2t - 1, -t + 4 \rangle, t = 3$

12 $\mathbf{r}(t) = \langle 5, t^3 \rangle, t = 2$

In Exercises 13–16 (a) find the domain of $\mathbf{r}$ and determine where $\mathbf{r}$ is continuous; (b) find $\mathbf{r}'(t)$ and $\mathbf{r}''(t)$.

13 $\mathbf{r}(t) = t^2\mathbf{i} + \tan t\mathbf{j} + 3\mathbf{k}$

14 $\mathbf{r}(t) = \sqrt[3]{t}\,\mathbf{i} + (1/t)\mathbf{j} + e^{-t}\mathbf{k}$

15 $\mathbf{r}(t) = \sqrt{t}\,\mathbf{i} + e^{2t}\mathbf{j} + t\mathbf{k}$

16 $\mathbf{r}(t) = \ln(1-t)\mathbf{i} + \sin t\mathbf{j} + t^2\mathbf{k}$

Parametric equations of a curve C are given in each of Exercises 17–20. Find parametric equations for the tangent line to C at the point P.

17 $x = 2t^3 - 1, y = -5t^2 + 3, z = 8t + 2; P(1, -2, 10)$

18 $x = 4\sqrt{t}, y = t^2 - 10, z = 4/t; P(8, 6, 1)$

19 $x = e^t, y = te^t, z = t^2 + 4; P(1, 0, 4)$

20 $x = t\sin t, y = t\cos t, z = t; P(\pi/2, 0, \pi/2)$

In Exercises 21 and 22 find two different unit tangent vectors to the given curve at the indicated point P.

21 $x = e^{2t}, y = e^{-t}, z = t^2 + 4; P(1, 1, 4)$

22 $x = 2 + \sin t, y = \cos t, z = t; P(2, 1, 0)$

Evaluate the integrals in Exercises 23–26.

23 $\int_0^2 (6t^2\mathbf{i} - 4t\mathbf{j} + 3\mathbf{k})\,dt$

24 $\int_{-1}^1 (-5t\mathbf{i} + 8t^3\mathbf{j} - 3t^2\mathbf{k})\,dt$

25 $\int_0^{\pi/4} (\sin t\mathbf{i} - \cos t\mathbf{j} + \tan t\mathbf{k})\,dt$

26 $\int_0^1 [te^{t^2}\mathbf{i} + \sqrt{t}\mathbf{j} + (t^2+1)^{-1}\mathbf{k}]\,dt$

27 If $\mathbf{u}'(t) = t^2\mathbf{i} + (6t+1)\mathbf{j} + 8t^3\mathbf{k}$ and $\mathbf{u}(0) = 2\mathbf{i} - 3\mathbf{j} + \mathbf{k}$, find $\mathbf{u}(t)$.

28 If $\mathbf{r}'(t) = 2\mathbf{i} - 4t^3\mathbf{j} + 6\sqrt{t}\mathbf{k}$ and $\mathbf{r}(0) = \mathbf{i} + 5\mathbf{j} + 3\mathbf{k}$, find $\mathbf{r}(t)$.

29 If $\mathbf{u}''(t) = 6t\mathbf{i} - 12t^2\mathbf{j} + \mathbf{k}$ and $\mathbf{u}'(0) = \mathbf{i} + 2\mathbf{j} - 3\mathbf{k}$, $\mathbf{u}(0) = 7\mathbf{i} + \mathbf{k}$, find $\mathbf{u}(t)$.

30 If $\mathbf{r}''(t) = 6t\mathbf{i} + 3\mathbf{j}$ and $\mathbf{r}'(0) = 4\mathbf{i} - \mathbf{j} + \mathbf{k}$, $\mathbf{r}(0) = 5\mathbf{j}$, find $\mathbf{r}(t)$.

31 If a curve C has a tangent vector $\mathbf{a}$ at a point P, then the **normal plane** to C at P is the plane through P with normal vector $\mathbf{a}$. Find an equation of the normal plane to the curve of Exercise 19 at the point $P(1, 0, 4)$.

32 Refer to Exercise 31. Find an equation for the normal plane to the curve of Exercise 20 at the point $P(\pi/2, 0, \pi/2)$.

33 If $\mathbf{r}$ and $\mathbf{s}$ are vector-valued functions which have limits as $t \to a$, prove the following.

(a) $\lim_{t\to a} [\mathbf{r}(t) + \mathbf{s}(t)] = \lim_{t\to a} \mathbf{r}(t) + \lim_{t\to a} \mathbf{s}(t)$

(b) $\lim_{t\to a} [\mathbf{r}(t) \cdot \mathbf{s}(t)] = \lim_{t\to a} \mathbf{r}(t) \cdot \lim_{t\to a} \mathbf{s}(t)$

(c) $\lim_{t\to a} c\mathbf{r}(t) = c \lim_{t\to a} \mathbf{r}(t)$ where c is a scalar

34 If a function f and a vector-valued function $\mathbf{u}$ have limits as $t \to a$, prove that

$$\lim_{t\to a} f(t)\mathbf{u}(t) = [\lim_{t\to a} f(t)][\lim_{t\to a} \mathbf{u}(t)].$$

35 Prove that $\lim_{t\to a} \mathbf{u}(t) = \mathbf{b}$ if and only if for every $\varepsilon > 0$ there is a $\delta > 0$ such that $|\mathbf{u}(t) - \mathbf{b}| < \varepsilon$ whenever $0 < |t - a| < \delta$. Interpret this fact geometrically.

36 If $\mathbf{u}$ and $\mathbf{v}$ are vector-valued functions which have limits as $t \to a$, prove that

$$\lim_{t\to a} [\mathbf{u}(t) \times \mathbf{v}(t)] = \lim_{t\to a} \mathbf{u}(t) \times \lim_{t\to a} \mathbf{v}(t).$$

37 Complete the proof of Theorem (15.8).

38 Prove (15.10).

39 Prove that if f and $\mathbf{u}$ are differentiable functions of t, then

$$D_t[f(t)\mathbf{u}(t)] = f(t)\mathbf{u}'(t) + f'(t)\mathbf{u}(t).$$

40 Prove the Chain Rule for vector-valued functions:

$$D_t\mathbf{u}(f(t)) = f'(t)\mathbf{u}'(f(t))$$

provided $\mathbf{u}$ and f are differentiable and the domains are suitably restricted.

41 Prove that if $\mathbf{u}$, $\mathbf{v}$, and $\mathbf{w}$ are differentiable vector-valued functions, then

$$D_t[\mathbf{u}(t) \cdot \mathbf{v}(t) \times \mathbf{w}(t)] = \mathbf{u}'(t) \cdot \mathbf{v}(t) \times \mathbf{w}(t) + \mathbf{u}(t) \cdot \mathbf{v}'(t) \times \mathbf{w}(t) + \mathbf{u}(t) \cdot \mathbf{v}(t) \times \mathbf{w}'(t).$$

42 Prove that if $\mathbf{u}'(t)$ and $\mathbf{u}''(t)$ exist, then

$$D_t[\mathbf{u}(t) \times \mathbf{u}'(t)] = \mathbf{u}(t) \times \mathbf{u}''(t).$$

43 If $\mathbf{u}$ and $\mathbf{v}$ are integrable on $[a, b]$ and if c is a scalar, prove the following.

(a) $\int_a^b [\mathbf{u}(t) + \mathbf{v}(t)]\, dt = \int_a^b \mathbf{u}(t)\, dt + \int_a^b \mathbf{v}(t)\, dt$

(b) $\int_a^b c\mathbf{u}(t)\, dt = c \int_a^b \mathbf{u}(t)\, dt$

44 If $\mathbf{u}$ is integrable on $[a, b]$ and $\mathbf{c}$ is any vector in V_3, prove that

$$\int_a^b \mathbf{c} \cdot \mathbf{u}(t)\, dt = \mathbf{c} \cdot \int_a^b \mathbf{u}(t)\, dt.$$

In Exercises 45 and 46, find $D_t[\mathbf{u}(t) \cdot \mathbf{v}(t)]$ and $D_t[\mathbf{u}(t) \times \mathbf{v}(t)]$.

45 $\mathbf{u}(t) = t\mathbf{i} + t^2\mathbf{j} + t^3\mathbf{k}$

$\mathbf{v}(t) = \sin t\mathbf{i} + \cos t\mathbf{j} + 2 \sin t\mathbf{k}$

46 $\mathbf{u}(t) = 2t\mathbf{i} + 6t\mathbf{j} + t^2\mathbf{k}$

$\mathbf{v}(t) = e^{-t}\mathbf{i} - e^{-t}\mathbf{j} + \mathbf{k}$

15.3 MOTION IN SPACE

In this section we shall introduce concepts needed for studying motion. If we assume that the mass of an object is concentrated at its center of gravity, then we can reduce our study to the motion of a point (or particle). Motion often takes place in a plane. For example, although the earth moves through space, its orbit lies in a plane. (This will be proved in Section 5.) Let us begin, therefore, by studying a particle moving in a coordinate plane. In order to analyze the behavior of the particle, it is essential to know its position (x, y) at every instant. Since at time t the particle has a unique position, it follows that the coordinates x and y are functions of t. Thus, we shall assume that at time t the particle is at the point $P(x, y)$, where $x = f(t)$ and $y = g(t)$ for certain functions f and g. If we let

$$\mathbf{r}(t) = \langle f(t), g(t) \rangle = f(t)\mathbf{i} + g(t)\mathbf{j}$$

then as t varies, the end-point of the position vector $\overrightarrow{OP}$ corresponding to $\mathbf{r}(t)$ traces the path C of the particle. From the discussion in the preceding section, the tangent line to C at P is parallel to $\mathbf{r}'(t)$ (see Figure 15.6). The magnitude of this vector is

$$|\mathbf{r}'(t)| = \sqrt{[f'(t)]^2 + [g'(t)]^2}. \tag{15.12}$$

If P_0 is the point on C corresponding to $t = t_0$, and if f and g are smooth functions, then by (13.17), the arc length $s(t)$ of C from P_0 to P is given by

$$s(t) = \int_{t_0}^{t} |\mathbf{r}'(t)|\, dt.$$

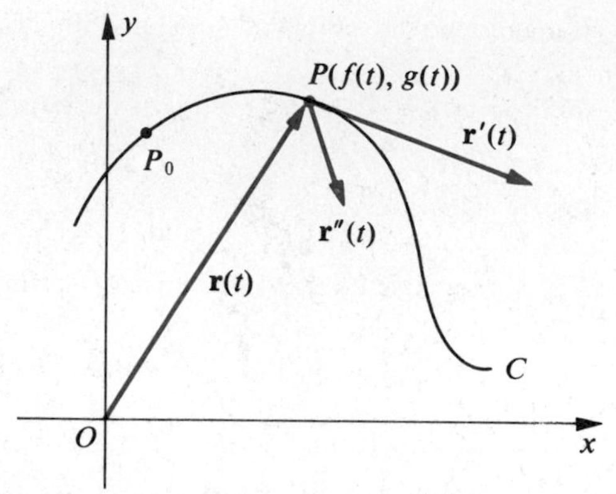

Figure 15.6

Applying (5.33) we obtain $s'(t) = |\mathbf{r}'(t)|$; that is, $|\mathbf{r}'(t)|$ *is the rate of change of arc length with respect to time.* For this reason we refer to $|\mathbf{r}'(t)|$ as the **speed** of the particle. The tangent vector $\mathbf{r}'(t)$, whose magnitude is the speed, is defined as the **velocity** of the particle at time t. In analogy with (4.31) we call the vector $\mathbf{r}''(t)$ the **acceleration** of the particle at time t and usually represent it by a geometric vector with initial point P. It can be shown that in most cases $\mathbf{r}''(t)$ is directed toward the concave side of C, as illustrated in Figure 15.6.

Example 1 The position of a particle moving in a plane is given by

$$\mathbf{r}(t) = (t^2 + t)\mathbf{i} + t^3\mathbf{j}$$

where $0 \le t \le 2$. Find the velocity and acceleration at time t. Sketch the path of the particle and represent $\mathbf{r}'(1)$ and $\mathbf{r}''(1)$ geometrically.

Solution The velocity and acceleration are

$$\mathbf{r}'(t) = (2t + 1)\mathbf{i} + 3t^2\mathbf{j} \quad \text{and} \quad \mathbf{r}''(t) = 2\mathbf{i} + 6t\mathbf{j}.$$

In particular, at $t = 1$

$$\mathbf{r}'(1) = 3\mathbf{i} + 3\mathbf{j} \quad \text{and} \quad \mathbf{r}''(1) = 2\mathbf{i} + 6\mathbf{j}.$$

These vectors and the path of the particle are sketched in Figure 15.7.

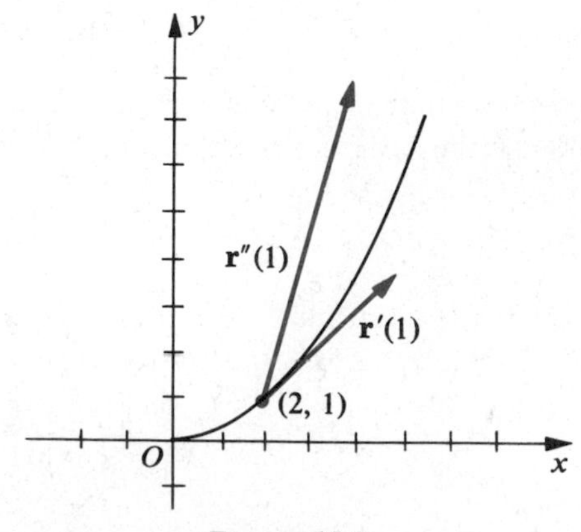

Figure 15.7

Example 2 Show that if a point P moves around a circle at a constant speed, then the acceleration is represented by a vector of constant magnitude directed from P toward the center of the circle. (This vector is called the *centripetal acceleration* vector.)

Solution There is no loss of generality in assuming that the center of the circle is at the origin O of a rectangular coordinate system. Suppose that the radius of the circle is a, the initial position of P is $A(a, 0)$, and θ is the angle generated by $\overrightarrow{OP}$ at the end of time t (see Figure 15.8). The fact that P moves around the circle at a constant speed is equivalent to stating that the rate at which the angle θ changes per unit time (called the **angular speed**) is a constant ω. This fact may be expressed by means of the differential equation $d\theta/dt = \omega$, or equivalently by $\theta = \omega t + c$ for some number c. Since $\theta = 0$ when $t = 0$, the number c is 0; that is $\theta = \omega t$. It follows that the coordinates (x, y) of P are

$$x = a \cos \omega t, \quad y = a \sin \omega t$$

and the motion of P is given by

$$\mathbf{r}(t) = a \cos \omega t\mathbf{i} + a \sin \omega t\mathbf{j}.$$

Consequently,

$$\mathbf{r}'(t) = -\omega a \sin \omega t\mathbf{i} + \omega a \cos \omega t\mathbf{j}$$
$$\mathbf{r}''(t) = -\omega^2 a \cos \omega t\mathbf{i} - \omega^2 a \sin \omega t\mathbf{j}$$

and hence $\mathbf{r}''(t) = -\omega^2\mathbf{r}(t)$. If $\overrightarrow{OP}$ is the position vector corresponding to $\mathbf{r}(t)$, then $-\omega^2\overrightarrow{OP}$ corresponds to $\mathbf{r}''(t)$. This shows that the direction of the acceleration vector is from P toward O. Since

$$|\mathbf{r}''(t)| = \sqrt{\omega^4 a^2(\cos^2 \omega t + \sin^2 \omega t)} = \omega^2 a$$

the magnitude of $\mathbf{r}''(t)$ is always $\omega^2 a$.

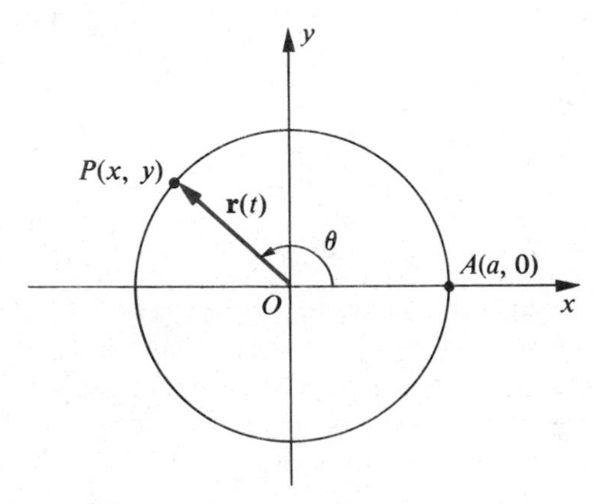

Figure 15.8

The preceding discussion may be extended easily to a particle moving in a three-dimensional coordinate system. Thus, suppose the position of a particle at time t is given by $P(t) = (f(t), g(t), h(t))$ where the functions f, g, and h are defined on an interval I. If we let

$$\mathbf{r}(t) = f(t)\mathbf{i} + g(t)\mathbf{j} + h(t)\mathbf{k}$$

then as t varies the end point of the position vector $\overrightarrow{OP}$ traces the path of the particle. As before, the derivative $\mathbf{r}'(t)$, if it exists, is called the **velocity** of the particle at time t and is a tangent vector to C at $P(t)$. The vector $\mathbf{r}''(t)$ is called the **acceleration** of the particle at time t. The **speed** of the particle is the magnitude $|\mathbf{r}'(t)|$ of the velocity and, as in two dimensions, equals the rate of change of arc length with respect to time.

Example 3 The position of a moving particle is given by $\mathbf{r}(t) = 2t\mathbf{i} + 3t^2\mathbf{j} + t^3\mathbf{k}$, where $0 \leq t \leq 2$. Find the velocity and acceleration at time t. Sketch the path of the particle and illustrate $\mathbf{r}'(1)$ and $\mathbf{r}''(1)$ geometrically.

Solution The velocity and acceleration are

$$\mathbf{r}'(t) = 2\mathbf{i} + 6t\mathbf{j} + 3t^2\mathbf{k} \quad \text{and} \quad \mathbf{r}''(t) = 6\mathbf{j} + 6t\mathbf{k}.$$

In particular, at $t = 1$ the particle is at the point $P(2, 3, 1)$ and

$$\mathbf{r}'(1) = 2\mathbf{i} + 6\mathbf{j} + 3\mathbf{k} \quad \text{and} \quad \mathbf{r}''(1) = 6\mathbf{j} + 6\mathbf{k}.$$

These vectors and the path of the particle are sketched in Figure 15.9.

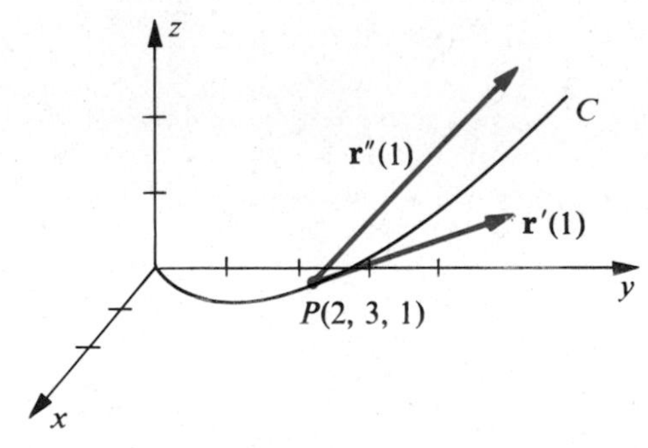

Figure 15.9

Suppose a particle P is rotating about the z-axis on a circle of radius a which lies in a plane parallel to the xy-plane, as illustrated in Figure 15.10. In addition, suppose the angular speed $d\theta/dt$ is a constant ω. The vector $\boldsymbol{\omega} = \omega\mathbf{k}$ directed along the z-axis and having magnitude ω is called the **angular velocity** of P. As in Example 2, the motion of P is given by

$$\mathbf{r}(t) = a\cos\omega t\mathbf{i} + a\sin\omega t\mathbf{j} + h\mathbf{k},$$

where h is the distance from the xy-plane to the particle. A direct computation shows that

$$\boldsymbol{\omega} \times \mathbf{r}(t) = -\omega a\sin\omega t\mathbf{i} + \omega a\cos\omega t\mathbf{j} = \mathbf{r}'(t).$$

Thus the velocity vector is the cross product of the angular velocity and the position vector of P. This fact can be generalized to the rotation of a particle P about any line.

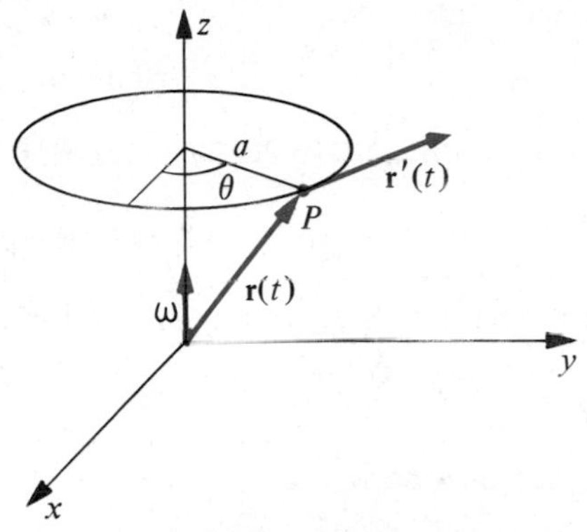

Figure 15.10

As a final illustration of how vectors may be applied to problems involving motion, suppose a projectile is fired from a cannon, with an angle of elevation α. If the initial speed is v_0, then we may introduce a coordinate plane as in Figure 15.11, where the muzzle of the cannon is at the origin O and the initial velocity $\mathbf{v}_0$ has direction α and magnitude v_0. Let $P(x, y)$ denote the position of the projectile after time t and let $\mathbf{r}(t)$ correspond to $\overrightarrow{OP}$. Our objective is to find an explicit form for $\mathbf{r}(t)$. We shall assume that air resistance is negligible and that the only force acting on the projectile is the force $\mathbf{g}$ of gravitational acceleration. Since $\mathbf{g}$ acts in the downward direction we may write $\mathbf{g} = -g\mathbf{j}$, where $|\mathbf{g}| = g \approx 32 \text{ ft sec}^2$.

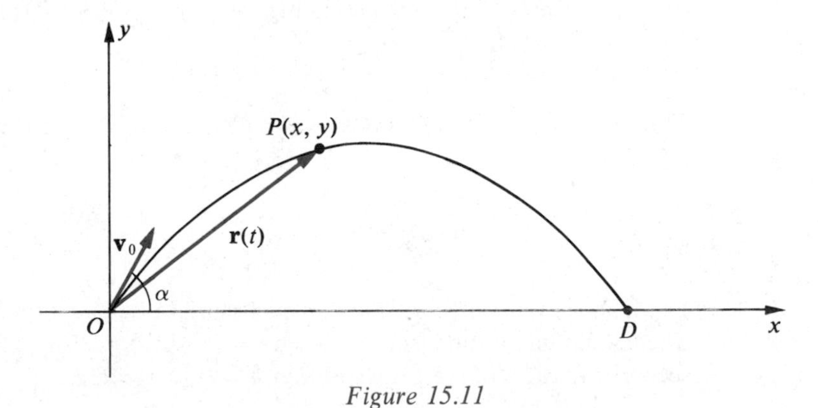

Figure 15.11

According to Newton's Second Law of Motion, the force $\mathbf{F}$ acting on an object of mass m is related to its acceleration $\mathbf{a}$ by the formula $\mathbf{F} = m\mathbf{a}$. In the present situation, $m\mathbf{a} = m\mathbf{g}$, which leads to the vector differential equation

$$\mathbf{r}''(t) = \mathbf{g}.$$

Indefinite integration gives us

$$\mathbf{r}'(t) = t\mathbf{g} + \mathbf{c}$$

where $\mathbf{c}$ is some vector. Since $\mathbf{r}'(t)$ is the velocity at time t,

$$\mathbf{v}_0 = \mathbf{r}'(0) = \mathbf{c}$$

and hence

$$\mathbf{r}'(t) = t\mathbf{g} + \mathbf{v}_0.$$

Integrating both sides of the last equation we obtain

$$\mathbf{r}(t) = \tfrac{1}{2}t^2\mathbf{g} + t\mathbf{v}_0 + \mathbf{d}$$

for some vector $\mathbf{d}$. Since $\mathbf{r}(0) = \mathbf{0}$, it follows that $\mathbf{d} = \mathbf{0}$. Consequently,

$$\mathbf{r}(t) = \tfrac{1}{2}t^2\mathbf{g} + t\mathbf{v}_0$$

which may also be written

$$x\mathbf{i} + y\mathbf{j} = -\tfrac{1}{2}t^2g\mathbf{j} + t(v_0 \cos \alpha\mathbf{i} + v_0 \sin \alpha\mathbf{j}).$$

Equating components we see that

$$\text{(15.13)} \qquad x = (v_0 \cos \alpha)t, \quad y = -\tfrac{1}{2}gt^2 + (v_0 \sin \alpha)t.$$

These are parametric equations for the path of the projectile. Eliminating the parameter gives us the rectangular equation

$$\text{(15.14)} \qquad y = \frac{-g}{2v_0^2 \cos^2 \alpha}x^2 + (\tan \alpha)x.$$

This shows that the path of the projectile is parabolic. To find the range, that is, the distance $d(O, D)$ from O to D in Figure 15.11, we let $y = 0$ in (15.13), obtaining

$$t(-\tfrac{1}{2}gt + v_0 \sin \alpha) = 0.$$

Thus the point D corresponds to $t = (2v_0 \sin \alpha)/g$. Using the first equation in (15.13) gives us

$$d(O, D) = (v_0 \cos \alpha)\left(\frac{2v_0 \sin \alpha}{g}\right) = \frac{v_0^2 \sin 2\alpha}{g}.$$

In particular, the range will have its maximum value v_0^2/g if $\sin 2\alpha = 1$, or $\alpha = 45°$. The maximum altitude h occurs when $t = (v_0 \sin \alpha)/g$. (Why?) Substituting in the second equation of (15.13) we obtain $h = (v_0^2 \sin^2 \alpha)/2g$.

EXERCISES 15.3

In each of Exercises 1–8 the position of a particle moving in a plane is given by $\mathbf{r}(t)$. Find its velocity, acceleration, and speed at time t. Sketch the path of the particle together with vectors corresponding to the velocity and acceleration at the indicated time t.

1 $\mathbf{r}(t) = 2t\mathbf{i} + (4t^2 + 1)\mathbf{j}, t = 1$

2 $\mathbf{r}(t) = (4 - 9t^2)\mathbf{i} + 3t\mathbf{j}, t = 1$

3 $\mathbf{r}(t) = (2/t)\mathbf{i} + (3/(t + 1))\mathbf{j}, t = 2$

4 $\mathbf{r}(t) = \sqrt{t}\mathbf{i} + (1 + \sqrt{t})\mathbf{j}, t = 4$

5 $\mathbf{r}(t) = \sin t\mathbf{i} + 4\cos 2t\mathbf{j}, t = \pi/6$

6 $\mathbf{r}(t) = \cos^2 t\mathbf{i} + 2\sin t\mathbf{j}, t = 3\pi/4$

7 $\mathbf{r}(t) = e^{2t}\mathbf{i} + e^{-t}\mathbf{j}, t = 0$

8 $\mathbf{r}(t) = 2t\mathbf{i} + e^{-t^2}\mathbf{j}, t = 1$

In Exercises 9–16 $\mathbf{r}(t)$ is the position vector of a particle moving in space. Find the velocity, acceleration, and speed at time t. Sketch the path of the particle together with vectors corresponding to $\mathbf{r}'(t)$ and $\mathbf{r}''(t)$ for the indicated values of t.

9 $\mathbf{r}(t) = \cos t\mathbf{i} + \sin t\mathbf{j} + t\mathbf{k};\ t = 0, \pi/4, \pi/2$

10 $\mathbf{r}(t) = t^2\mathbf{i} + t^3\mathbf{j} + t\mathbf{k};\ t = 0, 1, 2$

11 $\mathbf{r}(t) = t^2\mathbf{i} + 2\sqrt{t}\mathbf{j} + 4\sqrt{t^3}\mathbf{k};\ t - 1, 4, 9$

12 $\mathbf{r}(t) = 4\sin t\mathbf{i} + 2t\mathbf{j} + 9\cos t\mathbf{k};\ t = 0, \pi/2, 3\pi/4$

13 $\mathbf{r}(t) = e^t(\cos t\mathbf{i} + \sin t\mathbf{j} + \mathbf{k}), t = 0, \pi/4, \pi/2$

14 $\mathbf{r}(t) = t(\cos t\mathbf{i} + \sin t\mathbf{j} + t\mathbf{k}), t = 0, \pi/4, \pi/2$

15 $\mathbf{r}(t) = (1 + t)\mathbf{i} + 2t\mathbf{j} + (2 + 3t)\mathbf{k}, t = 0, 1, 2$

16 $\mathbf{r}(t) = 2t\mathbf{i} + \mathbf{j} + 9t^2\mathbf{k}, t = 0, 1, 2$

17 Prove that if a particle moves at a constant speed, then the velocity and acceleration vectors are orthogonal.

18 Prove that if the acceleration of a moving particle is always **0**, then the motion is rectilinear.

19 A projectile is fired with an initial speed of 1500 ft/sec and angle of elevation 30°. Find:
(a) the velocity at time t.
(b) the maximum altitude.
(c) the range.
(d) the speed at which the projectile strikes level ground.

20 Work Exercise 19 if the angle of elevation is 60°.

21 An outfielder on a baseball team threw a ball a distance of 250 feet. If he released the ball at an angle of 45° to the ground, what was the initial speed?

22 A projectile is fired horizontally with a velocity of 1800 ft/sec from an altitude of 1000 feet above level ground. When and where will it strike the ground?

15.4 CURVATURE

A plane curve may be represented parametrically in many different ways. Sometimes it is convenient to use arc length as the parameter. Thus, suppose a curve C in the xy-plane is given by

$$x - f(s), \quad y = g(s)$$

where the functions f and g are smooth on an interval I and where the parameter s is arc length measured along C from some fixed point A. Geometrically we have a situation similar to that illustrated in Figure 15.12 where for each s in I there corresponds a unique point $P(s) = (f(s), g(s))$ which is s units from A (measured along C). The positive direction along C is determined by increasing values of s.

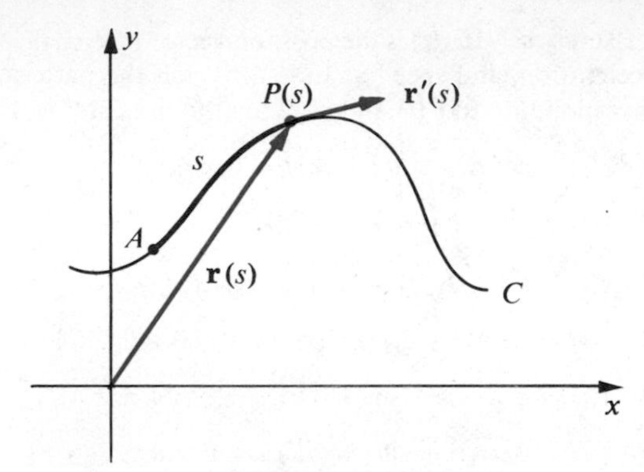

Figure 15.12

If $\mathbf{r}(s) = \langle f(s), g(s) \rangle$ is the position vector of $P(s)$, then the tangent vector $\mathbf{r}'(s)$ is always a unit vector. To establish this fact first note that by (13.17)

$$s = \int_0^s \sqrt{[f'(t)]^2 + [g'(t)]^2}\, dt.$$

Differentiating with respect to s gives us

$$1 = \sqrt{[f'(s)]^2 + [g'(s)]^2} = |\mathbf{r}'(s)|.$$

We shall denote the unit tangent vector $\mathbf{r}'(s)$ by $\mathbf{T}(s)$ and study its variation as $P(s)$ moves along C. For each s let θ be the least nonnegative angle between $\mathbf{T}(s)$ and $\mathbf{i}$, as illustrated in Figure 15.13. Note that θ is a function of s. The **curvature** K of C at the point $P(s)$ is defined by

(15.15) $$K = |D_s\theta|$$

provided $D_s\theta$ exists. Thus, K is the absolute value of the rate of change of θ with respect to s. In particular, if C is a straight line then θ is constant and the curvature is zero. The curvature is small for points such as R and S in Figure 15.13, since θ changes slowly as $P(s)$ moves along C. The curvature is large for the points Q and V. Thus, roughly speaking, K provides information about how sharp the curve is at various points.

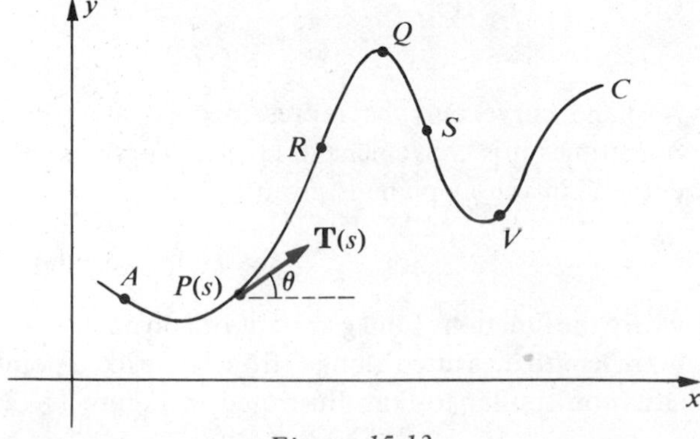

Figure 15.13

Suppose C can also be described by the graph of a rectangular equation $y = h(x)$, where the function h is smooth on some interval. Since $y' = D_x y$ is the slope of the tangent line at P we see, as in Figure 15.13, that

(15.16) $$\tan \theta = y' \quad \text{or} \quad \theta = \tan^{-1} y'.$$

As in Section 6.7, the arc length function s may be defined by

(15.17) $$s(x) = \int_a^x \sqrt{1 + (y')^2}\, dx$$

where a is the abscissa of the fixed point A on C. If y'' exists, then by the Chain Rule

$$\frac{d\theta}{dx} = \frac{d\theta}{ds}\frac{ds}{dx}, \quad \text{or} \quad D_x\theta = (D_s\theta)(D_x s).$$

Consequently

(15.18) $$K = |D_s\theta| = \frac{|D_x\theta|}{|D_x s|}.$$

Referring to (15.16) and (15.17),

$$D_x\theta = \frac{y''}{1 + (y')^2} \quad \text{and} \quad D_x s = \sqrt{1 + (y')^2}.$$

Substitution in (15.18) gives us

(15.19) $$K = \frac{|y''|}{[1 + (y')^2]^{3/2}}.$$

The preceding formula may be used to find K provided C is described in terms of a rectangular equation.

Example 1 Sketch the graph of $y = 1 - x^2$ and find the curvature at the following points: (x, y), $(0, 1)$, $(1, 0)$, $(2, -3)$.

Solution The graph (a parabola) is sketched in Figure 15.14. Since $y' = -2x$ and $y'' = -2$ we have, from (15.19),

$$K = \frac{2}{(1 + 4x^2)^{3/2}}.$$

In particular, at $(0, 1)$ we see that $K = 2$; that is, the direction of the tangent vector changes at the rate of 2 radians per unit change in arc length. At the point $(1, 0)$, $K = 2/(5)^{3/2} \approx 0.18$, which shows that the curve is less sharp here than at $(0, 1)$. This can also be seen from the graph. Finally, at $(2, -3)$, $K = 2/(17)^{3/2} \approx 0.03$. Observe that as x increases without bound the curvature approaches 0.

Example 2 Prove that the curvature at any point on a circle of radius a is $1/a$.

Solution There is no loss of generality in assuming that the circle has center at the origin and equation $x^2 + y^2 = a^2$. Implicit differentiation yields

$$2x + 2yy' = 0, \quad \text{or} \quad y' = -x/y.$$

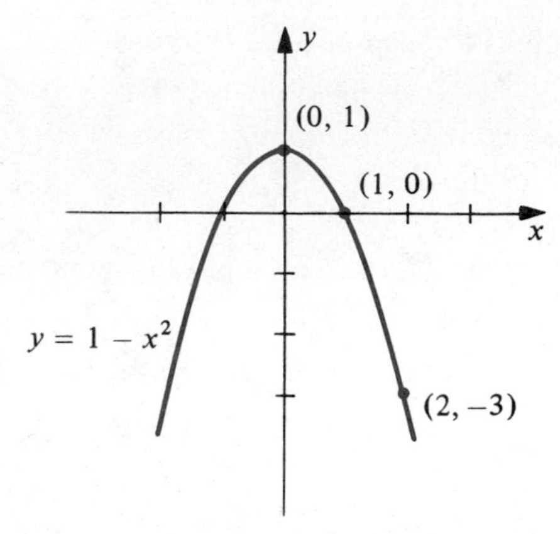

Figure 15.14

Consequently,

$$y'' = -\frac{y - xy'}{y^2} = -\frac{y + (x^2/y)}{y^2}$$
$$= -\frac{y^2 + x^2}{y^3} = -\frac{a^2}{y^3}.$$

Substituting in (15.19), and using the fact that $x^2 + y^2 = a^2$,

$$K = \frac{|a^2/y^3|}{[1 + x^2/y^2]^{3/2}} = \frac{|a^2/y^3|}{[(y^2 + x^2)/y^2]^{3/2}}$$
$$= \frac{|a^2/y^3|}{(a^2/y^2)^{3/2}} = \frac{1}{a}.$$

If the curvature of a curve C at P is not zero, then the circle of radius $\rho = 1/K$ whose center lies on the concave side of C and which has the same tangent line at P as C is called the **circle of curvature** of C at P. Its radius ρ and center are called the **radius of curvature** and **center of curvature**, respectively, of C at P. According to Example 2, the curvature of the circle of curvature is $1/\rho$, or K, and hence is the same as the curvature of C. For this reason the circle of curvature may be thought of as the circle which best coincides with C at P.

Example 3 Let C be the curve with parametric equations $x = t^2, y = t^3$. Find the curvature at the point P corresponding to $t = 1/2$. Sketch the graph of C and the circle of curvature at P.

Solution The point corresponding to $t = 1/2$ is $P(1/4, 1/8)$. We may eliminate the parameter in the given equations by noting that $t = y^{1/3}$ and, therefore, $x = t^2 = y^{2/3}$. The graph is sketched in Figure 15.15. The part above the x-axis is the graph of $y = x^{3/2}$. (Why?) Differentiating we obtain

$$y' = \tfrac{3}{2}x^{1/2} \quad \text{and} \quad y'' = \tfrac{3}{4}x^{-1/2}.$$

If $x > 0$ we may drop the absolute value sign in (15.19), obtaining

$$K = \frac{\frac{3}{4}x^{-1/2}}{[1 + (\frac{3}{2}x^{1/2})^2]^{3/2}}.$$

The reader may verify that this reduces to

$$K = \frac{6}{x^{1/2}(4 + 9x)^{3/2}}.$$

To find the curvature at P we let $x = 1/4$. This gives us

$$K = \frac{6}{(1/4)^{1/2}(25/4)^{3/2}} = \frac{6}{(1/2)(125/8)} = \frac{96}{125} \approx 0.77.$$

The radius of curvature ρ at P is given by

$$\rho = \frac{1}{K} = \frac{125}{96} \approx 1.3.$$

The circle of curvature is sketched in Figure 15.15.

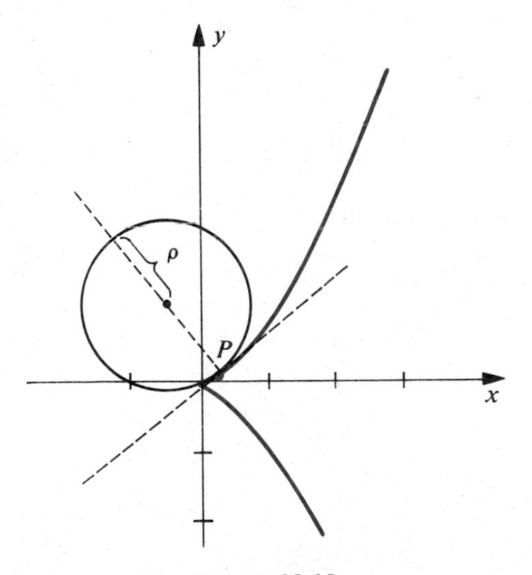

Figure 15.15

The definition of curvature $K = |D_s\theta|$ used in two dimensions has no immediate analogue in space and it is necessary to use an alternate approach to this concept. Of course, we must then show that if we specialize to vectors in a plane, the new definition agrees with the old. Let us proceed as follows.

Suppose a smooth space curve C is given by

$$x = f(s), \quad y = g(s), \quad z = h(s)$$

where s is arc length measured along C from a fixed point A to the point $P(s) = (f(s), g(s), h(s))$. As before, if $\mathbf{r}(s) = \langle f(s), g(s), h(s) \rangle$ is the position vector of $P(s)$, then $\mathbf{r}'(s)$ is a unit tangent vector to C at $P(s)$ which we shall denote by

$\mathbf{T}(s)$. Since $|\mathbf{T}(s)|$ is a constant, it follows that $\mathbf{T}(s)$ is orthogonal to $\mathbf{T}'(s)$ (see Example 4 of Section 2). If $\mathbf{T}'(s) \neq \mathbf{0}$, let

$$\mathbf{N}(s) = \frac{1}{|\mathbf{T}'(s)|}\mathbf{T}'(s). \tag{15.20}$$

The vector $\mathbf{N}(s)$ is a unit vector orthogonal to $\mathbf{T}(s)$ and is referred to as the **principal unit normal vector** to C at the point $P(s)$. From (15.20),

$$\mathbf{T}'(s) = |\mathbf{T}'(s)|\mathbf{N}(s).$$

The number $|\mathbf{T}'(s)|$ will be called the **curvature** of C at $P(s)$ and denoted by K. Thus

$$\mathbf{T}'(s) = K\mathbf{N}(s), \quad \text{where } K = |\mathbf{T}'(s)|. \tag{15.21}$$

In order to show that this definition of K reduces to $|D_s\theta|$ when all the vectors lie in a plane, we first note that in two dimensions,

$$\mathbf{T}(s) = \cos\theta\mathbf{i} + \sin\theta\mathbf{j}$$

where θ is the angle considered in (15.15). Regarding θ as a function of s and differentiating gives us

$$\mathbf{T}'(s) = -\sin\theta\, D_s\theta\mathbf{i} + \cos\theta\, D_s\theta\mathbf{j} = D_s\theta(-\sin\theta\mathbf{i} + \cos\theta\mathbf{j}).$$

Hence

$$|\mathbf{T}'(s)| = |D_s\theta|\,|-\sin\theta\mathbf{i} + \cos\theta\mathbf{j}| = |D_s\theta|$$

which is the same as (15.15).

In the next section we shall derive a formula for the curvature of a space curve.

EXERCISES 15.4

In Exercises 1–10 find the curvature of each given plane curve at the point P.

1 $y = 2 - x^3; P(1, 1)$

2 $y = x^4; P(1, 1)$

3 $y = e^{x^2}; P(0, 1)$

4 $y = \ln(x - 1); P(2, 0)$

5 $y = \sin x; P(\pi/2, 1)$

6 $y = \sec x; P(\pi/3, 2)$

7 $x = t - 1, y = \sqrt{t}; P(3, 2)$

8 $x = t + 1, y = t^2 + 4t + 3; P(1, 3)$

9 $x = 2\sin t, y = 3\cos t; P(1, 3\sqrt{3}/2)$

10 $x = t, y = \sqrt{t^2 - 9}; P(5, 4)$

For each of the plane curves described in Exercises 11–16, find the points at which the curvature is a maximum.

11 $y = e^{-x}$

12 $y = \cosh x$

13 $9x^2 + 4y^2 = 36$

14 $9x^2 - 4y^2 = 36$

15 $y = \ln x$

16 $y = \sin x$

In each of Exercises 17–20, find the points on the indicated curve at which the curvature is 0.

17 $y = x^4 - 12x^2$

18 $y = \tan x$

19 $y = \sinh x$

20 $y = e^{-x^2}$

15.5 TANGENTIAL AND NORMAL COMPONENTS OF ACCELERATION

The results of the previous section may be applied to the motion of a particle. Let us suppose that at time t the particle is at the point $P(x, y, z)$ on a curve C which is given parametrically by

$$x = f(t), \quad y = g(t), \quad z = h(t)$$

where f'', g'', and h'' exist. As usual, let $\mathbf{r}(t)$ represent the position vector $\overrightarrow{OP}$ and let s denote arc length measured along C. We shall also assume that s increases as t increases. The unit tangent vector $\mathbf{T}(s)$ introduced in the previous section may then be expressed as

$$\mathbf{T}(s) = \frac{1}{|\mathbf{r}'(t)|}\mathbf{r}'(t)$$

and hence

$$\mathbf{r}'(t) = |\mathbf{r}'(t)|\mathbf{T}(s) = \frac{ds}{dt}\mathbf{T}(s).$$

Differentiating with respect to t and using Exercises 39 and 40 of Section 15.2 gives us

$$\begin{aligned} \mathbf{r}''(t) &= \frac{d^2s}{dt^2}\mathbf{T}(s) + \frac{ds}{dt}\frac{d}{dt}\mathbf{T}(s) \\ &= \frac{d^2s}{dt^2}\mathbf{T}(s) + \frac{ds}{dt}\frac{ds}{dt}\mathbf{T}'(s). \end{aligned}$$

According to (15.21) this may also be written

$$\mathbf{r}''(t) = \frac{d^2s}{dt^2}\mathbf{T}(s) + \left(\frac{ds}{dt}\right)^2 K\mathbf{N}(s)$$

where K is the curvature of C and $\mathbf{N}(s)$ is the unit normal (15.20). If we denote the speed ds/dt by v and write $K = 1/\rho$, where ρ is the radius of curvature of C, then the last formula for $\mathbf{r}''(t)$ may be written

(15.22)
$$\mathbf{r}''(t) = \frac{dv}{dt}\mathbf{T}(s) + \frac{v^2}{\rho}\mathbf{N}(s).$$

This formula expresses the acceleration in terms of a **tangential component** dv/dt (the rate of change of speed with respect to time) and a **normal component** v^2/ρ. The sketch in Figure 15.16 illustrates one possibility which could occur. Note that the normal component depends only on the speed and the curvature (or radius of curvature) of the curve C. If the speed or curvature is large (or the radius of curvature is small), then the normal component of acceleration is large. This result, obtained theoretically, proves the well-known fact that an automobile driver should slow down when attempting to negotiate a sharp turn.

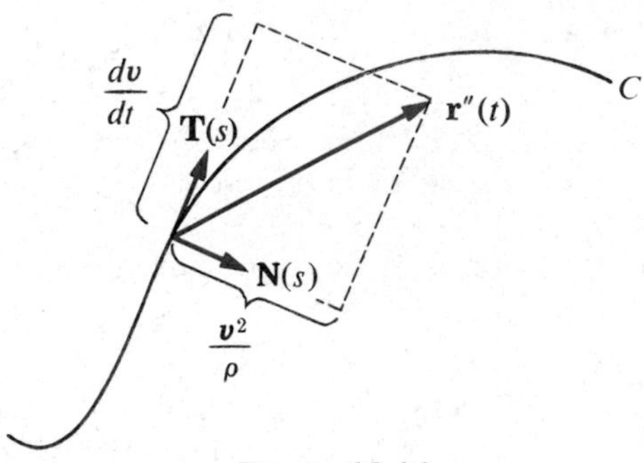

Figure 15.16

Let us next find formulas for the tangential and normal components of acceleration which depend only on $\mathbf{r}(t)$. First we recall that

$$\mathbf{r}'(t) = v\mathbf{T}(s).$$

Taking the dot product with (15.22) gives us

$$\mathbf{r}'(t) \cdot \mathbf{r}''(t) = \frac{dv}{dt} v\mathbf{T}(s) \cdot \mathbf{T}(s) + \frac{v^3}{\rho}\mathbf{T}(s) \cdot \mathbf{N}(s).$$

Since $\mathbf{T}(s)$ has magnitude 1 and is orthogonal to $\mathbf{N}(s)$, this reduces to

$$\mathbf{r}'(t) \cdot \mathbf{r}''(t) = \frac{dv}{dt} v = \frac{dv}{dt}|\mathbf{r}'(t)|.$$

Hence, if $|\mathbf{r}'(t)| \neq 0$, then

$$\text{(15.23)} \qquad \frac{dv}{dt} = \frac{\mathbf{r}'(t) \cdot \mathbf{r}''(t)}{|\mathbf{r}'(t)|} = \text{tangential component of acceleration.}$$

In like manner, if we take the cross product of $\mathbf{r}'(t)$ with (15.22), then

$$\begin{aligned} \mathbf{r}'(t) \times \mathbf{r}''(t) &= v\frac{dv}{dt}\mathbf{T}(s) \times \mathbf{T}(s) + \frac{v^3}{\rho}\mathbf{T}(s) \times \mathbf{N}(s) \\ &= \frac{v^3}{\rho}\mathbf{T}(s) \times \mathbf{N}(s). \end{aligned}$$

Since $\mathbf{T}(s)$ and $\mathbf{N}(s)$ are orthogonal unit vectors, $|\mathbf{T}(s) \times \mathbf{N}(s)| = 1$ and therefore

$$|\mathbf{r}'(t) \times \mathbf{r}''(t)| = \frac{v^3}{\rho}.$$

Consequently

$$\frac{v^2}{\rho} = \frac{|\mathbf{r}'(t) \times \mathbf{r}''(t)|}{|\mathbf{r}'(t)|} = \text{normal component of acceleration.} \tag{15.24}$$

Example 1 The position of a particle at time t is (t, t^2, t^3). Find the tangential and normal components of acceleration at time t.

Solution We have the following:

$$\begin{aligned}
\mathbf{r}(t) &= t\mathbf{i} + t^2\mathbf{j} + t^3\mathbf{k} \\
\mathbf{r}'(t) &= \mathbf{i} + 2t\mathbf{j} + 3t^2\mathbf{k} \\
\mathbf{r}''(t) &= 2\mathbf{j} + 6t\mathbf{k} \\
|\mathbf{r}'(t)| &= (1 + 4t^2 + 9t^4)^{1/2}.
\end{aligned}$$

Using (15.23), the tangential component of acceleration is

$$\frac{dv}{dt} = \frac{\mathbf{r}'(t) \cdot \mathbf{r}''(t)}{|\mathbf{r}'(t)|} = \frac{4t + 18t^3}{(1 + 4t^2 + 9t^4)^{1/2}}.$$

To find the normal component we first determine

$$\begin{aligned}
\mathbf{r}'(t) \times \mathbf{r}''(t) &= \begin{vmatrix} \mathbf{i} & \mathbf{j} & \mathbf{k} \\ 1 & 2t & 3t^2 \\ 0 & 2 & 6t \end{vmatrix} \\
&= 6t^2\mathbf{i} - 6t\mathbf{j} + 2\mathbf{k}.
\end{aligned}$$

Applying (15.24),

$$\frac{v^2}{\rho} = \frac{|\mathbf{r}'(t) \times \mathbf{r}''(t)|}{|\mathbf{r}'(t)|} = \frac{(36t^4 + 36t^2 + 4)^{1/2}}{(1 + 4t^2 + 9t^4)^{1/2}} = 2\left(\frac{9t^4 + 9t^2 + 1}{9t^4 + 4t^2 + 1}\right)^{1/2}.$$

Example 2 If a point P moves around a circle of radius a with a constant speed v, find the tangential and normal components of acceleration.

Solution Since $v = ds/dt$ is a constant, the tangential component dv/dt is 0. By Example 2 of the previous section, the curvature of the circle is $1/a$. Hence the normal component of acceleration (see (15.24)) is v^2/a. This shows that the acceleration is a vector of constant magnitude directed from P toward the center of the circle (see Example 2 in Section 15.3).

We may use (15.24) to obtain a formula for curvature. Specifically, if a curve is given parametrically by

$$x = f(t), \quad y = g(t), \quad z = h(t)$$

let

$$\mathbf{r}(t) = f(t)\mathbf{i} + g(t)\mathbf{j} + h(t)\mathbf{k}.$$

We may then regard C as the curve traced by the end-point of $\mathbf{r}(t)$ as t varies. Since

$v = |\mathbf{r}'(t)|$ and $K = 1/\rho$ we have, from (15.24),

$$K = \frac{|\mathbf{r}'(t) \times \mathbf{r}''(t)|}{|\mathbf{r}'(t)|^3}. \tag{15.25}$$

This formula may also be used for plane curves (see Exercises 23–27).

Example 3 Find the curvature of the twisted cubic $x = t$, $y = t^2$, $z = t^3$ and the point (x, y, z).

Solution If we let

$$\mathbf{r}(t) = t\mathbf{i} + t^2\mathbf{j} + t^3\mathbf{k}$$

then the curve is the same as that considered in Example 1. Substituting the expressions obtained there for $\mathbf{r}'(t)$ and $\mathbf{r}'(t) \times \mathbf{r}''(t)$ in (15.25) gives us

$$K = \frac{2(9t^4 + 9t^2 + 1)^{1/2}}{(1 + 4t^2 + 9t^4)^{3/2}}.$$

EXERCISES 15.5

In each of Exercises 1–8, $\mathbf{r}(t)$ is the position vector of a particle at time t. Find the tangential and normal components of acceleration at time t.

1 $\mathbf{r}(t) = t^2\mathbf{i} + (3t + 2)\mathbf{j}$

2 $\mathbf{r}(t) = (2t^2 - 1)\mathbf{i} + 5t\mathbf{j}$

3 $\mathbf{r}(t) = 3t\mathbf{i} + t^3\mathbf{j} + 3t^2\mathbf{k}$

4 $\mathbf{r}(t) = 4t\mathbf{i} + t^2\mathbf{j} + 2t^2\mathbf{k}$

5 $\mathbf{r}(t) = t(\cos t\mathbf{i} + \sin t\mathbf{j})$

6 $\mathbf{r}(t) = \cosh t\mathbf{i} + \sinh t\mathbf{j}$

7 $\mathbf{r}(t) = 4\cos t\mathbf{i} + 9\sin t\mathbf{j} + t\mathbf{k}$

8 $\mathbf{r}(t) = e^t(\sin t\mathbf{i} + \cos t\mathbf{j} + \mathbf{k})$

9–16 Use (15.25) to find the curvature, at the point corresponding to t, of the curve traced by the end-point of $\mathbf{r}(t)$ in each of Exercises 1–8.

17 A particle moves along the parabola $y = x^2$ such that the horizontal component of velocity is always 3. Find the tangential and normal components of acceleration at the point $P(1, 1)$. Sketch the path and illustrate the acceleration as a sum of two vectors as in (15.22).

18 Work Exercise 17 if the particle moves along the graph of $y = 2x^3 - x$.

19 Prove that if a particle moves along a curve C with a constant speed, then the acceleration is always normal to C.

20 Show that if a particle moves along the graph of $y = f(x)$, where $a \leq x \leq b$, then the normal component of acceleration is 0 at a point of inflection. Illustrate this fact by using $\mathbf{r}(t) = t\mathbf{i} + t^3\mathbf{j}$.

21 Show that the curvature at every point on the helix

$$x = a\cos t, \quad y = a\sin t, \quad z = bt$$

where $a > 0$, is given by $K = a/(a^2 + b^2)$.

22 Find the curvature, at (x, y, z), of the curve having parametric equations $x = a\cos t$, $y = b\sin t$, $z = ct$, where a, b, and c are positive.

In each of Exercises 23–26 use (15.25) to find the curvature of the given plane curve at the point P.

23 $x = t - t^2, y = 1 - t^3; P(0, 1)$

24 $x = t - \sin t, y = 1 - \cos t; P(\pi/2 - 1, 1)$

25 $x = 2\sin t, y = 3\cos t; P(1, 3\sqrt{3}/2)$

26 $x = \cos^3 t, y = \sin^3 t; P(\sqrt{2}/4, \sqrt{2}/4)$

27 If a plane curve is given parametrically by $x = f(t), y = g(t)$, where f'' and g'' exist, use (15.25) to prove that the curvature at the point K is given by

$$K = \frac{|f'(t)g''(t) - g'(t)f''(t)|}{[(f'(t))^2 + (g'(t))^2]^{3/2}}.$$

15.6 KEPLER'S LAWS

After many years of analyzing an enormous amount of empirical data, the German astronomer Johannes Kepler (1571–1630) formulated three laws that describe the motion of planets about the sun. These laws may be stated as follows.

Kepler's First Law. The orbit of each planet is an ellipse with the sun at one focus.

Kepler's Second Law. The vector from the sun to a moving planet sweeps out area at a constant rate.

Kepler's Third Law. If the time required for a planet to travel once around its elliptical orbit is T, and if the major axis of the ellipse is $2a$, then $T^2 = ka^3$ for some constant k.

Approximately 50 years later, Sir Isaac Newton (1642–1727) proved that Kepler's Laws were consequences of his Law of Universal Gravitation and Second Law of Motion. The achievements of both men were monumental, because these laws clarified all astronomical observations which had been made up to that time.

In this section, we shall prove Kepler's Laws by the use of vector techniques. Since the force of gravity which the sun exerts on a planet far exceeds that exerted by other celestial bodies, we shall neglect all other forces acting on a planet. From this point of view there are only two objects to consider: the sun and a planet revolving around it.

It is convenient to introduce a coordinate system with the center of mass of the sun at the origin, O, as illustrated in Figure 15.17. The point P represents the center of mass of the planet. To simplify the notation we shall denote the position vector of P by $\mathbf{r}$ instead of $\mathbf{r}(t)$, and use $\mathbf{v}$ and $\mathbf{a}$ to denote the velocity $\mathbf{r}'(t)$ and acceleration $\mathbf{r}''(t)$, respectively. Throughout our discussion we shall not distinguish between vectors represented by directed line segments and vectors represented by triples of real numbers.

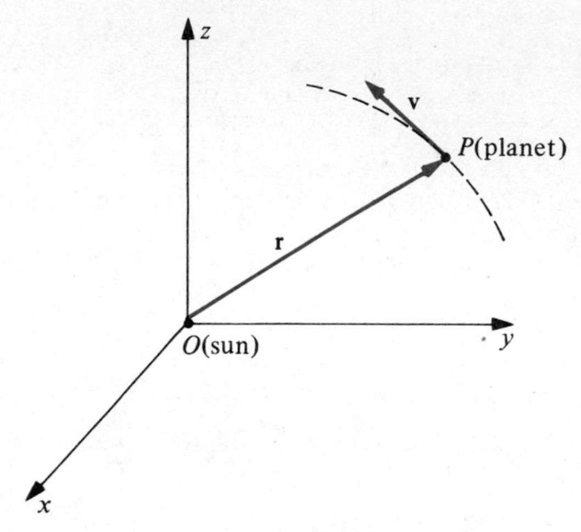

Figure 15.17

Before proving Kepler's Laws, let us show that the motion of the planet takes place in one plane. If we let $r = |\mathbf{r}|$, then $\mathbf{u} = (1/r)\mathbf{r}$ is a unit vector having the same directions as $\mathbf{r}$. According to Newton's Law of Gravitation, the force $\mathbf{F}$ of gravitational attraction on the planet is given by

$$\mathbf{F} = -G\frac{Mm}{r^2}\mathbf{u}$$

where M is the mass of the sun, m is the mass of the planet, and G is the gravitational constant. Newton's Second Law of Motion states that

$$\mathbf{F} = m\mathbf{a}.$$

If we equate these two expressions for $\mathbf{F}$ and solve for $\mathbf{a}$ we obtain

$$\mathbf{a} = -\frac{GM}{r^2}\mathbf{u}. \tag{15.26}$$

This shows that $\mathbf{a}$ is parallel to $\mathbf{r} = r\mathbf{u}$ and hence $\mathbf{r} \times \mathbf{a} = \mathbf{0}$. In addition, since $\mathbf{v} \times \mathbf{v} = \mathbf{0}$ we see that

$$\begin{aligned} \frac{d}{dt}(\mathbf{r} \times \mathbf{v}) &= \mathbf{r} \times \frac{d\mathbf{v}}{dt} + \frac{d\mathbf{r}}{dt} \times \mathbf{v} \\ &= \mathbf{r} \times \mathbf{a} + \mathbf{v} \times \mathbf{v} = \mathbf{0}. \end{aligned}$$

It follows that

$$\mathbf{r} \times \mathbf{v} = \mathbf{c} \tag{15.27}$$

where $\mathbf{c}$ is a constant vector. The vector $\mathbf{c}$ will play an important role in the proofs of Kepler's Laws.

Since $\mathbf{r} \times \mathbf{v} = \mathbf{c}$, the vector $\mathbf{r}$ is orthogonal to $\mathbf{c}$ for every value of t. This implies that the curve traced by P lies in one plane, that is, *the orbit of the planet is a plane curve.*

Let us now prove Kepler's First Law. There is no loss of generality in assuming that the motion of the planet takes place in the xy-plane. In this case the vector $\mathbf{c}$ is perpendicular to the xy-plane, and we may assume that $\mathbf{c}$ has the same direction as the positive z-axis, as illustrated in Figure 15.18.

Since $\mathbf{r} = r\mathbf{u}$ we see that

$$\mathbf{v} = \frac{d\mathbf{r}}{dt} = r\frac{d\mathbf{u}}{dt} + \frac{dr}{dt}\mathbf{u}.$$

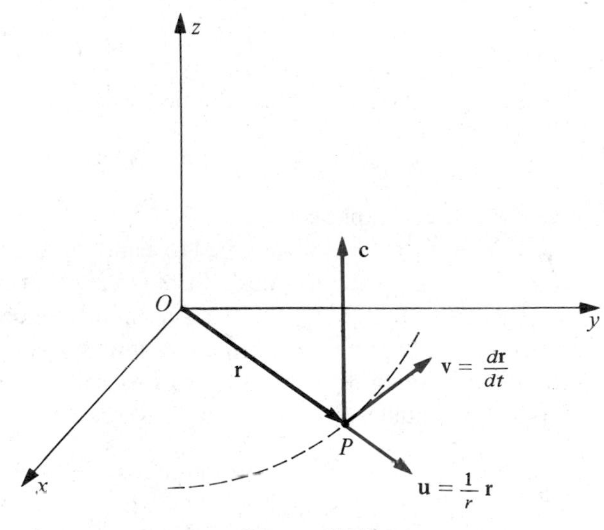

Figure 15.18

Substitution in $\mathbf{c} = \mathbf{r} \times \mathbf{v}$, and use of properties of vector products gives us

$$\begin{aligned}\mathbf{c} &= r\mathbf{u} \times \left(r\frac{d\mathbf{u}}{dt} + \frac{dr}{dt}\mathbf{u}\right)\\ &= r^2\left(\mathbf{u} \times \frac{d\mathbf{u}}{dt}\right) + r\frac{dr}{dt}(\mathbf{u} \times \mathbf{u}).\end{aligned}$$

Since $\mathbf{u} \times \mathbf{u} = \mathbf{0}$, this reduces to

$$\mathbf{c} = r^2\left(\mathbf{u} \times \frac{d\mathbf{u}}{dt}\right). \tag{15.28}$$

Using (15.28) and (15.26), together with (ii) and (vi) of Theorem 14.43, we see that

$$\begin{aligned}\mathbf{a} \times \mathbf{c} &= \left(-\frac{GM}{r^2}\mathbf{u}\right) \times \left[r^2\left(\mathbf{u} \times \frac{d\mathbf{u}}{dt}\right)\right]\\ &= -GM\left[\mathbf{u} \times \left(\mathbf{u} \times \frac{d\mathbf{u}}{dt}\right)\right]\\ &= -GM\left[\left(\mathbf{u} \cdot \frac{d\mathbf{u}}{dt}\right)\mathbf{u} - (\mathbf{u} \cdot \mathbf{u})\frac{d\mathbf{u}}{dt}\right].\end{aligned}$$

Since $|\mathbf{u}| = 1$ it follows from Example 4 of Section 15.2 that $\mathbf{u} \cdot (d\mathbf{u}/dt) = 0$. In addition, $\mathbf{u} \cdot \mathbf{u} = |\mathbf{u}|^2 = 1$, and hence the last formula for $\mathbf{a} \times \mathbf{c}$ reduces to

$$\mathbf{a} \times \mathbf{c} = GM\frac{d\mathbf{u}}{dt} = \frac{d}{dt}(GM\mathbf{u}).$$

We may also write

$$\mathbf{a} \times \mathbf{c} = \frac{d\mathbf{v}}{dt} \times \mathbf{c} = \frac{d}{dt}(\mathbf{v} \times \mathbf{c})$$

and consequently

$$\frac{d}{dt}(\mathbf{v} \times \mathbf{c}) = \frac{d}{dt}(GM\mathbf{u}).$$

Integrating both sides of this equation gives us

$$\mathbf{v} \times \mathbf{c} = GM\mathbf{u} + \mathbf{b} \tag{15.29}$$

where $\mathbf{b}$ is a constant vector.

The vector $\mathbf{v} \times \mathbf{c}$ is orthogonal to $\mathbf{c}$ and, therefore, is in the xy-plane. Since $\mathbf{u}$ is also in the xy-plane it follows from (15.29) that $\mathbf{b}$ is in the xy-plane.

Up to this point, our proof has been independent of the positions of the x- and y-axes. Let us now choose a coordinate system such that the positive x-axis has the direction of the constant vector $\mathbf{b}$, as illustrated in Figure 15.19. Let (r, θ) be polar coordinates for the point P, where $r = |\mathbf{r}|$. It follows that

$$\mathbf{u} \cdot \mathbf{b} = |\mathbf{u}|\,|\mathbf{b}| \cos\theta = b\cos\theta$$

where $b = |\mathbf{b}|$. If we let $c = |\mathbf{c}|$, then using (15.27) together with properties of the dot and vector products, and also (15.29),

$$\begin{aligned} c^2 &= \mathbf{c} \cdot \mathbf{c} = (\mathbf{r} \times \mathbf{v}) \cdot \mathbf{c} = \mathbf{r} \cdot (\mathbf{v} \times \mathbf{c}) \\ &= (r\mathbf{u}) \cdot (GM\mathbf{u} + \mathbf{b}) \\ &= rGM(\mathbf{u} \cdot \mathbf{u}) + r(\mathbf{u} \cdot \mathbf{b}) \\ &= rGM + rb\cos\theta. \end{aligned}$$

Solving the last equation for r gives us

$$r = \frac{c^2}{GM + b\cos\theta}.$$

Dividing numerator and denominator of this fraction by GM we obtain

$$r = \frac{p}{1 + e\cos\theta} \tag{15.30}$$

where $p = c^2/GM$ and $e = b/GM$. From Theorem (13.11), the graph of this polar equation is a conic with eccentricity e and focus at the origin. Since the orbit is a closed curve, it follows that $0 < e < 1$ and that the conic is an ellipse. This completes the proof of Kepler's First Law.

Let us next prove Kepler's Second Law. It may be assumed that the orbit of the planet is an ellipse in the xy-plane. Let $r = f(\theta)$ be a polar equation of the

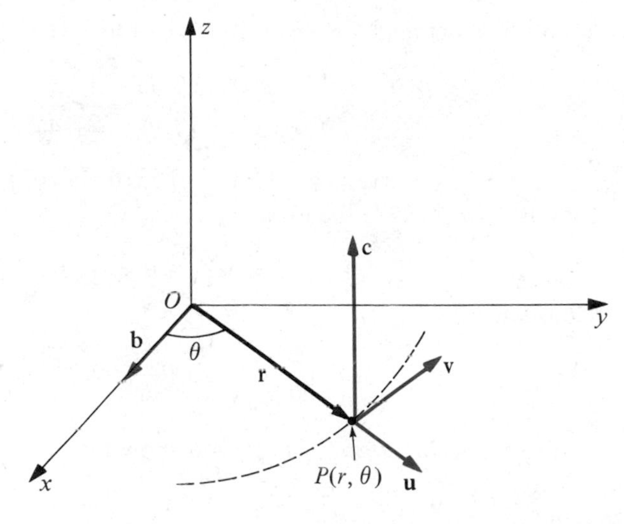

Figure 15.19

orbit, with the sun centered at the focus O. Let P_0 denote the position of the planet at time t_0, and P its position at any time $t \geq t_0$. As illustrated in Figure 15.20, θ_0 and θ will denote the angles measured from the positive x-axis to $\overrightarrow{OP_0}$ and $\overrightarrow{OP}$, respectively.

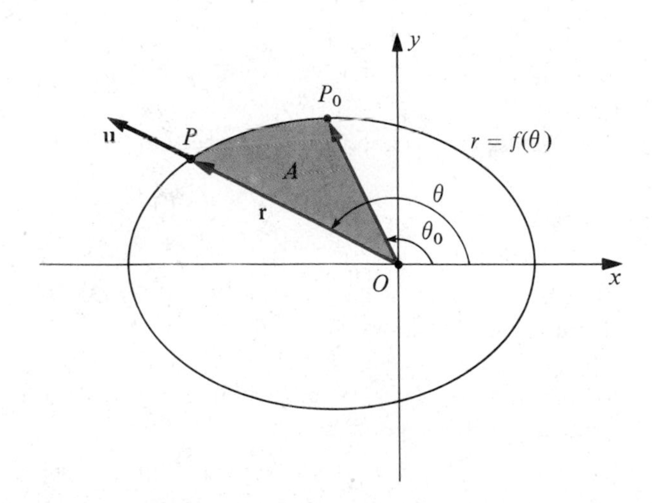

Figure 15.20

By (13.12), the area A swept out by $\overrightarrow{OP}$ in the time interval $[t_0, t]$ is given by

$$A = \int_{\theta_0}^{\theta} \frac{1}{2} r^2 \, d\theta$$

and hence

$$\frac{dA}{d\theta} = \frac{d}{d\theta} \int_{\theta_0}^{\theta} \frac{1}{2} r^2 \, d\theta = \frac{1}{2} r^2.$$

Using this fact and the Chain Rule gives us

$$\frac{dA}{dt} = \frac{dA}{d\theta}\frac{d\theta}{dt} = \frac{1}{2}r^2\frac{d\theta}{dt}. \tag{15.31}$$

Next we observe that since $\mathbf{r} = \langle r\cos\theta, r\sin\theta, 0\rangle$, the unit vector $\mathbf{u} = (1/r)\mathbf{r}$ may be expressed in the form

$$\mathbf{u} = \langle \cos\theta, \sin\theta, 0\rangle.$$

Consequently

$$\frac{d\mathbf{u}}{dt} = \left\langle -\sin\theta\frac{d\theta}{dt}, \cos\theta\frac{d\theta}{dt}, 0\right\rangle.$$

A direct calculation may be used to show that

$$\mathbf{u} \times \frac{d\mathbf{u}}{dt} = \frac{d\theta}{dt}\mathbf{k}.$$

If $\mathbf{c}$ is the vector obtained in the proof of Kepler's First Law, then by (15.28) and the last equation,

$$\mathbf{c} = r^2\left(\mathbf{u} \times \frac{d\mathbf{u}}{dt}\right) = r^2\frac{d\theta}{dt}\mathbf{k}$$

and hence

$$c = |\mathbf{c}| = r^2\frac{d\theta}{dt}. \tag{15.32}$$

Combining (15.31) and (15.32) we see that

$$\frac{dA}{dt} = \frac{1}{2}c \tag{15.33}$$

that is, the rate at which A is swept out by $\overrightarrow{OP}$ is a constant. This establishes Kepler's Second Law.

To prove Kepler's Third Law we shall retain the notation used in the proofs of the first two laws. In particular, we assume that a polar equation of the planetary orbit is given by

$$r = \frac{p}{1 + e\cos\theta}$$

where $p = c^2/GM$ and $e = b/GM$ (see (15.30)).

Let T denote the time required for the planet to make one complete revolution about the sun. By (15.33) the area swept out in the time interval $[0, T]$ is given by

$$A = \int_0^T \frac{dA}{dt}dt = \int_0^T \frac{1}{2}c\,dt = \frac{1}{2}cT.$$

This also equals the area of the plane region bounded by the ellipse. However, it was shown in Example 4 of Section 7.3 that $A = \pi ab$, where $2a$ and $2b$ are lengths of the major and minor axes, respectively, of the ellipse. Consequently

$$\frac{1}{2}cT = \pi ab, \quad \text{or} \quad T = \frac{2\pi ab}{c}.$$

From our work in Section 13.4, we know that

$$e = \frac{\sqrt{a^2 - b^2}}{a}$$

and hence

$$a^2e^2 = a^2 - b^2, \quad \text{or} \quad b^2 = a^2(1 - e^2).$$

Thus

$$T^2 = \frac{4\pi^2a^2b^2}{c^2} = \frac{4\pi^2a^4(1 - e^2)}{c^2}.$$

It was shown in the proof of Theorem 13.8 (where we had $p = de$) that

$$a^2 = \frac{p^2}{(1 - e^2)^2}, \quad \text{or} \quad a = \frac{p}{1 - e^2}$$

and hence

$$T^2 = \frac{4\pi^2a^4}{c^2}\left(\frac{p}{a}\right).$$

Since $p = c^2/GM$, this reduces to

$$T^2 = \frac{4\pi^2}{GM}a^3 = ka^3$$

where $k = 4\pi^2/GM$. This completes the proof.

In our proofs of Kepler's Laws, remember that we assumed that the only gravitational force acting on a planet was that of the sun. If forces exerted by other planets are taken into account, then irregularities in the elliptical orbits may occur. Indeed, it was because of observed irregularities in the motion of Uranus that the British astronomer J. Adams (1819–1892) and the French astronomer U. LeVerrier (1811–1877) were able to predict the orbit of an unknown planet that was causing the irregularities. Using their predictions, this planet, later named Neptune, was first observed with a telescope by the German astronomer J. Galle in 1846.

15.7 REVIEW

Concepts

Define or discuss each of the following.

1. Vector-valued function
2. Limit of a vector-valued function
3. Continuity of a vector-valued function
4. Derivative of a vector-valued function
5. Tangent vector to a curve
6. Velocity and acceleration
7. Differentiation rules for vector-valued functions
8. Integrals of vector-valued functions
9. Curvature of a plane curve
10. Curvature of a space curve
11. Circle of curvature
12. Radius of curvature
13. Tangential component of acceleration
14. Normal component of acceleration
15. Kepler's Laws

Exercises

1 If $\mathbf{r}(t) = t^2\mathbf{i} + (4t^2 - t^4)\mathbf{j}$,
(a) sketch the curve determined by the components of $\mathbf{r}(t)$.
(b) find $\mathbf{r}'(t)$ and $\mathbf{r}''(t)$.
(c) sketch geometric vectors corresponding to $\mathbf{r}'(t)$ and $\mathbf{r}''(t)$ if $t = 0$, $t = 1$, and $t = 2$.

2 The position of a particle moving in a plane is given by

$$\mathbf{r}(t) = (t - \sin t)\mathbf{i} + (1 - \cos t)\mathbf{j}.$$

Find its velocity, acceleration, and speed at time t. Sketch the path of the particle together with geometric vectors corresponding to the velocity and acceleration for the following values of t.
(a) $t = 0$ (b) $t = \pi/4$ (c) $t = \pi/2$
(d) $t = 3\pi/4$ (e) $t = \pi$ (f) $t = 5\pi/4$
(g) $t = 3\pi/2$ (h) $t = 7\pi/4$ (i) $t = 2\pi$

3 If the curve C is given by $x = e^t \sin t, y = e^t \cos t, z = e^t$, where $0 \leq t \leq 1$, find
(a) a unit tangent vector to C at the point P corresponding to $t = 0$.
(b) the length of C.

4 The position of a particle at time t is given by

$$\mathbf{r}(t) = 3t\mathbf{i} + t^3\mathbf{j} + t^4\mathbf{k}.$$

(a) Find the velocity and acceleration at time t.
(b) Sketch the path of the particle together with vectors corresponding to the velocity and acceleration at $t = 1$.
(c) Find the speed at $t = 1$.

5 If the curve C is given by $x = 3t^2 + 1, y = 4t, z = e^{t-1}$, find an equation of the tangent line at the point $P(4, 4, 1)$.

6 If $\mathbf{u}(t) = t^2\mathbf{i} + 6t\mathbf{j} + t\mathbf{k}$ and $\mathbf{v}(t) = t\mathbf{i} - 5t\mathbf{j} + 4t^2\mathbf{k}$ find
(a) the values of t for which $\mathbf{u}(t)$ and $\mathbf{v}'(t)$ are orthogonal.
(b) $D_t[\mathbf{u}(t) \times \mathbf{v}(t)]$.
(c) $D_t[\mathbf{u}(t) \cdot \mathbf{v}(t)]$.
(d) $\int \mathbf{u}(t)\,dt$.
(e) $\int_0^1 \mathbf{v}(t)\,dt$.

7 Find $\mathbf{u}(t)$ if $\mathbf{u}'(t) = e^{-t}\mathbf{i} - 4\sin 2t\mathbf{j} + 3\sqrt{t}\mathbf{k}$ and $\mathbf{u}(0) = -\mathbf{i} + 2\mathbf{j}$.

Verify the identities in Exercises 8 and 9 without using components.

8 $D_t|\mathbf{u}(t)|^2 = 2\mathbf{u}(t) \cdot \mathbf{u}'(t)$

9 $D_t(\mathbf{u}(t) \cdot \mathbf{u}'(t) \times \mathbf{u}''(t)) = \mathbf{u}(t) \cdot \mathbf{u}'(t) \times \mathbf{u}'''(t)$

In Exercises 10–12, find the curvature of the given curve at the point P.

10 $y = xe^x; P(0, 0)$

11 $x = 1/(1 + t), y = 1/(1 - t); P(2/3, 2)$

12 $x = 2t^2, y = t^4, z = 4t; P(x, y, z)$

In Exercises 13 and 14, find the tangential and normal components of acceleration at time t if the position vector of a particle is as indicated.

13 $\mathbf{r}(t) = \sin 2t\mathbf{i} + \cos t\mathbf{j}$

14 $\mathbf{r}(t) = 3t\mathbf{i} + t^3\mathbf{j} + t\mathbf{k}$

Partial Differentiation

In this chapter the concepts of limits, continuity, and derivatives are generalized to functions of more than one variable. Applications include finding rates of change, extrema, and approximations by differentials.

16.1 FUNCTIONS OF SEVERAL VARIABLES

Let us again recall that a function f is a correspondence that associates with each element of a set X a unique element of a set Y (see Definition (1.18)). If both X and Y are subsets of $\mathbb{R}$, then f is called a function of one (real) variable. If X is a subset of $\mathbb{R} \times \mathbb{R}$ and Y is a subset of $\mathbb{R}$, then f is called a *function of two (real) variables.* Another way of stating this is as follows.

(16.1) Definition

Let D be a set of ordered pairs of real numbers. A function f that associates with each pair (x, y) in D a unique real number, denoted by $f(x, y)$, is called a **function of two variables**. The set D is called the **domain** of f. The **range** of f consists of all real numbers $f(x, y)$, where (x, y) is in D.

In Chapter 1 the domain and range of a function of one variable were represented geometrically by points on a real line (see Figure 1.26). For functions of *two* variables we may represent the domain D by points in an xy-plane and the range by points on a real line, say a w-axis. This is illustrated in Figure 16.1, where several curved arrows are drawn from ordered pairs in D to the corresponding numbers in the range. To obtain a physical interpretation of this situation we could imagine a flat object having the shape of D. To each point (x, y) on the object there corresponds a temperature $f(x, y)$ which can be recorded on a thermometer, represented by the w-axis. As another illustration, we could regard D as the surface of a lake and let $f(x, y)$ denote the depth of the water under the point (x, y).

A function f of two variables is often defined by using an expression in x and y to specify the image $f(x, y)$. In this event, it is assumed that the domain is the set of all pairs (x, y) for which the given expression is meaningful, and we call f **a function of x and y**.

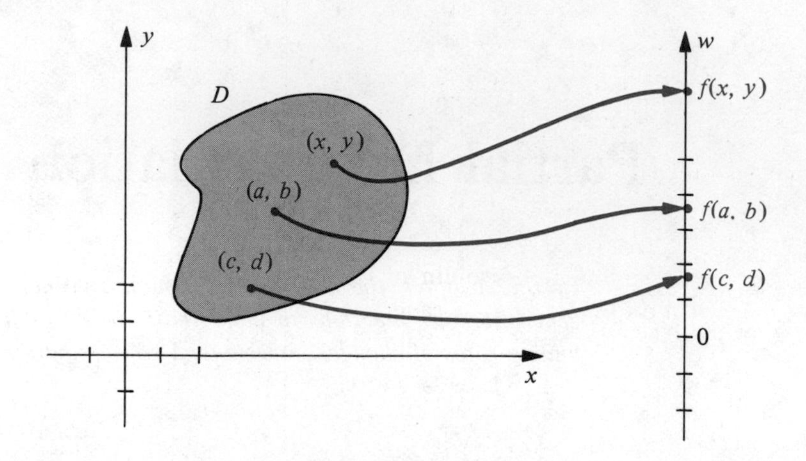

Figure 16.1

Example 1 If

$$f(x, y) = \frac{xy - 5}{2\sqrt{y - x^2}}$$

find the domain D of f. Illustrate D and the numbers $f(2, 5)$, $f(1, 2)$, and $f(-1, 2)$ in the manner of Figure 16.1.

Solution The domain D is the set of all pairs (x, y) such that $y - x^2$ is positive, that is, $y > x^2$. (Why?) Thus the graph of D is that part of the xy-plane which lies above the graph of $y = x^2$, as shown in Figure 16.2. By substitution,

$$f(2, 5) = \frac{(2)(5) - 5}{2\sqrt{5 - 4}} = \frac{5}{2}.$$

Similarly, $f(1, 2) = -3/2$ and $f(-1, 2) = -7/2$. These functional values are illustrated in Figure 16.2.

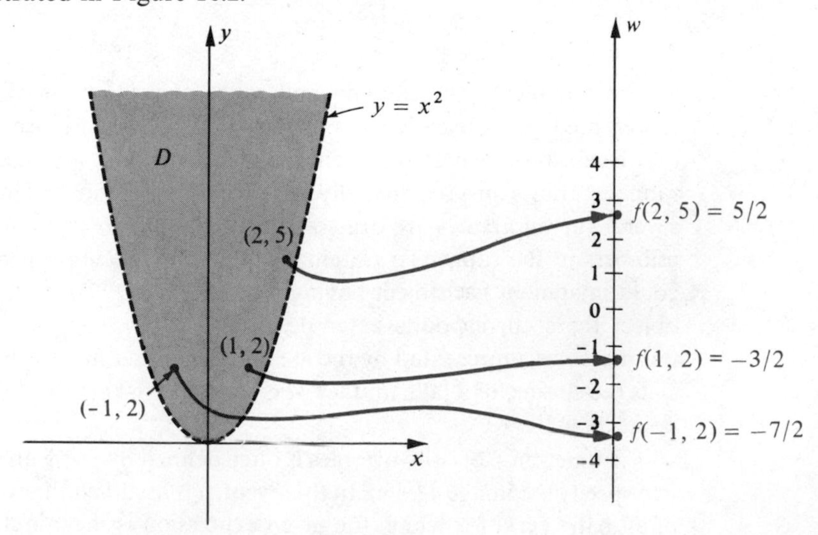

Figure 16.2

Formulas are sometimes used to define functions of two variables. For example, $V = \pi r^2 h$ expresses the volume V of a right circular cylinder as a function of the altitude h and base radius r. The symbols r and h are referred to as **independent variables** and V is the **dependent variable**.

A function f of three (real) variables is defined as in (16.1), except that the domain D is a subset of $\mathbb{R} \times \mathbb{R} \times \mathbb{R}$. In this case, with each ordered triple (x, y, z) in D there is associated a unique real number $f(x, y, z)$. If we represent D by a region in a three-dimensional rectangular coordinate system, then as illustrated in Figure 16.3, to each point (x, y, z) in D there corresponds a unique point on the w-axis with coordinate $f(x, y, z)$. To obtain a physical illustration we could regard D as a solid object and, as before, let $f(x, y, z)$ be the temperature at (x, y, z). As in the two-variable case, functions of three variables are often defined by means of expressions. For example,

$$f(x, y, z) = \frac{xe^y + \sqrt{z^2 + 1}}{xy \sin z}$$

determines a function of x, y, and z. Formulas such as $V = lwh$ for the volume of a rectangular parallelepiped of dimensions l, w, and h also illustrate how functions of three variables may arise.

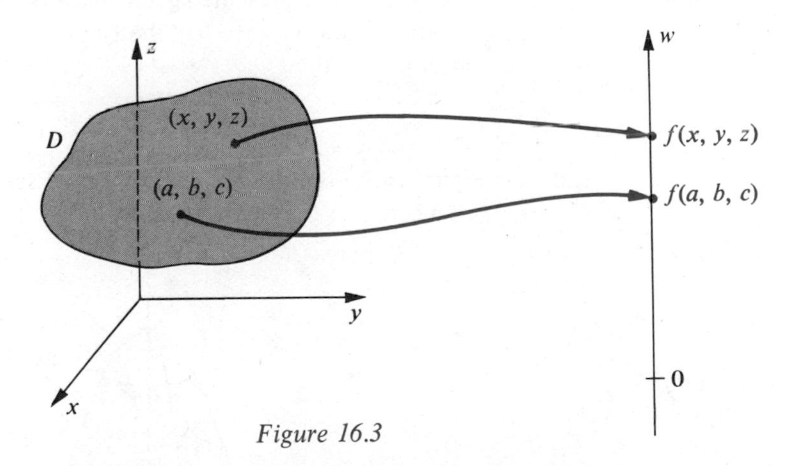

Figure 16.3

Functions of four, five, or *any* number of variables may be obtained in like manner. For example, if an object has the shape of D in Figure 16.3, we can let $f(x, y, z, t)$ denote the temperature at the point (x, y, z) at time t. This determines a function of four (independent) variables x, y, z, and t. As a final illustration, a manufacturer may find that the cost C of producing a certain item depends on material, labor, equipment, overhead, and maintenance charges. Thus C is a function of five variables.

Let us now return to a function f of two variables x and y. The **graph** of f is, by definition, the graph of the equation $z = f(x, y)$ in a three-dimensional rectangular coordinate system, and hence is usually a surface S of some type. If we represent D by a region in the xy-plane, then the pair (x, y) in D is represented by

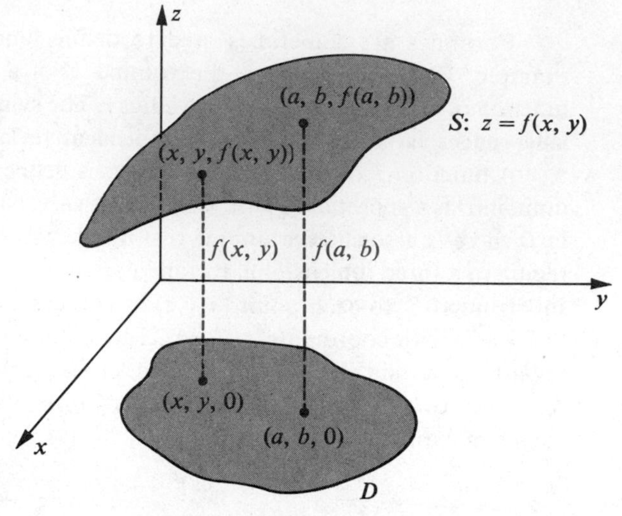

Figure 16.4

the point $(x, y, 0)$. Functional values $f(x, y)$ are the (directed) distances from the xy-plane to S as illustrated in Figure 16.4.

Example 2 Sketch the graph of f if $f(x, y) = 9 - x^2 - y^2$, where $D = \{(x, y): x^2 + y^2 \leq 9\}$.

Solution The domain of f may be represented geometrically by the collection of all points within and on the circle $x^2 + y^2 = 9$ in the xy-plane. The graph of f is the portion of the graph of

$$z = 9 - x^2 - y^2$$

shown in Figure 16.5. For future reference we have sketched circular traces on the planes $z = k$ where $k = 0, 3, 6, 8$, and 8.75.

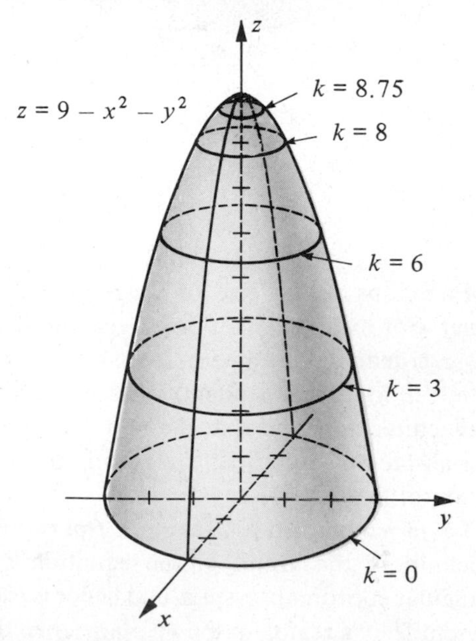

Figure 16.5. Circular traces are on the planes $z = k$.

There is another useful graphical technique for describing the behavior of a function f of two variables. The method consists of sketching, in an xy-plane, the graphs of the equations $f(x, y) = k$ for various values of k. The graphs obtained in this way are called the **level curves** associated with the function f. It is important to note that as a point (x, y) moves on a level curve, the values of the function are constant. This technique is often used in making **topographic maps** of rough terrain. For example, suppose $f(x, y)$ denotes the altitude (in feet) at a point (x, y) of latitude x and longitude y. On the hill pictured in (i) of Figure 16.6 curves (in three dimensions) corresponding to altitudes of $k = 50, 100, 200, 300$, and 400 feet are sketched. A person walking along one of these curves would always remain at the same altitude. In (ii) of the figure are the (two-dimensional) level curves corresponding to the same values of k. These are also referred to as **contour curves**. They represent the view obtained by looking down on the hill from an airplane. Similar configurations can be seen on weather maps. In this case $f(x, y)$ may represent atmospheric pressure at (x, y) and the level curves are called **isobars**.

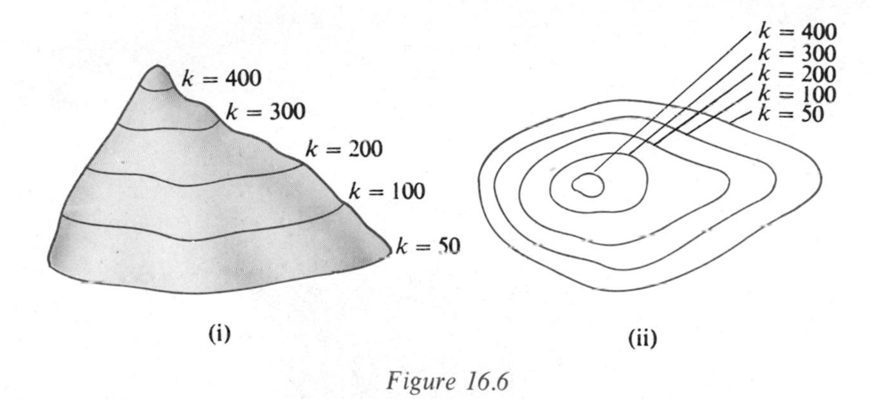

Figure 16.6

Example 3 Sketch some level curves associated with the function f of Example 2.

Solution The level curves are graphs of equations of the form $f(x, y) = k$, that is,

$$9 - x^2 - y^2 = k$$

or equivalently

$$x^2 + y^2 = 9 - k.$$

These are circles, provided $0 \leq k < 9$. In Figure 16.7 we have sketched the level curves corresponding to $k = 0, 3, 6, 8$, and 8.75. Note that the level curves correspond to traces of the graph of f on the planes $z = k$ shown in Figure 16.5.

If f is a function of three variables x, y, and z, then by definition the **level surfaces** associated with f are the graphs of $f(x, y, z) = k$ where k takes on suitable real values. If we let $k = w_0, w_1$, and w_2, there may result surfaces S_0, S_1, and S_2, respectively, as illustrated in Figure 16.8. For later use we have plotted a typical point P_0, P_1, and P_2 on each surface. As a point (x, y, z) moves along one of these surfaces, $f(x, y, z)$ does not change. In applications, if $f(x, y, z)$ is the temperature at (x, y, z), then the level surfaces are called **isothermal surfaces**. If $f(x, y, z)$ represents potential, they are called **equipotential surfaces**.

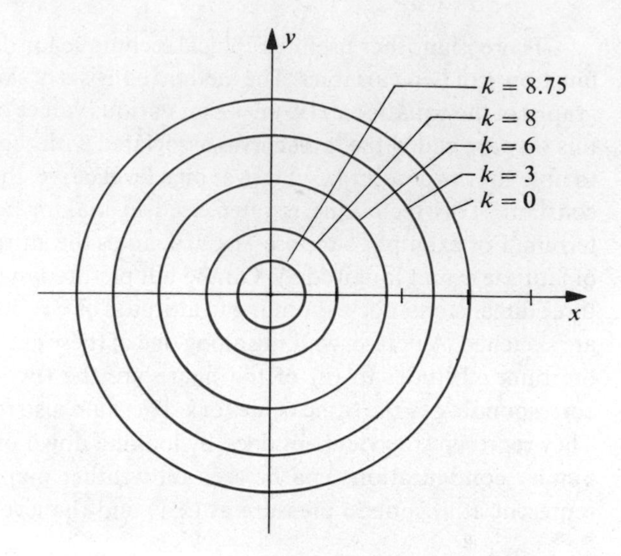

Figure 16.7

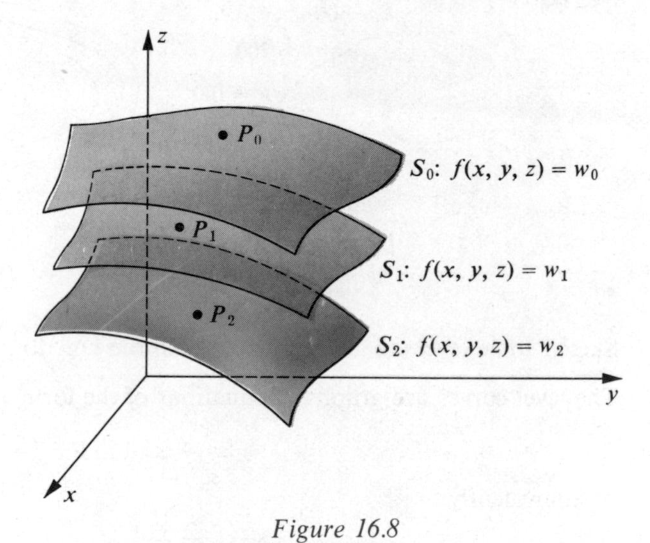

Figure 16.8

EXERCISES 16.1

In Exercises 1–8 determine the domain of f and the value of f at the indicated points.

1 $f(x, y) = 2x - y^2$; $(-2, 5), (5, -2), (0, -2)$

2 $f(x, y) = (y + 2)/x$; $(3, 1), (1, 3), (2, 0)$

3 $f(u, v) = uv/(u - 2v)$; $(2, 3), (-1, 4), (0, 1)$

4 $f(r, s) = \sqrt{1 - r} - e^{r/s}$; $(1, 1), (0, 4), (-3, 3)$

5 $f(x, y, z) = \sqrt{25 - x^2 - y^2 - z^2}$; $(1, -2, 2), (-3, 0, 2)$

6 $f(x, y, z) = 2 + \tan x + y \sin z; (\pi/4, 4, \pi/6), (0, 0, 0)$

7 $f(r, s, v, p) = rs^2 \tan v + 4sv \ln p; (3, -1, \pi/4, e)$

8 $f(x, y, u, v, w) = w \ln (x - y) - xue^{v/w}; (2, 1, 3, 4, -1)$

In Exercises 9–12 find

$$\frac{f(x + h, y) - f(x, y)}{h} \quad \text{and} \quad \frac{f(x, y + h) - f(x, y)}{h}$$

where $h \neq 0$.

9 $f(x, y) = x^2 + y^2$

10 $f(x, y) = y^2 + 3x$

11 $f(x, y) = xy^2 + 3x$

12 $f(x, y) = xy^3 + 4x^2 - 2$

In each of Exercises 13–18 describe the graph of f.

13 $f(x, y) = \sqrt{1 - x^2 - y^2}$

14 $f(x, y) = x^2 + y^2 - 1$

15 $f(x, y) = 6 - 2x - 3y$

16 $f(x, y) = \sqrt{x^2 + y^2}$

17 $f(x, y) = 5$

18 $f(x, y) = 4 - x$

In each of Exercises 19–24 sketch some level curves associated with f.

19 $f(x, y) = y^2 - x^2$

20 $f(x, y) = 3x - 2y$

21 $f(x, y) = x^2 - y$

22 $f(x, y) = xy$

23 $f(x, y) = y - \sin x$

24 $f(x, y) = 4x^2 + y^2$

In each of Exercises 25–30 describe the level surfaces associated with f.

25 $f(x, y, z) = x^2 + y^2 + z^2$

26 $f(x, y, z) = z - x^2 - y^2$

27 $f(x, y, z) = x + 2y + 3z$

28 $f(x, y, z) = x^2 + y^2 - z^2$

29 $f(x, y, z) = x^2 + y^2$

30 $f(x, y, z) = z$

16.2 LIMITS AND CONTINUITY

If f is a function of two variables, it is often important to study the variation of $f(x, y)$ as (x, y) varies through the domain D of f. As a physical illustration, consider a thin metal plate which has the shape of the region D in Figure 16.1. To each point (x, y) on the plate there corresponds a temperature $f(x, y)$ which is recorded on a thermometer represented by the w-axis. As the point (x, y) moves on the plate the temperature may increase, decrease, or remain stationary, and hence the point on the w-axis which corresponds to $f(x, y)$ will move in a positive direction, negative direction, or remain stationary, respectively. If the temperature $f(x, y)$ gets closer to a fixed value L as (x, y) gets closer and closer to a fixed point (a, b), we write

(16.2)
$$\lim_{(x, y) \to (a, b)} f(x, y) = L$$

and say that f *has the limit* L *as* (x, y) *approaches* (a, b). To make these remarks mathematically precise, let us employ a geometric device similar to that used for functions of one variable in Chapter 2. For any $\varepsilon > 0$, consider the open interval $(L - \varepsilon, L + \varepsilon)$ on the w-axis. If (16.2) is true, then as illustrated in Figure 16.9 there is a $\delta > 0$ such that for every point (x, y) inside the circle of radius δ with center (a, b), except possibly (a, b) itself, the number corresponding to $f(x, y)$ is in the interval $(L - \varepsilon, L + \varepsilon)$. This is equivalent to the statement

$$\text{if} \quad 0 < \sqrt{(x-a)^2 + (y-b)^2} < \delta, \quad \text{then} \quad |f(x, y) - L| < \varepsilon.$$

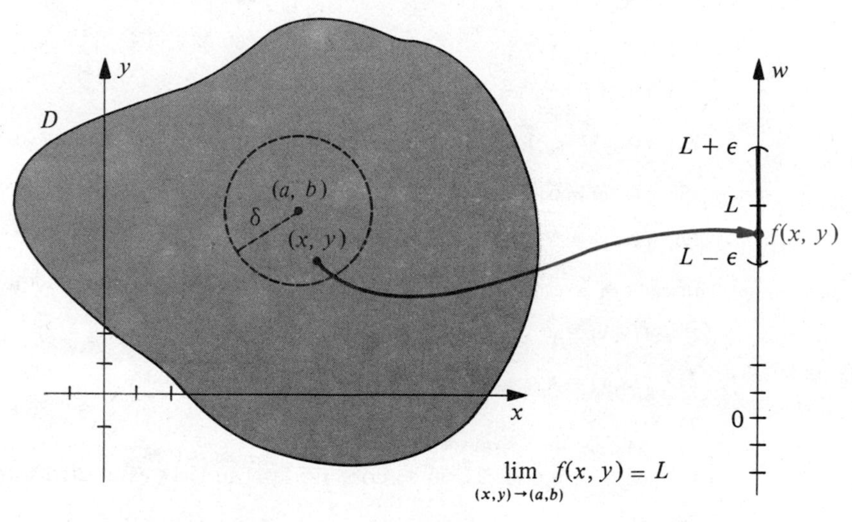

Figure 16.9

The preceding discussion is summarized in the next definition.

(16.3) Definition

Let a function f of two variables be defined throughout the interior of a circle with center (a, b), except possibly at (a, b) itself. The **limit of $f(x, y)$ as (x, y) approaches (a, b) is L**, written

$$\lim_{(x,y)\to(a,b)} f(x, y) = L$$

means that for every $\varepsilon > 0$ there corresponds a $\delta > 0$ such that

$$\text{if} \quad 0 < \sqrt{(x-a)^2 + (y-b)^2} < \delta, \quad \text{then} \quad |f(x, y) - L| < \varepsilon.$$

If we consider the graph of f as illustrated in Figure 16.4, then intuitively, Definition (16.3) means that as the point $(x, y, 0)$ approaches $(a, b, 0)$ on the xy-plane, the corresponding point $(x, y, f(x, y))$ on S approaches (a, b, L) (which may or may not be on S). It can be shown that if the limit L exists, it is unique.

If f and g are functions of two variables, then $f + g$, $f - g$, fg, and f/g are defined in the usual way, and Theorem (2.11) concerning limits of sums, products, and quotients can be extended. For example, if f and g have limits as (x, y) approaches (a, b), then it can be proved that

$$\lim_{(x,y)\to(a,b)} [f(x, y) + g(x, y)] = \lim_{(x,y)\to(a,b)} f(x, y) + \lim_{(x,y)\to(a,b)} g(x, y).$$

A function f of two variables is a **polynomial function** if $f(x, y)$ can be expressed as a sum of terms of the form $cx^m y^n$, where c is a real number and m and n are nonnegative integers. A **rational function** is a quotient of two polynomial functions. As in the single variable case, one can show that limits of such functions may be found by substituting directly for x and y. For example,

$$\begin{aligned}\lim_{(x,y)\to(2,-3)} (x^3 - 4xy^2 + 5y - 7) &= 2^3 - 4(2)(-3)^2 + 5(-3) - 7 \\ &= 8 - 72 - 15 - 7 = -86.\end{aligned}$$

The next example is an illustration of a function which has no limit as (x, y) approaches $(0, 0)$.

Example 1 Show that $\displaystyle\lim_{(x,y)\to(0,0)} \frac{x^2 - y^2}{x^2 + y^2}$ does not exist.

Solution If $f(x, y) = (x^2 - y^2)/(x^2 + y^2)$, then f is defined everywhere except $(0, 0)$. If we consider any point $(x, 0)$ on the x-axis, then $f(x, 0) = x^2/x^2 = 1$, provided $x \neq 0$. For points $(0, y)$ on the y-axis, $f(0, y) = -y^2/y^2 = -1$, if $y \neq 0$. Consequently, as illustrated in Figure 16.10, *every* circle with center $(0, 0)$ contains points at which the value of f is 1 and points at which the value of f is -1. It follows that the limit does not exist, for if we take $\varepsilon = 1$, there is no open interval $(L - \varepsilon, L + \varepsilon)$ on the w-axis containing both 1 and -1, and hence it is impossible to find a $\delta > 0$ which satisfies the conditions of Definition (16.3).

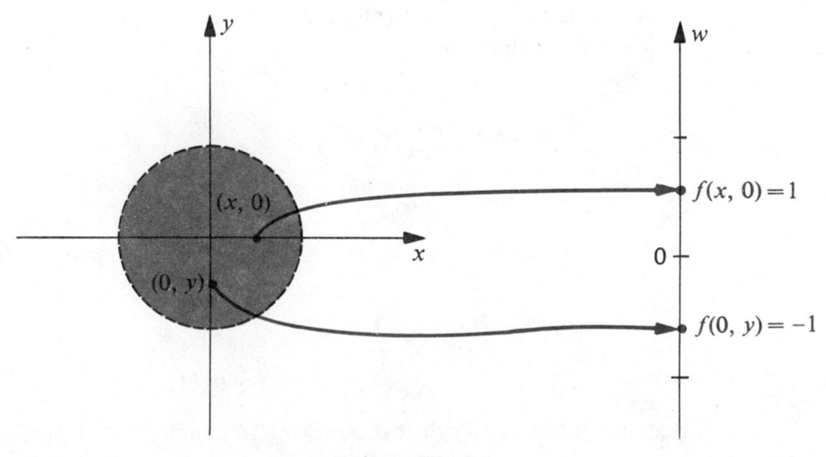

Figure 16.10

For certain functions in Chapter 2 we proved that $\lim_{x\to a} f(x)$ did not exist by showing that $\lim_{x\to a^-} f(x)$ and $\lim_{x\to a^+} f(x)$ were not equal. When considering such one-sided limits we may regard the point on the x-axis with coordinate x as "approaching" the point with coordinate a either from the left or from the right, respectively. The analogous situation for functions of two variables is more complicated, since in a coordinate plane there are an infinite number of different curves (which we shall call **paths**) along which (x, y) can approach (a, b). However, if the limit in Definition (16.3) exists, then $f(x, y)$ must have the limit L regardless of the path taken. This illustrates the following important rule for investigating certain limits.

If two different paths to a point $P(a, b)$ produce two different limiting values for f, then the limit of $f(x, y)$ as (x, y) approaches (a, b) does not exist.

Example 2 Rework Example 1 using the preceding rule.

Solution If, on the one hand, (x, y) approaches $(0, 0)$ along the x-axis, then the ordinate y is always zero and the expression $(x^2 - y^2)/(x^2 + y^2)$ reduces to x^2/x^2 or 1. Hence the limiting value is 1. On the other hand, if we let (x, y) approach $(0, 0)$ along the y-axis, then the abscissa x is 0 and $(x^2 - y^2)/(x^2 + y^2)$ reduces to $-y^2/y^2$ or -1. Since two different limits are obtained, the given limit does not exist. We could, of course, have chosen other paths to the origin $(0, 0)$. For example, if we let (x, y) approach $(0, 0)$ along the line $y = 2x$, then

$$\frac{x^2 - y^2}{x^2 + y^2} = \frac{x^2 - 4x^2}{x^2 + 4x^2} = \frac{-3x^2}{5x^2} = -\frac{3}{5}$$

and hence the limiting value would be $-3/5$.

Example 3 If $f(x, y) = \dfrac{x^2 y}{x^4 + y^2}$ find, if possible, $\displaystyle\lim_{(x,y)\to(0,0)} f(x, y)$.

Solution If we let (x, y) approach $(0, 0)$ along any line $y = mx$ which passes through the origin, we see that if $m \neq 0$,

$$\begin{aligned}\lim_{(x,y)\to(0,0)} \frac{x^2 y}{x^4 + y^2} &= \lim_{(x,y)\to(0,0)} \frac{x^2(mx)}{x^4 + (mx)^2} \\ &= \lim_{(x,y)\to(0,0)} \frac{mx^3}{x^4 + m^2x^2} \\ &= \lim_{(x,y)\to(0,0)} \frac{mx}{x^2 + m^2} \\ &= \frac{0}{0 + m^2} = 0.\end{aligned}$$

Because of this behavior it is tempting to conclude that $f(x, y)$ has the limit 0 as (x, y) approaches $(0, 0)$. However, if we let (x, y) approach $(0, 0)$ along the parabola

$y = x^2$, then

$$\lim_{(x,y)\to(0,0)} \frac{x^2y}{x^4 + y^2} = \lim_{(x,y)\to(0,0)} \frac{x^2(x^2)}{x^4 + (x^2)^2}$$

$$= \lim_{(x,y)\to(0,0)} \frac{x^4}{2x^4}$$

$$= \lim_{(x,y)\to(0,0)} \frac{1}{2} = \frac{1}{2}.$$

Thus not every path through $(0, 0)$ leads to the same limiting value of $f(x, y)$ and, therefore, the limit does not exist.

If R is a region in the xy-plane, then a point (a, b) is called an **interior point** of R if there exists a circle with center (a, b) which contains only points of R. An illustration of an interior point is shown in (i) of Figure 16.11. A point (a, b) is called a **boundary point** of R if *every* circle with center (a, b) contains points which are in R and points which are not in R, as illustrated in (ii) of Figure 16.11.

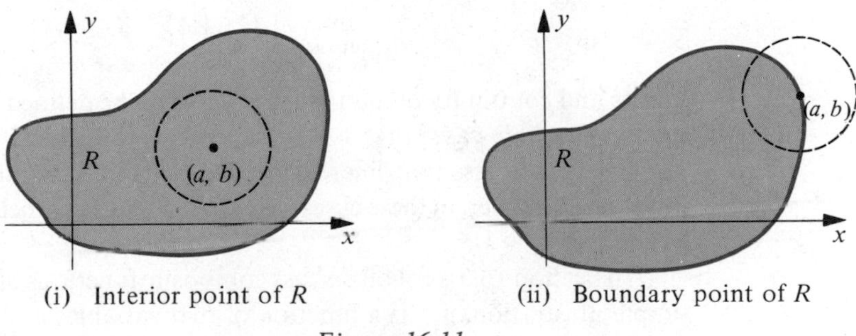

(i) Interior point of R (ii) Boundary point of R

Figure 16.11

Suppose a function f of two variables is defined for all (x, y) in a region R, except possibly at (a, b). If (a, b) is an interior point, then Definition (16.3) can be used to investigate the limit of f as (x, y) approaches (a, b). If (a, b) is a boundary point we shall use (16.3) with the added restriction that (x, y) must be both in R and in the circle of radius δ. Theorems on limits may then be extended to boundary points.

A function f of two variables is **continuous** at (a, b) if

$$\lim_{(x,y)\to(a,b)} f(x, y) = f(a, b)$$

where we apply the previous remarks if (a, b) is a boundary point of a region R. If R is in the domain D of f, then f is said to be **continuous on *R*** if it is continuous at every pair (a, b) in R. If f is continuous on R, then a small change in (x, y) produces a small change in $f(x, y)$. Referring to the graph S of f (see Figure 16.4), if (x, y) is close to (a, b), then the point $(x, y, f(x, y))$ on S is close to $(a, b, f(a, b))$. Thus there are no holes or vertical steps in the graph of a continuous function.

Theorems on continuity which are analogous to those for functions of one variable may be proved. In particular, polynomial functions are continuous everywhere and rational functions are continuous except at points where the denominator vanishes.

The preceding definitions and remarks on limits and continuity can be extended to functions of three or more variables. For example, if f is a function of three variables, then

$$\lim_{(x,y,z)\to(a,b,c)} f(x,y,z) = L$$

(16.4) means that for every $\varepsilon > 0$ there corresponds a $\delta > 0$ such that

$$\text{if} \quad 0 < \sqrt{(x-a)^2 + (y-b)^2 + (z-c)^2} < \delta, \quad \text{then} \quad |f(x,y,z) - L| < \varepsilon.$$

We may give a graphical interpretation for (16.4) which is similar to that illustrated in Figure 16.9. The difference here is that in place of the circle in the xy-plane, we use a sphere of radius δ with center at the point (a,b,c) in three dimensions (see Exercise 25).

A function f of three variables is **continuous** at (a,b,c) if

$$\lim_{(x,y,z)\to(a,b,c)} f(x,y,z) = f(a,b,c).$$

Limits and continuity on boundary points can be defined in a manner similar to the two-variable case.

There is no essential difference in defining limits for functions of four or more variables; however, in these cases we no longer give geometric interpretations (see Exercise 27).

In Section 16.5 we shall discuss composite functions of several variables. As a simple illustration, if f is a function of two variables and g is a function of one variable, then a function h of two variables may be defined by letting $h(x,y) = g(f(x,y))$, provided the range of f is in the domain of g.

Example 4 In each of the following, express $g(f(x,y))$ in terms of x and y, and find the domain of the resulting composite function.

(a) $f(x,y) = xe^y, g(t) = 3t^2 + t + 1$

(b) $f(x,y) = y - 4x^2, g(t) = \sin\sqrt{t}$

Solution In each case we substitute $f(x,y)$ for t. Thus:

(a) $g(f(x,y)) = g(xe^y) = 3x^2e^{2y} + xe^y + 1.$

(b) $g(f(x,y)) = g(y - 4x^2) = \sin\sqrt{y - 4x^2}.$

The domain in part (a) is $\mathbb{R} \times \mathbb{R}$, and the domain in part (b) consists of all ordered pairs (x,y) such that $y \geq 4x^2$.

The following analogue of (2.31) may be proved for functions of the type in Example 4.

(16.5) Theorem

> If a function f of two variables is continuous at (a, b) and a function g of one variable is continuous at $f(a, b)$, then the function h defined by $h(x, y) = g(f(x, y))$ is continuous at (a, b).

To illustrate, the function f in part (b) of Example 4 is continuous throughout $\mathbb{R} \times \mathbb{R}$ and g is continuous for all nonnegative real numbers. Hence, by Theorem (16.5), the composite function h, where

$$h(x, y) = g(f(x, y)) = \sin \sqrt{y - 4x^2}$$

is continuous at every pair (x, y) such that $y - 4x^2$ is nonnegative, or equivalently, $y \geq 4x^2$.

EXERCISES 16.2

In Exercises 1–10 find the limits, if they exist.

1 $\displaystyle\lim_{(x,y)\to(0,0)} \frac{x^2 - 2}{3 + xy}$

2 $\displaystyle\lim_{(x,y)\to(0,0)} \frac{x^3 - x^2y + xy^2 - y^3}{x^2 + y^2}$

3 $\displaystyle\lim_{(x,y)\to(0,0)} \frac{2x^2 - y^2}{x^2 + 2y^2}$

4 $\displaystyle\lim_{(x,y)\to(0,0)} \frac{x^2 - 2xy + 5y^2}{3x^2 + 4y^2}$

5 $\displaystyle\lim_{(x,y)\to(0,0)} \frac{4x^2y}{x^3 + y^3}$

6 $\displaystyle\lim_{(x,y)\to(0,0)} \frac{x^4 - y^4}{x^2 + y^2}$

7 $\displaystyle\lim_{(x,y)\to(1,2)} \frac{xy - 2x - y + 2}{x^2 + y^2 - 2x - 4y + 5}$

8 $\displaystyle\lim_{(x,y)\to(0,0)} \frac{3xy}{5x^4 + 2y^4}$

9 $\displaystyle\lim_{(x,y)\to(0,0)} \frac{x^3 - x^2y + y^2x - y^3}{x^2 + y^2}$

10 $\displaystyle\lim_{(x,y,z)\to(0,0,0)} \frac{xy + yz + xz}{x^2 + y^2 + z^2}$

In each of Exercises 11–16 discuss the continuity of the function f.

11 $f(x, y) = \ln(x + y - 1)$

12 $f(x, y) = xy/(x^2 - y^2)$

13 $f(x, y, z) = 1/(x^2 + y^2 - z^2)$

14 $f(x, y, z) = \sqrt{xy} \tan z$

15 $f(x, y) = \sqrt{x}e^{\sqrt{1-y^2}}$

16 $f(x, y) = \sqrt{25 - x^2 - y^2}$

In each of Exercises 17–20 find $g(f(x, y))$ and state the domain of the resulting composite function.

17 $f(x, y) = x^2 - y^2, g(t) = (t^2 - 4)/t$

18 $f(x, y) = 3x + 2y - 4, g(t) = \ln(t + 5)$

19 $f(x, y) = x + \tan y, g(z) = z^2 + 1$

20 $f(x, y) = y \ln x, g(w) = e^w$

21 If $f(x, y) = x^2 + 2y$, $g(t) = e^t$, and $h(t) = t^2 - 3t$, find $g(f(x, y))$, $h(f(x, y))$, and $f(g(t), h(t))$.

22 If $f(x, y, z) = 2x + ye^z$ and $g(t) = t^2$, find $g(f(x, y, z))$.

23 If $f(u, v) = uv - 3u + v$, $g(x, y) = x - 2y$, and $k(x, y) = 2x + y$, find $f(g(x, y), k(x, y))$.

24 If $f(x, y) = 2x + y$, find $f(f(x, y), f(x, y))$.

25 Give a graphical interpretation of (16.4).

26 Prove that if the limit L in Definition (16.3) exists, then it is unique.

27 (a) Define a function of four (real) variables.
(b) Extend Definition (16.3) to functions of four variables.

28 Prove that if f is a continuous function of two variables and $f(a, b) > 0$, then there is a circle C in the xy-plane with center (a, b) such that $f(x, y) > 0$ for all pairs (x, y) which are in the domain of f and within C.

29 Prove, directly from Definition (16.3), that

(a) $\lim_{(x,y)\to(a,b)} x = a$. (b) $\lim_{(x,y)\to(a,b)} y = b$.

30 If $\lim_{(x,y)\to(a,b)} f(x, y) = L$ and c is any real number, prove, directly from Definition (16.3), that

$$\lim_{(x,y)\to(a,b)} cf(x, y) = cL.$$

16.3 PARTIAL DERIVATIVES

In this section we begin the study of derivatives of functions of several variables.

(16.6) Definition

If f is a function of two variables, then the **first partial derivatives of f with respect to x and y** are the functions f_x and f_y defined as follows:

$$f_x(x, y) = \lim_{h\to 0} \frac{f(x + h, y) - f(x, y)}{h}$$

$$f_y(x, y) = \lim_{h\to 0} \frac{f(x, y + h) - f(x, y)}{h}$$

provided the limits exist.

Note that in the definition of $f_x(x, y)$, the variable y is fixed. It is for this reason that the notation for limits of functions of one variable is used instead of that introduced in the previous section. Indeed, if the symbol y in the definition of $f_x(x, y)$ is suppressed, then the limit in (3.6) results. It follows that to find $f_x(x, y)$ we may regard y as constant and differentiate $f(x, y)$ with respect to x in the usual way. Similarly, to find $f_y(x, y)$ the variable x is held constant and $f(x, y)$ is differentiated with respect to y.

Other common notations for partial derivatives are

$$f_x = \frac{\partial f}{\partial x} \quad \text{and} \quad f_y = \frac{\partial f}{\partial y}.$$

For brevity we often speak of $\partial f/\partial x$ or $\partial f/\partial y$ as *the partial of f with respect to x or y*, respectively.

If $w = f(x, y)$ we write

$$f_x(x, y) = \frac{\partial}{\partial x} f(x, y) = \frac{\partial w}{\partial x} = w_x$$

$$f_y(x, y) = \frac{\partial}{\partial y} f(x, y) = \frac{\partial w}{\partial y} = w_y.$$

Example 1 Find $f_x(x, y)$ and $f_y(x, y)$ if $f(x, y) = x^3y^2 - 2x^2y + 3x$.

Solution By the preceding remarks,

$$f_x(x, y) = 3x^2y^2 - 4xy + 3$$
$$f_y(x, y) = 2x^3y - 2x^2.$$

It is instructive to interpret Definition (16.6) by means of the illustration given at the beginning of the previous section, where $f(x, y)$ is the temperature of a hot plate at the point (x, y). Note that the points (x, y) and $(x + h, y)$ lie on a horizontal line, as indicated in Figure 16.12. The difference $f(x + h, y) - f(x, y)$ is the change in temperature as a point moves from (x, y) to $(x + h, y)$ and the ratio

(16.7)
$$\frac{f(x + h, y) - f(x, y)}{h}$$

is the *average* change in temperature. For example, if the temperature change is 2 degrees and h is 4, then this ratio is 1/2; that is, *on the average* the temperature changes at a rate of 1/2 degree per unit in distance. Taking the limit in (16.7) as h approaches 0 gives us the *rate of change* of temperature with respect to distance as the point (x, y) moves in a horizontal direction. Similarly, $f_y(x, y)$ measures the rate of change of $f(x, y)$ as (x, y) moves in a vertical direction. As another illustration, suppose D represents the surface of a lake and $f(x, y)$ is the depth of the water under the point (x, y) on the surface. In this case $f_x(x, y)$ is the rate at which the depth changes as a point moves away from (x, y) parallel to the x-axis, whereas $f_y(x, y)$ is the rate of change of the depth in the direction of the y-axis.

There is an interesting geometric interpretation for Definition (16.6) in terms of the graph of f. In order to illustrate $f_y(x, y)$, consider the points $M(x, y, 0)$ and $N(x, y + h, 0)$. The plane Γ, parallel to the yz-plane and passing through M and N, intersects S in a curve C. If we introduce a coordinate system on Γ, and if P and Q are the points on C which have projections M and N on the xy-plane, then as illustrated in Figure 16.13, $\overline{MP} = f(x, y)$, $\overline{NQ} = f(x, y + h)$, and $\overline{MN} = h$. It follows that the ratio $[f(x, y + h) - f(x, y)]/h$ is the slope of the secant line through P and Q in the plane Γ. The limit of this ratio as h approaches 0, namely

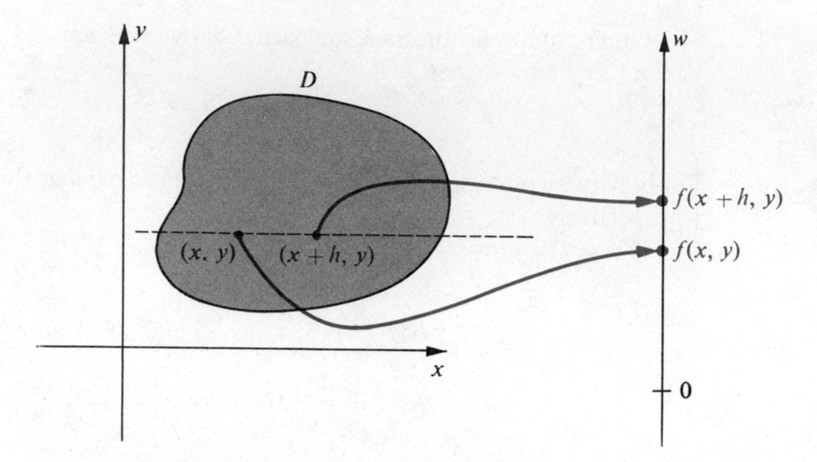

Figure 16.12

$f_y(x, y)$, is the slope of the tangent line l to C at P. Similarly, if C' is the trace of S on the plane Γ', parallel to the xz-plane and passing through M, then $f_x(x, y)$ is the slope of the tangent line to C' at P.

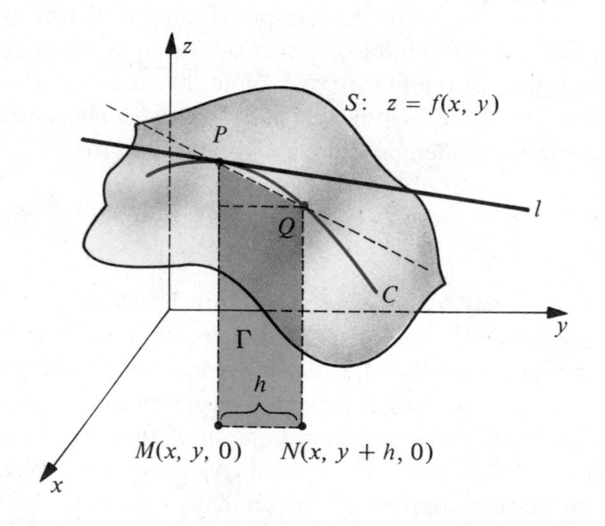

Figure 16.13

First partial derivatives of functions of three or more variables are defined in the same manner as in (16.6). Specifically, all variables except one are held constant and differentiation takes place with respect to the remaining variable. Thus, given $f(x, y, z)$ we may find f_x, f_y, and f_z (or equivalently $\partial f/\partial x$, $\partial f/\partial y$, and $\partial f/\partial z$). For example,

$$f_z(x, y, z) = \lim_{h \to 0} \frac{f(x, y, z + h) - f(x, y, z)}{h}.$$

In a manner similar to our discussion in two dimensions, if $f(x, y, z)$ is the temperature at the point $P(x, y, z)$, then $f_x(x, y, z)$ is the rate of change of temperature with respect to distance along a line through P which is parallel to the x-axis. The partial derivatives $f_y(x, y, z)$ and $f_z(x, y, z)$ give the rates of change in the directions of the y- and z-axes, respectively.

Example 2 If $w = x^2y^3 \sin z + e^{xz}$, find $\partial w/\partial x$, $\partial w/\partial y$, and $\partial w/\partial z$.

Solution

$$\frac{\partial w}{\partial x} = 2xy^3 \sin z + ze^{xz}$$

$$\frac{\partial w}{\partial y} = 3x^2y^2 \sin z$$

$$\frac{\partial w}{\partial z} = x^2y^3 \cos z + xe^{xz}$$

If f is a function of two variables x and y, then f_x and f_y are also functions of two variables and we may consider *their* first partial derivatives. The latter are called the **second partial derivatives** of f and are denoted as follows.

$$\frac{\partial}{\partial x} f_x = (f_x)_x = f_{xx} = \frac{\partial}{\partial x}\left(\frac{\partial f}{\partial x}\right) = \frac{\partial^2 f}{\partial x^2}$$

$$\frac{\partial}{\partial y} f_x = (f_x)_y = f_{xy} = \frac{\partial}{\partial y}\left(\frac{\partial f}{\partial x}\right) = \frac{\partial^2 f}{\partial y\, \partial x}$$

$$\frac{\partial}{\partial x} f_y = (f_y)_x = f_{yx} = \frac{\partial}{\partial x}\left(\frac{\partial f}{\partial y}\right) = \frac{\partial^2 f}{\partial x\, \partial y}$$

$$\frac{\partial}{\partial y} f_y = (f_y)_y = f_{yy} = \frac{\partial}{\partial y}\left(\frac{\partial f}{\partial y}\right) = \frac{\partial^2 f}{\partial y^2}$$

If $w = f(x, y)$ we write

$$\frac{\partial^2}{\partial x^2} f(x, y) = f_{xx}(x, y) = \frac{\partial^2 w}{\partial x^2} = w_{xx}$$

and likewise for the other second partial derivatives. It can be shown that if f has continuous second partial derivatives, then $f_{xy} = f_{yx}$. This will be true for most functions encountered in calculus and its applications. Similarly, if $w = f(x, y, z)$ and f has continuous second partial derivatives, then it can be proved that the order of differentiation is immaterial, that is,

$$\frac{\partial^2 w}{\partial y\, \partial x} = \frac{\partial^2 w}{\partial x\, \partial y}; \quad \frac{\partial^2 w}{\partial z\, \partial x} = \frac{\partial^2 w}{\partial x\, \partial z}; \quad \frac{\partial^2 w}{\partial z\, \partial y} = \frac{\partial^2 w}{\partial y\, \partial z}.$$

The proof is rather technical and will not be given in this book.

Example 3 Find the second partial derivatives of f if $f(x, y) = x^3y^2 - 2x^2y + 3x$.

Solution This function was considered in Example 1. Since $f_x(x, y) = 3x^2y^2 - 4xy + 3$ and $f_y(x, y) = 2x^3y - 2x^2$, we have

$$f_{xx}(x, y) = \frac{\partial}{\partial x}(3x^2y^2 - 4xy + 3) = 6xy^2 - 4y$$

$$f_{xy}(x, y) = \frac{\partial}{\partial y}(3x^2y^2 - 4xy + 3) = 6x^2y - 4x$$

$$f_{yx}(x, y) = \frac{\partial}{\partial x}(2x^3y - 2x^2) = 6x^2y - 4x$$

$$f_{yy}(x, y) = \frac{\partial}{\partial y}(2x^3y - 2x^2) = 2x^3.$$

Third and higher partial derivatives are defined in like manner. For example,

$$\frac{\partial}{\partial x} f_{xx} = f_{xxx} = \frac{\partial}{\partial x}\left(\frac{\partial^2 f}{\partial x^2}\right) = \frac{\partial^3 f}{\partial x^3}$$

$$\frac{\partial}{\partial x} f_{xy} = f_{xyx} = \frac{\partial}{\partial x}\left(\frac{\partial^2 f}{\partial y\, \partial x}\right) = \frac{\partial^3 f}{\partial x\, \partial y\, \partial x},$$

and so on. If third partial derivatives are continuous, then it can be shown that the order of differentiation is immaterial, that is,

$$f_{xyx} = f_{yxx} = f_{xxy} \quad \text{and} \quad f_{yxy} = f_{xyy} = f_{yyx}.$$

Similar notations and results apply to functions of more than two variables. Of course, letters other than x and y may be used. To illustrate, if f is a function of r and s, then symbols such as

$$f_r(r, s), \quad f_s(r, s), \quad f_{rs}, \quad \frac{\partial f}{\partial r}, \quad \frac{\partial^2 f}{\partial r^2}$$

are employed for partial derivatives.

EXERCISES 16.3

In Exercises 1–18 find the first partial derivatives of f.

1 $f(x, y) = 2x^4y^3 - xy^2 + 3y + 1$

2 $f(x, y) = (x^3 - y^2)^2$

3 $f(r, s) = \sqrt{r^2 + s^2}$

4 $f(s, t) = t/s - s/t$

5 $f(x, y) = xe^y + y \sin x$

6 $f(x, y) = e^x \ln xy$

7 $f(t, v) = \ln \sqrt{(t + v)/(t - v)}$

8 $f(u, w) = \arctan (u/w)$

9 $f(x, y) = x \cos \dfrac{x}{y}$

10 $f(x, y) = \sqrt{4x^2 - y^2} \sec x$

11 $f(r, s, t) = r^2 e^{2s} \cos t$

12 $f(x, y, t) = \dfrac{x^2 - t^2}{1 + \sin 3y}$

13 $f(x, y, z) = (y^2 + z^2)^x$

14 $f(r, s, v) = (2r + 3s)^{\cos v}$

15 $f(x, y, z) = xe^z - ye^x + ze^{-y}$

16 $f(r, s, v, p) = r^3 \tan s + \sqrt{s}e^{v^2} - v \cos 2p$

17 $f(q, v, w) = \sin^{-1} \sqrt{qv} + \sin vw$

18 $f(x, y, z) = xyze^{xyz}$

In Exercises 19–24 verify that $w_{xy} = w_{yx}$.

19 $w = xy^4 - 2x^2y^3 + 4x^2 - 3y$

20 $w = x^2/(x + y)$

21 $w = x^3e^{-2y} + y^{-2} \cos x$

22 $w = y^2e^{x^2} + 1/x^2y^3$

23 $w = x^2 \cosh \dfrac{z}{y}$

24 $w = \sqrt{x^2 + y^2 + z^2}$

25 If $w = 3x^2y^3z + 2xy^4z^2 - yz$, find w_{xyz}.

26 If $w = u^4vt^2 - 3uv^2t^3$, find w_{tut}.

27 If $u = v \sec rt$, find u_{rvr}.

28 If $v = y \ln (x^2 + z^4)$, find v_{zzy}.

29 If $w = \sin xyz$ find $\dfrac{\partial^3 w}{\partial z\, \partial y\, \partial x}$.

30 If $w = x^2/(y^2 + z^2)$, find $\dfrac{\partial^3 w}{\partial z\, \partial y^2}$.

31 If $w = r^4s^3t - 3s^2e^{rt}$, verify that $w_{rrs} = w_{rsr} = w_{srr}$.

32 If $w = \tan uv + 2 \ln (u + v)$, verify that $w_{uvv} = w_{vuv} = w_{vvu}$.

A function f of x and y is **harmonic** if $\partial^2 f/\partial x^2 + \partial^2 f/\partial y^2 = 0$. Prove that the functions defined in Exercises 33–36 are harmonic.

33 $f(x, y) = \ln \sqrt{x^2 + y^2}$

34 $f(x, y) = \arctan (y/x)$

35 $f(x, y) = \cos x \sinh y + \sin x \cosh y$

36 $f(x, y) = e^{-x} \cos y + e^{-y} \cos x$

37 If $w = \cos (x - y) + \ln (x + y)$, show that $\partial^2 w/\partial x^2 - \partial^2 w/\partial y^2 = 0$.

38 If $w = (y - 2x)^3 - \sqrt{y - 2x}$, show that $w_{xx} - 4w_{yy} = 0$.

39 If $w = e^{-c^2t} \sin cx$, show that $w_{xx} = w_t$ for every real number c.

40 The ideal gas law may be stated as $PV = cnT$, where n is the number of moles of gas, V is the volume, T is the temperature, P is the pressure, and c is a constant. Show that

$$\frac{\partial V}{\partial T} \frac{\partial T}{\partial P} \frac{\partial P}{\partial V} = -1.$$

In Exercises 41 and 42 show that v satisfies the **wave equation**

$$\frac{\partial^2 v}{\partial t^2} = a^2 \frac{\partial^2 v}{\partial x^2}.$$

41 $v = (\sin akx)(\sin kt)$

42 $v = (x - at)^4 + \cos (x + at)$

43 List all possible second partial derivatives of $w = f(x, y, z)$.

44 If $w = f(x, y, z, t, v)$, define w_t as a limit.

45 A hot metal plate is situated on an xy-plane such that the temperature T at the point (x, y) is given by $T = 10(x^2 + y^2)^2$. Find the rate of change of T with respect to distance at $P(1, 2)$ in the direction of (a) the x-axis and (b) the y-axis.

46 Suppose the electrical potential V at the point (x, y, z) is given by $V = 100/(x^2 + y^2 + z^2)$. Find the rate of change of V with respect to distance at $P(2, -1, 1)$ in the direction of (a) the x-axis, (b) the y-axis, and (c) the z-axis.

47 Let C be the trace of the paraboloid $z = 9 - x^2 - y^2$ on the plane $x = 1$. Find the equation of the tangent line l to C at the point $P(1, 2, 4)$. Sketch the paraboloid, C, and l.

48 Let C be the trace of the graph of $z = \sqrt{36 - 9x^2 - 4y^2}$ on the plane $y = 2$. Find the equation of the tangent line l to C at the point $(1, 2, \sqrt{11})$. Sketch the surface, C, and l.

16.4 INCREMENTS AND DIFFERENTIALS

If f is a function of two variables x and y, then the symbols Δx and Δy will denote increments of x and y. It should be observed that in the present situation Δy is an increment of the *independent* variable y and is not the same as that defined in (3.26), where y was a *dependent* variable. In terms of this increment notation, Definition (16.6) may be written

$$f_x(x, y) = \lim_{\Delta x \to 0} \frac{f(x + \Delta x, y) - f(x, y)}{\Delta x}$$

$$f_y(x, y) = \lim_{\Delta y \to 0} \frac{f(x, y + \Delta y) - f(x, y)}{\Delta y}.$$

If $w = f(x, y)$ and both x and y are given increments Δx and Δy, respectively, then Δw will denote the corresponding increment of the dependent variable, that is,

(16.8) $$\Delta w = f(x + \Delta x, y + \Delta y) - f(x, y).$$

Thus, Δw represents the change in the value of f if (x, y) changes to $(x + \Delta x, y + \Delta y)$.

Example 1 If $w = f(x, y) = 3x^2 - xy$, find Δw. What is the change in $f(x, y)$ if (x, y) changes from $(1, 2)$ to $(1.01, 1.98)$?

Solution From (16.8),

$$\begin{aligned}\Delta w &= [3(x + \Delta x)^2 - (x + \Delta x)(y + \Delta y)] - [3x^2 - xy] \\ &= [3x^2 + 6x\,\Delta x + 3(\Delta x)^2 - (xy + x\,\Delta y + y\,\Delta x + \Delta x\,\Delta y)] \\ &\quad - 3x^2 + xy \\ &= 6x\,\Delta x + 3(\Delta x)^2 - x\,\Delta y - y\,\Delta x - \Delta x\,\Delta y.\end{aligned}$$

To find the desired change in $f(x, y)$ we substitute $x = 1$, $y = 2$, $\Delta x = 0.01$, $\Delta y = -0.02$, and simplify, obtaining $\Delta w = 0.0605$. Of course, this number can also be found by calculating $f(1.01, 1.98) - f(1, 2)$.

The next theorem can be used to express Δw in terms of the first partial derivatives of f. In the statement of the theorem, η_1 and η_2 will denote functions of Δx and Δy; however, instead of using the symbols $\eta_1(\Delta x, \Delta y)$ and $\eta_2(\Delta x, \Delta y)$ for values, we shall use the simplified notation η_1 and η_2.

(16.9) Theorem

Let $w = f(x, y)$, where the function f is defined on a rectangular region $R = \{(x, y): a < x < b, c < y < d\}$, and suppose f_x and f_y exist in R and are continuous at the pair (x_0, y_0) in R. If $(x_0 + \Delta x, y_0 + \Delta y)$ is in R and $\Delta w = f(x_0 + \Delta x, y_0 + \Delta y) - f(x_0, y_0)$, then

$$\Delta w = f_x(x_0, y_0)\,\Delta x + f_y(x_0, y_0)\,\Delta y + \eta_1\,\Delta x + \eta_2\,\Delta y$$

where η_1 and η_2 are functions of Δx and Δy which have the limit 0 as $(\Delta x, \Delta y) \to (0, 0)$.

Proof. The graph of R and the pairs we wish to consider are illustrated in Figure 16.14. We first note that the increment Δw of (16.8) may be written

$$\Delta w = [f(x_0 + \Delta x, y_0 + \Delta y) - f(x_0, y_0 + \Delta y)] + [f(x_0, y_0 + \Delta y) - f(x_0, y_0)]. \tag{16.10}$$

If we next introduce a function g of *one* variable by defining $g(x) = f(x, y_0 + \Delta y)$, then $g'(x) = f_x(x, y_0 + \Delta y)$ for all x under consideration. Applying the Mean Value Theorem (4.12) to g we obtain

$$g(x_0 + \Delta x) - g(x_0) = g'(u)\,\Delta x$$

where u is between x_0 and $x_0 + \Delta x$. Using the definition of g, the last equation can be written

$$f(x_0 + \Delta x, y_0 + \Delta y) - f(x_0, y_0 + \Delta y) = f_x(u, y_0 + \Delta y)\,\Delta x.$$

Similarly, if we let $h(y) = f(x_0, y)$ and apply the Mean Value Theorem to the function h, we obtain

$$h(y_0 + \Delta y) - h(y_0) = h'(v)\,\Delta y$$

where v is between y_0 and $y_0 + \Delta y$, that is,

$$f(x_0, y_0 + \Delta y) - f(x_0, y_0) = f_y(x_0, v)\,\Delta y.$$

Substitution in (16.10) leads to

$$\Delta w = f_x(u, y_0 + \Delta y)\,\Delta x + f_y(x_0, v)\,\Delta y. \tag{16.11}$$

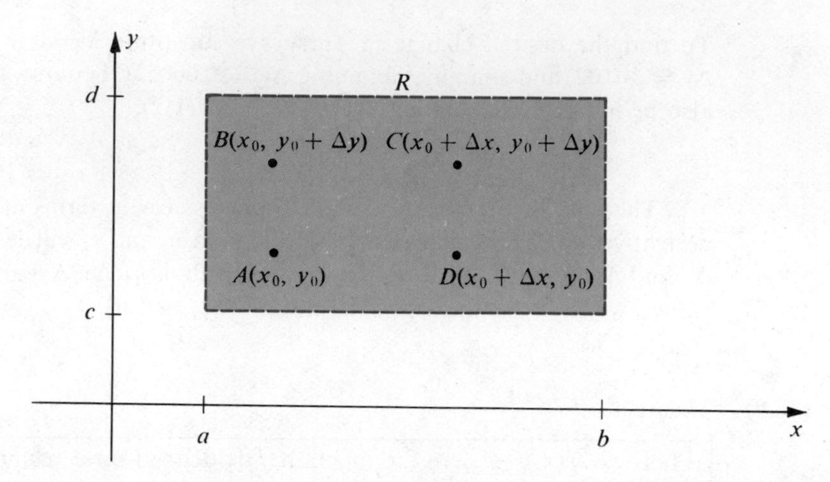

Figure 16.14

Let us define η_1 and η_2 by

(16.12)
$$\eta_1 = f_x(u, y_0 + \Delta y) - f_x(x_0, y_0)$$
$$\eta_2 = f_y(x_0, v) - f_y(x_0, y_0).$$

Using the fact that f_x and f_y are continuous and noting that $u \to x_0$ and $v \to y_0$ as $\Delta x \to 0$ and $\Delta y \to 0$, respectively, it follows that η_1 and η_2 have the limit 0 as $(\Delta x, \Delta y) \to (0, 0)$. Solving the equations in (16.12) for $f_x(u, y_0 + \Delta y)$ and $f_y(x_0, v)$, and substituting in (16.11), gives us

$$\Delta w = [f_x(x_0, y_0) + \eta_1]\,\Delta x + [f_y(x_0, y_0) + \eta_2]\,\Delta y$$

which leads to the conclusion of the theorem.

Example 2 If $w = 3x^2 - xy$, find values of η_1 and η_2 which satisfy the conclusion of Theorem (16.9).

Solution From the solution of Example 1,

$$\Delta w = 6x\,\Delta x + 3(\Delta x)^2 - x\,\Delta y - y\,\Delta x - \Delta x\,\Delta y$$

which may also be written

$$\Delta w = (6x - y)\,\Delta x + (-x)\,\Delta y + (3\,\Delta x)(\Delta x) + (-\Delta x)\,\Delta y.$$

If we define $\eta_1 = 3\,\Delta x$ and $\eta_2 = (-\Delta x)$, then the form in Theorem (16.9) is obtained. Note that η_1 and η_2 are not unique since we may also write

$$\Delta w = (6x - y)\,\Delta x + (-x)\,\Delta y + (3\,\Delta x - \Delta y)\,\Delta x + 0\,\Delta y$$

in which case $\eta_1 = 3\,\Delta x - \Delta y$ and $\eta_2 = 0$.

(16.13) Definition

If $w = f(x, y)$, the **differentials dx and dy of the independent variables** x and y are defined by

$$dx = \Delta x \quad \text{and} \quad dy = \Delta y$$

where Δx and Δy are increments of x and y. The **differential dw of the dependent variable** w is defined by

$$dw = f_x(x, y)\,dx + f_y(x, y)\,dy = \frac{\partial w}{\partial x}dx + \frac{\partial w}{\partial y}dy.$$

Using the conclusion of Theorem (16.9), with the pair (x_0, y_0) replaced by (x, y), we see that

$$\Delta w - dw = \eta_1\,\Delta x + \eta_2\,\Delta y$$

where η_1 and η_2 approach 0 as $(\Delta x, \Delta y) \to (0, 0)$. It follows that if Δx and Δy are small, then $\Delta w - dw \approx 0$, that is, $dw \approx \Delta w$. This fact may be used to approximate the change in w corresponding to small changes in x and y.

Example 3 If $w = 3x^2 - xy$, find dw and use it to approximate the change in w if (x, y) changes from $(1, 2)$ to $(1.01, 1.98)$. How does this compare with the exact change in w?

Solution Applying Definition (16.13),

$$\begin{aligned} dw &= \frac{\partial w}{\partial x}dx + \frac{\partial w}{\partial y}dy \\ &= (6x - y)\,dx + (-x)\,dy. \end{aligned}$$

Substituting $x = 1$, $y = 2$, $dx = \Delta x = 0.01$, and $dy = \Delta y = -0.02$, we obtain

$$dw = (6 - 2)(0.01) + (-1)(-0.02) = 0.06.$$

It was shown in Example 1 that $\Delta w = 0.0605$. Hence the error involved in using dw is 0.0005.

Example 4 The radius and altitude of a right circular cylinder are measured as 3 inches and 8 inches, respectively, with a possible error in measurement of ± 0.05 inches. Use differentials to approximate the maximum error in the calculated volume of the cylinder.

Solution The volume V of a cylinder of radius r and altitude h is $V = \pi r^2 h$. Consequently

$$dV = \frac{\partial V}{\partial r}dr + \frac{\partial V}{\partial h}dh = 2\pi rh\,dr + \pi r^2\,dh.$$

Substituting $r = 3$, $h = 8$, and $dr = dh = \pm 0.05$ gives us

$$dV = 48\pi(\pm 0.05) + 9\pi(\pm 0.05) = \pm 2.85\pi \approx \pm 8.95 \text{ in.}^3.$$

For a function of one variable, the term *differentiable* means that the derivative exists. For functions of two variables we use the much stronger condition stated in the next definition.

(16.14) Definition

If $w = f(x, y)$, then f is **differentiable** at (x_0, y_0) provided Δw can be expressed in the form

$$\Delta w = f_x(x_0, y_0)\,\Delta x + f_y(x_0, y_0)\,\Delta y + \eta_1\,\Delta x + \eta_2\,\Delta y$$

where η_1 and η_2 have the limit 0 as $(\Delta x, \Delta y) \to (0, 0)$.

According to Theorem (16.9), if f_x and f_y are continuous on a rectangular region R, then f is differentiable on R. The following theorem shows that a differentiable function is continuous.

(16.15) Theorem

If a function f of two variables is differentiable at (x_0, y_0), then f is continuous at (x_0, y_0).

Proof. We begin by writing the formula for Δw in Definition (16.14) as follows:

$$\Delta w = [f_x(x_0, y_0) + \eta_1]\,\Delta x + [f_y(x_0, y_0) + \eta_2]\,\Delta y.$$

If we let $x = x_0 + \Delta x$ and $y = y_0 + \Delta y$, then

$$\begin{aligned}\Delta w &= f(x, y) - f(x_0, y_0)\\ &= [f_x(x_0, y_0) + \eta_1](x - x_0) + [f_y(x_0, y_0) + \eta_2](y - y_0).\end{aligned}$$

It follows that

$$\lim_{(x,y)\to(x_0,y_0)} [f(x, y) - f(x_0, y_0)] = 0$$

or

$$\lim_{(x,y)\to(x_0,y_0)} f(x, y) = f(x_0, y_0).$$

Hence f is continuous at (x_0, y_0).

If f_x and f_y are continuous on a rectangular region, then f is differentiable and, therefore, continuous. It can be shown by means of examples that the mere *existence* of f_x and f_y is not enough to ensure continuity of f.

The discussion in this section can be extended to functions of more than two variables. For example, suppose $w = f(x, y, z)$ where f is defined in a suitable region R (such as a rectangular parallelepiped) and f_x, f_y, f_z exist in R and are continuous at (x, y, z). If x, y, and z are given increments Δx, Δy, and Δz, respectively, then the corresponding increment

$$\Delta w = f(x + \Delta x, y + \Delta y, z + \Delta z) - f(x, y, z)$$

can be written in the form

$$\Delta w = f_x(x, y, z)\,\Delta x + f_y(x, y, z)\,\Delta y + f_z(x, y, z)\,\Delta z + \eta_1\,\Delta x + \eta_2\,\Delta y + \eta_3\,\Delta z$$

where η_1, η_2, and η_3 are functions of $\Delta x, \Delta y$, and Δz which have the limit 0 as $(\Delta x, \Delta y, \Delta z) \to (0, 0, 0)$. The **differentials of the independent variables *x*, *y*, and *z*** are defined by

$$dx = \Delta x, \quad dy = \Delta y, \quad dz = \Delta z$$

and the **differential of the dependent variable *w*** is defined by

$$dw = \frac{\partial w}{\partial x}dx + \frac{\partial w}{\partial y}dy + \frac{\partial w}{\partial z}dz. \tag{16.16}$$

We can use dw to approximate Δw if the increments of x, y, and z are small. Differentiability is defined as before. The extension to four or more variables is made in like manner.

Example 5 Suppose the dimensions (in inches) of a rectangular parallelepiped change from 9, 6, and 4 to 9.02, 5.97, and 4.01, respectively. Use differentials to approximate the change in volume. What is the exact change in volume?

Solution If we denote the dimensions by x, y, and z, then the volume is $V = xyz$ and, by (16.16),

$$dV = yz\,dx + xz\,dy + xy\,dz.$$

Substituting $x = 9$, $y = 6$, $z = 4$, $dx = 0.02$, $dy = -0.03$, and $dz = 0.01$ gives us

$$\begin{aligned}\Delta V \approx dV &= 24(0.02) + 36(-0.03) + 54(0.01)\\ &= 0.48 - 1.08 + 0.54 = -0.06.\end{aligned}$$

Thus the volume decreases by approximately 0.06 cubic inches. The exact change in volume is

$$\Delta V = (9.02)(5.97)(4.01) - (9)(6)(4) = -0.063906.$$

EXERCISES 16.4

In Exercises 1–4 find values of η_1 and η_2 which satisfy Definition (16.14).

1 $f(x, y) = 4y^2 - 3xy + 2x$

2 $f(x, y) = (2x - y)^2$

3 $f(x, y) = x^3 + y^3$

4 $f(x, y) = 2x^2 - xy^2 + 3y$

In Exercises 5–16 find dw.

5 $w = x^3 - x^2y + 3y^2$

6 $w = 5x^2 + 4y - 3xy^3$

7 $w = x^2 \sin y + 2y^{3/2}$

8 $w = ye^{-2x} - 3x^4$

9 $w = x^2e^{xy} + (1/y^2)$

10 $w = \ln(x^2 + y^2) + x \tan^{-1} y$

11 $w = x^2 \ln(y^2 + z^2)$

12 $w = x^2y^3z + e^{-2z}$

13 $w = xyz/(x + y + z)$

14 $w = x^2e^{yz} + y \ln z$

15 $w = x^2z + 4yt^3 - xz^2t$

16 $w = x^2y^3zt^{-1}v^4$

17 Use differentials to approximate the change in $f(x, y) = x^2 - 3x^3y^2 + 4x - 2y^3 + 6$ if (x, y) changes from $(-2, 3)$ to $(-2.02, 3.01)$.

18 Use differentials to approximate the change in $f(x, y) = x^2 - 2xy + 3y$ if (x, y) changes from (1, 2) to (1.03, 1.99).

19 Use differentials to approximate the change in

$$f(x, y, z) = x^2z^3 - 3yz^2 + x^{-3} + 2y^{1/2}z$$

if (x, y, z) changes from (1, 4, 2) to (1.02, 3.97, 1.96).

20 Use differentials to approximate the change in $w = r^2 + 3sv + 2p^3$ if r changes from 1 to 1.02, s from 2 to 1.99, v from 4 to 4.01, and p from 3 to 2.97.

21 Use differentials to approximate $\sqrt[3]{26.98}\sqrt{36.04}$.

22 Use differentials to approximate $(32.03)^{2/5}/(1.95)^4$.

23 The dimensions of a closed rectangular box are measured as 3 ft, 4 ft, and 5 ft with a possible error of 1/16 in. Use differentials to approximate the maximum error in the following calculated values.
(a) The surface area of the box
(b) The volume of the box

24 The two shortest sides of a right triangle are measured as 3 in. and 4 in., respectively, with a possible error of 0.02 in. Use differentials to approximate the maximum error in the following calculated values.
(a) The hypotenuse
(b) The area of the triangle

25 An open cylindrical tin can has diameter 3 in. and altitude 4 in. Use differentials to approximate the amount of tin in the can if the tin is 0.015 in. thick.

26 The total resistance R of three resistances R_1, R_2, and R_3, connected in parallel, is given by

$$\frac{1}{R} = \frac{1}{R_1} + \frac{1}{R_2} + \frac{1}{R_3}.$$

If measurements of R_1, R_2, and R_3 are 100, 200, and 400 ohms, respectively, with a maximum error of 1% in each measurement, approximate the maximum error in the calculated value of R.

27 The specific gravity of an object is given by $s = A/(A - W)$, where A and W are the weights (in pounds) of the object in air and water, respectively. If measurements are $A = 12$ lb and $W = 5$ lb with maximum errors of 1/2 oz in air and 1 oz in water, what is the maximum error in the calculated value of s?

28 The equation relating the pressure P, volume V, and temperature T of a confined gas is $PV = kT$, where k is a constant. If $P = 0.5\ \text{lb/in.}^2$ when $V = 64\ \text{in.}^3$ and $T = 80°$, approximate the change in P if V and T change to $70\ \text{in.}^3$ and $76°$, respectively.

29 The temperature T at the point $P(x, y, z)$ in a rectangular coordinate system is given by $T = 8(2x^2 + 4y^2 + 9z^2)^{1/2}$. Use differentials to approximate the temperature difference between the points $(6, 3, 2)$ and $(6.1, 3.3, 1.98)$.

30 Approximate the change in area of an isosceles triangle if each of the two equal sides increases from 100 to 101 and the angle between them decreases from 120° to 119°.

31 If the cylinder described in Example 4 of this section has a closed top and bottom, use differentials to approximate the maximum error in the calculated total surface area.

32 Use differentials to approximate the change in surface area of the parallelepiped described in Example 5 of this section. What is the exact change?

16.5 THE CHAIN RULE

If f, g, and k are functions of two variables such that $w = f(u, v)$, $u = g(x, y)$, and $v = k(x, y)$, and if for each pair (x, y) in a subset D of $\mathbb{R} \times \mathbb{R}$ the corresponding pair (u, v) is in the domain of f, then $w = f(g(x, y), k(x, y))$ defines w as a (composite) function of x and y. For example, if

$$w = u^2 + u \sin v, \quad u = xe^{2y}, \quad \text{and} \quad v = xy$$

then

$$w = x^2 e^{4y} + xe^{2y} \sin xy.$$

The next theorem provides formulas for expressing $\partial w/\partial x$ and $\partial w/\partial y$ in terms of the first partial derivatives of the functions g, k, and f. In the statement of the theorem it is assumed that domains are chosen such that the composite function is defined on a suitable domain D. Each of the formulas stated in Theorem (16.17) is referred to as the **Chain Rule**.

(16.17) Theorem

If $w = f(u, v)$, $u = g(x, y)$, and $v = k(x, y)$ where f is differentiable and g and k have continuous first partial derivatives, then

$$\frac{\partial w}{\partial x} = \frac{\partial w}{\partial u}\frac{\partial u}{\partial x} + \frac{\partial w}{\partial v}\frac{\partial v}{\partial x}$$

$$\frac{\partial w}{\partial y} = \frac{\partial w}{\partial u}\frac{\partial u}{\partial y} + \frac{\partial w}{\partial v}\frac{\partial v}{\partial y}.$$

Proof. If x is given an increment Δx and y is held constant (that is, $\Delta y = 0$), then

$$\text{(a)} \qquad \Delta w = f(g(x + \Delta x, y), k(x + \Delta x, y)) - f(g(x, y), k(x, y)).$$

Also,

$$\text{(b)} \qquad \begin{aligned} \Delta u &= g(x + \Delta x, y) - g(x, y) \\ \Delta v &= k(x + \Delta x, y) - k(x, y) \end{aligned}$$

and hence

$$\begin{aligned} g(x + \Delta x, y) &= g(x, y) + \Delta u = u + \Delta u \\ k(x + \Delta x, y) &= k(x, y) + \Delta v = v + \Delta v. \end{aligned}$$

Substituting in equation (a),

$$\Delta w = f(u + \Delta u, v + \Delta v) - f(u, v).$$

Since f is differentiable, we may write

$$\text{(c)} \qquad \Delta w = \frac{\partial w}{\partial u}\Delta u + \frac{\partial w}{\partial v}\Delta v + \eta_1 \Delta u + \eta_2 \Delta v$$

where η_1 and η_2 are functions of Δu and Δv which have the limit 0 as $(\Delta u, \Delta v) \to (0, 0)$. Moreover, we may assume that both η_1 and η_2 are 0 if $(\Delta u, \Delta v) = (0, 0)$, for if they are not, they can be replaced by functions μ_1 and μ_2 which have this property and are equal to η_1 and η_2 elsewhere. With this agreement, the functions η_1 and η_2 in (c) are continuous at $(0, 0)$. Dividing both sides of that equation by Δx gives us

$$\text{(d)} \qquad \frac{\Delta w}{\Delta x} = \frac{\partial w}{\partial u}\frac{\Delta u}{\Delta x} + \frac{\partial w}{\partial v}\frac{\Delta v}{\Delta x} + \eta_1 \frac{\Delta u}{\Delta x} + \eta_2 \frac{\Delta v}{\Delta x}.$$

From equations (a) and (b)

$$\lim_{\Delta x \to 0} \frac{\Delta w}{\Delta x} = \frac{\partial w}{\partial x}, \quad \lim_{\Delta x \to 0} \frac{\Delta u}{\Delta x} = \frac{\partial u}{\partial x}, \quad \lim_{\Delta x \to 0} \frac{\Delta v}{\Delta x} = \frac{\partial v}{\partial x}.$$

If Δx approaches 0, then Δu and Δv also approach 0 and hence so do η_1 and η_2.

Consequently, if we take the limit in equation (d) as $\Delta x \to 0$, we obtain

$$\frac{\partial w}{\partial x} = \frac{\partial w}{\partial u}\frac{\partial u}{\partial x} + \frac{\partial w}{\partial v}\frac{\partial v}{\partial x}.$$

The second formula in the statement of the theorem is established in similar fashion.

Example 1 If $w = u^3 + e^{2v}$, $u = xy^2$, and $v = x^3 \sin y$, find $\partial w/\partial x$ and $\partial w/\partial y$.

Solution Applying Theorem (16.17),

$$\frac{\partial w}{\partial x} = (3u^2)(y^2) + (2e^{2v})(3x^2 \sin y) = 3u^2y^2 + 6e^{2v}x^2 \sin y$$

$$\frac{\partial w}{\partial y} = (3u^2)(2xy) + (2e^{2v})x^3 \cos y = 6u^2xy + 2e^{2v}x^3 \cos y.$$

If we desire to express these partial derivatives in terms of x and y alone, we may substitute xy^2 for u and $x^3 \sin y$ for v.

The Chain Rule can be extended to any number of variables, as stated in the following theorem.

(16.18) Theorem

If w is a differentiable function of n variables $u_1, u_2, \ldots, u_n$ where each u_i is a function of m variables $x_1, x_2, \ldots, x_m$ having continuous first partial derivatives, and if w is a (composite) function of $x_1, x_2, \ldots, x_m$, then

$$\frac{\partial w}{\partial x_i} = \frac{\partial w}{\partial u_1}\frac{\partial u_1}{\partial x_i} + \frac{\partial w}{\partial u_2}\frac{\partial u_2}{\partial x_i} + \cdots + \frac{\partial w}{\partial u_n}\frac{\partial u_n}{\partial x_i}$$

for $i = 1, 2, \ldots, m$.

Example 2 If $w = r^2s + v^3 \tan t$ where $r = x^3yz^2$, $s = xe^{yz}$, $v = x \cos y$, and $t = y \ln z$, find $\partial w/\partial z$.

Solution Applying Theorem (16.18),

$$\begin{aligned}\frac{\partial w}{\partial z} &= \frac{\partial w}{\partial r}\frac{\partial r}{\partial z} + \frac{\partial w}{\partial s}\frac{\partial s}{\partial z} + \frac{\partial w}{\partial v}\frac{\partial v}{\partial z} + \frac{\partial w}{\partial t}\frac{\partial t}{\partial z}\\ &= (2rs)(2x^3yz) + r^2(xye^{yz}) + (3v^2 \tan t)(0) + (v^3 \sec^2 t)(y/z)\\ &= 4rsx^3yz + r^2xye^{yz} + (v^3y/z)\sec^2 t.\end{aligned}$$

The following special case of the Chain Rule follows directly from Theorem (6.18) by letting $x_i = t$ for each i.

(16.19) Theorem

If w is a function of $u_1, u_2, \ldots, u_n$ and each u_i is a function of *one* variable t, then w is a function of t and

$$\frac{dw}{dt} = \frac{\partial w}{\partial u_1}\frac{du_1}{dt} + \frac{\partial w}{\partial u_2}\frac{du_2}{dt} + \cdots + \frac{\partial w}{\partial u_n}\frac{du_n}{dt}.$$

In Theorem (16.19) the derivative notations dw/dt and du_i/dt are used because w and each u_i is a function of *one* variable t. This situation could occur in applications where w is a function of several variables and each variable is a function of time t. In this case the derivative dw/dt would give us the rate of change of w with respect to time.

Example 3 If $w = x^3y^4$ and $x = t^2 - 2$, $y = 5t - 3$, find dw/dt at $t = 1$.

Solution This problem could be solved without using the Chain Rule by noting that

$$w = (t^2 - 2)^3(5t - 3)^4$$

and then finding dw/dt by means of the Product Rule (3.20). However, if we apply Theorem (16.19), then

$$\begin{aligned}\frac{dw}{dt} &= \frac{\partial w}{\partial x}\frac{dx}{dt} + \frac{\partial w}{\partial y}\frac{dy}{dt}\\ &= (3x^2y^4)(2t) + (4x^3y^3)(5).\end{aligned}$$

If $t = 1$, then $x = (1)^2 - 2 = -1$, $y = 5(1) - 3 = 2$, and consequently,

$$\frac{dw}{dt} = 3(-1)^2 2^4(2) + 4(-1)^3 2^3(5) = 96 - 160 = -64.$$

Partial derivatives can be used to find derivatives of functions which are defined implicitly. Suppose, as in Section 3.6, an equation $F(x, y) = 0$ defines a differentiable function f of one variable x such that $F(x, f(x)) = 0$ for all x in the domain D of f. If we let

$$w = F(u, y), \quad \text{where } u = x \text{ and } y = f(x)$$

then applying Theorem (16.19) and using the fact that $w = F(x, f(x)) = 0$ for all x in D,

$$\frac{dw}{dx} = \frac{\partial w}{\partial u}\frac{du}{dx} + \frac{\partial w}{\partial y}\frac{dy}{dx} = 0$$

that is,

$$\frac{\partial w}{\partial u}(1) + \frac{\partial w}{\partial y}f'(x) = 0.$$

If $\partial w/\partial y \neq 0$, then (since $u = x$)

$$f'(x) = -\left(\frac{\partial w}{\partial x}\right)\bigg/\left(\frac{\partial w}{\partial y}\right) = -\frac{F_x(x, y)}{F_y(x, y)}.$$

We may summarize this discussion as follows.

(16.20) Theorem

If an equation $F(x, y) = 0$ defines an implicit differentiable function f of one variable x such that $y = f(x)$, then

$$\frac{dy}{dx} = -\frac{F_x(x, y)}{F_y(x, y)}.$$

Example 4 If $y = f(x)$ satisfies the equation

$$y^4 + 3y - 4x^3 - 5x - 1 = 0,$$

find y'.

Solution If $F(x, y)$ is the expression on the left side of the given equation, then by Theorem (16.20),

$$y' = -\frac{-12x^2 - 5}{4y^3 + 3} = \frac{12x^2 + 5}{4y^3 + 3}.$$

The reader should compare this with Example 2 of Section 3.6, which was obtained using single-variable methods.

Given an equation such as

$$x^2 - 4y^3 + 2z - 7 = 0$$

we can solve for z, obtaining

$$z = \tfrac{1}{2}(-x^2 + 4y^3 + 7)$$

which is of the form

$$z = f(x, y).$$

In analogy with the single variable case, we say that the function f of two variables x and y is defined *implicitly* by the given equation. The next theorem gives us formulas for finding f_x and f_y or, equivalently, $\partial z/\partial x$ and $\partial z/\partial y$ without actually solving the equation for z.

(16.21) Theorem

If an equation $F(x, y, z) = 0$ defines an implicit differentiable function f of two variables x and y such that $z = f(x, y)$ for all (x, y) in the domain of f, then

$$\frac{\partial z}{\partial x} = -\frac{F_x(x, y, z)}{F_z(x, y, z)}, \qquad \frac{\partial z}{\partial y} = -\frac{F_y(x, y, z)}{F_z(x, y, z)}.$$

Proof. The statement that $F(x, y, z) = 0$ defines the function f, where $z = f(x, y)$, means that $F(x, y, f(x, y)) = 0$ for all (x, y) in the domain of f. If we let

$$w = F(u, v, z), \quad \text{where } u = x,\ v = y,\ z = f(x, y)$$

then by the Chain Rule and the fact that $w = 0$ for all (x, y) in D,

$$\frac{\partial w}{\partial x} = \frac{\partial w}{\partial u}\frac{\partial u}{\partial x} + \frac{\partial w}{\partial v}\frac{\partial v}{\partial x} + \frac{\partial w}{\partial z}\frac{\partial z}{\partial x} = 0.$$

Since $u = x$ and $v = y$, we have

$$\frac{\partial w}{\partial x}(1) + \frac{\partial w}{\partial y}(0) + \frac{\partial w}{\partial z}\frac{\partial z}{\partial x} = 0$$

and, if $\partial w/\partial z \neq 0$,

$$\frac{\partial z}{\partial x} = -\left(\frac{\partial w}{\partial x}\right)\bigg/\left(\frac{\partial w}{\partial z}\right) = -\frac{F_x(x, y, z)}{F_z(x, y, z)}.$$

The formula for $\partial z/\partial y$ may be obtained in similar fashion.

Example 5 If $z = f(x, y)$ satisfies the equation

$$x^2z^2 + xy^2 - z^3 + 4yz - 5 = 0$$

find $\partial z/\partial x$ and $\partial z/\partial y$.

Solution If we let $F(x, y, z)$ denote the expression on the left of the given equation, then by Theorem (16.21)

$$\frac{\partial z}{\partial x} = -\frac{2xz^2 + y^2}{2x^2z - 3z^2 + 4y}$$

$$\frac{\partial z}{\partial y} = -\frac{2xy + 4z}{2x^2z - 3z^2 + 4y}.$$

EXERCISES 16.5

In Exercises 1 and 2 find $\partial w/\partial x$ and $\partial w/\partial y$.

1 $w = u^2 \sin v,\ u = x^3 - 2y^3,\ v = xy^2$

2 $w = u^3 + u^2 v - 3v,\ u = \sin xy,\ v = y \ln x$

In Exercises 3 and 4 find $\partial w/\partial r$ and $\partial w/\partial s$.

3 $w = \sqrt{u^2 + v^2},\ u = re^{-s},\ v = s^2 e^{-r}$

4 $w = e^{t/v}, t = r^2 - s^2, v = r^3 + s^3$

In Exercises 5 and 6 find $\partial z/\partial x$ and $\partial z/\partial y$.

5 $z = (r + s)/v,\ r = x \cos y,\ s = y \sin x,\ v = 2x - y$

6 $z = uv^2 + v \ln w,\ u = 2x - y,\ v = x - 2y,\ w = -2x + 2y$

In Exercises 7 and 8 find $\partial r/\partial u$, $\partial r/\partial v$, and $\partial r/\partial t$.

7 $r = x^3 + 3y - xy^2,\ x = u + v \ln t,\ y = v^2 - \ln ut$

8 $r = x \cos y,\ x = u^2 - vt,\ y = v^2 - ut$

9 Find $\partial p/\partial r$ if $p = u^2 \cos vw,\ u = xye^{rs},\ v = s \ln xyr,\ w = ys + r \sin x$.

10 Find $\partial s/\partial y$ if $s = tr^2 \tan^{-1} uv,\ t = x + y^2 z,\ r = x^2 + 3yz,\ u = xz^2,\ v = xyz$.

In Exercises 11–14 find dw/dt.

11 $w = x^3 - y^3,\ x = 1/(t + 1),\ y = t/(t + 1)$

12 $w = \ln(u + v),\ u = e^{-2t},\ v = t^3 - t^2$

13 $w = r^2 - s \tan v,\ r = \sin^2 t,\ s = \cos t,\ v = 4t$

14 $w = x^2 y^3 z^4,\ x = 2t + 1,\ y = 3t - 2,\ z = 5t + 4$

In each of Exercises 15–18 find y' under the assumption that $y = f(x)$ satisfies the given equation.

15 $2x^3 + x^2 y + y^3 = 1$

16 $x^4 + 2x^2 y^2 - 3xy^3 + 2x = 0$

17 $6x + \sqrt{xy} = 3y - 4$

18 $x^{2/3} + y^{2/3} = 4$

In each of Exercises 19–22 find $\partial z/\partial x$ and $\partial z/\partial y$ under the assumption that $z = f(x, y)$ satisfies the given equation.

19 $2xz^3 - 3yz^2 + x^2 y^2 + 4z = 0$

20 $xz^2 + 2x^2 y - 4y^2 z + 3y - 2 = 0$

21 $xe^{yz} - 2ye^{xz} + 3ze^{xy} = 1$

22 $yx^2 + z^2 + \cos xyz = 4$

23 If $w = f(x, y)$, $x = r \cos \theta$, and $y = r \sin \theta$ show that

$$\left(\frac{\partial w}{\partial r}\right)^2 + \frac{1}{r^2}\left(\frac{\partial w}{\partial \theta}\right)^2 = \left(\frac{\partial w}{\partial x}\right)^2 + \left(\frac{\partial w}{\partial y}\right)^2.$$

24 If $w = f(x, y)$ and $x = e^r \cos \theta$, $y = e^r \sin \theta$, show that

$$\frac{\partial^2 w}{\partial x^2} + \frac{\partial^2 w}{\partial y^2} = e^{-2r}\left(\frac{\partial^2 w}{\partial r^2} + \frac{\partial^2 w}{\partial \theta^2}\right).$$

25 If $w = f(x, y)$ and $x = r\cos\theta$, $y = r\sin\theta$, show that

$$\frac{\partial^2 w}{\partial x^2} + \frac{\partial^2 w}{\partial y^2} = \frac{\partial^2 w}{\partial r^2} + \frac{1}{r^2}\frac{\partial^2 w}{\partial \theta^2} + \frac{1}{r}\frac{\partial w}{\partial r}.$$

26 If $v = f(x - at) + g(x + at)$ where f and g have second partial derivatives, show that v satisfies the wave equation

$$\frac{\partial^2 v}{\partial t^2} = a^2 \frac{\partial^2 v}{\partial x^2}.$$

(Compare with Exercise 42 of Section 16.3.)

27 If $w = \cos(x + y) + \cos(x - y)$, show, without using addition formulas, that $w_{xx} - w_{yy} = 0$.

28 If $w = f(x^2 + y^2)$, show that $y(\partial w/\partial x) - x(\partial w/\partial y) = 0$. (*Hint*: Let $u = x^2 + y^2$).

29 If $w = f(u, v)$, $u = g(x, y)$, and $v = k(x, y)$, show that

$$\frac{\partial^2 w}{\partial x^2} = \frac{\partial^2 w}{\partial u^2}\left(\frac{\partial u}{\partial x}\right)^2 + \left(\frac{\partial^2 w}{\partial v\,\partial u} + \frac{\partial^2 w}{\partial u\,\partial v}\right)\frac{\partial u}{\partial x}\frac{\partial v}{\partial x} + \frac{\partial^2 w}{\partial v^2}\left(\frac{\partial v}{\partial x}\right)^2$$
$$+ \frac{\partial w}{\partial u}\frac{\partial^2 u}{\partial x^2} + \frac{\partial w}{\partial v}\frac{\partial^2 v}{\partial x^2}.$$

30 Given w, u, and v as in Exercise 25, show that

$$\frac{\partial^2 w}{\partial y\,\partial x} = \frac{\partial^2 w}{\partial u^2}\frac{\partial u}{\partial x}\frac{\partial u}{\partial y} + \frac{\partial^2 w}{\partial v\,\partial u}\frac{\partial u}{\partial x}\frac{\partial v}{\partial y} + \frac{\partial^2 w}{\partial u\,\partial v}\frac{\partial u}{\partial y}\frac{\partial v}{\partial x}$$
$$+ \frac{\partial^2 w}{\partial v^2}\frac{\partial v}{\partial x}\frac{\partial v}{\partial y} + \frac{\partial w}{\partial u}\frac{\partial^2 u}{\partial y\,\partial x} + \frac{\partial w}{\partial v}\frac{\partial^2 v}{\partial y\,\partial x}.$$

31 If n resistances $R_1, R_2 \ldots, R_n$ are connected in parallel, then the total resistance R is given by

$$\frac{1}{R} = \sum_{i=1}^{n} \frac{1}{R_i}.$$

Prove that for $i = 1, 2, \ldots, n$,

$$\frac{\partial R}{\partial R_i} = \left(\frac{R}{R_i}\right)^2.$$

32 A function f of two variables is said to be **homogeneous of degree n** if

$$f(tx, ty) = t^n f(x, y)$$

for every t such that (tx, ty) is in the domain of f. Show that for such functions,

$$x f_x(x, y) + y f_y(x, y) = n f(x, y).$$

(*Hint*: Differentiate $f(tx, ty)$ with respect to t.)

33 The radius r and altitude h of a right circular cylinder are increasing at rates of 0.01 in./min and 0.02 in./min, respectively. Use the Chain Rule to find the rate at which the volume is increasing at the time when $r = 4$ in. and $h = 7$ in. At what rate is the curved surface area changing at this time?

34 The equal sides and included angle of an isosceles triangle are increasing at rates of 0.1 ft/hr and 2°/hr, respectively. Use the Chain Rule to find the rate at which the area of the triangle is increasing at the time when the length of the equal sides is 20 ft and the included angle is 60°.

35 The pressure p, volume V, and temperature T of a confined gas are related by $pV = cT$, where c is a constant. If p and V are changing at the rates dp/dt and dV/dt, respectively, use the Chain Rule to find a formula for dT/dt. Check your answer by using the product rule for functions of one variable.

36 If the base radius r and altitude h of a right circular cylinder are changing at the rates dr/dt and dh/dt, respectively, use the Chain Rule to find a formula for dV/dt, where V is the volume of the cylinder. Check your answer by using single variable methods.

16.6 DIRECTIONAL DERIVATIVES

Let us return to the illustration in which $w = f(x, y)$ is the temperature of a hot plate at the point $P(x, y)$. In Section 16.3 the partial derivatives $f_x(x, y)$ and $f_y(x, y)$ were interpreted as the rates of change of w with respect to distance as a point P moves horizontally or vertically, respectively. This can be generalized as follows. Let Q be any other point on the plate and let $R(x + \Delta x, y + \Delta y)$ be a point on the vector $\overrightarrow{PQ}$ as illustrated in Figure 16.15. The increment

$$\Delta w = f(x + \Delta x, y + \Delta y) - f(x, y)$$

is the difference in temperature between points P and R, and the ratio

$$\frac{\Delta w}{d(P, R)} = \frac{\Delta w}{\sqrt{(\Delta x)^2 + (\Delta y)^2}}$$

is the *average change in temperature* as point moves along $\overrightarrow{PQ}$ from P to R. For example, if the temperature at P is 50° and the temperature at R is 51.5°, then $\Delta w = 1.5°$. *If* $d(P, R) = 3$ inches, then the average rate of change in temperature as a point moves from P to R is

$$\frac{\Delta w}{d(P, R)} = 0.5 \text{ degrees per inch.}$$

If the ratio $\Delta w/d(P, R)$ has a limit as R approaches P along $\overrightarrow{PQ}$, that is, as $(\Delta x, \Delta y) \to (0, 0)$, then the limit is called the **rate of change of w with respect to distance in the direction of the vector $\overrightarrow{PQ}$**.

If f is any function of two variables, then the preceding discussion may be used to motivate the concept of the *directional derivative* of f at P in the direction of $\overrightarrow{PQ}$. If the direction of $\overrightarrow{PQ}$ is specified by θ (see Figure 16.16), then a unit vector having the same direction is $\mathbf{u} = \langle \cos\theta, \sin\theta \rangle$. If, as shown in the figure, $R(x_1, y_1)$ is any other point on the line l through P and Q, then $\overrightarrow{PR}$ is parallel to a geometric vector which represents $\mathbf{u}$. Since $\overrightarrow{PR}$ corresponds to $\langle x_1 - x, y_1 - y \rangle$ (see Theorem (14.2)), this is equivalent to the statement

$$\langle x_1 - x, y_1 - y \rangle = t\mathbf{u} = \langle t\cos\theta, t\sin\theta \rangle$$

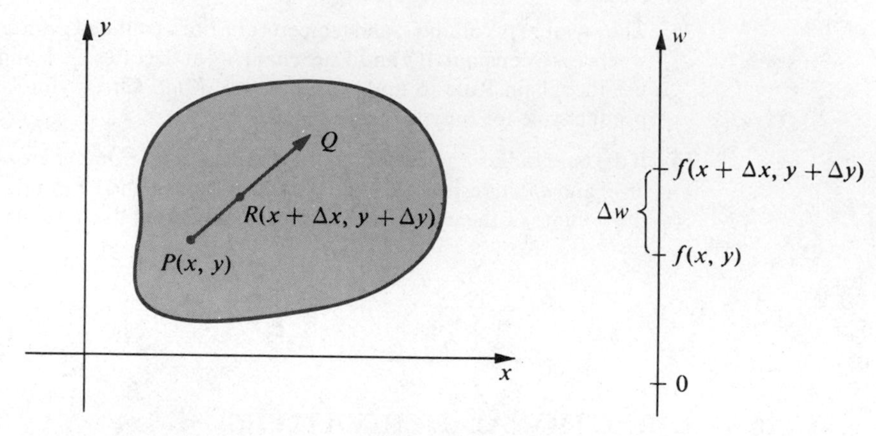

Figure 16.15

for some scalar t. Equating components and solving for x_1, y_1 we obtain

$$x_1 = x + t\cos\theta, \quad y_1 = y + t\sin\theta.$$

Thus every point R on l has coordinates $(x + t\cos\theta, y + t\sin\theta)$.

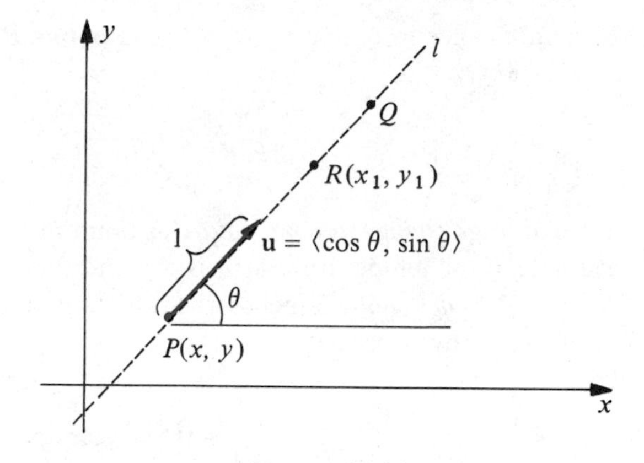

Figure 16.16

If we now let

$$\Delta x = t\cos\theta \quad \text{and} \quad \Delta y = t\sin\theta$$

then the change in $w = f(x, y)$ corresponding to the increments Δx and Δy is

$$\Delta w = f(x + t\cos\theta, y + t\sin\theta) - f(x, y).$$

If R is the point $(x + \Delta x, y + \Delta y)$ on l, then

$$d(P, R) = \sqrt{(\Delta x)^2 + (\Delta y)^2} = \sqrt{t^2\cos^2\theta + t^2\sin^2\theta} = \sqrt{t^2} = |t|.$$

Thus the average change in w is

$$\frac{\Delta w}{d(P, R)} = \frac{f(x + t\cos\theta, y + t\sin\theta) - f(x, y)}{|t|}.$$

Taking the limit as $t \to 0$ gives us the rate of change of f in the direction of $\overrightarrow{PQ}$. This discussion is summarized in the next definition.

(16.22) Definition

> If f is a function of x and y and if $\mathbf{u} = \langle \cos\theta, \sin\theta \rangle$ is a unit vector, then the **directional derivative of f in the direction of u**, denoted by $D_{\mathbf{u}} f(x, y)$, is given by
>
> $$D_{\mathbf{u}} f(x, y) = \lim_{t \to 0} \frac{f(x + t\cos\theta, y + t\sin\theta) - f(x, y)}{t}.$$

It is easy to see that the first partial derivatives of f are special cases of the directional derivative. As a matter of fact, if $\theta = 0$, then $\mathbf{u} = \langle 1, 0 \rangle = \mathbf{i}$, $\cos\theta = 1, \sin\theta = 0$, and the limit in Definition (16.22) reduces to

$$D_{\mathbf{i}} f(x, y) = \lim_{t \to 0} \frac{f(x + t, y) - f(x, y)}{t} = f_x(x, y).$$

Similarly, if $\theta = \pi/2$, then $\mathbf{u} = \langle 0, 1 \rangle = \mathbf{j}$, $\cos\theta = 0$, $\sin\theta = 1$, and we obtain

$$D_{\mathbf{j}} f(x, y) = \lim_{t \to 0} \frac{f(x, y + t) - f(x, y)}{t} = f_y(x, y).$$

The following theorem provides a formula for finding directional derivatives.

(16.23) Theorem

> If f is a differentiable function of two variables and $\mathbf{u} = \langle \cos\theta, \sin\theta \rangle$, then
>
> $$D_{\mathbf{u}} f(x, y) = f_x(x, y)\cos\theta + f_y(x, y)\sin\theta.$$

Proof. If g is the function of one variable t defined by

$$g(t) = f(x + t\cos\theta, y + t\sin\theta),$$

then from Theorem (3.7) and Definition (16.22)

$$g'(0) = \lim_{t \to 0} \frac{g(t) - g(0)}{t - 0} = D_{\mathbf{u}} f(x, y).$$

Since

$$g(t) = f(u, v), \quad \text{where } u = x + t\cos\theta,\ v = y + t\sin\theta,$$

it follows from Theorem (16.19) that

$$\begin{aligned} g'(t) &= f_u(u, v)\frac{du}{dt} + f_v(u, v)\frac{dv}{dt} \\ &= f_u(u, v)\cos\theta + f_v(u, v)\sin\theta. \end{aligned}$$

As we saw in the first part of the proof, the directional derivative equals $g'(0)$. However, if $t = 0$ then $u = x$, $v = y$ and we obtain the formula given in the statement of the theorem.

Example 1 If $f(x, y) = x^2 - 4xy$ and $\mathbf{u}$ is a unit vector with $\theta = \pi/3$, find $D_{\mathbf{u}} f(x, y)$. What is $D_{\mathbf{u}} f(-3, 2)$?

Solution Since $f_x(x, y) = 2x - 4y$, $f_y(x, y) = -4x$, and

$$\mathbf{u} = \langle \cos \pi/3, \sin \pi/3 \rangle = \langle 1/2, \sqrt{3}/2 \rangle$$

it follows from Theorem (16.23) that

$$\begin{aligned} D_{\mathbf{u}} f(x, y) &= (2x - 4y)(1/2) + (-4x)(\sqrt{3}/2) \\ &= (1 - 2\sqrt{3})x - 2y. \end{aligned}$$

In particular, at the point $P(-3, 2)$,

$$D_{\mathbf{u}} f(-3, 2) = (1 - 2\sqrt{3})(-3) - 2(2) = 6\sqrt{3} - 7 \approx 3.4.$$

We may use Theorem (16.23) to express the directional derivatives as a dot product of two vectors as follows:

$$D_{\mathbf{u}} f(x, y) = \langle f_x(x, y), f_y(x, y) \rangle \cdot \langle \cos\theta, \sin\theta \rangle = \langle f_x(x, y), f_y(x, y) \rangle \cdot \mathbf{u}.$$

The vector in V_2 whose components are the first partial derivatives of $f(x, y)$ is designated by the following special name and notation.

(16.24) Definition

If f is a function of two variables, then the **gradient of f** is defined by

$$\nabla f(x, y) = \langle f_x(x, y), f_y(x, y) \rangle = f_x(x, y)\mathbf{i} + f_y(x, y)\mathbf{j}.$$

In applications, the gradient $\nabla f(x, y)$ is sometimes denoted by grad $f(x, y)$. From the previous discussion we may write

(16.25) $$D_{\mathbf{u}} f(x, y) = \nabla f(x, y) \cdot \mathbf{u}$$

that is, *to find the directional derivative of f in the direction of the unit vector* $\mathbf{u}$, *dot the gradient of f with* $\mathbf{u}$. The symbol ∇ (called "del") is a **vector differential operator** and is symbolized by

$$\nabla = \mathbf{i}\frac{\partial}{\partial x} + \mathbf{j}\frac{\partial}{\partial y}.$$

It has properties similar to the operator d/dx. Standing alone it is meaningless; however, if it operates on $f(x, y)$ it produces (by definition) the two-dimensional vector function given by Definition (16.24).

Example 2 Find the directional derivative of $f(x, y) = x^3y^2$ at the point $P(-1, 2)$ in the direction of the vector $\mathbf{a} = 4\mathbf{i} - 3\mathbf{j}$.

Solution Using (16.24),

$$\nabla f(x, y) = 3x^2y^2\mathbf{i} + 2x^3y\mathbf{j}.$$

In particular, at $P(-1, 2)$ we have $\nabla f(-1, 2) = 12\mathbf{i} - 4\mathbf{j}$. A unit vector having the direction of $\mathbf{a}$ is

$$\mathbf{u} = \left(\frac{1}{|\mathbf{a}|}\right)\mathbf{a} = \frac{1}{5}(4\mathbf{i} - 3\mathbf{j}) = \frac{4}{5}\mathbf{i} - \frac{3}{5}\mathbf{j}.$$

Hence by (16.25)

$$D_{\mathbf{u}}f(-1, 2) = \nabla f(-1, 2) \cdot \mathbf{u} = (12\mathbf{i} - 4\mathbf{j}) \cdot \left(\frac{4}{5}\mathbf{i} - \frac{3}{5}\mathbf{j}\right) = \frac{48}{5} + \frac{12}{5} = 12.$$

Suppose $P(x, y)$ is a fixed point, and let us consider the directional derivative $D_{\mathbf{u}}f(x, y)$ as $\mathbf{u} = \langle \cos\theta, \sin\theta \rangle$ varies. For certain values of θ the directional derivative may be positive (that is, $f(x, y)$ may increase), or negative ($f(x, y)$ may decrease), or it may be 0. In many applications it is important to find the direction in which $f(x, y)$ increases most rapidly and also to find the maximum rate of change. The next theorem provides this information.

(16.26) Theorem

> The maximum value of $D_{\mathbf{u}}f(x, y)$ at the point $P(x, y)$ is $|\nabla f(x, y)|$ and it occurs in the direction of $\nabla f(x, y)$.

Proof. Let γ denote the angle between $\mathbf{u}$ and $\nabla f(x, y)$. Applying (16.25) and Theorem (14.18),

$$\begin{aligned} D_{\mathbf{u}}f(x, y) &= \nabla f(x, y) \cdot \mathbf{u} \\ &= |\nabla f(x, y)|\,|\mathbf{u}| \cos\gamma = |\nabla f(x, y)| \cos\gamma. \end{aligned}$$

The maximum value occurs if $\cos\gamma = 1$; that is, if $\gamma = 0$, in which case $\mathbf{u}$ has the same direction as $\nabla f(x, y)$. Moreover, in this direction, $D_{\mathbf{u}}f(x, y) = |\nabla f(x, y)|$. This completes the proof.

Directional derivatives in three dimensions are defined in a manner similar to (16.22). In this case we use direction angles α, β, and γ to specify the direction of the unit vector **u**, as in the next definition.

(16.27) Definition

If f is a function of three variables x, y, and z and if $\mathbf{u} = \langle \cos\alpha, \cos\beta, \cos\gamma \rangle$ is a unit vector having direction angles α, β, and γ, then the directional derivative of f in the direction of **u** is defined by

$$D_{\mathbf{u}} f(x, y, z) = \lim_{t \to 0} \frac{f(x + t\cos\alpha, y + t\cos\beta, z + t\cos\gamma) - f(x, y, z)}{t}.$$

We may use (16.27) to find the rate of change of f at (x, y, z) in the direction of **u**. As in Theorem (16.23) it can be shown that

(16.28)
$$D_{\mathbf{u}} f(x, y, z) = f_x(x, y, z)\cos\alpha + f_y(x, y, z)\cos\beta + f_z(x, y, z)\cos\gamma.$$

Moreover, if we define the **gradient of $f(x, y, z)$** by

(16.29)
$$\nabla f(x, y, z) = f_x(x, y, z)\mathbf{i} + f_y(x, y, z)\mathbf{j} + f_z(x, y, z)\mathbf{k}$$

then

(16.30)
$$D_{\mathbf{u}} f(x, y, z) = \nabla f(x, y, z) \cdot \mathbf{u}.$$

It is easy to show that the analogue of Theorem (16.26) is true in three dimensions.

Example 3 Find the directional derivative of $f(x, y, z) = x^2 - yz + z^2x$ at the point $P(1, -4, 3)$ in the direction from P to the point $Q(2, -1, 8)$. What is the maximum rate of increase of f at P?

Solution The vector in V_3 with geometric representation $\overrightarrow{PQ}$ is $\langle 2 - 1, -1 - (-4), 8 - 3 \rangle$, or $\langle 1, 3, 5 \rangle$. Consequently, a unit vector having the same direction as $\overrightarrow{PQ}$ is

$$\mathbf{u} = \frac{1}{\sqrt{35}}\langle 1, 3, 5 \rangle = \frac{1}{\sqrt{35}}(\mathbf{i} + 3\mathbf{j} + 5\mathbf{k}).$$

From (16.29),

$$\nabla f(x, y, z) = (2x + z^2)\mathbf{i} + (-z)\mathbf{j} + (-y + 2zx)\mathbf{k}$$

and, therefore,

$$\nabla f(1, -4, 3) = 11\mathbf{i} - 3\mathbf{j} + 10\mathbf{k}.$$

Applying (16.30)

$$\begin{aligned} D_{\mathbf{u}} f(1,-4,3) &= \nabla f(1,-4,3)\cdot \mathbf{u} \\ &= (11\mathbf{i} - 3\mathbf{j} + 10\mathbf{k})\cdot \frac{1}{\sqrt{35}}(\mathbf{i} + 3\mathbf{j} + 5\mathbf{k}) \\ &= \left(\frac{1}{\sqrt{35}}\right)(11 - 9 + 50) = \frac{52\sqrt{35}}{35} \approx 8.8. \end{aligned}$$

The maximum rate of increase of f at P is

$$|\nabla f(1,-4,3)| = \sqrt{121 + 9 + 100} = \sqrt{230} \approx 15.2.$$

Example 4 Suppose a rectangular coordinate system is located in space such that the temperature T at the point (x, y, z) is given by $T = 100/(x^2 + y^2 + z^2)$.

(a) Find the rate of change of T at the point $P(1, 3, -2)$ in the direction of the vector $\mathbf{a} = \mathbf{i} + \mathbf{j} + \mathbf{k}$.

(b) In what direction from P does T increase most rapidly? What is the maximum rate of change of T at P?

Solution (a) By definition, the gradient of T is

$$\nabla T = \frac{\partial T}{\partial x}\mathbf{i} + \frac{\partial T}{\partial y}\mathbf{j} + \frac{\partial T}{\partial z}\mathbf{k}.$$

Since

$$\frac{\partial T}{\partial x} = \frac{-200x}{(x^2 + y^2 + z^2)^2}, \quad \frac{\partial T}{\partial y} = \frac{-200y}{(x^2 + y^2 + z^2)^2}, \quad \frac{\partial T}{\partial z} = \frac{-200z}{(x^2 + y^2 + z^2)^2}$$

we have

$$\nabla T = \frac{-200}{(x^2 + y^2 + z^2)^2}(x\mathbf{i} + y\mathbf{j} + z\mathbf{k}).$$

If we let $\nabla T]_P$ denote the value of ∇T at the point $P(1, 3, -2)$, then

$$\nabla T]_P = \frac{-200}{196}(\mathbf{i} + 3\mathbf{j} - 2\mathbf{k}).$$

A unit vector $\mathbf{u}$ having the same direction as $\mathbf{a}$ is

$$\mathbf{u} = \frac{1}{\sqrt{3}}(\mathbf{i} + \mathbf{j} + \mathbf{k}).$$

Using (16.30), the rate of change of T at P in the direction of $\mathbf{a}$ is

$$D_{\mathbf{u}} T]_P = \nabla T]_P \cdot \mathbf{u} = \frac{-200}{196}\frac{(1 + 3 - 2)}{\sqrt{3}} \approx -1.18.$$

If, for example, T is measured in degrees and distance is in inches, then T is decreasing at a rate of 1.18 degrees per inch at P in the direction of **a**.

(b) The maximum rate of change of T at P occurs in the direction of the gradient, that is, in the direction of the vector $-\mathbf{i} - 3\mathbf{j} + 2\mathbf{k}$. The maximum rate of change equals the magnitude of the gradient, that is

$$|\nabla T]_P| = \frac{200}{196}\sqrt{1 + 9 + 4} \approx 3.8.$$

EXERCISES 16.6

In Exercises 1–14 find the directional derivative of f at the point P in the indicated direction.

1 $f(x, y) = x^2 - 5xy + 3y^2; P(3, -1), \theta = \pi/4$

2 $f(x, y) = x^3 - 3x^2y - y^3; P(1, -2), \theta = 2\pi/3$

3 $f(x, y) = \arctan y/x; P(4, -4), \mathbf{a} = 2\mathbf{i} - 3\mathbf{j}$

4 $f(x, y) = x^2 \ln y; P(5, 1), \mathbf{a} = -\mathbf{i} + 4\mathbf{j}$

5 $f(x, y) = \sqrt{9x^2 - 4y^2 - 1}; P(3, -2), \mathbf{a} = \mathbf{i} + 5\mathbf{j}$

6 $f(x, y) = (x - y)/(x + y); P(2, -1), \mathbf{a} = 3\mathbf{i} + 4\mathbf{j}$

7 $f(x, y) = x \cos^2 y; P(2, \pi/4), \mathbf{a} = \langle 5, 1 \rangle$

8 $f(x, y) = xe^{3y}; P(4, 0), \mathbf{a} = \langle -1, 3 \rangle$

9 $f(x, y, z) = xy^3z^2; P(2, -1, 4), \mathbf{a} = \mathbf{i} + 2\mathbf{j} - 3\mathbf{k}$

10 $f(x, y, z) = x^2 + 3yz + 4xy; P(1, 0, -5), \mathbf{a} = 2\mathbf{i} - 3\mathbf{j} + \mathbf{k}$

11 $f(x, y, z) = z^2e^{xy}; P(-1, 2, 3), \mathbf{a} = 3\mathbf{i} + \mathbf{j} - 5\mathbf{k}$

12 $f(x, y, z) = \sqrt{xy} \sin z; P(4, 9, \pi/4), \mathbf{a} = 2\mathbf{i} + 3\mathbf{j} - 2\mathbf{k}$

13 $f(x, y, z) = (x + y)(y + z); P(5, 7, 1), \mathbf{a} = \langle -3, 0, 1 \rangle$

14 $f(x, y, z) = z^2 \tan^{-1}(x + y); P(0, 0, 4), \mathbf{a} = \langle 6, 0, 1 \rangle$

In Exercises 15–18 find the directional derivative of f at P in the direction from P to Q. Also find the direction in which f increases most rapidly at P and find the maximum rate of increase.

15 $f(x, y) = x^2e^{-2y}; P(2, 0), Q(-3, 1)$

16 $f(x, y) = \sin(2x - y); P(-\pi/3, \pi/6), Q(0, 0)$

17 $f(x, y, z) = \sqrt{x^2 + y^2 + z^2}; P(-2, 3, 1), Q(0, -5, 4)$

18 $f(x, y, z) = x/y - y/z; P(0, -1, 2), Q(3, 1, -4)$

19 A hot metal plate is situated on an xy-plane such that the temperature T is inversely proportional to the distance from the origin. If the temperature at $P(3, 4)$ is 100°, find the rate of change of T at P in the direction of the vector $\mathbf{i} + \mathbf{j}$. In what direction does T increase most rapidly at P? In what direction is the rate of change 0?

20 The surface of a certain lake is represented by a region D in the xy-plane such that the depth (in feet) under the point corresponding to (x, y) is given by $f(x, y) = 300 - 2x^2 - 3y^2$. If a boy in the water is at the point $(4, 9)$, in what direction should he swim in order for the depth to decrease most rapidly? In what direction will the depth remain the same?

21 The electrical potential V at the point $P(x, y, z)$ in a rectangular coordinate system is given by $V = x^2 + 4y^2 + 9z^2$. Find the rate of change of V at $P(2, -1, 3)$ in the direction from P to the origin. Find the direction which produces the maximum rate of change of V at P. What is the maximum rate of change at P?

22 An object is situated in a rectangular coordinate system such that the temperature T at the point $P(x, y, z)$ is given by $T = 4x^2 - y^2 + 16z^2$. Find the rate of change of T at the point $P(4, -2, 1)$ in the direction of the vector $2\mathbf{i} + 6\mathbf{j} - 3\mathbf{k}$. In what direction does T increase most rapidly at P? What is this maximum rate of change?

23 Prove (16.28).

24 Prove that $D_{\mathbf{k}} f(x, y, z) = f_z(x, y, z)$.

If $u = f(x, y)$ and $v = g(x, y)$ where f and g are differentiable, prove each of the identities in Exercises 25–30.

25 $\nabla(cu) = c\nabla u$, where c is a constant

26 $\nabla(u + v) = \nabla u + \nabla v$

27 $\nabla(uv) = u\nabla v + v\nabla u$

28 $\nabla\left(\dfrac{u}{v}\right) = \dfrac{v\nabla u - u\nabla v}{v^2}, \quad v \neq 0$

29 $\nabla u^n = nu^{n-1}\nabla u$ for every real number n

30 If $w = h(u)$, then $\nabla w = \dfrac{dw}{du}\nabla u$.

16.7 TANGENT PLANES AND NORMAL LINES TO SURFACES

Suppose a surface S is the graph of an equation $F(x, y, z) = 0$, where F has continuous first partial derivatives. Let $P_0(x_0, y_0, z_0)$ be a point on S at which F_x, F_y, and F_z are not all zero. A **tangent line** to S at P_0 is, by definition, a tangent line to any curve C which lies on S and contains P_0. Suppose parametric equations for such a curve are $x = f(t)$, $y = g(t)$, $z = h(t)$ and let t_0 be the value of t which corresponds to P_0. For each t the point $(f(t), g(t), h(t))$ on C is also on S and, therefore,

$$F(f(t), g(t), h(t)) = 0.$$

If we let

$$w = F(x, y, z), \quad \text{where } x = f(t), y = g(t), z = h(t),$$

then using Theorem (16.19) and the fact that $w = 0$ for all t,

$$\frac{dw}{dt} = \frac{\partial w}{\partial x}\frac{dx}{dt} + \frac{\partial w}{\partial y}\frac{dy}{dt} + \frac{\partial w}{\partial z}\frac{dz}{dt} = 0$$

or

$$F_x(x, y, z)f'(t) + F_y(x, y, z)g'(t) + F_z(x, y, z)h'(t) = 0.$$

In particular, at the point P_0,

$$F_x(x_0, y_0, z_0)f'(t_0) + F_y(x_0, y_0, z_0)g'(t_0) + F_z(x_0, y_0, z_0)h'(t_0) = 0.$$

The preceding equation can be written compactly as

$$\nabla F]_{P_0} \cdot \mathbf{r}'(t_0) = 0,$$

where the symbol $\nabla F]_{P_0}$ denotes $\nabla F(x_0, y_0, z_0)$, and $\mathbf{r}(t) = \langle f(t), g(t), h(t) \rangle$. Since $\mathbf{r}'(t_0)$ is a tangent vector to C at P_0, this implies that the vector $\nabla F]_{P_0}$ is orthogonal to every tangent line l to S at P_0. The plane Γ through P_0 with normal vector $\nabla F]_{P_0}$ is called the *tangent plane to S at P_0*. We have shown that every tangent line l to S at P_0 lies in the tangent plane at P_0. A sketch illustrating the above concepts appears in Figure 16.17. The following theorem summarizes our discussion.

(16.31) Theorem

Suppose $F(x, y, z)$ has continuous first partial derivatives and P_0 is a point on the graph S of $F(x, y, z) = 0$. If F_x, F_y, and F_z are not all 0 at P_0, then $\nabla F]_{P_0}$ is normal to the tangent plane to S at P_0.

Applying Theorem (14.30) gives us the following corollary, where it is assumed that F_x, F_y, and F_z are not all 0 at (x_0, y_0, z_0).

(16.32) Corollary

An equation for the tangent plane to the graph of $F(x, y, z) = 0$ at the point (x_0, y_0, z_0) is

$$F_x(x_0, y_0, z_0)(x - x_0) + F_y(x_0, y_0, z_0)(y - y_0) + F_z(x_0, y_0, z_0)(z - z_0) = 0.$$

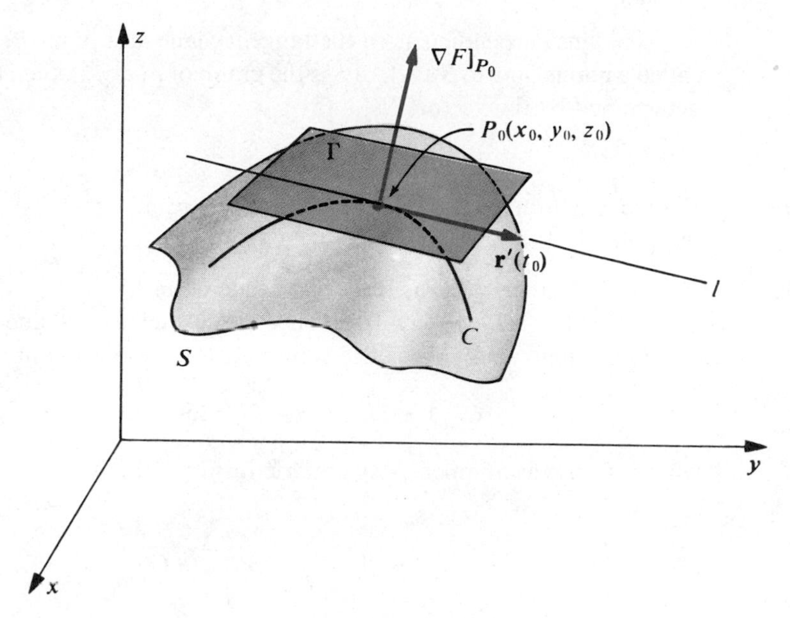

Figure 16.17

Example 1 Find an equation for the tangent plane to the ellipsoid $4x^2 + 9y^2 + z^2 - 49 = 0$ at the point $P_0(1, -2, 3)$.

Solution If $F(x, y, z)$ denotes the left side of the given equation, then

$$F_x(x, y, z) = 8x, \quad F_y(x, y, z) = 18y, \quad F_z(x, y, z) = 2z$$

and hence at P_0

$$F_x(1, -2, 3) = 8, \quad F_y(1, -2, 3) = -36, \quad F_z(1, -2, 3) = 6.$$

Applying Theorem (16.32) we obtain the equation

$$8(x - 1) - 36(y + 2) + 6(z - 3) = 0$$

which simplifies to

$$4x - 18y + 3z - 49 = 0.$$

If $z = f(x, y)$ is an equation for S, then letting $F(x, y, z) = f(x, y) - z$, the equation in Theorem (16.32) becomes

$$f_x(x_0, y_0)(x - x_0) + f_y(x_0, y_0)(y - y_0) + (-1)(z - z_0) = 0.$$

This shows that an equation of the tangent plane to the graph of $z = f(x, y)$ at the point (x_0, y_0, z_0) is

(16.33) $$z - z_0 = f_x(x_0, y_0)(x - x_0) + f_y(x_0, y_0)(y - y_0).$$

The line perpendicular to the tangent plane at a point P_0 on a surface S is called a **normal line** to S at P_0. If S is the graph of $F(x, y, z)$, then the normal line is determined by the vector $\nabla F]_{P_0}$.

Example 2 Find an equation of the normal line to the ellipsoid $4x^2 + 9y^2 + z^2 - 49 = 0$ at the point $P_0(1, -2, 3)$.

Solution This surface is the same as the one considered in Example 1. Consequently the vector $\nabla F]_{P_0} = \langle 8, -36, 6 \rangle$ determines the desired normal line. Using (14.29) we obtain the following parametric equations for the normal line:

$$x = 1 + 8t, \quad y = -2 - 36t, \quad z = 3 + 6t$$

where t is a real number. A symmetric form for the line (see (14.34)) is

$$\frac{x - 1}{8} = \frac{y + 2}{-36} = \frac{z - 3}{6}.$$

Suppose f is a function of x and y and S is the graph of the equation $z = f(x, y)$. The tangent plane Γ to S at $P_0(x_0, y_0, z_0)$ may be used to obtain a geometric interpretation for the differential

$$dz = f_x(x_0, y_0)\,\Delta x + f_y(x_0, y_0)\,\Delta y.$$

Since

$$\Delta z = f(x_0 + \Delta x, y_0 + \Delta y) - f(x_0, y_0)$$

the point $P(x_0 + \Delta x, y_0 + \Delta y, z_0 + \Delta z)$ is on S. Let $C(x_0 + \Delta x, y_0 + \Delta y, z)$ be on Γ and consider the points $A(x_0 + \Delta x, y_0 + \Delta y, 0)$ and $B(x_0 + \Delta x, y_0 + \Delta y, z_0)$ as illustrated in Figure 16.18. Since C is on Γ its coordinates satisfy equation (16.33), that is,

$$\begin{aligned} z - z_0 &= f_x(x_0, y_0)(x_0 + \Delta x - x_0) + f_y(x_0, y_0)(y_0 + \Delta y - y_0) \\ &= f_x(x_0, y_0)\,\Delta x + f_y(x_0, y_0)\,\Delta y = dz. \end{aligned}$$

Consequently, $\overline{BC} = dz$. Since $\overline{BP} = \Delta z$ we see that

$$\Delta z - dz = \overline{BP} - \overline{BC} = \overline{CP}.$$

It follows that when dz is used as an approximation to Δz, it is assumed that the surface S is almost the same as its tangent plane at points close to P_0.

As an application of the discussion in this section, suppose a solid, liquid, or gas is situated in a three-dimensional coordinate system such that $w = F(x, y, z)$ is the temperature at the point $P(x, y, z)$. If $P_0(x_0, y_0, z_0)$ is a fixed point, then the graph of the equation

$$F(x, y, z) = F(x_0, y_0, z_0)$$

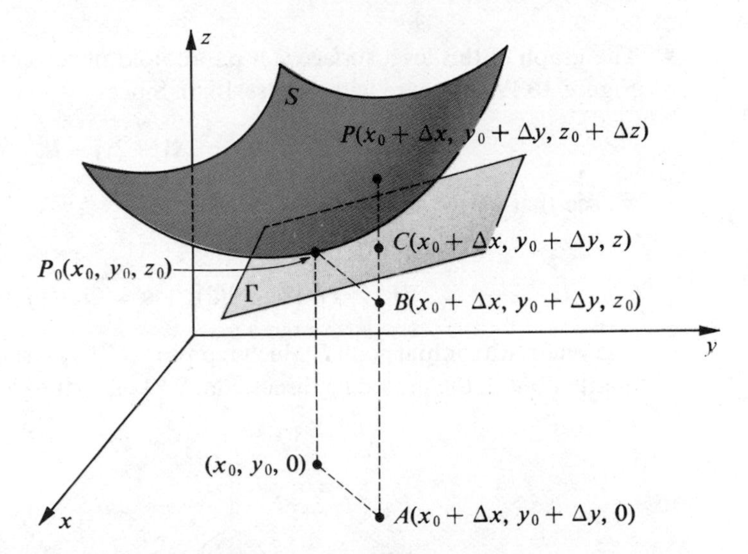

Figure 16.18

is the level (isothermal) surface S_0 which passes through P_0. Note that for every point on S_0 the temperature is $w_0 = F(x_0, y_0, z_0)$. It is convenient to visualize all possible level surfaces that can be obtained by choosing different points. Several surfaces S_0, S_1, and S_2 through P_0, P_1, and P_2, respectively, corresponding to temperatures w_0, w_1, and w_2, are sketched in Figure 16.8. By Theorem (16.31), $\nabla F]_{P_0}$ is a normal vector for the tangent plane Γ to S_0 at P_0. According to the work in the previous section, *the direction for the maximum rate of change of temperature at* P_0 *is orthogonal to the level surface through* P_0, that is, along a normal vector to Γ. An analogous situation occurs in electrical theory where $F(x, y, z)$ is the potential at $P(x, y, z)$. If a particle moves on one of the level (equipotential) surfaces the potential remains constant. As above, the potential changes most rapidly when a particle moves in a direction which is orthogonal to an equipotential surface.

Example 3 If $F(x, y, z) = x^2 + y^2 + z$, sketch the level surface for F which passes through the point $P(1, 2, 4)$ and draw a vector with initial point P which corresponds to $\nabla F]_P$.

Solution The level surfaces for F are graphs of equations of the form $F(x, y, z) = c$, where c is a constant. In particular, the level surface which passes through $P(1, 2, 4)$ is the graph of

$$F(x, y, z) = F(1, 2, 4).$$

Using the formula for $F(x, y, z)$ gives us

$$x^2 + y^2 + z = (1)^2 + (2)^2 + 4 = 9$$

or equivalently,

$$z = 9 - x^2 - y^2.$$

The graph of this level surface is a paraboloid of revolution and is sketched in Figure 16.19 (compare with Figure 16.5). Since

$$\nabla F(x, y, z) = 2x\mathbf{i} + 2y\mathbf{j} + \mathbf{k}$$

we see that

$$\nabla F]_P = 2(1)\mathbf{i} + 2(2)\mathbf{j} + \mathbf{k} = 2\mathbf{i} + 4\mathbf{j} + \mathbf{k}.$$

The vector with initial point P which represents $\nabla F]_P$ is shown in Figure 16.19. As pointed out in the preceding discussion, $\nabla F]_P$ is orthogonal to the level surface.

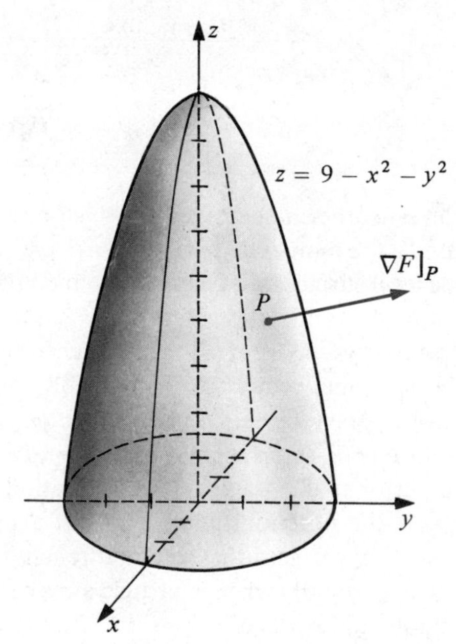

Figure 16.19

A similar result may be proved for functions of two variables. Specifically, given $f(x, y)$, we may show that the direction for the maximum rate of change of f at $P_0(x_0, y_0)$ is orthogonal to the level curve C through P_0, that is, orthogonal to the tangent line to C at P_0. For example, if $f(x, y) = 9 - x^2 - y^2$, the level curves are circles (see Figure 16.7) and hence the maximum rate of change of f takes place along lines through the origin. (Why?) This corresponds to going up or down the steepest part of the graph of f sketched in Figure 16.5.

Example 4 If $f(x, y) = x^2 + 2y^2$, sketch the level curve for f which passes through the point $P(3, 1)$ and draw a vector with initial point P which corresponds to $\nabla f]_P$.

Solution The level curve for f which passes through the point $P(3, 1)$ is given by $f(x, y) = f(3, 1)$, or

$$x^2 + 2y^2 = (3)^2 + 2(1)^2 = 11.$$

The graph is the ellipse sketched in Figure 16.20. The gradient of f is given by

$$\nabla f(x, y) = 2x\mathbf{i} + 4y\mathbf{j}$$

and hence

$$\nabla f]_P = 2(3)\mathbf{i} + 4(1)\mathbf{j} = 6\mathbf{i} + 4\mathbf{j}.$$

The vector with initial point P which represents $\nabla f]_P$ is shown in Figure 16.20. Note that this vector is orthogonal to the level curve through P.

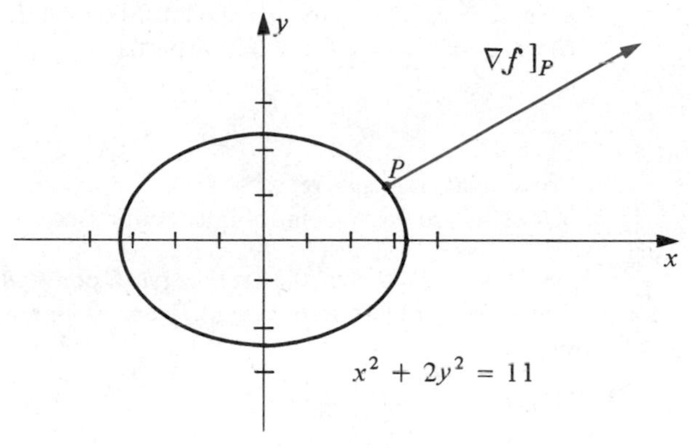

Figure 16.20

EXERCISES 16.7

In Exercises 1–10 find equations for the tangent plane and the normal line to the graph of the given equation at the indicated point P.

1 $4x^2 - y^2 + 3z^2 = 10$; $P(2, -3, 1)$

2 $9x^2 - 4y^2 - 25z^2 = 40$; $P(4, 1, -2)$

3 $z = 4x^2 + 9y^2$; $P(-2, -1, 25)$

4 $z = 4x^2 - y^2$; $P(5, -8, 36)$

5 $xy + 2yz - xz^2 + 10 = 0$; $P(-5, 5, 1)$

6 $x^3 - 2xy + z^3 + 7y + 6 = 0$; $P(1, 4, -3)$

7 $z = 2e^{-x} \cos y$; $P(0, \pi/3, 1)$

8 $z = \ln xy$; $P(1/2, 2, 0)$

9 $x = \ln (y/2z)$; $P(0, 2, 1)$

10 $xyz - 4xz^3 + y^3 = 10$; $P(-1, 2, 1)$

In each of Exercises 11–14, prove that an equation of the tangent plane to the given quadric surface at the point $P_0(x_0, y_0, z_0)$ may be written in the indicated form.

11 $\dfrac{x^2}{a^2} + \dfrac{y^2}{b^2} + \dfrac{z^2}{c^2} = 1$; $\dfrac{xx_0}{a^2} + \dfrac{yy_0}{b^2} + \dfrac{zz_0}{c^2} = 1$

12 $\dfrac{x^2}{a^2} - \dfrac{y^2}{b^2} + \dfrac{z^2}{c^2} = 1$; $\dfrac{xx_0}{a^2} - \dfrac{yy_0}{b^2} + \dfrac{zz_0}{c^2} = 1$

13 $\dfrac{x^2}{a^2} - \dfrac{y^2}{b^2} - \dfrac{z^2}{c^2} = 1$; $\dfrac{xx_0}{a^2} - \dfrac{yy_0}{b^2} - \dfrac{zz_0}{c^2} = 1$

14 $\dfrac{x^2}{a^2} + \dfrac{y^2}{b^2} = cz$; $\dfrac{2xx_0}{a^2} + \dfrac{2yy_0}{b^2} = c(z + z_0)$

15 Find the points on the hyperboloid $x^2 - 2y^2 - 4z^2 = 16$ at which the tangent plane is parallel to the plane $4x - 2y + 4z = 5$.

16 Prove that the sum of the squares of the x-, y-, and z-intercepts of every tangent plane to the graph of $x^{2/3} + y^{2/3} + z^{2/3} = a^{2/3}$ is a constant.

17 Prove that every normal line to a sphere passes through the center of the sphere.

18 Find the points on the paraboloid $z = 4x^2 + 9y^2$ at which the normal line is parallel to the line through $P(-2, 4, 3)$ and $Q(5, -1, 2)$.

19 Two surfaces are said to be **orthogonal** at a point of intersection $P(x, y, z)$ if their normal lines at P are orthogonal. Show that the graphs of $F(x, y, z) = 0$ and $G(x, y, z) = 0$ (where F and G have partial derivatives) are orthogonal at P if and only if

$$F_x G_x + F_y G_y + F_z G_z = 0.$$

20 Prove that the sphere $x^2 + y^2 + z^2 = a^2$ and the cone $x^2 + y^2 - z^2 = 0$ are orthogonal at every point of intersection (see Exercise 19).

In Exercises 21–24 sketch the level curve C for f which passes through the point P and draw a vector (with initial point P) corresponding to $\nabla f]_P$ (see Exercises 19–24 of Section 1).

21 $f(x, y) = y^2 - x^2$; $P(2, 1)$

22 $f(x, y) = 3x - 2y$; $P(-2, 1)$

23 $f(x, y) = x^2 - y$; $P(-3, 5)$

24 $f(x, y) = xy$; $P(3, 2)$

In Exercises 25–30 sketch the level surface S for F which passes through the point P and draw a vector (with initial point P) corresponding to $\nabla F]_P$ (see Exercises 25–30 of Section 1).

25 $F(x, y, z) = x^2 + y^2 + z^2$; $P(1, 5, 2)$

26 $F(x, y, z) = z - x^2 - y^2$; $P(2, -2, 1)$

27 $F(x, y, z) = x + 2y + 3z$; $P(3, 4, 1)$

28 $F(x, y, z) = x^2 + y^2 - z^2$; $P(3, -1, 1)$

29 $F(x, y, z) = x^2 + y^2$; $P(2, 0, 3)$

30 $F(x, y, z) = z$; $P(2, 3, 4)$

16.8 EXTREMA OF FUNCTIONS OF TWO VARIABLES

Throughout the remainder of this chapter, the term **rectangular region** will be used for points in a coordinate plane which lie inside a rectangle whose sides are parallel to the coordinate axes. If we wish to include the points on the boundary, the term **closed rectangular region** will be used.

In a manner analogous to the single variable case, a function f of two variables is said to have a **local maximum** at (a, b) if there is a rectangular region R containing (a, b) such that $f(x, y) \le f(a, b)$ for all other pairs (x, y) in R. Geometrically, if a surface S is the graph of f, then the local maxima correspond to the high points on S. If f_y exists, then as pointed out in Section 16.3, $f_y(a, b)$ is the slope of the tangent line to the trace C of S on the plane $x = a$ (see Figure 16.13). It follows that if $f(a, b)$ is a local maximum, then $f_y(a, b) = 0$. Similarly $f_x(a, b) = 0$.

The function f has a **local minimum** at (c, d) if there is a rectangular region R containing (c, d) such that $f(x, y) \geq f(c, d)$ for all other (x, y) in R. If f has first partial derivatives, then as above, they must be zero at (c, d). The local minima correspond to the low points on the graph of f.

It can be shown that if a function f of two variables is continuous on a closed rectangular region R, then f has an **absolute maximum** $f(a, b)$ and an **absolute minimum** $f(c, d)$ for some (a, b) and (c, d) in R. This means that

(16.34) $$f(c, d) \leq f(x, y) \leq f(a, b)$$

for all (x, y) in R. The proof may be found in more advanced books on calculus.

The local maxima and minima are called the **extrema** of f. If f has first partial derivatives, then from the preceding discussion, the pairs which give rise to extrema are solutions of *both* of the equations

(16.35) $$f_x(x, y) = 0 \quad \text{and} \quad f_y(x, y) = 0.$$

Consequently, to find the extrema we begin by solving equations (16.35) simultaneously and then, in some way, test each solution.

Example 1 If $f(x, y) = 1 + x^2 + y^2$, find the extrema of f.

Solution Since $f_x(x, y) = 2x$ and $f_y(x, y) = 2y$, equations (16.35) are

$$2x = 0 \quad \text{and} \quad 2y = 0.$$

The only pair which satisfies both equations is $(0, 0)$ and hence $f(0, 0) = 1$ is the only possible extremum. Moreover, since

$$f(x, y) = 1 + x^2 + y^2 > 1 \quad \text{if } (x, y) \neq (0, 0),$$

f has the local minimum 1 at $(0, 0)$. This fact may also be seen geometrically by referring to the graph of $z = f(x, y)$, sketched in Figure 16.21. The number $f(0, 0)$ is also the absolute minimum of f.

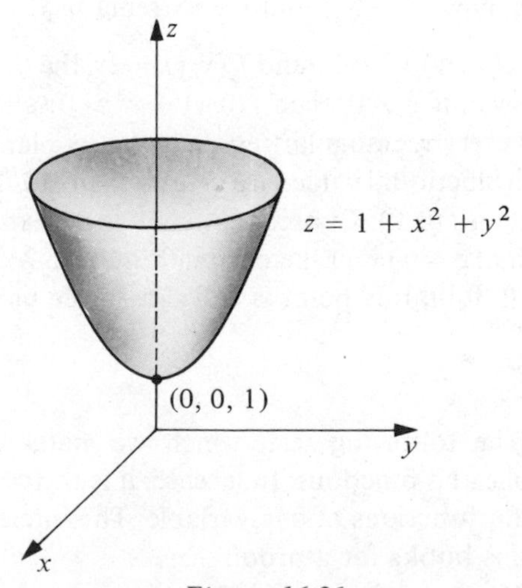

Figure 16.21

Example 2 If $f(x, y) = 4 - x^2 - y^2$, find the extrema of f.

Solution Since $f_x(x, y) = -2x$ and $f_y(x, y) = -2y$, equations (16.35) are

$$-2x = 0 \quad \text{and} \quad -2y = 0.$$

As in Example 1, the only pair which satisfies these equations is $(0, 0)$ and hence $f(0, 0) = 4$ is the only possible extremum. Moreover,

$$f(x, y) = 4 - (x^2 + y^2) \leq 4 \quad \text{if } (x, y) \neq (0, 0)$$

and, therefore, f has a local maximum 4 at $(0, 0)$. The number 4 is also the absolute maximum of f, as is evident from the graph of $z = f(x, y)$ sketched in Figure 16.22.

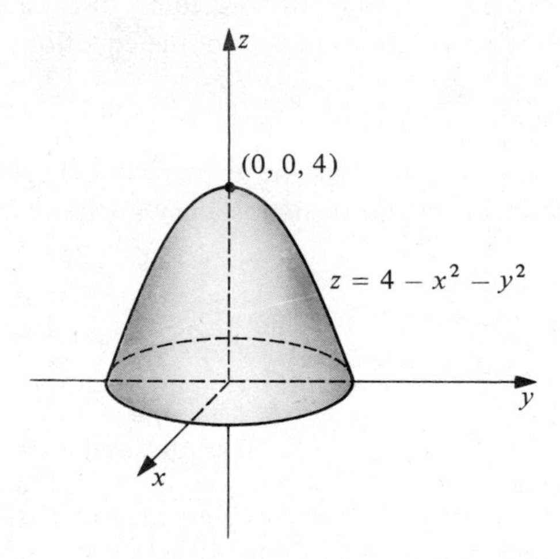

Figure 16.22

As illustrated in the next example, not every solution of (16.35) may lead to an extremum.

Example 3 If $f(x, y) = y^2 - x^2$, find the extrema of f.

Solution Since $f_x(x, y) = -2x$ and $f_y(x, y) = 2y$, the only possible extremum is $f(0, 0) = 0$. However, if $y \neq 0$, then $f(0, y) = y^2 > 0$; and if $x \neq 0$, then $f(x, 0) = -x^2 < 0$. Thus every rectangular region in the xy-plane containing $(0, 0)$ contains pairs at which functional values are greater than $f(0, 0)$ as well as pairs at which values are less than $f(0, 0)$. Consequently f has no extrema. This is also evident from the graph of $z = f(x, y)$ sketched in Figure 16.23. Because of the shape of the surface near $(0, 0, 0)$, this point is called a **saddle point** of the graph.

The following test, which we state without proof, is useful for more complicated functions. In a sense, it is the counterpart of the Second Derivative Test for functions of one variable. The interested reader may consult advanced calculus books for a proof.

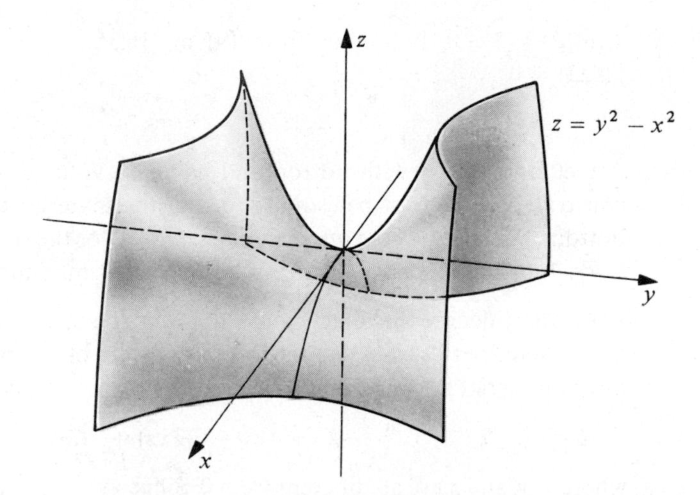

Figure 16.23

(16.36) Test for Extrema

Let f be a function of two variables which has continuous second partial derivatives on a rectangular region Q and let

$$g(x, y) = f_{xx}(x, y)f_{yy}(x, y) - [f_{xy}(x, y)]^2$$

for all (x, y) in Q. If (a, b) is in Q and $f_x(a, b) = 0$, $f_y(a, b) = 0$, then the following statements hold.

(i) $f(a, b)$ is a local maximum of f if $g(a, b) > 0$ and $f_{xx}(a, b) < 0$.

(ii) $f(a, b)$ is a local minimum of f if $g(a, b) > 0$ and $f_{xx}(a, b) > 0$.

(iii) $f(a, b)$ is not an extremum of f if $g(a, b) < 0$.

Note that if $g(a, b) > 0$, then $f_{xx}(a, b)f_{yy}(a, b)$ is positive and hence $f_{xx}(a, b)$ and $f_{yy}(a, b)$ agree in sign. Thus, we may replace $f_{xx}(a, b)$ in (i) and (ii) by $f_{yy}(a, b)$. If $g(a, b) = 0$ the test gives no information and in this case it is necessary to use a more direct approach such as that illustrated in Examples 1 and 2.

Example 4 If $f(x, y) = x^2 - 4xy + y^3 + 4y$, find the extrema of f.

Solution Since $f_x(x, y) = 2x - 4y$ and $f_y(x, y) = -4x + 3y^2 + 4$, equations (16.35) are

$$2x - 4y = 0 \quad \text{and} \quad -4x + 3y^2 + 4 = 0.$$

Solving these simultaneously gives us the pairs (4, 2) and (4/3, 2/3). The second partial derivatives of f are

$$f_{xx}(x, y) = 2, \quad f_{xy}(x, y) = -4, \quad \text{and} \quad f_{yy}(x, y) = 6y$$

and, therefore, $g(x, y) = 12y - 16$. Since $g(4/3, 2/3) = -8 < 0$, it follows from (iii) of (16.36) that $f(4/3, 2/3)$ is not an extremum of f. Since $g(4, 2) = 8 > 0$ and

$f_{xx}(4, 2) = 2 > 0$, it follows from (ii) of (16.36) that f has a local minimum $f(4, 2) = 0$.

Example 5 A rectangular box with no top and having a volume of 12 cubic feet is to be constructed. The cost per square foot of the material to be used is \$4 for the bottom, \$3 for two of the opposite sides, and \$2 for the remaining pair of opposite sides. Find the dimensions of the box that will minimize the cost.

Solution Let x and y denote the dimensions (in feet) of the base, and z the altitude (in feet). Since there are two sides of area xz and two sides of area yz, the cost C (in dollars) of the material is given by

$$C = 4xy + 3(2xz) + 2(2yz)$$

where x, y, and z are all different from 0. Since $xyz = 12$, it follows that $z = 12/xy$. Substituting in the formula for C and simplifying we obtain

$$C = 4xy + \frac{72}{y} + \frac{48}{x}.$$

To determine the possible extrema we must find the simultaneous solutions of the equations

$$C_x = 4y - \frac{48}{x^2} = 0 \quad \text{and} \quad C_y = 4x - \frac{72}{y^2} = 0.$$

These equations imply that

$$x^2y = 12 \quad \text{and} \quad xy^2 = 18.$$

We see from the first equation that $y = 12/x^2$. Substitution in the second equation gives us

$$x\left(\frac{144}{x^4}\right) = 18, \quad \text{or} \quad 144 = 18x^3.$$

Solving for x we obtain $x^3 = 8$, or $x = 2$. Since $y = 12/x^2$, the corresponding value of y is 12/4, or 3. Theorem 16.36 can be used to show that these values of x and y determine a minimum value of C. Finally, using $z = 12/xy$ we obtain $z = 12/(2 \cdot 3) = 2$. Thus the minimum cost occurs if the dimensions of the base are 3 ft by 2 ft and the altitude is 2 ft. The longer sides should be made out of the \$2 material and the shorter sides out of the \$3 material.

The material in this section can be generalized to functions of more than two variables. For example, given $f(x, y, z)$ we define local maxima and minima in a manner analogous to the two-variable case. If f has first partial derivatives, then a local extremum can occur only at a point where f_x, f_y, and f_z are simultaneously 0. It is difficult to obtain tests for determining whether such a point corresponds to a maximum, a minimum, or neither. However, in applications it is often possible to determine this by analyzing the physical nature of the problem.

EXERCISES 16.8

In Exercises 1–16, find the extrema of f.

1 $f(x, y) = x^2 + 2xy + 3y^2$

2 $f(x, y) = x^2 - 3xy - y^2 + 2y - 6x$

3 $f(x, y) = x^3 + 3xy - y^3$

4 $f(x, y) = 4x^3 - 2x^2y + y^2$

5 $f(x, y) = x^2 + 4y^2 - x + 2y$

6 $f(x, y) = 5 + 4x - 2x^2 + 3y - y^2$

7 $f(x, y) = x^4 + y^3 + 32x - 9y$

8 $f(x, y) = \cos x + \cos y$

9 $f(x, y) = e^x \sin y$

10 $f(x, y) = x \sin y$

11 $f(x, y) = (4y + x^2y^2 + 8x)/xy$

12 $f(x, y) = x/(x + y)$

13 $f(x, y) = \sin x + \sin y + \sin (x + y); 0 \leq x \leq 2\pi, 0 \leq y \leq 2\pi$

14 $f(x, y) = (x + y + 1)^2/(x^2 + y^2 + 1)$

15 $f(x, y) = yx^2 + y^3 - 4x^2$

16 $f(x, y) = x^2 - 6x \cos y + 9$

17 Find the shortest distance from the point $P(2, 1, -1)$ to the plane $4x - 3y + z = 5$.

18 Find the shortest distance between the planes $2x + 3y - z = 2$ and $2x + 3y - z = 4$.

19 Find the points on the graph of $xy^3z^2 = 16$ which are closest to the origin.

20 Find three positive real numbers whose sum is 1000 and whose product is a maximum.

21 If an open rectangular box is to have a fixed volume V, what relative dimensions will make the surface area a minimum?

22 If an open rectangular box is to have a fixed surface area A, what relative dimensions will make the volume a maximum?

23 Find the dimensions of the rectangular parallelepiped of maximum volume with faces parallel to the coordinate planes that can be inscribed in the ellipsoid $16x^2 + 4y^2 + 9z^2 = 144$.

24 Generalize Exercise 23 to any ellipsoid $x^2/a^2 + y^2/b^2 + z^2/c^2 = 1$.

25 Find the dimensions of the rectangular parallelepiped of maximum volume which has three of its faces in the coordinate planes, one vertex at the origin, and another vertex in the first octant on the plane $4x + 3y + z = 12$.

26 Generalize Exercise 25 to any plane $x/a + y/b + z/c = 1$ where a, b, and c are positive real numbers.

27 A company plans to manufacture closed rectangular boxes which have a volume of 8 cubic feet. If the material for the top and bottom costs twice as much as the material for the sides, find the dimensions that will minimize the cost.

28 A window has the shape of a rectangle surmounted by an isosceles triangle, as illustrated in Figure 16.24. If the perimeter of the window is 12 feet, what values of x, y, and θ will maximize the total area?

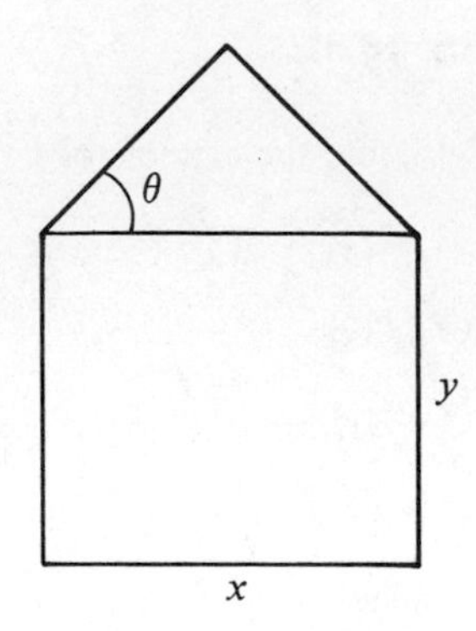

Figure 16.24

16.9 LAGRANGE MULTIPLIERS

The definitions of local extema given in the preceding section can be extended to a function f of many variables. Moreover, if $w = f(x, y, z, \ldots)$, where f has first partial derivatives, then it can be shown (see (16.35)) that a local maximum or minimum can occur at $(x, y, z, \ldots)$ only if

$$w_x = 0, \quad w_y = 0, \quad w_z = 0, \quad \ldots.$$

In applications it is often necessary to find the local extrema of f when the variables $x, y, z, \ldots$ are restricted in some manner. As an illustration, let us consider the following example.

Example 1 Find the volume of the largest rectangular box with faces parallel to the coordinate planes that can be inscribed in the ellipsoid $16x^2 + 4y^2 + 9z^2 = 144$.

Before proceeding to a formal solution, note that by symmetry it is sufficient to examine the part in the first octant, as illustrated in Figure 16.25. If $P(x, y, z)$ is the indicated vertex, then the volume V of the box is given by $V = 8xyz$. Our goal is to find the maximum value of V subject to the **constraint** (or **side condition**)

$$16x^2 + 4y^2 + 9z^2 - 144 = 0.$$

Solving this equation for z and substituting in the formula for V, we obtain

$$V = (8xy)\tfrac{1}{3}\sqrt{144 - 16x^2 - 4y^2}.$$

The extrema may then be found by means of (16.36); however, this method is cumbersome because of the manipulations involved in finding the partial derivatives of V. Another disadvantage of this technique is that in some cases it

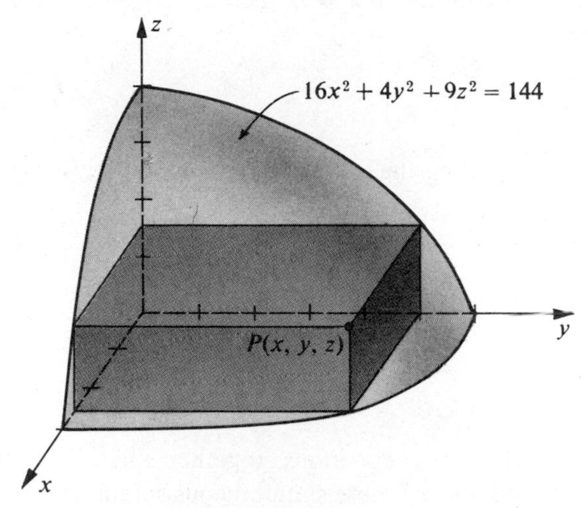

Figure 16.25

may be impossible to solve for z. For problems of this type it is usually simpler to employ the method of *Lagrange multipliers*, discovered by the great French mathematician Joseph Lagrange (1736–1813). We shall briefly, without proof, describe the method and then apply it to Example 1.

In general, suppose f and g are functions of x, y, and z and we wish to find the local extrema of $f(x, y, z)$ subject to the constraint $g(x, y, z) = 0$. Suppose further that we can solve the latter equation for z, obtaining $z = k(x, y)$, and that k, f, and g have continuous first partial derivatives throughout a suitable domain. We define a new function of *four* variables by

(16.37)
$$w = f(x, y, z) + \lambda g(x, y, z)$$

where the variable λ is called a **Lagrange multiplier**. Lagrange's result states that the values of x, y, and z which give the extrema of f are among the simultaneous solutions of the following four equations in four unknowns:

(16.38)
$$w_x = 0, \quad w_y = 0, \quad w_z = 0, \quad w_\lambda = g(x, y, z) = 0.$$

We shall give a partial proof of this fact. As an aid to simplifying the notation, let

$$u = f(x, y, z), \quad v = g(x, y, z) = 0, \quad z = k(x, y).$$

If we regard z as a function of the independent variables x and y and apply the Chain Rule, then at a local extremum (x, y, z) of f,

$$f_x(x, y, z) = u_x + u_z z_x = 0$$
$$f_y(x, y, z) = u_y + u_z z_y = 0.$$

Consequently, if $u_z \neq 0$, then

$$z_x = -\frac{u_x}{u_z}, \quad z_y = -\frac{u_y}{u_z}.$$

In addition, if $v_z \neq 0$, then by Theorem (16.21),

$$z_x = -\frac{v_x}{v_z}, \quad z_y = -\frac{v_y}{v_z}.$$

Equating the expressions for z_x and z_y we obtain

$$-\frac{u_x}{u_z} = -\frac{v_x}{v_z}, \quad -\frac{u_y}{u_z} = -\frac{v_y}{v_z}$$

and hence

$$u_x v_z = u_z v_x, \quad u_y v_z = u_z v_y. \tag{16.39}$$

These two equations, together with $v = 0$, provide three equations in three unknowns whose simultaneous solutions contain the values of x, y, and z which give the extrema of f.

Let us now take the Lagrange multiplier point of view. If as in (16.37) we write $w = u + \lambda v$, then by (16.38) the extrema are given by the solutions of the following four equations in four unknowns:

$$\begin{aligned} w_x &= u_x + \lambda v_x = 0 \\ w_y &= u_y + \lambda v_y = 0 \\ w_z &= u_z + \lambda v_z = 0 \\ w_\lambda &= \qquad v = 0. \end{aligned}$$

Assuming that the partial derivatives of v are not zero, we may solve each of the first three equations for λ, obtaining

$$\lambda = -\frac{u_x}{v_x} = -\frac{u_y}{v_y} = -\frac{u_z}{v_z}$$

which again gives us (16.39). Thus the Langrange multiplier technique also leads to the extrema of f.

***Solution* of Example 1** We wish to maximize

$$V = f(x, y, z) = 8xyz$$

subject to the constraint

$$g(x, y, z) = 16x^2 + 4y^2 + 9z^2 - 144 = 0.$$

We begin by considering equation (16.37):

$$w = 8xyz + \lambda(16x^2 + 4y^2 + 9z^2 - 144).$$

In this case equations (16.38) are

$$
\begin{aligned}
w_x &= 8yz + 32x\lambda = 0 \\
w_y &= 8xz + 8y\lambda = 0 \\
w_z &= 8xy + 18z\lambda = 0 \\
w_\lambda &= 16x^2 + 4y^2 + 9z^2 - 144 = 0.
\end{aligned}
\tag{16.40}
$$

Multiplying the first equation by x, the second by y, and the third by z and adding gives us

$$24xyz + 2\lambda(16x^2 + 4y^2 + 9z^2) = 0.$$

This, together with the fourth equation in (16.40), implies that

$$24xyz + 2\lambda(144) = 0, \quad \text{or} \quad xyz = -12\lambda.$$

The latter relation may be used to find x, y, and z. For example, starting with the first equation in (16.40) we obtain

$$
\begin{aligned}
8xyz + 32x^2\lambda &= 0 \\
8(-12\lambda) + 32x^2\lambda &= 0 \\
-32\lambda(3 - x^2) &= 0.
\end{aligned}
$$

Consequently, either $\lambda = 0$ or $x = \sqrt{3}$. We may reject $\lambda = 0$ since in this case $xyz = 0$, and hence $V = 8xyz = 0$. Thus the only possibility for x is $\sqrt{3}$.

Similarly, if we multiply both sides of the second equation in (16.40) by y we obtain

$$
\begin{aligned}
8xyz + 8y^2\lambda &= 0 \\
8(-12\lambda) + 8y^2\lambda &= 0 \\
8\lambda(-12 + y^2) &= 0
\end{aligned}
$$

and hence $y = \sqrt{12} = 2\sqrt{3}$.

Finally, multiplying the third equation in (16.40) by z and using the same technique results in $z = 4\sqrt{3}/3$. Hence the desired volume is

$$V = 8(\sqrt{3})(2\sqrt{3})\left(\frac{4\sqrt{3}}{3}\right) = 64\sqrt{3}.$$

As another illustration, suppose that $f(x, y, z)$ gives the temperature, mass, potential, or some other scalar quantity at the point (x, y, z) in a rectangular coordinate system. We might then wish to determine where, on a given surface, $f(x, y, z)$ attains a local maximum or minimum value. The mathematical counterpart of this type of problem is illustrated in the next example.

Example 2 If $f(x, y, z) = 4x^2 + y^2 + 5z^2$, find the point on the plane $2x + 3y + 4z = 12$ at which $f(x, y, z)$ has its least value.

Solution We wish to find the minimum value of $f(x, y, z)$ subject to the constraint $g(x, y, z) = 2x + 3y + 4z - 12 = 0$. If, as in (16.37), we let

$$w = 4x^2 + y^2 + 5z^2 + \lambda(2x + 3y + 4z - 12)$$

then equations (16.38) are

$$\begin{aligned} w_x &= 8x + 2\lambda = 0 \\ w_y &= 2y + 3\lambda = 0 \\ w_z &= 10z + 4\lambda = 0 \\ w_\lambda &= 2x + 3y + 4z - 12 = 0. \end{aligned}$$

Solving the first three equations for λ we obtain

$$\lambda = -4x = -\tfrac{2}{3}y = -\tfrac{5}{2}z.$$

Consequently, for a local extremum we need

$$y = 6x \quad \text{and} \quad z = \tfrac{8}{5}x.$$

Substituting in the equation $w_\lambda = 0$ we obtain

$$2x + 18x + \tfrac{32}{5}x - 12 = 0$$

or $x = 5/11$. Hence $y = 6(5/11) = 30/11$ and $z = (8/5)(5/11) = 8/11$. It follows that the minimum value occurs at the point $(5/11, 30/11, 8/11)$.

For some applications there may be more than one constraint. For example, consider the problem of finding the local extrema of $f(x, y, z)$ subject to the *two* side conditions

$$g(x, y, z) = 0 \quad \text{and} \quad h(x, y, z) = 0.$$

The method of Lagrange multipliers states that if we let

$$w = f(x, y, z) + \lambda g(x, y, z) + \mu h(x, y, z)$$

where λ and μ are variables, then the values of x, y, and z which yield the extrema of f are among the solutions of the following five equations in five unknowns:

$$w_x = 0, \quad w_y = 0, \quad w_z = 0, \quad w_\lambda = 0, \quad w_\mu = 0.$$

Lagrange's method can also be extended to functions of more than three variables and other numbers of side conditions. The reader is referred to texts on advanced calculus for discussions of the general case.

Example 3 Let C denote the first octant arc of the curve in which the paraboloid $2z = 16 - x^2 - y^2$ and the plane $x + y = 4$ intersect. Find the points on C which are closest to, and farthest from, the origin, Find the minimum and maximum distances from the origin to C.

Solution The arc C is sketched in Figure 16.26. If $P(x, y, z)$ is an arbitrary point on C, then we wish to find the largest and smallest values of $d(O, P) = \sqrt{x^2 + y^2 + z^2}$. This is equivalent to finding the extrema of

$$f(x, y, z) = x^2 + y^2 + z^2$$

subject to the following conditions:

$$g(x, y, z) = x^2 + y^2 + 2z - 16 = 0$$
$$h(x, y, z) = x + y - 4 = 0.$$

If we let

$$w = (x^2 + y^2 + z^2) + \lambda(x^2 + y^2 + 2z - 16) + \mu(x + y - 4)$$

then the desired values of $x, y,$ and z are solutions of

$$w_x = 2x + 2x\lambda + \mu = 0$$
$$w_y = 2y + 2y\lambda + \mu = 0$$
$$w_z = 2z + 2\lambda = 0.$$

From the first two equations we obtain

$$w_x - w_y = 2(x - y) + 2\lambda(x - y) = 0$$

and, therefore,

$$(x - y)(1 + \lambda) = 0.$$

It follows that either $\lambda = -1$ or $x = y$.

If $\lambda = -1$, then $w_z = 2z - 2 = 0$. Hence $z = 1$, and the first side condition gives us $x^2 + y^2 - 14 = 0$. Solving this equation simultaneously with $x + y - 4 = 0$ we see that either $x = 2 + \sqrt{3}$, $y = 2 - \sqrt{3}$ or $x = 2 - \sqrt{3}$, $y = 2 + \sqrt{3}$. Consequently, points on C which may lead to extrema are

$$P_1(2 + \sqrt{3}, 2 - \sqrt{3}, 1) \quad \text{and} \quad P_2(2 - \sqrt{3}, 2 + \sqrt{3}, 1).$$

The corresponding distances from O are

$$d(O, P_1) = \sqrt{15} = d(O, P_2).$$

If $y = x$, then using the side conditions we obtain $P_3(2, 2, 4)$ and $d(O, P_3) = 2\sqrt{6}$.

Referring to Figure 16.26, we may now make the following observations. As a point moves continuously along C from $A(4, 0, 0)$ to $B(0, 4, 0)$, its distance from the origin starts at $d(O, A) = 4$, decreases to a minimum value $\sqrt{15}$ at P_1 and then increases to a maximum value $2\sqrt{6}$ at P_3. The distance then decreases to $\sqrt{15}$ at P_2 and again increases to 4 at B.

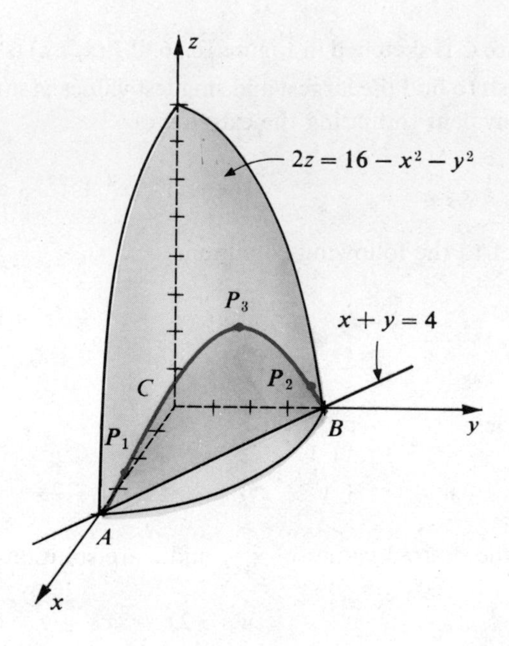

Figure 16.26

As a check on the solution, note that parametric equations for C are

$$x = 4 - t, \quad y = t, \quad z = 4t - t^2$$

where $0 \leq t \leq 4$. In this case

$$f(x, y, z) = (4 - t)^2 + t^2 + (4t - t^2)^2$$

and the extrema of f may be found using single variable methods. It is left to the reader to show that the same points are obtained.

EXERCISES 16.9

In Exercises 1–10 use Lagrange multipliers to find the local extrema of the given function f subject to the stated constraints.

1 $f(x, y) = y^2 - 4xy + 4x^2$; $x^2 + y^2 = 1$

2 $f(x, y) = 2x^2 + xy - y^2 + y$; $2x + 3y = 1$

3 $f(x, y, z) = x + y + z$; $x^2 + y^2 + z^2 = 25$

4 $f(x, y, z) = x^2 + y^2 + z^2$; $x + y + z = 25$

5 $f(x, y, z) = x^2 + y^2 + z^2$; $x - y + z = 1$

6 $f(x, y, z) = x + 2y - 3z$; $z = 4x^2 + y^2$

7 $f(x, y, z) = x^2 + y^2 + z^2$; $x - y = 1$; $y^2 - z^2 = 1$

8 $f(x, y, z) = z - x^2 - y^2$; $x + y + z = 1$; $x^2 + y^2 = 4$

9 $f(x, y, z, t) = xyzt$; $x - z = 2$; $y^2 + t = 4$

10 $f(x, y, z, t) = x^2 + y^2 + z^2 + t^2$; $x + y + 2z = 1$; $2x - z + t = 2$; $y + 3z + 2t = -1$

11 Find the point on the sphere $x^2 + y^2 + z^2 = 9$ that is closest to the point $(2, 3, 4)$.

12 Let C denote the line of intersection of the planes $3x + 2y + z = 6$ and $x - 4y + 2z = 8$. Find the point on C that is closest to the origin.

13 A closed rectangular box having a volume of 2 ft^3 is to be constructed. If the cost per square foot of the material for the sides, bottom, and top is \$1.00, \$2.00, and \$1.50, respectively, what dimensions will minimize the cost?

14 Find the volume of the largest rectangular box which has three of its vertices on the positive x-, y-, and z-axes, respectively, and a fourth vertex on the plane $2x + 3y + 4z = 12$.

15 A container with a closed top is to be constructed in the shape of a right circular cylinder. If the surface area has a fixed value S, what relative dimensions will maximize the volume?

16 Prove that the triangle of maximum area having a given perimeter p is equilateral. (*Hint:* If the sides are x, y, z and if $s = p/2$, then the area is $[s(s - x)(s - y)(s - z)]^{1/2}$.)

17–26 Work Exercises 17–26 of the previous section using Lagrange multipliers.

16.10 REVIEW

Concepts

Define or discuss each of the following.

1 Function of more than one variable

2 Level curves

3 Level surfaces

4 Limits of functions of more than one variable

5 Continuous function

6 First partial derivatives

7 Higher partial derivatives

8 Differentials

9 Differentiable function

10 The Chain Rule

11 Implicit differentiation

12 Directional derivative

13 Gradient of a function

14 Tangent plane to a surface

15 Extrema of functions of two variables

16 Lagrange multipliers

Exercises

In Exercises 1–4 determine the domain of the function f.

1 $f(x, y) = \sqrt{36 - 4x^2 + 9y^2}$ **2** $f(x, y) = \ln xy$

3 $f(x, y, z) = (z^2 - x^2 - y^2)^{-3/2}$ **4** $f(x, y, z) = \sec z/(x - y)$

In Exercises 5–10 find the first partial derivatives of f.

5 $f(x, y) = x^3 \cos y - y^2 + 4x$ **6** $f(r, s) = r^2 e^{rs}$

7 $f(x, y, z) = (x^2 + y^2)/(y^2 + z^2)$ **8** $f(u, v, t) = u \ln (v/t)$

9 $f(x, y, z, t) = x^2 z\sqrt{2y + t}$ **10** $f(v, w) = v^2 \cos w + w^2 \cos v$

In Exercises 11 and 12 find the second partial derivatives of f.

11 $f(x, y) = x^3y^2 - 3xy^3 + x^4 - 3y + 2$

12 $f(x, y, z) = x^2 e^{y^2 - z^2}$

13 If $u = (x^2 + y^2 + z^2)^{-1/2}$ prove that $\partial^2u/\partial x^2 + \partial^2u/\partial y^2 + \partial^2u/\partial z^2 = 0$.

14 (a) Find dw if $w = y^3 \tan^{-1} x^2 + 2x - y$. (*b*) Find dw if $w = x^2 \sin yz$.

15 Find Δw and dw if $w = x^2 + 3xy - y^2$. Use Δw to find the exact change and dw to find the approximate change in w if (x, y) changes from $(-1, 2)$ to $(-1.1, 2.1)$.

16 Ohm's Law states that $R = E/I$. If measurements are $E = 108$ volts and $I = 2$ amperes with a possible error of 0.2 volts and 0.01 amperes, respectively, use differentials to approximate the maximum error in the calculated value of R (in ohms).

Use the Chain Rule to find the solutions in Exercises 17–19.

17 If $s = u^2v + v^2w - w^3u$, $u = y \cos x$, $v = xe^{-y}$, and $w = y \ln x$, find $\partial s/\partial x$ and $\partial s/\partial y$.

18 If $z = \sqrt{x^2 + y^2}$, $x = re^{2st}$, and $y = tr^2e^{-s}$, find $\partial z/\partial r$, $\partial z/\partial s$, and $\partial z/\partial t$.

19 If $w = x \tan y + y \tan z$, $x = t^3$, $y = e^{-2t}$, and $z = 1/t^2$, find dw/dt.

20 Find the directional derivative of $f(x, y) = 3x^2 - y^2 + 5xy$ at the point $P(2, -1)$ in the direction of the vector $\mathbf{a} = -3\mathbf{i} - 4\mathbf{j}$. What is the maximum rate of increase of $f(x, y)$ at P?

21 The temperature at the point (x, y, z) is given by $T(x, y, z) = 3x^2 + 2y^2 - 4z$. Find the rate of change of T at the point $P(-1, -3, 2)$ in the direction from P to the point $Q(-4, 1, -2)$.

22 A curve C is given parametrically by $x = t$, $y = t^2$, $z = t^3$. If $f(x, y, z) = y^2 + xz$, find $D_{\mathbf{u}} f(2, 4, 8)$, where $\mathbf{u}$ is a unit tangent vector to C at $P(2, 4, 8)$.

23 Find equations of the tangent plane and normal line to the graph of $7z = 4x^2 - 2y^2$ at the point $P(-2, -1, 2)$.

24 Show that every plane tangent to the cone $x^2/a^2 - y^2/b^2 + z^2/c^2 = 0$ passes through the origin.

25 If $y = f(x)$ satisfies the equation $x^3 - 4xy^3 - 3y + x - 2 = 0$, use partial derivatives to find $f'(x)$.

26 If $z = f(x, y)$ satisfies the equation $x^2y + z \cos y - xz^3 = 0$, find $\partial z/\partial x$ and $\partial z/\partial y$.

27 Find the extrema of f if $f(x, y) = x^2 + 3y - y^3$.

28 The material for the bottom of a rectangular box costs twice as much per square inch as that for the sides and top. What relative dimensions will minimize the cost if the volume is fixed?

29 Given $f(x, y) = x^2/4 + y^2/25$, sketch several level curves associated with f and represent $\nabla f\,]_P$ by a vector for a point P on each curve.

30 Given $F(x, y, z) = z + 4x^2 + 9y^2$, sketch several level surfaces associated with F and represent $\nabla F\,]_P$ by a vector for a point P on each surface.

31 Find the local extrema of $f(x, y, z) = xyz$ subject to the constraint $x^2 + 4y^2 + 2z^2 = 8$.

32 Use Lagrange multipliers to find the local extrema of $f(x, y, z) = 4x^2 + y^2 + z^2$ subject to the constraints $2x - y + z = 4$ and $x + 2y - z = 1$. Check your answer using single variable methods.

33 Find the points on the graph of $1/x + 2/y + 3/z = 1$ which are closest to the origin.

34 A hopper in a grain elevator has the shape of a right circular cone of radius 2 ft, surmounted by a right circular cylinder. If the volume is $100\,\text{ft}^3$, find the altitudes h and k of the cylinder and cone, respectively, that will minimize the curved surface area.

Multiple Integrals

The concept of the definite integral which was introduced in Chapter 5 may be extended to functions of several variables. In this chapter we shall define double and triple integrals and discuss some of their fundamental properties and applications.

17.1 DOUBLE INTEGRALS

Of primary importance in the definition of double integrals are regions in the xy-plane of the types illustrated in Figure 17.1, where the functions g_1, g_2 and h_1, h_2 are continuous on the intervals $[a, b]$ and $[c, d]$, respectively, and where $g_1(x) \leq g_2(x)$ for all x in $[a, b]$ and $h_1(y) \leq h_2(y)$ for all y in $[c, d]$. A region such as that illustrated in (i) of Figure 17.1 will be called a **region of Type I**, whereas that illustrated in (ii) of the figure will be called a **region of Type II**.

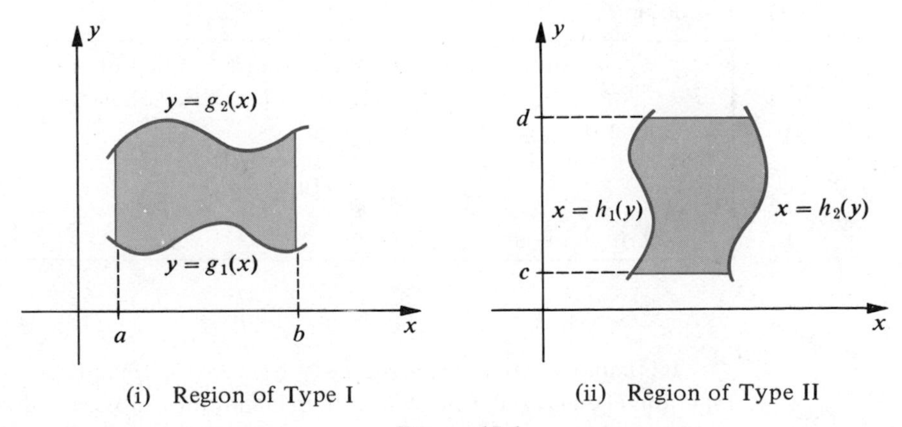

(i) Region of Type I (ii) Region of Type II

Figure 17.1

Throughout the remainder of this chapter, the symbol R will denote a region that can be decomposed into a finite number of subregions of Types I and II. Every such region R is a subset of a closed rectangular region W. If W is divided into smaller rectangles by means of a finite number of horizontal and vertical lines, then the totality of closed rectangular subregions which lie completely *within* R is called an **inner partition** of R. The shaded rectangles in Figure 17.2 illustrate an inner partition. If these rectangular subregions are labeled $R_1, R_2, \ldots, R_n$, then the inner partition is $P = \{R_i: i = 1, \ldots, n\}$. The length of the longest diagonal of the R_i will be denoted by $\|P\|$ and called the **norm of the partition** P. The symbol ΔA_i will be used for the area of R_i.

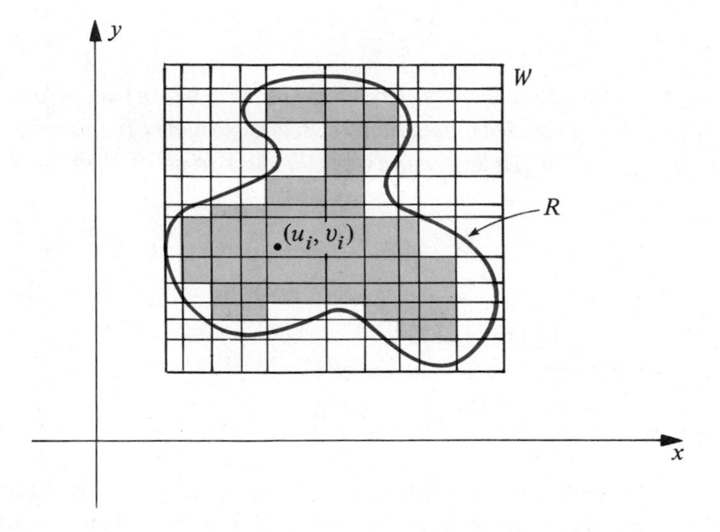

Figure 17.2. Inner partition

(17.1) Definition

> Let f be a function of two variables that is defined on a region R and let $P = \{R_i: i = 1, \ldots, n\}$ be an inner partition of R. A **Riemann sum** of f for P is any sum of the form
>
> $$\sum_{i=1}^{n} f(u_i, v_i)\, \Delta A_i$$
>
> where (u_i, v_i) is in R_i.

It is important to note the similarity between this definition and that given for functions of one variable in Chapter 5. In Definition (5.13) we evaluate a function f of *one* variable at a *number* w_i in the ith *subinterval* of a partition of $[a, b]$, multiply $f(w_i)$ by the *length* Δx_i of the subinterval, and then sum terms of the form $f(w_i)\, \Delta x_i$. In the present situation we evaluate a function f of *two* variables at a *pair* (u_i, v_i) in the ith *subregion* of a partition of R, multiply this number by the *area* ΔA_i of the subregion, and then sum all such terms.

Example Suppose $f(x, y) = 4x + 2y + 1$ and $R = \{(x, y): 0 \le x \le 4, 0 \le y \le (16 - x^2)/4\}$. Let $P = \{R_i: i = 1, \ldots, 7\}$ be the inner partition of R determined by vertical and horizontal lines with integer intercepts, as shown in Figure 17.3. If (u_i, v_i) is the centroid of R_i, calculate the Riemann sum (17.1).

Solution The points (u_i, v_i) are chosen as follows: $(1/2, 5/2)$ in R_1, $(3/2, 5/2)$ in R_2, $(1/2, 3/2)$ in R_3, $(3/2, 3/2)$ in R_4, $(1/2, 1/2)$ in R_5, $(3/2, 1/2)$ in R_6, and $(5/2, 1/2)$ in R_7. Since the area ΔA_i of each R_i is 1, the Riemann sum is

$$f(1/2, 5/2) \cdot 1 + f(3/2, 5/2) \cdot 1 + f(1/2, 3/2) \cdot 1 + f(3/2, 3/2) \cdot 1$$
$$+ f(1/2, 1/2) \cdot 1 + f(3/2, 1/2) \cdot 1 + f(5/2, 1/2) \cdot 1.$$

This simplifies to

$$8 + 12 + 6 + 10 + 4 + 8 + 12 = 60.$$

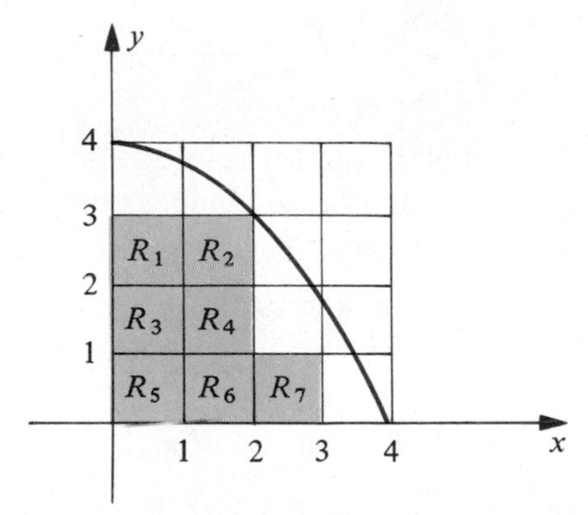

Figure 17.3

In the preceding example, the point (u_i, v_i) was the centroid of R_i. It is important to remember that when working with *arbitrary* Riemann sums, *any* pair (u_i, v_i) in R_i may be chosen (see Exercise 1). For brevity we shall often denote inner partitions of R by $P = \{R_i\}$ without specifying the domain of the summation variable, and the symbol Σ_i will signify that the summation takes place over all rectangles R_i in the partition. The next definition is patterned after Definition (5.15).

(17.2) Definition

> Let f be a function of two variables that is defined on a region R and let L be a real number. The statement
>
> $$\lim_{\|P\| \to 0} \sum_i f(u_i, v_i)\, \Delta A_i = L$$
>
> means that for every $\varepsilon > 0$ there exists a $\delta > 0$ such that if $P = \{R_i\}$ is an inner partition of R with $\|P\| < \delta$, then
>
> $$\left| \sum_i f(u_i, v_i)\, \Delta A_i - L \right| < \varepsilon$$
>
> for every choice of (u_i, v_i) in R_i.

Intuitively, (17.2) states that the Riemann sums of f get closer and closer to L as the norm $\|P\|$ approaches zero. The number L is called the **limit of a sum**. If a limit exists, it is unique. The next definition is analogous to (5.16).

(17.3) Definition

Let f be a function of two variables that is defined on a region R. The **double integral of f over R**, denoted by $\iint_R f(x, y)\, dA$, is given by

$$\iint_R f(x, y)\, dA = \lim_{\|P\| \to 0} \sum_i f(u_i, v_i)\, \Delta A_i$$

provided the limit exists.

If the double integral of f over R exists, then f is said to be **integrable over R**. It can be proved that if f is continuous on R, then f is integrable over R. If R is the union of two nonoverlapping regions R_1 and R_2 of the type under consideration, then

$$\iint_R f(x, y)\, dA = \iint_{R_1} f(x, y)\, dA + \iint_{R_2} f(x, y)\, dA. \tag{17.4}$$

The preceding formula may be extended to any finite number of regions.

There is a useful geometric interpretation for double integrals whenever f is continuous and $f(x, y) \geq 0$ throughout R. Let S denote the graph of f and T the solid which lies under S and over R, as illustrated in Figure 17.4. If $P_i(u_i, v_i, 0)$ is any

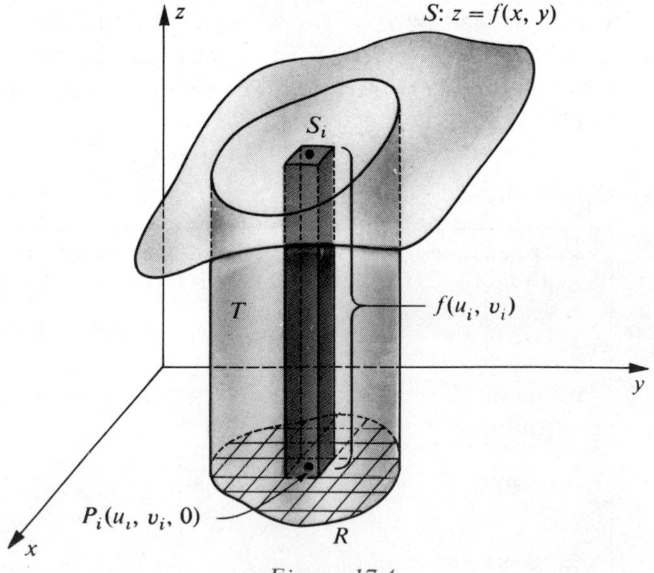

Figure 17.4

point in the subregion R_i of an inner partition P of R, then $f(u_i, v_i)$ is the distance from the xy-plane to the point S_i on S directly above P_i. The number $f(u_i, v_i)\,\Delta A_i$ is the volume of the prism with the rectangular base illustrated in Figure 17.4. The sum of all such volumes is an approximation to the volume V of T. Since it is evident that this approximation improves as $\|P\|$ approaches zero, we define V as the limit of a sum of the numbers $f(u_i, v_i)\,\Delta A_i$. Applying Definition (17.3) gives us

(17.5)
$$V = \iint_R f(x, y)\,dA.$$

Evidently, if $f(x, y) \leq 0$ throughout R, then the double integral of f over R is the *negative* of the volume of the solid which lies *over* the graph of f and *under* the region R.

We shall conclude this section by listing, without proof, some properties of double integrals which correspond to those given for definite integrals in Chapter 5. It is assumed that all regions and functions are suitably restricted so that the indicated integrals exist.

(17.6)
$$\iint_R cf(x, y)\,dA = c \iint_R f(x, y)\,dA, \quad \text{if } c \text{ is any real number}$$

(17.7)
$$\iint_R [f(x, y) + g(x, y)]\,dA = \iint_R f(x, y)\,dA + \iint_R g(x, y)\,dA$$

(17.8)
$$\text{If } f(x, y) \geq 0 \text{ throughout } R, \text{ then } \iint_R f(x, y)\,dA \geq 0.$$

EXERCISES 17.1

1 If f, R, and P are as in the Example, find the Riemann sum of f for P if (u_i, v_i) is as follows.
(a) The point in the lower left-hand corner of R_i
(b) The point in the upper right-hand corner of R_i

2 Suppose f and R are as in the Example and $P = \{R_i\}$ is the inner partition determined by horizontal and vertical lines having intercepts 0, 1/2, 3/2, 5/2, and 4. Find the Riemann sum of f for P if (u_i, v_i) is the centroid of R_i.

3 Let R be the triangular region with vertices $(0, 0)$, $(6, 0)$, $(6, 12)$ and let P be the inner partition of R determined by vertical lines with x-intercepts 0, 1, 3, 4, 6 and horizontal lines with y-intercepts 0, 2, 5, 7, 8, 12. If $f(x, y) = x^2y$, find the Riemann sum of f for P if (u_i, v_i) is the centroid of R_i.

4 Let $R = \{(x, y): 0 \leq x \leq 5, 0 \leq y \leq \sqrt{25 - x^2}\}$. Let P be the partition of R determined by vertical lines with x-intercepts 0, 2, 3, 4, 5 and by horizontal lines with y-intercepts 0, 2, 4, 5. If $f(x, y) = 4 - xy$, find the Riemann sum of f for P if (u_i, v_i) is the point in the lower right-hand corner of R_i.

5 Let R be the region bounded by the trapezoid with vertices $(0, 0)$, $(4, 4)$, $(8, 4)$, and $(12, 0)$. Let P be the partition of R determined by vertical lines with x-intercepts 0, 2, 4, 6, 8, 10, 12 and by horizontal lines with y-intercepts 0, 2, 4. If $f(x, y) = xy$, find the Riemann sum of f for P if (u_i, v_i) is the centroid of R_i.

6 Let R be the region bounded by the triangle with vertices $(-4, 0)$, $(0, 8)$, and $(4, 0)$. Let P be the partition of R determined by vertical lines with x-intercepts $-4, -2, 0, 1, 3, 4$ and by horizontal lines with y-intercepts 0, 2, 4, 6, 8. If $f(x, y) = x + y$, find the Riemann sum of f for P if (u_i, v_i) is the upper right-hand corner of R_i.

7 If $R = \{(x, y): a \le x \le b, c \le y \le d\}$ and $f(x, y) = k$ for all (x, y) in R, prove that the double integral of f over R equals $k(b - a)(d - c)$ by using (a) Definition (17.3) and (b) (17.5).

8 If the functions f and g are integrable over R and $f(x, y) \ge g(x, y)$ for all (x, y) in R, use properties of double integrals to prove that

$$\iint_R f(x, y)\, dA \ge \iint_R g(x, y)\, dA.$$

17.2 EVALUATION OF DOUBLE INTEGRALS

Except for very elementary cases, it is virtually impossible to find the value of a double integral directly from Definition (17.3). This section contains methods of evaluation which make use of the Fundamental Theorem of Calculus (5.31).

Suppose f is a function of two variables that is continuous on a closed rectangular region R of the type illustrated in Figure 17.5. We shall use the symbol $\int_c^d f(x, y)\, dy$ to mean that x is held fixed and the integration is performed with respect to y. This is sometimes called a **partial integration with respect to y**. To each x in the interval $[a, b]$ there corresponds a unique value for this integral and hence a function A is determined, where the value $A(x)$ is given by

(17.9) $$A(x) = \int_c^d f(x, y)\, dy.$$

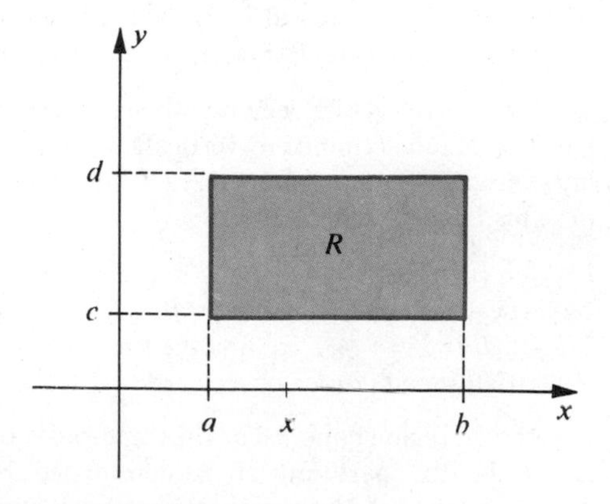

Figure 17.5

As an illustration of (17.9), if $f(x, y) = x^3 + 4y$, $c = 1$, and $d = 2$, then

$$\begin{aligned} A(x) &= \int_1^2 (x^3 + 4y)\, dy = x^3 y + 2y^2 \Big]_1^2 \\ &= (2x^3 + 8) - (x^3 + 2) \\ &= x^3 + 6. \end{aligned}$$

It can be proved that the function A defined by (17.9) is continuous on $[a, b]$ and hence has a definite integral which may be written

$$\int_a^b A(x)\, dx = \int_a^b \left[\int_c^d f(x, y)\, dy \right] dx.$$

The expression on the right side of this equation is called an **iterated (double) integral**. Ordinarily the notation is shortened by omitting the brackets. Thus, by definition,

(17.10) $$\int_a^b \int_c^d f(x, y)\, dy\, dx = \int_a^b \left[\int_c^d f(x, y)\, dy \right] dx.$$

Example 1 Evaluate $\displaystyle\int_1^4 \int_{-1}^2 (2x + 6x^2 y)\, dy\, dx.$

Solution As in (17.10), the integral equals

$$\begin{aligned} \int_1^4 \left[\int_{-1}^2 (2x + 6x^2 y)\, dy \right] dx &= \int_1^4 \left[2xy + 3x^2 y^2 \right]_{-1}^2 dx \\ &= \int_1^4 [(4x + 12x^2) - (-2x + 3x^2)]\, dx \\ &= \int_1^4 (6x + 9x^2)\, dx \\ &= 3x^2 + 3x^3 \Big]_1^4 = 234. \end{aligned}$$

Similarly, we may consider iterated integrals of the form

(17.11) $$\int_c^d \int_a^b f(x, y)\, dx\, dy = \int_c^d \left[\int_a^b f(x, y)\, dx \right] dy.$$

In this case we integrate first with respect to x (with y held fixed). After substituting the limits b and a for x in the usual way, the resulting expression is integrated with respect to y from c to d. It is proved in more advanced texts that the integrals in (17.10) and (17.11) are equal. This is illustrated in the next example.

Example 2 Evaluate $\int_{-1}^{2} \int_{1}^{4} (2x + 6x^2y)\,dx\,dy.$

Solution Applying (17.11),

$$\int_{-1}^{2} \left[\int_{1}^{4} (2x + 6x^2y)\,dx \right] dy = \int_{-1}^{2} \left[x^2 + 2x^3y \right]_{1}^{4} dy$$

$$= \int_{-1}^{2} [(16 + 128y) - (1 + 2y)]\,dy$$

$$= \int_{-1}^{2} (126y + 15)\,dy$$

$$= 63y^2 + 15y \Big]_{-1}^{2} = 234.$$

Iterated integrals may be defined over nonrectangular regions. In particular, if f is continuous on a region of the type illustrated in (i) of Figure 17.1, then by definition,

(17.12)
$$\int_{a}^{b} \int_{g_1(x)}^{g_2(x)} f(x, y)\,dy\,dx = \int_{a}^{b} \left[\int_{g_1(x)}^{g_2(x)} f(x, y)\,dy \right] dx.$$

In this case, after a partial integration with respect to y we substitute $g_2(x)$ and $g_1(x)$ for y in the usual way. The resulting expression in x is then integrated from a to b.

Similarly, if f is continuous on a region of the type illustrated in (ii) of Figure 17.1, then we define

(17.13)
$$\int_{c}^{d} \int_{h_1(y)}^{h_2(y)} f(x, y)\,dx\,dy = \int_{c}^{d} \left[\int_{h_1(y)}^{h_2(y)} f(x, y)\,dx \right] dy.$$

Example 3 Evaluate $\int_{0}^{2} \int_{x^2}^{2x} (x^3 + 4y)\,dy\,dx.$

Solution By (17.12) the integral equals

$$\int_{0}^{2} \left[\int_{x^2}^{2x} (x^3 + 4y)\,dy \right] dx = \int_{0}^{2} \left[x^3y + 2y^2 \right]_{x^2}^{2x} dx$$

$$= \int_{0}^{2} [(2x^4 + 8x^2) - (x^5 + 2x^4)]\,dx$$

$$= \frac{8}{3}x^3 - \frac{1}{6}x^6 \Big]_{0}^{2} = \frac{32}{3}.$$

Example 4 Evaluate $\int_{1}^{3} \int_{\pi/6}^{y^2} 2y \cos x\,dx\,dy.$

Solution By (17.13) the integral equals

$$\int_1^3 \left[\int_{\pi/6}^{y^2} 2y \cos x \, dx \right] dy = \int_1^3 \left[2y \sin x \right]_{\pi/6}^{y^2} dy$$
$$= \int_1^3 (2y \sin y^2 - y) \, dy$$
$$= -\cos y^2 - \frac{1}{2}y^2 \Big]_1^3$$
$$= \left(-\cos 9 - \frac{9}{2} \right) - \left(-\cos 1 - \frac{1}{2} \right)$$
$$= \cos 1 - \cos 9 - 4 \approx -2.55$$

The next theorem indicates that the double integral defined in Section 1 may often be evaluated by means of iterated integrals.

(17.14) Evaluation Theorem for Double Integrals

(i) Let R be a region of Type I which lies between the graphs of $y = g_1(x)$ and $y = g_2(x)$, where g_1 and g_2 are continuous on $[a, b]$. If f is continuous on R, then

$$\iint_R f(x, y) \, dA = \int_a^b \int_{g_1(x)}^{g_2(x)} f(x, y) \, dy \, dx.$$

(ii) Let R be a region of Type II which lies between the graphs of $x = h_1(y)$ and $x = h_2(y)$, where h_1 and h_2 are continuous on $[c, d]$. If f is continuous on R, then

$$\iint_R f(x, y) \, dA = \int_c^d \int_{h_1(y)}^{h_2(y)} f(x, y) \, dx \, dy.$$

It is important to note that in order to use (i) of Theorem 17.14, the region R must have a **lower boundary**, consisting of the graph of an equation $y = g_1(x)$ and an **upper boundary** consisting of the graph of an equation $y = g_2(x)$. To use (ii), R must have a **left-hand boundary** consisting of the graph of an equation $x = h_1(y)$ and a **right-hand boundary** consisting of the graph of an equation $x = h_2(y)$. If the region is more complicated, then it is often possible to divide R into subregions of the required type and then apply (i) or (ii) to each subregion.

Instead of giving a rigorous proof of Theorem (17.14), we shall present an intuitive discussion which makes the result plausible, at least for nonnegative-valued functions. Suppose $f(x, y) \geq 0$ throughout the region R of Type I illustrated in (i) of Figure 17.1. Let S denote the graph of f, T the solid which lies under S and over R, and V the volume of T. We shall begin by considering the plane Γ which is parallel to the yz-plane and intersects the x-axis at the point $(x, 0, 0)$, where $a \leq x \leq b$. As illustrated in Figure 17.6, C will denote the trace of S on Γ. The points $P(x, g_1(x), 0)$ and $Q(x, g_2(x), 0)$ indicate where Γ intersects the boundaries $y = g_1(x)$ and $y = g_2(x)$ of R. The shaded portion in the figure is that part of Γ which lies above the segment PQ and under the surface S. From our work in Chapter 6, the area $A(x)$ of the shaded portion is given by

(17.15)
$$A(x) = \int_{g_1(x)}^{g_2(x)} f(x, y)\, dy.$$

Since $A(x)$ is the area of a typical cross section of T, it follows from (6.11) that

(17.16)
$$V = \int_a^b A(x)\, dx = \int_a^b \int_{g_1(x)}^{g_2(x)} f(x, y)\, dy\, dx.$$

Using the fact that V is also given by (17.5) we obtain the evaluation formula in part (i) of Theorem (17.14). A similar discussion can be given for part (ii).

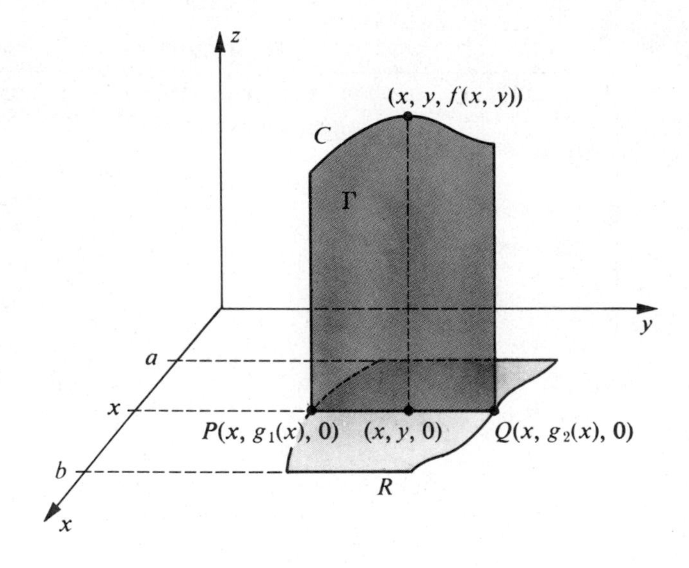

Figure 17.6

Example 5 Evaluate $\iint_R (x^3 + 4y)\, dA$, where R is the region in the xy-plane bounded by the graphs of the equations $y = x^2$ and $y = 2x$.

Solution The graph of R is sketched in Figure 17.7. Note that R is a region both of Type I and Type II. If we regard R as a region of Type I with lower boundary $y = x^2$ and upper boundary $y = 2x$, where $0 \leq x \leq 2$, then by (i) of Theorem 17.14,

$$\iint_R (x^3 + 4y)\, dA = \int_0^2 \int_{x^2}^{2x} (x^3 + 4y)\, dy\, dx.$$

From Example 3 we know that the last integral equals 32/3.

If we regard R as a region of Type II, then the left-hand boundary is the graph of $x = 1/2y$ and the right-hand boundary is the graph of $x = \sqrt{y}$, where $0 \leq y \leq 4$. Hence by (ii) of Theorem 17.14,

$$\iint_R f(x, y)\, dA = \int_0^4 \int_{(1/2)y}^{\sqrt{y}} (x^3 + 4y)\, dx\, dy$$

$$= \int_0^4 \left[\frac{1}{4}x^4 + 4yx\right]_{(1/2)y}^{\sqrt{y}} dy$$

$$= \int_0^4 \left[\left(\frac{1}{4}y^2 + 4y^{3/2}\right) - \left(\frac{1}{64}y^4 + 2y^2\right)\right] dy = \frac{32}{3}.$$

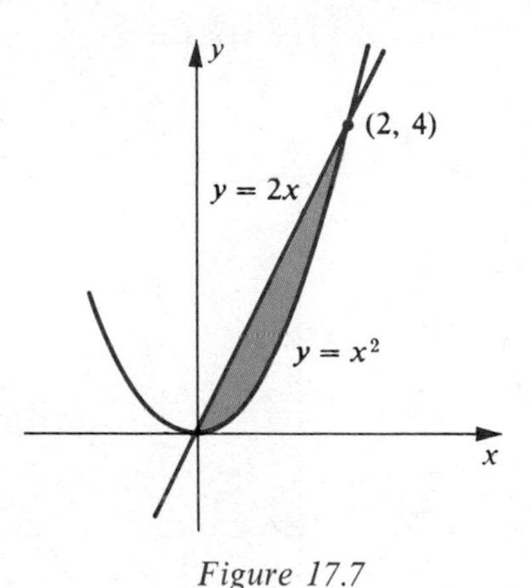

Figure 17.7

Example 6 Let R be the region bounded by the graphs of $y = \sqrt{x}$, $y = \sqrt{3x - 18}$ and $y = 0$. If f is continuous on R, express $\iint_R f(x, y)\, dA$ in terms of iterated integrals by

(a) using only part (i) of Theorem 17.14.
(b) using only part (ii) of Theorem 17.14.

Solution The graphs of $y = \sqrt{x}$ and $y = \sqrt{3x - 18}$ are the top halves of the parabolas $y^2 = x$ and $y^2 = 3x - 18$. The region R is sketched in Figure 17.8.

(a) If we wish to use only part (i) of Theorem 17.14, then it is necessary to employ two iterated integrals, because if $0 \leq x \leq 6$, then the lower boundary of the region is the graph of $y = 0$, whereas if $6 \leq x \leq 9$, the lower boundary is the graph of $y = \sqrt{3x - 18}$. If R_1 denotes the part of the region R which lies between $x = 0$ and $x = 6$, and if R_2 denotes the part between $x = 6$ and $x = 9$, then both R_1 and

R_2 are regions of Type I. Hence by (17.4) and (i) of Theorem (17.14),

$$\iint_R f(x, y)\, dA = \iint_{R_1} f(x, y)\, dA + \iint_{R_2} f(x, y)\, dA$$
$$= \int_0^6 \int_0^{\sqrt{x}} f(x, y)\, dy\, dx + \int_6^9 \int_{\sqrt{3x-18}}^{\sqrt{x}} f(x, y)\, dy\, dx.$$

(b) To use (ii) of Theorem 17.14 we must solve each of the given equations for x in terms of y, obtaining

$$x = y^2 \quad \text{and} \quad x = \frac{y^2 + 18}{3} = \frac{1}{3}y^2 + 6.$$

Only one iterated integral is required in this case since R is a region of Type II. Thus

$$\iint_R f(x, y)\, dA = \int_0^3 \int_{y^2}^{(1/3)y^2+6} f(x, y)\, dx\, dy.$$

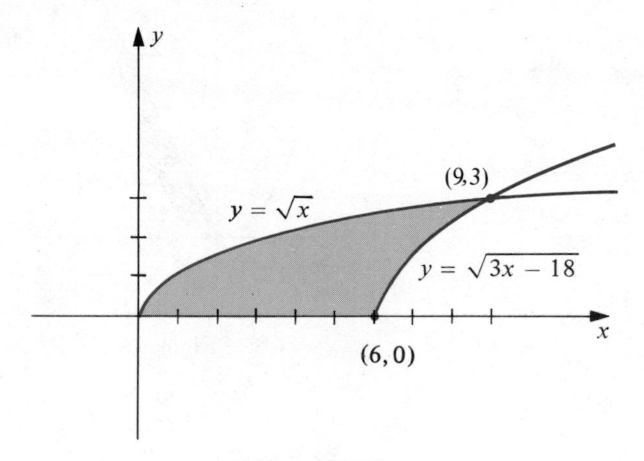

Figure 17.8

Examples 5 and 6 indicate how certain double integrals may be evaluated by using either (i) or (ii) of Theorem 17.14. Generally, the choice of the *order of integration* $dy\,dx$ or $dx\,dy$ often depends on the form of $f(x, y)$ and the region R. Sometimes it is extremely difficult, or even impossible, to evaluate a given iterated double integral. However, by **reversing the order of integration** from $dy\,dx$ to $dx\,dy$, or vice versa, it may be possible to find an equivalent iterated double integral which is easy to evaluate. This technique is illustrated in the next example.

Example 7 Given $\int_0^4 \int_{\sqrt{y}}^2 y \cos x^5\, dx\, dy$, reverse the order of integration and evaluate the resulting integral.

Solution Since the given order of integration is $dx\,dy$, the region R is of Type II. As illustrated in Figure 17.9, the left-hand and right-hand boundaries are the graphs of $x = \sqrt{y}$ and $x = 2$, respectively, and $0 \le y \le 4$. Note that R is also a region of Type I whose lower and upper boundaries are given by $y = 0$ and $y = x^2$, respectively, and where $0 \le x \le 2$. Hence by Theorem 17.14,

$$\begin{aligned}
\int_0^4 \int_{\sqrt{y}}^2 y \cos x^5 \, dx \, dy &= \iint_R y \cos x^5 \, dA \\
&= \int_0^2 \int_0^{x^2} y \cos x^5 \, dy \, dx \\
&= \int_0^2 \frac{y^2}{2} \cos x^5 \Big]_0^{x^2} dx \\
&= \int_0^2 \frac{x^4}{2} \cos x^5 \, dx \\
&= \frac{1}{10} \int_0^2 \cos x^5 (5x^4) \, dx \\
&= \frac{1}{10} \sin x^5 \Big]_0^2 = \frac{1}{10} \sin 32 \approx 0.055.
\end{aligned}$$

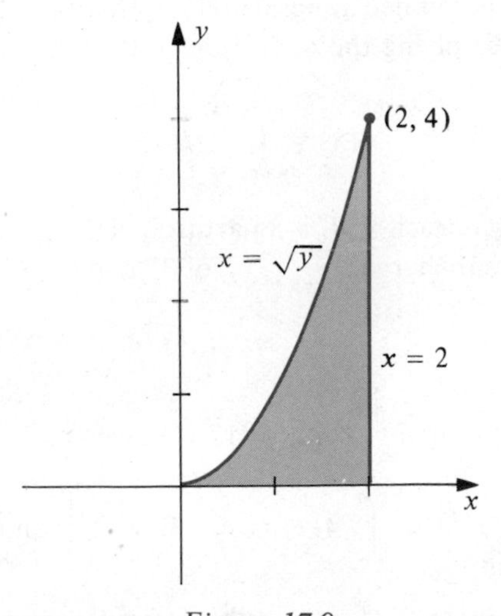

Figure 17.9

It is often useful to regard the first integral in (17.16) as a limit of a sum. By (6.11),

$$V = \int_a^b A(x) \, dx = \lim_{\|P'\| \to 0} \sum_i A(u_i) \, \Delta x_i \tag{17.17}$$

where P' is a partition of the interval $[a, b]$, u_i is any number in the ith subinterval $[x_{i-1}, x_i]$ of P', and $\Delta x_i = x_i - x_{i-1}$. As illustrated in Figure 17.10, $A(u_i) \, \Delta x_i$ is the volume of a lamina L_i whose face is parallel to the yz-plane and which has as its base a strip of width Δx_i in the xy-plane. Consequently, the volume V may be regarded as a limit of a sum of volumes of such laminas.

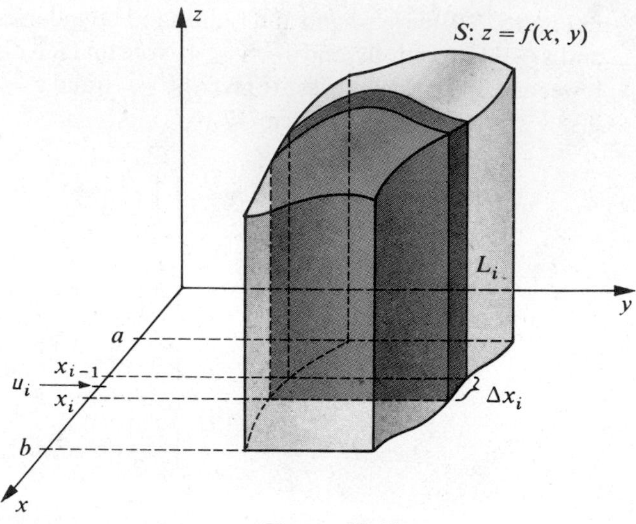

Figure 17.10

The iterated integral in (17.16) may also be expressed in terms of limits of sums. Thus, applying the definition of the definite integral to (17.15),

$$A(x) = \int_{g_1(x)}^{g_2(x)} f(x, y)\,dy = \lim_{\|P''\| \to 0} \sum_j f(x, v_j)\,\Delta y_j$$

where, for each x, P'' is a partition of the y-interval $[g_1(x), g_2(x)]$, v_j is a number in the jth subinterval $[y_{j-1}, y_j]$ of P'', and $\Delta y_j = y_j - y_{j-1}$. Hence, for each u_i in $[a, b]$,

$$A(u_i) = \lim_{\|P''\| \to 0} \sum_j f(u_i, v_j)\,\Delta y_j.$$

Substituting for $A(u_i)$ in (17.17) gives us

$$V = \int_a^b A(x)\,dx = \lim_{\|P'\| \to 0} \sum_i \left[\lim_{\|P''\| \to 0} \sum_j f(u_i, v_j)\,\Delta y_j \right] \Delta x_i.$$

If we ignore the limits in this formula, there results what is called a **double sum**, written

$$\sum_i \sum_j f(u_i, v_j)\,\Delta y_j\,\Delta x_i.$$

Referring to Figure 17.11, we see that the expression $f(u_i, v_j)\,\Delta y_j\,\Delta x_i$ may be regarded as the volume of a prism having base area $\Delta y_j\,\Delta x_i$ and altitude $f(u_i, v_j)$. The double sum may then be interpreted as follows. *Holding x fixed* we sum the volumes of the prisms in the direction of the y-axis, thereby obtaining the volume of the lamina illustrated in Figure 17.10. The volumes of these laminas are then added by summing in the direction of the x-axis. In a certain sense, the iterated integral does this and at the same time takes limits as the bases of the prisms shrink to points. This interpretation is often useful in visualizing applications of the double integral. To emphasize this point of view we sometimes write

$$V = \lim_{\|\Delta\| \to 0} \sum_i \sum_j f(u_i, v_j)\,\Delta y_j\,\Delta x_i = \int_a^b \int_{g_1(x)}^{g_2(x)} f(x, y)\,dy\,dx \tag{17.18}$$

where the symbol $\|\Delta\| \to 0$ is used to signify that all the Δx_i and Δy_j approach 0.

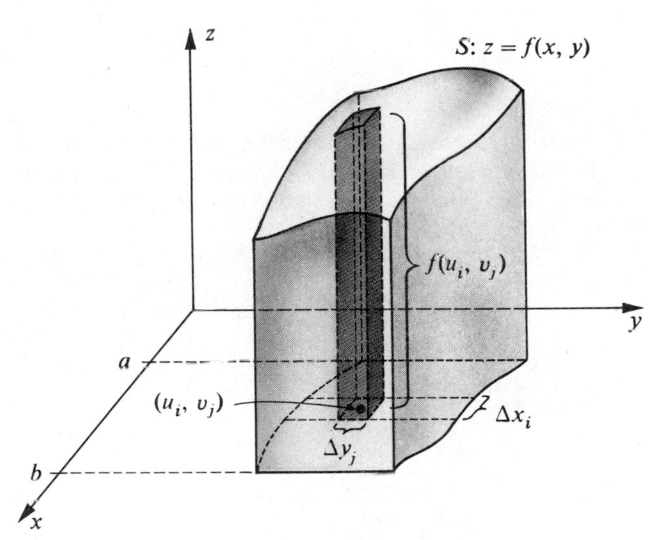

Figure 17.11

EXERCISES 17.2

Using the values given in Exercises 1 and 2, verify that

$$\int_a^b \int_c^d f(x, y)\,dy\,dx = \int_c^d \int_a^b f(x, y)\,dx\,dy.$$

1 $a = 1, b = 2, c = -1, d = 2;\ f(x, y) = 12xy^2 - 8x^3$

2 $a = -2, b = -1, c = 0, d = 3;\ f(x, y) = 4xy^3 + y$

Evaluate the integrals in Exercises 3–12.

3 $\displaystyle\int_1^2 \int_{1-x}^{\sqrt{x}} x^2 y\,dy\,dx$

4 $\displaystyle\int_{-1}^1 \int_{x^3}^{x+1} (3x + 2y)\,dy\,dx$

5 $\displaystyle\int_0^2 \int_{y^2}^{2y} (4x - y)\,dx\,dy$

6 $\displaystyle\int_0^1 \int_{-y-1}^{y-1} (x^2 + y^2)\,dx\,dy$

7 $\displaystyle\int_1^2 \int_{x^3}^{x} e^{y/x}\,dy\,dx$

8 $\displaystyle\int_0^{\pi/6} \int_0^{\pi/2} (x \cos y - y \cos x)\,dy\,dx$

9 $\displaystyle\int_1^e \int_0^x \ln x\,dy\,dx$

10 $\displaystyle\int_0^1 \int_y^1 \frac{1}{1 + y^2}\,dx\,dy$

11 $\int_{\pi/6}^{\pi/4} \int_{\tan x}^{\sec x} (y + \sin x)\, dy\, dx$

12 $\int_{\pi/6}^{\pi/4} \int_{0}^{\sin x} e^y \cos x\, dy\, dx$

In Exercises 13–18 sketch the region R bounded by the graphs of the equations and express $\iint_R f(x, y)\, dA$ as an iterated integral (a) using part (i) of Theorem (17.14); (b) using part (ii) of Theorem (17.14).

13 $y = \sqrt{x}, x = 4, y = 0$

14 $y = \sqrt{x}, x = 0, y = 2$

15 $y = x^3, x = 0, y = 8$

16 $y = x^3, x = 2, y = 0$

17 $y = \sqrt{x}, y = x^3$

18 $y = \sqrt{1 - x^2}, y = 0$

In each of Exercises 19–24 sketch the region R bounded by the graphs of the given equations. If f is continuous on R, use (17.4) to express $\iint_R f(x, y)\, dA$ as a sum of two iterated integrals of the type used in (a) part (i) of Theorem (17.14); (b) part (ii) of Theorem (17.14).

19 $8y = x^3,\ y - x = 4,\ 4x + y = 9$

20 $x = 2\sqrt{y},\ \sqrt{3}x = \sqrt{y},\ y = 2x + 5$

21 $x = \sqrt{3 - y},\ y = 2x,\ x + y + 3 = 0$

22 $x + 2y = 5, x - y = 2, 2x + y = -2$

23 $y = e^x, y = \ln x, x + y = 1, x + y = 1 + e$

24 $y = \sin x, \pi y = 2x$

In Exercises 25–30, sketch the region of integration for the iterated integral.

25 $\int_{-1}^{2} \int_{-\sqrt{4-x^2}}^{4-x^2} f(x, y)\, dy\, dx$

26 $\int_{-1}^{2} \int_{x^2-4}^{x-2} f(x, y)\, dy\, dx$

27 $\int_{0}^{1} \int_{\sqrt{y}}^{\sqrt[3]{y}} f(x, y)\, dx\, dy$

28 $\int_{-2}^{-1} \int_{3y}^{2y} f(x, y)\, dx\, dy$

29 $\int_{-3}^{1} \int_{\tan^{-1} x}^{e^x} f(x, y)\, dy\, dx$

30 $\int_{-1}^{2} \int_{\sinh x}^{\cosh x} f(x, y)\, dy\, dx$

In Exercises 31–36 reverse the order of integration and evaluate the resulting integral.

31 $\int_{0}^{1} \int_{2x}^{2} e^{y^2}\, dy\, dx$

32 $\int_{0}^{9} \int_{\sqrt{y}}^{3} \sin x^3\, dx\, dy$

33 $\int_{0}^{2} \int_{y^2}^{4} y \cos x^2\, dx\, dy$

34 $\int_{1}^{e} \int_{0}^{\ln x} y\, dy\, dx$

35 $\int_{0}^{8} \int_{\sqrt[3]{y}}^{2} \frac{y}{\sqrt{16 + x^7}}\, dx\, dy$

36 $\int_{0}^{1} \int_{x}^{1} \frac{1}{y} \sin y \cos \frac{x}{y}\, dy\, dx$

37 Interpret the iterated integral in part (ii) of Theorem (17.14) as a limit of a double sum in a manner analogous to (17.18).

17.3 AREAS AND VOLUMES

It is customary to denote the double integral $\iint_R 1\,dA$ by $\iint_R dA$. If R is a region of Type I, as illustrated in (i) of Figure 17.1, then by Theorem (17.14),

$$\iint_R dA = \int_a^b \int_{g_1(x)}^{g_2(x)} 1\,dy\,dx = \int_a^b y\Big]_{g_1(x)}^{g_2(x)} dx$$

$$= \int_a^b [g_2(x) - g_1(x)]\,dx$$

which, according to (6.1), equals the area A of R. The same is true if R is a region of Type II, as illustrated in (ii) of Figure 17.1. These facts are also evident from the definition of the double integral, for if $f(x, y) = 1$ throughout R, then the Riemann sum in Definition (17.3) is a sum of areas of rectangles in an inner partition P of R. As the norm $\|P\|$ approaches zero, these rectangles cover more of R and it is apparent that the limit equals the area of R.

If an iterated integral is used to find an area it may be regarded as a limit of a double sum in a manner similar to that used for volumes at the end of the previous section. Specifically, as in (17.18), with $f(x, y) = 1$, we write

$$A = \iint_R dA = \int_a^b \int_{g_1(x)}^{g_2(x)} dy\,dx = \lim_{\|\Delta\| \to 0} \sum_i \sum_j \Delta y_j\,\Delta x_i.$$

The double sum may be interpreted as follows. First, *holding x fixed* we sum over rectangles of areas $\Delta y_j\,\Delta x_i$ in the direction of the y-axis, from the graph of $y = g_1(x)$ to the graph of $y = g_2(x)$. This gives us the area of a rectangular strip which is parallel to the y-axis, as illustrated in (i) of Figure 17.12. Second, we sweep out the region R by summing over all such strips from $x = a$ to $x = b$. The limit of the double sum (that is, the iterated integral) does this and, at the same time, forces all the Δx_i and Δy_j to approach 0.

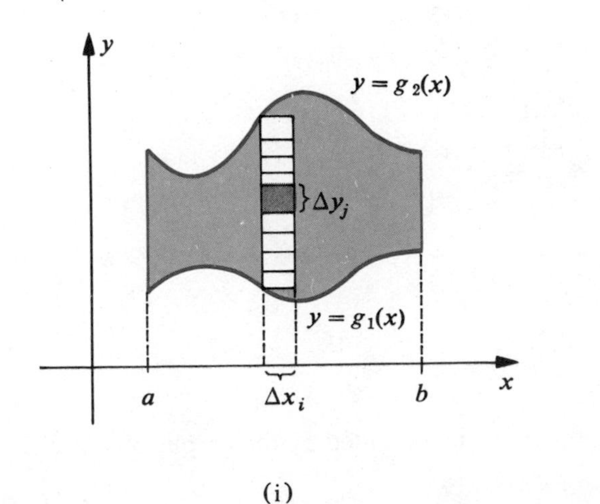

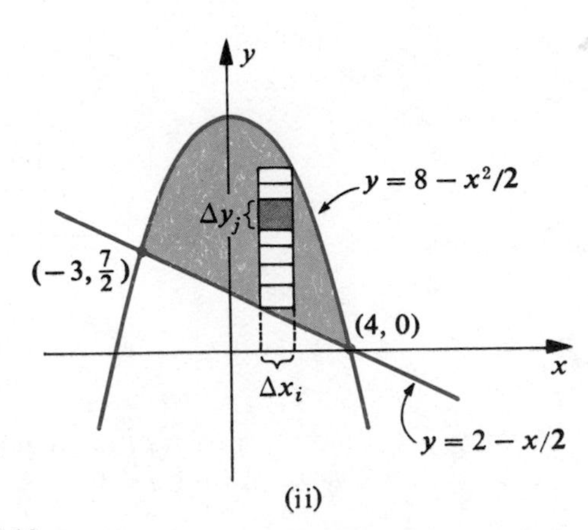

Figure 17.12

Example 1 Find the area A of the region bounded by the graphs of $2y = 16 - x^2$ and $x + 2y - 4 = 0$.

Solution The region lies under the parabola $y = 8 - x^2/2$ and over the line $y = 2 - x/2$, as illustrated in (ii) of Figure 17.12. Also shown in the figure is a typical rectangle of area $\Delta y_j\, \Delta x_i$ and the strip obtained by summing in the direction of the y-axis. Using an iterated integral we have

$$\begin{aligned} A &= \int_{-3}^{4} \int_{2-(x/2)}^{8-(x^2/2)} dy\, dx \\ &= \int_{-3}^{4} \left[\left(8 - \frac{x^2}{2}\right) - \left(2 - \frac{x}{2}\right) \right] dx \\ &= 6x - \frac{x^3}{6} + \frac{x^2}{4} \Big]_{-3}^{4} = \frac{343}{12}. \end{aligned}$$

If part (ii) of Theorem (17.14) is used to find areas then

$$A = \iint_R dA = \int_c^d \int_{h_1(y)}^{h_2(y)} dx\, dy = \lim_{\|\Delta\| \to 0} \sum_j \sum_i \Delta x_i\, \Delta y_j.$$

The preceding double sum may be interpreted as follows. First, *holding y fixed* we sum over rectangles of areas $\Delta x_i\, \Delta y_j$ in the direction of the x-axis, from the graph of $x = h_1(y)$ to the graph of $x = h_2(y)$. This gives us the area of a rectangular strip which is parallel to the x-axis, as illustrated in Figure 17.13. Second, we sweep out the region R by summing over all such strips from $y = c$ to $y = d$. As before, the iterated integral accomplishes this and simultaneously forces all the Δx_i and Δy_j to approach 0.

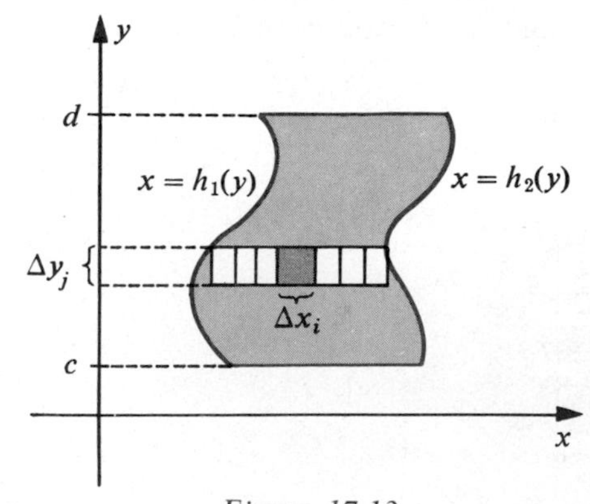

Figure 17.13

Example 2 Find the area A of the region in the xy-plane bounded by the graphs of $x = y^3$, $x + y = 2$, and $y = 0$.

Solution The region is sketched in Figure 17.14 together with a typical rectangle of area

$\Delta x_i \, \Delta y_j$ and a strip obtained by summing in the direction of the x-axis. Using an iterated integral gives us

$$A = \iint_R dA = \int_0^1 \int_{y^3}^{2-y} dx\, dy$$

$$= \int_0^1 x\Big]_{y^3}^{2-y} dy$$

$$= \int_0^1 (2 - y - y^3)\, dy = 5/4.$$

The area can also be found by employing part (i) of Theorem (17.14); however, in this case it is necessary to divide R into two parts by means of the vertical line through (1, 1). We then have

$$A = \int_0^1 \int_0^{\sqrt[3]{x}} dy\, dx + \int_1^2 \int_0^{2-x} dy\, dx.$$

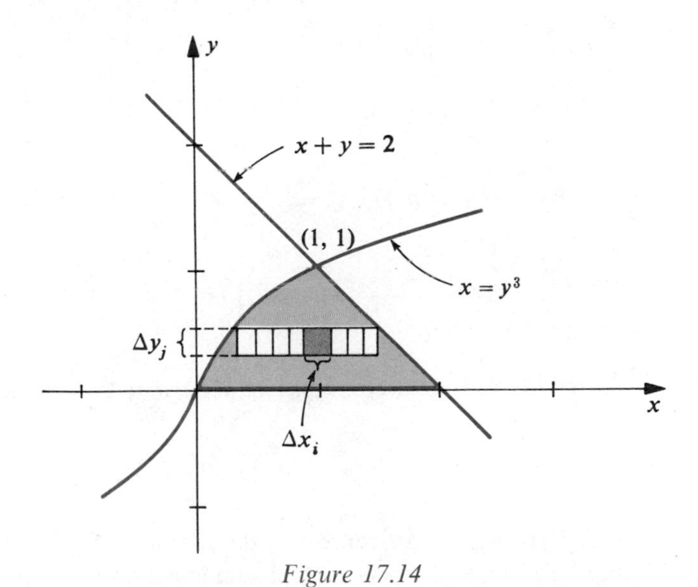

Figure 17.14

We can use formula (17.5) to express volumes of certain solids in terms of double integrals which may then be evaluated by means of iterated integrals. This is illustrated in the next example.

Example 3 Find the volume V of the solid in the first octant bounded by the coordinate planes and the graphs of the equations $z = x^2 + y^2 + 1$ and $2x + y = 2$.

Solution As illustrated in Figure 17.15, the solid lies under the paraboloid $z = x^2 + y^2 + 1$ and over the triangular region R in the xy-plane bounded by the coordinate axes and the line $y = 2 - 2x$.

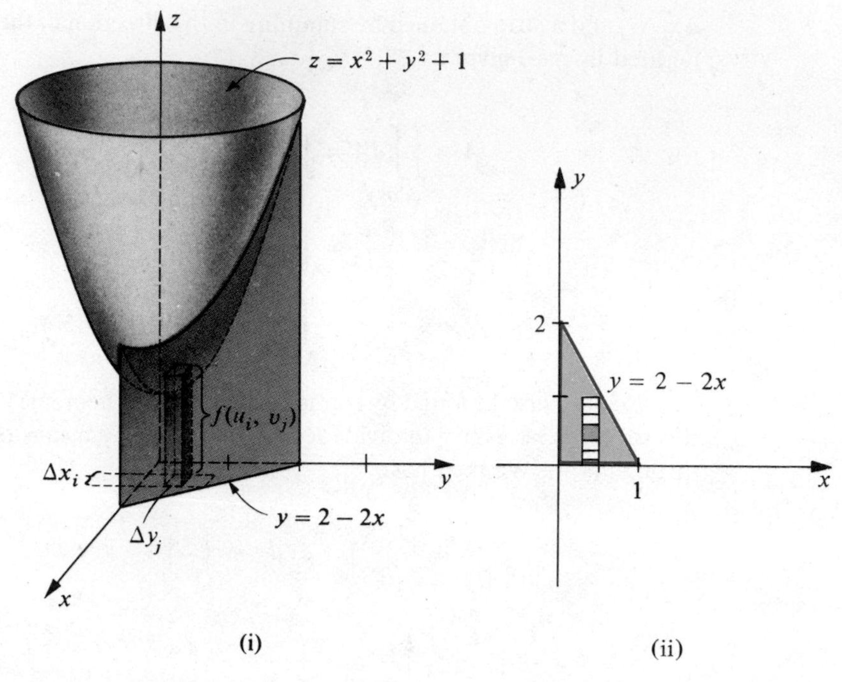

Figure 17.15

By (17.5), with $f(x, y) = x^2 + y^2 + 1$,

$$V = \iint_R (x^2 + y^2 + 1)\, dA.$$

As in (17.18), this integral may be interpreted as the limit of a double sum

$$\sum_i \sum_j f(u_i, v_j)\, \Delta y_j\, \Delta x_i$$

where $f(u_i, v_j)\, \Delta y_j\, \Delta x_i$ represents the volume of a prism of the type shown in (i) of Figure 17.15, and where we first sum in the direction of the y-axis from $y = 0$ to $y = 2 - 2x$. The sketch in (ii) of Figure 17.15 shows the region R in the xy-plane and a strip corresponding to the first summation. This gives us the volume of a lamina whose face is parallel to the yz-plane. We then sum these laminar volumes in the direction of the x-axis from $x = 0$ to $x = 1$. Using an iterated integral we have

$$\begin{aligned} V &= \int_0^1 \int_0^{2-2x} (x^2 + y^2 + 1)\, dy\, dx \\ &= \int_0^1 \left[x^2 y + \frac{1}{3} y^3 + y \right]_0^{2-2x} dx \\ &= \int_0^1 \left(-\frac{14}{3} x^3 + 10x^2 - 10x + \frac{14}{3} \right) dx = \frac{11}{6}. \end{aligned}$$

We may also find V by integrating first with respect to x. The reader should verify that in this case the iterated integral has the form

$$V = \int_0^2 \int_0^{(2-y)/2} (x^2 + y^2 + 1)\, dx\, dy.$$

Example 4 Find the volume V of the solid bounded by the graphs of $x^2 + y^2 = 9$ and $y^2 + z^2 = 9$.

Solution The graphs are cylinders of radius 3 whose axes intersect orthogonally. That portion of the solid which lies in the first octant is illustrated in (i) of Figure 17.16. By symmetry it is sufficient to find the volume of this portion and then multiply the result by 8. Referring to (i) of the figure we see that the solid lies under the graph of $z = \sqrt{9 - y^2}$ and over the region R bounded by the x- and y-axes and the graph of $x^2 + y^2 = 9$. Thus, by (17.5),

$$V = 8 \iint_R (9 - y^2)^{1/2}\, dA.$$

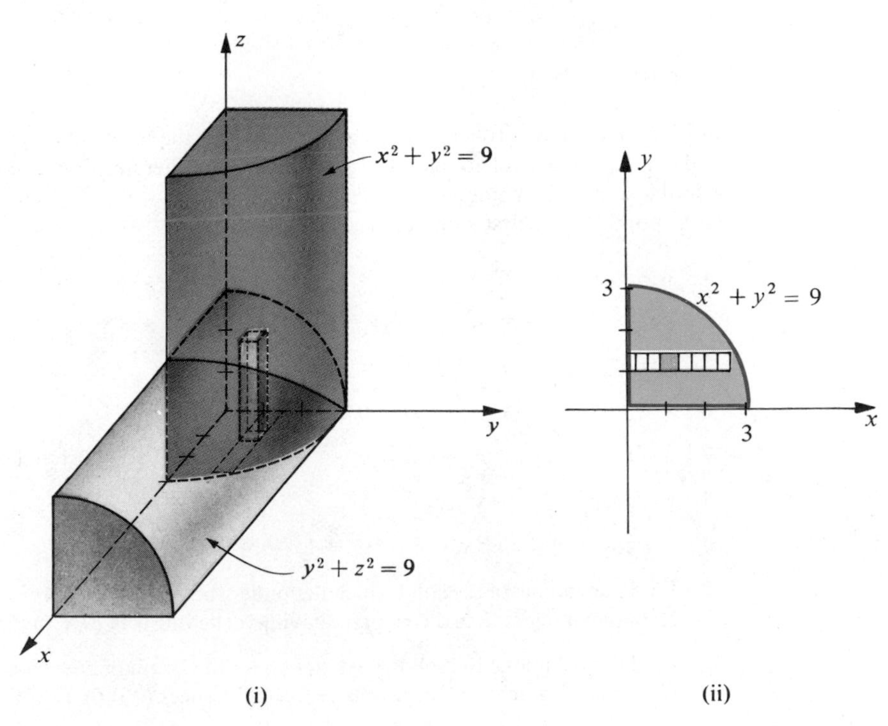

Figure 17.16

We shall evaluate the double integral by using (ii) of (17.14). In this case we regard the integral as a limit of a double sum of volumes of prisms of the type shown in (i) of Figure 17.16, where we first sum in the direction of the x-axis from $x = 0$ to

$x = (9 - y^2)^{1/2}$. The sketch in (ii) of Figure 17.16 indicates the region R in the xy-plane and a strip corresponding to this first summation. This gives us the volume of a lamina whose face is parallel to the xz-plane. We then sum these laminar volumes in the direction of the y-axis from $y = 0$ to $y = 3$. Using an iterated integral we have

$$\begin{aligned} V &= 8\int_0^3 \int_0^{(9-y^2)^{1/2}} (9 - y^2)^{1/2}\,dx\,dy \\ &= 8\int_0^3 \Big[(9 - y^2)^{1/2}x\Big]_0^{(9-y^2)^{1/2}} dy \\ &= 8\int_0^3 (9 - y^2)\,dy \\ &= 8\Big[9y - \frac{1}{3}y^3\Big]_0^3 = 144. \end{aligned}$$

The double integral could also be evaluated by integrating first with respect to y; however, this order of integration is much more complicated than the one discussed above.

EXERCISES 17.3

In each of Exercises 1–10 sketch the region bounded by the graphs of the given equations and find its area by means of double integrals. Interpret each iterated double integral as a limit of a double sum by sketching a typical rectangle of area $\Delta y_j\,\Delta x_i$ and the rectangular strip which corresponds to the first summation of the double sum.

1 $y = 1/x^2, y = -x^2, x = 1, x = 2$

2 $y = \sqrt{x}, y = -x, x = 1, x = 4$

3 $y^2 = -x, x - y = 4, y = -1, y = 2$

4 $x = y^2, y - x = 2, y = -2, y = 3$

5 $y = x, y = 3x, x + y = 4$

6 $x - y + 1 = 0, 7x - y - 17 = 0, 2x + y + 2 = 0$

7 $y = e^x, y = \sin x, x = -\pi, x = \pi$

8 $y = \ln|x|, y = 0, y = 1$

9 $y = x^2, y = 1/(1 + x^2)$

10 $y = \cosh x, y = \sinh x, x = -1, x = 1$

11 Find the volume of the solid which lies under the graph of $z = 4x^2 + y^2$ and over the rectangular region R in the xy-plane having vertices $(0, 0, 0), (0, 1, 0), (2, 0, 0)$, and $(2, 1, 0)$.

12 Find the volume of the solid which lies under the graph of $z = x^2 + 4y^2$ and over the triangular region R in the xy-plane having vertices $(0, 0, 0)$, $(1, 0, 0)$, and $(1, 2, 0)$.

In each of Exercises 13–20 sketch the solid in the first octant which is bounded by the graphs of the given equations and find its volume.

13 $x^2 + z^2 = 9, y = 2x, y = 0, z = 0$

14 $z = 4 - x^2, x + y = 2, x = 0, y = 0, z = 0$

15 $2x + y + z = 4, x = 0, y = 0, z = 0$

16 $y^2 = z, y = x, x = 4, z = 0$

17 $z = x^2 + y^2, 2x + 3y = 6, x = 0, y = 0, z = 0$

18 $x + 2y + 5z = 10, x = 0, y = 0, z = 0$

19 $z = x^3, y = x, y = 1, z = 0$

20 $x^2 + y^2 = 16, x = z, y = 0, z = 0$

In Exercises 21 and 22 find the volume of the solid bounded by the graphs of the given equations.

21 $z = x^2 + 4, y = 4 - x^2, x + y = 2, z = 0$

22 $y = x^3, y = x^4, z - x - y = 4, z = 0$

23 Find the volume of the solid bounded by the graphs of $x^2 + z^2 = a^2$ and $x^2 + y^2 = a^2$, where $a > 0$.

24 Find the volume of the solid bounded by the graph of $x^{2/3} + y^{2/3} + z^{2/3} = a^{2/3}$, where $a > 0$.

The iterated integral in each of Exercises 25–30 represents the volume of a solid. Describe the solid.

25 $\int_{-2}^{1} \int_{x-1}^{1-x^2} (x^2 + y^2)\, dy\, dx$

26 $\int_{0}^{1} \int_{3-x}^{3-x^2} \sqrt{25 - x^2 - y^2}\, dy\, dx$

27 $\int_{0}^{4} \int_{y/4}^{\sqrt{y}} (x + y)\, dx\, dy$

28 $\int_{0}^{1} \int_{y^{1/2}}^{y^{1/3}} \sqrt{x^2 + y^2}\, dx\, dy$

29 $\int_{0}^{4} \int_{-1}^{2} 3\, dy\, dx$

30 $\int_{0}^{\pi} \int_{0}^{\sin x} dy\, dx$

17.4 MOMENTS AND CENTER OF MASS

The first moment and center of mass of a homogeneous lamina were discussed in Chapter 10. We shall now employ double integrals to extend these concepts to laminas in which the density is not constant.

Let T denote a lamina which has the shape of the region R shown in Figure 17.2 and suppose that the area density, that is, the density per unit area, at the point (x, y) is given by $\rho(x, y)$, where ρ is a continuous function on R. If $P = \{R_i\}$ is an inner partition of R, then (u_i, v_i) will denote a point in R_i and the symbol T_i will represent the part of T which corresponds to R_i. Since ρ is continuous, a small change in (x, y) produces a small change in the density $\rho(x, y)$, that is, ρ is almost constant on R_i. Hence, if $\|P\| \approx 0$, then the mass of T_i may be approximated by $\rho(u_i, v_i)\,\Delta A_i$, where ΔA_i is the area of R_i. The sum $\Sigma_i\, \rho(u_i, v_i)\,\Delta A_i$ is an approximation to the mass of the lamina T. The mass M of T is defined as the limit of such sums as $\|P\| \to 0$. This gives us

(17.19)
$$M = \lim_{\|P\| \to 0} \sum_i \rho(u_i, v_i)\,\Delta A_i = \iint_R \rho(x, y)\, dA.$$

If the mass of T_i is assumed to be concentrated at (u_i, v_i), then as in Chapter 10, the moment of T_i with respect to the x-axis is the product $v_i\rho(u_i, v_i)\,\Delta A_i$. The moment M_x of T with respect to the x-axis is defined as the limit of such sums, that is,

(17.20)
$$M_x = \lim_{\|P\| \to 0} \sum_i v_i\rho(u_i, v_i)\,\Delta A_i = \iint_R y\rho(x, y)\,dA.$$

Similarly, the moment M_y of T with respect to the y-axis is

(17.21)
$$M_y = \lim_{\|P\| \to 0} \sum_i u_i\rho(u_i, v_i)\,\Delta A_i = \iint_R x\rho(x, y)\,dA.$$

As usual, the center of mass of the lamina is the point $(\bar{x}, \bar{y})$ such that

(17.22)
$$\bar{x} = \frac{M_y}{M}, \quad \bar{y} = \frac{M_x}{M}.$$

Example 1 A lamina T has the shape of an isosceles right triangle with equal sides of length a. Find the center of mass if the density at a point Q is directly proportional to the square of the distance from Q to the vertex V which is opposite the hypotenuse.

Solution It is convenient to introduce a coordinate system as illustrated in Figure 17.17, with V at the origin and the hypotenuse of the triangle along the line $x + y = a$. Shown in the figure is a typical rectangle of an inner partition P, where (u_i, v_i) is a point in the rectangle.

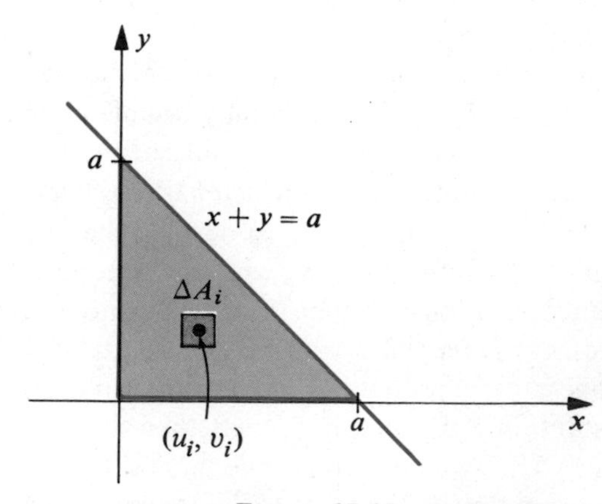

Figure 17.17

By hypothesis, the density at $Q(x, y)$ is given by $\rho(x, y) = k(x^2 + y^2)$ for some constant k. Applying (17.19) and Theorem (17.14), the mass of T is

$$M = \lim_{\|P\| \to 0} \sum_i k(u_i^2 + v_i^2)\,\Delta A_i$$

$$= \iint_R k(x^2 + y^2)\,dA$$

$$= \int_0^a \int_0^{a-x} k(x^2 + y^2)\,dy\,dx$$

$$= k\int_0^a \left[x^2 y + \frac{1}{3}y^3\right]_0^{a-x} dx$$

$$= k\int_0^a \left[x^2(a - x) + \frac{1}{3}(a - x)^3\right] dx.$$

The reader may verify that the last integral equals $ka^4/6$. Similarly, by (17.21),

$$M_y = \lim_{\|P\| \to 0} \sum_i u_i k(u_i^2 + v_i^2)\,\Delta A_i$$

$$= \iint_R xk(x^2 + y^2)\,dA$$

$$= \int_0^a \int_0^{a-x} xk(x^2 + y^2)\,dy\,dx$$

which may be shown to equal $ka^5/15$. From (17.22),

$$\bar{x} = \frac{ka^5/15}{ka^4/6} = \frac{2a}{5}.$$

By symmetry $\bar{y} = 2a/5$.

Example 2 A lamina has the shape of the region R in the xy-plane bounded by the graphs of $x = y^2$ and $x = 4$. Find the center of mass if the density at the point $P(x, y)$ is directly proportional to the distance from the y-axis to P.

Solution The region is sketched in Figure 17.18 where, by hypothesis, $\rho(x, y) = kx$. It follows from the form of ρ and the symmetry of the region that the center of mass is on the x-axis, that is, $\bar{y} = 0$. Using (17.19) and then integrating first with respect to x as indicated by the strip in Figure 17.18,

$$M = \iint_R kx\,dA = k\int_{-2}^{2} \int_{y^2}^{4} x\,dx\,dy$$

$$= k\int_{-2}^{2} \left.\frac{x^2}{2}\right]_{y^2}^{4} dy = \frac{k}{2}\int_{-2}^{2} (16 - y^4)\,dy = \frac{128}{5}k.$$

By (17.21),

$$M_y = \iint_R x(kx)\,dA = k\int_{-2}^{2}\int_{y^2}^{4} x^2\,dx\,dy$$

$$= k\int_{-2}^{2} \frac{x^3}{3}\bigg]_{y^2}^{4} dy = \frac{k}{3}\int_{-2}^{2} (64 - y^6)\,dy = \frac{512}{7}k.$$

Consequently

$$\bar{x} = \frac{M_y}{M} = \frac{512k}{7}\cdot\frac{5}{128k} = \frac{20}{7}$$

and hence the centroid is $(20/7, 0)$.

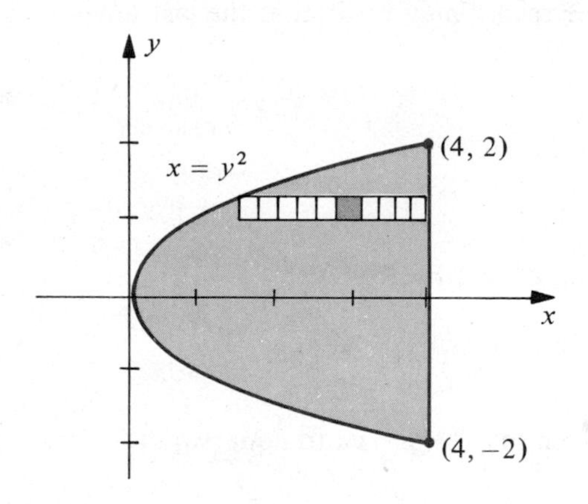

Figure 17.18

If n particles of masses $m_1, m_2, \ldots, m_n$ are located at points (x_1, y_1), $(x_2, y_2), \ldots, (x_n, y_n)$, respectively, then the moments of the system with respect to the x- and y-axes were defined in (10.10) by

$$M_x = \sum_{i=1}^{n} y_i m_i \quad \text{and} \quad M_y = \sum_{i=1}^{n} x_i m_i.$$

These are also called the *first moments* of the system with respect to the coordinate axes. If we use the *squares* of the distances from the coordinate axes we obtain the *second moments*, or **moments of inertia** I_x and I_y of the system with respect to the x-axis and y-axis, respectively. Thus, by definition,

$$I_x = \sum_{i=1}^{n} y_i^2 m_i \quad \text{and} \quad I_y = \sum_{i=1}^{n} x_i^2 m_i.$$

This concept may be extended to laminas by employing the limiting processes used for double integrals. In particular, let us consider a lamina T of the type used at the beginning of this section (see Figure 17.2). If the density at (x, y) is $\rho(x, y)$

where ρ is continuous, then it is natural to define the **moment of inertia I_x of T with respect to the x-axis** by replacing v_i in (17.20) by the square of v_i. Thus

(17.23)
$$I_x = \lim_{\|P\| \to 0} \sum_i v_i^2 \rho(u_i, v_i) \Delta A_i = \iint_R y^2 \rho(x, y) \, dA.$$

In like manner, the **moment of inertia I_y of T with respect to the y-axis** is given by

(17.24)
$$I_y = \lim_{\|P\| \to 0} \sum_i u_i^2 \rho(u_i, v_i) \Delta A_i = \iint_R x^2 \rho(x, y) \, dA.$$

If we multiply $\rho(u_i, v_i) \Delta A_i$ by the square $u_i^2 + v_i^2$ of the distance from the origin to (u_i, v_i) and take the limit of a sum of such terms, we obtain the **moment of inertia I_O of T with respect to the origin**. Thus

(17.25)
$$I_O = \iint_R (x^2 + y^2) \rho(x, y) \, dA.$$

The number I_O is also called the **polar moment of inertia** of T. Observe that $I_O = I_x + I_y$.

Example 3 A lamina T has the semicircular shape illustrated in (i) of Figure 17.19. If the density at a point P is directly proportional to the distance from the diameter AB to P, find the moment of inertia of T with respect to the line through A and B.

Solution If we introduce a coordinate system as in (ii) of Figure 17.19, then the density at (x, y) is $\rho(x, y) = ky$ and by (17.23) the desired moment of inertia is

$$\begin{aligned} I_x &= \int_{-a}^{a} \int_{0}^{\sqrt{a^2 - x^2}} y^2(ky) \, dy \, dx \\ &= k \int_{-a}^{a} \left[\frac{1}{4} y^4 \right]_0^{\sqrt{a^2 - x^2}} dx \\ &= \frac{k}{4} \int_{-a}^{a} (a^4 - 2a^2x^2 + x^4) \, dx. \end{aligned}$$

Carrying out the integration yields $I_x = 4ka^5/15$.

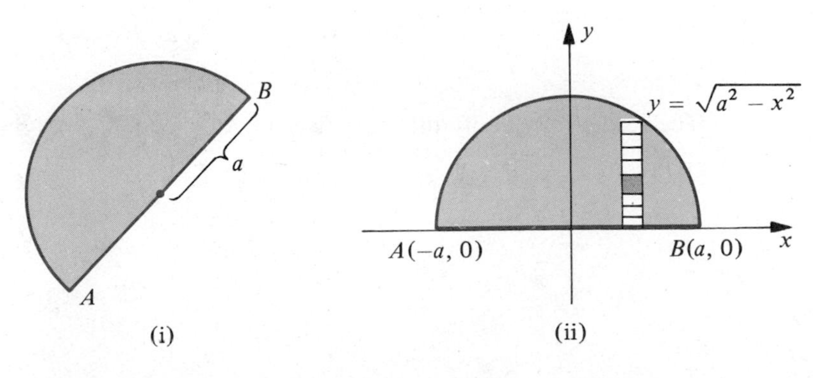

Figure 17.19

Moments of inertia are useful in problems which involve the rotation of an object about a fixed axis. For example, suppose a wheel is rotating about an axle. If a particle P of mass m on the wheel is a distance a from the axis of rotation (see Figure 15.10), then the moment of inertia I of P with respect to the axis is ma^2. If the angular speed is ω, then the speed v of the particle, that is, the distance traveled per unit time, is $a\omega$. By definition, the **kinetic energy** K.E. of P is given by

$$\text{K.E.} = \tfrac{1}{2}mv^2.$$

Consequently

(17.26)
$$\text{K.E.} = \tfrac{1}{2}ma^2\omega^2 = \tfrac{1}{2}I\omega^2.$$

If we represent the wheel by a flat disc and introduce a limit of a sum, then the same formula may be derived for the wheel. The formula can also be extended to laminas having noncircular shapes. The kinetic energy of a rotating object tells the engineer or physicist the amount of work necessary to bring the object to rest. Since (17.26) states that the kinetic energy of a rotating lamina is directly proportional to the moment of inertia, it follows that for a fixed ω, the larger the moment of inertia, the larger the amount of work required to stop the rotation.

If a lamina T of mass M has moments of inertia I_x and I_y, then the **radius of gyration $\bar{y}$ of T with respect to the x-axis**, and the **radius of gyration $\bar{x}$ of T with respect to the y-axis** are defined by the equations

(17.27)
$$\bar{y}^2 M = I_x \quad \text{and} \quad \bar{x}^2 M = I_y.$$

These formulas imply that $(\bar{x}, \bar{y})$ is the point at which all the mass can be concentrated without changing the moments of inertia of T with respect to the coordinate axes.

Example 4 Find the radius of gyration with respect to the x-axis of the lamina described in Example 3.

Solution In Example 3 we found that $I_x = 4ka^5/15$. The mass of T is

$$M = \int_{-a}^{a} \int_{0}^{\sqrt{a^2 - x^2}} ky\, dy\, dx.$$

The reader may verify that this gives us $M = 2ka^3/3$. Applying (17.27),

$$\begin{aligned}\bar{y} &= \sqrt{I_x/M} \\ &= \sqrt{\frac{4ka^5/15}{2ka^3/3}} \\ &= \sqrt{\frac{2a^2}{5}} = \frac{a\sqrt{10}}{5}.\end{aligned}$$

EXERCISES 17.4

In each of Exercises 1–14 find the mass and the center of mass of the lamina which has the shape of the region bounded by the graphs of the given equations and having the indicated density.

1 $y = \sqrt{x}, x = 9, y = 0; \rho(x, y) = x + y$ **2** $y = \sqrt[3]{x}, x = 8, y = 0; \rho(x, y) = y^2$

3 $y = x^2, y = 4$; density at the point $P(x, y)$ is directly proportional to the distance from the y-axis to P

4 $y = x^3, y = 2x$; density at the point $P(x, y)$ is directly proportional to the distance from the x-axis to P

5 $y = e^{-x^2}, y = 0, x = -1, x = 1; \rho(x, y) = |xy|$

6 $y = \sin x, y = 0, x = 0, x = \pi; \rho(x, y) = y$

7 $x = y^2, y - x = 2, y = -2, y = 3; \rho(x, y) = 1$

8 $y = x, y = 3x, x + y = 4; \rho(x, y) = 2$

9 $y = \sec x, x = -\pi/4, x = \pi/4, y = 1/2; \rho(x, y) = 4$

10 $xy^2 = 1, y = 1, y = 2; \rho(x, y) = x^2 + y^2$

11 $y = 1/x, y = x, x = 2, y = 0; \rho(x, y) = x$

12 $x = e^y, y = -1, x = 2; \rho(x, y) = 1$

13 $y = e^{-x}, y = 0, x = 0, x = 1; \rho(x, y) = y^2$

14 $y = \ln x, y = 0, x = 2; \rho(x, y) = 1/x$

15 Find I_x, I_y, and I_O for the lamina of Exercise 1.

16 Find I_x, I_y, and I_O for the lamina of Exercise 2.

17 Find I_x, I_y, and I_O for the lamina of Exercise 3.

18 Find I_x, I_y, and I_O for the lamina of Exercise 4.

19 A homogeneous lamina has the shape of a square of side a. Find the moment of inertia with respect to the following.
(a) A side (b) A diagonal (c) The center of mass

20 A homogeneous lamina has the shape of an equilateral triangle of side a. Find the moment of inertia with respect to the following.
(a) An altitude (b) A side (c) A vertex

21 Find the radius of gyration in (a) of Exercise 19.

22 Find the radius of gyration in (a) of Exercise 20.

23 Find the moment of inertia and radius of gyration of a homogeneous circular disc of radius a with respect to a line along a diameter.

24 A lamina T has the shape of an isosceles right triangle with equal sides of length a. The density at a point P is directly proportional to the square of the distance from P to the vertex V which is opposite the hypotenuse. Find the moment of inertia and the radius of gyration with respect to a line along one of the equal sides (see Example 1).

17.5 DOUBLE INTEGRALS IN POLAR COORDINATES

A region in a polar coordinate plane of the type illustrated in Figure 17.20, bounded by arcs of circles of radii r_1 and r_2 with centers at the pole and by two rays from the origin, will be called an **elementary polar region**. If $\Delta\theta$ denotes the radian measure of the angle between the rays and $\Delta r = r_2 - r_1$, then the area ΔA of the region is given by

$$\Delta A = \tfrac{1}{2}r_2^2\,\Delta\theta - \tfrac{1}{2}r_1^2\,\Delta\theta.$$

This formula may also be written

$$\Delta A = \tfrac{1}{2}(r_2^2 - r_1^2)\,\Delta\theta = \tfrac{1}{2}(r_2 + r_1)(r_2 - r_1)\,\Delta\theta.$$

If we denote the **average radius** $\frac{1}{2}(r_2 + r_1)$ by $\bar{r}$, then

(17.28) $$\Delta A = \bar{r}\,\Delta r\,\Delta\theta.$$

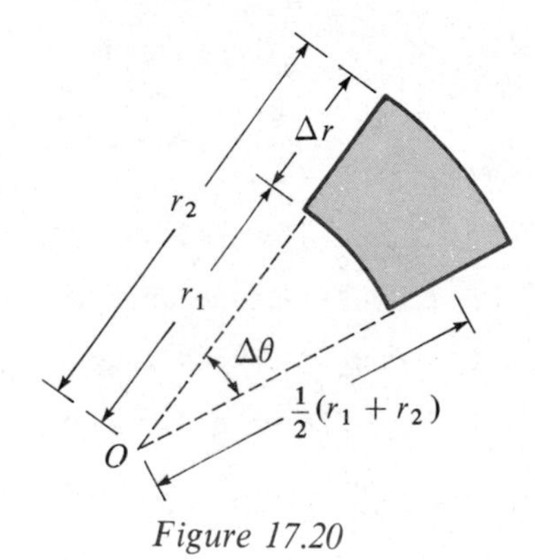

Figure 17.20

Next, consider a region R of the type illustrated in (i) of Figure 17.21, bounded by two rays which make positive angles α and β with the polar axis and by the graphs of two polar equations $r = g_1(\theta)$ and $r = g_2(\theta)$, where the functions g_1 and g_2 are continuous and $g_1(\theta) \le g_2(\theta)$ for all θ in the interval $[\alpha, \beta]$. If R is subdivided by means of circular arcs and rays as shown in (ii) of Figure 17.21, then the collection of elementary polar regions $R_1, R_2, \ldots, R_n$ which lie completely within R is called an **inner polar partition** P of R. The norm $\|P\|$ of P is the length of the longest diagonal of the R_i. If we choose a point (r_i, θ_i) in R_i such that r_i is the average radius, then by (17.28) the area ΔA_i of R_i is $r_i\,\Delta r_i\,\Delta\theta_i$. If f is a continuous function of the polar variables r and θ, then it can be proved that

(17.29) $$\lim_{\|P\|\to 0} \sum f(r_i, \theta_i) r_i\,\Delta r_i\,\Delta\theta_i = \iint_R f(r, \theta)\,dA$$

where the double integral may be evaluated as follows:

(17.30)
$$\iint_R f(r,\theta)\,dA = \int_\alpha^\beta \int_{g_1(\theta)}^{g_2(\theta)} f(r,\theta) r\,dr\,d\theta.$$

It is important to note that the integrand on the right is the product of $f(r,\theta)$ and r. This is true because ΔA_i equals $r_i\,\Delta r_i\,\Delta\theta_i$.

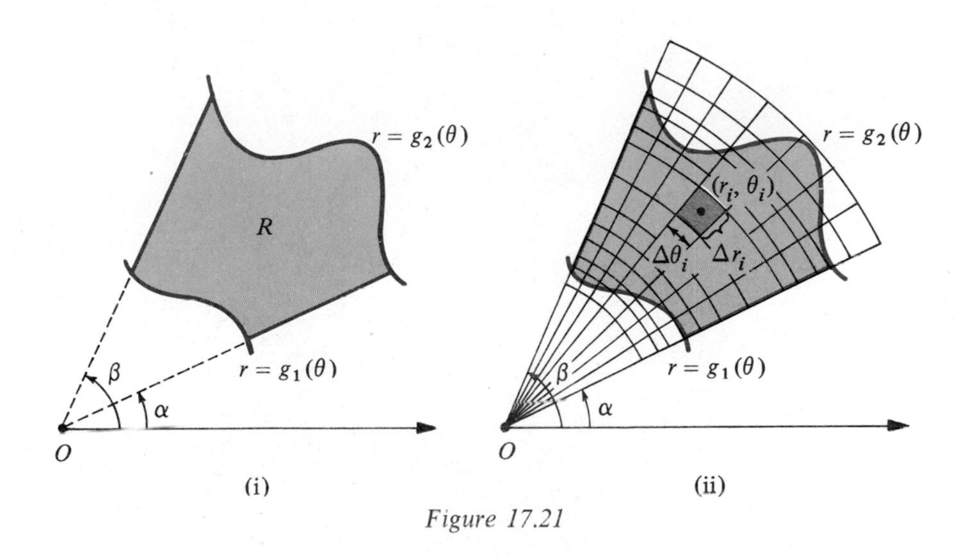

Figure 17.21

It is sometimes convenient to regard the iterated integral in (17.30) in terms of limits of double sums in a manner similar to our discussion for rectangular coordinates at the end of Section 17.2. In the present situation, we first hold θ fixed and sum along one of the wedge-shaped regions shown in (ii) of Figure 17.21, from the graph of g_1 to the graph of g_2. For the second summation we sweep out the region by letting θ vary from α to β. Intuitively, the iterated integral does this and simultaneously takes the limit as $\|P\|$ approaches zero.

If $f(r,\theta) = 1$ throughout R, then the integral in (17.30) equals the area of R. This also follows from our work in Chapter 13 since after the partial integration with respect to r, the formula given in (13.14) is obtained.

Example 1 A lamina has the shape of the region R which lies outside the graph of $r = a$ and inside the graph of $r = 2a\sin\theta$. Find the mass if the density at a point $P(r,\theta)$ is inversely proportional to the distance from the pole to P.

Solution The region is sketched in Figure 17.22, together with a typical strip of elementary polar regions involved in summing from one boundary of R to the other. By hypothesis, the density $f(r,\theta)$ at the point (r,θ) is k/r, where k is some real number.

The mass M of the lamina is

$$M = \iint_R f(r, \theta)\, dA = \int_{\pi/6}^{5\pi/6} \int_a^{2a\sin\theta} (k/r) r\, dr\, d\theta$$

$$= k \int_{\pi/6}^{5\pi/6} \Big] r \Big]_a^{2a\sin\theta} d\theta$$

$$= k \int_{\pi/6}^{5\pi/6} (2a\sin\theta - a)\, d\theta$$

$$= ka \Big[-2\cos\theta - \theta \Big]_{\pi/6}^{5\pi/6}$$

$$= ka[(\sqrt{3} - 5\pi/6) - (-\sqrt{3} - \pi/6)]$$

$$= ka(2\sqrt{3} - 2\pi/3).$$

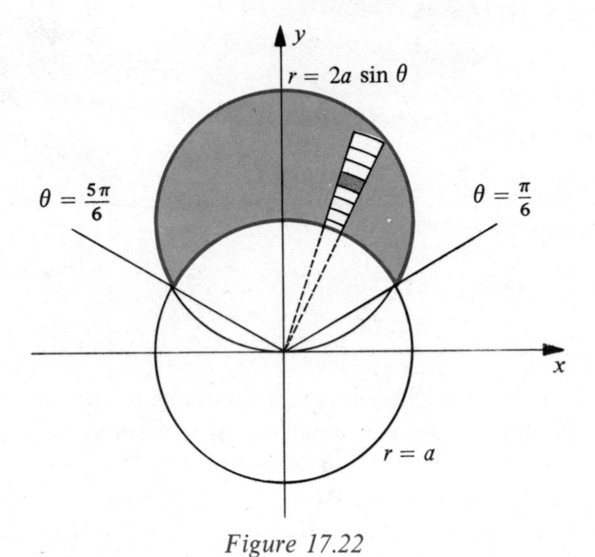

Figure 17.22

It can be shown that under suitable conditions an iterated double integral in rectangular coordinates can be transformed into a double integral in polar coordinates. First, the variables x and y in the integrand are replaced by $r\cos\theta$ and $r\sin\theta$. Next, $dy\,dx$ (or $dx\,dy$) is replaced by $r\,dr\,d\theta$ (or $r\,d\theta\,dr$) and the limits of integration are changed to polar coordinates. As an illustration, formula (17.24) for the moment of inertia with respect to the y-axis (or polar axis) of a lamina having the shape of (i) in Figure 17.21 is

(17.31)
$$I_y = \int_\alpha^\beta \int_{g_1(\theta)}^{g_2(\theta)} (r\cos\theta)^2 \rho(r\cos\theta, r\sin\theta) r\, dr\, d\theta,$$

where $\rho(x, y)$ is the density at (x, y). For the special case of the lamina in Example 1, (17.31) takes on the form

$$I_y = \int_{\pi/6}^{5\pi/6} \int_{a}^{2a\sin\theta} r^2 \cos^2 \theta (k/r) r \, dr \, d\theta.$$

Another illustration is given in the following example.

Example 2 Use polar coordinates to evaluate

$$\int_{-a}^{a} \int_{0}^{\sqrt{a^2 - x^2}} (x^2 + y^2)^{3/2} \, dy \, dx.$$

Solution Evidently the region of integration is bounded by the graphs of $y = \sqrt{a^2 - x^2}$ and $y = 0$ illustrated in (ii) of Figure 17.19. Replacing $x^2 + y^2$ in the integrand by r^2, $dy\,dx$ by $r\,dr\,d\theta$, and changing the limits, we have

$$\begin{aligned} \int_{-a}^{a} \int_{0}^{\sqrt{a^2 - x^2}} (x^2 + y^2)^{3/2} \, dy \, dx &= \int_{0}^{\pi} \int_{0}^{a} r^3 r \, dr \, d\theta \\ &= \frac{a^5}{5} \int_{0}^{\pi} d\theta = \frac{\pi a^5}{5}. \end{aligned}$$

Polar coordinates can also be used for double integrals over a region R of the type illustrated in (i) of Figure 17.23. In this case R is bounded by the arcs of two circles of radii a and b and by the graphs of polar equations $\theta = h_1(r)$ and $\theta = h_2(r)$, where the functions h_1 and h_2 are continuous and $h_1(r) \le h_2(r)$ for all r in the interval $[a, b]$. If f is a function of r and θ which is continuous on R, then the limit in (17.29) exists and the corresponding evaluation theorem is

(17.32)
$$\iint_R f(r, \theta) \, dA = \int_{a}^{b} \int_{h_1(r)}^{h_2(r)} f(r, \theta) r \, d\theta \, dr.$$

The iterated integral in (17.32) may be interpreted in terms of limits of double sums. In this case we first hold r fixed and sum along a circular arc, as illustrated by the darker of the shaded portions in (ii) of Figure 17.23. Secondly, we sweep out R by summing terms which correspond to such ring-shaped regions.

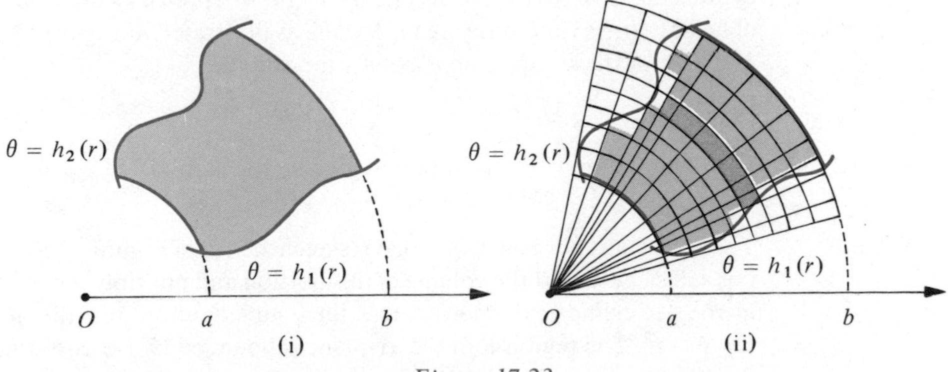

Figure 17.23

Example 3 Find the polar moment of inertia of a homogeneous lamina which has the shape of the smaller of the regions bounded by the polar axis, the graphs of $r = 1, r = 2$, and the part of the spiral $r\theta = 1$ from $\theta = \frac{1}{2}$ to $\theta = 1$.

Solution The region, together with some elementary polar regions, is sketched in Figure 17.24. By hypothesis, the density at every point (r, θ) is a constant k. Applying (17.25) and the previous remarks,

$$\begin{aligned} I_O &= \iint_R (x^2 + y^2)\rho(x, y)\, dA \\ &= \int_1^2 \int_0^{1/r} r^2 kr\, d\theta\, dr \\ &= k \int_1^2 r^3\theta \Big]_0^{1/r} dr \\ &= k \int_1^2 r^2\, dr = \frac{7k}{3}. \end{aligned}$$

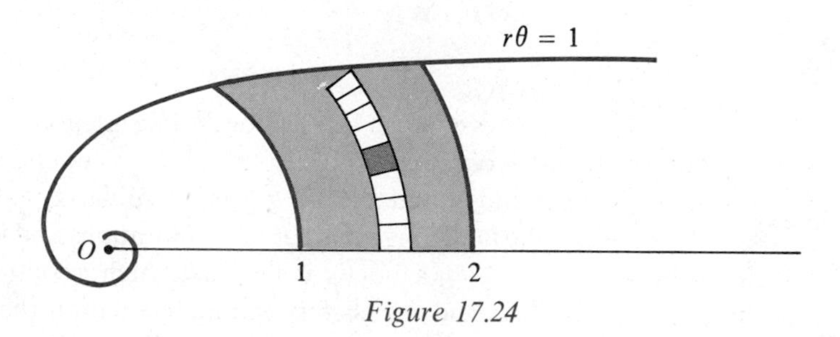

Figure 17.24

If $f(r, \theta) \geq 0$ throughout a polar region R, then the double integral $\iint_R f(r, \theta)\, dA$ in (17.29) may be regarded as the volume of a solid in a manner completely analogous to our formulation of (17.5) or (17.18) in rectangular coordinates. The principal difference in the present situation is that we consider the graph S of $z = f(r, \theta)$ in *cylindrical* coordinates. The desired solid lies under the graph of S and over the region R. If we use a polar partition of R, then the expression $f(r_i, \theta_i)r_i\, \Delta r_i\, \Delta\theta_i$ in (17.29) may be interpreted as the volume of a prism of height $f(r_i, \theta_i)$ and base area $r_i\, \Delta r_i\, \Delta\theta_i$, as illustrated in Figure 17.25. The limit of a sum of these expressions equals the volume.

Example 4 Find the volume V of the solid bounded by the paraboloid $z = 4 - x^2 - y^2$ and the xy-plane.

Solution The first octant portion of the solid is sketched in (i) of Figure 17.26. By symmetry, it is sufficient to find the volume of this portion and multiply the result by 4. If we introduce cylindrical coordinates, then an equation for the paraboloid is $z = 4 - r^2$. The region R in the xy-plane is bounded by the coordinate axes and one-fourth of the circle $r = 2$. This is illustrated in (ii) of Figure 17.26, together

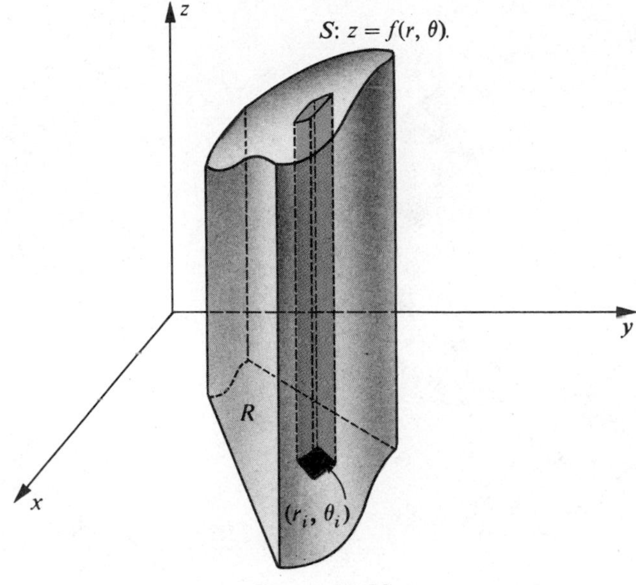

Figure 17.25

with a wedge of elements of a polar partition. By the preceding discussion, with $f(r, \theta) = 4 - r^2$, we have

$$\begin{aligned} V &= 4 \iint_R (4 - r^2)\, dA \\ &= 4 \int_0^{\pi/2} \int_0^2 (4 - r^2) r\, dr\, d\theta \\ &= 4 \int_0^{\pi/2} \left[2r^2 - \frac{r^4}{4} \right]_0^2 d\theta \\ &= 4 \int_0^{\pi/2} 4\, d\theta = 16\theta \Big]_0^{\pi/2} = 8\pi. \end{aligned}$$

The same problem, using rectangular coordinates, leads to the following double integral:

$$\begin{aligned} V &= 4 \iint_R (4 - x^2 - y^2)\, dA \\ &= 4 \int_0^2 \int_0^{\sqrt{4 - x^2}} (4 - x^2 - y^2)\, dy\, dx. \end{aligned}$$

After evaluating the last integral, the reader should be thoroughly convinced of the advantage of using cylindrical (or polar) coordinates in certain problems.

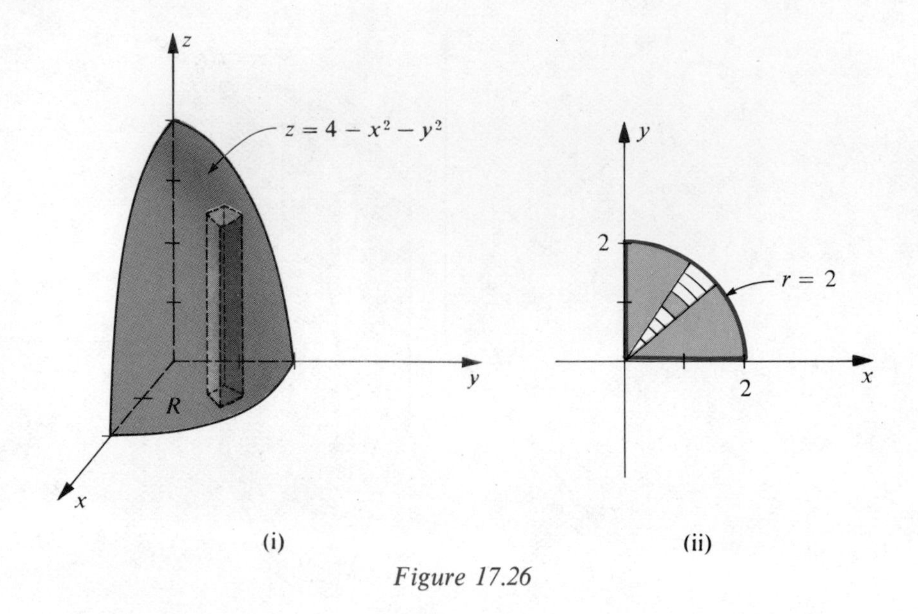

Figure 17.26

EXERCISES 17.5

In Exercises 1–4 use a double integral to find the area of the indicated region.

1 One loop of $r^2 = 9 \sin 2\theta$

2 One loop of $r = 2 \cos 4\theta$

3 Inside $r = 2 - 2\cos\theta$ and outside $r = 3$

4 Bounded by $r = 3 + 2\sin\theta$

In each of Exercises 5–8, find the mass and center of mass of the lamina which has the indicated shape and density.

5 Bounded by $r = 2\cos\theta$; density at the point $P(r, \theta)$ is directly proportional to the distance from the pole to P.

6 Bounded by the loop of $r^2 = a\cos 2\theta$ from $\theta = -\pi/4$ to $\theta = \pi/4$; density at the point $P(r, \theta)$ is directly proportional to the square of the distance from the pole to P

7 Inside the graph of $r = 3\sin\theta$ and outside the graph of $r = 1 + \sin\theta$; density at the point $P(r, \theta)$ is inversely proportional to the distance from the pole to P.

8 Inside the graph of $r = 4\cos\theta$ and outside the graph of $r = 2$; density at the point $P(r, \theta)$ is directly proportional to the distance from the polar axis to P.

9 Find the polar moment of inertia of a homogeneous lamina which has the shape of the region bounded by $r^2 = a \sin 2\theta$.

10 A homogeneous lamina has the shape of the region $R = \{(r, \theta): r^{-1} \le \theta \le 2r^{-1}, 1 \le r \le 2\}$. Find the polar moment of inertia.

11 Find the moment of inertia and radius of gyration of a homogeneous circular lamina of radius a with respect to a line through the center.

12 Find the moment of inertia and radius of gyration of a homogeneous circular lamina of radius a with respect to a tangent line.

Evaluate the integrals in Exercises 13–16 by changing to polar coordinates.

13 $\displaystyle\int_{-a}^{a}\int_{0}^{\sqrt{a^2-x^2}} e^{-(x^2+y^2)}\,dy\,dx$

14 $\displaystyle\int_{0}^{a}\int_{0}^{\sqrt{a^2-x^2}} (x^2+y^2)^{3/2}\,dy\,dx$

15 $\displaystyle\int_{1}^{2}\int_{0}^{x} \frac{1}{\sqrt{x^2+y^2}}\,dy\,dx$

16 $\displaystyle\int_{0}^{1}\int_{0}^{\sqrt{1-x^2}} e^{\sqrt{x^2+y^2}}\,dy\,dx$

17 Find the volume of the solid which lies inside the sphere $x^2 + y^2 + z^2 = 25$ and outside the cylinder $x^2 + y^2 = 9$.

18 Find the volume of the solid which is cut out of the ellipsoid $4x^2 + 4y^2 + z^2 = 16$ by the cylinder $x^2 + y^2 = 1$.

19 Find the volume of the solid bounded by the cone $z = r$ and the cylinder $r = 2\cos\theta$.

20 Find the volume of the solid bounded by the paraboloid $z = 4r^2$, the cylinder $r = 3\sin\theta$, and the plane $z = 0$.

17.6 TRIPLE INTEGRALS

It is possible to define *triple integrals* for a function f of *three* variables x, y, and z in a manner similar to that used in Section 17.2 for functions of two variables. The simplest case occurs when f is continuous throughout a region of the form

$$Q = \{(x, y, z): a \le x \le b, c \le y \le d, k \le z \le l\}.$$

If Q is divided into subregions $Q_1, Q_2, \ldots, Q_n$ by means of planes parallel to the three coordinate planes, then the collection $\{Q_i : i = 1, \ldots, n\}$ is called a **partition** P of Q. The **norm** $\|P\|$ of the partition is the length of the longest diagonal of all the Q_i. As illustrated in Figure 17.27, if Δy_i, Δy_i, and Δz_i are the dimensions of Q_i, then

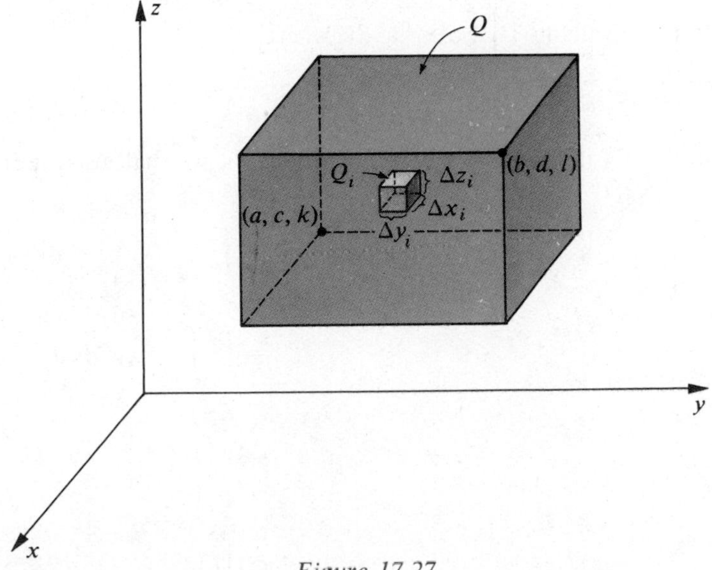

Figure 17.27

the volume ΔV_i of Q_i is the product $\Delta x_i \, \Delta y_i \, \Delta z_i$. If (u_i, v_i, w_i) is a point in Q_i, then the sum

(17.33)
$$\sum_{i=1}^{n} f(u_i, v_i, w_i) \, \Delta V_i$$

is called a **Riemann sum** of f for P.

The concept of a limit of Riemann sums for a function of three variables is defined in a manner similar to that given in (17.2) for functions of two variables. If this limit exists it is called the **triple integral of f over Q** denoted by

(17.34)
$$\iiint_Q f(x, y, z) \, dV = \lim_{\|P\| \to 0} \sum_i f(u_i, v_i, w_i) \, \Delta V_i.$$

Moreover, it can be shown that

$$\iiint_Q f(x, y, z) \, dV = \int_k^l \int_c^d \int_a^b f(x, y, z) \, dx \, dy \, dz,$$

where the **iterated integral** on the right means that the first integration is with respect to x (with y and z fixed), the second is with respect to y (with z fixed), and the third is with respect to z. There are five other iterated integrals which equal the triple integral of f over the parallelepiped Q. For example, if we integrate in the order y, z, x, then

$$\iiint_Q f(x, y, z) \, dV = \int_a^b \int_k^l \int_c^d f(x, y, z) \, dy \, dz \, dx.$$

Example 1 Evaluate $\iiint_Q 3xy^3z^2 \, dV$ where

$$Q = \{(x, y, z)\colon -1 \le x \le 3, 1 \le y \le 4, 0 \le z \le 2\}.$$

Solution Of the six possible iterated integrals, we shall use the following:

$$\int_1^4 \int_{-1}^3 \int_0^2 3xy^3z^2 \, dz \, dx \, dy = \int_1^4 \int_{-1}^3 \Big]_0^2 xy^3z^3 \, dx \, dy$$

$$= \int_1^4 \int_{-1}^3 8xy^3 \, dx \, dy$$

$$= \int_1^4 \Big]_{-1}^3 4x^2y^3 \, dy$$

$$= \int_1^4 (36y^3 - 4y^3) \, dy = 8y^4 \Big]_1^4 = 2040.$$

Triple integrals may be defined over regions other than those bounded by parallelepipeds. Suppose, for example, that R is a region in the xy-plane which can be divided into subregions of Types I and II, and that Q is the region in three dimensions defined by

$$Q = \{(x, y, z) : (x, y) \text{ is in } R \text{ and } k_1(x, y) \le z \le k_2(x, y)\}$$

where the functions k_1 and k_2 have continuous first partial derivatives throughout R. Geometrically, Q lies between the graphs of $z = k_1(x, y)$ and $z = k_2(x, y)$ and over or under the region R. If Q is subdivided by means of planes parallel to the three coordinate planes, then the resulting parallelepipeds $Q_1, Q_2, \ldots, Q_n$ which lie completely within Q form an **inner partition** P of Q. The sketch in Figure 17.28 shows a region Q of the type under consideration together with a typical element Q_i of an inner partition.

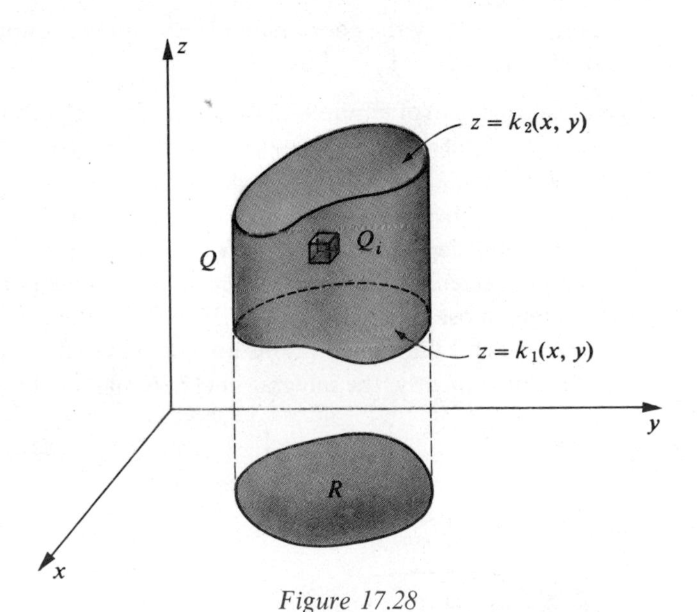

Figure 17.28

A **Riemann sum** of f for P is any sum of the form (17.33), where (u_i, v_i, w_i) is an arbitrary point in Q_i. The triple integral of f over Q is again defined as the limit (17.34). It can be shown that if f is continuous throughout Q, then

(17.35)
$$\iiint_Q f(x, y, z)\, dV = \iint_R \left[\int_{k_1(x, y)}^{k_2(x, y)} f(x, y, z)\, dz \right] dA.$$

In particular, if R is of Type I, as in (i) of Figure 17.1, then

(17.36)
$$\iiint_Q f(x, y, z)\, dV = \int_a^b \int_{g_1(x)}^{g_2(x)} \int_{k_1(x, y)}^{k_2(x, y)} f(x, y, z)\, dz\, dy\, dx$$

where the symbol on the right is called an **iterated triple integral**. It is evaluated by means of partial integrations of $f(x, y, z)$ in the order z, y, and x, substituting the indicated limits in the usual way. Similarly, if R is of Type II as in (ii) of Figure 17.1, then

(17.37)
$$\iiint_Q f(x, y, z)\, dV = \int_c^d \int_{h_1(y)}^{h_2(y)} \int_{k_1(x, y)}^{k_2(x, y)} f(x, y, z)\, dz\, dx\, dy.$$

It is convenient to regard the iterated integral in (17.36) or (17.37) as a limit of a *triple sum* of terms of the form $f(u_i, v_j, w_k)\, \Delta z_k\, \Delta y_j\, \Delta x_i$, where we first sum along a column of parallelepipeds Q_n in the direction of the z-axis from the lower surface (with equation $z = k_1(x, y)$) to the upper surface (with equation $z = k_2(x, y)$). The resulting double integral is then evaluated over R using the techniques discussed in Section 17.2.

Example 2 Express $\iiint_Q f(x, y, z)\, dV$ as an iterated integral, where Q is the region in the first octant bounded by the coordinate planes and the graphs of $z - 2 = x^2 + y^2/4$ and $x^2 + y^2 = 1$.

Solution As illustrated in (i) of Figure 17.29, Q lies under the graph of $z = 2 + x^2 + y^2/4$ (a paraboloid) and over the graph of $z = 0$ (the xy-plane). The region R in the xy-plane is bounded by the coordinate axes and the graph of $y = \sqrt{1 - x^2}$. Also shown in (i) of the figure is a column which corresponds to a first summation over the parallelepipeds of volumes $\Delta z_k\, \Delta y_j\, \Delta x_i$ in the direction of the z-axis. Since the column extends from the xy-plane ($z = 0$) to the paraboloid, the lower limit on the innermost integral sign in (17.36) is 0 and the upper limit is $2 + x^2 + y^2/4$. The second and third integrations are taken over the region R (see (ii) of Figure 17.29). Consequently, the integral in (17.36) has the form

$$\int_0^1 \int_0^{\sqrt{1 - x^2}} \int_0^{2 + x^2 + y^2/4} f(x, y, z)\, dz\, dy\, dx.$$

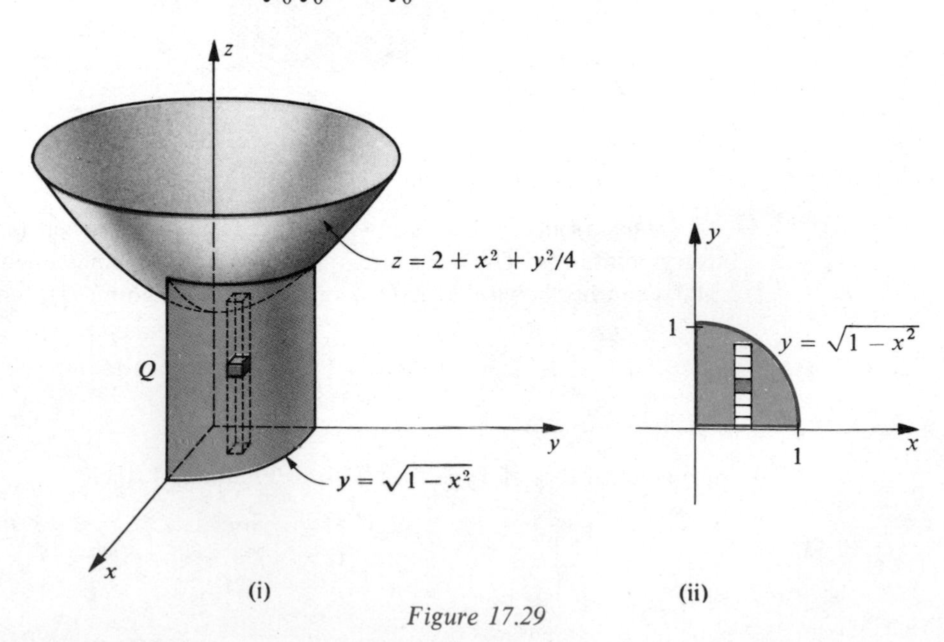

Figure 17.29

If $f(x, y, z) = 1$ throughout Q, then the triple integral of f over Q is written $\iiint_Q dV$ and its value is the volume of the region Q. It follows from Example 2 that the value of the integral

$$\int_0^1 \int_0^{\sqrt{1-x^2}} \int_0^{2+x^2+y^2/4} dz\, dy\, dx$$

is the volume of the region Q shown in (i) of Figure 17.29.

Example 3 Find the volume of the solid in the first octant which is bounded by the plane $y + z = 4$, the cylinder $y = x^2$, and the xy- and yz-planes.

Solution The solid is sketched in (i) of Figure 17.30 together with a column corresponding to a first summation in the z-direction. Note that the column extends from $z = 0$ to $z = 4 - y$. The region R in the xy-plane is shown in (ii) of the figure, together with a strip corresponding to the first (of a double) integration with respect to y. Applying (17.36) with $f(x, y, z) = 1$,

$$\begin{aligned} V &= \int_0^2 \int_{x^2}^4 \int_0^{4-y} dz\, dy\, dx \\ &= \int_0^2 \int_{x^2}^4 (4 - y)\, dy\, dx \\ &= \int_0^2 \left[4y - \frac{1}{2}y^2 \right]_{x^2}^4 dx \\ &= \int_0^2 \left(8 - 4x^2 + \frac{1}{2}x^4 \right) dx \\ &= 8x - \frac{4}{3}x^3 + \frac{1}{10}x^5 \Big]_0^2 = \frac{128}{15}. \end{aligned}$$

If we use (17.37), then the strip in (ii) of Figure 17.30 would be horizontal and

$$V = \int_0^4 \int_0^{\sqrt{y}} \int_0^{4-y} dz\, dx\, dy.$$

The reader may verify that the value of this integral is also 128/15.

For certain regions, triple integrals may be evaluated by means of an iterated triple integral in which the first integration is with respect to y or x. Thus, consider

$$Q = \{(x, y, z): a \le x \le b, h_1(x) \le z \le h_2(x), k_1(x, z) \le y \le k_2(x, z)\}$$

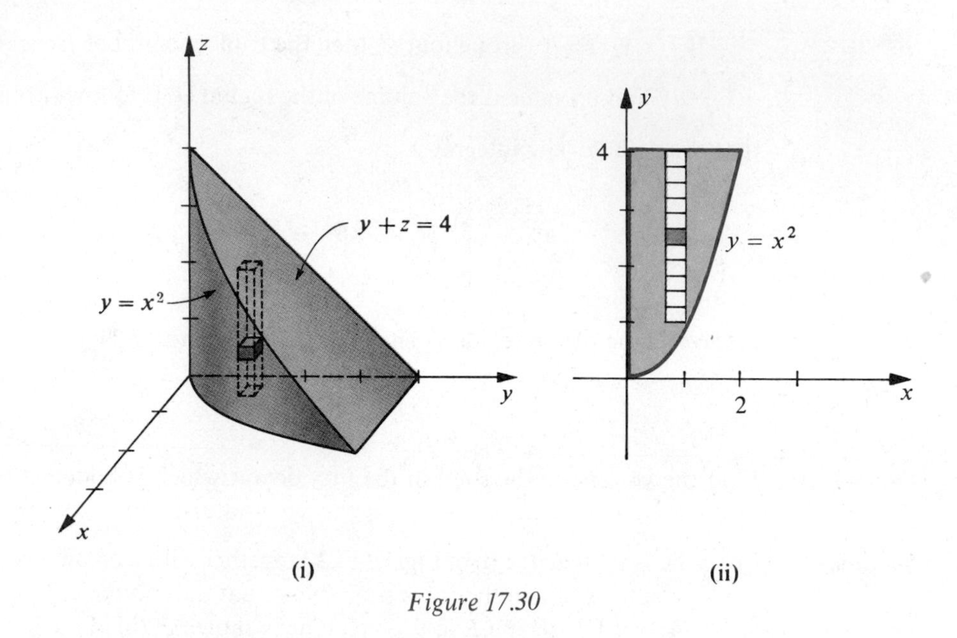

Figure 17.30

where the given functions are sufficiently well behaved. The graph of such a region is illustrated in Figure 17.31. In this case it can be shown that

(17.38)
$$\iiint_Q f(x, y, z)\, dV = \int_a^b \int_{h_1(x)}^{h_2(x)} \int_{k_1(x,z)}^{k_2(x,z)} f(x, y, z)\, dy\, dz\, dx.$$

This iterated integral should be interpreted as a limit of a triple sum obtained by first summing over a row of parallelepipeds in the direction of the y-axis from the left surface (with equation $y = k_1(x, z)$) to the right surface (with equation $y = k_2(x, z)$), as indicated in Figure 17.31. The resulting double integral is then evaluated over the region R in the xz-plane.

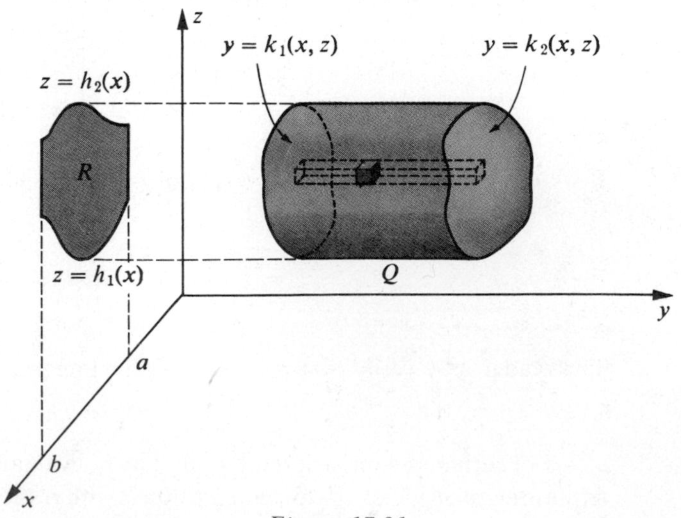

Figure 17.31

Example 4 Find the volume of the region Q bounded by the graphs of $z = 3x^2$, $z = 4 - x^2$, $y = 0$, and $z + y = 6$.

Solution As illustrated in (i) of Figure 17.32, Q lies under the cylinder $z = 4 - x^2$, over the cylinder $z = 3x^2$, to the right of the xz-plane, and to the left of the plane $z + y = 6$. Hence Q is a region of the type illustrated in Figure 17.31, where $k_1(x, z) = 0$ and $k_2(x, z) = 6 - z$. The region R in the xz-plane is sketched in (ii) of Figure 17.32. Applying (17.38),

$$V = \iiint_Q dV = \int_{-1}^{1} \int_{3x^2}^{4-x^2} \int_{0}^{6-z} dy\, dz\, dx$$

$$= \int_{-1}^{1} \int_{3x^2}^{4-x^2} (6 - z)\, dz\, dx$$

$$= \int_{-1}^{1} \left[6z - \frac{1}{2}z^2 \right]_{3x^2}^{4-x^2} dx$$

$$= \int_{-1}^{1} (16 - 20x^2 + 4x^4)\, dx = \frac{304}{15}.$$

If a different order of integration is used it is necessary to use several triple integrals. (Why?)

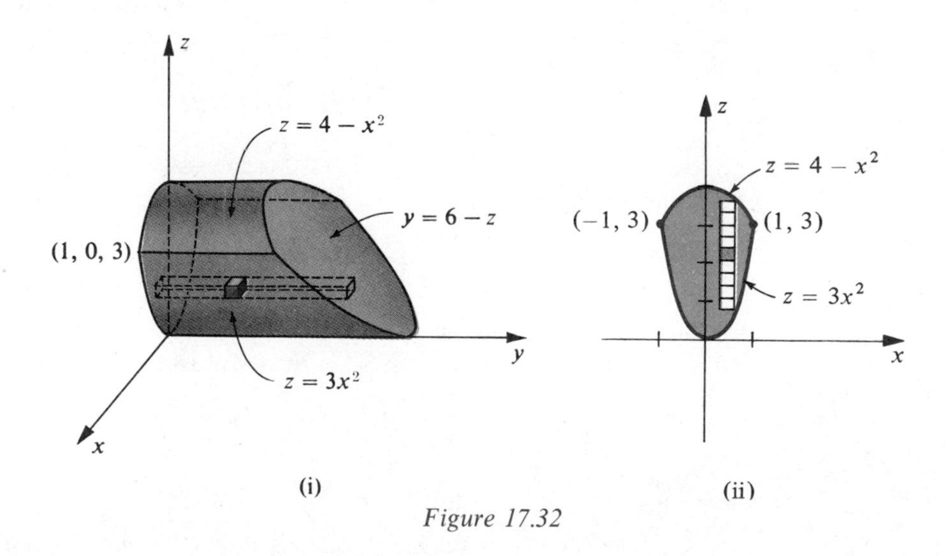

Figure 17.32

Finally, if Q is a region of the type illustrated in Figure 17.33, where the functions l_1 and l_2 have continuous first partial derivatives on a suitable region R in the yz-plane, then

$$\iiint_Q f(x, y, z)\, dV = \iint_R \left[\int_{l_1(y,z)}^{l_2(y,z)} f(x, y, z)\, dx \right] dA.$$

In the final iterated double integral, dA will be replaced by $dz\, dy$ or $dy\, dz$.

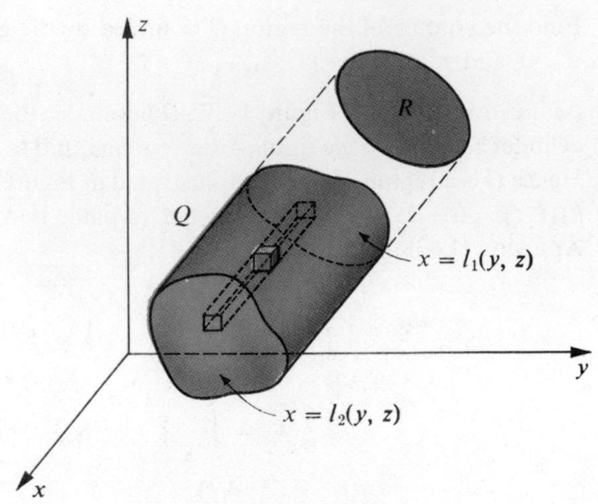

Figure 17.33

Example 5 Rework Example 3 by integrating first with respect to x.

Solution Referring to (i) of Figure 17.30, a first integration with respect to x would correspond to a summation along a row of parallelepipeds extending from the yz-plane ($x = 0$) to the cylinder ($x = \sqrt{y}$). The region R in the yz-plane is bounded by the y- and z-axes and the line $y + z = 4$. Consequently

$$V = \int_0^4 \int_0^{4-z} \int_0^{\sqrt{y}} dx\, dy\, dz.$$

The reader may verify that the value of this integral is 128/15. If we take the second integration with respect to z instead of y, then the iterated integral is

$$V = \int_0^4 \int_0^{4-y} \int_0^{\sqrt{y}} dx\, dz\, dy.$$

It is not difficult to extend the definition of multiple integrals to functions of *any* number of variables. In Exercise 28 the reader is asked to define

$$\iiiint_U f(x, y, z, t)\, dW \tag{17.39}$$

where

$$U = \{(x, y, z, t): a \le x \le b, c \le y \le d, m \le z \le n, p \le t \le q\}.$$

EXERCISES 17.6

1 Evaluate the integral of Example 1 using five orders of integration different from that used in the solution.

2 If $Q = \{(x, y, z): 1 \le x \le 2,\ -1 \le y \le 0,\ 0 \le z \le 3\}$, evaluate

$$\iiint_Q (x + 2y + 4z)\, dV$$

in six different ways.

Evaluate the integrals in Exercises 3–6.

3 $\int_0^1 \int_{1+x}^{2x} \int_z^{x+z} x\,dy\,dz\,dx$

4 $\int_1^2 \int_0^{z^2} \int_{x+z}^{x-z} z\,dy\,dx\,dz$

5 $\int_{-1}^2 \int_1^{x^2} \int_0^{x+y} 2x^2y\,dz\,dy\,dx$

6 $\int_2^3 \int_0^{3y} \int_1^{yz} (2x + y + z)\,dx\,dz\,dy$

In each of Exercises 7–10 sketch the region Q bounded by the graphs of the given equations and express $\iiint_Q f(x, y, z)\,dV$ as an iterated integral in six different ways.

7 $x + 2y + 3z = 6; x = 0, y = 0, z = 0$

8 $x^2 + y^2 = 9, z = 0, z = 2$

9 $z = 9 - 4x^2 - y^2,\ z = 0$

10 $36x^2 + 9y^2 + 4z^2 = 36$

In Exercises 11–20 sketch the region bounded by the graphs of the given equations and use a triple integral to find its volume.

11 $z + x^2 = 4,\ y + z = 4,\ y = 0,\ z = 0$

12 $x^2 + z^2 = 4, y^2 + z^2 = 4$

13 $y = 2 - z^2, y = z^2, x + z = 4, x = 0$

14 $z = 4y^2, z = 2, x = 2, x = 0$

15 $y^2 + z^2 = 1, x + y + z = 2, x = 0$

16 $z = x^2 + y^2, y + z = 2$

17 $z = 9 - x^2,\ z = 0,\ y = -1,\ y = 2$

18 $z = e^{x+y}, y = 3x, x = 2, y = 0, z = 0$

19 $z = x^2,\ z = x^3,\ y = z^2,\ y = 0$

20 $y = x^2 + z^2, z = x^2, z = 4, y = 0$

Each of the iterated integrals in Exercises 21–26 represents the volume of a region Q. Describe Q.

21 $\int_0^1 \int_{\sqrt{1-z}}^{\sqrt{4-z}} \int_2^3 dx\,dy\,dz$

22 $\int_0^1 \int_{z^3}^{\sqrt{z}} \int_0^{4-x} dy\,dx\,dz$

23 $\int_0^2 \int_{x^2}^{2x} \int_0^{x+y} dz\,dy\,dx$

24 $\int_0^1 \int_x^{3x} \int_0^{xy} dz\,dy\,dx$

25 $\int_1^2 \int_{-\sqrt{z}}^{\sqrt{z}} \int_{-\sqrt{z-x^2}}^{\sqrt{z-x^2}} dy\,dx\,dz$

26 $\int_1^4 \int_{-z}^{z} \int_{-\sqrt{z^2-y^2}}^{\sqrt{z^2-y^2}} dx\,dy\,dz$

27 Define the limit of Riemann sums for functions of three variables (see (17.2)).

28 Define (17.39).

17.7 APPLICATIONS OF TRIPLE INTEGRALS

If a solid has the shape of a three-dimensional region Q, and the density at (x, y, z) is $\rho(x, y, z)$, where ρ is continuous throughout Q, then as in (17.19) the mass is given by

(17.40)
$$M = \iiint_Q \rho(x, y, z)\, dV.$$

If a particle of mass m is at the point (x, y, z), then its moments with respect to the xy-, xz-, and yz-planes are defined as zm, ym, and xm, respectively. By employing the usual techniques involving limits of sums, we are led to define the moments of the *solid* with respect to the coordinate planes as

(17.41)
$$M_{xy} = \iiint_Q z\rho(x, y, z)\, dV$$
$$M_{xz} = \iiint_Q y\rho(x, y, z)\, dV$$
$$M_{yz} = \iiint_Q x\rho(x, y, z)\, dV.$$

For example, if $\{Q_i\}$ is a partition of Q and if T_i is the part of the solid corresponding to Q_i, then the mass of T_i may be approximated by $\rho(x_i, y_i, z_i)\,\Delta V_i$ where (x_i, y_i, z_i) is a point in Q_i. The moment of T_i with respect to the xy-plane may be approximated by $z_i\rho(x_i, y_i, z_i)\,\Delta V_i$. Summing all such moments and taking the limit leads to the first integral in (17.41). The other formulas may be motivated in similar fashion. The **center of mass** is the point $(\bar{x}, \bar{y}, \bar{z})$, where

(17.42)
$$\bar{x} = \frac{M_{yz}}{M}, \quad \bar{y} = \frac{M_{xz}}{M}, \quad \bar{z} = \frac{M_{xy}}{M}.$$

If Q is homogeneous, then the function ρ is constant and hence cancels in (17.42). Consequently the center of mass of a homogeneous solid depends only on the shape of Q. As in two dimensions, the corresponding point for geometric solids is called the **centroid**.

Example 1 A solid has the shape of a right circular cylinder of base radius a and height h. Find the center of mass if the density at a point P is directly proportional to the distance from one of the bases to P.

Solution If we introduce a coordinate system as in Figure 17.34, then the solid is bounded by the graphs of $x^2 + y^2 = a^2$, $z = 0$, and $z = h$. By hypothesis we may assume that the density at (x, y, z) is $\rho(x, y, z) = kz$ for some number k. Evidently the center of mass is on the z-axis and, therefore, it is sufficient to find $\bar{z} = M_{xy}/M$.

Moreover, by the form of ρ and the symmetry of the solid, we may calculate M and M_{xy} for the first octant portion and multiply by 4. Using (17.40),

$$\begin{aligned} M &= 4\int_0^a \int_0^{\sqrt{a^2-x^2}} \int_0^h kz\,dz\,dy\,dx \\ &= 4k\int_0^a \int_0^{\sqrt{a^2-x^2}} \frac{h^2}{2}\,dy\,dx \\ &= 2kh^2\int_0^a \sqrt{a^2-x^2}\,dx = 2kh^2\left(\frac{\pi a^2}{4}\right) = \frac{k\pi h^2 a^2}{2}. \end{aligned}$$

Next using (17.41),

$$\begin{aligned} M_{xy} &= 4\int_0^a \int_0^{\sqrt{a^2-x^2}} \int_0^h z(kz)\,dz\,dy\,dx \\ &= 4k\int_0^a \int_0^{\sqrt{a^2-x^2}} \frac{h^3}{3}\,dy\,dx \\ &= \frac{4kh^3}{3}\int_0^a \sqrt{a^2-x^2}\,dx \\ &= \frac{4kh^3}{3}\left(\frac{\pi a^2}{4}\right) = \frac{k\pi h^3 a^2}{3}. \end{aligned}$$

Finally, by (17.42),

$$\bar{z} = \frac{M_{xy}}{M} = \frac{k\pi h^3 a^2}{3}\cdot\frac{2}{k\pi h^2 a^2} = \frac{2h}{3}.$$

Hence the center of mass is on the axis of the cylinder, 2/3 of the way from the lower base.

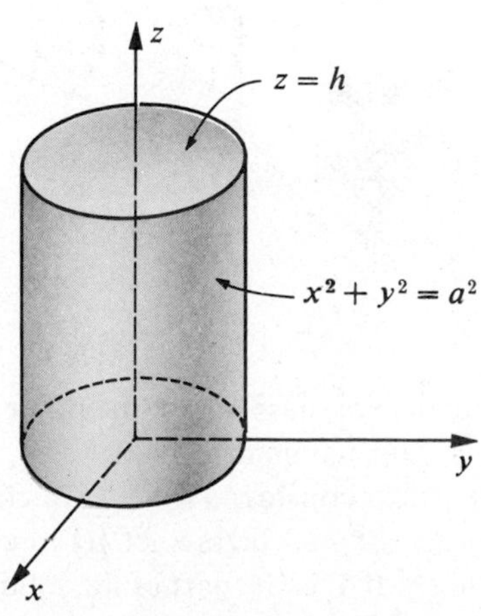

Figure 17.34

In the following example we shall merely *set up* the integrals necessary for the solution; that is, the integrals will be expressed in iterated form, but not evaluated. Indeed, the evaluations would be an almost impossible task.

Example 2 A solid has the shape of the region in the first octant bounded by the graphs of $4 - z = 9x^2 + y^2$, $y = 4x$, $z = 0$, and $y = 0$. If the density at the point $P(x, y, z)$ is proportional to the distance from the origin to P, set up the integrals needed to find $\bar{x}$.

Solution The region is sketched in Figure 17.35. By hypothesis, the density at (x, y, z) is $\rho(x, y, z) = k(x^2 + y^2 + z^2)^{1/2}$ for some k. Using (17.40) and (17.41), M and M_{yz} are given by

$$M = \int_0^{8/5} \int_{y/4}^{\sqrt{4-y^2}/3} \int_0^{4-9x^2-y^2} k(x^2 + y^2 + z^2)^{1/2}\, dz\, dx\, dy$$

$$M_{yz} = \int_0^{8/5} \int_{y/4}^{\sqrt{4-y^2}/3} \int_0^{4-9x^2-y^2} xk(x^2 + y^2 + z^2)^{1/2}\, dz\, dx\, dy.$$

By (17.42), $\bar{x} = M_{yz}/M$.

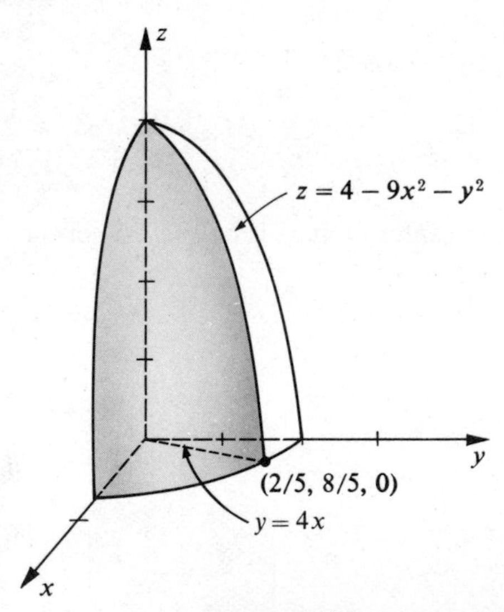

Figure 17.35

If a particle of mass m is at the point (x, y, z), then its distance to the z-axis is $(x^2 + y^2)^{1/2}$ and its **moment of inertia I_z with respect to the z-axis** is defined as $(x^2 + y^2)m$. Next consider a solid T which has the shape of a region Q and density $\rho(x, y, z)$ at (x, y, z). As before, let $\{Q_i\}$ be a partition of Q and choose any point (x_i, y_i, z_i) in Q_i. If T_i is the part of the solid corresponding to Q_i, then the moment of inertia of T_i with respect to the z-axis may be approximated by

$(x_i^2 + y_i^2)\rho(x_i, y_i, z_i)\,\Delta V_i$. Taking the limit of a sum of such terms leads to the following definition of the moment of inertia of T with respect to the z-axis:

(17.43)
$$I_z = \iiint\limits_Q (x^2 + y^2)\rho(x, y, z)\,dV.$$

Similarly, the moments of inertia with respect to the x- and y-axis, respectively, are

(17.44)
$$I_x = \iiint\limits_Q (y^2 + z^2)\rho(x, y, z)\,dV, \quad I_y = \iiint\limits_Q (x^2 + z^2)\rho(x, y, z)\,dV.$$

If a solid of mass M has moment of inertia I with respect to a line, then the **radius of gyration** is, by definition, the number d such that $I = Md^2$. This formula implies that the radius of gyration is the distance from the line at which all the mass could be concentrated without changing the moment of inertia of the solid.

Example 3 Find the moment of inertia and radius of gyration, with respect to the axis of symmetry, of the solid described in Example 1.

Solution The solid is sketched in Figure 17.34, where $\rho(x, y, z) = kz$. Using (17.43) and symmetry,

$$\begin{aligned}
I_z &= 4\int_0^a \int_0^{\sqrt{a^2 - x^2}} \int_0^h (x^2 + y^2)kz\,dz\,dy\,dx \\
&= 4k\int_0^a \int_0^{\sqrt{a^2 - x^2}} (x^2 + y^2)\frac{h^2}{2}\,dy\,dx \\
&= 2kh^2\int_0^a \left[x^2 y + \frac{y^3}{3}\right]_0^{\sqrt{a^2 - x^2}} dx \\
&= 2kh^2\int_0^a [x^2\sqrt{a^2 - x^2} + \tfrac{1}{3}\sqrt{(a^2 - x^2)^3}]\,dx.
\end{aligned}$$

The last integral may be evaluated by using a trigonometric substitution or a table of integrals. This results in $I_z = k\pi h^2 a^4/4$. If d is the radius of gyration, then $d^2 = I_z/M$. Using the value for M found in Example 1 we obtain

$$d^2 = \frac{k\pi h^2 a^4}{4} \cdot \frac{2}{k\pi h^2 a^2} = \frac{a^2}{2}.$$

Hence $d = a/\sqrt{2} \approx 0.7a$; that is, the radius of gyration is a distance from the axis of the cylinder which is approximately 7/10 of the radius of the cylinder.

Example 4 A homogeneous solid has the shape of the region Q bounded by the graphs of $4x^2 - y^2 + z^2 = 0$ and $y = 3$. Set up a triple integral for finding the moment of inertia with respect to the y-axis.

Solution The region is part of an elliptic cone and is sketched in Figure 17.36. Note that the trace of the cone on the xy-plane is the pair of lines $x = \pm y/2$. If we denote the constant density by k, then applying (17.44),

$$I_y = \int_0^3 \int_{-y/2}^{y/2} \int_{-\sqrt{y^2-4x^2}}^{\sqrt{y^2-4x^2}} (x^2 + z^2)k\, dz\, dx\, dy.$$

We could also find I_y by multiplying by 4 the moment of inertia of that part of the solid which lies in the first octant. Thus

$$I_y = 4\int_0^3 \int_0^{y/2} \int_0^{\sqrt{y^2-4x^2}} (x^2 + z^2)k\, dz\, dx\, dy.$$

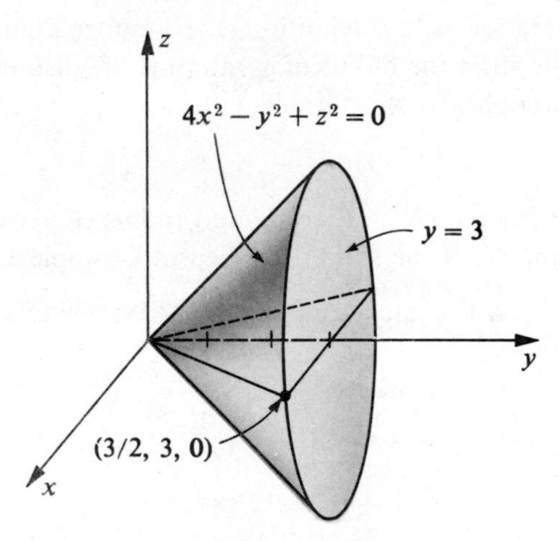

Figure 17.36

EXERCISES 17.7

In each of Exercises 1–4 set up, but do not evaluate, the integrals needed to find the centroid of the given solid.

1 The solid in the first octant bounded by the coordinate planes and the graphs of $z = 9 - x^2$ and $2x + y = 6$

2 The solid bounded by the graphs of $z = x^2$, $y = x^2$, $y = x^3$, and $z = 0$

3 The solid bounded by the graphs of $y = 4x^2 + 9z^2$, $z = x^2$, $z = 1$, and $y = 0$

4 The solid in the first octant bounded by the coordinate planes and the graphs of $x^2 + z^2 = 9$ and $y^2 + z^2 = 9$

5 The density at a point P in a cubical solid of edge a is directly proportional to the square of the distance from P to a fixed corner of the cube. Find the center of mass.

6 Let Q be the tetrahedron bounded by the coordinate planes and the plane $2x + 5y + z = 10$. Find the center of mass if the density at the point $P(x, y, z)$ is directly proportional to the distance from the xz-plane to P.

7 Let Q be the solid bounded by the paraboloid $x = y^2 + 4z^2$ and the plane $x = 4$. If the density at (x, y, z) is given by $\rho(x, y, z) = x^2 + z^2$, set up, but do not evaluate, the integrals necessary for finding the center of mass.

8 Let Q be the solid bounded by the hyperboloid $y^2 - x^2 - z^2 = 1$ and the plane $y = 2$. If the density at (x, y, z) is given by $\rho(x, y, z) = x^2y^2z^2$, set up, but do not evaluate, the integrals necessary for finding the center of mass.

9 Find the moment of inertia with respect to a line through one of its edges of a homogeneous cube of volume a^3. What is the radius of gyration?

10 Find the moment of inertia with respect to the z-axis of the tetrahedron described in Exercise 6. What is the radius of gyration?

In each of Exercises 11–16 set up the integral for the moment of inertia with respect to the z-axis of the indicated solid.

11 The solid bounded by the graphs of $z = 36 - 4x^2 - 9y^2$ and $z = 0$; $\rho(x, y, z) = z$

12 The cylinder bounded by the graphs of $4x^2 + y^2 = a^2$, $z = 0$, and $z = h$; $\rho(x, y, z) = 1 + z$

13 A sphere of radius a with center at the origin; $\rho(x, y, z) = x^2 + y^2 + z^2$

14 The cone bounded by the graphs of $x^2 + 9y^2 - z^2 = 0$ and $z = 36$; $\rho(x, y, z) = x^2 + y^2$

15 The homogeneous tetrahedron bounded by the coordinate planes and the graph of $x/a + y/b + z/c = 1$, where a, b, and c are positive

16 The homogeneous solid bounded by the ellipsoid $\dfrac{x^2}{a^2} + \dfrac{y^2}{b^2} + \dfrac{z^2}{c^2} = 1$

17.8 TRIPLE INTEGRALS IN CYLINDRICAL AND SPHERICAL COORDINATES

Triple integrals may sometimes be expressed in terms of cylindrical coordinates. The simplest case occurs if a function of r, θ, and z is continuous throughout a region of the form

$$Q = \{(r, \theta, z) : a \le r \le b, c \le \theta \le d, k \le z \le l\}.$$

We first divide Q into subregions $Q_1, Q_2, \ldots, Q_n$ which have the same shape as Q by using graphs of equations $r = a_i$, $\theta = c_i$, and $z = k_i$, where a_i, c_i, and k_i are numbers in the intervals $[a, b]$, $[c, d]$, and $[k, l]$, respectively. A typical subregion Q_i, appropriately labeled, is sketched in Figure 17.37, where $\bar{r}_i$ is the average radius of the base of Q_i. The volume ΔV_i of Q_i is the product of the area of the base $\bar{r}_i \Delta r_i \Delta\theta_i$ and the altitude Δz_i. Thus

$$\Delta V_i = \bar{r}_i \Delta r_i \Delta\theta_i \Delta z_i.$$

If (r_i, θ_i, z_i) is any point in Q_i, then the triple integral of f over Q is given by

$$\iiint_Q f(r, \theta, z)\, dV = \lim_{\|P\| \to 0} \sum_i f(r_i, \theta_i, z_i)\, \Delta V_i$$

where the norm $\|P\|$ is the length of the longest diagonal of the Q_i. Furthermore, it can be proved that

$$\iiint\limits_Q f(r, \theta, z)\, dV = \int_k^l \int_c^d \int_a^b f(r, \theta, z) r\, dr\, d\theta\, dz.$$

There are five other possible orders of integration for this iterated integral.

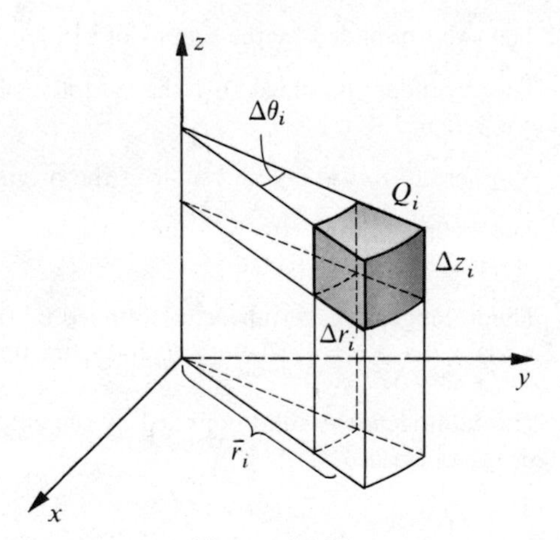

Figure 17.37

Triple integrals in cylindrical coordinates may be defined over more complicated regions by employing inner partitions. For example, it can be shown that if R is a polar region of the type discussed in Section 17.5 and if

$$Q = \{(r, \theta, z) : (r, \theta) \text{ is in } R \text{ and } k_1(r, \theta) \leq z \leq k_2(r, \theta)\}$$

where the functions k_1 and k_2 have continuous first partial derivatives throughout R, then

$$\iiint\limits_Q f(r, \theta, z)\, dV = \iint\limits_R \left[\int_{k_1(r, \theta)}^{k_2(r, \theta)} f(r, \theta, z)\, dz \right] dA.$$

In particular, if R is of the type illustrated in Figure 17.21, then

(17.45) $$\iiint\limits_Q f(r, \theta, z)\, dV = \int_\alpha^\beta \int_{g_1(\theta)}^{g_2(\theta)} \int_{k_1(r, \theta)}^{k_2(r, \theta)} f(r, \theta, z)\, r\, dz\, dr\, d\theta.$$

The iterated integral in (17.45) may be interpreted as a limit of a triple sum, in which the first summation takes place along a column of the subregions Q_i from the lower surface (with equation $z = k_1(r, \theta)$) to the upper surface (with equation $z = k_2(r, \theta)$) (see Figure 17.38). The second summation extends over a row of these columns, where θ is held fixed and r is allowed to vary. At this stage we have summed over a slice of Q as illustrated in Figure 17.38. Finally, we sweep out Q by letting θ vary from α to β.

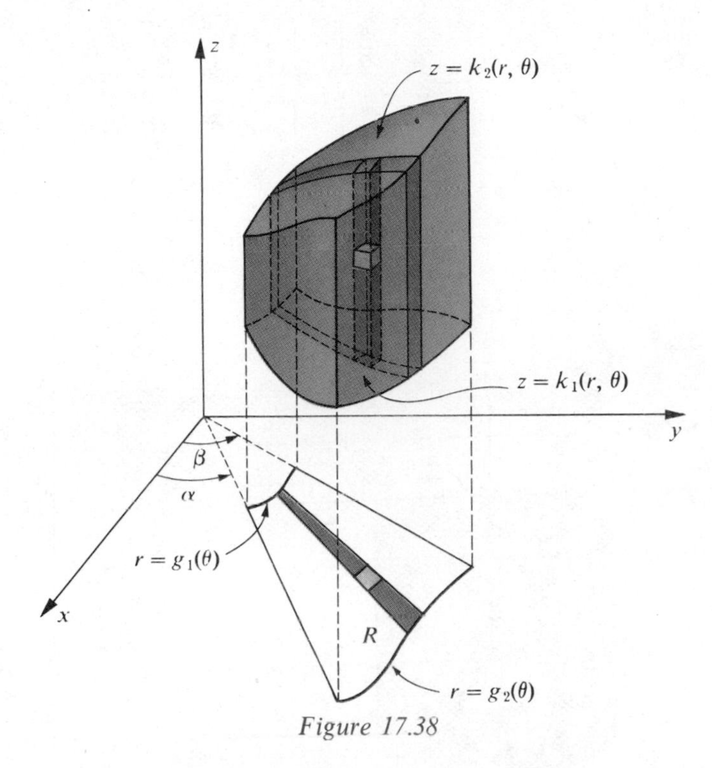

Figure 17.38

Example 1 A solid has the shape of the region Q which lies inside the cylinder $r = a$, within the sphere $r^2 + z^2 = 4a^2$ and above the xy-plane. Find the center of mass and the moment of inertia I_z if the density at a point P is directly proportional to the distance from the xy-plane to P.

Solution The region Q is sketched in Figure 17.39 together with a column which corresponds to a first summation in the direction of the z-axis. Since the density at the point $P(r, \theta, z)$ is given by kz for some (fixed) number k, the mass M may be found by applying (17.45) with $f(r, \theta, z) = kz$. Thus

$$M = \int_0^{2\pi} \int_0^a \int_0^{\sqrt{4a^2 - r^2}} kzr\, dz\, dr\, d\theta$$

$$= \frac{k}{2} \int_0^{2\pi} \int_0^a z^2 r \Big]_0^{\sqrt{4a^2 - r^2}} dr\, d\theta$$

$$= \frac{k}{2} \int_0^{2\pi} \int_0^a (4a^2 r - r^3)\, dr\, d\theta$$

$$= \frac{k}{2} \int_0^{2\pi} 2a^2 r^2 - \frac{1}{4} r^4 \Big]_0^a d\theta$$

$$= \frac{7a^4 k}{8} \int_0^{2\pi} d\theta = \frac{7a^4 \pi k}{4}.$$

To find I we employ (17.43) and use cylindrical coordinates, obtaining

$$I_z = \int_0^{2\pi}\int_0^a\int_0^{\sqrt{4a^2-r^2}} r^2kzr\,dz\,dr\,d\theta$$

$$= \frac{k}{2}\int_0^{2\pi}\int_0^a r^3z^2\Big]_0^{\sqrt{4a^2-r^2}}\,dr\,d\theta$$

$$= \frac{k}{2}\int_0^{2\pi}\int_0^a (4a^2r^3 - r^5)\,dr\,d\theta = \frac{5a^6\pi k}{6}.$$

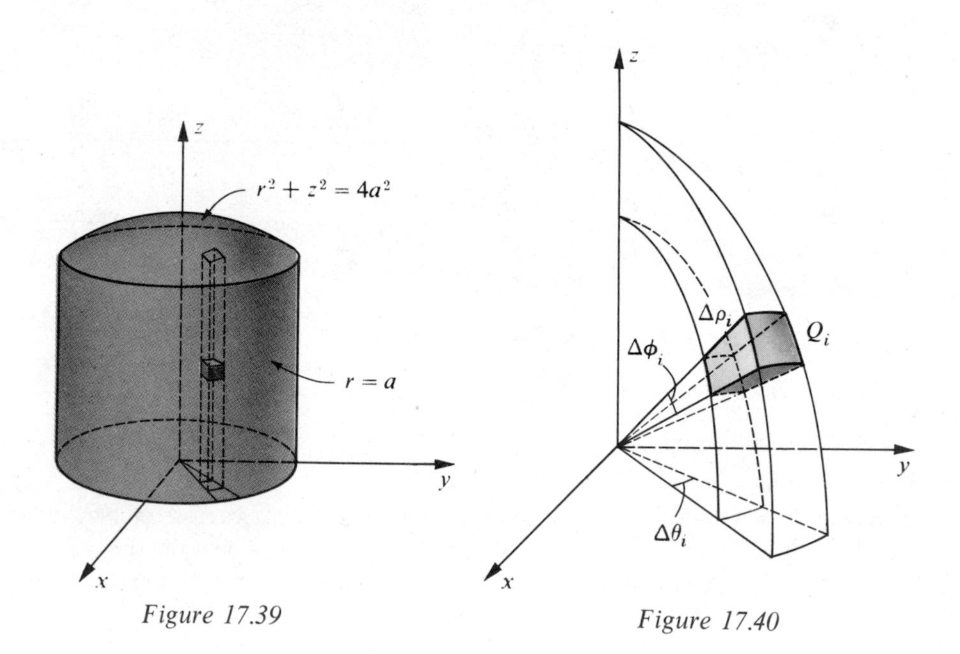

Figure 17.39

Figure 17.40

Triple integrals in spherical coordinates may also be considered. Suppose a function f of ρ, θ, and ϕ is continuous throughout a region of the form

$$Q = \{(\rho, \theta, \phi): a \le \rho \le b, c \le \theta \le d, k \le \phi \le l\}.$$

We again divide Q into subregions $Q_1, Q_2, \ldots, Q_n$ having the same shape as Q by means of graphs of equations $\rho = a_i, \theta = c_i$, and $\phi = k_i$. A typical subregion Q_i is sketched in Figure 17.40. It is left as an exercise to show that if ΔV_i is the volume of Q_i, then

(17.46) $$\Delta V_i \approx \rho_i^2 \sin\phi_i \Delta\rho_i \Delta\phi_i \Delta\theta_i$$

where $(\rho_i, \theta_i, \phi_i)$ is any point in Q_i and the increments are those illustrated in Figure 17.40. If the triple integral of f over Q is defined as a limit of a sum in the usual way, then it can be proved that

(17.47)
$$\iiint_Q f(\rho, \theta, \phi)\, dV = \int_c^d \int_k^l \int_a^b f(\rho, \theta, \phi)\rho^2 \sin\phi \, d\rho \, d\phi \, d\theta.$$

There are five other possible orders of integration. Spherical coordinates may also be employed over more complicated regions by using inner partitions. In this case the limits of integration in (17.47) must be suitably adjusted.

Example 2 Find the volume and the centroid of the region Q which is bounded above by the sphere $\rho = a$ and below by the cone $\phi = m$ where $0 < m < \pi/2$.

Solution The region Q is sketched in Figure 17.41. The volume V is given by

$$\begin{aligned} V &= \int_0^{2\pi} \int_0^m \int_0^a \rho^2 \sin\phi \, d\rho \, d\phi \, d\theta \\ &= \int_0^{2\pi} \int_0^m \frac{a^3}{3} \sin\phi \, d\phi \, d\theta \\ &= \frac{a^3}{3} \int_0^{2\pi} \Big[-\cos\phi \Big]_0^m d\theta \\ &= \frac{a^3}{3} \int_0^{2\pi} (1 - \cos m)\, d\theta \\ &= \frac{2\pi a^3}{3}(1 - \cos m). \end{aligned}$$

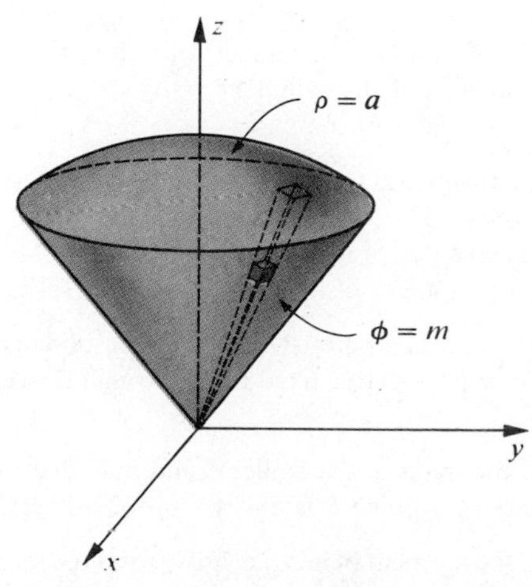

Figure 17.41

By symmetry, the centroid is on the z-axis. If (x, y, z) are rectangular coordinates of a point, then by (14.52) $z = \rho \cos \phi$. Consequently,

$$\begin{aligned}
M_{xy} &= \iiint_Q z\, dV = \int_0^{2\pi} \int_0^m \int_0^a \rho \cos \phi \rho^2 \sin \phi \, d\rho \, d\phi \, d\theta \\
&= \int_0^{2\pi} \int_0^m \frac{a^4}{4} \sin \phi \cos \phi \, d\phi \, d\theta \\
&= \frac{a^4}{4} \int_0^{2\pi} \frac{1}{2} \sin^2 \phi \Big]_0^m d\theta \\
&= \frac{a^4}{8} \sin^2 m \int_0^{2\pi} d\theta = \frac{\pi a^4}{4} \sin^2 m.
\end{aligned}$$

The centroid, in rectangular coordinates, is $(0, 0, \bar{z})$, where

$$\bar{z} = \frac{M_{xy}}{V} = \frac{3}{8} a(1 + \cos m).$$

The letter ρ used in spherical coordinates should not be confused with the density function. For problems which involve spherical coordinates *and* a density function, a different symbol should be used to denote density.

EXERCISES 17.8

Use cylindrical coordinates in Exercises 1–10.

1 Find the volume and the centroid of the solid bounded by the graphs of $z = x^2 + y^2$, $x^2 + y^2 = 4$, and $z = 0$.

2 Find the volume and the centroid of the solid bounded by the graphs of $x^2 + y^2 - z^2 = 0$ and $x^2 + y^2 = 4$.

3 Find the moment of inertia of a homogeneous right circular cylinder of altitude h and radius of base a with respect to each of the following.
(a) The axis of the cylinder
(b) A diameter of the base

4 A homogeneous solid is bounded by the graphs of $z = r$ and $z = r^2$. Find the following.
(a) The center of mass
(b) The moment of inertia with respect to the z-axis

5 The density at a point P of a spherical solid of radius a is directly proportional to the distance from P to a fixed line l through the center of the solid. Find the mass of the solid.

6 Find the mass of the conical solid bounded by the graphs of $z = r$ and $z = 4$ if the density at a point P is directly proportional to the distance from the z-axis to P.

7 Find the moment of inertia with respect to l for the solid described in Exercise 5. What is the radius of gyration?

8 Find the moment of inertia with respect to the z-axis for the solid described in Exercise 6. What is the radius of gyration?

9 Let $Q = \{(x, y, z): 1 \le z \le 5 - x^2 - y^2, x^2 + y^2 \ge 1\}$. If the density at the point $P(x, y, z)$ is directly proportional to the distance from the xy-plane to P, find the mass and the center of mass of Q.

10 Find the mass and center of mass of the solid which lies inside both the cylinder $x^2 + y^2 - 2y = 0$ and the sphere $x^2 + y^2 + z^2 = 4$ if the density at the point $P(x, y, z)$ is directly proportional to the distance from the xy-plane to P.

Use spherical coordinates in Exercises 11–18.

11 Find the mass and center of mass of a solid hemisphere of radius a if the density at a point P is directly proportional to the distance from the center of the base to P.

12 Rework Exercise 2.

13 Find the moment of inertia with respect to the axis for the hemisphere in Exercise 11.

14 Find the moment of inertia with respect to a diameter of the base of a homogeneous solid hemisphere of radius a.

15 Find the volume of the solid which lies above the cone $z^2 = x^2 + y^2$ and inside the sphere $x^2 + y^2 + z^2 = 4z$.

16 Find the volume of the solid which lies outside the cone $z^2 = x^2 + y^2$ and inside the sphere $x^2 + y^2 + z^2 = 1$.

17 Find the mass of the solid which lies outside the sphere $\rho = 1$ and inside the sphere $\rho = 2$ if the density at a point P is directly proportional to the square of the distance from the center of the spheres to P.

18 A homogeneous spherical shell has inner radius a and outer radius b. Find its moment of inertia with respect to a line through the center.

19 Verify (17.46) by treating Q_i as a rectangular parallelepiped.

20 Interpret (17.47) as a limit of a triple sum.

17.9 SURFACE AREA

In Chapter 13 we derived formulas for the area of a surface of revolution (see (13.22) and (13.25)). We shall now discuss a method for finding areas of surfaces of the type illustrated in Figure 17.4. Suppose that $f(x, y) \ge 0$ throughout a region R in the xy-plane and that f has continuous first partial derivatives on R. Let S denote the part of the graph of f whose projection on the xy-plane is R. To simplify the discussion we shall assume that no normal vector to S is parallel to the xy-plane. Our objective is to define the area A of S and to find a formula which can be used to calculate A.

Let $P = \{R_i\}$ be an inner partition of R, where the dimensions of the rectangle R_i are Δx_i and Δy_i. We choose a point $(x_i, y_i, 0)$ in each R_i and let $B_i(x_i, y_i, f(x_i, y_i))$ be the corresponding point on S. Next consider the tangent plane to S at B_i. Let ΔT_i and ΔS_i be the areas of the regions on the tangent plane and S, respectively, obtained by projecting R_i vertically upward (see Figure 17.42). If $\|P\|$ is small, then ΔT_i is an approximation to ΔS_i, and $\Sigma_i \Delta T_i$ is an approximation to the area A of S. Since this approximation should improve as $\|P\|$ approaches 0, we define A by

(17.48)
$$A = \lim_{\|P\| \to 0} \sum_i \Delta T_i.$$

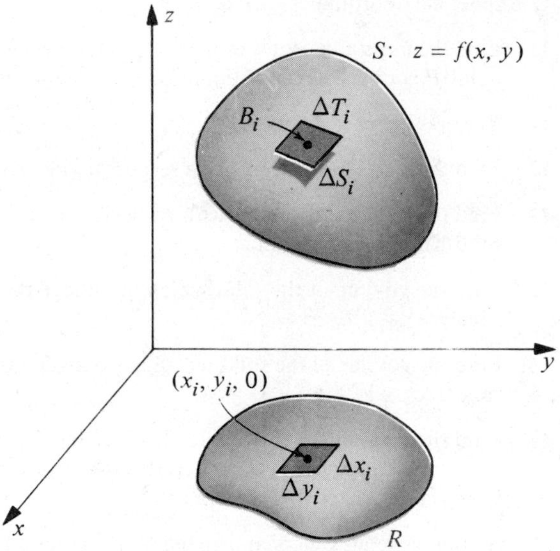

Figure 17.42

In order to find an integral formula for A, let us choose $(x_i, y_i, 0)$ as the corner of R_i closest to the origin. Let $\mathbf{a}$ and $\mathbf{b}$ be the vectors with initial point $B_i(x_i, y_i, f(x_i, y_i))$ which are tangent to the traces of S on the planes $y = y_i$ and $x = x_i$, respectively, as illustrated in the isolated view shown in Figure 17.43. Using the geometric interpretation for partial derivatives given in Section 16.3, the slopes of the lines determined by $\mathbf{a}$ and $\mathbf{b}$ in these planes are $f_x(x_i, y_i)$ and $f_y(x_i, y_i)$, respectively. It follows that

(17.49)
$$\begin{aligned} \mathbf{a} &= \Delta x_i \mathbf{i} + f_x(x_i, y_i)\,\Delta x_i \mathbf{k} \\ \mathbf{b} &= \Delta y_i \mathbf{j} + f_y(x_i, y_i)\,\Delta y_i \mathbf{k}. \end{aligned}$$

The area ΔT_i of the parallelogram determined by $\mathbf{a}$ and $\mathbf{b}$ is $|\mathbf{a} \times \mathbf{b}|$. Since

$$\mathbf{a} \times \mathbf{b} = \begin{vmatrix} \mathbf{i} & \mathbf{j} & \mathbf{k} \\ \Delta x_i & 0 & f_x(x_i, y_i)\,\Delta x_i \\ 0 & \Delta y_i & f_y(x_i, y_i)\,\Delta y_i \end{vmatrix}$$

we have

(17.50)
$$\mathbf{a} \times \mathbf{b} = -f_x(x_i, y_i)\,\Delta x_i \Delta y_i \mathbf{i} - f_y(x_i, y_i)\,\Delta x_i \Delta y_i \mathbf{j} + \Delta x_i \Delta y_i \mathbf{k}.$$

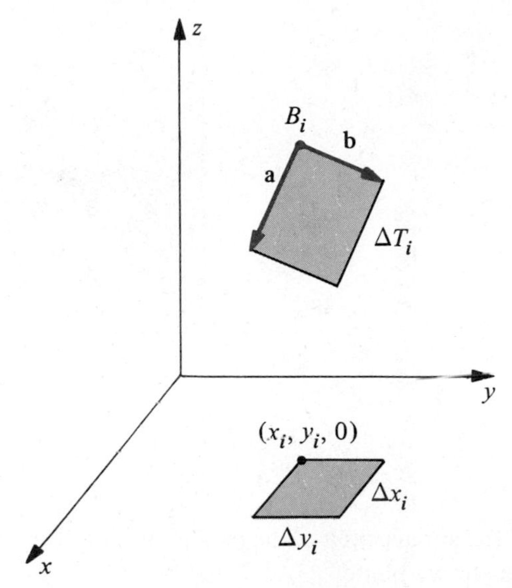

Figure 17.43

Consequently

(17.51)
$$\Delta T_i = |\mathbf{a} \times \mathbf{b}| = \sqrt{(f_x(x_i, y_i))^2 + (f_y(x_i, y_i))^2 + 1}\ \Delta x_i \Delta y_i$$

where $\Delta x_i \Delta y_i = \Delta A_i$. If, as in (17.48), we take the limit of a sum of the ΔT_i and apply the definition of the double integral, we obtain

(17.52)
$$A = \iint_R \sqrt{(f_x(x, y))^2 + (f_y(x, y))^2 + 1}\, dA.$$

This formula may also be used if $f(x, y) \leq 0$ on R.

Example 1 Let R be the triangular region in the xy-plane with vertices $(0, 0, 0)$, $(0, 1, 0)$, and $(1, 1, 0)$. Find the surface area of that part of the graph of $z = 3x + y^2$ which lies over R.

Solution The region R is bounded by the graphs of $y = x$, $x = 0$, and $y = 1$ (see Figure 17.44). Letting $f(x, y) = 3x + y^2$ and applying (17.52) gives us

$$\begin{aligned} A &= \iint_R \sqrt{3^2 + (2y)^2 + 1}\, dA \\ &= \int_0^1 \int_0^y (10 + 4y^2)^{1/2}\, dx\, dy \\ &= \int_0^1 (10 + 4y^2)^{1/2} x \Big]_0^y\, dy \\ &= \int_0^1 (10 + 4y^2)^{1/2} y\, dy \\ &= \frac{1}{12}(10 + 4y^2)^{3/2} \Big]_0^1 = \frac{14^{3/2} - 10^{3/2}}{12}. \end{aligned}$$

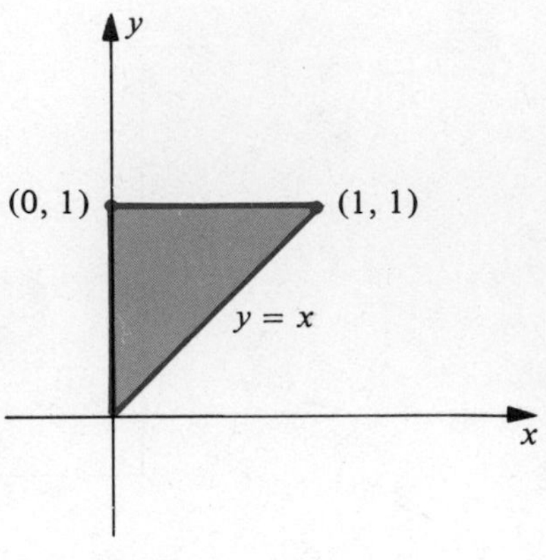

Figure 17.44

Example 2 Find the surface area of the part of the graph of $z = 9 - x^2 - y^2$ which lies above or on the xy-plane.

Solution The graph is sketched in Figure 16.5. By (17.52),

$$A = \iint_R \sqrt{4x^2 + 4y^2 + 1}\, dx\, dy$$

where R is the region in the xy-plane bounded by the circle $x^2 + y^2 = 9$. If we use polar coordinates to evaluate the double integral, then

$$A = \int_0^{2\pi} \int_0^3 (4r^2 + 1)^{1/2} r\, dr\, d\theta.$$

It is left to the reader to show that $A = \pi(37^{3/2} - 1)/6$.

Formulas similar to (17.52) may be stated if the surface S has suitable projections on the yz- or xz-planes. Thus, if S is the graph of an equation $y = g(x, z)$ and if the projection on the xz-plane is R_1, then

$$A = \iint_{R_1} \sqrt{(g_x(x, z))^2 + (g_z(x, z))^2 + 1}\, dx\, dz.$$

Another formula of this type may be stated if S is given by $x = h(y, z)$.

EXERCISES 17.9

1 Find the surface area of that part of the graph of $z = y + x^2/2$ which lies over the square region in the xy-plane having vertices $(0, 0, 0)$, $(1, 0, 0)$, $(1, 1, 0)$, and $(0, 1, 0)$.

2 Find the area of the region on the plane $z = y + 1$ which lies inside the cylinder $x^2 + y^2 = 1$.

3 Find the area of the region on the plane $x/a + y/b + z/c = 1$ cut out by the cylinder $x^2 + y^2 = d^2$, where a, b, c, and d are positive.

4 Set up an integral for finding the surface area of that part of the sphere $x^2 + y^2 + z^2 = 4$ which lies over the square region in the xy-plane having vertices $(1, 1, 0)$, $(1, -1, 0)$, $(-1, 1, 0)$, and $(-1, -1, 0)$.

5 Find the surface area of that part of the sphere $x^2 + y^2 + z^2 = a^2$ which lies inside the cylinder $x^2 + y^2 - ay = 0$.

6 Find the surface area of that part of the cylinder $y^2 + z^2 = a^2$ which lies inside the cylinder $x^2 + y^2 = a^2$.

7 Find the surface area of the part of the paraboloid $z = x^2 + y^2$ which is cut off by the plane $z = 1$.

8 Let R be the triangular region in the xy-plane with vertices $(0, 0, 0)$, $(0, 2, 0)$, and $(2, 2, 0)$. Find the surface area of that part of the graph of $z = y^2$ which lies over R.

9 Find the surface area of the part of the graph of $z = xy$ which is inside the cylinder $x^2 + y^2 = 4$.

10 Find the surface area of that part of the graph of $z = x^2 - y^2$ which lies in the first octant and is inside the cylinder $x^2 + y^2 = 1$.

17.10 REVIEW

Concepts

Define or discuss each of the following

1 Inner partitions

2 Riemann sums

3 Double integral

4 Iterated double integral

5 Evaluation of double integrals

6 Areas and volumes using double integrals

7 Center of mass of a lamina

8 Moments of inertia of a lamina

9 Radius of gyration

10 Double integrals in polar coordinates

11 Triple integral

12 Iterated triple integral

13 Evaluation of triple integrals

14 Applications of triple integrals

15 Triple integrals in cylindrical and spherical coordinates

16 Surface area

Exercises

Evaluate the integrals in Exercises 1–6.

1 $\int_{-1}^{0} \int_{x+1}^{x^3} (x^2 - 2y)\, dy\, dx$

2 $\int_{1}^{2} \int_{1}^{\ln y} \frac{1}{y}\, dx\, dy$

3 $\int_{0}^{3} \int_{r}^{r^2+1} r\, d\theta\, dr$

4 $\int_{2}^{0} \int_{0}^{z^2} \int_{x}^{z} (x + z)\, dy\, dx\, dz$

5 $\int_{0}^{2} \int_{\sqrt{y}}^{1} \int_{z^2}^{y} xy^2z^3\, dx\, dz\, dy$

6 $\int_{0}^{\pi/2} \int_{0}^{\pi/4} \int_{0}^{a\cos\phi} \rho^2 \sin\phi\, d\rho\, d\phi\, d\theta$

In each of Exercises 7–10 express $\iint_R f(x, y)\, dA$ as an iterated integral if R is the region bounded by the graphs of the given equations.

7 $x^2 - y^2 = 4, x = 4$

8 $x^2 - y^2 = 4, y = 4, y = 0$

9 $y^2 = 4 + x, y^2 = 4 - x$

10 $y = -x^2 + 4, y = 3x^2$

Each of the integrals in Exercises 11 and 12 represents the area of a region R in the xy-plane. Describe R.

11 $\int_{-1}^{1} \int_{e^y}^{y^3} dx\, dy$

12 $\int_{-1}^{0} \int_{x}^{-x^2} dy\, dx$

In Exercises 13 and 14 reverse the order of integration and evaluate the resulting integral.

13 $\int_{0}^{3} \int_{y^2}^{9} ye^{-x^2}\, dx\, dy$

14 $\int_{0}^{1} \int_{x}^{\sqrt{x}} e^{x/y}\, dy\, dx$

In Exercises 15 and 16 find the mass and the center of mass of the lamina which has the shape of the region bounded by the graphs of the given equations and having the indicated density.

15 $y = x, y = 2x, x = 3$; density at the point $P(x, y)$ is directly proportional to the distance from the y-axis to P.

16 $y^2 = x, x = 4$; density at the point $P(x, y)$ is directly proportional to the distance from the line with equation $x = -1$ to P.

17 A lamina has the shape of the region which lies inside the graph of $r = 2 + \sin\theta$ and outside the graph of $r = 1$. Find the mass if the density at the point $P(r, \theta)$ is inversely proportional to the distance from the pole to P.

18 Find the area of the region bounded by the polar axis and the graphs of $r = e^\theta$ and $r = 2$ from $\theta = 0$ to $\theta = \ln 2$.

19 Use polar coordinates to evaluate

$$\int_{-a}^{0} \int_{-\sqrt{a^2 - x^2}}^{0} \sqrt{x^2 + y^2}\, dy\, dx.$$

20 Find I_x, I_y, and I_O for the lamina bounded by the graphs of $y = x^2$ and $y = x^3$ if the density at the point $P(x, y)$ is directly proportional to the distance from the y-axis to P.

21 A homogeneous lamina has the shape of a right triangle with sides of lengths a, b, and $\sqrt{a^2 + b^2}$. Find the moment of inertia and the radius of gyration with respect to a line along the side of length a.

22 A lamina has the shape of the region between concentric circles of radii a and b, where $a < b$. If the density at a point P is directly proportional to the distance from the center to P, use polar coordinates to find the moment of inertia with respect to a line through the center.

23 Find the volume of the solid which lies under the graph of $z = xy^2$ and over the rectangle in the xy-plane with vertices $(1, 1, 0)$, $(2, 1, 0)$, $(1, 3, 0)$, and $(2, 3, 0)$.

24 Express $\iiint_Q f(x, y, z)\, dV$ as an iterated integral in six different ways, where Q is bounded by the graphs of $y = x^2 + 4z^2$ and $y = 4$.

25 Use triple integrals to find the volume and centroid of the solid bounded by the graphs of $z = x^2$, $z = 4$, $y = 0$, and $y + z = 4$.

26 Set up a triple integral for the moment of inertia with respect to the z-axis of the solid bounded by the graphs of $z = 9x^2 + y^2$ and $z = 9$ if the density at the point $P(x, y, z)$ is inversely proportional to the square of the distance from the point $(0, 0, -1)$ to P.

27 Set up a triple integral for the moment of inertia with respect to the y-axis of the solid bounded by the graphs of $x^2 - y^2 + z^2 = 1$, $y = 0$, and $y = 4$ if the density at the point $P(x, y, z)$ is directly proportional to the distance from the y-axis to P.

28 A homogeneous solid is bounded by the graphs of $z = 9 - x^2 - y^2$, $x^2 + y^2 = 4$, and $z = 0$. Use cylindrical coordinates to find the following.
(a) The mass
(b) The center of mass
(c) The moment of inertia with respect to the z-axis

29 A solid has the shape of a sphere of radius a. Use spherical coordinates to find the mass if the density at a point P is directly proportional to the distance from the center to P.

30 Find the surface area of that part of the cone $z = (x^2 + y^2)^{1/2}$ which is inside the cylinder $x^2 + y^2 = 4x$.

Topics in Vector Calculus

The concepts introduced in this chapter have many applications in the physical sciences. After discussing vector fields *and defining* line integrals, *we consider evaluation theorems and determine conditions for independence of path.* Green's Theorem *is then proved for elementary plane regions. This important result provides a connection between line integrals and double integrals. We next turn our attention to the* divergence *and* curl *of a vector field and introduce the notion of* surface integral. *We then discuss two of the major theorems in vector calculus: the* Divergence Theorem *and* Stokes' Theorem. *The chapter ends with a study of* transformations of coordinates, Jacobians, *and* change of variables in multiple integrals.

18.1 VECTOR FIELDS

In many applications a unique vector having initial point K is associated with each point K in some region. The totality of such vectors is called a **vector field**. The diagram in (i) of Figure 18.1 illustrates a vector field determined by a wheel rotating about a fixed point. To each point on the wheel there corresponds a velocity vector. A vector field of this type is called a **velocity field**. Other examples of velocity fields are those determined by the flow of water or of wind. For example, (ii) of Figure 18.1 could indicate the velocity vectors associated with fluid particles moving in a stream. Sketches of this type show only a few vectors of the field. It is important to remember that a vector is associated with *every* point in the region under consideration. Other common vector fields are **force fields** which arise in the study of mechanics or electricity. In the illustrations given above it was assumed that the vectors were independent of time. These are called **steady vector fields** and are the only types we shall consider in this chapter.

If a rectangular coordinate system is introduced, then the vector associated with the point $K(x, y, z)$ may be denoted by $\mathbf{F}(x, y, z)$. Since its components depend on the coordinates of K we have

$$\mathbf{F}(x, y, z) = M(x, y, z)\mathbf{i} + N(x, y, z)\mathbf{j} + P(x, y, z)\mathbf{k} \tag{18.1}$$

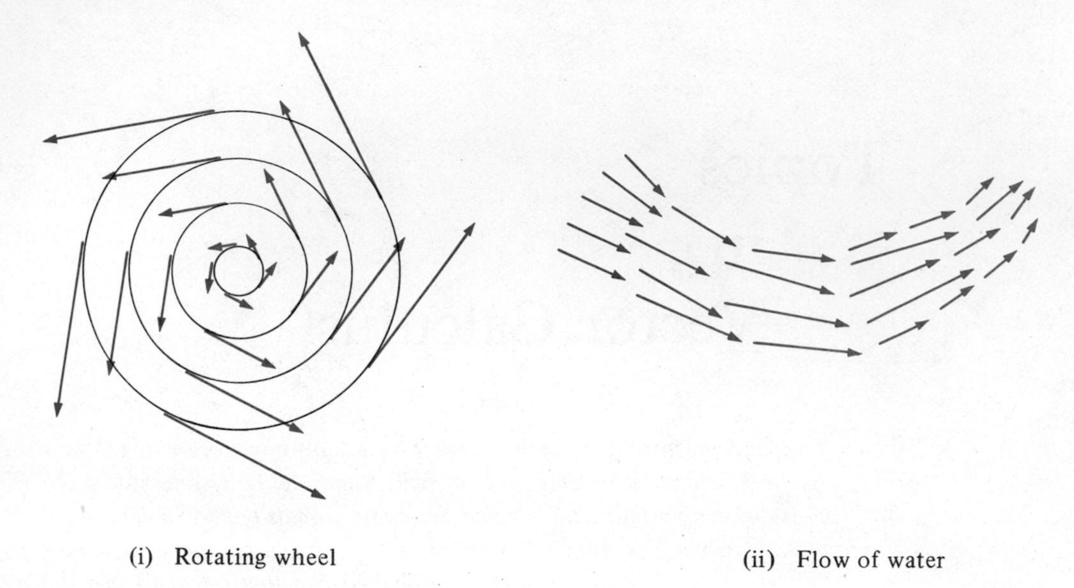

Figure 18.1. Velocity fields

where M, N, and P are functions of x, y, and z. Conversely, every equation of this type determines a vector field. It follows that a vector field may be defined as a **vector function F** whose domain D is a subset of $\mathbb{R} \times \mathbb{R} \times \mathbb{R}$ and whose range is a subset of V_3. As a special case, if the domain corresponds to a region in a plane and the range is a subset of V_2, then a vector field (in two dimensions) is given by

(18.2)
$$\mathbf{F}(x, y) = M(x, y)\mathbf{i} + N(x, y)\mathbf{j}$$

where M and N are functions of x and y. For convenience we shall use the terms "vector field" and "vector function" interchangeably. The functions discussed in Chapter 16, which associate a *number* with each point K, will be called **scalar functions**.

Example 1 Describe the vector field **F** if

$$\mathbf{F}(x, y) = -y\mathbf{i} + x\mathbf{j}.$$

Solution If $\mathbf{r} = x\mathbf{i} + y\mathbf{j}$ is the position vector of the point $K(x, y)$, then

$$\mathbf{r} \cdot \mathbf{F}(x, y) = -xy + yx = 0.$$

Thus $\mathbf{F}(x, y)$ is always orthogonal to the position vector **r** and, therefore, is tangent to the circle of radius $|\mathbf{r}|$ with center at the origin. Moreover,

$$|\mathbf{F}(x, y)| = \sqrt{y^2 + x^2} = |\mathbf{r}|$$

and hence the magnitude of $\mathbf{F}(x, y)$ equals the radius of this circle. It follows that the vectors $\mathbf{F}(x, y)$ have the appearance of the velocity field illustrated in Figure 18.1, where the origin is at the point of rotation.

Example 2 If $\mathbf{r} = x\mathbf{i} + y\mathbf{j} + z\mathbf{k}$ is the position vector of a point $K(x, y, z)$, then a vector field $\mathbf{F}$ is called an **inverse square field** if

$$\mathbf{F}(x, y, z) = \frac{c}{|\mathbf{r}|^3}\mathbf{r} = \frac{c}{|\mathbf{r}|^2}\mathbf{u}$$

where c is a scalar, $\mathbf{r} \neq \mathbf{0}$, and $\mathbf{u} = (1/|\mathbf{r}|)\mathbf{r}$ is a unit vector having the same direction as $\mathbf{r}$. Describe $\mathbf{F}$ if $c < 0$.

Solution If we let $\mathbf{r} = x\mathbf{i} + y\mathbf{j} + z\mathbf{k}$ we obtain

$$\mathbf{F}(x, y, z) = \frac{c}{(x^2 + y^2 + z^2)^{3/2}}(x\mathbf{i} + y\mathbf{j} + z\mathbf{k})$$

which may then be expressed in form (18.1). It is simpler, however, to analyze the vectors in the field by using the given formula. Since $\mathbf{F}(x, y, z)$ is a negative scalar multiple of $\mathbf{r}$, the direction of $\mathbf{F}(x, y, z)$ is toward the origin O. Also,

$$|\mathbf{F}(x, y, z)| = \frac{|c|}{|\mathbf{r}|^3}|\mathbf{r}| = \frac{|c|}{|\mathbf{r}|^2}$$

and hence the magnitude of $\mathbf{F}(x, y, z)$ is inversely proportional to the square of the distance from O to the point (x, y, z). This means that as $K(x, y, z)$ moves away from the origin, the length of the associated vector $\mathbf{F}(x, y, z)$ decreases. Some typical vectors in the field are sketched in Figure 18.2.

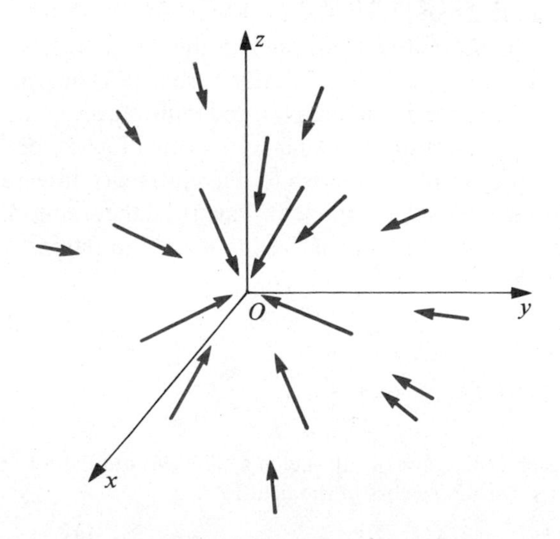

Figure 18.2. Inverse square field

Inverse square fields (see Example 2) occur frequently in applications. For example, if a particle of mass M is located at the origin, then the force of gravitational attraction exerted on a particle of mass m located at $K(x, y, z)$ is given by

$$\mathbf{F}(x, y, z) = -\frac{GMm}{|\mathbf{r}|^3}\mathbf{r} = -G\frac{Mm}{|\mathbf{r}|^2}\mathbf{u} \tag{18.3}$$

where G is a gravitational constant, $\mathbf{r}$ is the position vector of K, and $\mathbf{u} = (1/|\mathbf{r}|)\mathbf{r}$. Inverse square fields also occur in electrical theory.

Important types of vector fields are obtained by using the gradient of a scalar function f of two or three variables. For example, if $w = f(x, y, z)$, then as in (16.29)

$$\nabla w = f_x(x, y, z)\mathbf{i} + f_y(x, y, z)\mathbf{j} + f_z(x, y, z)\mathbf{k}.$$

We know from our work in Chapter 16 that the direction of ∇w at any point K is orthogonal to the level surface associated with f which contains K. Moreover, the magnitude of ∇w equals the maximum rate of change of f at K. If a vector field $\mathbf{F}$ is the gradient of a scalar function, that is, if

$$\mathbf{F}(x, y, z) = \nabla f(x, y, z)$$

for some f, then $\mathbf{F}$ is called a **conservative vector field** and $f(x, y, z)$ is called the **potential** at K. The function f is called a **potential function** for $\mathbf{F}$. Inverse square fields are conservative. Indeed, if $\mathbf{F}$ is the field given in Example 2, then

$$\mathbf{F}(x, y, z) = \nabla\left(\frac{-c}{r}\right)$$

where $r = |\mathbf{r}| = (x^2 + y^2 + z^2)^{1/2}$. The verification of this fact is left to the reader (see Exercise 11). In applications, the potential function of a conservative vector field $\mathbf{F}$ is the function p such that $\mathbf{F}(x, y, z) = -\nabla p(x, y, z)$. In this event, the field in Example 2 would be expressed as $\mathbf{F}(x, y, z) = \nabla(c/r)$. We shall have more to say about conservative fields later in this chapter.

If a vector field $\mathbf{F}$ is described as in (18.1) or (18.2), then we may define limits, continuity, partial derivatives, and multiple integrals by using the components of $\mathbf{F}(x, y, z)$ in a fashion similar to that used for vector-valued functions in Chapter 15. For example, if we wish to differentiate or integrate $\mathbf{F}(x, y, z)$, we differentiate or integrate each component. The usual theorems may be established. Thus, $\mathbf{F}$ is continuous if and only if the component functions M, N, and P are continuous, and so on.

EXERCISES 18.1

In each of Exercises 1–10 sketch a sufficient number of vectors $\mathbf{F}(x, y, z)$ to illustrate the pattern of the vectors in the field $\mathbf{F}$.

1 $\mathbf{F}(x, y) = x\mathbf{i} - y\mathbf{j}$

2 $\mathbf{F}(x, y) = -x\mathbf{i} + y\mathbf{j}$

3 $\mathbf{F}(x, y) = 2x\mathbf{i} + 3y\mathbf{j}$

4 $\mathbf{F}(x, y) = 3\mathbf{i} + x\mathbf{j}$

5 $\mathbf{F}(x, y) = (x^2 + y^2)^{-1/2}(x\mathbf{i} + y\mathbf{j})$

6 $\mathbf{F}(x, y, z) = x\mathbf{i} + z\mathbf{k}$

7 $\mathbf{F}(x, y, z) = -x\mathbf{i} - y\mathbf{j} - z\mathbf{k}$

8 $\mathbf{F}(x, y, z) = x\mathbf{i} + y\mathbf{j} + z\mathbf{k}$

9 $\mathbf{F}(x, y, z) = \mathbf{i} + \mathbf{j} + \mathbf{k}$

10 $\mathbf{F}(x, y, z) = 2\mathbf{k}$

11 If $\mathbf{F}$ is the vector field of Example 2, prove that $\mathbf{F}(x, y, z) = \nabla(-c/r)$, where $r = |\mathbf{r}|$.

12 Find a potential function for the vector field given by (18.3).

In each of Exercises 13–20 find a conservative vector field which has the given potential.

13 $f(x, y, z) = x^2 - 3y^2 + 4z^2$

14 $f(x, y, z) = \sin(x^2 + y^2 + z^2)$

15 $f(x, y, z) = xe^{2y} \cos z$

16 $f(x, y, z) = \ln xyz$

17 $f(x, y) = \arctan(xy)$

18 $f(x, y) = y^2 e^{-3x}$

19 $f(x, y, z) = \frac{1}{2}(x^2 + y^2 + z^2)$

20 $f(x, y) = x + y$

21 If **F** is given by (18.1), define

$$\lim_{(x,y,z)\to(x_0,y_0,z_0)} \mathbf{F}(x, y, z) = \mathbf{a}$$

in a manner analogous to Definition (15.3). Give an ε-δ definition for this limit. What is the geometric significance of the limit?

22 If **F** is given by (18.1), define the notion of *continuity* at (x_0, y_0, z_0) in a manner similar to Definition (15.4). What is the geometric significance of a continuous vector function?

18.2 LINE INTEGRALS

The concept of *line integral* is a natural generalization of the definite integral $\int_a^b f(x)\,dx$ defined in Chapter 5. Recall that in the latter case we began by dividing the interval $[a, b]$ into n subintervals of lengths $\Delta x_1, \Delta x_2, \ldots, \Delta x_n$. We then chose a number w_i in each subinterval and took the limit of the Riemann sum $\Sigma_i f(w_i)\Delta x_i$ as all the Δx_i approached 0. A similar process can be carried out for functions of several variables. For example, if $f(x, y)$ is defined on a finite plane curve C, we could begin by dividing C into n subarcs of lengths $\Delta s_1, \Delta s_2, \ldots, \Delta s_n$. After choosing a point (u_i, v_i) in each subarc we could consider the limit of the sum $\Sigma_i f(u_i, v_i)\Delta s_i$ as all the Δs_i approach 0. The same process could be carried out for a function of three variables defined on a space curve. Let us now make these remarks more precise.

Suppose a plane curve C is given parametrically by

$$x = g(t), \quad y = h(t), \quad \text{where } a \le t \le b \tag{18.4}$$

and where the functions g and h are smooth on the interval $[a, b]$. Consider $f(x, y)$ where f is continuous on a region D containing C. Let A and B be the points on C determined by the parameter values a and b, respectively. We shall assign the positive direction along C as that determined by increasing values of t. Let us partition the parameter interval $[a, b]$ by choosing

$$a = t_0 < t_1 < t_2 < \cdots < t_n = b.$$

This leads to a subdivision of C into subarcs $\widehat{P_{i-1}P_i}$, where $P_i(x_i, y_i)$ is the point on C corresponding to t_i. As in Figure 18.3, let $\Delta x_i = x_i - x_{i-1}$, $\Delta y_i = y_i - y_{i-1}$, and let Δs_i denote the length of the subarc $\widehat{P_{i-1}P_i}$. The **norm** $\|\Delta\|$ of the subdivision of C is, by definition, the largest of the Δs_i.

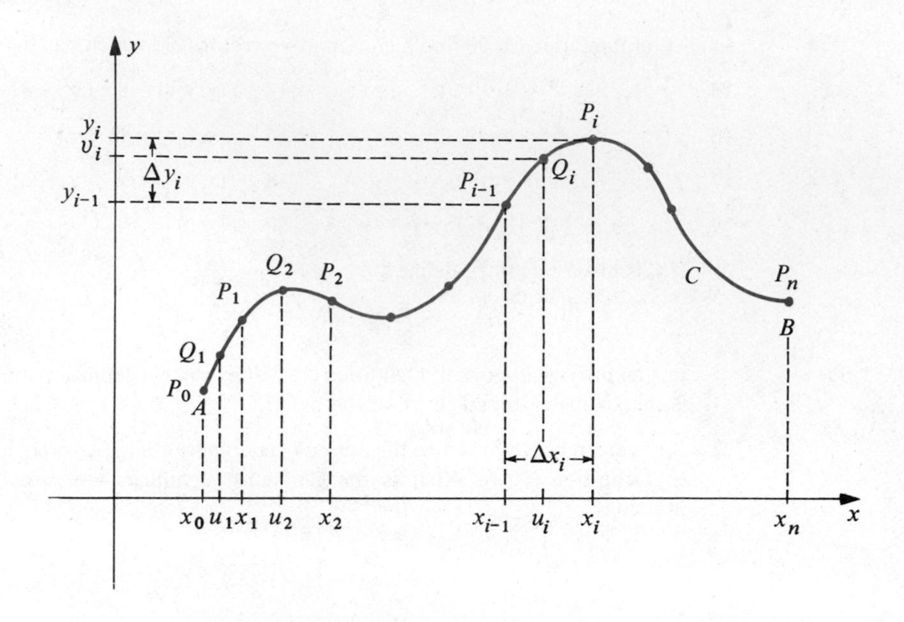

Figure 18.3

We next choose a point $Q_i(u_i, v_i)$ in each $\widehat{P_{i-1}P_i}$, as illustrated in Figure 18.3. For each i, we evaluate the function f at (u_i, v_i), multiply this number by Δs_i, and form the sum

(18.5)
$$\sum_{i=1}^{n} f(u_i, v_i)\,\Delta s_i.$$

If, in a manner similar to Definition (5.15), this sum has a limit L as $n \to \infty$ and $\|\Delta\| \to 0$ which is independent of the partition of $[a, b]$ and the points Q_i, then L is called the **line integral of f along C from A to B** and is denoted by

(18.6)
$$\int_C f(x, y)\,ds = \lim_{\|\Delta\| \to 0} \sum_i f(u_i, v_i)\,\Delta s_i.$$

The terminology *line integral* is a misnomer. It would be more descriptive to use *curve* integral for (18.6).

If we wish to specify the end-points A and B of the curve C, then instead of (18.6) we write

$$\int_A^B f(x, y)\,ds.$$

When this notation is employed, the curve C from A to B should be kept in mind, except for certain cases to be considered later, where the line integral has the same value for *all* curves joining A to B.

It can be shown that if f is continuous on D, then the limit in (18.6) exists and is the same for all parametric representations of C. Moreover, the integral may be evaluated as follows:

$$\int_C f(x,y)\,ds = \int_a^b f(g(t),h(t))\sqrt{[g'(t)]^2 + [h'(t)]^2}\,dt \tag{18.7}$$

that is, we use the parametric equations $x = g(t)$, $y = h(t)$ of C to substitute for x and y, and replace ds by the indicated expression involving the radical. The form of the radical is a consequence of the discussion of arc length in Chapter 13 (see (13.17)). It is worth noting that if C is an interval on the x-axis, then $\Delta s_i = \Delta x_i$, all $v_i = 0$, and (18.6) reduces to a definite integral of the typc considered in Chapter 5.

Definition (18.6) can be extended to more complicated curves. In particular, suppose C is a **piecewise smooth curve**, in the sense that it can be expressed as the union of a finite number of curves $C_1, C_2, \ldots, C_n$ of the type given in (18.4), where the terminal point B_i of C_i is the initial point A_{i+1} of C_{i+1} for $i = 1, 2, \ldots, n-1$. In this case the line integral of f along C is defined as the sum of the line integrals along the individual curves.

Example 1 Evaluate $\int_C xy^2\,ds$ if C has parametric equations $x = \cos t$, $y = \sin t$ where $0 \le t \le \pi/2$.

Solution The curve C is that part of the unit circle with center O which lies in the first quadrant. Applying (18.7),

$$\begin{aligned}\int_C xy^2\,ds &= \int_0^{\pi/2} \cos t \sin^2 t\sqrt{\sin^2 t + \cos^2 t}\,dt\\ &= \int_0^{\pi/2} \sin^2 t \cos t\,dt\\ &= \frac{1}{3}\sin^3 t\Big]_0^{\pi/2} = \frac{1}{3}.\end{aligned}$$

Properties of line integrals may be proved which are similar to those obtained for the definite integrals of Chapter 5. For example, it can be shown that reversing the direction of integration changes the sign of the integral, that the integral of a sum of two functions is the sum of the integrals of the individual functions, and so on.

To obtain a geometric interpretation for (18.6), suppose $f(x,y) \ge 0$ throughout D and consider the graph S of the equation $z = f(x,y)$ in a three-dimensional rectangular coordinate system. As illustrated in Figure 18.4, $f(u_i, v_i)\,\Delta s_i$ is the area of a strip with base $\overset{\frown}{P_{i-1}P_i}$ in the xy-plane and altitude $f(u_i, v_i)$. The limit of a sum of such terms is the area of that part of a cylinder with directrix C and rulings parallel to the z-axis which lies between the xy-plane and S.

An elementary physical application of the line integral (18.6) may be obtained by regarding the curve as a thin wire of variable density. If the wire is represented by the curve C in Figure 18.3, and if the density at the point (x, y) is

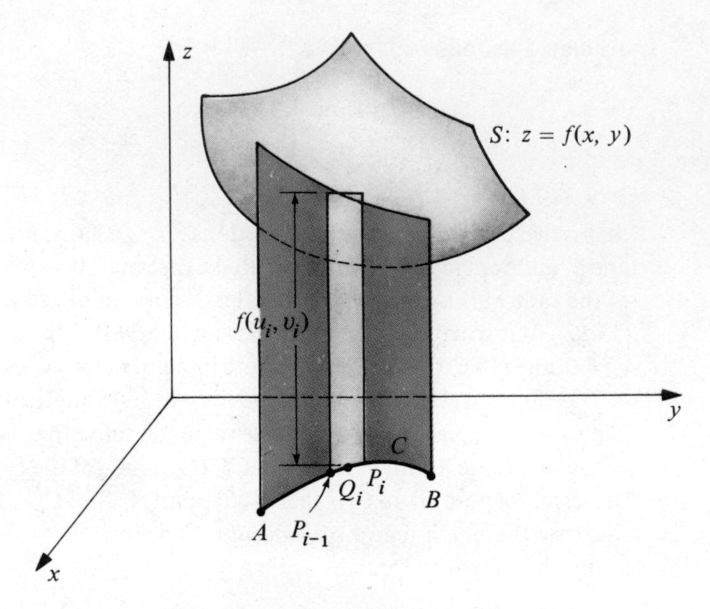

Figure 18.4

given by $f(x, y)$, then $f(u_i, v_i)\,\Delta s_i$ is an approximation to the mass Δm_i of the part of the wire between P_{i-1} and P_i. The sum

$$\sum_i \Delta m_i = \sum_i f(u_i, v_i)\,\Delta s_i$$

is, therefore, an approximation to the total mass M of the wire. To define M we take the limit of this sum, obtaining

$$M = \int_C f(x, y)\,ds.$$

The moments M_x and M_y of the wire with respect to the x- and y-axes, respectively, are defined in a manner analogous to the moments of laminas in (17.20) and (17.21). Specifically,

(18.8)

$$M_x = \lim_{\|\Delta\| \to 0} \sum_i v_i f(u_i, v_i)\,\Delta s_i = \int_C y f(x, y)\,ds$$

$$M_y = \lim_{\|\Delta\| \to 0} \sum_i u_i f(u_i, v_i)\,\Delta s_i = \int_C x f(x, y)\,ds.$$

As in (17.22), the center of mass $(\bar{x}, \bar{y})$ of the wire is given by

(18.9)

$$\bar{x} = \frac{M_y}{M}, \quad \bar{y} = \frac{M_x}{M}.$$

Moments of inertia can also be defined in the usual way (see Exercise 25).

Example 2 A thin wire is bent into the shape of a semicircle of radius a. If the density at a point P is directly proportional to its distance from the line through the end-points, find the center of mass of the wire.

Solution If we introduce a coordinate system such that the shape of the wire coincides with the upper half of a circle of radius a and center O, then parametric equations for C are

$$x = a\cos t, \quad y = a\sin t; \quad 0 \le t \le \pi.$$

By hypothesis, the density function f is given by $f(x, y) = ky$ for some constant k. Hence the mass of the wire is

$$\begin{aligned} M = \int_C ky\,ds &= \int_0^{\pi} ka\sin t\sqrt{a^2\sin^2 t + a^2\cos^2 t}\,dt \\ &= \int_0^{\pi} ka(\sin t)a\,dt \\ &= -ka^2\cos t\Big]_0^{\pi} = 2ka^2. \end{aligned}$$

From (18.8),

$$\begin{aligned} M_x = \int_C yky\,ds &= \int_0^{\pi} ka^3\sin^2 t\,dt \\ &= ka^3\int_0^{\pi}\frac{1-\cos 2t}{2}\,dt \\ &= ka^3\left[\frac{1}{2}t - \frac{1}{4}\sin 2t\right]_0^{\pi} = \frac{1}{2}\pi ka^3. \end{aligned}$$

Consequently, by (18.9),

$$\bar{y} = \frac{M_x}{M} = \frac{\frac{1}{2}\pi ka^3}{2ka^2} = \frac{1}{4}\pi a \approx 0.8a.$$

Since it is evident that $\bar{x} = 0$, we see that the center of mass is on the y-axis, approximately 4/5 of the way from the x-axis to the top of the semicircle.

Two other types of line integrals can be obtained by using Δx_i and Δy_i in place of Δs_i in (18.6). They are called the **line integrals of f along C with respect to x and y**, respectively. Thus, by definition,

(18.10)

$$\begin{aligned} \int_C f(x, y)\,dx &= \lim_{\|\Delta\|\to 0}\sum_i f(u_i, v_i)\,\Delta x_i \\ \int_C f(x, y)\,dy &= \lim_{\|\Delta\|\to 0}\sum_i f(u_i, v_i)\,\Delta y_i. \end{aligned}$$

If we wish to specify the end-points A and B of C we shall write these line integrals as

$$\int_A^B f(x,y)\,dx \quad \text{and} \quad \int_A^B f(x,y)\,dy.$$

If C is given parametrically by $x = g(t)$, $y = h(t)$, where $a \le t \le b$, then these integrals may be evaluated by substituting $g(t)$ and $h(t)$ for x and y, respectively, letting $dx = g'(t)\,dt$, $dy = h'(t)\,dt$, and using a and b for the limits of integration.

Example 3 Evaluate $\int_C f(x,y)\,dx$ and $\int_C f(x,y)\,dy$ if $f(x,y) = xy^2$ and C is the part of the parabola $y = x^2$ from $A(0,0)$ to $B(2,4)$.

Solution Parametric equations for C are $x = t$, $y = t^2$, where $0 \le t \le 2$. Hence $dx = dt$, $dy = 2t\,dt$, and

$$\int_C xy^2\,dx = \int_0^2 t^5\,dt = \frac{1}{6}t^6\bigg]_0^2 = \frac{32}{3},$$

$$\int_C xy^2\,dy = \int_0^2 t^5 \cdot 2t\,dt = \frac{2}{7}t^7\bigg]_0^2 = \frac{256}{7}.$$

If C is the graph of a rectangular equation $y = g(x)$ where $a \le x \le b$, then parametric equations for C can be found by letting

$$x = t, \quad y = g(t) \quad \text{where} \quad a \le t \le b.$$

The line integrals in (18.10) may then be evaluated as follows:

$$\int_C f(x,y)\,dx = \int_a^b f(t, g(t))\,dt = \int_a^b f(x, g(x))\,dx$$

$$\int_C f(x,y)\,dy = \int_a^b f(t, g(t))g'(t)\,dt = \int_a^b f(x, g(x))g'(x)\,dx.$$

This shows that for curves given in the rectangular form $y = g(x)$ where $a \le x \le b$, we may bypass the parametric equations by substituting $y = g(x)$, $dy = g'(x)\,dx$, and then using a and b for the limits of integration. To illustrate, in Example 3, where $y = x^2$, we could have written

$$\int_C xy^2\,dx = \int_0^2 x(x^4)\,dx = \frac{32}{3}$$

$$\int_C xy^2\,dy = \int_0^2 x(x^4)2x\,dx = \frac{256}{7}.$$

In application, line integrals often occur in the combination

$$\int_C M(x,y)\,dx + \int_C N(x,y)\,dy$$

where M and N are continuous functions on a domain D containing C. This sum is usually abbreviated by writing

$$\int_C M(x, y)\, dx + N(x, y)\, dy.$$

If C is given parametrically by $x = g(t)$, $y = h(t)$, where $a \le t \le b$, then the integral may be evaluated by substituting for x, y, dx, and dy as was done previously.

Example 4 Evaluate $\int_C xy\, dx + x^2\, dy$ if:

(a) C consists of line segments from $(2, 1)$ to $(4, 1)$ and from $(4, 1)$ to $(4, 5)$.

(b) C is the line segment from $(2, 1)$ to $(4, 5)$.

(c) parametric equations for C are $x = 3t - 1$, $y = 3t^2 - 2t$; $1 \le t \le 5/3$.

Solution (a) If C is subdivided into two parts C_1 and C_2 as shown in (i) of Figure 18.5, then parametric equations for these curves are

$$C_1: x = t, y = 1; \quad 2 \le t \le 4$$
$$C_2: x = 4, y = t; \quad 1 \le t \le 5.$$

The line integral along C may be expressed as a sum of two line integrals, the first along C_1 and the second along C_2. On C_1 we have $dy = 0$, $dx = dt$, and hence

$$\int_{C_1} xy\, dx + x^2\, dy = \int_2^4 t(1)\, dt + 0 = \frac{1}{2}t^2\bigg]_2^4 = 6.$$

On C_2 we have $dx = 0$, $dy = dt$, and therefore

$$\int_{C_2} xy\, dx + x^2\, dy = \int_1^5 0 + 16\, dt = 16t\bigg]_1^5 = 64.$$

Consequently the line integral along C equals $6 + 64$ or 70.

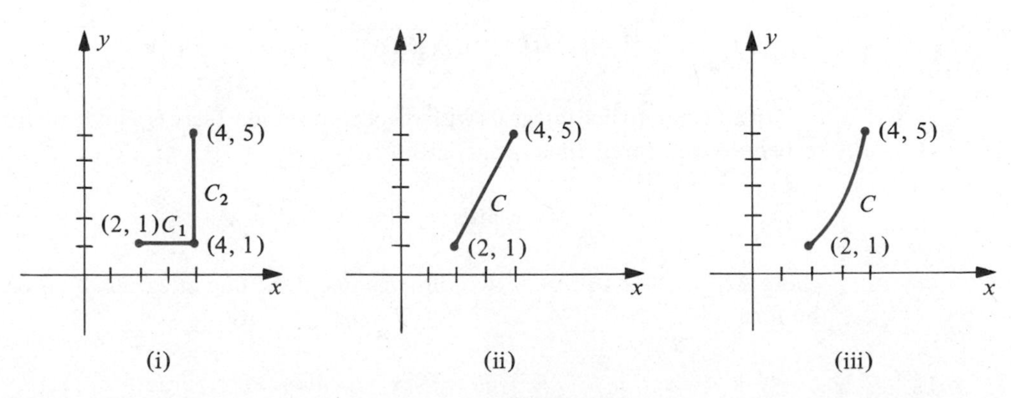

Figure 18.5

(b) The graph of C is sketched in (ii) of Figure 18.5. Since a rectangular equation for C is $y = 2x - 3$ where $2 \leq x \leq 4$, we may write

$$\int_C xy\,dx + x^2\,dy = \int_2^4 x(2x - 3)\,dx + x^2 2\,dx = \int_2^4 (4x^2 - 3x)\,dx = \frac{170}{3} = 56\frac{2}{3}.$$

(c) The graph of C is part of a parabola (see (iii) of Figure 18.5). Since $dx = 3\,dt$ and $dy = (6t - 2)\,dt$, the given integral equals

$$\int_1^{5/3} (3t - 1)(3t^2 - 2t)3\,dt + (3t - 1)^2(6t - 2)\,dt.$$

It is left to the reader to show that the value is 58. Another method of solution is to use the rectangular equation $y = \frac{1}{3}(x^2 - 1)$ for the parabola, where $2 \leq x \leq 4$. In this event we obtain the following integral:

$$\int_C xy\,dx + x^2\,dy = \int_2^4 x\,\frac{1}{3}(x^2 - 1)\,dx + x^2\left(\frac{2}{3}x\right)dx = 58.$$

In the preceding example we obtained three different values for the line integral along three different paths from (2, 1) to (4, 5). In Section 3 we shall consider line integrals having the same value along every curve joining two points A and B. For such integrals we use the phrase **independent of path**.

If a curve C in three dimensions is given parametrically by

$$x = g(t), \quad y = h(t), \quad z = k(t)$$

where the functions g, h, and k are smooth on an interval $[a, b]$, then line integrals of a function f of *three* variables are defined in a manner similar to those for two variables. In this case, instead of using (x_i, y_i) and (u_i, v_i) as the coordinates of P_i and Q_i on C (see Figure 18.3), we use (x_i, y_i, z_i) and (u_i, v_i, w_i), respectively. Instead of (18.6) we now have

$$\int_C f(x, y, z)\,ds = \lim_{\|\Delta\| \to 0} \sum_i f(u_i, v_i, w_i)\,\Delta s_i.$$

This integral may be evaluated by using the formula

$$\int_a^b f(g(t), h(t), k(t))\sqrt{(g'(t))^2 + (h'(t))^2 + (k'(t))^2}\,dt.$$

In addition to line integrals with respect to x and y there is a line integral with respect to z in three dimensions, given by

$$\int_C f(x, y, z)\,dz = \lim_{\|\Delta\| \to 0} \sum_i f(u_i, v_i, w_i)\,\Delta z_i$$

where $\Delta z_i = z_i - z_{i-1}$. As in two dimensions, these line integrals often occur in the form

$$\int_C M(x, y, z)\,dx + N(x, y, z)\,dy + P(x, y, z)\,dz$$

where M, N, and P are functions of x, y, and z which are continuous throughout a region containing C. If C is given parametrically, then this line integral may be evaluated by substituting for x, y, and z in the same manner as in the two-variable case.

Example 5 Evaluate

$$\int_C yz\,dx + xz\,dy + xy\,dz$$

where C is the twisted cubic

$$x = t, \quad y = t^2, \quad z = t^3; \quad 0 \le t \le 2.$$

Solution Substituting for x, y, and z and using $dx = dt$, $dy = 2t\,dt$, $dz = 3t^2\,dt$, we obtain

$$\int_0^2 t^5\,dt + 2t^5\,dt + 3t^5\,dt = \int_0^2 6t^5\,dt = t^6\Big]_0^2 = 64.$$

One of the most important physical applications of line integrals has to do with force fields. Let us suppose that the force acting at a point (x, y, z) is given by the vector field

$$\mathbf{F}(x, y, z) = M(x, y, z)\mathbf{i} + N(x, y, z)\mathbf{j} + P(x, y, z)\mathbf{k}$$

where the functions M, N, and P are continuous on a suitable domain. Our objective is to find the work done as the point of application of $\mathbf{F}(x, y, z)$ moves along a curve C, where C is given parametrically by $x = g(t)$, $y = h(t)$, $z = k(t)$; $a \le t \le b$.

Let us begin by subdividing C as illustrated in Figure 18.6 where for each i, P_i and Q_i have coordinates (x_i, y_i, z_i) and (u_i, v_i, w_i), respectively. If the norm $\|\Delta\|$ is small, then the work done by $\mathbf{F}(x, y, z)$ along the arc $\widehat{P_{i-1}P_i}$ can be approximated by the work Δw_i done by the constant force $\mathbf{F}(u_i, v_i, w_i)$ as its point of application moves along $\overrightarrow{P_{i-1}P_i}$. If $\Delta x_i = x_i - x_{i-1}$, $\Delta y_i = y_i - y_{i-1}$, and $\Delta z_i = z_i - z_{i-1}$, then $\overrightarrow{P_{i-1}P_i}$ corresponds to the vector $\Delta x_i\mathbf{i} + \Delta y_i\mathbf{j} + \Delta z_i\mathbf{k}$ of V_3. By Definition (14.26) the work Δw_i is given by

$$\begin{aligned}\Delta w_i &= \mathbf{F}(u_i, v_i, w_i)\cdot(\Delta x_i\mathbf{i} + \Delta y_i\mathbf{j} + \Delta z_i\mathbf{k})\\ &= M(u_i, v_i, w_i)\,\Delta x_i + N(u_i, v_i, w_i)\,\Delta y_i + P(u_i, v_i, w_i)\,\Delta z_i.\end{aligned}$$

The work W done by $\mathbf{F}$ along C is, by definition,

$$W = \lim_{\|\Delta\|\to 0} \sum_i \Delta w_i$$

that is,

(18.11)
$$W = \int_C M(x, y, z)\,dx + N(x, y, z)\,dy + P(x, y, z)\,dz.$$

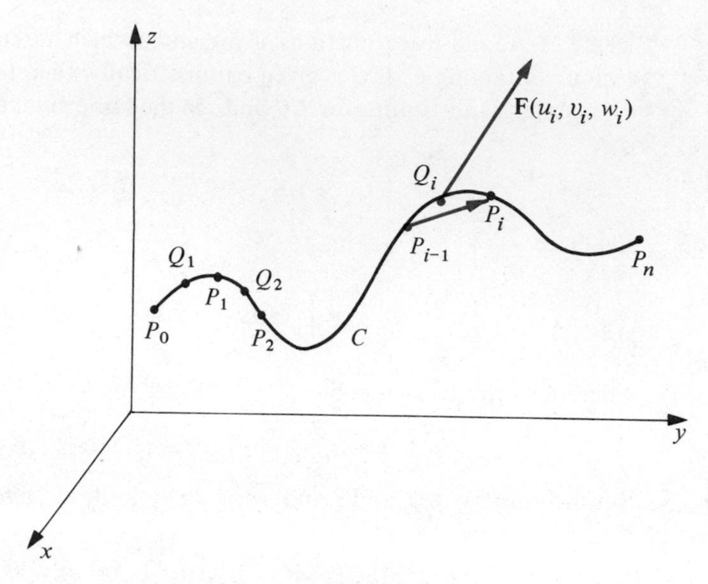

Figure 18.6

A similar situation exists in two dimensions. In this case if a force field is given by

$$\mathbf{F}(x, y) = M(x, y)\mathbf{i} + N(x, y)\mathbf{j}$$

and C is a finite piecewise smooth plane curve, then the work done as the point of application of $\mathbf{F}(x, y)$ moves along C is

(18.12)
$$W = \int_C M(x, y)\,dx + N(x, y)\,dy.$$

It is of interest to interpret (18.11) vectorially. If we let

$$\mathbf{r}(t) = x\mathbf{i} + y\mathbf{j} + z\mathbf{k}$$

where $x = g(t)$, $y = h(t)$, and $z = k(t)$, then $\mathbf{r}(t)$ is the position vector of the point $Q(x, y, z)$ on C. If s denotes arc length measured along C, then as in Section 15.4, a unit tangent vector $\mathbf{T}(s)$ to C at Q is given by

$$\mathbf{T}(s) = \frac{d}{ds}\mathbf{r}(t) = \frac{dx}{ds}\mathbf{i} + \frac{dy}{ds}\mathbf{j} + \frac{dz}{ds}\mathbf{k}.$$

The vectors $\mathbf{r}(t)$, $\mathbf{F}(x, y, z)$, and $\mathbf{T}(s)$ are illustrated in Figure 18.7.

The **tangential component of F at Q** is

$$\mathbf{F}(x, y, z) \cdot \mathbf{T}(s) = M(x, y, z)\frac{dx}{ds} + N(x, y, z)\frac{dy}{ds} + P(x, y, z)\frac{dz}{ds}.$$

Formula (18.11) may then be rewritten

(18.13)
$$W = \int_C \mathbf{F}(x, y, z) \cdot \mathbf{T}(s)\,ds$$

that is, *the work done as the point of application of* $\mathbf{F}(x, y, z)$ *moves along C equals the line integral of the tangential component of* **F** *along C.* Sometimes it is

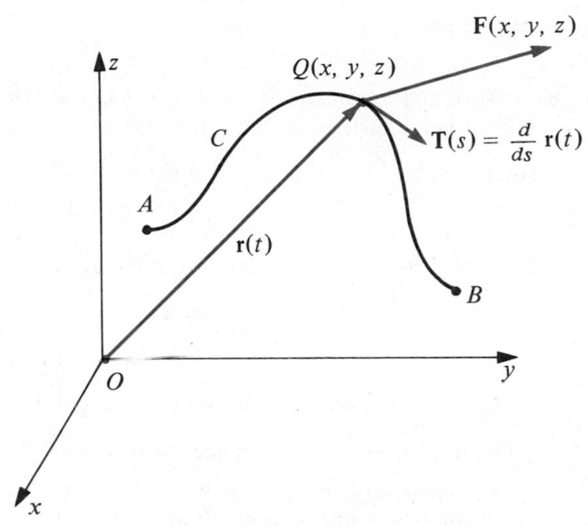

Figure 18.7

convenient to abbreviate $\mathbf{F}(x, y, z)$ by $\mathbf{F}$, $\mathbf{T}(s)$ by $\mathbf{T}$, and formally let

$$d\mathbf{r} = dx\,\mathbf{i} + dy\,\mathbf{j} + dz\,\mathbf{k} = \mathbf{T}\,ds.$$

With this agreement on notation, (18.11) and (18.13) may be written

(18.14) $$W = \int_C \mathbf{F} \cdot d\mathbf{r} = \int_C \mathbf{F} \cdot \mathbf{T}\,ds.$$

The same formula may be used for (18.12). Intuitively, we may regard $\mathbf{F} \cdot d\mathbf{r}$ as representing the work done as the point of application of $\mathbf{F}$ moves along the tangent vector $d\mathbf{r}$ to C. The integral sign represents the limit of a sum of such elements of work.

Example 6 If an inverse force field $\mathbf{F}$ is given by

$$\mathbf{F}(x, y, z) = \frac{k}{|\mathbf{r}|^3}\mathbf{r}$$

find the work done by $\mathbf{F}$ as its point of application moves along the x-axis from $A(1, 0, 0)$ to $B(2, 0, 0)$.

Solution Let C denote the line segment AB. Since

$$\mathbf{F}(x, y, z) = \frac{k}{(x^2 + y^2 + z^2)^{3/2}}(x\mathbf{i} + y\mathbf{j} + z\mathbf{k}),$$

it follows from (18.14) that

$$W = \int_C \mathbf{F} \cdot d\mathbf{r} = \int_C \frac{k}{(x^2 + y^2 + z^2)^{3/2}}(x\,dx + y\,dy + z\,dz),$$

where C is given parametrically by $x = t$, $y = 0$, $z = 0$ for $1 \leq t \leq 2$. Substituting in the preceding integral and simplifying,

$$W = \int_1^2 \frac{k}{(t^2)^{3/2}} t\,dt = \int_1^2 \frac{k}{t^2}\,dt = -\frac{k}{t}\bigg]_1^2 = \frac{k}{2}.$$

The units for W depend on those for distance and $|\mathbf{F}(x, y, z)|$.

EXERCISES 18.2

In Exercises 1 and 2 evaluate the line integrals $\int_C f(x,y)\,ds$, $\int_C f(x,y)\,dx$, and $\int_C f(x,y)\,dy$ if C is defined parametrically as indicated.

1 $f(x,y) = x^3 + y$; $x = 3t$, $y = t^3$; $0 \le t \le 1$

2 $f(x,y) = xy^{2/5}$; $x = \frac{1}{2}t$, $y = t^{5/2}$; $0 \le t \le 1$

3 Evaluate $\int_C 6x^2y\,dx + xy\,dy$ where C is the graph of $y = x^3 + 1$ from $(-1,0)$ to $(1,2)$.

4 Evaluate $\int_C y\,dx + (x+y)\,dy$ where C is the graph of $y = x^2 + 2x$ from $(0,0)$ to $(2,8)$.

5 Evaluate $\int_C (x-y)\,dx + x\,dy$ where C is the graph of $y^2 = x$ from $(4,-2)$ to $(4,2)$.

6 Evaluate $\int_C xy\,dx + x^2y^3\,dy$ where C is the graph of $x = y^3$ from $(0,0)$ to $(1,1)$.

7 Evaluate $\int_C xy\,dx + (x+y)\,dy$ for the following.

(a) C consists of line segments from $(0,0)$ to $(1,0)$ and from $(1,0)$ to $(1,3)$.
(b) C consists of line segments from $(0,0)$ to $(0,3)$ and from $(0,3)$ to $(1,3)$.
(c) C is the line segment from $(0,0)$ to $(1,3)$.
(d) C is the part of the parabola $y = 3x^2$ from $(0,0)$ to $(1,3)$.

8 Evaluate $\int_C (x^2+y^2)\,dx + 2x\,dy$ for the following.

(a) C consists of line segments from $(1,2)$ to $(1,8)$ and from $(1,8)$ to $(-2,8)$.
(b) C consists of line segments from $(1,2)$ to $(-2,2)$ and from $(-2,2)$ to $(-2,8)$.
(c) C is the line segment from $(1,2)$ to $(-2,8)$.
(d) C is the graph of $y = 2x^2$ from $(1,2)$ to $(-2,8)$.

9 Evaluate $\int_C xz\,dx + (y+z)\,dy + x\,dz$ if C is given by $x = e^t$, $y = e^{-t}$, $z = e^{2t}$; $0 \le t \le 1$.

10 Evaluate $\int_C y\,dx + z\,dy + x\,dz$ if C is given by $x = \sin t$, $y = 2\sin t$, $z = \sin^2 t$; $0 \le t \le \pi/2$.

11 Evaluate $\int_{(0,0,0)}^{(2,3,4)} (x+y+z)\,dx + (x-2y+3z)\,dy + (2x+y-z)\,dz$ for each of the following curves C from $(0,0,0)$ to $(2,3,4)$.

(a) C consists of three line segments, the first parallel to the x-axis, the second parallel to the y-axis, and the third parallel to the z-axis.

(b) C consists of three line segments, the first parallel to the z-axis, the second parallel to the x-axis, and the third parallel to the y-axis.

(c) C is a line segment.

12 Evaluate $\int_{(1,-2,3)}^{(-4,5,2)} (x-y)\,dx + (y-z)\,dy + x\,dz$ where the curve C from $(1,-2,3)$ to $(-4, 5, 2)$ is of the type described in parts (a)–(c) of Exercise 11.

13 Evaluate $\int_C xyz\,ds$, if C is the line segment from $(0,0,0)$ to $(1,2,3)$.

14 Evaluate $\int_C (xy+z)\,ds$ if C is the helix $x = a\cos t$, $y = a\sin t$, $z = bt$, $0 \le t \le 2\pi$.

15 A thin wire is situated in a coordinate plane such that its shape coincides with the part of the parabola $y = 4 - x^2$ between $(-2,0)$ and $(2,0)$. Find the mass and center of mass if the density at the point (x,y) is directly proportional to its distance from the y-axis.

16 The shape of a thin wire in a coordinate plane coincides with the graph of $y = \sqrt{1-x^2}$ from $(-1,0)$ to $(0,1)$. Find the mass and center of mass if the density at the point (x,y) is directly proportional to its distance from the x-axis.

17 Extend the definitions of mass and center of mass of a wire to three dimensions.

18 A wire of constant density is bent into the shape of the helix $x = a\cos t$, $y = a\sin t$, $z = bt$, where $0 \le t \le 3\pi$. Find the mass and center of mass of the wire (see Exercise 17).

19 If $\mathbf{F}(x, y) = xy^2\mathbf{i} + x^2y\mathbf{j}$, evaluate $\int_C \mathbf{F} \cdot d\mathbf{r}$ along the curves described in parts (a)–(d) of Exercise 7.

20 If $\mathbf{F}(x, y) = (2x + y)\mathbf{i} + (x + 2y)\mathbf{j}$, evaluate $\int_C \mathbf{F} \cdot d\mathbf{r}$ along the curves described in parts (a)–(d) of Exercise 8.

21 The force acting at a point (x, y) in a coordinate plane is given by

$$\mathbf{F}(x, y) = \frac{4\mathbf{r}}{|\mathbf{r}|^3}, \quad \text{where } \mathbf{r} = x\mathbf{i} + y\mathbf{j}.$$

Find the work done by $\mathbf{F}$ as its point of application moves along the upper half of the circle $x^2 + y^2 = a^2$ from $(-a, 0)$ to $(a, 0)$.

22 The force at a point (x, y) in a coordinate plane is given by

$$\mathbf{F}(x, y) = (x^2 + y^2)\mathbf{i} + xy\mathbf{j}.$$

Find the work done by $\mathbf{F}(x, y)$ as its point of application moves along the graph of $y = x^2$ from $(0, 0)$ to $(2, 4)$.

23 The force at a point (x, y, z) in three dimensions is $\mathbf{F}(x, y, z) = y\mathbf{i} + z\mathbf{j} + x\mathbf{k}$. Find the work done by $\mathbf{F}(x, y, z)$ as its point of application moves along the curve $x = t$, $y = t^2$, $z = t^3$ from $(0, 0, 0)$ to $(2, 4, 8)$.

24 Solve Exercise 23 if $\mathbf{F}(x, y, z) = e^x\mathbf{i} + e^y\mathbf{j} + e^z\mathbf{k}$.

25 If a thin wire of variable density is represented by a curve C in the xy-plane, define the moments of inertia I_x and I_y with respect to the x- and y-axes, respectively. Find I_x and I_y for the wire in Exercise 16.

26 Find the moments of inertia I_x and I_y for the wire in Exercise 15.

27 If a thin wire of variable density is represented by a curve in three dimensions, define the moments of inertia with respect to the x-, y-, and z-axes.

28 Given the wire in Exercise 18, find the moment of inertia with respect to the z-axis (see also Exercise 27).

29 If an object moves through a force field $\mathbf{F}$ such that at each point (x, y, z) its velocity vector is orthogonal to $\mathbf{F}(x, y, z)$, show that the work done by $\mathbf{F}$ on the object is 0.

30 If a constant force $\mathbf{c}$ acts on a moving object as it travels once around a circle, show that the work done by $\mathbf{c}$ on the object is 0.

18.3 INDEPENDENCE OF PATH

A piecewise smooth curve which connects two points A and B is sometimes called a **path** from A to B. We shall now obtain conditions under which a line integral is **independent of path** in a region, in the sense that if A and B are arbitrary points, then the same value is obtained for *every* path from A to B. The results will be established for line integrals in two dimensions. Proofs for the three-dimensional case are similar and will be omitted.

It will be assumed throughout this section that all regions are **connected**. This means that any two points in a region can be joined by a piecewise smooth curve which lies in the region. We shall also assume that for any point A in a plane region D, there exists a circle with center A which lies completely in D. A region of this type is called an **open region**. The next theorem gives us the fundamental result that if a vector function $\mathbf{F}$ is continuous on D, then the line integral $\int_C \mathbf{F} \cdot d\mathbf{r}$ *is independent of path if and only if* $\mathbf{F}$ *is conservative.*

(18.15) Theorem

> If $\mathbf{F}(x, y) = M(x, y)\mathbf{i} + N(x, y)\mathbf{j}$ is continuous on an open connected region, then the line integral $\int_C \mathbf{F} \cdot d\mathbf{r}$ is independent of path if and only if $\mathbf{F}(x, y) = \nabla f(x, y)$ for some function f.

Proof. Suppose the integral is independent of path. If (x_0, y_0) is a fixed point in D, let f be defined by

$$f(x, y) = \int_{(x_0, y_0)}^{(x, y)} \mathbf{F} \cdot d\mathbf{r}$$

where (x, y) is arbitrary in D. Since the integral is independent of path, f depends only on x and y, and not on the path C from (x_0, y_0) to (x, y). Choose a circle in D with center (x, y) and let (x_1, y) be a point within the circle such that $x_1 \neq x$. Let C_1 be any path from (x_0, y_0) to (x_1, y) and let C_2 be the horizontal segment from (x_1, y) to (x, y), as illustrated in (i) of Figure 18.8.

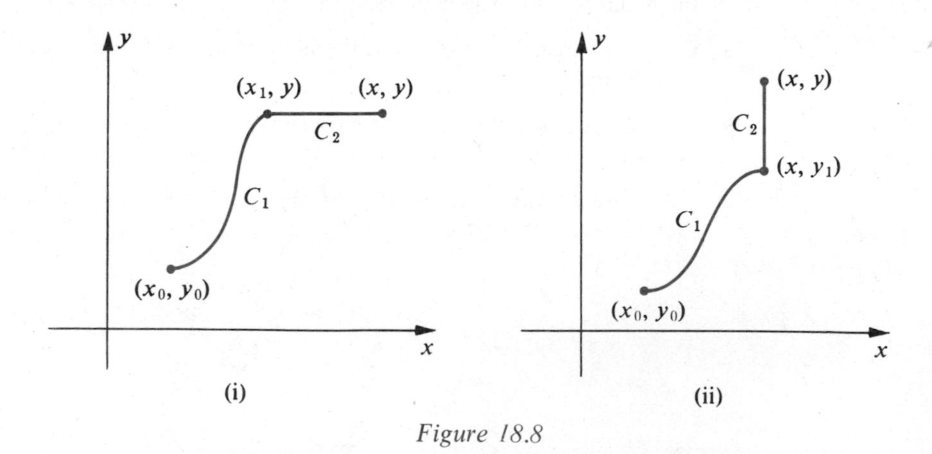

Figure 18.8

We may, therefore, write

$$f(x, y) = \int_{C_1} \mathbf{F} \cdot d\mathbf{r} + \int_{C_2} \mathbf{F} \cdot d\mathbf{r} = \int_{(x_0, y_0)}^{(x_1, y)} \mathbf{F} \cdot d\mathbf{r} + \int_{(x_1, y)}^{(x, y)} \mathbf{F} \cdot d\mathbf{r}.$$

Since the first integral does not depend on x,

$$\frac{\partial}{\partial x} f(x, y) = 0 + \frac{\partial}{\partial x} \int_{(x_1, y)}^{(x, y)} \mathbf{F} \cdot d\mathbf{r}.$$

Writing $\mathbf{F} \cdot d\mathbf{r} = M(x, y)\,dx + N(x, y)\,dy$ and using the fact that $dy = 0$ on C_2 gives us

$$\frac{\partial}{\partial x} f(x, y) = \frac{\partial}{\partial x} \int_{(x_1, y)}^{(x, y)} M(x, y)\,dx.$$

Since y is fixed in this partial differentiation we may regard the integrand as containing only the variable x. Applying (5.33) we obtain

$$\frac{\partial}{\partial x} f(x, y) = M(x, y).$$

Similarly, if we choose the path shown in (ii) of Figure 18.8 and differentiate with respect to y, there results

$$\frac{\partial}{\partial y} f(x, y) = N(x, y).$$

This proves that $\nabla f(x, y) = \mathbf{F}(x, y)$.

Conversely, if there exists a function f such that $\mathbf{F}(x, y) = \nabla f(x, y)$, then

$$M(x, y)\mathbf{i} + N(x, y)\mathbf{j} = f_x(x, y)\mathbf{i} + f_y(x, y)\mathbf{j}$$

and hence

$$M(x, y) = f_x(x, y) \quad \text{and} \quad N(x, y) = f_y(x, y).$$

Consequently, if $A(x_1, y_1)$ and $B(x_2, y_2)$ are in D, then

$$\int_C \mathbf{F} \cdot d\mathbf{r} = \int_C f_x(x, y)\,dx + f_y(x, y)\,dy$$

where C is any path from A to B. If C is smooth and is given parametrically by $x = g(t)$, $y = h(t)$ where $t_1 \le t \le t_2$, then substituting in the integrand and using Theorem (16.19),

$$\int_C \mathbf{F} \cdot d\mathbf{r} = \int_{t_1}^{t_2} [f_x(g(t), h(t))g'(t) + f_y(g(t), h(t))h'(t)]\,dt.$$

Applying (16.19) and the Fundamental Theorem of Calculus,

$$\begin{aligned}\int_C \mathbf{F} \cdot d\mathbf{r} &= \int_{t_1}^{t_2} \frac{d}{dt}[f(g(t), h(t))]\,dt \\ &= f(g(t_2), h(t_2)) - f(g(t_1), h(t_1)) \\ &= f(x_2, y_2) - f(x_1, y_1).\end{aligned}$$

Thus the line integral depends only on the coordinates of A and B, not on the path C, which is what we wished to prove. The proof can be extended to piecewise smooth curves by subdividing C into a finite number of smooth curves.

Theorem (18.15) can be extended to line integrals in three dimensions. The proof is left as an exercise. It is convenient to restate (18.15) in the following nonvector form.

(18.16) Theorem

If M and N are continuous in an open connected region, then the line integral

$$\int_A^B M(x,y)\,dx + N(x,y)\,dy$$

is independent of path if and only if there exists a function f such that

$$\frac{\partial f}{\partial x} = M \quad \text{and} \quad \frac{\partial f}{\partial y} = N.$$

If the function f indicated in (18.16) can be found, then

$$df = \frac{\partial f}{\partial x}dx + \frac{\partial f}{\partial y}dy = M(x,y)\,dx + N(x,y)\,dy,$$

and our previous work allows us to write

$$\text{(18.17)} \qquad \int_{(x_1,y_1)}^{(x_2,y_2)} M(x,y)\,dx + N(x,y)\,dy = \int_{(x_1,y_1)}^{(x_2,y_2)} df = f(x_2,y_2) - f(x_1,y_1)$$

where (x_1, y_1) and (x_2, y_2) are coordinates of A and B, respectively. This formula is analogous to the Fundamental Theorem of Calculus (5.31).

In terms of conservative force fields, (18.17) implies that the work done in going along any path C from A to B equals the difference in the potentials between A and B. It follows that if C is a closed curve, that is, if $A = B$, then the work done in traversing C is 0. It can be shown that a converse of this result is also true, namely, if $\int_C \mathbf{F} \cdot d\mathbf{r}$ is 0 for every simple closed curve C, then the line integral is independent of path and hence the field is conservative. These facts are very important in applications, since many of the vector fields which occur in nature (such as inverse square fields) are conservative. In physical terms, if a unit particle moves completely around a closed curve in a conservative force field, and if there is no loss of energy due to friction, then the work done is 0. This is also a consequence of the law of conservation of energy.

Suppose $\mathbf{F}$ is a conservative vector field with potential function f, that is, $\mathbf{F}(x, y) = \nabla f(x, y)$. If $A(x_1, y_1)$ is a point at which the potential is 0 and if $B(x, y)$ is any other point, then as in (18.17),

$$\int_A^B \mathbf{F} \cdot d\mathbf{r} = f(x, y) - f(x_1, y_1) = f(x, y).$$

This shows that the potential at any point B is the work done in going from a point of zero potential to the point B. This is analogous to the classical physical description of **potential energy** as the type of energy that a body has by virtue of its position.

The following result provides additional information about independence of path.

(18.18) Theorem

If $M(x, y)$ and $N(x, y)$ have continuous first partial derivatives on an open connected region D, and if

$$\int_C M(x, y)\,dx + N(x, y)\,dy$$

is independent of path in D, then

$$\frac{\partial M}{\partial y} = \frac{\partial N}{\partial x}.$$

Proof. By Theorem (18.16) there is a function f such that

$$M = \frac{\partial f}{\partial x} \quad \text{and} \quad N = \frac{\partial f}{\partial y}.$$

Consequently

$$\frac{\partial M}{\partial y} = \frac{\partial^2 f}{\partial y\,\partial x} = \frac{\partial^2 f}{\partial x\,\partial y} = \frac{\partial N}{\partial x}.$$

It follows from Theorem (18.18) that if $\partial M/\partial y \neq \partial N/\partial x$, then the indicated line integral is *not* independent of path.

The converse of (18.18) is false. However, it can be proved that if D is a **simply connected region**, in the sense that every simple closed curve C in D encloses only points in D (that is, there are no "holes" in the region), then the condition $\partial M/\partial y = \partial N/\partial x$ implies that the line integral is independent of path.

Example 1 If $\mathbf{F}(x, y) = (2x + y^3)\mathbf{i} + (3xy^2 + 4)\mathbf{j}$, show that $\int_C \mathbf{F} \cdot d\mathbf{r}$ is independent of path and evaluate

$$\int_{(0,1)}^{(2,3)} \mathbf{F} \cdot d\mathbf{r}.$$

Solution By Theorem (18.15) or (18.16) the integral is independent of path if there exists a (potential) function f such that

$$f_x(x, y) = 2x + y^3 \quad \text{and} \quad f_y(x, y) = 3xy^2 + 4.$$

If we (partially) integrate $f_x(x, y)$ with respect to x we obtain

$$f(x, y) = x^2 + xy^3 + k(y)$$

where k is a function of y alone. Differentiating with respect to y and comparing with $f_y(x, y) = 3xy^2 + 4$ gives us

$$f_y(x, y) = 3xy^2 + k'(y) = 3xy^2 + 4.$$

Consequently $k'(y) = 4$, or $k(y) = 4y + c$ for some constant c. Thus

$$f(x, y) = x^2 + xy^3 + 4y + c$$

defines a function of the desired type. Applying (18.17),

$$\begin{aligned}\int_{(0,1)}^{(2,3)} (2x + y^3)\,dx + (3xy^2 + 4)\,dy &= \int_{(0,1)}^{(2,3)} d(x^2 + xy^3 + 4y) \\ &= x^2 + xy^3 + 4y\Big]_{(0,1)}^{(2,3)} \\ &= (4 + 54 + 12) - 4 = 66.\end{aligned}$$

Example 2 Prove that $\int_A^B x^2y\,dx + 3xy^2\,dy$ is not independent of path.

Solution If we let $M = x^2y$ and $N = 3xy^2$, then

$$\frac{\partial M}{\partial y} = x^2 \quad \text{and} \quad \frac{\partial N}{\partial x} = 3y^2.$$

Since $\partial M/\partial y \neq \partial N/\partial x$, the integral is not independent of path (see (18.18)).

In many applications it is essential to find a potential function f for a conservative vector field $\mathbf{F}$. In Example 1 we determined f for a two-dimensional field. The solution of the next example illustrates a technique that can be used in three dimensions.

Example 3 If $\mathbf{F}(x, y, z) = y^2 \cos x\mathbf{i} + (2y \sin x + e^{2z})\mathbf{j} + 2ye^{2z}\mathbf{k}$, prove that $\int_C \mathbf{F} \cdot d\mathbf{r}$ is independent of path and find a potential function f for $\mathbf{F}$.

Solution The integral is independent of path if there exists a differentiable function f of x, y, and z such that $\nabla f(x, y, z) = \mathbf{F}(x, y, z)$; that is, if

$$\text{(a)} \quad \begin{aligned} f_x(x, y, z) &= y^2 \cos x \\ f_y(x, y, z) &= 2y \sin x + e^{2z} \\ f_z(x, y, z) &= 2ye^{2z}. \end{aligned}$$

If we (partially) integrate $f_x(x, y, z)$ with respect to x there results

(b) $$f(x, y, z) = y^2 \sin x + g(y, z)$$

for some function g of y and z. Differentiating with respect to y and comparing with the equation for f_y in (a) gives us

$$f_y(x, y, z) = 2y \sin x + g_y(y, z) = 2y \sin x + e^{2z}$$

and hence

$$g_y(y, z) = e^{2z}.$$

Integrating with respect to y we obtain

$$g(y, z) = ye^{2z} + k(z)$$

where k is a function of z alone. Consequently, from equation (b)

$$f(x, y, z) = y^2 \sin x + ye^{2z} + k(z).$$

Differentiating with respect to z, and using the equation for f_z in (a),

$$f_z(x, y, z) = 2ye^{2z} + k'(z) = 2ye^{2z}.$$

It follows that $k'(z) = 0$ or $k(z) = c$ for some constant c. Thus

$$f(x, y, z) = y^2 \sin x + ye^{2z} + c$$

defines a potential function for **F**.

EXERCISES 18.3

In Exercises 1–10 determine whether or not $\int_C \mathbf{F} \cdot d\mathbf{r}$ is independent of path. If it is, find a potential function f for **F**.

1 $\mathbf{F}(x, y) = (3x^2y + 2)\mathbf{i} + (x^3 + 4y^3)\mathbf{j}$

2 $\mathbf{F}(x, y) = (6x^2 - 2xy^2)\mathbf{i} + (2x^2y + 5)\mathbf{j}$

3 $\mathbf{F}(x, y) = e^x\mathbf{i} + (3 - e^x \sin y)\mathbf{j}$

4 $\mathbf{F}(x, y) = (x^2 + y^2)^{-1/2}(-y\mathbf{i} + x\mathbf{j})$

5 $\mathbf{F}(x, y) = 4xy^3\mathbf{i} + 2xy^3\mathbf{j}$

6 $\mathbf{F}(x, y) = y^3 \cos x\mathbf{i} - 3y^2 \sin x\mathbf{j}$

7 $\mathbf{F}(x, y, z) = (y \sec^2 x - ze^x)\mathbf{i} + \tan x\mathbf{j} - e^x\mathbf{k}$

8 $\mathbf{F}(x, y, z) = (y + z)\mathbf{i} + (x + z)\mathbf{j} + (x + y)\mathbf{k}$

9 $\mathbf{F}(x, y, z) = 8xz\mathbf{i} + (1 - 6yz^3)\mathbf{j} + (4x^2 - 9y^2z^2)\mathbf{k}$

10 $\mathbf{F}(x, y, z) = e^{-z}\mathbf{i} + 2y\mathbf{j} + xe^{-z}\mathbf{k}$

In each of Exercises 11–14, prove that the line integral is independent of path and find its value.

11 $\int_{(-1,2)}^{(3,1)} (y^2 + 2xy)\,dx + (x^2 + 2xy)\,dy$ **12** $\int_{(0,0)}^{(1,\pi/2)} e^x \sin y\,dx + e^x \cos y\,dy$

13 $\int_{(1,0,2)}^{(-2,1,3)} (6xy^3 + 2z^2)\,dx + 9x^2y^2\,dy + (4xz + 1)\,dz$

14 $\int_{(4,0,3)}^{(-1,1,2)} (yz + 1)\,dx + (xz + 1)\,dy + (xy + 1)\,dz$

15 Suppose a force $\mathbf{F}(x, y, z)$ is directed toward the origin with a magnitude that is inversely proportional to the distance from the origin. Prove that $\mathbf{F}$ is conservative and find a potential function for $\mathbf{F}$.

16 Suppose a force $\mathbf{F}(x, y, z)$ is directed away from the origin with a magnitude that is directly proportional to the distance from the origin. Prove that $\mathbf{F}$ is conservative and find a potential function for $\mathbf{F}$.

17 If $\int_C M(x, y, z)\,dx + N(x, y, z)\,dy + P(x, y, z)\,dz$ is independent of path and M, N, and P have continuous first partial derivatives, prove that

$$\frac{\partial M}{\partial y} = \frac{\partial N}{\partial x}, \quad \frac{\partial M}{\partial z} = \frac{\partial P}{\partial x}, \quad \frac{\partial N}{\partial z} = \frac{\partial P}{\partial y}.$$

18 Prove the analogue of Theorem (18.15) for line integrals in three dimensions.

Use Exercise 17 to prove that the following line integrals are not independent of path.

19 $\int_{(0,0,0)}^{(1,2,3)} 2xy\,dx + (x^2 + z^2)\,dy + yz\,dz$

20 $\int_{(1,0,0)}^{(0,1,\pi)} e^y \cos z\,dx + xe^y \cos z\,dy + xe^y \sin z\,dz$

18.4 GREEN'S THEOREM

Recall that a plane curve C is termed **smooth** if it can be represented parametrically by $x = g(t)$, $y = h(t)$ where g' and h' are continuous on an interval $[a, b]$, and are not simultaneously zero, except possibly at a or b. If

$$A = (g(a), h(a)) = (g(b), h(b)) = B,$$

then C is said to be a **smooth closed curve**. If, in addition, $(g(t_1), h(t_1)) \neq (g(t_2), h(t_2))$ for all other numbers t_1, t_2 in $[a, b]$, then C does not cross itself between A and B and it is referred to as **simple**. Some common examples of smooth simple closed curves are circles and ellipses. A **piecewise smooth simple closed curve** consists of a finite union of smooth curves C_i such that as t varies from a to b, the point $P(t)$ obtained from parametric representations of the C_i traces C exactly one time, with the exception that $P(a) = P(b)$. A curve of this type forms the boundary of a region R in the plane and, by definition, the positive direction along C is such that R is on the left as $P(t)$ traces C. This is illustrated in Figure 18.9, where the arrows indicate the positive direction along C. The symbol

$$\oint_C M(x, y)\,dx + N(x, y)\,dy$$

will denote a line integral along C once in the positive direction.

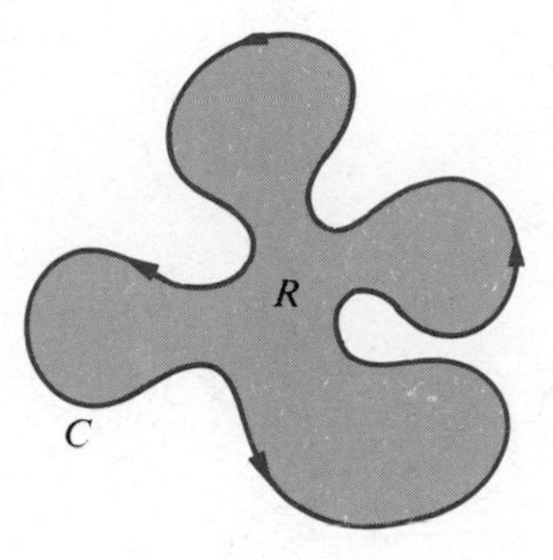

Figure 18.9

The next theorem, named after the English mathematical physicist George Green (1793–1841), indicates the relationship between the line integral around C and a double integral over R. To simplify the statement, the symbols M, N, $\partial M/\partial x$, and $\partial N/\partial y$ are used in the integrands to denote the values of these functions at (x, y).

(18.19) Green's Theorem

Let C be a piecewise smooth simple closed curve and let R be the region consisting of C and its interior. If M and N are functions that are continuous and have continuous first partial derivatives throughout an open region D containing R, then

$$\oint_C M\,dx + N\,dy = \iint_R \left(\frac{\partial N}{\partial x} - \frac{\partial M}{\partial y}\right) dA.$$

Partial proof. We shall prove the theorem for a region R which is both of Type I and of Type II. Thus we may write

$$R = \{(x, y): a \le x \le b, g_1(x) \le y \le g_2(x)\} \quad \text{and}$$
$$R = \{(x, y): c \le y \le d, h_1(y) \le x \le h_2(y)\}$$

where the functions g_1, g_2, h_1, and h_2 are smooth. An illustration of a region of this type appears in Figure 18.10. It is sufficient to show that each of the following is true

(a) $$\oint_C M\,dx = -\iint_R \frac{\partial M}{\partial y}\,dA$$

(b) $$\oint_C N\,dy = \iint_R \frac{\partial N}{\partial x}\,dA$$

since addition of these integrals produces the desired conclusion.

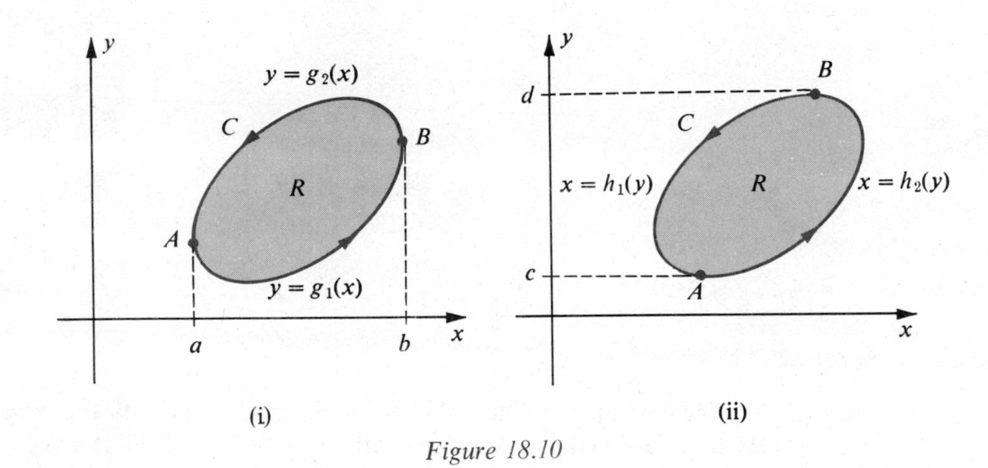

Figure 18.10

To prove (a) we refer to (i) of Figure 18.10 and note that C consists of two curves C_1 and C_2 which have equations $y = g_1(x)$ and $y = g_2(x)$, respectively. The line integral $\oint_C M\,dx$ may be written

$$\begin{aligned}\oint_C M\,dx &= \int_{C_1} M(x, y)\,dx + \int_{C_2} M(x, y)\,dx\\ &= \int_a^b M(x, g_1(x))\,dx + \int_b^a M(x, g_2(x))\,dx\\ &= \int_a^b M(x, g_1(x))\,dx - \int_a^b M(x, g_2(x))\,dx.\end{aligned}$$

Applying (i) of (17.14) gives us

$$\begin{aligned}\iint_R \frac{\partial M}{\partial y}\,dA &= \int_a^b \int_{g_1(x)}^{g_2(x)} \frac{\partial M}{\partial y}\,dy\,dx\\ &= \int_a^b M(x, y)\Big]_{g_1(x)}^{g_2(x)}\,dx\\ &= \int_a^b [M(x, g_2(x)) - M(x, g_1(x))]\,dx.\end{aligned}$$

Comparing the last expression with that obtained for $\oint_C M\,dx$ gives us (a). The formula in (b) may be established in similar fashion by referring to (ii) of Figure 18.10. This is left as an exercise.

Although we shall not prove it, Green's Theorem is true for regions of the types illustrated in Figure 17.1, where part of the boundary consists of horizontal or vertical line segments. The theorem may then be extended to the case where R is a finite union of such regions. For example if, as illustrated in Figure 18.11, $R = R_1 \cup R_2$, where the boundary of R_1 is $C_1 \cup C_1'$ and the boundary of R_2 is $C_2 \cup C_2'$, then

$$\iint_{R_1} \left(\frac{\partial N}{\partial x} - \frac{\partial M}{\partial y}\right) dy\,dx = \oint_{C_1 \cup C_1'} M\,dx + N\,dy$$

$$\iint_{R_2} \left(\frac{\partial N}{\partial x} - \frac{\partial M}{\partial y}\right) dy\,dx = \oint_{C_2 \cup C_2'} M\,dx + N\,dy.$$

The sum of the two double integrals above equals a double integral over R. We next observe that a line integral along C_1' in the direction indicated in Figure 18.11 is the negative of that along C_2' since the curve is the same but the directions are opposite one another. It follows that the sum of the two line integrals reduces to a line integral along $C_1 \cup C_2$, which is the boundary C of R. Thus

$$\iint_{R} \left(\frac{\partial N}{\partial x} - \frac{\partial M}{\partial y}\right) dy\,dx = \oint_{C} M\,dx + N\,dy.$$

We may now proceed, by mathematical induction, to any finite union. The proof of Green's Theorem for the most general case is beyond the scope of this book.

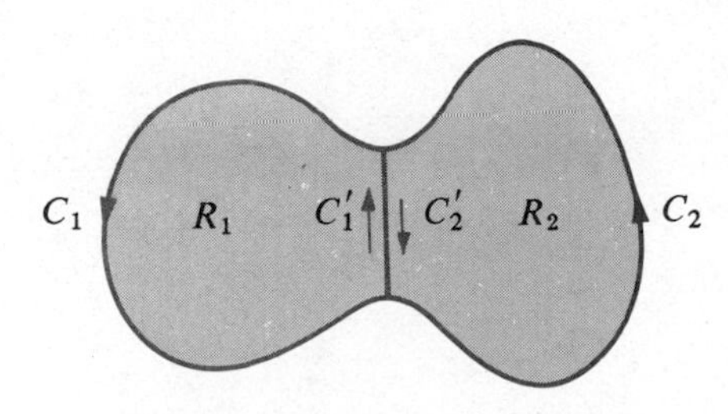

Figure 18.11

Example 1 Use Green's Theorem to evaluate $\oint_C 5xy\,dx + x^3\,dy$, where C is the closed curve consisting of the graphs of $y = x^2$ and $y = 2x$ between the points $(0, 0)$ and $(2, 4)$.

Solution The region R bounded by C is illustrated in Figure 17.7. Applying (18.19) with $M(x, y) = 5xy$ and $N(x, y) = x^3$,

$$\oint_C 5xy\,dx + x^3\,dy = \iint_R \left[\frac{\partial}{\partial x}(x^3) - \frac{\partial}{\partial y}(5xy)\right] dA$$

$$= \int_0^2 \int_{x^2}^{2x} (3x^2 - 5x)\,dy\,dx$$

$$= \int_0^2 \left[3x^2y - 5xy\right]_{x^2}^{2x} dx$$

$$= \int_0^2 (11x^3 - 10x^2 - 3x^4)\,dx = -\frac{28}{15}.$$

Of course, the line integral can also be evaluated directly.

Example 2 Use Green's Theorem to evaluate the line integral

$$\oint_C 2xy\,dx + (x^2 + y^2)\,dy$$

where C is the ellipse $4x^2 + 9y^2 = 36$.

Solution If R is the region bounded by C, then applying (18.19) with $M(x, y) = 2xy$ and $N(x, y) = x^2 + y^2$,

$$\oint_C 2xy\,dx + (x^2 + y^2)\,dy = \iint_R (2x - 2x)\,dA$$

$$= \iint_R 0\,dA = 0.$$

Example 3 Evaluate

$$\oint_C (e^{x^2} + y)\,dx + (x^2 + \tan^{-1}\sqrt{y})\,dy$$

where C is the boundary of the rectangle having vertices $(1, 2)$, $(5, 2)$, $(5, 4)$, and $(1, 4)$.

Solution The region R bounded by C is sketched in Figure 18.12. Applying Green's Theorem, the line integral equals

$$\iint_R (2x - 1)\,dA = \int_2^4 \int_1^5 (2x - 1)\,dx\,dy$$

$$= \int_2^4 \Big[x^2 - x \Big]_1^5 dy$$

$$= \int_2^4 20\,dy$$

$$= 20y \Big]_2^4 = 40.$$

The reader will gain a deeper appreciation of Green's Theorem by attempting to evaluate the line integral in this example directly.

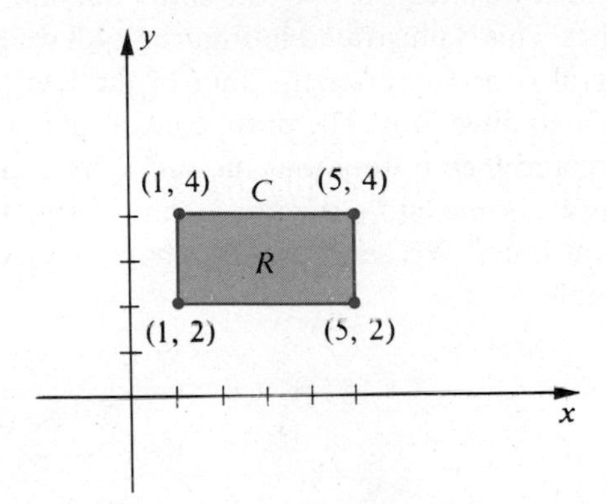

Figure 18.12

Green's Theorem may be used to derive an interesting and useful formula for finding the area A of a region R which is bounded by a piecewise smooth simple closed curve C. On the one hand, if we let $M = 0$ and $N = x$ in (18.19) we obtain

$$A = \iint_R dA = \oint_C x\,dy.$$

On the other hand, if we let $M = -y$ and $N = 0$ there results

$$A = \iint_R dA = -\oint_C y\,dx.$$

We may combine these two formulas for A by adding both sides of the equations and dividing by 2. This gives us

(18.20) $$A = \tfrac{1}{2}\int_C x\,dy - y\,dx.$$

Example 4 Use (18.20) to find the area of the ellipse $\dfrac{x^2}{a^2} + \dfrac{y^2}{b^2} = 1$.

Solution Parametric equations for the ellipse are $x = a\cos t$, $y = b\sin t$, where $0 \le t \le 2\pi$. Substituting in (18.20) we obtain

$$\begin{aligned} A &= \frac{1}{2}\int_0^{2\pi} (a\cos t)(b\cos t)\,dt - (b\sin t)(-a\sin t)\,dt \\ &= \frac{1}{2}\int_0^{2\pi} ab(\cos^2 t + \sin^2 t)\,dt \\ &= \frac{ab}{2}\int_0^{2\pi} dt = \frac{ab}{2}(2\pi) = \pi ab. \end{aligned}$$

Green's Theorem can be extended to a region R which contains holes, provided we integrate over the entire boundary and always keep the region R to the left. This is illustrated in Figure 18.13, where it can be shown that the double integral over R equals the sum of the line integrals along C_1 and C_2 in the indicated directions. The proof consists of making a slit in R as illustrated in the figure and then noting that the sum of two line integrals in opposite directions along the same curve is zero. A similar argument can be used if the region has several holes. We shall use the above observation in the solution of the next example.

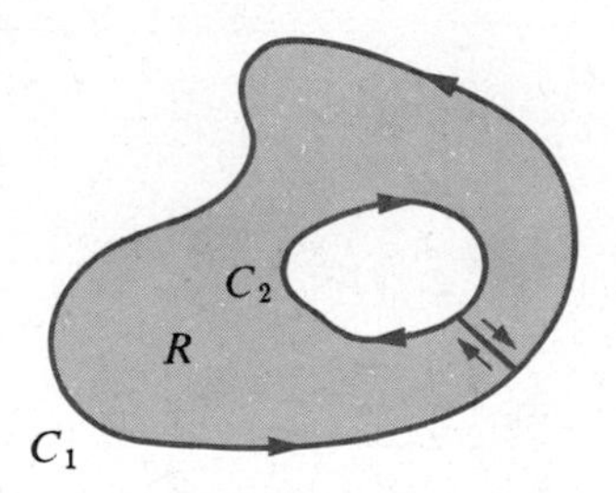

Figure 18.13

Example 5 Let C_1 and C_2 be two nonintersecting piecewise smooth simple closed curves having the origin O as an interior point. If $M = -y/(x^2 + y^2)$ and $N = x/(x^2 + y^2)$, prove that

$$\oint_{C_1} M\,dx + N\,dy = \oint_{C_2} M\,dx + N\,dy.$$

Solution If R denotes the region between C_1 and C_2, then we have a situation similar to that illustrated in Figure 18.13, where O is inside of C_2. By the remarks preceding this example,

$$\oint_{C_1} M\,dx + N\,dy + \oint_{C_2} M\,dx + N\,dy = \iint_R \left(\frac{\partial N}{\partial x} - \frac{\partial M}{\partial y}\right) dA,$$

where $\circlearrowleft$ indicates the positive direction along C_2 *with respect to* R. Since

$$\frac{\partial N}{\partial x} = \frac{(x^2 + y^2)(1) - x(2x)}{(x^2 + y^2)^2} = \frac{y^2 - x^2}{(x^2 + y^2)^2} = \frac{\partial M}{\partial y}$$

the double integral over R is zero. Consequently

$$\oint_{C_1} M\,dx + N\,dy = -\oint_{C_2} M\,dx + N\,dy.$$

Since the positive direction along C_2 *with respect to the region* R_2 *in the interior of* C_2 is opposite to that indicated in the last integral we have

$$\oint_{C_1} M\,dx + N\,dy = \oint_{C_2} M\,dx + N\,dy$$

which is what we wished to prove.

Example 6 If

$$\mathbf{F}(x, y) = \frac{1}{x^2 + y^2}(-y\mathbf{i} + x\mathbf{j})$$

prove that

$$\oint_C \mathbf{F} \cdot d\mathbf{r} = 2\pi$$

for every piecewise smooth simple closed curve C having the origin in its interior.

Solution The line integral has the same form as those considered in Example 5. Hence, if we choose a circle C_1 of radius a with center at the origin which lies entirely within C, then by Example 5,

$$\oint_C \mathbf{F} \cdot d\mathbf{r} = \oint_{C_1} \mathbf{F} \cdot d\mathbf{r}.$$

Since parametric equations for C_1 are

$$x = a\cos t, \quad y = a\sin t; \quad 0 \le t \le 2\pi,$$

we have

$$\begin{aligned}\oint_C \mathbf{F} \cdot d\mathbf{r} &= \oint_{C_1} \frac{-y}{x^2 + y^2}\,dx + \frac{x}{x^2 + y^2}\,dy \\ &= \int_0^{2\pi} \frac{-a\sin t}{a^2}(-a\sin t)\,dt + \frac{a\cos t}{a^2}(a\cos t)\,dt \\ &= \int_0^{2\pi} (\sin^2 t + \cos^2 t)\,dt = \int_0^{2\pi} dt = 2\pi.\end{aligned}$$

Green's Theorem has far-reaching consequences in advanced mathematics. This is especially true in the area known as *complex variables*, a subject which is fundamental for many important applications in the physical sciences and engineering.

EXERCISES 18.4

Use Green's Theorem to evaluate the line integrals in Exercises 1–10.

1 $\oint_C (x^2 + y)\,dx + xy^2\,dy$, where C is the closed curve determined by $y^2 = x$ and $y = -x$ from $(0, 0)$ to $(1, -1)$

2 $\oint_C (x + y^2)\,dx + (1 + x^2)\,dy$, where C is the closed curve determined by $y = x^3$ and $y = x^2$ from $(0, 0)$ to $(1, 1)$

3 $\oint_C x^2y^2\,dx + (x^2 - y^2)\,dy$ where C is the square with vertices $(0, 0)$, $(1, 0)$, $(1, 1)$, and $(0, 1)$

4 $\oint_C \sqrt{y}\,dx + \sqrt{x}\,dy$ where C is the triangle with vertices $(1, 1)$, $(3, 1)$, and $(2, 2)$

5 $\oint_C xy\,dx + (y + x)\,dy$ where C is the unit circle with center at $(0, 0)$

6 $\oint_C e^x \sin y\,dx + e^x \cos y\,dy$ where C is the ellipse $3x^2 + 8y^2 = 24$

7 $\oint_C xy\,dx + \sin y\,dy$ where C is the triangle with vertices $(1, 1)$, $(2, 2)$, and $(3, 0)$

8 $\oint_C \tan^{-1} x\,dx + 3x\,dy$ where C is the rectangle with vertices $(1, 0)$, $(0, 1)$, $(2, 3)$, and $(3, 2)$

9 $\oint_C y^2(1 + x^2)^{-1}\,dx + 2y \tan^{-1} x\,dy$ where C is the graph of $x^{2/3} + y^{2/3} = 1$

10 $\oint_C (x^2 + y^2)\,dx + 2xy\,dy$ where C is the boundary of the region bounded by the graphs of $y = \sqrt{x}$, $y = 0$, and $x = 4$

In Exercises 11 and 12 use (18.20) to find the area of the region bounded by the curve C with the given parametric equations.

11 $x = a\cos^3 t, y = a\sin^3 t; 0 \le t \le 2\pi$

12 $x = a\cos t, y = a\sin t; 0 \le t \le 2\pi$

In Exercises 13 and 14 use (18.20) to find the area of the region bounded by the graphs of the given equations.

13 $y = 4x^2, y = 16x$

14 $y = x^3, y^2 = x$

15 Prove formula (b) in the partial proof of Green's Theorem.

16 If $\mathbf{F}(x, y)$ is a two-dimensional vector field and $\int_A^B \mathbf{F} \cdot d\mathbf{r}$ is independent of path in a region D, use Green's Theorem to prove that $\oint_C \mathbf{F} \cdot d\mathbf{r} = 0$ for every piecewise smooth simple closed curve in D.

17 Let R be the region bounded by a piecewise smooth simple closed curve C in the xy-plane. If the area of R is A, use Green's Theorem to prove that the centroid $(\bar{x}, \bar{y})$ is given by

$$\bar{x} = \frac{1}{2A}\oint_C x^2\,dy, \quad \bar{y} = -\frac{1}{2A}\oint_C y^2\,dx.$$

18 Suppose a homogeneous lamina of density k has the shape of a region in the xy-plane which is bounded by a piecewise smooth simple closed curve C. Prove that the moments of inertia with respect to the x- and y-axes are given by

$$I_x = -\frac{k}{3}\oint_C y^3\,dx, \quad I_y = \frac{k}{3}\oint_C x^3\,dy.$$

19 Use Exercise 17 to find the centroid of a semicircular region of radius a.

20 Use Exercise 18 to find the moment of inertia of a homogeneous circular disc of radius a with respect to a diameter.

18.5 DIVERGENCE AND CURL

The vector differential operator ∇ in three dimensions is defined by

$$\nabla = \mathbf{i}\frac{\partial}{\partial x} + \mathbf{j}\frac{\partial}{\partial y} + \mathbf{k}\frac{\partial}{\partial z}.$$

It has no practical use standing alone; however, if it operates on a scalar function f it produces the **gradient** of f, as given by

$$\text{grad}\, f = \nabla f = \frac{\partial f}{\partial x}\mathbf{i} + \frac{\partial f}{\partial y}\mathbf{j} + \frac{\partial f}{\partial z}\mathbf{k}.$$

We shall now use ∇ as an operator on *vector* functions. Most of the work in this section will be manipulative. The physical significance of the functions to be defined will be discussed later in this chapter.

Suppose a vector function $\mathbf{F}$ in three dimensions is given by

$$\mathbf{F}(x, y, z) = M(x, y, z)\mathbf{i} + N(x, y, z)\mathbf{j} + P(x, y, z)\mathbf{k}$$

where M, N, and P have partial derivatives in some region. The **curl** of $\mathbf{F}$, denoted by curl $\mathbf{F}$ or $\nabla \times \mathbf{F}$, may be defined as the determinant "expansion" (by the first row) of the following expression:

(18.21)
$$\text{curl}\,\mathbf{F} = \nabla \times \mathbf{F} = \begin{vmatrix} \mathbf{i} & \mathbf{j} & \mathbf{k} \\ \dfrac{\partial}{\partial x} & \dfrac{\partial}{\partial y} & \dfrac{\partial}{\partial z} \\ M & N & P \end{vmatrix}.$$

The above determinant notation is ambiguous, since the first row is made up of vectors, the second row of partial derivative operators, and the third row of scalar functions; however, it is an extremely useful device for remembering the cumbersome formula

(18.22)
$$\nabla \times \mathbf{F} = \left(\frac{\partial P}{\partial y} - \frac{\partial N}{\partial z}\right)\mathbf{i} + \left(\frac{\partial M}{\partial z} - \frac{\partial P}{\partial x}\right)\mathbf{j} + \left(\frac{\partial N}{\partial x} - \frac{\partial M}{\partial y}\right)\mathbf{k}.$$

The curl of **F** determines a vector field in three dimensions which has important applications in the physical sciences.

The **divergence** of **F**, denoted by div **F** or $\nabla \cdot \mathbf{F}$, is defined by

(18.23)
$$\operatorname{div} \mathbf{F} = \nabla \cdot \mathbf{F} = \frac{\partial M}{\partial x} + \frac{\partial N}{\partial y} + \frac{\partial P}{\partial z}.$$

The reason for using the symbol $\nabla \cdot \mathbf{F}$ is evident from the fact that the formula for div **F** may be obtained by taking the formal inner product of ∇ and **F**. Note that div **F** is a *scalar* function.

Example 1 Find $\nabla \times \mathbf{F}$ and $\nabla \cdot \mathbf{F}$ if

$$\mathbf{F}(x, y, z) = xy^2z^4\mathbf{i} + (2x^2y + z)\mathbf{j} + y^3z^2\mathbf{k}.$$

Solution For simplicity we use the symbols $\nabla \times \mathbf{F}$ and $\nabla \cdot \mathbf{F}$ to denote the values of these functions at (x, y, z). Applying (18.21)–(18.23),

$$\nabla \times \mathbf{F} = \begin{vmatrix} \mathbf{i} & \mathbf{j} & \mathbf{k} \\ \dfrac{\partial}{\partial x} & \dfrac{\partial}{\partial y} & \dfrac{\partial}{\partial z} \\ xy^2z^4 & (2x^2y + z) & y^3z^2 \end{vmatrix}$$
$$= (3y^2z^2 - 1)\mathbf{i} + 4xy^2z^3\mathbf{j} + (4xy - 2xyz^4)\mathbf{k}.$$

$$\nabla \cdot \mathbf{F} = \frac{\partial}{\partial x}(xy^2z^4) + \frac{\partial}{\partial y}(2x^2y + z) + \frac{\partial}{\partial z}(y^3z^2)$$
$$= y^2z^4 + 2x^2 + 2y^3z.$$

A number of interesting algebraic properties can be established for the curl and divergence of vector functions. A property which is analogous to the product rule for derivatives is given in the next example. Several other properties are stated in exercises at the end of this section.

Example 2 If f and **F** are scalar and vector functions, respectively, which possess partial derivatives, show that

$$\nabla \cdot [f\mathbf{F}] = f[\nabla \cdot \mathbf{F}] + [\nabla f] \cdot \mathbf{F}.$$

Solution If we write $\mathbf{F} = M\mathbf{i} + N\mathbf{j} + P\mathbf{k}$, where M, N, and P are functions of x, y, and z, then

$$f\mathbf{F} = fM\mathbf{i} + fN\mathbf{j} + fP\mathbf{k}.$$

Applying (18.23),

$$\nabla \cdot [f\mathbf{F}] = \frac{\partial}{\partial x}(fM) + \frac{\partial}{\partial y}(fN) + \frac{\partial}{\partial z}(fP)$$

$$= f\frac{\partial M}{\partial x} + \frac{\partial f}{\partial x}M + f\frac{\partial N}{\partial y} + \frac{\partial f}{\partial y}N + f\frac{\partial P}{\partial z} + \frac{\partial f}{\partial z}P.$$

Rearranging terms gives us

$$\nabla \cdot [f\mathbf{F}] = f\left[\frac{\partial M}{\partial x} + \frac{\partial N}{\partial y} + \frac{\partial P}{\partial z}\right] + \left[\frac{\partial f}{\partial x}M + \frac{\partial f}{\partial y}N + \frac{\partial f}{\partial z}P\right]$$

$$= f[\nabla \cdot \mathbf{F}] + [\nabla f] \cdot \mathbf{F}.$$

We see from (18.22) that the coefficient of **k** in the formula for $\nabla \times \mathbf{F}$ has the same form as the integrand of the double integral in the statement of Green's Theorem (18.19). Consequently, if we let

$$\mathbf{F}(x, y) = M(x, y)\mathbf{i} + N(x, y)\mathbf{j} + 0\mathbf{k}$$

and consider the tangent vector

$$\mathbf{T} = \frac{dx}{ds}\mathbf{i} + \frac{dy}{ds}\mathbf{j} + \frac{dz}{ds}\mathbf{k}$$

to C, where s represents arc length, then the conclusion of Green's Theorem (18.19) may be written

(18.24) $$\oint_C \mathbf{F} \cdot \mathbf{T}\, ds = \iint_R (\nabla \times \mathbf{F}) \cdot \mathbf{k}\, dA.$$

Since $(\nabla \times \mathbf{F}) \cdot \mathbf{k}$ is the component of curl **F** in the direction of the z-axis we shall refer to it as the **normal component** (to R) of curl **F**. In words, (18.24) may be phrased as follows: *the line integral of the tangential component of* **F** *taken along C once in the positive direction is equal to the double integral over R of the normal component of* curl **F**. The three-dimensional analogue of this result is *Stokes' Theorem*, which will be discussed in Section 8.

As a final remark on notation, the differential operator Lap is denoted by

$$\nabla^2 = \nabla \cdot \nabla = \frac{\partial^2}{\partial x^2} + \frac{\partial^2}{\partial y^2} + \frac{\partial^2}{\partial z^2}.$$

If it operates on $f(x, y, z)$ it produces a scalar function called the **Laplacian** of f. Thus, by definition,

(18.25) $$\text{Lap}\, f = \nabla^2 f = \frac{\partial^2 f}{\partial x^2} + \frac{\partial^2 f}{\partial y^2} + \frac{\partial^2 f}{\partial z^2}.$$

Functions which satisfy **Laplace's equation** $\nabla^2 f = 0$ are called *harmonic* and are very important in physical applications.

EXERCISES 18.5

In Exercises 1–4 find $\nabla \times \mathbf{F}$ and $\nabla \cdot \mathbf{F}$.

1 $\mathbf{F}(x, y, z) = x^2 z\mathbf{i} + y^2 x\mathbf{j} + (y + 2z)\mathbf{k}$

2 $\mathbf{F}(x, y, z) = (3x + y)\mathbf{i} + xy^2 z\mathbf{j} + xz^2\mathbf{k}$

3 $\mathbf{F}(x, y, z) = 3xyz^2\mathbf{i} + y^2 \sin z\mathbf{j} + xe^{2z}\mathbf{k}$

4 $\mathbf{F}(x, y, z) = x^3 \ln z\mathbf{i} + xe^{-y}\mathbf{j} - (y^2 + 2z)\mathbf{k}$

Verify the identities in Exercises 5–8.

5 $\nabla \times (\mathbf{F} + \mathbf{G}) = \nabla \times \mathbf{F} + \nabla \times \mathbf{G}$

6 $\nabla \cdot (\mathbf{F} + \mathbf{G}) = \nabla \cdot \mathbf{F} + \nabla \cdot \mathbf{G}$

7 $\nabla \times (f\mathbf{F}) = f(\nabla \times \mathbf{F}) + (\nabla f) \times \mathbf{F}$

8 $\nabla \cdot (\mathbf{F} \times \mathbf{G}) = (\nabla \times \mathbf{F}) \cdot \mathbf{G} - (\nabla \times \mathbf{G}) \cdot \mathbf{F}$

If f and $\mathbf{F}$ have continuous second partial derivatives, verify the identities in Exercises 9–12.

9 $\operatorname{curl} \operatorname{grad} f = \mathbf{0}$

10 $\operatorname{div} \operatorname{curl} \mathbf{F} = 0$

11 $\operatorname{curl}(\operatorname{grad} f + \operatorname{curl} \mathbf{F}) = \operatorname{curl} \operatorname{curl} \mathbf{F}$

12 $\operatorname{div} \operatorname{grad} f = \operatorname{Lap} f$

13 If $\mathbf{r} = x\mathbf{i} + y\mathbf{j} + z\mathbf{k}$ prove that $\nabla \cdot \mathbf{r} = 3, \nabla \times \mathbf{r} = \mathbf{0}$, and $\nabla|\mathbf{r}| = \mathbf{r}/|\mathbf{r}|$.

14 If $\mathbf{r} = x\mathbf{i} + y\mathbf{j} + z\mathbf{k}$ and $\mathbf{a}$ is a constant vector, prove that $\operatorname{curl}(\mathbf{a} \times \mathbf{r}) = 2\mathbf{a}$.

15 Prove that both the curl and the divergence of an inverse square vector field are zero.

16 Let $\mathbf{r} = x\mathbf{i} + y\mathbf{j} + z\mathbf{k}$ and $r = |\mathbf{r}|$. If $\mathbf{F}(x, y, z) = (c/r^k)\mathbf{r}$, where c is a constant and k is any positive real number, prove that the curl of $\mathbf{F}$ is $\mathbf{0}$. (*Hint:* Use Exercise 7.)

17 If a vector field $\mathbf{F}$ is conservative and has continuous partial derivatives, prove that $\operatorname{curl} \mathbf{F} = \mathbf{0}$.

18 If f and g are scalar functions which have second partial derivatives, prove that

$$\operatorname{Lap}(fg) = f \operatorname{Lap} g + g \operatorname{Lap} f + 2(\operatorname{grad} f) \cdot (\operatorname{grad} g).$$

Prove that the functions defined in Exercises 19 and 20 satisfy Laplace's equation.

19 $f(x, y, z) = (x^2 + y^2 + z^2)^{-1/2}$

20 $f(x, y, z) = ax^2 + by^2 + cz^2$, where $a + b + c = 0$.

18.6 SURFACE INTEGRALS

Line integrals are evaluated along curves. Double and triple integrals are defined on regions in two and three dimensions, respectively. It is also possible to consider an integral of a function over a surface. For simplicity we shall restrict our discussion to rather well-behaved surfaces, and our demonstrations will be at an intuitive level. Rigorous treatments may be found in advanced texts on calculus.

If the projection of a surface S on a coordinate plane is a region of the type we considered for double integrals, then S is said to have a **regular projection** on the coordinate plane. Suppose that S is the graph of $z = f(x, y)$ where S has a regular projection R on the xy-plane, and that f has continuous first partial derivatives on R. In the previous chapter we defined the area A of S. A similar technique may be used to define an integral of $g(x, y, z)$ over the surface S, where the function g is continuous throughout a region containing S. We shall employ the notation used in Section 17.9. In particular, as illustrated in Figure 17.42, ΔS_i and ΔT_i denote areas of parts of S and the tangent plane to S at $B_i(x_i, y_i, z_i)$, respectively, which project onto the rectangle R_i of an inner partition P of R. In a manner analogous to the definitions of all previous integrals, we evaluate g at B_i for each i and form the sum $\Sigma_i\, g(x_i, y_i, z_i)\,\Delta T_i$. By definition, the **surface integral of g over S** is given by

(18.26)
$$\iint\limits_S g(x, y, z)\,dS = \lim_{\|P\|\to 0} \sum_i g(x_i, y_i, z_i)\,\Delta T_i$$

where the limit of the sum is defined in the usual way. In a manner similar to the development of (17.52), this integral may be evaluated by means of the formula

(18.27)
$$\iint\limits_S g(x, y, z)\,dS = \iint\limits_R g(x, y, f(x, y))\sqrt{(f_x(x, y))^2 + (f_y(x, y))^2 + 1}\,dA.$$

Observe that if $g(x, y, z) = 1$ for all (x, y, z), then (18.27) reduces to (17.52) and hence the surface integral equals the surface area of S.

An elementary physical application may be obtained by considering a thin metal sheet which has the shape of S. If the area density at (x, y, z) is $g(x, y, z)$, then (18.27) is the mass of the sheet. The center of gravity and moments of inertia may be obtained by employing the methods used for solids in the preceding chapter.

Example 1 Evaluate $\iint_S x^2 z\,dS$ where S is the portion of the cone $z^2 = x^2 + y^2$ which lies between the planes $z = 1$ and $z = 4$.

Solution As shown in Figure 18.14, the projection R of S onto the xy-plane is the annular region bounded by circles of radii 1 and 4 with centers at the origin. If we write the equation for S in the form

$$z = (x^2 + y^2)^{1/2} = f(x, y)$$

then

$$f_x(x, y) = \frac{x}{(x^2 + y^2)^{1/2}} \quad \text{and} \quad f_y(x, y) = \frac{y}{(x^2 + y^2)^{1/2}}.$$

Applying (18.27) and noting that the radical reduces to $\sqrt{2}$ we obtain

$$\iint\limits_S x^2 z\,dS = \iint\limits_R x^2(x^2 + y^2)^{1/2}\sqrt{2}\,dx\,dy.$$

Using polar coordinates to evaluate the double integral,

$$\iint_S x^2 z\, dS = \int_0^{2\pi}\int_1^4 (r^2\cos^2\theta) r\,\sqrt{2} r\, dr\, d\theta$$

$$= \sqrt{2}\int_0^{2\pi} \cos^2\theta \frac{r^5}{5}\Big]_1^4\, d\theta$$

$$= \frac{1023\sqrt{2}}{5}\int_0^{2\pi} \frac{1+\cos 2\theta}{2}\, d\theta$$

$$= \frac{1023\sqrt{2}}{10}\Big[\theta + \frac{1}{2}\sin 2\theta\Big]_0^{2\pi}$$

$$= \frac{1023\sqrt{2}\pi}{5} \approx 909.0.$$

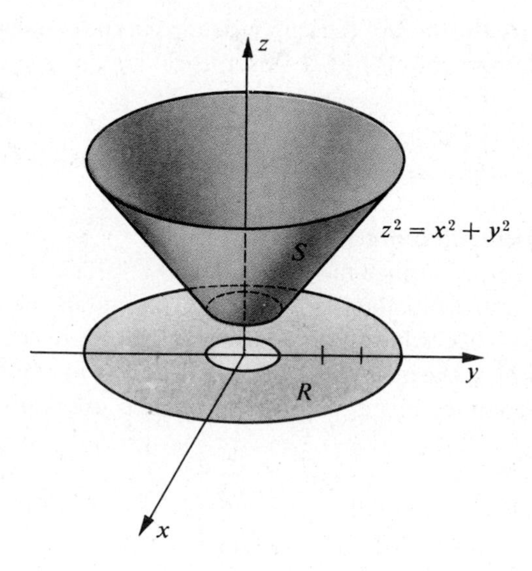

Figure 18.14

As before, suppose S is the graph of $z = f(x, y)$, where S has a regular projection R on the xy-plane, and f has continuous first partial derivatives on R. Let us consider a unit normal vector

(18.28) $$\mathbf{n} = \cos\alpha\mathbf{i} + \cos\beta\mathbf{j} + \cos\gamma\mathbf{k}$$

to S where α, β, γ are the direction cosines of $\mathbf{n}$. As illustrated in Figure 18.15 we shall choose the **upper normal** to S, in the sense that $0° \leq \gamma \leq 90°$. If B_i, $\mathbf{a}$, and $\mathbf{b}$ are as shown in Figure 17.43, then

$$\mathbf{n} = \frac{1}{|\mathbf{a}\times\mathbf{b}|}\mathbf{a}\times\mathbf{b}.$$

Applying (17.50) and (17.51) and abbreviating $f_x(x_i, y_i)$ and $f_y(x_i, y_i)$ by f_x and f_y gives us

$$\mathbf{n} = \frac{1}{\sqrt{f_x^2 + f_y^2 + 1}}(-f_x\mathbf{i} - f_y\mathbf{j} + \mathbf{k}). \tag{18.29}$$

Comparing this formula for $\mathbf{n}$ with (18.28) we see that

$$\cos\gamma = \frac{1}{\sqrt{f_x^2 + f_y^2 + 1}}$$

and hence, if γ is acute, (18.27) may be written

$$\iint_S g(x, y, z)\,dS = \iint_R g(x, y, f(x, y)) \sec\gamma\,dx\,dy. \tag{18.30}$$

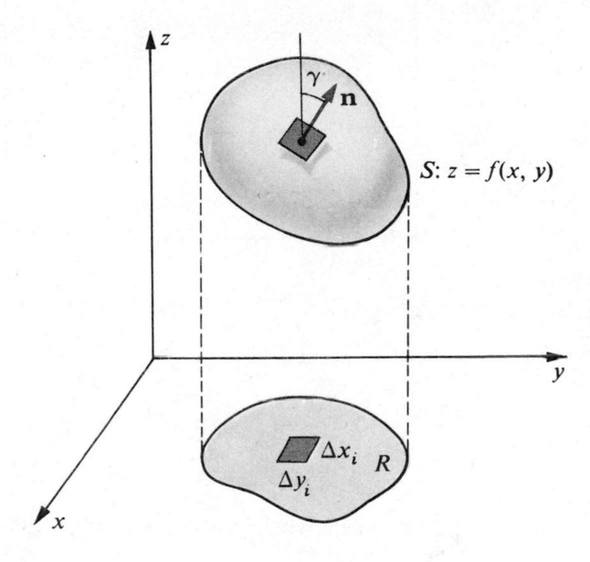

Figure 18.15

In like manner, if an equation for S is $y = h(x, z)$ where h has continuous first partial derivatives, and S has a regular projection R_1 on the xz-plane (see (i) of Figure 18.16), then

$$\iint_S g(x, y, z)\,dS = \iint_{R_1} g(x, h(x, z), z) \sec\beta\,dx\,dz.$$

This integral can be evaluated by means of a formula similar to (18.27), using h in place of f and z in place of y as follows:

$$\iint_S g(x, y, z)\,dS = \iint_{R_1} g(x, h(x, z), z)\sqrt{(h_x(x, z))^2 + (h_z(x, z))^2 + 1}\,dx\,dz.$$

Similarly, if S is given by $x = k(y, z)$ where k has continuous first partial derivatives and if S has a regular projection R_2 on the yz-plane (see (ii) of Figure 18.16), then

$$\iint_S g(x, y, z)\, dS = \iint_{R_2} g(k(y, z), y, z) \sec \alpha \, dy\, dz.$$

This integral may be evaluated as follows:

$$\iint_S g(x, y, z)\, dS = \iint_{R_2} g(k(y, z), y, z)\sqrt{(k_y(y, z))^2 + (k_z(y, z))^2 + 1}\, dy\, dz.$$

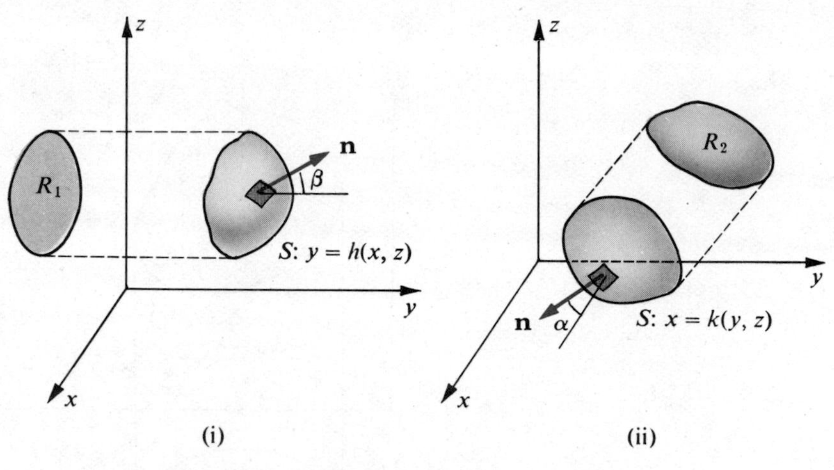

Figure 18.16

Example 2 Evaluate $\iint_S (xz/y)\, dS$, where S is the part of the cylinder $x = y^2$ which lies in the first octant between the planes $z = 0$, $z = 5$, $y = 1$, and $y = 4$.

Solution The surface S is sketched in Figure 18.17, where for clarity we have used a different unit of length on the x-axis. Since the projection of S on the yz-plane is the rectangle with vertices $(0, 1, 0)$, $(0, 4, 0)$, $(0, 4, 5)$, and $(0, 1, 5)$, we may use the formula which precedes this example, with $k(y, z) = y^2$, as follows:

$$\begin{aligned}\iint_S \frac{xz}{y}\, dS &= \int_1^4 \int_0^5 \frac{y^2 z}{y} \sqrt{(2y)^2 + 0^2 + 1}\, dz\, dy \\ &= \int_1^4 \int_0^5 yz\sqrt{4y^2 + 1}\, dz\, dy \\ &= \int_1^4 y\sqrt{4y^2 + 1}\left[\frac{z^2}{2}\right]_0^5 dy\end{aligned}$$

$$= \frac{25}{2}\int_1^4 y\sqrt{4y^2+1}\,dy$$

$$= \frac{25}{24}(4y^2+1)^{3/2}\Big]_1^4$$

$$= \frac{25}{24}[65^{3/2} - 5^{3/2}] \approx 534.2.$$

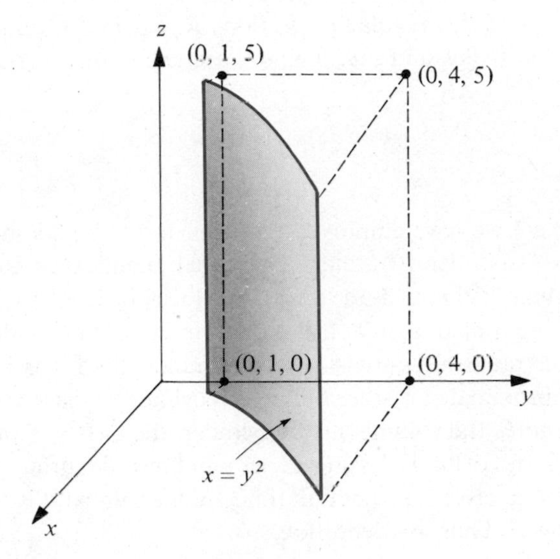

Figure 18.17

From (17.51) we see that

$$\Delta T_i = \sec\gamma\,\Delta x_i\,\Delta y_i$$

that is, the area on the tangent plane can be found by multiplying the area of R_i by $\sec\gamma$. If we represent the area of R_i by $dx\,dy$, then we are motivated to define the **differential dS of surface area** by

$$dS = \sec\gamma\,dx\,dy.$$

In applications we may regard dS as the element of surface area on S which projects onto a rectangular region in the xy-plane of area $dx\,dy$. Similarly, for the surface integrals corresponding to (i) and (ii) in Figure 18.16 we write

$$dS = \sec\beta\,dx\,dz \quad \text{and} \quad dS = \sec\alpha\,dy\,dz$$

respectively. In certain cases we may consider surface integrals which involve a vector function $\mathbf{F}$, where

$$\mathbf{F}(x, y, z) = M(x, y, z)\mathbf{i} + N(x, y, z)\mathbf{j} + P(x, y, z)\mathbf{k}$$

and M, N, and P are continuous functions. If $\mathbf{n}$ is the unit normal (18.28), then $\mathbf{F}(x, y, z) \cdot \mathbf{n}$ is called the **normal component** of $\mathbf{F}$ at the point (x, y, z). By definition, the **surface integral of the normal component of F over *S*** is given by

(18.31)
$$\iint_S \mathbf{F} \cdot \mathbf{n}\, dS = \iint_S (M \cos \alpha + N \cos \beta + P \cos \gamma)\, dS$$

where we have used the symbols for the functions to denote their values at (x, y, z).

If S has regular projections R, R_1, and R_2 on the xy-, xz-, and yz-planes, respectively, and if α, β, and γ are acute, then (18.31) may be written

(18.32)
$$\iint_S \mathbf{F} \cdot \mathbf{n}\, dS = \iint_{R_2} M\, dy\, dz + \iint_{R_1} N\, dx\, dz + \iint_{R} P\, dx\, dy$$

where we have employed the three forms for dS introduced previously.

In order to attach a physical significance to (18.31), suppose that S is submerged in a fluid having a velocity field $\mathbf{F}(x, y, z)$. Let dS represent a small element of area on S. If $\mathbf{F}$ is continuous, then it is almost constant on dS and, as illustrated in Figure 18.18, the amount of fluid crossing dS in a unit of time may be approximated by the volume of a cylinder of base area dS and altitude $\mathbf{F} \cdot \mathbf{n}$. If dV denotes the volume of this cylinder, then $dV = \mathbf{F} \cdot \mathbf{n}\, dS$. Since dV represents the amount of fluid crossing dS per unit time, the surface integral (18.31) is the volume of fluid crossing S per unit time. In this context it is called the *flux* of $\mathbf{F}$ *through* or *over* S. Thus, by definition,

(18.33)
$$\textbf{the flux of F through } S = \iint_S \mathbf{F} \cdot \mathbf{n}\, dS.$$

This terminology is also used for arbitrary vector fields.

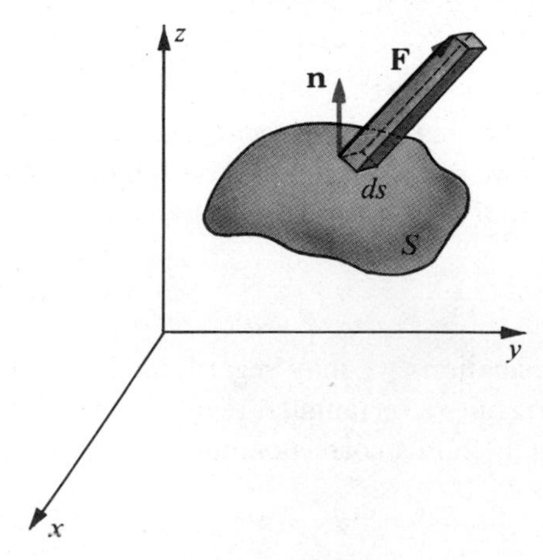

Figure 18.18

Surface integrals may be defined over closed surfaces, such as spheres, ellipsoids, and so on. In this case if we choose $\mathbf{n}$ as the *outer normal*, then the flux measures the *net outward flow* per unit time. If the integral in (18.33) is positive, then the flow out of S exceeds the flow into S and we say there is a **source** of $\mathbf{F}$ within S. If the integral is negative, then the flow into S exceeds the flow out of S and we say there is a **sink** within S. If the integral is 0, then the flow into and the flow out of S are equal, that is, the sources and sinks balance one another.

Example 3 Let S be the part of the graph of $z = 9 - x^2 - y^2$ such that $z \geq 0$. If $\mathbf{F}(x, y, z) = 3x\mathbf{i} + 3y\mathbf{j} + z\mathbf{k}$, find the flux of $\mathbf{F}$ through S.

Solution The graph is sketched in Figure 16.5. According to (18.29), a unit upper normal to S at the point (x, y, z) is

$$\mathbf{n} = \frac{2x\mathbf{i} + 2y\mathbf{j} + \mathbf{k}}{\sqrt{4x^2 + 4y^2 + 1}}.$$

Hence the flux of $\mathbf{F}$ through S is

$$\iint_S \mathbf{F} \cdot \mathbf{n}\, dS = \iint_S \frac{6x^2 + 6y^2 + z}{\sqrt{4x^2 + 4y^2 + 1}}\, dS.$$

Applying (18.27)

$$\begin{aligned}\iint_S \mathbf{F} \cdot \mathbf{n}\, dS &= \iint_R (6x^2 + 6y^2 + 9 - x^2 - y^2)\, dx\, dy \\ &= \iint_R (5x^2 + 5y^2 + 9)\, dx\, dy\end{aligned}$$

where R is the circular region in the xy-plane bounded by the graph of $x^2 + y^2 = 9$. Using polar coordinates gives us

$$\iint_S \mathbf{F} \cdot \mathbf{n}\, dS = \int_0^{2\pi} \int_0^3 (5r^2 + 9)r\, dr\, d\theta = \frac{567\pi}{2} \approx 890.6.$$

EXERCISES 18.6

In Exercises 1–4, evaluate $\iint_S f(x, y, z)\, dS$.

1 $f(x, y, z) = x^2$; S the upper half of the sphere $x^2 + y^2 + z^2 = a^2$

2 $f(x, y, z) = x^2 + y^2 + z^2$; S the part of the plane $z = y + 4$ that is inside the cylinder $x^2 + y^2 = 4$

3 $f(x, y, z) = x + y$; S the first octant portion of the plane $2x + 3y + z = 6$

4 $f(x, y, z) = (x^2 + y^2 + 1)^{1/2}$; S the part of the paraboloid $2z = x^2 + y^2$ that lies inside the cylinder $x^2 + y^2 = 2y$

In each of Exercises 5–8 set up, but do not evaluate, the given surface integral by using a projection of S on (a) the yz-plane; (b) the xz-plane.

5 $\iint_S xy^2z^3\,dS$; S the part of the plane $2x + 3y + 4z = 12$ that is in the first octant.

6 $\iint_S (xz + 2y)\,dS$; S the part of the graph of $y = x^3$ between the planes $y = 0$, $y = 8$, $z = 2$, and $z = 0$.

7 $\iint_S (x^2 - 2y + z)\,dS$; S the part of the graph of $4x + y = 8$ bounded by the coordinate planes and the plane $z = 6$.

8 $\iint_S (x^2 + y^2 + z^2)\,dS$; S the first octant part of the graph of $x^2 + y^2 = 4$ bounded by the coordinate planes and the plane $x + z = 2$.

9 Interpret $\iint_S f(x, y, z)\,dS$ geometrically if f is the constant function $f(x, y, z) = c$, where $c > 0$, and S has a regular projection on the xy-plane.

10 Show that a double integral $\iint_R f(x, y)\,dA$ of the type considered in Theorem 17.14 is a special case of a surface integral.

In Exercises 11–14 find $\iint_S \mathbf{F} \cdot \mathbf{n}\,dS$, where $\mathbf{n}$ is a unit upper normal to S.

11 $\mathbf{F} = x\mathbf{i} + y\mathbf{j} + z\mathbf{k}$; S the upper half of the sphere $x^2 + y^2 + z^2 = a^2$

12 $\mathbf{F} = x\mathbf{i} - y\mathbf{j}$; S the part of the sphere $x^2 + y^2 + z^2 = a^2$ that lies in the first octant

13 $\mathbf{F} = 2\mathbf{i} + 5\mathbf{j} + 3\mathbf{k}$; S the part of the cone $z = (x^2 + y^2)^{1/2}$ that is inside the cylinder $x^2 + y^2 = 1$

14 $\mathbf{F} = x\mathbf{i} + y\mathbf{j} + z\mathbf{k}$; S the part of the plane $3x + 2y + z = 12$ cut out by the planes $x = 0$, $y = 0$, $x = 1$, and $y = 2$

15 Prove that if S is given by $z = f(x, y)$, then (18.31) can be written

$$\iint_S \mathbf{F} \cdot \mathbf{n}\,dS = \iint_R (-Mf_x - Nf_y + P)\,dx\,dy.$$

16 Suppose a metal funnel has the shape of the surface S described in Example 1 (see Figure 18.14). If the unit of length is centimeters, and the area density at the point (x, y, z) is z^2 gm/cm^2, find the mass of the funnel. Use methods analogous to those employed for solids to find the center of gravity of the funnel. Find the moment of inertia of the funnel with respect to the z-axis.

18.7 THE DIVERGENCE THEOREM

One of the most important theorems in applications of vector calculus is the *Divergence Theorem.* It is also called *Gauss' Theorem* in honor of one of the greatest mathematicians of all time, Karl Friedrich Gauss (1777–1855). The theorem has to do with a surface S which forms the complete boundary of a closed and bounded region Q. In this section the vector $\mathbf{n}$ in (18.28) will denote a unit outer normal to S. Illustrations of a surface S of the type to be considered and some typical positions for $\mathbf{n}$ appear in Figure 18.19.

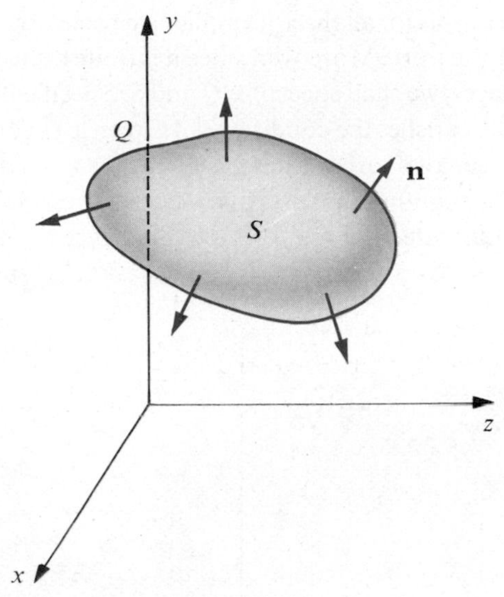

Figure 18.19

If a vector function $\mathbf{F}$ is given by

$$\mathbf{F}(x, y, z) = M(x, y, z)\mathbf{i} + N(x, y, z)\mathbf{j} + P(x, y, z)\mathbf{k}$$

where M, N, and P have continuous first partial derivatives, then the following can be proved.

(18.34) The Divergence Theorem

$$\iint_S \mathbf{F} \cdot \mathbf{n}\, dS = \iiint_Q \nabla \cdot \mathbf{F}\, dV$$

In physical applications this formula states that *the flux of* $\mathbf{F}$ *over S equals the triple integral of the divergence of* $\mathbf{F}$ *over Q*. Using (18.31) and (18.23), the nonvector form of (18.34) may be written

(18.35) $$\iint_S (M\cos\alpha + N\cos\beta + P\cos\gamma)\, dS = \iiint_Q \left(\frac{\partial M}{\partial x} + \frac{\partial N}{\partial y} + \frac{\partial P}{\partial z}\right) dV.$$

In order to prove (18.35) it is sufficient to show that

$$\iint_S M\cos\alpha\, dS = \iiint_Q \frac{\partial M}{\partial x}\, dV$$

$$\iint_S N\cos\beta\, dS = \iiint_Q \frac{\partial N}{\partial y}\, dV$$

$$\iint_S P\cos\gamma\, dS = \iiint_Q \frac{\partial P}{\partial z}\, dV.$$

The proofs for all three formulas are basically similar and therefore we shall prove only the third. Moreover, since it is quite difficult to prove the theorem for general surfaces, we shall specialize Q and S. Specifically, let R be a region in the xy-plane which satisfies the conditions of Green's Theorem, and suppose Q lies over R and between two surfaces having equations $z = f_1(x, y)$ and $z = f_2(x, y)$, where f_1 and f_2 have continuous first partial derivatives and $f_1(x, y) \leq f_2(x, y)$ for all (x, y) in R. As illustrated in Figure 18.20, the surface S which bounds Q consists of a bottom surface S_1, a top surface S_2, and a lateral surface S_3.

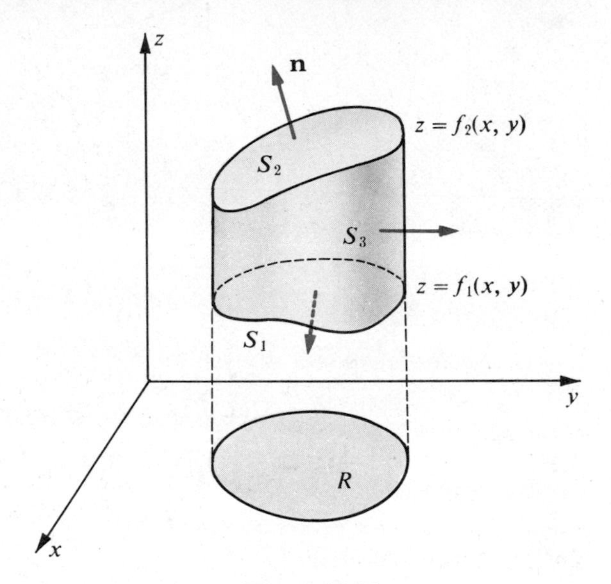

Figure 18.20

On S_3 we have $\gamma = 90°$ and, therefore, $\cos \gamma = 0$. Consequently

$$\iint_S P \cos \gamma \, dS = \iint_{S_1} P \cos \gamma \, dS + \iint_{S_2} P \cos \gamma \, dS.$$

Applying (18.30),

$$\iint_{S_2} P \cos \gamma \, dS = \iint_R P(x, y, f_2(x, y)) \, dx \, dy.$$

Since the formulas for surface integrals in Section 6 were derived under the assumption that **n** is always an *upper* normal, and since the outer normal **n** on S_1 is a *lower* normal, it is necessary to change the sign on the right side of (18.30) when working with S_1. Indeed, a formula for **n** on S_1 may be found by taking the negative of the vector in (18.29). With this in mind, we see that

$$\iint_{S_1} P \cos \gamma \, dS = -\iint_R P(x, y, f_1(x, y)) \, dx \, dy$$

and hence

$$\iint_S P \cos \gamma \, dS = \iint_R [P(x, y, f_2(x, y)) - P(x, y, f_1(x, y))] \, dx \, dy$$

$$= \iint_R \left[\int_{f_1(x,y)}^{f_2(x,y)} \frac{\partial P}{\partial z} dz \right] dx\, dy$$

$$= \iiint_Q \frac{\partial P}{\partial z} dV$$

which is what we wished to prove. It is not difficult to extend our proof to finite unions of regions of the above type. The proofs of the formulas for $\iint_S M \cos \alpha\, dS$ and $\iint_S N \cos \beta\, dS$ are done in like manner, provided Q and S are suitably restricted.

Example 1 Let Q be the region bounded by the graphs of $x^2 + y^2 = 4$, $z = 0$, and $z = 3$. Let S denote the surface of Q, and let $\mathbf{n}$ be a unit outer normal to S. If $\mathbf{F}(x, y, z) = x^3\mathbf{i} + y^3\mathbf{j} + z^3\mathbf{k}$, use the Divergence Theorem to evaluate $\iint_S \mathbf{F} \cdot \mathbf{n}\, dS$.

Solution The surface S and some typical positions of $\mathbf{n}$ are sketched in Figure 18.21. Since

$$\nabla \cdot \mathbf{F} = 3x^2 + 3y^2 + 3z^2 = 3(x^2 + y^2 + z^2)$$

we have, by the Divergence Theorem (18.34),

$$\iint_S \mathbf{F} \cdot \mathbf{n}\, dS = 3 \iiint_Q (x^2 + y^2 + z^2)\, dV.$$

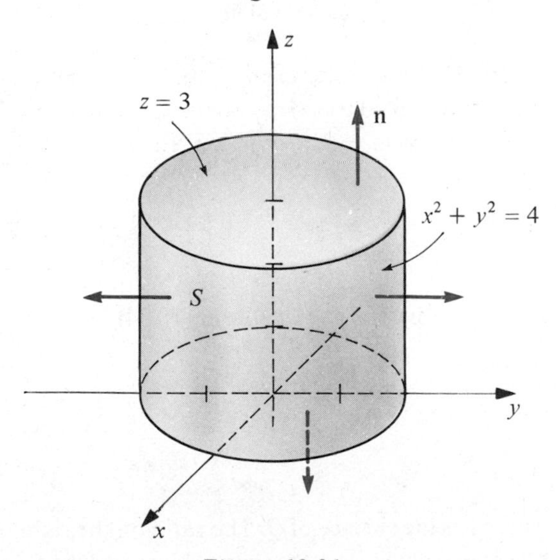

Figure 18.21

It is convenient to use cylindrical coordinates to evaluate the triple integral. Thus

$$\iint_S \mathbf{F} \cdot \mathbf{n}\, dS = 3 \int_0^{2\pi} \int_0^2 \int_0^3 (r^2 + z^2) r\, dz\, dr\, d\theta$$

$$= 3 \int_0^{2\pi} \int_0^2 \left[r^2 z + \frac{1}{3} z^3 \right]_0^3 r\, dr\, d\theta$$

$$= 3\int_0^{2\pi}\int_0^2 (3r^2 + 9)r\,dr\,d\theta$$
$$= 3\int_0^{2\pi}\left[\frac{3}{4}r^4 + \frac{9}{2}r^2\right]_0^2 d\theta$$
$$= 3\int_0^{2\pi} 30\,d\theta = 90\theta\Big]_0^{2\pi} = 180\pi.$$

We may use the Divergence Theorem to obtain a physical interpretation for the divergence of a vector field. Let us begin by recalling from the Mean Value Theorem (5.29) that if a function f of one variable is continuous on a closed interval $[a, b]$, then there is a number c in (a, b) such that

$$\int_a^b f(x)\,dx = f(c)L$$

where $L = b - a$ is the length of $[a, b]$. An analogous result may be proved for triple integrals. Specifically, if a function f of three variables is continuous throughout a spherical region Q, then there is a point $A(u, v, w)$ in the interior of Q such that

$$\iiint_Q f(x, y, z)\,dV = f(x, y, z)]_A V$$

where V is the volume of Q and $f(x, y, z)]_A$ denotes $f(u, v, w)$. This result is sometimes called the **Mean Value Theorem for Triple Integrals**. It follows that if $\mathbf{F}$ is a continuous vector function, then

$$\iiint_Q \nabla\cdot\mathbf{F}\,dV = \nabla\cdot\mathbf{F}]_A V$$

and hence by Gauss' Theorem (18.34)

$$\nabla\cdot\mathbf{F}]_A = \frac{\iint_S \mathbf{F}\cdot\mathbf{n}\,dS}{V}$$

where S is the surface of Q. The ratio on the right may be interpreted as *the flux of* $\mathbf{F}$ *per unit volume over the sphere.* Next let P be an arbitrary point and suppose $\mathbf{F}$ is continuous throughout a region containing P in its interior. Let S_k be the surface of a sphere of radius k with center at P. From the previous discussion, for each k there is a point P_k within S_k such that

$$\nabla\cdot\mathbf{F}]_{P_k} = \frac{\iint_{S_k} \mathbf{F}\cdot\mathbf{n}\,dS}{V_k}$$

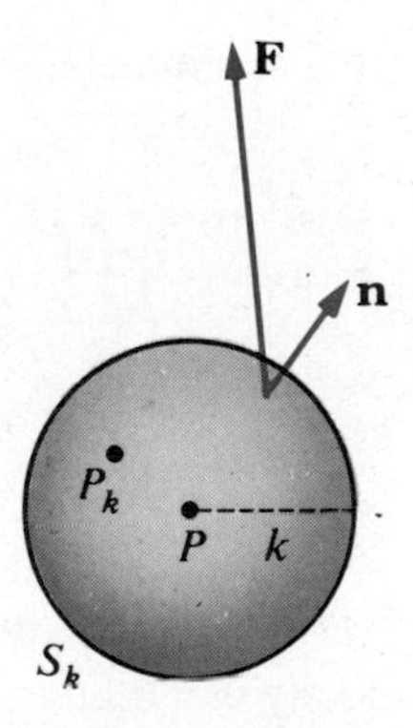

Figure 18.22

where V_k is the volume of the sphere (see Figure 18.22).

If we let $k \to 0$, then $P_k \to P$ and we obtain

(18.36)
$$\operatorname{div} \mathbf{F}]_P = \lim_{k \to 0} \frac{\displaystyle\iint_{S_k} \mathbf{F} \cdot \mathbf{n}\, dS}{V_k};$$

that is, *the divergence of* **F** *at* P *is the limiting value of the flux per unit volume over a sphere with center* P, *as the radius of the sphere approaches* 0. In particular, if **F** represents the velocity of a fluid, then $\operatorname{div} \mathbf{F}]_P$ can be interpreted as the rate of loss or gain of fluid per unit volume at P. It follows from the remarks at the end of Section 6 that there is a source or sink at P if $\operatorname{div} \mathbf{F}]_P > 0$ or $\operatorname{div} \mathbf{F}]_P < 0$, respectively. If the fluid is incompressible and no sources or sinks are present, then there can be no gain or loss within the volume element V_k and hence at every point P

$$\operatorname{div} \mathbf{F} = \frac{\partial M}{\partial x} + \frac{\partial N}{\partial y} + \frac{\partial P}{\partial z} = 0.$$

This formula is called the **equation of continuity** for incompressible fluids. The limit formulation of divergence given in (18.36) may also be applied to physical concepts such as magnetic or electric flux, since these entities have many characteristics which are similar to those of fluids.

Example 2 Let Q be the rectangular parallelepiped bounded by the coordinate planes and the graphs of $x = 1$, $y = 2$, and $z = 3$. If

$$\mathbf{F}(x, y, z) = 2e^{-x}z\mathbf{i} + xy^2\mathbf{j} + z^3\mathbf{k},$$

use the Divergence Theorem to find the flux of **F** over the surface S of Q.

Solution By the Divergence Theorem (18.34)

$$\iint_S \mathbf{F}\cdot\mathbf{n}\,dS = \iiint_Q (-2e^{-x}z + 2xy + 3z^2)\,dV$$

$$= \int_0^1\int_0^2\int_0^3 (-2e^{-x}z + 2xy + 3z^2)\,dz\,dy\,dx.$$

The reader may verify that the value of this integral is $60 + 18e^{-1} \approx 66.6$. If **F** represents the velocity of a fluid or gas, then the fact that the integral is positive tells us that the flow out of Q exceeds the flow into Q. Hence the sources within Q exceed the sinks.

Example 3 Suppose a surface S forms the complete boundary of a region Q of the type considered in the proof of the Divergence Theorem, and suppose that the origin O is an interior point of Q. If an inverse square field is given by $\mathbf{F} = (q/r^3)\mathbf{r}$ where q is a constant, $\mathbf{r} = x\mathbf{i} + y\mathbf{j} + z\mathbf{k}$, and $r = |\mathbf{r}|$, prove that the flux of **F** over S is $4\pi q$ regardless of the shape of Q.

Solution Since **F** is not continuous at O the Divergence Theorem cannot be applied directly; however, we may proceed as follows. Let S_1 be a sphere of radius a and center O which lies completely within S (see Figure 18.23), and let Q_1 denote the region which lies outside of S_1 and inside of S. Since **F** is continuous throughout Q_1 we may apply (18.34), obtaining

$$\iiint_{Q_1} \nabla\cdot\mathbf{F}\,dV = \iint_S \mathbf{F}\cdot\mathbf{n}\,dS + \iint_{S_1} \mathbf{F}\cdot\mathbf{n}\,dS.$$

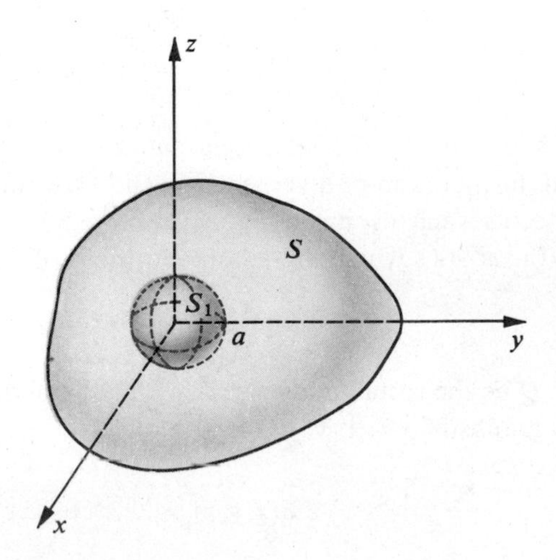

Figure 18.23

It can be shown that if $\mathbf{F}$ is an inverse square field, then $\nabla \cdot \mathbf{F} = 0$ (see Exercise 15 in Section 5) and, therefore,

$$\iint_S \mathbf{F} \cdot \mathbf{n}\, dS = -\iint_{S_1} \mathbf{F} \cdot \mathbf{n}\, dS.$$

Since the outer normal to the surface S_1 is

$$\mathbf{n} = -\frac{1}{r}\mathbf{r}$$

and $r = a$ on S_1 we have

$$\begin{aligned}
\iint_S \mathbf{F} \cdot \mathbf{n}\, dS &= -\iint_{S_1} \left(\frac{q}{r^3}\right)\mathbf{r} \cdot \left(-\frac{1}{r}\mathbf{r}\right) dS \\
&= \iint_{S_1} \frac{q}{r^4}(\mathbf{r} \cdot \mathbf{r})\, dS \\
&= \iint_{S_1} \frac{q}{r^2}\, dS \\
&= \frac{q}{a^2} \iint_{S_1} dS \\
&= \frac{q}{a^2}(4\pi a^2) = 4\pi q.
\end{aligned}$$

Example 4 Let Q be the region bounded by the cylinder $z = 4 - x^2$, the plane $y + z = 5$, and the xy- and xz-planes. If $\mathbf{n} = \langle \cos\alpha, \cos\beta, \cos\gamma \rangle$ is a unit outer normal to the surface S of Q, evaluate the surface integral

$$\iint_S [(x^3 + e^{-y} \sin z)\cos\alpha + (x^2 y + \tan^{-1} z)\cos\beta + \sqrt{y}\sec x \cos\gamma]\, dS.$$

Solution The region Q is illustrated in Figure 18.24. It would be an incredibly difficult job to evaluate the integral directly; however, using form (18.35) of the Divergence Theorem we can obtain the value from the triple intregral

$$\iiint_Q (3x^2 + x^2)\, dV.$$

Referring to Figure 18.24 and using (17.38), we see that this integral equals

$$\begin{aligned}\int_{-2}^{2}\int_{0}^{4-x^2}\int_{0}^{5-z} 4x^2\,dy\,dz\,dx &= \int_{-2}^{2}\int_{0}^{4-x^2} 4x^2(5-z)\,dz\,dx \\ &= \int_{-2}^{2} \Big[20x^2z - 2x^2z^2\Big]_0^{4-x^2} dx \\ &= \int_{-2}^{2} (48x^2 - 4x^4 - 2x^6)\,dx \\ &= \frac{4608}{35} \approx 131.7.\end{aligned}$$

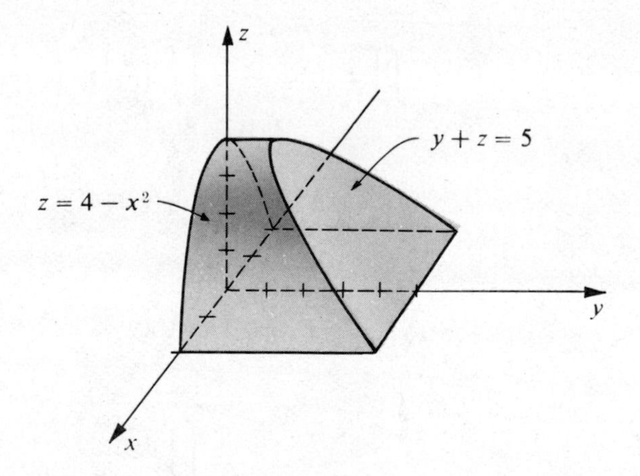

Figure 18.24

EXERCISES 18.7

In Exercises 1–4, use the Divergence Theorem to find $\iint_S \mathbf{F}\cdot\mathbf{n}\,dS$, where $\mathbf{n}$ is a unit outer normal to S.

1 $\mathbf{F} = y\sin x\mathbf{i} + y^2z\mathbf{j} + (x + 3z)\mathbf{k}$; S the surface of the region bounded by the planes $x = \pm 1$, $y = \pm 1$, $z = \pm 1$

2 $\mathbf{F} = y^3e^z\mathbf{i} - xy\mathbf{j} + x\arctan y\mathbf{k}$; S the surface of the region bounded by the coordinate planes and the plane $x + y + z = 1$

3 $\mathbf{F} = (x^2 + \sin yz)\mathbf{i} + (y - xe^{-z})\mathbf{j} + z^2\mathbf{k}$; S the surface of the region bounded by the graphs of $x^2 + y^2 = 4$, $x + z = 2$, $z = 0$

4 $\mathbf{F} = 2xy\mathbf{i} + z\cosh x\mathbf{j} + (z^2 + y\sin^{-1}x)\mathbf{k}$; S the surface of the region bounded by the graphs of $z = x^2 + y^2$ and $z = 9$

In Exercises 5–8 verify (18.34) for the given $\mathbf{F}$ and S.

5 $\mathbf{F} = x\mathbf{i} + y\mathbf{j} + z\mathbf{k}$; S the graph of $x^2 + y^2 + z^2 = a^2$

6 $\mathbf{F} = x^2\mathbf{i} + y^2\mathbf{j} + z^2\mathbf{k}$; S the surface of the cube bounded by the coordinate planes and the planes $x = a$, $y = a$, $z = a$

7 $\mathbf{F} = (x + z)\mathbf{i} + (y + z)\mathbf{j} + (x + y)\mathbf{k}$; S the surface of the region

$$Q = \{(x, y, z): 0 \le y^2 + z^2 \le 1, 0 \le x \le 2\}$$

8 $\mathbf{F} = |\mathbf{r}|^2\mathbf{r}$, where $\mathbf{r} = x\mathbf{i} + y\mathbf{j} + z\mathbf{k}$; S the graph of $x^2 + y^2 + z^2 = a^2$

9 If $\iint_S \mathbf{F} \cdot \mathbf{n}\, dS = 0$ for every closed surface of the type considered in the Divergence Theorem, prove that $\operatorname{div} \mathbf{F} = 0$.

10 Use the Divergence Theorem to prove that if a scalar function f has continuous second partial derivatives, then

$$\iiint_Q \nabla^2 f\, dV = \iint_S D_{\mathbf{n}} f\, dS$$

where $D_{\mathbf{n}} f$ is the directional derivative of f in the direction of an outer normal $\mathbf{n}$ to S.

In Exercises 11–16 assume that S and Q satisfy the conditions of the Divergence Theorem.

11 If f and g are scalar functions which have continuous second partial derivatives, prove that

$$\iiint_Q (f\nabla^2 g + \nabla f \cdot \nabla g)\, dV = \iint_S (f\nabla g) \cdot \mathbf{n}\, dS.$$

(*Hint*: Let $\mathbf{F} = f\nabla g$ in (18.34).)

12 If f and g are scalar functions which have continuous second partial derivatives, prove that

$$\iiint_Q (f\nabla^2 g - g\nabla^2 f)\, dV = \iint_S (f\nabla g - g\nabla f) \cdot \mathbf{n}\, dS.$$

(*Hint*: Use the identity in Exercise 11 together with that obtained by interchanging f and g.)

13 If $\mathbf{F}$ is a conservative vector field with potential f, and if $\operatorname{div} \mathbf{F} = 0$, prove that

$$\iiint_Q \mathbf{F} \cdot \mathbf{F}\, dV = \iint_S f\mathbf{F} \cdot \mathbf{n}\, dS.$$

14 If $\mathbf{r} = x\mathbf{i} + y\mathbf{j} + z\mathbf{k}$ and V is the volume of Q, prove that

$$V = \frac{1}{3}\iint_S \mathbf{r} \cdot \mathbf{n}\, dS$$

15 If $\mathbf{F}$ has continuous second partial derivatives prove that

$$\iint_S \operatorname{curl} \mathbf{F} \cdot \mathbf{n}\, dS = 0.$$

16 If $\mathbf{a}$ is a constant vector prove that

$$\iint_S \mathbf{a} \cdot \mathbf{n}\, dS = 0.$$

A surface (triple) integral of a vector function is defined as the sum of the surface (triple) integrals of each component of the function. With this understanding, establish the results in Exercises 17 and 18, where S and Q satisfy the conditions of the Divergence Theorem.

17 $\iint_S \mathbf{F} \times \mathbf{n}\, dS = -\iiint_Q \nabla \times \mathbf{F}\, dV$ (*Hint*: Apply the Divergence Theorem and Exercise 8 of Section 18.5 to $\mathbf{c} \times \mathbf{F}$, where $\mathbf{c}$ is an arbitrary constant vector.)

18 $\iint_S f\mathbf{n}\, dS = \iiint_Q \nabla f\, dV$ (*Hint*: Apply the Divergence Theorem to $f\mathbf{c}$ where $\mathbf{c}$ is an arbitrary constant vector.)

19 If $\mathbf{F}$ is orthogonal to S at each point (x, y, z), prove that

$$\iiint_Q \operatorname{curl} \mathbf{F}\, dV = 0.$$

20 If $\mathbf{r}$ is the position vector of (x, y, z) and $r = |\mathbf{r}|$, prove that

$$\iiint_Q \mathbf{r}\, dV = \iint_S (\operatorname{grad} r^2)\mathbf{n}\, dS.$$

18.8 STOKES' THEOREM

In (18.24) we stated the following vector form for Green's Theorem:

$$\oint_C \mathbf{F} \cdot \mathbf{T}\, ds = \iint_R \operatorname{curl} \mathbf{F} \cdot \mathbf{k}\, dA$$

where the plane curve C is the boundary of R. This result may be extended to a piecewise smooth simple closed curve C in three dimensions which forms the boundary of an open surface S. The sketch in Figure 18.25 illustrates a special case in which S is the graph of $z = f(x, y)$, where f has continuous first partial derivatives and the projection C_1 of C on the xy-plane is a curve which bounds a region R of the type considered in Green's Theorem. In the figure, $\mathbf{n}$ represents a unit upper normal to S. We take the positive direction along C as that which corresponds to the positive direction along C_1. The symbol $\mathbf{T}$ denotes a unit tangent vector to C. If $\mathbf{F}$ is a vector function which has continuous partial derivatives in a region containing S, then we have the following theorem, named after the English mathematical physicist G. Stokes (1819–1903).

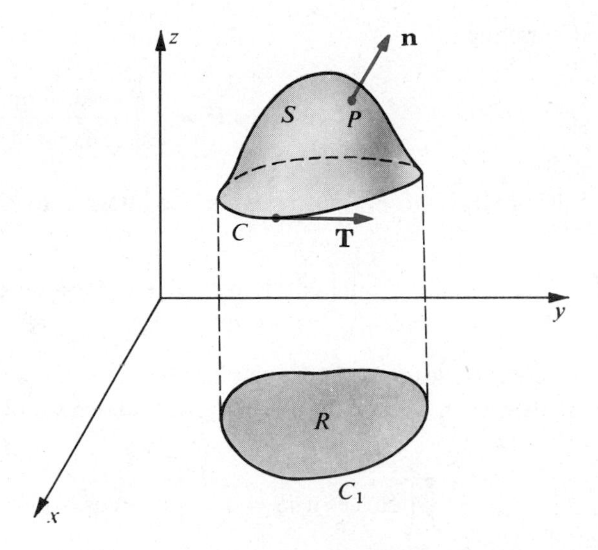

Figure 18.25

(18.37) Stokes' Theorem

$$\oint_C \mathbf{F} \cdot \mathbf{T}\, ds = \iint_S \operatorname{curl} \mathbf{F} \cdot \mathbf{n}\, dS$$

This may be phrased as follows: *the line integral of the tangential component of* **F** *taken along C once in the positive direction equals the surface integral of the normal component of curl* **F** *over S*. Of course, the line integral may also be written $\oint_C \mathbf{F} \cdot d\mathbf{r}$, where $\mathbf{r}$ is the position vector of the point (x, y, z) on C. In order to consider more general open surfaces than the one pictured in Figure 18.25 it is necessary to consider what is called an *oriented surface S* and define the positive direction along C in a suitable manner. The proof of Stokes' Theorem is difficult and is usually not considered in a first course in calculus. The interested reader should consult more advanced texts for a proof.

Example 1 Let S be the part of the graph of $z = 9 - x^2 - y^2$ where $z \geq 0$, and let C be the trace of S on the xy-plane. Verify Stokes' Theorem if $\mathbf{F} = 3z\mathbf{i} + 4x\mathbf{j} + 2y\mathbf{k}$.

Solution We wish to show that the two integrals in (18.37) have the same value.

The surface is the same as that considered in Example 3 of Section 6 (see also Figure 16.5), where we found that

$$\mathbf{n} = \frac{2x\mathbf{i} + 2y\mathbf{j} + \mathbf{k}}{\sqrt{4x^2 + 4y^2 + 1}}.$$

By (18.21)

$$\operatorname{curl} \mathbf{F} = \begin{vmatrix} \mathbf{i} & \mathbf{j} & \mathbf{k} \\ \dfrac{\partial}{\partial x} & \dfrac{\partial}{\partial y} & \dfrac{\partial}{\partial z} \\ 3z & 4x & 2y \end{vmatrix} = 2\mathbf{i} + 3\mathbf{j} + 4\mathbf{k}.$$

Consequently

$$\iint_S \operatorname{curl} \mathbf{F} \cdot \mathbf{n} \, dS = \iint_S \frac{4x + 6y + 4}{\sqrt{4x^2 + 4y^2 + 1}} \, dS.$$

Using (18.27) to evaluate this surface integral gives us

$$\iint_S \operatorname{curl} \mathbf{F} \cdot \mathbf{n} \, dS = \iint_R (4x + 6y + 4) \, dx \, dy$$

where R is the region in the xy-plane bounded by the circle of radius 3 with center at the origin. Changing to polar coordinates we obtain

$$\begin{aligned}
\iint_S \operatorname{curl} \mathbf{F} \cdot \mathbf{n} \, dS &= \int_0^{2\pi} \int_0^3 (4r \cos\theta + 6r \sin\theta + 4) r \, dr \, d\theta \\
&= \int_0^{2\pi} \int_0^3 [r^2(4\cos\theta + 6\sin\theta) + 4r] \, dr \, d\theta \\
&= \int_0^{2\pi} (36\cos\theta + 54\sin\theta + 18) \, d\theta \\
&= \Big[36\sin\theta - 54\cos\theta + 18\theta \Big]_0^{2\pi} = 36\pi.
\end{aligned}$$

The line integral in (18.37) may be written

$$\oint \mathbf{F} \cdot \mathbf{T} \, ds = \oint_C \mathbf{F} \cdot d\mathbf{r} = \oint_C 3z \, dx + 4x \, dy + 2y \, dz$$

where C is the circle in the xy-plane having equation $x^2 + y^2 = 9$. Since $z = 0$ on C, this reduces to

$$\oint_C \mathbf{F} \cdot d\mathbf{r} = \oint_C 4x \, dy.$$

Using the parametric equations $x = 3\cos t$, $y = 3\sin t$ for C, we obtain

$$\begin{aligned}
\oint_C \mathbf{F} \cdot d\mathbf{r} &= \int_0^{2\pi} 36\cos^2 dt \\
&= 18 \int_0^{2\pi} (1 + \cos 2t) \, dt = 36\pi
\end{aligned}$$

which is the same as the value of the surface integral.

We may use Stokes' Theorem to obtain a physical interpretation for curl $\mathbf{F}$ in a manner similar to that in which the Divergence Theorem was used for div $\mathbf{F}$ in (18.36). If P is any point, let S_k be a circular disc of radius k with center at P and let C_k denote the circumference of S_k (see Figure 18.26).

Applying Stokes' Theorem (18.37) and a mean value theorem for double integrals leads to

(18.38)
$$\int_{C_k} \mathbf{F} \cdot \mathbf{T}\, ds = \iint_{S_k} \operatorname{curl} \mathbf{F} \cdot \mathbf{n}\, dS = [\operatorname{curl} \mathbf{F} \cdot \mathbf{n}]_{P_k} (\pi k^2)$$

where P_k is some point in S_k. Thus

$$[\operatorname{curl} \mathbf{F} \cdot \mathbf{n}]_{P_k} = \frac{1}{\pi k^2} \int_{C_k} \mathbf{F} \cdot \mathbf{T}\, ds.$$

If we let $k \to 0$, then $P_k \to P$ and hence

(18.39)
$$[\operatorname{curl} \mathbf{F} \cdot \mathbf{n}]_P = \lim_{k \to 0} \frac{1}{\pi k^2} \int_{C_k} \mathbf{F} \cdot \mathbf{T}\, ds.$$

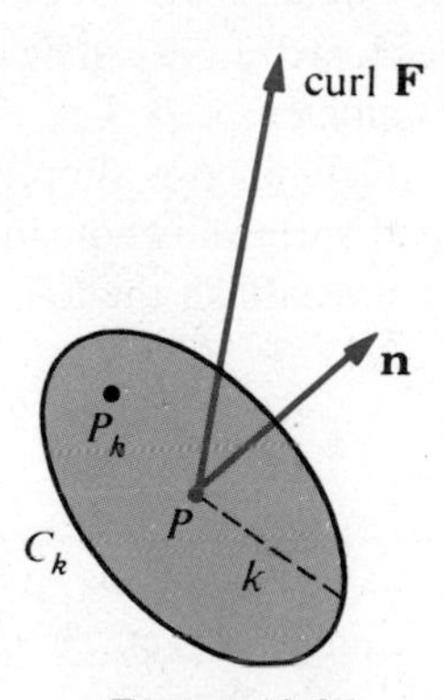

Figure 18.26

If $\mathbf{F}$ is the velocity field of a fluid, then the line integral in (18.38) is called the **circulation around C_k**. It measures the average tendency of the fluid to move or *circulate* around the curve. We see from (18.39) that $[\operatorname{curl} \mathbf{F} \cdot \mathbf{n}]_P$ provides information about the motion of the fluid around the circumference of a circular disc which is perpendicular to $\mathbf{n}$, as the disc shrinks to a point. Since $[\operatorname{curl} \mathbf{F} \cdot \mathbf{n}]_P$ will have its maximum value when $\mathbf{n}$ is parallel to $\operatorname{curl} \mathbf{F}$, it follows that the direction of curl $\mathbf{F}$ at P is that for which the circulation around the boundary of a disc perpendicular to curl $\mathbf{F}$ will have its maximum value as the disc shrinks to a point. Occasionally curl $\mathbf{F} \cdot \mathbf{n}$ is called the **rotation of F about n** and is denoted by **rot F**, since it has to do with the tendency of the vector field to *rotate* about P.

In order to bring out another connection between curl $\mathbf{F}$ and rotational aspects of motion, let us consider the special case in which a fluid is rotating uniformly about an axis, as if it were a rigid body. If we concentrate on a single fluid particle P, then we have a situation similar to that illustrated in Figure 15.9, where $\mathbf{r} = x\mathbf{i} + y\mathbf{j} + z\mathbf{k}$ is the position vector of P and $\boldsymbol{\omega} = \omega_1\mathbf{i} + \omega_2\mathbf{j} + \omega_3\mathbf{k}$ is the (constant) angular velocity. If we denote the velocity $d\mathbf{r}/dt$ by $\mathbf{F}$, then as in Section 15.3,

$$\mathbf{F} = \boldsymbol{\omega} \times \mathbf{r} = (\omega_2 z - \omega_3 y)\mathbf{i} + (\omega_3 x - \omega_1 z)\mathbf{j} + (\omega_1 y - \omega_2 x)\mathbf{k}.$$

By (18.21),

$$\operatorname{curl}\mathbf{F} = \begin{vmatrix} \mathbf{i} & \mathbf{j} & \mathbf{k} \\ \dfrac{\partial}{\partial x} & \dfrac{\partial}{\partial y} & \dfrac{\partial}{\partial z} \\ \omega_2 z - \omega_3 y & \omega_3 x - \omega_1 z & \omega_1 y - \omega_2 x \end{vmatrix}.$$

Using the fact that $\boldsymbol{\omega}$ is a constant vector it may be verified that

$$\operatorname{curl}\mathbf{F} = 2\omega_1\mathbf{i} + 2\omega_2\mathbf{j} + 2\omega_3\mathbf{k} = 2\boldsymbol{\omega}.$$

This shows that the magnitude of curl **F** is twice the angular velocity, and the direction of curl **F** is along the axis of rotation.

The concept of a simply connected region can be extended to three dimensions; however, the usual definition requires properties of curves and surfaces studied in more advanced courses. For our purposes a region D will be called **simply connected** if every simple closed curve C in D is the boundary of a surface S which satisfies the conditions of Stokes' Theorem. In regions of this type any simple closed curve can be continuously shrunk to a point in D without crossing the boundary of D. For example, the region inside of a sphere or a rectangular parallelepiped is simply connected. The region inside of a torus (a doughnut-shaped surface) is not simply connected. Using this rather restrictive definition we can establish the following result.

(18.40) **Theorem**

> If $\mathbf{F}(x, y, z)$ has continuous partial derivatives throughout a simply connected region D, then $\operatorname{curl}\mathbf{F} = \mathbf{0}$ in D if and only if $\oint_C \mathbf{F}\cdot d\mathbf{r} = 0$ for every simple closed curve C in D.

Proof. If $\operatorname{curl}\mathbf{F} = \mathbf{0}$, then by Stokes' Theorem (18.37),

$$\oint_C \mathbf{F}\cdot d\mathbf{r} = \iint_S \operatorname{curl}\mathbf{F}\cdot\mathbf{n}\, dS = 0.$$

Conversely, suppose the indicated line integral is 0 for every C. If $\operatorname{curl}\mathbf{F} \neq \mathbf{0}$ at some point P, then by continuity there is a subregion of D containing P throughout which $\operatorname{curl}\mathbf{F} \neq \mathbf{0}$. If, in this subregion, we choose a circular disc of the type illustrated in Figure 18.26, where $\mathbf{n}$ is parallel to curl **F**, then

$$\oint_{C_k} \mathbf{F}\cdot d\mathbf{r} = \iint_{S_k} \operatorname{curl}\mathbf{F}\cdot\mathbf{n}\, dS > 0,$$

a contradiction. Consequently, $\operatorname{curl}\mathbf{F} = \mathbf{0}$ throughout D.

If $\operatorname{curl}\mathbf{F} = \mathbf{0}$ in D, then the circulation around every closed curve C is 0 (see (18.38)). This also means that the rotation of **F** about any unit vector $\mathbf{n}$ is 0. For this reason vector fields of this type are often called **irrotational**.

It was pointed out earlier that if **F** is continuous on a suitable region, then the vanishing of the line integral $\oint \mathbf{F} \cdot d\mathbf{r}$ around every simple closed curve is equivalent to independence of path. Moreover, as in Theorem (18.15), independence of path is equivalent to $\mathbf{F} = \nabla f$ for some function f (that is, **F** is conservative). Combining these facts with Theorem (18.40), we obtain the following *equivalent conditions for a vector field to be conservative, provided* **F** *has continuous first partial derivatives and the region is simply connected.*

(18.41)

(a) $$\mathbf{F} = \nabla f$$

(b) $$\int_C \mathbf{F} \cdot d\mathbf{r} \text{ is independent of path}$$

(c) $$\int_C \mathbf{F} \cdot d\mathbf{r} = 0 \text{ for every simple closed curve } C$$

(d) $$\nabla \times \mathbf{F} = \mathbf{0}$$

Example 2 Prove that if

$$\mathbf{F}(x, y, z) = (3x^2 + y^2)\mathbf{i} + 2xy\mathbf{j} - 3z^2\mathbf{k},$$

then **F** is conservative.

Solution The function **F** has continuous partial derivatives throughout $\mathbb{R} \times \mathbb{R} \times \mathbb{R}$. Since

$$\begin{aligned} \nabla \times \mathbf{F} &= \begin{vmatrix} \mathbf{i} & \mathbf{j} & \mathbf{k} \\ \dfrac{\partial}{\partial x} & \dfrac{\partial}{\partial y} & \dfrac{\partial}{\partial z} \\ 3x^2 + y^2 & 2xy & -3z^2 \end{vmatrix} \\ &= (0 - 0)\mathbf{i} + (0 - 0)\mathbf{j} + (2y - 2y)\mathbf{k} = \mathbf{0}, \end{aligned}$$

F is conservative by (d) of (18.41).

In advanced mathematics and applications it is useful to study pairs of vector fields **F** and **G** which have the form

$$\begin{aligned} \mathbf{F}(x, y) &= v(x, y)\mathbf{i} + u(x, y)\mathbf{j} \\ \mathbf{G}(x, y) &= u(x, y)\mathbf{i} - v(x, y)\mathbf{j} \end{aligned}$$

where the functions u and v have continuous partial derivatives. Note that $\mathbf{F}(x, y)$ and $\mathbf{G}(x, y)$ are orthogonal at every point since

$$\mathbf{F} \cdot \mathbf{G} = vu - uv = 0.$$

Moreover, under suitable restrictions (see Theorem 18.18) it can be shown that **F** and **G** are both conservative if and only if

(18.42) $$\frac{\partial u}{\partial x} = \frac{\partial v}{\partial y} \quad \text{and} \quad \frac{\partial u}{\partial y} = -\frac{\partial v}{\partial x}.$$

These conditions on u and v are the celebrated **Cauchy–Riemann equations** which are exploited extensively in the study of complex variables. We see from (18.42) that

$$\frac{\partial^2 u}{\partial x^2} = \frac{\partial^2 v}{\partial x\,\partial y} = \frac{\partial^2 v}{\partial y\,\partial x} = -\frac{\partial^2 u}{\partial y^2}$$

and hence

$$\nabla^2 u = \frac{\partial^2 u}{\partial x^2} + \frac{\partial^2 u}{\partial y^2} = 0.$$

Similarly, we can show that

$$\nabla^2 v = \frac{\partial^2 v}{\partial x^2} + \frac{\partial^2 v}{\partial y^2} = 0.$$

Consequently u and v satisfy Laplace's equation and hence are harmonic (see (18.25)).

EXERCISES 18.8

In each of Exercises 1–4, verify Stokes' Theorem for the given $\mathbf{F}$ and S.

1 $\mathbf{F} = y^2\mathbf{i} + z^2\mathbf{j} + x^2\mathbf{k}$; S the first octant portion of the plane $x + y + z = 1$

2 $\mathbf{F} = 2y\mathbf{i} - z\mathbf{j} + 3\mathbf{k}$; S the part of the paraboloid $z = 4 - x^2 - y^2$ which lies inside the cylinder $x^2 + y^2 = 1$

3 $\mathbf{F} = z\mathbf{i} + x\mathbf{j} + y\mathbf{k}$; S the hemisphere $z = (a^2 - x^2 - y^2)^{1/2}$

4 $\mathbf{F} = x^2\mathbf{i} + y^2\mathbf{j} + z^2\mathbf{k}$; S the part of the cone $z = (x^2 + y^2)^{1/2}$ cut off by the plane $z = 1$

5 If $\mathbf{F} = (3z - \sin x)\mathbf{i} + (x^2 + e^y)\mathbf{j} + (y^3 - \cos z)\mathbf{k}$, use Stokes' Theorem to evaluate $\oint_C \mathbf{F}\cdot d\mathbf{r}$ where C is the curve $x = \cos t$, $y = \sin t$, $z = 1$; $0 \le t \le 2\pi$.

6 If $\mathbf{F} = yz\mathbf{i} + xy\mathbf{j} + xz\mathbf{k}$, use Stokes' Theorem to evaluate $\oint_C \mathbf{F}\cdot d\mathbf{r}$, where C is the square with vertices $(0, 0, 2)$, $(1, 0, 2)$, $(1, 1, 2)$, $(0, 1, 2)$.

7 If $\mathbf{F} = 2y\mathbf{i} + e^z\mathbf{j} - \arctan x\mathbf{k}$, use Stokes' Theorem to evaluate $\iint_S \operatorname{curl}\mathbf{F}\cdot\mathbf{n}\,dS$, where S is the part of the paraboloid $z = 4 - x^2 - y^2$ cut off by the xy-plane.

8 If $\mathbf{F} = xy^2\mathbf{i} + yz^2\mathbf{j} + zx^2\mathbf{k}$, use Stokes' Theorem to evaluate $\iint_S \operatorname{curl}\mathbf{F}\cdot\mathbf{n}\,dS$, where S is the triangle with vertices $(1, 0, 0)$, $(0, 2, 0)$, $(0, 0, 3)$.

Establish the identities in Exercises 9–11 under the assumption that C and S satisfy the conditions of Stokes' Theorem.

9 $\oint_C f\nabla g\cdot d\mathbf{r} = \iint_S (\nabla f \times \nabla g)\cdot\mathbf{n}\,dS$, where f and g are scalar functions

10 $\oint_C \mathbf{a}\times\mathbf{r}\cdot d\mathbf{r} = 2\mathbf{a}\cdot\iint_S \mathbf{n}\,dS$, where $\mathbf{a}$ is a constant vector

11 $\oint_C f\,d\mathbf{r} = \iint_S \mathbf{n}\times\nabla f\,dS$, where f is a scalar function (*Hint*: Compare with Exercise 18 of Section 15.7.)

12 If $\mathbf{F}(x, y, z) = y\mathbf{i} + (x + e^z)\mathbf{j} + (1 + ye^z)\mathbf{k}$, use (b) of (18.41) to prove that $\mathbf{F}$ is conservative.

13 If u and v have continuous partial derivatives and satisfy the Cauchy-Riemann equations (18.42), prove that v satisfies Laplace's equation.

14 Let $\mathbf{n}$ denote the unit outer normal at any point P on the surface of a sphere S. If $\mathbf{F}$ has continuous first partial derivatives within and on S, prove that $\iint_S \text{curl}\,\mathbf{F} \cdot \mathbf{n}\, dS = 0$ by using (a) the Divergence Theorem and (b) Stokes' Theorem.

18.9 TRANSFORMATIONS OF COORDINATES

In previous sections of this text we studied many different types of scalar and vector functions. Let us now consider a function T whose domain and range are subsets of $\mathbb{R} \times \mathbb{R}$. Thus, to each ordered pair (x, y) in the domain there corresponds a unique ordered pair (u, v) in the range such that $T(x, y) = (u, v)$. We may represent T geometrically as illustrated in Figure 18.27, where the point (u, v) in the uv-plane corresponds to the point (x, y) in the xy-plane. As usual, (u, v) is referred to as the **image** of (x, y) under T.

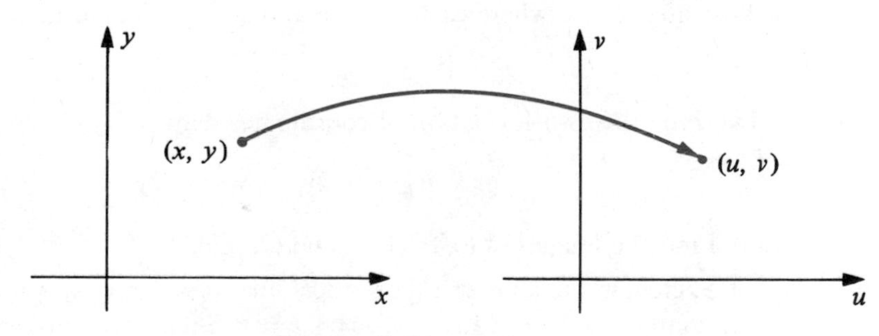

Figure 18.27. $T(x, y) = (u, v)$

Since each pair (u, v) is determined by (x, y), it follows that u and v are functions of x and y; that is,

$$u = f(x, y), \quad v = g(x, y) \tag{18.43}$$

where f and g have the same domain as T. We shall refer to equations (18.43), or to the function T, as a **transformation of coordinates** from the xy-plane to the uv-plane.

Given the transformation of coordinates (18.43), let us partition a region in the uv-plane by means of vertical lines $u = c_1, u = c_2, u = c_3, \ldots$, and horizontal lines $v = d_1, v = d_2, v = d_3, \ldots$. The corresponding level curves for f and g, that is

$$u = f(x, y) = c_i, \quad v = g(x, y) = d_j$$

where $i = 1, 2, 3, \ldots$ and $j = 1, 2, 3, \ldots$ determine a corresponding "curvilinear" partition of a region in the xy-plane. Figure 18.28 illustrates four curves of each type associated with a certain transformation T.

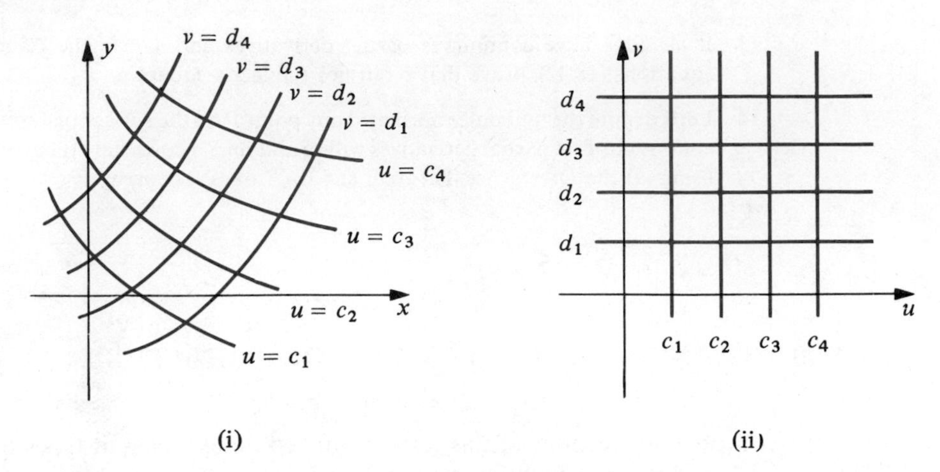

Figure 18.28

We shall refer to the curves $u = f(x, y) = c_i$ and $v = g(x, y) = d_j$ in the xy-plane as ***u*-curves** and ***v*-curves**, respectively. Of course, the types of curves obtained depend on the nature of the functions f and g. The next example illustrates a case where each u-curve and v-curve is a straight line.

Example 1 Let T be the transformation of coordinates defined by

$$u = x + 2y, \quad v = x - 2y.$$

(a) Find the images of $(0, 1)$, $(1, 2)$, and $(2, -3)$.

(b) Sketch, in the uv-plane, the vertical lines $u = 2$, $u = 4$, $u = 6$, $u = 8$, and the horizontal lines $v = -1$, $v = 1$, $v = 3$, $v = 5$. Sketch the corresponding u-curves and v-curves in the xy-plane.

Solution (a) To find the images of the given ordered pairs we may use the formula

$$T(x, y) = (u, v) = (x + 2y, x - 2y).$$

This gives us

$$T(0, 1) = (2, -2), \quad T(1, 2) = (5, -3), \quad T(2, -3) = (-4, 8).$$

(b) The desired vertical and horizontal lines in the uv-plane are sketched in (ii) of Figure 18.29. The u-curves in the xy-plane are the lines

$$x + 2y = 2, \quad x + 2y = 4, \quad x + 2y = 6, \quad x + 2y = 8$$

and the v-curves are the lines

$$x - 2y = -1, \quad x - 2y = 1, \quad x - 2y = 3, \quad x - 2y = 5$$

as illustrated in (i) of the figure.

Dear: Cuong

What is your address and when is my birth-day? I shall close but I shall write again.

Much

Love

To

You

I

/

Love

You

Cuong

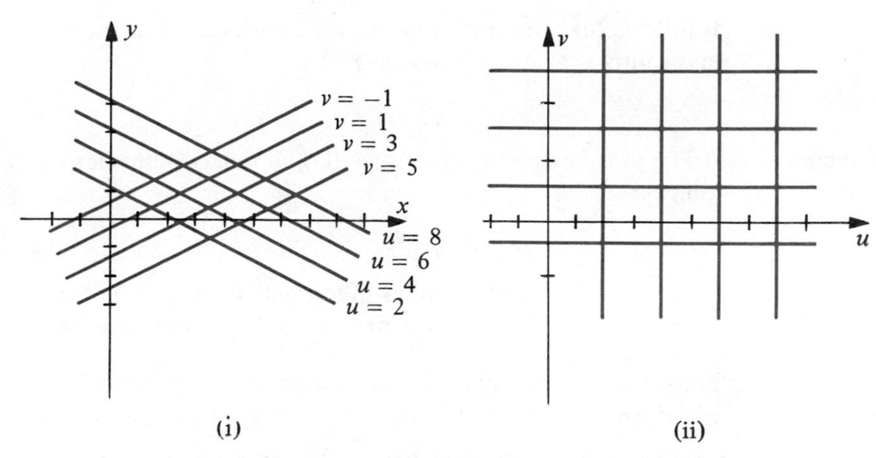

Figure 18.29

The transformation T in Example 1 is one-to-one, in the sense that different ordered pairs in the xy-plane have different images in the uv-plane (see Exercise 13). In general, if T is a one-to-one transformation of coordinates, then by reversing the correspondence we obtain a transformation T^{-1} from the uv-plane to the xy-plane called the **inverse** of T. We may specify T^{-1} by means of equations of the form

$$x = \phi(u, v), \quad y = \psi(u, v).$$

Note that T, followed by T^{-1}, or vice versa, is an identity function.

Example 2 (a) Find the inverse of the transformation T defined in Example 1.

(b) Find the curve in the uv-plane which maps into the ellipse $x^2 + 4y^2 = 1$ under T^{-1}.

Solution (a) The transformation T is given by

$$u = x + 2y, \quad v = x - 2y.$$

If we add corresponding sides of these equations we obtain $u + v = 2x$. Subtracting corresponding sides leads to $u - v = 4y$. Thus T^{-1} is given by

$$x = \tfrac{1}{2}(u + v), \quad y = \tfrac{1}{4}(u - v).$$

(b) Since x and y are related by means of the last two equations in part (a), the points (u, v) which map into $x^2 + 4y^2 = 1$ must satisfy the equation

$$[\tfrac{1}{2}(u + v)]^2 + 4[\tfrac{1}{4}(u - v)]^2 = 1.$$

This simplifies to

$$u^2 + v^2 = 2.$$

It follows that the circle of radius $\sqrt{2}$ with center at the origin in the uv-plane maps into $x^2 + 4y^2 = 1$ under T^{-1}.

Example 3 If $P(x, y)$ is any point other than $(0, 0)$ in the xy-plane, let r and θ be defined as follows:

$$r = \sqrt{x^2 + y^2}$$

(18.44)

θ = the smallest nonnegative angle from the positive x-axis to the position vector $\overrightarrow{OP}$.

Thus, r and θ are polar coordinates for P. The statements in (18.44) define a transformation of coordinates T from the xy-plane to the $r\theta$-plane.

(a) Sketch, in the $r\theta$-plane, the graphs of $r = 1, r = 2, r = 3$ and $\theta = 1/2, \theta = 3/2, \theta = 5/2$. Sketch the corresponding r-curves and θ-curves in the xy-plane.

(b) Find T^{-1}.

Solution (a) Note that we must exclude points (r, θ) in the $r\theta$-plane such that $r \leq 0, \theta < 0$, or $\theta \geq 2\pi$. (Why?) The desired graphs are sketched in Figure 18.30.

(b) It is not difficult to show that T is one-to-one (see Exercise 14). The familiar polar coordinate formulas

$$x = r\cos\theta, \quad y = r\sin\theta$$

define the inverse transformation T^{-1}.

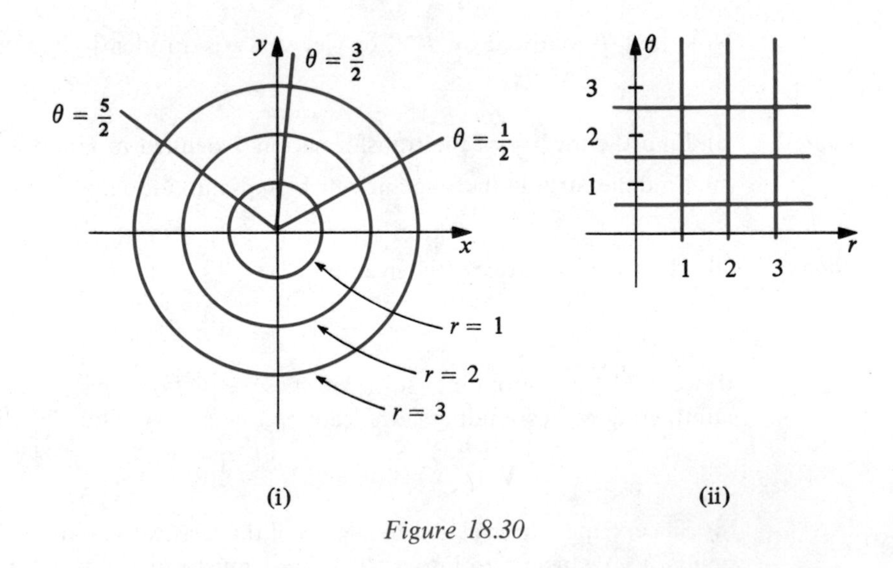

Figure 18.30

The discussion in this section can be extended to three dimensions. Thus, a transformation of coordinates T from an xyz-coordinate system to a uvw-coordinate system is determined by three equations of the form

$$u = f(x, y, z), \quad v = g(x, y, z), \quad w = h(x, y, z)$$

where the functions f, g, and h are defined on some region R. In this case the planes $u = c_i$, $v = d_j$, $z = e_k$, where c_i, d_j, e_k are real numbers, determine ***u*-surfaces, *v*-surfaces** and ***w*-surfaces** in the xyz-coordinate system (see Exercises 15 and 16).

EXERCISES 18.9

In each of Exercises 1–10 let T be the transformation from the xy-plane to the uv-plane defined by the given equations.

(a) Describe the u-curves and the v-curves.

(b) If T is one-to-one, find equations $x = \phi(u, v)$, $y = \psi(u, v)$ which specify T^{-1}.

1 $u = 3x, v = 5y$

2 $u = e^x$, $v = e^y$

3 $u = x - y$, $v = 3y + 2x$

4 $u = 5x - 4y$, $v = 2x + 3y$

5 $u = 2x + y$, $v = xy$

6 $u = x^3$, $v = x + y$

7 $u = x^2 + 4y^2$, $v = 4x^2 - y^2$

8 $u = x + y$, $v = x^2 - y$

9 $u = x^2 - y^2$, $v = 2xy$

10 $u = y/(x^2 + y^2)$, $v = x/(x^2 + y^2)$, $x^2 + y^2 \neq 0$

11 If T is the transformation defined in Exercise 1, find the image of the rectangle with vertices (0, 0), (0, 1), (2, 1), and (2, 0). What is the image of the unit circle $x^2 + y^2 = 1$?

12 If T is the transformation defined in Exercise 3, find the image of the triangle with vertices (0, 0), (0, 1), (2, 0). What is the image of the line $x + 2y = 1$?

13 Prove that the transformation T of Example 1 is one-to-one.

14 Prove that the transformation T of Example 3 is one-to-one.

In Exercises 15 and 16 let T be the indicated transformation from the xyz-coordinate system to the uvw-coordinate system. Find the u-surfaces, the v-surfaces, and the w-surfaces. If T is one-to-one, determine T^{-1}.

15 $u = 2x + y$, $v = y + z$, $w = x + y + z$

16 $u = x^2 + y^2 + z^2$, $v = x^2 + y^2 - z^2$, $w = x^2 - y^2 - z^2$

18.10 CHANGE OF VARIABLES IN MULTIPLE INTEGRALS

In Section 5.6 we developed a technique for making a change of variables in a definite integral $\int_a^b f(x)\,dx$. Under suitable conditions, if we let $x = g(u)$, then $dx = g'(u)\,du$ and

$$\int_a^b f(x)\,dx = \int_c^d f(g(u))g'(u)\,du$$

where $a = g(c)$ and $b = g(d)$. In this section we shall obtain an analogous formula for a change of variables in a double integral $\iint_R F(x, y)\,dA$. In particular, suppose we let

(18.45)
$$x = f(u, v), \quad y = g(u, v)$$

where f and g have continuous second partial derivatives. These equations define a transformation of coordinates T from the uv-plane to the xy-plane. If we use (18.45) to substitute for x and y in the double integral, then the integrand becomes a function of u and v. Our objective is to find a region S in the uv-plane which maps onto R under T, as illustrated in Figure 18.31, and such that

$$\iint_R F(x, y)\,dA = \iint_S F(f(u, v), g(u, v))\,dA.$$

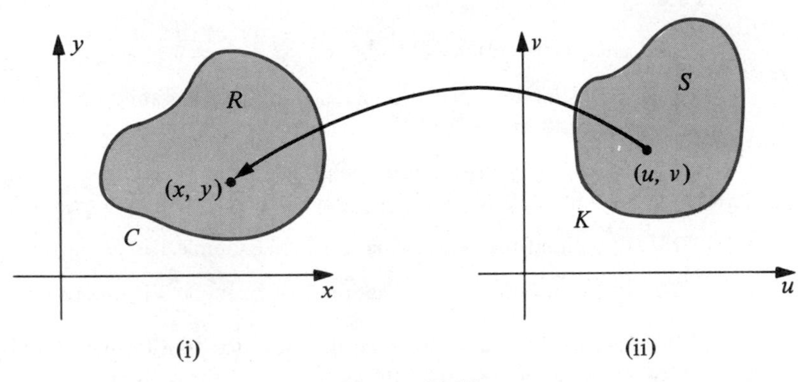

Figure 18.31

In order to change variables in a double intregral it is necessary to impose suitable restrictions on the regions and functions which occur. Specifically, we shall assume that R consists of all points which are either inside or on a piecewise smooth simple closed curve C, and that F has continuous first partial derivatives throughout an open region containing R. We shall also require that T maps a region S of the uv-plane in a one-to-one manner onto R, and that S is bounded by a piecewise smooth simple closed curve K which T maps onto C. Finally, we impose the condition that as (u, v) traces K once in the positive direction, the corresponding point (x, y) traces C once, either in the positive or negative direction. It is possible to weaken these conditions; however, that is beyond the scope of our work.

The function of u and v introduced in the next definition will be used in the change of variable process. It is named after the mathematician C. G. Jacobi (1804–51).

(18.46) Definition

If $x = f(u, v)$ and $y = g(u, v)$, then the **Jacobian** of x and y with respect to u and v, denoted by $\partial(x, y)/\partial(u, v)$, is given by

$$\frac{\partial(x,y)}{\partial(u,v)} = \begin{vmatrix} \dfrac{\partial x}{\partial u} & \dfrac{\partial x}{\partial v} \\ \dfrac{\partial y}{\partial u} & \dfrac{\partial y}{\partial v} \end{vmatrix} = \frac{\partial x}{\partial u}\frac{\partial y}{\partial v} - \frac{\partial y}{\partial u}\frac{\partial x}{\partial v}.$$

In the next theorem all symbols have the same meanings as before, and it is assumed that the regions and functions satisfy the conditions we have discussed. In the statement of the theorem the notations $dx\,dy$ and $du\,dv$ are used in place of dA so that there is no misunderstanding about the region of integration.

(18.47) Theorem

If $x = f(u, v)$, $y = g(u, v)$ is a transformation of coordinates, then

$$\iint_R F(x, y)\,dx\,dy = \pm \iint_S F(f(u, v), g(u, v))\frac{\partial(x, y)}{\partial(u, v)}\,du\,dv.$$

The $+$ sign or the $-$ sign is chosen if, as (u, v) traces the boundary K of S once in the positive direction, the corresponding point (x, y) traces the boundary C of R once in the positive or negative direction, respectively.

Proof. Let us begin by choosing $G(x, y)$ such that $\partial G/\partial x = F$. Applying Green's Theorem (18.19) with $G = N$ gives us

$$\iint_R F(x, y)\,dx\,dy = \iint_R \frac{\partial}{\partial x}[G(x, y)]\,dx\,dy = \oint_C G(x, y)\,dy. \tag{18.48}$$

Suppose the curve K in the uv-plane is given parametrically by

$$u = \phi(t), v = \psi(t), \quad \text{where } a \leq t \leq b.$$

From our assumptions on the transformation T, parametric equations for the curve C in the xy-plane are

$$\begin{aligned} x &= f(u, v) = f(\phi(t), \psi(t)) \\ y &= g(u, v) = g(\phi(t), \psi(t)) \end{aligned} \tag{18.49}$$

where $a \leq t \leq b$. We may, therefore, evaluate the line integral $\oint_C G(x, y)\,dy$ in (18.48) by formal substitutions for x and y. To simplify the notation, let

$$G(t) = G[f(\phi(t), \psi(t)), g(\phi(t), \psi(t))].$$

Applying the Chain Rule to y in (18.49) gives us

$$\frac{dy}{dt} = \frac{\partial y}{\partial u}\frac{du}{dt} + \frac{\partial y}{\partial v}\frac{dv}{dt} = \frac{\partial y}{\partial u}\,\phi'(t) + \frac{\partial y}{\partial v}\psi'(t).$$

Consequently,

$$\begin{aligned}\oint_C G(x, y)\,dy &= \oint_C G(t)\frac{dy}{dt}\,dt \\ &= \int_a^b G(t)\left[\frac{\partial y}{\partial u}\phi'(t) + \frac{\partial y}{\partial v}\psi'(t)\right]dt.\end{aligned}$$

Since $du = \varphi'(t)\,dt$ and $dv = \psi'(t)\,dt$, we may regard the last line integral as a line integral around the curve K in the uv-plane. Thus

$$\oint_C G(x, y)\,dy = \pm \oint_K G\frac{\partial y}{\partial u}\,du + G\frac{\partial y}{\partial v}\,dv \tag{18.50}$$

where for simplicity we have used G as an abbreviation for $G(f(u, v), g(u, v))$. The choice of the $+$ sign or the $-$ sign is determined by letting t vary from a to b and noting whether (x, y) traces C in the same or opposite direction, respectively, as (u, v) traces K.

The line integral on the right in (18.50) has the form

$$\oint_K M\,du + N\,dv$$

where

$$M = G\frac{\partial y}{\partial u} \quad \text{and} \quad N = G\frac{\partial y}{\partial v}.$$

Applying Green's Theorem (18.19) we obtain

$$\begin{aligned}\oint_K M\,du + N\,dv &= \iint_S \left(\frac{\partial N}{\partial u} - \frac{\partial M}{\partial v}\right)du\,dv \\ &= \iint_S \left(G\frac{\partial^2 y}{\partial u\,\partial v} + \frac{\partial G}{\partial u}\frac{\partial y}{\partial v} - G\frac{\partial^2 y}{\partial v\,\partial u} - \frac{\partial G}{\partial v}\frac{\partial y}{\partial u}\right)du\,dv \\ &= \iint_S \left[\left(\frac{\partial G}{\partial x}\frac{\partial x}{\partial u} + \frac{\partial G}{\partial y}\frac{\partial y}{\partial u}\right)\frac{\partial y}{\partial v} - \left(\frac{\partial G}{\partial x}\frac{\partial x}{\partial v} + \frac{\partial G}{\partial y}\frac{\partial y}{\partial v}\right)\frac{\partial y}{\partial u}\right]du\,dv \\ &= \iint_S \frac{\partial G}{\partial x}\left(\frac{\partial x}{\partial u}\frac{\partial y}{\partial v} - \frac{\partial y}{\partial u}\frac{\partial x}{\partial v}\right)du\,dv.\end{aligned}$$

Using the fact that $\partial G/\partial x = F(x, y)$, together with the definition of Jacobian (18.46), gives us

$$\oint_K M\,du + N\,dv = \iint_S F(f(u, v), g(u, v))\frac{\partial(x, y)}{\partial(u, v)}\,du\,dv.$$

Combining this formula with (18.48) and (18.50) leads to the desired result.

Example 1 Evaluate

$$\iint_R e^{(y-x)/(y+x)}\,dx\,dy$$

where R is the region in the xy-plane bounded by the trapezoid with vertices (0, 1), (0, 2), (2, 0), and (1, 0),

Solution Since $y - x$ and $y + x$ appear in the integrand, it is convenient to let

$$u = y - x, \quad v = y + x.$$

These equations define a transformation T from the xy-plane to the uv-plane. In order to apply Theorem (18.47), it is necessary to use the inverse transformation T^{-1}. To find T^{-1} we solve the preceding equations for x and y in terms of u and v, obtaining

$$x = \tfrac{1}{2}(v - u), \quad y = \tfrac{1}{2}(v + u).$$

The Jacobian of this transformation from the uv-plane to the xy-plane is

$$\frac{\partial(x, y)}{\partial(u, v)} = \begin{vmatrix} -\frac{1}{2} & \frac{1}{2} \\ \frac{1}{2} & \frac{1}{2} \end{vmatrix} = -\frac{1}{4} - \frac{1}{4} = -\frac{1}{2}.$$

The region R is sketched in (i) of Figure 18.32.

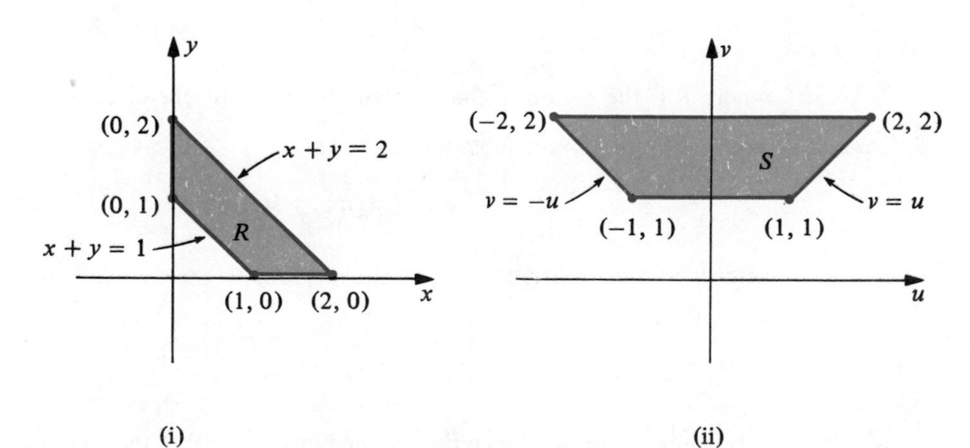

Figure 18.32

To determine the region S in the uv-plane which corresponds to R, we note that the sides of R lie on the lines

$$x = 0, \quad y = 0, \quad x + y = 1, \quad x + y = 2.$$

Using the equations for the transformations T and T^{-1} we see that the corresponding curves in the uv-plane are

$$v = u, \quad v = -u, \quad v = 1, \quad v = 2,$$

respectively. These lines form the boundary of the trapezoidal region S shown in (ii) of Figure 18.32. It is left to the reader to verify that interior points of S correspond to interior points of R, and that as (u, v) traces the boundary of S once in the *positive* (counterclockwise) direction, the corresponding point (x, y) traces the boundary of R once in the *negative* (clockwise) direction. Applying Theorem 18.47,

$$\begin{aligned}\iint_R e^{(y-x)/(y+x)}\,dx\,dy &= -\int_1^2\int_{-v}^{v} e^{u/v}\left(-\frac{1}{2}\right)du\,dv\\ &= \frac{1}{2}\int_1^2\left[ve^{(u/v)}\right]_{-v}^{v} dv\\ &= \frac{1}{2}\int_1^2 v(e - e^{-1})\,dv\\ &= \frac{1}{2}(e - e^{-1})\left[\frac{v^2}{2}\right]_1^2\\ &= \frac{3}{4}(e - e^{-1}).\end{aligned}$$

Example 2 Evaluate

$$\iint_R e^{-(x^2+y^2)}\,dx\,dy$$

where R is the region in the xy-plane bounded by the circle $x^2 + y^2 = a^2$.

Solution The polar coordinate substitutions

$$x = r\cos\theta, \quad y = r\sin\theta$$

determine a transformation from the $r\theta$-plane to the xy-plane, where

$$\frac{\partial(x, y)}{\partial(r, \theta)} = \begin{vmatrix}\cos\theta & -r\sin\theta\\ \sin\theta & r\cos\theta\end{vmatrix} = r.$$

A transformation similar to that defined in (18.44) maps the rectangle S in the $r\theta$-plane bounded by $r = 0$, $r = a$, $\theta = 0$, and $\theta = 2\pi$ onto R. Moreover, as (r, θ)

traces the boundary of S once in the positive direction, the corresponding point (x, y) traces R once in the positive direction. Hence, from Theorem (18.47),

$$\iint_R e^{-(x^2+y^2)}\,dx\,dy = \int_0^{2\pi}\int_0^a e^{-r^2}\,r\,dr\,d\theta$$

$$= \int_0^{2\pi} \left[-\frac{1}{2}e^{-r^2}\right]_0^a d\theta$$

$$= -\frac{1}{2}(e^{-a^2} - 1)2\pi$$

$$= \pi(1 - e^{-a^2}).$$

The theory we have developed can be extended to triple integrals. Given a transformation

$$x = f(u, v, w), \quad y = g(u, v, w), \quad z = h(u, v, w)$$

from a uvw-coordinate system to an xyz-coordinate system, we define the **Jacobian** $\partial(x, y, z)/\partial(u, v, w)$ of the transformation by

$$\frac{\partial(x, y, z)}{\partial(u, v, w)} = \begin{vmatrix} \dfrac{\partial x}{\partial u} & \dfrac{\partial x}{\partial v} & \dfrac{\partial x}{\partial w} \\ \dfrac{\partial y}{\partial u} & \dfrac{\partial y}{\partial v} & \dfrac{\partial y}{\partial w} \\ \dfrac{\partial z}{\partial u} & \dfrac{\partial z}{\partial v} & \dfrac{\partial z}{\partial w} \end{vmatrix}.$$

If R and S are corresponding regions in the xyz- and uvw-coordinate systems, then under suitable restrictions,

(18.51)
$$\iiint_R F(x, y, z)\,dx\,dy\,dz = \pm \iiint_S G(u, v, w)\frac{\partial(x, y, z)}{\partial(u, v, w)}\,du\,dv\,dw$$

where $G(u, v, w)$ is the expression obtained by substituting for $x, y,$ and z in $F(x, y, z)$.

To illustrate, if we use the spherical coordinate formulas (14.52), then

$$x = \rho \sin\phi \cos\theta$$
$$y = \rho \sin\phi \sin\theta$$
$$z = \rho \cos\phi.$$

It can be shown (see Exercise 17), that

(18.52)
$$\frac{\partial(x, y, z)}{\partial(\rho, \phi, \theta)} = \rho^2 \sin\phi.$$

Hence (18.51) takes on the form

$$\iiint_R F(x, y, z)\, dx\, dy\, dz = \pm \iiint_S G(\rho, \theta, \phi)\rho^2 \sin\phi \; d\rho\, d\phi\, d\theta \tag{18.53}$$

which is in agreement with formula (17.47), obtained in an intuitive manner.

EXERCISES 18.10

In Exercises 1–4 find the Jacobian $\partial(x, y)/\partial(u, v)$.

1 $x = u^2 - v^2,\ y = 2uv$

2 $x = e^u \sin v,\ y = e^u \cos v$

3 $x = ve^{-2u},\ y = u^2e^{-v}$

4 $x = u/(u^2 + v^2),\ y = v/(u^2 + v^2)$

In Exercises 5 and 6 find $\partial(x, y, z)/\partial(u, v, w)$.

5 $x = 2u + 3v - w,\ y = v - 5w,\ z = u + 4w$

6 $x = u^2 + vw,\ y = 2v + u^2w,\ z = uvw$

In each of Exercises 7–10 use the indicated change of variables to express the integral as a double integral over a region S in the uv-plane. (Do not evaluate.)

7 $\iint_R (y - x)\, dx\, dy$; R the region bounded by $y = 2x,\ y = 0,\ x = 2$; $x = u + v,\ y = 2v$

8 $\iint_R (3x - 4y)\, dx\, dy$; R the region bounded by $y = 3x,\ 2y = x,\ x = 4$; $x = u - 2v$, $y = 3u - v$

9 $\displaystyle\iint_R \left(\frac{x^2}{4} + \frac{y^2}{9}\right) dx\, dy$; $R = \left\{(x, y): \dfrac{x^2}{4} + \dfrac{y^2}{9} \le 1\right\}$; $x = 2u,\ y = 3v$

10 $\iint_R xy\, dx\, dy$; R the region bounded by $y = 2\sqrt{1 - x},\ x = 0,\ y = 0$; $x = u^2 - v^2$, $y = 2uv$

In each of Exercises 11–16 evaluate the given integral by making the indicated change of variable.

11 $\iint_R (x - y)^2 \cos^2 (x + y)\, dx\, dy$; R the region bounded by the square with vertices $(0, 1)$, $(1, 2)$, $(2, 1)$, $(1, 0)$; $u = x - y,\ v = x + y$

12 $\displaystyle\iint_R \sin\frac{y - x}{y + x}\, dx\, dy$; R the trapezoid with vertices $(1, 1)$, $(2, 2)$, $(4, 0)$, $(2, 0)$; $u = y - x$, $v = y + x$

13 $\iint_R (x^2 + 2y^2)\, dx\, dy$; R the region in the first quadrant bounded by the graphs of $xy = 1,\ xy = 2,\ y = x,\ y = 2x$; $x = u/v,\ y = v$

14 $\iint_R (4x - 4y + 1)^{-2}\, dx\, dy$; R the region bounded by $x = \sqrt{-y},\ x = y$, $x = 1$; $x = u + v, y = v - u^2$

15 $\displaystyle\iint_R \frac{2y + x}{y - 2x}\, dx\, dy$; R the trapezoid with vertices $(-1, 0)$, $(-2, 0)$, $(0, 4)$, $(0, 2)$; $u = y - 2x,\ v = 2y + x$

16 $\iint_R \left(\sqrt{x - 2y} + \frac{y^2}{4}\right) dx\,dy$; R the triangle with vertices $(0,0)$, $(4,0)$, $(4,2)$; $u = y/2$, $v = x - 2y$

17 Verify (18.52).

18 Use Theorem (18.47) to derive a formula for $\iiint_R F(x, y, z)\,dx\,dy\,dz$ for a transformation from rectangular to cylindrical coordinates.

19 If the transformation of coordinates $x = f(u, v)$, $y = g(u, v)$ is one-to-one, show that

$$\frac{\partial(x, y)}{\partial(u, v)} \frac{\partial(u, v)}{\partial(x, y)} = 1.$$

(*Hint*: Use the following property of determinants:

$$\begin{vmatrix} a & b \\ c & d \end{vmatrix} \begin{vmatrix} p & q \\ r & s \end{vmatrix} = \begin{vmatrix} ap + br & aq + bs \\ cp + dr & cq + ds \end{vmatrix}.)$$

20 Given the transformation $x = f(u, v)$, $y = g(u, v)$ and the transformation $u = h(r, s)$, $v = k(r, s)$ show that

$$\frac{\partial(x, y)}{\partial(u, v)} \frac{\partial(u, v)}{\partial(r, s)} = \frac{\partial(x, y)}{\partial(r, s)}.$$

(*Hint*: Use the hint given in Exercise 19).

18.11 REVIEW

Concepts

Define or discuss each of the following.

1. Vector field
2. Vector function
3. Scalar function
4. Inverse square field
5. Conservative vector field
6. Line integral
7. Applications of line integrals
8. Independence of path
9. Green's Theorem
10. Divergence

11 Curl

12 Surface integrals

13 The Divergence Theorem (Gauss' Theorem)

14 Stokes' Theorem

15 Transformations of coordinates

16 Jacobians

17 Change of variables in multiple integrals

Exercises

In Exercises 1–4 give a geometric description of the vector field **F**.

1 $\mathbf{F}(x, y) = 2x\mathbf{i} + y\mathbf{j}$

2 $\mathbf{F}(x, y, z) = x\mathbf{i} + y\mathbf{j} + \mathbf{k}$

3 $\mathbf{F}(x, y, z) = -\mathbf{k}$

4 $\mathbf{F}(x, y, z) = \nabla(x^2 + y^2 + z^2)^{-1/2}$

5 Find a two-dimensional conservative vector field which has the potential function $f(x, y) = y^2 \tan x$.

6 Find a three-dimensional conservative vector field which has the potential function $f(x, y, z) = \ln(x + y + z)$.

Given the points $A(1, 0)$ and $B(-1, 4)$, evaluate the line integral

$$\int_C y^2\, dx + xy\, dy$$

if C is the curve described in each of Exercises 7–10.

7 C is given parametrically by $x = 1 - t$, $y = t^2$; $0 \le t \le 2$.

8 C consists of the line segments from A to the point $D(-1, 0)$, and from D to B.

9 C is the line segment from A to B.

10 C is the graph of $y = 2 - 2x^3$ from A to B.

11 Evaluate $\int_C xy\, ds$, where C is the part of the graph of $y = x^4$ between $(-1, 1)$ and $(2, 16)$.

Given $A(0, 0, 0)$ and $B(2, 4, 8)$ evaluate $\int_C x\, dx + (x + y)\, dy + (x + y + z)\, dz$ if C is the curve described in each of Exercises 12–14.

12 C consists of three line segments, the first parallel to the z-axis, the second parallel to the x-axis, and the third parallel to the y-axis.

13 C consists of the line segment from A to B.

14 C is given parametrically by $x = t$, $y = t^2$, $z = t^3$.

15 If $\mathbf{F}(x, y) = (x + y)\mathbf{i} + (x - y)\mathbf{j}$, evaluate $\int_C \mathbf{F} \cdot d\mathbf{r}$, where C has parametric equations $x = \cos t$, $y = \sin t$, $-\pi \le t \le 0$.

16 The force **F** at a point (x, y, z) in three dimensions is given by $\mathbf{F}(x, y, z) = xy\mathbf{i} + y^2z\mathbf{j} + xz^2\mathbf{k}$. If C is the square with vertices $A_1(1, 1, 1)$, $A_2(-1, 1, 1)$, $A_3(-1, -1, 1)$, and $A_4(1, -1, 1)$, find the work done by **F** as its point of application moves around C once in the direction determined by increasing subscripts on the A_i.

Prove that the line integrals in Exercises 17 and 18 are independent of path and find their values.

17 $\int_{(1,-1)}^{(2,3)} (x + y)\,dx + (x + y)\,dy$

18 $\int_{(0,0,0)}^{(2,1,3)} (8x^3 + z^2)\,dx - 3z\,dy + (2xz - 3y)\,dz$

19 If $\mathbf{F}(x, y, z) = 2xe^{2y}\mathbf{i} + 2(x^2e^{2y} + y\cot z)\mathbf{j} - y^2\csc^2 z\mathbf{k}$, prove that $\int_A^B \mathbf{F}\cdot d\mathbf{r}$ is independent of path and find a potential function f for $\mathbf{F}$.

Use Green's Theorem to evaluate the line integral $\oint_C xy\,dx + (x^2 + y^2)\,dy$ if C is the curve described in each of Exercises 20–22.

20 C is the closed curve determined by $y = x^2$ and $y - x = 2$ from $(-1, 1)$ to $(2, 4)$.

21 C is the triangle with vertices $(0, 0)$, $(1, 0)$, and $(0, 1)$.

22 C is the circle with equation $x^2 + y^2 = 1$.

23 Find $\operatorname{div}\mathbf{F}$ and $\operatorname{curl}\mathbf{F}$ if $\mathbf{F}(x, y, z) = x^3z^4\mathbf{i} + xyz^2\mathbf{j} + x^2y^2\mathbf{k}$.

24 If f and g are scalar functions of three variables prove that

$$\nabla\cdot(f\nabla g) = f\nabla^2 g + \nabla f\cdot\nabla g.$$

25 Evaluate $\iint_S xyz\,dS$, where S is the part of the plane $z = x + y$ which lies over the triangular region in the xy-plane having vertices $(0, 0, 0)$, $(1, 0, 0)$, and $(0, 2, 0)$.

26 Evaluate $\iint_S x^2z^2\,dS$ where S is the top half of the cylinder $y^2 + z^2 = 4$ between $x = 0$ and $x = 1$.

27 Let Q be the region bounded by the cylinder $x^2 + y^2 = 1$ and the planes $z = 0$ and $z = 1$. If $\mathbf{F} = x^3\mathbf{i} + y^3\mathbf{j} + z^3\mathbf{k}$, use the Divergence Theorem to find $\iint_S \mathbf{F}\cdot\mathbf{n}\,dS$, where S is the surface of Q and $\mathbf{n}$ is the unit outer normal to S.

28 Verify the Divergence Theorem if $\mathbf{F} = 2x\mathbf{i} + y\mathbf{j} - z\mathbf{k}$ and S is the surface of the parallelepiped bounded by the planes $x = \pm 1$, $y = \pm 2$, $z = \pm 3$.

Verify Stokes' Theorem for the given $\mathbf{F}$ and S in Exercises 29 and 30.

29 $\mathbf{F} = y^2\mathbf{i} + 2x\mathbf{j} + 5y\mathbf{k}$ and S is the hemisphere $z = (4 - x^2 - y^2)^{1/2}$

30 $\mathbf{F} = (x + y)\mathbf{i} + (y + z)\mathbf{j} + (x + z)\mathbf{k}$ and S is the region bounded by the triangle with vertices $(1, 0, 0)$, $(0, 1, 0)$, and $(0, 0, 1)$.

31 If a transformation T of the xy-plane to the uv-plane is given by $u = 2x + 5y$, $v = 3x - 4y$, find:

(a) the u-curves. (b) the v-curves.
(c) T^{-1}. (d) $\partial(x, y)/\partial(u, v)$.
(e) the image of the line $ax + by + c = 0$.
(f) the image of the circle $x^2 + y^2 = a^2$.

32 Evaluate the integral

$$\int_0^1 \int_y^{2-y} e^{(x-y)/(x+y)}\,dx\,dy$$

by means of the change of variables $u = x - y$, $v = x + y$.

Differential Equations

A differential equation *is an equation which involves derivatives or differentials. If only derivatives of a function of one variable occur, then a differential equation is called* ordinary. *A* partial differential equation *contains partial derivatives. The primary objective of this chapter is to develop techniques for solving certain basic types of ordinary differential equations. The discussion is not intended to be a treatise on the subject, but rather to serve as an introduction to this vast and important area of mathematics.*

19.1 INTRODUCTION

An equation of the form

$$F(x, y, y', y'', \ldots, y^{(n)}) = 0 \tag{19.1}$$

where F is a function of $n + 2$ variables, y is a function of x, and $y^{(k)}$ denotes the kth derivative of y with respect to x, is called an **ordinary differential equation of order *n***. The following are examples of ordinary differential equations of orders 1, 2, 3, and 4, respectively.

$$y' = 2x$$

$$\frac{d^2y}{dx^2} + x^2\left(\frac{dy}{dx}\right)^3 - 15y = 0$$

$$(y''')^4 - x^2(y'')^5 + 4xy = xe^x$$

$$\left(\frac{d^4y}{dx^4}\right)^2 - 1 = x^3\frac{dy}{dx}$$

If, in (19.1), F is a polynomial function, then by definition the **degree** of the differential equation is the greatest exponent associated with the highest order derivative $y^{(n)}$. The degrees of the preceding differential equations are 1, 1, 4, and 2, respectively.

If a function f has the property that when $f(x)$ is substituted for y in a differential equation, the resulting expression is an identity for all x in some interval, then $f(x)$ (or f) is called a **solution** of the differential equation. For example, if C is any real number, then a solution of $y' = 2x$ is

$$f(x) = x^2 + C$$

because substitution of $f(x)$ for y leads to the identity $2x = 2x$. We call $x^2 + C$ the **general solution** of $y' = 2x$, since every solution has this form.

Example 1 If C_1 and C_2 are any real numbers, prove that

$$f(x) = C_1e^{5x} + C_2e^{-5x}$$

is a solution of $y'' - 25y = 0$.

Solution Since

$$f'(x) = 5C_1e^{5x} - 5C_2e^{-5x} \quad \text{and}$$
$$f''(x) = 25C_1e^{5x} + 25C_2e^{-5x} = 25f(x)$$

we have $f''(x) - 25f(x) = 0$. This shows that $f(x)$ is a solution of $y'' - 25y = 0$.

The solution given in Example 1 is called the general solution of $y'' - 25y = 0$. Observe that the differential equation is of order 2 and the general solution contains two arbitrary parameters C_1 and C_2. The precise definition of general solution involves the concept of **independent parameters** and is left for more advanced courses. It can be shown that general solutions of nth order differential equations contain n independent parameters $C_1, C_2, \ldots, C_n$. A **particular solution** is obtained by assigning specific values to the parameters. Some differential equations have solutions which are not special cases of the general solution. These so-called **singular solutions** will not be discussed in this book.

Example 2 Find the particular solution of $y' = 2x$ which satisfies the condition that $y = 5$ if $x = 2$.

Solution Let us write the general solution of $y' = 2x$ in the form $y = x^2 + C$. If the given condition is to be satisfied, then necessarily $5 = 4 + C$ or $C = 1$. Hence the desired particular solution is $y = x^2 + 1$.

Conditions of the type stated in Example 2 are called **boundary conditions** for the differential equation. If the general solution contains one parameter, one boundary condition is sufficient to find a particular solution. If, as in Example 1, the general solution contains two parameters, then two boundary conditions are needed for particular solutions. Similar statements hold if more than two parameters are involved.

The solutions we have considered express y explicitly in terms of x. Solutions of certain differential equations are stated implicitly. In this case, implicit differentiation is used to check the solution, as illustrated in the following example.

Example 3 Show that $x^3 + x^2y - 2y^3 = C$ is an implicit solution of

$$(x^2 - 6y^2)y' + 3x^2 + 2xy = 0.$$

Solution If the first equation is satisfied by $y = f(x)$, then differentiating implicitly,

$$3x^2 + 2xy + x^2y' - 6y^2y' = 0 \quad \text{or}$$

$$(x^2 - 6y^2)y' + 3x^2 + 2xy = 0.$$

Thus $y = f(x)$ satisfies the differential equation.

The simplest type of differential equation is one which can be written in the form

(19.2) $$M(x) + N(y)y' = 0$$

where M and N are continuous functions. If $y = f(x)$ is a solution, then

$$M(x) + N(f(x))f'(x) = 0.$$

If $f'(x)$ is continuous, then indefinite integration leads to

$$\int M(x)\,dx + \int N(f(x))f'(x)\,dx = C$$

or, equivalently,

$$\int M(x)\,dx + \int N(y)\,dy = C.$$

The last equation is an (implicit) solution of (19.2). A device which is useful for remembering this method of solution is to write (19.2) as

$$M(x) + N(y)\frac{dy}{dx} = 0$$

and then change to the following *differential form*:

(19.3) $$M(x)\,dx + N(y)\,dy = 0.$$

The solution is then found by formally integrating each term. The differential equation (19.2) is said to be **separable**, since the variables x and y may be separated as indicated in (19.3).

Example 4 Solve the differential equation

$$2xy + 6x + (x^2 - 4)y' = 0.$$

Solution The given equation may be written

$$2x(y + 3) + (x^2 - 4)\frac{dy}{dx} = 0.$$

Assuming that $(y + 3)(x^2 - 4) \neq 0$, we may divide both sides by this product and express the equation in the differential form (19.3) as follows:

$$\frac{2x}{x^2 - 4}\,dx + \frac{1}{y + 3}\,dy = 0.$$

Thus the given equation is separable, and integration gives us

$$\ln|x^2 - 4| + \ln|y + 3| = C_1 \quad \text{or}$$
$$\ln|(x^2 - 4)(y + 3)| = C_1.$$

This may also be written

$$|(x^2 - 4)(y + 3)| = e^{C_1} = C_2.$$

It can be shown, furthermore, that the solution is

$$(x^2 - 4)(y + 3) = C$$

for every nonzero C. In explicit form,

$$y = \frac{C}{x^2 - 4} - 3.$$

Since it was assumed that $(x^2 - 4)(y + 3) \neq 0$, the cases $x = \pm 2$ and $y = -3$ require special attention. Direct substitution shows that $y = -3$ is a solution of the given differential equation; however, y' is undefined if $x = \pm 2$. Consequently $y = C/(x^2 - 4) - 3$ is the solution if C is any real number.

Since general solutions of differential equations contain arbitrary parameters, the graphs of these solutions are families of curves. For example, the solution $y = x^2 + C$ of $y' = 2x$ leads to a family of parabolas of the type illustrated in Figure 19.1. Conversely, if we *begin* with the equation of a family of curves, then implicit differentiation leads to a differential equation which is called a **differential equation for the family**.

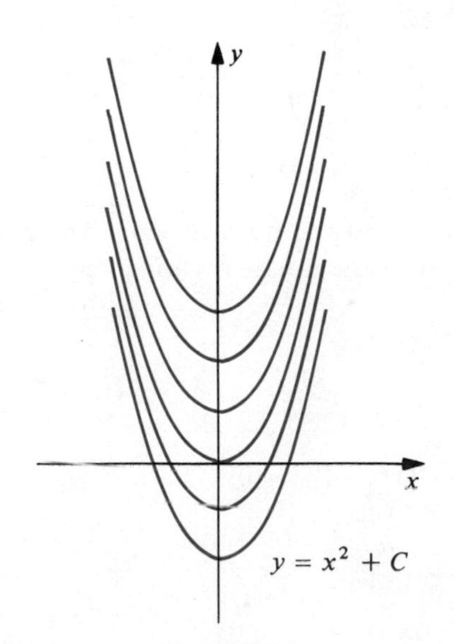

Figure 19.1

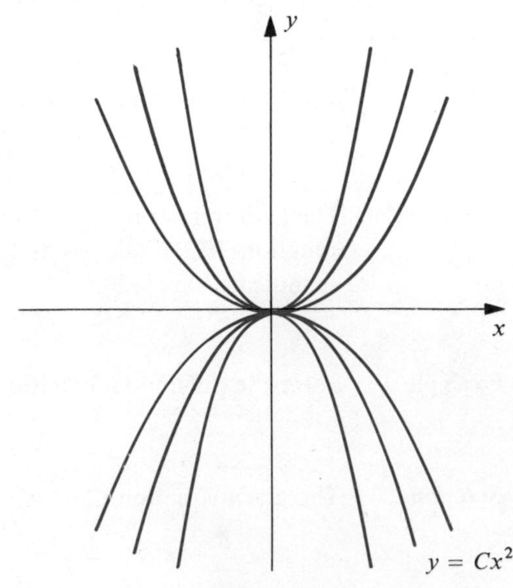

Figure 19.2

Example 5 Find a differential equation for the family of parabolas defined by $y = Cx^2$ where C is a nonzero real number.

Solution The family consists of all parabolas with vertex at the origin and focus on the y-axis as illustrated in Figure 19.2. Differentiating the given equation we obtain $y' = 2Cx$ and therefore

$$y = Cx^2 = (Cx)x = (\tfrac{1}{2}y')x.$$

Hence a differential equation for the family is

$$2y - y'x = 0.$$

Letting $y' = dy/dx$ we obtain the differential form

$$2y\,dx - x\,dy = 0.$$

An equation in x and y which contains two arbitrary constants is said to define a **two-parameter family of curves**. Similar terminology is used if there are more than two constants. In order to find a differential equation when several parameters are involved, it is necessary to differentiate the given equation several times.

Example 6 Find a differential equation for the two-parameter family of curves

$$y = C_1 \cos x + C_2 \sin x.$$

Solution Differentiating twice we obtain

$$y' = -C_1 \sin x + C_2 \cos x$$
$$y'' = -C_1 \cos x - C_2 \sin x.$$

Comparing with the given equation we see that

$$y'' = -y, \quad \text{or} \quad y'' + y = 0.$$

This is a differential equation for the family.

EXERCISES 19.1

In each of Exercises 1–8 prove that y is a solution of the indicated differential equation.

1 $y = C_1e^x + C_2e^{2x}$; $y'' - 3y' + 2y = 0$

2 $y = Ce^{-3x}$; $y' + 3y = 0$

3 $y = e^x(C_1 \cos x + C_2 \sin x)$; $y'' - 2y' + 2y = 0$

4 $y = (C_1 + C_2x)e^{-2x}$; $y'' + 4y' + 4y = 0$

5 $y = Cx^{-2/3}$; $2xy^3\,dx + 3x^2y^2\,dy = 0$

6 $y = Cx^3$; $x^3y''' + x^2y'' - 3xy' - 3y = 0$

7 $y^2 - x^2 - xy = C$; $(x - 2y)y' + 2x + y = 0$

8 $x^2 - y^2 = C$; $yy' = x$

In each of Exercises 9–14 find a differential equation which has the given general solution.

9 $y^2 = x^3 + C$

10 $x^2 + y^2 = C^2$

11 $y = e^{Cx}$

12 $y = C_1x^3 + C_2x$

13 $y = C_1e^x + C_2xe^x$

14 $y = C_1e^x + C_2xe^x + C_3e^{-x}$

In each of Exercises 15–18, find a differential equation for the indicated family of curves.

15 All nonvertical lines through the origin

16 All circles with centers at the origin

17 All circles with centers on the x-axis

18 All parabolas having focus and vertex on the x-axis

Solve the differential equations in Exercises 19–32.

19 $x\,dy - y\,dx = 0$

20 $(1 + y^2)\,dx + (1 + x^2)\,dy = 0$

21 $2y\,dx + (xy + 5x)\,dy = 0$

22 $(xy - 4x)\,dx + (x^2y + y)\,dy = 0$

23 $y' = x - 1 + xy - y$

24 $(y + yx^2)\,dy + (x + xy^2)\,dx = 0$

25 $e^{x+2y}\,dx - e^{2x-y}\,dy = 0$

26 $\cos x\,dy - y\,dx = 0$

27 $y(1 + x^3)y' + x^2(1 + y^2) = 0$

28 $x^2y' - yx^2 = y$

29 $x \tan y - y' \sec x = 0$

30 $xy + y'e^{x^2} \ln y = 0$

31 $e^y \sin x\,dx - \cos^2 x\,dy = 0$

32 $\sin y \cos x\,dx + (1 + \sin^2 x)\,dy = 0$

In each of Exercises 33–40, find the particular solution of the differential equation which satisfies the given boundary conditions.

33 $2y^2y' = 3y - y'$; $x = 3$, $y = 1$

34 $\sqrt{x}y' - \sqrt{y} = x\sqrt{y}$; $x = 9$, $y = 4$

35 $x\,dy - (2x + 1)e^{-y}\,dx = 0$; $x = 1$, $y = 2$

36 $\sec 2y\,dx - \cos^2 x\,dy = 0$; $x = \pi/4$, $y = \pi/6$

37 $(xy + x)\,dx + \sqrt{4 + x^2}\,dy = 0$; $x = 0$, $y = 1$

38 $x\,dy - \sqrt{1 - y^2}\,dx = 0$; $x = 1$, $y = \frac{1}{2}$

39 $\cot x\,dy - (1 + y^2)\,dx = 0$; $x = 0$, $y = 1$

40 $\csc y\,dx - e^x\,dy = 0$; $x = 0$, $y = 0$

19.2 EXACT DIFFERENTIAL EQUATIONS

The separable differential equations considered in the previous section are special cases of the first-order equation

$$P(x, y) + Q(x, y)\frac{dy}{dx} = 0$$

where P and Q are functions of x and y. We often express this equation in terms of differentials as

$$P(x, y)\,dx + Q(x, y)\,dy = 0.$$

In this section we shall restrict our attention to differential equations which are *exact*, in the sense of the following definition.

(19.4) Definition

If P and Q have continuous first partial derivatives, then

$$P(x, y) + Q(x, y)\frac{dy}{dx} = 0, \quad \text{or} \quad P(x, y)\,dx + Q(x, y)\,dy = 0$$

is an **exact differential equation** if and only if

$$\frac{\partial P}{\partial y} = \frac{\partial Q}{\partial x}.$$

Example 1 Show that the differential equation

$$(3x^2y - 2y^3 + 3)\,dx + (x^3 - 6xy^2 + 2y)\,dy = 0$$

is exact.

Solution Setting $P(x, y) = 3x^2y - 2y^3 + 3$ and $Q(x, y) = x^3 - 6xy^2 + 2y$ we see that

$$P_y(x, y) = 3x^2 - 6y^2 = Q_x(x, y).$$

Hence the equation is exact by Definition (19.4).

(19.5) Theorem

If $P(x, y) + Q(x, y)y' = 0$ is an exact differential equation, then there exists a function F of x and y such that

$$\frac{\partial F}{\partial x} = P \quad \text{and} \quad \frac{\partial F}{\partial y} = Q.$$

Proof. If (x_1, y_1) is in the domain of P and Q, let us define the function F by

$$F(x, y) = \int_{x_1}^{x} P(t, y)\,dt + \int_{y_1}^{y} Q(x_1, t)\,dt.$$

It is left as an exercise to prove that $F_x(x, y) = P(x, y)$ and $F_y(x, y) = Q(x, y)$.

The next result is the fundamental existence theorem for solutions of exact differential equations.

(19.6) Theorem

> An exact differential equation $P(x, y) + Q(x, y)y' = 0$ has a solution of the form $F(x, y) = C$, where C is a constant and F is any function of x and y such that $F_x = P$ and $F_y = Q$.

Proof. By Theorem (19.5) there exists at least one function F such that $F_x = P$ and $F_y = Q$. We shall complete the proof by showing that $y = f(x)$ is a solution of the exact differential equation if and only if $F(x, f(x)) = C$ for some constant C.

By the Chain Rule,

$$D_x F(x, f(x)) = F_x(x, f(x)) + F_y(x, f(x))f'(x)$$

or equivalently

$$D_x F(x, f(x)) = P(x, f(x)) + Q(x, f(x))f'(x).$$

If $y = f(x)$ is a solution of the differential equation, we see that $D_x F(x, f(x)) = 0$ and, therefore, $F(x, f(x)) = C$ for some C. Conversely, if $F(x, f(x)) = C$, then $D_x F(x, f(x)) = 0$, that is,

$$0 = P(x, f(x)) + Q(x, f(x))f'(x)$$

which means that $f(x)$ is a solution of the given differential equation.

The following examples illustrate a technique for finding the solution $F(x, y) = C$ in Theorem (19.6).

Example 2 Solve the differential equation

$$(3x^2y - 2y^3 + 3)\,dx + (x^3 - 6xy^2 + 2y)\,dy = 0.$$

Solution It was shown in Example 1 that the equation is exact. Consequently, by (19.6), there is a function F such that

$$F_x(x, y) = 3x^2y - 2y^3 + 3$$
$$F_y(x, y) = x^3 - 6xy^2 + 2y.$$

Integrating $F_x(x, y)$ with respect to x gives us

$$F(x, y) = x^3y - 2xy^3 + 3x + g(y)$$

where g is a function of y. It follows that

$$F_y(x, y) = x^3 - 6xy^2 + g'(y).$$

Comparing this equation with the previous formula for $F_y(x, y)$ we see that $g'(y) = 2y$, and hence $g(y) = y^2 + C_1$. Substituting for $g(y)$ in the formula for $F(x, y)$ gives us

$$F(x, y) = x^3y - 2xy^3 + 3x + y^2 + C_1.$$

Hence, by Theorem (19.6), a solution of the differential equation is

$$x^3y - 2xy^3 + 3x + y^2 = C$$

where the constant C_1 has been absorbed by the constant C. This solution may be checked by implicit differentiation.

Example 3 Solve the differential equation

$$2x \sin y - y \sin x + (x^2 \cos y + \cos x)y' = 0$$

subject to the boundary conditions $y = \pi/6$ if $x = \pi/2$.

Solution Setting $P(x, y) = 2x \sin y - y \sin x$ and $Q(x, y) = x^2 \cos y + \cos x$ we see that

$$P_y(x, y) = 2x \cos y - \sin x = Q_x(x, y)$$

and hence the differential equation is exact. By Theorem (19.5) there is a function F such that

$$F_x(x, y) = 2x \sin y - y \sin x$$
$$F_y(x, y) = x^2 \cos y + \cos x.$$

Integrating $F_x(x, y)$ with respect to x gives us

$$F(x, y) = x^2 \sin y + y \cos x + g(y)$$

where g is a function of y. Consequently

$$F_y(x, y) = x^2 \cos y + \cos x + g'(y).$$

Comparing with the other form of $F_y(x, y)$ we see that $g'(y) = 0$ and therefore $g(y) = C_1$. Applying Theorem (19.6), a solution of the given differential equation is

$$x^2 \sin y + y \cos x = C$$

where again C_1 has been absorbed by C. Finally, the boundary conditions imply that

$$\left(\frac{\pi}{2}\right)^2\left(\frac{1}{2}\right) + \left(\frac{\pi}{6}\right) \cdot 0 = C$$

or $C = \pi^2/8$. Thus the required particular solution is

$$x^2 \sin y + y \cos x = \frac{\pi^2}{8}.$$

EXERCISES 19.2

Solve the differential equations in Exercises 1–14.

1 $(2x + y)\,dx + (2y + x)\,dy = 0$

2 $(2xy + y^3)\,dx + (x^2 + 3y^2x)\,dy = 0$

3 $(1 - 2xy + 3x^2y^2)\,dx - (x^2 + 3y^2 - 2x^3y)\,dy = 0$

4 $(2x - 3x^2y)\,dx + (1 + 2y - x^3)\,dy = 0$

5 $(x^2 \cos y + 4y)y' + 2x \sin y = -5$

6 $y^2e^x + 4y + (2ye^x + 4x)y' = 0$

7 $e^{xy}(xy + 1)\,dx + (x^2e^{xy} + 2y)\,dy = 0$

8 $(v \ln v + 2u \ln u)\,du + (u \ln v + u)\,dv = 0$

9 $(\sin^2 \theta + 2r \cos \theta)\,dr + r(\sin 2\theta - r \sin \theta)\,d\theta = 0$

10 $(x^2 + 2xy - y^2)y' = -y^2 - 2xy + x^2$

11 $(e^{-x} \cos y - 2x)\,dx + e^{-x} \sin y\,dy = 0$

12 $y \cos x\,dx + (\sin x - \sin y)\,dy = 0$

13 $(x - yx^{-1} + \ln y) + (xy^{-1} - \ln x)y' = 0$

14 $\left(\dfrac{x}{1 + y^2} + y^3 - 1\right)y' + \left(\dfrac{1}{1 + x^2} + \tan^{-1} y\right) = 0$

In Exercises 15–18 find a solution of the differential equation which satisfies the given boundary conditions.

15 $(2xy^3 + 8x)\,dx + (3x^2y^2 + 5)\,dy = 0$; $y = -1$ when $x = 2$

16 $(x^2e^y + 3e^x)y' + (2xe^y + 3ye^x) = 0$; $y = 0$ when $x = 1$

17 $\left(ye^{2x} + \dfrac{y}{1 + 4y^2}\right)dy + (y^2e^{2x} - 1)\,dx = 0$; $x = 0$, $y = 1/2$

18 $(x \cos x + \sin x + y)\,dx + (\sin y + x)\,dy = 0$; $x = \pi/6$, $y = \pi/3$

19 Complete the proof of Theorem (19.5).

20 Prove that the separable differential equation (19.2) is a special case of

$$P(x, y) + Q(x, y)\frac{dy}{dx} = 0.$$

19.3 HOMOGENEOUS DIFFERENTIAL EQUATIONS

A function f of two variables is said to be **homogeneous of degree *n*** if

(19.7)
$$f(tx, ty) = t^n f(x, y)$$

for every $t > 0$ such that (tx, ty) is in the domain of f. For example, if

$$f(x, y) = 2x^4 - x^2y^2 + 5xy^3$$

then f is homogeneous of degree 4, since

$$\begin{aligned} f(tx, ty) &= 2(tx)^4 - (tx)^2(ty)^2 + 5(tx)(ty)^3 \\ &= t^4(2x^4 - x^2y^2 + 5xy^3) = t^4 f(x, y). \end{aligned}$$

Similarly, if

$$f(x, y) = \frac{1}{x^2 + y^2} e^{x/y}$$

then f is homogeneous of degree -2, since

$$f(tx, ty) = \frac{1}{t^2x^2 + t^2y^2} e^{tx/ty} = t^{-2} f(x, y).$$

A **homogeneous differential equation** is an equation which can be written in the form

(19.8)
$$P(x, y)\,dx + Q(x, y)\,dy = 0$$

where P and Q are homogeneous functions of the same degree. Equations of this type can be transformed into separable equations by means of the substitution

(19.9)
$$y = xv, \quad \text{where } v = g(x)$$

for some function g. To verify this fact we note that from (19.9),

$$dy = v\,dx + x\,dv.$$

Thus, substitution of xv for y in (19.8) yields

$$P(x, xv)\,dx + Q(x, xv)(v\,dx + x\,dv) = 0.$$

If P and Q are homogeneous of degree n, then

$$P(x, xv) = x^n P(1, v) \quad \text{and} \quad Q(x, xv) = x^n Q(1, v).$$

Substituting in the preceding differential equation and dividing both sides by x^n we obtain

$$P(1, v)\,dx + Q(1, v)(v\,dx + x\,dv) = 0.$$

This equation may be written in the separable form

$$\frac{1}{x}\,dx + \frac{Q(1,v)}{P(1,v) + vQ(1,v)}\,dv = 0, \tag{19.10}$$

provided no zero denominators occur. We have proved that if $y = xv$ is a solution of (19.8), then v is a solution of (19.10). Conversely, if v is a solution of (19.10), then reversing our argument shows that $y = xv$ is a solution of (19.8). It is not advisable to memorize the final form (19.10). Instead, remember the substitution $y = xv$ which is used to simplify the homogeneous equation. It is also possible to use the substitution $x = vy$ to solve the equation.

Example 1 Solve the differential equation

$$(y^2 - xy)\,dx + x^2 dy = 0.$$

Solution If $P(x, y) = y^2 - xy$ and $Q(x, y) = x^2$, then the functions P and Q are both homogeneous of degree 2. Hence the differential equation is homogeneous and we substitute

$$y = xv, \quad dy = v\,dx + x\,dv.$$

This leads to the following chain of equations.

$$\begin{aligned}(x^2v^2 - x^2v)\,dx + x^2(v\,dx + x\,dv) &= 0\\ x^2(v^2 - v)\,dx + x^2(v\,dx + x\,dv) &= 0\\ (v^2 - v)\,dx + v\,dx + x\,dv &= 0\\ v^2\,dx + x\,dv &= 0\\ \frac{1}{x}\,dx + \frac{1}{v^2}\,dv &= 0\end{aligned}$$

Integrating each term gives us

$$\ln|x| - \frac{1}{v} = C_1.$$

Since $v = y/x$, the last equation is equivalent to

$$\ln|x| - \frac{x}{y} = C_1.$$

If we let $C_1 = \ln|C|$, then

$$\ln|x| - \ln|C| = \frac{x}{y}, \quad \text{or} \quad \ln\left|\frac{x}{C}\right| = \frac{x}{y}.$$

This may be written $x/C = e^{x/y}$ or $x = Ce^{x/y}$. An explicit form for the solution may also be obtained.

Example 2 Solve the differential equation

$$\left(y - x\cot\frac{x}{y}\right)dy + y\cot\frac{x}{y}\,dx = 0.$$

Solution The equation is homogeneous of degree 1. It is convenient to let $x = vy$, since in this case $\cot(x/y)$ reduces to $\cot v$. Thus we substitute

$$x = vy, \quad dx = v\,dy + y\,dv$$

obtaining

$$(y - vy\cot v)\,dy + y\cot v(v\,dy + y\,dv) = 0.$$

Dividing both sides by y and simplifying leads to

$$\begin{aligned}(1 - v\cot v)\,dy + \cot v(v\,dy + y\,dv) &= 0\\ dy + y\cot v\,dv &= 0.\end{aligned}$$

Separating the variables and integrating we obtain

$$\begin{aligned}\frac{1}{y}dy + \cot v\,dv &= 0\\ \ln|y| + \ln|\sin v| = C_1 &= \ln|C|\\ \ln|y\sin v| &= \ln|C|\\ y\sin v &= C.\end{aligned}$$

Since $v = x/y$, a solution to the given equation is

$$y\sin\frac{x}{y} = C.$$

EXERCISES 19.3

Solve the differential equations in Exercises 1–10.

1 $(x + 3y)\,dx + x\,dy = 0$

2 $(2x - y)\,dx + (x + 2y)\,dy = 0$

3 $\left(y\sin\frac{y}{x} + x\cos\frac{y}{x}\right)dx - x\sin\frac{y}{x}\,dy = 0$

4 $x^2\,dy + (y^2 - xy)\,dx = 0$

5 $(x^2 + y^2)\,dx - x^2\,dy = 0$

6 $(y^2 - xy + x^2)\,dx + xy\,dy = 0$

7 $y' = x^3/(4x^3 - 3x^2y)$

8 $y' = x^{-1}y - \cos x^{-1}y$

9 $(y + \sqrt{x^2 - y^2})\,dx = x\,dy$

10 $(x + y)\,dx = (x - y)\,dy$

Solve the differential equations in Exercises 11–14 by using the substitution $x = vy$.

11 $2xy\,dx + (y^2 - x^2)\,dy = 0$

12 $y^2\,dx + x(x - y)\,dy = 0$

13 $3y\,dx + (x + 2y)\,dy = 0$

14 $dx - (y^{-1}x + \tan y^{-1}x)\,dy = 0$

In Exercises 15 and 16 show that the differential equation is both homogeneous and exact and solve it by the corresponding method.

15 $(x + 3y)\,dx + (3x - 2y)\,dy = 0$ **16** $(x^2 + y^2)\,dx + 2xy\,dy = 0$

19.4 LINEAR DIFFERENTIAL EQUATIONS OF THE FIRST ORDER

A **first-order linear differential equation** is an equation of the form

$$y' + P(x)y = Q(x) \tag{19.11}$$

where P and Q are continuous functions. If $Q(x) = 0$ for all x, then (19.11) is separable and we may write

$$\frac{1}{y}y' = -P(x)$$

provided $y \neq 0$. Integrating we obtain

$$\ln|y| = -\int P(x)\,dx + \ln|C|.$$

We have expressed the constant of integration as $\ln|C|$ in order to change the form of the last equation as follows:

$$\begin{aligned}
\ln|y| - \ln|C| &= -\int P(x)\,dx \\
\ln\left|\frac{y}{C}\right| &= -\int P(x)\,dx \\
\frac{y}{C} &= e^{-\int P(x)\,dx} \\
ye^{\int P(x)\,dx} &= C.
\end{aligned}$$

We next observe that

$$\begin{aligned}
D_x\left[ye^{\int P(x)\,dx}\right] &= y'e^{\int P(x)\,dx} + P(x)ye^{\int P(x)\,dx} \\
&= e^{\int P(x)\,dx}\left[y' + P(x)y\right].
\end{aligned}$$

Consequently, if we multiply both sides of (19.11) by $e^{\int P(x)\,dx}$, then the resulting equation may be written

$$D_x\left[ye^{\int P(x)\,dx}\right] = Q(x)e^{\int P(x)\,dx}.$$

This gives us the following (implicit) solution of (19.11):

$$ye^{\int P(x)\,dx} = \int Q(x)e^{\int P(x)\,dx}\,dx + D. \tag{19.12}$$

Solving this equation for y leads to an explicit solution. The expression $e^{\int P(x)dx}$ is called an **integrating factor** of (19.11). We have shown that multiplication of both sides of (19.11) by this expression leads to an equation which has the solution (19.12).

Example 1 Solve the differential equation $x^2y' + 5xy + 3x^5 = 0$ where $x \neq 0$.

Solution In order to find an integrating factor we begin by expressing the given differential equation in the "standardized" form (19.11), where the coefficient of y' is 1. Thus, dividing both sides by x^2 we obtain

$$y' + \frac{5}{x}y = -3x^3$$

which has the form (19.11) with $P(x) = 5/x$ and $Q(x) = -3x^3$. From the preceding discussion, the required integrating factor is

$$e^{\int P(x)\,dx} = e^{5\ln|x|} = e^{\ln|x|^5} = |x|^5.$$

If $x > 0$ then $|x|^5 = x^5$, whereas if $x < 0$ then $|x|^5 = -x^5$. In either case, multiplying both sides of the standardized form by $|x|^5$ gives us

$$x^5y' + 5x^4y = -3x^8 \quad \text{or} \quad D_x(x^5y) = -3x^8.$$

Thus a solution is

$$x^5y = -\frac{x^9}{3} + C$$

or

$$y = -\frac{x^4}{3} + \frac{C}{x^5}.$$

A generalization of (19.11) is the **Bernoulli equation**

(19.13) $$y' + P(x)y = Q(x)y^n,$$

where $n \neq 0$. Evidently $y = 0$ is a solution. If $y \neq 0$ we may divide both sides by y^n, obtaining

(19.14) $$y^{-n}y' + P(x)y^{1-n} = Q(x).$$

If we let $w = y^{1-n}$, then

$$w' = D_xw = (1-n)y^{-n}y'$$

and hence

$$y^{-n}y' = \frac{1}{1-n}w'.$$

Replacing $y^{-n}y'$ in (19.14) by the last expression gives us

$$\frac{1}{1-n}w' + P(x)w = Q(x).$$

This first-order linear differential equation may be solved for w using the integrating factor technique. After w has been found, the solution of (19.13) is given by $y^{1-n} = w$ (and $y = 0$).

Example 2 Solve the differential equation

$$y' + 2x^{-1}y = x^6y^3.$$

Solution The equation has the Bernoulli form (19.13) with $n = 3$. If, as in the preceding discussion, we multiply both sides by y^{-3} and substitute $w = y^{1-n} = y^{-2}$ we obtain

$$\begin{aligned} y^{-3}y' + 2x^{-1}y^{-2} &= x^6 \\ -\tfrac{1}{2}w' + 2x^{-1}w &= x^6 \\ w' + 4x^{-1}w &= -2x^6. \end{aligned}$$

Since the integrating factor for the last equation is

$$e^{\int(-4/x)dx} = e^{-4\ln|x|} = e^{\ln|x|^{-4}} = |x|^{-4} = x^{-4}$$

we write

$$x^{-4}w' - 4x^{-5}w = -2x^2.$$

Consequently

$$x^{-4}w = -\tfrac{2}{3}x^3 + C$$

or

$$w = -\tfrac{2}{3}x^7 + Cx^4.$$

Finally, since $w = y^{-2}$, the solution of the given equation is

$$y^{-2} = -\tfrac{2}{3}x^7 + Cx^4$$

or

$$(-\tfrac{2}{3}x^7 + Cx^4)y^2 = 1.$$

EXERCISES 19.4

Solve the differential equations in Exercises 1–26.

1 $y' + 2y = e^{2x}$

2 $y' - 3y = 2$

3 $xy' - 3y = x^5$

4 $y' + y\cot x = \csc x$

5 $xy' + y + x = e^x$

6 $xy' + (1 + x)y = 5$

7 $x^2\,dy + (2xy - e^x)\,dx = 0$

8 $x^2\,dy + (x - 3xy + 1)\,dx = 0$

9 $y' + y\cot x = 4x^2 \csc x$

10 $y' + y\tan x = \sin x$

11 $(y\sin x - 2)\,dx + \cos x\,dy = 0$

12 $(x^2y - 1)\,dx + x^3\,dy = 0$

13 $(x^2\cos x + y)\,dx - x\,dy = 0$

14 $y' + y = \sin x$

15 $xy' + (2 + 3x)y = xe^{-3x}$

16 $(x + 4)y' + 5y = x^2 + 8x + 16$

17 $xy' - y = x^3y^4$

18 $xy' + 2y = 4x^4y^4$

19 $(2xy - x^2y^2)\,dx + (1 + x^2)\,dy = 0$

20 $y' + y = y^2e^x$

21 $x^{-1}y' + 2y = 3$

22 $y' - 5y = e^{5x}$

23 $\tan x\,dy + (y - \sin x)\,dx = 0$

24 $\cos x\,dy - (y\sin x + e^{-x})\,dx = 0$

25 $y' + 3x^2y = x^2 + e^{-x^3}$

26 $y' + y\tan x = \cos^3 x$

In Exercises 27–30 find the particular solution of the differential equation which satisfies the given boundary conditions.

27 $xy' - y = x^2 + x; x = 1, y = 2$

28 $y' + 2y = e^{-3x}; x = 0, y = 2$

29 $xy' + y + xy = e^{-x}; x = 1, y = 0$

30 $y' + 2xy - e^{-x^2} = x; x = 0, y = 1$

19.5 APPLICATIONS

Differential equations are indispensable in the physical sciences. Earlier in the text we considered elementary applications involving laws of exponential growth, motion, geometry, economics, and several other concepts. As problems become more difficult the corresponding differential equations become more complicated, and computer approximations are often employed.

In this section we shall give several more illustrations of how differential equations may be applied. The interested student will find additional examples in texts on differential equations or in advanced books in engineering or physics.

We have previously used antidifferentiation to derive laws of motion for falling bodies, assuming that air resistance could be neglected (see Example 5 in Section 4.9). This is a valid assumption for small objects near the surface of the earth; however, in many cases air resistance must be taken into account. Indeed, this frictional force often increases as the speed of the object increases. In the following example we shall derive the law of motion for a falling body under the assumption that the resistance due to the air is directly proportional to the speed of the body.

Example 1 An object of mass m is released from a balloon. Find the distance it falls in t seconds, if the force of resistance due to the air is directly proportional to the speed of the object.

Solution Let us introduce a vertical axis with positive direction downward and origin at the point of release as illustrated in Figure 19.3. We wish to find the distance $s(t)$ from the origin to the object at time t.

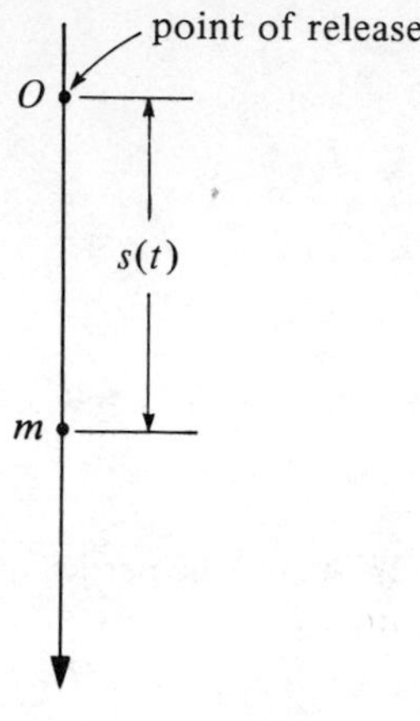

Figure 19.3

From our earlier work, the speed of the object is $v = s'(t)$ and the magnitude of the acceleration is $a = dv/dt = s''(t)$. If g is the gravitational constant, then the object is attracted toward the earth with a force of magnitude mg. By hypothesis, the force of resistance due to the air is kv for some constant k, and this force is directed opposite to the motion. It follows that the downward force F on the object is $mg - kv$. Since Newton's Second Law of Motion states that $F = ma = m(dv/dt)$, we arrive at the following differential equation:

$$m\frac{dv}{dt} = mg - kv \tag{19.15}$$

or equivalently,

$$dv = \left(g - \frac{k}{m}v\right) dt.$$

If we denote the constant k/m by c, then the last equation may be written in the separable form

$$\frac{dv}{g - cv} = dt.$$

Integrating both sides we obtain

$$-\frac{1}{c}\ln(g - cv) = t + D$$

for some constant D. This may also be written as

$$\ln(g - cv) = -c(t + D)$$

or

$$g - cv = e^{-c(t+D)} = e^{-cD}e^{-ct}.$$

If we let $t = 0$, then $v = 0$ and hence $g = e^{-cD}$. Consequently

$$g - cv = ge^{-ct}.$$

Solving for v gives us

$$v = \frac{g - ge^{-ct}}{c}$$

or, since $v = s'(t)$,

$$s'(t) = \frac{g}{c} - \frac{g}{c}e^{-ct}.$$

Integrating both sides of this equation we see that

$$s(t) = \frac{g}{c}t + \frac{g}{c^2}e^{-ct} + E.$$

The constant E may be found by letting $t = 0$. Since $s(0) = 0$ we obtain

$$0 = 0 + \frac{g}{c^2} + E, \quad \text{or} \quad E = -\frac{g}{c^2}.$$

Consequently, the distance the object falls in t seconds is

$$s(t) = \frac{g}{c}t + \frac{g}{c^2}e^{-ct} - \frac{g}{c^2}.$$

It is interesting to compare this formula for $s(t)$ with that obtained when the air resistance is neglected. In the latter case (19.15) reduces to $dv/dt = g$ and it follows that $s'(t) = v = gt$. Integrating both sides leads to the much simpler formula $s(t) = \frac{1}{2}gt^2$.

Example 2 A simple electrical circuit consists of a resistance R and an inductance L connected in series, as illustrated schematically in Figure 19.4, where E is a constant electromotive force. If the switch S is closed at $t = 0$, then it follows from electrical laws that for $t > 0$, the current i satisfies the differential equation

$$L\frac{di}{dt} + Ri = E.$$

Express i as a function of t.

Solution The given equation may be written

$$\frac{di}{dt} + \frac{R}{L}i = \frac{E}{L}$$

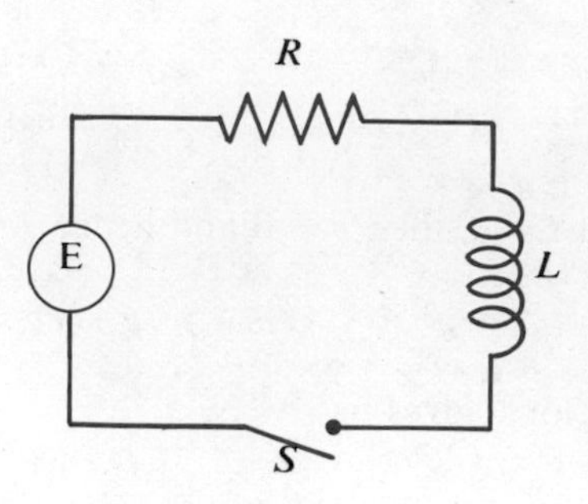

Figure 19.4

which is of the form (19.11). Multiplying both sides by the integrating factor

$$e^{\int (R/L)dt} = e^{(R/L)t}$$

gives us

$$e^{(R/L)t}\frac{di}{dt} + \frac{R}{L}e^{(R/L)t}i = \frac{E}{L}e^{(R/L)t}.$$

Consequently, as in the previous section,

$$ie^{(R/L)t} = \int \frac{E}{L}e^{(R/L)t}\,dt = \frac{E}{R}e^{(R/L)t} + C.$$

Since $i = 0$ when $t = 0$, it follows that $C = -E/R$. Substituting for C we obtain

$$ie^{(R/L)t} = \frac{E}{R}e^{(R/L)t} - \frac{E}{R}.$$

Finally, multiplying both sides by $e^{-(R/L)t}$ gives us

$$i = \frac{E}{R}[1 - e^{-(R/L)t}].$$

Observe that as t increases without bound i approaches E/R, which is the current when no inductance is present. In Exercise 11 the problem is to find a formula for i if the electromotive force is given by $E \sin \omega t$ where ω is a constant.

An **orthogonal trajectory** of a family of curves is a curve which intersects each curve of the family orthogonally. We shall restrict our discussion to curves in a coordinate plane. To illustrate, given the family $y = 2x + b$ of all lines of slope 2, every line $y = (-1/2)x + c$ of slope $-1/2$ is an orthogonal trajectory. We sometimes call two such families of curves **mutually orthogonal**. As another example, the family of all lines through the origin and the family of all concentric circles with centers at the origin are mutually orthogonal.

Pairs of mutually orthogonal families of curves occur frequently in physical applications of mathematics. In the theory of electricity and magnetism, the lines of force associated with a given field are orthogonal trajectories of the corresponding equipotential curves. Similarly, the stream lines studied in aerodynamics and hydrodynamics are orthogonal trajectories of the so-called *velocity-equipotential* curves. As a final illustration, in the study of thermodynamics the flow of heat across a plane surface is shown to be orthogonal to the isothermal curves.

The next example illustrates a technique for finding the orthogonal trajectories of a family of curves.

Example 3 Find the orthogonal trajectories of the family of ellipses $x^2 + 3y^2 = c$ and sketch several members of each family.

Solution Differentiating the given equation implicitly we obtain

$$2x + 6yy' = 0, \quad \text{or} \quad y' = -\frac{x}{3y}.$$

Hence the slope of the tangent line at any point (x, y) on one of the ellipses is $y' = -x/3y$. If dy/dx is the slope of the tangent line on a corresponding orthogonal trajectory, then it must equal the negative reciprocal of y'. This gives us the following differential equation for the family of orthogonal trajectories:

$$\frac{dy}{dx} = \frac{3y}{x}.$$

Separating the variables,

$$\frac{dy}{y} = 3\frac{dx}{x}.$$

Integrating, and writing the constant of integration as $\ln |k|$ gives us

$$\ln |y| = 3 \ln |x| + \ln |k| = \ln |kx^3|.$$

It follows that $y = kx^3$ is an equation for the family of orthogonal trajectories. Several members of the given family and corresponding orthogonal trajectories are sketched (with dashes) in Figure 19.5.

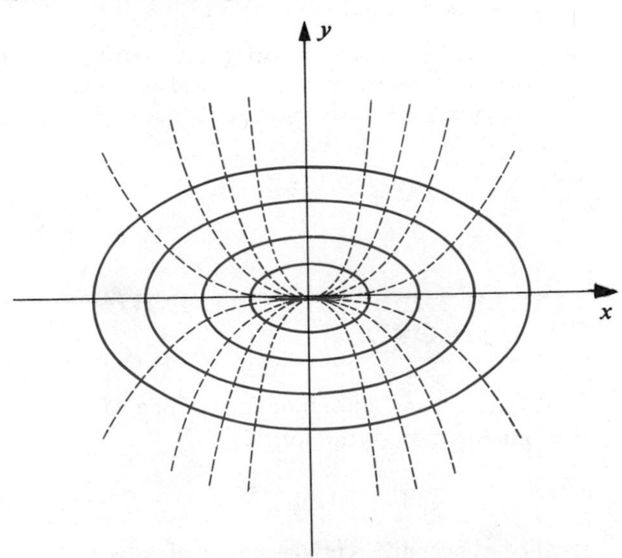

Figure 19.5. Orthogonal trajectories

EXERCISES 19.5

Find the orthogonal trajectories of the families of curves in Exercises 1–8 and sketch several members of each family.

1 $x^2 - y^2 = c$

2 $xy = c$

3 $y^2 = cx$

4 $y = cx^2$

5 $y^2 = cx^3$

6 $y = ce^{-x}$

7 $x^2 + y^2 = 2cx$

8 $x^2 + y^2 - 2cy = 0$

9 The differential equation $p\,dv + cv\,dp = 0$ describes the adiabatic change of state of air, where p and v are the pressure and volume, respectively, and c is a constant. Solve for p as a function of v.

10 The equation

$$R\frac{dQ}{dt} + \frac{Q}{C} = E$$

describes the charge Q on a condenser of capacity C during a charging process involving a resistance R and electromotive force E. If the charge is 0 when $t = 0$, express Q as a function of t.

11 Solve Example 2 for the case where the electromotive force is $E \sin \omega t$, where E and ω are constants.

12 The equation

$$R\frac{di}{dt} + \frac{i}{C} = \frac{dE}{dt}$$

describes an electrical circuit consisting of a resistance R and capacity C in series. Express i as a function of t if $i = i_0$ when $t = 0$, and (a) E is a constant; (b) $E = E_0 \sin \omega t$ where E_0 and ω are constants.

13 Solve Example 1 if the force of resistance due to the air is directly proportional to the square of the speed of the object.

14 An object of mass m is moving rectilinearly subject to a force given by $F(t) = e^{-t}$, where t is time. The motion is resisted by a frictional force that is numerically equal to twice the speed of the object. If $v = 0$ at $t = 0$, find a formula for v at any time $t > 0$.

19.6 LINEAR DIFFERENTIAL EQUATIONS OF THE SECOND ORDER

If $f_1, f_2, \ldots, f_n$ and k are functions of one variable which have the same domain, then an equation of the form

$$y^{(n)} + f_1(x)y^{(n-1)} + \cdots + f_{n-1}(x)y' + f_n(x)y = k(x) \tag{19.16}$$

is called a **linear differential equation of order n.** If $k(x) = 0$ for all x, the equation is said to be **homogeneous.** Notice that this meaning of the word *homogeneous* is different from that in Section 3. If $k(x) \neq 0$ for some x, then (19.16) is said to be **nonhomogeneous**. A thorough analysis of (19.16) will be found in textbooks on differential equations. We shall restrict our work to second-order equations in which f_1 and f_2 are constant functions. In this section the homogeneous case is considered. Nonhomogeneous equations will be discussed in the next section.

The general second-order homogeneous linear differential equation with constant coefficients has the form

(19.17)
$$y'' + by' + cy = 0$$

where b and c are constants. Before attempting to find particular solutions let us establish the following result.

(19.18) Theorem

If $y = f(x)$ and $y = g(x)$ are solutions of $y'' + by' + cy = 0$, then

$$y = C_1 f(x) + C_2 g(x)$$

is a solution for all real numbers C_1 and C_2.

Proof. By hypothesis,

$$f''(x) + bf'(x) + cf(x) = 0$$
$$g''(x) + bg'(x) + cg(x) = 0.$$

If we multiply the first of these equations by C_1, the second by C_2, and add, the result is

$$[C_1 f''(x) + C_2 g''(x)] + b[C_1 f'(x) + C_2 g'(x)] + c[C_1 f(x) + C_2 g(x)] = 0.$$

Thus $C_1 f(x) + C_2 g(x)$ is a solution.

It can be shown that if the solutions f and g in Theorem (19.18) have the property that $f(x) \neq Cg(x)$ for all real numbers C, and if $g(x)$ is not identically 0, then $y = C_1 f(x) + C_2 g(x)$ is a general solution of $y'' + by' + cy = 0$. Thus, to determine the general solution it is sufficient to find two such functions f and g and employ (19.18).

In our search for a solution of (19.17) we shall use $y = e^{mx}$ as a trial solution. Since $y' = me^{mx}$ and $y'' = m^2 e^{mx}$, it follows that $y = e^{mx}$ is a solution of $y'' + by' + cy = 0$ if and only if

$$m^2 e^{mx} + bme^{mx} + ce^{mx} = 0$$

or, since $e^{mx} \neq 0$, if and only if

(19.19)
$$m^2 + bm + c = 0.$$

The equation in (19.19) is called the **auxiliary equation** of $y'' + by' + cy = 0$. It can be obtained from this differential equation by replacing y'' by m^2, y' by m, and

y by 1. In simple cases the roots of the auxiliary equation (19.19) can be found by factoring. If a factorization is not evident, then applying the quadratic formula, we see that the roots of the auxiliary equation are

(19.20)
$$m = \frac{-b \pm \sqrt{b^2 - 4c}}{2}.$$

Thus the auxiliary equation has unequal real roots m_1 and m_2, a double real root m, or two complex conjugate roots according as $b^2 - 4c$ is positive, zero, or negative, respectively. The next theorem is a consequence of the remark following the proof of Theorem (19.18).

(19.21) Theorem

If the roots m_1, m_2 of the auxiliary equation are real and unequal, then the general solution of $y'' + by' + cy = 0$ is

$$y = C_1 e^{m_1 x} + C_2 e^{m_2 x}.$$

Example 1 Solve the differential equation $y'' - 3y' - 10y = 0$.

Solution The auxiliary equation is $m^2 - 3m - 10 = 0$, or $(m - 5)(m + 2) = 0$. Since the roots $m_1 = 5$ and $m_2 = -2$ are real and unequal, it follows from Theorem (19.21) that the general solution is

$$y = C_1 e^{5x} + C_2 e^{-2x}.$$

(19.22) Theorem

If the auxiliary equation has a double root m, then the general solution of $y'' + by' + cy = 0$ is

$$y = C_1 e^{mx} + C_2 x e^{mx}.$$

Proof. Using (19.20), with $b^2 - 4c = 0$, we obtain $m = -b/2$ or $2m + b = 0$. Since m satisfies the auxiliary equation, $y = e^{mx}$ is a solution of the differential equation. According to the remark following the proof of Theorem (19.18), it is sufficient to show that $y = xe^{mx}$ is also a solution. Substitution of xe^{mx} for y in $y'' + by' + cy = 0$ gives us

$$\begin{aligned}(2me^{mx} + m^2xe^{mx}) &+ b(mxe^{mx} + e^{mx}) + cxe^{mx} \\ &= (m^2 + bm + c)xe^{mx} + (2m + b)e^{mx} \\ &= 0xe^{mx} + 0e^{mx} = 0\end{aligned}$$

which is what we wished to show.

Example 2 Solve the differential equation

$$y'' - 6y' + 9y = 0.$$

Solution The auxiliary equation $m^2 - 6m + 9 = 0$, or equivalently $(m - 3)^2 = 0$, has a double root 3. Hence, by Theorem (19.22), the general solution is

$$y = C_1e^{3x} + C_2xe^{3x} = e^{3x}(C_1 + C_2x).$$

We may also consider second-order differential equations of the form

$$ay'' + by' + cy = 0$$

where $a \neq 1$. It is possible to obtain form (19.17) by dividing both sides by a; however, it is usually simpler to employ the auxiliary equation

$$am^2 + bm + c = 0$$

as illustrated in the next example.

Example 3 Solve the differential equation $6y'' - 7y' + 2y = 0$.

Solution The auxiliary equation $6m^2 - 7m + 2 = 0$ can be factored as follows:

$$(2m - 1)(3m - 2) = 0.$$

Hence the roots are $m_1 = 1/2$ and $m_2 = 2/3$. Applying Theorem (19.21), the general solution of the given equation is

$$y = C_1e^{x/2} + C_2e^{2x/3}.$$

The final case to consider is that in which the roots of the auxiliary equation $m^2 + bm + c = 0$ of $y'' + by' + cy = 0$ are complex numbers. Recall that complex numbers may be represented by expressions of the form $a + bi$, where a and b are real numbers, and i is a symbol which may be manipulated in the same manner as a real number, but has the additional property that $i^2 = -1$. Two complex numbers $a + bi$ and $c + di$ are said to be **equal**, and we write $a + bi = c + di$, if and only if $a = c$ and $b = d$. Operations of addition, subtraction, multiplication, and division are defined *just as though* all letters denote real numbers, with the additional stipulation that whenever i^2 occurs, it may be replaced by -1. For example, the formulas for addition and multiplication of two complex numbers $a + bi$ and $c + di$ are

$$(a + bi) + (c + di) = (a + c) + (b + d)i$$
$$(a + bi)(c + di) = (ac - bd) + (ad + bc)i.$$

We may regard the real numbers as a subset of the complex numbers by identifying the real number a with the complex number $a + 0i$. A complex number of the form $0 + bi$ is abbreviated bi.

Complex numbers are often required for solving equations of the form $f(x) = 0$, where $f(x)$ is a polynomial. For example, if only real numbers are allowed, then the equation $x^2 = -4$ has no solutions. However, if complex numbers are available, then the equation has a solution $2i$, since

$$(2i)^2 = 2^2 i^2 = 4(-1) = -4.$$

Similarly, $-2i$ is a solution of $x^2 = -4$.

Since $i^2 = -1$, we sometimes use the symbol $\sqrt{-1}$ in place of i and write

$$\sqrt{-13} = \sqrt{13}i, \quad 2 + \sqrt{-25} = 2 + \sqrt{25}i = 2 + 5i$$

and so on. A *quadratic equation* $ax^2 + bx + c = 0$, where a, b, c are real numbers and $a \neq 0$, has roots given by the **quadratic formula**

$$x = \frac{-b \pm \sqrt{b^2 - 4ac}}{2a}.$$

If $b^2 - 4ac < 0$, then the roots are complex numbers. To illustrate, if we apply the quadratic formula to the equation $x^2 - 4x + 13 = 0$ we obtain

$$x = \frac{4 \pm \sqrt{16 - 52}}{2} = \frac{4 \pm \sqrt{-36}}{2} = \frac{4 \pm 6i}{2} = 2 \pm 3i.$$

Thus the equation has the two complex roots $2 + 3i$ and $2 - 3i$.

The complex number $a - bi$ is called the **conjugate** of the complex number $a + bi$. We see from the quadratic formula that if a quadratic equation with real coefficients has complex roots, then they are necessarily conjugates of one another.

It follows from the preceding discussion that if the auxiliary equation $m^2 + bm + c = 0$ in (19.19) has complex roots, then they are of the form

$$z_1 = s + ti \quad \text{and} \quad z_2 = s - ti$$

where s and t are real numbers. We may anticipate, from Theorem (19.21), that the general solution of the differential equation $y'' + by' + cy = 0$ is

$$y = C_1 e^{z_1 x} + C_2 e^{z_2 x}$$

that is,

(19.23) $$y = C_1 e^{(s+ti)x} + C_2 e^{(s-ti)x}.$$

In order to handle such complex exponents it is necessary to extend some of the concepts of calculus to include functions whose domains include complex numbers. Since a complete development is beyond the scope of our work, we shall merely outline the main ideas.

In Chapter 12 we discussed how certain functions can be represented by power series. In particular, from (12.43)-(12.45),

$$e^x = 1 + x + \frac{x^2}{2!} + \cdots + \frac{x^n}{n!} + \cdots$$

$$\sin x = x - \frac{x^3}{3!} + \frac{x^5}{5!} - \cdots + (-1)^n \frac{x^{2n+1}}{(2n+1)!} + \cdots$$

$$\cos x = 1 - \frac{x^2}{2!} + \frac{x^4}{4!} - \cdots + (-1)^n \frac{x^{2n}}{(2n)!} + \cdots$$

for every real number x. It is not difficult to extend the definitions and theorems of Chapter 12 to infinite series which involve complex numbers. Since this is true, we *define* e^z, $\sin z$, and $\cos z$ for every complex number z as follows:

(19.24)

$$e^z = 1 + z + \frac{z^2}{2!} + \cdots + \frac{z^n}{n!} + \cdots$$

$$\sin z = z - \frac{z^3}{3!} + \frac{z^5}{5!} - \cdots + (-1)^n \frac{z^{2n+1}}{(2n+1)!} + \cdots$$

$$\cos z = 1 - \frac{z^2}{2!} + \frac{z^4}{4!} - \cdots + (-1)^n \frac{z^{2n}}{(2n)!} + \cdots.$$

Using the first formula in (19.24) gives us

$$e^{iz} = 1 + (iz) + \frac{(iz)^2}{2!} + \frac{(iz)^3}{3!} + \frac{(iz)^4}{4!} + \frac{(iz)^5}{5!} + \cdots$$

$$= 1 + iz + i^2\frac{z^2}{2!} + i^3\frac{z^3}{3!} + i^4\frac{z^4}{4!} + i^5\frac{z^5}{5!} + \cdots$$

Since $i^2 = -1, i^3 = -i, i^4 = 1, i^5 = i$, and so on, we see that

$$e^{iz} = 1 + iz - \frac{z^2}{2!} - i\frac{z^3}{3!} + \frac{z^4}{4!} + i\frac{z^5}{5!} - \cdots$$

which may also be written in the form

$$e^{iz} = \left(1 - \frac{z^2}{2!} + \frac{z^4}{4!} - \cdots\right) + i\left(z - \frac{z^3}{3!} + \frac{z^5}{5!} - \cdots\right).$$

If we now use the formulas for $\cos z$ and $\sin z$ in (19.24), we obtain the following important result.

(19.25) Euler's Formula

For every complex number z,

$$e^{iz} = \cos z + i \sin z.$$

It can be shown that the Laws of Exponents are true for complex numbers. In addition, formulas for derivatives developed earlier in this text can be extended to functions of a *complex* variable z. One such formula is $D_z e^{kz} = ke^{kz}$, where k is a complex number. It can then be proved that the general solution of $y'' + by' + c = 0$, where the roots of the auxiliary equation are the complex numbers $s \pm ti$, is given by (19.23). The form of this solution may be changed as follows:

$$\begin{aligned} y &= C_1 e^{(s+ti)x} + C_2 e^{(s-ti)x} \\ &= C_1 e^{sx+txi} + C_2 e^{sx-txi} \\ &= C_1 e^{sx} e^{txi} + C_2 e^{sx} e^{-txi} \end{aligned}$$

or equivalently,

(19.26)
$$y = e^{sx}(C_1 e^{itx} + C_2 e^{-itx}).$$

This can be further simplified by using Euler's Formula. Specifically, we see from (19.25) that

$$\begin{aligned} e^{itx} &= \cos tx + i \sin tx, \\ e^{-itx} &= \cos tx - i \sin tx \end{aligned}$$

from which it follows that

(19.27)
$$\cos tx = \frac{e^{itx} + e^{-itx}}{2}, \qquad \sin tx = \frac{e^{itx} - e^{-itx}}{2i}.$$

If we let $C_1 = C_2 = 1/2$ in (19.26) and use (19.27) we obtain the particular solution $y = e^{sx} \cos tx$ of $y'' + by' + cy = 0$. Letting $C_1 = -C_2 = i/2$ gives us the particular solution $y = e^{sx} \sin tx$. This is a partial proof of the next theorem.

(19.28) Theorem

If the auxiliary equation $m^2 + bm + c = 0$ has distinct complex roots $s \pm ti$, then the general solution of $y'' + by' + cy = 0$ is

$$y = e^{sx}(C_1 \cos tx + C_2 \sin tx).$$

Example 4 Solve the differential equation

$$y'' - 10y' + 41y = 0.$$

Solution The roots of the auxiliary equation $m^2 - 10m + 41 = 0$ are

$$m = \frac{10 \pm \sqrt{100 - 164}}{2} = \frac{10 \pm \sqrt{-64}}{2} = \frac{10 \pm 8i}{2} = 5 \pm 4i.$$

Hence by Theorem (19.28), the general solution of the differential equation is

$$y = e^{5x}(C_1 \cos 4x + C_2 \sin 4x).$$

EXERCISES 19.6

Solve the differential equations in Exercises 1–22.

1 $y'' - 5y' + 6y = 0$

2 $y'' - y' - 2y = 0$

3 $y'' - 3y' = 0$

4 $y'' + 6y' + 8y = 0$

5 $y'' + 4y' + 4y = 0$

6 $y'' - 4y' + 4y = 0$

7 $y'' - 4y' + y = 0$

8 $6y'' - 7y' - 3y = 0$

9 $y'' + 2\sqrt{2}y' + 2y = 0$

10 $4y'' + 20y' + 25y = 0$

11 $8y'' + 2y' - 15y = 0$

12 $y'' + 4y' + y = 0$

13 $9y'' - 24y' + 16y = 0$

14 $4y'' - 8y' + 7y = 0$

15 $2y'' - 4y' + y = 0$

16 $2y'' + 7y' = 0$

17 $y'' - 2y' + 2y = 0$

18 $y'' - 2y' + 5y = 0$

19 $y'' - 4y' + 13y = 0$

20 $y'' + 4 = 0$

21 $\dfrac{d^2y}{dx^2} + 6\dfrac{dy}{dx} + 2y = 0$

22 $\dfrac{d^2y}{dx^2} + 2\dfrac{dy}{dx} + 6y = 0$

In each of Exercises 23–30 find the particular solution of the differential equation which satisfies the stated boundary conditions.

23 $y'' - 3y' + 2y = 0$; $y = 0$ and $y' = 2$ when $x = 0$

24 $y'' - 2y' + y = 0$; $y = 1$ and $y' = 2$ when $x = 0$

25 $y'' + y = 0$; $y = 1$ and $y' = 2$ when $x = 0$

26 $y'' - y' - 6y = 0$; $y = 0$ and $y' = 1$ when $x = 0$

27 $y'' + 8y' + 16y = 0$; $y = 2$ and $y' = 1$ when $x = 0$

28 $y'' + 5y = 0$; $y = 4$ and $y' = 2$ when $x = 0$

29 $\dfrac{d^2y}{dx^2} - 2\dfrac{dy}{dx} + 5y = 0$; $y = 0$ and $\dfrac{dy}{dx} = 1$ when $x = 0$

30 $\dfrac{d^2y}{dx^2} - 6\dfrac{dy}{dx} + 13y = 0$; $y = 2$ and $\dfrac{dy}{dx} = 3$ when $x = 0$

19.7 NONHOMOGENEOUS LINEAR DIFFERENTIAL EQUATIONS

In this section we shall consider second-order nonhomogeneous linear differential equations with constant coefficients, that is, equations of the form

$$y'' + by' + cy = k(x) \qquad (19.29)$$

where b and c are constants and the function k is continuous.

It is convenient to use the differential operator symbols D and D^2 where if $y = f(x)$, then

$$Dy = y' = f'(x) \quad \text{and} \quad D^2y = y'' = f''(x).$$

We shall also employ the **linear differential operator**

$$L = D^2 + bD + c$$

where by definition,

$$L(y) = (D^2 + bD + c)y = D^2y + bDy + cy = y'' + by' + cy.$$

Using this notation, (19.29) can be written in the compact form $L(y) = k(x)$. It is easy to verify that for every real number C,

(19.30) $$L(Cy) = CL(y).$$

Also, if $y_1 = f_1(x)$ and $y_2 = f_2(x)$, then it can be shown that

(19.31) $$L(y_1 \pm y_2) = L(y_1) \pm L(y_2).$$

Given the differential equation (19.29), that is, $L(y) = k(x)$, the corresponding homogeneous equation $L(y) = 0$ is called the **complementary equation**. Suppose that y_p is a particular solution of $L(y) = k(x)$ and y_c is any solution of the complementary equation. Since $L(y_p) = k(x)$ and $L(y_c) = 0$,

$$L(y_p + y_c) = L(y_p) + L(y_c) = k(x) + 0 = k(x)$$

which means that $y_p + y_c$ is a solution of (19.29). Moreover, if $y = f(x)$ is any other solution of $L(y) = k(x)$, then

$$L(y - y_p) = L(y) - L(y_p) = k(x) - k(x) = 0.$$

Consequently $y - y_p$ is a solution of the complementary equation, that is, $y - y_p = y_c$ for some y_c. This proves the next theorem.

(19.32) Theorem

If y_p is a particular solution of the differential equation $L(y) = k(x)$ and if y_c is the general solution of the complementary equation $L(y) = 0$, then the general solution of $L(y) = k(x)$ is $y = y_p + y_c$.

If we use the results of the previous section to find the general solution y_c of $L(y) = 0$, then according to Theorem (19.32) all that is needed to determine the general solution of $L(y) = k(x)$ is *one* particular solution y_p.

Example 1 Solve the differential equation $y'' - 4y = 6x - 4x^3$.

Solution We see by inspection that $y_p = x^3$ is a particular solution of the given equation. The complementary equation is $y'' - 4y = 0$ which by (19.21), has general solution

$$y_c = C_1 e^{2x} + C_2 e^{-2x}.$$

Applying Theorem (19.32), the general solution of the given nonhomogeneous equation is

$$y = C_1 e^{2x} + C_2 e^{-2x} + x^3.$$

In most cases a particular solution of (19.29) cannot be found by inspection, as was done in Example 1. The following technique, called **variation of parameters**, may then be employed. Given the differential equation $L(y) = k(x)$, let y_1 and y_2 be the expressions that appear in the general solution $y = C_1 y_1 + C_2 y_2$ of the complementary equation $L(y) = 0$. For example we might have, as in (19.21), $y_1 = e^{m_1 x}$ and $y_2 = e^{m_2 x}$. Let us now attempt to find a particular solution of $L(y) = k(x)$ which has the form

(19.33)
$$y_p = uy_1 + vy_2$$

where $u = g(x)$ and $v = h(x)$ for some functions g and h. The first and second derivatives of y_p are

$$y_p' = (uy_1' + vy_2') + (u'y_1 + v'y_2)$$
$$y_p'' = (uy_1'' + vy_2'') + (u'y_1' + v'y_2') + (u'y_1 + v'y_2)'.$$

Substituting these in $L(y_p) = y_p'' + by_p' + cy_p$ and rearranging terms we obtain

(19.34)
$$L(y_p) = u(y_1'' + by_1' + cy_1) + v(y_2'' + by_2' + cy_2) + b(u'y_1 + v'y_2) + (u'y_1 + v'y_2)' + (u'y_1' + v'y_2').$$

Since y_1 and y_2 are solutions of $y'' + by' + cy = 0$, the first two terms on the right in (19.34) are 0. Hence, in order to obtain $L(y_p) = k(x)$, it is sufficient to choose u and v such that

(19.35)
$$u'y_1 + v'y_2 = 0$$
$$u'y_1' + v'y_2' = k(x).$$

It can be shown that this system of equations always has a unique solution u' and v'. We may then determine u and v by integration and use (19.33) to find y_p.

Example 2 Solve the differential equation $y'' + y = \cot x$.

Solution The complementary equation is $y'' + y = 0$. Since the auxiliary equation $m^2 + 1 = 0$ has roots $\pm i$, we see from Theorem (19.28) that the general solution of the homogeneous equation $y'' + y = 0$ is $y = C_1 \cos x + C_2 \sin x$. As in the preceding discussion, we let $y_1 = \cos x$ and $y_2 = \sin x$. The system (19.35) is, therefore,

$$u' \cos x + v' \sin x = 0$$
$$-u' \sin x + v' \cos x = \cot x.$$

Solving for u' and v' gives us

$$u' = -\cos x, \quad v' = \csc x - \sin x.$$

If we integrate each of these expressions (and drop the constants of integration) we obtain

$$u = -\sin x, \quad v = \ln |\csc x - \cot x| + \cos x.$$

Applying (19.33), a particular solution of the given equation is

$$y_p = -\sin x \cos x + \sin x \ln |\csc x - \cot x| + \sin x \cos x$$

or

$$y_p = \sin x \ln |\csc x - \cot x|.$$

Finally, by (19.32), the general solution of $y'' + y = \cot x$ is

$$y = C_1 \cos x + C_2 \sin x + \sin x \ln |\csc x - \cot x|.$$

Given the differential equation

$$L(y) = y'' + by' + cy = e^{nx}$$

where e^{nx} is not a solution of $L(y) = 0$, it is reasonable to expect that there exists a particular solution of the form $y_p = Ae^{nx}$, since e^{nx} is the result of finding $y'' + by' + cy$. This suggests that we use Ae^{nx} as a trial solution in the given equation and attempt to find the value of the coefficient A. This technique is called the **method of undetermined coefficients**, and is illustrated in the next example.

Example 3 Solve the differential equation $y'' + 2y' - 8y = e^{3x}$.

Solution Since the auxiliary equation $m^2 + 2m - 8 = 0$ of $y'' + 2y' - 8y = 0$ has roots 2 and -4 it follows from the previous section that the general solution of the complementary equation is

$$y_c = C_1 e^{2x} + C_2 e^{-4x}.$$

From the preceding remarks we seek a particular solution of the form $y_p = Ae^{3x}$. Since $y_p' = 3Ae^{3x}$ and $y_p'' = 9Ae^{3x}$, substitution in the given equation leads to

$$9Ae^{3x} + 6Ae^{3x} - 8Ae^{3x} = e^{3x}.$$

Dividing both sides by e^{3x} we obtain

$$9A + 6A - 8A = 1, \quad \text{or} \quad A = \tfrac{1}{7}.$$

Thus $y_p = (1/7)e^{3x}$ and by Theorem (19.32), the general solution is

$$y = C_1 e^{2x} + C_2 e^{-4x} + \tfrac{1}{7} e^{3x}.$$

Three rules for arriving at trial solutions to second-order nonhomogeneous differential equations with constant coefficients are stated without proof in the next theorem. The reader is referred to texts on differential equations for a more extensive treatment of this topic.

(19.36) Theorem

(i) If $y'' + by' + cy = e^{nx}$, and n is not a root of the auxiliary equation $m^2 + bm + c = 0$, then there is a particular solution of the form $y_p = Ae^{nx}$.

(ii) If $y'' + by' + cy = xe^{nx}$, and n is not a solution of the auxiliary equation $m^2 + bm + c = 0$, then there is a particular solution of the form $y_p = (A + Bx)e^{nx}$.

(iii) If either

$$y'' + by' + cy = e^{sx} \sin tx$$

or

$$y'' + by' + cy = e^{sx} \cos tx$$

and $s + ti$ is not a solution of the auxiliary equation $m^2 + bm + c = 0$, then there is a particular solution of the form

$$y_p = Ae^{sx} \cos tx + Be^{sx} \sin tx.$$

Rule (i) of Theorem (19.36) was used in the solution of Example 3. The use of rules (ii) and (iii) is illustrated in the following examples.

Example 4 Solve $y'' - 3y' - 18y = xe^{4x}$.

Solution Since the auxiliary equation $m^2 - 3m - 18 = 0$ has roots 6 and -3 it follows from the preceding section that the general solution of $y'' - 3y' - 18y = 0$ is

$$y = C_1e^{6x} + C_2e^{-3x}.$$

Since 4 is not a root of the auxiliary equation we see from (ii) of (19.36) that there is a particular solution of the form

$$y_p = (A + Bx)e^{4x}.$$

Differentiating we obtain

$$y_p' = (4A + 4Bx + B)e^{4x}$$
$$y_p'' = (16A + 16Bx + 8B)e^{4x}.$$

Substitution in the given differential equation produces

$$(16A + 16Bx + 8B)e^{4x} - 3(4A + 4Bx + B)e^{4x} - 18(A + Bx)e^{4x} = xe^{4x}$$

which reduces to

$$-14A + 5B - 14Bx = x.$$

Thus y_p is a solution provided

$$-14A + 5B = 0 \quad \text{and} \quad -14B = 1.$$

This gives us $B = -1/14$ and $A = -5/196$. Consequently

$$y_p = \left(-\frac{5}{196} - \frac{1}{14}x\right)e^{4x} = -\frac{1}{196}(5 + 14x)e^{4x}.$$

Applying (19.32), the general solution is

$$y = C_1e^{6x} + C_2e^{-3x} - \tfrac{1}{196}(5 + 14x)e^{4x}.$$

Example 5 Solve $y'' - 10y' + 41y = \sin x$.

Solution The general solution $y_c = e^{5x}(C_1 \cos 4x + C_2 \sin 4x)$ of the complementary equation was found in Example 4 of the preceding section. Referring to (iii) of (19.36) with $s = 0$ and $t = 1$, we seek a particular solution of the form

$$y_p = A\cos x + B\sin x.$$

Since

$$y_p' = -A\sin x + B\cos x \quad \text{and} \quad y_p'' = -A\cos x - B\sin x,$$

substitution in the given equation produces

$$\begin{aligned} -A\cos x - B\sin x + 10A\sin x - 10B\cos x \\ + 41A\cos x + 41B\sin x = \sin x \end{aligned}$$

which can be written

$$(40A - 10B)\cos x + (10A + 40B)\sin x = \sin x.$$

Consequently, y_p is a solution provided

$$40A - 10B = 0 \quad \text{and} \quad 10A + 40B = 1.$$

The solution of this system of equations is $A = 1/170$ and $B = 4/170$. Hence

$$y_p = \frac{1}{170}\cos x + \frac{4}{170}\sin x = \frac{1}{170}(\cos x + 4\sin x)$$

and the general solution is

$$y = e^{5x}(C_1 \cos 4x + C_2 \sin 4x) + \frac{1}{170}(\cos x + 4\sin x).$$

EXERCISES 19.7

Solve the differential equations in Exercises 1–10 by using the method of variation of parameters.

1 $y'' + y = \tan x$

2 $y'' + y = \sec x$

3 $y'' - 6y' + 9y = x^2e^{3x}$

4 $y'' + 3y' = e^{-3x}$

5 $y'' - y = e^x \cos x$

6 $y'' - 4y' + 4y = x^{-2}e^{2x}$

7 $y'' - 9y = e^{3x}$

8 $y'' + y = \sin x$

9 $\dfrac{d^2y}{dx^2} - 3\dfrac{dy}{dx} - 4y = 2$

10 $\dfrac{d^2y}{dx^2} - \dfrac{dy}{dx} = x + 1$

Solve the differential equations in Exercises 11–18 by using undetermined coefficients.

11 $y'' - 3y' + 2y = 4e^{-x}$

12 $y'' + 6y' + 9y = e^{2x}$

13 $y'' + 2y' = \cos 2x$

14 $y'' + y = 5 \sin x$

15 $y'' - y = xe^{2x}$

16 $y'' + 3y' - 4y = xe^{-x}$

17 $\dfrac{d^2y}{dx^2} - 6\dfrac{dy}{dx} + 13y = e^x \cos x$

18 $\dfrac{d^2y}{dx^2} - 2\dfrac{dy}{dx} + 2y = e^{-x} \sin 2x$

19 Prove (19.30).

20 Prove (19.31).

19.8 SERIES SOLUTIONS OF DIFFERENTIAL EQUATIONS

As shown in Chapter 12, a power series $\Sigma\, a_n x^n$ determines a function f, such that

$$y = f(x) = a_0 + a_1x + a_2x^2 + a_3x^3 + a_4x^4 + \cdots \tag{19.37}$$

for every x in the interval of convergence. Moreover, series representations for the derivatives of f may be obtained by differentiating each term of (19.37). Thus

$$\begin{aligned} y' &= a_1 + 2a_2x + 3a_3x^2 + 4a_4x^3 + \cdots = \sum_{n=1}^{\infty} na_nx^{n-1} \\ y'' &= 2a_2 + 3\cdot 2a_3x + 4\cdot 3a_4x^2 + \cdots = \sum_{n=2}^{\infty} n(n-1)a_nx^{n-2} \end{aligned} \tag{19.38}$$

and so on. Power series may be used to solve certain differential equations. In this event, the solution is often expressed as an infinite series, and is called a **series solution** of the differential equation.

Example 1 Find a series solution of the differential equation $y' = 2xy$.

Solution If the solution is given by $y = \Sigma\, a_n x^n$, then $y' = \Sigma\, n a_n x^{n-1}$, and substitution in the differential equation gives us

$$\sum_{n=1}^{\infty} n a_n x^{n-1} = 2x \sum_{n=0}^{\infty} a_n x^n = \sum_{n=0}^{\infty} 2a_n x^{n+1}.$$

It is convenient to change the summation on the left so that the same power of x appears as in the summation on the right. This may be accomplished by replacing n by $n + 2$ and beginning the summation at $n = -1$. Thus

$$\sum_{n=-1}^{\infty} (n+2)a_{n+2}x^{n+1} = \sum_{n=0}^{\infty} 2a_n x^{n+1}$$

or

$$a_1 + 2a_2 x + \cdots + (n+2)a_{n+2}x^{n+1} + \cdots = 2a_0 x + \cdots + 2a_n x^{n+1} + \cdots.$$

Comparing coefficients we see that $a_1 = 0$ and $(n+2)a_{n+2} = 2a_n$ if $n \geq 0$. Consequently, the coefficients are given by

$$a_1 = 0 \quad \text{and} \quad a_{n+2} = \frac{2}{n+2}a_n \quad \text{if} \quad n \geq 0.$$

In particular,

$$a_1 = 0,\ a_2 = a_0,\ a_3 = \tfrac{2}{3}a_1 = 0,\ a_4 = \tfrac{1}{2}a_2 = \tfrac{1}{2}a_0,\ a_5 = \tfrac{2}{5}a_3 = 0,$$

$$a_6 = \tfrac{1}{3}a_4 = \frac{1}{2\cdot 3}a_0,\ a_7 = \tfrac{2}{7}a_5 = 0,\ a_8 = \tfrac{1}{4}a_6 = \frac{1}{2\cdot 3\cdot 4}a_0,$$

and so on. It can be shown that if n is odd, then $a_n = 0$, whereas $a_{2n} = (1/n!)a_0$ for every positive integer n. The series solution is, therefore,

$$y = \sum_{n=0}^{\infty} a_n x^n = a_0\left(1 + x^2 + \frac{1}{2!}x^4 + \cdots + \frac{1}{n!}x^{2n} + \cdots\right).$$

It follows from (12.45) that the series solution in Example 1 can be written as $y = a_0 e^{x^2}$. Indeed, this form may be found directly from $y' = 2xy$ by using the separation of variables technique discussed in Section 1. The objective in Example 1, however, was to illustrate series solutions and not to find the solution in the simplest manner. In many instances it is impossible to find the sum of $\Sigma\, a_n x^n$ and the solution must be left in series form.

Example 2 Solve the differential equation $y'' - xy' - 2y = 0$.

Solution Substituting for y, y', and y'' from (19.37) and (19.38) we obtain

$$\sum_{n=2}^{\infty} n(n-1)a_n x^{n-2} - x\sum_{n=1}^{\infty} n a_n x^{n-1} - 2\sum_{n=0}^{\infty} a_n x^n = 0$$

or, equivalently,

$$\sum_{n=2}^{\infty} n(n-1)a_n x^{n-2} = \sum_{n=0}^{\infty} na_n x^n + \sum_{n=0}^{\infty} 2a_n x^n.$$

We next adjust the summation on the left so that the power x^n appears instead of x^{n-2}. This can be accomplished by replacing n by $n+2$ and starting the summation at $n = 0$. This gives us

$$\sum_{n=0}^{\infty} (n+2)(n+1)a_{n+2} x^n = \sum_{n=0}^{\infty} (n+2)a_n x^n.$$

Comparing coefficients we see that $(n+2)(n+1)a_{n+2} = (n+2)a_n$, that is,

$$a_{n+2} = \frac{1}{n+1} a_n.$$

Letting $n = 0, 1, 2, \ldots, 7$ leads to the following form for the coefficients a_k:

$$a_2 = a_0 \qquad a_3 = \frac{1}{2}a_1$$

$$a_4 = \frac{1}{3}a_2 = \frac{1}{3}a_0 \qquad a_5 = \frac{1}{4}a_3 = \frac{1}{2\cdot 4}a_1$$

$$a_6 = \frac{1}{5}a_4 = \frac{1}{3\cdot 5}a_0 \qquad a_7 = \frac{1}{6}a_5 = \frac{1}{2\cdot 4\cdot 6}a_1$$

$$a_8 = \frac{1}{7}a_6 = \frac{1}{3\cdot 5\cdot 7}a_0 \qquad a_9 = \frac{1}{8}a_7 = \frac{1}{2\cdot 4\cdot 6\cdot 8}a_1$$

In general,

$$a_{2n} = \frac{1}{1\cdot 3\cdots(2n-1)}a_0, \quad a_{2n+1} = \frac{1}{2\cdot 4\cdots(2n)}a_1 = \frac{1}{2^n n!}a_1.$$

The solution $y = \Sigma\, a_n x^n$ may, therefore, be expressed as a sum of two infinite series:

$$y = a_0\left[1 + \sum_{n=1}^{\infty} \frac{1}{1\cdot 3\cdots(2n-1)} x^{2n}\right] + a_1 \sum_{n=0}^{\infty} \frac{1}{2^n n!} x^{2n+1}.$$

EXERCISES 19.8

Find series solutions for the differential equations in Exercises 1–12.

1 $y'' + y = 0$

2 $y'' - 4y = 0$

3 $y'' - 2xy = 0$

4 $y'' + 2xy' + y = 0$

5 $\dfrac{d^2y}{dx^2} - x\dfrac{dy}{dx} + 2y = 0$

6 $\dfrac{d^2y}{dx^2} + x^2y = 0$

7 $(x+1)y' = 3y$

8 $y' = 4x^3y$

9 $y'' - y = 5x$

10 $y'' - xy = x^4$

11 $(x^2-1)y'' + 6xy' + 4y = -4$

12 $y'' + y = e^x$

19.9 REVIEW

Concepts

Discuss methods for solving the following types of differential equations.

1 Separable

2 Exact

3 Homogeneous

4 First-order linear

5 Bernoulli

6 Second-order linear

Exercises

Solve the differential equations in Exercises 1–40

1 $xe^y\,dx - \csc x\,dy = 0$

2 $(2xy-1)\,dx + (x^2+2y)\,dy = 0$

3 $(3x-y)\,dx + (x+y)\,dy = 0$

4 $y' + 4y = e^{-x}$

5 $y^2 - ye^{-x} + (e^{-x} + 2xy + 3)y' = 0$

6 $(x^2y + x^2)\,dy + y\,dx = 0$

7 $y\tan x + y' = 2\sec x$

8 $(x^2+y^2) - xyy' = 0$

9 $y\sqrt{1-x^2}y' = \sqrt{1-y^2}$

10 $(2y + x^3)\,dx - x\,dy = 0$

11 $\left(2x\sin\dfrac{y}{x} - y\cos\dfrac{y}{x}\right)dx + x\cos\dfrac{y}{x}\,dy = 0$

12 $(y\cos x - 2x) + (\sin x + 2y)y' = 0$

13 $xy' - 2y = x^3y^3$

14 $y'' + y' - 6y = 0$

15 $y'' - 8y' + 16y = 0$

16 $y'' - 6y' + 25y = 0$

17 $\dfrac{d^2y}{dx^2} - 2\dfrac{dy}{dx} = 0$

18 $\dfrac{d^2y}{dx^2} = y + \sin x$

19 $y'' - y = e^x \sin x$

20 $y'' - y' - 6y = e^{2x}$

21 $\sec^2 y\,dx = \sqrt{1-x^2}\,dy - x\sec^2 y\,dx$

22 $(2x - yx^{-1} + \ln y)\,dx + (xy^{-1} - \ln x + 1)\,dy = 0$

23 $y' + y = e^{4x}$

24 $y'' + 2y' = 0$

25 $y'' - 3y' + 2y = e^{5x}$

26 $xe^y\,dx - (x+1)y\,dy = 0$

27 $xy' + y = (x - 2)^2$

28 $(3x^2 - 2xy^2 + 1)\,dx + (y^2 - 2x^2y)\,dy = 0$

29 $y'' - y' - 20y = xe^{-x}$

30 $(x^2 - y^2)y' + 3xy = 0$

31 $\dfrac{d^2y}{dx^2} + 5\dfrac{dy}{dx} + 7y = 0$

32 $\dfrac{d^2y}{dx^2} + y = \csc x$

33 $e^{x+y}\,dx - \csc x\,dy = 0$

34 $y'' + 10y' + 25y = 0$

35 $\cot x\,dy = (y - \cos x)\,dx$

36 $y'' + y' + y = e^x \cos x$

37 $(y - 2e^{-2x} \sin y)\,dx + (e^{-2x} \cos y + x)\,dy = 0$

38 $y''' = 0$

39 $y' + y \csc x = \tan x$

40 $xy^2y' = x^3 + y^3$

Mathematical Induction

The method of proof known as **mathematical induction** may be used to show that certain statements or formulas are true for all positive integers. For example, if n is a positive integer, let P_n denote the statement

$$(xy)^n = x^n y^n$$

where x and y are real numbers. Thus, P_1 represents the statement $(xy)^1 = x^1 y^1$, P_2 denotes $(xy)^2 = x^2 y^2$, P_3 is $(xy)^3 = x^3 y^3$, and so on. It is easy to show that P_1, P_2, and P_3 are *true* statements. However, since the set of positive integers is infinite, it is impossible to check the validity of P_n for every positive integer n. In order to give a proof, the method provided by (I.1) is required. This method is based on the following fundamental axiom.

(I.1) Axiom of Mathematical Induction

> Suppose a set S of positive integers has the following two properties:
> (i) S contains the integer 1.
> (ii) Whenever S contains a positive integer k, S also contains $k + 1$.
> Then S contains every positive integer.

The reader should have little reluctance about accepting (I.1). If S is a set of positive integers satisfying property (ii), then whenever S contains an arbitrary positive integer k, it must also contain the next positive integer, $k + 1$. If S also satisfies property (i), then S contains 1 and hence by (ii), S contains $1 + 1$, or 2. Applying (ii) again, we see that S contains $2 + 1$, or 3. Once again, S must contain $3 + 1$, or 4. If we continue in this manner, it can be argued that if n is any *specific* positive integer, then n is in S, since we can proceed a step at a time as above, eventually reaching n. Although this argument does not *prove* (I.1), it certainly makes it plausible.

We shall use (I.1) to establish the following fundamental principle.

(I.2) Principle of Mathematical Induction

If with each positive integer n there is associated a statement P_n, then all the statements P_n are true provided the following two conditions hold:

(i) P_1 is true.

(ii) Whenever k is a positive integer such that P_k is true, then P_{k+1} is also true.

Proof. Assume that (i) and (ii) of (I.2) hold and let S denote the set of all positive integers n such that P_n is true. By assumption, P_1 is true and consequently 1 is in S. Thus S satisfies property (i) of (I.1). Whenever S contains a positive integer k, then by the definition of S, P_k is true and hence from assumption (ii) of (I.2), P_{k+1} is also true. This means that S contains $k + 1$. We have shown that whenever S contains a positive integer k, then S also contains $k + 1$. Consequently, property (ii) of (I.1) is true. Hence by (I.1), S contains every positive integer; that is, P_n is true for every positive integer n.

There are other variations of the principle of mathematical induction. One variation appears in (I.9). In most of our work the statement P_n will usually be given in the form of an equation involving the arbitrary positive integer n, as in our illustration $(xy)^n = x^n y^n$.

When applying (I.2), the following two steps should always be followed:

(I.3)

Step (i) Prove that P_1 is true.

Step (ii) Assume that P_k is true and prove that P_{k+1} is true.

Step (ii) is usually the most confusing for the beginning student. We do not *prove* that P_k is true (except for $k = 1$). Instead, we show that *if* P_k is true, then the statement P_{k+1} is true. That is all that is necessary according to (I.2). The assumption that P_k is true is referred to as the **induction hypothesis**.

Example 1 Prove that for every positive integer n, the sum of the first n positive integers is $n(n + 1)/2$.

Solution For any positive integer n, let P_n denote the statement

$$1 + 2 + 3 + \cdots + n = \frac{n(n+1)}{2} \tag{I.4}$$

where by convention, when $n \leq 4$, the left side is adjusted so that there are precisely n terms in the sum. The following are some special cases of P_n:

If $n = 2$, then P_2 is

$$1 + 2 = \frac{2(2+1)}{2}, \quad \text{or} \quad 3 = 3.$$

If $n = 3$, then P_3 is

$$1 + 2 + 3 = \frac{3(3 + 1)}{2}, \quad \text{or} \quad 6 = 6.$$

If $n = 5$, then P_5 is

$$1 + 2 + 3 + 4 + 5 = \frac{5(5 + 1)}{2}, \quad \text{or} \quad 15 = 15.$$

We wish to show that P_n is true for every n. Although it is instructive to check (I.4) for several values of n as we did above, it is unnecessary to do so. We need only follow steps (i) and (ii) of (I.3).

(i) If we substitute $n = 1$ in (I.4), then by convention the left side collapses to 1 and the right side is $\frac{1(1 + 1)}{2}$, which also equals 1. This proves that P_1 is true.

(ii) *Assume* that P_k is true. Thus the induction hypothesis is

$$1 + 2 + 3 + \cdots + k = \frac{k(k + 1)}{2}. \tag{I.5}$$

Our goal is to prove that P_{k+1} is true, that is,

$$1 + 2 + 3 + \cdots + (k + 1) = \frac{(k + 1)[(k + 1) + 1]}{2}. \tag{I.6}$$

By the induction hypothesis we already have a formula for the sum of the first k positive integers. Hence a formula for the sum of the first $k + 1$ positive integers may be found simply by adding $(k + 1)$ to both sides of (I.5). Doing so and simplifying, we obtain

$$\begin{aligned} 1 + 2 + 3 + \cdots + k + (k + 1) &= \frac{k(k + 1)}{2} + (k + 1) \\ &= \frac{k(k + 1) + 2(k + 1)}{2} \\ &= \frac{k^2 + 3k + 2}{2} \\ &= \frac{(k + 1)(k + 2)}{2} \\ &= \frac{(k + 1)[(k + 1) + 1]}{2}. \end{aligned}$$

We have shown that P_{k+1} is true, and therefore the proof by mathematical induction is complete.

The Laws of Exponents can be proved by mathematical induction. In order to apply (I.3), we shall use the following definition of exponents.

(I.7) Definition

If x is any real number, then
(i) $x^1 = x$
(ii) whenever k is a positive integer for which x^k is defined, let $x^{k+1} = x^k \cdot x$.

A definition such as (I.7) is called a **recursive definition**. In general, if a concept is defined for every positive integer n in such a way that the case corresponding to $n = 1$ is given, and if it is also stated how any case after the first is obtained from the preceding one, then the definition is a recursive definition. For example, by (i) of (I.7) we have $x^1 = x$. Next, applying (ii) of (I.7) we obtain

$$x^2 = x^{1+1} = x^1 \cdot x = x \cdot x.$$

Since x^2 is now defined, we may employ (ii) again (with $k = 2$), obtaining

$$x^3 = x^{2+1} = x^2 \cdot x = (x \cdot x) \cdot x.$$

This defines x^3, and hence (ii) of (I.7) may be used again to obtain x^4. Thus,

$$x^4 = x^{3+1} = x^3 \cdot x = [(x \cdot x) \cdot x] \cdot x.$$

Observe that this agrees with the formulation of x^n as a product of x by itself n times. It can be shown (by mathematical induction) that (I.7) defines x^n for every positive integer n.

Example 3 If x is a real number, prove that $x^m \cdot x^n = x^{m+n}$ for all positive integers m and n.

Solution Let m be an arbitrary positive integer. For each positive integer n, let P_n denote the statement

$$x^m \cdot x^n = x^{m+n}. \tag{I.8}$$

We shall use (I.3) to prove that P_n is true for every positive integer n.

(i) To show that P_1 is true we may use (i) and (ii) of (I.7) as follows:

$$\begin{aligned} x^m \cdot x^1 &= x^m \cdot x \\ &= x^{m+1} \end{aligned}$$

which is formula (I.8) with $n = 1$. Hence P_1 is true.

(ii) Assume that P_k is true. Thus the induction hypothesis is

$$x^m \cdot x^k = x^{m+k}.$$

We wish to prove that P_{k+1} is true, that is,

$$x^m \cdot x^{k+1} = x^{m+(k+1)}.$$

The proof may be arranged as follows, where reasons are stated to the right of each step.

$$\begin{aligned} x^m \cdot x^{k+1} &= x^m \cdot (x^k \cdot x) && \text{(ii) of (I.7)} \\ &= (x^m \cdot x^k) \cdot x && \text{(associative law in } \mathbb{R}) \\ &= x^{m+k} \cdot x && \text{(induction hypothesis)} \\ &= x^{(m+k)+1} && \text{(ii) of (I.7)} \\ &= x^{m+(k+1)} && \text{(associative law for integers).} \end{aligned}$$

By (I.3) the proof by induction is complete.

Consider a positive integer j and suppose that with each integer $n \geq j$ there is associated a statement P_n. For example, if $j = 6$, then the statements are numbered P_6, P_7, P_8, and so on. The principle of mathematical induction may be extended to cover this situation. Just as before, two steps are used. Specifically, to prove that the statements S_n are true for $n \geq j$, we use the following two steps.

(I.9)

> (i′) Prove that S_j is true,
>
> (ii′) *Assume* that S_k is true for $k \geq j$ and *prove* that S_{k+1} is true.

Example 4 Let a be a nonzero real number such that $a > -1$. Prove that $(1 + a)^n > 1 + na$ for every integer $n \geq 2$.

Solution For each positive integer n, let P_n denote the inequality $(1 + a)^n > 1 + na$. Note that P_1 is *false*, since $(1 + a)^1 = 1 + (1)(a)$. However, we can show that P_n is true for $n \geq 2$ by using (I.9) with $j = 2$.

(i′) We first note that $(1 + a)^2 = 1 + 2a + a^2$. Since $a \neq 0$, we have $a^2 > 0$ and therefore $1 + 2a + a^2 > 1 + 2a$. This gives us $(1 + a)^2 > 1 + 2a$, and hence P_2 is true.

(ii′) Assume that P_k is true for $k \geq 2$. Thus the induction hypothesis is

$$(1 + a)^k > 1 + ka.$$

We wish to show that P_{k+1} is true, that is,

$$(1 + a)^{k+1} > 1 + (k + 1)a.$$

Since $a > -1$, we have $a + 1 > 0$, and hence multiplying both sides of the induction hypothesis by $1 + a$ will not change the inequality sign. Consequently,

$$(1 + a)^k(1 + a) > (1 + ka)(1 + a)$$

which may be rewritten as

$$(1 + a)^{k+1} > 1 + ka + a + ka^2$$

or as

$$(1 + a)^{k+1} > 1 + (k + 1)a + ka^2.$$

Since $ka^2 > 0$, we have

$$1 + (k + 1)a + ka^2 > 1 + (k + 1)a$$

and therefore

$$(1 + a)^{k+1} > 1 + (k + 1)a.$$

Thus, P_{k+1} is true and the proof is complete.

EXERCISES

In each of Exercises 1–18, prove that the given formula is true for every positive integer n.

1 $2 + 4 + 6 + \cdots + 2n = n(n + 1)$

2 $1 + 4 + 7 + \cdots + (3n - 2) = \dfrac{n(3n - 1)}{2}$

3 $1 + 3 + 5 + \cdots + (2n - 1) = n^2$

4 $3 + 9 + 15 + \cdots + (6n - 3) = 3n^2$

5 $2 + 7 + 12 + \cdots + (5n - 3) = \dfrac{n}{2}(5n - 1)$

6 $2 + 6 + 18 + \cdots + 2 \cdot 3^{n-1} = 3^n - 1$

7 $1 + 2 \cdot 2 + 3 \cdot 2^2 + 4 \cdot 2^3 + \cdots + n \cdot 2^{n-1} = 1 + (n - 1) \cdot 2^n$

8 $(-1)^1 + (-1)^2 + (-1)^3 + \cdots + (-1)^n = \dfrac{(-1)^n - 1}{2}$

9 $1^2 + 2^2 + 3^2 + \cdots + n^2 = \dfrac{n(n + 1)(2n + 1)}{6}$

10 $1^3 + 2^3 + 3^3 + \cdots + n^3 = \left[\dfrac{n(n + 1)}{2}\right]^2$

11 $\dfrac{1}{1 \cdot 2} + \dfrac{1}{2 \cdot 3} + \dfrac{1}{3 \cdot 4} + \cdots + \dfrac{1}{n(n + 1)} = \dfrac{n}{n + 1}$

12 $\dfrac{1}{1 \cdot 2 \cdot 3} + \dfrac{1}{2 \cdot 3 \cdot 4} + \dfrac{1}{3 \cdot 4 \cdot 5} + \cdots + \dfrac{1}{n(n + 1)(n + 2)} = \dfrac{n(n + 3)}{4(n + 1)(n + 2)}$

13 $3 + 3^2 + 3^3 + \cdots + 3^n = \frac{3}{2}(3^n - 1)$

14 $1^3 + 3^3 + 5^3 + \cdots + (2n - 1)^3 = n^2(2n^2 - 1)$

15 $n < 2^n$

16 $1 + 2n \le 3^n$

17 $1 + 2 + 3 + \cdots + n < \frac{1}{8}(2n + 1)^2$

18 If $0 < a < b$, then $\left(\dfrac{a}{b}\right)^{n+1} < \left(\dfrac{a}{b}\right)^n$.

Prove that the statements in Exercises 19–22 are true for every positive integer n.

19 3 is a factor of $n^3 - n + 3$.

20 2 is a factor of $n^2 + n$.

21 4 is a factor of $5^n - 1$.

22 9 is a factor of $10^{n+1} + 3 \cdot 10^n + 5$.

23 Use mathematical induction to prove that if a is any real number greater than 1, then $a^n > 1$ for every positive integer n.

24 If $a \neq 1$, prove that

$$1 + a + a^2 + \cdots + a^{n-1} = \frac{a^n - 1}{a - 1}$$

for every positive integer n.

25 If a and b are real numbers, use mathematical induction to prove that $(ab)^n = a^n b^n$ for every positive integer n.

26 If a is a real number, prove that $(a^m)^n = a^{mn}$ for all positive integers m and n.

27 Use mathematical induction to prove that $a - b$ is a factor of $a^n - b^n$ for every positive integer n. (*Hint:* $a^{k+1} - b^{k+1} = a^k(a - b) + (a^k - b^k)b$.)

28 Prove that $a + b$ is a factor of $a^{2n-1} + b^{2n-1}$ for every positive integer n.

29 If z is a complex number and $\bar{z}$ is its conjugate, prove that $\overline{z^n} = \bar{z}^n$ for every positive integer n.

30 Prove that for every positive integer n, if $z_1, z_2, \ldots, z_n$ are complex numbers, then $\overline{z_1 z_2 \ldots z_n} = \bar{z}_1 \bar{z}_2 \ldots \bar{z}_n$

31 Prove that

$$\log(a_1 a_2 \ldots a_n) = \log a_1 + \log a_2 + \cdots + \log a_n$$

for all $n \geq 2$, where each a_i is a positive real number.

32 Prove the **Generalized Distributive Law**

$$a(b_1 + b_2 + \cdots + b_n) = ab_1 + ab_2 + \cdots + ab_n$$

for all $n \geq 2$, where a and each b_i are real numbers.

33 Prove that

$$a + ar + ar^2 + \cdots + ar^{n-1} = \frac{a(1 - r^n)}{1 - r}$$

where n is any positive integer and a and r are real numbers with $r \neq 1$.

34 Prove that

$$a + (a + d) + (a + 2d) + \cdots + [a + (n - 1)d] = (n/2)[2a + (n - 1)d]$$

where n is any positive integer and a and d are real numbers.

35 If a and b are real numbers and n is any positive integer, prove that

$$(a - b)(a^{n-1} + a^{n-2}b + \cdots + ab^{n-2} + b^{n-1}) = a^n - b^n.$$

Theorems on Limits and Definite Integrals

This appendix contains proofs for some theorems stated in Chapters 2, 3, and 5. Part of the numbering system corresponds to that given in those chapters.

II.0 Uniqueness Theorem for Limits

> If $f(x)$ has a limit as x approaches a, then that limit is unique.

Proof. Suppose that $\lim_{x\to a} f(x) = L_1$ and $\lim_{x\to a} f(x) = L_2$ where $L_1 \neq L_2$. It may be assumed, without loss of generality, that $L_1 < L_2$. Choose ε such that $\varepsilon < \frac{1}{2}(L_2 - L_1)$ and consider the open intervals $(L_1 - \varepsilon, L_1 + \varepsilon)$ and $(L_2 - \varepsilon, L_2 + \varepsilon)$ on the coordinate line l' (see Figure II.1). Since $\varepsilon < \frac{1}{2}(L_2 - L_1)$ these two intervals do not intersect. By (2.7) there is a $\delta_1 > 0$ such that whenever x is in $(a - \delta_1, a + \delta_1)$, but $x \neq a$, then $f(x)$ is in $(L_1 - \varepsilon, L_1 + \varepsilon)$. Similarly, there is a $\delta_2 > 0$ such that whenever x is in $(a - \delta_2, a + \delta_2)$, but $x \neq a$, then $f(x)$ is in $(L_2 - \varepsilon, L_2 + \varepsilon)$. This is illustrated in Figure II.1, where the case $\delta_1 < \delta_2$ is shown. If an x is selected which is in *both* $(a - \delta_1, a + \delta_1)$ and $(a - \delta_2, a + \delta_2)$, then $f(x)$ is in $(L_1 - \varepsilon, L_1 + \varepsilon)$ and also in $(L_2 - \varepsilon, L_2 + \varepsilon)$, contrary to the fact that these two intervals do not intersect. Hence our original supposition is false and consequently $L_1 = L_2$. It is also possible to give a strictly algebraic proof of this result.

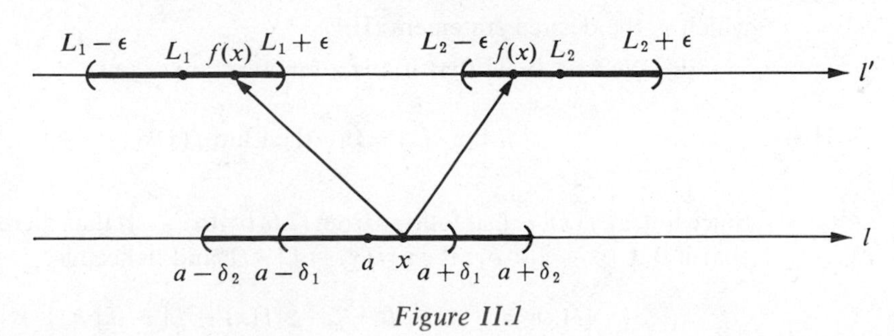

Figure II.1

(2.11) Theorem

If $\lim_{x\to a} f(x) = L$ and $\lim_{x\to a} g(x) = M$, then

(i) $\lim_{x\to a} [f(x) + g(x)] = L + M$

(ii) $\lim_{x\to a} [f(x) \cdot g(x)] = L \cdot M$

(iii) $\lim_{x\to a} \dfrac{f(x)}{g(x)} = \dfrac{L}{M}$, provided $M \neq 0$.

Proof. (i) According to (2.6) we must show that for every $\varepsilon > 0$ there corresponds a $\delta > 0$ such that

$$\text{if} \quad 0 < |x - a| < \delta, \quad \text{then} \quad |f(x) + g(x) - (L + M)| < \varepsilon. \tag{II.1}$$

We begin by writing

$$|f(x) + g(x) - (L + M)| = |(f(x) - L) + (g(x) - M)|. \tag{II.2}$$

Employing the triangle inequality (1.5),

$$|(f(x) - L) + (g(x) - M)| \leq |f(x) - L| + |g(x) - M|.$$

Combining the last inequality with (II.2) gives us

$$|f(x) + g(x) - (L + M)| \leq |f(x) - L| + |g(x) - M|. \tag{II.3}$$

Since $\lim_{x\to a} f(x) = L$ and $\lim_{x\to a} g(x) = M$, the numbers $|f(x) - L|$ and $|g(x) - M|$ can be made arbitrarily small by choosing x sufficiently close to a. In particular, they can be made less than $\varepsilon/2$. Thus there exist $\delta_1 > 0$ and $\delta_2 > 0$ such that

$$\begin{aligned} &\text{if} \quad 0 < |x - a| < \delta_1 \quad \text{then} \quad |f(x) - L| < \varepsilon/2, \text{ and} \\ &\text{if} \quad 0 < |x - a| < \delta_2 \quad \text{then} \quad |g(x) - M| < \varepsilon/2. \end{aligned} \tag{II.4}$$

If δ denotes the *smaller* of δ_1 and δ_2, then whenever $0 < |x - a| < \delta$, the inequalities in (II.4) involving $f(x)$ and $g(x)$ are both true. Consequently, if $0 < |x - a| < \delta$, then from (II.4) and (II.3),

$$|f(x) + g(x) - (L + M)| < \varepsilon/2 + \varepsilon/2 = \varepsilon$$

which is the desired statement (II.1).

(ii) We first show that if k is a function and

$$\text{if } \lim_{x\to a} k(x) = 0, \quad \text{then } \lim_{x\to a} f(x)k(x) = 0. \tag{II.5}$$

Since $\lim_{x\to a} f(x) = L$, it follows from (2.6) (with $\varepsilon = 1$) that there is a $\delta_1 > 0$ such that if $0 < |x - a| < \delta_1$, then $|f(x) - L| < 1$ and hence also

$$|f(x)| = |f(x) - L + L| \leq |f(x) - L| + |L| < 1 + |L|.$$

Consequently

$$\text{(II.6)} \qquad \text{if} \quad 0 < |x - a| < \delta_1, \quad \text{then} \quad |f(x)k(x)| < (1 + |L|)\,|k(x)|.$$

Since $\lim_{x \to a} k(x) = 0$, for every $\varepsilon > 0$ there corresponds a $\delta_2 > 0$ such that

$$\text{(II.7)} \qquad \text{if} \quad 0 < |x - a| < \delta_2, \quad \text{then} \quad |k(x) - 0| < \frac{\varepsilon}{1 + |L|}.$$

If δ denotes the smaller of δ_1 and δ_2, then whenever $0 < |x - a| < \delta$ both inequalities (II.6) and (II.7) are true and consequently

$$|f(x)k(x)| < (1 + |L|) \cdot \frac{\varepsilon}{1 + |L|}.$$

Therefore

$$\text{if} \quad 0 < |x - a| < \delta, \quad \text{then} \quad |f(x)k(x) - 0| < \varepsilon$$

which proves (II.5).

Next consider the identity

$$\text{(II.8)} \qquad f(x)g(x) - LM = f(x)[g(x) - M] + M[f(x) - L].$$

Since $\lim_{x \to a} [g(x) - M] = 0$ it follows from (II.5), with $k(x) = g(x) - M$, that $\lim_{x \to a} f(x)[g(x) - M] = 0$. In addition, $\lim_{x \to a} M[f(x) - L] = 0$ and hence, from (II.8), $\lim_{x \to a} [f(x)g(x) - LM] = 0$; that is, $\lim_{x \to a} f(x)g(x) = LM$.

(iii) It is sufficient to show that $\lim_{x \to a} 1/(g(x)) = 1/M$, for once this is done, the desired result may be obtained by applying (ii) to the product $f(x) \cdot 1/(g(x))$. Consider

$$\text{(II.9)} \qquad \left|\frac{1}{g(x)} - \frac{1}{M}\right| = \left|\frac{M - g(x)}{g(x)M}\right| = \frac{1}{|M|\,|g(x)|}|g(x) - M|.$$

Since $\lim_{x \to a} g(x) = M$, there exists a $\delta_1 > 0$ such that if $0 < |x - a| < \delta_1$ then $|g(x) - M| < |M|/2$. Consequently, for all such x,

$$\begin{aligned} |M| &= |g(x) + (M - g(x))| \\ &\le |g(x)| + |M - g(x)| \\ &< |g(x)| + |M|/2 \end{aligned}$$

and, therefore,

$$\frac{|M|}{2} < |g(x)| \quad \text{or} \quad \frac{1}{|g(x)|} < \frac{2}{|M|}.$$

Substitution in (II.9) leads to

$$\text{(II.10)} \qquad \left|\frac{1}{g(x)} - \frac{1}{M}\right| < \frac{2}{|M|^2}|g(x) - M|, \quad \text{provided } 0 < |x - a| < \delta_1.$$

Again using the fact that $\lim_{x\to a} g(x) = M$, it follows that for every $\varepsilon < 0$ there corresponds a $\delta_2 > 0$ such that

$$\text{(II.11)} \qquad \text{if} \quad 0 < |x - a| < \delta_2, \quad \text{then} \quad |g(x) - M| < \frac{|M|^2}{2}\varepsilon.$$

If δ denotes the smaller of δ_1 and δ_2, then both inequalities (II.10) and (II.11) are true. Thus

$$\text{if} \quad 0 < |x - a| < \delta, \quad \text{then} \quad \left|\frac{1}{g(x)} - \frac{1}{M}\right| < \varepsilon$$

which means that $\lim_{x\to a} 1/(g(x)) = 1/M$.

(2.20) Theorem

> If $a > 0$ and n is a positive integer, or if $a < 0$ and n is an odd positive integer, then $\lim_{x\to a} \sqrt[n]{x} = \sqrt[n]{a}$.

Proof. Suppose $a > 0$ and n is any positive integer. It must be shown that for every $\varepsilon > 0$ there corresponds a $\delta > 0$ such that

$$\text{if} \quad 0 < |x - a| < \delta, \quad \text{then} \quad |\sqrt[n]{x} - \sqrt[n]{a}| < \varepsilon$$

or, equivalently,

$$\text{(II.12)} \qquad \text{if} \quad -\delta < x - a < \delta, x \neq a, \quad \text{then} \quad -\varepsilon < \sqrt[n]{x} - \sqrt[n]{a} < \varepsilon.$$

It is sufficient to prove (II.12) if $\varepsilon < \sqrt[n]{a}$, for if a δ exists under this condition then the same δ can be used for any *larger* value of ε. Thus, in the remainder of the proof $\sqrt[n]{a} - \varepsilon$ is considered as a positive number less than ε. The inequalities in the following list are all equivalent:

$$-\varepsilon < \sqrt[n]{x} - \sqrt[n]{a} < \varepsilon$$

$$\sqrt[n]{a} - \varepsilon < \sqrt[n]{x} < \sqrt[n]{a} + \varepsilon$$

$$(\sqrt[n]{a} - \varepsilon)^n < x < (\sqrt[n]{a} + \varepsilon)^n$$

$$(\sqrt[n]{a} - \varepsilon)^n - a < x - a < (\sqrt[n]{a} + \varepsilon)^n - a$$

$$-[a - (\sqrt[n]{a} - \varepsilon)^n] < x - a < (\sqrt[n]{a} + \varepsilon)^n - a.$$

If δ denotes the smaller of the two positive numbers $a - (\sqrt[n]{a} - \varepsilon)^n$ and $(\sqrt[n]{a} + \varepsilon)^n - a$, then whenever $-\delta < x - a < \delta$ the last inequality in the list is true and hence so is the first. This gives us (II.12).

Next suppose $a < 0$ and n is an odd positive integer. In this case $-a$ and $\sqrt[n]{-a}$ are positive and, by the first part of the proof, we may write

$$\lim_{-x\to -a} \sqrt[n]{-x} = \sqrt[n]{-a}.$$

Thus for every $\varepsilon > 0$, there corresponds a $\delta > 0$ such that

$$\text{if} \quad 0 < |-x - (-a)| < \delta, \quad \text{then} \quad |\sqrt[n]{-x} - \sqrt[n]{-a}| < \varepsilon$$

or equivalently

$$\text{if} \quad 0 < |x - a| < \delta, \quad \text{then} \quad |\sqrt[n]{x} - \sqrt[n]{a}| < \varepsilon.$$

The latter inequalities imply that $\lim_{x \to a} \sqrt[n]{x} = \sqrt[n]{a}$.

(2.23) The Sandwich Theorem

> If $f(x) \leq h(x) \leq g(x)$ for all x in an open interval containing a, except possibly at a, and if $\lim_{x \to a} f(x) = L = \lim_{x \to a} g(x)$, then $\lim_{x \to a} h(x) = L$.

Proof. For every $\varepsilon > 0$, there correspond $\delta_1 > 0$ and $\delta_2 > 0$ such that

$$\text{(II.13)} \qquad \begin{aligned} &\text{if} \quad 0 < |x - a| < \delta_1, \quad \text{then} \quad |f(x) - L| < \varepsilon, \\ &\text{if} \quad 0 < |x - a| < \delta_2, \quad \text{then} \quad |g(x) - L| < \varepsilon. \end{aligned}$$

If δ denotes the smaller of δ_1 and δ_2, then whenever $0 < |x - a| < \delta$, both inequalities in (II.13) which involve ε are true, that is,

$$-\varepsilon < f(x) - L < \varepsilon \quad \text{and} \quad -\varepsilon < g(x) - L < \varepsilon.$$

Consequently, if $0 < |x - a| < \delta$, then $L - \varepsilon < f(x)$ and $g(x) < L + \varepsilon$. Since $f(x) \leq h(x) \leq g(x)$ it follows that if $0 < |x - a| < \delta$, then $L - \varepsilon < h(x) < L + \varepsilon$ or, equivalently, $|h(x) - L| < \varepsilon$, which is what we wished to prove.

(3.7) Theorem

> If f is defined on an open interval containing a, then
>
> $$f'(a) = \lim_{x \to a} \frac{f(x) - f(a)}{x - a}$$
>
> provided the limit exists.

Proof. Suppose

$$\lim_{x \to a} \frac{f(x) - f(a)}{x - a} = L$$

for some number L. According to the definition of limit (2.6), this means that for every $\varepsilon > 0$ there exists a $\delta > 0$ such that

$$\text{if} \quad 0 < |x - a| < \delta, \quad \text{then} \quad \left| \frac{f(x) - f(a)}{x - a} - L \right| < \varepsilon.$$

If we let $h = x - a$, then $x = a + h$ and the last statement may be written

$$\text{if} \quad 0 < |h| < \delta, \quad \text{then} \quad \left| \frac{f(a+h) - f(a)}{h} - L \right| < \varepsilon$$

which means that

$$\lim_{h \to 0} \frac{f(a+h) - f(a)}{h} = L.$$

However, by Definition (3.4) this limit must equal $f'(a)$, and consequently $L = f'(a)$. This gives us the formula in the statement of the theorem. Conversely, if $f'(a)$ exists then by reversing the steps in the previous proof we arrive at the desired limit.

(3.36) The Chain Rule

> If $y = f(u)$, $u = g(x)$, and the derivatives $D_u y$ and $D_x u$ both exist, then the composite function defined by $y = f(g(x))$ has a derivative given by
>
> $$D_x y = (D_u y)(D_x u) = f'(u)g'(x).$$

Proof. Let us begin by using (3.28), which states that if $y = f(x)$ and $\Delta x \approx 0$, then the difference between the derivative $f'(x)$ and the ratio $\Delta y/\Delta x$ is numerically small. Since this difference depends on the size of Δx, we shall represent it by means of the functional notation $\eta(\Delta x)$. Thus, for each $\Delta x \neq 0$,

$$\text{(a)} \qquad \eta(\Delta x) = \frac{\Delta y}{\Delta x} - f'(x).$$

It should be noted that $\eta(\Delta x)$ does *not* represent the product of η and Δx, but rather that *η is a function of Δx*, whose values are given by (a). Moreover, applying (3.27) we see that

$$\text{(b)} \qquad \lim_{\Delta x \to 0} \eta(\Delta x) = \lim_{\Delta x \to 0} \left[\frac{\Delta y}{\Delta x} - f'(x) \right] = 0.$$

The function η has been defined only for nonzero values of Δx. It is convenient to extend the definition of η to include $\Delta x = 0$ by letting $\eta(0) = 0$. It then follows from (b) that *η is continuous at* 0.

Multiplying both sides of (a) by Δx and rearranging terms gives us

$$\text{(c)} \qquad \Delta y = f'(x)\,\Delta x + \eta(\Delta x) \cdot \Delta x$$

which is true whether $\Delta x \neq 0$ or $\Delta x = 0$. Since $f'(x)\,\Delta x = dy$, it follows from (c) that

$$\text{(d)} \qquad \Delta y - dy = \eta(\Delta x) \cdot \Delta x.$$

Let us now consider the situation stated in the hypothesis of the theorem, where

$$y = f(u) \quad \text{and} \quad u = g(x).$$

If $g(x)$ is in the domain of f, then we may write

$$y = f(u) = f(g(x)),$$

that is, y is a function of x. If we give x an increment Δx there corresponds an increment Δu in u and, in turn, an increment Δy in $y = f(u)$. Thus

$$\Delta u = g(x + \Delta x) - g(x)$$
$$\Delta y = f(u + \Delta u) - f(u).$$

Since $D_u y$ exists we may use (c) with u as the independent variable to write

$$\text{(e)} \qquad \Delta y = f'(u)\,\Delta u + \eta(\Delta u) \cdot \Delta u$$

where η is a function of Δu and where, by (b),

$$\text{(f)} \qquad \lim_{\Delta u \to 0} \eta(\Delta u) = 0.$$

Moreover, η is continuous at $\Delta u = 0$ and (e) is true if $\Delta u = 0$. Dividing both sides of (e) by Δx gives us

$$\frac{\Delta y}{\Delta x} = f'(u)\frac{\Delta u}{\Delta x} + \eta(\Delta u) \cdot \frac{\Delta u}{\Delta x}.$$

If we now take the limit as Δx approaches zero and use the fact that

$$\lim_{\Delta x \to 0} \frac{\Delta y}{\Delta x} = D_x y \quad \text{and} \quad \lim_{\Delta x \to 0} \frac{\Delta u}{\Delta x} = D_x u$$

we see that

$$D_x y = f'(u) D_x u + \lim_{\Delta x \to 0} \eta(\Delta u) \cdot D_x u.$$

Since $f'(u) = D_u y$, we may complete the proof by showing that the limit indicated in the last equation is 0. To accomplish this we first observe that since g is differentiable it is continuous, and hence

$$\lim_{\Delta x \to 0} [g(x + \Delta x) - g(x)] = 0$$

or equivalently

$$\lim_{\Delta x \to 0} \Delta u = 0.$$

In other words, Δu approaches 0 as Δx approaches 0. Using this fact, together with (f), gives us

$$\text{(g)} \qquad \lim_{\Delta x \to 0} \eta(\Delta u) = \lim_{\Delta u \to 0} \eta(\Delta u) = 0$$

and the theorem is proved. (The fact that $\lim_{\Delta x \to 0} \eta(\Delta u) = 0$ can also be established by means of an $\varepsilon - \delta$ argument using (2.6).)

(5.22) Theorem

> If f is integrable on $[a,b]$ and k is any number, then kf is integrable on $[a,b]$ and
>
> $$\int_a^b kf(x)\,dx = k\int_a^b f(x)\,dx.$$

Proof. If $k = 0$ the result follows from (5.21). Assume, therefore, that $k \neq 0$. Since f is integrable, $\int_a^b f(x)\,dx = I$ for some number I. If P is a partition of $[a,b]$, then every Riemann sum R_P for the function kf has the form $\Sigma_i kf(w_i)\,\Delta x_i$,where for each i, w_i is in the ith subinterval $[x_{i-1}, x_i]$ of P. We wish to show that for every $\varepsilon > 0$ there corresponds a $\delta > 0$ such that whenever $\|P\| < \delta$, then

$$\left|\sum_i kf(w_i)\,\Delta x_i - kI\right| < \varepsilon \tag{II.14}$$

for all w_i in $[x_{i-1}, x_i]$. If we let $\varepsilon' = \varepsilon/|k|)$, then since f is integrable there exists a $\delta > 0$ such that whenever $\|P\| < \delta$,

$$\left|\sum_i f(w_i)\,\Delta x_i - I\right| < \varepsilon' = \varepsilon/|k|.$$

Multiplying both sides of this inequality leads to (II.14). Hence

$$\lim_{\|P\|\to 0} \sum_i kf(w_i)\,\Delta x_i = kI = k\int_a^b f(x)\,dx.$$

(5.23) Theorem

> If f and g are integrable on $[a,b]$, then $f + g$ is integrable on $[a,b]$ and
>
> $$\int_a^b [f(x) + g(x)]\,dx = \int_a^b f(x)\,dx + \int_a^b g(x)\,dx.$$

Proof. By hypothesis there exist real numbers I_1 and I_2 such that

$$\int_a^b f(x)\,dx = I_1 \quad \text{and} \quad \int_a^b g(x)\,dx = I_2.$$

Let P denote a partition of $[a,b]$ and let R_P denote an arbitrary Riemann sum for $f + g$ associated with P, that is,

$$R_P = \sum_i [f(w_i) + g(w_i)]\,\Delta x_i \tag{II.15}$$

where w_i is in $[x_{i-1}, x_i]$ for each i. We wish to show that given any $\varepsilon > 0$ there corresponds a $\delta > 0$ such that whenever $\|P\| < \delta$, then $|R_P - (I_1 + I_2)| < \varepsilon$. Using (i) of (5.2), we may write (II.15) in the form

$$R_P = \sum_i f(w_i)\,\Delta x_i + \sum_i g(w_i)\,\Delta x_i.$$

Rearranging terms and using the triangle inequality (1.5),

$$\text{(II.16)}\qquad \begin{aligned} |R_P - (I_1 + I_2)| &= \left|\left(\sum_i f(w_i)\,\Delta x_i - I_1\right) + \left(\sum_i g(w_i)\,\Delta x_i - I_2\right)\right| \\ &\le \left|\sum_i f(w_i)\,\Delta x_i - I_1\right| + \left|\sum_i g(w_i)\,\Delta x_i - I_2\right|. \end{aligned}$$

By the integrability of f and g, if $\varepsilon' = \varepsilon/2$, then there exist $\delta_1 > 0$ and $\delta_2 > 0$ such that whenever $\|P\| < \delta_1$ and $\|P\| < \delta_2$,

$$\text{(II.17)}\qquad \begin{aligned} &\left|\sum_i f(w_i)\,\Delta x_i - I_1\right| < \varepsilon' = \varepsilon/2, \quad \text{and} \\ &\left|\sum_i g(w_i)\,\Delta x_i - I_2\right| < \varepsilon' = \varepsilon/2 \end{aligned}$$

for all w_i in $[x_{i-1}, x_i]$. If δ denotes the smaller of δ_1 and δ_2, then whenever $\|P\| < \delta$, both inequalities in (II.17) are true and hence, from (II.16),

$$|R_P - (I_1 + I_2)| < \varepsilon/2 + \varepsilon/2 = \varepsilon$$

which is what we wished to prove.

(5.25) Theorem

> If $a < c < b$ and f is integrable on both $[a, c]$ and $[c, b]$, then f is integrable on $[a, b]$ and
>
> $$\int_a^b f(x)\,dx = \int_a^c f(x)\,dx + \int_c^b f(x)\,dx.$$

Proof. By hypothesis there exist real numbers I_1 and I_2 such that

$$\text{(II.18)}\qquad \int_a^c f(x)\,dx = I_1 \quad \text{and} \quad \int_c^b f(x)\,dx = I_2.$$

Let us denote a partition of $[a, c]$ by P_1, of $[c, b]$ by P_2, and of $[a, b]$ by P. Arbitrary Riemann sums associated with P_1, P_2, and P will be denoted by R_{P_1}, R_{P_2}, and R_P, respectively. It must be shown that for every $\varepsilon > 0$ there corresponds a $\delta > 0$ such that if $\|P\| < \delta$, then $|R_P - (I_1 + I_2)| < \varepsilon$.

If we let $\varepsilon' = \varepsilon/4$, then by (II.18) there exist positive numbers δ_1 and δ_2 such that if $\|P_1\| < \delta_1$ and $\|P_2\| < \delta_2$, then

$$\text{(II.19)}\qquad |R_{P_1} - I_1| < \varepsilon' = \varepsilon/4 \quad \text{and} \quad |R_{P_2} - I_2| < \varepsilon' = \varepsilon/4.$$

If δ denotes the smaller of δ_1 and δ_2, then both inequalities in (II.19) are true whenever $\|P\| < \delta$. Moreover, since f is integrable on $[a, c]$ and $[c, b]$ it is bounded on both intervals and hence there exists a number M such that $|f(x)| \le M$ for all x in $[a, b]$. We shall now assume that δ has been chosen so that in addition to the previous requirement we also have $\delta < \varepsilon/4M$.

Let P be a partition of $[a, b]$ such that $\|P\| < \delta$. If the numbers which determine P are

$$a = x_0, x_1, x_2, \ldots, x_n = b,$$

then there is a unique half-open interval of the form $(x_{k-1}, x_k]$ which contains c. If $R_P = \Sigma_{i=1}^n f(w_i)\,\Delta x_i$ we may write

$$R_P = \sum_{i=1}^{k-1} f(w_i)\,\Delta x_i + f(w_k)\,\Delta x_k + \sum_{i=k+1}^{n} f(w_i)\,\Delta x_i. \tag{II.20}$$

Let P_1 denote the partition of $[a, c]$ determined by $\{a, x_1, \ldots, x_{k-1}, c\}$, let P_2 denote the partition of $[c, b]$ determined by $\{c, x_k, \ldots, x_{n-1}, b\}$, and consider the Riemann sums

$$\begin{aligned} R_{P_1} &= \sum_{i=1}^{k-1} f(w_i)\,\Delta x_i + f(c)(c - x_{k-1}) \\ R_{P_2} &= f(c)(x_k - c) + \sum_{i=k+1}^{n} f(w_i)\,\Delta x_i. \end{aligned} \tag{II.21}$$

Using the triangle inequality and (II.19),

$$\begin{aligned} |(R_{P_1} + R_{P_2}) - (I_1 + I_2)| &= |(R_{P_1} - I_1) + (R_{P_2} - I_2)| \\ &\leq |R_{P_1} - I_1| + |R_{P_2} - I_2| \\ &< \varepsilon/4 + \varepsilon/4 = \varepsilon/2. \end{aligned} \tag{II.22}$$

It follows from (II.20) and (II.21) that

$$|R_P - (R_{P_1} + R_{P_2})| = |f(w_k) - f(c)|\,\Delta x_k.$$

Employing the triangle inequality and the choice of δ gives us

$$\begin{aligned} |R_P - (R_{P_1} + R_{P_2})| &\leq \{|f(w_k)| + |f(c)|\}\,\Delta x_k \\ &\leq (M + M) \cdot \varepsilon/4M = \varepsilon/2 \end{aligned} \tag{II.23}$$

provided $\|P\| < \delta$. If we write

$$\begin{aligned} |R_P - (I_1 + I_2)| &= |R_P - (R_{P_1} + R_{P_2}) + (R_{P_1} + R_{P_2}) - (I_1 + I_2)| \\ &\leq |R_P - (R_{P_1} + R_{P_2})| + |(R_{P_1} + R_{P_2}) - (I_1 + I_2)| \end{aligned}$$

then it follows from (II.23) and (II.22) that whenever $\|P\| < \delta$

$$|R_P - (I_1 + I_2)| < \varepsilon/2 + \varepsilon/2 = \varepsilon$$

for every Riemann sum R_P. This completes the proof.

(5.27) Theorem

> If f is integrable on $[a, b]$ and if $f(x) \geq 0$ for all x in $[a, b]$, then
>
> $$\int_a^b f(x)\,dx \geq 0.$$

Proof. We shall give an indirect proof. Thus let $\int_a^b f(x)\,dx = I$ and suppose that $I < 0$. Consider any partition P of $[a, b]$ and let $R_P = \Sigma_i f(w_i)\,\Delta x_i$ be an arbitrary Riemann sum associated with P. Since $f(w_i) \geq 0$ for all w_i in $[x_{i-1}, x_i]$ it follows that $R_P \geq 0$. If we let $\varepsilon = -(I/2)$, then according to (5.15), whenever $\|P\|$ is sufficiently small,

$$|R_P - I| < \varepsilon = -(I/2).$$

It follows that $R_P < I - (I/2) = I/2 < 0$, a contradiction. Therefore, the supposition $I < 0$ is false and hence $I \geq 0$.

The Trigonometric Functions

Let U be a unit circle, that is, a circle of radius 1, with center at the origin of a rectangular coordinate system, and let A be the point with coordinates $(1, 0)$. If t is any number between 0 and 2π, then as illustrated in Figure III.1 there is precisely one point $P(x, y)$ on U such that the length of arc $\widehat{AP}$ measured in the counterclockwise direction from A is t. More generally, if t is *any* nonnegative real number, then a unique point P is obtained by measuring a distance t from A in the counterclockwise direction along U. Of course, if $t > 2\pi$, more than one revolution of U is necessary in order to arrive at P. For example, if $t = 3\pi$, then P has coordinates $(-1, 0)$. Similarly, with each negative real number t, a point P may be obtained by measuring a distance $|t|$ from A in the clockwise direction along U. The point P which is located in this manner will be called *the point on U corresponding to the real number t*. Since the length of U is 2π, the point corresponding to t is the same as the point corresponding to $t + 2\pi$, $t + 4\pi$, $t - 2\pi$, and in general to $t + 2\pi n$, where n is any integer.

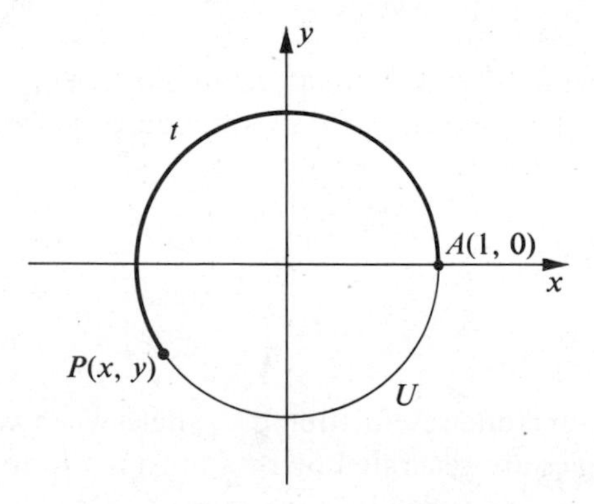

Figure III.1

The unit circle U may be used to define the trigonometric functions. These functions are referred to as the **sine**, **cosine**, **tangent**, **cosecant**, **secant**, and **cotangent** functions and are denoted by sin, cos, tan, csc, sec, and cot, respectively. If t is a real number, then the number which the sine function associates with t is denoted by either $\sin(t)$ or $\sin t$. Similar notation is used for the other five functions.

(III.1) Definition

If t is any real number and $P(x, y)$ is the point on the unit circle U corresponding to t, then

$$\sin t = y \qquad \csc t = \frac{1}{y} \quad (\text{if } y \neq 0)$$

$$\cos t = x \qquad \sec t = \frac{1}{x} \quad (\text{if } x \neq 0)$$

$$\tan t = \frac{y}{x} \quad (\text{if } x \neq 0) \qquad \cot t = \frac{x}{y} \quad (\text{if } y \neq 0)$$

Since the same point $P(x, y)$ is obtained for $t + 2\pi n$, where n is any integer, the values of the trigonometric functions repeat in successive intervals of length 2π. A function f with domain X is said to be **periodic** if there exists a positive real number k such that $f(t + k) = f(t)$ for every t in X. Geometrically, this means that the graph of f repeats itself as abscissas of points vary over successive intervals of length k. If a least such positive real number k exists, it is called the **period** of f. It can be shown that the sine, cosine, cosecant, and secant functions have period 2π, whereas the tangent and cotangent functions have period π.

It follows from (III.1) that

(III.2)
$$\csc t = \frac{1}{\sin t}, \quad \sec t = \frac{1}{\cos t}, \quad \cot t = \frac{1}{\tan t},$$
$$\tan t = \frac{\sin t}{\cos t}, \quad \cot t = \frac{\cos t}{\sin t}$$

provided the denominators are not zero.

The following formulas are consequences of the fact that if $P(x, y)$ is on U, then $y^2 + x^2 = 1$.

(III.3) $$\sin^2 t + \cos^2 t = 1$$

(III.4) $$\tan^2 t + 1 = \sec^2 t$$

(III.5) $$\cot^2 t + 1 = \csc^2 t$$

It is often useful to employ angles when working with trigonometric functions. Angles are generated by rotating a ray l_1 about its end point O, in a plane, to a position specified by another ray l_2. Rays l_1 and l_2 are called the **initial side** and **terminal side**, respectively, and O is called the **vertex** of the angle. The **standard position** of angle θ is obtained by taking the vertex as the origin of a rectangular coordinate system and the initial side as the positive x-axis. In this case θ is generated by rotating the positive x-axis about O in the xy-plane. If the rotation is counterclockwise, then θ is considered a **positive angle**, whereas a clockwise rotation produces a **negative angle**. If U is the unit circle with center O, then as the x-axis rotates to its terminal position, its point of intersection with U travels a certain distance t before arriving at its final position P, as illustrated in Figure

(III.15) Product Formulas

$$\sin u \cos v = \tfrac{1}{2}[\sin(u+v) + \sin(u-v)]$$
$$\cos u \sin v = \tfrac{1}{2}[\sin(u+v) - \sin(u-v)]$$
$$\cos u \cos v = \tfrac{1}{2}[\cos(u+v) + \cos(u-v)]$$
$$\sin u \sin v = \tfrac{1}{2}[\cos(u-v) - \cos(u+v)]$$

(III.16) Factoring Formulas

$$\sin u \pm \sin v = 2\cos\frac{u \mp v}{2}\sin\frac{u \pm v}{2}$$
$$\cos u + \cos v = 2\cos\frac{u+v}{2}\cos\frac{u-v}{2}$$
$$\cos u - \cos v = 2\sin\frac{v+u}{2}\sin\frac{v-u}{2}$$

EXERCISES

1 Verify the entries in the table of radians and degrees preceding (III.8).

2 Verify (III.4) and (III.5).

3 Find the quadrant containing θ if
(a) $\sec\theta < 0$ and $\sin\theta > 0$.
(b) $\cot\theta > 0$ and $\csc\theta < 0$.
(c) $\cos\theta > 0$ and $\tan\theta < 0$.

4 Find the values of the remaining trigonometric functions if
(a) $\sin t = -4/5$ and $\cos t = 3/5$.
(b) $\csc t = \sqrt{13}/2$ and $\cot t = -3/2$.

5 Without the use of tables, find the values of the trigonometric functions corresponding to each of the following real numbers.
(a) $9\pi/2$ (b) $-5\pi/4$ (c) 0 (d) $11\pi/6$

6 Find the radian measures that correspond to the following degree measures: $330°$, $405°$, $-150°$, $240°$, $36°$.

7 Find the degree measures that correspond to the following radian measures: $9\pi/2$, $-2\pi/3$, $7\pi/4$, 5π, $\pi/5$.

8 A central angle θ subtends an arc 20 cm long on a circle of radius 2 meters. What is the radian measure of θ?

9 Find the values of the six trigonometric functions of θ if θ is in standard position and satisfies the stated condition.
(a) The point $(30, -40)$ is on the terminal side of θ.
(b) The terminal side of θ is in quadrant II and is parallel to the line $2x + 3y + 6 = 0$.
(c) $\theta = -90°$.

10 Find each of the following without the use of tables.
(a) $\cos 225°$ (b) $\tan 150°$ (c) $\sin(-\pi/6)$
(d) $\sec(4\pi/3)$ (e) $\cot(7\pi/4)$ (f) $\csc(300°)$

III.2. According to the preceding discussion, P is the point corresponding to the real number t. A natural way of assigning a measure to θ is to use t. If this is done, then θ is referred to as an **angle of t radians** and we write $\theta = t$. In particular, if $\theta = 1$, then θ is an angle that subtends an arc of unit length on the unit circle U.

The radian measure of an angle can be found by using a circle of any radius. Specifically, if θ is a central angle of a circle of radius r and if θ subtends an arc of length s, where $0 \le s < 2\pi r$, then it follows from plane geometry that the radian measure of θ is given by

(III.6) $$\theta = \frac{s}{r}.$$

It can also be shown that the area A of a sector of a circle of radius r, determined by a central angle of radian measure θ, is

(III.7) $$A = \tfrac{1}{2}r^2\theta.$$

If the angle θ is generated by one-half of a complete counterclockwise rotation, then $\theta = 180°$. The radian measure of θ, however, is π. This gives us the basic relation $180° = \pi$ radians. Equivalent formulas are

$$1° = \pi/180 \text{ radians} \quad \text{and} \quad 1 \text{ radian} = (180/\pi)°.$$

By long division we obtain the fact that $1° \approx 0.01745$ radians, and 1 radian $\approx$ $57.296°$.

The following table gives the relationship between the radian and degree measure of several common angles.

Radians	$\pi/6$	$\pi/4$	$\pi/3$	$\pi/2$	$2\pi/3$	$3\pi/4$	$5\pi/6$	π
Degrees	30°	45°	60°	90°	120°	135°	150°	180°

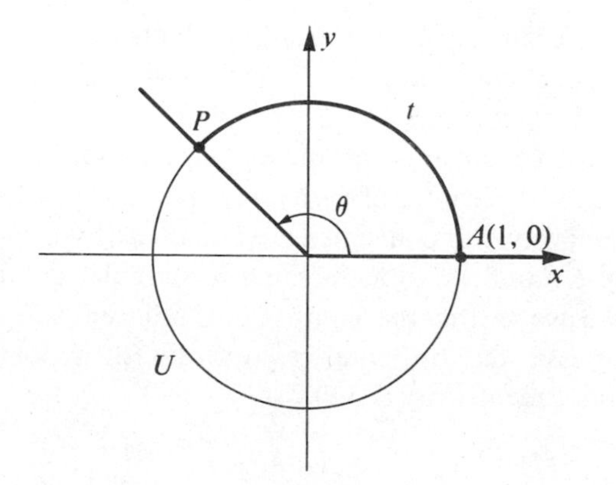

Figure III.2

In certain applications it is convenient to change the domains of the trigonometric functions from a subset of $\mathbb{R}$ to the set of angles. This may be accomplished by using the following definition.

(III.8) Definition

If θ is an angle and if the radian measure of θ is t, then the value of each trigonometric function at θ is its value at the real number t.

It follows from (III.8) that $\sin\theta = \sin t$, $\cos\theta = \cos t$, and so on, where t is the radian measure of θ. For convenience, the terminology *trigonometric functions* will be used regardless of whether angles or real numbers are employed for the domain.

The values of the trigonometric functions at an angle θ may be determined by means of an arbitrary point on the terminal side of θ as indicated in the next theorem.*

(III.9) Theorem

Let θ be an angle in standard position on a rectangular coordinate system and let $P(x, y)$ be any point other than O on the terminal side of θ (see (i) of Figure III.3). If $d(O, P) = r$, then the following relations hold.

$$\sin\theta = \frac{y}{r} \qquad \csc\theta = \frac{r}{y} \quad (\text{if } y \neq 0)$$
$$\cos\theta = \frac{x}{r} \qquad \sec\theta = \frac{r}{x} \quad (\text{if } x \neq 0)$$
$$\tan\theta = \frac{y}{x} \quad (\text{if } x \neq 0) \qquad \cot\theta = \frac{x}{y} \quad (\text{if } y \neq 0)$$

It can be shown, by using similar triangles, that the formulas given in (III.9) are independent of the point $P(x, y)$ that is chosen on the terminal side of θ. Note that if $r = 1$, then (III.9) reduces to (III.1).

An acute angle θ can be regarded as an angle of a right triangle and we may refer to the lengths of the **hypotenuse**, the **opposite side**, and the **adjacent side** in the usual way. For convenience we shall use hyp, opp, and adj, respectively, to denote these numbers. Introducing a rectangular coordinate system as in (ii) of Figure III.3, we see that the lengths of the adjacent side and the opposite side for θ are the abscissa and ordinate, respectively, of a point P on the terminal side of θ. Consequently, by (III.9),

(III.10)
$$\sin\theta = \frac{\text{opp}}{\text{hyp}}, \quad \csc\theta = \frac{\text{hyp}}{\text{opp}}, \quad \tan\theta = \frac{\text{opp}}{\text{adj}},$$
$$\cos\theta = \frac{\text{adj}}{\text{hyp}}, \quad \sec\theta = \frac{\text{hyp}}{\text{adj}}, \quad \cot\theta = \frac{\text{adj}}{\text{opp}}.$$

*See E. W. Swokowski, *Fundamentals of Algebra and Trigonometry*, Fourth Edition (Boston: Prindle, Weber & Schmidt Inc., 1978) page 256.

The formulas given in (III.10) are useful in problems dealing with triangles.

Formulas (III.2)–(III.5) are called the **fundamental identities**. Some other identities involving trigonometric functions are the following.

(III.11) Negative Angle Formulas

$$\sin(-u) = -\sin u, \quad \cos(-u) = \cos u, \quad \tan(-u) = -\tan u$$
$$\csc(-u) = -\csc u, \quad \sec(-u) = \sec u, \quad \cot(-u) = -\cot u$$

(III.12) Addition Formulas

$$\sin(u \pm v) = \sin u \cos v \pm \cos u \sin v$$
$$\cos(u \pm v) = \cos u \cos v \mp \sin u \sin v$$
$$\tan(u \pm v) = \frac{\tan u \pm \tan v}{1 \mp \tan u \tan v}$$

(III.13) Double Angle Formulas

$$\sin 2u = 2 \sin u \cos u$$
$$\cos 2u = \cos^2 u - \sin^2 u = 1 - 2\sin^2 u = 2\cos^2 u - 1$$
$$\tan 2u = \frac{2\tan u}{1 - \tan^2 u}$$

(III.14) Half-Angle Formulas

$$\sin^2\frac{u}{2} = \frac{1 - \cos u}{2} \qquad \cos^2\frac{u}{2} = \frac{1 + \cos u}{2}$$
$$\tan\frac{u}{2} = \frac{1 - \cos u}{\sin u} = \frac{\sin u}{1 + \cos u}$$

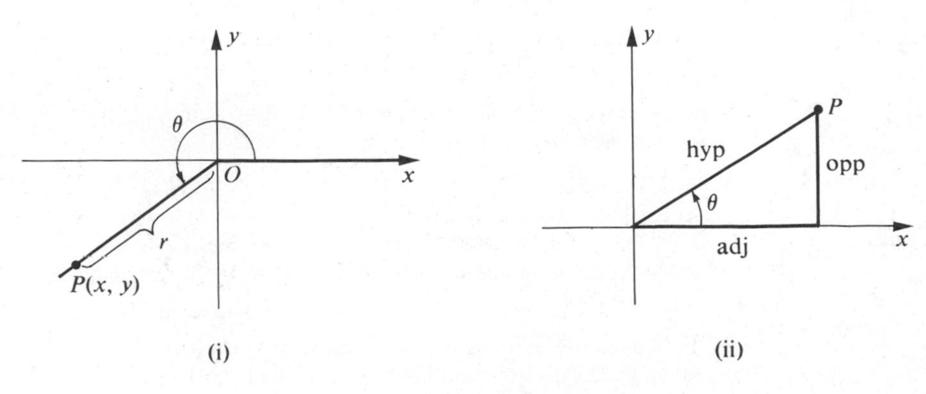

Figure III.3

Verify the identities in Exercises 11–36.

11 $\cos\theta\sec\theta = 1$

12 $\tan\alpha\cot\alpha = 1$

13 $\sin\theta\sec\theta = \tan\theta$

14 $\sin\alpha\cot\alpha = \cos\alpha$

15 $\dfrac{\csc x}{\sec x} = \cot x$

16 $\cot\beta\sec\beta = \csc\beta$

17 $(1 + \cos\alpha)(1 - \cos\alpha) = \sin^2\alpha$

18 $\cos^2 x(\sec^2 x - 1) = \sin^2 x$

19 $\cos^2 t - \sin^2 t = 2\cos^2 t - 1$

20 $(\tan\theta + \cot\theta)\tan\theta = \sec^2\theta$

21 $\dfrac{\sin t}{\csc t} + \dfrac{\cos t}{\sec t} = 1$

22 $1 - 2\sin^2 x = 2\cos^2 x - 1$

23 $(1 + \sin\alpha)(1 - \sin\alpha) = \dfrac{1}{\sec^2\alpha}$

24 $(1 - \sin^2 t)(1 + \tan^2 t) = 1$

25 $\sec\beta - \cos\beta = \tan\beta\sin\beta$

26 $\dfrac{\sin w + \cos w}{\cos w} = 1 + \tan w$

27 $\dfrac{\csc^2\theta}{1 + \tan^2\theta} = \cot^2\theta$

28 $\sin x + \cos x\cot x = \csc x$

29 $\sin t(\csc t - \sin t) = \cos^2 t$

30 $\cot t + \tan t = \csc t\sec t$

31 $\ln e^{\tan t} = \tan t$

32 $e^{\ln|\sin t|} = |\sin t|$

33 $\ln\cot x = -\ln\tan x$

34 $\ln\sec\theta = -\ln\cos\theta$

35 $-\ln|\sec\theta - \tan\theta| = \ln|\sec\theta + \tan\theta|$

36 $\ln|\csc x - \cot x| = -\ln|\csc x + \cot x|$

In each of Exercises 37–48 find the solutions of the given equation which are in the interval $[0, 2\pi)$, and also find the degree measure of each solution.

37 $2\cos^3\theta - \cos\theta = 0$

38 $2\cos\alpha + \tan\alpha = \sec\alpha$

39 $\sin\theta = \tan\theta$

40 $\csc^5\theta - 4\csc\theta = 0$

41 $2\cos^3 t + \cos^2 t - 2\cos t - 1 = 0$

42 $\cos x\cot^2 x = \cos x$

43 $\sin\beta + 2\cos^2\beta = 1$

44 $\cos 2x + 3\cos x + 2 = 0$

45 $2\sec u\sin u + 2 = 4\sin u + \sec u$

46 $\sin 2u = \sin u$

47 $2\cos^2\frac{1}{2}\theta - 3\cos\theta = 0$

48 $\sec 2x\csc 2x = 2\csc 2x$

In Exercises 49–52 find the exact values without the use of tables or calculators.

49 $\cos 75°$ **50** $\tan 285°$ **51** $\sin 195°$ **52** $\csc \pi/8$

If θ and ϕ are acute angles such that $\csc\theta = 5/3$ and $\cos\phi = 8/17$, find the numbers in Exercises 53–61.

53 $\sin(\theta + \phi)$ **54** $\cos(\theta + \phi)$ **55** $\tan(\theta - \phi)$

56 $\sin(\phi - \theta)$ **57** $\sin 2\phi$ **58** $\cos 2\phi$

59 $\tan 2\theta$ **60** $\sin\theta/2$ **61** $\tan\theta/2$

62 Express $\cos(\alpha + \beta + \gamma)$ in terms of functions of α, β, and γ.

63 Express each of the following products as a sum or difference.
(a) $\sin 7t \sin 4t$ (b) $\cos(u/4)\cos(-u/6)$ (c) $6\cos 5x \sin 3x$

64 Express each of the following as a product.
(a) $\sin 8u + \sin 2u$ (b) $\cos 3\theta - \cos 8\theta$ (c) $\sin(t/4) - \sin(t/5)$

Tables

TABLE 1. TRIGONOMETRIC FUNCTIONS

Degrees	Radians	Sin	Tan	Cot	Cos		
0	.000	.000	.000		1.000	1.571	90
1	.017	.017	.017	57.29	1.000	1.553	89
2	.035	.035	.035	28.64	.999	1.536	88
3	.052	.052	.052	19.081	.999	1.518	87
4	.070	.070	.070	14.301	.998	1.501	86
5	.087	.087	.087	11.430	.996	1.484	85
6	.105	.105	.105	9.514	.995	1.466	84
7	.122	.122	.123	8.144	.993	1.449	83
8	.140	.139	.141	7.115	.990	1.431	82
9	.157	.156	.158	6.314	.988	1.414	81
10	.175	.174	.176	5.671	.985	1.396	80
11	.192	.191	.194	5.145	.982	1.379	79
12	.209	.208	.213	4.705	.978	1.361	78
13	.227	.225	.231	4.331	.974	1.344	77
14	.244	.242	.249	4.011	.970	1.326	76
15	.262	.259	.268	3.732	.966	1.309	75
16	.279	.276	.287	3.487	.961	1.292	74
17	.297	.292	.306	3.271	.956	1.274	73
18	.314	.309	.325	3.078	.951	1.257	72
19	.332	.326	.344	2.904	.946	1.239	71
20	.349	.342	.364	2.747	.940	1.222	70
21	.367	.358	.384	2.605	.934	1.204	69
22	.384	.375	.404	2.475	.927	1.187	68
23	.401	.391	.424	2.356	.921	1.169	67
24	.419	.407	.445	2.246	.914	1.152	66
		Cos	Cot	Tan	Sin	Radians	Degrees

(Table I is continued on the next page.)

Table I. Trigonometric Functions (continued)

Degrees	Radians	Sin	Tan	Cot	Cos		
25	.436	.423	.466	2.144	.906	1.134	65
26	.454	.438	.488	2.050	.899	1.117	64
27	.471	.454	.510	1.963	.891	1.100	63
28	.489	.469	.532	1.881	.883	1.082	62
29	.506	.485	.554	1.804	.875	1.065	61
30	.524	.500	.577	1.732	.866	1.047	60
31	.541	.515	.601	1.664	.857	1.030	59
32	.559	.530	.625	1.600	.848	1.012	58
33	.576	.545	.649	1.540	.839	.995	57
34	.593	.559	.675	1.483	.829	.977	56
35	.611	.574	.700	1.428	.819	.960	55
36	.628	.588	.727	1.376	.809	.942	54
37	.646	.602	.754	1.327	.799	.925	53
38	.663	.616	.781	1.280	.788	.908	52
39	.681	.629	.810	1.235	.777	.890	51
40	.698	.643	.839	1.192	.766	.873	50
41	.716	.656	.869	1.150	.755	.855	49
42	.733	.669	.900	1.111	.743	.838	48
43	.750	.682	.933	1.072	.731	.820	47
44	.768	.695	.966	1.036	.719	.803	46
45	.785	.707	1.000	1.000	.707	.785	45
		Cos	Cot	Tan	Sin	Radians	Degrees

TABLE II. EXPONENTIAL FUNCTIONS

x	e^x	e^{-x}	x	e^x	e^{-x}
0.00	1.0000	1.0000	2.50	12.182	0.0821
0.05	1.0513	0.9512	2.60	13.464	0.0743
0.10	1.1052	0.9048	2.70	14.880	0.0672
0.15	1.1618	0.8607	2.80	16.445	0.0608
0.20	1.2214	0.8187	2.90	18.174	0.0550
0.25	1.2840	0.7788	3.00	20.086	0.0498
0.30	1.3499	0.7408	3.10	22.198	0.0450
0.35	1.4191	0.7047	3.20	24.533	0.0408
0.40	1.4918	0.6703	3.30	27.113	0.0369
0.45	1.5683	0.6376	3.40	29.964	0.0334
0.50	1.6487	0.6065	3.50	33.115	0.0302
0.55	1.7333	0.5769	3.60	36.598	0.0273
0.60	1.8221	0.5488	3.70	40.447	0.0247
0.65	1.9155	0.5220	3.80	44.701	0.0224
0.70	2.0138	0.4966	3.90	49.402	0.0202
0.75	2.1170	0.4724	4.00	54.598	0.0183
0.80	2.2255	0.4493	4.10	60.340	0.0166
0.85	2.3396	0.4274	4.20	66.686	0.0150
0.90	2.4596	0.4066	4.30	73.700	0.0136
0.95	2.5857	0.3867	4.40	81.451	0.0123
1.00	2.7183	0.3679	4.50	90.017	0.0111
1.10	3.0042	0.3329	4.60	99.484	0.0101
1.20	3.3201	0.3012	4.70	109.95	0.0091
1.30	3.6693	0.2725	4.80	121.51	0.0082
1.40	4.0552	0.2466	4.90	134.29	0.0074
1.50	4.4817	0.2231	5.00	148.41	0.0067
1.60	4.9530	0.2019	6.00	403.43	0.0025
1.70	5.4739	0.1827	7.00	1096.6	0.0009
1.80	6.0496	0.1653	8.00	2981.0	0.0003
1.90	6.6859	0.1496	9.00	8103.1	0.0001
2.00	7.3891	0.1353	10.00	22026.0	0.00005
2.10	8.1662	0.1225			
2.20	9.0250	0.1108			
2.30	9.9742	0.1003			
2.40	11.0232	0.0907			

TABLE III. NATURAL LOGARITHMS

n	.0	.1	.2	.3	.4	.5	.6	.7	.8	.9
0*		7.697	8.391	8.796	9.084	9.307	9.489	9.643	9.777	9.895
1	0.000	0.095	0.182	0.262	0.336	0.405	0.470	0.531	0.588	0.642
2	0.693	0.742	0.788	0.833	0.875	0.916	0.956	0.993	1.030	1.065
3	1.099	1.131	1.163	1.194	1.224	1.253	1.281	1.308	1.335	1.361
4	1.386	1.411	1.435	1.459	1.482	1.504	1.526	1.548	1.569	1.589
5	1.609	1.629	1.649	1.668	1.686	1.705	1.723	1.740	1.758	1.775
6	1.792	1.808	1.825	1.841	1.856	1.872	1.887	1.902	1.917	1.932
7	1.946	1.960	1.974	1.988	2.001	2.015	2.028	2.041	2.054	2.067
8	2.079	2.092	2.104	2.116	2.128	2.140	2.152	2.163	2.175	2.186
9	2.197	2.208	2.219	2.230	2.241	2.251	2.262	2.272	2.282	2.293
10	2.303	2.313	2.322	2.332	2.342	2.351	2.361	2.370	2.380	2.389

*Subtract 10 if $n < 1$; for example, $\ln 0.3 \approx 8.796 - 10 = -1.204$.

Formulas from Geometry

The most frequently used notation is as follows.

r = radius

h = altitude

b (or a) = length of base

A = area

C = circumference

V = volume

S = curved surface area

B = area of base

Formulas for areas, circumference of a circle, volumes, and curved surface area are as follows.

TRIANGLE	$A = \frac{1}{2}bh$
CIRCLE	$A = \pi r^2, \quad C = 2\pi r$
PARALLELOGRAM	$A = bh$
TRAPEZOID	$A = \frac{1}{2}(a + b)h$
RIGHT CIRCULAR CYLINDER	$V = \pi r^2 h, \quad S = 2\pi rh$
RIGHT CIRCULAR CONE	$V = \frac{1}{3}\pi r^2 h, \quad S = \pi r\sqrt{r^2 + h^2}$
SPHERE	$V = \frac{4}{3}\pi r^3, \quad S = 4\pi r^2$
PRISM	$V = Bh$
PYRAMID	$V = \frac{1}{3}Bh$

Answers to Odd-Numbered Exercises

CHAPTER 1

Exercises 1.1, page 9

1 (a) $>$ (b) $<$ (c) $=$ (d) $>$ (e) $=$ (f) $<$
3 (a) 3 (b) 7 (c) 7 (d) 3 (e) $22/7 - \pi$ (f) -1 (g) 0 (h) 9
5 (a) 4 (b) 8 (c) 8 (d) 12 **7** $(17/5, \infty)$ **9** $[-2, \infty)$ **11** $(-\infty, -3) \cup (2, \infty)$
13 $(5, \infty)$ **15** $(-4/5, 3]$ **17** $[-3, 1)$ **19** $(-\infty, 7/2)$ **21** $(9.7, 10.3)$ **23** $[5/3, 3]$
25 $(-\infty, 1/25) \cup (3/5, \infty)$ **27** $(-2, 1/3)$ **29** $(-\infty, -4] \cup [-1/2, \infty)$
31 $(-\infty, -1/10) \cup (1/10, \infty)$ **33** $[-2/3, 7/2)$
35 $|a| = |(a - b) + b| \leqslant |a - b| + |b|$. Hence $|a| - |b| \leqslant |a - b|$.

Exercises 1.2, page 19

1 5, $(4, -1/2)$ **3** $\sqrt{26}, (-1/2, -9/2)$ **5** 5, $(-11/2, -2)$ **7** 35

9

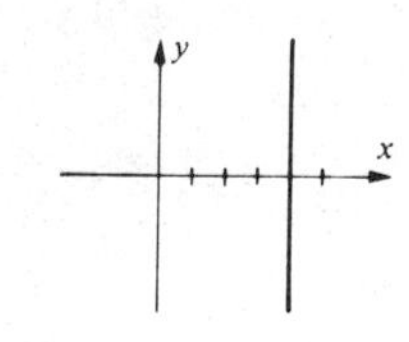

11

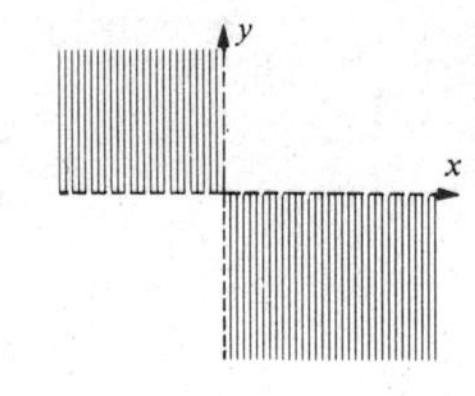

13

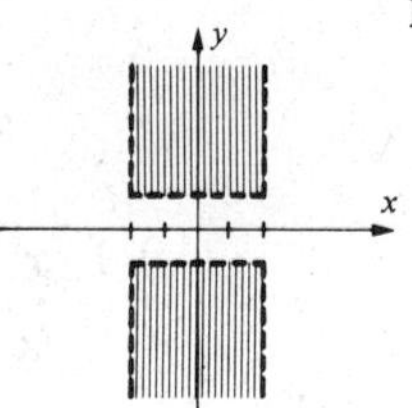

15

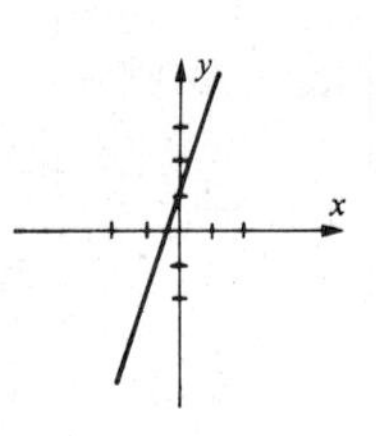

17

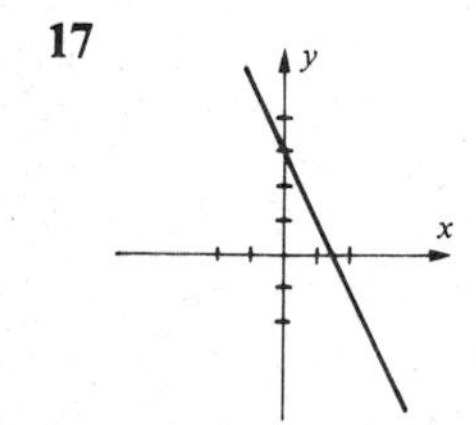

19

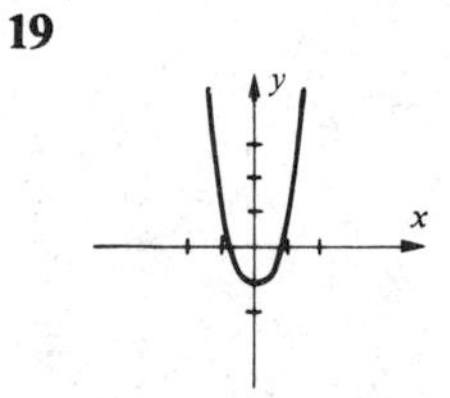

21

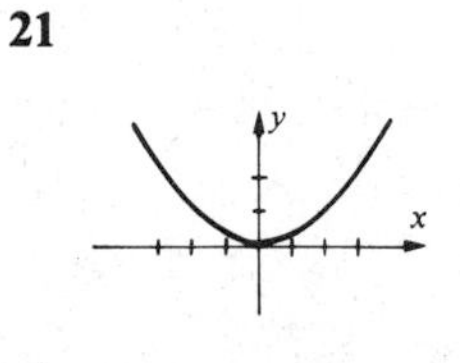

23

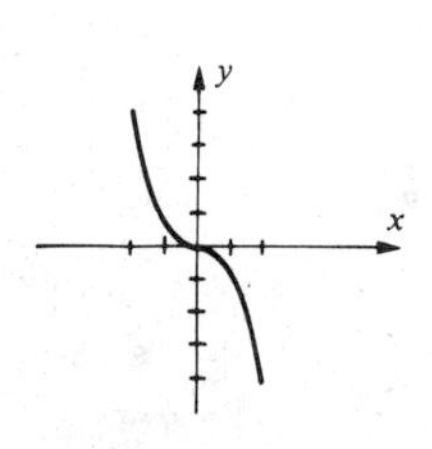

25

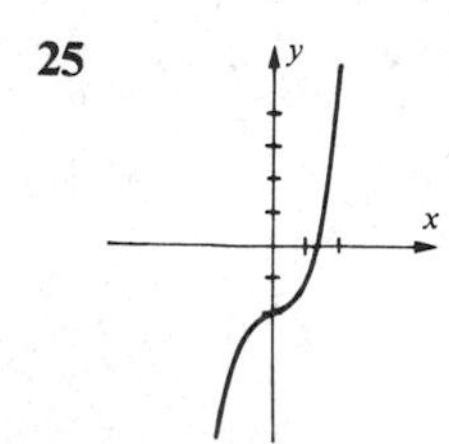

27

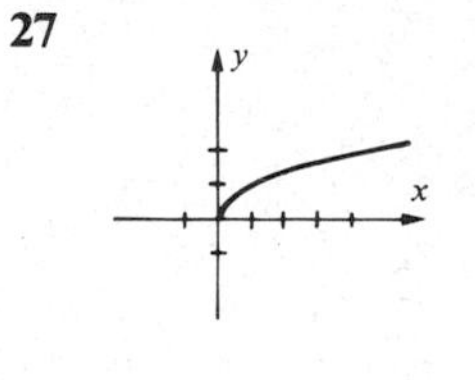

29

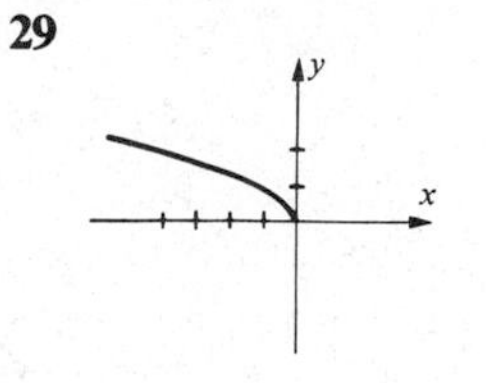

31 Circle of radius 4, center at the origin **33** $(x-3)^2+(y+2)^2=16$
35 $x^2+y^2=34$ **37** $(x+4)^2+(y-2)^2=4$ **39** $(x-1)^2+(y-2)^2=34$
41 $(-2,3)$, 3 **43** $(-3,0)$, 3 **45** $(1/4,-1/4)$, $\sqrt{26}/4$

Exercises 1.3, page 28

1 4 **3** The slope does not exist. **5** The slopes of opposite sides are equal.
7 *Hint*: Show that opposite sides are parallel and two adjacent sides are perpendicular.
9 $(-12,0)$ **11** $x-2y-14=0$ **13** $3x-8y-41=0$ **15** $x-8y-24=0$
17 (a) $x=10$ (b) $y=-6$ **19** $5x+2y-29=0$ **21** $5x-7y+15=0$
23 $x+6y-9=0$; $3x-5y+5=0$; $4x+y-4=0$; $(15/23, 32/23)$
25 $m=3/4$, $b=2$ **27** $m=-1/2$, $b=0$ **29** $m=0$, $b=4$ **31** $m=-5/4$, $b=5$

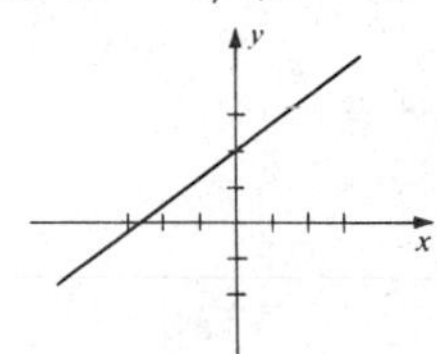

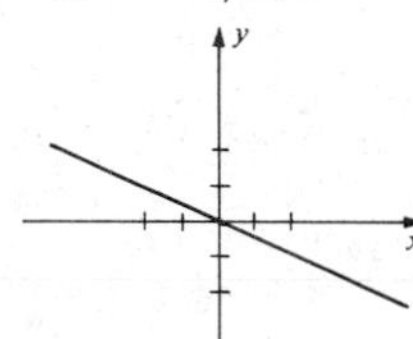

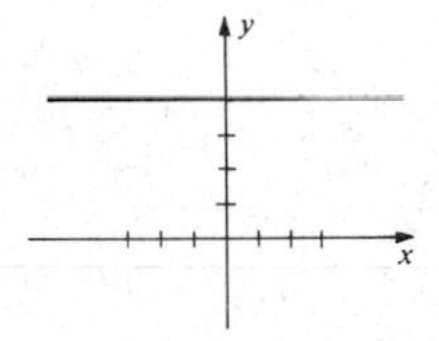

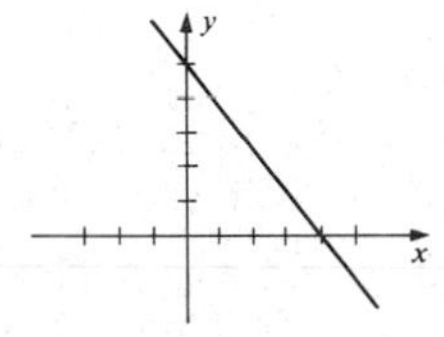

33 $m=1/3$, $b=-7/3$ **35** $k=-3$ **37** $x/(3/2)+y/(-3)=1$ **39** $r<1$ or $r>2$

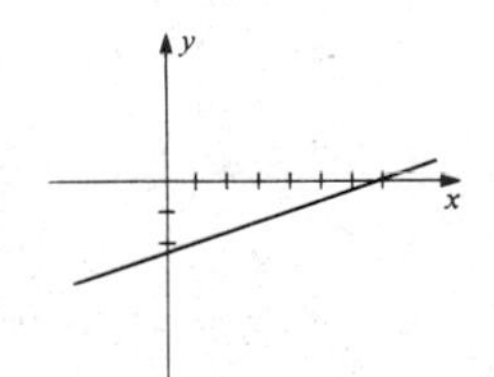

Exercises 1.4, page 37

1 $2, -8, -3, 6\sqrt{2}-3$
3 (a) $3a^2-a+2$ (b) $3a^2+a+2$ (c) $-3a^2+a-2$ (d) $3a^2+6ah+3h^2-a-h+2$
(e) $3a^2+3h^2-a-h+4$ (f) $6a+3h-1$
5 (a) $a^2/(1+4a^2)$ (b) a^2+4 (c) $1/(a^4+4)$ (d) $1/(a^2+4)^2$ (e) $1/(a+4)$ (f) $1/\sqrt{a^2+4}$
7 $[5/3,\infty)$ **9** $[-2,2]$ **11** All real numbers except 0, 3, and -3
13 $9/7$, $(a+5)/7$, $\mathbb{R}$ **15** 19, a^2+3, all nonnegative real numbers
17 $\sqrt[3]{4}$, $\sqrt[3]{a}$, $\mathbb{R}$ **19** Yes **21** No **23** Yes **25** No
27 Odd **29** Even **31** Even **33** Neither **35** Neither
37 $(-\infty,\infty)$; $(-\infty,\infty)$ **39** $(-\infty,\infty)$; $\{-3\}$ **41** $(-\infty,\infty)$; $(-\infty,4]$

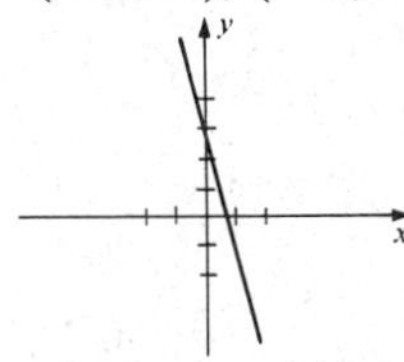

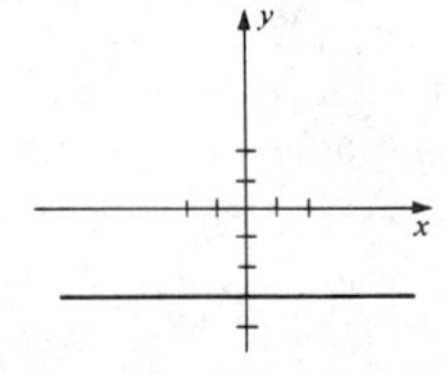

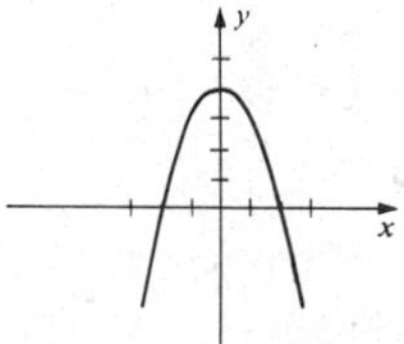

43 $[-2,2]$; $[0,2]$ **45** $(-\infty,4)\cup(4,\infty)$; $(-\infty,0)\cup(0,\infty)$ **47** $(-\infty,4)\cup(4,\infty)$; $(0,\infty)$

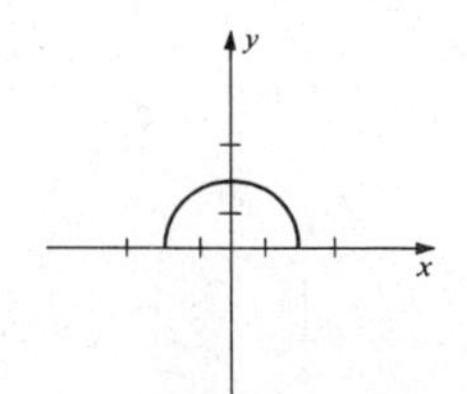

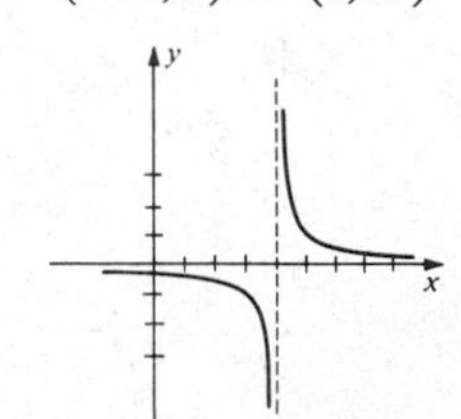

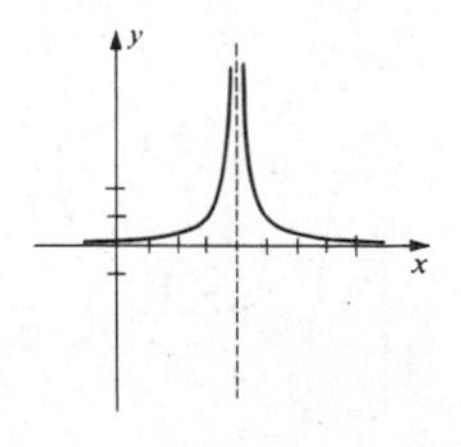

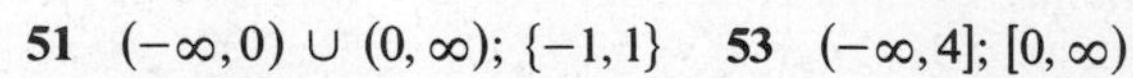

49 $(-\infty, \infty)$; $[0, \infty)$

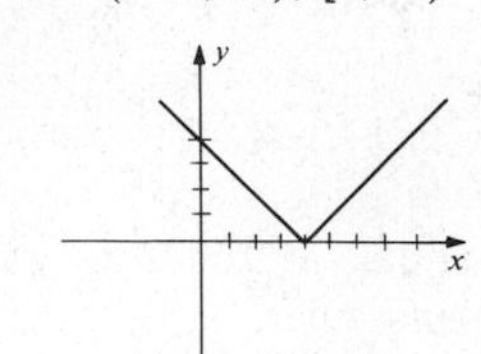

51 $(-\infty, 0) \cup (0, \infty)$; $\{-1, 1\}$

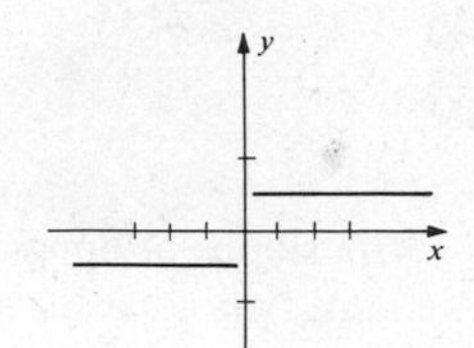

53 $(-\infty, 4]$; $[0, \infty)$

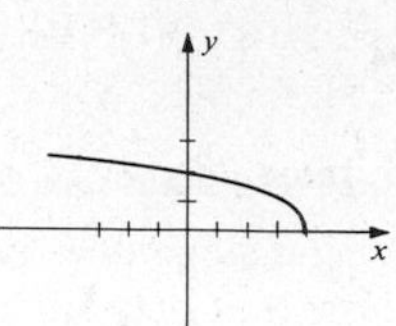

55 $(-\infty, \infty)$; $\{1, -1\}$

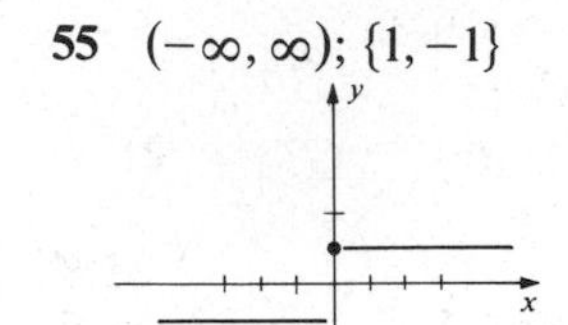

57 $(-\infty, \infty)$; $[-5, 5]$

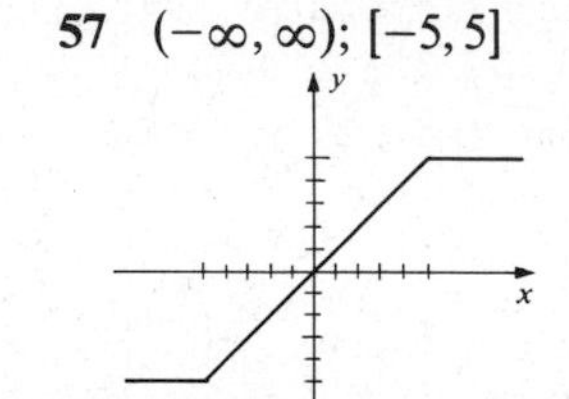

59

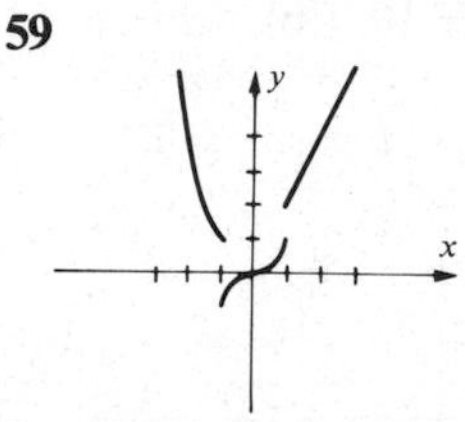

61 If $-1 < x < 1$, then there are two different points on the graph with abscissa x.

Exercises 1.5, page 42

1 $3x^2 + 1/(2x-3)$; $3x^2 - 1/(2x-3)$; $3x^2/(2x-3)$; $3x^2(2x-3)$
3 $2x$; $2/x$, $x^2 - 1/x^2$, $(x^2+1)/(x^2-1)$ 5 $2x^3 + x^2 + 7$; $2x^3 - x^2 - 2x + 3$; $2x^5 + 2x^4 + 3x^3 + 4x^2 + 3x + 10$, $(2x^3 - x + 5)/(x^2 + x + 2)$
7 $98x^2 - 112x + 37$; $-14x^2 - 31$ 9 $(x+1)^3$; $x^3 + 1$
11 $3/(3x^2+2)^2 + 2$; $1/(27x^4 + 36x^2 + 14)$ 13 $\sqrt{2x^2+7}$; $2x + 4$ 15 5, -5
17 $1/x^4$, $1/x^4$ 19 x, x 21 No; the coefficients of x^5 may be the negatives of one another.
25 *Hint*: Consider $f(x) = \frac{1}{2}[f(x) + f(-x)] + \frac{1}{2}[f(x) - f(-x)]$.

Exercises 1.6, page 47

5 $f^{-1}(x) = (x-8)/11$ 7 $f^{-1}(x) = \sqrt{6-x}$, where $0 \leqslant x \leqslant 6$
9 $f^{-1}(x) = (x^2+2)/7$, where $x \geqslant 0$ 11 $f^{-1}(x) = \sqrt[3]{(7-x)/3}$
13 $f^{-1}(x) = (x^{1/5} - 8)^{1/3}$ 15

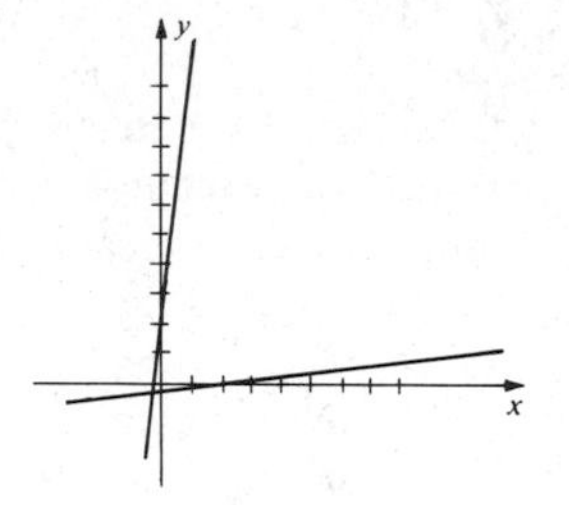

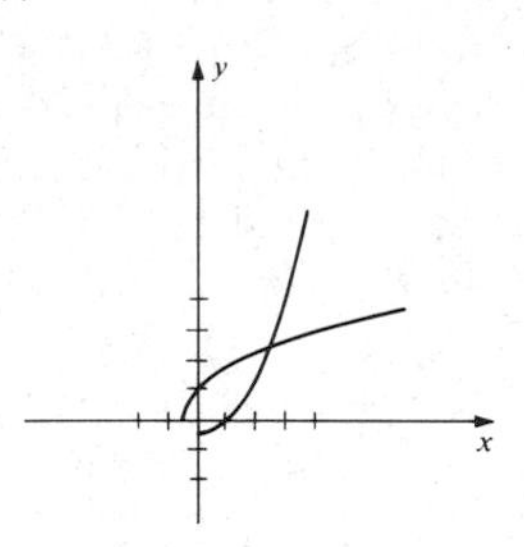

17 $f^{-1}(x) = (x-b)/a$; a constant function has no inverse. The identity function is its own inverse.
19 If $f(x) = x^2$, then f is not one-to-one and hence has no inverse function.

1.7 Review Exercises, page 48

1 $(-\infty, -3/5)$ 3 $[3.495, 3.505]$ 5 $(1, 3/2)$ 7 $(-5, 1/3) \cup (7/5, \infty)$
9 (a) 12 (b) $(1/2, 5/2)$ (c) 7

11

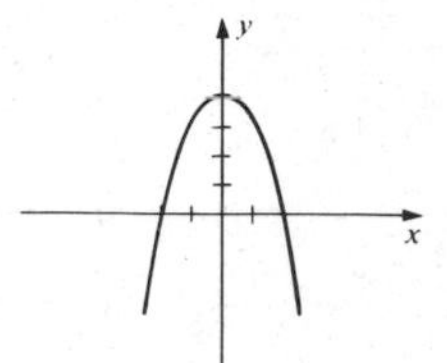

13

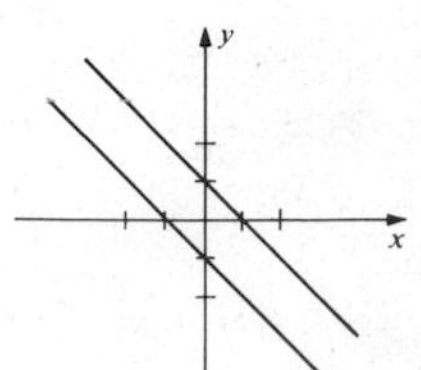

15

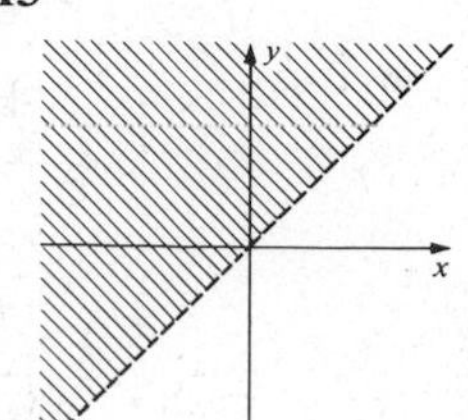

17

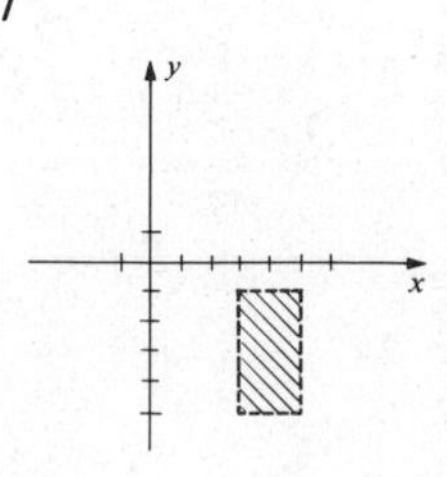

19 $(x + 4)^2 + (y + 3)^2 = 81$ **21** $(5, -7)$; 9 **23** $6x - 7y + 24 = 0$ **25** $x = -4$
27 $(-\infty, 0) \cup (0, 1) \cup (1, \infty)$
29 The open interval (5, 7) **31** (a) $1/\sqrt{2}$ (b) $1/2$
(c) 1 (d) $1/\sqrt[4]{2}$ (e) $1/\sqrt{1 - x}$ (f) $-1/\sqrt{x + 1}$ (g) $1/\sqrt{x^2 + 1}$ (h) $1/(x + 1)$.
33 **35**

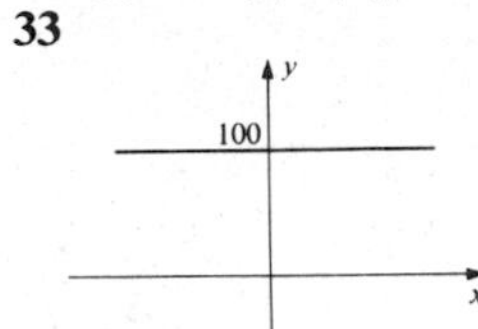

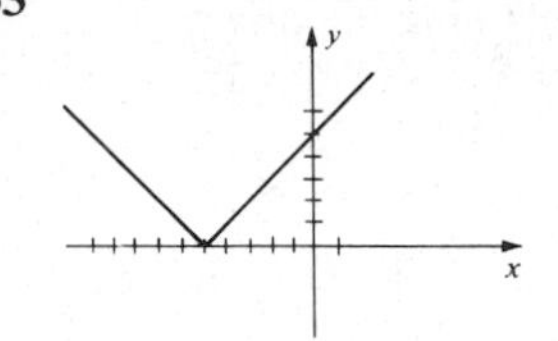

37 $x^2 + 4 + \sqrt{2x + 5}$; $x^2 + 4 - \sqrt{2x + 5}$; $(x^2 + 4)\sqrt{2x + 5}$; $(x^2 + 4)/\sqrt{2x + 5}$; $2x + 9$; $\sqrt{2x^2 + 13}$ **39** $f^{-1}(x) = (5 - x)/7$

CHAPTER 2

Exercises 2.1, page 56

1 4 **3** 1/9 **5** 17/13 **7** 32 **9** $2x$ **11** 12 **13** Does not exist
15 Does not exist **17** (a) $10a - 4$ (b) $y = 16x - 20$ **19** (a) $3a^2$ (b) $y = 12x - 16$
21 3; $y = 3x + 2$ **23** $-1/a^2$; $x + 4y - 4 = 0$ **29** (3,9)

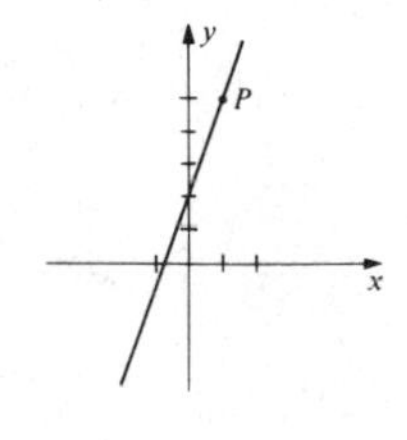

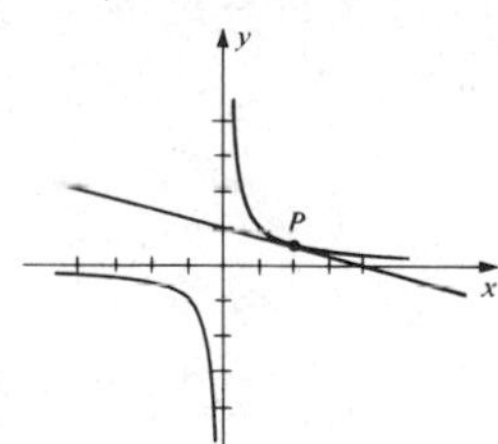

Exercises 2.2, page 62

1 Given any ϵ, choose $\delta \leqslant \epsilon/5$. **3** Given any ϵ, choose $\delta \leqslant \epsilon/9$.
5 Given any ϵ, choose $\delta \leqslant 4\epsilon$.
7 Given any ϵ, let δ be any positive number.
9 Given any ϵ, choose $\delta \leqslant \epsilon$.
11 Every interval $(3 - \delta, 3 + \delta)$ contains numbers for which the quotient equals 1, and other numbers for which the quotient equals -1.
13 $1/(x + 5)$ can be made as large as desired by choosing x sufficiently close to -5.
15 There are many examples. One is $f(x) = (x^2 - 1)/(x - 1)$ if $x \neq 1$ and $f(1) = 3$.
17 Every interval $(a - \delta, a + \delta)$ contains numbers such that $f(x) = 0$ and other numbers such that $f(x) = 1$.

Exercises 2.3, page 70

1 -13 **3** $5\sqrt{2} - 20$ **5** -2 **7** 0
9 15 **11** -7 **13** 1/12 **15** 8 **17** -23 **19** 2 **21** 72/7
23 -2 **25** $-1/8$ **27** 0 **29** -810 **31** -64 **33** -108 **35** $-1/4$
37 If $a < 0$ and $r = m/n$ is in lowest terms, then n must be an odd integer.

Exercises 2.4, page 75

1 4 **3** -6 **5** 3 **7** -1 **9** 1 **11** 1/8 **13** 1 **15** Does not exist
17 6; 4 **19** -1; 1/11; does not exist **21** $(-1)^{n-1}$; $(-1)^n$ **23** 0; 0 **25** 1; 0

Exercises 2.5, page 83

11 $\{x: x \neq 3/2, x \neq -1\}$ **13** $[3/2, \infty)$ **15** $(-\infty, -1) \cup (1, \infty)$
17 $\{x: x \neq -9\}$ **19** $\{x: x \neq 0, x \neq 1\}$
21 $[-5, -3] \cup [3, 4) \cup (4, 5]$ **23, 25, 27** Discontinuous at each integer n
29 f is discontinuous on any open interval containing the origin.
31 No. $\lim_{x \to 3} f(x)$ does not exist. **35** $c = \sqrt[3]{w-1}$ **37** $c = \sqrt{w} - 2$

2.6 Review Exercises, page 85

1 13 **3** $-4 - \sqrt{14}$ **5** $7/8$ **7** $32/3$ **9** Does not exist **11** 3 **13** -1
15 $4a^3$ **17** $-3/16$ **19** -1 **21** -6 **25** $\mathbb{R}$ **27** $[-3, -2) \cup (-2, 2) \cup (2, 3]$
29 Discontinuous at 4 and -4 **31** Discontinuous at 0 and 2 **33** $c = 1/\sqrt{w}$

CHAPTER 3

Exercises 3.1, page 92

1 $-3a^2$ **3** $1/2\sqrt{a}$
5 (a) 11.8, 11.4, 11.04, 11.004 cm/sec (b) 11 cm/sec (c) $(-3/8, \infty)$ (d) $(-\infty, -3/8)$
7 48 ft/sec, 16 ft/sec, -16 ft/sec. Maximum height attained at $t = 7/2$ sec. Ground struck at $t = 7$ sec with velocity -112 ft/sec. **9** If $f(t) = at + b$, then $v(t) = a$.

Exercises 3.2, page 98

1 0 **3** 9 **5** $8 - 10x$ **7** $-1/(x-2)^2$ **9** $3/2\sqrt{3x+1}$ **11** $-7/2\sqrt{x^3}$
13 $6x^2 - 4$ **15** $2a$ **17** $-12/a^3$ **19** $-1/(a+5)^2$
23 *Hint:* Use (3.7).
25 If $f(x) = ax + b$, then $f'(x) = a$ has degree 0. If $f(x)$ has degree 2, then $f'(x)$ has degree 1. If $f(x)$ has degree 3, then $f'(x)$ has degree 2.

Exercises 3.3, page 106

1 $f'(x) = 20x + 9$ **3** $f'(s) = -1 + 8s - 20s^3$ **5** $g'(x) = 10x^4 + 9x^2 - 28x$
7 $h'(r) = 18r^5 - 21r^2 + 4r$ **9** $f'(x) = 23/(3x+2)^2$
11 $h'(z) = (-27z^2 + 12z + 70)/(2 - 9z)^2$ **13** $f'(x) = 9x^2 - 4x + 4$
15 $F'(t) = 2t - 2/t^3$ **17** $g'(x) = 416x^3 - 195x^2 + 64x - 20$ **19** $G'(v) = 6v^2/(v^3+1)^2$
21 $f'(x) = -(1 + 2x + 3x^2)/(1 + x + x^2 + x^3)^2$ **23** $g'(z) = 72z^5 - 64z^3 - 18z^2 - 70z - 7$
25 $K'(s) = (-4/81)s^{-5}$ **27** $h'(x) = 10(5x - 4)$
29 $f'(t) = (6 - 20t - 21t^2)/5(2 + 7t^2)^2$ where $t \neq 0$ **31** $M'(x) = 2 - 4/x^2 - 6/x^3$
33 $y' = (-3x + 2)/x^3$ **35** $y' = 120x - 49$
39 $y' = (8x-1)(x^2+4x+7)(3x^2) + (8x-1)(2x+4)(x^3-5) + 8(x^2+4x+7)(x^3-5)$
41 (a) $y = 5$ (b) $5x + 2y - 10 = 0$ (c) $4x - 5y + 13 = 0$ **43** (a) $2/3, -2$ (b) $0, -4/3$
45 (a) $(-\infty, -5)$ and $(2, \infty)$ (b) $(-5, 2)$. The velocity is 0 at $t = -5$ and $t = 2$.
47 (a) 1 (b) -3 (c) -4 (d) 11 (e) $-1/25$ **49** (0,1); (1,2); (3,22)
51 $16y - 40x + 99 = 0$ **53** $y - 9 = 2(x - 5)$, point of tangency is (1, 1); $y - 9 = 18(x - 5)$, point of tangency is (9, 81).

Exercises 3.4, page 113

1 -0.72 **3** $-1.89/98.01 \approx -0.019$
5 (a) $\Delta y = (6x+5)\Delta x + 3(\Delta x)^2$ (b) $dy = (6x+5)\,dx$ (c) $dy - \Delta y = -3(\Delta x)^2$
7 (a) $\Delta y = -\Delta x/x(x + \Delta x)$ (b) $dy = -dx/x^2$ (c) $dy - \Delta y = -(\Delta x)^2/x^2(x + \Delta x)$
9 (a) $-9\Delta x$ (b) $-9\Delta x$ (c) 0
11 (a) $3x^2\Delta x + 3x(\Delta x)^2 + (\Delta x)^3$ (b) $3x^2\Delta x$ (c) $-3x(\Delta x)^2 - (\Delta x)^3$

13 0.06 **15** $dw = (3z^2 - 6z + 2)dz$, $\Delta w \approx -1.30$ **17** 0.96π; 0.015; 1.5%
19 30 in.^{3}; 30.301 in.^{3} **21** Area $\approx$ 3301.661; maximum error $\approx$ 11.459; average error $\approx$ 0.00347; percentage error $\approx$ 0.347% **23** $1/50\pi \approx 0.00637$ **25** -1
27 0.92; 0.92236816 **29** dA is the shaded region.

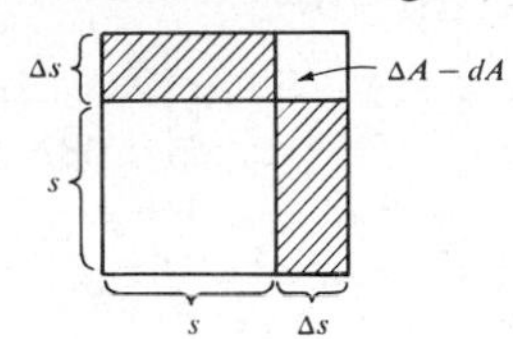

Exercises 3.5, page 119

1 $3(x^2 - 3x + 8)^2(2x - 3)$ **3** $-40(8x - 7)^{-6}$ **5** $-(7x^2 + 1)/(x^2 - 1)^5$
7 $f'(x) = 5(8x^3 - 2x^2 + x - 7)^4(24x^2 - 4x + 1)$ **9** $F'(v) = 17{,}000\,(17v - 5)^{999}$
11 $s'(t) = -2(4t^5 - 3t^3 + 2t)^{-3}(20t^4 - 9t^2 + 2)$
13 $N'(x) = 32x(6x - 7)^3(8x^2 + 9) + 18(8x^2 + 9)^2(6x - 7)^2$ which reduces to $(6x - 7)^2(8x^2 + 9)(336x^2 - 224x + 162)$ **15** $g'(z) = 6(z^2 - 1/z^2)^5(2z + 2/z^3)$
17 $k'(u) = -20(u^2 + 1)^3/(4u - 5)^6 + 6u(u^2 + 1)^2/(4u - 5)^5$ which reduces to $(u^2 + 1)^2(4u^2 - 30u - 20)/(4u - 5)^6$ **19** $f'(x) = 124x(3x^2 - 5)/(2x^2 + 7)^3$
21 $G'(s) = -6(s^{-4} + 3s^{-2} + 2)^{-7}(-4s^{-5} - 6s^{-3})$
23 $h'(x) = 200(2x + 1)^9[(2x + 1)^{10} + 1]^9$ **25** $F'(t) = 2(2t + 1)(2t + 3)^2(24t^2 + 26t + 3)$
27 (a) $y - 81 = 864(x - 2)$ (b) (1, 1), (1/2, 0), (3/2, 0)
29 (a) $y - 1 = 20(x - 1)$ (b) (1/2, 0)
31 $dy = 10(x^4 - 3x^2 + 1)^9(4x^3 - 6x)dx$, $\Delta y \approx 0.2$ **33** $\dfrac{dw}{ds} = \dfrac{dw}{dz}\dfrac{dz}{ds}$
35 $dw/ds = (3z^2 + 2/z^2)5(s^2 + 1)^4 2s$ **37** -0.02 **39** 15 **45** $x + 2y - 5 = 0$
47 $4x - 2y - 3 = 0$ **49** $64x - 16y - 7 = 0$ **51** $(4x^3 + 4x)/(x^4 + 2x^2 + 5)$

Exercises 3.6, page 125

1 $f(x) = -(2/5)x^2 + 2x + 4/5$, $\mathbb{R}$ **3** $f(x) = \sqrt{16 - x^2}$, $[-4, 4]$. There are other answers.
5 $f(x) = x + \sqrt{x}$, $[0, \infty)$ **7** $f(x) = 1 - 2\sqrt{x} + x$, $[0, 1]$ **9** $y' = -8x/y$
11 $y' = -(6x^2 + 2xy)/(x^2 + 3y^2)$ **13** $y' = (10x - y)/(x + 8y)$ **15** $y' = -y^3/x^3$
17 $y' = -(2xy^3 + 4y + 1)/(3x^2y^2 + 4x - 6)$ **19** $y' = (4x^2 + 3x - 1)(8x + 3)/4y(y^2 - 9)^3$
21 $4x - y + 16 = 0$ **23** $y + 3 = (-36/23)(x - 2)$
25 If it did, then $[f(x)]^2 + x^2 = -1$, an impossibility.
27 (a) Infinitely many (b) One, $f(x) = 0$, with domain $x = 0$ **29** 0.09

Exercises 3.7, page 128

1 $f'(x) = (2/3)x^{-1/3} + 6x^{1/2}$ **3** $k'(r) = 8r^2(8r^3 + 27)^{-2/3}$
5 $F'(v) = -5v^4(v^5 - 32)^{-6/5}$ **7** $f'(x) = 1/\sqrt{2x}$ **9** $F'(z) = 15\sqrt{z} - 1/z\sqrt[3]{z}$
11 $g'(w) = (w^2 + 4w - 9)/2w^{5/2}$ **13** $M'(x) = (8x - 7)/2\sqrt{4x^2 - 7x + 4}$
15 $f'(t) = -48t/(9t^2 + 16)^{5/3}$ **17** $H'(u) = -1/2\sqrt{(3u + 8)(2u + 5)^3}$
19 $k'(s) = 16(s^2 + 9)^{1/4}(4s + 5)^3 + \frac{1}{2}s(4s + 5)^4(s^2 + 9)^{-3/4}$
21 $h'(x) = \dfrac{9x^2}{5}(x^2 + 4)^{5/3}(x^3 + 1)^{-2/5} + \dfrac{10x}{3}(x^2 + 4)^{2/3}(x^3 + 1)^{3/5}$
23 $g'(z) = -3\sqrt[3]{2z + 3}/2\sqrt{(3z + 2)^3} + 2/(3\sqrt{3z + 2}\sqrt[3]{(2z + 3)^2})$
25 $f'(x) = 6(7x + \sqrt{x^2 + 3})^5(7 + x/\sqrt{x^2 + 3})$ **27** $2x + \sqrt{3}y - 1 = 0$
29 $y - 5 = (-33/24)(x + 8)$ **31** (4, 2) **33** $y' = -\sqrt{y/x}$
35 $y' = (12\sqrt{xy} + y)/(6\sqrt{xy} - x)$ **37** $y' = (18x^{5/3}y^{2/3} + y)/(12x^{2/3}y^{5/3} - x)$
39 $dy = 2(6x + 11)^{-2/3}dx$ **41** $4 + (1/48) \approx 4.02$ **43** -0.24
45 60π cm^2; maximum error $\approx$ 1.508; percentage error $\approx$ 0.8%

Exercises 3.8, page 131

1 $f'(x) = 12x^3 - 8x + 1, f''(x) = 36x^2 - 8$
3 $H'(s) = 1/3\sqrt[3]{s^2} - 4/s^3, H''(s) = -2/9\sqrt[3]{s^5} + 12/s^4$
5 $g'(z) = 3/2\sqrt{3z+1}, g''(z) = -9/4\sqrt{(3z+1)^3}$
7 $k'(r) = 20(4r+7)^4, k''(r) = 320(4r+7)^3$ **9** $f'(x) = x(x^2+4)^{-1/2}, f''(x) = 4(x^2+4)^{-3/2}$
11 $D_x^3 y = 120x^2 + 18$ **13** $D_x^3 y = 594/(3x+1)^4$ **15** $D_x^3 y = -270(2-9x)^{-8/3}$
17 $y'' = (2xy^3 - 2x^4)/y^5$ **19** $y'' = 10(y^2 - 3xy + x^2)/(2y - 3x)^3$
21 $f'(x) = 6x^5 - 8x^3 + 9x^2 - 1, f''(x) = 30x^4 - 24x^2 + 18x, f'''(x) = 120x^3 - 48x + 18,$
$f^{(4)}(x) = 360x^2 - 48, f^{(5)}(x) = 720x, f^{(6)}(x) = 720$
23 $f^{(n)}(x) = (-1)^n n!/x^{n+1}, f^{(n)}(1) = (-1)^n n!$
25 The degree of f' is $n-1$, of f'' is $n-2, \ldots,$ of $f^{(n)}$ is 0. Since $f^{(n)}$ is a constant function, all higher derivatives are zero. **27** $f''(1) = 18, f''(-2) = -18$
29 0.14 **31** $D_x^2 y = f''(g(x))(g'(x))^2 + f'(g(x))g''(x)$

3.9 Review Exercises, page 132

1 $f'(x) = -24x/(3x^2+2)^2$ **3** $f'(x) = 6x^2 - 7$ **5** $g'(t) = 3/\sqrt{6t+5}$
7 $\frac{1}{3}(7z^2 - 4z + 3)^{-2/3}(14z - 4)$ **9** $-144x/(3x^2-1)^5$ **11** $-2(y^2 - y^{-2})^{-3}(2y + 2y^{-3})$
13 $(12/5)(3x+2)^{-1/5}$ **15** $4(8s^2-4)^3(72s^4 - 108s^2 + 16s)/(1-9s^3)^5$
17 $(x^6+1)^4(3x+2)^2(99x^6 + 60x^5 + 9)$
19 $\frac{4}{3}(7y-2)^{-2}(2y+1)^{-1/3} - 14(7y-2)^{-3}(2y+1)^{2/3}$
21 $2x[(x^2+2)(x^2+3) + (x^2+1)(x^2+3) + (x^2+1)(x^2+2)]$
23 $\dfrac{1}{2\sqrt{x+\sqrt{x+\sqrt{x}}}}\left(1 + \dfrac{2\sqrt{x}+1}{4\sqrt{x}\sqrt{x+\sqrt{x}}}\right)$
25 $[\frac{2}{3}(3x+2)^{1/2}(2x+3)^{-2/3} - \frac{3}{2}(2x+3)^{1/3}(3x+2)^{-1/2}]/(3x+2)$
27 $g'(z) = 3(9z^{5/3} - 5z^{3/5})^2(15z^{2/3} - 3z^{-2/5})$ **29** $k'(s) = (9s-1)^3(108s^2 - 139s + 39)$
31 $f'(w) = -53/2\sqrt{(2w+5)(7w-9)^3}$ **33** $y' = (4xy^2 - 15x^2)/(12y^2 - 4x^2y)$
35 $y' = 1/\sqrt{x}(3\sqrt{y}+2)$ **37** $9x - 4y - 12 = 0$ **39** $x - y + 4 = 0$
41 $y' = -2x^{-3} - x^{-2}; y'' = 6x^{-4} + 2x^{-3}; y''' = -24x^{-5} - 6x^{-4}$
43 $y'' = 5(y^2 - 4xy - x^2)/(y-2x)^3$ **45** $f^{(n)}(x) = n!/(1-x)^{n+1}$ **47** $0.06\sqrt{3}, 0.015$
49 -0.57

CHAPTER 4

Exercises 4.1, page 142

1 5; -3 **3** 1; -3
5 (a) Since $f'(x) = 1/(3x^{2/3})$, $f'(0)$ does not exist. If $a \neq 0$, then $f'(a) \neq 0$. Hence 0 is the only critical number of f. The number $f(0) = 0$ is not a local extremum since $f(x) < 0$ if $x < 0$ and $f(x) > 0$ if $x > 0$.
(b) The facts that 0 is the only critical number and that there is a vertical tangent line at (0, 0) follow as in part (a). The number $f(0) = 0$ is a local minimum since $f(x) > 0$ if $x \neq 0$.
7 There is a critical number 0, but $f(0)$ is not a local extremum since $f(x) < f(0)$ if $x < 0$ and $f(x) > f(0)$ if $x > 0$. The function is continuous at every number a since $\lim_{x \to a} f(x) = f(a)$. If $0 < x_1 < x_2 < 1$, then $f(x_1) < f(x_2)$ and hence there is neither a maximum nor minimum on (0, 1). This does not contradict (4.4) because the interval (0, 1) is open. **9** 3/8
11 5/3 and -2 **13** 2 **15** 4 and -4 (not 0) **17** 0, 15/7, and 5/2 **19** None
21 If $f(x) = cx + d$ and $c \neq 0$, then $f'(x) = c \neq 0$. Hence there are no critical numbers. On $[a, b]$ the function has absolute extrema at a and b.
23 If $x = n$ is an integer, then $f'(n)$ does not exist. Otherwise, $f'(x) = 0$ for all $x \neq n$.

25 If $f(x) = ax^2 + bx + c$ and $a \neq 0$, then $f'(x) = 2ax + b$. Hence $-b/2a$ is the only critical number of f.

27 Since $f'(x) = nx^{n-1}$, the only possible critical number is $x = 0$, and $f(0) = 0$. If n is even then $f(x) > 0$ if $x \neq 0$ and hence 0 is a local minimum. If n is odd, then 0 is not an extremum since $f(x) < 0$ if $x < 0$ and $f(x) > 0$ if $x > 0$.

Exercises 4.2, page 146

1 f is not differentiable at the number 0 in the interval $(-1, 1)$.

3 f is not continuous on the interval $[-1, 4]$. **5** $c = 2$ **7** $c = 0$ **9** $c = 2$

11 $c = 2$ **13** f is not differentiable at $x = 0$. **15** $c = 2$ **17** $c = (2 - \sqrt{7})/3$

19 If $f(x) = cx + d$, then $f'(x) = c$ for all x. Moreover $f(b) - f(a) = (cb + d) - (ca + d) = c(b - a) = f'(x)(b - a)$.

21 If f has degree 3, then $f'(x)$ is a polynomial of degree 2. Consequently the equation $f(b) - f(a) = f'(x)(b - a)$ has at most two solutions x_1 and x_2. If f has degree 4, there are at most three such solutions. If f has degree n, there are at most $n - 1$ solutions.

Exercises 4.3, page 153

1 Max: $f(-7/8) = 129/16$; increasing on $(-\infty, -7/8]$; decreasing on $[-7/8, \infty)$

3 Max: $f(-2) = 29$; min: $f(5/3) = -548/27$; increasing on $(-\infty, -2]$ and $[5/3, \infty)$; decreasing on $[-2, 5/3]$

5 Max: $f(0) = 1$; min; $f(-2) = -15$; min: $f(2) = -15$; increasing on $[-2, 0]$ and $[2, \infty)$; decreasing on $(-\infty, -2]$ and $[0, 2]$

7 Min: $f(-1) = -3$; increasing on $[-1, \infty)$; decreasing on $(-\infty, -1]$

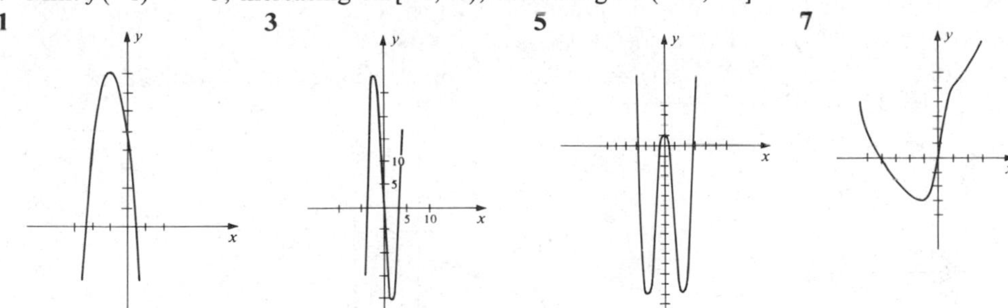

9 Max: $f(0) = 0$; min: $f(-\sqrt{3}) = f(\sqrt{3}) = -3$; increasing on $[-\sqrt{3}, 0]$ and $[\sqrt{3}, \infty)$; decreasing on $(-\infty, -\sqrt{3}]$ and $[0, \sqrt{3}]$

11 Max: $f(7/4) \approx 42$; min: $f(0) = 2$; min: $f(7) = 2$; increasing on $[0, 7/4]$ and $[7, \infty)$; decreasing on $(-\infty, 0]$ and $[7/4, 7]$

13 Max: $f(-1) = -4$; min: $f(1) = 4$; increasing on $(-\infty, -1]$ and $[1, \infty)$; decreasing on $[-1, 0)$ and $(0, 1]$.

15 Max: $f(3/5) \approx 0.346$; min: $f(1) = 0$; increasing on $(-\infty, 3/5]$ and $[1, \infty)$; decreasing on $[3/5, 1]$

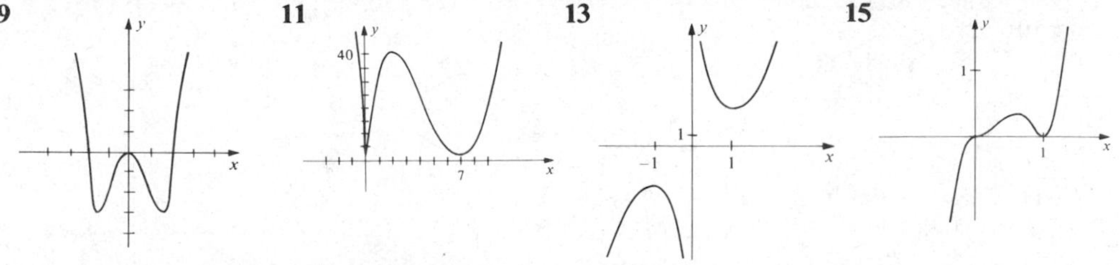

17 Max: $f(-\sqrt{3}) = (6\sqrt{3})^{1/3}$; min: $f(\sqrt{3}) = -(6\sqrt{3})^{1/3}$

19 Max: $f(-1) = 0$; min: $f(5/7) = -9^3 12^4/7^7$ **21** None

In Exercises 23 and 25 the absolute maximum is given first and the absolute minimum second.

23 (a) $f(-7/8) = 129/16$; $f(1) = -6$ (b) $f(-7/8) = 129/16$; $f(-4) = -31$ (c) $f(0) = 5$; $f(5) = -130$ **25** (a) $f(-1) = 20$; $f(1) = -16$ (b) $f(-2) = 29$; $f(-4) = -31$ (c) $f(5) = 176$; $f(5/3) = -548/27$ **27** (*art below*) **29** Max: $f'(-1) = 8$; min: $f'(1) = -8$; f' is increasing on $(-\infty, -1]$ and $[1, \infty)$; decreasing on $[-1, 1]$.

27

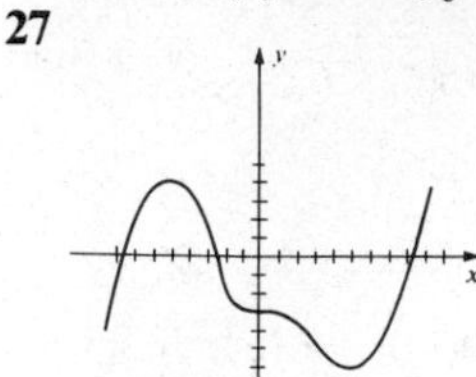

29

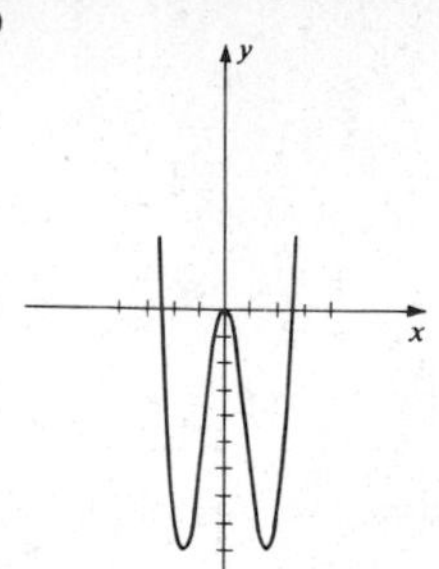

31 $a = 3/4$, $b = 0$, $c = -9/4$, $d = 1/2$

Exercises 4.4, page 162

In Exercises 1-11 the notations CU and CD mean that the graph is concave upward or downward, respectively, in the interval which follows. PI denotes point(s) of inflection.

1 Max: $f(1/3) = 31/27$; min: $f(1) = 1$. CD on $(-\infty, 2/3)$; CU on $(2/3, \infty)$; abscissa of PI is $2/3$.

3 Min: $f(1) = 5$; CU on $(-\infty, 0)$ and $(2/3, \infty)$; CD on $(0, 2/3)$; abscissas of PI are 0 and $2/3$.

5 Max: $f(0) = 0$ (by first derivative test); min: $f(-\sqrt{2}) = f(\sqrt{2}) = -8$. CU on $(-\infty, -\sqrt{6/5})$ and $(\sqrt{6/5}, \infty)$; CD on $(-\sqrt{6/5}, \sqrt{6/5})$; abscissas of PI are $\pm\sqrt{6/5}$.

7 Max: $f(0) = 1$; min: $f(-1) = f(1) = 0$. CU on $(-\infty, -1/\sqrt{3})$ and $(1/\sqrt{3}, \infty)$; CD on $(-1/\sqrt{3}, 1/\sqrt{3})$; abscissas of PI are $\pm 1/\sqrt{3}$.

1

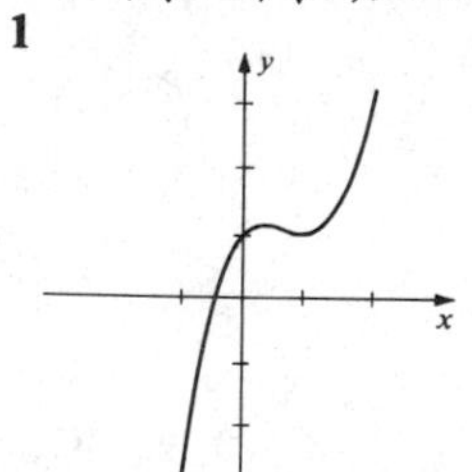

3

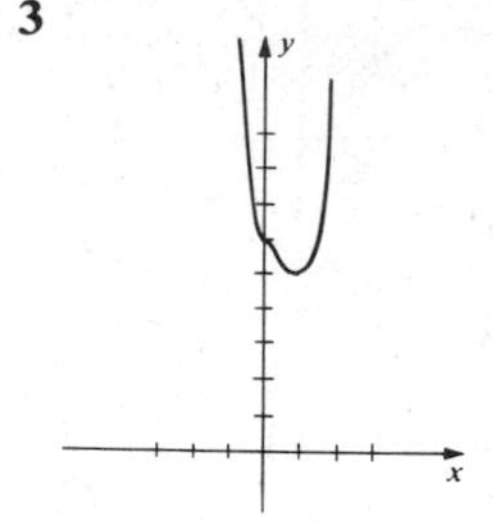

5

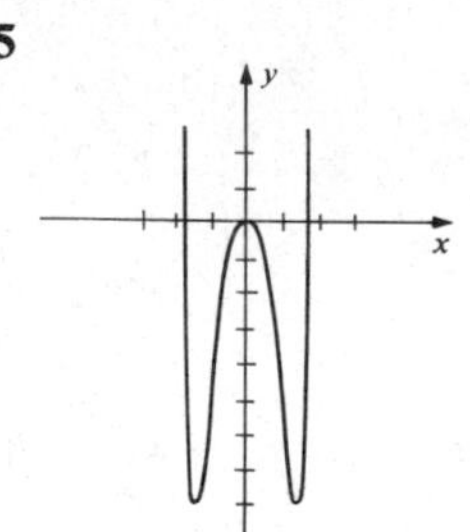

7

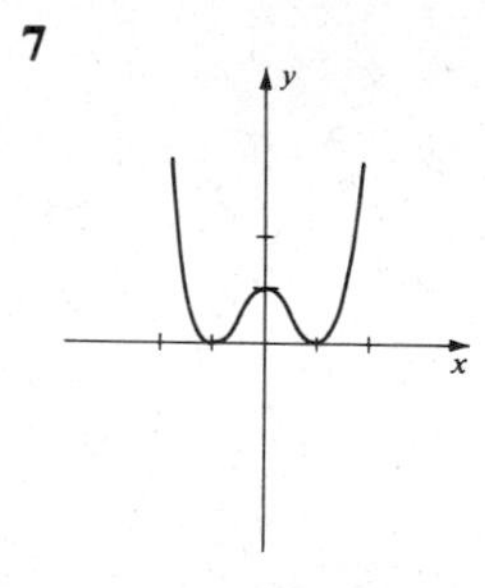

9 No local extrema; CU on $(-\infty, 0)$; CD on $(0, \infty)$; PI $(0, -1)$

11 No max or min. CU on $(-\infty, -3)$ and $(3, \infty)$; CD on $(-3, 0)$ and $(0, 3)$; abscissas of PI are ± 3.

13 Min: $f(-1) = -1/2$; max: $f(1) = 1/2$. CU on $(-\sqrt{3}, 0)$ and $(\sqrt{3}, \infty)$; CD on $(-\infty, -\sqrt{3})$ and $(0, \sqrt{3})$; abscissas of PI are 0, $\pm\sqrt{3}$.

15 Max: $f(-4/3) \approx 7.27$; min: $f(0) = 0$; CD on $(-\infty, 0)$ and $(0, 2/3)$; CU on $(2/3, \infty)$; PI is $(2/3, 10\sqrt[3]{12}/3)$.

9

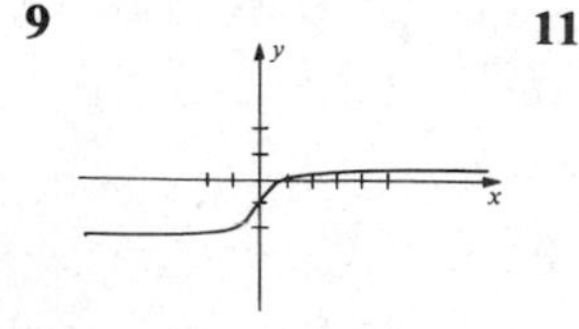

11

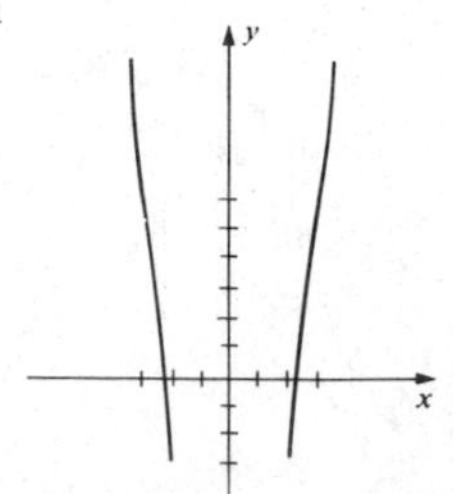

13

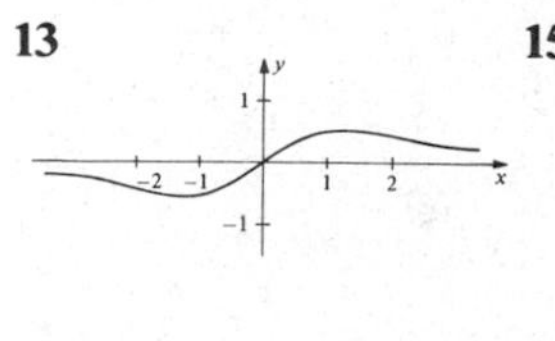

15

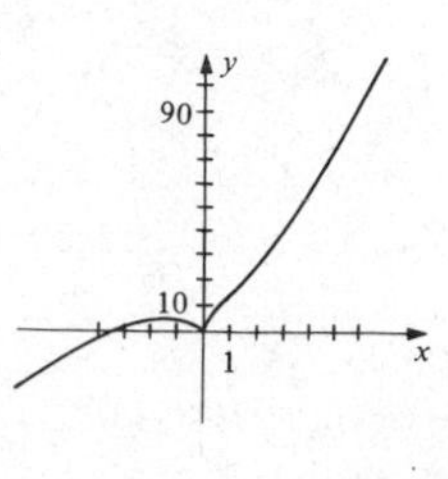

17 Min: $f(-2) \approx -7.55$. CU on $(-\infty, 0)$ and $(4, \infty)$; CD on $(0, 4)$; abscissas of PI are 0 and 4.

17

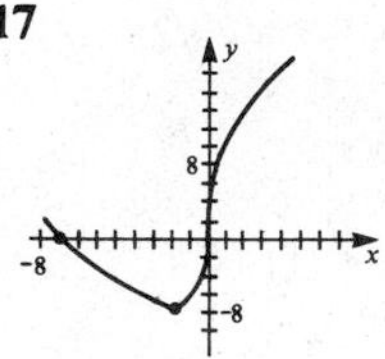

19

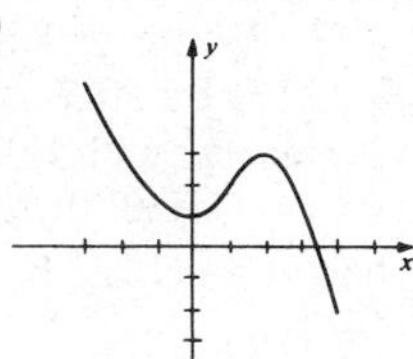

21

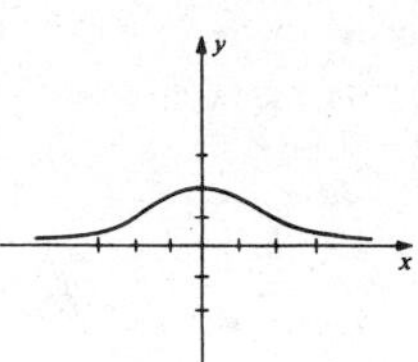

23

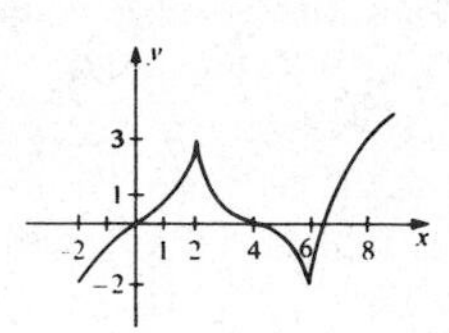

25

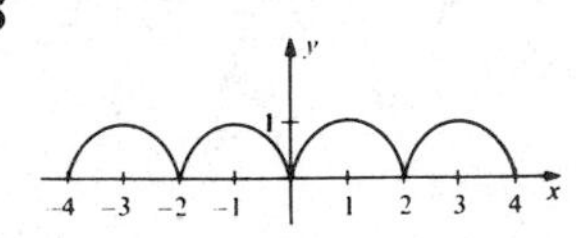

27 If $f(x) = ax^2 + bx + c$, then $f''(x) = 2a$. (a) CU if $a > 0$ (b) CD if $a < 0$

29 If f has degree n, then f'' has degree $n - 2$.

Exercises 4.5, page 175

1 5/2 **3** $-7/3$ **5** 1 **7** 0 **9** 5/2 **11** -1 **13** $y = 4$

15 Both limits equal the quotient of the coefficients of x^n in $f(x)$ and $g(x)$. If f has lower degree than g, then both limits are 0.

21 ∞; $-\infty$; $x = 4$; $y = 0$

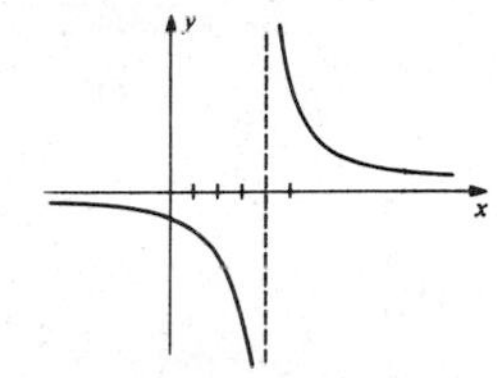

23 ∞; $-\infty$; $x = -5/2$; $y = 0$

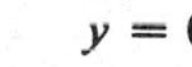

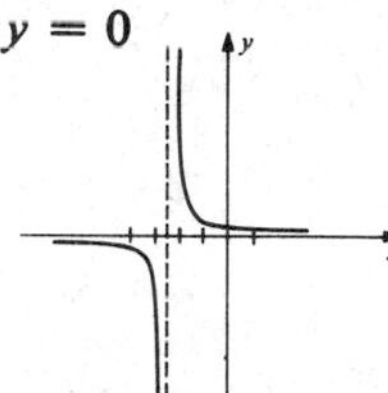

25 $-\infty$; $-\infty$; $x = -8$; $y = 0$

27 $-\infty, \infty$ for $a = -1$; $\infty, -\infty$ for $a = 2$; $x = -1$, $x = 2$, $y = 2$

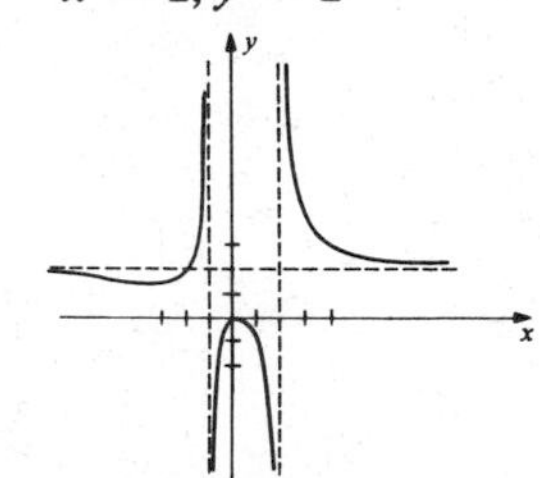

29 $\infty, -\infty$ for $a = 0$; ∞, ∞ for $a = 3$; $x = 0$, $x = 3$, $y = 0$

31 $x = -3$, $x = 1$, $y = 1$

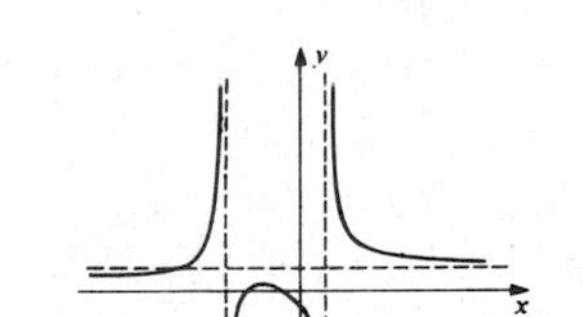

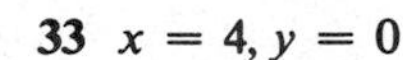

33 $x = 4$, $y = 0$

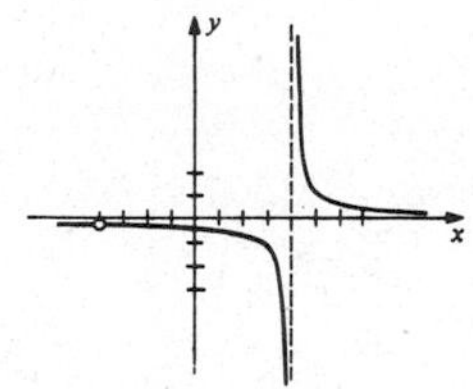

35 ∞ or $-\infty$ depending on whether the leading coefficients of f and g have the same sign or opposite signs, respectively. **37** $(-2, 0)$ **39** $(4/3, 3)$ and $(-4/3, 3)$

Exercises 4.6, page 185

1 20, -20 **3** Side of base = 2 ft; height = 1 ft **5** $5\sqrt{5}$ ft ≈ 11.2 ft

7 Length of base $= \sqrt{2}a$, height $= a/\sqrt{2}$ **9** $32\pi a^3/81$
11 Radius of base $=$ height $= 1/\sqrt[3]{\pi}$ **13** 3 in. **17** Width $= 2a/\sqrt{3}$, depth $= 2\sqrt{2}a/\sqrt{3}$
19 (1,2) **21** 500 **23** $d\sqrt[3]{s_1}/(\sqrt[3]{s_1} + \sqrt[3]{s_2})$ **25** 25 ft by 50/7 ft
27 There can be no cylindrical part. The tank must be spherical with radius $\sqrt[3]{75/\pi}$.
29 (a) Use $36\sqrt{3}/(2 + \sqrt{3}) \approx 16.71$ cm for the rectangle. (b) Use all the wire for the rectangle.
31 17.20 ft **33** width $= 12/(6 - \sqrt{3}) \approx 2.81$ ft; height $= (18 - 6\sqrt{3})/(6 - \sqrt{3}) \approx 1.78$ ft
35 37 **37** 16 in., 16 in., 32 in.

Exercises 4.7, page 193

1 (a) $16^{-2/3}$ cm/min (b) 36π cm^3/min (c) $(6)4^{1/3}\pi$ cm^2/min
3 The following are (beats/min)/sec: (a) 7 (b) 15 (c) 23
5 (a) 3200π cm^2/sec (b) 6400π cm^2/sec (c) 9600π cm^2/sec

In Exercises 7-15 the notation $[a, b]$ denotes the interval of time from $t = a$ to $t = b$.

7 $v(t) = 6t - 12$, $a(t) = 6$. The point moves to the left in $[0, 2]$ and to the right in $[2, 5]$.
9 $v(t) = 3t^2 - 9$, $a(t) = 6t$; to the right in $[-3, -\sqrt{3}]$; to the left in $[-\sqrt{3}, \sqrt{3}]$; to the right in $[\sqrt{3}, 3]$
11 $v(t) = 1 - 4/t^2$, $a(t) = 8/t^3$; to the left in $[1, 2]$; to the right in $[2, 4]$
13 $v(t) = 8t^3 - 12t$, $a(t) = 24t^2 - 12$; to the left in $[-2, -\sqrt{3/2}]$; to the right in $[-\sqrt{3/2}, 0]$; to the left in $[0, \sqrt{3/2}]$; to the right in $[\sqrt{3/2}, 2]$
15 $v(t) = 3t^2$, $a(t) = 6t$; to the right in $[0, 4]$
17 $v(t) = 144 - 32t$, $a(t) = -32$; $v(3) = 48$, $a(3) = -32$. Maximum height is 324 ft. Strikes ground at $t = 9$. **19** $a(2) = -6$, $a(3) = 6$; $v(10/3) = 44/3$
21 -0.12 units/ft **23** $dr/dc = 1/2\pi$ **25** 9/5
27 $dT/dl = C/\sqrt{l}$, where C is a constant
29 $V = 2400s - 200s^2 + 4s^3$; $dV/ds = 2400 - 400s + 12s^2$

Exercises 4.8, page 198

1 $-3\sqrt{336}/8 \approx -6.9$ ft/sec **3** $20/9\pi \approx 0.71$ ft/min **5** 64/11 ft/sec, 20/11 ft/sec
7 $-7442\pi \approx -23{,}368$ in.^{3}/hr **9** 10/3 ft/sec **11** Increasing at a rate of 5 in.^{3}/min
13 $15\sqrt{3}/32 \approx 0.8$ ft/min **15** 10 units/sec **17** $-4\sqrt{\sqrt{3}/600} \approx -0.215$ cm/min
19 π m/sec **21** 11/1600 ohms/sec **23** $13.37/112\pi \approx 0.38$ft/min **25** 64ft/sec
27 $(6 + \sqrt{2})180/\sqrt{10 + 3\sqrt{2}} \approx 353.6$ mph **29** $-27/25\pi \approx 0.344$ in./hr

Exercises 4.9, page 206

1 $3x^3 - 2x^2 + 3x + C$ **3** $x^4/2 - x^3/3 + 3x^2/2 - 7x + C$ **5** $-(1/2x^2) + 3/x + C$
7 $2x^{3/2} + 2x^{1/2} + C$ **9** $9x^{2/3} - (1/8)x^{4/3} + 7x + C$
11 $(8/9)x^{9/4} + (24/5)x^{5/4} - x^{-3} + C$ **13** $3x^3 - 3x^2 + x + C$
15 $(24/5)x^{5/3} - (15/2)x^{2/3} + C$ **17** $(10/9)x^{9/5} + C$ **19** $x^3/3 + x^2/2 + x + C, (x \neq 1)$
21 $f(x) = 4x^3 - 3x^2 + x + 3$ **23** $f(x) = (2/3)x^3 - x^2/2 - 8x + 65/6$
25 $s(t) = -t^3 + t^2 - 5t + 4$ **27** $s(t) = -16t^2 + 1600t$. Maximum height is $s(50) = 40{,}000$.
29 (a) $s(t) = -16t^2 - 16t + 96$ (b) $t = 2$ (c) 80 ft/sec **33** $a(t) = 10$ ft/sec^2
35 $F = (9/5)C + 32$ **37** $V = 2t^{3/2} + (1/8)t^2 + 2$

Exercises 4.10, page 213

1 (a) 806 (b) $c(x) = (800/x) + 0.04 + 0.0002x$; $C'(x) = 0.04 + 0.0004x$; $c(100) = 8.06$; $C'(100) = 0.08$ (c) 0.84 (d) $x = 0$
3 (a) 11,250 (b) $c(x) = (250/x) + 100 + 0.001x^2$; $C'(x) = 100 + 0.003x^2$; $c(100) = 112.50$; $C'(100) = 130$ (c) 107.5 (d) $x = 0$ **5** 16

Answers for 7-13 indicate functional values at x.

7 $(800/3) - (4/3)x$; $-4/3$; $(800/3)x - (4/3)x^2$; $(800/3) - (8/3)x$; 100 units at 133.33
9 $4 - \sqrt{x}$; $-1/2\sqrt{x}$; $4x - x^{3/2}$; $4 - (3/2)\sqrt{x}$; 7 units at 1.35
11 (a) $(800 - 3x)/\sqrt{400 - x}$ (b) 6158.39 (c) $2\sqrt{400 - x}$
13 (a) $300 - 3x^2$ (b) 2000 (c) $300 - x^2$
15 (a) $-1/10$ (b) $50x - (x^2/10)$ (c) $48x - (x^2/10) - 10$ (d) $48 - (x/5)$ (e) 5750 (f) 2
17 (a) $1800x - 2x^2$ (b) $1799x - 2.01x^2 - 1000$ (c) 100 (d) \$158,800
19 3990 units, \$15,420.10 **21** $p(x) = 150 - (x/10)$; $R(x) = 150x - (x^2/10)$
25 $C(x) = 20x - (0.015/2)x^2 + 5.0075$; 986.26 **27** $R(x) = x^3/3 - 3x^2 + 15x$, $g'(x) = (2x/3) - 3$

4.11 Review Exercises, page 216

1 Tangent: $y + 3 = -(30/23)(x - 2)$; normal: $y + 3 = (23/30)(x - 2)$
3 Horizontal at $x = -3/2$; vertical at $x = -1$, $x = -2$ **5** $(\sqrt{61} - 1)/3$
7 Local maximum $f(0) = 1$; increasing on $(-\infty, 0]$, decreasing on $[0, \infty)$
9 Local maximum $f((4 + \sqrt{7})/3)$ and local minimum $f((4 - \sqrt{7})/3)$; increasing on $[(4 - \sqrt{7})/3, (4 + \sqrt{7})/3]$, decreasing on $(-\infty, (4 - \sqrt{7})/3]$ and $[(4 + \sqrt{7})/3, \infty)$; abscissa of point of inflection is $x = 4/3$; concave upward on $(-\infty, 4/3)$ and concave downward on $(4/3, \infty)$.
11 Local maximum $f(\sqrt[3]{20}) = 400$; increasing on $(-\infty, \sqrt[3]{20}]$, decreasing on $[\sqrt[3]{20}, \infty)$. Abscissas of points of inflection are 0 and 2. Concave upward on (0, 2); concave downward on $(-\infty, 0)$ and $(2, \infty)$.

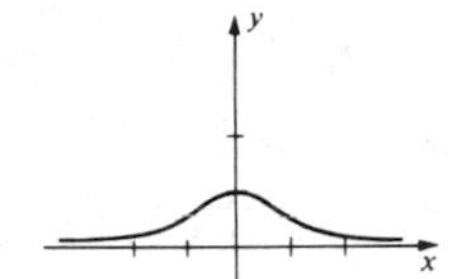

9

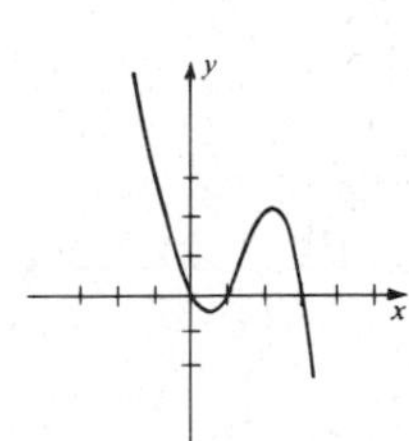

11

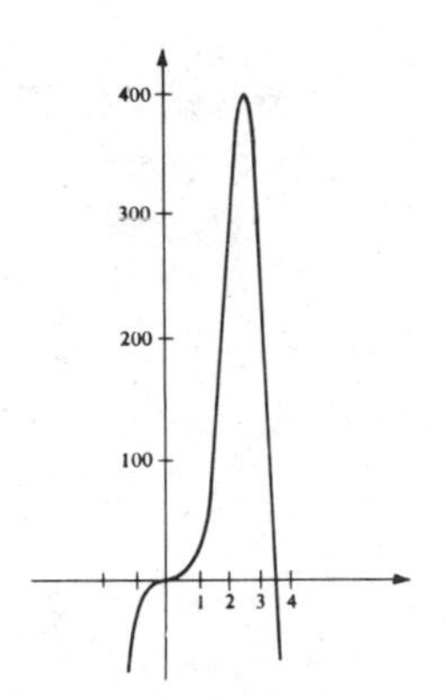

13 $s(t) = -16t^2 - 30t + 900$; $v(5) = -190$ ft/sec; $t = 15(-1 + \sqrt{65})/16 \approx 6.6$ sec
15 $(x^2/2)(x^4 + x^2 - 1) + C$ **17** $(10/13)x^{13/5} - (10/21)x^{21/10} + C$
19 $f(x) = (9/28)x^{7/3} - (5/2)x^2 + (25/4)x - 169/14$ **21** $\sqrt{2}a$
23 Radius of semicircle is $1/8\pi$ mi, length of rectangle is 1/8 mi. **25** $-18/5\pi \approx -1.15$ ft/min
27 $20\sqrt{5} \approx 44.7$ mi/hr **29** $dp/dv = -p/v$ **31** 3/2 **33** 0 **35** $-\infty$ **37** $-\infty$
39 $y = 1/3$, $x = 5/3$, $x = -5/3$

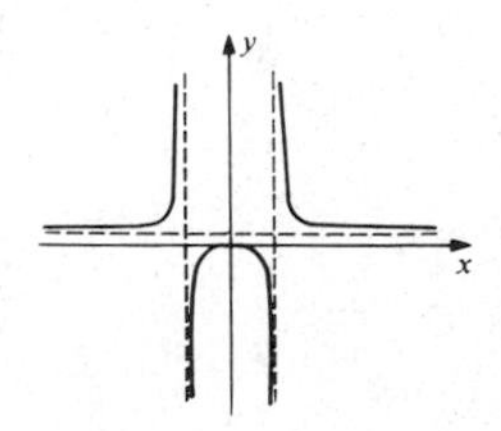

CHAPTER 5

Exercises 5.1, page 227

1 -5 **3** 34 **5** 40 **7** 510 **9** 500 **13** $(2n^3 + 12n^2 + 40n)/6$
15 $(4n^3 - 12n^2 + 11n)/3$ **17** 28 **19** 125/3 **21** 78 **23** 18 **25** 19/4

Exercises 5.2, page 234

1 1.1, 1.5, 1.1, 0.4, 0.9; $\|P\| = 1.5$ **3** 0.3, 1.7, 1.4, 0.5, 0.1; $\|P\| = 1.7$
5 (a) 40 (b) 32 (c) 36 **7** 49/4 **9** 79 **11** $\int_{-1}^{2} (3x^2 - 2x + 5)\,dx$
13 $\int_0^4 2\pi x(1 + x^3)\,dx$ **15** $-14/3$ **17** 30 **19** 25 **21** $9\pi/4$ **23** 625/4
25 Any unbounded function. For example, $f(x) = 1/x$, $g(x) = 1/\sqrt{1 - x}$, $h(x) = \csc x$. There is no contradiction since the interval in (5.19) is closed.

Exercises 5.3, page 240

1 30 **3** -12 **5** 2 **7** 78 **9** $-291/2$ **11** $-14\sqrt{5}/3$ **13** 215/6
19 $\int_{-3}^{1} f(x)\,dx$ **21** $\int_h^{c+h} f(x)\,dx$ **29** *Hint*: $-|f(x)| \leqslant f(x) \leqslant |f(x)|$

Exercises 5.4, page 244

1 $\sqrt{3}$ **3** 3 **5** $\sqrt[3]{15/4}$ **7** $\sqrt{50} - 3$ **9** $a\sqrt{16 - \pi^2}/4$ **11** If $f(x) = k$ on $[a, b]$, then $\int_a^b f(x)\,dx = k(b - a) = f(c)(b - a)$ for every c in $[a, b]$, since $f(c) = k$.

Exercises 5.5, page 250

1 -18 **3** 265/2 **5** 5 **7** 31/256 **9** 20/3 **11** 352/5 **13** $-37/6$
15 13/3 **17** $-7/2$ **19** 0 **21** 10/3 **23** 53/2 **25** $8\sqrt{3} + 16$ **29** $1/x$
31 6 **37** $z = 16/9$ **39** $\sqrt[3]{5/4}$ **41** (a) 3/2

Exercises 5.6, page 257

1 $(3x + 1)^5/15 + C$ **3** $\frac{2}{9}(t^3 - 1)^{3/2} + C$ **5** $-1/(4(x^2 - 4x + 3)^2) + C$
7 $-\frac{3}{8}(1 - 2s^2)^{2/3} + C$ **9** $\frac{2}{5}(\sqrt{u} + 3)^5 + C$ **11** 14/3 **13** 0 **15** 1/3
17 $(1/7)x^7 + (3/5)x^5 + x^3 + x + C$ **19** $(5/12)(8x + 5)^{3/2} + C$ **21** 5/36
23 (a) $(x + 4)^3/3 + C_1$ (b) $x^3/3 + 4x^2 + 16x + C_2$, where $C_2 = C_1 + 64/3$
25 $(2/3)(\sqrt{x} + 3)^3 + C_1$, $(2/3)x^{3/2} + 6x + 18x^{1/2} + C_2$, $18 + C_1 = C_2$
29 $1/\sqrt{x^3 + x + 5}$ **31** 1 **33** 14/3 **35** $z = \sqrt{3}$ **37** 544/225

Exercises 5.7, page 265

1 (a) 1.41 (b) 1.39 **3** (a) 0.88 (b) 0.88 **5** (a) 0.39 (b) 0.39
9 (a) 8.65 (b) 8.59

5.8 Review Exercises, page 266

1 70 **3** 11/4 **5** -10 **7** 3/5 **9** 1/6
11 $-(1/16)(1 - 2x^2)^4 + C$ **13** $\sqrt{8} - \sqrt{3} \approx 1.10$ **15** $-2/(1 + \sqrt{x}) + C$
17 $3x - x^2 - (5/4)x^4 + C$ **19** 52/9 **21** $(1/6)(4t^2 + 2t - 7)^3 + C$
23 $-x^{-2} + 3x^{-1} + C$ **25** $\sqrt[5]{y^4 + 2y^2 + 1} + C$ **27** 0 **29** 416/15

CHAPTER 6

Exercises 6.1, page 278

1 17/6

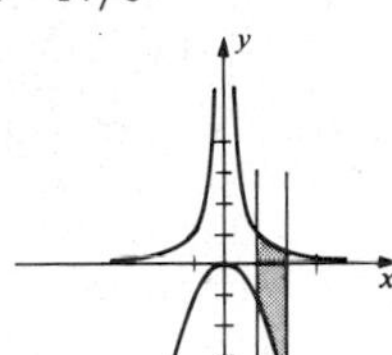

3 33/2

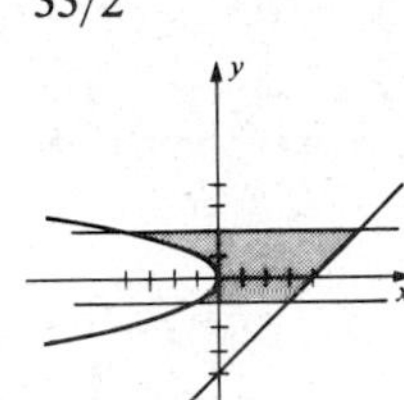

5 32/3

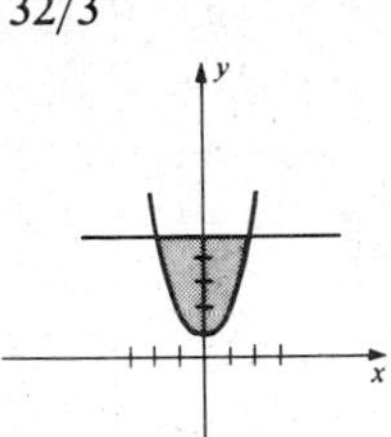

7 32/3

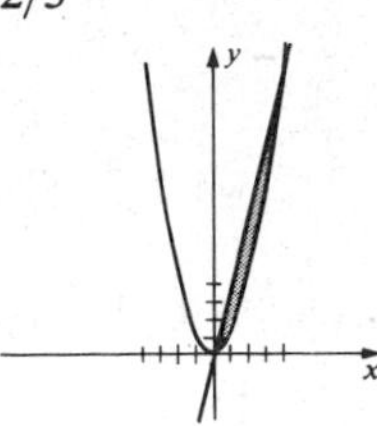

9 9/2

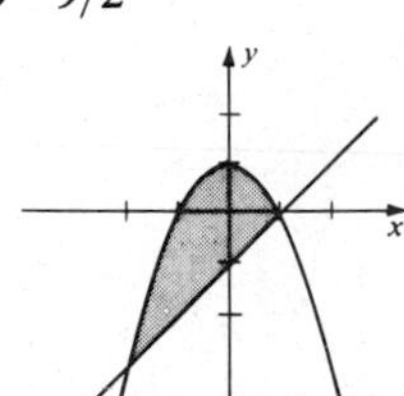

11 $8\sqrt{3}$

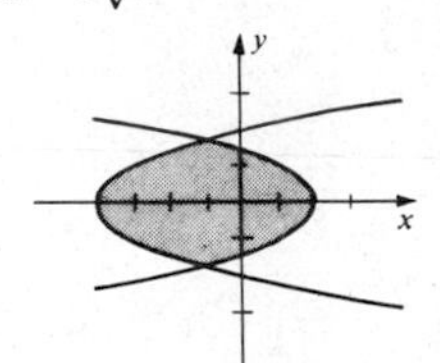

13 2

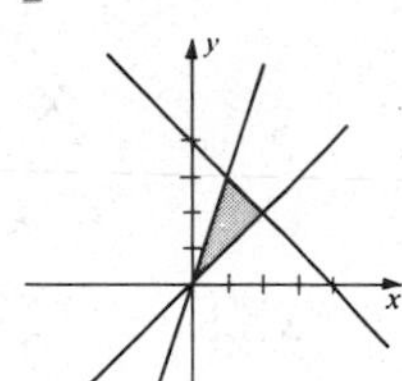

15 1/2

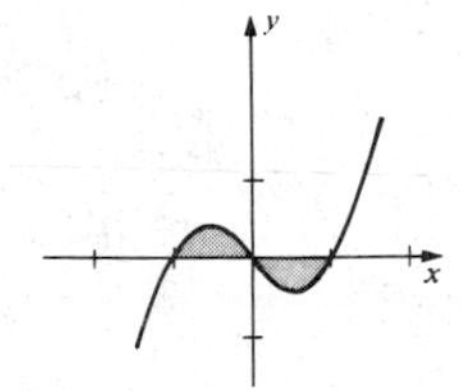

17 8

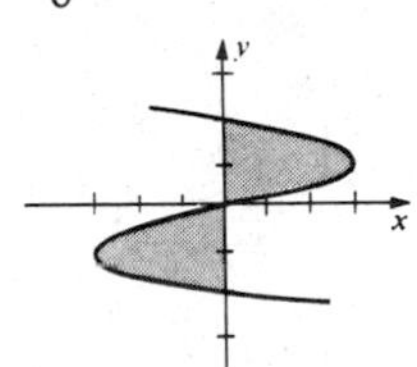

19 16/3

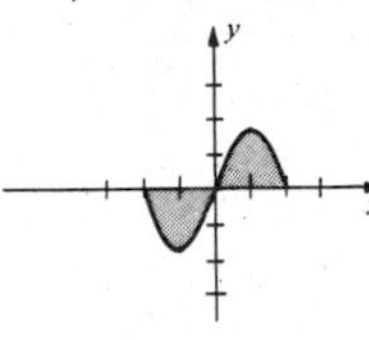

21 Let $f(x) = \sqrt{9 - (x - 4)^2}$ on [1, 7]. Then $A = \lim_{\|P\| \to 0} \sum_i 2f(w_i)\Delta x_i$. Since the region is bounded by a circle of radius 3, $A = 9\pi$.

23 The limit equals the area of the region under the graph of $y = 4x + 1$ from 0 to 1. $A = 3$.

25 The limit equals the area of the region to the left of the graph of $x = 4 - y^2$ and to the right of the y-axis from $y = 0$ to $y = 1$. $A = 11/3$.

27 The area A of $\{(x,y)\colon 2 \leqslant x \leqslant 5, 0 \leqslant y \leqslant x(x^2 + 1)^{-2}\}$, $A = 21/260$

29 The area A of $\{(x,y)\colon 1 \leqslant y \leqslant 4, 0 \leqslant x \leqslant (5 + \sqrt{y})/\sqrt{y}\}$, $A = 13$

Exercises 6.2, page 287

1 $2\pi/3$

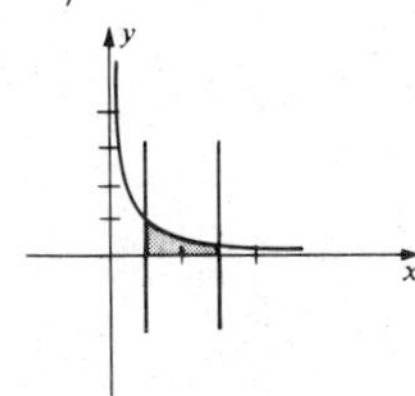

3 2π

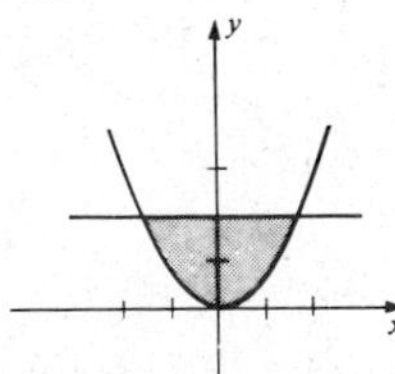

5 $512\pi/15$

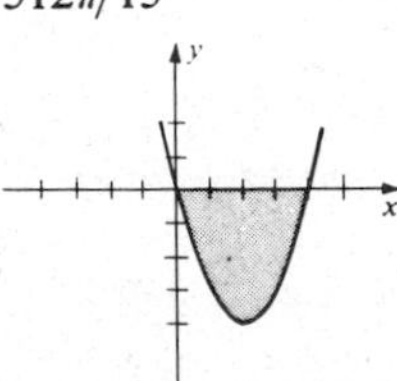

7 $64\pi/15$

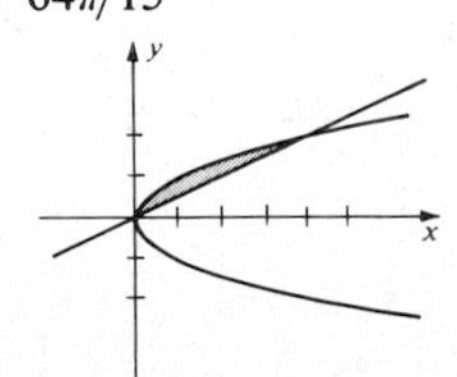

9 $64\sqrt{2}\,\pi/3$

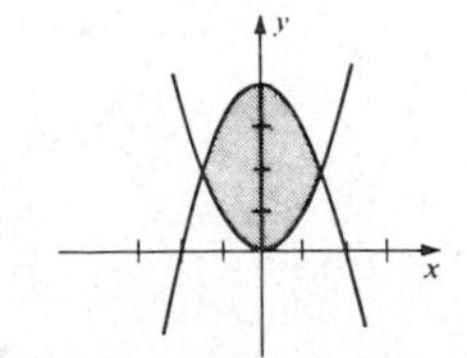

11 $72\pi/5$

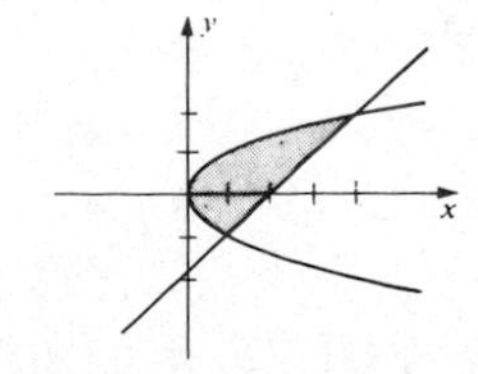

13 (a) $512\pi/15$ (b) $832\pi/15$ (c) $128\pi/3$
15 $V = \pi\int_{-2}^{0}[(8-4x)^2 - (8-x^3)^2]\,dx + \pi\int_0^2[(8-x^3)^2 - (8-4x)^2]\,dx$
17 $V = \pi\int_2^3[(y-1)^2 - (2-\sqrt{3-y})^2]\,dy$
19 $V = \pi\int_{-1}^{1}[(5+\sqrt{1-y^2})^2 - (5-\sqrt{1-y^2})^2]\,dy = 20\pi\int_{-1}^{1}\sqrt{1-y^2}\,dy$
21 $V = \frac{1}{3}\pi r^2 h$ **23** $V = \frac{1}{3}\pi h(r_1^2 + r_2^2 + r_1 r_2)$
25 The limit equals the volume of the solid obtained by revolving the region between $y = x^2$ and $y = x^3$, $0 \leqslant x \leqslant 1$, about the x-axis. $V = 2\pi/35$.

Exercises 6.3, page 293

1 $128\pi/5$

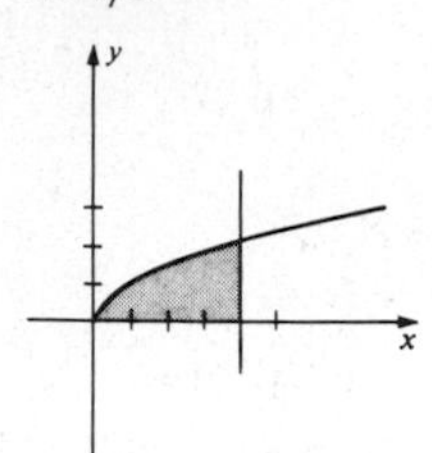

3 $24\pi/5$

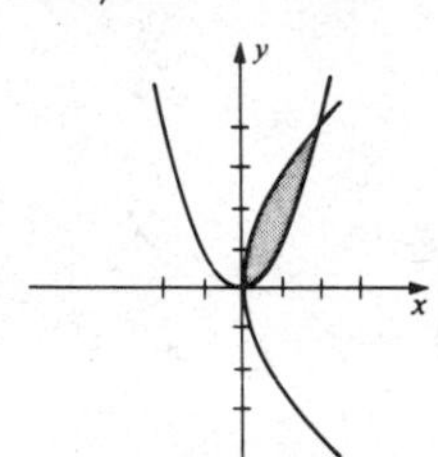

5 $135\pi/2$

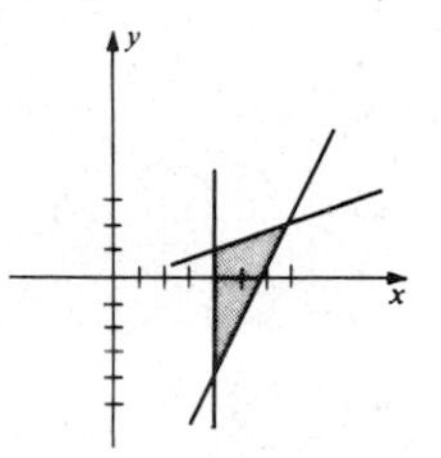

7 $512\pi/5$

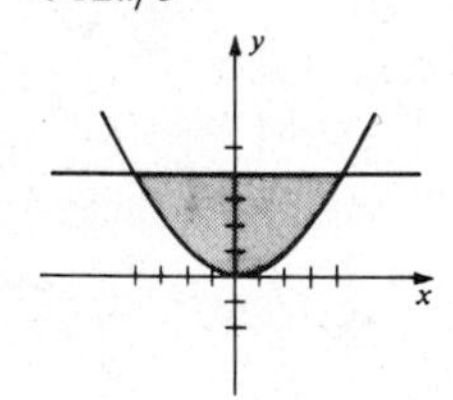

9 72π

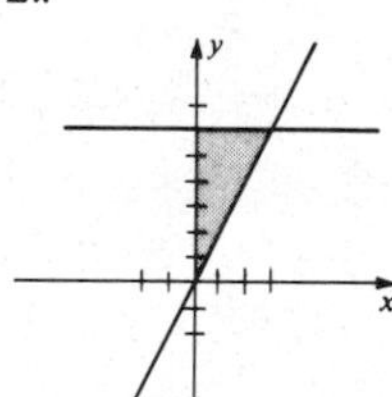

11 (a) 16π (b) $64\pi/3$ **15** $V = 2\pi\int_{-8}^{0}(8-y)(y/4 - y^{1/3})\,dy + 2\pi\int_0^8(8-y)(y^{1/3} - y/4)\,dy$
17 $V = 2\pi\int_0^1(2-x)(x-x^2)\,dx$ **19** $V = 4\pi\int_{-1}^{1}(5-x)\sqrt{1-x^2}\,dx$
25 The limit of the sum equals the volume of the solid obtained by revolving the region between $y = x$ and $y = x^2$, $0 \leqslant x \leqslant 1$, about the y-axis. The volume is $\pi/6$.

Exercises 6.4, page 297

1 $16a^3/3$ **3** $128/15$ **5** $2a^2h/3$ **7** $128\pi/15$ **9** $2a^3/3$ **11** $\pi a^2 b/2$ **13** 4

Exercises 6.5, page 304

1 (a) 128/3 in.-lb (b) 64/3 in.-lb **3** $F_2 = 3F_1$ **5** 2250 ft-lb **7** 31,050 ft-lb
9 (a) $81(62.5)\pi/2 \approx 7952$ ft-lb (b) $(189)(62.5)\pi/2 \approx 18{,}555$ ft-lb **11** 500 ft-lb
13 (a) $3k/10$ dyne-cm (b) $9k/40$ dyne-cm (k a constant) **15** $575(1/2 - 1/\sqrt[5]{40}) \approx 12.55$
17 $W = gm_1 m_2 h/4000(4000+h)$ **19** 36.85

Exercises 6.6, page 309

1 (a) 31.25 lb (b) 93.75 lb **3** (a) $62.5/\sqrt{3}$ lb (b) $62.5\sqrt{3}/24$ lb **5** 320 lb
7 303,356.25 lb **9** $(592)(62.5)/3$ lb $\approx 12{,}333.3$ lb **11** 3200/3 lb **13** 4500 lb

Exercises 6.7, page 316

1 $(4 + (16/81))^{3/2} - (1 + (16/81))^{3/2} \approx 7.29$ **3** $(8/27)(10^{3/2} - (13^{3/2}/8)) \approx 7.63$
7 13/12 **9** 353/240 **11** $s = \int_0^2\sqrt{(53/4) - 21y^2 + 9y^4}\,dy$ **13** 6
15 $s = (x^{2/3} + \frac{4}{9})^{3/2} - (1 + \frac{4}{9})^{3/2}$; $\Delta s = (1/27)[(9(1.1)^{2/3} + 4)^{3/2} - 13^{3/2}] \approx 0.1196$;
$ds = \sqrt{13}/30 \approx 0.1202$ **17** $ds = \sqrt{17}(0.1) \approx 0.412$, $d(A,B) = \sqrt{0.1781} \approx 0.422$
19 8.61

Exercises 6.8, page 322

1 (a) \$49.54 (b) \$96.30 (c) \$137.50 (d) \$170.37
3 $10^{3/5}$ yr $\approx$ 3.98 yr $\approx$ 4 yr (to the nearest month)
5 (a) $9((601)^{2/3} - 1) \approx 632$ min (b) $9((301)^{2/3} - 1) \approx 395$ min
7 Minutes: (a) 18.16 (b) 66.22 (c) 115.24 (d) 197.12

6.9 Review Exercises, page 323

1 $64/3$ 3 $5\sqrt{5}/6$ 5 10π 7 $3\pi/5$

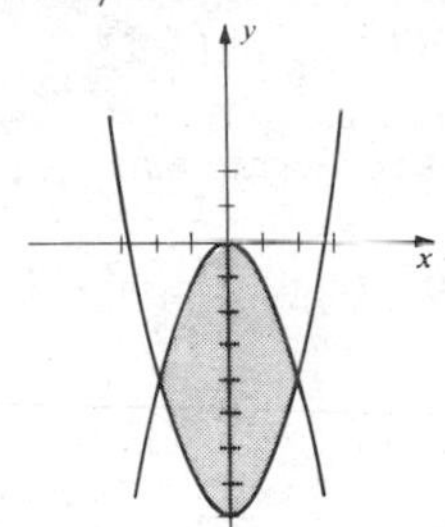

9 (a) $1152\pi/5$ (b) 54π (c) $1728\pi/5$ 11 $(37^{3/2} - 10^{3/2})/27 \approx 7.16$
13 $432(62.5)\pi$ ft-lb $\approx$ 84,823 ft-lb 15 6,000 lb 17 $\pi/5$
19 Two possibilities exist. The solid could be obtained by revolving the region under $y = x^2$ from $x = 0$ to $x = 1$ around the x-axis, or by revolving the region under $y = \frac{1}{2}x^3$ from $x = 0$ to $x = 1$ around the y-axis.

CHAPTER 7

Exercises 7.2, page 333

1 $V(0,0)$; $F(0,-3)$; $y = 3$ 3 $V(0,0)$; $F(-3/8,0)$; $x = 3/8$
5 $V(0,0)$; $F(0,1/32)$; $y = -1/32$ 7 $V(-1,0)$; $F(2,0)$; $x = -4$
9 $V(2,-2)$; $F(2,-7/4)$; $y = -9/4$ 11 $V(-4,2)$; $F(-7/2,2)$; $x = -9/2$
13 $V(-5,-6)$; $F(-5,-97/16)$; $y = -95/16$ 15 $V(0,1/2)$; $F(0,-9/2)$; $y = 11/2$
17 Let $y' = 2ax + b = 0$ to obtain the abscissa $x = -b/2a$ of the vertex. Given $x = ay^2 + by + c$ let $2ay + b = 0$ to obtain the ordinate $-b/2a$ of the vertex.
19 $y^2 = 8x$ 21 $(x - 6)^2 = 12(y - 1)$ 23 $3x^2 = -4y$ 25 9/16 ft from the vertex
27 $y = 2x^2 - 3x + 1$ 31 (a) 8/3 (b) 2π (c) $16\pi/5$ 33 200/3 pounds

Exercises 7.3, page 339

1 $V(\pm 3,0)$; $F(\pm\sqrt{5},0)$ 3 $V(0,\pm 4)$; $F(0,\pm 2\sqrt{3})$ 5 $V(0,\pm\sqrt{5})$; $F(0,\pm\sqrt{3})$
7 $V(\pm 1/2,0)$; $F(\pm\sqrt{21}/10,0)$ 9 $x^2/64 + y^2/39 = 1$ 11 $4x^2/9 + y^2/25 = 1$
13 $8x^2/81 + y^2/36 = 1$ 15 $\{(2,2),(4,1)\}$ 17 $2\sqrt{21}$ feet 19 $y - 3 = (5/6)(x + 2)$
21 $(\pm\sqrt{3},3/2)$ 23 $(4/3)\pi ab^2$ 25 (a) 864 (b) $216\sqrt{3}$ 27 $2a/\sqrt{2}$ and $2b/\sqrt{2}$

Exercises 7.4, page 345

1 $V(\pm 3,0)$, $F(\pm\sqrt{13},0)$, $y = \pm 2x/3$ 3 $V(0,\pm 3)$, $F(0,\pm\sqrt{13})$, $y = \pm 3x/2$
5 $V(0,\pm 4)$, $F(0,\pm 2\sqrt{5})$, $y = \pm 2x$ 7 $V(\pm 1,0)$, $F(\pm\sqrt{2},0)$, $y = \pm x$
9 $V(\pm 5,0)$, $F(\pm\sqrt{30},0)$, $y = \pm(\sqrt{5}/5)x$ 11 $V(0,\pm\sqrt{3})$, $F(0,\pm 2)$, $y = \pm\sqrt{3}x$
13 $15y^2 - x^2 = 15$
15 $x^2/9 - y^2/16 = 1$ 17 $y^2/21 - x^2/4 = 1$ 19 $x^2/9 - y^2/36 = 1$
21 $\{(0,4),(8/3,20/3)\}$ 23 Conjugate hyperbolas have the same asymptotes.
25 $(y - 1) = (-4/5)(x + 2)$
27 $(\pm 2\sqrt{2},-6)$ 29 $5x - 6y - 16 = 0$ 31 $\pi b^2[\sqrt{a^2 + b^2}(b^2 - 2a^2) + 2a^3]/3a^2$

Exercises 7.5, page 351

1 Parabola; vertex $(-1, 5)$, focus $(1, 5)$
3 Ellipse; center $(4, 2)$, vertices $(1, 2)$ and $(7, 2)$, end points of minor axis $(4, 4)$ and $(4, 0)$
5 Hyperbola; center $(-5, 1)$, vertices $(-9, 1)$ and $(-1, 1)$, end points of conjugate axis $(-5, 6)$ and $(-5, -4)$
7 Ellipse; center $(-5, 3)$, vertices $(-5, 2)$ and $(-5, 4)$, end points of minor axis $(-11/2, 3)$ and $(-9/2, 3)$
9 Hyperbola; center $(5, 0)$, vertices $(5, 4)$ and $(5, -4)$, end points of conjugate axis $(-5, 0)$ and $(15, 0)$
11 Ellipse; center $(-3, 1)$, vertices $(-7, 1)$ and $(1, 1)$, end points of minor axis $(-3, 4)$ and $(-3, -2)$
13 Ellipse; center $(5, 2)$, vertices $(5, 7)$ and $(5, -3)$, end points of minor axis $(3, 2)$ and $(7, 2)$
15 Parabola; vertex $(-1/2, -25/16)$, focus $(-1/2, 7/16)$
17 Hyperbola; center $(-2, -5)$, vertices $(-2, -2)$ and $(-2, -8)$, end points of conjugate axis $(4, -5)$ and $(-8, -5)$
19 Hyperbola; center $(6, 2)$, vertices $(6, 4)$ and $(6, 0)$, end points of conjugate axis $(0, 2)$ and $(12, 2)$

21

(5, −4)

23

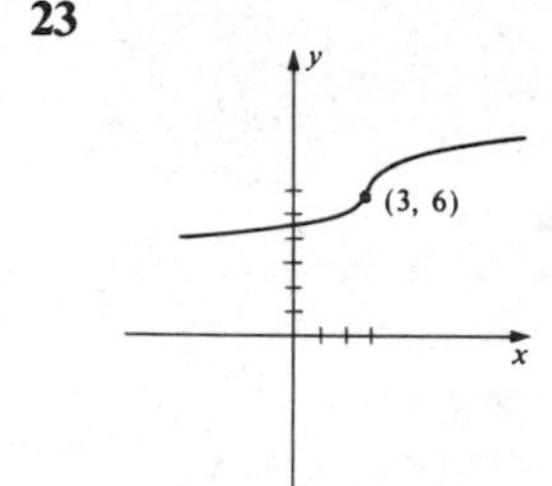

25

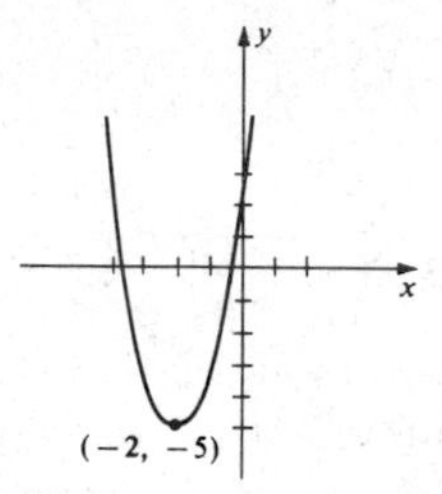

27 $(x-h)^2/a^2 - (y-k)^2/b^2 = 1$ **29** $8x - y - 26 = 0$

Exercises 7.6, page 356

The following answers contain equations in x' and y' resulting from a rotation of axis.

1 Ellipse, $(x')^2 + 16(y')^2 = 16$ **3** Hyperbola, $4(x')^2 - (y')^2 = 1$
5 Ellipse, $(x')^2 + 9(y')^2 = 9$ **7** Parabola, $(y')^2 = 4(x' - 1)$
9 Hyperbola, $2(x')^2 - (y')^2 - 4y' - 3 = 0$ **11** Parabola, $(x')^2 - 6x' - 6y' + 9 = 0$
13 Ellipse, $(x')^2 + 4(y')^2 - 4x' = 0$
15 Sketch of proof: It can be shown that in (7.28) and (7.29), $B^2 - 4AC = B'^2 - 4A'C'$. For a suitable rotation of axes we obtain $B' = 0$ and (7.29) has the form $A'x'^2 + C'y'^2 + D'x' + E'y' + F' = 0$. Except for degenerate cases, the graph of the latter equation is an ellipse, hyperbola, or parabola if $A'C' > 0$, $A'C' < 0$, or $A'C' = 0$, respectively. However, if $B' = 0$, then $B^2 - 4AC = -4A'C'$ and hence the graph is an ellipse, hyperbola, or parabola if $B^2 - 4AC < 0$, $B^2 - 4AC > 0$, or $B^2 - 4AC = 0$, respectively.

7.7 Review Exercises, page 357

1 Parabola; $F(16, 0)$; $V(0, 0)$ **3** Ellipse; $F(0, \pm\sqrt{7})$; $V(0, \pm 4)$
5 Hyperbola; $F(\pm 2\sqrt{2}, 0)$; $V(\pm 2, 0)$ **7** Parabola; $F(0, -9/4)$; $V(0, 4)$
9 Hyperbola: vertices $(-5,5)$ and $(-3,5)$; foci $(-4 \pm \sqrt{10}/3, 5)$ **11** $9y^2 - 49x^2 = 441$
13 $x^2 = -40y$ **15** $4x^2 + 3y^2 = 300$ **17** $y^2 - 81x^2 = 36$
19 Ellipse; center $(-3, 2)$, vertices $(-6, 2)$ and $(0, 2)$, end points of minor axis $(-3, 0)$ and $(-3, 4)$
21 Parabola; vertex $(2, -4)$, focus $(4, -4)$
23 Hyperbola; center $(-4, 0)$, vertices $(-7, 0)$ and $(-1, 0)$, asymptotes $y = \pm(x+4)/3$
25 $2 \pm 2\sqrt{2}$ **27** $64\pi/3$
29 Parabola; $(y')^2 - 3x' = 0$ is an equation after a rotation of axis.

CHAPTER 8

Exercises 8.1, page 367

1 $9/(9x+4)$ **3** $-15/(2-3x)$ **5** $-3x^2/(7-2x^3)$
7 $(6x-2)/(3x^2-2x+1)$ **9** $(8x+7)/3(4x^2+7x)$ **11** $1+\ln x$
13 $\dfrac{1}{2x}\left(1+\dfrac{1}{\sqrt{\ln x}}\right)$ **15** $-\dfrac{1}{x}\left(\dfrac{1}{(\ln x)^2}+1\right)$ **17** $\dfrac{20}{5x-7}+\dfrac{6}{2x+3}$
19 $\dfrac{x}{x^2+1}-\dfrac{18}{9x-4}$ **21** $\dfrac{2x}{3(x^2-1)}-\dfrac{2x}{3(x^2+1)}$ **23** $\dfrac{1}{\sqrt{x^2-1}}$ **25** $\dfrac{(2x^2-1)y}{x(3y+1)}$
27 $(y^2-xy\ln y)/(x^2-xy\ln x)$ **29** $y=8x-15$
31 The graphs coincide if $x>0$; however, the graph of $y=\ln(x^2)$ contains points with negative abscissas.
33 $(1, 1), (2, 4+4\ln 2)$ **35** 0.6982
37 $v(t)=2t-4/(t+1)$; $a(t)=2+4/(t+1)^2$. Moves to the left in $[0, 1]$ and to the right in $[1, 4]$.
39 1, 1/5, 1/10, 1/100, 1/1000; the slope approaches 0; the slope increases without bound.

Exercises 8.2, page 374

1 $-5e^{-5x}$ **3** $6xe^{3x^2}$ **5** $e^{2x}/\sqrt{1+e^{2x}}$ **7** $e^{\sqrt{x+1}}/(2\sqrt{x+1})$
9 $(-2x^2+2x)e^{-2x}$ **11** $\dfrac{e^x(x^2+1)-2xe^x}{(x^2+1)^2}$ **13** $12(e^{4x}-5)^2e^{4x}$
15 $-e^{1/x}/x^2-e^{-x}$ **17** $\dfrac{(e^x+e^{-x})^2-(e^x-e^{-x})^2}{(e^x+e^{-x})^2}$ or $4/(e^x+e^{-x})^2$
19 $e^{2x}(\frac{1}{x}-2\ln x)$ **21** $\dfrac{e^x}{e^x+1}-\dfrac{e^x}{e^x-1}$ **23** $\dfrac{e^{2x}-e^{-2x}}{(e^{2x}+e^{-2x})\sqrt{\ln(e^{2x}+e^{-2x})}}$
25 1 (for $x>0$) **27** $\dfrac{3x^2-ye^{xy}}{xe^{xy}+6y}$ **29** $\dfrac{6x-e^y}{3y^2+xe^y}$
31 $y=(e+3)x-e-1$ **33** $c=\ln((e^b-e^a)/(b-a))$ **35** $(1/2, e)$
37 Min: $f(-1)=-e^{-1}$; decreasing on $(-\infty,-1)$, increasing on $(-1,\infty)$; CU on $(-2,\infty)$, CD on $(-\infty,-2)$; PI at $(-2,-2e^{-2})$
39 No local extrema; decreasing on $(-\infty,\infty)$; CU on $(-\infty,\infty)$, no PI

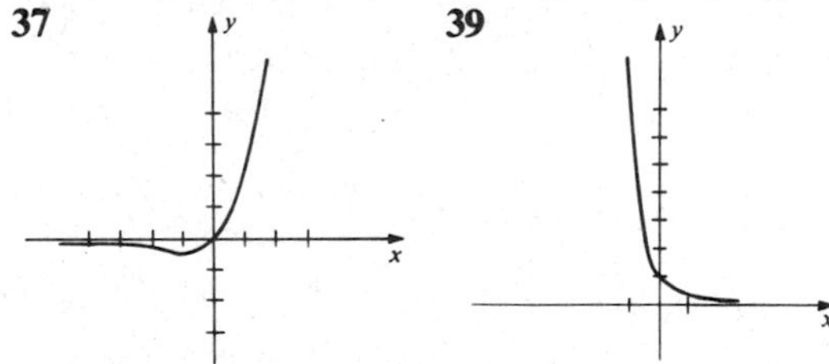

41 $0.03\,e\approx 0.082$, 2.800 **43** 1, 2 **45** 1, 2, −2

Exercises 8.3, page 381

1 $(1/2)\ln(x^2+1)+C$ **3** $(-1/5)\ln|7-5x|+C$ **5** $(1/2)\ln|x^2-4x+9|+C$
7 $(1/3)\ln|x^3+1|+C$ **9** $(1/2)(\ln 9-\ln 3)$, or $\ln\sqrt{3}$ **11** $4(\ln 6-\ln 5)$, or $4\ln(6/5)$
13 $(1/2)x^2+(1/5)e^{5x}+C$ **15** $(1/2)(\ln x)^2+C$ **17** $(-1/4)(e^{-12}-e^{-4})$
19 $2e^{\sqrt{x}}+C$ **21** $e^x+2x-e^{-x}+C$ **23** $\ln(e^x+e^{-x})+C$ **25** $-1/(x+1)+C$
27 $x^2+x-4\ln|x-3|+C$ **29** 4 **31** $\ln 2+e^{-2}-e^{-1}\approx 0.46$ **33** $\pi(1-e^{-1})$
35 $(1-e^{-6})/3$ **37** (a) $\int_0^1\sqrt{1+e^{2x}}\,dx$ (b) If $f(x)=\sqrt{1+e^{2x}}$, then $L\approx(1/10)[f(0)+2f(0.2)+2f(0.4)+2f(0.6)+2f(0.8)+f(1)]$ (c) $L\approx 2.0096$
39 $(5x+2)^2(6x+1)(150x+39)$ **41** $\dfrac{(19x^2+20x-3)(x^2+3)^4}{2(x+1)^{3/2}}$

43 $\left[\dfrac{2}{3(2x+1)}+\dfrac{8}{(4x-1)}+\dfrac{12}{(3x+5)}\right]\sqrt[3]{2x+1}(4x-1)^2(3x+5)^4$

45 $\left[\dfrac{3x}{3x^2+2}+\dfrac{3}{2(6x-7)}\right]\sqrt{(3x^2+2)\sqrt{6x-7}}$ **47** $-2/(3-2x)$ **49** $\dfrac{6e^{-2x}}{1-e^{-2x}}$

Exercises 8.4, page 389

1 $7^x \ln 7$ **3** $8^{x^2+1}(2x \ln 8)$ **5** $(4x^3+6x)/\ln 10(x^4+3x^2+1)$

7 $5^{3x-4}3 \ln 5$ **9** $\dfrac{-(x^2+1)10^{1/x}\ln 10}{x^2}+2x10^{1/x}$ **11** $\dfrac{2x^3(\ln 7)7^{\sqrt{x^4+9}}}{\sqrt{x^4+9}}$

13 $\dfrac{30x}{(3x^2+2)\ln 10}$ **15** $\left(\dfrac{6}{6x+4}-\dfrac{2}{2x-3}\right)/\ln 5$ **17** $1/(x \ln x \ln 10)$ **19** $ex^{e-1}+e^x$

21 $(x+1)^x\left(\dfrac{x}{x+1}+\ln(x+1)\right)$ **23** $(1/(3 \ln 10))10^{3x}+C$ **25** $(-1/(2 \ln 3))3^{-x^2}+C$

27 $(1/\ln 2)\ln(2^x+1)+C$ **29** $\dfrac{5^{-2}-5^{-4}}{2 \ln 5}$ **31** $\dfrac{2^{x^3}}{3 \ln 2}+C$

33 $\ln 10 \ln|\log_{10} x|+C$ **35** $1/\ln 2 - 1/2$

37 Tangent: $y-3=3(1+\ln 3)(x-1)$; normal: $y-3=-(3(1+\ln 3))^{-1}(x-1)$

41 They are symmetric with respect to the line $y=x$.

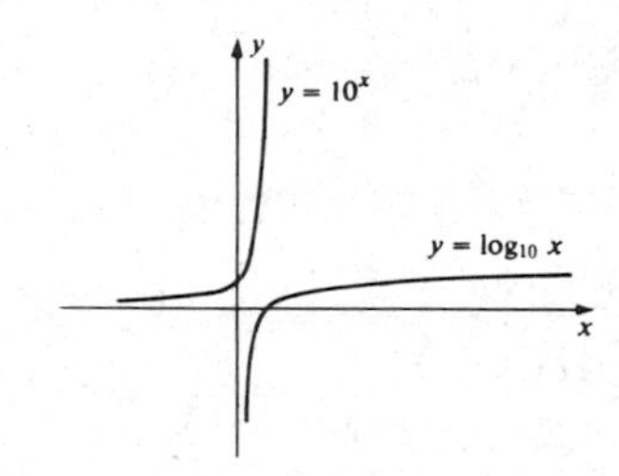

Exercises 8.5, page 396

1 $q(t)=5{,}000\,(3)^{t/10}$; 45,000; $10 \ln 10/\ln 3 \approx 21$ hr **3** $30(29/30)^5 \approx 25.3$

5 In May of 2090 **7** $5 \ln 6/\ln 3 \approx 8.2$ min

9 We must determine i such that at $t=0$, $Ri+L(di/dt)=0$ and $i=I$. Proceeding in a manner similar to the solution of Example 3 we obtain $(1/i)di=(-R/L)\,dt$; $\ln i=(-R/L)t+C$ and $i=e^c e^{(-R/L)t}$. Since $i=I$ at $t=0$, $i=Ie^{-Rt/L}$

11 Write $P(1+r/m)^{mt}=P((1+r/m)^{m/r})^{rt}$. If we let $h=r/m$, then $h \to 0$ as $m \to \infty$ and hence

$$\lim_{m\to\infty} P((1+r/m)^{m/r})^{rt}=\lim_{h\to 0} P((1+h)^{1/h})^{rt}=Pe^{rt}.$$

17 (a) $-\ln 2/100$ (b) $-\ln 2/1000$ (c) $-\ln 2/c$ **19** \$14,310.84

Exercises 8.6, page 402

1 $(x^2-3)/2$, $[\sqrt{5},5]$; x **3** $\sqrt{4-x}$, $[-45,4]$; $-1/2\sqrt{4-x}$ **5** $1/x$, $(0,\infty)$; $-1/x^2$

7 $\sqrt{-\ln x}$, $(0,1]$; $-1/2x\sqrt{-\ln x}$ **9** $\ln(x+\sqrt{x^2+4})-\ln 2$; $\mathbb{R}$; $1/\sqrt{x^2+4}$

11 f is increasing since $f'(x)=5x^4+9x^2+2 \geqslant 2 > 0$. Slope is 1/16.

13 f is increasing since $f'(x)=4e^{2x}/(e^{2x}+1)^2 > 0$. Slope is 1.

15 Since f decreases on $(-\infty,0]$ and increases on $[0,\infty)$ there is no inverse function. If the domain is restricted to a subset of one of these intervals, then f will have an inverse function.

17 f increases on $(-\infty,0]$ and decreases on $[0,\infty)$. If the domain is restricted to a subset of one of these intervals, then f^{-1} will exist.

8.7 Review Exercises, page 403

1 $-2(1 + \ln|1 - 2x|)$ **3** $12/(3x + 2) + 3/(6x - 5) - 8/(8x - 7)$
5 $-4x/(2x^2 + 3)[\ln(2x^2 + 3)]^2$ **7** $2x$ **9** $(\ln 10)10^x \log x + 10^x/x \ln 10$
11 $(1/x)(2 \ln x)x^{\ln x}$ **13** $2x(1 - x^2)e^{1-x^2}$ **15** $2^{-1/x}[(x^3 + 4)\ln 2 - 3x^4]/x^2(x^3 + 4)^2$
17 $e(1 + \sqrt{x})^{e-1}/2\sqrt{x}$ **19** $(10^{\ln x}\ln 10)/x$ **21** $y(e^{xy} - 1)/x(1 - e^{xy})$
25 $4e^2 + 12$ cm **27** $-(1/2)e^{-2x} - 2e^{-x} + x + C$ **29** $2(e^{-1} - e^{-2})$
31 $x^2/6 - 2x/9 + (4/27)\ln|3x + 2| + C$ **33** $3/2 \ln 4$ **35** $2 \ln 10\sqrt{\log x} + C$
37 $(1/2)x^2 - x + 2 \ln|x + 1| + C$ **39** $(5e)^x/(1 + \ln 5) + C$ **41** $x^{e+1}/(e + 1) + C$
43 $y - e = -2(1 + e)(x - 1)$ **45** $(\pi/8)(e^{-16} - e^{-24})$
47 (a) $f'(x) > 0$, so that f is increasing and f^{-1} exists with domain $[4, \infty)$.
(b) $\ln(\sqrt{x} - 1)$, $1/2\sqrt{x}(\sqrt{x} - 1)$ (c) 4, 1/4
49 The amount in solution at any time t hours past 1:00 P.M. is $10(1 - 2^{-t/3})$.
(a) 2.21 hours (approximately 6:14 P.M.) (b) $10(1 - 2^{-7/3}) \approx 8.016$ lb

CHAPTER 9

Exercises 9.1, page 411

1 1 **3** 0 **5** 2 **7** 1 **9** 0 **11** 3/7 **13** 1 **15** Does not exist.
17 Does not exist. **19** Does not exist.

Exercises 9.2, page 419

3 $8 \cos(8x + 3)$ **5** $(1/2\sqrt{x-1}) \sec(\sqrt{x-1}) \tan(\sqrt{x-1})$ **7** $-(3x^2 - 2)\csc^2(x^3 - 2x)$
9 $-6x \sin 3x^2$ **11** $-6 \cos 3x \sin 3x$ **13** $-5x^2 \csc 5x \cot 5x + 2x \csc 5x$
15 $3 \tan^3 x \sec^3 x + 2 \tan x \sec^5 x$ **17** $5(\sin 5x - \cos 5x)^4(5 \cos 5x + 5 \sin 5x)$
19 $-9 \cot^2(3x + 1)\csc^2(3x + 1)$ **21** $\dfrac{-4(1 - \sin 4x)\sin 4x + 4 \cos^2 4x}{(1 - \sin 4x)^2}$, or $\dfrac{4}{1 - \sin 4x}$
23 $\dfrac{-\csc x \cot x - \csc^2 x}{\csc x + \cot x}$, or $-\csc x$ **25** $\dfrac{e^{-3x}\sec^2\sqrt{x}}{2\sqrt{x}} - 3e^{-3x}\tan\sqrt{x}$ **27** $\dfrac{2 \tan 2x}{\ln \sec 2x}$
29 $6 \tan^2 2x \sec^2 2x - 6 \tan 2x \sec^3 2x$ **31** $-3 \csc 3x[(x^3 + 1)\cot 3x + x^2]/(x^3 + 1)^2$
33 $x^{\sin x}[\cos x \ln x + (1/x)\sin x]$
35 $y' = 6 \sec^2 3x \tan 3x$, $y'' = 18 \sec^4 3x + 36 \sec^2 3x \tan^2 3x$
37 $y' = x \sin x$, $y'' = x \cos x + \sin x$ **39** $y' = \dfrac{\sec^2 x}{2\sqrt{\tan x}}$, $y'' = \sec^2 x\sqrt{\tan x} - \dfrac{1}{4}\dfrac{\sec^4 x}{\sqrt{\tan^3 x}}$
41 $\sin y/(1 - x \cos y)$ **43** $(e^x \cos y - e^y)/(e^x \sin y + xe^y)$
45 Max: $f(\pi/6) = f(\pi/6 + 2\pi) = 3\sqrt{3}/2$; min: $f(5\pi/6) = f(5\pi/6 + 2\pi) = -3\sqrt{3}/2$
47 Max: $f(\pi/4) = -e^{-\pi/4}/\sqrt{2} \approx 0.32$, $f(\pi/4 + 2\pi) = e^{-9\pi/4}/\sqrt{2} \approx 6 \times 10^{-4}$;
min: $f(5\pi/4) = -e^{-5\pi/4}/\sqrt{2} \approx -0.014$, $f(5\pi/4 + 2\pi) = e^{-3.25\pi}/\sqrt{2} \approx -3 \times 10^{-5}$

45

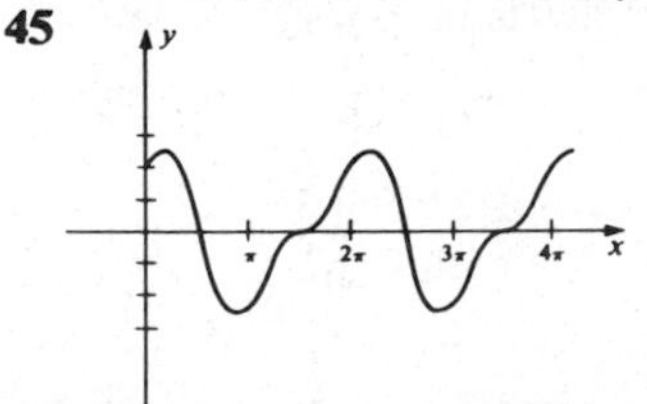

47

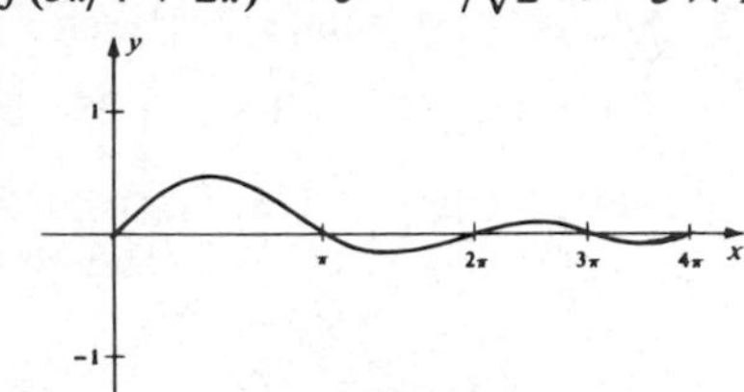

49 Tangent: $y - 1 = 3\sqrt{3}(x - \pi/6)$, normal: $y - 1 = (-1/3\sqrt{3})(x - \pi/6)$ **51** $\pm\pi/9$
53 $10{,}000\ \pi/135 \approx 232.7$ ft/sec
55 $s'(t) = -6 \sin 2t$, $s''(t) = -12 \cos 2t$. The point oscillates between 3 and -3, completing one such cycle in each π units of time. **57** $2\pi(1 - \sqrt{2/3})$.

Exercises 9.3, page 425

1 $(-1/4)\cos 4x + C$ **3** $(1/3)\sec 3x + C$ **5** $(1/3)(\ln|\sec 3x| + \ln|\sec 3x + \tan 3x|) + C$
7 $(1/2)\ln|\sec 2x + \tan 2x| + C$ **9** $(-1/2)\cot(x^2 + 1) + C$ **11** $(1/6)\sin 6x + C$
13 $(1/2)\tan^2 x]_0^{\pi/4} = 1/2$ **15** $(1/2)(\ln|\sec 2x + \tan 2x| - \sin 2x) + C$
17 $(1/2)\sin^2 x]_{\pi/6}^{\pi} = -1/8$ **19** $\ln|x + \cos x| + C$ **21** $\tan x + \sec x]_{\pi/4}^{\pi/3} = \sqrt{3} - \sqrt{2} + 1$
23 $e^x + \sin e^x + C$ **25** $-e^{\cos x} + C$ **27** $(1/2)\ln|2\tan x + 1| + C$ **29** 2
31 $\ln\left(\dfrac{\sqrt{2}+1}{\sqrt{2}-1}\right) = \ln(3 + 2\sqrt{2}) = 2\ln(1 + \sqrt{2})$ **33** $2\pi\sqrt{3}$
39 (a) $L = \int_0^{\pi/2} \sqrt{1 + \cos^2 x}\,dx$ (b) If $f(x) = \sqrt{1 + \cos^2 x}$, then
$L \approx (\pi/24)[f(0) + 4f(\pi/8) + 2f(\pi/4) + 4f(3\pi/8) + f(\pi/2)]$ (c) $L \approx 1.91$

Exercises 9.4, page 430

1 (a) $\pi/3$ (b) $-\pi/3$ **3** (a) $\pi/4$ (b) $3\pi/4$ **5** (a) $\pi/3$ (b) $-\pi/3$ **7** $1/2$ **9** $4/5$
11 $\pi - \sqrt{5}$ **13** 0 **15** Undefined **17** $-24/25$ **19** $x/\sqrt{x^2 + 1}$
21 $\sqrt{2 + 2x}/2$ **31** $\cot^{-1} x = y$ if and only if $\cot y = x$, where $0 < y < \pi$

Exercises 9.5, page 436

1 $\dfrac{3}{9x^2 - 30x + 26}$ **3** $\dfrac{1}{2\sqrt{x}\sqrt{1-x}}$ **5** $\dfrac{-e^{-x}}{\sqrt{e^{-2x} - 1}} - e^{-x}\,\text{arcsec}\, e^{-x}$
7 $\dfrac{2x^3}{1 + x^4} + 2x \arctan x^2$ **9** $-\dfrac{9(1 + \cos^{-1} 3x)^2}{\sqrt{1 - 9x^2}}$ **11** $\dfrac{2x}{(1 + x^4)\arctan x^2}$
13 $\dfrac{-1}{(\sin^{-1} x)^2\sqrt{1 - x^2}}$ **15** $\dfrac{x}{(x^2 - 1)\sqrt{x^2 - 2}}$ **17** $\dfrac{1 - 2x\arctan x}{(x^2 + 1)^2}$
19 $(1/2\sqrt{x})\sec^{-1}\sqrt{x} + 1/2\sqrt{x}\sqrt{x - 1}$ **21** $(1 - x^6)^{-1/2}(3\ln 3)x^2 3^{\arcsin x^3}$
23 $(\tan x)^{\arctan x}[\cot x \sec^2 x \arctan x + (\ln\tan x)/(1 + x^2)]$
25 $(ye^x - \sin^{-1} y - 2x)/\left[\dfrac{x}{\sqrt{1 - y^2}} - e^x\right]$ **27** $\pi/16$ **29** $\pi/12$
31 $-\arctan(\cos x) + C$ **33** $2\arctan\sqrt{x} + C$ **35** $\sin^{-1}(e^x/4) + C$
37 $(1/6)\sec^{-1}(x^3/2) + C$ **39** $(1/2)\ln(x^2 + 9) + C$ **41** $(1/5)\sec^{-1}(e^x/5) + C$
43 $4\pi/3$ **45** $\pm 7/3576 \approx \pm 0.002$ **47** $-25/1044$ rad/sec **49** $40\sqrt{3}$
53 Tangent: $y - \pi/6 = (2/\sqrt{3})(x - 3/2)$, normal: $y - \pi/6 = (-\sqrt{3}/2)(x - 3/2)$
55 CU on $(-\infty, 0)$; CD on $(0, \infty)$ **57** $(\pi e^2/2) - (\pi^2/4) - (\pi/2) \approx 7.57$
59 $2\pi/27$ mi/sec ≈ 0.233 mi/sec

Exercises 9.6, page 444

15 $5\cosh 5x$ **17** $(1/2\sqrt{x})(\sqrt{x}\,\text{sech}^2\sqrt{x} + \tanh\sqrt{x})$
19 $-2x\,\text{sech}\, x^2[(x^2 + 1)\tanh x^2 + 1]/(x^2 + 1)^2$ **21** $3x^2 \sinh x^3$ **23** $3\cosh^2 x \sinh x$
25 $2\coth 2x$ **27** $-e^{3x}\,\text{sech}\, x \tanh x + 3e^{3x}\,\text{sech}\, x$ **29** $-\text{sech}^2 x/(\tanh x + 1)^2$
31 $\dfrac{y(e^x - \cosh xy)}{(x\cosh xy - e^x)}$ **33** $2\cosh\sqrt{x} + C$ **35** $\ln|\sinh x| + C$
37 $(1/2)\sinh^2 x + C$ (or $(1/2)\cosh^2 x + C$, or $(1/4)\sinh 2x + C$) **39** $(-1/3)\text{sech}\, 3x + C$
41 $(1/9)\tanh^3 3x + C$ **43** $(-1/2)\ln|1 - 2\tanh x| + C$ **45** $(-1 + \cosh 3)/3$
47 $(\ln(2 + \sqrt{3}), \sqrt{3})$, $(\ln(2 - \sqrt{3}), -\sqrt{3})$ **51**

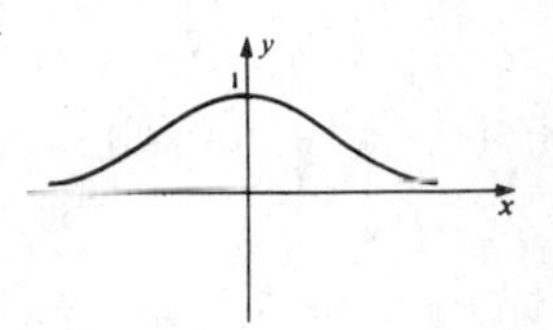

Exercises 9.7, page 448

7 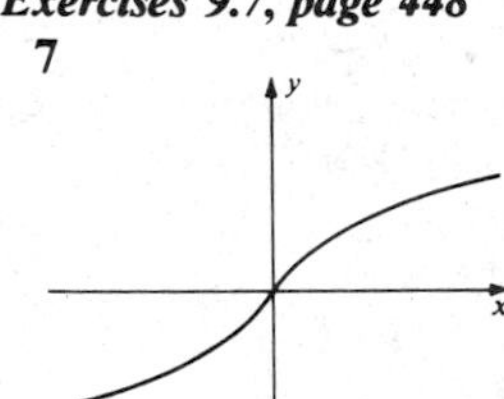

9 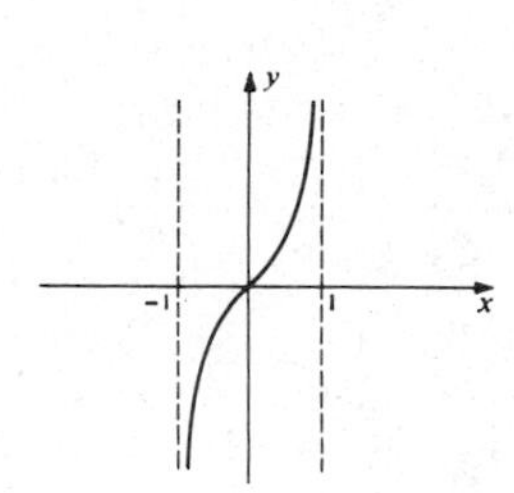

11 $5/\sqrt{25x^2+1}$ **13** $1/(2\sqrt{x}\sqrt{x-1})$ **15** $2x/(2x^2-x^4)$
17 $-|x|/x\sqrt{x^2+1}+\sinh^{-1}(1/x)$ **19** $4/(\sqrt{16x^2-1}\cosh^{-1}4x)$ **21** $(1/3)\sinh^{-1}(3/5)x+C$
23 $(1/14)\tanh^{-1}(2/7)x+C$ **25** $\cosh^{-1}e^x/4+C$ **27** $(-1/6)\,\text{sech}^{-1}(x^2/3)+C$

9.8 Review Exercises, page 449

1 $-(6x+1)\sin\sqrt{3x^2+x}/2\sqrt{3x^2+x}$ **3** $5(\sec x+\tan x)^4(\sec x\tan x+\sec^2 x)$
5 $2x\,\text{arcsec}\,x^2+2x/\sqrt{x^4-1}$
7 $[12(3x+7)^3\sin^{-1}5x-5(3x+7)^4(1-25x^2)^{-1/2}]/(\sin^{-1}5x)^2$
9 $(\cos x)^{x+1}[\ln\cos x-(x+1)\tan x]$ **11** $(2x-2\,\text{sech}\,4x\tanh 4x)/\sqrt{2x^2+\text{sech}\,4x}$
13 $-6\cot 2x$ **15** $-(2+2\sec^2 x\tan x)/(2x+\sec^2 x)^2$
17 $-\sin x e^{\cos x}-e\sin x(\cos x)^{e-1}$ **19** $-5e^{-5x}\sinh e^{-5x}$ **21** $(1/3)x^{-2/3}$
23 $2^{\arctan 2x}(2\ln 2)/(1+4x^2)$
25 $-6e^{-2x}\sin^2 e^{-2x}\cos e^{-2x}$ **27** $-2xe^{-x^2}(\csc^2 x^2+\cot x^2)$
29 $[-(\cot x+1)\csc x\cot x+\csc^2 x(\csc x+1)]/(\cot x+1)^2$ **31** $-x/\sqrt{x^2(1-x^2)}$
33 $3\sec^2(\sin 3x)\cos 3x$ **35** $4(\tan x+\tan^{-1}x)^3[\sec^2 x+1/(1+x^2)]$
37 $1/(1+x^2)[1+(\tan^{-1}x)^2]$ **39** $-e^{-x}(e^{-x}\cosh e^{-x}+\sinh e^{-x})$
41 $(\cosh x-\sinh x)^{-2}$, or e^{2x} **43** $2x/\sqrt{x^4+1}$ **45** $(1/5)\cos(3-5x)+C$
47 $2\tan\sqrt{x}+C$ **49** $(1/9)\ln|\sin 9x|+(1/9)\ln|\csc 9x-\cot 9x|+C$
51 $-\ln|\cos e^x|+C$ **53** $-(1/3)\cot 3x+(2/3)\ln|\csc 3x-\cot 3x|+x+C$
55 $(1/4)\sin 4x+C$ **57** $(1/18)\ln(4+9x^2)+C$ **59** $-\sqrt{1-e^{2x}}+C$
61 $(1/2)\sinh x^2+C$ **63** $\pi/3$ **65** $(1/3)(1+\tan x)^3+C$ **67** $-\ln|2+\cot x|+C$
69 $\cosh(\ln x)+C$ **71** $(1/2)\sin^{-1}(2x/3)+C$ **73** $-(1/3)\text{sech}^{-1}|2x/3|+C$
75 $(1/25)\sqrt{25x^2+36}+C$ **77** Max: $f(\pi/2)=f(5\pi/2)=3$ and $f(3\pi/2)$ $=f(7\pi/2)=-1$; min: $f(7\pi/6)=f(19\pi/6)=f(11\pi/6)=f(23\pi/6)=-3/2$; tangent: $y-(1/2)=2\sqrt{3}(x-\pi/6)$; normal: $y-(1/2)=(-1/2\sqrt{3})(x-\pi/6)$

77 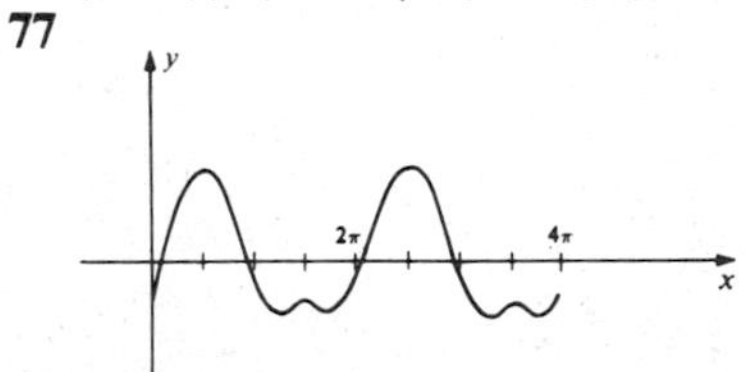

79 $(4/15,\sin^{-1}(4/5)),\ (-4/15,\sin^{-1}(-4/5))$ **81** π **83** $(\pi/4)-0.01(1+\pi)$
85 CU on $(-1,1)$ **87** $-1/2\sqrt{3}$ rad/sec

CHAPTER 10

Exercises 10.1, page 458

1 $-(x+1)e^{-x}+C$ **3** $e^{3x}[(x^2/3)-(2x/9)+(2/27)]+C$
5 $(1/25)\cos 5x+(1/5)x\sin 5x+C$ **7** $x\sec x-\ln|\sec x+\tan x|+C$

9 $(x^2 - 2)\sin x + 2x \cos x + C$
11 $x \tan^{-1} x - (1/2)\ln(1 + x^2) + C$ **13** $(2/9)x^{3/2}(3 \ln x - 2) + C$
15 $-x \cot x + \ln|\sin x| + C$ **17** $-\frac{1}{2}e^{-x}(\cos x + \sin x) + C$
19 $\cos x(1 - \ln \cos x) + C$ **21** $\frac{1}{2}(\sec x \tan x + \ln|\sec x + \tan x|) + C$
23 $(1/3)(2 - \sqrt{2})$ **25** $\pi/4$ **27** $(1/40400)(2x + 3)^{100}(200x - 3) + C$
29 $(e^{4x}/41)(4 \sin 5x - 5 \cos 5x) + C$ **31** $x(\ln x)^2 - 2x \ln x + 2x + C$
33 $x^3 \cosh x - 3x^2 \sinh x + 6x \cosh x - 6 \sinh x + C$ **35** $2 \cos \sqrt{x} + 2\sqrt{x} \sin \sqrt{x} + C$
37 $x \cos^{-1} x - \sqrt{1 - x^2} + C$ **43** $e^x(x^5 - 5x^4 + 20x^3 - 60x^2 + 120x - 120) + C$
45 2π **47** $(\pi/2)(e^2 + 1)$ **49** $(62.5\pi)/4 = 125\pi/8$

Exercises 10.2, page 464

1 $\sin x - (1/3)\sin^3 x + C$ **3** $(1/8)x - (1/32)\sin 4x + C$
5 $(1/5)\cos^5 x - (1/3)\cos^3 x + C$ **7** $(1/8)(\frac{5}{2}x - 2 \sin 2x + \frac{3}{8} \sin 4x + \frac{1}{6} \sin^3 2x) + C$
9 $\frac{1}{4} \tan^4 x + \frac{1}{6} \tan^6 x + C$ **11** $\frac{1}{5} \sec^5 x - \frac{1}{3} \sec^3 x + C$
13 $\frac{1}{5} \tan^5 x - \frac{1}{3} \tan^3 x + \tan x - x + C$ **15** $\frac{2}{3} \sin^{3/2} x - \frac{2}{7} \sin^{7/2} x + C$
17 $\tan x - \cot x + C$ **19** $2/3 - 5/6\sqrt{2}$ **21** $\frac{1}{4} \sin 2x - \frac{1}{16} \sin 8x + C$ **23** $3/5$
25 $(-1/5)\cot^5 x - (1/7)\cot^7 x + C$ **27** $-\ln|2 - \sin x| + C$
29 $-1/(1 + \tan x) + C$ **31** $\pi^2/2$ **33** $5/2$

Exercises 10.3, page 470

1 $2 \sin^{-1}(x/2) - \frac{1}{2}x\sqrt{4 - x^2} + C$ **3** $\frac{1}{3} \ln\left|\frac{\sqrt{9 + x^2} - 3}{x}\right| + C$ **5** $\sqrt{x^2 - 25}/25x + C$
7 $-\sqrt{4 - x^2} + C$ **9** $-x/\sqrt{x^2 - 1} + C$ **11** $\frac{1}{432}\left(\tan^{-1}\frac{x}{6} + \frac{6x}{36 + x^2}\right) + C$
13 $\frac{9}{4} \sin^{-1}(2x/3) + (x/2)\sqrt{9 - 4x^2} + C$ **15** $1/(2(16 - x^2)) + C$
17 $(1/243)(9x^2 + 49)^{3/2} - (49/81)(9x^2 + 49)^{1/2} + C$ **19** $(3 + 2x^2)\sqrt{x^2 - 3}/27x^3 + C$
21 $(1/2)x^2 + 8 \ln|x| - 8x^{-2} + C$ **29** $25\pi(\sqrt{2} - \ln(\sqrt{2} + 1)) \approx 41.849$
31 $\sqrt{5} + (1/2)\ln(2 + \sqrt{5}) \approx 2.96$ **33** πab
35 $f(x) = \sqrt{x^2 - 16} - 4 \tan^{-1}(\sqrt{x^2 - 16}/4)$
37 $-\sqrt{25 + x^2}/25x + C$ **39** $-\sqrt{1 - x^2}/x + C$

Exercises 10.4, page 476

1 $3 \ln|x| + 2 \ln|x - 4| + C$, or $\ln|x|^3 (x - 4)^2 + C$
3 $4 \ln|x + 1| - 5 \ln|x - 2| + \ln|x - 3| + C$, or $\ln \frac{(x + 1)^4|x - 3|}{|x - 2|^5} + C$
5 $6 \ln|x - 1| + 5/(x - 1) + C$ **7** $3 \ln|x - 2| - 2 \ln|x + 4| + C$, or $\ln \frac{|x - 2|^3}{(x + 4)^2} + C$
9 $2 \ln|x| - \ln|x - 2| + 4 \ln|x + 2| + C$, or $\ln \frac{x^2(x + 2)^4}{|x - 2|} + C$
11 $5 \ln|x + 1| - \frac{1}{x + 1} - 3 \ln|x - 5| + C$, or $\ln \frac{|x + 1|^5}{|x - 5|^3} - \frac{1}{x + 1} + C$
13 $5 \ln|x| - \frac{2}{x} + \frac{3}{2x^2} - \frac{1}{3x^3} + 4 \ln|x + 3| + C$
15 $\frac{1}{6} \ln|x - 3| - \frac{7}{2(x - 3)} + (5/6)\ln|x + 3| - \frac{3}{2(x + 3)} + C$
17 $3 \ln|x + 5| + \ln(x^2 + 4) + (1/2)\tan^{-1}(x/2) + C$, or $\ln(x^2 + 4)|x + 5|^3 + (1/2)\tan^{-1}(x/2) + C$
19 $\ln \sqrt{\frac{x^2 + 1}{x^2 + 4}} + \frac{1}{2} \tan^{-1}(x/2) + C$ **21** $\ln(x^2 + 1) - 4/(x^2 + 1) + C$
23 $(x^2/2) + x + 2 \ln|x| + 2 \ln|x - 1| + C$, or $(x^2 + 2x)/2 + \ln(x^2 - x)^2 + C$
25 $(x^3/3) - 9x - (1/9x) - (1/2)\ln(x^2 + 9) + (728/27)\tan^{-1}(x/3) + C$
27 $2 \ln|x + 4| + 6(x + 4)^{-1} - 5(x - 3)^{-1} + C$

29 $2\ln|x-4| + 2\ln|x+1| - (3/2)(x+1)^{-2} + C$
31 $(3/2)\ln(x^2+1) + \ln|x-1| + x^2 + C$
37 $(1/2)\ln 3$ **39** $(\pi/27)(4\ln 2 + 3) \approx 0.672$

Exercises 10.5, page 480

1 $(1/2)\tan^{-1}(x-2)/2 + C$ **3** $\sin^{-1}(x-2)/2 + C$
5 $-2\sqrt{9-8x-x^2} - 5\sin^{-1}(x+4)/5 + C$
7 $(1/3)\ln|x-1| - (1/6)\ln(x^2+x+1) - (1/\sqrt{3})\tan^{-1}(2x+1)/\sqrt{3} + C$
9 $(1/2)[\tan^{-1}(x+2) + (x+2)/(x^2+4x+5)] + C$
11 $(x+3)/4\sqrt{x^2+6x+13} + C$ **13** $(2/3\sqrt{7})\tan^{-1}(4x-3)/3\sqrt{7} + C$
15 $(1/3)\ln 2 + 2\pi/6\sqrt{3} \approx 0.8356$ **17** $\pi\ln(1.8) + (2\pi/3)(\tan^{-1}\frac{1}{3} - \pi/4) \approx 0.8755$

Exercises 10.6, page 484

1 $(3/7)(x+9)^{7/3} - (27/4)(x+9)^{4/3} + C$ **3** $(5/81)(3x+2)^{9/5} - (5/18)(3x+2)^{4/5} + C$
5 $2\sqrt{x} - 8\ln(\sqrt{x}+4) + C$ **7** $\frac{6}{7}x^{7/6} - \frac{6}{5}x^{5/6} + 2x^{1/2} - 6x^{1/6} + 6\arctan x^{1/6} + C$
9 $\frac{2}{\sqrt{3}}\tan^{-1}\sqrt{\frac{x-2}{3}} + C$ **11** $(5/9)(x+4)^{4/5}(x-5) + C$
13 $(2/7)(1+e^x)^{7/2} - (4/5)(1+e^x)^{5/2} + (2/3)(1+e^x)^{3/2} + C$
15 $e^x - 4\ln(e^x+4) + C$ **17** $2\sin\sqrt{x+4} - 2\sqrt{x+4}\cos\sqrt{x+4} + C$
19 $-1/4(x-1)^4 - 1/5(x-1)^5 + C$, or $(1-5x)/20(x-1)^5 + C$
21 $\frac{2}{\sqrt{3}}\tan^{-1}\left(\frac{2\tan(x/2)+1}{\sqrt{3}}\right) + C$ **23** $\ln|1+\tan(x/2)| + C$
25 $\frac{1}{5}\ln\left|\frac{\tan(x/2)+2}{2\tan(x/2)-1}\right| + C$

Exercises 10.7, page 486

1 $\sqrt{4+9x^2} - 2\ln\left|\frac{2+\sqrt{4+9x^2}}{3x}\right| + C$
3 $-(x/8)(2x^2-80)\sqrt{16-x^2} + 96\sin^{-1}(x/4) + C$
5 $-(2/135)(9x+4)(2-3x)^{3/2} + C$
7 $(-1/18)\sin^5 3x\cos 3x - (5/72)\sin^3 3x\cos 3x - (5/48)\sin 3x\cos 3x + 5x/16 + C$
9 $(-1/3)\cot x\csc^2 x - (2/3)\cot x + C$
11 $(1/2)x^2\sin^{-1}x + (1/4)x\sqrt{1-x^2} - (1/4)\sin^{-1}x + C$
13 $(1/13)e^{-3x}(-3\sin 2x - 2\cos 2x) + C$
15 $\sqrt{5x-9x^2} + (5/6)\cos^{-1}(5-18x)/5 + C$
17 $\frac{1}{4\sqrt{15}}\ln\left|\frac{\sqrt{5}x^2-\sqrt{3}}{\sqrt{5}x^2+\sqrt{3}}\right| + C$ **19** $(1/4)(2e^{2x}-1)\cos^{-1}e^x - (1/4)e^x\sqrt{1-e^{2x}} + C$

Exercises 10.8, page 495

1 $-27, -46, \bar{x} = -23/7, \bar{y} = -27/14$
3 $\bar{x} = 4/5, \bar{y} = 2/7$ **5** $\bar{x} = \pi/2, \bar{y} = \pi/8$ **7** $\bar{x} = -1/2, \bar{y} = -3/5$

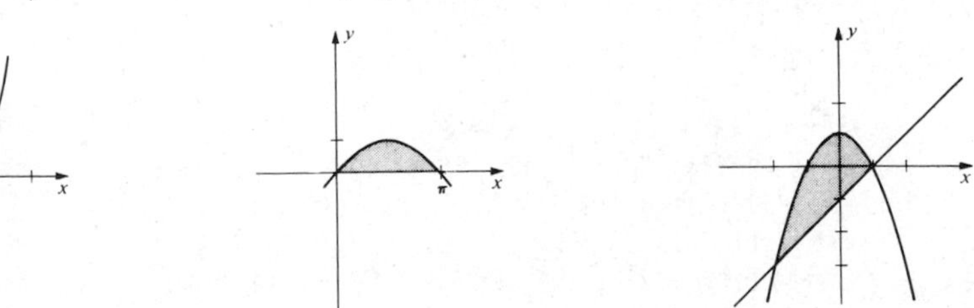

9 $\bar{x} = 1/\ln 2, \bar{y} = \tan^{-1}(3/4)/8 \ln 2$ **11** $\bar{x} = (3e^{-2} - 1)/2(1 - e^{-2})$, $\bar{y} = (1 - e^{-4})/4(1 - e^{-2})$

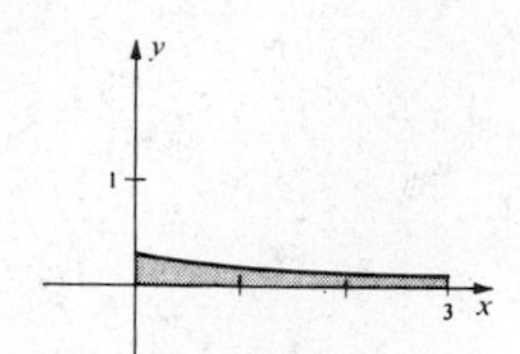

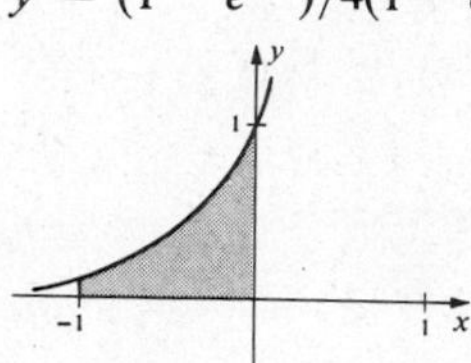

15 $\bar{x} = 0, \bar{y} = 4a/3\pi$ if the region is bounded by the x-axis and the upper half of the circle centered at (0, 0).
17 $\bar{x} = 0, \bar{y} = -20a/3(8 + \pi)$. The figure is positioned vertically with the origin at the center of the circle. **19** $\bar{x} = 0, \bar{y} = 4b/3\pi$

Exercises 10.9, page 501

1 $\bar{x} = 2 \ln 2$ **3** $\bar{x} = (e^2 + 1)/2(e^2 - 1)$ **5** $\bar{x} = 25/48$ **7** $\bar{y} = 27/8$
9 $\bar{y} = (7e^8 + 1)/8(3e^4 + 1)$ **11** $\bar{y} = 1/(100 \ln 2)$ **13** $\bar{y} = 3b/8$
15 $\bar{y} = (\pi + 2)/16$ **17** (a) 576π (b) 288π **19** $(4a/3\pi, 4a/3\pi)$

10.10 Review Exercises, page 502

1 $(x^2/2)\sin^{-1}x - (1/4)\sin^{-1}x + (x/4)\sqrt{1 - x^2} + C$ **3** $2 \ln 2 - 1$
5 $(1/6)\sin^3 2x - (1/10)\sin^5 2x + C$ **7** $(1/5)\sec^5 x + C$ **9** $x/25\sqrt{x^2 + 25} + C$
11 $2 \ln\left|\dfrac{2 - \sqrt{4 - x^2}}{x}\right| + \sqrt{4 - x^2} + C$ **13** $2 \ln|x - 1| - \ln|x| - \dfrac{x}{(x - 1)^2} + C$
15 $\ln \dfrac{(x + 3)^2(x^2 + 9)^2}{|x - 3|^5} + \dfrac{1}{3} \tan^{-1}\dfrac{x}{3} + C$ **17** $-\sqrt{4 + 4x - x^2} + 2 \sin^{-1}\dfrac{x - 2}{\sqrt{8}} + C$
19 $3(x + 8)^{1/3} + \ln[(x + 8)^{1/3} - 2]^2 - \ln[(x + 8)^{2/3} + 2(x + 8)^{1/3} + 4]$
$-\dfrac{6}{\sqrt{3}} \tan^{-1}\dfrac{(x + 8)^{1/3} + 1}{\sqrt{3}} + C$
21 $(1/13)e^{2x}(2 \sin 3x - 3 \cos 3x) + C$ **23** $\frac{1}{4} \sin^4 x - \frac{1}{6} \sin^6 x + C$ **25** $-\sqrt{4 - x^2} + C$
27 $\dfrac{1}{3}x^3 - x^2 + 3x - \dfrac{1}{2x} - \dfrac{1}{4} \ln|x| - \dfrac{23}{4} \ln|x + 2| + C$ **29** $2 \tan^{-1}(x^{1/2}) + C$
31 $\ln|\sec e^x + \tan e^x| + C$ **33** $\frac{1}{125}[10x \sin 5x - (25x^2 - 2)\cos 5x] + C$
35 $\frac{2}{7} \cos^{7/2} x - \frac{2}{3} \cos^{3/2} x + C$ **37** $\frac{2}{3}(1 + e^x)^{3/2} + C$
39 $(1/16)[2x\sqrt{4x^2 + 25} - 25 \ln(2x + \sqrt{4x^2 + 25})] + C$ **41** $(1/3)\tan^3 x + C$
43 $-x \csc x + \ln|\csc x - \cot x| + C$ **45** $-(1/4)(8 - x^3)^{4/3} + C$
47 $2 \cos \sqrt{x} + 2\sqrt{x} \sin \sqrt{x} + C$ **49** $(1/2)e^{2x} - e^x + \ln(1 + e^x) + C$
51 $(2/5)x^{5/2} - (8/3)x^{3/2} + 6x^{1/2} + C$ **53** $(1/3)(16 - x^2)^{3/2} - 16(16 - x^2)^{1/2} + C$
55 $(11/2)\ln|x + 5| - (15/2)\ln|x + 7| + C$ **57** $x \tan^{-1} 5x - (1/10)\ln|1 + 25x^2| + C$
59 $e^{\tan x} + C$ **61** $(1/\sqrt{5})\ln|\sqrt{5}x + \sqrt{7 + 5x^2}| + C$
63 $(-1/5)\cot^5 x - (1/3)\cot^3 x - \cot x - x + C$
65 $(1/5)(x^2 - 25)^{5/2} + (25/3)(x^2 - 25)^{3/2} + C$ **67** $x^3/3 - (1/4)\tanh 4x + C$
69 $(-1/4)x^2e^{-4x} - (1/8)xe^{-4x} - (1/32)e^{-4x} + C$
71 $3 \sin^{-1}(x + 5)/6 + C$ **73** $(-1/7)\cos 7x + C$
75 $18 \ln|x - 2| - 9 \ln|x - 1| - 5 \ln|x - 3| + C$
77 $x^3 \sin x + 3x^2 \cos x - 6x \sin x - 6 \cos x + \sin x + C$
79 $(-1/x)\sqrt{9 - 4x^2} - 2 \sin^{-1}(2x/3) + C$
81 $24x - (10/3)\ln|\sin 3x| - (1/3)\cot 3x + C$
83 $-\ln x - 4/\sqrt[4]{x} + 4 \ln(1 + \sqrt[4]{x}) + C$ **85** $-2\sqrt{1 + \cos x} + C$
87 $-x/2(25 + x^2) + (1/10)\tan^{-1}(x/5) + C$ **89** $(1/3)\sec^3 x - \sec x + C$
91 $\ln(x^2 + 4) - (3/2)\tan^{-1}(x/2) + (7/\sqrt{5})\tan^{-1}(x/\sqrt{5}) + C$
93 $(1/4)x^4 - 2x^2 + 4 \ln|x| + C$ **95** $(2/5)x^{5/2}\ln x - (4/25)x^{5/2} + C$

97 $(3/64)(2x+3)^{8/3} - (9/20)(2x+3)^{5/3} + (27/16)(2x+3)^{2/3} + C$
99 $(1/2)e^{x^2}(x^2-1)+C$ **101** $2\pi^2$ **103** $\ln(2+\sqrt{3})$
105 $\bar{x}=3/5, \bar{y}=12/35$ **107** $\bar{x}=8/3$ **109** $\bar{y}=(1-7e^{-6})/8(1-4e^{-3})$

CHAPTER 11

Exercises 11.1, page 512

1 1/2 **3** 1/40 **5** 3/13 **7** 0 **9** −1/2 **11** −1/2 **13** 1/6
15 ∞ **17** 1/3 **19** ∞ **21** 1 **23** 0 **25** Does not exist
27 2/5 **29** ∞ **31** 0 **33** ∞ **35** 2 **37** Does not exist **39** 3/5
41 Does not exist **43** Does not exist **45** Does not exist **47** Does not exist **49** 2

Exercises 11.2, page 516

1 0 **3** 0 **5** 0 **7** 0 **9** 1 **11** 0 **13** e^5 **15** 1 **17** 1
19 Does not exist **21** e^2 **23** 2 **25** 0 **27** 1 **29** Does not exist **31** 1/2
33 Does not exist **35** e **37** Does not exist **39** 1/3

Exercises 11.3, page 521

1 3 **3** Diverges **5** Diverges **7** 1/2 **9** −1/2 **11** Diverges
13 Diverges **15** 0 **17** Diverges **19** Diverges **21** π **23** ln 2
25 (a) Does not exist (b) π **27** (a) 1 (b) $\pi/32$ **29** (a) $n<-1$ (b) $n \geqslant -1$
31 There are many possible answers; e.g. $f(x)=x, f(x)=\sin x, f(x)=x^7+x$.
33 $4\cdot 10^5$ mi/lb **35** $1/s, s>0$ **37** $s/(s^2+1), s>0$ **39** $1/(s-a), s>a$

Exercises 11.4, page 528

1 6 **3** Diverges **5** Diverges **7** Diverges **9** $3\sqrt[3]{4}$ **11** Diverges
13 $\pi/2$ **15** Diverges **17** −1/4 **19** Diverges **21** Diverges **23** Diverges
25 0 **27** (a) 2 (b) Value cannot be assigned.
29 Values cannot be assigned to either the area or the volume. **31** $n>-1$

Exercises 11.5, page 538

1 $\sin x = 1 - \frac{1}{2}\left(x-\frac{\pi}{2}\right)^2 + \frac{\sin z}{4!}\left(x-\frac{\pi}{2}\right)^4$, where z is between x and $\pi/2$.

3 $\sqrt{x} = 2 + \frac{1}{4}(x-4) - \frac{1}{64}(x-4)^2 + \frac{1}{512}(x-4)^3 - \frac{5}{128}z^{-7/2}(x-4)^4$, where z is between x and 4.

5 $\tan x = 1 + 2\left(x-\frac{\pi}{4}\right) + 2\left(x-\frac{\pi}{4}\right)^2 + \frac{8}{3}\left(x-\frac{\pi}{4}\right)^3 + \frac{10}{3}\left(x-\frac{\pi}{4}\right)^4 + \frac{g(z)}{5!}\left(x-\frac{\pi}{4}\right)^5$,
where z is between x and $\pi/4$ and $g(z) = 16\sec^6 z + 88\sec^4 z\tan^2 z + 16\sec^2 z\tan^4 z$

7 $1/x = (-1/2) - (1/4)(x+2) - (1/8)(x+2)^2 - (1/16)(x+2)^3 - (1/32)(x+2)^4 - (1/64)(x+2)^5 + z^{-7}(x+2)^6$, where z is between x and -2

9 $\tan^{-1} x = \frac{\pi}{4} + \frac{1}{2}(x-1) - \frac{1}{4}(x-1)^2 + \frac{3z^2-1}{3(1+z^2)^3}(x-1)^3$, where z is between 1 and x

11 $xe^x = -\frac{1}{e} + \frac{1}{2e}(x+1)^2 + \frac{1}{3e}(x+1)^3 + \frac{1}{8e}(x+1)^4 + \frac{ze^z+5e^z}{120}(x+1)^5$, where z is between x and -1

13 $\ln(x+1) = x - \frac{1}{2}x^2 + \frac{1}{3}x^3 - \frac{1}{4}x^4 + \frac{x^5}{5(z+1)^5}$, where z is between 0 and x

15 $\cos x = 1 - \frac{x^2}{2!} + \frac{x^4}{4!} - \frac{x^6}{6!} + \frac{x^8}{8!} - \frac{\sin z}{9!}x^9$, where z is between x and 0

17 $e^{2x} = 1 + 2x + 2x^2 + \frac{4}{3}x^3 + \frac{2}{3}x^4 + \frac{4}{15}x^5 + \frac{4}{45}e^{2z}x^6$, where z is between x and 0

19 $1/(x-1)^2 = 1 + 2x + 3x^2 + 4x^3 + 5x^4 + 6x^5 + 7x^6/(z-1)^8$, where z is between 0 and x

21 $\arcsin x = x + \dfrac{1+2z^2}{6(1-z^2)^{5/2}}x^3$, where z is between 0 and x

23 $f(x) = 7 - 3x + x^2 - 5x^3 + 2x^4$ **25** 0.9998 **27** 2.0075
29 0.454545, error $\leqslant$ 0.0000005 **31** 0.223, error $\leqslant$ 0.0002
33 0.8660254, error $\leqslant$ $(8.1)(10^{-9})$
35 Five decimal places since $|R_3(x)| \leqslant 4.2 \times 10^{-6}$
37 Three decimal places **39** Four decimal places

Exercises 11.6, page 542

1 1.2599 **3** 0.5641 **5** 1.3315 **7** 1.2927 **9** 1.5572 **11** ± 3.34
13 -1, 1.35 **15** -0.64 **17** 2.71

11.7 Review Exercises, page 543

1 $(1/2)\ln 2$ **3** ∞ **5** 8/3 **7** 0 **9** $-\infty$ **11** e^8 **13** e
15 Diverges **17** Diverges **19** $-9/2$ **21** Diverges **23** $\pi/2$ **25** Diverges

27 (a) $\ln\cos x = \ln\dfrac{\sqrt{3}}{2} - \dfrac{1}{\sqrt{3}}\left(x - \dfrac{\pi}{6}\right) - \dfrac{2}{3}\left(x-\dfrac{\pi}{6}\right)^2 - \dfrac{4}{9\sqrt{3}}\left(x - \dfrac{\pi}{6}\right)^3$
$-\dfrac{1}{12}(\sec^4 z + 2\sec^2 z\tan^2 z)\left(x - \dfrac{\pi}{6}\right)^4$, where z is between x and $\pi/6$

(b) $\sqrt{x-1} = 1 + \frac{1}{2}(x-2) - \frac{1}{8}(x-2)^2 + \frac{1}{16}(x-2)^3 - \frac{5}{128}(x-2)^4 + \frac{7}{256}(z-1)^{-9/2}(x-2)^5$, where z is between x and 2

29 0.4651 (with $n = 3$, $|R_3(x)| \leqslant 1.6 \times 10^{-6}$) **31** 2.5170

CHAPTER 12

Exercises 12.1, page 553

1 1/5, 1/4, 3/11, 2/7; 1/3 **3** 3/5, $-9/11$, $-29/21$, $-57/35$; -2
5 $-5, -5, -5, -5$; -5 **7** 2, 7/3, 25/14, 7/5; 0 **9** $2/\sqrt{10}, 2/\sqrt{13}, 2/\sqrt{18}, 2/5$; 0
11 3/10, $-6/17$, 9/26, $-12/37$; 0 **13** 1.1, 1.01, 1.001, 1.0001; 1
15 2, 0, 2, 0; the limit does not exist. **17** 0 **19** $\pi/2$ **21** Does not exist
23 0 **25** Does not exist **27** Does not exist **29** e **31** 0 **33** 1/2
35 Two of many possible answers are:

$$a_n = \begin{cases} 2n \text{ if } 1 \leqslant n \leqslant 4 \\ (n-4)a \text{ if } n \geqslant 5 \end{cases}; \qquad a_n = 2n + (n-1)(n-2)(n-3)(n-4)(a-10)/24.$$

Exercises 12.2, page 561

1 Converges, 4 **3** Converges, $\sqrt{5}/(\sqrt{5}+1)$ **5** Converges, 37/99 **7** Diverges
9 Diverges **11** Converges **13** Converges **15** Diverges **17** Diverges
19 Diverges **21** Converges **23** Converges **25** Diverges **27** Diverges
29 1/2 **35** 23/99 **37** 16181/4995 **39** 20 meters

Exercises 12.3, page 571

1 Converges **3** Diverges **5** Diverges **7** Diverges **9** Converges
11 Converges **13** Converges **15** Diverges **17** Converges **19** Converges
21 Diverges **23** Diverges **25** Converges **27** Converges **29** Converges

31 Converges **33** Converges **35** Diverges **37** Converges **39** Converges
41 Converges **43** Converges **45** Converges
47 Converges if $k > 1$, diverges if $k \leqslant 1$

Exercises 12.4, page 575

1 Converges **3** Converges **5** Diverges **7** Converges **9** Diverges
11 Converges **13** 0.368 **15** 0.305 **17** 316 **19** 7

Exercises 12.5, page 581

1 Conditionally convergent **3** Absolutely convergent **5** Absolutely convergent
7 Absolutely convergent **9** Divergent **11** Absolutely convergent
13 Absolutely convergent **15** Conditionally convergent **17** Absolutely convergent
19 Absolutely convergent **21** Absolutely convergent **23** Divergent **25** Divergent
27 Absolutely convergent **29** Absolutely convergent **31** Divergent **33** Convergent
35 Absolutely convergent **37** Divergent

Exercises 12.6, page 587

1 $[-1, 1)$ **3** $(-2, 2)$ **5** $(-1, 1]$ **7** $[-1, 1)$ **9** $[-1, 1]$ **11** $(-6, 14)$
13 Converges only for $x = 0$ **15** $(-2, 2)$ **17** $(-\infty, \infty)$ **19** $[17/9, 19/9)$
21 $(-12, 4)$ **23** $(0, 2e)$ **25** $(-5/2, 7/2]$ **27** $r = 3/2$ **29** $r = 1/e$

Exercises 12.7, page 593

1 $\sum_{n=0}^{\infty} x^n$, $-1 < x < 1$ **3** $\sum_{n=1}^{\infty} nx^{n-1}$, $-1 < x < 1$ **5** $\sum_{n=0}^{\infty} x^{2(n+1)}$, $-1 < x < 1$

7 $\displaystyle\sum_{n=0}^{\infty} \frac{3^n}{2^{n+1}} x^{n+1}$, $-2/3 < x < 2/3$ **9** $-1 - x - 2\sum_{n=2}^{\infty} x^n$, $-1 < x < 1$

11 0.182 **13** $\displaystyle\sum_{n=0}^{\infty} \frac{(-1)^n}{n!} x^n$ **15** $\displaystyle\sum_{n=0}^{\infty} \frac{1}{(2n)!} x^{2n}$ **17** $\displaystyle\sum_{n=0}^{\infty} \frac{3^n}{n!} x^{n+1}$ **19** 0.3333

21 0.0999 **23** 0.9677 **25** $\sum_{n=1}^{\infty} 2nx^{2n-1}$ **27** $\displaystyle\sum_{n=1}^{\infty} \frac{(-1)^n}{2n(n!)} x^{2n}$ **29** $\displaystyle -\sum_{n=1}^{\infty} \frac{1}{n^2} x^n$

Exercises 12.8, page 600

1 $\displaystyle\sum_{n=0}^{\infty} \frac{(-1)^n}{(2n)!} x^{2n}$ **3** $\displaystyle\sum_{n=0}^{\infty} \frac{2^n}{n!} x^n$ **5** $\displaystyle\sum_{n=0}^{\infty} \frac{1}{n!} x^{n+2}, \infty$ **7** $\displaystyle\sum_{n=0}^{\infty} \frac{1}{(2n+1)!} x^{2n+1}, \infty$

9 $\displaystyle\sum_{n=0}^{\infty} \frac{(-1)^n 3^{2n+1}}{(2n+1)!} x^{2n+2}, \infty$ **11** $\displaystyle 1 + \sum_{n=1}^{\infty} \frac{(-1)^n 2^{2n-1}}{(2n!)} x^{2n}, \infty$

13 $\displaystyle\sum_{n=0}^{\infty} \frac{(-1)^n}{\sqrt{2}(2n+1)!}\left(x - \frac{\pi}{4}\right)^{2n+1} + \sum_{n=0}^{\infty} \frac{(-1)^n}{\sqrt{2}(2n)!}\left(x - \frac{\pi}{4}\right)^{2n}$

15 $\displaystyle\sum_{n=0}^{\infty} \frac{(-1)^n}{2^{n+1}}(x-2)^n$ **17** $\displaystyle\sum_{n=0}^{\infty} \frac{(\ln 10)^n}{n!} x^n$ **19** $\displaystyle\sum_{n=0}^{\infty} \frac{e^{-2}2^n}{n!}(x+1)^n$

21 $\displaystyle 2 + 2\sqrt{3}\left(x - \frac{\pi}{3}\right) + 7\left(x - \frac{\pi}{3}\right)^2 + \frac{23\sqrt{3}}{3}\left(x - \frac{\pi}{3}\right)^3 + \cdots$

23 $\displaystyle \frac{\pi}{6} + \frac{2}{\sqrt{3}}(x - \tfrac{1}{2}) + \frac{2}{3\sqrt{3}}(x - \tfrac{1}{2})^2 + \frac{8}{9\sqrt{3}}(x - \tfrac{1}{2})^3$

25 $-(1/e) + (1/2e)(x+1)^2 + (1/3e)(x+1)^3 + (1/8e)(x+1)^4$ **27** 1.6487 **29** 0.7468
31 0.4864 **33** 0.4484 **35** 0.1689

Exercises 12.9, page 603

1 (a) $1 + \frac{1}{2}x + \sum_{n=2}^{\infty} (-1)^{n-1} \frac{1 \cdot 3 \cdot 5 \cdots (2n-3)}{2^n n!} x^n$, 1

(b) $1 - \frac{1}{2}x^3 - \sum_{n=2}^{\infty} \frac{1 \cdot 3 \cdot 5 \cdots (2n-3)}{2^n n!} x^{3n}$, 1

3 $1 - \sum_{n=1}^{\infty} (-1)^n \frac{1}{2}(n+1)(n+2)x^n$, 1 **5** $2 + \frac{1}{12}x + 2\sum_{n=2}^{\infty} (-1)^{n-1} \frac{2 \cdot 5 \cdot 8 \cdots (3n-4)}{3^n 8^n n!} x^n$, 8

7 $x + \sum_{n=1}^{\infty} \frac{1 \cdot 3 \cdot 5 \cdots (2n-1)}{2^n n!\,(2n+1)} x^{2n+1}$, 1 **9** 0.508

12.10 Review Exercises, page 605

1 0 **3** Does not exist **5** 5 **7** Divergent **9** Convergent **11** Divergent
13 Divergent **15** Convergent **17** Divergent **19** Divergent
21 Absolutely convergent **23** Conditionally convergent **25** Absolutely convergent
27 Convergent **29** Convergent **31** Conditionally convergent **33** Convergent
35 Convergent **37** Divergent **39** 0.159 **41** $(-3, 3)$ **43** $[-12, -8)$ **45** 1/4

47 $\sum_{n=1}^{\infty} \frac{(-1)^{n+1}}{(2n)!} x^{2n-1}$, ∞ **49** $\sum_{n=0}^{\infty} \frac{(-1)^n 2^{2n}}{(2n+1)!} x^{2n+1}$, ∞

51 $1 + \frac{2}{3}x + 2\sum_{n=2}^{\infty} (-1)^{n-1} \frac{1 \cdot 4 \cdot 7 \cdots (3n-5)}{3^n n!} x^n$, 1 **53** $e^2 \sum_{n=0}^{\infty} \frac{(-1)^n}{n!}(x+2)^n$

55 $2 + \frac{x-4}{4} + \sum_{n=2}^{\infty} (-1)^{n-1} \frac{1 \cdot 3 \cdot 5 \cdots (2n-3)}{2^{3n-1} n!}(x-4)^n$ **57** 0.189 **59** 1.002

CHAPTER 13

Exercises 13.1, page 615

1 $y = 2x + 7$ **3** $y = x - 2$ **5** $x = y^2 - 6y + 4$ **7** $y = 1/x^2$

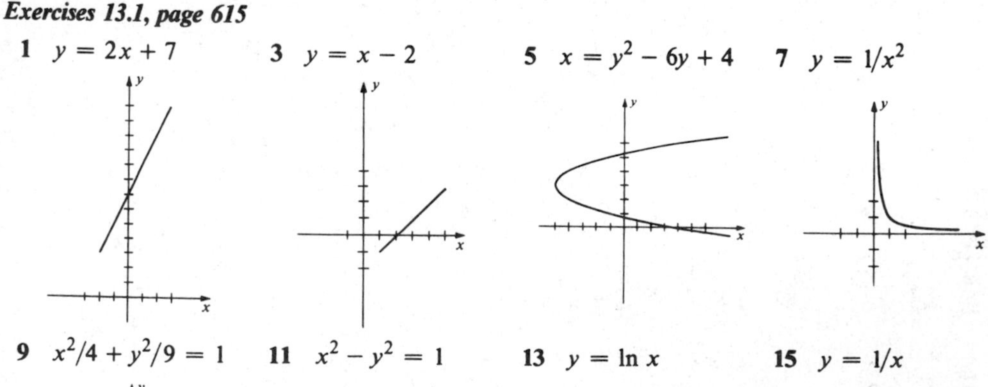

9 $x^2/4 + y^2/9 = 1$ **11** $x^2 - y^2 = 1$ **13** $y = \ln x$ **15** $y = 1/x$

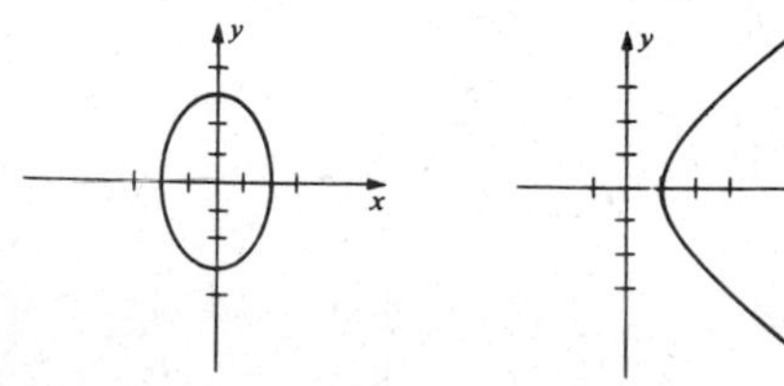

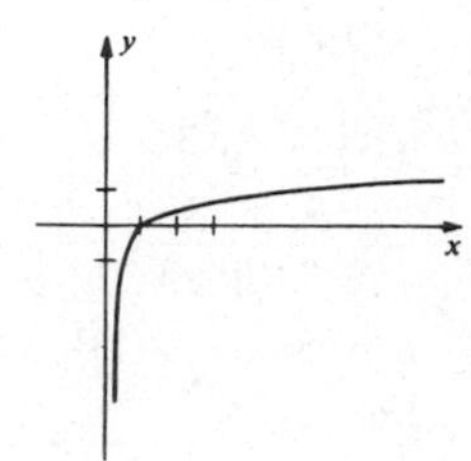

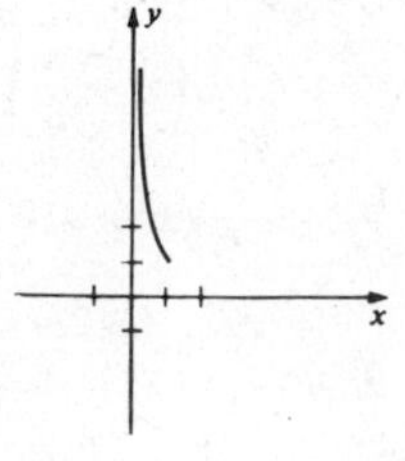

17 $x^2 - y^2 = 1$

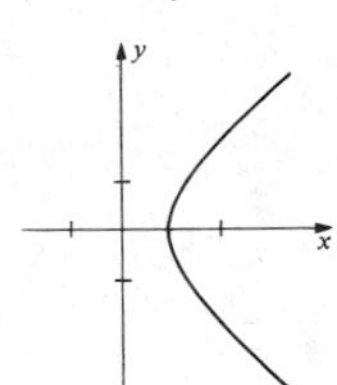

19 $y = \sqrt{x^2 - 1}$
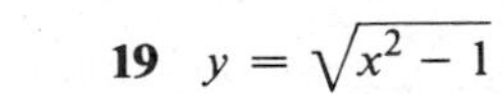

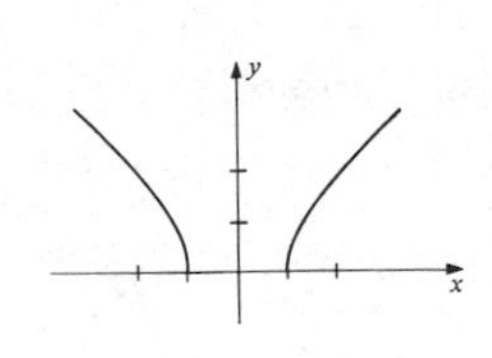

21 Piecewise smooth

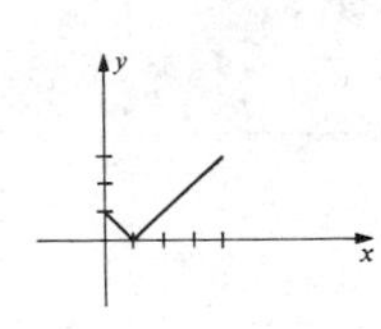

23 Smooth

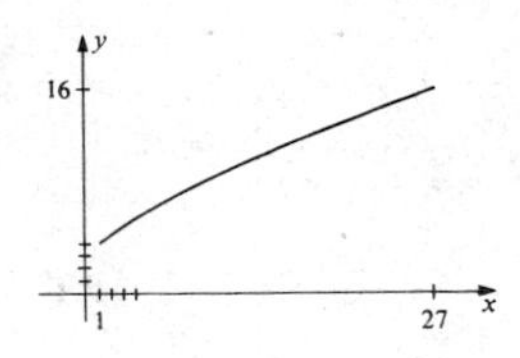

25

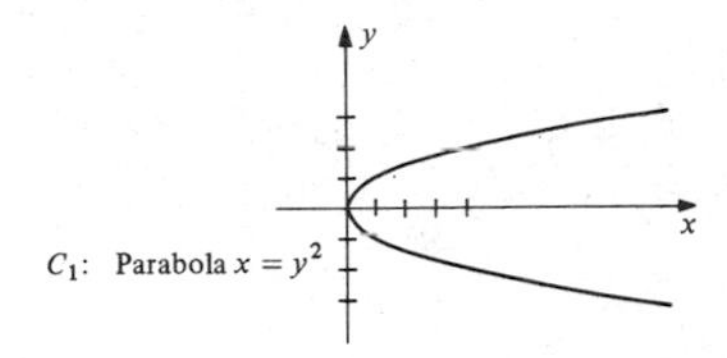

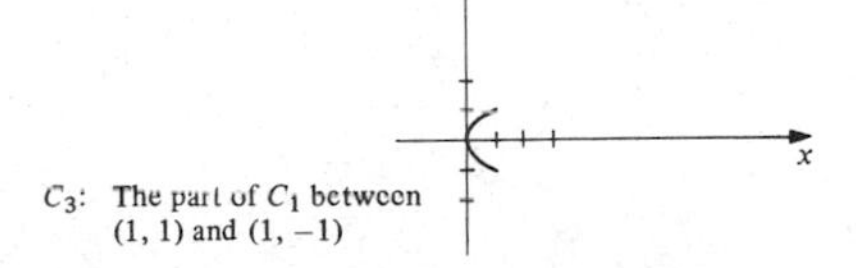

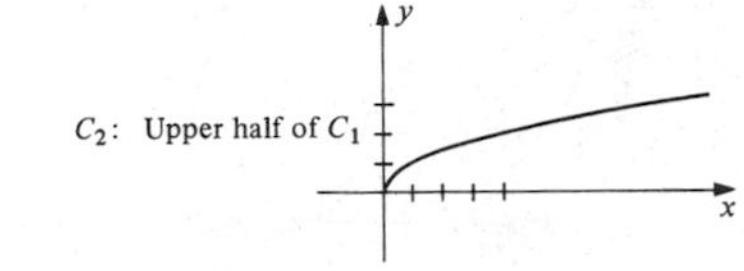

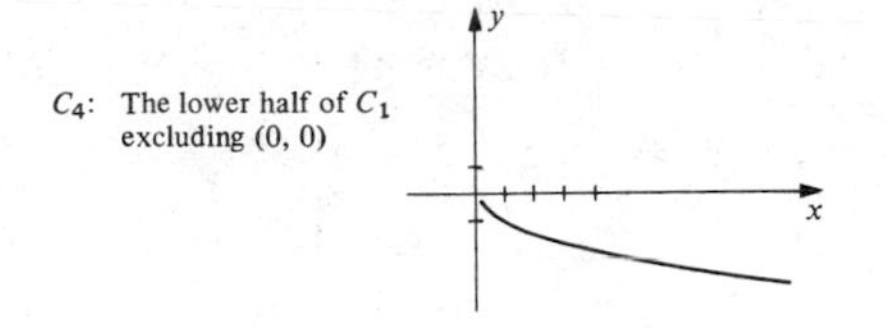

27 (a) P moves counterclockwise along the circle $x^2 + y^2 = 1$ from (1, 0) to $(-1, 0)$. (b) P moves clockwise along the circle $x^2 + y^2 = 1$ from (0, 1) to $(0, -1)$. (c) P moves clockwise along the circle $x^2 + y^2 = 1$ from $(-1, 0)$ to (1, 0).

29 $x = (x_2 - x_1)t^n + x_1$, $y = (y_2 - y_1)t^n + y_1$, are parametric equations for l if n is any odd positive integer.

Exercises 13.2, page 619

1 2 **3** 1 **5** 1/4 **7** $-2e^{-3}$ **9** $-\frac{3}{2}\tan 1$ **11** $4x - y = 49$

13 $y - 1 = m(x - 4)$ where $m = 1/(12 \pm 8\sqrt{2})$

15 Horizontal at $(16, -16)$ and (16, 16); vertical at (0, 0); $d^2y/dx^2 = (3t^2 + 12)/64t^3$

17 Vertical at (0, 0) and $(-3, 1)$; no horizontal; $d^2y/dx^2 = (1 - 3t)/144t^{3/2}(t - 1)^3$

19 $\sin t/(1 - \cos t)$; horizontal when $t = (2n + 1)\pi$; vertical when $t = 2n\pi$; slope 1 when $t = \pi/2 + 2n\pi$, where n is an integer

Exercises 13.3, page 625

1

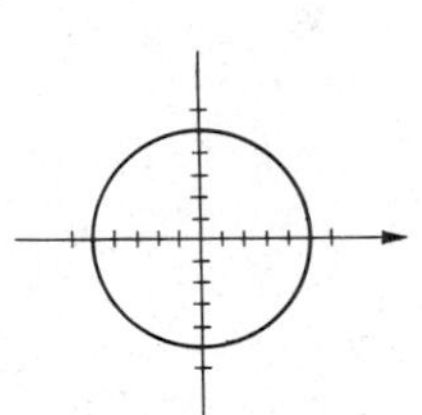

3

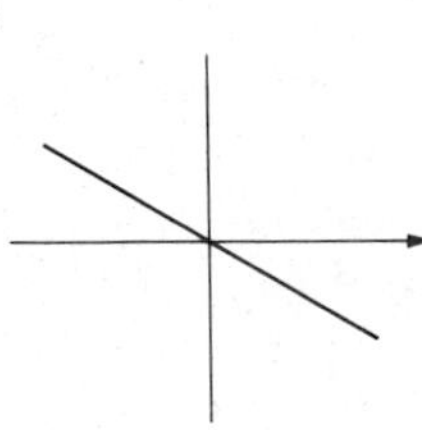

5

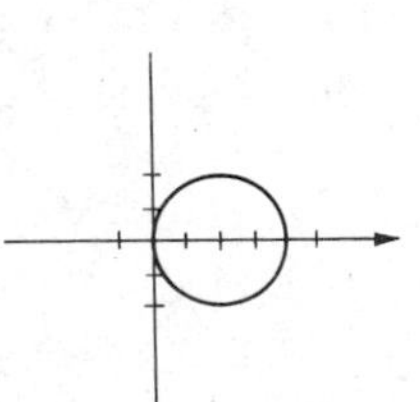

7

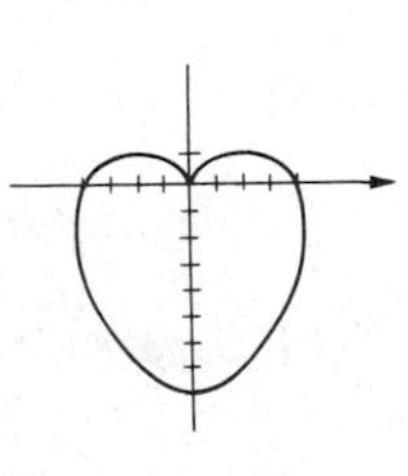

9

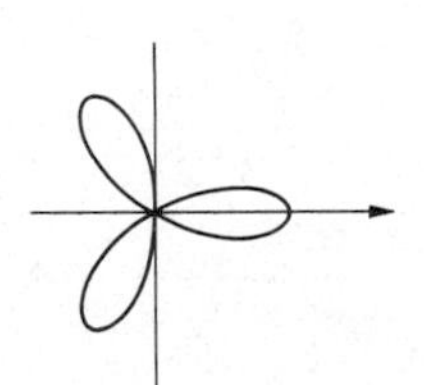

11

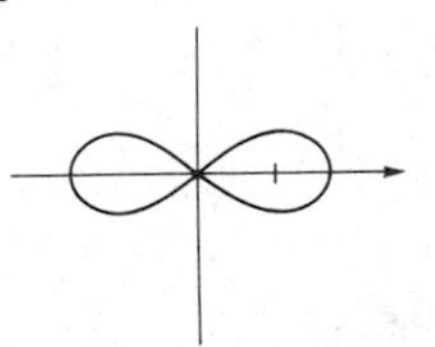

13

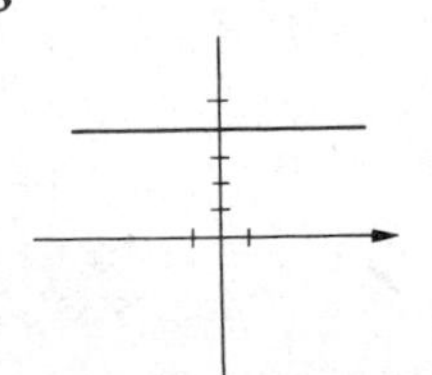

15

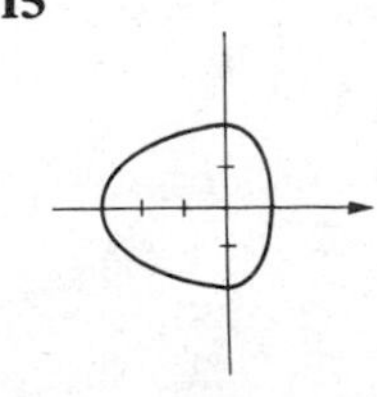

17

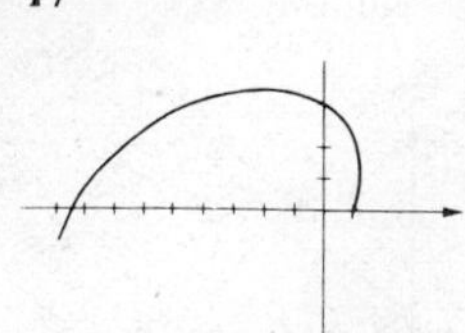

19

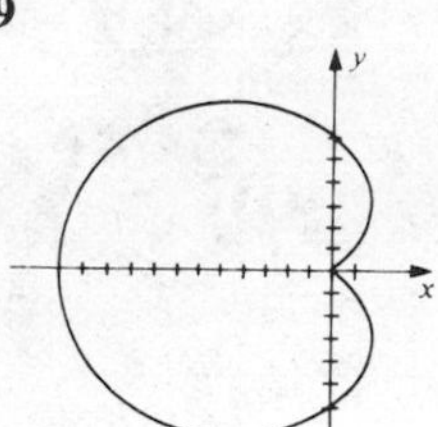

21

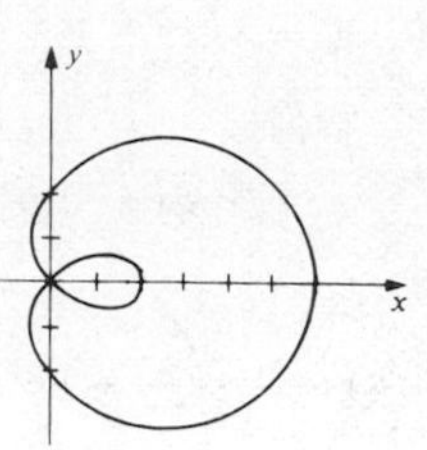

23

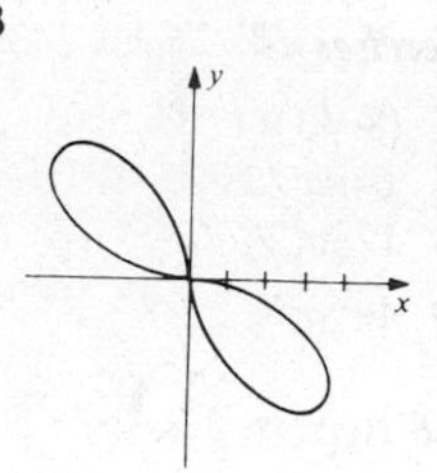

25 $r = -3 \sec \theta$ **27** $r = 4$ **29** $r = 6 \csc \theta$ **31** $r^2 = 16 \sec 2\theta$ **33** $x = 5$
35 $x^2 + y^2 - 6y = 0$ **37** $x^2 + y^2 = 4$ **39** $y^2 = x^4/(1 - x^2)$
41 $y^2/9 - x^2/4 = 1$ **43** $x^2 - y^2 = 1$ **45** $y - 2x = 6$ **47** $y^2 = 1 - 2x$

33

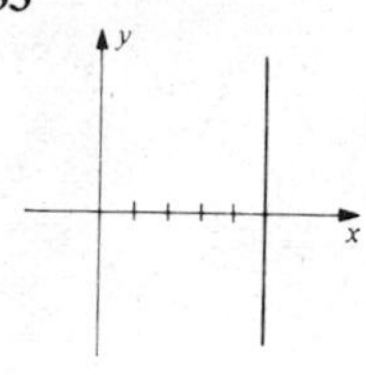

35

37

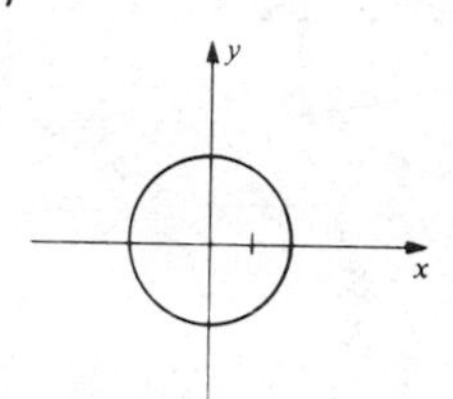

39

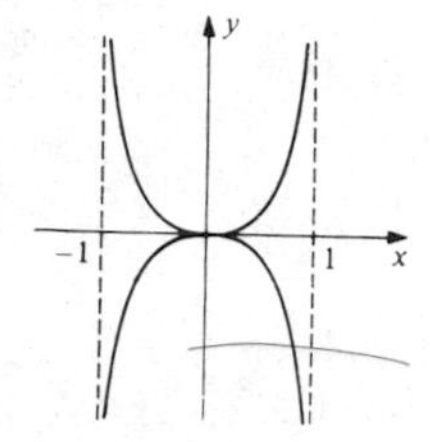

41

43

45

47

Exercises 13.4, page 632

1 Ellipse; vertices $(3/2, \pi/2)$ and $(3, 3\pi/2)$, foci $(0, 0)$ and $(3/2, 3\pi/2)$
3 Hyperbola; vertices $(-3, 0)$ and $(3/2, \pi)$, foci $(0, 0)$ and $(-9/2, 0)$
5 Parabola; $V(3/4, 0)$, $F(0, 0)$
7 Ellipse; vertices $(-4, 0)$ and $(-4/3, \pi)$, foci $(0, 0)$ and $(-8/3, 0)$
9 Hyperbola (except for the points $(\pm 3, 0)$); vertices $(6/5, \pi/2)$ and $(-6, 3\pi/2)$, foci $(0, 0)$ and $(-36/5, 3\pi/2)$
11 $9x^2 + 8y^2 + 12y - 36 = 0$ **13** $y^2 - 8x^2 - 36x - 36 = 0$ **15** $4y^2 = 9 - 12x$
17 $3x^2 + 4y^2 + 8x - 16 = 0$ **19** $4x^2 - 5y^2 + 36y - 36 = 0,\ y \neq 0$ **21** $r = 2/(3 + \cos \theta)$
23 $r = 12/(1 - 4 \sin \theta)$ **25** $r = 5/(1 + \cos \theta)$ **27** $r = 8/(1 + \sin \theta)$
31 $r = 2/(1 + \cos \theta)$, $e = 1$, $r = 2 \sec \theta$ **33** $r = 4/(1 + 2 \sin \theta)$, $e = 2$, $r = 2 \csc \theta$
35 $r = 2/(3 + \cos \theta)$, $e = 1/3$, $r = 2 \sec \theta$

Exercises 13.5, page 636

1 π **3** $3\pi/2$ **5** $\pi/2$ **7** $33\pi/2$ **9** $(e^{\pi} - 1)/4$ **11** 2 **13** $9\pi/20$
15 $2\pi + 9\sqrt{3}/2$ **17** $4\sqrt{3} - 4\pi/3$ **19** $5\pi/24 - \sqrt{3}/4$
21 $11 \sin^{-1}(1/4) + 3\pi/4 - \sqrt{15}/4$ **23** $4/\sqrt{3}$ **25** $64\sqrt{2}/3$

Exercises 13.6, page 640

1 $\frac{2}{27}[34^{3/2} - 125]$ **3** $\sqrt{2}(e^{\pi/2} - 1)$ **5** $\ln|2 + \sqrt{3}|$ **7** 483/32 **9** $\sqrt{2}(1 - e^{-2\pi})$
11 2 **13** $10\sqrt{26} + 2 \ln(5 + \sqrt{26})$ **15** $\pi^2/8$
19 $\sqrt{2} - \sqrt{5}/2 + \ln(2 + \sqrt{5}) - \ln(1 + \sqrt{2})$

Exercises 13.7, page 645

1 $(8\pi/3)(17^{3/2} - 1)$ **3** $11\pi/9$ **5** $(\pi/27)(145^{3/2} - 10^{3/2})$
7 $64\pi a^3/3$ **9** $536\pi/5$ **11** $2\sqrt{2}\,\pi(2e^{\pi} + 1)/5$
13 $128\pi(2\sqrt{2} - 1)/3$ **15** $\pi[2\sqrt{5} + \ln(2 + \sqrt{5}) - \sqrt{2} - \ln(1 + \sqrt{2})]$ **17** $128\pi/5$
19 $4\pi^2 a^2$ **21** 10π **25** $4\pi^2 ab$

13.8 Review Exercises, page 646

1 $y = 2(x - 1) - 1/(x - 1), 3$ **3** $y = e^{-x^2}, -2e^{-1}$

5

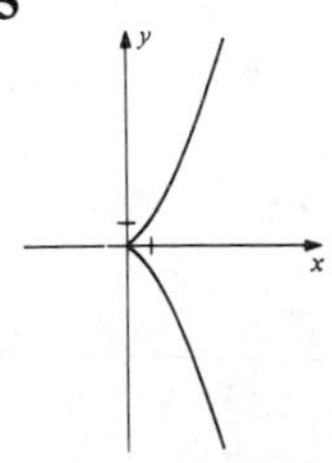

(a)

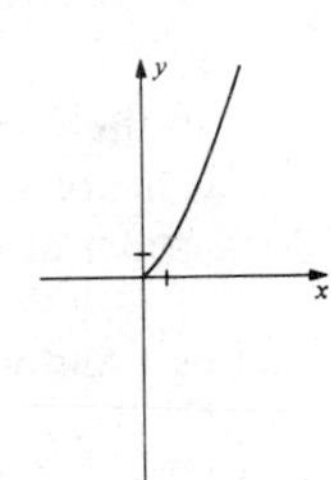

(b)

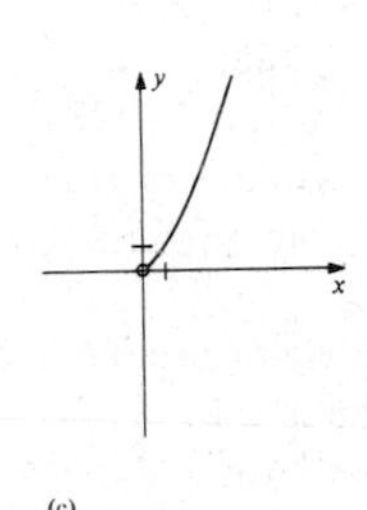

(c)

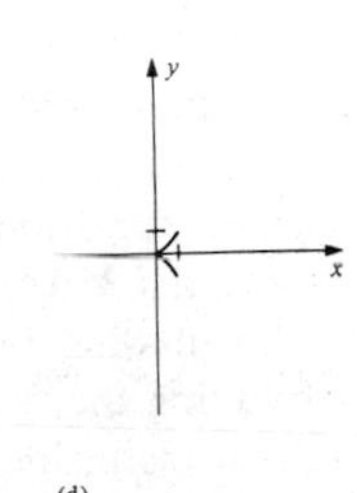

(d)

7 $(x^2 + y^2)^{3/2} = 6(x^2 - y^2)$

9 $(x^2 + y^2)^2 + 8xy = 0$ **11** $3x - 2y = 6$ **13** $x^2 - y^2 = 1$

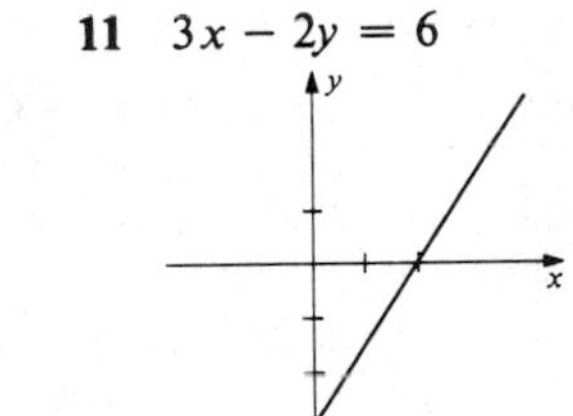

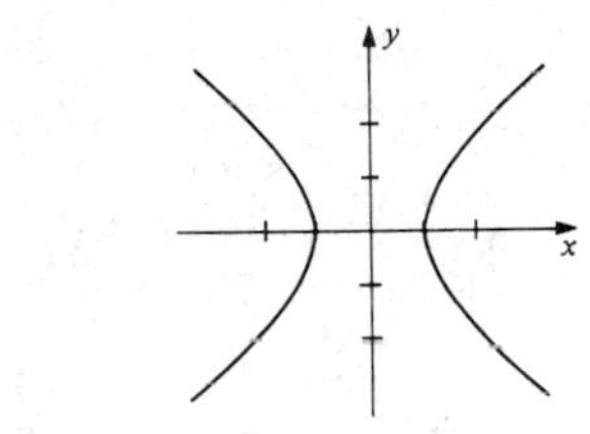

15 $x^2 + y^2 = 2(\sqrt{x^2 + y^2} + x)$ **17** $x^2 + y^2 = 1$ **19** $8x^2 + 9y^2 + 16x - 64 = 0$

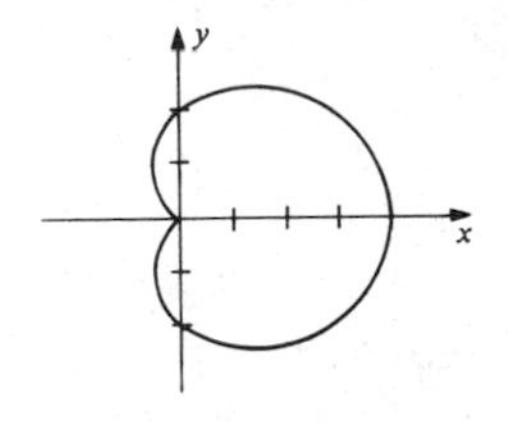

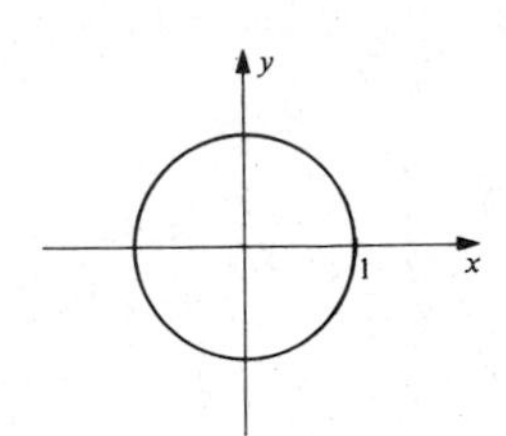

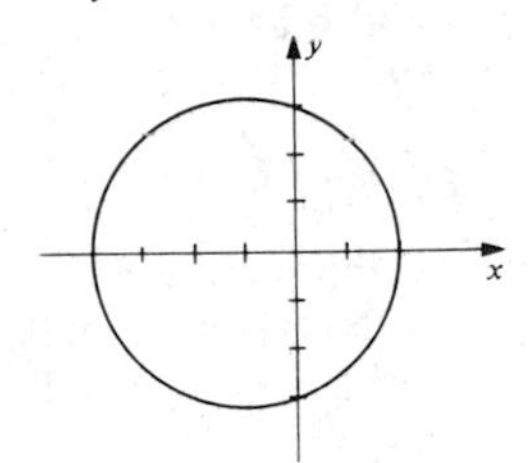

21 $r = 2\cos\theta \sec 2\theta$ **23** 2 **25** $\sqrt{2} + \ln(1 + \sqrt{2})$
27 $(\pi/2)(\sinh 2 + 2)$ **29** $((2\sqrt{2}\,\pi)/5)[e^2(2\cos 1 + \sin 1) - 2]$

CHAPTER 14

Exercises 14.1, page 659

11 $\langle 3, 1\rangle, \langle 1, -7\rangle, \langle 13, 8\rangle, \langle 3, -32\rangle$ **13** $\langle -15, 6\rangle, \langle 1, -2\rangle, \langle -68, 28\rangle, \langle 12, -12\rangle$
15 $4\mathbf{i} - 3\mathbf{j}, -2\mathbf{i} + 7\mathbf{j}, 19\mathbf{i} - 17\mathbf{j}, -11\mathbf{i} + 33\mathbf{j}$ **17** $-2\mathbf{i} - 5\mathbf{j}, -6\mathbf{i} + 7\mathbf{j}, -6\mathbf{i} - 26\mathbf{j}, -26\mathbf{i} + 34\mathbf{j}$
19 $-3\mathbf{i} + 2\mathbf{j}, 3\mathbf{i} + 2\mathbf{j}, -15\mathbf{i} + 8\mathbf{j}, 15\mathbf{i} + 8\mathbf{j}$ **21** $\langle 4, 7\rangle$ **23** $\langle -6, 0\rangle$ **25** $\langle 9, -3\rangle$
31 $3\sqrt{2}$ **33** 5 **35** $\sqrt{41}$ **37** 18
39 (a) $(-8/17)\mathbf{i} + (15/17)\mathbf{j}$ (b) $(8/17)\mathbf{i} - (15/17)\mathbf{j}$
41 (a) $\langle 2/\sqrt{29}, -5/\sqrt{29}\rangle$ (b) $\langle -2/\sqrt{29}, 5/\sqrt{29}\rangle$
43 (a) $\langle -12, 6\rangle$ (b) $\langle -3, 1.5\rangle$ **45** $(24/\sqrt{65})\mathbf{i} - (42/\sqrt{65})\mathbf{j}$

Exercises 14.2, page 665

1 (a) $\sqrt{104}$ (b) $(3, 1, -1)$ **3** (a) $\sqrt{53}$ (b) $(-1/2, -1, 1)$ **5** (a) $\sqrt{3}$ (b) $(1/2, 1/2, 1/2)$
7 $(2, 5, 1), (-4, 2, -3), (-4, 5, 1), (2, 2, -3), (-4, 5, -3), (2, 2, 1)$
9 *Hint*: Show that $d(A,B)^2 + d(A,C)^2 = d(B,C)^2$; $3\sqrt{2}/2$
11 $x^2 + y^2 + z^2 - 6x + 2y - 4z + 5 = 0$ **13** $4x^2 + 4y^2 + 4z^2 + 40x - 8z + 103 = 0$
15 (a) $x^2 + y^2 + z^2 + 4x - 8y + 12z + 52 = 0$
(b) $x^2 + y^2 + z^2 + 4x - 8y + 12z + 40 = 0$
(c) $x^2 + y^2 + z^2 + 4x - 8y + 12z + 20 = 0$
17 $C(-2, 1, -1)$; 2 **19** $C(4, 0, -4)$; 4 **21** $C(0, -2, 0)$; 2
23 $6x - 12y + 4z + 13 = 0$; a plane
25 (a) A plane parallel to the *xy*-plane (b) A plane parallel to the *xz*-plane (c) The *yz*-plane
27 All points within and on a sphere of radius 1 with center at the origin
29 All points within and on a rectangular parallelepiped with center at the origin and having edges of lengths 2, 4, and 6
31 All points inside and on a right circular cylinder with radius 1 and vertical axis the *z*-axis, excluding the points on the *z*-axis.
33 All points inside the sphere with center (2, 1, 0) and radius 2.

Exercises 14.3, page 675

3 $\langle 1, 3, 0\rangle, \langle -22, 42, 9\rangle, -25, \sqrt{41}, 3\sqrt{41}, 3\sqrt{41}, \langle -5, 9, 2\rangle, \sqrt{110}$
5 $4\mathbf{i} - 2\mathbf{j} - 3\mathbf{k}, 11\mathbf{i} - 28\mathbf{j} + 30\mathbf{k}, -15, \sqrt{29}, 3\sqrt{29}, 3\sqrt{29}, 2\mathbf{i} - 6\mathbf{j} + 7\mathbf{k}, \sqrt{89}$
7 $\mathbf{i} + \mathbf{k}, 5\mathbf{i} + 9\mathbf{j} - 4\mathbf{k}, -1, \sqrt{2}, 3\sqrt{2}, 3\sqrt{2}, \mathbf{i} + 2\mathbf{j} - \mathbf{k}, \sqrt{6}$ **11** 3 **13** -12
15 -159 **17** $-3/\sqrt{30}$ **19** 0 **21** $-3/\sqrt{534}$ **23** $6/13$ **25** 74
27 $\cos^{-1}(37/\sqrt{3081}) \approx 48.20°$ **29** $-82/\sqrt{126}$
31 $-2(x - 8) + 4(y + 3) - 12(z - 5) = 0$. A plane through P. **33** 1
35 $-12/\sqrt{3}$ **37** $3/10$
39 $\dfrac{1}{\sqrt{1589}}\langle 32, 23, 6\rangle$ **41** (a) $2\mathbf{a} = 28\mathbf{i} - 30\mathbf{j} + 12\mathbf{k}$ (b) $-\frac{1}{3}\mathbf{a} = -(\frac{14}{3})\mathbf{i} + 5\mathbf{j} - 2\mathbf{k}$
43 (a) If **a** and **b** have the same or opposite directions (b) If **a** and **b** have the same direction

Exercises 14.4, page 679

1 $-2/\sqrt{30}, 1/\sqrt{30}, 5/\sqrt{30}$ for **a** and 3**a**; $2/\sqrt{30}, -1/\sqrt{30}, -5/\sqrt{30}$ for -2**a**
3 $0, 4/5, -3/5$ for **a** and 3**a**; $0, -4/5, 3/5$ for $-$**a**
5 For **i**: $0, \pi/2, \pi/2$; 1, 0, 0. For **j**: $\pi/2, 0, \pi/2$; 0, 1, 0. For **k**: $\pi/2, \pi/2, 0$; 0, 0, 1.
7 $-4/\sqrt{57}, 4/\sqrt{57}, -5/\sqrt{57}$ **9** $1/\sqrt{26}, 0, -5/\sqrt{26}$
11 $\langle 2/\sqrt{3}, 2/\sqrt{3}, 2/\sqrt{3}\rangle$ **13** $\sqrt{7}/3$

Exercises 14.5, page 682

1 $x = 5 - 3t, y = -2 + 8t, z = 4 - 3t$; $(1, 26/3, 0), (17/4, 0, 13/4), (0, 34/3, -1)$
3 $x = 2 - 8t, y = 0, z = 5 - 2t$; $(-18, 0, 0), (0, 0, 9/2)$, (line lies on the *xz*-plane)
5 $x = 4 + (1/3)t, y = 2 + 2t, z = -3 + (1/2)t$ **7** $x = 0, y = t, z = 0$
9 $x = -6 - 3s, y = 4 + s, z = -3 + 9s$ **11** $\cos^{-1}(15/2\sqrt{171}) \approx 55°$ **13** $(5, -7, 3)$
15 Do not intersect **17** Intersect at $(2, -4, -1)$ **19** $\sqrt{5411}/3\sqrt{10}$

Exercises 14.6, page 688

9 (a) $z = 4$ (b) $x = 6$ (c) $y = -7$ **11** $6x - 5y - z + 84 = 0$
13 $3x - y + 2z + 11 = 0$ **15** $20x - 5y + 2z = 0$ **17** $11x - 2y - 6z - 69 = 0$
19 $(x - 5)/3 = (y + 2)/(-8) = (z - 4)/3$ **21** $(x - 4)/(-7) = (z + 3)/8$; $y = 2$
23 (a) No (b) Yes (c) No **25** (a) No (b) No (c) Yes, at (8, 0, 1) **27** $7/\sqrt{59}$
29 3 **31** $6x + 11y + 4z = 38$ **33** $71x - 15y + 17z = -10$

Exercises 14.7, page 698

1 $\langle 5, 10, 5 \rangle$ **3** $-6\mathbf{i} - 8\mathbf{j} + 18\mathbf{k}$ **5** $\langle -4, 2, -1 \rangle$ **7** $-40\mathbf{i} + 15\mathbf{k}$ **9** $\mathbf{0}$
11 $\langle 12, -14, 24 \rangle$, $\langle 16, -2, -5 \rangle$ **13** $(\mathbf{a} \times \mathbf{b}) \cdot \mathbf{b} = \mathbf{a} \cdot (\mathbf{b} \times \mathbf{b}) = \mathbf{a} \cdot \mathbf{0} = 0$
17 No. $\mathbf{a} \times (\mathbf{b} - \mathbf{c}) = \mathbf{0}$ implies that $\mathbf{b} - \mathbf{c}$ is parallel to $\mathbf{a}$ **19** (a) $13\mathbf{i} + 7\mathbf{j} + 5\mathbf{k}$ (b) $13x + 7y + 5z - 16 = 0$ (c) $9\sqrt{3}/2$
21 (a) $c(5\mathbf{i} + 4\mathbf{j} + 10\mathbf{k})$ where $c \neq 0$ (b) $5x + 4y + 10z = 20$ (c) $\sqrt{141}$ **25** 16
27 $89/\sqrt{521}$ **29** $\sqrt{474/17}$

Exercises 14.8, page 704

1

3

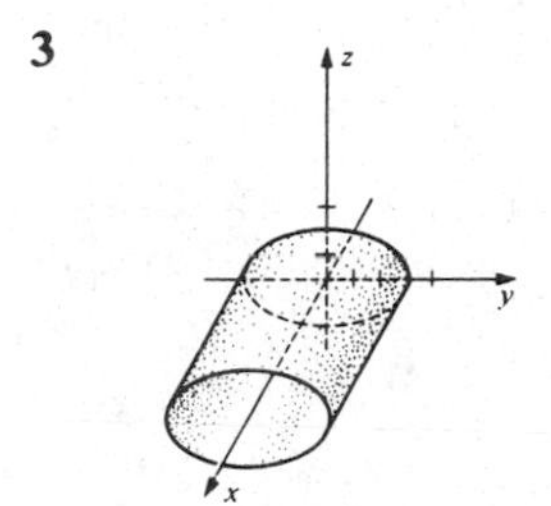

5

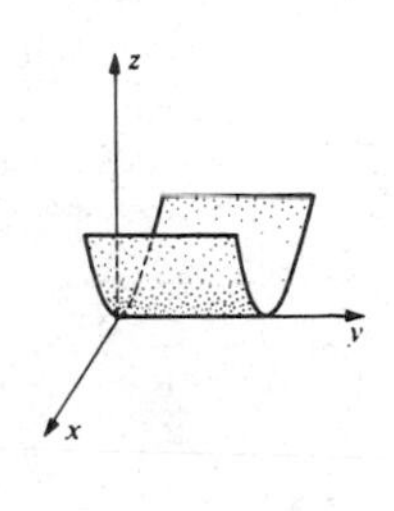

7

9

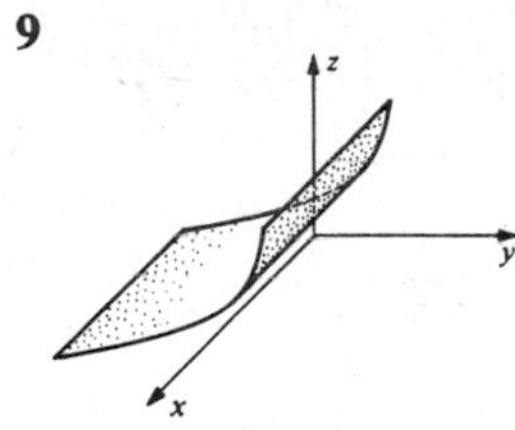

11 $x^2 + z^2 + 4y^2 = 16$ **13** $z = 4 - x^2 - y^2$ **15** $y^2 + z^2 - x^2 = 1$

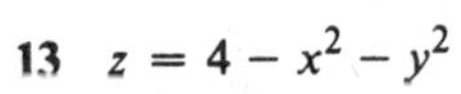

17

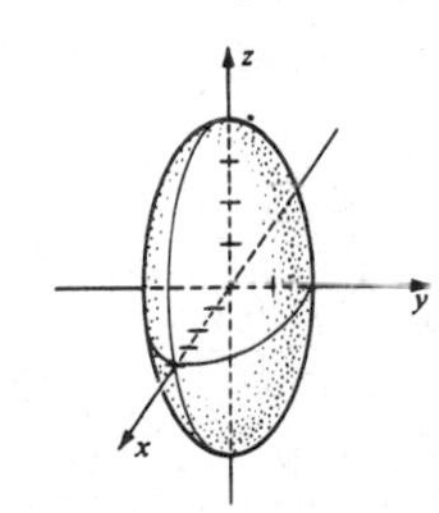

19

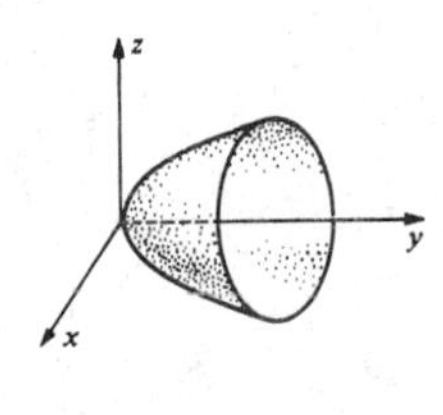

Exercises 14.9, page 710

1 Paraboloid; axis along the z-axis **3** Hyperboloid of one sheet; axis along the z-axis
5 Hyperboloid of two sheets; axis along the x-axis **7** Cone; axis along the x-axis
9 Ellipsoid **11** Hyperboloid of one sheet; axis along the x-axis
13 Hyperboloid of two sheets; axis along the y-axis **15** Paraboloid; axis along the x-axis
17 Cone; axis along the y-axis **19** Hyperbolic paraboloid

Exercises 14.10, page 713

1 (a) $(0, 5, 3)$ (b) $(3, 3\sqrt{3}, -5)$ **3** (a) $(\sqrt{10}, \pi/4, \cos^{-1}(-2/\sqrt{5}))$ (b) $(2, \pi/3, \pi/2)$
5 (a) $(2\sqrt{3}, \pi/3, 2)$ (b) $(1, \pi/4, -\sqrt{3})$ **7** $x^2 + y^2 = 16$. Circular cylinder
9 $z = \sqrt{3x^2 + 3y^2}$. Half cone **11** $x^2 + y^2 = 4x$. Circular cylinder
13 $x^2 + y^2 + z^2 = 4z$. Sphere **15** The z-axis
17 A cylinder with a lemniscate as its directrix and rulings parallel to the z-axis
19 Spheres of radii 1 and 2 with centers at the origin
21 $r^2 + z^2 = 4$, $\rho = 2$
23 $3r\cos\theta + r\sin\theta - 4z = 12$, $\rho(3\sin\phi\cos\theta + \sin\phi\sin\theta - 4\cos\phi) = 12$
25 $r^2\sin^2\theta + z^2 = 9$, $\rho^2(\sin^2\phi\sin^2\theta + \cos^2\phi) = 9$

14.11 Review Exercises, page 714

1 $12\mathbf{i} + 19\mathbf{j}$ **3** -8 **5** $\sqrt{29} - \sqrt{17}$ **7** $\tan^{-1}(5/2)$ **9** $(1/\sqrt{29})(5\mathbf{i} - 2\mathbf{j})$
11 (a) $\sqrt{38}$ (b) $(2, -7/2, 5/2)$ (c) $x^2 + y^2 + z^2 + 2x + 8y - 6z + 10 = 0$ (d) $y = -4$
(e) $x = 5 + 6t, y = -3 + t, z = 2 - t$ (f) $6x + y - z - 25 = 0$
13 $6x - 15y + 5z = 30$ **15** $x^2/64 + y^2/9 + z^2 = 1$
17 Sphere with center $(7, -3, 4)$ and radius 8
19 Plane; x-, y-, and z-intercepts $10/3$, -2, and 5, respectively
21 Elliptic cylinder with rulings parallel to the y-axis
23 Hyperboloid of one sheet; axis along the x-axis
25 Hyperboloid of two sheets; axis along the z-axis **27** Hyperbolic paraboloid
29 36 **31** $\sqrt{33} + \sqrt{37}$ **33** $3/\sqrt{26}, -1/\sqrt{26}, -4/\sqrt{26}$ **35** $22\mathbf{i} - 2\mathbf{j} + 17\mathbf{k}$
37 $9/\sqrt{33}$ **39** 26 **41** **0** **43** $-4\mathbf{i} - 10\mathbf{j} + 4\mathbf{k}$
45 (a) $1/\sqrt{6}, -2/\sqrt{6}, 1/\sqrt{6}$ (b) $(1/\sqrt{66})\langle 1, 4, 7\rangle$ (c) $x + 4y + 7z - 5 = 0$
(d) $x = 2 + 7t, y = -1 - 7t, z = 1 + 3t$ (e) 59 (f) $\cos^{-1}(59/\sqrt{3745}) \approx 15.4°$ (g) $\sqrt{66}$
47 $x = 3 + 2t, y = -1 - 4t, z = 5 + 8t$; $x = -1 + 7t, y = 6 - 2t, z = (-7/2) - 2t$
49 $\cos^{-1}(-25/\sqrt{2295}) \approx 121.46°$ **51** $(2\sqrt{2}, -\pi/4, 1), (3, -\pi/4, \cos^{-1}(1/3))$
53 $z = -\sqrt{(x^2 + y^2 + z^2)/2}$. Bottom half of a cone **55** $x = 1$. Plane
57 $r = 1, \rho^2 \sin^2\phi = 1$ **59** $r^2 + z^2 - 2z = 0, \rho - 2\cos\phi = 0$

CHAPTER 15

Exercises 15.1, page 720

1 Elliptic helix

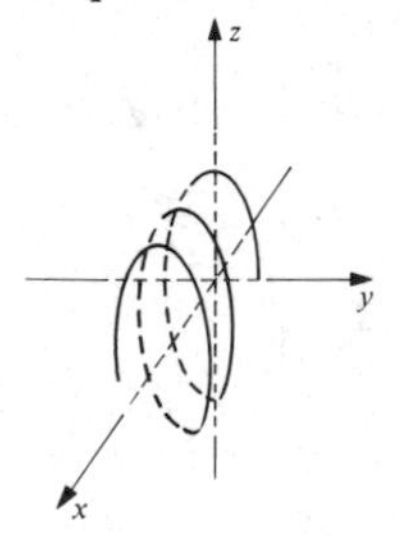

3 Twisted cubic

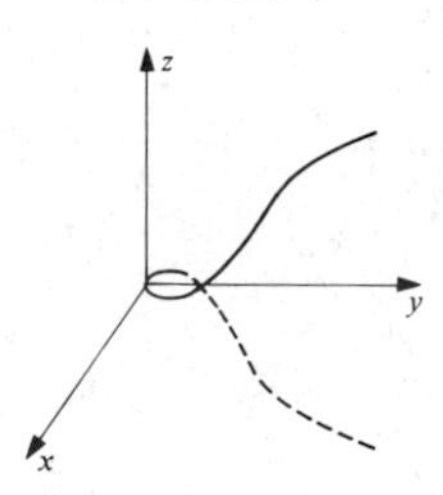

5 Parabola on the plane $z = 3$

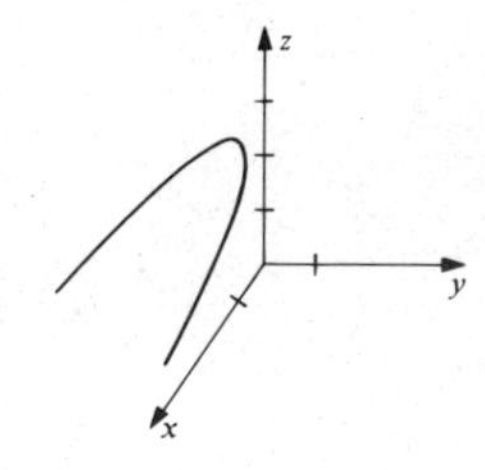

7

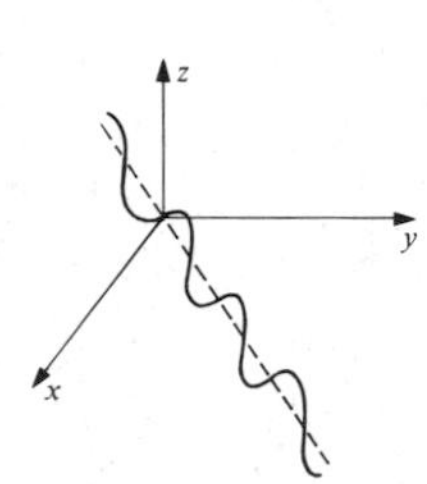

9

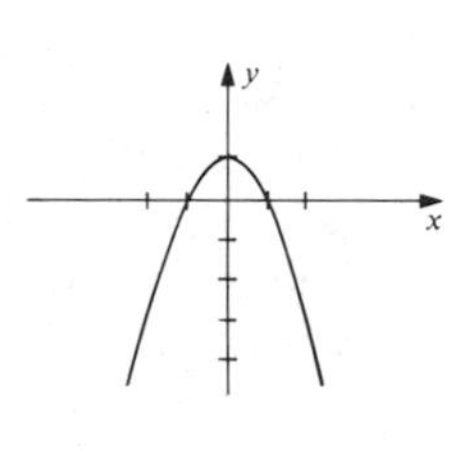

11

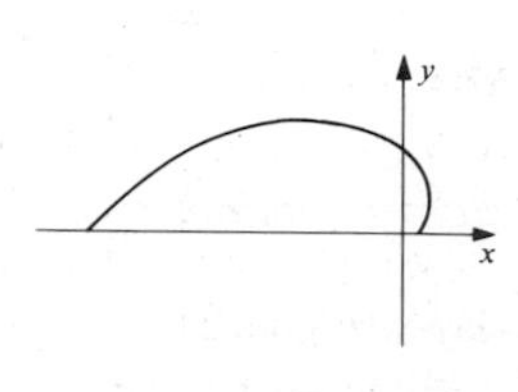

13 $5[4\sqrt{17} + \ln(4 + \sqrt{17})]/4 \approx 23.23$ **15** $\sqrt{3}(e^{2\pi} - 1)$ **17** 7

Exercises 15.2, page 726

1 (a) Continuous throughout its domain [1, 2]
(b) $\mathbf{r}'(t) = (1/2)(t-1)^{-1/2}\mathbf{i} - (1/2)(2-t)^{-1/2}\mathbf{j}$, $\mathbf{r}''(t) = (-1/4)(t-1)^{-3/2}\mathbf{i} - (1/4)(2-t)^{-3/2}\mathbf{j}$
3 (a) Continuous throughout its domain $\{t: t \neq \pi/2 + n\pi\}$
(b) $\mathbf{r}'(t) = \langle \sec^2 t, 2t + 8\rangle$, $\mathbf{r}''(t) = \langle 2\sec^2 t \tan t, 2\rangle$
5 (b) $\mathbf{r}'(t) = \langle -t^3, 2t\rangle$, $\mathbf{r}(2) = \langle -4, 4\rangle$, $\mathbf{r}'(2) = \langle -8, 4\rangle$
7 (b) $\mathbf{r}'(t) = -4\sin t\,\mathbf{i} + 2\cos t\,\mathbf{j}$, $\mathbf{r}(3\pi/4) = -2\sqrt{2}\mathbf{i} + \sqrt{2}\mathbf{j}$, $\mathbf{r}'(3\pi/4) = -2\sqrt{2}\mathbf{i} - \sqrt{2}\mathbf{j}$
9 (b) $\mathbf{r}'(t) = 3t^2\mathbf{i} - 3t^{-4}\mathbf{j}$; $\mathbf{r}(1) = \mathbf{i} + \mathbf{j}$, $\mathbf{r}'(1) = 3\mathbf{i} - 3\mathbf{j}$ **11** (b) $\mathbf{r}'(t) = 2\mathbf{i} - \mathbf{j}$; $\mathbf{r}(3) = 5\mathbf{i} + \mathbf{j}$

5 **7**

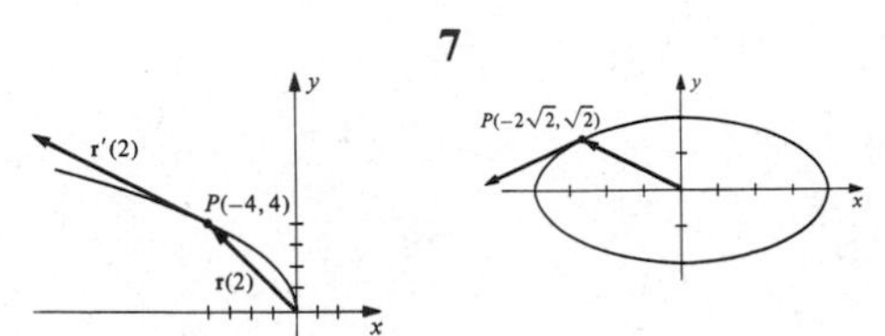

9

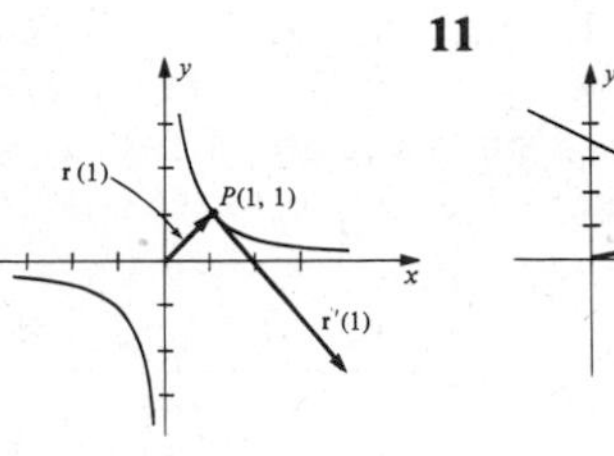

11

13 (a) Domain $= \{t\colon t \neq \pi/2 + k\pi$, k any integer$\}$, $\mathbf{r}$ is continuous on D.
(b) $\mathbf{r}'(t) = 2t\mathbf{i} + \sec^2 t\mathbf{j}$; $\mathbf{r}''(t) = 2\mathbf{i} + 2\sec^2 t \tan t\mathbf{j}$
15 (a) $t \geqslant 0$, $t \geqslant 0$ (b) $(1/2\sqrt{t})\mathbf{i} + 2e^{2t}\mathbf{j} + \mathbf{k}$, $(-1/4t\sqrt{t})\mathbf{i} + 4e^{2t}\mathbf{j}$
17 $x = 1 + 6t$, $y = -2 - 10t$, $z = 10 + 8t$ **19** $x = 1 + s$, $y = s$, $z = 4$
21 $\pm(1/\sqrt{5})\langle 2, -1, 0\rangle$ **23** $16\mathbf{i} - 8\mathbf{j} + 6\mathbf{k}$ **25** $(1 - 1/\sqrt{2})\mathbf{i} - (1/\sqrt{2})\mathbf{j} + (1/2)\ln 2\mathbf{k}$
27 $\left(\dfrac{t^3}{3} + 2\right)\mathbf{i} + (3t^2 + t - 3)\mathbf{j} + (2t^4 + 1)\mathbf{k}$
29 $(t^3 + t + 7)\mathbf{i} - (t^4 - 2t)\mathbf{j} + (t^2/2 - 3t + 1)\mathbf{k}$ **31** $x + y - 1 = 0$
45 $(1 + 5t^2)\sin t + (3t + 2t^3)\cos t$,
$[(t^3 + 4t)\sin t - t^2 \cos t]\mathbf{i} + [(3t^2 - 2)\sin t + (t^3 - 2t)\cos t]\mathbf{j} + [-3t \sin t + (1 - t^2)\cos t]\mathbf{k}$

Exercises 15.3, page 734

Answers for Exercises 1-7 are in the order, $\mathbf{r}'(t)$, $\mathbf{r}''(t)$, $|\mathbf{r}'(t)|$, $\mathbf{r}'(a)$, $\mathbf{r}''(a)$ where a is the indicated time.

1 $2\mathbf{i} + 8t\mathbf{j}$, $8\mathbf{j}$, $2\sqrt{1 + 16t^2}$, $2\mathbf{i} + 8\mathbf{j}$, $8\mathbf{j}$
3 $\dfrac{-2}{t^2}\mathbf{i} - \dfrac{3}{(t+1)^2}\mathbf{j}$, $\dfrac{4}{t^3}\mathbf{i} + \dfrac{6}{(t+1)^3}\mathbf{j}$, $\sqrt{\dfrac{4}{t^4} + \dfrac{9}{(t+1)^4}}$, $-\frac{1}{2}\mathbf{i} - \frac{1}{3}\mathbf{j}$, $\frac{1}{2}\mathbf{i} + \frac{2}{9}\mathbf{j}$
5 $\cos t\mathbf{i} - 8\sin 2t\mathbf{j}$, $-\sin t\mathbf{i} - 16\cos 2t\mathbf{j}$, $\sqrt{\cos^2 t + 64\sin^2 2t}$, $(\sqrt{3}/2)\mathbf{i} - 4\sqrt{3}\mathbf{j}$, $-\frac{1}{2}\mathbf{i} - 8\mathbf{j}$
7 $2e^{2t}\mathbf{i} - e^{-t}\mathbf{j}$, $4e^{2t}\mathbf{i} + e^{-t}\mathbf{j}$, $\sqrt{4e^{4t} + e^{-2t}}$, $2\mathbf{i} - \mathbf{j}$, $4\mathbf{i} + \mathbf{j}$
9 $-\sin t\mathbf{i} + \cos t\mathbf{j} + \mathbf{k}$, $-\cos t\mathbf{i} - \sin t\mathbf{j}$, $\sqrt{2}$
11 $2t\mathbf{i} + (1/\sqrt{t})\mathbf{j} + 6\sqrt{t}\mathbf{k}$, $2\mathbf{i} - (1/2\sqrt{t^3})\mathbf{j} + (3/\sqrt{t})\mathbf{k}$, $\sqrt{4t^2 + (1/t) + 36t}$
13 $e^t(\cos t - \sin t)\mathbf{i} + e^t(\cos t + \sin t)\mathbf{j} + e^t\mathbf{k}$, $-2e^t \sin t\mathbf{i} + 2e^t \cos t\mathbf{j} + e^t\mathbf{k}$, $\sqrt{3}\,e^t$
15 $\mathbf{i} + 2\mathbf{j} + 3\mathbf{k}$, $\mathbf{0}$, $\sqrt{14}$
19 (a) $750\sqrt{3}\mathbf{i} + (-gt + 750)\mathbf{j}$ (b) $(1500)^2/8g \approx 8{,}789$ ft (c) $(1500)^2\sqrt{3}/2g \approx 60{,}892$ ft
(d) 1500 ft/sec **21** $40\sqrt{5} \approx 89.4$ ft/sec

Exercises 15.4, page 740

1 $6/10^{3/2}$ **3** 2 **5** 1 **7** $2/17^{3/2}$ **9** $48/(21)^{3/2}$ **11** $(\ln\sqrt{2}, 1/\sqrt{2})$
13 $(0, \pm 3)$ **15** $(\sqrt{2}/2, -(1/2)\ln 2)$ **17** $(\pm\sqrt{2}, -20)$ **19** $(0, 0)$

Exercises 15.5, page 744

1 $4t/(4t^2 + 9)^{1/2}$, $6/(4t^2 + 9)^{1/2}$
3 $6t(t^2 + 2)/(t^4 + 4t^2 + 1)^{1/2}$, $6(t^4 + t^2 + 1)^{1/2}/(t^4 + 4t^2 + 1)^{1/2}$
5 $t/(1 + t^2)^{1/2}$, $(2 + t^2)/(1 + t^2)^{1/2}$
7 $-65 \sin t \cos t/(16\sin^2 t + 81\cos^2 t + 1)^{1/2}$,
$(81\sin^2 t + 16\cos^2 t + 1296)^{1/2}/(16\sin^2 t + 81\cos^2 t + 1)^{1/2}$ **9** $6/(4t + 9)^{3/2}$
11 $2(t^4 + t^2 + 1)^{1/2}/3(t^4 + 4t^2 + 1)^{3/2}$
13 $(2 + t^2)/(1 + t^2)^{3/2}$ **15** $(81\sin^2 t + 16\cos^2 t + 1296)^{1/2}/(16\sin^2 t + 81\cos^2 t + 1)^{3/2}$
17 $36/\sqrt{5}$, $18/\sqrt{5}$ **23** 0 **25** $48/(21)^{3/2}$

15.7 Review Exercises, page 752

1 $\mathbf{r}'(t) = 2t\mathbf{i} + (8t - 4t^3)\mathbf{j}$, $\mathbf{r}''(t) = 2\mathbf{i} + (8 - 12t^2)\mathbf{j}$ **3** (a) $(1/\sqrt{3})(\mathbf{i} + \mathbf{j} + \mathbf{k})$

(b) $\sqrt{3}(e-1) \approx 2.98$ **5** $x = 4 + 6t, y = 4 + 4t, z = 1 + t$
7 $-e^{-t}\mathbf{i} + 2\cos 2t\mathbf{j} + 2t^{3/2}\mathbf{k}$ **11** $108/82^{3/2}$
13 $(\sin t \cos t - 8 \sin 2t \cos 2t)/(4\cos^2 2t + \sin^2 t)^{1/2}$,
$(2|\cos 2t \cos t + 2 \sin 2t \sin t|)/(4 \cos^2 2t + \sin^2 t)^{1/2}$

CHAPTER 16

Exercises 16.1, page 760

1 $\mathbb{R} \times \mathbb{R}$; $-29, 6, -4$ **3** $\{(u, v)\colon u \neq 2v\}$; $-3/2, 4/9, 0$
5 $\{(x, y, z)\colon x^2 + y^2 + z^2 \leqslant 25\}$; $4, 2\sqrt{3}$ **7** $\{(r, s, v, p)\colon v \neq \pi/2 + n\pi, p > 0\}$; $3 - \pi$
9 $2x + h, 2y + h$ **11** $y^2 + 3, 2xy + xh$ **13** The top half of the sphere $x^2 + y^2 + z^2 = 1$
15 The plane with x-, y-, and z-intercepts 3, 2, and 6, respectively **17** The plane $z = 5$
19

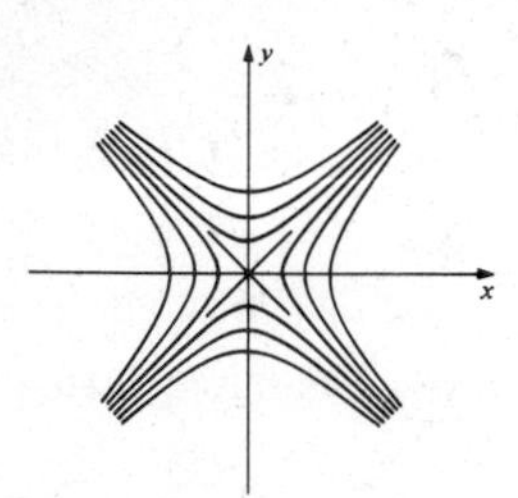

21

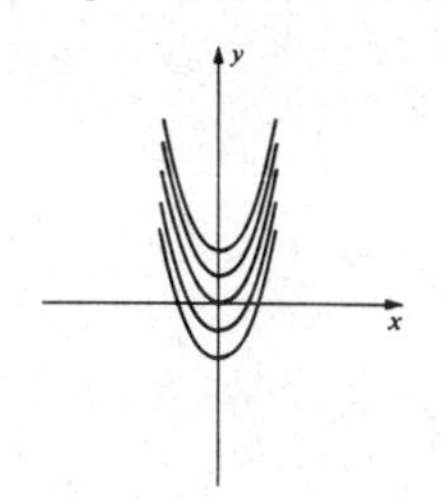

23

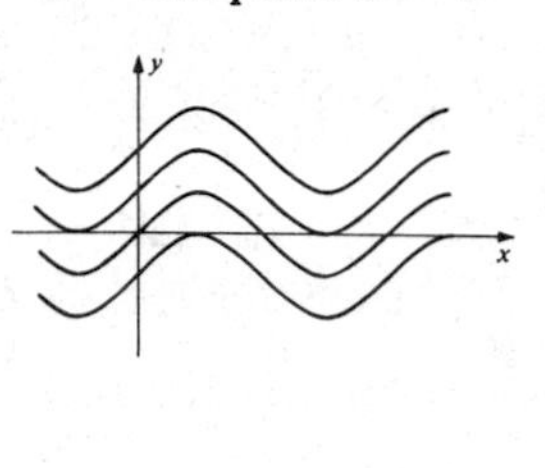

25 The origin and all spheres with center at the origin **27** All planes with normal vector $\langle 1, 2, 3\rangle$
29 The z-axis and all circular cylinders having axis along the z-axis.

Exercises 16.2, page 767

1 $-2/3$ **3** Does not exist **5** Does not exist **7** Does not exist **9** 0
11 Continuous on $\{(x, y)\colon x + y > 1\}$ **13** Continuous at (x, y, z) if $x^2 + y^2 \neq z^2$
15 Continuous on $\{(x, y)\colon x > 0, -1 < y < 1\}$
17 $(x^4 - 2x^2y^2 + y^4 - 4)/(x^2 - y^2)$; $\{(x, y)\colon y \neq \pm x\}$
19 $x^2 + 2x \tan y + \tan^2 y + 1$; $\{(x, y)\colon y \neq \pi/2 + k\pi, k \text{ any integer}\}$
21 $e^{x^2+2y}, (x^2 + 2y)(x^2 + 2y - 3), e^{2t} + 2t^2 - 6t$
23 $(x - 2y)(2x + y) - 3(x - 2y) + (2x + y)$
27 (a) Replace the ordered pairs by ordered quadruples (x, y, z, w), where $x, y, z, w \in \mathbb{R}$.
(b) $\lim_{(x,y,z,w)\to(a,b,c,d)} f(x, y, z, w) = L$ means that for every $\epsilon > 0$ there corresponds a $\delta > 0$ such that if $0 < \sqrt{(x-a)^2 + (y-b)^2 + (z-c)^2 + (w-d)^2} < \delta$, then $|f(x, y, z, w) - L| < \epsilon$.

Exercises 16.3, page 772

1 $f_x(x, y) = 8x^3y^3 - y^2, f_y(x, y) = 6x^4y^2 - 2xy + 3$
3 $f_r(r, s) = r/\sqrt{r^2 + s^2}, f_s(r, s) = s/\sqrt{r^2 + s^2}$
5 $f_x(x, y) = e^y + y \cos x, f_y(x, y) = xe^y + \sin x$
7 $f_t(t, v) = -v/(t^2 - v^2), f_v(t, v) = t/(t^2 - v^2)$
9 $f_x(x, y) = \cos(x/y) - (x/y)\sin(x/y), f_y(x, y) = (x/y)^2 \sin(x/y)$
11 $f_r(r, s, t) = 2re^{2s} \cot t, f_s(r, s, t) = 2r^2e^{2s} \cot t, f_t(r, s, t) = -r^2e^{2s} \csc^2 t$
13 $f_x(x, y, z) = (y^2 + z^2)^x \ln(y^2 + z^2), f_y(x, y, z) = 2xy(y^2 + z^2)^{x-1}, f_z(x, y, z) = 2xz(y^2 + z^2)^{x-1}$
15 $f_x(x, y, z) = e^z - ye^x, f_y(x, y, z) = -e^x - ze^{-y}, f_z(x, y, z) = xe^z + e^{-y}$
17 $f_q(q, v, w) = v/2\sqrt{qv}\sqrt{1 - qv}, f_v(q, v, w) = (q/2\sqrt{qv}\sqrt{1 - qv}) + w \cos vw, f_w(q, v, w) = v \cos vw$
25 $18xy^2 + 16y^3z$ **27** $t^2 \sec rt(\sec^2 rt + \tan^2 rt)$ **29** $(1 - x^2y^2z^2)\cos xyz - 3xyz \sin xyz$
43 $w_{xx}, w_{xy}, w_{xz}, w_{yx}, w_{yy}, w_{yz}, w_{zx}, w_{zy}, w_{zz}$ **45** 200, 400 **47** $x = 1, y = t, z = -4t + 12$

Exercises 16.4, page 780

There are other correct answers to 1 and 3 (see Example 2).

1 $\eta_1 = -3\Delta y,\ \eta_2 = 4\Delta y$ **3** $\eta_1 = 3x\Delta x + (\Delta x)^2,\ \eta_2 = 3y\Delta y + (\Delta y)^2$
5 $dw = (3x^2 - 2xy)\,dx + (6y - x^2)\,dy$ **7** $dw = 2x \sin y\,dx + (x^2 \cos y + 3y^{1/2})\,dy$
9 $dw = (x^2y + 2x)e^{xy}\,dx + (x^3e^{xy} - 2y^{-3})\,dy$
11 $dw = 2x \ln(y^2 + z^2)\,dx + (2x^2y/(y^2 + z^2))\,dy + (2x^2z/(y^2 + z^2))\,dz$
13 $dw = \dfrac{yz(y + z)}{(x + y + z)^2}dx + \dfrac{xz(x + z)}{(x + y + z)^2}dy + \dfrac{xy(x + y)}{(x + y + z)^2}dz$
15 $dw = (2xz - z^2t)\,dx + 4t^3\,dy + (x^2 - 2xzt)\,dz + (12yt^2 - xz^2)\,dt$ **17** Approximately 7.38
19 Approximately 1.87 **21** Approximately 18.006 **23** (a) $1/4\ \text{ft}^2$ (b) $47/192\ \text{ft}^3$
25 $57\pi(0.015)/4 \approx 0.67\ \text{in.}^3$ **27** 0.0185 **29** 2.96 **31** $1.7\pi\ \text{in.}^2$

Exercises 16.5, page 787

1 $\partial w/\partial x = 2u(\sin v)3x^2 + u^2(\cos v)y^2,\ \partial w/\partial y = -12u(\sin v)y^2 + 2u^2(\cos v)xy$
3 $\partial w/\partial r = (u/\sqrt{u^2 + v^2})e^{-s} - (v/\sqrt{u^2 + v^2})s^2e^{-r}$,
$\partial w/\partial s = -(u/\sqrt{u^2 + v^2})re^{-s} + (v/\sqrt{u^2 + v^2})2se^{-r}$
5 $\partial z/\partial x = (1/v)\cos y + (1/v)y \cos x - 2(r + s)/v^2$,
$\partial z/\partial y = (1/v)(-x \sin y) + (1/v)\sin x + (r + s)/v^2$
7 $\partial r/\partial u = (3x^2 - y^2) + (3 - 2xy)(-1/u),\ \partial r/\partial v = (3x^2 - y^2)\ln t + (3 - 2xy)(2v)$,
$\partial r/\partial t = (3x^2 - y^2)(v/t) + (3 - 2xy)(-1/t)$
9 $\partial p/\partial r = (2u \cos vw)(sxye^{rs}) - (u^2w \sin vw)(s/r) - (vu^2 \sin vw)\sin x$
11 $dw/dt = [-3x^2/(t + 1)^2] - [3y^2/(t + 1)^2]$
13 $dw/dt = 4r \sin t \cos t + \tan v \sin t - 4s \sec^2 v$ **15** $y' = -(6x^2 + 2xy)/(x^2 + 3y^2)$
17 $y' = -\left(6 + \dfrac{\sqrt{y}}{2\sqrt{x}}\right)\Big/\left(\dfrac{\sqrt{x}}{2\sqrt{y}} - 3\right) = \dfrac{12\sqrt{xy} + y}{6\sqrt{xy} - x}$
19 $\partial z/\partial x = -(2z^3 + 2xy^2)/(6xz^2 - 6yz + 4),\ \partial z/\partial y = -(2x^2y - 3z^2)/(6xz^2 - 6yz + 4)$
21 $\partial z/\partial x = -(e^{yz} - 2yze^{xz} + 3yze^{xy})/(xye^{yz} - 2xye^{xz} + 3e^{xy})$
$\partial z/\partial y = -(xze^{yz} - 2e^{xz} + 3xze^{xy})/(xye^{yz} - 2xye^{xz} + 3e^{xy})$
33 $0.88\pi \approx 2.76\ \text{in.}^3/\text{min},\ 0.46\pi \approx 1.44\ \text{in.}^2/\text{min}$ **35** $dT/dt = (v/c)(dp/dt) + (p/c)(dv/dt)$

Exercises 16.6, page 796

1 $-10/\sqrt{2}$ **3** $-1/8\sqrt{13}$ **5** $67/8\sqrt{26}$ **7** $1/2\sqrt{26}$ **9** $16\sqrt{14}$ **11** $15e^{-2}/\sqrt{35}$
13 $-12/\sqrt{10}$ **15** $-28\sqrt{26},\ 4\mathbf{i} - 8\mathbf{j},\ \sqrt{80}$ **17** $-25/\sqrt{14}\sqrt{77},\ (1/\sqrt{14})(-2\mathbf{i} + 3\mathbf{j} + \mathbf{k}),\ 1$
19 $-28/\sqrt{2},\ -12\mathbf{i} - 16\mathbf{j},\ c(4\mathbf{i} - 3\mathbf{j})$ for any $c \neq 0$
21 $-178/\sqrt{14},\ 4\mathbf{i} - 8\mathbf{j} + 54\mathbf{k},\ \sqrt{2996} \approx 54.8$

Exercises 16.7, page 803

1 $16(x - 2) + 6(y + 3) + 6(z - 1) = 0;\ x = 2 + 16t,\ y = -3 + 6t,\ z = 1 + 6t$
3 $16(x + 2) + 18(y + 1) + (z - 25) = 0;\ x = -2 + 16t,\ y = -1 + 18t,\ z = 25 + t$.
5 $4(x + 5) - 3(y - 5) + 20(z - 1) = 0;\ x = -5 + 4t,\ y = 5 - 3t,\ z = 1 + 20t$
7 $x + \sqrt{3}(y - \pi/3) + (z - 1) = 0;\ x = t,\ y = (\pi/3) + \sqrt{3}t,\ z = 1 + t$
9 $-x + (1/2)(y - 2) - (z - 1) = 0;\ x = -t,\ y = 2 + (1/2)t,\ z = 1 - t$
15 $(8\sqrt{2}/\sqrt{5}, 2\sqrt{2}/\sqrt{5}, -2\sqrt{2}/\sqrt{5}),\ (-8\sqrt{2}/\sqrt{5}, -2\sqrt{2}/\sqrt{5}, 2\sqrt{2}/\sqrt{5})$

21

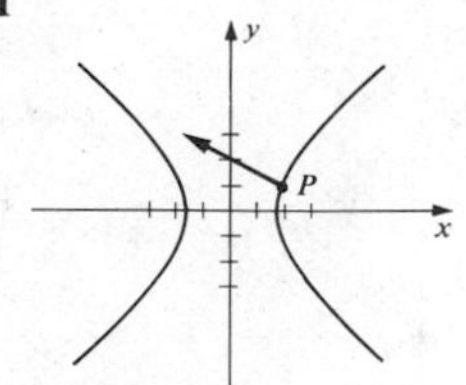

23

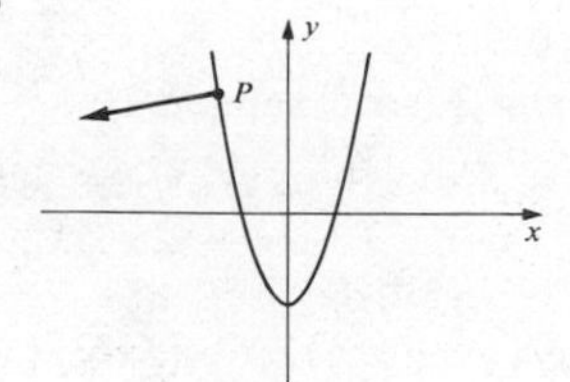

25

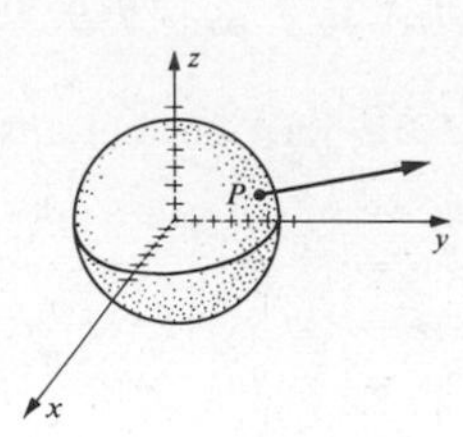

27

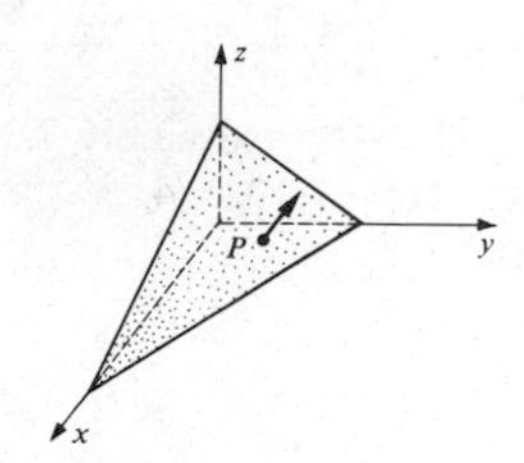

29

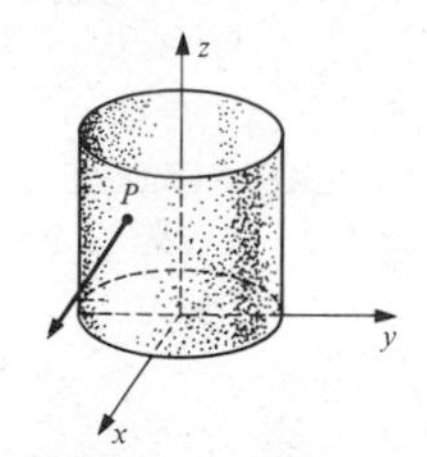

Exercises 16.8, page 809

The following answers refer to local maxima and minima.

1 Min: $f(0,0) = 0$ **3** Min: $f(1,-1) = -1$ **5** Min: $f(1/2,-1/4) = 1/4$
7 Min: $f(-2,\sqrt{3}) = -48 - 6\sqrt{3}$ **9** No extrema **11** Min: $f(\sqrt[3]{2}, 2\sqrt[3]{2}) = 12/\sqrt[3]{2}$
13 Max: $f(\pi/3,\pi/3) = 3\sqrt{3}/2$; min: $f(5\pi/3, 5\pi/3) = -3\sqrt{3}/2$
15 At the point (0, 0) f attains neither a maximum nor a minimum. **17** $1/\sqrt{26}$
19 $(\pm 2/\sqrt[4]{12}, \pm\sqrt[4]{12}, \pm 2\sqrt{2}/\sqrt[4]{12})$
21 Square base; altitude one-half the length of the side of the base **23** $6/\sqrt{3}$, $12/\sqrt{3}$, $8/\sqrt{3}$
25 1, 4/3, 4 **27** Square base of side $\sqrt[3]{4}$ ft, height $2\sqrt[3]{4}$ ft

Exercises 16.9, page 816

1 Min: $f(1/\sqrt{5}, 2/\sqrt{5}) = 0 = f(-1/\sqrt{5}, -2/\sqrt{5})$; max: $f(2/\sqrt{5}, -1/\sqrt{5}) = 5 = f(-2/\sqrt{5}, 1/\sqrt{5})$
3 Min: $f(-5/\sqrt{3}, -5/\sqrt{3}, -5/\sqrt{3}) = -5/\sqrt{3}$; max: $f(5/\sqrt{3}, 5/\sqrt{3}, 5/\sqrt{3}) = 5/\sqrt{3}$
5 Min: $f(1/3, -1/3, 1/3) = 1/3$ **7** Min: $f(0,-1,0) = 1$
9 Min: $f(1, 2/\sqrt{3}, -1, 8/3) = -16/3\sqrt{3}$; max: $f(1, -2/\sqrt{3}, -1, 8/3) = 16/3\sqrt{3}$
11 $(6/\sqrt{29}, 9/\sqrt{29}, 12/\sqrt{29})$ **13** Square base $2/\sqrt[3]{7}$ by $2/\sqrt[3]{7}$, height $7/2\sqrt[3]{7}$
15 Height = twice the radius

16.10 Review Exercises, page 818

1 $\{(x,y): 4x^2 - 9y^2 \leqslant 36\}$ **3** $\{(x,y): z^2 > x^2 + y^2\}$
5 $f_x(x,y) = 3x^2 \cos y + 4$, $f_y(x,y) = -x^3 \sin y - 2y$
7 $f_x(x,y,z) = 2x/(y^2 + z^2)$; $f_y(x,y,z) = 2y(z^2 - x^2)/(y^2 + z^2)^2$;
$f_z(x,y,z) = -(x^2 + y^2)2z/(y^2 + z^2)^2$
9 $f_x(x,y,z,t) = 2xz\sqrt{2y+t}$, $f_y(x,y,z,t) = x^2 z/\sqrt{2y+t}$, $f_z(x,y,z,t) = x^2\sqrt{2y+t}$, $f_t(x,y,z,t)$
$= x^2 z/2\sqrt{2y+t}$
11 $f_{xx}(x,y) = 6xy^2 + 12x^2$, $f_{yy}(x,y) = 2x^3 - 18xy$, $f_{xy}(x,y) = 6x^2 y - 9y^2$
15 $\Delta w = (2x + 3y)\Delta x + (3x - 2y)\Delta y + (\Delta x)^2 + 3\Delta x\Delta y - (\Delta y)^2$;
$dw = (2x + 3y)\,dx + (3x - 2y)\,dy$; $\Delta w = -1.13$, $dw = -1.1$
17 $\partial s/\partial x = (2uv - w^3)(-y \sin x) + (u^2 + 2vw)e^{-y} + (v^2 - 3w^2 u)(y/x)$,
$\partial s/\partial y = (2uv - w^3)\cos x + (u^2 + 2vw)(-xe^{-y}) + (v^2 - 3w^2 u)\ln x$
19 $dw/dt = (\tan y)3t^2 + (x \sec^2 y + \tan z)(-2e^{-2t}) + y \sec^2 z(-2/t^3)$ **21** $-14/\sqrt{41}$
23 $-16(x + 2) + 4(y + 1) - 7(z - 2) = 0$; $x = -2 - 16t$, $y = -1 + 4t$, $z = 2 - 7t$
25 $f'(x) = (3x^2 - 4y^3 + 1)/(12xy^2 + 3)$ **27** Min: $f(0,-1) = -2$

29

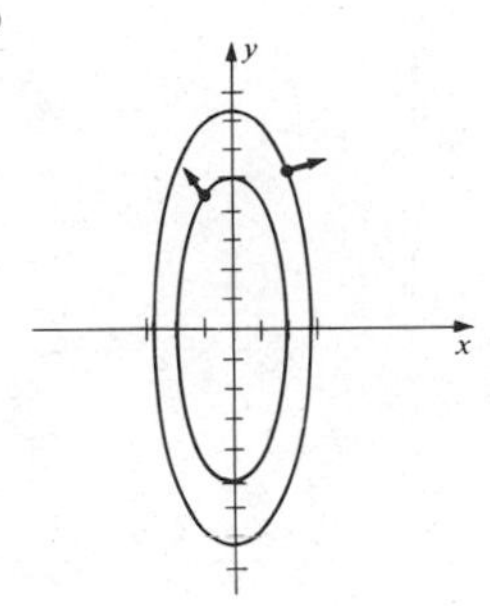

31 Min: $f(-\sqrt{8/3}, -\sqrt{2/3}, -\sqrt{4/3}) = f(-\sqrt{8/3}, \sqrt{2/3}, \sqrt{4/3}) = f(\sqrt{8/3}, -\sqrt{2/3}, \sqrt{4/3})$
$= f(\sqrt{8/3}, \sqrt{2/3}, -\sqrt{4/3}) = -8\sqrt{3}/3$; max: $f(\sqrt{8/3}, \sqrt{2/3}, \sqrt{4/3})$
$= f(-\sqrt{8/3}, -\sqrt{2/3}, \sqrt{4/3} = f(-\sqrt{8/3}, \sqrt{2/3}, -\sqrt{4/3})$
$= f(\sqrt{8/3}, -\sqrt{2/3}, -\sqrt{4/3}) = 8/3\sqrt{3}$

33 If $a = 1 + 2^{2/3} + 3^{2/3}$, the point is $(a, 2^{1/3}a, 3^{1/3}a)$.

CHAPTER 17

Exercises 17.1, page 825

1 (a) 39 (b) 81 **3** 1769.13 **5** 240

Exercises 17.2, page 835

1 Both are -36 **3** $163/120$ **5** $36/5$ **7** $(4e - e^4)/2$ **9** $(e^2 + 1)/4$

11 $\pi/24 + \ln[3/\sqrt{2}(\sqrt{2} + 1)] + 1/\sqrt{2} - 1/2 \approx 0.2087$

13 (a) $\int_0^4 \int_0^{\sqrt{x}} f(x,y)\,dy\,dx$ (b) $\int_0^2 \int_{y^2}^4 f(x,y)\,dx\,dy$

15 (a) $\int_0^2 \int_{x^3}^8 f(x,y)\,dy\,dx$ (b) $\int_0^8 \int_0^{y^{1/3}} f(x,y)\,dx\,dy$

17 (a) $\int_0^1 \int_{x^3}^{x^{1/2}} f(x,y)\,dy\,dx$ (b) $\int_0^1 \int_{y^2}^{y^{1/3}} f(x,y)\,dx\,dy$

13

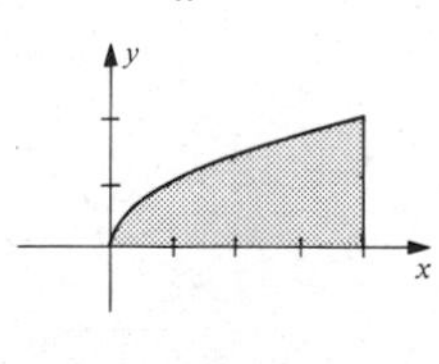

15

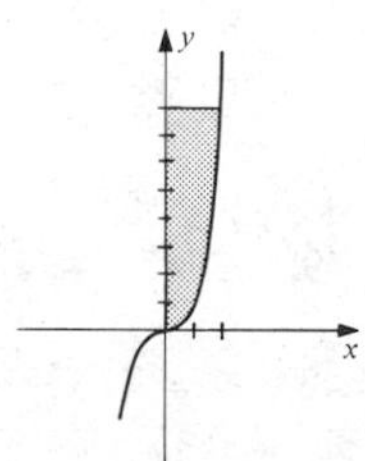

17

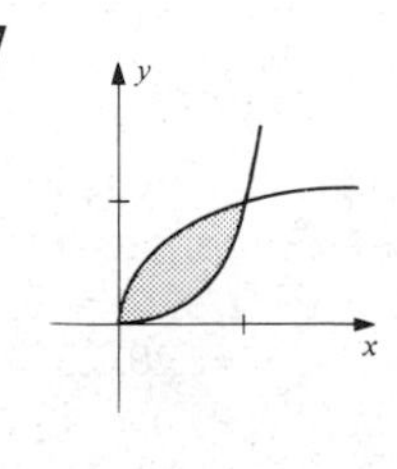

19 (a) $\int_1^2 \int_{9-4x}^{4+x} f(x,y)\,dy\,dx + \int_2^4 \int_{x^3/8}^{4+x} f(x,y)\,dy\,dx$

(b) $\int_1^5 \int_{(9-y)/4}^{2y^{1/3}} f(x,y)\,dx\,dy + \int_5^8 \int_{y-4}^{2y^{1/3}} f(x,y)\,dx\,dy$

21 (a) $\int_{-1}^1 \int_{-x-3}^{2x} f(x,y)\,dy\,dx + \int_1^3 \int_{-x-3}^{3-x^2} f(x,y)\,dy\,dx$

(b) $\int_{-6}^{-2} \int_{-y-3}^{\sqrt{3-y}} f(x,y)\,dx\,dy + \int_{-2}^2 \int_{y/2}^{\sqrt{3-y}} f(x,y)\,dx\,dy$

23 (a) $\int_0^1 \int_{1-x}^{e^x} f(x,y)\,dy\,dx + \int_1^e \int_{\ln x}^{1+e-x} f(x,y)\,dy\,dx$

(b) $\int_0^1 \int_{1-y}^{e^y} f(x,y)\,dx\,dy + \int_1^e \int_{\ln y}^{1+e-y} f(x,y)\,dx\,dy$

19

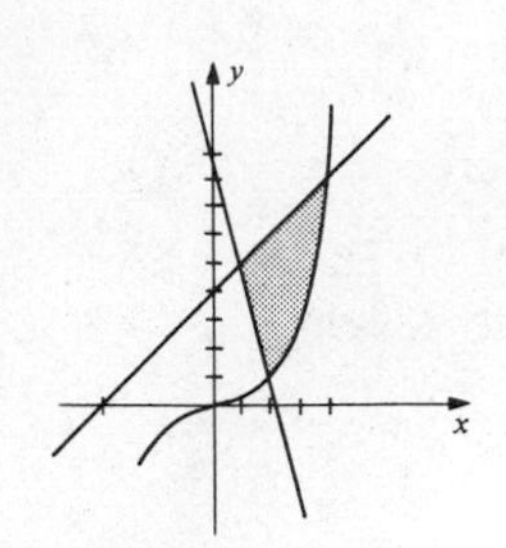

21

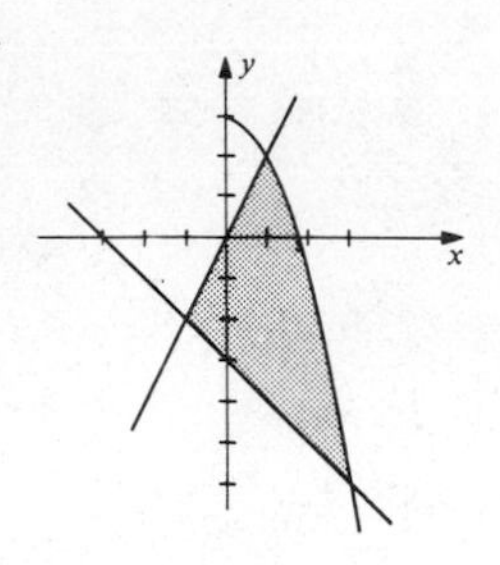

23

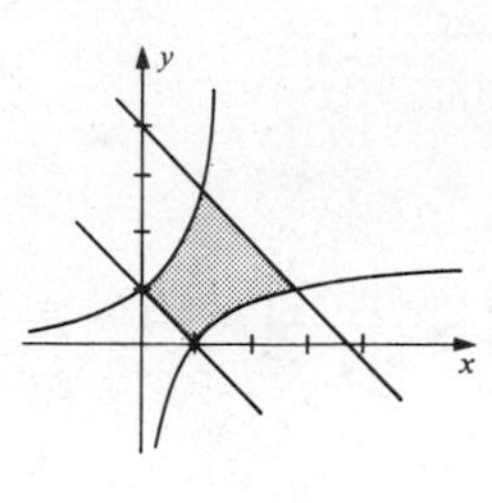

25

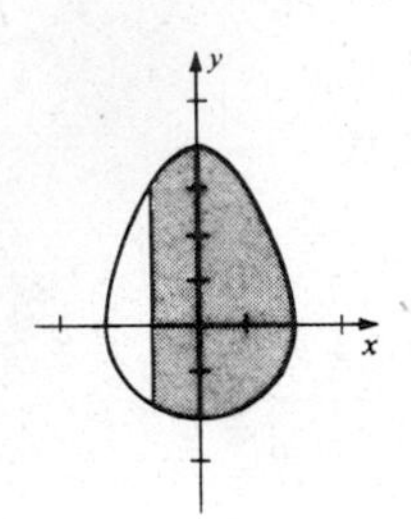

27

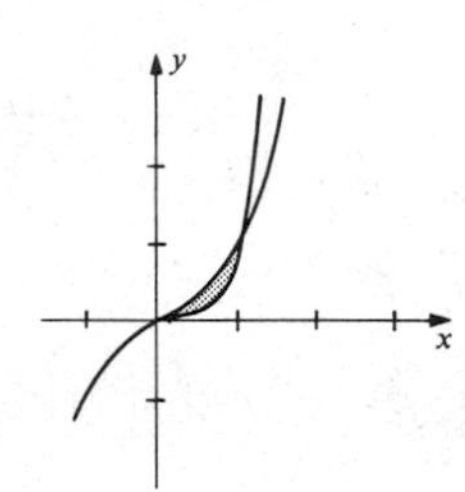

29

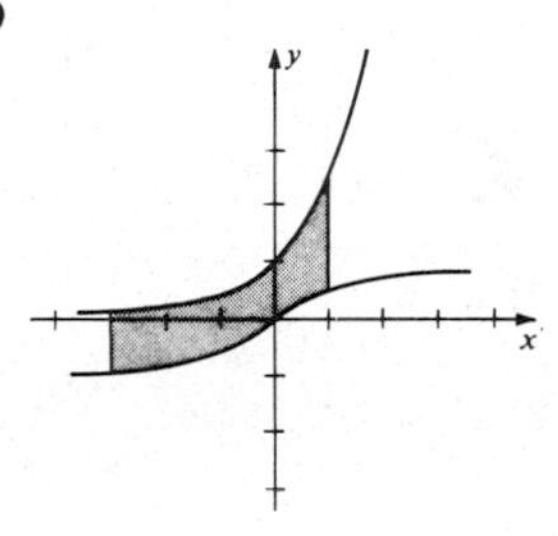

31 $\int_0^2 \int_0^{y/2} e^{y^2}\,dx\,dy = (e^4 - 1)/4$ **33** $\int_0^4 \int_0^{\sqrt{x}} y \cos x^2\,dy\,dx = (1/4)\sin 16$

35 $\int_0^2 \int_0^{x^3} \dfrac{y}{\sqrt{16 + x^7}}\,dy\,dx = 8/7$

Exercises 17.3, page 842

1 17/6 **3** 33/2 **5** 2 **7** $e^{\pi} - e^{-\pi}$
9 $2(\tan^{-1} a^{1/2} - a^{3/2}/3)$ where $a = (-1 + \sqrt{5})/2$ **11** 34/3 **13** 18 **15** 16/3
17 13/2 **19** 1/20 **21** 423/20 **23** $16a^3/3$
25 The solid lies under the paraboloid $z = x^2 + y^2$ and over the region in the xy-plane bounded by the graphs of $y = x - 1$, $y = 1 - x^2$, $x = -2$, and $x = 1$.
27 The solid lies under the plane $z = x + y$ and over the region in the xy-plane bounded by the graphs of $x = y/4$, $x = \sqrt{y}$, $y = 0$, and $y = 4$.
29 The rectangular parallelepiped bounded by the planes $z = 0$, $z = 3$, $x = 0$, $x = 4$, $y = -1$, $y = 2$

Exercises 17.4, page 849

1 $M = 2349/20$, $\bar{x} = 1290/203$, $\bar{y} = 38/29$
3 $M = 8k$ (k a proportionality constant), $\bar{x} = 0$, $\bar{y} = 8/3$
5 $M = (1 - e^{-2})/4$, $\bar{x} = 0$, $\bar{y} = 4(1 - e^{-3})/9(1 - e^{-2}) \approx 0.46$
7 $M = 115/6$, $\bar{x} = 20/23$, $\bar{y} = 1/2$
9 $M = 4 \ln(\sqrt{2} + 1) - 4 \ln(\sqrt{2} - 1) - \pi \approx 3.9$, $\bar{y} = (16 - \pi)/4M \approx 0.8$, $\bar{x} = 0$
11 $M = 4/3$, $\bar{x} = 21/16$, $\bar{y} = (3 + 12 \ln 2)/32 \approx 0.35$
13 $M = (1 - e^{-3})/9 \approx 0.106$, $\bar{x} = (1 - 4e^{-3})/3(1 - e^{-3}) \approx 0.28$, $\bar{y} = 9(1 - e^{-4})/16(1 - e^{-3}) \approx 0.58$
15 $I_x = 3^5(31/28)$, $I_y = 3^7(19/8)$, $I_0 = 3^5(1259/56)$ **17** $I_x = 64k$, $I_y = 32k/3$, $I_0 = 224k/3$
19 (a) $\rho a^4/3$ (b) $\rho a^4/12$ (c) $\rho a^4/6$ (ρ = density) **21** $a/\sqrt{3}$ **23** $\rho a^4\pi/4$, $a/2$

Exercises 17.5, page 856

1 9/2 **3** $9\sqrt{3}/2 - \pi$ **5** $M = 32k/9$, $\bar{x} = 6/5$, $\bar{y} = 0$
7 $M = 2k(\sqrt{3} - \pi/3)$, $\bar{x} = 0$, $\bar{y} = (27\sqrt{3} - 4\pi)/(24\sqrt{3} - 8\pi) \approx 2.1$ **9** $\rho a^2\pi/8$
11 $\rho a^4\pi/4$, $a/2$ **13** $(\pi/2)(1 - e^{-a^2})$ **15** $\ln(\sqrt{2} + 1)$ **17** $256\,\pi/3$ **19** 64/9

Exercises 17.6, page 864

3 $-1/12$ **5** $513/8$

7 $\int_0^6 \int_0^{(6-x)/2} \int_0^{(6-x-2y)/3} f(x,y,z)\,dz\,dy\,dx \quad \int_0^3 \int_0^{6-2y} \int_0^{(6-x-2y)/3} f(x,y,z)\,dz\,dx\,dy$

$\int_0^6 \int_0^{(6-x)/3} \int_0^{(6-x-3z)/2} f(x,y,z)\,dy\,dz\,dx \quad \int_0^2 \int_0^{6-3z} \int_0^{(6-x-3z)/2} f(x,y,z)\,dy\,dx\,dz$

$\int_0^2 \int_0^{(6-3z)/2} \int_0^{6-2y-3z} f(x,y,z)\,dx\,dy\,dz \quad \int_0^3 \int_0^{(6-2y)/3} \int_0^{6-2y-3z} f(x,y,z)\,dx\,dz\,dy$

9 $\int_{-3/2}^{3/2} \int_{-\sqrt{9-4x^2}}^{\sqrt{9-4x^2}} \int_0^{9-4x^2-y^2} f(x,y,z)\,dz\,dy\,dx \quad \int_{-3}^{3} \int_{-\frac{1}{2}\sqrt{9-y^2}}^{\frac{1}{2}\sqrt{9-y^2}} \int_0^{9-4x^2-y^2} f(x,y,z)\,dz\,dx\,dy$

$\int_{-3/2}^{3/2} \int_0^{9-4x^2} \int_{-\sqrt{9-z-4x^2}}^{\sqrt{9-z-4x^2}} f(x,y,z)\,dy\,dz\,dx \quad \int_0^9 \int_{-\frac{1}{2}\sqrt{9-z}}^{\frac{1}{2}\sqrt{9-z}} \int_{-\sqrt{9-z-4x^2}}^{\sqrt{9-z-4x^2}} f(x,y,z)\,dy\,dx\,dz$

$\int_{-3}^{3} \int_0^{9-y^2} \int_{-\frac{1}{2}\sqrt{9-z-y^2}}^{\frac{1}{2}\sqrt{9-z-y^2}} f(x,y,z)\,dx\,dz\,dy \quad \int_0^9 \int_{-\sqrt{9-z}}^{\sqrt{9-z}} \int_{-\frac{1}{2}\sqrt{9-z-y^2}}^{\frac{1}{2}\sqrt{9-z-y^2}} f(x,y,z)\,dx\,dy\,dz$

11 $128/5$ **13** $32/3$ **15** 2π **17** 108 **19** $1/70$

21 The region above the rectangle with vertices (2,0,0), (2,0,1), (3,0,0), (3,0,1) and between the graphs of $y = \sqrt{1-z}$ and $y = \sqrt{4-z}$

23 Q is the region under the plane $z = x + y$ and over the region in the xy-plane bounded by the graphs of $y = x^2$, $y = 2x$, $x = 0$, and $x = 2$.

25 Q is the region bounded by the paraboloid $z = x^2 + y^2$ and the planes $z = 1$ and $z = 2$.

Exercises 17.7, page 870

1 Let $I = \int_0^3 \int_0^{6-2x} \int_0^{9-x^2} f(x,y,z)\,dz\,dy\,dx$. To find M, M_{xy}, M_{xz}, M_{yz}, let $f(x,y,z)$ equal 1, z, y, and x, respectively.

3 Let $I = \int_{-1}^{1} \int_{x^2}^{1} \int_0^{4x^2+9z^2} f(x,y,z)\,dz\,dy\,dx$ and use the method given in the answer to Exercise 1.

5 $\bar{x} = \bar{y} = \bar{z} = 7a/12$

7 $M = \int_0^4 \int_{-\sqrt{x}/2}^{\sqrt{x}/2} \int_{-\sqrt{x-4z^2}}^{\sqrt{x-4z^2}} (x^2+z^2)\,dy\,dz\,dx$, $\bar{x} = \int_0^4 \int_{-\sqrt{x}/2}^{\sqrt{x}/2} \int_{-\sqrt{x-4z^2}}^{\sqrt{x-4z^2}} x(x^2+z^2)\,dy\,dz\,dx/M$. The $\bar{y}$ and $\bar{z}$ integrals have the same limits but the integrands are $y(x^2+z^2)$ and $z(x^2+z^2)$, respectively.

9 $2\rho a^5/3$, $(\sqrt{2/3})a$ **11** $I_z = \int_{-3}^{3} \int_{-\frac{1}{3}\sqrt{36-4x^2}}^{\frac{1}{3}\sqrt{36-4x^2}} \int_0^{36-4x^2-9y^2} (x^2+y^2)z\,dz\,dy\,dx$

13 $I_z = \int_{-a}^{a} \int_{-\sqrt{a^2-x^2}}^{\sqrt{a^2-x^2}} \int_{-\sqrt{a^2-x^2-y^2}}^{\sqrt{a^2-x^2-y^2}} (x^2+y^2)(x^2+y^2+z^2)\,dz\,dy\,dx$

15 $I_z = \int_0^a \int_0^{b(1-x/a)} \int_0^{c(1-x/a-y/b)} (x^2+y^2)\rho\,dz\,dy\,dx$

Exercises 17.8, page 876

1 8π; $\bar{x} = 0, \bar{y} = 0, \bar{z} = 4/3$ **3** (a) $h\pi a^4 k/2$ (b) $h\pi a^2 k\left(\frac{a^2}{4} + \frac{h^2}{3}\right)$ (k = density)

5 $k\pi^2 a^4/4$ **7** $k\pi^2 a^6/8$, $a/\sqrt{2}$ **9** $M = 9\pi k$, $\bar{x} = \bar{y} = 0$, $\bar{z} = 9/4$

11 $ka^4\pi/2$ (where k is the proportionality constant) center of mass is $2a/5$ from base along the axis of symmetry **13** $2\pi a^6 k/9$ **15** 8π **17** $124\pi k/5$

Exercises 17.9, page 880

1 $(\sqrt{3} + 2\ln(1+\sqrt{3}) - \ln 2)/2$ **3** $\pi c d^2 \sqrt{\frac{1}{a^2} + \frac{1}{b^2} + \frac{1}{c^2}}$ **5** $2a^2(\pi - 2)$

7 $\pi(5^{3/2} - 1)/6$ **9** $2\pi(5^{3/2} - 1)/3$

17.10 Review Exercises, page 882

1 $-5/84$ **3** $63/4$ **5** $-107/210$ **7** $\int_2^4 \int_{-\sqrt{x^2-4}}^{\sqrt{x^2-4}} f(x,y)\,dy\,dx$

9 $\int_{-2}^{2} \int_{y^2-4}^{-y^2+4} f(x,y)\,dx\,dy$

11 R is bounded by the graphs of $x = e^y$, $x = y^3$, $y = -1$, and $y = 1$.

13 $\int_0^9 \int_0^{\sqrt{x}} ye^{-x^2}\,dy\,dx = (1 - e^{-81})/4$

15 $M = 9k$ (k the constant of proportionality), $\bar{x} = 9/4$, $\bar{y} = 27/8$ **17** $2\pi k$ **19** $\pi a^3/6$

21 $kab^3/12$ (k = density), $b/\sqrt{6}$ **23** 13 **25** $V = 256/15$; $\bar{x} = 0$, $\bar{y} = 8/7$, $\bar{z} = 12/7$

27 $\int_0^4 \int_{-\sqrt{1+y^2}}^{\sqrt{1+y^2}} \int_{-\sqrt{1+y^2-x^2}}^{\sqrt{1+y^2-x^2}} k(x^2+z^2)^{3/2}\,dz\,dx\,dy$ **29** $M = k\pi a^4$

CHAPTER 18

Exercises 18.1, page 888

1

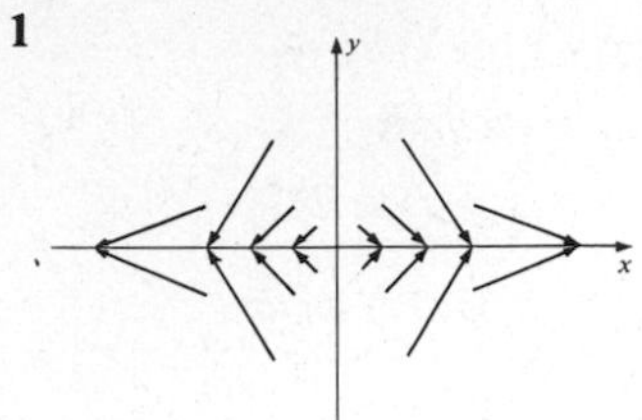

3

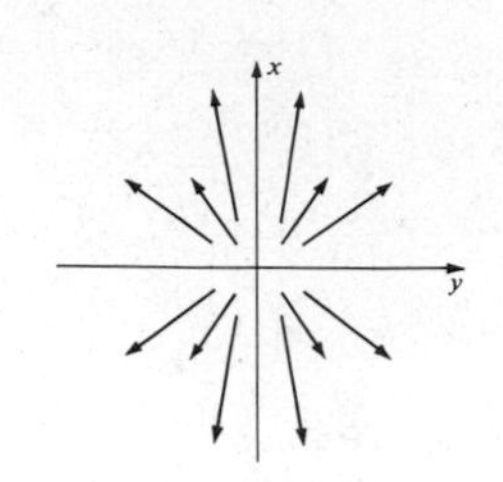

5

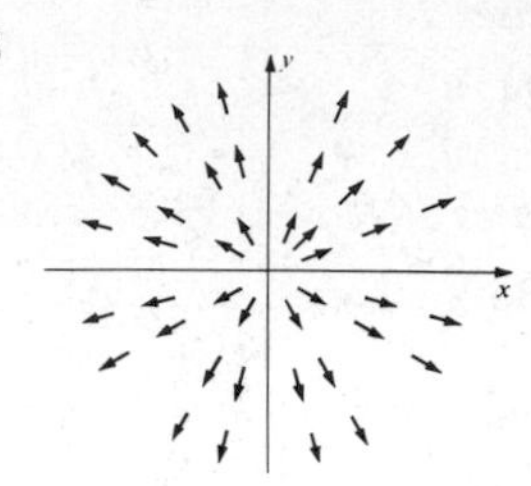

7

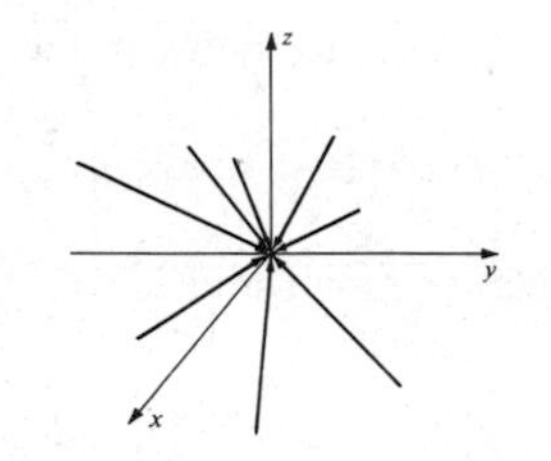

9

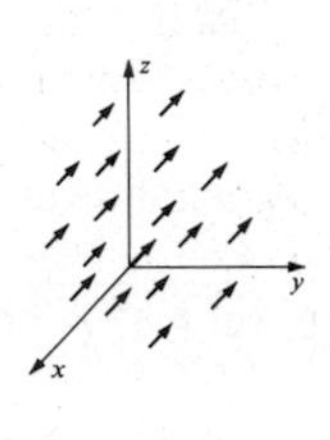

13 $\nabla f(x,y,z) = 2x\mathbf{i} - 6y\mathbf{j} + 8z\mathbf{k}$ **15** $\mathbf{F}(x,y,z) = e^{2y}\cos z\mathbf{i} + 2xe^{2y}\cos z\mathbf{j} - xe^{2y}\sin z\mathbf{k}$
17 $\mathbf{F}(x,y) = (1 + x^2y^2)^{-1}(y\mathbf{i} + x\mathbf{j})$ **19** $\mathbf{F}(x,y,z) = x\mathbf{i} + y\mathbf{j} + z\mathbf{k}$

Exercises 18.2, page 900

1 21, 14, $14(2\sqrt{2} - 1)$ **3** 34/7 **5** $-16/3$ **7** (a) 15/2 (b) 6 (c) 7 (d) 29/4
9 $(3e^4 + 6e^{-2} - 12e + 8e^3 - 5)/12$ **11** (a) 19 (b) 35 (c) 27 **13** $3\sqrt{14}/2$
15 $M = k(17^{3/2} - 1)/6$; $\bar{x} = 0, \bar{y} = (17^{5/2} - 41)/(10(17^{3/2} - 1)) \approx 1.67$ **19** 9/2 (for all paths) **21** 0 **23** 412/15 **25** $I_x = 4k/3, I_y = 2k/3$
27 If the density at (x,y,z) is $f(x,y,z)$, then $I_x = \int_C (y^2 + z^2)f(x,y,z)\,ds$, $I_y = \int_C (x^2 + z^2)f(x,y,z)\,ds$, $I_z = \int_C (x^2 + y^2)f(x,y,z)\,ds$.

Exercises 18.3, page 907

1 Yes, $f(x,y) = x^3y + 2x + y^4 + c$ **3** No **5** No
7 Yes, $f(x,y,z) = y\tan x - ze^x + c$ **9** Yes, $f(x,y,z) = 4x^2z + y - 3y^2z^3 + c$ **11** 14
13 -31
15 If $\mathbf{F}(x,y,z) = (-c/|\mathbf{r}|^2)\mathbf{r}$, where $c > 0$, then $f(x,y,z) = -c\ln r = -(c/2)\ln(x^2 + y^2 + z^2)$.

Exercises 18.4, page 916

1 $-7/60$ **3** 2/3 **5** π **7** 3 **9** 0 **11** $3\pi a^2/8$ **13** 128/3
19 $\bar{x} = 0, \bar{y} = 4a/3\pi$

Exercises 18.5, page 920

1 $\mathbf{i} + x^2\mathbf{j} + y^2\mathbf{k}$, $2xz + 2yx + 2$
3 $-y^2\cos z\mathbf{i} + (6xyz - e^{2z})\mathbf{j} - 3xz^3\mathbf{k}$, $3yz^2 + 2y\sin z + 2xe^{2z}$

Exercises 18.6, page 927

1 $2\pi a^4/3$ **3** 5
5 (a) $\int_0^4 \int_0^{3-3y/4} (6 - 3y/2 - 2z)y^2z^3(\sqrt{29}/2)\,dz\,dy$
(b) $\int_0^3 \int_0^{6-2z} x(4 - 2x/3 - 4z/3)^2(\sqrt{29}/3)\,dx\,dz$
7 (a) $\int_0^8 \int_0^6 (4 - 3y + y^2/16 + z)\sqrt{17}/4\,dz\,dy$
(b) $\int_0^2 \int_0^6 (x^2 + 8x - 16 + z)\sqrt{17}\,dz\,dx$

9 The value of the integral equals the volume of a cylinder of altitude c, rulings parallel to the z-axis, and whose base is the projection of S on the xy-plane. **11** $2\pi a^3$ **13** 3π

Exercises 18.7, page 936

1 24 **3** 20π **5** Both integrals equal $4\pi a^3$. **7** Both integrals equal 4π.

Exercises 18.8, page 944

1 Both integrals equal -1. **3** Both integrals equal πa^2. **5** 0 **7** -8π

Exercises 18.9, page 949

1 (a) Vertical lines; horizontal lines (b) $x = u/3,\ y = v/5$
3 (a) Straight lines $x - y = c$ and $3y + 2x = d$, where c and d are arbitrary constants
(b) $x = \frac{3}{5}u + \frac{1}{5}v,\ y = -\frac{2}{5}u + \frac{1}{5}v$
5 (a) Straight lines $2x + y = c$ and hyperbolas $xy = d$ (b) Not one-to-one
7 (a) Ellipses $x^2 + 4y^2 = c$ and hyperbolas $4x^2 - y^2 = d$ (b) Not one-to-one
9 (a) Hyperbolas $x^2 - y^2 = c$ and $xy = d$ (b) Not one-to-one
11 Rectangle with vertices (0,0), (6,0), (6,5), (0,5); the ellipse $u^2/9 + v^2/25 = 1$
15 Planes having equations of the form
$2x + y = c,\ y + z = d,\ x + y + z = k;\ x = w - v,\ y = u + 2v - 2w,\ z = -u - v + 2w$

Exercises 18.10, page 956

1 $4u^2 + 4v^2$ **3** $2u(vu - 1)e^{-(2u+v)}$ **5** -6 **7** $\int_0^2 \int_0^{2-u} (v - u)2\, dv\, du$
9 $\int_{-1}^{1} \int_{-\sqrt{1-v^2}}^{\sqrt{1-v^2}} (u^2 + v^2)\, 6\, du\, dv$ **11** $\int_{-1}^{1} \int_1^3 \frac{1}{2}u^2 \cos^2 v\, dv\, du = \frac{1}{3} + \frac{1}{12}\sin 6 - \frac{1}{12}\sin 2$
13 $\int_1^2 \int_{\sqrt{u}}^{\sqrt{2u}} (u^2 v^{-3} + 2v)\, dv\, du = 15/8$ **15** $\int_2^4 \int_{-u/2}^{2u} \frac{v}{u}\frac{1}{5}\, dv\, du = 9/4$

18.11 Review Exercises, page 958

1 **3**

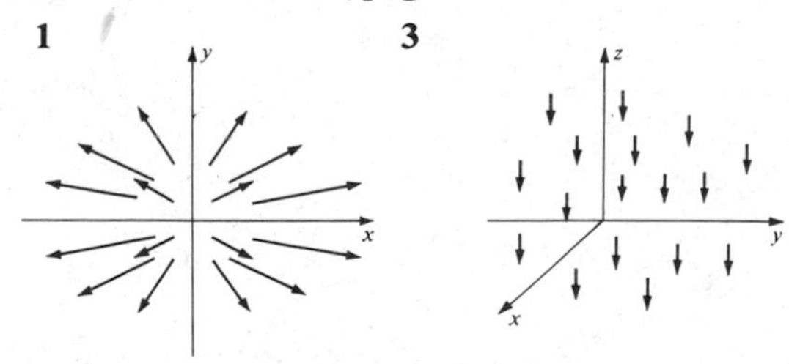

5 $\mathbf{F}(x,y) = y^2 \sec^2 x\mathbf{i} + 2y \tan x\mathbf{j}$ **7** $-56/5$ **9** $-40/3$
11 $(1025\sqrt{1025} - 17\sqrt{17})/144$ **13** 70 **15** 0 **17** 25/2
19 $f(x,y,z) = x^2e^{2y} + y^2 \cot z + c$ **21** 1/6
23 $3x^2z^4 + xz^2,\ (2x^2y - 2xyz)\mathbf{i} + (4x^3z^3 - 2xy^2)\mathbf{j} + yz^2\mathbf{k}$ **25** $\sqrt{3}/5$ **27** $5\pi/2$
29 Both integrals equal 8π. **31** (a) The lines $2x + 5y = c$ (b) The lines $3x - 4y = d$
(c) $x = (4/23)u + (5/23)v,\ y = (3/23)u - (2/23)v$ (d) $-1/23$
(e) $(4a + 3b)u + (5a - 2b)v + 23c = 0$ (f) $25u^2 + 28uv + 29v^2 = 529a^2$

CHAPTER 19

Exercises 19.1, page 965

9 $2yy' = 3x^2$ **11** $xy' = y \ln y$ **13** $y'' - 2y' + y = 0$ **15** $xy' - y = 0$
17 $yy'' + y'^2 + 1 = 0$ **19** $y = Cx$ **21** $x^2y^5e^y = C,\ C \neq 0$ and $y = 0$

23 $y = -1 + Ce^{x^2/2-x}$ **25** $y = -(1/3)\ln(3C + 3e^{-x})$ **27** $y^2 = C(1 + x^3)^{-2/3} - 1$
29 $\cos x + x \sin x - \ln|\sin y| = C$ **31** $\sec x + e^{-y} = C$ **33** $y^2 + \ln y = 3x - 8$
35 $y = \ln(2x + \ln x + e^2 - 2)$ **37** $y = 2e^{2-\sqrt{4+x^2}} - 1$ **39** $\tan^{-1}y - \ln|\sec x| = \pi/4$

Exercises 19.2, page 970

1 $x^2 + xy + y^2 = C$ **3** $x - x^2y + x^3y^2 - y^3 = C$ **5** $x^2 \sin y + 5x + 2y^2 = C$
7 $xe^{xy} + y^2 = C$ **9** $r \sin^2\theta + r^2 \cos\theta = C$ **11** $e^{-x}\cos y + x^2 = C$
13 $x^2/2 - y \ln x + x \ln y = C$ **15** $x^2y^3 + 4x^2 + 5y = 7$
17 $4y^2e^{2x} - 8x + \ln(1 + 4y^2) = 1 + \ln 2$

Exercises 19.3, page 973

1 $x^4(1 + 4yx^{-1}) = C$ **3** $x \cos(y/x) = C$ **5** $\ln x = (2/\sqrt{3})\tan^{-1}((2y - x)/\sqrt{3}x) + C$
7 $\ln((y/x) - 1) - 3 \ln((3y/x) - 1) = \ln x^2 + C$ **9** $\ln x = \sin^{-1}(y/x) + C$
11 $y(1 + x^2y^{-2}) = C$ **13** $\ln y^2 + (3/2)\ln((2x/y) + 1) = C$ **15** $x^2/2 + 3xy - y^2 = C$

Exercises 19.4, page 976

1 $y = (1/4)e^{2x} + Ce^{-2x}$ **3** $y = (x^5/2) + Cx^3$ **5** $y = (1/x)e^x - (1/2)x + C/x$
7 $y = (e^x + C)/x^2$ **9** $y = (4/3)x^3 \csc x + C \csc x$ **11** $y = 2 \sin x + C \cos x$
13 $y = x \sin x + Cx$ **15** $y = \left(\frac{x}{3} + \frac{C}{x^2}\right)e^{-3x}$ **17** $y = \left(-\frac{x^3}{2} + \frac{C}{x^3}\right)^{-1/3}$
19 $y = [-\frac{1}{2}(1 + x^2)\tan^{-1}x + x/2 + C(1 + x^2)]^{-1}$ **21** $y = 3/2 + Ce^{-x^2}$
23 $y = (1/2)\sin x + C/\sin x$ **25** $y = 1/3 + (x + C)e^{-x^3}$ **27** $y = x(x + \ln x + 1)$
29 $y = e^{-x}(1 - x^{-1})$

Exercises 19.5, page 982

1 $y = k/x$ **3** $2x^2 + y^2 = k$ **5** $2x^2 + 3y^2 = k$ **7** $x^2 + y^2 = ky$ **9** $p = kv^{-1/c}$
11 $i = [E/(R^2 + \omega^2L^2)](R \sin \omega t - L\omega \cos \omega t + L\omega e^{-(R/L)t})$
13 $s = (1/c)\ln(e^{2\sqrt{gc}\,t} + 1) - (\sqrt{g/c})t - (\ln 2)/c$, where $c = k/m$

Exercises 19.6, page 989

1 $y = C_1e^{2x} + C_2e^{3x}$ **3** $y = C_1 + C_2e^{3x}$ **5** $y = C_1e^{-2x} + C_2xe^{-2x}$
7 $y = C_1e^{(2+\sqrt{3})x} + C_2e^{(2-\sqrt{3})x}$ **9** $y = C_1e^{-\sqrt{2}x} + C_2xe^{-\sqrt{2}x}$
11 $y = C_1e^{-5x/4} + C_2e^{3x/2}$ **13** $y = C_1e^{4x/3} + C_2xe^{4x/3}$
15 $y = C_1e^{(2+\sqrt{2})x/2} + C_2e^{(2-\sqrt{2})x/2}$
17 $y = C_1e^x \cos x + C_2e^x \sin x$ **19** $y = C_1e^{2x} \sin 3x + C_2e^{2x} \cos 3x$
21 $y = C_1e^{(-3+\sqrt{7})x} + C_2e^{(-3-\sqrt{7})x}$ **23** $y = 2e^{2x} - 2e^x$
25 $y = \cos x + 2 \sin x$ **27** $y = (2 + 9x)e^{-4x}$ **29** $y = e^x \sin 2x/2$

Exercises 19.7, page 995

1 $y = C_1 \sin x + C_2 \cos x - \cos x \ln|\sec x + \tan x|$ **3** $y = (C_1 + C_2x + x^4/12)e^{3x}$
5 $y = C_1e^x + C_2e^{-x} + (2/5)e^x \sin x - (1/5)e^x \cos x$ **7** $y = (C_1 + x/6)e^{3x} + C_2e^{-3x}$
9 $y = C_1e^{4x} + C_2e^{-x} - (1/2)$ **11** $y = C_1e^x + C_2e^{2x} + (2/3)e^{-x}$
13 $y = C_1 + C_2e^{-2x} + (1/8)\sin 2x - (1/8)\cos 2x$
15 $y = C_1e^x + C_2e^{-x} + (1/9)(3x - 4)e^{2x}$
17 $y = e^{3x}(C_1 \cos 2x + C_2 \sin 2x) + (1/65)(7 \cos x - 4 \sin x)e^x$

Exercises 19.8, page 997

1 $y = a_0 \sum_{n=0}^{\infty} \frac{(-1)^n}{(2n)!} x^{2n} + a_1 \sum_{n=0}^{\infty} \frac{(-1)^n}{(2n+1)!} x^{2n+1} (= a_0 \cos x + a_1 \sin x)$

3 $y = a_0\left[1 + \sum_{n=1}^{\infty} \frac{2^n(3n-2)(3n-5)\cdots 7 \cdot 4 \cdot 1}{(3n)!} x^{3n}\right]$
$+a_1\left[x + \sum_{n=1}^{\infty} \frac{2^n(3n-1)(3n-4)\cdots 8 \cdot 5 \cdot 2}{(3n+1)!} x^{3n+1}\right]$

5 $y = a_0(1 - x^2) + a_1\left[x - \sum_{n=1}^{\infty} \frac{(2n-3)(2n-5)\cdots 5 \cdot 3 \cdot 1}{(2n+1)!} x^{2n+1}\right]$

7 $y = a_0(x+1)^3$

9 $y = -5x + a_0 \sum_{n=0}^{\infty} \frac{x^n}{n!} + a_1 \sum_{n=0}^{\infty} \frac{(-x)^n}{n!} = -5x + a_0 e^x + a_1 e^{-x}$

11 $y = a_0 \sum_{n=0}^{\infty} (n+1)x^{2n} + a_1 \sum_{n=0}^{\infty} \left(\frac{2n+3}{3}\right) x^{2n+1} + \sum_{n=1}^{\infty} (n+1)x^{2n}$

19.9 Review Exercises, page 998

1 $\sin x - x \cos x + e^{-y} = C$ **3** $(1/\sqrt{3})\tan^{-1}(y/\sqrt{3}x) + (1/2)\ln(3x^2 + y^2) = C$
5 $xy^2 + ye^{-x} + 3y = C$ **7** $y = 2 \sin x + C \cos x$ **9** $\sqrt{1 - y^2} + \sin^{-1} x = C$
11 $x^2 \sin(y/x) = C$ **13** $y = (Cx^{-4} - 2x^3/7)^{-1/2}$ **15** $y = (C_1 + C_2 x)e^{4x}$
17 $y = C_1 + C_2 e^{2x}$ **19** $y = C_1 e^x + C_2 e^{-x} - (1/5)e^x(\sin x + 2 \cos x)$
21 $\sin^{-1} x - \sqrt{1 - x^2} - y/2 - (\sin 2y)/4 = C$ **23** $y = Ce^{-x} + e^{4x}/5$
25 $y = C_1 e^x + C_2 e^{2x} + e^{5x}/12$ **27** $y = (x-2)^3/3x + C/x$
29 $y = C_1 e^{5x} + C_2 e^{-4x} + ((1/108) - (1/18)x)e^{-x}$
31 $y = e^{-(5/2)x}(C_1 \cos \sqrt{3}x/2 + C_2 \sin \sqrt{3}x/2)$ **33** $e^x(\sin x - \cos x) + 2e^{-y} = C$
35 $y = \cos x/2 + C \sec x$ **37** $xy + e^{-2x} \sin y = C$
39 $y = (\ln|\sec x + \tan x| - x + C)/(\csc x - \cot x)$

APPENDIX III

Exercises, page A26

3 (a) II (b) III (c) IV

The order in Exercises 5 and 11 is sin, cos, tan, csc, sec, cot.

5 (a) 1, 0,—, 1,—, 0 (b) $\sqrt{2}/2, -\sqrt{2}/2, -1, \sqrt{2}, -\sqrt{2}, -1$
(c) 0, 1, 0,—, 1,— (d) $-1/2, \sqrt{3}/2, -\sqrt{3}/3, -2, 2\sqrt{3}/3, -\sqrt{3}$
7 810°, −120°, 315°, 900°, 36°
9 (a) −4/5, 3/5, −4/3, −5/4, 5/3, −3/4
(b) $2\sqrt{13}/13, -3\sqrt{13}/13, -2/3, \sqrt{13}/2, -\sqrt{13}/3, -3/2$ (c) −1, 0, –, −1, –, 0
37 $\pi/2, 3\pi/2, \pi/4, 3\pi/4, 5\pi/4, 7\pi/4$; 90°, 270°, 45°, 135°, 225°, 315°
39 $0, \pi$; 0°, 180° **41** $0, \pi, 2\pi/3, 4\pi/3$; 0°, 180°, 120°, 240°
43 $\pi/2, 7\pi/6, 11\pi/6$; 90°, 210°, 330° **45** $\pi/6, 5\pi/6, \pi/3, 5\pi/3$; 30°, 150°, 60°, 300°
47 $\pi/3, 5\pi/3$; 60°, 300° **49** $\sqrt{2 - \sqrt{3}}/2$ or $(\sqrt{6} - \sqrt{2})/4$
51 $-\sqrt{2 - \sqrt{3}}/2$ or $(\sqrt{2} - \sqrt{6})/4$ **53** 84/85 **55** −36/77 **57** 240/289
59 24/7 **61** 1/3
63 (a) $(1/2)\cos 3t - (1/2)\cos 11t$
(b) $(1/2)\cos(u/12) + (1/2)\cos(5u/12)$
(c) $3 \sin 8x - 3 \sin 2x$

Index

75 $\int \tan^n u\,du = \frac{1}{n-1}\tan^{n-1} u - \int \tan^{n-2} u\,du$

76 $\int \cot^n u\,du = \frac{-1}{n-1}\cot^{n-1} u - \int \cot^{n-2} u\,du$

77 $\int \sec^n u\,du = \frac{1}{n-1}\tan u \sec^{n-2} u + \frac{n-2}{n-1}\int \sec^{n-2} u\,du$

78 $\int \csc^n u\,du = \frac{-1}{n-1}\cot u \csc^{n-2} u + \frac{n-2}{n-1}\int \csc^{n-2} u\,du$

79 $\int \sin au \sin bu\,du = \frac{\sin (a-b)u}{2(a-b)} - \frac{\sin (a+b)u}{2(a+b)} + C$

80 $\int \cos au \cos bu\,du = \frac{\sin (a-b)u}{2(a-b)} + \frac{\sin (a+b)u}{2(a+b)} + C$

81 $\int \sin au \cos bu\,du = -\frac{\cos (a-b)u}{2(a-b)} - \frac{\cos (a+b)u}{2(a+b)} + C$

82 $\int u \sin u\,du = \sin u - u \cos u + C$

83 $\int u \cos u\,du = \cos u + u \sin u + C$

84 $\int u^n \sin u\,du = -u^n \cos u + n \int u^{n-1} \cos u\,du$

85 $\int u^n \cos u\,du = u^n \sin u - n \int u^{n-1} \sin u\,du$

86 $\int \sin^n u \cos^m u\,du = -\frac{\sin^{n-1} u \cos^{m+1} u}{n+m} + \frac{n-1}{n+m}\int \sin^{n-2} u \cos^m u\,du$

$= \frac{\sin^{n+1} u \cos^{m-1} u}{n+m} + \frac{m-1}{n+m}\int \sin^n u \cos^{m-2} u\,du$

Inverse Trigonometric Forms

87 $\int \sin^{-1} u\,du = u \sin^{-1} u + \sqrt{1-u^2} + C$

88 $\int \cos^{-1} u\,du = u \cos^{-1} u - \sqrt{1-u^2} + C$

89 $\int \tan^{-1} u\,du = u \tan^{-1} u - \frac{1}{2}\ln (1+u^2) + C$

90 $\int u \sin^{-1} u\,du = \frac{2u^2-1}{4}\sin^{-1} u + \frac{u\sqrt{1-u^2}}{4} + C$

91 $\int u \cos^{-1} u\,du = \frac{2u^2-1}{4}\cos^{-1} u - \frac{u\sqrt{1-u^2}}{4} + C$

92 $\int u \tan^{-1} u\,du = \frac{u^2+1}{2}\tan^{-1} u - \frac{u}{2} + C$

93 $\int u^n \sin^{-1} u\,du = \frac{1}{n+1}\left[u^{n+1}\sin^{-1} u - \int \frac{u^{n+1}\,du}{\sqrt{1-u^2}}\right], \quad n \neq -1$

94 $\int u^n \cos^{-1} u\,du = \frac{1}{n+1}\left[u^{n+1}\cos^{-1} u + \int \frac{u^{n+1}\,du}{\sqrt{1-u^2}}\right], \quad n \neq -1$

95 $\int u^n \tan^{-1} u\,du = \frac{1}{n+1}\left[u^{n+1}\tan^{-1} u - \int \frac{u^{n+1}\,du}{1+u^2}\right], \quad n \neq -1$

Exponential and Logarithmic Forms

96 $\int ue^{au}\,du = \frac{1}{a^2}(au-1)e^{au} + C$

97 $\int u^n e^{au}\,du = \frac{1}{a}u^n e^{au} - \frac{n}{a}\int u^{n-1}e^{au}\,du$

98 $\int e^{au} \sin bu\,du = \frac{e^{au}}{a^2+b^2}(a \sin bu - b \cos bu) + C$

99 $\int e^{au} \cos bu\,du = \frac{e^{au}}{a^2+b^2}(a \cos bu + b \sin bu) + C$

100 $\int \ln u\,du = u \ln u - u + C$

101 $\int u^n \ln u\,du = \frac{u^{n+1}}{(n+1)^2}[(n+1)\ln u - 1] + C$

102 $\int \frac{1}{u \ln u}\,du = \ln |\ln u| + C$

Hyperbolic Forms

103 $\int \sinh u\,du = \cosh u + C$

104 $\int \cosh u\,du = \sinh u + C$

105 $\int \tanh u\,du = \ln \cosh u + C$

106 $\int \coth u\,du = \ln |\sinh u| + C$

107 $\int \operatorname{sech} u\,du = \tan^{-1} |\sinh u| + C$

108 $\int \operatorname{csch} u\,du = \ln |\tanh \frac{1}{2}u| + C$

109 $\int \operatorname{sech}^2 u\,du = \tanh u + C$

110 $\int \operatorname{csch}^2 u\,du = -\coth u + C$

111 $\int \operatorname{sech} u \tanh u\,du = -\operatorname{sech} u + C$

112 $\int \operatorname{csch} u \coth u\,du = -\operatorname{csch} u + C$